C000112193

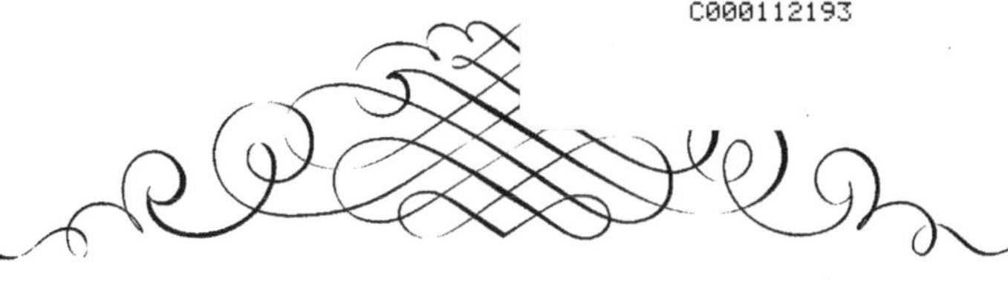

ISBN 978-1-5278-7976-8
PIBN 10894898

TC
D

A HANDBOOK

OF THE

PETROLEUM INDUSTRY

BY

DAVID T. DAY, Ph.D.
EDITOR-IN-CHIEF

In Collaboration with the Following Contributors and Associate Editors

FREDERICK G. CLAPP
FREDERICK H. LAHEE
LOUIS C. SANDS
ANNE B. COONS
FORREST M. TOWL
T. G. DELBRIDGE
H. C. COOPER
REGINALD G. SMITH
DAVID ELIOT DAY
A. D. SMITH
ROLAND B. DAY
W. N. BEST
ARTHUR H. GOLDINGHAM
JOHN D. GILL
FRANK NEWMAN SPELLER

IN TWO VOLUMES

VOLUME I

NEW YORK
JOHN WILEY & SONS, INC.
LONDON: CHAPMAN & HALL, LIMITED
1922

PRESS OF
BRAUNWORTH & CO.
BOOK MANUFACTURERS
BROOKLYN, N. Y.

PREFACE

For years the impending shortage in the supply of petroleum has been seen well enough by the petroleum producers and refiners; it has required, however, the experiences of the past few years to give general recognition to this condition. Repeated appeals have been made to petroleum investigators for greater energy in searching for new supplies, as well as for more careful utilization of the stores which have already been found.

Consultation with many engineers has shown that the utilization of the oil, as well as the development of new supplies, can best be aided, in so far as books are concerned, by presenting a handbook of the industry. Such a book is here offered. It has omitted all historical matter in so far as is consistent with clear presentation of present-day conditions. The authors have not, however, feared to point out advances which are likely, though still in the future. Wherever possible, information has been reduced to tabular form.

In the hope of profiting by all possible criticism of a constructive nature, the editor calls attention to the fact that this is the first edition.

In 1921, the oil-fields of the United States yielded 470 million (almost half a billion) barrels of crude petroleum. This was over 62 per cent of the world's production in that year. The amount held in storage in reserve tanks in this country is not one-third of a year's supply; the amount held in storage in the earth in the United States is reckoned at 5 to 7 billion barrels. The lower figure would mean total exhaustion in less than fifteen years at the present rate, and the higher one, total exhaustion in less than twenty years. Of course, the necessary producing conditions make it impossible for the production to come to an abrupt stop. The supply must dwindle away at a rate of decline which will be rapid at first and then very slow, as has been the case in the Pennsylvania field. The important consideration is to determine when the daily output will fall below necessary requirement. In 1921 the oil consumed amounted to 525 million barrels, and in 1920 more oil was consumed than was produced. If all refineries were operated to capacity the present production would not nearly suffice. The United States is already dependent upon a considerable contribution from foreign fields to fill the demand.

This condition of our resources has brought about another new feature of the industry, and this book signalizes a great change in the attitude of both the producer and the refiner. Formerly, both were in the habit of gaining commercial advantage by keeping their methods of work as secret as possible. Now that the necessity for more production and for more refining has become paramount, there is an obvious gain to every element of the industry in the contribution of information to that end. This has been evinced by the hearty response of authorities in each specialty to the request that they contribute their special knowledge to this volume, in the hope that more oil may be found and better utilization be given it.

The handbook is written for the public, but with especial reference to the engineers who produce and refine oil. First is given a summation of the conditions of occurrence of petroleum, including such a statement of the geological associations as will aid in finding new fields, as well as in bringing neglected fields into effective use. A chapter is devoted to the field methods used in finding oil and in measuring oil resources. This is followed by a statement of the methods of producing oil and the conditions of transportation and storage. The next chapter gives the statistics of petroleum produced, refined, exported and imported. A chapter on the characteristics of the crude oils of the world is followed by a chapter on methods of testing oil products according to most authoritative practice. In addition to the chapter on the occurrence of petroleum and natural gas, a chapter is devoted to the production of gasoline from natural gas, and another to the occurrence and utilization of asphalt. The solution of the problem of diminishing supply lies in the eventual extraction of oil from shales, bituminous sandstones and limestones, and from coals, which subject is next treated.

Undoubtedly the feature of the industry which most needs exposition is modern practice in refinery construction, to which considerable space has been devoted. The discussion of refining methods includes a description of the chief products for which refining is carried out, their characteristics and uses. A special chapter is devoted to the branch of refining called "cracking," because of its importance to the gasoline supply. Refining practice has produced certain cast-off products known generally as fuel oils. The low market price of these commodities has resulted in great abuse in the methods of burning them as fuels, and in the spirit of reform a chapter is devoted to a study of the engineering features involved in their more intelligent utilization. The outlook in this direction is especially towards their use in internal combustion engines, a subject which is brought up to date. The study of lubrication as practiced by mechanical engineers has reached a stage where a clear statement of the fundamental principles involved is needed. It has been given, therefore, with the expectation that it may simplify, to some extent, the use of lubricants. The use of pipes in the petroleum industry has until recently been mainly confined to the transfer of oils from place to place. Lately, however, pipes have filled a very much more serious function—that of tubular retorts and stills of many forms. Therefore a special section is given to their manufacture and treatment. It seems proper for any handbook to include such general tables of the properties of matter as refer especially to the main subject treated and to which ready reference must continually be had in working out the various engineering problems of the industry. In order that terminology in the petroleum industry may be more generally understood, and to the end that various terms, such as gasoline, may receive better definition, a glossary is appended.

The names of the contributors of chapters and subchapters are given at the head of the respective sections, but acknowledgment must be here made of the great service rendered by other authorities in making the handbook. Many of the tables useful to the petroleum industry have been found in "Kent," while others have been taken from the publications of the National Tube Company. The Bureau of Standards has published a table giving the relation of Baumé to specific gravity, calculated according to a very slightly different modulus from the customary Tagliabue tables. With heavy oils the differences between the two tables are negligible, but they are unfortunately so considerable with gasoline as to make it necessary for the oil fraternity to have ready reference to both sets of tables. The Bureau of Standards has also furnished many standardized methods of testing oil products, the use of which is gratefully acknowledged. The statistics of production of crude petroleum are taken from the published reports of the U. S. Geological Survey, which has developed an accuracy that is generally recognized as authoritative. The same is true of the refining statistics of the Bureau of Mines. Helpful use has also been made of the

Bureau's many publications intended as aids in efficiency in oil production. A great debt is hereby acknowledged to many oil companies who have contributed analyses of crude petroleum, field maps and refining data, as well as advice concerning information which the volume should contain. Still more highly appreciated is the enthusiastic service given by a host of helpers, among whom especial acknowledgment is offered to David Eliot Day, who, in addition to contributing the chapter on oil-shales, most helpfully served as associate editor. He was succeeded by Miss Helen Stone, who also edited and revised the illustrative material. Her devoted service has greatly increased the accuracy of the handbook and has advanced the date of publication. Miss Anne B. Coons has aided in many lines from the inception of this work, in addition to preparing the chapter on production statistics. Miss Altha T. Coons likewise has given general assistance on all chapters, as well as directing the compilation of the index. Grateful acknowledgment is given to the work of Roland B. Day, whose advice and inspiration have aided in shaping the entire work, as well as in selecting and arranging the tables. He also wrote the chapter on cracking processes. All of these have contributed largely without compensation. Grateful acknowledgment is also offered to the members of the regular staff, who have given loyal support to this enterprise.

THE EDITOR-IN-CHIEF.

CONTENTS

VOLUME I

PAGE

The Occurrence of Petroleum, by Frederick G. Clapp, Consulting Geologist and Petroleum Engineer, New York, N. Y.; formerly Geologist, U. S. Geological Survey; Chief Geologist, The Associated Petroleum Engineers.

Introduction 1
Essential Conditions for Oil Occurrence 1
Relation of Occurrence to Surface Indications 2
Occurrence and Metamorphism 9
Stratigraphy and Occurrence 11
Lithologic Character and Occurrence 41
Geological Structure 44
Relation of Movements of Oil to Occurrence 73
Geographic Occurrence of Petroleum 75

Field Methods in Petroleum Geology, by Frederic H. Lahee, Ph.D., Chief Geologist, the Sun Company, Dallas, Texas, and the Twin State Oil Company; formerly of the Geological Department, Harvard University, and Massachusetts Institute of Technology.

Geological Observations 167
Types of Geological Maps 171
Mapping 172
Instruments Used in Field Work 178
Methods Employed in Surface Mapping 182

Oil-field Development and Petroleum Production, by Louis C. Sands, Vice-President and General Manager, Oil Well Supply Company, Pittsburg, Pa.

Introduction 201
Methods or Systems of Drilling 201
- The Standard or Cable-tool System 203
- The Rotary System 251
- Fishing and Fishing Tools 274

Miscellaneous Factors in Oil-field Development 290
Tubular Goods for Oil-country Use 300
Methods of Producing Oil 309

Statistics of Petroleum and Natural Gas Production, by Anne B. Coons, Assistant Statistician, U. S. Geological Survey.

Sources of Information 321
World's Production of Petroleum 322
United States Production 328
Stocks 345
Deliveries 347

PAGE

Imports and Exports 349
Wells 351
Natural-gas Gasoline 353
Natural Gas 353

TRANSPORTATION OF PETROLEUM PRODUCTS, by Forrest M. Towl, Petroleum Engineer, New York, N. Y.; formerly Pipe-line Engineer, Standard Oil Company; Honorary Lecturer in Mechanical Engineering at Columbia University.

Introduction 359
Railroad Transportation 359
Boat Transportation, by N. J. Pluymert 370
The Tank Steamer 371
Ship Classification Rules 376
Pipe-line Transportation 379
Flow of Oil through Pipes 380
Pipe-line Computations 396
Pipe-line Construction 406
Pipe-line Operation 424
Gathering Oil Regulations in West Virginia 428
Transportation of Natural Gas, by Thomas R. Weymouth and Forrest M. Towl 432
Meters 432
Flow of Gas at Low Pressures and with Small Pressure Drop 446
Compression of Natural Gas 447
Pipe-line Formula Development 438
Problems with Single Line of Several Diameters of Pipe 442
Problems with Complex Line System in General 444
Effect of Position of Pipes of Different Diameters 445
Pipe-lines 452

CHARACTERISTICS OF PETROLEUM, by David T. Day, Ph.D., Petroleum Chemist, Washington, D. C.; formerly Chief of the Division of Mineral Resources, U. S. Geological Survey, and formerly Consulting Chemist, U. S. Bureau of Mines.

Composition 457
The Hydrocarbons in Petroleum 459
Oxygen Compounds 524
Sulphur 525
Nitrogen 528
Physical Properties 531
Analyses of Petroleum 538

PETROLEUM TESTING METHODS, by T. G. Delbridge, Ph.D., Chief Chemist, Atlantic Refining Company, Philadelphia, Pa.; formerly of the Chemical Department, Cornell University.

Introduction 587
Gravity 594
Flash and Fire Points 623
Viscosity 640
Cold Tests 657
Cloud Point 658
Pour Point 660

PAGE
Color 662
Gravimetric Determination of "B. S." 670
End-point Distillation of Gasoline, Naphtha and Kerosene 673
Modified Assay Distillation 678
Melting Point 679
"Doctor Test" for Gasoline, Naphtha and Illuminating Oil 689
Acid Heat Test 690
Floc Test for Illuminating Oils 691
Lamp-burning Tests 691
Emulsification Tests 695
The R. E. Test 696
The Conradson Steam Test 701
Herschel Demulsibility Test 701
Carbon Residue 704
Heat Test for Lubricating and Insulating Oils 705
Corrosion of Metals 706
Sulphur by Lamp Method 707
Sulphur by Bomb Method 709
Acidity of Petroleum Products 710
Fatty-oil Content 711
Gasoline Precipitation Test for Lubricating Oils 713
Tar Tests 714
Solubility 715
Evaporation Tests 716
Penetration of Asphalts 719
Asphalt Content 720
Ductility 721
Float Test 722
Fixed Carbon and Ash 723
Oil and Moisture in Paraffin Wax 724
Approximate Vapor Pressure of Gasoline 726
Conclusion 728

NATURAL-GAS GASOLINE, by H. C. Cooper, Chief Engineer, Hope Natural Gas Company, Clarksburg, W. Va.; In charge of recovering gasoline from natural gas by the absorption method.
Physical Chemistry of Natural Gas and Natural-gas Gasoline 729
Compression and Cooling Method for Separating Gasoline from Natural Gas 735
Absorption Method for Separating Gasoline from Natural Gas 738
Process of Separation of Gasoline Constituents from Natural Gas by Means of Solid Substances 744
Physical Characteristics of Natural Gas and Natural-gas Gasoline 751
Methods of Blending, Shipping, Storing and Transferring Gasoline 757
Testing Natural Gas for Gasoline Content 759
Construction of Gasoline Extraction Plants 764

ASPHALT, by Reginald G. Smith, Chemical Engineer, San Francisco, Calif., Engineer in Charge of the Production and Utilization of Asphalt in the Standard Oil Company of California.
Introduction 787
Classification 787
The Refining of Asphalt 787
Native Asphalts 790

PAGE

Production of Asphalt........ 795
Characteristics of Asphalt........ 798
Characteristics of Petroleum Asphalt........ 809
Packages........ 814
Uses of Asphalt........ 819

OIL SHALE, by David Eliot Day, Managing Engineer, The Day Company, Consulting Petroleum Engineers., San Francisco, Calif.

General........ 831
Occurrence........ 835
Commercial Considerations........ 849
Mining Methods........ 854
Crushing........ 857
Retorting Methods........ 861
Refining........ 870
The Known Shale Areas of the United States—Occurrence, Geology and Characteristics........ 871

INDEX........ 895

THE OCCURRENCE OF PETROLEUM

BY

FREDERICK G. CLAPP[1]

INTRODUCTION

Wherever petroleum occurs in nature, natural gas is generally found in the same locality or region. Gas, however, is less restricted in its distribution than is oil, and numerous gas pools exist where no petroleum occurs in close proximity. Because of the relationship of natural gas to petroleum, the two substances must, to a considerable extent, be considered together. Both petroleum and natural gas are among the great industrial resources of the world and the countries which possess them in large quantities are exceedingly fortunate. The fact that these substances have not yet been developed in a particular locality is rarely proof of their absence, as no year passes without the discovery of new and often unsuspected fields or pools.

The occurrence of oil and gas is widespread, no stratified rock being so young, and scarcely any so old, as absolutely to preclude their existence. Consequently, any country in which geological conditions are favorable is warranted in exerting every effort and undertaking great expense to investigate and determine its resources in petroleum and natural gas.

In this outline of the occurrence of petroleum the numerous references made use of are given credit where possible. In addition to those specifically mentioned there are certain general sources of information, such as the report of S. F. Peckham entitled, "The Production, Technology and Uses of Petroleum and its Products," in volume 10 of the Tenth Census of the United States, 1880, pages 1–319; "A Treatise on Petroleum" by Sir Boverton Redwood, London, 1913, three volumes; "Petroleum Mining and Oil Field Development," by Arthur Beeby Thompson, London, 1916, 648 pages; the annual statistical reports in the Mineral Resources series of the United States Geological Survey, by David T. Day, S. H. Shotwell, Joseph D. Weeks, F. H. Oliphant, W. T. Griswold and John D. Northrop, and many other Geological Survey references of a general nature.

In this chapter the word oil is frequently used as being sufficiently synonymous with petroleum for the subject matter in hand.

ESSENTIAL CONDITIONS FOR PETROLEUM OCCURRENCE

The old statement that the requisites for occurrence of petroleum are substantially (*a*) a porous rock to contain the oil, (*b*) an impervious cover to prevent it from escaping, (*c*) geological structure to permit the accumulation of oil from relatively large areas into restricted area, and (*d*) a source from which oil has been derived, has become somewhat

[1] Petroleum Engineer, 30 Church St., New York, N. Y.

time-worn; yet to the layman this expresses the essential conditions about as clearly as possible. It should be further understood, however, that many parts of the world are unsuitable for oil occurrence for a variety of reasons. For instance, no petroleum will be found in the immense regions occupied at the surface by rocks of Archean or Algonkian age or in those of later periods which are entirely of an igneous or intensely metamorphosed character. Thus perhaps half of the earth's surface is ruled out at the start. We must also understand that metamorphism has removed oil from great additional areas of stratified rocks which, by a casual observer, are frequently thought suitable for its occurrence; but since the exact line of demarcation between the too highly metamorphosed areas and those in which petroleum may still remain can be learned only by a close study of advanced petroleum geology, one must leave this phase of the subject with a much briefer analysis than it really warrants.

In an effort to delineate the areas in which petroleum may occur, Woodruff [1] has recently prepared a paper on "Petroliferous Provinces." While other geologists do not absolutely concur in his mapping it outlines the essential particulars, viz., that there are large areas of the earth's surface in which the presence of oil is impossible, and still others where its occurrence is improbable, owing to the absence of some condition essential to oil accumulation.

Various surface indications of the occurrence of petroleum are found throughout the world, in the form of oil and gas exudations, mud volcanoes, bituminous dikes and even lakes of asphalt; but surface indications are not essential, and in the last analysis oil occurrence is found to be independent of the existence or non-existence of surface indications, even though these may lead to its discovery in some regions.

The conditions essential to the occurrence of petroleum may now be stated in the order of their importance in the following revised form:

Conditions Essential to Occurrence of Oil

A. Presence of rocks of sedimentary origin, or in a few instances of those in proximity to rocks of sedimentary origin.
B. Absence of intense metamorphism.
C. Presence of sandstones, limestones, sands or other strata sufficiently porous to hold oil.
D. Some source from which the oil may have been formed.
E. Such water conditions as do not prohibit the accumulation of oil in pools.
F. Suitable cover to prevent the oil from seeping away or being pushed to the surface of the earth by underground waters.
G. Suitable structure or attitude of the strata to cause oil to be collected locally into pools, with the assistance of such other factors as water, gravitation, rock pressure, etc.

The occurrence of oil will be described under the following headings:

A. Relation to surface indications.
B. Relation to metamorphism.
C. Relation to stratigraphy.
D. Relation to lithologic character of the strata.
E. Relation to structure or attitude of the strata.
F. Relation to water.
G. Relation of movements of oil to its occurrence.
H. Geographic occurrence.

[1] Woodruff, E. G.: Bull. 150, Am. Inst. Min. and Met. Engrs., 1919, p. 907, et seq.

RELATION OF PETROLEUM OCCURRENCE TO SURFACE INDICATIONS

Classification of surface indications.—While the evidences of oil occurrence which are commonly noted by the geologist or petroleum engineer are not such as are ordinarily seen and comprehended by the layman, there are, however, in most oil-fields certain evideuces which have a definite or relative bearing on the existence of oil, either in that particular locality or at a distance. The principal so-called "surface evidences" may be classified as follows:

(1) Oil seepages or oil springs.
(2) Natural gas springs.
(3) Outcrops of sandstones or limestones impregnated with petroleum or bitumen.
(4) Bituminous lakes or other bituminous seepages.
(5) Bituminous dikes.
(6) Mud volcanoes.
(7) Burnt clays.
(8) Occurrence of salt.
(9) Occurrence of sulphur.

Any one of these frequently has some association with oil, but oil may occur at a great distance from the point where the evidences appear on the surface. To illustrate, it is a fact that a formation which reaches the surface of the ground at a particular point, and from which seepages of oil, gas or asphalt are seen to emerge, may descend at such an angle that the structurally favorable locality for the accumulation of oil lies many miles from the exposed outcrop. For this reason, it is generally poor policy—based on a lack of geological knowledge—to drill on or near seepages unless evidence exists that the main deposit of oil occurs directly below.

1. Oil seepages or oil springs.—Seepages of oil may emerge in either one of two ways: first, where the outcrop of an oil-bearing bed reaches the surface; and, second, where there is a crevice or fault through which the oil has risen to the surface from some depth. Seepages are commonly found in lowlands, in swamps, or along small streams; sometimes appearing merely as iridescent scums on the water, but one case has been observed where oil and asphalt run down the side of a small basaltic hill from crevices 100 feet or more above its base. As a rule the upper part of an outcrop of oil sand will be found to have lost its signs of oil owing to weathering; but sometimes oil will still be found in that stratum at water level.

It is very important for the novice to learn to distinguish between scums of oil and other scums which sometimes appear on surface-water. For instance, iron scums have, times without number, been mistaken for oil. On the strength of such supposed evidence, persons have paid thousands of dollars for expert examination of territory, only to learn upon the arrival of the expert, that the scum was not oil, and sometimes that the geological conditions of the locality were absolutely unsuitable for the occurrence of oil.

As a rule a film of oil can be distinguished from a film of iron by its odor. A person experienced in the business will always be able to distinguish between the two substances; and often a novice can distinguish between them by remembering that a drop of oil on a water surface will expand as a thin film enlarging about a common center and giving a beautiful concentric iridescence. A small stick thrust into iron scum will break it into separate patches; while an oil scum is so thin and cohesive that it will still retain its oily and unbroken appearance after being thus disturbed.

As a rule, oil seepages exist as very faint scums of oil on the surface of rivers or lakes. This was true in certain of the Pennsylvania, Ohio and West Virginia oil-fields at the time of their discovery. In other cases, as in Mexico, the tarry oil emerges from the earth in very large quantities. In certain Biblical localities it is said to have gushed out in great streams. Rivers of oil are reported to have emerged in past times, from beneath the Caspian Sea, and natural explosions of burning oil are said to have taken place, throwing clay and stones into the air, uplifting portions of the sea bottom and giving rise to small islands.[1]

In seepages of considerable size the green or black oil occurs in drops or patches instead of in films. These patches are frequently associated with tarry or asphaltic substances, and gas sometimes bubbles through the water at the same point. While oil seepages are generally very small, an instance is mentioned by Craig,[2] where as much as 20 barrels of oil per day flow into a stream, from a certain outcrop. A few cases are known where oil has risen through the waters of the sea and appeared on the surface, such instances being reported in the Gulf of Mexico, the Caspian Sea, and off the coasts of South America and Trinidad.

The student will naturally ask why oil seepages are so seldom known in many productive oil-fields, such as Pennsylvania, Ohio, West Virginia, Oklahoma, and Illinois. The reason is that the beds are so slightly tilted as to have remained unbroken, and there has not been enough fissuring to permit the oil to reach the surface. Moreover, where the petroleum formations actually reach the surface in a distant locality, the oil has frequently leaked away, owing to the fact that the productive strata in the abovementioned states are of Carboniferous and earlier ages, and the oil has had plenty of time to disappear. In California, Trinidad and Mexico, however, where seepages are so abundant, the predominant formations are of more recent age, so that even though the sands are tilted and eroded at the surface, the oil has not yet entirely disappeared. A similar statement may be made in respect to natural gas exudations.

In 1915 a classification of seepages or oil springs was proposed by DeGolyer,[3] as follows:

I. Seepages associated with igneous intrusions.
   (a) At contact zones of volcanic plugs and sedimentary rocks.
   (b) At contact zones of dikes and sedimentary rocks.
   (c) Through cracks and fissures in the igneous rock itself.
   (d) As intrusions in the igneous rock.
   (e) From metamorphosed rock above an intrusion which does not outcrop.
II. Seepages not associated with intrusions.
   (a) At crest of domes or anticlines.
   (b) Along marked fault or fissure planes.
   (c) From steeply dipping strata.
   (d) Isolated occurrences of uncertain relationship.

2. Natural gas springs.—In some places in Ohio and West Virginia, bubbles of gas rise in minute quantities through the water to the surface of streams. In other localities, like Baku on the Apsheron Peninsula in Russia, gas has been actually burning for thousands of years, the exudations having been frequented by a peculiar sect of people known as "Fire Worshipers." Gas springs, while not constituting actual proof of the existence of oil in a locality, show that conditions are favorable, for where natural gas is found, we may expect petroleum also within a lateral range of the same formations.

3. Outcrops of sands, sandstones or limestones impregnated with tar or bitumen.—Outcrops of "sands" impregnated with tar or bitumen are not common, but they exist

[1] de Lapparent, A.: Traité de Geologie, 1883, p. 490.
[2] Cunningham-Craig, E. H.: Oil Finding, London, 1912, p. 90.
[3] DeGolyer, E.: Econ. Geol. Vol. X, 1915, p. 654.

in some parts of the world. Perhaps the best known outcrops of this kind are the "Tar sands" of the Athabasca and other rivers in northern Alberta in Canada. Here the Dakota formation is impregnated with tar for many scores of miles along the rivers, leading to the supposition that oil and gas will be found in great quantity in those portions of the sands which are under great cover and which have the requisite structure.

The outcrop of an oil sand is seldom visibly petroliferous, on account of the oxidation and evaporation that is continually taking place. In the case of oils having an asphalt base, however, the outcrop may be distinctly bituminous or may have a brownish residue. Sometimes enough rock can be broken out with the aid of a pick to show that the formation is oily and of a lighter color. In other cases, where oil sands outcrop along

Fig. 1.—Surface of the "Pitch Lake," in Trinidad, W. I., one of the world's largest seepages.

the coast and are cut by the waves they are still soft and sticky. Limestones frequently contain little globules of oil, and if oil be present in the vicinity of the outcrop, its presence can be detected by the smell after the rock has been broken with a hammer.

4. Bituminous lakes or other bituminous seepages.—Asphalt and related bitumens may occur as dikes or veins in the rock, or they may appear at the surface in the form of lakes or rivers. In Mexico, there are large numbers of seepages, some of small size, and many others covering thousands of square feet. These seepages, frequently called "pitch" or "chapapote," also occur extensively in California, Venezuela, Trinidad, West Africa, Borneo, Russia, etc. The bitumen is generally in the form of pools or lakes; but sometimes, according to Bosworth,[1] it forms cones some of which are 50 feet

[1] Bosworth, T. C.: Pet. Review, Apr. 6, 1912, p. 204.

high. The best known example of a lake of asphalt is the Pitch Lake of Trinidad, frequently described in literature.

5. Bituminous dikes.—In the case of dikes of asphalt and other bitumens, the relationship is not so apparent, since these substances are frequently solidified, and some varieties are as hard and compact as coal. These have been so deceptive, that in one case, at least, namely, in the Albertite Mines in Albert County, New Brunswick, the courts have decided that the material shall be legally known as coal. In reality, however, it is an entirely different bitumen. The grahamite dikes of the Ritchie Mines, West Virginia, and those of Oklahoma are noteworthy examples.

Fig 1a.—An oil seepage at Saladero, Mexico.

6. Mud volcanoes.—A mud volcano is an outburst of water, sometimes heated, discharging mud into the air, occasionally with a rumbling noise, and sometimes forming a conspicuous cone many feet in height. The value of mud volcanoes as an evidence of petroleum has been frequently questioned, and perhaps not yet adequately established, as they occur in parts of the world where petroleum is unknown. It may be said that they certainly constitute an evidence of gas and that this fact leads us to infer the association of petroleum also.

Sir Boverton Redwood states that "in Russia and in India the relationship between petroleum and mud volcanoes is very noticeable"; and he goes on to say that mud volcanoes are generally considered as a favorable indication. Mud volcanoes are found in the vicinity of the oil-fields of the Apsheron, Crimean and Taman Peninsulas of Russia, in Venezuela and Colombia, at Minbu and on the Island of Ramri in Burma, and also to some extent in Rumania and Poland. At Baku on the Apsheron Peninsula, they have been known for centuries, and have been associated with large quantities of natural gas and petroleum. The height of the mud volcanoes near Baku is reported to be as much as 1300 feet in some places, while those of Transylvania in Rumania are seldom over 30 feet in height.

Natural mud volcanoes are not known to exist in the United States, but certain gas wells of great volume in the Caddo field of Louisiana and in some parts of Oklahoma and Texas, where the formations are comparatively soft and of a clayey texture, have broken loose, and run wild for years, throwing water and mud into the air. It is easy to conjecture that these wells may be of similar internal structure to natural mud volcanoes but they are evidences of gas and not of oil. It is probable that mud volcanoes are not necessarily associated either with naural gas or oil in quantity, but that they may occur

Fig. 2.—Gas bubbling up through water, Egbell (Gbèly), Czecho-Slovakia. Petroleum has been discovered near by.

even where only small amounts of any kind of gas are imprisoned in or below muddy strata.

7. Burnt clays.—In a few places, as in Trinidad and California, there are beds of shale which reach a thickness of 10 feet, and which have been burnt into hard red rock resembling brick, by the natural combustion of the bituminous material in or associated with them.

8. Presence of **salt water or massive** salt **deposits**—Salt is frequently associated with petroleum, as crystalline masses, as salt water or as saliferous strata. Water which accompanies petroleum is generally more saline than that normally present in sedimentary rocks. On the other hand, extensive masses of salt often exist entirely unassociated with oil, this being the case in Poland, England, Kansas, Ohio, and Ontario. Consequently, recent theories of the origin of oil have in the main ignored any relationship with salt as a factor in the process, assuming first that brine, where found in an oil-field during drilling, is a normal constituent of deeply buried formations of marine origin. In many localities in Ohio, Louisiana and Rumania, oil is associated with salt in such a manner that there appears to be a direct relationship, as German, Hungarian and Rumanian geologists and chemists have assumed. Perhaps the idea that salt and oil are necessarily associated is derived in part from the fact that the latter was originally discovered in the United States during the search for brine, which was at one time obtained from the wells in Pennsylvania and West Virginia. While the connection of oil with dome-shaped masses of salt, as in Texas and Louisiana, may be due merely to the fact that oil is concentrated against the salt mass as a suitable structure, nevertheless the frequent association of the two substances seems to indicate some possible relationship in origin.

FIG. 3.—A mud volcano, Buzd, Transylvania, Rumania.

9. Occurrence of sulphur.—In Louisiana and Texas, oil is associated with sulphur in more than one instance, and calcium sulphate or gypsum is a common mineral in certain oil fields, leading geologists to infer some relationship between oil and sulphur. Many persons have supposed that the occurrence of hydrogen sulphide is favorable to petroleum, but there seems to be no certainty of this.

Summary.—Summarizing the foregoing statements, it must be acknowledged that surface indications are not essential to the existence of an oil field, since many fields show an almost entire lack of surface indications, petroleum having been discovered only when tapped accidentally by the drill or located on favorable geologic structure. We can say, however, that where petroleum seepages exist, they indicate oil below the surface in greater or smaller quantity. Similarly, since it is known that asphaltic deposits are derived from petroleum, oil must have existed at some time in localities where such deposits exist and the geologist must decide whether or not the conditions are such that petroleum is still to be expected there. In many cases where asphaltic deposits are very old geologically, the petroleum seems to have completely disappeared. Where natural gas exists without petroleum, there is no positive evidence of the latter; but since both are believed to be derived from a common source, it is important to look for the geological evidence of petroleum somewhere in the gas-producing region.

Asphaltic deposits constitute a direct evidence that oil does exist or has existed in the vicinity, for these substances are the desiccated remains of heavy oils which have emerged from the earth in the past. In many cases it is possible for a geologist to locate the field from which the asphalt has escaped. Consequently, it is important not to neglect the surface evidences; but, in studying them, one must remember that certain structural relations hold true; and consequently that oil pools will very seldom be found directly underneath the points of emergence of the substances.

It is of course necessary for the expert to take account of all such circumstances as may prove the general region favorable; but he must be careful not to give undue weight to the chances of finding oil in the particular locality where a "surface evidence" is seen. As an additional surface indication, we may mention the great deposits of bituminous shale which exist in some parts of the world and which, although in most cases not yet commercially workable, prove that petroleum does exist in greater or smaller quantities.

Limited value of "oil lines."—One of the greatest mistakes made throughout the early development of the petroleum industry in the United States was the supposition on the part of a majority of prospectors that the fields run universally in a given direction. With that idea in mind, the pioneers drilled thousands of wells on lines of various bearings extending for miles through the territory. Underlying their belief was the fact that the general trend of the Appalachian fields, and certain other fields is from northeast to southwest. Consequently, by drilling in a direction parallel to the mountains one is somewhat more likely to strike petroleum than by drilling toward the southeast or northwest; but, on the other hand, the above statement in regard to the trend of the fields is only true in the very broad sense, and an individual pool is just as likely to disappear toward the southwest as to be continued. The only class of lines which are of real value in petroleum geology are the "structure contour lines" which are drawn by the geologist upon his maps and which, when extended from an initial well, will furnish some clue to the probable direction of extension of the pool.

RELATION OF PETROLEUM OCCURRENCE TO METAMORPHISM

Since oil appears to be a product of sedimentary conditions or of secondary causes arising in sedimentary rocks, it is not commonly found in rocks of igneous or metamorphic origin. As has been stated, we can rule out from possible oil-bearing territory practically all those portions of the world underlain by rocks of Archean and Algonkian age, which are igneous and in which oil never existed or are intensely metamorphosed rocks from which all oil has been removed in past ages by natural distillation.

Oil has been removed by natural means not only from such highly metamorphosed rocks, but also from those of even moderate metamorphism. For instance, the strata of the greater part of the Appalachian Mountain system, and other mountain masses formed previous to Mesozoic time, are folded, faulted and broken, and to an oil geologist are obviously unfavorable for oil, the latter having been removed during the mountain-making process. Moreover, there are no rocks known to contain oil in commercial quantities within about 100 miles to the west of the Allegheny Front in Pennsylvania, though some rocks within this limit contain natural gas. Similar conditions prevail in other mountain regions where the folding on a broad scale dates from pre-Mesozoic time, though in the case of more recent strata the time has not generally been sufficient to remove the oil, even in regions of high folding.

Until a few years ago, geologists were unable to explain the absence of oil from the regions adjacent to some of the principal mountain systems of the world; and no guide had been discovered by which we could delimit the regions where oil may be expected from those where it does not exist. In 1915, however, White [1] published a paper showing that in regions where coal-beds exist, the percentage of fixed carbon decreases in a regular

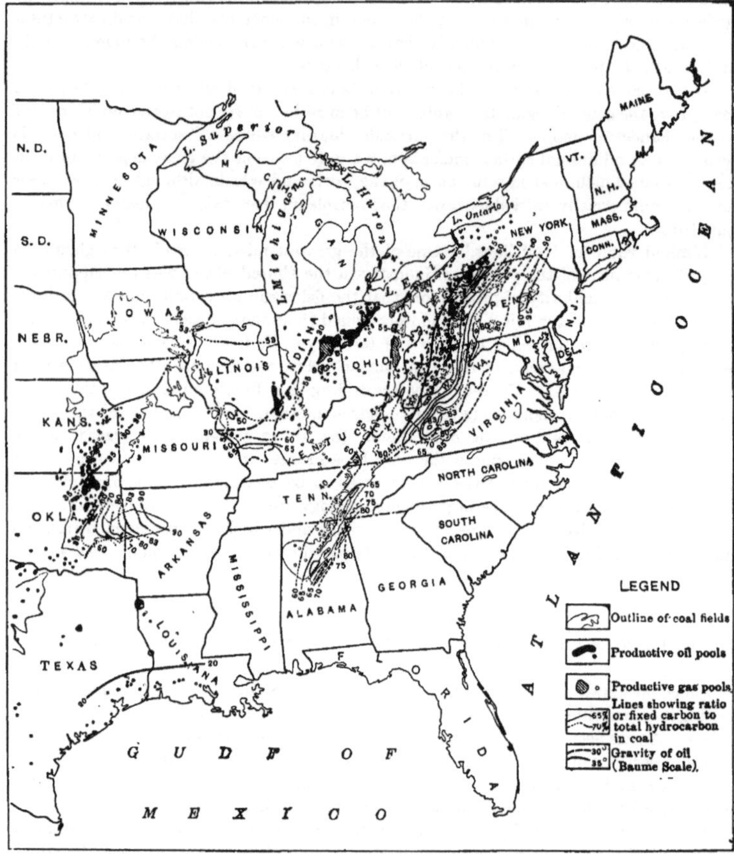

FIG. 4. Map showing relation between regional metamorphism of coals and that of petroleums in Eastern United States. After David White.

manner from the anthracite regions or regions of highly bituminous coals to the softer bituminous coal regions, and that in traveling from soft to harder coal regions this ratio passes a line beyond which no commercial oil fields may be expected. Fig. 4 shows the isovols or lines of equal volatile matter, as they were worked out for a portion of the United States. This branch of the science is still in its infancy, though it has been used

[1] White, David: Some relations in origin between coal and petroleum. Jour. Wash. Acad. Sci., Vol. V, No. 6, 1915, pp. 189–212.

to great advantage by The Associated Petroleum Engineers, the geologists of the Sun Company [1] and others,[2] who have prepared isovol maps of several states. There is no doubt that they can be similarly worked out for all countries where bituminous coal occurs. The percentage of impurities must be taken into account.

Where isovols or carbon ratios have been studied and used it has been found that oil fields of commercial importance seldom exist beyond the 65 per cent line and that commercial gas fields seldom exist beyond the 70 per cent line. The carbon ratio may also vary with increasing depth in any locality; but this variation has been less studied than that of the surface rocks.

RELATION OF PETROLEUM OCCURRENCE TO STRATIGRAPHY

Although petroleum occurs in greater or smaller quantities in formations of nearly all geologic ages, it has been found in commercial quantity only in those of the Cambrian and later ages. The most important fields have been found in formations of the Ordovician, Silurian, Devonian, Carboniferous, Cretaceous and Tertiary systems. Tertiary rocks are believed to yield over 50 per cent of the world's production of oil.

Considering fields throughout the world, the distribution of oil in the various geologic systems may be represented in part as follows:

Stratigraphic Occurrence of Petroleum in Various Parts of the World.

Period	System	Country
Cenozoic	Quaternary	United States
	Tertiary	United States, Mexico, Trinidad, Russia, Rumania, Poland, Ukraine, East Indies, Peru, Japan, Italy, Colombia, Venezuela, Germany, Persia
Mesozoic	Cretaceous	United States, Canada, Mexico, Persia, Poland, Ukraine, Colombia, Venezuela, Ecuador, Argentina
	Jurassic	United States, Germany, Argentina
	Triassic	United States
Paleozoic	Carboniferous	United States, Canada, China, England
	Devonian	United States, Canada
	Silurian	United States, Canada
	Ordovician	United States, Canada
	Cambrian	United States, Canada

In North America the principal occurrences of petroleum by geologic series are as follows:

[1] Fuller, M. L.: Relation of oil to carbon ratios of Pennsylvanian coals in North Texas, Econ. Geol., Vol. XIV, No. 7, 1919, pp. 536–542; Carbon ratios in carboniferous coals of Oklahoma, and their relation to petroleum. Econ. Geo., Vol. XV, No. 3, 1920, pp. 225–235.

[2] Semmes, D. R.: Mining and Met., Bull. No. 159, Sec. 5, Am. Inst. Min. and Met. Engrs., Mch. 1920, p. 2.

Productive Strata of Petroleum and Natural Gas in North America

(NOTE.—s. = sand (drillers' usage), S. = sandstone, L. = limestone, F. = formation)

Period	System	Series or other subdivisions		Mexico	California	Montana	Wyoming	Texas	Louisiana	Arkansas	Oklahoma	Kansas	Western Canada
Cenozoic	Quaternary	Pleistocene			Alluvium			Alluvium	Alluvian				
	Tertiary	Pline series			Fernando F.								
		Miocene series			Various formations								
		Oligocene series			Sespe F.								
		Eocene series			Tejon F.								
Mesozoic	Cretaceous	Laramie series			Chico F.								
		Upper Cretaceous	Montana			Eagle S.	Shannon s.	Nacatoch s.	Nacatoch s.				Virgelle S.
			Colorado			(Frontier group of sands)	Frontier group of sands	Wood-bine s.	Selma chalk Blossom s. Woodbine s.	Eldorado s. Blossom s. Woodbine s. Dixie s.			
		Comanche or Lower Cretaceous		Tamasopo L.		Kootenai s.	Dakota S.	Edwards L.			Trinity s.		Dakota S.
	Jurassic						Sundance F.						
	Triassic						Chugwater F.						

Paleozoic	Carboniferous	Permian series			Embar S.	Embar S.	Several sands			Many sands		
		Pennsylvanian series			Tensleep S. and Amsden F.	Tensleep S.	Several sands		Various sands (gas only)	Many sands	Many sands	
		Mississippian series					Bend lime-stone			Missis-sippi s. and L.	Missis-sippi s. and L.	
	Devonian	Chemung and Catskill								Wilcox s.		Macken-zi valley sands
		Corniferous or Onondaga										
	Silurian	Salina										
		Niagara										
		Clinton										
		Medina										
	Ordovician	Trenton										
	Cambrian	Potsdam										

Productive Strata of Petroleum and Natural Gas in North America—Continued

(NOTE.—s. = sand, S. = sandstone, L. = limestone, F. = formation)

Period	System	Series or Other subdivisions		Kentucky	Illinois	Indiana	Ohio	West Virginia	Pennsylvania	New York	Ontario	Colorado	Quebec	New Brunswick
Cenozoic	Quaternary	[illegible]												
		Pliocene series												
		Miocene series												
	Tertiary	Oligocene series												
		Eocene series												
Mesozoic	Cretaceous	Laramie series												
		Upper Cretaceous	Montana									Pierre F.		
			Colorado									Mancos shale		
		Comanche or Lower Cretaceous												
	Jurassic													
	Triassic													

Paleozoic	Carboniferous	Permian series											
		Pennsylvanian series	Several sands			Several sands	Several sands	Dunkard s. "Gas s." Salt s.					
		Mississippian series	Several sands	Several sands	Several sands	Big Lime (Maxville L.) Keener s. Big Injun s. Berea s.	Big Lime (Greenbrier L.) Keener s. Big Injun s. Berea s.	Big Injun and Murraysville sands					Several sands
	Devonian	Chemung and Catskill					Many sands	Many sands					
		Corniferous or Onondaga	Several sands		Corniferous L.				Corniferous L.	Corniferous L.			
	Silurian	Salina											
		Niagara	Niagara L.			Newburg s.			Guelph L.	Guelph L.			
		Clinton								Clinton L.			
		Medina				"Clinton sand"			Medina red s. Medina white s.	Medina red s. Medina white s.			
	Ordovician	Trenton	Trenton L.	Trenton L.	Trenton L.	Trenton L.			Trenton L.	Trenton L.			
	Cambrian[1]	Potsdam							Potsdam S.	Potsdam S.		Quebec group	

[1] The Cambrian production is everywhere small and unimportant.

In a strictly geological treatise, the oil-bearing formations enumerated in the preceding tables should be considered from below upwards; but from the standpoint of the operators in any particular field it is simpler to have the uppermost strata considered first. For this reason the usual geological procedure is reversed. No attempt will be made in the following stratigraphic outlines to include oil-fields outside of North America; and the reader is referred to the pages describing geographic occurrence for such information as has been abstracted relative to the stratigraphy of foreign fields.

Quaternary System

Although petroleum and natural gas are found in formations of probable Quaternary age in Louisiana and Texas, the most important Quaternary localities, from the standpoint of oil indications, are in California. In some places the Quaternary deposits of that state are 1000 feet in thickness, although the majority are not more than 100 feet thick. The San Pedro formation of this age in Los Angeles and Ventura Counties is extremely rich in fossils, mostly of a type still living in the warmer waters of southern California; and a great asphalt seepage on Rancho la Brea, in the Salt Lake field near Los Angeles, contains one of the most important deposits of extinct vertebrate fossils in the world, including varieties from the fox to the mastodon, as well as birds and insects, all of which have been caught in past ages in the liquid asphalt which emerges from the ground.

Tertiary System

As already stated, the most important geologic system, from the standpoint of the occurrence of petroleum throughout the world, is the Tertiary; but there are comparatively few parts of the United States where the Tertiary is productive. In the California fields, however, Tertiary strata yield immense quantities of petroleum.

Pliocene and Miocene Series

Fernando formation.—While many of the most important Pliocene sources of petroleum are in foreign countries, the California fields contain a great group of strata, 3000 to 10,000 feet in thickness, extending downward from the Quaternary into the Miocene. This is known as the Fernando formation, and its basal portion contains the sands from which the most important fields of the state derive their oil. The Fernando is unconformable on the underlying Monterey shale, which is believed to be the principal source of oil in the state; and the lower part of the Fernando includes the horizon of the McKittrick and underlying Santa Margarita formations, from which great quantities of oil are derived. The Fernando formation consists of conglomerates, sands, shales and clays. Some of the fields drawing oil from it are the McKittrick, Belridge, Lost Hills, Midway, Sunset, Kern River, Summerland, and Whittier. The greater part of the oil produced from the Fernando formation is a fuel-oil grade of 11° to 19° Bé., though some exceptions are known.

Monterey formation.—Underlying the Fernando and Santa Margarita formations of Upper Miocene age in California is the Monterey shale, a deposit sometimes 7000 feet in thickness, believed by California geologists to constitute the source of most of the oil produced in that state, even when the oil is actually drawn from sands in overlying formations. The principal characteristic of the Monterey is its content of diatoms, on account of which it is known as a diatomaceous shale. Diatoms are fossil microscopic plants which are found in this shale in immense quantities. The upper part of the Monterey is productive of oil in the Santa Clara Vallev, and in the Whittier, Fullerton and Puente Hills fields. The upper portions of the shale formation contain limestone and sandstone, and there are frequent flinty layers near the base. The oil of the Santa Maria and Lompoc fields is derived from this lower section, where it occurs in interstices and crevices in fractured flinty shale and limestone.

Vaqueros formation.—Underlying the Monterey shale in California, and overlying the Sespe, is a formation of extreme Lower Miocene age which is likewise petroliferous. The Vaqueros formation is continuous from north to south throughout the state, but ranges from 200 to 5000 feet in thickness. It constitutes one of the important oil-bearing formations of California, yielding the principal production of the Coalinga field, and it is an important producer in the Puente Hills and Santa Clara fields. The oil-bearing beds at Coalinga are prominent sands from 30 to 100 feet in thickness, near the base of the Vaqueros formation, while in the Puente Hills and the Santa Clara Valley the oil occurs in sands and associated beds close to the top of the Vaqueros and hence directly beneath the Monterey.

Oligocene Series

Sespe formation.—In other parts of the world a number of oil fields exist in the Oligocene series. In this country such fields are mainly confined to California, where the Sespe formation of this age varies from a few hundred to over 4000 feet in thickness. The Sespe is characterized by reddish-brown or green sand or sands, and lies directly beneath the Vaqueros formation. At several places in California it contains commercial deposits of oil, both in its upper and lower portions. The oil occurs in hard sandstones and shales and sometimes in well-developed sands. Most of the Sespe oil is of high quality, ranging from 25° to 36° Bé.

Eocene Series

Tejon formation.—Notwithstanding the fact that the Tejon formation, which is 5000 feet thick in some places, is mentioned as one of the producing formations of California, it is an unimportant producer. It carries diatomaceous shales; and since some oil in commercial quantities has been obtained from it, the California geologists expect it to be of future value as a producer.

Cretaceous System

Upper Cretaceous Series

Chico formation.—At the top of the Upper Cretaceous series in California is a formation known as the Chico, which sometimes attains a thickness of 6000 feet, and consists of purple petroliferous shales, coarse concretionary sandstones (its most characteristic phase) and, at its base, coarse conglomerate. The formation is occasionally productive of oil.

Shannon sandstone.—With the possible exception of some of the sands of the Chico formation of California, the stratigraphically highest important productive oil sand in the United States is the Shannon sandstone, lying in and near the base of the Pierre formation of the Montana group. The Shannon wells are situated at the north end of the Salt Creek field in Natrona County, Wyoming. The daily output of each well was from 5 to 15 barrels of oil. The Shannon sand lies about 2000 feet above the Wall Creek sand, and consists of two members, separated by about 30 feet of shale, the upper member being about 40 feet in thickness and the lower one 50 feet.

Wall Creek sandstone.—One of the most important sources of production in Wyoming is the Wall Creek sandstone, which constitutes one of the Frontier group of sands, and lies near the top of the Benton shale in the middle of the Colorado group. It varies in color from a dirty gray to white, and is sometimes considerably over 100 feet in thickness, frequently containing layers of sandy shale a fraction of an inch, or more, in thickness. This sand, although widely productive in Wyoming, is best known from its great wells in Salt Creek field of Natrona County. It is probably equivalent to the Torchlight sandstone of northwestern Wyoming.

Second or Lower Wall Creek sandstone.—The Lower Wall Creek sand was recognized in wells in 1915, on the western border of the Salt Creek field in Wyoming, although it had previously been known to geologists from its outcrop. It has since been found widely productive in that state. This sand, like the Wall Creek sand, is a member of

Fig. 5.—Cliff of Wall Creek sandstone on Poison Spider Creek, Natrona County, Wyoming.

the Frontier group of sands in the Benton shale. In character it is a massive sandstone, similar to the Wall Creek, but considerably less thick, and frequently distinguished by a dirty brownish color on its outcrop. This sand may be equivalent to the Peay sandstone, which is productive in northwestern Wyoming.

Nacatoch sand.—Below the Arkadelphia clay, which is 300 to 400 feet in thickness and lies at the top of the Upper Cretaceous or Gulf series in northwestern Louisiana, lies a persistent sand, 50 to 200 feet thick, known geologically as the Nacatoch, and to well-drillers as the "Caddo," "Vivian" or "Shreveport" gas sand. It supplies great quantities of natural gas throughout the northwestern Louisiana fields, and a portion of the oil of the Homer pool in Claiborne Parish. The Nacatoch consists of fine-grained, light gray to greenish sand, with some layers of hard sandstone and some of clay, and it is locally calcareous. Scattered grains of glauconite are common in it, giving it a greenish color. The sand is very porous, so that where it is not a gas- or oil-bearing sand it is generally filled with a great volume of salt water.

Annona chalk.—Throughout Louisiana and eastern Texas extends a great chalk or chalky formation, of the Austin group of Upper Cretaceous age, known in Louisiana as the Annona chalk and in Alabama as the Selma chalk, and constituting one of the best-known formations in that part of the country. It grades upward into the Marlbrook marl and downward into the Brownstone marl; and although in places it may be a great chalk formation 1000 feet in thickness, in others it is largely composed of shale or "gumbo." At a rough distance of 150 feet above the Annona, is a bed of chalk 20 feet in thickness known as the "Saratoga chalk." This appears very persistently in

Fig. 6.—Dakota sandstone with high dip, outcropping near crest of Oil Mountain, Natrona County, Wyoming.

well logs from Texas to Florida. Although the Annona is not, strictly speaking, an oil sand, some wells of great volume, though generally of short life, have been struck in it in the northwestern Louisiana fields. The Annona is an important key rock in the Gulf fields, but must be carefully distinguished from the Saratoga chalk above mentioned.

Blossom sand.—Lying at the top of the Eagle Ford clay and 200 to 400 feet above the Woodbine sand (described below), in northwestern Louisiana, is a prominent sand that commonly contains an abundance of salt water. Other designations of the Blossom sand are "sub-Clarksville," "second gas sand," "1800-foot sand," "the sandrock," etc. It comprises a fairly homogeneous sandstone or sand, contains some lenses of clay, and generally shows innumerable minute dark-colored or greenish specks of a glauconitic mineral. In character the Blossom sand is similar, and in fact indistinguishable, from the Woodbine sand lower down, except for its stratigraphic position and perhaps through the fact that the common thickness of the Blossom is about 30 feet, while that of the Woodbine may be several hundred feet.

Woodbine sand.—The lowest formation in the Upper Cretaceous or Gulf series in northeastern Texas and northwestern Louisiana is the Woodbine sand, which consists of ferruginous and argillaceous sands and some bituminous clays, and contains occasional plant fossils. It also contains numerous dark or greenish specks of pyrite or glauconite, which give the formation a yellow or brown color on its weathered outcrop, and which sometimes form a cementing material and bind the sand into a compact sandstone. The Woodbine is, at present, the lowest producing sand of the northwest Louisiana fields with the exception of a small pool of Lower Cretaceous production on the Texas border south of the main northwestern Louisiana fields. The Woodbine may be several hundred feet in thickness.

In the early part of 1921 the Woodbine sand was found productive in the Mexia field of east central Texas and wells of immense yield have been drilled.

Lower Cretaceous or Comanche Series

Tamasopo limestone.—Throughout the northern part of the State of Vera Cruz, in which the great Mexican oil-fields occur, lies a limestone formation thousands of feet in thickness. It is supposed that the oil found in the overlying strata has migrated from this formation, which is regarded as the original source of all oil produced in the Mexican fields. The Tamasopo is the formation from which the famous Dos Bocas, Juan Casiano, Potrero del Llano, Cerro Azul, Panuco, and other great gushers produced their oil. In character the Tamasopo varies greatly, being white and finely crystalline in places, with almost unrecognizable fossils, and having chert distributed throughout it. Elsewhere it is less compact and ranges in color from dark blue almost to black. In places it is dolomitized and has traces of petroleum on its outcrop.

Trinity sand.—What is known as the Trinity sand exists in Texas and southern Oklahoma at the base of the Lower Cretaceous or Comanche series. On its outcrop it is a fine clean sand, containing pebbles and occasional quartz boulders, sometimes with fine lenses of clay and limestone and fragments of silicified wood and occasionally with other fossils. The Trinity sand is productive in the Madill pool of Marshall County, Oklahoma; but some authorities suppose the oil there to be a seepage from underlying Carboniferous sediments.

Dakota sand.—In the States of Wyoming, Montana and Colorado, and extending northward through western Canada, is a persistent bed of hard generally conglomeratic sand known as the Dakota sandstone, which though commonly only 50 to 100 feet thick in Wyoming, is several hundred feet thick at many places under the Great Plains, where it furnishes great quantities of artesian water. At many points it carries some oil and is believed equivalent to a part of the Kootenai formation, containing some of the Cat Creek sand of Montana.

Morrison formation.—Although geologists are not agreed as to whether it belongs in the Cretaceous system or in the underlying Jurassic, the formation known as the Morrison underlies the Dakota formation in Wyoming, being 250 feet thick in places, and consisting of shales with several hard interbedded sandstones. The Morrison is characterized by its variegated colors. Oil seepages emerge from several of its constituent sands in the Powder River field.

Jurassic System

Sundance formation.—Oil is found in the Jurassic in only a few localities in the United States. The principal known region of its occurrence is Wyoming, where seepages emerge from the Sundance formation in the Powder River field. The formation is 150 feet thick on its outcrop and consists of gray shales with some thin beds of argillaceous fossiliferous limestone.

Carboniferous System

PERMIAN SERIES

Most of the known Permian oil in the United States comes from southwestern Oklahoma and from Wichita, Wilbarger, and Clay counties in northern Texas; but natural gas in enormous volume is derived from Permian strata in the Amarillo fields of northwestern Texas and oil is found in them in Wyoming.

The Permian is divided geologically into several formations in the respective States where it occurs, and the lowest formation is the one most commonly productive. In other words, the Permian oil and gas in southwestern Oklahoma are found in what are believed to be the lowermost 3000 feet of the series, whereas the higher-lying Permian beds in the center of the deep basins have been found barren to date.

Permian rocks or "Red beds" form the surface throughout large areas in the Appalacbian basin of southwestern Pennsylvania and western West Virginia, but little or no oil or gas has been found in them. The entire region formed by the western halves of Oklahoma, Kansas, and Nebraska has Permian rocks at or under the surface, generally to a depth of several thousand feet. The same rocks reach a depth of at least 7000 feet in parts of western Texas, and similar thicknesses prevail in extreme southeastern New Mexico. In western Texas, however, the series is largely made up of limestones, instead of the normal alternation of irregularly bedded red rocks found elsewhere at the same horizons. In Wyoming, oil is found in the Embar formation of Permian age. In Russia and China, the Permian is very extensive and has thicknesses comparable to those of the United States, generally with a similar lack of oil indications.

During the first forty years of oil production in the United States, most geologists believed that no oil would be found in the Permian, on account of the characteristic red color of the beds, as this is indicative of subaerial or shallow fresh water origin and popular hypotheses precluded the existence of petroleum in beds of such origin. This view seemed countenanced by the fact that up to that time no oil had been found in the Permian in test wells sunk in Oklahoma, Kansas, and Texas. With the discovery of the Electra and other Wichita and Clay County fields in the Permian of northern Texas, this view had to be revised so as to explain the existence of oil in sandstones interstratified in the red beds. This was done through the assumption that, since no oil could have originated in formations not deposited in the ocean, it must have seeped up into them from underlying Pennsylvanian rocks. This is now the common view.

We should acknowledge that the whole problem of Permian oil has not yet passed the theoretical stage, and that consequently it is not safe to rule out a region as regards oil possibilities simply because the underlying Pennsylvanian is absent. Moreover, one cannot say that oil may not be found in the thousands of feet of Permian limestones which abound in western Texas and perhaps elsewhere, and which clearly must have been water-deposited and may contain oil if other essential conditions prevail.

Blackwell sand.—This prominent sand lies in the Garrison formation of Lower Permian age in the Blackwell, Billings and Ponca City pools in the north-central part of Oklahoma; and it is the principal sand of Permian age which is productive in that general region. The top of the Blackwell sand lies about 2650 feet above the top of the Wheeler sand or Oswego limestone.

Sands in the Permian series of southwest Oklahoma.—In the region surrounding the Wichita and Arbuckle Mountains in southwestern Oklahoma and adjoining portions of northern Texas, the Permian strata average less than 3000 feet in thickness; though it must be acknowledged that geologists are not agreed as to exactly what constitutes the base of the Permian there. The Wichita and Arbuckle Mountains are composed of older formations, about the base of which the Mississippian and Pennsylvanian rocks are concealed where present; and the Permian sandstones and shales, cross-bedded and largely red in color, overlap against the pre-Carboniferous rocks. Seepages of asphalt of considerable size are common in this border zone.

The Permian of southern Oklahoma has been productive for years in Carter County and on the edge of Stephens County, but the great oil development of the Permian series of this region may properly be said to have commenced in 1920 when the great Stephens County fields were opened up in earnest and became equally important with some of the large fields of the Pennsylvanian system. Some of the Stephens County oil comes from the Pennsylvanian.

The most important characteristic of the Permian beds in that region is their irregularity. Sometimes a red shale hundreds of feet in thickness grades laterally within a few hundred feet into a blue shale or even into a massive sandstone. Similarly the sands, which range from a few feet to several hundred feet in thickness, are irregularly stratified between the shales and hence are very "spotted," so that a single producing sand seldom holds for an entire pool. The Fortuna sand, in the Cement field of Caddo County, was thought to be an exception, for the reason that similar characteristics and similar production had been found in wells 5 miles apart. It is now known, however, that several intermediate wells produce from shallower sands and not from the Fortuna.

In searching for oil in the Permian it must be remembered that prejudice against that formation was largely based on its red color and inferences as to its origin. However, since the Permian contains sands of excellent texture and cover, and since favorable types of structures are abundant in it, there is now no reason for not giving the series as much attention as other subdivisions of the Carboniferous system.

Sands in the Permian series of northern Texas.—The productive sands of the Permian series in the Electra and Petrolia pools of Wichita and Clay counties, Texas, have been given no definite names by which they might possibly be correlated with sands in the same series elsewhere; but the assumption is that in northern Texas, as in most other localities the Permian sands are of too local a character for such correlation. They are contained in the Wichita formation, which is 1000 to 1500 feet in thickness, and is composed of red, bluish and gray sandstones and some shales. The dominant color is red, as in southern Oklahoma, but in counties farther west and south the red color disappears, and bluish-black shales and limestones prevail. The pools above mentioned are the principal ones producing from the Permian series, although the Burkburnett pool may derive some of its oil from this source.

Embar sand.—The name Embar was first used for strata of Lower Permian age outcropping with a thickness of 250 feet in central Wyoming. Sandstone of this age has since been found productive of heavy oil at Lander and some other localities in that state. The Embar in Wyoming overlies the Tensleep sandstone of Pennsylvanian age and underlies the Chugwater red beds of Triassic age. At present the Embar is not of great importance from the standpoint of oil production, but it is a part of the Quadrant group, in which great production has been found in Big Horn County, Montana.

PENNSYLVANIAN SERIES

The rocks of the Pennsylvanian series are sometimes called the Carboniferous proper. They are the strata in which the greatest coal-beds of Carboniferous age occur and which are mined so extensively in Pennsylvania, West Virginia, Ohio, Kentucky, Missouri, Indiana, Illinois, Kansas, Oklahoma, and Texas, as well as in many foreign countries, as the British Isles, China, and the Province of Nova Scotia. Pennsylvanian strata occupy the surface throughout large areas in the oil fields of the states above mentioned, and in immense areas elsewhere throughout the world. As a rule, where the Permian series is present, the Pennsylvanian exists below it.

Pennsylvanian rocks, while they contain coal beds, are made up largely of shales, sandstones, and limestones. In some places these rocks show frequent alternations while elsewhere the same characteristics persist for thousands of feet.

Pennsylvanian Series of Northern Appalachian Fields

Pittsburgh coal.—Although not an oil or gas sand, the Pittsburgh coal-bed is important on account of its being a persistent stratum throughout southwestern Pennsylvania,

Fig. 7.—Ames limestone 3 feet in thickness, near middle of picture, a persistent and prominent "key rock" in western Pennsylvania and southeastern Ohio, outcropping in railroad cut near Clarksburg, Indiana County, Pennsylvania.

northern West Virginia and southeastern Ohio. It is therefore useful to well drillers as a datum for calculating the positions of oil sands encountered in wells. The Pittsburgh coal is so well known and of such unusual thickness as to be generally distinguishable from all other coals in the region.

Cow Run sand.—The position of the First Cow Run or stratigraphically highest producing sand in the Appalachian fields is 280 to 330 feet below the Pittsburgh coal-bed, and from 60 to 100 feet below the Ames or "Crinoidal" limestone—the persistent key bed in the middle of the Conemaugh formation in Ohio and Pennsylvania (see Fig. 7). The Cow Run is an important sand in Washington and Morgan counties,

Ohio, but is non-persistent and patchy in character. It ranges from shaly to conglomeratic in texture and its thickness varies from 5 to 150 feet. It is often missing, though present in many places throughout southeastern Ohio, West Virginia, and southwestern Pennsylvania. At Macksburg, Ohio, this sand is known as the 140-foot sand. It is named from Cow Run, Ohio, where it was first encountered in drilling in 1861.

Dunkard sand.—The sand known to drillers in southwestern Pennsylvania and adjoining portions of West Virginia as the Dunkard is 30 to 150 feet thick, and lies near the base of the Conemaugh formation (i.e., above the Upper Freeport coal). Sometimes the sand is composed of two members, known respectively from above downwards as the Little Dunkard and Big Dunkard, and 350 to 550 feet respectively below

FIG. 8.—Saltsburg sandstone, outcropping at Saltsburg, Pennsylvania, to illustrate the type of Cow Run sand at the outcrop.

the Pittsburgh coal. The Big Dunkard is sometimes known as the Hurry-up sand. The Little Dunkard is generally considered to be equivalent to the Saltsburg sandstone (see Fig. 8) which outcrops farther east and north, whereas the Big Dunkard is the equivalent of the Mahoning sandstone. These sands were named from Dunkard Creek, in extreme southern Pennsylvania, near the mouth of which oil was discovered in 1861.

Upper Freeport coal.—Many prominent coal-beds exist in western Pennsylvania, West Virginia, and southeastern Ohio, and they are not always distinguishable from each other. The Upper Freeport is one which is rather persistent and important as a marker, in that it lies from 400 to 600 feet below the Pittsburgh coal. It is the next bed below the Dunkard sand (where present), is of considerable economic importance, and is, furthermore, the top of the Allegheny formation.

Macksburg 500-foot sand.—Below the Kittanning coal beds in the Allegheny formation of Pennsylvanian age lies a patchy sand which is important only in the vicinity of Macksburg, Ohio. Its position is 580 feet below the Pittsburgh coal and 300 feet below the First Cow Run sand.

Second **Cow Run sand.**—Near the base of the Allegheny formation, at an average distance of 770 feet below the Pittsburgh coal, the Second Cow Run sand is sometimes found. It is not of great importance, but produces oil at Cow Run in Washington County, Ohio, and elsewhere. This sand is occasionally mistaken for various other sands below the First Cow Run, so that well logs which mention it are sometimes questionable. The ordinary interval between the First and Second Cow Run is about 400 feet.

Gas sand.—"Gas sands" of Pennsylvanian age are many in number, but the sand generally distinguished by this name without other qualifications lies in the Allegheny or Pottsville formation, and variously coincides with the Kittanning, Clarion, or Homewood sandstone, according to locality. As these are not synonymous, even the true "Gas sand" is somewhat different in identity according to locality. In a general way, it is a name given to a sand 800 to 900 feet below the Pittsburgh coal.

Salt sands.—The sandstones generally known as the "Salt sands" in Ohio, West Virginia, and southwest Pennsylvania apparently lie entirely in the Pottsville formation, below the Second Cow Run and above the Maxton sand, and commonly 900 to 1000 feet below the Pittsburgh coal. These sands should not be confused with what is locally known as "Salt sand" in Carroll and certain other counties in eastern Ohio, as the latter is in reality a salt water bearing phase of the Berea sand. The true Salt sands are also an important source of salt water, as indicated by the numerous salt works in Ohio and adjoining states during the first half of the last century.

Maxton sand.—Where present, the Maxton rests on the Big Lime (Maxville limestone) and hence lies at the base of the Pennsylvanian series. Its representative in northern Pennsylvania was called by the drillers the First Mountain sand. Its important localities are in Washington and Monroe counties, Ohio, although it is present in Pennsylvania and West Virginia. Its thickness ranges from 8 to 100 feet. It is not very important as a producer of oil or gas.

Pennsylvanian Series of Kentucky Fields

Beaver, Horton, and Pike sands.—In Kentucky the Pottsville conglomerate of Lower Pennsylvanian age contains three sands, not necessarily correlated with the Salt and Maxton sands of Pennsylvania. They are known as the Beaver, Horton and Pike sands and are all petroliferous. Their thickness is variable, and any one of them may be 50 to 250 feet thick. The Beaver, or uppermost, is usually the thickest.

Pennsylvanian Series of Mid-Continent Fields

Newkirk sand.—Near the top of the Pennsylvanian series in northern Oklahoma, with its top 650 to 750 feet below the top of the Blackwell sand and 1900 feet above the top of the Wheeler sand, is a stratum known at its outcrop as the Elgin sandstone. Where present below ground, it is known as the Newkirk sand. It is productive in the Billings, Ponca City, and Newkirk fields.

Ponca City sand.—The principal oil sand of the Ponca City field of Oklahoma, believed by some to correlate with the deep sand of the Billings pool, is the Ponca City sand, which lies about 1300 feet below the top of the Blackwell sand and an equal distance above the Oswego limestone or Wheeler sand.

Layton sand.—The most important locality of the Layton sand is the Cushing field of Oklahoma, and the sand appears to be confined to that state. Its maximum thickness is about 100 feet, and the average for the limited area in which it is prominent is about 50 feet. Its horizon is in the Coffeyville formation of upper Pennsylvanian age, and its top is approximately 1000 feet above the Oswego or Fort Scott limestone or Wheeler sand.

Jones sand.—This is a local name, used in the Cushing field of northeastern Oklahoma, for a slightly productive sand 15 to 35 feet in thickness, forming a part of the

Fig. 9.—Pottsville sandstone outcropping in the railroad cut, 3 miles east of Tunnelton, West Virginia.

Nowata formation. Its top is about 500 feet above the Oswego or Fort Scott limestone or Wheeler sand.

Big Lime (of **Oklahoma**).—The Big lime of the Oklahoma and Kansas fields is quite different from any of the various limestones of the same designation found in Ohio, Pennsylvania, Kentucky and West Virginia. The Big lime of Oklahoma and Kansas is identical with the Oologah limestone of Middle Pennsylvanian age, which outcrops at Oologah and other points in northeastern Oklahoma, and is easily recognizable in well logs. It is not an oil or gas sand, but is important as a key rock useful in correlations between wells, and in calculating the positions of the expected sands below.

Cleveland sand.—In the Cushing field and in Osage and Pawnee Counties, Oklahoma, the Cleveland sand is an important producer. Its top lies 200 to 300 feet above the Wheeler sand (Oswego limestone), and about 450 feet below the top of the Layton sand.

Peru sand.—The sand known as the Peru in Washington and Osage counties, Oklahoma, and in southern Kansas, lies in the Labette formation of Middle Pennsylvanian age, intermediate between the positions of the Cleveland sand (above) and the Oswego limestone, or Wheeler sand (below). Another Peru sand, or "Markham," as it is sometimes called, lies about 200 feet *below* the top of the Oswego in Osage County, instead of *above* it, and is therefore not the same as the Peru of Kansas, though of considerable importance as an oil and gas producer.

Fig. 10.—Overhanging cliff of Pottsville sandstone, Rockville, Preston County, West Virginia, to illustrate its common extremely massive type of outcrop.

Oswego lime.—The formation known as the Oswego lime, which is equivalent to the Fort Scott limestone of Middle Pennsylvanian age, outcropping at the surface in eastern Kansas and northeastern Oklahoma, is important, not only for the natural gas that it contains, but also because it is used by the drillers as a datum in calculating the depths at which to expect the various sands. The top of this limestone ranges from 150 to 250 feet below the Big lime. Where it is a limestone the Oswego is 50 to 100 feet thick, interlaminated with a few bands of black shale. In Tulsa County and southward the Oswego is broken into two members separated by over 120 feet of shale; hence the total distance from top to bottom of the inclusive Oswego limes is 200 feet.

Wheeler or Bixler sand.—Directly underneath the Oswego lime in Oklahoma there is sometimes a sand from 10 to 40 feet in thickness called the Wheeler or Bixler. It appears to be rather persistent, producing oil in Washington and Nowata counties, although commonly containing salt water in considerable quantity. In the Cushing field it is the most uniform sand and appears to correlate with the Oswego lime. It is a great producer of oil and gas in the Cushing field.

Squirrel sand.—This sand lies near the top of the Cherokee shales and about 130 feet below the top of the Wheeler sand or Oswego lime, in widely scattered localities in Osage and adjoining counties in northeastern Oklahoma. It sometimes produces gas, but is not an important sand.

Skinner sand.—About 150 feet below the top of the Wheeler is a sand that is locally productive in the Cushing field in northeastern Oklahoma but is seldom recognized elsewhere.

Red Fork sand.—The Red Fork sand is only known to be productive in Tulsa County, Oklahoma, where it lies 100 to 150 feet above the Bartlesville sand, high in the Cherokee shales of Pennsylvanian age. Its position is therefore not far from that of the Skinner sand mentioned above.

Bartlesville sand.—Although the Bartlesville, which is in the middle of the Cherokee shales formation, is the most widespread and one of the most productive sands of the Pennsylvanian series in northeastern Oklahoma, it is also one of the most irregular. Its top lies 250 to 450 feet below the top of the Oswego limestone or Wheeler sand, and it ranges in thickness from a few inches to over 200 feet and in porosity from shale to coarse sandstone. This sand has been immensely productive in over thirty pools in northeastern Oklahoma. It may be the same as the "gas sand" of the Morris pool, but is not the Glenn sand of that pool, which lies 600 to 800 feet deeper and near the base of the Cherokee shales. The sand may be of a double nature, consisting of an upper member or gas sand, and a lower, or oil sand. In northeastern Oklahoma the Bartlesville is in reality a composite of several lenses of sandstone rather than a single or double bed.

Glenn sand.—Notwithstanding the fact that the Glenn sand of eastern Oklahoma is generally spoken of as equivalent to the Bartlesville sand, this identity is by no means certain. In fact, the evidence seems to indicate that the Glenn lies 200 to 300 feet below the top of the Red Fork sand, which, in turn, is somewhat below the horizon of the Bartlesville; and furthermore the Glenn is 700 feet below the top of the Oswego, instead of only 400, as is the Bartlesville where recognized in Tulsa County.

Pennsylvanian Series of Northern and Central Texas

Cisco formation.—The highest formation recognized in the Pennsylvanian of northern and central Texas is the Cisco, which is 900 feet in thickness and is productive in Clay, Wichita, Shackelford and Taylor Counties. Probably some of the deep sands of the Electra and Burkburnett fields belong in this formation. In character the strata are blue shales, sandstones, conglomerates, limestones, and important coals.

Canyon formation.—In the middle of the Pennsylvanian series is what is known as the Canyon formation, productive in Brown, McCulloch, Wichita, and Taylor counties. This formation is 600 to 800 feet in thickness and consists of limestones, blue shales, sandstones, and conglomerate. It furnishes the principal producing sand of the Burkburnett field and some deep sands of the Electra field, and the shallow sand at Brownwood is productive from it.

Strawn formation.—The Strawn formation is 1700 feet in thickness and lies at the base of the Pennsylvanian series, yielding oil at Strawn and Moran, in Palo Pinto and Shackelford counties respectively. It consists of sandstones and blue shales, with some conglomerate, coals and limestones.

MISSISSIPPIAN SERIES

Sandstones of Mississippian age are productive in Illinois, West Virginia, Ohio, and Kentucky, while Misssisippian limestones produce in Oklahoma, Kansas, Ohio and West Virginia. Rocks of Mississippian age outcrop in northern Pennsylvania, central West Virginia, central Ohio, Michigan, Kentucky, Tennessee, Alabama, Indiana, Missouri, Iowa, Arkansas and Oklahoma; and they underlie the whole extent of the Pennsylvanian sediments in the basins of western Pennsylvania, West Virginia, southeastern Ohio, southwestern Indiana, eastern Illinois, Kansas and northern Oklahoma. In character the series ranges from the Mauch Chunk red shales underlain by the Greenbrier or Maxville limestone (Big lime), 60 feet or less in thickness in southwest Pennsylvania, to the Mansfield sandstones of eastern Illinois, and, farther west, to the well-known Mississippi lime of the Oklahoma well drillers, which is several hundred feet in thickness and frequently so hard and cherty that it was once thought to be impenetrable.

Mississippian Series of Northern Mid-Continent Fields

Mississippi lime sand or Hogshooter **sand.**—At the top of the Mississippian series in northeastern Oklahoma and southern Kansas, 100 to 250 feet below the Bartlesville sand and 420 to 560 feet below the top of the Oswego lime, exists a porous zone known to well-drillers as the Mississippi lime, Mississippi sand or simply the "Mississippi." It constitutes one of the most productive horizons of Oklahoma; but, on account of an unconformity between it and the overlying Cherokee shale, the pools are not always situated exactly beneath the domes in the surface formations.

The character of the Mississippi sand or "lime" appears to be as variable as its productivity; and, on account of the erosion that took place before the deposition of the Pennsylvanian series, it is certain that the productive horizon of the uppermost Mississippian, as encountered in wells, does not always lie in absolutely the same stratigraphic position. Another reason for this discrepancy is that the Pitkin limestone formation, (the geological name for the upper part of the Mississippian series in northeastern Oklahoma), lenses out toward the south, so that one new sand after another enters in that direction, until in southeastern Oklahoma the limestone nature of the series has entirely disappeared.

The thickness of the Hogshooter sand at its type locality in Oklahoma is 20 to 100 feet. A prominent characteristic of the sand is its salt water content, which is very great in all except the extreme anticlinal and domal portions of the stratum. For this reason the Hogshooter and other great gas pools in northeastern Oklahoma are gradually encroached on by water until the gas finally disappears.

Mississippi lime.—A book could be written on what is known, and another on what is not known, concerning the Mississippian limestone, or Mississippi lime as it is called by the well drillers. A few of its eccentricities in Oklahoma alone are as follows:

(*a*) It appears to be 1000 feet thick, according to some of the well logs in southeastern Kansas, whereas the total thickness of limestone at this horizon in southern Tulsa County, Oklahoma, is not over 300 feet, and toward the south it gives way entirely to sandstone.

(*b*) Its relation to the immediately underlying Chattanooga shale appears to be as constant in northeastern Oklahoma as it is in Tennessee, over 400 miles away.

(*c*) Whereas the top of the Oswego lime is only about 400 feet above the Mississippi lime in southeastern Kansas, in Tulsa County, Oklahoma, it is 1000 feet above the Mississippi lime.

(*d*) Similarly, the distance from the top of the Bartlesville sand ranges from 200 feet in central Washington County, Oklahoma, to nearly 700 feet, 40 miles to the south.

(*e*) The Hogshooter sand, which rests on the Mississippi lime in central Washington County, is not the same as the sand which rests on it in Tulsa County and southward.

In a general way it may be said that the Mississippian of Oklahoma consists of two members—the Pitkin limestone above and the Boone chert below, separated by the Fayetteville formation, which consists largely of shale. The Boone chert, or Lower Mississippian, rests on the Chattanooga shale.

The first and second "breaks" in the lime.—The name Mississippi lime has been loosely used in the oil business, and is commonly used in referring to all limestone encountered below the Pennsylvanian series. Hence, the intervals between different limestone formations of Mississippian and lower age are generally spoken of as "breaks" in the lime. Recently oil has been found in Osage County, Oklahoma, as low as the "second break" in the lime, which geologists believe to be well below the true Mississippian. The sands described immediately below are therefore in the southward extension of the "breaks" referred to.

Tucker, Burgess, Meadows or Booch sand.—The second sand of the Mississippian series in northeastern Oklahoma is the Burgess; but it is often erroneously classed as Pennsylvanian because the top of the Mississippi lime dies out toward the south. The interval from the top of the Hogshooter to the top of the Burgess is about 100 feet where the two sands are present in the Collinsville field.

Lying 150 to 300 feet below the top of the Bartlesville sand, 500 to 600 feet below the top of the Oswego or Wheeler and near the base of the Pennsylvanian series, farther south in northeastern Oklahoma, is the Tucker sand. Although some doubts are expressed as to the true correlations, its stratigraphic position is about the same as that of the Burgess sand of the extreme northern part of northeastern Oklahoma and the Booch sand of Okmulgee and Muskogee counties. The Tucker is productive in widespread localities. In character it is medium-grained, and in color bluish-green.

Second Booch sand.—In Tulsa, Wagoner, and Muskogee counties, a productive sand is found about 100 feet below the top of the Booch, and has been called the Second Booch, Dutcher or Colbert sand.

Mounds sand.—The Mounds sand, which lies 400 to 500 feet below the top of the Booch sand, is productive in the vicinity of Mounds in northern Okmulgee and southern Tulsa counties, Oklahoma, but is seldom important elsewhere. It may be more important in the future.

Morris or Sapulpa sand.—About 500 or 600 feet below the top of the Booch sand, in Okmulgee and portions of adjoining counties, is the Morris sand, named from the field of the same name in Okmulgee County. The Sapulpa sand appears to be the equivalent of it.

Leidecker sand.—The Leidecker sand is only known to be productive in Muskogee County, where it lies about 550 feet below the top of the Booch sand, near the base of the formation overlying the Mississippi lime, perhaps Pennsylvanian, but supposed to be the upper sandy representative of the Mississippian in that region.

Muskogee or Boynton sand.—Principally productive at Muskogee and jointly with the Leidecker sand in the Boynton pool of western Muskogee County, the Muskogee sand also supplies gas in the Morris pool of eastern Okmulgee County. This sand lies

700 to 850 feet below the top of the Booch sand, and is the lowest sand known in the Carboniferous system of Oklahoma.

Mississippian Series of Southern Mid-Continent Fields

Bend limestone.—The Bend limestone is the great producing formation of the central Texas fields, resembling, in this respect, the Trenton limestone of the Lima-Indiana fields and the Tamasopo limestone in Mexico. The Bend, however, is quite different in age from either the Trenton or the Tamasopo. It is considered here as of Upper Mississippian age, although some geologists place it in the base of the Pennsylvanian, this question being merely a technicality. The Bend has been immensely productive throughout the so-called Ranger, Desdemona, Caddo (Stephens County), and Breckenridge fields, in Eastland, Comanche, Stephens, Callahan and Coleman counties. In some places it yields oil as far as 4000 feet below the surface, being therefore one of the deepest producing formations in the world. In the Bend series the Duke, Gordon, Jones and Veale sands are recognized, all being interstratified in the Marble Falls limestone member of the Bend.

The latest great discovery in the Bend consists of the Young County fields of Texas, comprising an extension of the Eastland and Stephens Counties area.

Mississippian Series of the Appalachian Fields

Mauch Chunk red shales.—Not important as an oil producer, but decidedly so as key beds, are the Mauch Chunk shales, which extend from Pennsylvania to Kentucky and are generally characterized by their red color. They are supposed to occupy the horizon of the Chester group of the Mississippi Valley. The thickness of the Mauch Chunk varies from 90 to 800 feet, on account of an erosion unconformity at its top. In Kentucky and West Virginia a white sand productive of some gas occurs in the red shales and is known as the Maxon sand; but it can not be positively identified with the Maxton sand which is supposed to lie at the base of the Pottsville formation in southern Pennsylvania, northern West Virginia and southeastern Ohio.

Big lime (of Kentucky, West Virginia, southern Pennsylvania and southeastern Ohio).—The geologic formation which is the Maxville limestone in southeastern Ohio and the Greenbrier limestone in West Virginia, and which may be continuous with the Ste. Genevieve and St. Louis limestones of Kentucky, must not be confused with the Big lime of the Central Ohio fields, which is of Devonian and Silurian age and not of Mississippian. The Big lime which is now being discussed is a formation which is generally productive in southeastern Ohio, often in the same areas in which the underlying Keener is productive. In Martin, Floyd and Knott counties, Kentucky, it is important from the standpoint of natural gas, which is contained in a thin lens of tan sand lying in the midst of the limestone. The thickness of the Big lime in Ohio and West Virginia ranges from a few feet up to 100 feet or more; but in Kentucky it is sometimes over 1000 feet in thickness.

Keener sand.—One of the most productive sands in southeastern Ohio and the contiguous portions of West Virginia, is the Keener, named from the Keener farm, near Sistersville, West Virginia. It is an important producer in Washington and Monroe counties, Ohio. Its thickness ranges from 25 to 60 feet. It is generally regarded as a member of the Big Injun sand series of Pocono or Lower Mississippian age, but is separated from the Big Injun by shale.

Big Injun sand.—About 1200 feet below the Pittsburgh coal is the Big Injun—one of the thickest sands in the Appalachian basin, ranging from 50 to more than 400 feet, though not as important a producer as some other sands. It attains its best development in West Virginia and Pennsylvania, although it exists in eastern Kentucky and Ohio. It is productive at many places in southeastern Ohio. It is rather coarse in texture, and in many places contains pebbly layers which constitute the "pays." The Big Injun is generally water-bearing, and wells in it may be drowned out by the influx of water. In some places it is separated by shales into several sands. It belongs in the Mountain sand group of northern Pennsylvania; at Washington, Pennsylvania, it is called the Manifold sand, while its geologic equivalent is the Burgoon sandstone of Pocono age.

Squaw sand.—In many localities in southeastern Ohio as well as in West Virginia, a non-persistent sand, sometimes reaching a thickness of 150 feet occurs a short distance

Fig. 11.—Quarry of Greenbrier limestone ("Big lime" of West Virginia, Ohio and southwest Pennsylvania), near Terra Alta, West Virginia.

below the Big Injun, and is known as the Squaw. In a few cases, a still smaller related sand has been facetiously called the Papoose. Still lower are the Weir sands.

Berea sand.—The Berea grit, or Berea sand, as the oil men call it, is one of the most important sands of the Appalachian fields, and extends throughout western West Virginia, southeastern and central Ohio, into Kentucky, and northward into the western edge of Pennsylvania. It lies above the Bedford and Ohio shales, directly below the black Sunbury shale and, in Ohio, averages 1375 feet below the Ames limestone and 1450 to 1750 feet below the Pittsburgh coal, ranging in thickness from 5 to 170 feet. Its type locality is found at its outcrop at Berea, Ohio, where it is quarried. In texture it is rather variable, being fine-grained and shaly in some localities, and elsewhere a pure white sand with several "pays."

The Berea is the most persistent and best-known producing horizon of southeastern Ohio, and it is also productive over a wide area in West Virginia and at some places in Kentucky. In many parts of southeastern Ohio, and especially along a curved line passing through Carroll, Guernsey, Morgan, Athens, and Meigs counties, the Berea is composed of two sands, separated by a "break" of shale. In such cases the upper layer is sometimes known as the Salt sand (this must not be confused with the Salt sand of Lower Pennsylvanian age in Pennsylvania and West Virginia and localities in Ohio) because it is full of salt water, on account of the dying out of the upper sandstone phase southeastward and locally. The second of the two sandstone members is thinner and less persistent, but is more productive.

Murraysville or Butler 30-foot sand.—No certainty exists that the Murraysville sand, which is a conspicuous gas producer in the counties east and north of Pittsburgh, is identical with the Berea sand found farther west, though the similarity of its stratigraphic position is suggestive. The Murraysville directly overlies the Bedford "red

FIG. 12.—Tunnel through Pocono sandstone (Big Injun, Squaw and Keener Sands) on Licking River, near Blackband, eastern Licking County, Ohio.

rock." In places it is 150 feet thick. Other names given to it are "the Butler gas sand" and simply "the 30-foot sand." It lies 300 to 450 feet below the base of the Big Injun. In northern Pennsylvania its horizon is marked by the Third Mountain sand, also called there the Pithole grit.

Ohio shale gas wells.—In Ashtabula and Lake counties, and elsewhere in northeastern Ohio, a great number of shallow wells derive gas from the Ohio shale, a formation 800 to 1500 feet in thickness, underlying the Berea grit. The gas has low pressure but is very persistent in occurrence. The wells are long-lived, and are sufficiently productive, in most cases, to supply the gas used by one or two families for cooking and lighting purposes. The rock pressure never exceeds 100 pounds per square incn, and is seldom over 40 pounds. The daily capacity of individual wells occasionally rises to 100,000 cubic feet, but is rarely above 20,000 cubic feet.

Stray, Mt. Pisgah, Beaver, Otter, Cooper, and Slickford sands.—Below the Berea sand, the Waverly group of the Mississippian series in Kentucky contains lower sands, known, from the top downwards, as the Stray, Mt. Pisgah, Beaver, Otter, Cooper, and Slickford. These sands are locally productive in Wayne and adjoining counties in that state.

Venango First sand, Butler County Second sand, or 100-foot sand.—Situated at an average distance of 1900 feet below the Pittsburgh coal are sands which are believed to be approximately identical in their position, and in southern Pennsylvania are supposed to form the basal sand of the Pocono formation, being thus at the base of the Carboniferous system. In southwestern Pennsylvania the 100-foot sand, as the group is there called, consists of two members—the Gantz sand above and the 50-foot below. These two, together with the intervening "break" of shale, may be 50 to 200 feet in thickness, the 50-foot sand member being the larger of the two. Both members are prominent oil and gas sands, white or gray in color, of medium grain and hardness, containing occasional lenses of coarse pebbly sand or "pays." The determination of the identity of the 100-foot sand throughout southwestern Pennsylvania is greatly simplified by the fact that it is about 100 feet below a thin band of red rock known as the Bedford shale. In northern Pennsylvania, the Venango First sand is supposed to lie in the Catskill group of Upper Devonian age; but the boundary between the Mississippian and Devonian is indefinite.

Devonian System

Chattanooga shale.—Although seldom a productive oil formation in itself, the Chattanooga shale of Upper Devonian age is important as a key rock. Moreover, it has a high oil content that may some day render it of commercial importance for distillation purposes. This shale, which is the apparent equivalent of the New Albany shale of Indiana, the Ohio shales of Ohio, and the Genesee shale of New York, extends, with practically identical characteristics, from eastern Tennessee westward throughout Tennessee and Kentucky and into Oklahoma. Its thickness is from 20 to 200 feet or more; and it is a dense black, crisp and brittle bituminous shale, quite different from others above and below. It is overlain in Tennessee by the Ridgetop shale of Lower Mississippian age, and in Oklahoma by the Boone chert or its representative; and it is underlain by limestones or by a sand which is known in eastern Oklahoma as the Claremore. In a few wells in Meade, Floyd, Barren and Allen counties, Kentucky, gas is found in commercial quantities in the Chattanooga shale. From 10 to 25 gallons of oily substance per ton have been distilled from this shale.

Claremore sand.—The sand in which "radium water" exists in the vicinity of Claremore, eastern Oklahoma, and which contains oil and gas at some localities in Tulsa County and farther south in the same state, lies directly below the Chattanooga black shale. In northeastern Oklahoma it is 400 to 500 feet below the top of the Mississippi lime; but the limestone practically dies out toward the south, so that the interval does not prevail except locally.

Wilcox sand.—The original locality of the Wilcox sand is near Beggs, in northern Okmulgee County, Oklahoma; and the sand is not known to be productive except in neighboring portions of Okmulgee and Creek counties, as in the Youngstown pool, where it is the principal producer of oil. The exact geologic horizon of the Wilcox sand has not been absolutely determined, but it is known to be beneath the Chattanooga shale and therefore may be a part of the Devonian system.

Catskill Red Shales and Venango Sand Group

Underneath the Mississippian strata in southwestern Pennsylvania is a thickness of 300 to 400 feet (inclusive of interstratified sands) of red shales in which is contained the

Venango sand group. These sands were first discovered at Oil Creek, Pennsylvania, near Oil City, although the red shales have long been known from their outcrops in the Catskill Mountains of New York. The Venango group includes all sands from the top of the " First " oil sand (" Second " sand of Butler County) to the bottom of the "Third" sand (called the " Fourth " in Butler, Armstrong, and Clarion counties and the "Fifth " sand in some parts of Venango County and in southwestern Pennsylvania). The group comprises a number of variable sandstones, sometimes conglomeratic, separated by the irregular beds of red shales. The " First " and " Second " sands of this group are exposed in the gorges of the Allegheny River and Dennis Run in Warren County, Pennsylvania.

These sands, as passed through in the original drilling on Oil Creek, are related somewhat as follows:

Partial section of Venango sand group on Oil Creek, Pennsylvania

	Feet
First sand	40
Interval	105
Second sand	25
Interval	110
Third sand	35
Total	315

Sometimes a " Stray " Third sand is also present, 15 to 20 feet above the Third; it is 12 to 25 feet in thickness and sometimes pebbly. This sand has generally been found to contain black oil, so that at Pleasantville it was called the Fourth or " black oil sand." The other members of the group carry green oil.

Although the Second sand of the group exists as a single mass on Oil Creek, toward the southeast, at Pleasantville, Pithole, Cashup, and Fagundus it is split into two sands with an interval of 15 to 30 feet of shale between them. These few introductory remarks will explain something of the irregularity of the Venango sand group and the difficulty in making correlations at a distance.

Gray, Boulder, Snee or Blue Monday sand.—Several names have been used interchangeably and rather loosely, to designate from one to three sands, occurring at the top of the Catskill red beds in southwestern Pennsylvania, usually 40 to 100 feet above the Gordon sand, and approximately 2100 feet below the Pittsburgh coal. It is seldom that any one of these sands exceeds 25 feet in thickness, although they are as thick as 60 feet in some places. They are generally white to gray in color, and sometimes soft and pebbly. In general, the name " Boulder sand " is used towards the north, and " Snee " and " Blue Monday " to the south, in northern Allegheny County. No certainty exists that the sands are ever absolutely identical, although they are similarly situated stratigraphically.

Nineveh 30-foot sand.—In Greene County, Pennsylvania, there is a sand of which the exact correlation to the north is uncertain, but which lies 50 to 100 feet below the 50-foot sand, and averages 2050 feet below the Pittsburgh coal. By some geologists it has been correlated with the Boulder sand to the north. It is a hard gray sandstone 10 to 20 feet thick, sometimes with a pebbly " pay " streak in the middle. In extreme southwestern Pennsylvania the Nineveh is considered the uppermost member of the Catskill series.

Venango Second, Butler Third, or Gordon sand (and the accompanying Butler Third Stray or Gordon Stray).—The Gordon sand is fairly persistent, though it is most easily recognized when considered in connection with the accompanying " Stray sand." The combined thickness of the two may be about 75 feet. The sands lie in the midst of the red shales, about 2100 to 2200 feet below the Pittsburgh coal. They are generally soft, white and pebbly, and form exceptionally productive oil reservoirs. The "Stray" sand is not always present.

Venango Third, Butler Fourth, or Fourth sand.—The next sand below the Gordon in southwest Pennsylvania is called the Fourth, and is situated 40 to 80 feet below the Gordon, and approximately 2200 feet below the Pittsburgh coal. The Fourth sand ranges from 10 feet to more than 50 feet in thickness, and is usually a soft, white, porous sand which has been a great producer of oil.

Fifth sand.—Near the base of the Catskill red beds, from 40 to 100 feet below the Fourth sand and approximately 2200 to 2300 feet below the Pittsburgh coal in southwestern Pennsylvania and northern West Virginia, lies the Fifth sand, the interval being filled with prevailingly red to green shales. The thickness of the Fifth is seldom over 50 feet and frequently not over 10 feet; but the sand is very persistent and a great producer of oil. It is commonly a fine-grained, hard, white to gray, quartz sandstone, similar in character to the Fourth, but more persistent. It is variable in texture, containing occasional embedded lenses of pebbles constituting the "pay" streaks. In southwestern Pennsylvania the fifth sand is sometimes called the McDonald sand.

Chemung Formation

Bayard or Sixth sand.—At an average distance of 2400 feet below the Pittsburgh coal and 100 feet below the top of the Fifth sand in southwestern Pennsylvania and northern West Virginia, lies the Bayard or Sixth, which is often a great gas producer. It may attain a thickness as great as 50 feet.

Elizabeth sand.—In extreme southwestern Pennsylvania, there is a thin sand usually not over 7 to 20 feet in thickness, which lies 100 to 200 feet below the top of the Bayard sand, and at an average distance of 2500 feet below the Pittsburgh coal. The name is derived from the deep well drilled in 1904 at West Elizabeth, Pennsylvania.

Warren oil sand group.—The Warren group of oil sands is about 300 feet in thickness and is supposed to lie a similar distance above the Speechley sand and about 400 to 600 feet below the Gordon. The group extends from the top of the North Warren shale to the bottom of the Stoneham sandstone and includes the Second, Third and Fourth sands of the Warren group. It is not persistent toward the south. Where these sands are present in extreme southwestern Pennsylvania, the Warren First sand, 30 feet in thickness, is recorded as 2700 feet, and the Warren Second sand 2750 feet below the Pittsburgh coal.

Speechley sand.—The top of the Speechley sand is from 100 to 150 feet above the top of the Tiona sand and it ranges in thickness from a few inches to nearly 100 feet. It is a hard, close-grained sandstone, and has been found productive of gas in western Pennsylvania as far south as southern Westmoreland County. This sand may average about 2900 feet below the Pittsburgh coal.

Tiona sand.—There is some discrepancy in the statements of drillers as to the identity of the Tiona sand, some of them placing it above the Speechley sand. It is here considered as underlying the latter by 100 to 150 feet, and underlying the Pittsburgh coal by an average distance of 3000 to 3100 feet. It appears to be limited to western Pennsylvania, and has been found productive of gas as far south as northern Westmoreland County. It generally lies about 200 feet above the First Bradford sand and ranges in thickness from a few inches to 100 feet. Ordinarily it is close-grained and hard, but in places it is open enough to yield good production.

Bradford sands.—Although questions of correlation exist, the Bradford group of sands is supposed to average somewhere near 3300 feet below the Pittsburgh coal. These sands are rather fine-grained, light to dark brown sandstones, sometimes containing pebbles, fossil shells and fish bones. The type locality is the Bradford oil-field, in northern Pennsylvania, although various members of the Bradford group are recognized at different places in western Pennsylvania. The position of the Bradford sand is supposedly about 400 feet below the Warren oil group, although satisfactory correlations seem never to have been made.

Silurian-Devonian Strata in Ohio, Kentucky, Etc.

Big lime (of Central Ohio fields).—Various " Big limes " are recognized throughout Ohio, West Virginia, Pennsylvania, Texas and Oklahoma, and they must not be confused. The Big lime of central Ohio is a name given by the drillers to a composite of the

Fig. 13.—Alternating sandstones and red shales of Catskill age, outcropping 1 mile west of Terra Alta, West Virginia. These beds contain the Nineveh, Gordon, Fourth and Fifth sands.

Delaware, Columbus, Monroe, Salina, and Niagara formations of Devonian and Silurian age, lying directly below the Ohio shale group. Although these formations are largely composed of limestones, they also contain numerous beds of gypsum, dolomite, and sometimes sandstone; and in northeastern Ohio extensive beds of rock salt are found in them. The Salina formation consists not only of rock salt, but also of gypsum, anhydrite, and limestone, irregularly interbedded. From one to four beds of salt are reported by drillers in northern Ohio.

The lower portion of the Big lime is of Niagara age and consists of light gray magnesian limestone. An important characteristic of the Big lime is that it contains a large amount of salt water, which must be shut off before the well is drilled below it to the Newburg or the Clinton sand, as the case may be.

Corniferous or Onondaga limestone or Irvine or Ragland sand.—Directly below the Chattanooga black shale at the base of the Devonian system in Kentucky is a limestone which has become known to drillers in that state as the Corniferous, or, in places, as the Irvine sand. It is geologically correlated with the Onondaga, or perhaps with the Hamilton and Onondaga, of farther north. Its thickness in eastern Kentucky is 25 to 45 feet; but in Ontario it ranges from 200 to 400 feet. It is a thick-bedded, massive, magnesian limestone, generally characterized on the outcrop by cherty inclusions, which produce an unequal surface on weathering and cause the Onondaga to be called a " hornstone." It has a tendency toward minute porosity due to solution and dolomitization, rendering it capable of acting as a reservoir for oil. The Corniferous

Fig. 14.—Outcrop of Clinton limestone near Hamilton, Ontario.

crops out in Ontario and New York near the outlet of Lake Erie, and dips south and west, so that in northern Pennsylvania it is estimated to be 2000 feet below the base of the Bradford Third sand. Many pools in Kentucky derive their oil from the Corniferous; and it is productive to some extent in Indiana, New York and Ontario.

Silurian System

Niagara formation.—Certain strata included in the Big lime of Ohio are oil-bearing in places. In Kentucky, the Niagara, which takes on a sandy character with high porosity, ranks second to the Corniferous in importance as a producer.

Newburg or Stadler sand.—The Newburg sand lies in the lower part of the Big lime, and is really a porous phase of the Niagara limestone which lies 275 to 360 feet above the "Clinton sand" in Ohio. The Newburg is chocolate-brown in color, and oily, and is thin and pockety in character. Its chief productive localities are the Newburg pool and other points near Cleveland, Ohio.

Clinton formation.—As will be explained below, the so-called "Clinton" sand of the Central Ohio fields is believed to be really of Upper Medina age and not as was once supposed equivalent to the Clinton limestone of southwestern Ohio. The Clinton is somewhat petroliferous in Kentucky, where it sometimes reaches a thickness of 10 or 20 feet; and it is reported sometimes to contain oil in commercial quantities. In Ontario a persistent bed of dolomite or limestone appears in the Clinton shales, and may be the equivalent of the Clinton limestone of southwestern Ohio. It is frequently productive of oil and gas in Ontario.

"Clinton" sand.—The principal sand in the Central Ohio fields is the "Clinton," which is widely productive of gas in the counties of Fairfield, Licking, Knox, Richland, Ashland, Wayne, Medina and Cuyahoga. It also contains several good oil pools in Wayne, Muskingum, Licking, Perry, Hocking, and Fairfield counties. The Clinton is a very persistent bed, averaging 30 feet in thickness throughout central Ohio, but increasing to several hundred feet in the Allegheny Mountains of central Pennsylvania, while in the other direction it runs out to a feather edge in west-central Ohio, along a line bounding the great gas fields on the west. The "Clinton" sand lies from 6 to 300 feet below the base of the Clinton limestone, the thinnest part being in northern Ohio and the thickest in Morgan County in the southeast. The "Clinton" is credited with being absolutely free of water, although no evidence exists that it does not contain water somewhere farther east than the present limits of production.

Correlations throughout Ontario, northwestern Pennsylvania and northern Ohio lead to the belief that one of the Medina sands, which lies a short distance below the Clinton of Ontario, is the real Ontario representative of the so-called "Clinton" of Ohio.

Medina sands.—As stated above, the "Clinton" sand of the Central Ohio fields is thought to be, in reality, of Upper Medina age; and it may be identical with one of the Medina sands which are productive at many places in southern Ontario. In that province there are two sands, a few feet apart, known as the White Medina and the Red Medina, and both may be productive in some places.

Ordovician System

Outcrops of Ordovician strata are approximately as numerous, and their buried portions are probably as extensive, as those of the Cambrian, while the Ordovician system is far more productive of gas and oil than is the Cambrian. The Ordovician might be called a typically favorable oil system, since it contains the exceedingly productive Trenton limestone near its middle, and, overlying this, the Utica, Hudson River and Cincinnati shales, which form an excellent cover. Ordovician rocks are also widely distributed in Europe; but, unlike the Ordovician of North America, those of Europe contain relatively little limestone.

Caney and Upper Sunnybrook sands.—It is only in Kentucky that oil is produced in commercial quantities from the Cincinnati series, which occupies the interval of 500 to 700 feet below the Medina series, and constitutes the Upper Ordovician. Although the Cincinnati series generally consists of limestones, blue shales, and thin calcareous sandstones, in Wolfe and Morgan counties, Kentucky, it contains the Caney sand, and in Wayne County of the same state the Upper Sunnybrook sand, both of which are productive. Natural gas is derived from the Cincinnati shales at a number of localities in Oldham, Bourbon, Clinton, and Barren counties.

Trenton limestone or "Trenton rock."—One of the most widely distributed and best known geologic formations is the Trenton, which extends under various names from New Jersey and New York westward through Ontario and the Middle West, where it forms the reservoir rock of the famous Lima-Indiana fields, thence into Minnesota, probably underlying a great part of the Mississippi Valley and portions of the Great Plains.

At its type locality, at Trenton Falls, New York, this formation is a dark blue, almost black limestone, massive and rather evenly bedded, with the beds sometimes separated by layers of black shale. At that place its uppermost few feet contain *Crinoidal* fossils, and the associated shales are also fossiliferous. The color and texture change somewhat throughout the lateral extent of the formation but these fossils have enabled geologists to identify it throughout a wide stretch of this and other countries. Where exposed at the surface in quarries at Point Pleasant, Ohio, on the Ohio River, 20 miles above Cincinnati, the Trenton is a light bluish-gray limestone, crystalline, massive, and fossiliferous.

In Kentucky the Trenton proper is a series of gray, granular and sometimes crystalline limestones less than 300 feet thick, known as the Lexington formation. Oil was found in the Trenton of that state in 1829, and, since then, considerable oil has been produced from it in McCreary and Wayne counties. Below the Lexington in Kentucky lie at least 600 feet of thick-bedded and compact limestone, known as the High Bridge formation, and yielding some production in Barren, Clinton and Cumberland counties.

Unlike sandstone, the Trenton limestone is distinguished by a peculiar chemical composition and character in regions where it is productive of oil, on account of being locally more magnesian in such regions. That is, whereas the normal composition of pure limestone is calcium carbonate ($CaCO_3$), the top or productive portion of the Trenton in the oil-fields contains an average of only 50 per cent of this constituent, the greater part of the balance being replaced by magnesium carbonate ($MgCO_3$) as explained on page 41.

The Trenton is the sole producing formation of the northwestern Ohio and eastern Indiana fields, and its thickness under those states is believed to be great.

Knox dolomite.—Although practically continuous with the overlying Trenton and generally considered part of the same geologic system, the Knox dolomite of the Kentucky fields is a distinct formation. It is about 250 feet in thickness, and consists of dark and light-colored dolomitic limestones, of which the so-called "deep sand" of Wayne County is an example. Oil is said to have occurred in the Knox dolomite, but its importance as a producer has not yet been demonstrated and some of the oil credited to it in Kentucky may in reality have come from the overlying High Bridge formation. Small seepages of oil and good showings of natural gas in wells have been found in several localities in northwestern Alabama. The Knox is sometimes referred to as belonging to the "Cambro-Ordovician" system.

Cambrian System

Although outcrops of Cambrian sediments are comparatively limited in North America, these sediments are supposed to underlie enormous areas, probably including the entire Appalachian basin, Cincinnati anticline, Mississippi Valley, the Great Plains of Oklahoma, Kansas, Nebraska, eastern Colorado, eastern Wyoming, eastern New Mexico, and the greater part of Texas. Cambrian sediments frequently consist of sandstones of a type which is likely to contain oil, and various rocks which form an excellent cover. The fact that Cambrian rocks are so seldom productive may be explained by their metamorphism, which is found to be more and more complete as the age of the strata increases; and as the Cambrian is the oldest system of rocks in the Paleozoic Period, great metamorphism has taken place in it throughout wide expanses of country. Another reason for the common absence of oil from the Cambrian may be that this system lies at the base of the Paleozoic, directly upon still older and more highly metamorphosed sedimentary or crystalline rocks; hence the highly porous sandstones of the Cambrian form a reservoir for all waters accumulating from higher altitudes in the Cambrian and overlying strata, as well as those descending in the stratum from distant mountain outcrops. On account of their great age, the waters of Cambrian sediments, where found in the midst of great basins, are frequently highly mineralized.

In the past the Cambrian appears to have been productive of oil to a very limited extent in the old Pincher Creek district of southwestern Alberta and southeastern British Columbia, as well as in the Gaspé Peninsula at the northeastern extremity of the Province of Quebec. It contains natural gas in small quantities at some places in Ontario and New York.

RELATION OF PETROLEUM OCCURRENCE TO LITHOLOGIC CHARACTER OF THE STRATA

Under this heading we must consider the character of the individual strata in which oil occurs, since a suitable lithologic character is essential in order to assure a reservoir, except where there is some substitute for porosity, either in the cavernous nature of the stratum or in the presence of joint-cracks or cleavage-planes. Ordinarily the productive stratum consists of sand or sandstone, in which the interstices between individual grains furnish the pores. The presence of oil in limestone, however, may be due to natural porosity of the rock, to solution along well-defined channels, or to secondary chemical changes resulting in the formation of dolomite, which, since it occupies less space than does calcium carbonate, holds oil in the resulting interstices. It is not always realized that the exact spot of occurrence of limestone oil in any generally favorable locality is determined, in the last analysis, by the internal characteristics of the stratum; yet, as a rule, the favorable surface structure is also the most favorable from a strictly lithologic standpoint. Since the upper portions of a limestone are the ones altered from calcium carbonate to dolomite, the highest position on the anticline or dome is the most favorable where limestone is the producing stratum as it is in the case of a sandstone or sand.

In the case of sandstone, the location of an accumulation is determined not only by structure, but by the degree of continuity of the stratum, positions of old shore lines,[1] etc. Not all sandstone throughout an oil region contains oil, even when other conditions are favorable, one essential requirement being that the individual grains shall be sufficiently rounded to make the bed porous, and that they shall be comparatively uniform in size, rather than an indiscriminate mixture of grains of all sizes, as in the case of many conglomerates. Experimental tests on oil sands and other rocks have been made,[2] which showed that productive sands seldom have a porosity of less than 10 per cent and they may be as high as 30 per cent. Non-producing sands, however, may have a very low pore space. Well-drillers recognize the internal variations of a rock when they speak of a sand as "open" or "close," "soft" or "hard," "good" or "poor" in character, etc. All porous strata are commonly termed "oil sands" by the drillers.

[1] Jones, W. F.: The relation of oil pools to ancient shore-lines. Econ. Geology, Vol. XV, No. 1, 1920, pp. 81–87.

[2] Melcher, A. F.: Determination of pore space of oil and gas sands, Min. and Met., No. 160, Sect. 5, Apr., 1920, 22 pp.

The per cent of pore space in some well known producing sands is shown by the following figures:[1]

Pore Space of Typical Oil Sands

Name of sand	Locality	Per cent of pore space
Berea	Woodsfield, Ohio	11.2
Big Injun	Lewisville, Ohio	13.1
Cabin Creek	Dawes, W. Va	19.3
Gas sand	Petrolia, Tex	26.6
Wall Creek	Salt Creek, Wyo	25.8
Bartlesville	Bartlesville, Okla	16.6
Bradford	Custer City, Pa	18.0
Woodbine	Shreveport, La	17.4

As a rule the porosity of a gas sand is somewhat greater than that of one productive of oil, but this is not always true. It must not be supposed that the figures mentioned above are anything like uniform for a particular sand, since the porosity varies greatly within short distances. The table is given merely to show what are some of the maximum porosities for the respective typical sands.

Ordinarily the bed immediately overlying an " oil sand " is known as the " cap-rock"; but a shale or other formation which is hundreds of feet thick and prevents the oil from seeping upward and disappearing, may more appropriately be termed the " cover." The " cap-rock," in its ordinary sense, is generally very hard and may consist of a limestone or shale bed only a few feet in thickness. One of the most common covers for oil-bearing strata is the Utica shale, which overlies the Trenton limestone in the Ohio-Indiana fields; but many shales of Carboniferous and Devonian ages also form excellent covers. The Clinton sand of the central Ohio fields is overlain by the Clinton shale, and the Corniferous limestone of Kentucky lies just below the Chattanooga shale. In many fields of Tertiary and Cretaceous ages, the cover is a great thickness of clay, while in certain parts of the world an intrusive igneous bed furnishes the cover.

The relations of the porous bed, cap-rock, cover and overlying and underlying strata are shown in Fig. 15, the vertical scale frequently being as much as 1000 feet to 1 inch and the horizontal scale about 1 mile to an inch.

It has been shown by Schuchert[2] that while all the above mentioned conditions have an important bearing on the occurrence of petroleum, nevertheless paleography, or the geography of former lands and seas, is also an essential factor, as it accounts for such conditions as the porosity of strata.

Some references on the character of oil-bearing rocks are as follows:

References

Lauer, A. W.: The petrology of reservoir rocks and its influence on the accumulation of petroleum, Econ. Geology, Vol. XII, No. 5, 1917, pp. 435–465.

Lewis, James O.: Discussion of " Petrology of reservoir rocks," by A. W. Lauer, Econ. Geology, Vol. XIII, No. 1, 1918, pp. 65–9.

[1] Melcher, A. F.: loc. cit.

[2] Schuchert, Charles: Discussion of "Petroliferous Provinces," by E. G. Woodruff, Min. and Met. Bull. No. 155, Am. Inst. Min. Engrs., 1919, pp. 3058–3070.

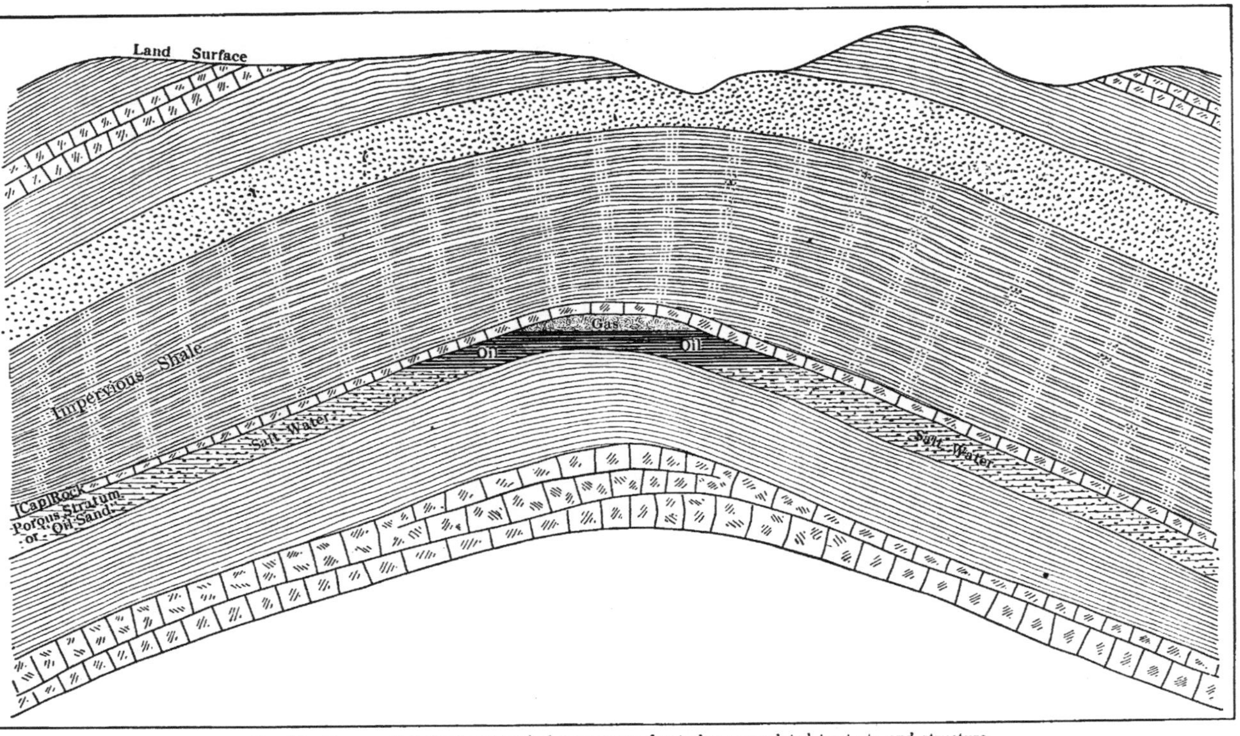

Fig. 15.—Cross-section showing a typical occurrence of petroleum as related to strata and structure.

Rogers, G. Sherburne: Discussion of " Petrology of reservoir rocks and its influence on the accumulation of petroleum " by A. W. Lauer, Econ. Geology, Vol. XIII, No. 4, 1918, pp. 316–324.

Shaw, E. W.: Crevices and cavities in oil sands, Econ. Geology, Vol. XIII, No. 3, 1918, 207–222.

RELATION OF PETROLEUM OCCURRENCE TO GEOLOGIC STRUCTURE

Introduction

The relation between geologic structure and oil occurrence has been realized for several decades by the exponents of the " anticlinal " theory, which was first suggested by Hunt [1] and later investigated and advocated by Andrews,[2] Winchell,[3] Stevenson,[4] Minshall,[5] Newberry,[6] Höfer,[7] and others. The theory was not definitely formulated, however, until 1885, when White [8] worked out its details and made the first practical application of it in the location of a number of pools in West Virginia and Pennsylvania. Dr. White's theory was strictly applicable only to rocks saturated with water. Orton [9] deserves great credit for deciphering the detailed structure of structural terraces along similar lines of research. The various theories for the accumulation of oil up to 1911 were ably summarized by Campbell.[10]

A marked increase in the value of geology to oil development was due to the evolution of the " structural theory " for oil occurrence, proposed in 1910 [11] as an offshoot from the anticlinal theory, and revised in 1916.[12] The structural theory, as understood by the writer, is as follows:

Through some means, by organic or inorganic agency or agencies, the petroleum and gas have come into or been generated in the porous formations in which they are found. The deposits may have originated through the decomposition of plant or animal remains on an ancient sea-bottom, as the adherents of the organic theory claim; or they may be the product of chemical action taking place deep in the earth, as is believed by the adherents of the inorganic theory; or some petroleum deposits may be of organic and others of inorganic origin. The theory of the origin of oil and gas is not an essential part of the structural theory. Whatever their true origin may be, the oil, gas and water in the formations (assumed to have been approximately horizontal at the time these sub-

[1] Hunt, T. Sterry: Notes on the history of petroleum or rock oil, Can. Nat., Vol. VI, Aug., 1861, pp. 241–255; Canada Geol. Survey, 17th Rept. of Progress, 1863–1866, p. 233.

[2] Andrews, E. Benjamin: Rock oil, its geologic relations and distribution, Am. Jour. Sci., 2d ser., Vol. XXXII, 1861, pp. 85–93.

[3] Winchell, Alexander: On the oil formation in Michigan and elsewhere. Am. Jour. Sci., 2d ser., Vol. XXXIX, 1865, p. 352.

[4] Stevenson, J. J.: 2d Geol. Survey Pa., Vol. H, 1875, pp. 394–395.

[5] Minshall, F. W.: In letters to the State Journal, Parkersburg, W. Va., in 1881.

[6] Newberry, J. S.: Ohio Geol. Survey, Vol. I, 1873, p. 160.

[7] Höfer, Hans: Das Erdöl und seine Verwandten, 3d edition, p. 166; Geologie des Erdöls, p. 18.

[8] White, I. C.: Science, Vol. VI, June 26, 1885; Bull. Geol. Soc. Am., Vol. III, 1892, pp. 187–216; W. Va. Geol. Survey, Vol. 1-A, 1904, pp. 48–64.

[9] Orton, Edward: Ohio Geol. Survey, Vol. VI, 1886, pp. 21 and 94.

[10] Campbell, M. R.: Historical review of theories advanced by American geologists to account for the origin and accumulation of oil. Econ. Geology, Vol VI, No. 4, 1911, pp. 362–386.

[11] Clapp, F. G.: A proposed classification of petroleum and natural gas fields based on structure. Read before Geol. Soc. Washington, March 9, 1910; Abstract Science, Vol. XXXI, No. 801, May 6, 1910, pp. 718–719. Published in full in Econ. Geol., Vol. V, No. 6, 1910, pp, 503–521. The use of geological science in the petroleum and natural gas business. Proc. Engrs. Soc. W. Pa., Vol. XXVI, No. 4, 1910, pp. 87–120. Read before that Society, April 19, 1910.

[12] Clapp, F. G.: Revision of the structural classification of petroleum and natural gas fields. Bull. Geol. Soc. Am., Vol. XXVIII, 1917, pp. 553–602.

stances entered them) were at first widely diffused in the porous formations or contiguous strata. They have remained in their diffused condition to the present time in many parts of the world, where only small quantities of oil and gas, too slight for profitable development, have been found, and where the dip of the rocks is very slight.

Where beds have been folded, however, as in most of the oil fields of the world, the oil, gas and water have been allowed to separate and arrange themselves according to their relative specific gravities. The separation and concentration may have been assisted by rock-pressure, diastrophism, hydraulic pressure, seepage, capillarity, molecular attraction, internal heat, metamorphism or other causes; but whatever causes may account for the movement of the oil, gas and water, the law of gravitation, being ever operative, was of the greatest importance in determining their arrangement; hence the accumulation was in the order of the densities of the substances.

The structural hypothesis agrees with the anticlinal theory, of which it is an outgrowth, in acknowledging that, on a given anticlinal, monoclinal, or quaquaversal strue-ture, gas lies nearest the crest and oil lower down, while the salt water, when present, lies still lower on the flanks of the uplift. Whether the pools occur at the exact crest of the anticlines, lower on their slopes, or in the synclines, is determined by factors of **secondary importance.**

The Structural Classification

The various structures in which oil occurs may be divided into classes and sub-classes, each sub-class including a special type of structure, and all having certain features in common. The classification, as revised by the writer in 1916, is as follows:

Classification of Oil and Gas Structures

I. Aclinal or sub-aclinal structure.
II. Anticlinal and synclinal structures.
  (*a*) Strong anticlines standing alone.
  (*b*) Well-defined alternating anticlines and synclines.
  (c) Broad geanticlinal folds.
  (*d*) Overturned folds.
  (*e*) Lenticular sands.
III. Monoclinal structure.
  (*a*) Monoclinal noses.
  (*v*) Monoclinal ravines.
  (*c*) Structural terraces or "arrested anticlines."
  (*d*) Lenticular sands.
IV. Quaquaversal structures or "domes."
  (*a*) Anticlinal bulges or "cross anticlines."
  (*b*) Monoclinal bulges.
  (*c*) Closed saline domes.
  (*d*) Quaquaversal structure caused by volcanic plugs.
  (*e*) Perforated saline domes.
V. Contact of sedimentary and igneous rocks.
  (*a*) Contact of sedimentaries with volcanic plugs.
  (*b*) Contact of sedimentaries with dikes.
  (*c*) Contact of sedimentaries with intrusive beds.
  (*d*) Contact of sedimentaries with older igneous rocks.
VI. Strata dipping unconformably away from old shore lines.
VII. Crevices of igneous rocks.

VIII. Crevices of sedimentary rocks.
IX. Faults.
  (a) Upthrow side.
  (b) Downthrow side.
  (c) Overthrusts.
X. Oil sands sealed in by bituminous deposits.

Class I.—Fields in Aclinal or Sub-Aclinal Structure

True aclinal formations are rare, though they are proposed by Johnson and Huntley [1] as a part of a summarized classification. The term aclinal is defined by Webster's

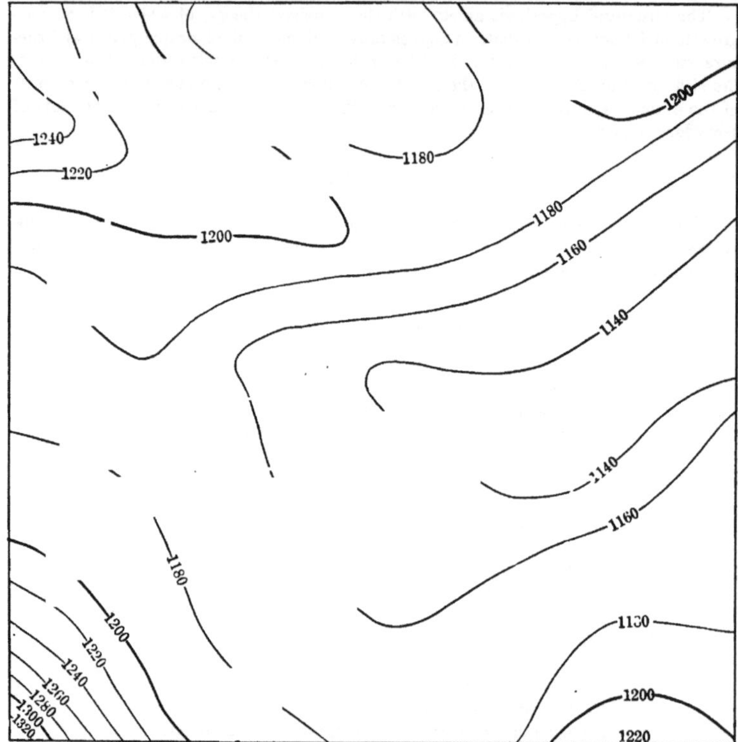

Fig. 16.—Sketch map showing an example of sub-aclinal structure (Class I) in Stephens County, Oklahoma. (Geology by Clyde T. Griswold for The Associated Petoleum Engineers.) Contour interval 20 feet.

Dictionary as "without inclination; horizontal." Properly speaking, we may define sub-aclinal beds as those which are approximately flat, sometimes dipping as little as 10

[1] Johnson, R. H., and Huntley, L. G.: Principles of oil and gas production. New York, 1916, p. 63.

or 20 feet per mile, too slight a dip to separate fully the oil and gas from the accompanying water. Occurrences of gas or oil in such regions are generally mere showings, encouraging to a prospector but seldom resulting in real production. The Electra pool in Texas appears to be an example of sub-aclinal structure, the maximum dip being only 15 feet per mile. A non-productive structure of this class, mapped by The Associated Petroleum Engineers in Stephens County, Oklahoma, is shown in Fig. 16.

Class II.—Fields Associated with Anticlinal and Synclinal Structures

General discussion.—This is perhaps the most common class of oil accumulation, occurring frequently in the Appalachian, Ohio-Indiana, Illinois, Mid-Continent, Wyoming, northern Louisiana, and California fields in this country, in many South American and Mexican fields and in the fields of Russia, Galicia, Burma, and Borneo. Class II is divided into five sub-classes, in order to distinguish between various structural conditions under which oil is found associated with anticlines and synclines.

The original examples of Classes II and III come from Ohio, West Virginia and Pennsylvania. Therefore a generalized cross-section of these fields from west to east is shown in Fig. 17. The pools of Sub-class II (*c*) occur in the Trenton limestone on the crest of the Cincinnati anticline in northwestern Ohio and eastern Indiana.

A good example of a symmetrical anticline is, according to Thompson,[1] the Yenangyaung oil field of Burma, which has yielded the main oil supply in that country. The Bibi-Eibat field of Russia is mentioned by the same writer as another symmetrical anticline modified by doming and faulting. Asymmetrical anticlines are, however, most prevalent in oil fields, examples being the Grosny field of Russia, the Yenangyat field of Burma, and the Campina field of Rumania. A cross-section of the Grosny field is given by Kalitsky,[2] of the Yenangyat field by Pascoe[3] (see Fig. 20), and the Campina field by Mrazek[4] (see Fig. 21). Many of the fields of Poland are of this nature, a good example being the Boryslav-Tustanowice, or principal field of that country.[5]

Sub-class II (*a*)—Where strong anticlines stand alone.—In this division are included fields that bear a direct relation to very pronounced uplifts, easily recognizable, and constituting a marked geologic feature of the region. A prominent example is the Eureka-Volcano-Burning Springs anticline in West Virginia, which is over 50 miles in length, with a flat crest an eighth of a mile to half a mile broad, and side dips of from 20 to 60 degrees. It was one of the first anticlines to be recognized in this country, having been described by White,[6] Andrews[7] and Evans,[8] and extensively drilled. A map of a typical section of it is shown in Fig. 22. The pools of the La Salle anticline in Illinois and some of those in California belong to this class, as do some pools in Oklahoma, Kansas and Texas.

Sub-class II (*b*)—Where well-defined alternating anticlines and synclines exist.—This may be considered as a composite of Sub-class II. It includes the important pools of the Appalachian region in Pennsylvania and West Virginia, some of the pools in southern Indiana and Illinois, certain Oklahoma fields, the Caddo field of Louisiana, and certain ones in Wyoming. The anticlinal crests belonging to this sub-class may be from 2 miles to 200 miles apart.

[1] Thompson, A. Beeby: Trans. Inst. Min. and Met., Vol. XX, 1910–1911, p. 219 (1911).
[2] Kalitsky, K.: Mem. Geol. Com., St. Petersburg, No. 24, 1906.
[3] Pascoe, E. H.: Records Geol. Survey India, Vol. XXXIV, 1906.
[4] Mrazek, L. L.: Congress International du Petrole, 1907.
[5] Grybowski, J.: Bull, Acad. Sci., Cracow, 1907.
[6] White, I. C.: Bull. Geol. Soc. Am., Vol. X., 1899, pp. 277–284.
[7] Andrews, E. B.: Am. Jour. Sci., 2d ser., Vol. XXXII, 1861, pp. 85–93.
[8] Evans, E. W.: Am. Jour. Sci., 2d ser., Vol. XLII, 1866, pp. 334–343.

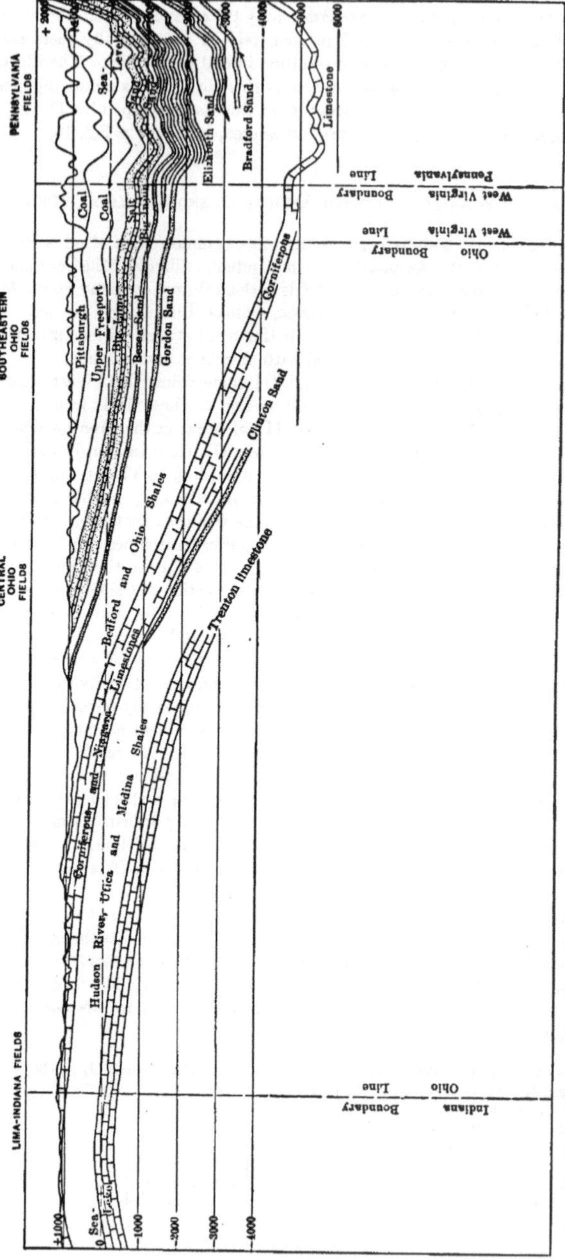

Fig. 17.—Generalized section from Cincinnati anticline to Allegheny Mountains; showing relative positions and differences in geologic structure of the Ohio and Pennsylvania fields. Scale: Horizontal, 50 miles = 1 inch; vertical, 5000 feet = 1 inch.

FIG. 18.—Looking along an anticline in Upper Cretaceous series in Natrona County, Wyoming, showing prominent structural ridge and dips north and south of it.

FIG. 19.—Beds of the Sarmatian formation of Upper Miocene age dipping away from saline dome, near Gezés, Transylvania, Rumania.

The strata in the fields of Sub-class II (*b*) are folded into alternating anticlines and synclines, having dips which are seldom as much as 30° from the horizontal. This is the sub-class to which the anticlinal theory was originally applied. It is illustrated in

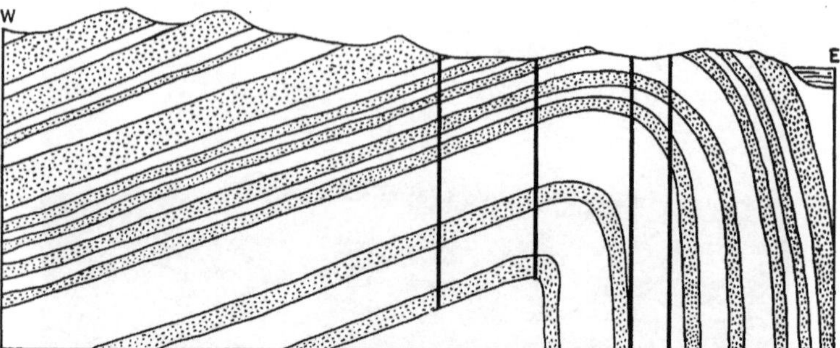

Fig. 20.—Cross-section of an asymmetrical anticline in the Yenangyat oil field, Burma (after Pascoe, Records Geol. Survey, India, Vol. XXXIV, 1905).

Fig. 23, where the productive sand is, or has been, wet; but in regions where the sands are practically dry the oil field occupies the syncline. Several of the California fields

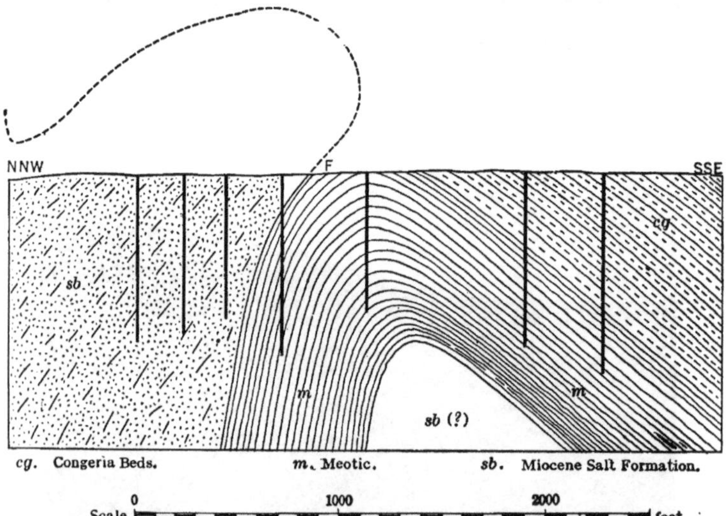

Fig. 21.—Cross-section of an asymmetrical anticline in the Campina oil field, Rumania (after Mrazek, Congress International du Petrole, 1907).

also belong in this sub-class, among these being the Coalinga and Los Angeles fields.[1] The Baku and Surakhany fields of Russia and the Yenangyaung field in Burma appear

[1] Eldridge, Geo. H.: Bull. 213, U. S. Geol. Survey, 1902, pp. 306–321. Arnold, Ralph, and Anderson, Robert: Bull. 357, U. S. Geol. Survey, 1908, pp. 70–71.

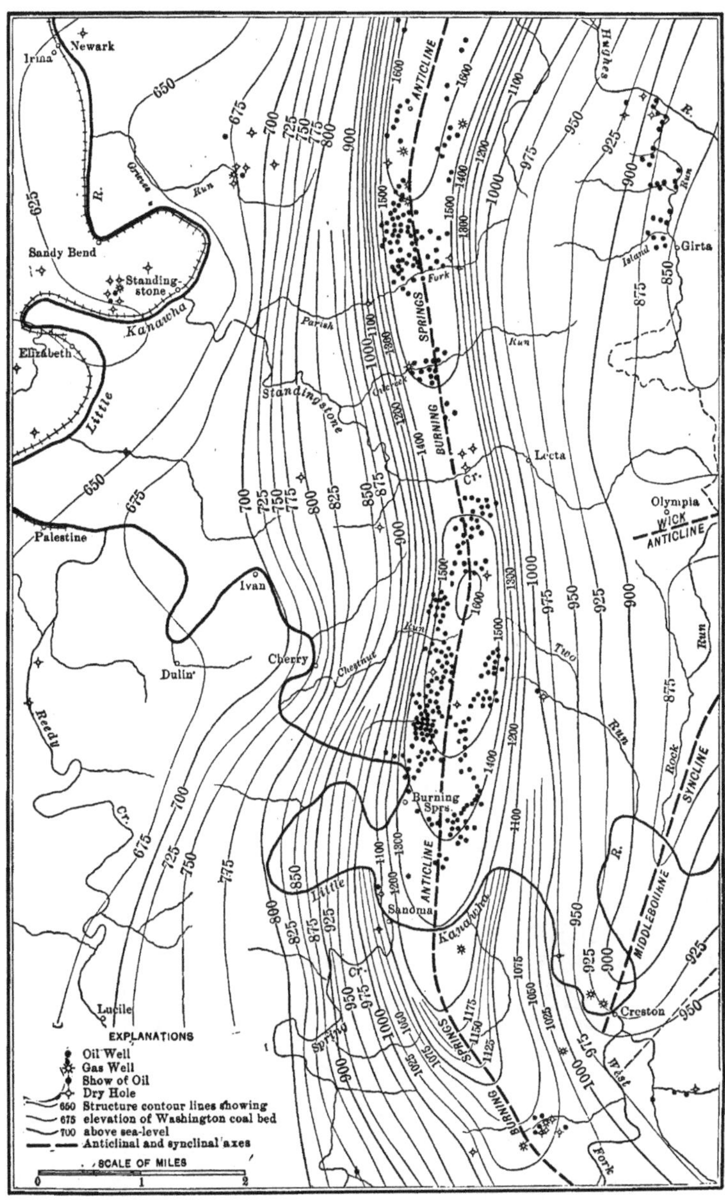

Fig. 22.—Sketch map showing geologic structure of a portion of the Eureka-Volcano-Burning Springs anticline, in Wirt County, West Virginia. to illustrate fields of Sub-class (II *a*), bounded on both sides by unproductive sub-aclinal structure of Class I. Contour interval 25 feet. (After Ray V. Hennen: County report and map of Roane, Wirt and Calhoun Counties, W. Va. Geol. Survey, 1911.)

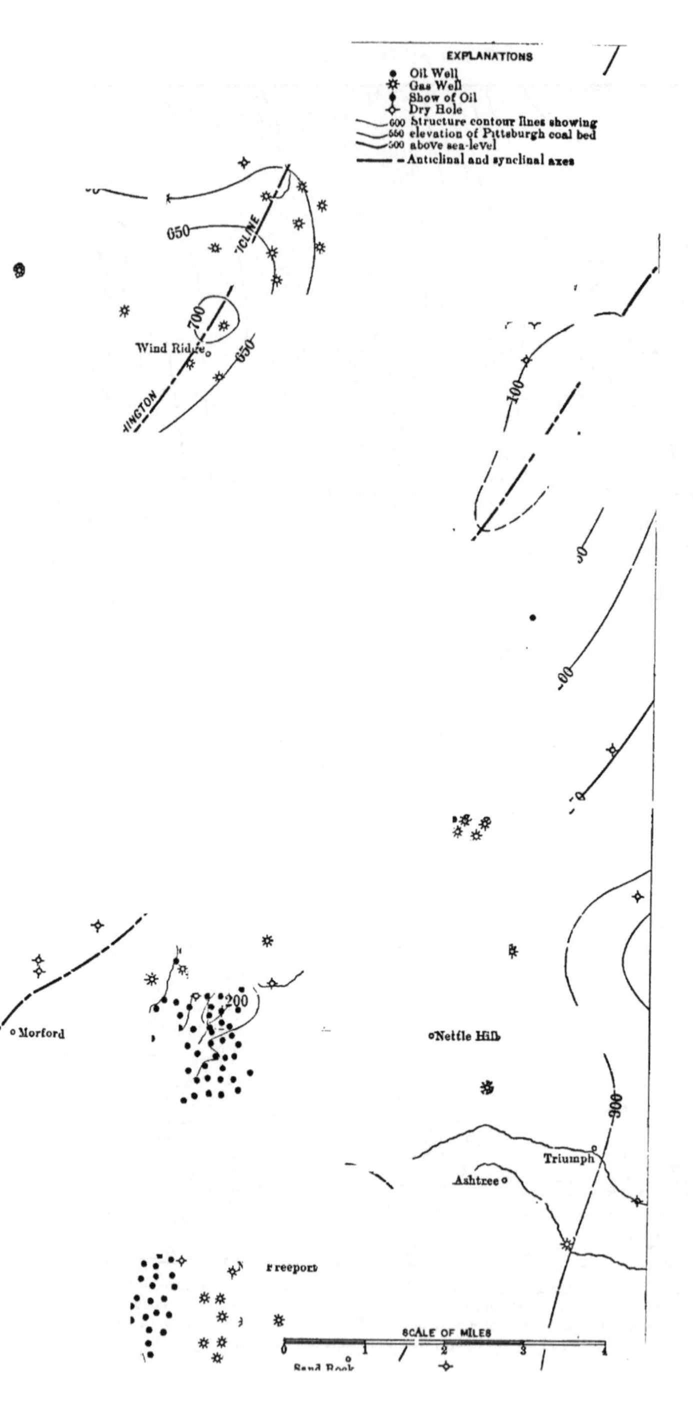
EXPLANATIONS
Oil Well
Gas Well
Show of Oil
Dry Hole
600 Structure contour lines showing
550 elevation of Pittsburgh coal bed
500 above sea-level
Anticlinal and synclinal axes
650
700
Wind Ridge
650
100
Morford
200
Nettle Hill
300
Triumph
Ashtree
Freeport
SCALE OF MILES
0
1
2
3
4

to belong here. The Negritos and Lobitos fields in Peru are reported to lie on the eastern flanks of a series of anticlinal structures, the axes of which are almost parallel with the shore of the Pacific Ocean.

Sub-class II (*c*).—**Broad geanticlinal** folds.—This is an extreme type of II (*a*). By a geanticline is meant an anticline so extremely long and broad that it constitutes more than a local feature and extends over thousands or tens of thousands of square miles. One of the best examples in this country is the Cincinnati anticline, a cross-section of which appears in Fig. 17. Another great geanticline, which is important for natural gas development, is that in southwestern Canada, extending north from the International Boundary.[1]

Sub-class II (*d*).—**Overturned** folds.—Examples of oil and gas occurring in connection with overturned folds are not common, but conspicuous instances are found in California, as shown by Arnold and Johnson (Fig. 24).[2] Other instances are said to occur in Poland and Rumania.

Sub-class II (*e*).—**Lenticular sands.**—In all types of structure there are numerous instances where sandstones or other porous beds are too hard or close-grained in certain places to hold the oil, or else they pinch out laterally between shales. A typical example is shown in Fig. 25, where the sand pinches out toward the anticlinal crest, causing the gas to collect on the side of the anticline and the oil still lower on its flank. The occurrence of this type in nature is illustrated in Fig. 26.

Class III.—Monoclinal Structure

The term monocline was proposed by W. B. and H. D. Rogers in 1842. It is defined by Anderson and Pack,[3] " in conformity with general usage," as " a succession of beds dipping in one direction"; but the word "homocline" has been used by some writers to express this idea.[4] A monoclinal dip is seldom, if ever, perfectly uniform for a distance of many miles, and it usually shows many changes of dip in a short distance. The steeper dips are generally confined within small areas, while the gentler ones are often uniformly continuous for many miles. On monoclinal dips the oil which was once widely disseminated through the porous strata has finally been accumulated at favorable positions where local interruptions break the regularity of the dip. In such cases gas has collected either on the up-dip side, where the sand is interrupted by pinching out, as in Sub-class III (*d*) or by local flattening as in Sub-class III (*c*), while oil has collected on the down-dip side of the structural interruption. In the Bremen field of Ohio the most productive oil wells are situated at the points of greatest change in rate of dip. Since the sand in that field appears to be dry, the accumulations are presumably due to catchment of the descending oil by these interruptions, during the lowering of the original water-level in the sand.

On a monoclinal dip there seem to be two main types of interruptions; (1) those due to longitudinal warping, parallel with the direction of strike of the sand, and (2) those due to lateral warping, parallel with the direction of the dip. The second type produces structural " noses," as illustrated in Fig. 32. The first type of interruption produces a " ravine " or " notch " in the monoclinal slope, such warpings being common in the Ohio fields. The ravine-like structure, being a conspicuous type of abnormality in monoclinal dips, is favorable for oil occurrence where the sands are dry. Actual

[1] Clapp, F. G., and Huntley, L. G.: Petroleum and natural gas resources of Canada, by F. G. Clapp and others, Vol. II, 1915, pp. 271–272.

[2] Arnold, Ralph, and Johnson, Harry W.: Bull. 406, U. S. Geol. Survey, 1901, p. 97.

[3] Anderson, Robert, and Pack, Robert W.: Bull, 603, U. S. Geol. Survey, 1915, p. 109.

[4] Johnson, Roswell H.: Science, N. S., Vol. XLII, pp. 450–452.

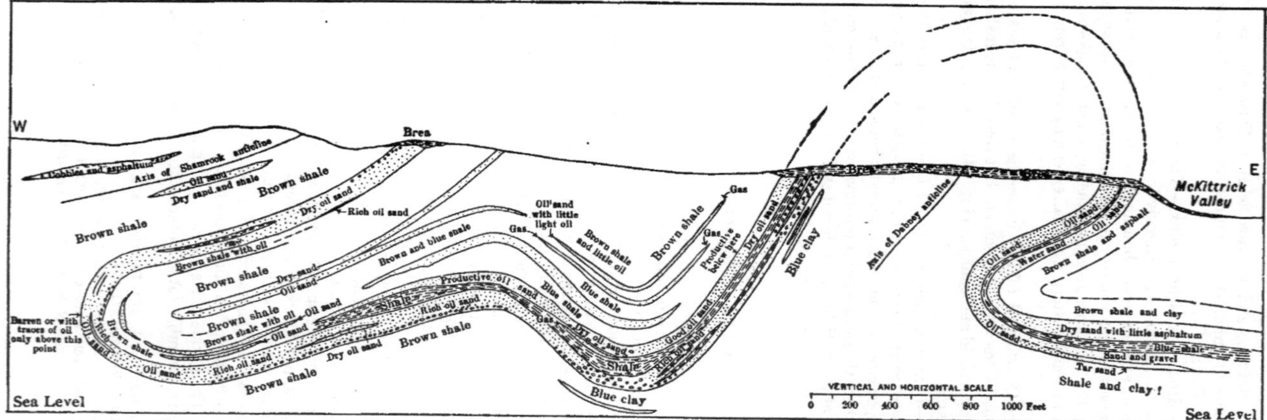

FIG. 24.—Hypothetical section across the south end of the McKittrick field, California, to illustrate occurrence of petroleum in structures of Sub-class II (*d*) and III (*d*), etc. (After Ralph Arnold and Harry R. Johnson, Bull. 406, U. S. Geol. Survey, Plate V, 1910.)

determinations of the structure of the Clinton sand under extensive areas have indicated that where types (1) and (2) intersect, as in many localities in the central Ohio pools, the largest accumulations of oil are found.

It is evident that fields of commercial importance are no more likely to occur in regions of plane monoclines than in absolute aclines, since no factors of separation exist in either case.

It must be remembered, of course, that if the sands contain water, as most sands do at some locality or other, an oil pool is likely to rest upon it. However, some degree of inclination is necessary to separate the oil from the water; and it has been found in practice that where the dip is less than half a degree the separation is so incomplete as to cause few if any commercial pools. Manifestly, the only way to locate an oil pool on a plane monocline is to drill for it, since the surface structure will afford no aid.

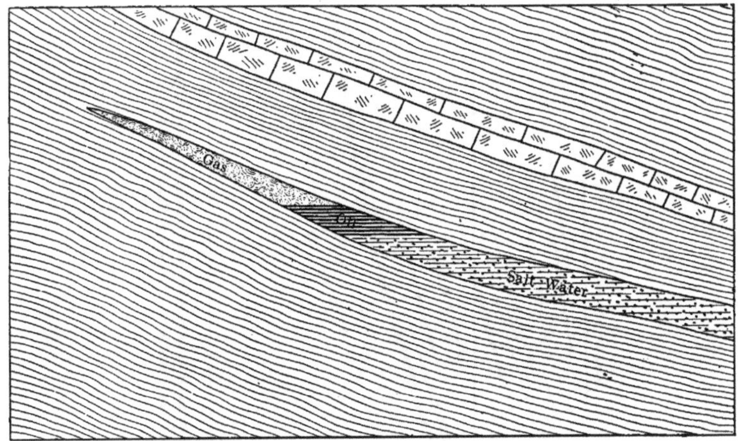

Fig. 25.—Ideal section of a sand. Showing the relations of gas, oil and water in structures of Sub-classes II (*e*) and III (*d*).

Sub-class III (*a*).—Monoclinal noses.—Attention was first called to the monoclinal nose type, although no particular name was given to it, by the present writer in 1910.[1] This type of structure is common in the gas-fields of central Ohio and in the north-central Texas fields. It may be considered a less prominent form of Sub-class III (c), if the terrace is not well defined. While it has been noticed by the writer mostly in Ohio and Oklahoma, a number of examples are known in Kentucky.

Sub-class III (*b*).—**Monoclinal ravines.**—The term "structural ravine" was perhaps first used by the writer in 1911.[2] It was applied to a structure bearing the same relation to an inclined sand as a topographic ravine bears to a sloping hillside. In the revised classification the term is changed to "monoclinal ravine," which is somewhat more specific.

Sub-class III (*c*).—Structural terraces or "**arrested anticlines.**"—Terrace structure was first described by Orton in 1866.[3] The terraces described by him were in the Find-

[1] Clapp, F. G.: Econ. Geology, Vol. V, No. 6, 1910, p. 508, Fig. 53.
[2] Clapp, F. G.: Econ. Geology, Vol. VI, No. 1, p. 10.
[3] Orton, Edward: Science, Vol. VII, p. 563.

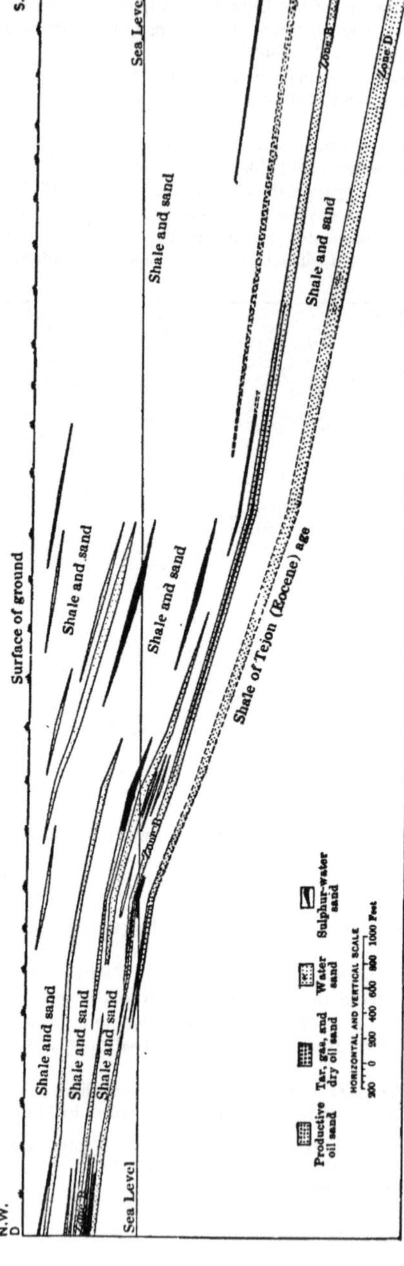

FIG. 26.—Hypothetical section through the Coalinga oil-field in California to illustrate the occurrence of oil in structures of Class III. (After Ralph Arnold and Robert Anderson, Bull. 398, Plate XX, U. S. Geol. Survey, 1908.)

lay field of northwestern Ohio, where the oil and gas occurred in two terraces, separated by a short, sharp monocline. The upper terrace yielded dry gas, the lower one oil and water.

While structural terraces might be described under Class II, they are more properly a variety of monoclinal structure and an extreme case of Sub-classes III (*a*) and (*b*). They were named by Orton[1] "arrested anticlines"; and the Macksburg field of southern Ohio was cited by him as an example of their occurrence, while the terrace structure of the Macksburg field was first described by Minshall[2] in the same volume. Other good examples of terrace structure in connection with oil occurrence have been found in Jefferson County, Ohio,[3] and described by Griswold and Munn. Fig. 27 is an illustration of this class of structure, taken from their report.

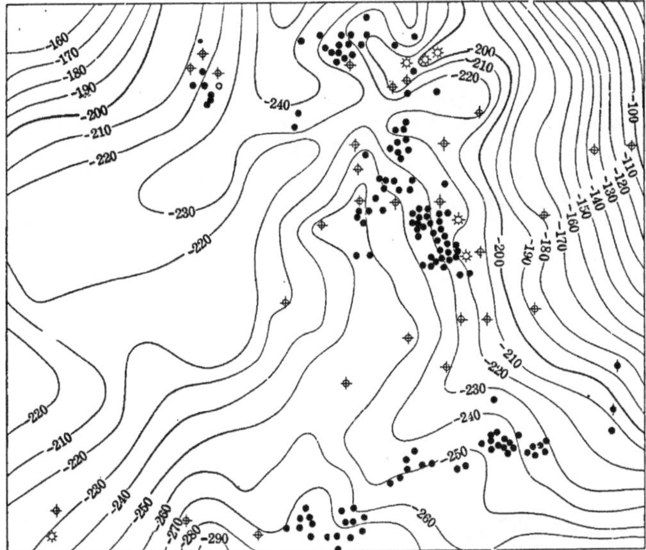

FIG. 27.—Occurrence of petroleum on structural terraces of Sub-class III (*c*) in southeastern Ohio. (After W. T. Griswold and M. J. Munn, Bull. 318, U. S. Geol. Survey, 1907.) For explanations see Fig. 22. Contour interval, 10 feet. Figures are distances below sea-level.

During the past two decades, hundreds of structural terraces have been discovered in Kansas, Texas, southeastern Ohio, and other states, and most of them are favorable for oil and gas development. Generally, though not always, the structure can be practically determined from the appearance of the surface, so that one is not obliged to bore until he is ready to make his test.

Sub-class III (*d*).—**Lenticular sands.—Accumulations on monoclines due to thinning out or change in texture of the sand.**—While it has been said that a change in texture or the dying out of the sand is not necessarily responsible for the exact positions and limits of oil pools on monocline dips, there are cases in which accumulations are due to

[1] Orton, Edward: Geology of Ohio, Vol. VI, 1888, p. 94.

[2] Minshall, F. W.: Geology of Ohio, Vol. VI, 1888, p. 95.

[3] Griswold, W. T. and Munn, M. J.: Bull. 318, U. S. Geol. Survey, 1907.

FIG. 28.—View in the Spindle Top pool, Jefferson County, Texas, ten years after the discovery of the field, showing the absence of the surface topography which normally overlies saline domes.

FIG. 29.—A basaltic cone in Mexico, to illustrate the type of topography frequently prevailing in fields of the volcanic plug type. Seepages exist and pools have been developed contiguous to such hills.

these causes. In the Louisiana fields, some of the oil occurs in lenticular sands, which thin out or grade laterally into shale.[1] This appears to be more frequently the case in Kansas, where the sands diminish in importance as they extend northward, than it is in Oklahoma, where they are more persistent. The relations of oil, gas and water are commonly similar to their general relations in any other monocline, except that the outlines of the pools are defined by the extent of the sands (Fig. 25). Similar lenticular sands are abundant in the California fields (Fig. 26). Doubtless a great number of examples of this type exist; but the most typical one known to the writer is the so-called " Clinton sand " of Ohio, which rises westward from the Appalachian basin and gradually grows thinner as it passes through central Ohio so that it never reaches the surface. The feather edge is enclosed by shale. This formation furnishes an ideal substitute for an anticline and is the repository of one of the greatest gas fields in the world, while on its lower border are the Bremen, Junction City, Wooster, New Straitsville and several other oil fields. A cross-section of the west side of the Appalachian basin, illustrating Sub-class III (*d*), is shown in Fig. 17.

Class IV.—Quaquaversal Structures or "Domes"

General discussion.—In the classification of oil pools, the subdivision entitled " Quaquaversal structures " is considered as including those structures in which the sand descends in all directions from a central point, such as the saline domes of Louisiana, certain domes in Oklahoma and West Virginia, the basalt plugs of Mexico, and the perforated and non-perforated salt domes of Rumania.

Sub-class IV (*a*).—**Anticlinal** bulges or " **cross-anticlines.**"—This type of structure merges into those described in Sub-classes II (*a*) and II (*b*) of the classification, since practically all anticlines consist of alternate contractions and bulges, where their crests are respectively depressed or elevated. Anticlinal bulges are frequent in Wyoming and are found in Pennsylvania and West Virginia, and occasionally in Ohio. The fact that the Wheeler and Healdton pools in southern Oklahoma and the Petrolia pool in northern Texas owe their position to distinct doming of the strata seems to have been first mentioned by Gardner;[2] and the structure of the Petrolia pool was originally deciphered by Udden and Phillips.[3] Many of these domes have since been mapped by the United States Geological Survey and by the writer of this chapter, as well as by many geologists who have had occasion to study them in the course of private work.

Sub-class IV (*b*).—**Monoclinal** bulges.—This type is frequently confused with the anticlinal bulge type, but is quite distinct in structure. Anticlinal bulges are expansions and elevations in the crests of definite anticlines or continuous folds, while monoclinal bulges are domes that rise with apparently irregular spacing on a monoclinal slope in which structures of Class III also exist. In Sub-class IV (*b*) the monoclinal structure gives place locally to a quaquaversal structure. Few domes are known in the Clinton sand, on the great monocline of central Ohio; and there are long stretches in the Berea where none are found. In Kansas and Oklahoma, however, monoclinal bulges are among the commonest types of structure. Since in the last mentioned state the sands are commonly saturated with water, the oil and gas occur high on the dome.

An example of the monoclinal bulge type is shown in Fig. 32.

Sub-class IV (*c*).—Closed **saline domes.**—The credit of having discovered oil in domes of this form is due largely to Capt. Lucas,[4] who drilled in 1901 at Spindle Top,

[1] Harris, G. D.: Bull. 429, U. S. Geol. Survey, 1910, pp. 128–129.

[2] Gardner, James H.: Econ. Geology, Vol. X, No. 5, 1915, pp. 422–434.

[3] Udden, J. A., and Phillips, D. McN.: Tex. Univ. Bull., 246, 1912.

[4] Lucas, A. F.: The dome theory of the Coastal Plain, Science, N. S. Vol. XXXV, No. 912, June 21, 1912, pp. 961–964.

Texas, and opened a famous well. As early as 1894, he had made diamond-drill borings at Jefferson Island, Belle Isle, Weeks Island, and Anse La Butte, Louisiana, and had discovered salt masses of limited area but of great depth. Lucas'[1] paper on this subject appeared in 1899 and his discoveries were confirmed by a paper published by Hill in 1902.[2] This type of quaquaversal structure was described by Hayes and Kennedy in 1903[3] and more fully by Fenneman in 1906.[4] The saline domes of Louisiana were described by Harris in 1908,[5] 1909[6] and 1910.[7] The structure is typical of most of the

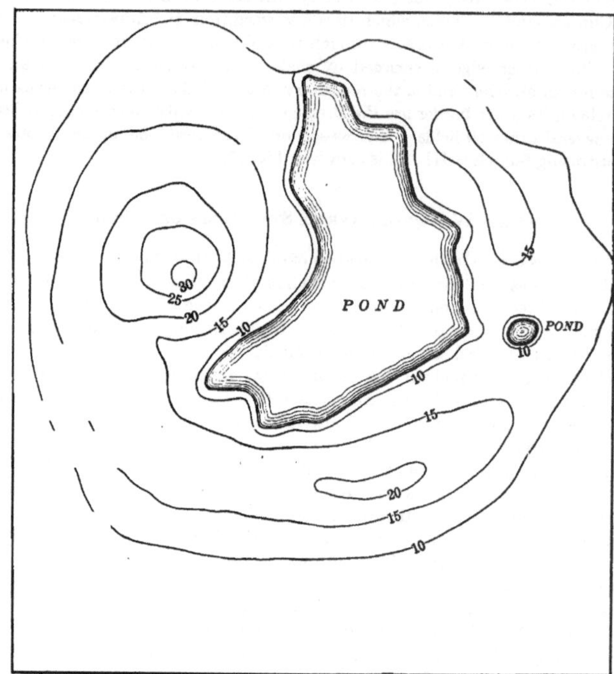

FIG. 30.—Sketch-map showing surface topography of the Vinton dome, an oil pool of Sub-class IV (c) in Southern Louisiana.

Gulf Coast fields in Louisiana and Texas, or those situated within 100 miles of the Gulf of Mexico. The Caddo and Corsicana fields, Mexia field, the North and Central Texas fields and the West Texas fields are not of this type.

[1] Lucas, A. F.: Rock-salt in Louisiana: Trans. Am. Inst. Min. Eng., 1899; also Journ. Ind. and Eng. Chemistry, Vol. iv, No. 2, Feb., 1912.

[2] Hill, Robert T.: Journ. Franklin Inst., Vol. cliv, Aug. and Oct., 1902, pp. 143, 225, 263.

[3] Hayes, C. W., and Kennedy, William,: Oil fields of the Texas-Louisiana Gulf Coastal Plain, Bull. 212, U. S. Geol. Survey.

[4] Fenneman, N. M.: Oil-fields of the Texas-Louisiana Gulf Coastal Plain, Bull. 282, U. S. Geol. Survey.

[5] Harris, G. D.: Rock salt, Bull. No 7, Rept. of 1907, La. Geol. Survey.

[6] Harris, G. D.: Geological occurrence of rock salt in Louisiana and East Texas. Econ. Geology, Vol. iv, No. 1, 1909, pp. 12-34.

[7] Harris, G. D.: Oil and gas in Louisiana. Bull. 429, U. S. Geol. Survey.

In southern Louisiana there are five elevations known as the "Five Islands," or "South Islands," which constitute the most prominent landmarks for hundreds of miles along the coast of the Gulf of Mexico.[1] These are salt domes. It must not be supposed, however, that the presence of the geologic structure known as a saline dome is always indicated on the surface by a topographic dome, as many saline-dome pools are situated where the surface is practically flat. In the other cases the surface is irregular, or even depressed, as in Fig. 30, illustrating the topography of the surface of the Vinton dome in Louisiana. In the case of the Spindle Top dome, the bulge on the surface is not over 10 feet in height.

Whether or not there is any particular surface topography indicative of a dome in a locality of this class, there is a very marked upward bending of the strata as they

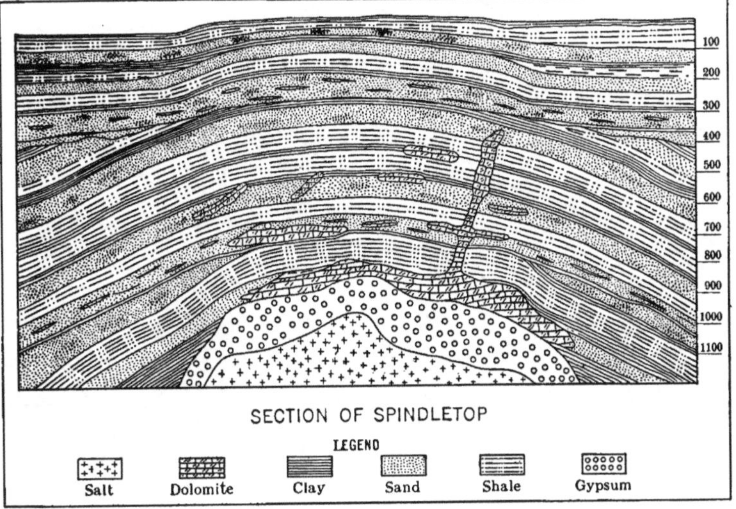

FIG. 31.—Hypothetical cross-section of the Spindle Top oil-pool in Texas, to illustrate Sub-class IV (c). (After G. D. Harris, Bull. 7, La. Geol. Survey, 1907, Fig. 19.)

approach the edge of the dome, so that they may be practically vertical at its circumference. An uplift of several thousand feet within a distance of 1 mile is not uncommon (see Fig. 31). While no Cretaceous beds of normal structure reach the surface in Louisiana, there are several saline domes in which these formations have been uplifted to the surface in limited areas; and extensive deposits of rock salt, sulphur, gypsum, and other minerals lie beneath the Cretaceous beds and are interlaminated with them in the domes.

Practically all the known saline domes except those discovered within the past twelve years have been described and mapped by Veatch [2] and Harris;[3] but not all of them contain oil.

In Texas and Louisiana, as in the Transylvanian fields of Rumania, the saline domes occur at the intersection of probable fault-lines. This fact was first mentioned by

[1] Lucas, A. F.: Trans. Am. Inst. Min. Engrs., Vol. XXIX, p. 464.
[2] Veatch, A. C.: La. Geol. Survey, Rept. 1902, pp. 41–100.
[3] Harris, G. D.: Bull. 429, U. S. Geol. Survey, 1910.

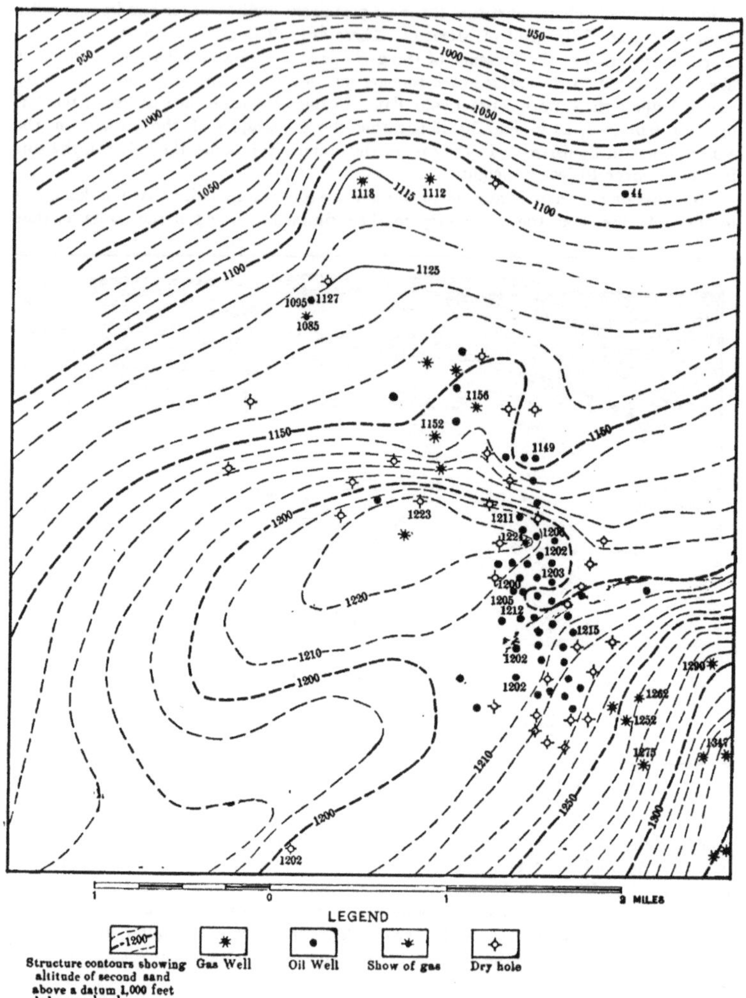

FIG. 32.—Map showing geologic structure in the vicinity of Strawn, Texas, to illustrate the monoclinal-bulge type of dome structure, Sub-class IV (*b*). Contour interval 10 feet. (After E. W. Shaw, Bull. 629, U. S. Geol. Survey, Plate III.)

Lucas,[1] and was also perceived by Hayes and Kennedy,[2] who published a map showing possible lines of flexures or faults. The alignment was mapped in greater detail by Harris.[3]

Theories of the origin of the salt domes are given in the articles listed in the footnotes.[4]

Sub-class IV (*d*).—**Volcanic plugs.**—The best known examples of the volcanic-neck type of quaquaversal structure are found in Mexico, where there are oil fields of various types of geologic structure. The type in question consists of plugs or necks of basalt which rise through the Cretaceous and Tertiary sediments in the Coastal Plain to elevations of several hundred feet above the surface. At the base of the upheavals and in close proximity to them the Tamasopo limestone and overlying formations make pockets in which large deposits of oil have accumulated. In the Tamasopo limestone and the San Felipe beds these oil deposits are presumably the result of concentration from surrounding portions of the same strata.

The presence of oil accumulations surrounding the plugs is sometimes, although not always, indicated by large seepages of oil. In certain cases the lower beds actually reach the surface and a true quaquaversal structure exists. Doubt has been expressed as to whether this is commonly the case;[5] but it is certain that there is definite doming around some of the plugs. At any rate, where the plugs are present, pockets of oil have accumulated, and the plugs themselves may be considered as quaquaversal structures.

It would appear that large deposits of oil might be expected in the vicinity of such intrusive masses, in cases where porous sands are overlain by a suitable cover, so as to prevent the escape of oil. Where the impervious covering or cap-rock is unusually thick or without fractures, seepages may be absent, although they exist in the vicinity of most of the plugs.

The presence of igneous rock in saline domes has never been definitely established; but Capt. Lucas thought he had igneous rock beneath the salt in a 3300-foot well at Belle Isle, Louisiana.[6] That the volcanic-neck type is probably more common than we know is suggested by the fact that on the southern edge of the Transylvanian basin in Rumania, where saline domes, arranged in slightly curved linear series similar to those of Louisiana, are the prevailing type, a plug of basalt rises through the plain, at the very point where the saline dome might be expected.

Sub-class IV (*e*).—**Perforated saline domes.**—In Rumania the saline-dome type of structure frequently reaches an exaggerated phase, owing to the fact that the dome-shaped salt masses have reached the surface of the earth, and the surrounding strata have been compressed outward to such an extent that they are vertical, or in some cases even overturned, in a narrow belt surrounding the dome.[7] In Rumania, large oil fields are found in such structures, which were originally described by Professor Mrazek as "diapir structures," or perforated domes. Perforated domes may be considered as an intermediate class between Sub-classes IV (*c*) and IV (*d*). So far as we know, the oil-

[1] Lucas, A. F.: Trans. Am. Inst. Min. Eng., Vol. XXIX, 1899, p. 463, Fig. 1.

[2] Hayes, C. W. and Kennedy, Wm.: Bull. 212, U. S. Geol. Survey, 1903, p. 144.

[3] Harris, G. D.: Bull. 429, U. S. Geol. Survey, 1910.

[4] Hager, Lee: Eng. and Min. Jour., Vol. LXXVIII, 1904, pp. 137-139 and 180-183. Washburne, C. W.: Trans. Am. Inst. Min. Eng., Vol. XLVIII, 1914, p. 691. Norton, E. G.: Origin of the Louisiana and east Texas salines. Am. Inst. Min. Eng., Bull. No. 97, Jan. 1915, p. 93. Dumble, E. T.: The occurrences of petroleum in eastern Mexico as contrasted with those of Texas and Louisiana; Fuel-Oil Journal, Oct. 1915, p. 86. Garfias, V. R.: The effect of igneous intrusions on the accumulation of oil in northeastern Mexico; Jour. Geol., Vol. XX, No. 7, 1912, p. 666.

[5] DeGolyer, E.: The effect of igneous intrusions on the accumulation of oil in the Tampico-Tuxpan region, Mexico, Econ. Geology, Vol. X, 1915, pp. 651–662.

[6] Lucas, A. F.: Trans. Am. Inst. Min. Eng., Vol. XLVIII, p. 693.

[7] Bosworth, T. O.: Pet. Review, March 23, 1912, p. 172. Thompson, A. Beeby: Trans. Inst. Min. Met., Vol. XX, 1910–1911, p. 223, Fig. 40.

bearing domes of this type are limited to Rumania, although it is probable that some saline domes of the Gulf Coastal Plain of the United States may be of the perforated type. Fig. 33 is an illustration of a typical occurrence.

Class V.—Contact of Sedimentary and Igneous Rocks

Sub-class V (*a*).—Contact of sedimentaries with volcanic plugs.—The importance of this class is attested by the fact that the close association of seepages with the volcanic plugs of Cerro de la Pez and Cerro de la Dicha was the direct cause of the discovery of the Ebano field in Mexico. A large number of seepages also occur surrounding volcanic plugs at Cerros Chapapote and Las Borrachas near Juan Felipe, at Cerros Palma Real and Cacalote near Potrero del Llano, at Cerro Pelon near Solis, at Mata de Chapapote, at Caracol and Apachiltepec on Tlacolula, and many other places in Mexico. The principal function of igneous intrusions in the accumulation of oil is believed by DeGol-

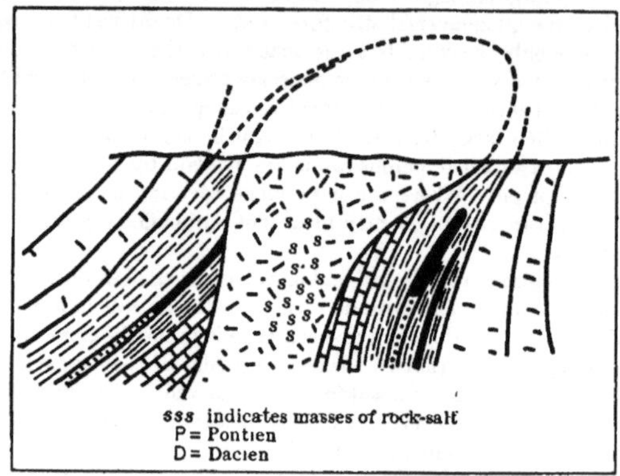

Fig. 33.—Cross-section of oil-field at Baicoi, Rumania, illustrating occurrence of oil in structures belonging to Sub-class IV (*e*). (After Bosworth, Pet. Review, March 23, 1912.)

yer[1] to have been the formation of channels through which the oil has been able to migrate into the overlying formations and even to reach the surface. A secondary and relatively unimportant function of intrusion is believed to have been the formation by brecciation and metamorphism of reservoirs capable of containing oil.

Sub-class V (*b*).—Contact of sedimentaries with dikes.—Some of the Mexican seepages occur along dikes of basaltic rock in the Tertiary sediments. The Tampalachi seepage near Panuco occurs on a concealed dike; and there are seepages along dikes at Tamijuin, Acala, Chapapote in the San José de las Rusias hacienda, and 1 mile southeast of Cervantes. This last is illustrated in Fig. 34.

Sub-class V (*c*).—Contact of sedimentaries with intrusive beds or laccoliths.—There are only a few places in the world where oil is positively known to occur below intrusive beds. Fig. 35 illustrates this type of structure, which undoubtedly exists in Mexico and is believed by some to occur in Cuba. The Furbero field is reported by DeGolyer[2] to overlie and underlie an altered laccolith of gabbro in the baked and broken Mendez

[1] DeGolyer, E.: Econ. Geology, Vol. X, 1915, p. 661.
[2] Ibid, p. 653.

shales (Fig. 35). Thompson [1] supposed that the structure of the field on the island of Tcheleken in the Caspian Sea might be due to an underlying laccolith.

Sub-class V (*d*).—**Contact of sedimentaries with older igneous rocks.**—While no deposits of oil are positively known to occur in structures of this type, a limited amount

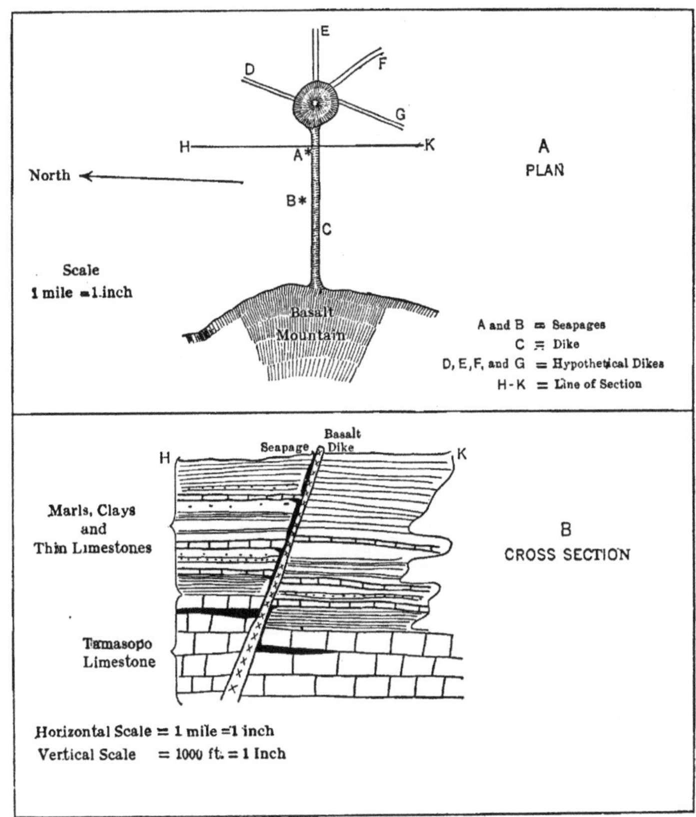

FIG. 34.—Plan and hypothetical cross-section of a dike in Mexico, showing occurrence of oil seepage in structure belonging to Sub-class V (*b*).

of gas has been found under such conditions in the provinces of Quebec and Ontario, Canada, and in northern New York State.

CLASS VI.—STRATA DIPPING UNCONFORMABLY AWAY FROM OLD SHORE-LINES

A good example of oil in an unconformity on a monocline is found in the Maikop field of Russia,[2] where strata of Upper Oligocene and Lower Neocene age, dipping 7 to

[1] Thompson, A. Beeby: Trans. Inst. Min. and Met., Vol. XX, 1910–1911, p. 230.

[2] Ibid, p. 229.

10 degrees, rest unconformably on, or overlap, Cretaceous strata, as shown in Fig. 36. The unconformity has sealed up the upper end of the Neocene sands. The structural

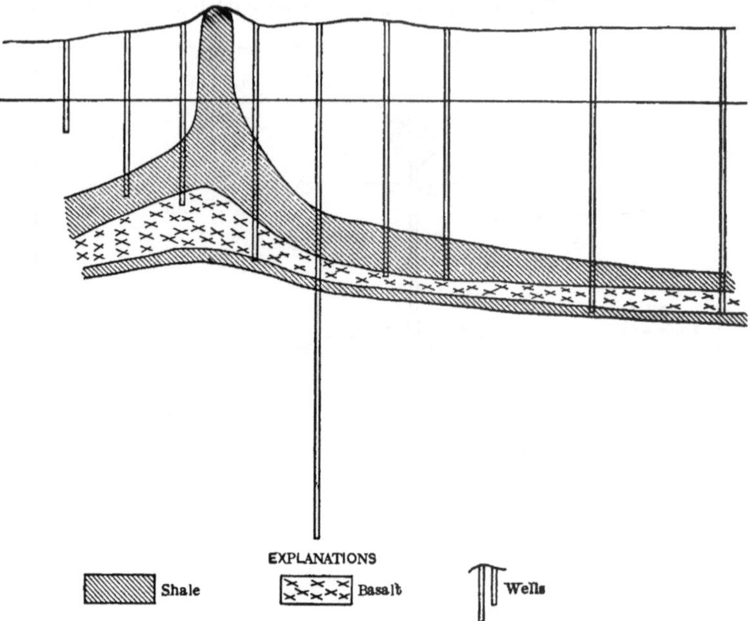

Fig. 35.—Cross-section of the Furbero field, Mexico, to illustrate occurrence of oil in structures belonging to Sub-classes V (c), V (d) and VII. Igneous rock is indicated by crosses in pattern. (After DeGolyer, Trans. Am. Inst. Min. Eng., Vol. LII, 1915, Fig. 2.)

relations of certain pools in the vicinity of the Arbuckle and Wichita Mountains in southern Oklahoma are also of this class.

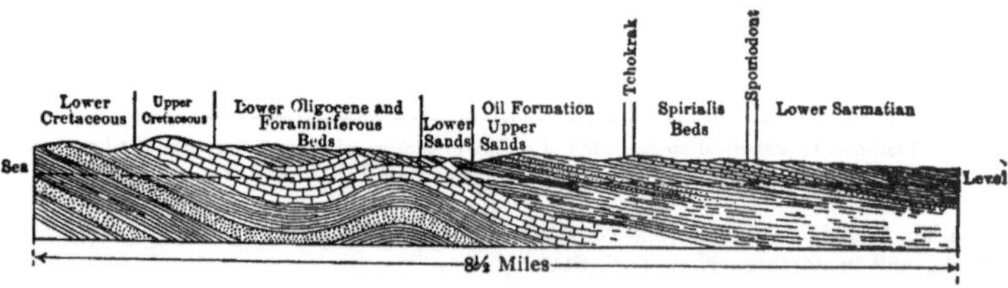

Fig. 36.—Geological section through Shirvanskaya Wells, Maikop field, Russia, showing occurrence of petroleum according to Sub-class VI. (After A. Beeby Thompson.)

Class VII.—Crevices of Igneous Rocks

Petroleum and solid bitumen in small quantities have been noticed by various observers in traps, basalts, and other igneous rocks. An interesting instance was mentioned by Logan [1] in a greenstone dike at Tar Point, Gaspé, in the Province of Quebec. Another occurrence, mentioned by Dr. David T. Day, is a boulder of vesicular basalt in Colorado, in which the vesicles were filled with oil. In the vicinity of Binny Craig, Scotland, a volcanic neck, or pipe, of trap was encountered in an oil shale working. According to Cadell [2] this dike contains cavities in which mineral wax, pitch, or paraffin have been found. In his opinion, however, these phenomena do not indicate igneous origin of the oil, but are due to the intrusion of the volcanic rock from below into the sedimentary formations which contained the oil, and to the subsequent absorption by the volcanic rock of large quantities of oil from the surrounding formations.

Rateau [3] states that at Roczk, Galicia, a trachitic rock is impregnated with petroleum. Near Pagosa Springs, in Archuleta County, Colorado,[4] there are dikes of igneous rock, reported to be more or less saturated with petroleum, which is associated with hot sulphur water. DeGolyer [5] mentions probable examples of oil seepages from dike fissures at Cerros de la Pez and de la Dicha, at Ebano, Mexico. The best example known to the writer is at Cerro de Chapapote, between Tepezintla and Pierre Labrada, Mexico, where oil is visibly seeping out of a basalt plug at a point about 60 feet above its base. Washburne [6] mentions the occurrence of small amounts of oil in porous basalt on the Johnson Ranch, on the North Fork of Siuslaw River, in the western part of Lane County, Oregon.

A somewhat different type of occurrence is found in granite and associated crystalline rocks on Copper Mountain [7] in the northeastern part of Fremont County, Wyoming. For many years, asphalt, oil tar, or "brea," as it is frequently called, was gathered for fuel from various points on the granite mountain. A detailed study of the geology of this region has been made by Darton,[8] who found that Copper Mountain is a dome, and that its granite core was at one time overlain by stratified rocks, which have been removed by erosion. The brea and deposits of heavy oil have accumulated in hollows in the upper part of the mountain, and oil has been encountered in shafts and tunnels high up in the granite. The oil first accumulated in the Embar sandstone of Permian age which overlay the dome, and when faulting occurred the residual petroleum settled into the crevices of the granite as far down as water level. The downward migration is supposed to have covered a distance of 2000 feet, but its direction was probably lateral, as it took place during tilting of the rocks to form the dome.

Class VIII.—Crevices of Sedimentary Rocks

The presence of oil in the Florence field of Colorado is shown by Washburne [9] to be due to joint cracks and fissures in shale. The oil does not follow any particular beds or series of beds, and the oil zone contains no sandstone or other porous formations. In northeastern Ohio, southern Texas, and in some localities in Wyoming, small oil and gas wells are frequently located in shales, where there are no known domes, anticlines, or other favorable structures. The oil obtained from these wells is called "shale oil," or "crevice oil," to distinguish it from the product of normal structural types. Crevice oil is believed by many to be indigenous to the beds in which it is found.

[1] Logan, Sir William: Geology of Canada, 1863, pp. 400–789.

[2] Cadell, Henry M.: Oil-shale holdings of the Lothians. Trans. Inst. Min. Engrs., Vol. XXII, pp. 347–353.

[3] Rateau, M. A.: Annales des Mines, 8th ser., Vol. XI, p. 152.

[4] Oliphant, F. H.: Mineral resources of United States for 1910. U. S. Geol. Survey (1912), p. 561.

[5] DeGolyer, E.: Econ. Geology, Vol. X, 1915, p. 655.

[6] Washburne, C. W.: U. S. Geol. Survey, Bull. 590.

[7] Trumbull, L. W.: Bull. No. 1, Scientific Series, State of Wyoming Geologist's office, 1916, pp. 5–16.

[8] Darton, N. H.: Prof. paper 51, U. S. Geol. Survey.

[9] Washburne, C. W.: Bull. 381, U. S. Geol. Survey, 1910, pp. 521–523.

Class IX.—Oil Associated with Closed Faults

General discussion.—Known examples of this class are found in the Los Angeles field and in the Lompoc field in California, as described by Arnold.[1] In these cases the steeply inclined oil sands are cut off abruptly by faults, which seal in the oil and prevent its escape to the surface. In regard to the probable frequency of such occurrences, it may be worth while to state that oil springs occur along fault-lines in British Columbia, the Province of Quebec, Wyoming and Mexico. This type of structure is illustrated in Figs. 37 and 38.

Sub-class IX (*a*).—**Oil on** the upthrow side.—Examples of oil on the upthrow side of faults are frequent in Oklahoma (Fig. 38) and they may exist elsewhere.

Sub-class IX (*b*).—**Oil on the** downthrow side.—A good example of oil along the downthrow side of faults in the Coalinga field in California is illustrated in Fig. 37. Numerous other examples doubtless exist.

Oil on both sides of **the fault.**—In many of the Oklahoma fields, oil is found on both the upthrow and the downthrow sides. In the Bibi-Eibat field of Russia, oil occurs on both sides of the normal faults which cut the crest of the anticline, as shown by D. Golubiatnikoff,[2] and, according to Thompson,[3] the faults in the Bibi-Eibat field (Fig. 39) exercise an important influence on the production of the wells. The faults are inclined from the vertical, and the wells are drilled so as to strike them at great depth. In some cases, the wells on one side of the fault are much more productive than those on the other.

Sub-class **IX** (*c*).—**Oil** along overthrust faults.—Examples of oil-fields along overthrust faults are well known in Rumania. A cross-section of the Bustenari field (after Bosworth [4]) is shown in Fig. 40. A less important instance is found in the Pincher Creek pool of southern Alberta. One of the best examples is described by Arnold and Johnson [5] and occurs in the McKittrick field in California (Fig. 41), where the shales of the Monterey and Santa Margarita formations of Middle Miocene age are supposed to have been overthrust along a slightly inclined plane, on top of the McKittrick gravels, clays, and oil sands of Upper Miocene age. In the Whittier pool the oil sands are found in a vertical position on the downthrow side of the overthrust fault (see Fig. 42).

Class X.—Oil Sands Sealed in by Bituminous Deposits at Outcrop

The pitch lakes of Trinidad and Venezuela are the best known examples of this type of oil occurrence. In some California fields [6] the outcrop of the sands is believed to be closed by brea. A portion of the oil found near the vein of grahamite at Ritchie Mines, West Virginia,[7] may belong to this class, although the accumulation here is also due to anticlinal and synclinal structures belonging to Sub-class II (*b*). The source of the West Virginia grahamite dike is believed to have been the Cairo oil sand, which lies at a depth of about 1300 feet from the surface; and there is no doubt that a part of the oil is held in by grahamite, or has been so held in at some time in the past.

The albertite dike in Albert County, New Brunswick, is due to oil which has intruded

[1] Eldridge, Geo. H. and Arnold, Ralph: The Santa Clara Valley, Puente Hills and Los Angeles oil districts, southern California, Bull. 309, U. S. Geol. Survey, 1907.

[2] Golubiatnikoff, D.: Bull. Geol. Com., St. Petersburg, Vol. XXIII, 1904.

[3] Thompson, A. Beeby: Petroleum mining and oil-field development, New York and London, 1910, p. 57.

[4] Bosworth, T. O.: Pet. Review, March 23, 1912, p. 172.

[5] Arnold, Ralph, and Johnson, Harry R.: Bull. 406, U. S. Geol. Survey, 1910, pp. 97–99.

[6] Ibid., pl. V.

[7] White, I. C.: Bull. Geol. Soc. Am., Vol. XIX, 1899, pp. 277–284.

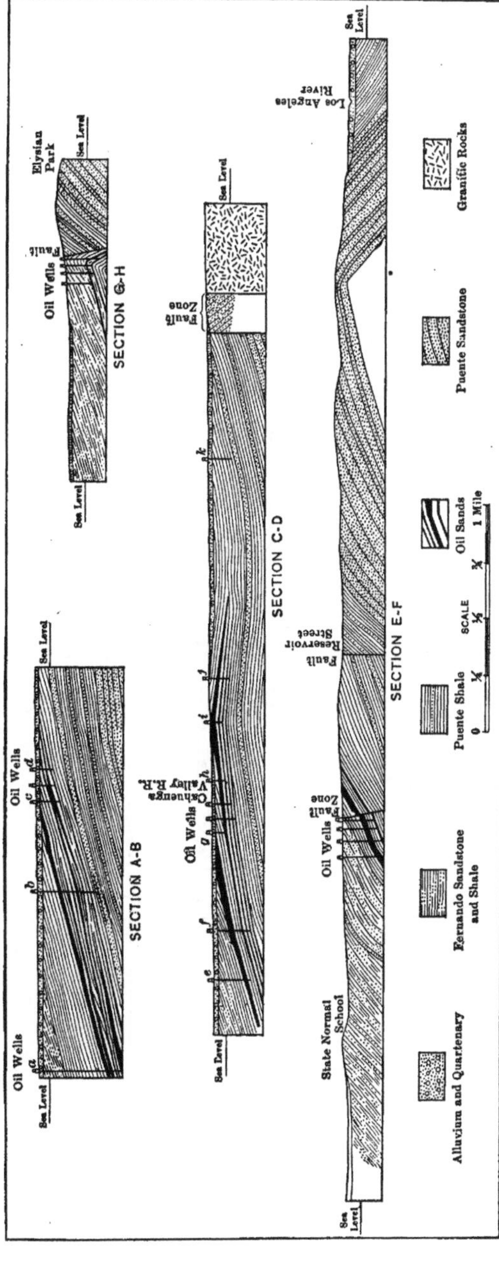

FIG. 37.—Geological sections through the Los Angeles oil-fields, California, to illustrate occurrences of petroleum in structures belonging to Sub-classes II (*b*), II (*c*) and IX (*b*). (After Eldridge and Arnold, Bull. 309, U. S. Geol. Survey, Plate XX.)

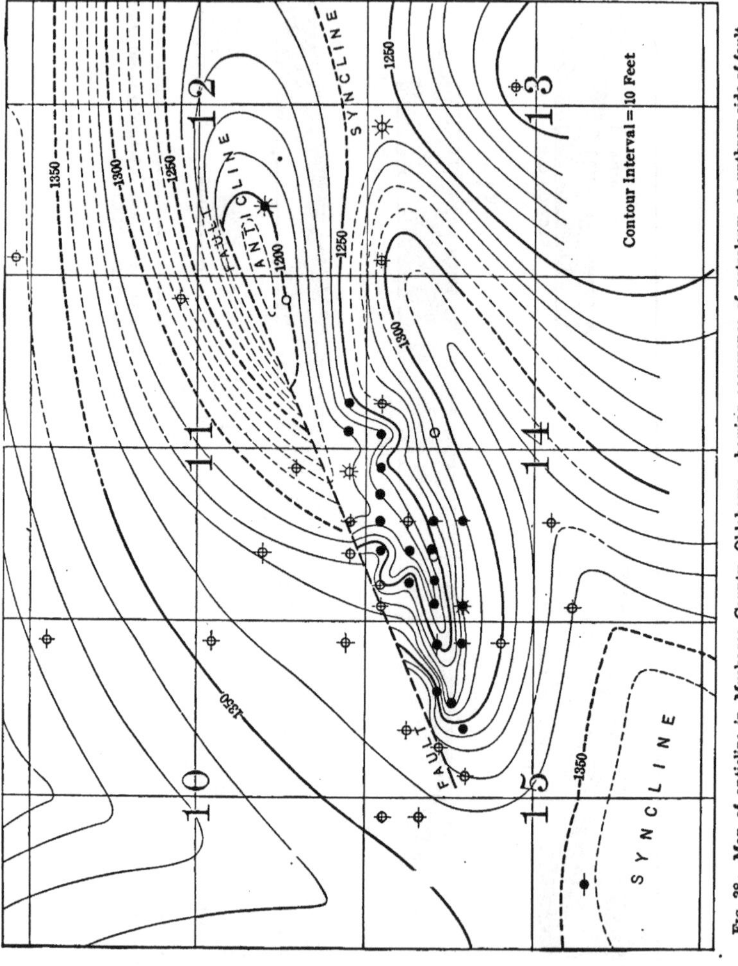

FIG. 38.—Map of anticline in Muskogee County, Oklahoma, showing occurrence of petroleum on upthrow side of fault, belonging to Sub-class IX (a). (Geology by F. G. Clapp.) Contour interval 10 ft.

from petroliferous strata and filled a large vertical fissure in the Albert shales [1] of Mississippian or Devonian age. Dikes of grahamite or similar bitumens exist in Stephens, Carter, Pushmataha, and other counties in southern Oklahoma. The uintaite (gilsonite) of Utah has been shown by Eldridge to occupy a fractured zone in the central Uinta synclinal basin. Oil will not be found in proximity to all such dikes, as some of

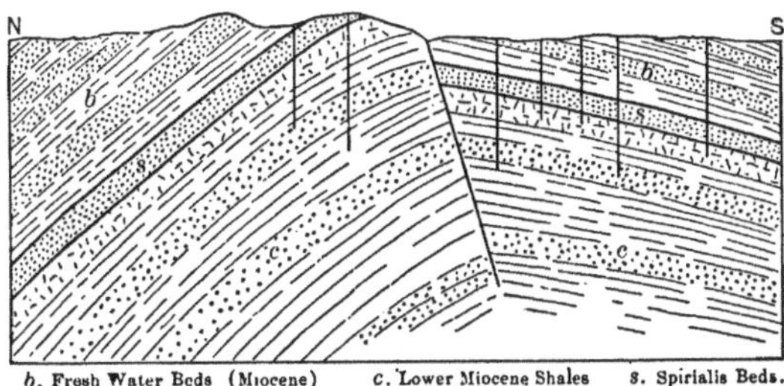

Fig. 39.—Geological section through Bibi-Eibat field of Russia, showing occurrence of petroleum on both sides of faults, belonging to Sub-classes IX (a) and (b) and also falling in Sub-class II (b). (After Thompson.)

the bitumens show by their composition that the locality has suffered too great metamorphism. In some cases, however, the deposits furnish indications of oil.

To illustrate the importance of bituminous dikes as indications of the former presence of petroleum and natural gas, it may be said that the grahamite dike of West Virginia stands in the center of one of the greatest oil and gas regions in the world; that the

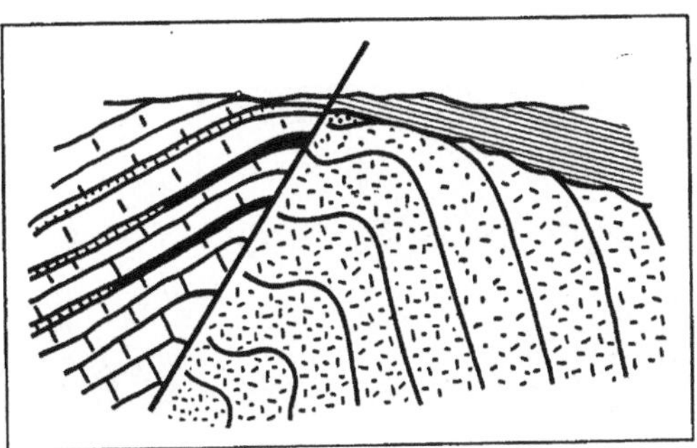

Fig. 40.—Cross-section of the Bustenari field, Rumania, to illustrate occurrence of oil on overthrust fault, belonging to Sub-class IX (c). (After Bosworth, Pet. Review, March 23, 1912.)

albertite of New Brunswick is only a few miles from the Stony Creek gas field; that the grahamite dikes of Oklahoma intersect known oil fields in several cases; and that the uintaite dikes of Utah lead, in a general way, toward the oil deposits which are found across the boundary in Colorado.

[1] Bailey, L. W. and Ells, R. W.: Canada Geol. Survey, 1876–1877, p. 354, et seq.

Structural "Habits" Peculiar to Individual Fields

The deposits of oil in a given field generally exhibit certain peculiarities or "habits," which seem to prevail throughout that field. Although the relations of the substances involved depend upon structural and other principles of general application,

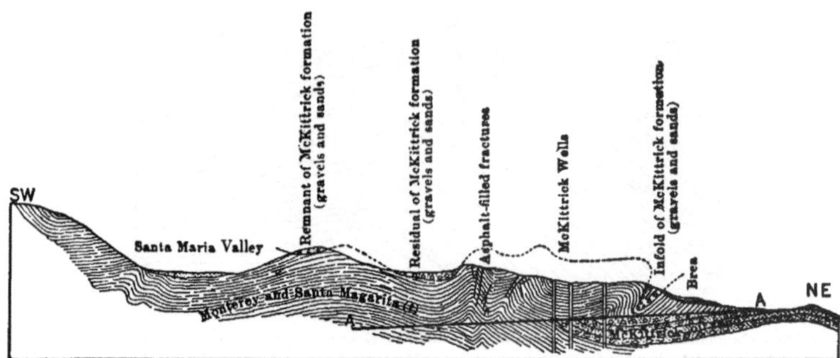

Fig. 41.—Geological cross-section in McKittrick field, California, illustrating occurrence of oil below overthrust fault, Sub-class IX (*c*). (After Arnold and Anderson, Bull. 406, U. S. Geol. Survey, 1910, Fig. 2.)

there are modifying conditions which cause certain peculiarities to run entirely through a field. The central Ohio fields, for instance, owe their monoclinal structure to the Cincinnati uplift; they are too far from the Alleghany Mountains to be subject to the prominent folds produced in Pennsylvania and West Virginia, but they all show

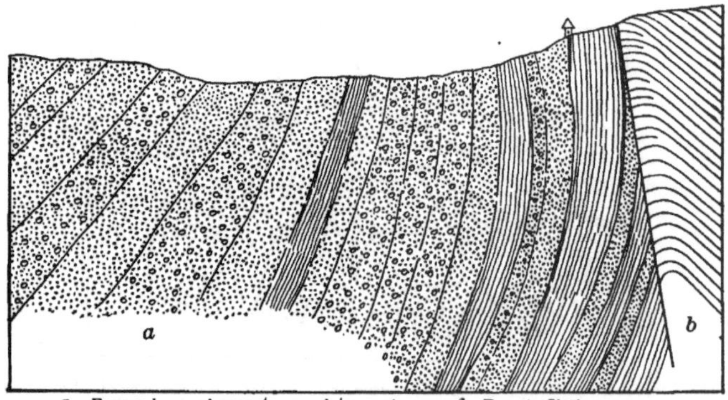

Fig. 42.—Cross-section in Whittier field, California, to illustrate occurrence of oil in vertical strata adjacent to overthrust faults, Sub-class IX (*c*). (After Eldridge and Arnold, Bull. 309, U. S. Geol. Survey, 1907, Plate XI.)

certain tendencies toward anticlinal structure, as exhibited by monoclinal noses, ravines and terraces and by changes in rate of dip within short distances. On the other hand, in the Oklahoma and Kansas fields, situated on the great monocline between the Ozark Mountains and the Great Plains geosyncline, there are numerous domes, due to the pressure of some forces other than those which elevated the mountains on the east or west of the fields.

RELATION OF PETROLEUM OCCURRENCE TO WATER

It has been said on page 45 that where gas, oil and water are all present in a definitely inclined stratum of any particular structure, they are arranged in the order named, gas being found at the top and water at the bottom; i.e., that gas most commonly occurs on the top of the anticline or dome, oil on its flanks, and water lower in the synclines or basins. In cases where the strata are saturated with water, the oil is often at the crest of the structure, in close association with the gas. Where the water-level is lower, however, the oil may be halfway down the flank of the anticline, and where the rocks are dry, the synclines furnish the resting place for the oil. Oil does not always exist in the synclines in dry sands, because all stratified rocks, with a few exceptions, have been at some time saturated with water, which, if present at the time of the tilting of the s'rata, must have helped to arrange the substances in the order of their specific gravities as described; and with the disappearance of water during geological time, the oil has settled into its natural position on the structure. Thus in the Clinton sand of central Ohio, where the structure is mainly monoclinal, interrupted by noses, ravines, terraces, etc., the oil has settled on terraces or in the monoclinal ravines. On the other hand, in West Virginia, where the sands likewise have become fairly dry, the oil has descended from the crest of the anticline to the bottom of the syncline in some cases, while in others it follows the flank of the anticline, or even forms an irregular belt from top to bottom of the anticline along one of its flanks. This may be due to the varying texture of the sands in which it occurs.

In the Oklahoma fields, however, on account of the prevalence of water, oil most commonly lies near the crest of the dome, and in many cases one can not predict from the surface whether a well at the crest of the dome will come in as a gas well or as an oil well.

References on Relationship to Water

Johnson, Roswell H.: Water displacement in oil and gas sands, Min. and Met., Bull. 157, Sect. 7, Am. Inst. Min. and Met Eng., Jan. 1920, 4 pp.

Reeves, Frank: Origin of the natural brines of oil-fields, Johns Hopkins Univ. Circ., N. S., 1917, No. 3; Absence of water in certain sandstones of the Appalachian field, Econ. Geology, Vol. XII, 1917, pp. 354–378.

RELATION OF MIGRATIONS OF PETROLEUM IN THE STRATA TO ITS OCCURRENCE

Statement of anticlinal occurrence.—Since the strata are generally arched into folds, we infer that as oil has migrated up the slope of an inclined stratum in the past, it has generally encountered a more or less well developed anticline. If the fold is an incipient one it may have the form of an "arrested anticline," "structural terrace" or any one of a variety of irregular forms which tend to produce oil pools by forming barriers which prevent the water from forcing the oil higher. If the anticline is a perfect one and lies entirely below the ground-water level, the oil, with suitable conditions of porosity, will migrate to the anticlinal crest where its movement will be arrested by the dipping of the bed in the opposite direction. In such folds, therefore, the oil or gas has a tendency to occupy the highest part of the crest of the fold, on top of the water which has forced them into that position (see Fig. 15). This is the accepted version of the anticlinal theory.

Effect of declining water-level on accumulation of oil.—If a pool existed on the crest of an anticline, or directly above water-level along its flanks, at a time when the previously existing water-level in the stratum was subsiding, it will be found that the oil has not remained exactly at the point to which the water forced it, but has gradually returned to the lower level under the influence of gravity, moving rather slowly on account of lithologic and structural resistance, and becoming diffused through the porous stratum without collecting in large quantity at any particular point. Under such conditions every point of slight structural interruption constitutes a potential accumulation, and many pools of small production are found to exist.

Effect of low water-level on position of oil.—Most oil pools owe their present position to combined structural and water conditions, which have caused an upward migration of oil along a monoclinal slope, or up the flank of an anticline from the bottom of a synclinal basin. During this migration oil has accumulated in any minor folds or wrinkles which may have existed on the flanks of major structures below water-level. In the absence of such structural features the oil tends to accumulate at water-level on the anticlinal flank which forms the monoclinal slope towards the center of the adjacent basin. For this reason, a broad basin and long monoclinal slope, constituting a large gathering ground, are more favorable for the formation of an oil pool of considerable magnitude than is a short slope.

Effect of time on migration of oil.—In regions of extremely complex geology, like California, Mexico, Rumania, and Louisiana, many amplifications of the structural theory are necessary; but the underlying principles of migration and accumulation are just as operative in such localities as they are in the Appalachian region. In comparing conditions in California or Mexico with those in the Appalachian fields, it must be understood that time is an important element. So much time has elapsed since the formation of pools in Paleozoic strata that any avenue of escape, however small, has been sufficient in most cases to allow the oil to disappear; whereas, in California, a comparatively short time has passed since the deposition of the strata, so that even where the oil sands are tilted, eroded and exposed at the surface, only a part of the oil has disappeared. It is undoubtedly true, however, that if Californian or Mexican oil were left in the ground, in the course of future ages it would sooner or later disappear into the air.

It must not be supposed that this failure of the oil to escape through the eroded edges of the porous beds is due entirely to lack of time, although time is probably the most important element in the process. The tendency of thick asphaltic oils, like those of California, to part with their lighter constituents after being exposed to the air, and to clog the sands with tarry residue, has undoubtedly played a prominent part in preventing the escape of much of the oil.

Effect of impervious covers on upward migrations.—In the Florence field of Colorado, the oil is held in fissures in shale, in a broad open syncline, and the impervious layer that prevents its escape until it is reached by the drill appears to be a zone of water-saturated shale which extends a certain distance below the surface. Day [1] has shown that oil will diffuse readily through dry fullers' earth, but that moistened earth is absolutely impervious. From this it is seen that the impervious cover over an oil reservoir may be in the nature of fine-grained clay or shale so cemented as to prevent the passage of the oil through it, or it may consist merely of fine material saturated with water.

[1] Day, David T.: The conditions of accumulation of petroleum in the earth, Trans. Am. Inst. Min. Eng., Vol. XLI, 1911, pp. 212–224.

References on Movements and Migration

Johnson, Roswell H.: The accumulation of oil and gas in sandstone, Science, N. S., Vol. XXXV, 1912, pp. 458–459.

Washburne, Chester W.: The capillary concentration of gas, Trans. Am. Inst. Min. Eng., Vol. L, 1914, pp. 829–842.

McCoy, A. W.: Some effects of capillarity on oil accumulation, Jour. Geology, Vol. XXIV, 1916, pp. 798–805.

Washburne, Chester W.: Discussion of "Some effects of capillarity on oil accumulation" by A. W. McCoy, Jour. Geol., Vol. XXV, 1917, pp. 584–586.

Ziegler, Victor: The movements of oil and gas through rocks. Econ. Geology, Vol. XIII, No. 5, 1918, pp. 335–348.

Johnson, Roswell H.: Water displacement in oil and gas sands, Min. and Met. Bull. 157, Sect. 7, Am. Inst. Min. and Met. Eng., Jan., 1920, 4 pp.

McCoy, Alexander W.: Notes on principles of oil accumulation, Jour. Geology, Vol. XXVII, No. 4, 1919, pp. 252–262.

GEOGRAPHIC OCCURRENCE OF PETROLEUM

INTRODUCTION

All five continents of the world contain petroleum in commercial quantities. Their order of importance, as regards present development, is as follows:

1. North America.
2. Europe.
3. Asia.
4. South America.
5. Africa.

The East and West Indies are also producers, and oil is known to exist in the Arctic regions and in many islands. There are many indications that South America may rise, in the course of a few years, to second place in this list, although, in the opinion of the writer, the order of importance of the continents as regards ultimate oil production, when fully developed, is likely to be as follows:

1. North America.
2. Asia.
3. South America.
4. Europe.
5. Africa.

It is possible that North America may ultimately be surpassed in production by Asia or South America. Some idea of the known distribution of petroleum may be gained by the following figures, representing the total quantity of oil which has been produced in the various countries:

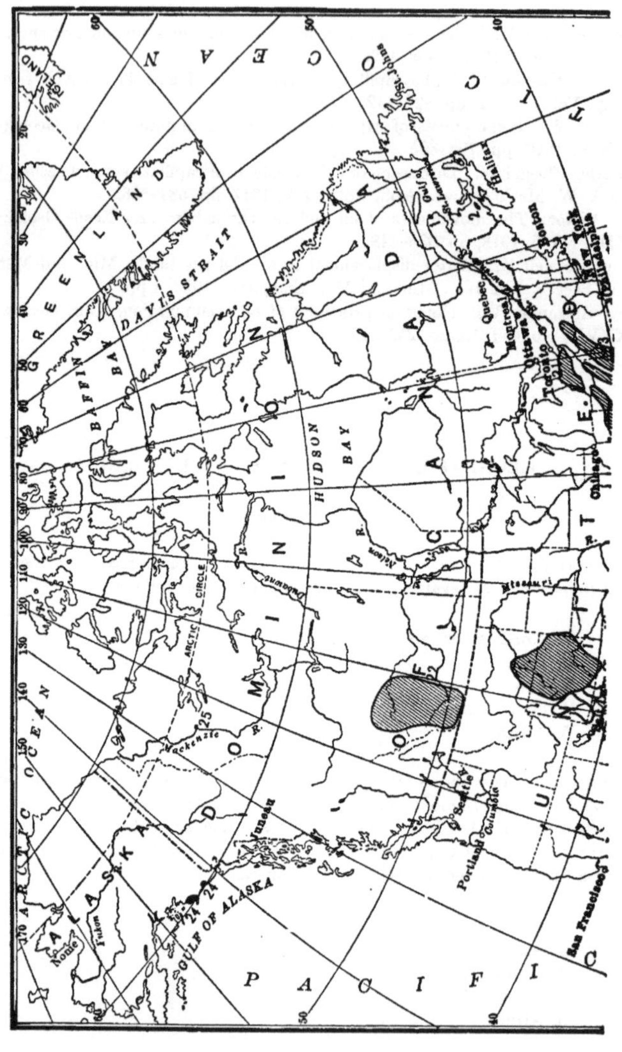
GREENLAND
BAFFIN BAY
DAVIS STRAIT
HUDSON BAY
ARCTIC CIRCLE
ARCTIC OCEAN
ALASKA
GULF OF ALASKA
PACIFIC
Mackenzie R.
Juneau
Nome
Yukon
Seattle
Portland
Columbia
San Francisco
Missouri
Nelson R.
Chicago
Toronto
Ottawa
Montreal
Quebec
St. Lawrence
Gulf of St. Lawrence
Halifax
Boston
New York
Philadelphia
St. Johns

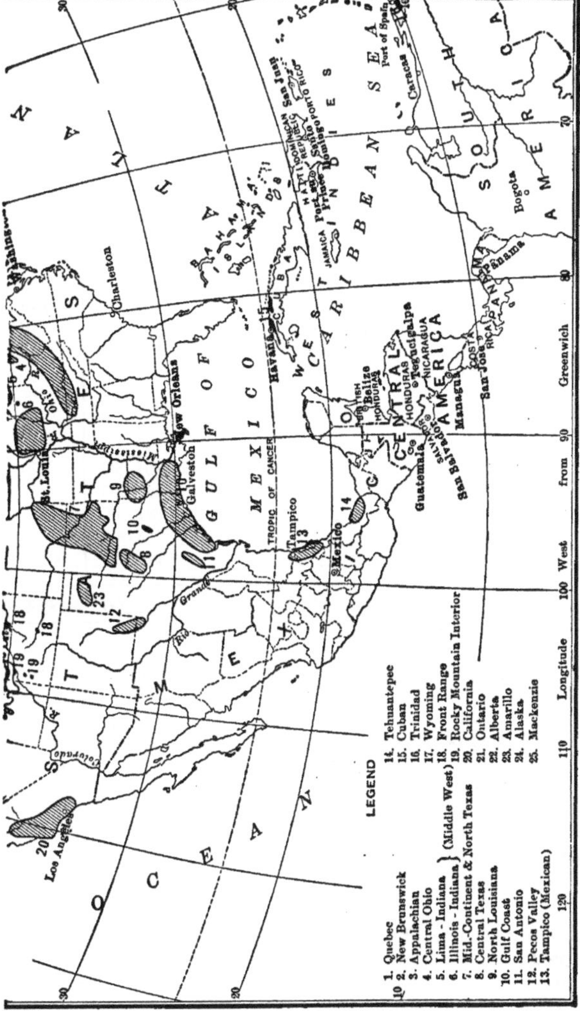

FIG. 43.—Map showing occurrence and distribution of oil-fields in North America and vicinity.

World Production of Petroleum *

Total Production, 1857–1920

Country.	Barrels of 42 U. S. Gallons.	Metric Tons.
United States	5,429,693,000	729,640,000
Mexico	536,524,000	80,047,000
Russia	1,904,021,000	252,072,000
Dutch East Indies	219,584,000	29,690,000
Persia	48,070,000	6,558,000
India	122,583,000	16,343,000
Rumania	165,462,000	23,013,000
Poland (Galicia)	171,263,000	23,700,000
Peru	29,797,000	3,968,000
Japan and Formosa	42,810,000	5,708,000
Trinidad	11,356,000	1,580,000
Argentina	7,225,000	1,043,000
Egypt	6,990,000	1,017,000
British Borneo (Sarawak)	4,052,000	584,000
Venezuela	1,335,000	203,000
France (Alsace)	723,000	102,000
Germany	17,120,000	2,318,000
Canada	24,864,000	3,315,000
Italy	1,042,000	148,000
Algeria	37,000	6,000
England	5,000	56,000
Other	416,000	
	8,744,972,000	1,181,111,000

* G. B. Richardson, U. S. Geological Survey.

Although the preceding table shows the relative importance of the countries of the world as regards the production of oil, it is not an absolute index of its occurrence, as some of the most prolific sources of supply have not yet been opened up, especially in Asia and South America. On account of the backwardness of these continents in physical development, their resources have been rather imperfectly realized; and this, in addition to the lack of transportation facilities and the lack of market in many cases, has prevented their development. The following account of the occurrence of oil in the various countries is necessarily greatly abridged, but, even so, it is an essential part of a chapter on the occurrence of petroleum.

North America

North America is at present first among the five continents of the world in oil production, the productive countries being, in the order of their importance, the United States, Mexico, and Canada. The principal producing regions of North America are commonly classified as Appalachian, California, Mid-Continent, Lima-Indiana, Illinois, Gulf, Rocky Mountain, Mexico, New Brunswick, Ontario, Western Canada and Alaska fields, though there are others which are of less importance at the present time. They are outlined in Fig. 43. The various oil regions are differentiated structurally and stratigraphically as well as geographically. The oil fields of each country will be described in turn, in the order of importance of the countries in which they occur.

United States

General Discussion

The first seven of the oil districts mentioned above are contained wholly within the United States and are, in order of importance in 1919, and excluding Alaska, as follows:

1. Mid-Continent.
2. California.
3. Appalachian.
4. Gulf.
5. Rocky Mountain.
6. Illinois.
7. Lima-Indiana.

Their ultimate importance and order of production may differ considerably from this, however. The following outline gives a general idea of the modes of occurrence of petroleum in the different fields:

Occurrence of Petroleum in United States Fields

Name of oil district	Predominant type of structure	Geologic age	Character of oil-bearing rock
Appalachian	Anticlines, synclines, terraces, monoclines	Cambrian to Carboniferous	Mostly sandstone, some limestone
Lima-Indiana	Cincinnati geanticline	Ordovician	Limestone
Illinois	Anticlines and domes	Ordovician to Carboniferous	Sandstones and limestones
Mid-Continent	Anticlines, domes and monoclines	Ordovician to Cretaceous	Sandstones, limestones and true sands
Rocky Mountain	Anticlines, domes and joint-cracks	Carboniferous to Tertiary	Sandstones and shales
Gulf	Anticlines and saline domes	Cretaceous and Tertiary	Limestones, sands, and clays
California	Anticlines, domes and faults	Tertiary and Quaternary	Sands, sandstones, shales and conglomcrates

Alabama

No wells have yet been known to produce oil in quantity in Alabama, though gas wells exist in Fayette County,[1] at Huntsville [2] and other localities in the northwestern

[1] Munn, M. J.: The Fayette gas-field, Alabama, Bull. 471-A-2, U. S. Geol. Survey, 1912, pp. 26–51.
[2] Smith, E. A.: Concerning Oil and Gas in Alabama, Circ. No. 3, Geol. Survey of Ala., 1917.

part of the state, the productive formations being of Carboniferous and Cambrian age; and there are some showings of oil, in addition to seepages and other surface indications. Showings and seepages of oil and gas also exist [1] in formations of Cretaceous age in southwestern Alabama, notably in Washington and Choctaw Counties.[2] Favorable geologic structure occurs in several places [3] and in general the state is believed to be promising for oil.[4]

Alaska

Martin [5] describes petroleum seepages on the Pacific Coast near Katalla, and 60 miles east, near Yakataga. At Katalla a small amount of oil has been produced from several wells, the first of which was drilled in 1901.[6] The formations exposed on the surface in these localities are Tertiary sandstones and shales, faulted and closely folded; but the oil is presumed to be derived from deeply buried formations.

Petroleum seepages also occur on the west side of Cook Inlet near Iniskin Bay and at Cold Bay, 160 miles to the southwest, from Jurassic shales and sandstones, broadly folded but faulted. Very little drilling has been done.

Traces of oil residue have been found near the south end of Smith Bay, 60 miles east of Point Barrow.

Arizona

Only traces of oil are known in Arizona. They are reported to have been found at Goodridge in the San Juan Valley and at Seven Lakes on the Chaco Plateau.[7]

Arkansas

Petroleum was discovered in Arkansas in large quantities in 1920, in counties adjoining the Louisiana fields and from similar formations. The principal field developed to date is at El Dorado in Union County. Much drilling has been done, and it is expected that other fields will be opened up. Small quantities of oil are reported to have been produced in Washington County in shale of Mississippian age; and asphalt occurs in Madison, Scott, Pike and Sevier Counties. Since 1902 natural gas has been produced in Sebastian County,[8] and good gas fields now exist in western Arkansas.

California

California has been the largest oil producer in the United States, though its annual production has been exceeded by Oklahoma since 1914. The oil-producing counties are Los Angeles,[9] Orange, Ventura, Santa Barbara, Monterey,[10] San Luis Obispo,[11] Santa Clara, Fresno, Kern, and Contra Costa,[12] although Monterey and Contra Costa can

[1] Smith, E. A.: Historical Sketch of Oil and Gas Development in Alabama, Oil Trade Journal, Vol. IX, April, 1918, p. 133.

[2] Hopkins, O. B.: Oil and gas possibilities of the Hatchetigbee anticline, Alabama, Bull. 661, U. S. Geol. Survey, 1917.

[3] Hager, Dorsey: Possible oil and gas fields in the Cretaceous beds of Alabama, Trans. Am. Inst. Min. Eng., Vol. LIX, 1918, pp. 424–434.

[4] Semmes, D. R.: Oil possibilities in Alabama, Min. and Met., No. 159, Am. Inst. Min. and Met. Eng., March, 1920, 10 pp.

[5] Martin, George C.: The petroleum fields of the Pacific Coast of Alaska, etc., Bull. 250, U. S. Geol. Survey, 1905, 64 pp.

[6] Oliphant, F. H.: Min. Res. of U. S. for 1902, pp. 582–584.

[7] Gregory, Herbert E.: Prof. Paper 92, U. S. Geol. Survey, 1917, p. 145.

[8] Smith, C. D.: Bull. 541-B, U. S. Geol. Survey, 1913, pp. 3–13

[9] Arnold, Ralph: Bull. 309, U. S. Geol. Survey, 1907, pp. 138–198.

[10] Angel, M.: 10th Ann. Rept., State Mineralogist, 1890, p. 345.

[11] Ibid., p. 567.

[12] Arnold, Ralph: The Miner Ranch oil field, Contra Costa County, Cal., Bull. 340, U. S. Geol. Survey, 1908, pp. 339–342.

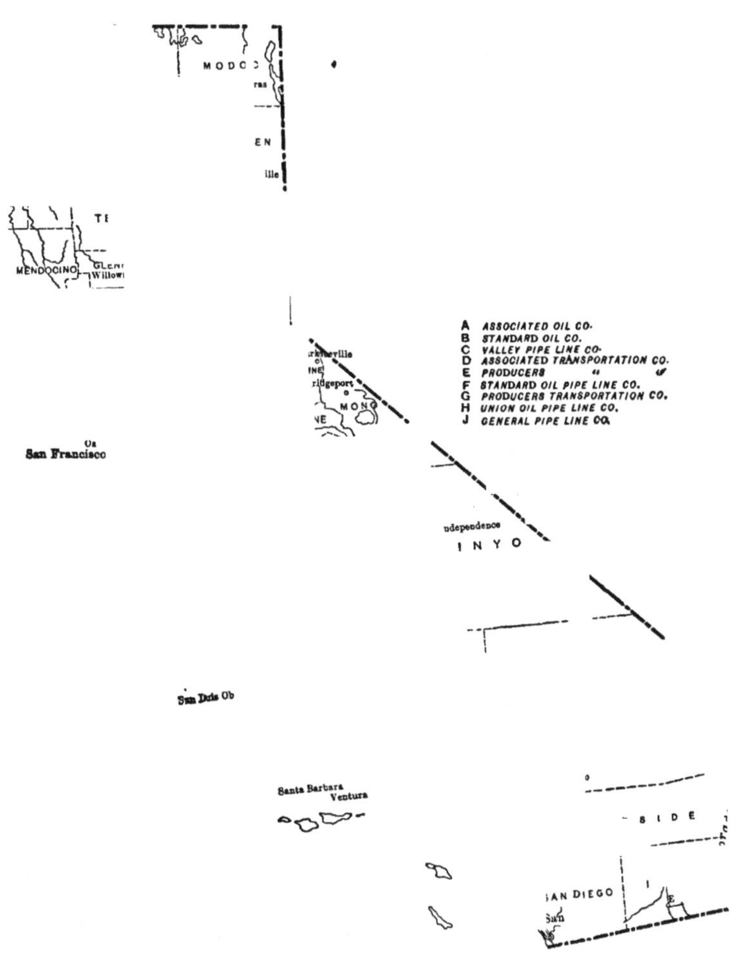

FIG. 44.—Principal oil pools and pipe lines of California.

Los Angeles, Sour Lake
Puente Hills, including Whittier, Coyote Hills and Fullerton Fields
Ventura Fields, including Nordhoff, Santa Paula, Fillmore, Piru, Pico, Newhall, Wiley Canyon Fields
4. Santa Maria, Cox Canyon and Lompoc
5. Midway, Sunset
6. McKittrick
7. Kern River (Bakersfield)
8. Bellridge
9. Lost Hills
10. Devil's Den
11. Coalinga
12. Elk Hills
13. Vallecitos
14. Summerland
15. Monte Bello
16. Edna

hardly be considered commercially productive now. The principal fields [1] in the order of their importance (see Fig. 44) are the Midway, Coalinga,[2] Whittier-Fullerton, Kern River,[3] Sunset, Lost Hills-Belridge, Lompoc-Santa Maria,[4] McKittrick,[5] Salt Lake,[6] Santa Paula-Newhall, Montebello, Los Angeles City, Summerland,[7] and Watsonville. The oil production in 1919 amounted to 101,538,000 barrels.

Oil indications of greater or less importance also occur in other parts of southern California, and there are some traces of oil in northwestern California.[8]

The oil-bearing formations in California are mainly of Tertiary age and are folded and faulted to a much greater degree than those of other fields in the United States (see Figs. 24, 26, 37 and 41). The California sands are generally of greater thickness, and the average life of the wells is longer. The sands are very numerous and are distributed throughout the thousands of feet of sediments represented by the table on the following page.

References on California Oil Occurrence

Eldridge, Geo. H., and Arnold, Ralph: The Santa Clara Valley, Puente Hills and Los Angeles oil districts, southern California, Bull. 309, U. S. Geol. Survey, 1907, 218 pp.

Arnold, Ralph, and Anderson, Robert: Preliminary report on the Santa Maria oil district, Santa Barbara County, California, Bull. 317, U. S. Geol. Survey, 1907, 161 pp.

Arnold, Ralph: Geology and oil resources of the Summerland district, Santa Barbara County, California, Bull. 321, U. S. Geol. Survey, 1907, 93 pp.

Arnold, Ralph, and Anderson, Robert: Preliminary report on the Coalinga oil district, Fresno and Kings Counties, California, Bull. 357, 1908, 354 pp.

Arnold, Ralph, and Johnson, Harry R.: Preliminary report on the McKittrick-Sunset oil region, Kern and San Luis Obispo Counties, California, Bull. 406, U. S. Geol. Survey, 1910, 225 pp.

Anderson, Robert: Preliminary report on the geology and oil property of the Cantua-Panoche region, California, Bull. 431, U. S. Geol. Survey, 1911, pp. 58–87.

Anderson, Robert: Preliminary report on the geology and possible oil resources of the south end of the San Joaquin Valley, California, Bull. 471-A, U. S. Geol. Survey, 1912, pp. 102–132.

Allen, Irving C., and Jacobs, W. A.: Physical and chemical properties of the petroleums, Bull. 19, U. S. Bureau of Mines, 1912, 60 pp.

Prutzman, Paul W.: Petroleum in Southern California, 1913, Cal. State Min. Bur., Bull. 63, 1913, 419 pp.

McLaughlin, R. P., and Waring, C. A.: Petroleum industry of California, Cal. State Min. Bur., Bull. 69, 1914, 519 pp.

Arnold, Ralph, and Garfias, V. R.: Geology and technology of the California oil fields, Trans. Am. Inst. Min. Eng., February, 1914, pp. 383–469.

Pack, Robert W. and English, Walter A.: Geology and oil prospects in Waltham, Priest, Bitterwater and Peachtree Valleys, California, Bull. 581-D, U. S. Geol. Survey, 1914, pp. 119–160.

[1] Northrop, John D.: Min. Res. of U. S, 1917, Part II, p. 846.
[2] Young, W. G.: Eng. and Min. Jour., Vol. LXXI, 1901, p. 403.
[3] Angel, M.: 10th Ann. Rept., State Mineralogist, 1890, p. 219.
[4] Arnold, Ralph: Compt. rendu, Cong. Internat. Petrole, Sess. 2, 1910, Vol. II, p. 365.
[5] Young, W. G.: Eng. and Min. Jour., Vol. LXXI, 1901, p. 30.
[6] Arnold, Ralph: Bull. 285, U. S. Geol. Survey, 1906, pp. 357–361.
[7] Arnold, Ralph: Bull. 321, U. S. Geol. Survey, 1907.
[8] Weber, A. H.: 7th Ann. Rept., State Mineralogist Bur., No. 3, 1894; also Min. and Sci. Press, Vol. LXXIX, 1899, pp. 144 and 172.

Producing Formations in California Fields

Period	System	Series	Formation	Productive districts	Maximum thickness feet
Cenozoic	Quaternary	Recent and Pleistocene	Alluvium and San Pedro	La Brea asphalt deposit, Salt Lake field	1000
	Tertiary	Pliocene	Fernando		1000
		Upper Miocene	Fernando: McKittrick	KernRiver, Sunset Midway, McKittrick Belridge, Lost Hills	7000
			Fernando: Santa Margarita	Whittier, Fullerton, Puente Hills	2000
		Lower Miocene	Monterey shale	Santa Maria, Santa Clara Valley	7000
				Santa Maria	
			Vaqueros	Coalinga, Santa Maria, Puente Hills, Santa Clara Valley	3000
		Oligocene	Sespe	Ventura County	4000
		Eocene	Tejon	Coalinga, Midway, Santa Clara Valley	5000
			Martinez		4000
Mesozoic	Cretaceous	Upper Cretaceous	Chico	Coalinga	6000
			Generally unproductive		

Anderson, Robert and Pack, Robert W.: Geology and oil resources of the West Border of the San Joaquin Valley north of Coalinga. California, Bull. 603, U. S. Geol. Survey, 1915, 220 pp.

English, Walter A.: Geology and oil prospects of Cayuma Valley, California, Bull. 621–M, U. S. Geol. Survey, 1916, pp. 191–215.

English, Walter A.: Geology and oil prospects of the Salinas Valley-Parkfield area, California, Bull. 691-H, U. S. Geol. Survey, 1918, pp. 219–250.

Kew, William S. W.: Structure and oil resources of the Simi Valley, southern California, Bull. 691–M, U. S. Geol. Survey, 1919, pp. 323–355.

Rogers, G. Sherburne: The Sunset-Midway oil field, California, Prof. Papers, 116 and 117, U. S. Geol. Survey, 1919, (Parts I and II).

Colorado

Petroleum is commercially [1] produced in less than half a dozen counties in Colorado, the principal pools (see Fig. 45) in the order of their importance, being the Florence,[2] which produces about 95 per cent of the total and the Boulder [3] field which produces most of the balance. Some oil is also credited to the Rangeley [4] field, and oil from the Yampa and De Beque fields is used in the vicinity of these fields. There are also numerous indications and oil prospects elsewhere in Colorado [5] and much drilling has been done.

Oil shales of Eocene age are very abundant throughout northwestern Colorado,[6] and will be of great commercial value in the future. The total oil production of Colorado in 1920 was only 110,000 barrels.

References

Fenneman, N. M.: Geology of the Boulder district, Colorado, Bull. 265, U. S. Geol. Survey, 1905, 101 pp.

Gale, Hoyt S.: Geology of the Rangeley oil district, Rio Blanco County, Colorado, Bull. 350, U. S. Geol. Survey, 1908, 61 pp.

Washburne, C. W.: The Florence oil field, Colorado, Bull. 381, U. S. Geol. Survey, 1910, pp. 517–544

Woodruff, E. G.: Geology and petroleum resources of the De Beque oil field, Colorado, Bull. 531-C, U. S. Geol. Survey, 1913, pp. 3–16.

Woodruff, E. G., and Day, David T.: Oil shale of northwestern Colorado and northeastern Utah, Bull. 581-A, U. S. Geol. Survey, 1914, pp. 1–20.

Winchester, Dean E.: Oil shale in northwestern Colorado and adjacent areas, Bull. 641-F, U. S. Geol. Survey, 1916, pp. 139–198

Florida

Oil has not yet been discovered in commercial quantities in Florida, but showings have been found in shales of Eocene or Upper Cretaceous age in a well which was started in 1920 in Washington County. Showings had previously been reported in Escambia and Marion Counties.

Reference

Matson, G. C. and Clapp, F. G.: A preliminary report on the geology of Florida, with special reference to the stratigraphy. Second Ann. Rept. Fla. Geol. Survey, 1908–9, pp. 28–162,

Georgia

No commercial deposits of oil have yet been discovered in Georgia, but small seepages occur at Scotland in Telfair County.[7] Several wells have been drilled in various parts of the state, without success.

[1] Lakes, Arthur: Mines and Minerals, Vol. XXIII, 1903, p. 399.

[2] Fenneman, N. M.: Bull. 260, U. S. Geol. Survey, 1905, pp. 436–440.

[3] Fenneman, N. M.: Bull. 213, U. S. Geol. Survey, 1903, pp. 322–332; Bull. 225, U. S. Geol. Survey, 1904, pp. 383–391.

[4] Gale, Hoyt S.: Bull. 350, U. S. Geol. Survey, 1908.

[5] Lakes, Arthur: Mines and Minerals, Vol. XIX, p. 477; Vol. XXI, p. 981; Vol. XXII, p. 150; XXIII, p. 107; Bull. Colorado School of Mines, Vol. I, 1901, p. 221.

[6] Woodruff, E. G., and Day, David T.: Bull. 581-A, U. S. Geol. Survey, 1914, pp. 1–21. Winchester, Dean E.: Bull. 641-F, U. S. Geol. Survey, 1916, pp. 139–198.

[7] Hull, J. P. D. and Teas, L. P.: A preliminary report on the oil prospect near Scotland, Telfair County Georgia, Ga. Geol. Survey, 1919, 23 pp.

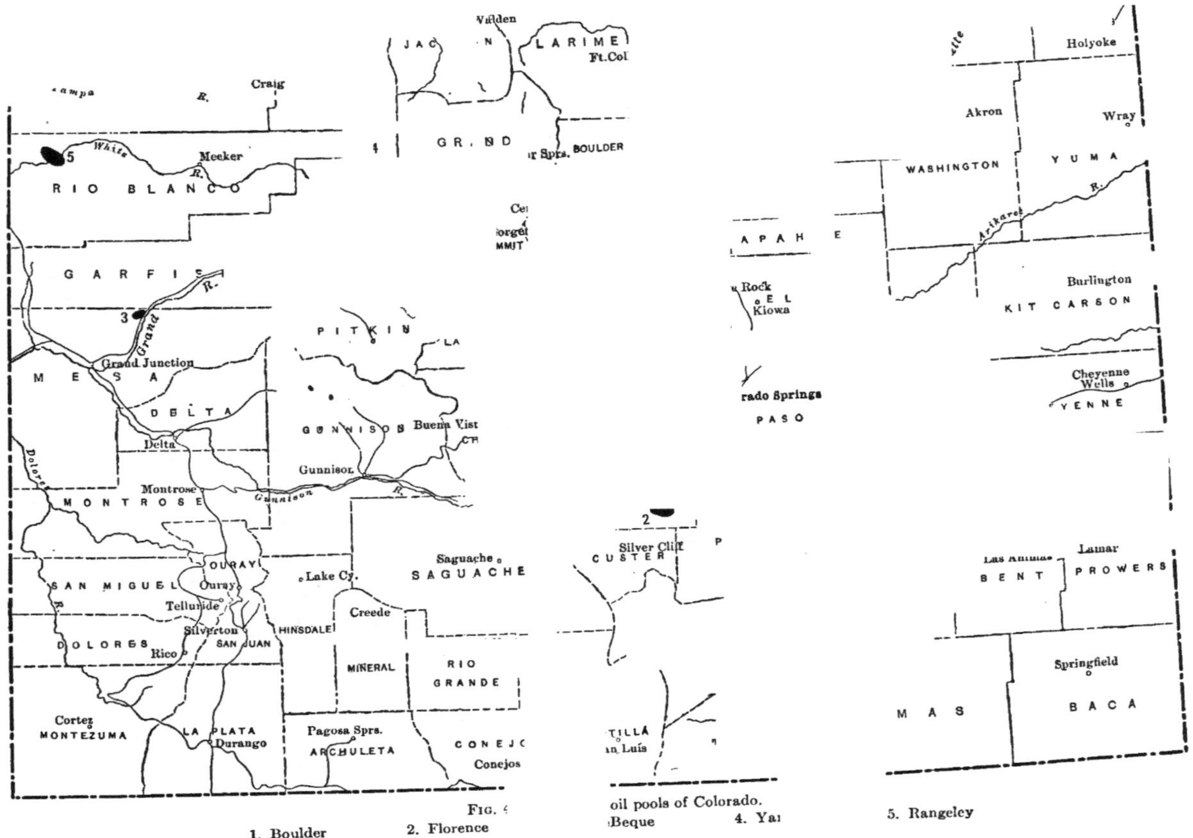

FIG. oil pools of Colorado.
1. Boulder 2. Florence Beque 4. Ya 5. Rangeley

Reference

Veatch, Otto and Stephenson, L. W.: A preliminary report on the Coastal Plain of Georgia, Bull. 26, Ga. Geol. Survey, 1911, 456 pp.

Idaho

Traces of oil and natural gas have been found in Tertiary deposits near Payette,[1] but no production has been secured anywhere in the state.

Illinois

The principal fields of Illinois,[2] as shown by Fig. 46, are situated along the La Salle anticline, in Coles, Edgar, Clark, Cumberland, Crawford, Lawrence, and Wabash Counties. Those toward the north in the first four counties mentioned, constitute the shallow field from which the sands descend southward through the fields of Crawford and Lawrence Counties. Lawrence County is the richest oil-producing county in the state, having seven sands from 400 to 2000 feet below the surface. Elsewhere in the southern half of the state, are a number of small pools, of which the Sandoval, Carlyle,[3] and Carbonville pools are the only important ones. In 1919 Illinois ranked fourth among the oil-producing states, producing 12,535,000 barrels.

Until 1919, practically all the oil produced in the state was derived from the Pennsylvanian and Mississippian series; but in that year a good well was found in the Trenton limestone at the north end of the shallow field, which is now being actively developed for that formation.

Some oil has also been found in Bond and Montgomery counties and in 1921 a pool was opened in Monroe County.

References

Blatchley, R. S.: Oil resources of Illinois, Bull. 16, Ill. State Geol. Survey, 1910, pp. 42–176.

Shaw, E. W.: The Carlyle oil field and surrounding territory, Bull. 20, Geol. Survey Ill., 1910, pp. 7–50.

Blatchley, R. S.: The oil fields of Crawford and Lawrence Counties, Bull. 22, Ill. Geol. Survey, 1913, 442 pp. and maps.

Blatchley, R. S.: The Plymouth oil field, Ill. Geol. Survey, Bull. 23, 1914, pp. 1–7.

Blatchley, R. S.: Oil and gas in Bond, Macoupin and Montgomery Counties, Illinois, Ill. Geol. Survey, Bull. 28, 1914, 51 pp.

Hinds, Henry: Oil and gas in the Colchester and Macomb quadrangles, Bull. 23, Ill. Geol. Survey, 1914, pp. 8–12.

Morse, Wm. C. and Kay, F. H.: Area south of the Colmar oil field, Bull. 31, Ill. Geol. Survey, 1915, pp. 1–35.

Morse, Wm. C. and Kay, F. H.: The Colmar oil field—a re-study, Bull. 31, Ill. Geol. Survey, 1915, pp. 36–55.

[1] Washburne, C. W.: Gas and oil prospects near Vale, Oregon and Payette, Idaho, Bull. 431, U. S. Geol. Survey, 1911, pp. 22–55.

[2] Blatchley, R. S.: Petroleum industry of Southeastern Illinois, Bull. 2, Geol. Survey, Ill., 1906, 109 pp.

[3] Shaw, E. W.: In a report issued by University of Illinois in 1912; Wheeler, H. A.: Eng. and Min. Jour., Vol. XCII, 1911, p. 63; DeWolf, F. W. and Mylius, L. A.: A new Trenton field in Illinois. Bull. Am. Assoc. Pet. Geologists, Vol. IV, No. 1, 1920, pp. 43–46.

Fig. 46.—Principal oil and gas pools and oil pipe lines of Illinois.

1. Siggins 6. Litchfield 11. Plymouth

Rich, J. L.: The Allendale oil field, Bull. 31, Ill. Geol. Survey, 1915, pp. 57–68.

Lee, Wallace: Oil and gas in the Gillespie and Mt. Olive quadrangles, Illinois, Bull. 31, Ill. Geol. Survey, 1915, pp. 71-107.

Kay, F. H.: Oil fields of Illinois, Bull. Geol. Soc. Am., Vol. XXVIII, 1917, pp. 655–656.

Kay, F. H. and others: Oil investigations in Illinois in 1916, Bull. 35, Ill. Geol. Survey, 1917, 80 pp.

Barrett, N. O. and others: Oil investigations in Illinois in 1917 and 1918, Bull. 30, Ill. Geol. Survey, 1919, 144 pp.

Mylius, L. A.: A re-study of the Staunton gas pool, Bull. 44, Ill. Geol. Survey, 1919, pp. 1-23.

Indiana

This state has two entirely separate oil districts; viz., (1) the Lima-Indiana or Trenton rock field of the eastern part of the state, in Adams, Wells, Huntington, Grant, Blackford, Jay, Delaware, Randolph and Madison Counties, and (2) a group of small fields of Mississippian and Corniferous ages in Gibson,[1] Vigo, Pike,[2] and Sullivan[3] Counties in the southwestern part of the state. Small deposits have also been opened in Marion, Fulton, and Jasper Counties. The pools are outlined in Fig. 47. In many other parts of the state showings of oil are found in wells, and gas has been abundant in the past.

In the Trenton rock fields the wells average 1000 to 1200 feet in depth, while in the southwestern fields they are as deep as 1500 feet. The production in 1918 was 877,558 barrels, Indiana being twelfth in rank among the oil-producing states.

The Huron sandstone of Mississippian age is the productive formation of the Princeton field of Gibson County, and of a few abandoned wells at Loogootee in Martin County.

Corniferous oil has been found at Terre Haute in Vigo County, in small quantities near Birdseye in Dubois County, neår Salem in Washington County, and northwest of Medaryville in Jasper County. Most of the oil found in the Corniferous in Indiana has come from a single well, drilled in 1889 at Terre Haute and credited with being the best-paying well ever drilled in the state.[4]

In eastern Indiana the oil is found in the Trenton limestone of Ordovician age, which descends from 1000 feet below the surface near the Ohio line to a probable depth of 3000 feet in southwestern Indiana. The oil is found in the porous and more magnesian portions of this rock, along the crest of the Cincinnati anticline (see Fig. 17). The Lima-Indiana field passed its zenith over twenty years ago.

References

Bownocker, J. A.: Petroleum in Ohio and Indiana, Bull. Geol. Soc. Am., Vol. XXVIII, 1917, pp. 667–676.

Bownocker, J. A.: Rise and decline in production of petroleum in Ohio and Indiana, Min. and Met., Bull. No. 158, Sec. 22, Am. Inst. Min. and Met. Eng., Feb., 1920, 12 pp.

Fuller, M. L.: Asphalt, oil and gas in southwestern Indiana, Bull. 213, U. S. Geol. Survey, 1903, pp. 333–335.

[1] Blatchley, W. S.: 31st Ann. Rept., 1906, Dept. Geol. and Nat. Res., pp. 559–593.

[2] Blatchley, Ralph F.: The Oakland City, Indiana, oil field in 1910, 35th Ann. Rept., Dept. Geol. and Nat. Res., 1911, pp. 81–142.

[3] Barrett, Edward: The Sullivan County oil-field, 38th Ann. Rept., Dept. Geol. and Nat. Res., 1914, pp. 9–40.

[4] Blatchley, W. S.: 25th Ann. Rept., Dept. Geol. and Nat. Res. of Ind., 1900, p. 517; 33d Ann. Rept., 1908, p. 373.

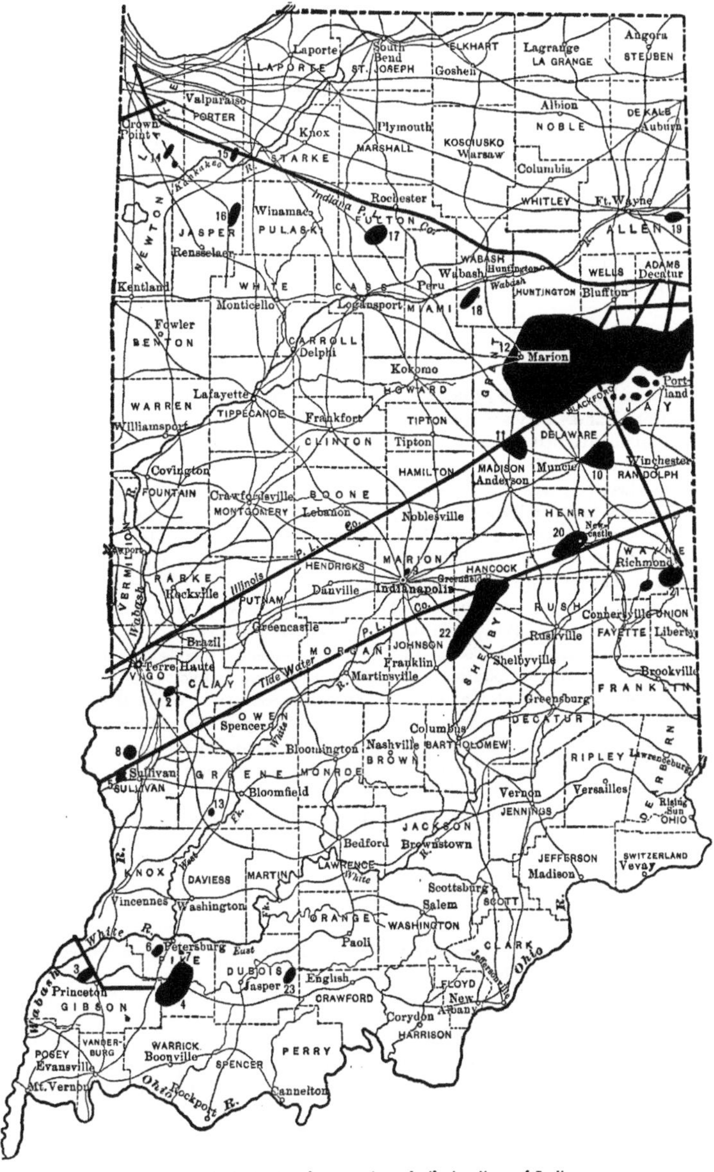

Fig. 47.—Principal oil and gas pools and oil pipe lines of Indiana.

1. Terra Haute
2. Riley
3. Princeton
4. Oakland City
5. Sullivan
8. Shelburn
9. Broad Ripple
10. Muncie
11. Alexandria
12. Marion
14. Lake County
15. Porter County
16. Jasper County
17. Rochester
18. Wabash
20. Newcastle
21. Richmond
22. Shelbyville
23. Jasper

Iowa

Petroleum is not known to occur in Iowa in commercial quantity, though traces are reported in wells near Fort Lee, Madison County. Traces of natural gas are said to have been found in Dallas, Hamilton, Louisa, Muscatine, Polk, Sac and Story Counties.

Reference

Calvin, Samuel: 11th Ann. Rept., Iowa Geol. Survey, 1901, pp. 9–30; 12th Ann. Rept., 1902, pp. 11–27.

Kansas

As shown in Fig. 48, the oil fields of Kansas lie in the southeastern part of the state, extending from Kansas City southwest to the Oklahoma line, and forming a continuation of the productive belt of eastern Oklahoma. The Kansas fields lie at the extreme northern end of the Mid-Continent field. The first well in the state was drilled in 1860, near Paola, where seepages had been observed.

The oil occurs in several sands of Pennsylvanian age and in the underlying Mississippian limestone. These formations lie close to the surface near the Missouri line but descend to a depth of more than 3000 feet in central Kansas, hence the shallow fields are situated in the east and the deep fields toward the west. Those figuring most conspicuously in the oil news lie along various anticlines in Cowley, Butler, and Marion Counties in east-central Kansas, while the older and shallower fields are farther east in Chautauqua, Montgomery, Labette, Elk, Wilson, Neosho, Woodson, Greenwood, Allen, Franklin, and Miami Counties. In 1921 oil was found in Lyon County. Showings of oil and small wells also occur in many other counties. While the primary structure in the deeper fields is anticlinal and domal, many of the shallow fields owe their position to terraces or to variations in texture of sands on the monoclinal dip, some sands being lenticular in form. In 1920 Kansas was fourth in rank among the oil-producing states, yielding 38,501,000 barrels of oil. The Kansas fields are described in detail in many published reports and the facts concerning them have been summarized by able writers.

References

Haworth, Erasmus and others: Special report on oil and gas, Univ. Geol. Survey of Kansas, Vol. IX, 1908, 586 pp.

Gardner, James H.: The Mid-Continent oil-fields, Bull. Geol. Soc. Am., Vol. XXVIII, 1917, pp. 685–720.

Moore, Raymond C.: Petroleum resources of Kansas, Min. and Met., Bull. Am. Inst. Min. and Met. Eng., No. 158, Sect. 29, 1920, 11 pp.

Snider, L. C.: Oil and gas in the Mid-Continent fields, Oklahoma City, 1920, pp. 155–171.

Moore, R. C. and Haynes, W. I.: Oil and gas resources of Kansas, Bull. 3, Kansas Geol. Survey, 1917, 391 pp.

Perrine, Irving: Geological conditions in central Kansas, Bull. Am. Assoc. Pet. Geologists, Vol. II, 1918, pp. 70–97.

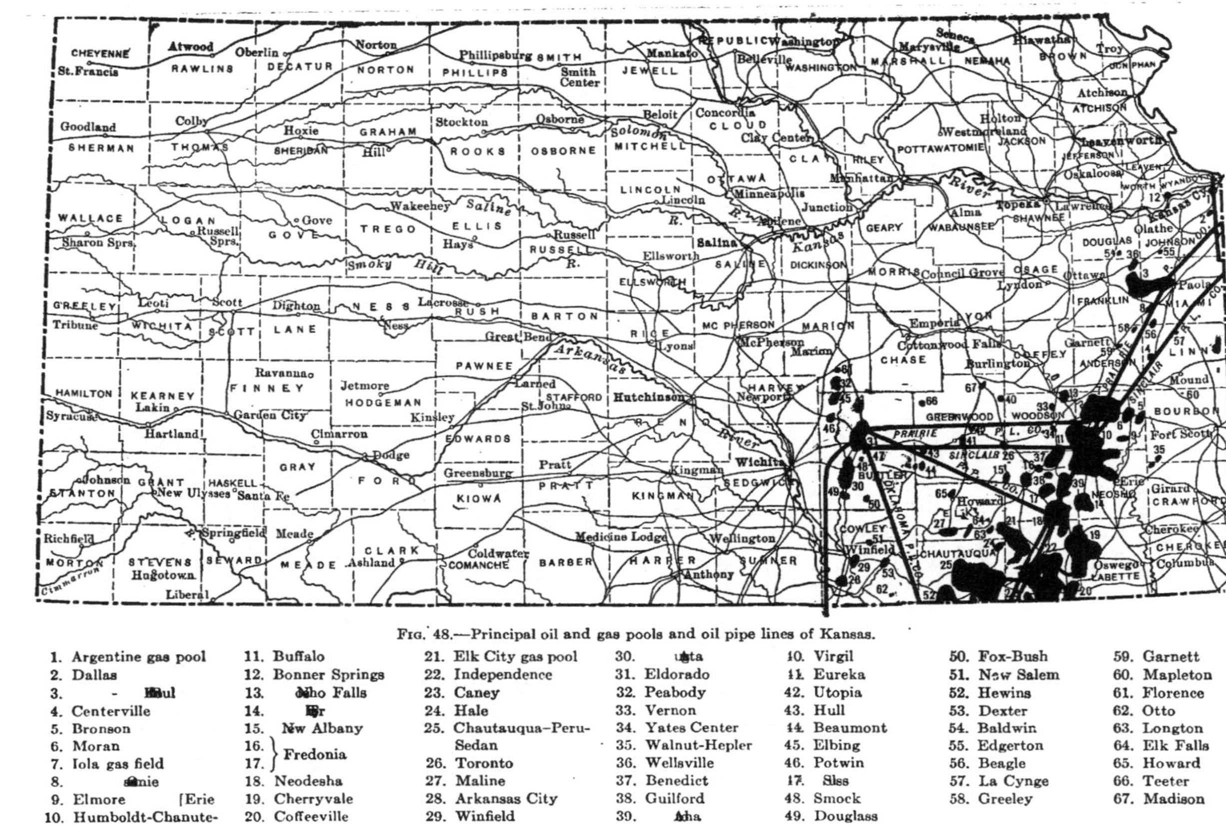

Fig. 48.—Principal oil and gas pools and oil pipe lines of Kansas.

1. Argentine gas pool
2. Dallas
3. [illegible] - [illegible]ul
4. Centerville
5. Bronson
6. Moran
7. Iola gas field
8. [illegible]nie
9. Elmore [Erie
10. Humboldt-Chanute-
11. Buffalo
12. Bonner Springs
13. [illegible]ho Falls
14. [illegible]r
15. New Albany
16. } Fredonia
17. }
18. Neodesha
19. Cherryvale
20. Coffeeville
21. Elk City gas pool
22. Independence
23. Caney
24. Hale
25. Chautauqua-Peru-Sedan
26. Toronto
27. Maline
28. Arkansas City
29. Winfield
30. [illegible]ta
31. Eldorado
32. Peabody
33. Vernon
34. Yates Center
35. Walnut-Hepler
36. Wellsville
37. Benedict
38. Guilford
39. [illegible]ha
40. Virgil
41. Eureka
42. Utopia
43. Hull
44. Beaumont
45. Elbing
46. Potwin
47. [illegible]
48. Smock
49. Douglass
50. Fox-Bush
51. New Salem
52. Hewins
53. Dexter
54. Baldwin
55. Edgerton
56. Beagle
57. La Cynge
58. Greeley
59. Garnett
60. Mapleton
61. Florence
62. Otto
63. Longton
64. Elk Falls
65. Howard
66. Teeter
67. Madison

Kentucky

The oil fields of Kentucky mapped in Fig. 49 are fairly well distributed throughout the eastern and south-central parts of the state, where oil occurs in many formations [1] ranging from Ordovician to Pennsylvanian age. The producing counties, as shown by the following table, are Lee, Allen, Barren,[2] Bath,[2] Rowan,[3] Menifee, Morgan, Wolfe,[4] Powell, Estill, Breathitt, Knott, Knox, Wayne,[5] McCreary,[5] Lawrence, Mason, Johnson, Magoffin, Warren, Lincoln, Grayson, and Ohio. Showings of oil and gas are also found in many other counties, and the productive regions are susceptible of extension.[6]

The structure of the fields is sometimes monoclinal, sometimes domal and anticlinal. The state was eighth in rank among oil-producing states in 1920, producing 8,679,900 barrels of oil. The total production for 1919 was 9,346,593 barrels, and the Lee County field was credited with two-thirds of the production of the state.

The oil produced [7] in Floyd and Knott Counties comes from the Beaver, Horton, Pike, and Salt sands and that of Knox County from the Wages, Jones, and Epperson sands of Carboniferous age; the oil of Lawrence County comes from the Berea sand,* and that of Wayne and McCreary Counties from the Stray and Beaver Creek sands of Mississippian age; while the Corniferous (or Onondaga) limestone of Devonian age is the most important producer of the state, being productive in Allen, Warren, and Ohio Counties and in the Olympia, Ragland, Cannel City, Stillwater, Campton, Irvine,[8] Big Sinking Creek, Ross Creek, Station Camp Creek, Lanhart, Buck Creek, Miller's Creek and Heidelberg fields. The Corniferous limestone yields about 96 per cent of the oil produced in Kentucky. The producing sands are shown in the table on page 94.

References

Hoeing, J. B.: The oil and gas sands of Kentucky, Bull. 1, Ky. Geol. Survey, 1905, 233 pp.

Fuller, M. L.: Appalachian oil-fields, Bull. Geol. Soc. Am., Vol. XXVIII, 1917, pp. 617–654.

Shaw, E. W. and Mather, K. F.: The oil-fields of Allen County, Kentucky, Bull. 688, U. S. Geol. Survey, 1919, 126 pp.

Jillson, W. R.: The oil and gas resources of Kentucky, Ky. Dept. Geol. and Forestry, 1919, 630 pp.

Louisiana

This state has two producing districts which may be classed as the North and South Louisiana fields,[9] the principal pools of which are mapped in Fig. 50. The northern fields produce entirely from strata of Cretaceous age, and are in anticlines and domes superposed on the Sabine uplift, which underlies the northwestern part

[1] Hoeing, J. B.: Bull. No. 1, Ky. Geol. Survey, 1905.

[2] Fischer, M.: Eng. and Min. Jour., Vol. XLIX, 1890, p. 197.

[3] Munn, M. J.: Bull. 531, U. S. Geol. Survey, 1913.

[4] Munn, M. J.: Bull. 471, U. S. Geol. Survey, 1910.

[5] Munn, M. J.: Bull. 579, U. S. Geol. Survey, 1914.

[6] Fohs, F. Julius: Bull. 99, Am. Inst. Min. Eng., 1915, p. 621.

[7] Glenn, L. C.: Oil fields of Kentucky and Tennessee, Min. and Met., Bull. Am. Inst. Min. and Met. Eng., No. 157, Sect. 5, 1920, 12 pp.

[8] St. Clair, Stuart: Irvine oil district, Kentucky, Bull. Am. Inst. Min. and Met. Eng., Sept., 1919, pp. 1079–1089.

[9] Harris, G. D.: Oil and gas in Louisiana, Bull. 429, U. S. Geol. Survey, 1910, 192 pp.

* Oil has recently been discovered in considerable quantity in the Wier sand in Lawrence and Johnson counties.

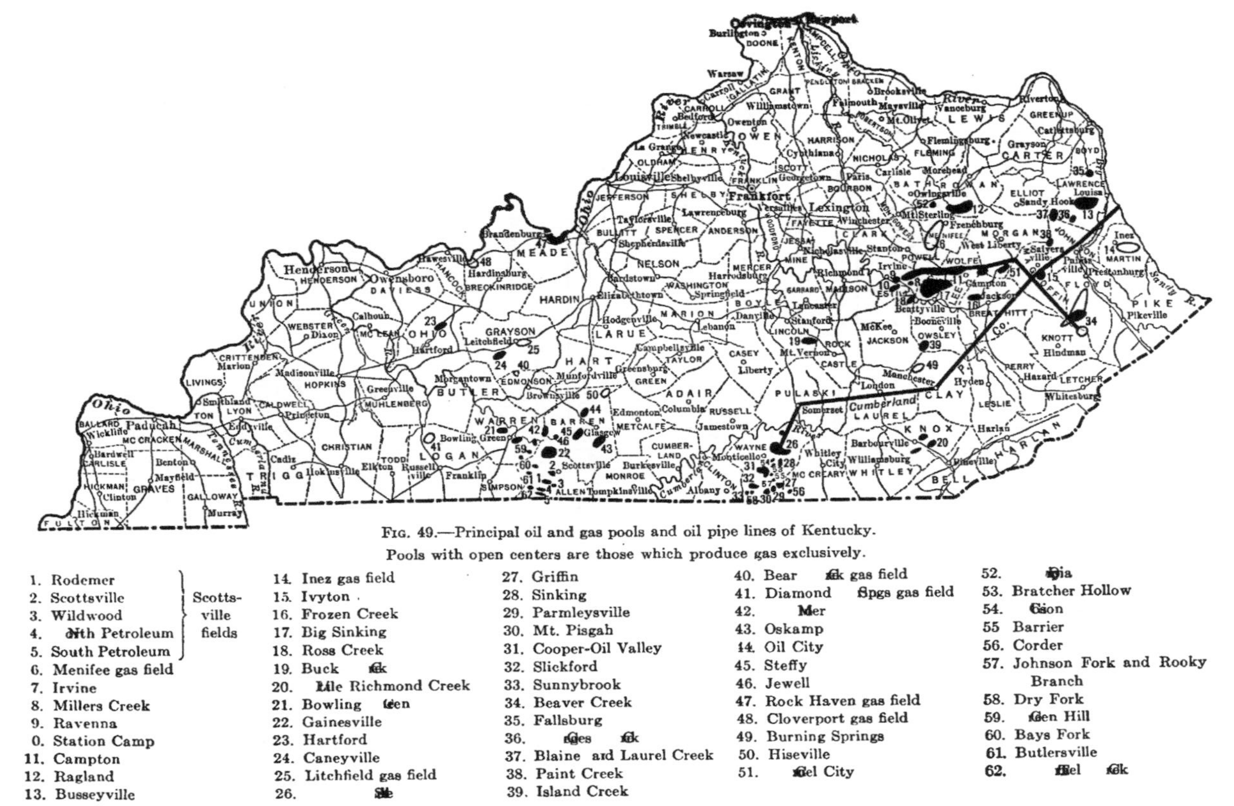

FIG. 49.—Principal oil and gas pools and oil pipe lines of Kentucky.
Pools with open centers are those which produce gas exclusively.

1. Rodemer } Scottsville fields
2. Scottsville } Scottsville fields
3. Wildwood } Scottsville fields
4. [illegible] Petroleum } Scottsville fields
5. South Petroleum } Scottsville fields
6. Menifee gas field
7. Irvine
8. Millers Creek
9. Ravenna
0. Station Camp
11. Campton
12. Ragland
13. Busseyville
14. Inez gas field
15. Ivyton
16. Frozen Creek
17. Big Sinking
18. Ross Creek
19. Buck [illegible]
20. [illegible] Richmond Creek
21. Bowling [illegible]
22. Gainesville
23. Hartford
24. Caneyville
25. Litchfield gas field
26. [illegible]
27. Griffin
28. Sinking
29. Parmleysville
30. Mt. Pisgah
31. Cooper-Oil Valley
32. Slickford
33. Sunnybrook
34. Beaver Creek
35. Fallsburg
36. [illegible]
37. Blaine and Laurel Creek
38. Paint Creek
39. Island Creek
40. Bear [illegible] gas field
41. Diamond [illegible] gas field
42. [illegible]
43. Oskamp
44. Oil City
45. Steffy
46. Jewell
47. Rock Haven gas field
48. Cloverport gas field
49. Burning Springs
50. Hiseville
51. [illegible] City
52. [illegible]
53. Bratcher Hollow
54. [illegible]
55. Barrier
56. Corder
57. Johnson Fork and Rooky Branch
58. Dry Fork
59. [illegible] Hill
60. Bays Fork
61. Butlersville
62. [illegible]

Producing Sands of Kentucky

System	Series	Formation or group	Sands	Region productive	Thickness (feet)
Carboniferous	Pennsylvanian	Pottsville	Beaver Horton Pike	Eastern	60–1000
	Mississippian	Mauch Chunk	Maxon	Eastern	30–800
		Ste. Genevieve and St. Louis	"Tan sand"	Martin, Floyd and Knott Counties	25–1000
		Waverly	Keener Big Injun Squaw Wier Berea	Eastern	200–600
			Stray Mt. Pisgah Beaver Otter Cooper Slickford	Wayne County	
Devonian		Chattanooga shale	Stray sand	Unproductive	20–240
		Hamilton and Onondaga	Corniferous Irvine Ragland Campton	Many fields	9–45
Silurian	Niagaran	Niagara	Niagaran	Western	50–250
		Clinton	Clinton	Various	5–20
Ordovician	Cincinnatian	Cincinnati	Caney Upper Sunnybrook Barren County "deep" Cumberland "shallow"	Wolfe, Morgan, Wayne, Cumberland and Warren Counties	500–700
	Trenton	Lexington	Upper Trenton	Wayne and McCreary Counties	300
		High Bridge	Lower Trenton	Barren, Clinton and Cumberland Counties	600
		Knox dolomite	Wayne County "deep"	Wayne County	250

of the state. These fields are situated in Caddo, Claiborne, Bossier, De Soto, Red River, Ouachita and Morehouse Parishes. The following are the strata from which oil is obtained:

Producing Sands of Northern Louisiana

<table>
<tr><th>System</th><th>Series</th><th>Group</th><th>Formation</th><th>Producing sands</th></tr>
<tr><td>Tertiary</td><td></td><td></td><td></td><td></td></tr>
<tr><td rowspan="8">Cretaceous</td><td rowspan="7">Gulf or Upper Cretaceous</td><td rowspan="3"></td><td>Arkadelphia</td><td></td></tr>
<tr><td>Nacatoch sand</td><td>Nacatoch sand</td></tr>
<tr><td>Marlbrook marl</td><td></td></tr>
<tr><td rowspan="2">Austin</td><td>Annona chalk</td><td>Annona chalk</td></tr>
<tr><td>Brownstone marl</td><td></td></tr>
<tr><td rowspan="2"></td><td>Eagle Ford clay</td><td>Blossom sand</td></tr>
<tr><td>Woodbine sand</td><td>Woodbine sand</td></tr>
<tr><td>Comanche or Lower Cretaceous</td><td></td><td></td><td></td></tr>
</table>

The South Louisiana fields are all of the salt-dome type (see Fig. 31 and page 59) and are mostly situated in Calcasieu, Jefferson Davis, Acadia, and St. Martin parishes, the oil occurring in strata of Tertiary and Cretaceous age. Domes in other parishes contain smaller amounts of oil. The structure of the gas fields of Terrebonne and La Fourche parishes has never been determined.

References

Hayes, C. W., and Kennedy, William: Oil fields of the Texas-Louisiana Gulf Coastal Plain, Bull. 212, U. S. Geol. Survey, 1903, 146 pp.

Matson, G. C.: The Caddo oil and gas field, Louisiana and Texas, Bull. 619, U. S. Geol. Survey, 1916, 62 pp.

Matson, G. C., and Hopkins, O. B.: The De Soto-Red River oil and gas field, Louisiana, Bull. 661-C, U. S. Geol. Survey, 1917, pp. 101–140.

Matteson, W. G.: Principles and problems of oil prospecting in the Gulf Coast Country, Trans. Am. Inst. Min. Eng., Vol. LIX, 1918, pp. 435–491.

Lucas, A. F.: Possible existence of deep-seated oil deposits on the Gulf Coast, Bull. Am. Inst. Min. Eng., Vol. LXI, 1919, pp. 501–519.

Michigan

There has been little development in Michigan, though small wells in St. Clair County have been producing a little oil for some years, and in 1920 some small wells were drilled in Monroe County. Showings have also been found at Saginaw and in Sanilac County. Michigan is thought by geologists to have some oil prospects.

References

Lane, A. C.: Ann. Rept. Mich. Geol. Survey, 1901, pp. 211–237.

Smith, A. R.: The occurrence of oil and gas in Michigan, Pub. 14, Mich. Geol. and Biol. Survey, 1914, 281 pp.

Minnesota

Traces of oil have been reported in Minnesota, but no pool of any economic importance has been found. Some natural gas occurs, but this is also in non-commercial quantities.[1]

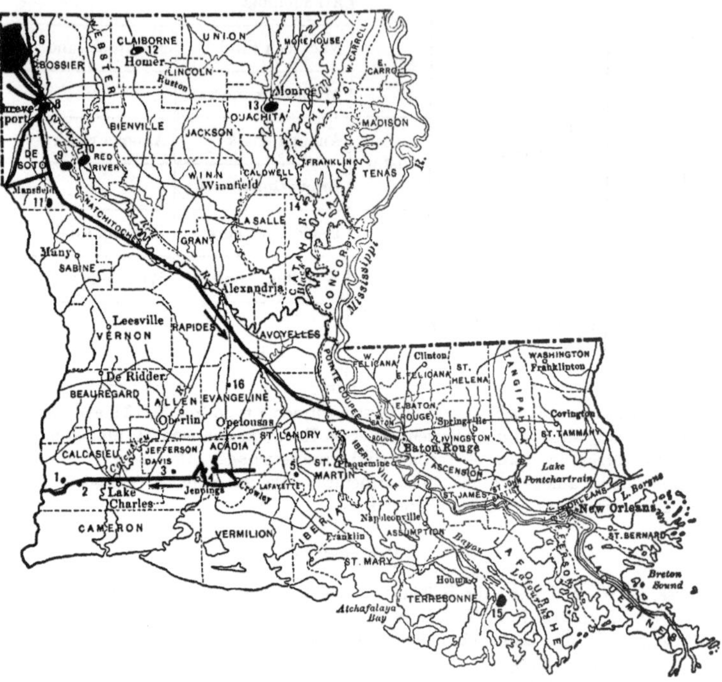

Fig. 50.—Principal oil and gas pools and oil pipe lines of Louisiana.

1. Vinton
2. Edgerly
3. Welsh
4. Jennings
5. Anse la Butte
6. Caddo
7. } Shreveport gas field
8. }
9. } De Soto-Red River
10. }
11. Pelican
12. Homer
13. Monroe gas field
14. Sicily Island
15. Terrebonne gas field
16. Pine Prairie

Mississippi

The State of Mississippi is not yet in the list of oil producers, though good showings of natural gas have been found at a number of widely scattered localities, and several deep test wells have been drilled at various points.

[1] Winchell, N. H.: Bull. Minn. Geol. Survey, 1889, No. 5.

Reference

Hopkins, O. B.: Structure of the Vicksburg-Jackson area, Mississippi, with special reference to oil and gas, Bull. 641-D, U. S. Geol. Survey, 1916, pp. 93–120.

Lowe E. N.: Oil and gas prospecting in Mississippi, Bull. 15, Miss. Geol. Survey, 1919, 80 pp.

Missouri

Although described many years ago,[1] the commercial production of oil in Missouri has been limited to Cass and Jackson Counties, from both of which only small amounts have been obtained. Traces have been found, however, in drilled wells in Bates, Vernon and Adair Counties.

Montana

Although there was some drilling in Montana as long ago as 1900, the first commercial success was in the discovery in 1904 of natural gas on the Glendive or Cedar Creek anticline in Dawson County in the eastern part of the state and the piping of the gas to Glendive. Commercial quantities of gas were secured later in other localities, but production was on a small scale. Petroleum in commercial quantity was first discovered in 1915, at Elk Basin on the line between the State of Wyoming and Carbon County, Montana. The field is a good producer, but is generally considered part of the Wyoming field. The oil is derived from the Frontier group of sands of the Upper Cretaceous series.

Following the discovery of oil in Devil's Basin in northern Musselshell County, in 1919, the Cat Creek field was opened up in 1920 and 1921 in Fergus County, in the central part of the state. In 1921 the Soap Creek pool was discovered in Big Horn County. So far as known, the prospective and producing sands of Montana are similar to those of Wyoming and Alberta, so that a bright future may be expected for Montana as an oil-producing state. The position of the producing sands in the geologic column is shown on page 98.

References

Bowen, C. F.: Possibilities of oil in the Porcupine dome, Rosebud County, Montana, Bull. 621-F, U. S. Geol. Survey, 1915, pp. 61–70.

Rowe, J. P.: Probable oil and gas in Montana, Eng. and Min. Jour., Vol. XCIX, Apr. 10, 1915, pp. 647–649.

Collier, A. J.: The Bowdoin dome, Montana, a possible reservoir of oil or gas, Bull. 661-E, U. S. Geol. Survey, 1917, pp. 193–209.

Hancock, E. T.: Geology, and oil and gas prospects of the Lake Basin field, Montana, Bull. 601–D, U. S. Geol. Survey 1918, pp. 101–147.

Stebinger, Eugene: Oil and gas geology of the Birch-Creek-Sun River area, northwestern Montana, 691-E, U. S. Geol. Survey, 1918, pp. 149–184.

Rowe, J. P.: Oil and gas in Montana, Eng. and Min. Jour., Vol. CX, No. 9, Aug. 28, 1920, pp. 412–417.

Nebraska

Little in the way of oil indication has ever been found in Nebraska, though oil is reported in very small quantities in Brown and Rock Counties.[2]

[1] Shumard, B. F.: Trans. Acad Sci, St. Louis, Vol. II, 1866, p. 263; Robinson,—. —.: Eng. and Min. Jour., Vol. IV, p. 297, and Vol. V, p. 261, 1868; Broadhead, G. C.: Trans. Acad. Sci., St. Louis, Vol. III, 1875, pp. 224–226.

[2] Barbour, E. H.: Neb. Geol. Survey, Vol. I, 1903.

Productive Sands in Montana Fields

System	Series	Group	Formation	Productive sands	Representative thickness (feet)
Cretaceous	Upper Cretaceous	Montana	Pierre	Virgelle sandstone	3200
		Colorado	Niobrara shale		1500
			Frontier	Torchlight sandstone Peay sandstone Mosby sand	400
			Mowry shale		250
			Thermopolis shale	Cat Creek	200
	Lower Cretaceous	Dakota	Kootenai	Lupton	800
Jurassic	Upper Jurassic		Ellis		300
GREAT UNCONFORMITY					
Carboniferous	Permian		Quadrant	Embar	Thin
	Pennsylvanian			Tensleep and Amsden	1200

Nevada

Nevada is not considered a promising field for oil production, though bitumen and asphalt exist in otherwise unfavorable localities, and rich oil shales occur in the vicinity of Elko.

REFERENCES

Anderson, Robert: An occurrence of asphaltite in northeastern Nevada, Bull. 380, U. S. Geol. Survey, 1909, pp. 283–285.

Anderson, Robert: Geology and oil prospects of the Reno Region, Nevada, Bull. 381, 1910, pp. 475–493.

Macfarland, Ira: Development of petroleum in Nevada, Proc. Am. Min. Cong., Vol. XII, 1909, p. 418.

New Jersey

Although several wells have been drilled for oil and the writer has inspected some very small showings of oil and gas, which might appear to indicate the occurrence of petroleum, geological conditions are not such that the state is considered promising.

New Mexico

Although great development work is going on in New Mexico, the producing wells are small and few in number, and the occurrences are confined mainly to the vicinity of Artesia in Chaves County, Raton in Colfax County, Dayton in Eddy County,[1] the Seven Lakes field in McKinley County and to Bernadillo County. In Eddy County development work has been conducted with only moderate success; and a number of wells elsewhere throughout the state have been failures. Many companies are at present prospecting, hoping to strike paying production and believing that the conditions are more or less favorable.

References

Howard, G. G.: A general discussion of the geology of New Mexico with special reference to the production of petroleum. Privately published in Los Angeles, 1919, 30 pp.

Winchester, Dean E.: Geology of Alamosa Creek Valley, Socorro County, New Mexico, with special reference to the occurrence of oil and gas, Bull. 716-A, U. S. Geol. Survey, 1920, pp. 1–15.

Knox, John K.: Geology of New Mexico as an index to probable oil resources, Bull. Am. Assoc. Pet. Geologists, Vol. IV, No. 1, 1920, pp. 95–112.

Merritt, J. W.: Structure in western Chaves County, New Mexico, Bull. Am. Assoc. Pet. Geologists, Vol. IV, No. 1, 1920, pp. 53–58.

Garrett, Dan L.: Stratigraphy and structure of northeastern New Mexico, Bull. Am. Assoc. Pet. Geologists, Vol. IV, No. 1, 1920, pp. 73–82.

Hager, Dorsey, and Robitaille, A. Edmond: Geological report on oil possibilities in eastern New Mexico, privately published, 1919, 115 pp.

New York

New York State's oil production is from three counties, Allegany, Cattaraugus and Steuben, and their relative importance is indicated by the order in which they are named. The total number of producing wells now remaining in the three counties is about 12,000. The great majority of the wells in Allegany and Cattaraugus Counties have produced for over thirty years, but the development of Steuben County, which is of little importance, was begun fifteen years ago, at which time the Rexville pool was discovered. Many counties in the western part of the state have produced natural gas.

References

Ashburner, C. A.: Trans. Am. Inst. Min. Eng., Vol. XIV, 1886, p. 419; Vol. XVI, 1888, p. 906.

Van Ingen, D. A.: Bull. N. Y. State Mus., Vol. III, 1896, p. 558.

Orton, Edward: Petroleum and natural gas in New York, Bull. N. Y. State Mus., Vol. VI, 1899, pp. 395–526.

Bishop, I. P.: Petroleum and natural gas in western New York, Ann. Rept., N. Y. State Mus., Vol. LI, 1899, pp. 9–63; Oil and gas in southwestern New York, Vol. LIII, 1901, pp. 105–134.

[1] Richardson, G. B.: Petroleum near Dayton, New Mexico, Bull. 541-D, U. S. Geol. Survey, 1912, pp. 22–27.

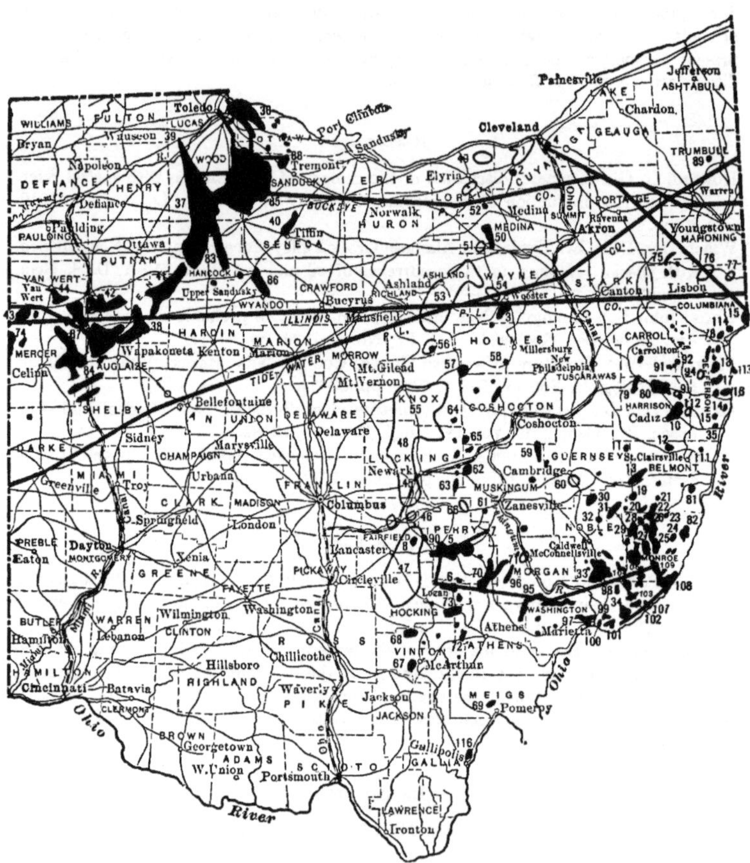

FIG. 51.—Principal oil and gas pools and oil pipe lines of Ohio. (Oil pools in solid black, gas pools with open centers)

Key to oil and gas pools of Ohio

1. Cleveland gas field
2. Berea
3. Wooster
4. Brooklyn
5. Bremen-Junction City
6. New Straitsville
7. Crooksville
8. Jewett
10. (Cadiz)-Maxwell
11. Oakgrove
12. Flushing
13. (Barnesville)
14. Gould
15. McIntyre
16. Bluck
17. Island Creek
18. Knoxville
19. Temperanceville
20. Malaga-Brushy Creek-Harper
21. Stumptown
22. Scott-Griffith
23. Beallsville
24. Schriver
25. Woodsfield Berea Grit-Jackson Ridge
26. Hoffman-Moore-Burkhart
27. Denbow-Wahl-Black-Whitterbrook
28. Egger-Monroefield-Cooper
29. Lewisville-Decker-Fisher
30. Chaseville
31. Summerfield gas pool
32. Salt Run
33. Macksburg
34. Cow Run-Newells Run-Bosworth
35. Mingo
36. East Toledo
37. North Baltimore
38. Lima
39. Dunbridge (Dowling)
40. Tiffin
41. Beaver Dam
42. Elida
43. Menden
44. Venedocia
45. Newark gas field
46. Thurston gas field
47. Sugar Grove-Gibsonville gas field
48. Homer gas field
49. Elyria gas field
50. Medina
51. Lodi gas pool
52. Litchfield
53. Ashland
54. Wooster gas pool
55. Mt. Vernon
56. Butler
57. Brinkhaven
58. Kilbuck
59. Otsego
60. Cambridge gas pools
61. White Cottage
62. Nashport
63. Hopewell
64. Bladensburg
65. Frazeysburg
66. Mt. Perry
67. McArthur
68. Locust Grove
69. Thomas Fork
70. Corning
71. McConnellsville gas field
72. Athens
73. Starr
74. Chattanooga
75. Homeworth
76. Lisbon
77. New Waterford
78. Wellsville
79. Philadelphia Road-Plumb Run-Bowerston
80. Scio
81. Armstrong mills
82. Cameron
83. Findlay
84. St. Marys
85. Gibsonburg-Helena
86. Upper Sandusky
87. Spencerville
88. Fremont
89. Mecca
90. Pleasantville
91. Kilgore
92. Amsterdam
94. Richmond
95. Chester Hill
96. Buck Run
97. Moores Junction
98. Fifteen and Wingett P.O.
99. Goose Run and Mitchell
100. Hendershot
101. Sand Hill
102. Archers Fork
103. Elk Run
104. Flints Mills
105. Hohman
106. Rinards Mills-Wilson Run-Whitacre-Clift-Sycamore
107. Sheets Run
108. Sistersville
109. Graysville
110. Moose Ridge (No. not given, lies N. of 108)
111. Colerain
112. Bricker-Snyder
113. Toronto gas pool
114. East Liverpool gas pool
115. Klondike
116. Gallipolis

North Carolina

No oil or gas in commercial quantities has ever been discovered in North Carolina, though petroliferous beds of Mesozoic and probably also of Triassic age occur in several localities. They appear to be unimportant.

North Dakota

In Lamoure and Bottineau Counties natural gas has been obtained from Cretaceous beds, but there has been no commercial production of oil or gas.

Reference

Leonard, A. G.: Natural gas in North Dakota, Bull. 431, U. S. Geol. Survey, 1911, pp. 7–10.

Ohio

The Ohio fields may be divided into three groups, those of the Lima-Indiana district in the northwestern part of the state, those of the Appalachian district in southeastern Ohio, and an intermediate district in the central part of the state. Their distribution is shown in Fig. 51. Production in the northwestern Ohio fields is from the Trenton limestone of Ordovician age, and that in the southeastern Ohio fields from sands of Pennsylvanian and Mississippian age, while that of the intermediate field is from the so-called "Clinton" sand in the Medina formation of Silurian age. The relationship between these three fields appears in Fig. 17, which shows that while the northwestern Ohio fields lie on the crest of the Cincinnati anticline, and the southeastern fields on terraces, domes, and minor structures (see Fig. 27) subsidiary to the Appalachian geosyncline, the Central Ohio field, on the other hand, lies on a great monocline connecting the two distinct regions. The Clinton sand is destitute of water, and the oil occurs at points of minor interruption, such as monoclinal ravines and terraces, while the great Central Ohio gas-field owes its position mainly to the fact that the Clinton sand rises and dies out toward the west (Figs. 17 and 25). The range in depth of sands in the southeastern Ohio fields is 12 to 2200 feet, while the depth of the Trenton limestone in the Lima fields is from 1000 to 1500 feet, and the Clinton sand, in its productive area, ranges from 2300 to 3400 feet from the surface.

The industry began in Ohio with the discovery of natural gas at Findlay in 1884, while oil was found a year later. Oil is obtained from the Clinton sand in the counties of Hocking, Perry, Fairfield, Licking, Muskingum, Knox, and Wayne, and gas in several other Counties.

Oil derived from the Mississippian and Pennsylvanian rocks is produced in Washington, Monroe, Noble, Harrison, Guernsey, Morgan, Belmont, Jefferson, Carroll, Mahoning, Perry, Muskingum, and Medina Counties.

The Trenton limestone production is mainly from the Counties of Wood, Hancock, Allen, Sandusky, Auglaize, Mercer, Van Wert, Ottawa, Seneca, and Wyandot.

The principal producing sands of Ohio are given on page 103.

References

Fuller, M. L.: Appalachian oil-fields, Bull. Geol. Soc. Am., Vol. XXVIII, 1917, pp. 617–654.

Bownocker, J. A.: Petroleum in Ohio and Indiana, Bull. Geol. Soc. Am., Vol. XXVIII, 1917, pp. 667–676.

Bownocker, J. A.: Rise and decline in production of petroleum in Ohio and Indiana, Min. and Met., Bull. No. 158, Sec. 22, Am. Inst. Min. and Met. Eng., Feb., 1920, 12 pp.
Griswold, W. T.: The Berea grit oil sand in the Cadiz quadrangle, Ohio, Bull. 198, U. S. Geol. Survey, 1902, 43 pp.
Bownocker, J. A.: The occurrence and exploitation of petroleum and natural gas in Ohio, Bull. 1, Geol. Survey Ohio, 4th Series, 1903, 320 pp.
Griswold, W. T.: Structure of the Berea oil sand in the Flushing quadrangle, Harrison, Belmont and Guernsey Counties, Ohio, Bull. 346, U. S. Geol. Survey, 1908, 30 pp.
Bownocker, J. A.: The Bremen oil field, Bull. 12, Geol. Survey Ohio, 4th Series, 1910, 68 pp.

Principal Producing Sands of Ohio

Age	Name of sand	Productive region
Pennsylvanian	Cow Run or 140-foot Macksburg 300-foot Macksburg 500-foot Macksburg 800-foot Salt Maxton	Southeastern Ohio
Mississippian	Big lime (Maxville limestone) Keener Big Injun	Southeastern Ohio
	Berea	Southeastern and Central Ohio
Silurian	Niagara formation	City of Cleveland
	"Clinton"	Central and Northern Ohio
Ordovician	Trenton limestone	Northwestern Ohio

Oklahoma

The Oklahoma fields are the most important of the Mid-Continent district, and, although so numerous and productive that Oklahoma is first in rank among the oil-producing states, they are relatively simple in geology as compared with those of some other productive states. They form a connecting link between the Kansas and the North Texas fields, to which they are very similar in age and structure. The individual pools in which oil is developed are very numerous, being no less than one hundred in number, as shown by Fig. 52, and are situated in the counties of Nowata, Washington, Osage, Kay, Noble, Garfield, Rogers, Muskogee, Wagoner, Tulsa, Okmulgee, McIntosh, Okfuskee, Creek, Payne, Pawnee, Pontotoc, Carter, Stephens, Caddo, Grady, Comanche, and Cotton, while small pools and good showings occur in other counties. The productive formations are of Permian, Pennsylvanian, and Mississippian age, except in Marshall County, where the age is Cretaceous, and in 1920 the Wilcox sand of supposed Devonian age was discovered widely productive in Okmulgee and Creek Counties.

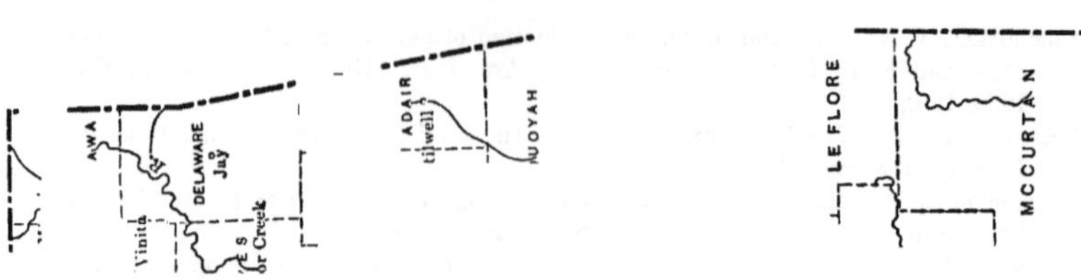

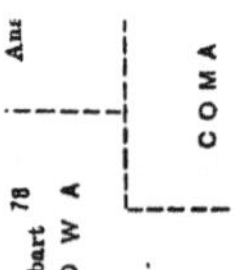

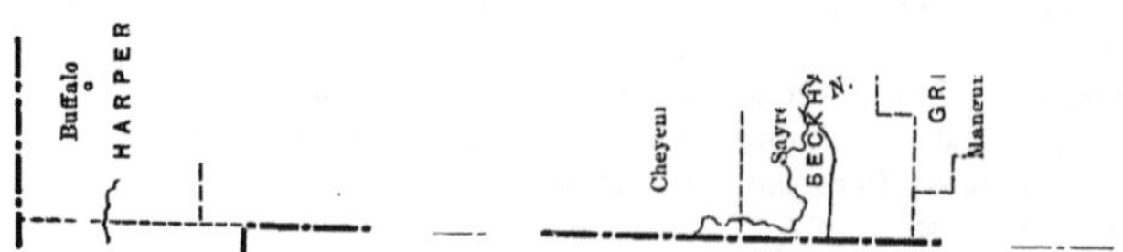

FIG. 52.—Principal oil and gas pools and oil pipe lines of Oklahoma.

Oil and gas pools shown in Fig. 52.

1. Blackwell
2. Newkirk
3. Ponca City
4. Hickory Creek-Pond Creek
5. Caney
6. Canary-Copan
7. Wann
8. California Creek
9. Delaware-Childers
10. Adair
11. Coodys Bluff-Alluwe
12. Nowata-Claggett
13. Hogshooter
14. Glenoak
15. Dewey-Bartlesville
16. Ochelata-Ramona
17. Avant
18. Bigheart
19. Wynona
20. Pawhuska
21. Little Hominy
22. Boston
23. Cleveland
24. Chelsea
25. Bird Creek-Flat Rock-Skiatook
26. Collinsville
27. Owasso
28. Claremore
29. Cushing-Oilton-Drumright-Shamrock
30. Taneha-Jenks-Redfork-Glen Pool
31. Bixby
32. Leonard
33. Catoosa
34. Inola
35. Broken Arrow
36. Coweta
37. Wagoner
38. Stone Bluff
39. Haskell
40. Beggs
41. Bald Hill
42. Preston
43. Hamilton Switch
44. Yarhola
45. Muskogee
46. Chicken Farm
47. Boynton
48. Brushy Mountain gas field
49. Shepard
50. Morris
51. Tiger Mountain gas field
52. Tiger Flats
53. Youngstown
54. Schulter
55. Deaner
56. Poteau gas field
57. Ada gas field
58. Allen
59. Coal County gas pool
60. Bristow
61. Madill
62. Healdton
63. Hewitt
64. Wheeler
65. Graham-Fox
66. Pooleville
67. Loco
68. Homer
69. "Duncan gas field" (Cruce)
70. Velma
71. Kilgore
72. Comanche
73. Walters
74. Lawton
75. Cement
76. Billings
77. Garber
78. Gotebo
79. Granite
80. Vera
81. Weimar
82. Pearson
83. Meyers
84. Buck Creek
85. Okesa
86. South Elgin
87. Barnes
88. Pershing
89. Hominy
90. Hominy Townsite
91. Osage Junction
92. Jennings
93. Hallett
94. Quay-Yale
95. Casey
96. Cushing Townsite
97. Ingalls
98. Ripley
99. Kelleyville
100. Porter
101. Paden
102. Stroud
103. Morse
104. Yeager
105. Maud
106. Francis gas field
107. Empire
108. Three-four
109. Okemah
110. Quinton gas pool

Throughout the eastern part of the state, the strata show a generally moderate westward dip, interrupted by domes and terraces of monoclinal type, so that the beds which occupy the surface in northeastern Oklahoma lie at a depth of several thousand feet in the western part. Some of the most pronounced interruptions in the dip are in the counties between and surrounding the Wichita and Arbuckle Mountains in the southern part of the state.

Development began in the northeastern part of Oklahoma and gradually extended southwest, though intervening pools are being continually developed.

References

Taff, J. A., and Reed, W. J.: The Madill oil pool, Oklahoma, Bull. 381, U. S. Geol. Survey, 1910, pp. 504–513.

Ohern, D. W., and Garrett, R. E.: The Ponca City oil and gas field, Bull. 16, Okla. Geol. Survey, 1912, 30 pp.

Smith, C. D.: The Glenn oil and gas pool and vicinity, Bull. 541-B, U. S. Geol. Survey, 1913, pp. 14–28.

Snider, L. C.: Petroleum and natural gas in Oklahoma, Oklahoma City, 1913, 196 pp.

Buttram, Frank: The Cushing oil and gas field, Oklahoma, Bull. 18, Okla. Geol. Survey, 1914, 64 pp. and tables.

Gardner, James H.: The oil pools of southern Oklahoma and Northern Texas, Econ. Geology, Vol. X, 1915, pp. 422–434.

Wood, Robert H.: Oil and gas development in north-central Oklahoma, Bull. 531-B, U. S. Geol. Survey, 1915, pp. 27–53.

Wegemann, C. H., and Heald, K. C.: The Healdton oil field, Carter County, Oklahoma, Bull. 621-B, U. S. Geol. Survey, 1915, pp. 13–30.

Wegemann, C. H., and Howell, R. W.: The Lawton oil and gas field, Oklahoma, Bull. 621-G, U. S. Geol. Survey, 1915, pp. 70–85.

Oklahoma Geological Survey: Petroleum and natural gas in Oklahoma, Parts I and II, Bulls. 18 and 19, 1915 and 1917.

Heald, K. C.: The oil and gas geology of the Foraker quadrangle, Osage County, Oklahoma, Bull. 641-B, U. S. Geol. Survey, 1916, pp. 17–47.

Beal, Carl H.: Geologic structure in the Cushing oil and gas field, Oklahoma, Bull. 658, U. S. Geol. Survey, 1917, 64 pp. and maps.

Oil sands are numerous, the principal ones being shown in the accompanying table:

Principal Oil Sands of Oklahoma

System	Series	Name of sand	Productive region
Cretaceous	Lower Cretaceous	Madill	Marshall County only
Carboniferous	Permian	Many sands	Southern Oklahoma
	Pennsylvanian	Blackwell Newkirk Layton Cleveland Peru Wheeler Squirrel	Northern Oklahoma
		Bartlesville Tucker Booch Dutcher	Northeastern Oklahoma
	Mississippian	Burgess Mississippi or Hogshooter Mounds Muskogee	Northeastern Oklahoma
Devonian		Wilcox	Okmulgee County and vicinity

Fath, A. E.: Structure of the northern part of the Bristow quadrangle, Creek County. Oklahoma, with reference to petroleum and natural gas, Bull. 661-B, U. S. Geol, Survey, 1917, pp. 69–99.

Gardner, James H.: The Mid-Continent oil fields, Bull. Geol. Soc. Am., Vol. XXVIII, 1917, pp. 685–720.

Bloesch, Edward: Unconformities in Oklahoma and their importance in petroleum geology, Bull. Am. Assoc. Pet. Geologists, 1919, pp. 253–285.

White, David, and others: Structure and oil and gas resources of the Osage Reservation, Oklahoma, Bull. 686, U. S. Geol. Survey, 1919.

Swigart, T. E.: Underground problems in the Comanche oil and gas field, Stephens County, Oklahoma, Bur. of Mines, Rept. for Sept., 1920, 42 pp.

Oregon

Petroleum has been reported in Malheur and Crook Counties, but there has been no commercial production in the state.

References

Washburne, C. W.: Gas and oil prospects near Vale, Oregon, and Payette, Idaho, Bull. 431, U. S. Geol. Survey, 1911, pp. 26–55.

Washburne, C. W.: Reconnaissance of the geology and oil prospects of northwestern Oregon, Bull. 590, U. S. Geol. Survey, 1914, 111 pp.

Pennsylvania

The earliest known reference to petroleum in North America occurs in a letter written in the year 1632 by a Franciscan missionary, who mentions springs in what is now Allegany County, New York. The oils from these springs and from Oil Creek, Pennsylvania, were used by the Indians for medicinal purposes and sold under the name of Seneca oil. Between 1790 and 1820, numerous wells were drilled west of the Alleghany Mountains in the search for brine for salt manufacture. In 1854 the Pennsylvania Rock-Oil Company was organized to procure petroleum on Oil Creek near Titusville, Venango County, Pennsylvania. After encountering various difficulties, some of the stockholders employed Edwin L. Drake to drill an artesian well, and on August 23, 1859, he struck oil at a depth of 69 feet from the surface. The success of this well marked the beginning of the oil industry in the United States. Prospectors, investors and speculators came from all directions, and, in the years which followed, thousands of wells were drilled. The depth of wells ranges from 60 to 3500 feet, and the age of the producing formations from Devonian to Pennsylvanian; or, more specifically, oil may be found anywhere from the upper part of the Conemaugh formation to the base of the Chemung, the principal sands being as shown in the table on page 108.

The oil fields lie directly above water-level in the sands following the flanks of anticlines which parallel the Appalachian Mountains, becoming deeper toward the geosyncline to the west. (See Figs. 17 and 53*a*.) Natural gas occupies the crests of the anticlines. In general, the gas fields lie east of Pittsburgh and nearer the mountains while the oil fields lie toward the west; but, as we travel northward, we find the oil-producing region extending farther and farther toward the east. The distribution of Pennsylvanian pools is shown in Fig. 53*a*.

References

Fuller, M. L.: The Gaines oil field in northern Pennsylvania, 22nd Ann. Rept., U. S. Geol. Survey, 1900–1901, pp. 579–627.

Clapp, F. G.: Economic Geology of the Amity quadrangle, eastern Washington County, Pennsylvania, Bull. 300, U. S. Geol. Survey, 1907, 145 pp.

Stone, R. W., and Clapp, F. G.: Oil and gas fields of Greene County, Pa., Bull. 304, U. S. Geol. Survey, 1907, 110 pp.

Griswold, W. T., and Munn, M. J.: Geology of oil and gas fields in Steubenville, Burgettstown and Claysville quadrangles, Bull. 318, U. S. Geol. Survey, 1907, 196 pp.

Munn, M. J.: Geology of the oil and gas fields of the Sewickley quadrangle, Rept. No. 1, Top. and Geol. Survey, Pa., 1910, 170 pp.

Munn, M. J.: Geology of the oil and gas fields of the Clarion quadrangle, Rept. No. 3, Top. and Geol. Survey Pa., 1910, 111 pp.

Shaw, E. W., and Munn, M. J.: Coal, oil and gas of the Foxburg quadrangle, Pennsylvania, Bull. 454, U. S. Geol. Survey, 1911, 85 pp.

Hice, R. R.: Oil and gas map of Southwestern Penna., 1915, Top. and Geol. Survey, Pa. (1916), 22 pp. and map.

Fuller, M. L.: Appalachian oil-field, Bull. Geol. Soc. Am., Vol. XXVIII, 1917, pp. 617–654.

Principal Producing Sands of Pennsylvania

Geologic series	Formation	Name of sand	Productive region
Pennsylvanian	Conemaugh	Cow Run Dunkard	Southwestern
	Allegheny		
	Pottsville	Gas sand, Salt or 40-foot, Maxton	Southwestern
Mississippian	Mauch Chunk	Big Lime	
	Pocono	Big Injun or Mountain	Central, western and southwestern
		Butler gas sand, Murraysville or Berea	Central western and extreme western
		Gantz and 50-foot | 100-foot	Southwestern
Devonian	Catskill	Nineveh 30-foot Gordon Fourth Fifth Sixth or Bayard	Southwestern
	Chemung	Elizabeth	Southwestern
		Warren First and Second	Northwestern
		Speechley	Northwestern and central western
		Tiona Bradford Kane	Northwestern

PRINCIPAL OIL AND GAS POOLS OF PENNSYLVANIA

1. Washington—Taylorstown
2. Fonner or Dumas Station
3. Zollarsville
4. Ross and Somerset Twps.
5. Linden
6. Canonsburg
7. Venice
8. McDonald
9. Cecil
10. Mawhinney
11. McMurry
12. Finleyville
13. Clifton or Sodom
14. Castle Shannon
15. Bridgeville
16. Cuddy
17. Lickskillet
18. Hopper
19. Imperial
20. Moon
21. Aten
22. McCormick
23. Ewings Mill
24. Moon Run and Neville Island
25. Chartiers
26. Bellevue and Avalon
27. Woodville or Heidelberg
28. McCurdy
29. Coraopolis and Haysville
30. Shannopin
31. Leetsdale
32. Phillips and Sevins

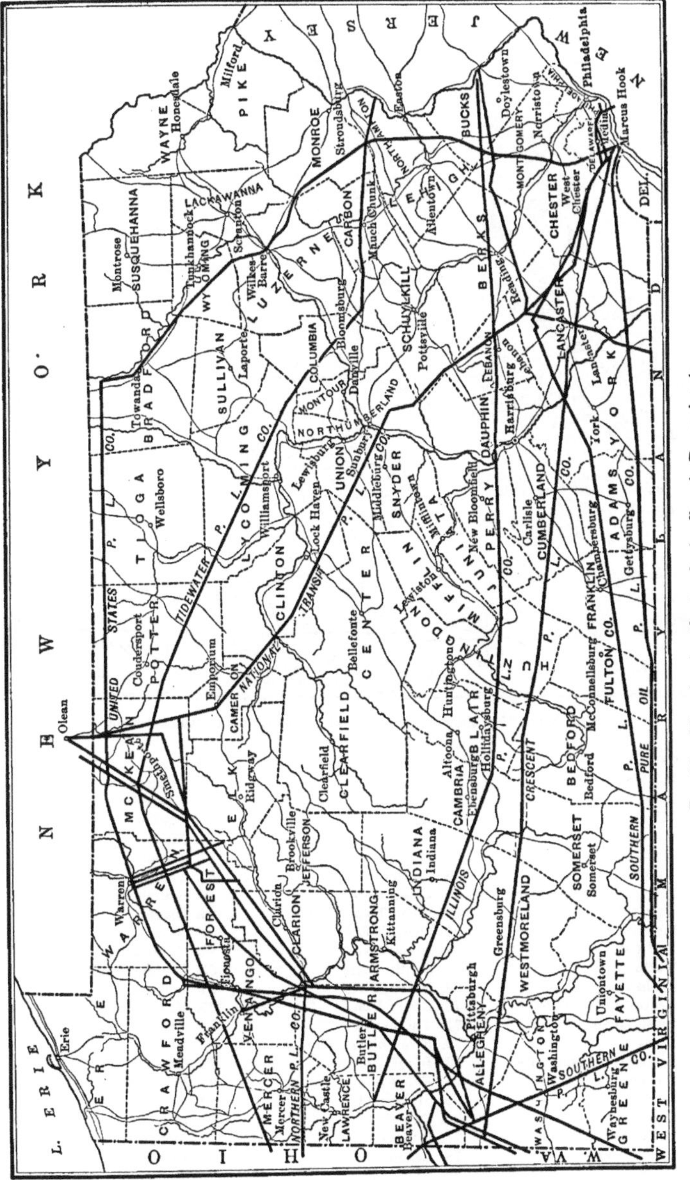

FIG. 53.—Map showing the trunk pipe lines in Pennsylvania.

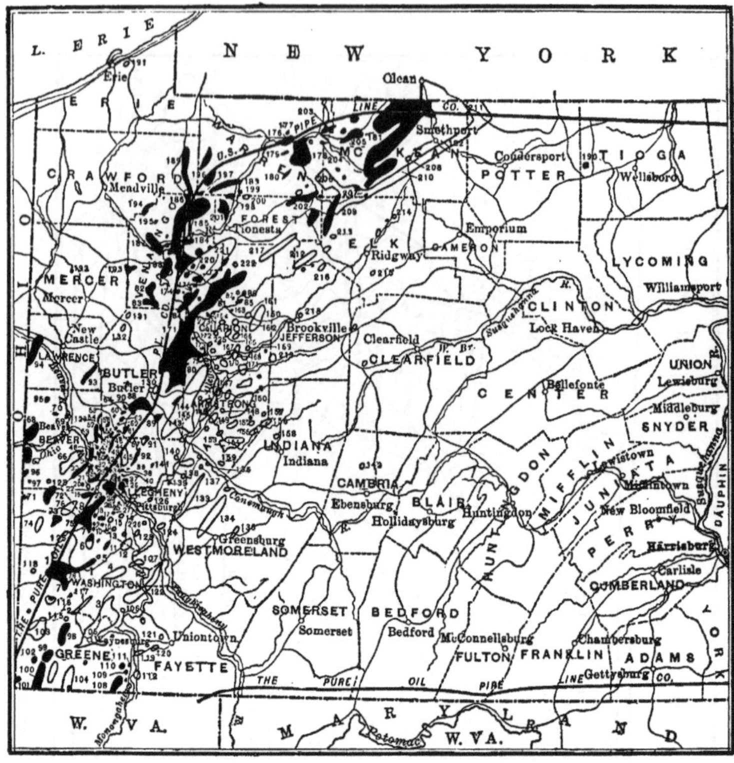

FIG. 53a.—Map showing principal oil and gas pools of Pennsylvania.

*33. Grubbs
34. Mt. Nebo
*35. Glenfield
36. Dixmont
37. West View
38. Sandle
39. Wildwood
40. Perrysville or Hammerschmitt
41. Ingomar
42. Duff City
43. Wexford
44. English
45. Keown or Fourth Sand
46. Brush Creek
47. Zimar
48. Economy—Legionville—Harmony
*49. Davis
50. Craig
51. Cookson
52. Crows Run
53. Dunn
54. Brenner
55. Dumbaugh
56. Garvin
57. Hendersonville
58. Crider
59. Duncan
60. Ramsey
61. Mars
62. Lockwood
63. Callery
64. Evans City
65. Moon Twp.
66. New Sheffield
67. Hookstown
68. Smiths Ferry
69. Beaver Falls
70. Brady Run
71. Turkeyfoot
72. Five Points—Florence
73. Burgettstown
74. Eldersville
75. McDonald
76. Candor
77. Buffalo
78. Lagonda
79. Bradys Bend
80. Karns City
81. Elk City including Edenberg and Triangle pools
82. Bullion
83. Clintonville
84. Emlenton—Ritchie Run
85. Clarion
86. Miola
87. Manor
88. Thorn Creek
89. Saxonburg
90. Glade
91. Bakerstown
92. Gibsonia
93. Slippery Rock
94. Bessemer
95. New Gallilee
96. Carson
97. Kendall
98. Grays Park
99. Bristoria
100. New Freeport
101. Board Tree
102. Aleppo
103. Richhill
104. Lantz
105. Waynesburg
106. Centerville (Washington Co.)

107. Forward
108. Mt. Morris
109. Whiteley Creek
110. Garrison
111. Blackshire
112. New Geneva
115. Nineveh
116. Dague
117. Point Lookout
118. Mahaffey
119. Masontown
120. Fayette
121. Hadenville
122. Belleveron
123. Jefferson
124. McKeesport
125. Hickory
126. Homewood
127. Crafton
128. Murdocksville
129. Lovi
130. Butler
131. Harrisville
132. Centerville (Butler Co.)
133. Murrysville
134. Grapeville
135. Latrobe
136. Apollo
137. Pine Run
138. Leechburg
139. Plum Creek
140. Tarentum
141. Dorseyville
142. Carrolltown
143. Lardintown
144. Winfield
145. Slatelick
146. Glade Run
147. Limestone or Limestone Run
148. Garrett Run
149. Ford City
150. Cowanshannock
151. Rockville
152. Say
153. Shellhammer
154. Girty
155. Plum Run
156. Willetts
157. Atwood
158. Creekside
159. Roaring Run
160. Frogtown-Greenville
161. George
162. Kifer
163. Shamburg
164. Sligo
165. Bittenbender
166. Piolett
167. Blair
168. Leatherwood
169. New Bethlehem
170. Madison
171. Petrolia (including Rosenberry, Sucker Rod, and Cross Belt pools)
172. Rattlesnake
173. Armstrong Run
174. Black Run
175. Beall
176. Warren
177. North Warren
178. Clarendon and Glade
179. Morison Run
180. Cherry Grove
181. Bradford
182. Smethport
183. Tidioute
184. Oil City
185. Petroleum Center
186. Titusville
187. Franklin
188. Venango
189. Centerville (Crawford Co.)
190. Gaines
191. Erie
192. Mercer
193. Raymilton
194. Lake Creek
195. Sugar Creek
196. Grand Valley
197. Enterprise
198. Fagundus
199. Dennis Run
200. Endeavor
201. West Hickory
202. Balltown
203. Cornplanter
204. Dewdrop
205. Sugar Run
206. Sheffield
207. Kane
208. Ormsby
209. Nansen
210. Hazelhurst
211. Shinglehouse
212. Halltown
213. Bear Creek
214. Glenhazel
215. Ridgway
216. Millstone
217. Pine Grove
218. Brookville
219. New Salem
220. Sandy Creek
221. Kossuth
222. Fryburg
223. Fiddlers Run
224. Catfish Run
225. Rimersburg
226. Mifflin

* Nos. 33, 35 and 49 are omitted on map for lack of space.

South Dakota

Small natural gas wells have been obtained in South Dakota, and there have been shows of oil, but not in commercial quantities.

Tennessee

Oil was found in Tennessee as early as 1820 to 1840, in wells which were being drilled for brine, but drilling was not resumed until 1892. Oil has been found in the Spurrier and Riverton districts in Pickett County, though there was no production in the state until 1915, when oil was discovered at Oneida in Scott County, in the St. Louis limestone of Mississippian age. It has also been found in this formation at Glenmary [1] in the same county, and, in small quantities, in rocks of Corniferous age in Dickson, Sumner, and Robertson Counties, while mere showings have been found in Cretaceous or Tertiary strata in the vicinity of Memphis. The wells range from less than 1000 feet to more than 3000 feet deep. There is little commercial production in the state, but the conditions are promising in several localities.

References

Glenn, L. C.: Oil fields of Kentucky and Tennessee, Min. and Met., Bull. Am. Inst. Min. and Met. Eng., No. 157, Sect. 5, Jan. 1920, pp. 1–3.

[1] Glenn, L. C.: The Glenmary oil field, Res. of Tenn., Vol. VIII, No. 3, July, 1918, pp. 211–219.

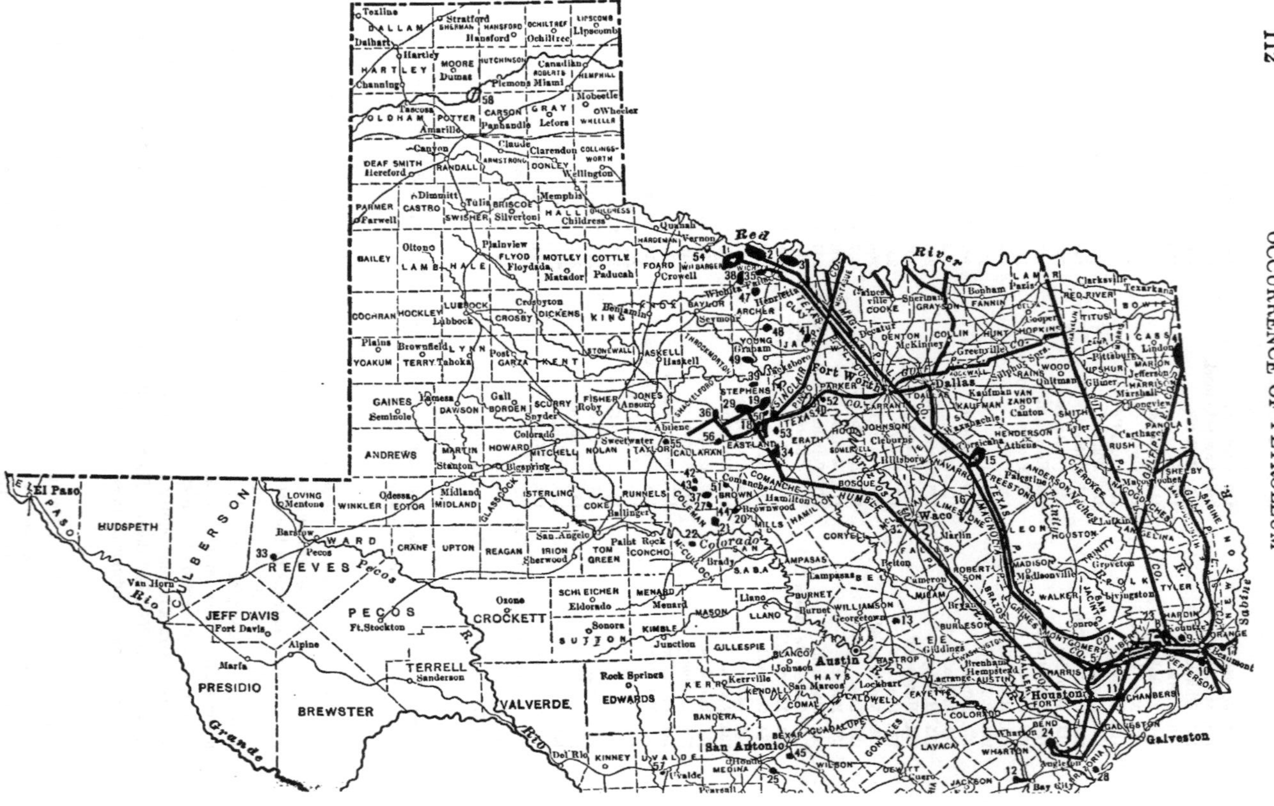
Red
River
El Paso
HUDSPETH
CULBERSON
REEVES
Pecos
JEFF DAVIS
PECOS
PRESIDIO
BREWSTER
TERRELL
CROCKETT
VALVERDE
EDWARDS
Rio
Grande
Colorado
ANDREWS
Amarillo
Lubbock
Fort Worth
Dallas
Waco
Austin
San Antonio
Houston
Galveston
Beaumont
Sabine

Fig. 54.—Principal oil and gas pools and oil pipe lines of Texas.

Key to oil and gas pools of Texas.

1. Electra
2. Burkburnett
3. Petrolia
4. Caddo (La.)
5. Humble
6. Dayton
7. Batson
8. Saratoga
9. Sourlake
10. Spindle Top
11. Goose Creek
12. Markham
13. Thrall
14. Strawn
15. Corsicana
16. Mexia
17. Orange
18. Ranger
19. Caddo (Stephens Co.)
20. Brownwood
21. Trickham
22. Lohn
23. Hull
24. Damons Mound
25. Somerset
26. White Point gas field
27. Santa Anna
28. Hoskins Mound
29. Breckenridge (Parks)
30. Reiser gas field
31. Jennings gas field
32. South Bosque
33. Toyah
34. Desdemona
35. Iowa Park
36. Moran
37. Burkett
38. Culbertson
39. Black Ranch
40. Mineral Wells gas field
41. Avis
42. Gray
43. Morris
44. Bangs
45. Alta Vista
46. Piedras Pintas
47. Holliday
48. Westfork
49. Graham
50. Veale, etc.
51. Bailey
52. Millsap
53. Allen
54. Vernon
55. Abilene
56. Putnam
57. Uvalde
58. Amarillo gas field

Texas

Texas is favored with such a complexity of conditions, and its territory is so vast, that it has a great variety of oil-fields. Their distribution is shown in Fig. 54. For convenience they may be divided into seven groups: (1) North Texas (2) Central Texas, (3) Gulf-Coast, (4) Caddo (extension from northern Louisiana), (5) Pecos Valley, (6) fields along the east side of the Balcones Fault zone, embracing the Corsicana, Mexia, San Antonio, Uvalde and Webb County fields, and (7) Amarillo fields. Each of these groups must be described separately.

The North and Central fields are generally described together, but as they are somewhat different they deserve separate mention. In the North Texas fields, or those of Wichita, Clay, Wilbarger, Archer, and Jack Counties, the producing formations are of Permian and Pennsylvanian age, and the wells range in depth from 500 to 2000 feet, while the structures consist of nearly flat strata of slight dip, or arranged in gentle anticlines paralleling the Wichita Mountains which lie to the north. These fields were opened in 1904 and attained their principal development at Burkburnett in 1919.

The Central Texas fields, or Ranger fields, as they are frequently called, lie in Eastland, Stephens, Palo Pinto, Comanche, Callahan, Coleman, Brown, and McCulloch Counties, their positions being due to terrace structures, low domes (see Fig. 32) or lithologic changes on a great structural nose extending north for 100 miles from the Llano-Burnett uplift, in the direction of the Wichita Mountains, and having a breadth averaging as much as 40 miles. The wells in this region vary greatly in depth, the deepest being toward the north, where some reach a depth of 4000 feet. They probably average less than 3400 feet and are only a few hundred feet deep in McCulloch County. The producing formation is mainly the Bend limestone, supposedly of Pennsylvanian age but called Mississippian by some geologists. Some oil is also found in shallower sandstone of Pennsylvanian age. The Bend fields were discovered in 1917 and attained intensive development during the three succeeding years.

The Gulf-Coast fields are of the salt-dome type, and are numerous but of small lateral extent. The producing strata are believed to vary in age from late Tertiary to Cretaceous, and may be Quaternary in some cases. They vary greatly in depth. These fields gained their notoriety in 1901, when the Spindle Top dome was discovered by Capt. Lucas. The Gulf-Coast fields may be said to lie in Jefferson, Orange, Hardin, Liberty, Harris, Brazoria, Fort Bend, Matagorda and San Patricio Counties, though it is probable that similar structures will be found to be productive in other counties.

A few small wells have been successful in Pecos Valley in Reeves and adjoining counties in strata of Permian age and rather shallow in depth. No considerable field has been developed.

The Amarillo fields are limited in production to natural gas at the time this report is written, though some of the most prominent anticlinal and domal structures in Texas exist in that locality, and one oil well was discovered in Carson County in these fields in 1921. Some of the largest gas wells in the country exist in these fields.

The Corsicana field, together with several smaller adjoining fields in Navarro County, is of Upper Cretaceous age and at one time was of great importance for its oil production, while the Mexia field was a notable gas field, until the development of great oil wells in 1921. The structures are anticlinal. Farther southwest in Williamson County, in the same belt, the Thrall field is a dome which owes its position to a circular intrusive plug or bed of basalt, while the San Antonio and Somerset fields in Bexar and Atascosa Counties are rather complicated and little known as regards their detailed structure. The same may be said of the Laredo gas field in Webb County, where some oil has been found, and of the Piedras Pintas field in Duval County.

References

Adams, G. I.: Oil and gas fields of the western interior and northern Texas coal measures, Bull. 184, U. S. Geol. Survey, 1901, 64 pp.
Hayes, C. W., and Kennedy, William: Oil fields of the Texas-Louisiana Gulf Coastal Plain, Bull. 212, U. S. Geol. Survey, 1903, 146 pp.
Udden, J. A., and Phillipps, D. M.: A reconnaissance report on the geology of the oil and gas fields of Wichita and Clay Counties, Texas, Bull. No. 246, Univ. of Texas, 1912, 303 pp.
Gardner, James H.: The Mid-Continent oil-fields, Bull. Geol. Soc. Am., Vol. XXVIII, 1917, pp. 685–720.
Matson, G. C., and Hopkins, O. B.: The Corsicana oil and gas field, Texas, Bull. 661-F, U. S. Geol. Survey, 1917, pp. 211–251.
Matteson, W. G.: A review of the development of the new central Texas oil fields during 1918, Econ. Geology, Vol. XIV, No. 2, 1919, pp. 95–146.
Riddle, R. A.: The Marathon fold and its influence on petroleum accumulation, Univ. of Tex., Bull. 1847, 1918, pp. 9–16.
Fuller, M. L.: Relation of oil to carbon ratios of Pennsylvanian coals in North Texas, Econ. Geology, Vol. XIV, No. 7, Nov., 1919, pp. 536–542.
Pratt, W. E.: Geologic structures and producing areas in North Texas petroleum fields, Bull. Am. Assoc. Pet. Geologists, Vol. III, 1919, pp. 44–70.
Udden, J. A.: Oil-bearing formations in Texas, Bull. Am. Assoc., Pet. Geologists, Vol. III, 1919, pp. 82–98.
Sellards, E. H.: Structural conditions in the oil fields of Bexar County, Texas, Bull. Am. Assoc. Pet. Geologists, 1919, pp. 299–309.
Hager, Dorsey: Geology of oil fields of North Central Texas, Trans. Am. Inst. Min. Eng., Vol. LXI, 1919, pp. 520–531.

Utah

Productive oil wells have been obtained in San Juan, Uinta and Washington Counties, Utah, and showings have been found in the Counties of Emery, Grand, Wayne, Sanpete and Summit, but there has been no commercial production. Conditions in Utah are believed to be promising.

References

Richardson, G. B.: Petroleum in southern Utah, Bull. 340, U. S. Geol. Survey, 1908, pp. 343–347.
Woodruff, E. G.: Geology of the San Juan oil field, Utah, Bull. 471, U. S. Geol. Survey, 1912.
Lupton, C. T.: Oil and gas near Green River, Grand County, Utah, Bull. 541-D, U. S. Geol. Survey, pp. 3–21.

Virginia

Only insignificant deposits of petroliferous character exist in Virginia, and no oil production should be expected.

Washington

Seepages and showings of oil and gas have been found in the Olympic Peninsula, where the substances exude from Cretaceous strata. Favorable structures are reported, but little success is known to have been achieved in drilling.

REFERENCES

Lupton, C. T.: Oil and gas in the western Olympic Peninsula, Washington, Bull. 581-B, U. S. Geol. Survey, 1914, pp. 23-81.

Weaver, C. E.: The possible occurrence of oil and gas fields in Washington, Trans. Am. Inst. Min. Eng., Vol. LII, 1915, pp. 239-249.

West Virginia

The oil-fields of West Virginia, outlined in Fig. 55, are among the most important in the United States, the state having ranked second in production in the year 1900 and seventh in 1916. The fields are distributed throughout the western part of the state and are intimately related to natural gas fields; although, in general, gas may be said to lie toward the east and the principal oil fields toward the west. To this statement there are important exceptions.

The oil-producing counties, beginning at the north, are Hancock, Brooke, Ohio, Marshall, Wetzel, Monongalia, Marion, Tyler, Harrison, Doddridge, Pleasants, Ritchie, Wood, Lewis, Gilmer, Calhoun, Roane, Clay, Jackson, Wirt, Kanawha, Lincoln, and Wayne. Gas occurs in these counties and also in Taylor, Barbour, Upshur, Braxton, Boone, Logan, Mingo, Cabell, Putnam, and Raleigh, in several of which it is not of great importance.

In its oil horizons West Virginia ranges from the Chemung and Catskill formations and upward through the Pennsylvanian. The principal producing sands are as follows:

Principal Producing Sands of West Virginia

Geologic series	Formation	Name of sand
Pennsylvanian	Conemaugh	Cow Run and other sands
	Allegheny	
	Pottsville	Salt Maxton
Mississippian	Mauch Chunk	Big Lime (Greenbrier), (same as Maxville of Ohio)
	Pocono	Keener Big Injun Squaw Berea Gantz (50-foot and 100-foot)
Devonian	Catskill	Nineveh Gordon Fourth Fifth Sixth or Bayard
	Chemung	Elizabeth

The oil-bearing structures consist of alternating anticlines and synclines. In the latter, the oil commonly lies on the flanks or in the bottom, on account of the dryness of the sands in this state, especially in the deeper formations. However, on the Eureka-

Volcano-Burning Springs anticline, situated in Pleasants, Wood, and Wirt Counties, the oil lies on the crest of an anticline, which is in many respects a rather remarkable one. (See Fig 22.)

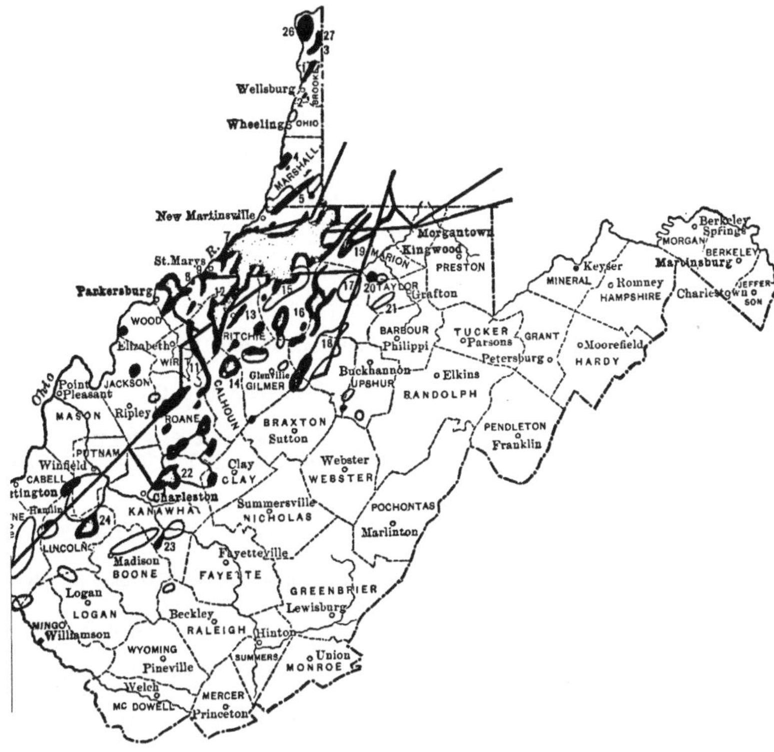

FIG. 55.—Principal oil and gas pools and oil pipe lines of West Virginia.
(Pools mainly oil are solid black. Gas pools with open centers.)

1. Hollidays Cove
2. Wellsburg
3. Turkeyfoot
4. Moundsville
5. Cameron
6. Boardtree
7. Sistersville
8. Eureka
9. St. Marys
10. Volcano
11. Burning Springs
12. Cairo
13. Harrisville
14. Grantsville
15. West Union gas field
16. Salem
17. Clarksville gas field
18. Weston gas field
19. Mannington
20. Shinnston
21. Bridgeport gas field
22. Blue Creek
23. Cabin Creek
24. Griffithsville
25. Glenns Run
26. Moscow
27. Carson

REFERENCES

White, I. C.: Petroleum and natural gas, W. Va. Geol. Survey, Vol. I, 1899, pp. 123–378, Vol. I-A, 1904, pp. 1–557; also various County Reports.

oger, David B.: Recent oil and gas developments in West Virginia, Bull. Am. Assoc. Pet. Geologists, Vol. IV, No. 1, 1920, pp. 27–32.

Wyoming

As shown in Fig. 56, petroleum is widely distributed throughout the state. The most notable producer is the Salt Creek field in Natrona County, and other important fields are Grass Creek, Big Muddy, Basin, Elk Basin, Little Buffalo Basin, Lander, Greybull, Pilot Butte, Oregon Basin, Byron, Moorcroft, Upton-Thornton, Lance Creek, Douglas, Powder River, Poison Spider, Iron Creek, etc.

The oil occurs largely in the Wall Creek and associated members of the Frontier sands in the Colorado group of Upper Cretaceous age, but is present to some extent in Lower Cretaceous, Jurassic, and Pennsylvanian rocks, as well as in the Montana group overlying the Colorado. The structures consist mainly of domes and anticlines, some of the former being among the most remarkable and easily decipherable in the world; and immense quantities of oil have been taken from them.

The stratigraphic distribution of the oil production is shown on page 119.

References

Washburne, C. W.: Gas fields of the Bighorn Basin, Wyoming, Bull. 340, U. S. Geol. Survey, 1908, pp. 348–363.

Schultz, A. R.: The Labarge oil field, Central Uinta County, Wyo., Bull. 340, U. S. Geol. Survey, 1908, pp. 364–373.

Wegemann, C. H.: The Powder River oil field, Bull. 471-A, U. S. Geol. Survey, 1912, pp. 52–71.

Hewitt, D. F., and Lupton, C. T.: Anticlines in the southern part of the Big Horn Basin, Wyoming, a preliminary report on the occurrence of oil, Bull. 656, U. S. Geol. Survey, 1917, 192 pp.

Trumbull, L. W.: Petroleum Geology of Wyoming, Cheyenne, 1917, 78 pp.

Wegemann, C. H.: The Salt Creek oil field, Wyoming, Bull. 670, U. S. Geol. Survey, 1918, 52 pp.

Collier, A. J.: Oil in the Warm Springs and Hamilton domes, near Thermopolis, Wyoming, Bull. 711-D, U. S. Geol. Survey, 1920, 73 pp.

Jamison, C. E.: The Douglas oil field, Converse County, Wyoming, Bull. 3-A, State Geologist's office, 1913, 24 pp.

Jamison, C. E.: The Muddy Creek oil field, Carbon County, Wyoming, Bull. 3-B, State Geologist's office, 1913, 7 pp.

Hintze, F. F. Jr.: The Basin and Greybull oil and gas fields, Bull. 10, State Geologist's office, 1915, 62 pp.

Jamison, C. E.: The Little Buffalo Basin oil and gas field Bull. 11, State Geologist office, 1915, Pt. I, pp. 67–90.

Jamison, C. E.: The Grass Creek oil and gas field Bull. 11, State Geologist's office, 1915, Pt. II, pp. 91–120.

Trumbull, L. W.: Light oil fields of Wyoming, Bull. 12, State Geologist's office, 1916, pp. 123–134.

Ziegler, Victor: The Pilot Butte oil field, Fremont County, Bull. 13, State Geologist's office, 1916, pp. 141–176.

Ziegler, Victor: The Byron oil and gas field, Bull. 14, State Geologist's office, 1917, pp. 181–207.

Ziegler, Victor: The Oregon Basin gas and oil field, Bull. 15, State Geologist's office, 1917, pp. 211–242.

Barnett, V. H.: The Douglas oil and gas field, Converse County, Wyoming, Bull. 541-C, U. S. Geol. Survey, 1914, pp. 51–88.

Barnett, V. H.: The Moorcroft oil field Crook County, Wyoming, Bull. 581-C, U. S. Geol. Survey, 1913, pp. 83–104.

Lupton, C. T.: Oil and gas near Basin, Big Horn County, Wyoming, Bull. 621-L, U. S. Geol. Survey, 1916, pp. 157–190.

Producing Sands in Wyoming Fields

System	Series	Group	Formations	Productive sands	Representative thickness (feet)
Cretaceous	Upper Cretaceous	Montana	Pierre		4500
				Shannon	
		Colorado	Niobrara shale		700
			Benton shale	Wall Creek	700
				Second Wall Creek	
					250
				Mowry shale	300
					200
	Lower Cretaceous	Dakota sandstone		Cloverly Dakota	150
Cretaceous?			Morrison	Morrison	250
Jurassic	Upper Jurassic		Sundance	Sundance	150
Triassic			Chugwater		
Carboniferous	Permian		Embar	Embar	

The oils of the Lander and Warm Springs pools in Fremont and Hot Springs Counties respectfully, are from the Embar sand. In character they are heavy and asphaltic, dark brown in color, and have a disagreeable odor. Triassic oil, which is found in the Hamilton pool in Hot Springs County, is of a better grade, while the Cretaceous oils are of still better quality. The depth of successful Wyoming wells runs from 60 to 4000 feet.

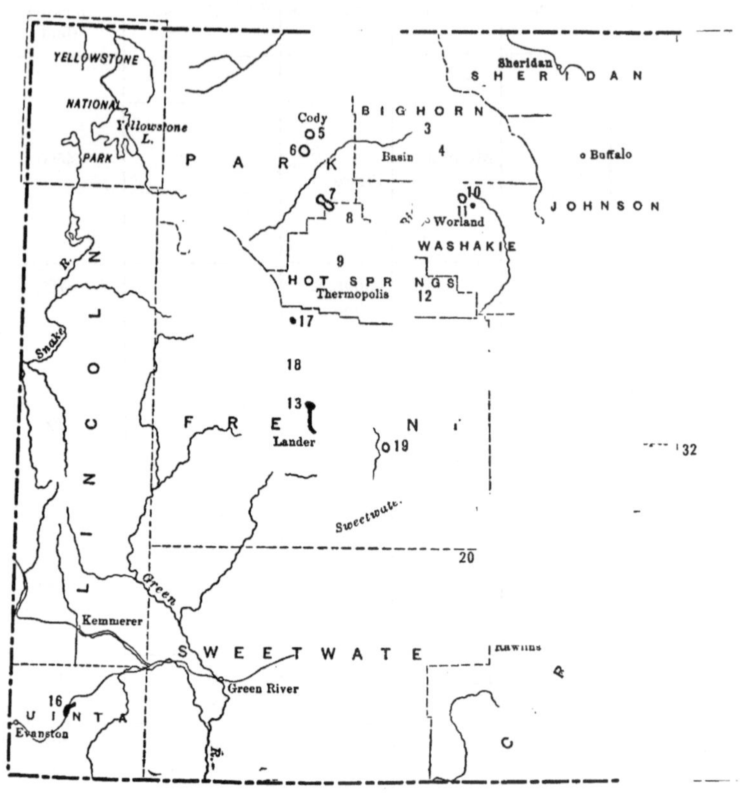

Fig. 56.—Principal oil pools and oil pipe lines of Wyoming. (From Ernest Marquardt)

1. Elk Basin
2. Byron
3. Greybull
4. Basin
5. North Oregon Basin gas pool
6. South Oregon Basin gas pool
7. Buffalo Basin gas pool
8. Grass Creek
9. Hamilton (Cottonwood) dome
10. Hidden Dome gas pool
11. Sherad
12. Warm Springs
13. Sage Creek
14. Hudson
15. Dallas
16. Spring Valley
17. Maverick Springs
18. Pilot Butte
19. Sand Draw gas pool
20. Lost Soldier
21. Ferris
22. Mahoney gas pool
23. Shannon
24. Salt Creek
25. Boone Dome pool
26. Pine Dome gas pool
27. Poison Spider
28. Oil Mountain
29. Iron Creek

Dominion of Canada

References to the occurrence of oil, asphalt and natural gas in Canada date back to its earliest history. Sir Alexander Mackenzie noted the tar springs of the Athabaska region in the " Voyages through North America to the Frozen and Pacific Ocean," 1789 and 1793; and since that time these tar springs have been commented upon by all visitors to that region.

As far back as 1830 residents of Enniskillen in Lambton County, Ontario, noted the presence of oil on swamps of that region, and in 1857 a shallow well was dug near Enniskillen, at a place which became known as Oil Springs. In fact, not only does the actual use of oil in Canada antedate the drilling of the Drake well in 1859, in Titusville, Pennsylvania, but in 1857, a Mr. Williams drilled a deep well in Ontario, with successful results.

Drake's discovery caused great excitement in the Oil Springs region in Ontario. By 1860 hundreds of derricks had been erected, and wells were sunk 100 feet or so at Black Creek in the township of Enniskillen, the first flowing well being struck on January 11, 1862.

Shortly after the excitement caused by the development of Oil Springs, another large pool was struck at Bothwell, in Kent County, about 30 miles southeast; and in 1865 Petrolia, 7 miles north of Oil Springs, developed a larger field.

As far back as 1844, Sir William Logan noted the presence of several oil springs near the extremity of Gaspé Peninsula in the Province of Quebec, but drilling produced no important result.

The principal producing formations of Canada are noted in the following table.

Producing Sands of Eastern Canada

Geologic series	Oil-bearing formations	Productive region
Mississippian	Several sands of Pocono age	Albert County, N. B.
Devonian	Hamilton limestone	Petrolia and Oil Springs, Ontario
	Onondaga limestone	Oil Springs and Petrolia
	Oriskany sandstone	Euphemia, Lambton County, Ontario
Silurian	Guelph limestone	Essex and Welland counties, Ontario
	Clinton limestone	Welland County, Ontario
	Red Medina sand	Welland County, Ontario
	White Medina sands	Welland and Brant Counties, Ontario
Ordovician	Trenton limestone	Western Ontario (and showings of gas in Welland County).
Cambrian (unimportant)	Potsdam sandstone	St. Catharines, Ontario
	Quebec group	Province of Quebec

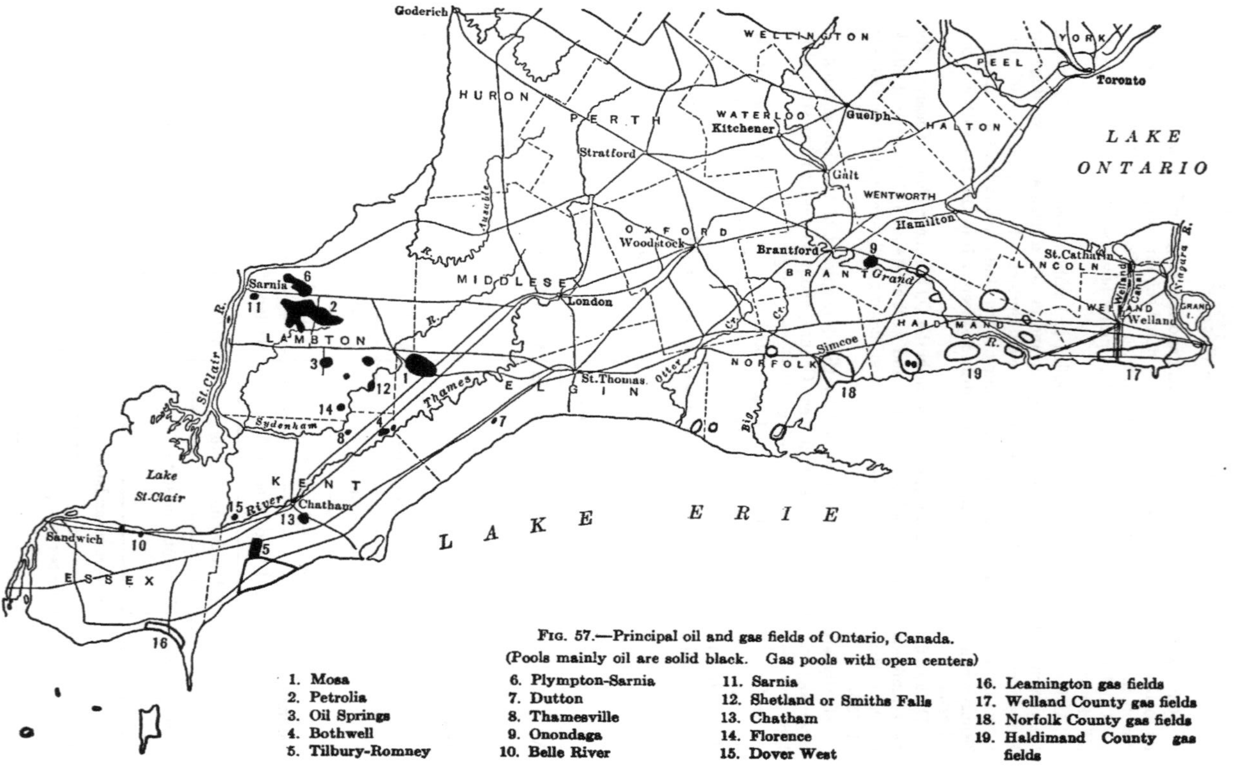

Fig. 57.—Principal oil and gas fields of Ontario, Canada.
(Pools mainly oil are solid black. Gas pools with open centers)

1. Mosa
2. Petrolia
3. Oil Springs
4. Bothwell
5. Tilbury-Romney
6. Plympton-Sarnia
7. Dutton
8. Thamesville
9. Onondaga
10. Belle River
11. Sarnia
12. Shetland or Smiths Falls
13. Chatham
14. Florence
15. Dover West
16. Leamington gas fields
17. Welland County gas fields
18. Norfolk County gas fields
19. Haldimand County gas fields

The known fields of Canada comprise the Ontario, New Brunswick and Western Canada fields. Up to the present, the greater part of the Dominion's production has come from Ontario, which in 1917 produced 202,991 barrels, or 98.8 per cent of the whole production of the country. Fig. 57 shows in detailed form the relative positions of the oil pools of Ontario. Their relative importance in 1918 was as follows;

1. Mosa.
2. Petrolia.
3. Oil Springs.
4. Bothwell.
5. Tilbury-Romney.
6. Plympton-Sarnia.
7. Dutton.
8. Thamesville.
9. Onondaga.
10. Belle River.

These fields are situated in the southwestern part of Ontario in the counties of Lambton, Kent, Essex and Brant, but only the first mentioned was ever of great importance. All except the first three fields are mainly gas fields, and gas is also produced commercially in Norfolk, Haldimand, Lincoln, Welland, and Elgin Counties. The oil lies along domes subsidiary to the Cincinnati anticline in the extreme southwestern part of Ontario and in minor structures farther east. Tests in other parts of Ontario have not resulted in much success, nor has commercial production been secured in the Province of Quebec. In New Brunswick, however, a productive gas field exists in the vicinity of Moncton, and some oil has been found in the same field which occurs on a definite anticline in Mississippian or Upper Devonian sands.

The greater part of the prospecting now going on is in western Canada. For many years gas has been obtained from an immense field on the geanticline occupying the plains of eastern Alberta, and some small wells along the base of the Rockies and on the International Boundary have produced oil; but not much success in oil drilling has been attained. Small wells have resulted from recent testing in the Peace River region, and wells have also been drilled, though with little success, on the Athabasca, along which the famous outcrops of "tar sands" occur. All oil and gas found in western Canada has been in the Dakota sand and other formations of Cretaceous age.

The great oil excitement in the Mackenzie River region had little result in 1921.

Canada can hardly be said to be an important producer of oil at present, being eighteenth in importance among oil-producing countries; but in the aggregate it has produced 24,864,000 barrels and is ninth in importance as regards total production.

References

Clapp, F. G., and others: Petroleum and Natural Gas Resources of Canada, Dept. of Mines, Mines Branch, 2 vols., 1914 and 1915, 802 pp.

Huntley, L. G.: Oil, gas and water content of Dakota sand in Canada and United States, Trans. Am. Inst. Min. Eng., Vol. LII, 1915, pp. 329–352.

Dowling, D. B.: Correlation and geologic structure of the Alberta oil fields, Trans. Am. Inst. Min. Eng., Vol. LII, 1915, pp. 353–362.

Dowling, D. B., Slipper, S. E., and McLearn, F. H.: Investigations in the gas and oil fields of Alberta, Saskatchewan and Manitoba, Can. Dept. Mines, Mines Branch, Mem. 116, 1919, Pub. No. 1722, 89 pp.

Miller, W. G.: Petroleum in Canada, Bull. Geol. Soc. Am., Vol. XXVIII, 1917, pp. 721–726.

Newfoundland

Petroleum has been found in non-commercial quantities at Parsons Pond, St.Pauls Inlet and at Port-au-Port on the northwestern and western coast of Newfoundland. Some little production has been obtained from about three wells by pumping, but there is little prospect of paying fields.

Mexico

When Capt. Lucas discovered oil in 1901 at Spindle Top, Texas, E. L. Doheny conceived the idea that the large asphalt deposits in Mexico represented dried-up oil seepages, and with this idea he drilled at Ebano, about 50 miles from Tampico on the railroad line leading to San Luis Potosi. The wells were very successful, and production from them has continued until recently.

A little later S. Pearson & Son developed oil near Minatitlan in southern Vera Cruz, building a refinery at that place and organizing a general campaign for oil throughout Vera Cruz. This resulted in the drilling of the oil wells at San Cristobal, the famous gusher at Dos Bocas, and one of the largest wells in the world at Potrero del Llano, about 40 miles north of the Tuxpan River.

Oil was first shipped from the Mexican fields about 1904, and in 1920 Mexico produced 163,540,000 barrels of oil, which was 23.5 per cent of the world's production, and placed Mexico second in the list of oil-producing countries. The total production of Mexico, inclusive of that date, was 536,524,000 barrels, or 6.1 per cent of the world's yield.

The known oil fields of Mexico are included within two great regions,—Tampico-Tuxpan and Tehuantepec-Tabasco—both of which are segments of the Gulf Coastal Plain. They include about a dozen producing fields, covering areas of over 20,000 square miles, of which only a very small portion is productive. The principal oil-yielding rocks are the Tamasopo limestone of Cretaceous age and the immediately overlying Eocene strata, although some oil is obtained locally from later Tertiaries. Wells range in depth from 800 to 3000 feet, the usual depth being somewhat over 2000 feet.

The most important fields are those of the Tuxpan-Tampico region, situated in the northern part of the State of Vera Cruz and extending into the southern edge of Tamaulipas. Some of the greatest wells that the world has ever known have been drilled in this district. The Tampico-Tuxpan region includes ten recognized fields, the Panuco, Ebano-Chijol, Topila, and San Pedro fields in the Panuco valley, and the Potrero del Llano-Alazan, Cerro Azul, Tepetate-Casiano, Alamo-Chapapote, Furbero, and Tanhuijo-San Marco fields farther south. The Zacamixtle field was opened in 1920.

The twelve recognized fields of the country are shown in Fig. 58. They are named below.

1. Potrero del Llano-Alazan.
2. Panuco.
3. Cerro Azul.
4. Tepetate-Casiano-Los Naranjos.
5. Alamo-Chapapote.
6. Ebano-Chijol.
7. Topila.
8. Furbero.
9. Tehuantepec-Tabasco.
10. Dos Bocas.
11. Zacamixtle.
12. Tanhuijo-San Marco.

The Dos Bocas field was a world wonder but it became exhausted years ago.

The production of individual Mexican wells is abnormally large, and consequently about 95 per cent of the production of the country has come from less than 100 wells, the daily output of which has ranged between 2000 and 150,000 barrels per well.

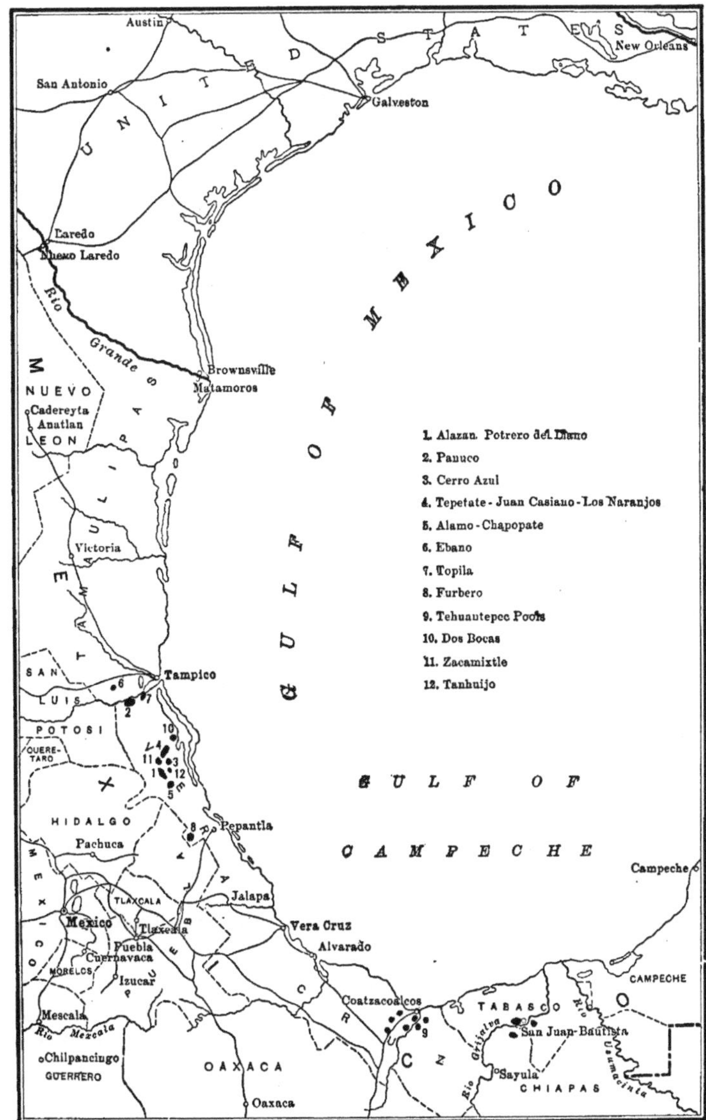

Fig. 58.—Principal oil fields of Mexico.

One of the earliest known pools of Mexico and the southernmost field of northern Vera Cruz is that at Furbero, situated 42 miles south of the Alamo field. The Furbero field is of a different geologic type from the other Mexican oil fields, the occurrence of the oil being in part dependent on an intrusive igneous bed and laccolith.[1] The oil occurs both in the igneous and in the contiguous metamorphosed rocks. (See Fig. 35.)

The oil of northern Vera Cruz is found mainly in the great Tamasopo limestone of Lower Cretaceous age, though there are a few wells in the overlying San Felipe shales. The geological structure consists of large domes in anticlines of great continuity, sometimes faulted and frequently cut by dikes and plugs. Most of the fields mentioned above are on these domes.

The great difficulty in the Mexican fields is the encroachment of salt water, which saturates all but the crests of the domes or anticlines. In practically all of the old productive fields it became so serious in 1920 and 1921 as to menace their existence. The question of the ultimate life of these fields is problematical, but geologists have already shown that the ultimate failure of the known Mexican fields is coming soon.

References

DeGolyer, E.: The effect of igneous intrusion on the accumulation of oil in the Tampico-Tuxpan region, Mexico, Econ. Geology, Vol. X, 1915, pp. 651–662.

Dumble, E. T.: The occurrences of petroleum in eastern Mexico as contrasted with those of Texas and Louisiana, Trans. Am. Inst. Min. Eng., Vol. LII, 1915, pp. 250–267.

Huntley, L. G.: The Mexican oil fields, Trans. Am. Inst. Min. Eng., Vol. LII, 1915, pp. 281–321.

DeGolyer, E.: The significance of certain Mexican oil-field temperatures, Econ. Geology, Vol. XIII, No. 4, 1918, pp. 275–301.

DeGolyer, E.: The petroleum industry of Mexico, Evening Post Oil-Industry Supplement, New York, Aug. 31, 1918, p. 17.

Ordoñez, Ezequiel: Oil in southern Tamaulipas, Mexico, Trans. Am. Inst. Min. Eng., Sept., 1918, pp. 1,001–1,008.

Central America

Indications of petroleum exist at many places in Central America, but there is as yet no commercial development. The following brief notes may be of some value:

British Honduras

Indications of petroleum are reported, but nothing definite is known of them.

Costa Rica

Some explorations have been made by an American Company, without known success. The greater part of the country is unfavorable.

Honduras

In the Republic of Honduras, indications of petroleum are reported in limestone, presumably of Neocomian (Lower Cretaceous) age in the Guare Mountains near Comayagua.

Panama

Oil is reported in the Province of Los Santos and elsewhere in Panama but, so far as known, the occurrences are unimportant.

[1] DeGolyer, E.: The Furbero oil field, Mexico, Trans. Am. Inst. Min. Eng., Vol. LII, 1915, pp. 268–280.

EUROPE

General Discussion

Europe ranks second among the five continents of the world as a present oil producer, and will probably maintain its position for a few years if the search for new fields is prosecuted vigorously. Of the various countries in Europe, Russia is the chief producer, standing next to the United States, and followed, in the order of importance, by Poland, Rumania, France, Czecho-Slovakia and Italy. The total production of Europe from 1857–1917 was only about half as great as the production of the United States and far greater than that of Asia, the nearest competing continent. The occurrence of oil in Europe is considered by countries in the following pages.

Russia

The producing fields of Russia are the Apsheron Peninsula or Baku fields, comprising Sabunchi, Surakany, Bibi-Eibat, Balakhani, Romani, and Binagadi. These fields furnish 75 per cent of the oil produced in Russia and are named in the order of present importance. Outside the Apsheron Peninsula, the fields are, in the order of importance, Grosny, Emba, Isle of Sviatoi, Tcheleken, Maikop and Ferghana, the last mentioned being in Asia.

With the exception of Mexico, the one important rival of the United States, in the actual production of petroleum, is Russia. Oil has been known at Baku since the earliest times, and, up to the date of the Saracen conquest in A. D. 636, that city was the principal point of pilgrimage of Persian and Hindoo fire-worshipers, being visited every year by thousands. In the thirteenth century the oil and gas exudations of Baku were fully described by Marco Polo;[1] and Olearius, who accompanied a German embassy to Persia in 1656, saw over thirty petroleum springs near what is now Schemakha, west of Baku.

As late as 1880, the Temple of Surakhani, which had been the seat of the Sacred Fire for centuries, was still visited by priests from India. Machine drilling was begun in Russia in 1871, and since then many gushers have been struck, the maximum production of the Baku fields being that of 1901, when 80,977,638 barrels were produced. The wells were originally 400 to 800 feet deep, but may have been deepened.

The Baku district occupies an area of only 9 square miles, on the Apsheron Peninsula. The oil is derived chiefly from the Oligocene series, consisting here of alternate beds of shaly marls and fine-grained calcareous sands or sandstone. Their upturned edges are covered with Pliocene or later deposits, also somewhat disturbed, and sometimes bearing oil derived from the underlying Oligocene.

The Grosny field lies north of the Caucasus Mountains on a sharp anticline, in Miocene beds about 500 miles northwest of Baku (Lat. 43° 30′ North, Long. 44° 45′ East). It was worked by means of pits, as early as 1823, but its modern development began in 1893, with the first drilled well. The production has increased quite rapidly, and the total for 1917 was 14,760,000 barrels.

The Maikop field lies on the north flank of the Caucasus Mountains in Kuban province, northeast of the Black Sea. The rocks are Miocene, and the oil-bearing series, which is about 1000 feet in thickness, consists of dolomitic limestone grading downward into shales and sandstone. The dip is commonly as much as 45° to 60° from the horizontal, but shows considerable variation. Oil in large quantities was first produced in 1910, and in 1912 the annual output had risen to 1,104,442 barrels. In 1916 it had declined to 240,096 barrels.

[1] The Book of Ser Marco Polo the Venetian, Ed. by Col. Yule, London, 1871, Vol. I, p. 4.

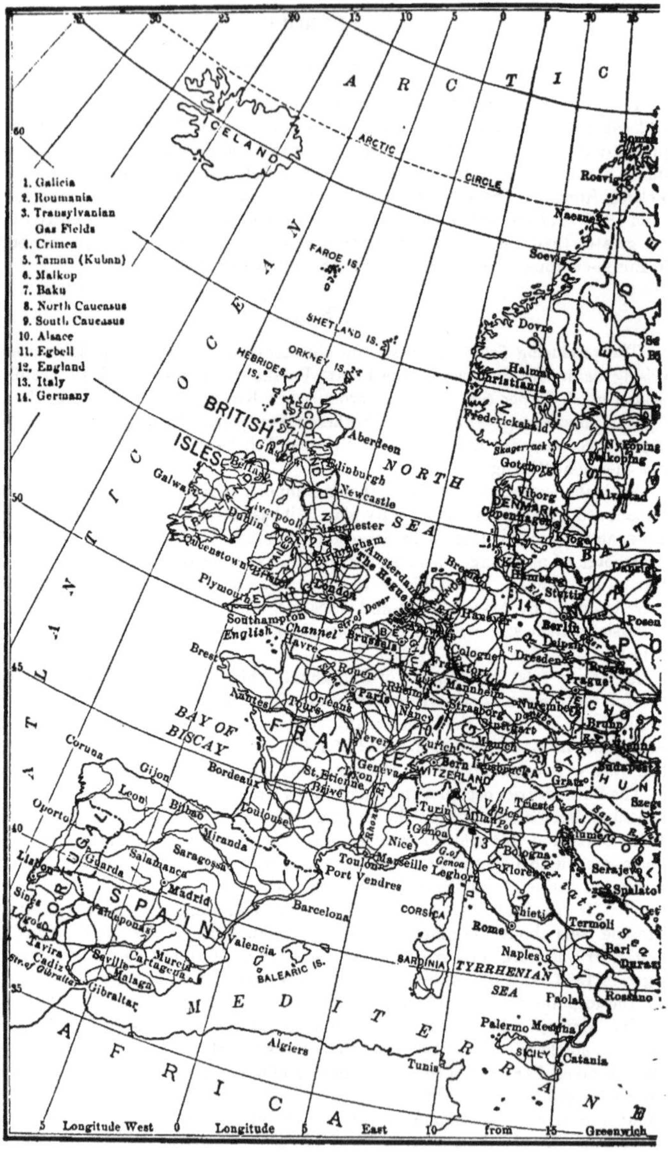

Fig. 59.—Principal oil

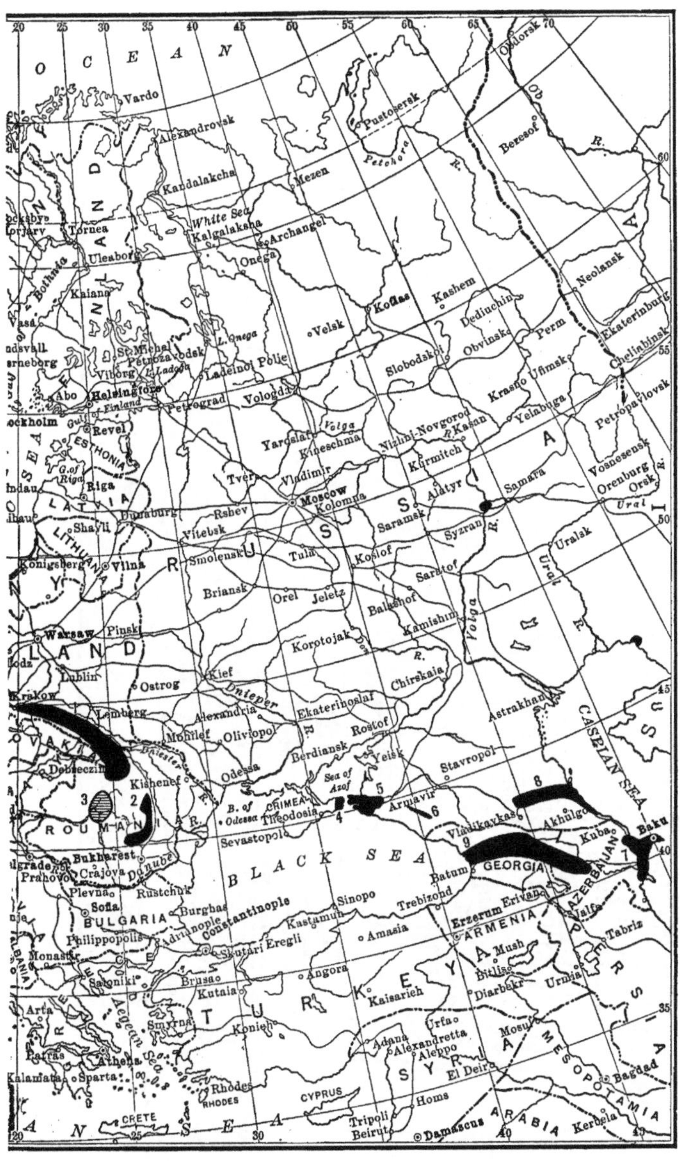

and gas fields of Europe.

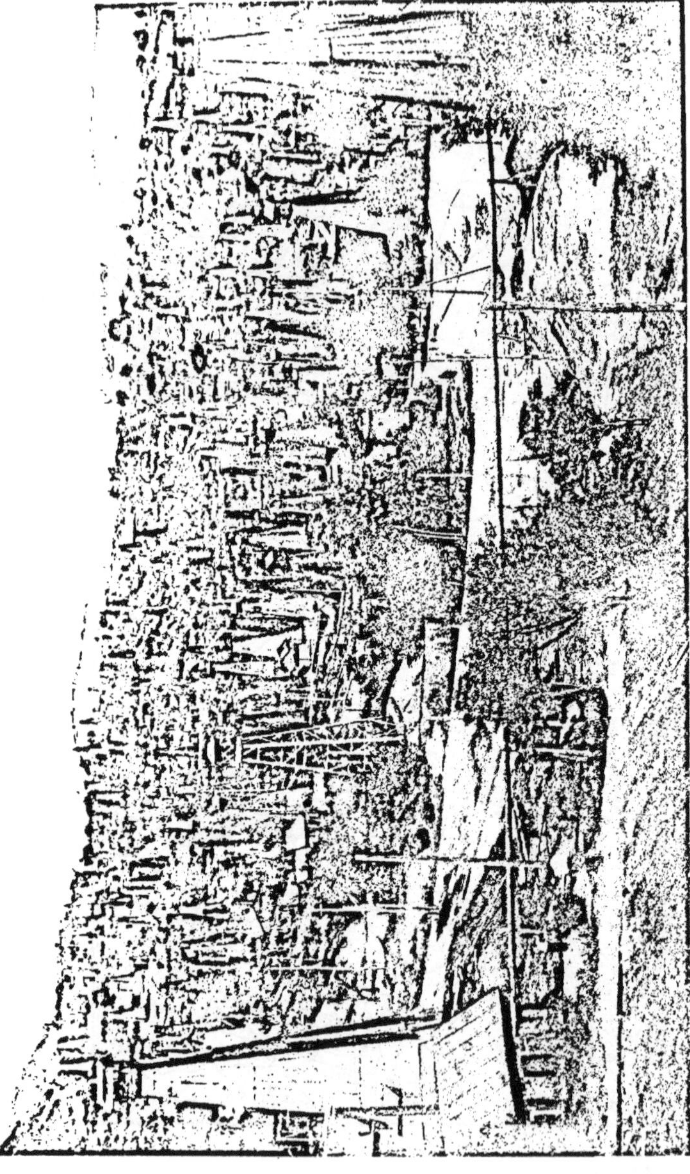

The Emba field was opened in 1913, and in 1914 produced 2,001,801 barrels of oil. In 1917 its production had declined to about 1,800,000 barrels.

Oil show.ngs exist in many other parts of Russia, including Finland, Vologda Province, the central and southern provinces, parts of Caucasia, the vicinity of Kertch in the Crimea, Taman peninsula, Terek, Daghestan territories, Kutais, Tiflis, etc. Several of these districts are commercially productive. Extensive oil deposits are believed to exist on the Ukhta River, in Archangel, and in Nova Zembla. The aggregate yield of all Russian fields in 1910 (the best year) was 70,336,574 barrels, and the estimate for 1917 was 69,960,000 barrels.

Rumania

Petroleum has long been known in Rumania, as indicated by the frequent occurrence of the village name Pacureti, derived from the Wallachian word pacura, meaning petroleum. As early as 1750, travelers reported that it was used in treating diseases of cattle, for lighting courtyards, and for wagon grease. Modern development began in 1866; but for fifteen years the wells were dug by hand. The years 1880–1887 witnessed the first successful machine drilling, and thereafter progress was steadily made.

The Rumanian oil district is practically continuous with that of Galicia, as it lies along the flanks of the Carpathian Mountains. The principal productive districts are Prahova, Dambovitza Bacau, and Buzeu. Most of the oil is derived from Miocene and Pliocene strata, with less important production from the Eocene and Oligocene and possibly from the Cretaceous. The rocks are mainly shales, sands and sandy clays, with some limestone, gypsum, and salt, the whole compressed into a succession of sharp, narrow, and irregular domes and anticlines, associated with some faulting.

In Bukowina, oil has been found along three main anticlines in the Cretaceous series, as well as in several minor folds.

In Transylvania, numerous gas springs and mud volcanoes emerge from the Mediterranean and Sarmatian formations of Miocene age, and gas wells of great volume have been obtained by drilling, while some gas appears to exist in the Pliocene and Eocene series. The gas occurs in closed domes similar to those illustrated in Figs. 31 and 33 overlying masses of rock salt. Petroleum is reported in the Eocene at Kovacs and Monastir and also from Neocomian (Lower Cretaceous) strata. Producing wells have been secured in either Oligocene or Cretaceous at Sosmezo in the Ojtoz.

The increase in yield has been fairly steady from small beginnings to 13,544,768 barrels in 1913, but on account of the destruction of wells during the German invasion, only 3,720,760 barrels (estimated) were produced in 1917. In 1919 and 1920 a rapid revival took place. Of the maximum production 86 per cent came from Prahova, 6 per cent each from Dambovitza and Buzeu and 2 per cent from Bacau.

Poland

The most important part of the former Austrian Province of Galicia, from the standard of petroleum production, is that now incorporated in Poland and known as "Little Poland"; hence it is proper to consider this district in conjunction with the adjoining one of the Ukraine. Although "earth-balsam" was used medicinally, as early as 1506, the first systematic development in Galicia occurred in the Bobrka field in 1853, at which time considerable quantities were collected from trenches and shallow wells. The introduction of drilling in 1870 was accompanied by a rapid increase in production, and after 1885 the development of other Galician fields followed in rapid succession, several distinct districts being recognized. Many wells were remarkable producers, and yields of from 1000 to 3000 barrels per day were not uncommon.

The principal fields of East Galicia are Tustanowice, Boryslaw, Mraznica, Schodnica, and Urycz, in the order of their importance in 1917, while those of West Galicia are Potok, Rogi, Rowne, Krosno, Tarnawa-Wielopole-Zagorz, Kobylanka, Kyrg, Zalawie, Lipinki, and Libusza. The fields of West Galicia produce less than 15 per cent of the oil produced by Galicia or by Poland and the Ukraine combined.

The Galician field forms a belt 50 miles wide and 227 miles long on the north flank of the Carpathian Mountains from the Raba River as far as Czerenosz. The geology is complex and the rocks characterized by sharply compressed irregular overthrust folds. Lithologically the rocks are largely shales and sandstones, with some conglomerate and, more rarely, limestone. The oil-bearing formations appear to include the Lower and Upper Cretaceous, Eocene, Oligocene and Miocene series. The Eocene strata, which are very persistent, are likewise the most productive. Beds of identical character exist at different horizons, adding greatly to the complexity of the geological problems.

The production of Polish and Ukranian Galicia increased from 149,837 barrels in 1874 to 14,932,799 barrels in 1909, but the following year saw the beginning of a marked decline, and in 1915 the production was only 4,158,899 barrels. In 1917 Galicia was seventh in importance among the oil-producing districts of the world. The production in 1919 is given as 6,600,000 barrels.

Ukraine

Several distinct producing districts are recognized. The production of Ukraine comes entirely from Galicia, and on account of the geographic, geologic and historic associations the reader is referred to the description under Poland.

Italy

Petroleum was used for illuminating purposes at Agrigentum, Sicily, before the beginning of the Christian era. Its use is described by Strabo and Pliny, who call it "Sicilian oil." The presence of burning gas springs in northern Italy led in 1226 to the adoption, by the town of Salsomaggiore, of a salamander surrounded by flames, as its official emblem. In 1400 a concession was granted for the collection of petroleum in Miano di Medesano. Petroleum was discovered in 1640 at Modena and used for lighting, medicinal and other purposes, while oil was collected early in the eighteenth century from Montechiaro near Piacenza and at Montechino. In 1802 Genoa was lighted by petroleum from the wells of Miano, and the famous Theater at Reggio was lighted with oil from wells at Monte Gibbio.

In 1868 a treatise was published by E. St. John Fairman on the petroleum zones of Italy.

Oil has been developed in commercial quantity in Eocene, Miocene and Pliocene strata of complicated structure in the Emilian provinces northeast of the Gulf of Genoa, in Eocene beds in the Valley of Pescara, in the province of Chieti in central Italy and in Eocene shales in the valley of the Liri near San Giovanni Incarico, midway between Rome and Naples. The Emilia production, after fluctuating for twenty years from 15 to 200 barrels annually, began to increase in 1880, and reached a maximum of nearly 25,000 barrels in 1895. The yield of the Pescara district has been very irregular, varying from nothing to over 2500 barrels. The San Giovanni Incarico field reached a maximum of over 4000 in 1878, but was practically exhausted by 1890. The total production of Italy for 1860–1920 inclusive was only 1,042,000 barrels. Recent years have witnessed renewed activity in Italy, and the production is increasing, although in 1920 the yield was still only 34,180 barrels for the country as a whole.

References

Camerana, E.: L'Industrie dea hydrocarbures en Italie, read before the Third International Petroleum Congress, Bucharest, 1907.
Chamber, D. M.: Italy as an oil-producing country. Petroleum World, Nov., 1912, pp. 467–470.
La production de pétrole en Italie: Journal du pétrole, Vol. XVII, No. 10, Oct., 1917, p. 5.

France

Many evidences of petroleum exist in France, in the form of seepages of oil adjacent to gas springs, and certain structural and stratigraphic conditions and indications of bitumen have long been known. The oil springs near Walsbron, Lorraine, were famous in the Middle Ages and were even known to the Romans. The gas of the "burning fountain," at La Gua, near Grenoble, Department of Isère, was described in 1618. At this date, or even earlier, petroleum was collected from a spring at Gabian, Department of Hérault, on the Gulf of Lyons, and used for medicinal purposes. Small seepages are reported from gray Miocene marls, limestones, and sandstones at depths of 125 to 425 feet at Clermont-Ferrand, Puy de la Poix, Malintrat and Coeur, in the plain of Limagne, Puy-de-Dôme, between the ranges of Puy-de-Dôme and Forrests in southeastern France, and at Chatillon on the flanks of the Alps in Savoy. A deep test made in 1896 at Macholle, near Riom, produced only a few gallons of oil mixed with salt water at a depth of nearly 3500 feet, and other borings have been equally unsuccessful.

Oil is found at Lobsann and Schivabweiler in Lower Alsace, at Olhungen and Woerth, and in Upper Alsace at Altkirch near Basle. The first tests for oil in Lower Alsace were made in 1735 by French miners. In 1745 a small refinery was set up, so that the refining industry in Péchelbronn may be said to date back 175 years, making it one of the oldest in the world. The production was negligible until machine drilling was introduced in 1881. Even now, Péchelbronn is the only petroleum mine worked to a depth of over 800 feet by means of adits through which oil is brought to the surface. Its extraction is as dangerous as it is difficult, as the sand in which it is found is saturated with inflammable gas.

Reference

Petroleum Times, Dec. 27, 1919.
Petroleum World: Oil industry of France and Alsace, Pet. World, Sept., 1919, pp. 370–374.

Cyprus

Signs of oil have been reported in Cyprus, but, according to Cadman,[1] "Indications of petroleum appeared to be confined to a dark bearing phase of the crystalline limestone of probable Cretaceous age, and it seemed doubtful that oil in commercial quantities would be discovered."

Germany

Petroleum was medicinally used in Germany as early as 1436, under the name "St. Quirinus Oil," and was derived from the Tegernsee district of Bavaria. In the sixteenth century petroleum indications were found in Hannover, Prussia, Hildesheim,

[1] Cadman, Sir John: Petroleum Times, June 12, 1920.

Luneburg and elsewhere. Modern commercial developments date back less than half a century and include (1) the North German field, and (2) Bavaria, (Tegernsee district). The North German or Prussian field is situated in a belt lying between the Weser and Elbe rivers north of the Hartz Mountains and including the Wietze, Steinford, Oelheim, and other pools of Hannover, etc. Production on a commercial scale dates from 1880. The oil is said to come chiefly from limestones and sandstones of Upper Jurassic age, or from transitional beds between the Jurassic and Cretaceous series, but that of the Bavarian field comes from Eocene marls and sandstones. The production is very light. The locality of the Limmer asphalt is near Hannover.

Some oil shale was worked prior to the war in the Rhine Provinces and near Reutlingen and Messel; at the latter place 32 gallons of crude oil, 71 gallons of ammonia water and 1900 cubic feet of gas are now obtained.

The total estimated production for Germany in 1917 was only 995,764 barrels, of which one-fifth is credited to Alsace; and the total production credited to Germany (including Alsace) from 1880 to 1918 inclusive was only 16,664,121 barrels.

Czecho-Slovakia

Oil is found in two regions: viz., (1) in Moravia on the west, where it is found at Ratiskovice near the town of Hodonin, and at Bohuslavice on the Vlara River, where some light oil is obtained from Oligocene strata, and (2) in Slovakia on the east, where the Gbely (or Egbell) field is situated. Many wells were drilled in the last-mentioned field between 1913 and 1920. The petroleum, which is of a heavy grade, is found in the Sarmatian formation of Miocene age. The average depth of the wells is between 800 and 900 feet. The production of the Gbely field, which is the only real one in the country, was only 56,000 barrels in 1918.

Jugo-Slavia

In Croatia and Slavonia oil springs have been noted in Pliocene shales at Ludberg, about 35 miles northeast of Agram and at Lepavina, 9 miles south of that place. Wells have obtained small productions at Ribejak, and natural gas has been found near Ivanich. A thick tarry oil exudes from Miocene marls near Kutina in the Moslalvina Hills, at Bacindol, and at Petrovoszelo. Petroleum associated with salt water and gas has been reported seeping from Triassic shales at Bukowik, southwest of the head of Lake Scutari in Montenegro.

Oil has been found in small quantity in Pliocene shales in the hills near Styria and Zala, from both Miocene and Pliocene strata at Szelnica and Peklenica on the Murakoz peninsula, in ryolite-tuffs at Recsk, Heves, and in complexly folded Cretaceous, Eocene, and Oligocene beds at various localities in the Carpathian Mountains.

European Turkey

Petroleum is known in Miocene strata bordering the Sea of Marmora for several miles between Sarkani and Gamos, and petroleum and bitumen occur on the island of Koraka opposite Salagora in the Gulf of Arta. In tests made 40 miles south of Keshan, near the junction of Deli Osman River and Milos Brah, some oil and much salt water were found. Rather indefinite reports also exist of a Miocene oil occurrence near Feredzik on the Maritza, and in rocks of the same age under the Jewish quarter of Constantinople. Submarine springs in the last-mentioned vicinity were recorded in 1681 by Count Maraigli.

References

English, T.: Quart. Jour. Geol. Soc. Vol. LVIII, (1902), p. 157.
Redwood, Sir Boverton: A Treatise on Petroleum, Vol. I, 1913, p. 135.
Dominion, Leon: Trans. Am. Inst. Min. Eng., Vol. LVI, 1916, p. 256.

England

As long ago as 1667 Shirley [1] described gas in a spring near Wigan (Lancashire), some of which he successfully collected and burned, while in 1739 Clayton[2] described a ditch in the same locality, the water of which " would seemingly burn like brandy, the flame being so fierce that several strangers boiled eggs over it." In 1864 naphtha or liquid bitumen was described by Plot [3] on the surface of a spring near Pitchford (Shropshire). The petroleum of this locality later became better known and was used medicinally.

Most of the indications in England have been found in rocks of Carboniferous age. The localities include Whitehaven (Cumberland), Clowne and Alfreton (Derbyshire), Worsley (Lancashire), Langton (North Staffordshire), and Coalbrookdale, Coalport, and Pitchford (Shropshire). From 70 to 100 gallons of oil per day were pumped, with water, from a depth of 960 feet in the Southgate colliery at Clowne. Other petroliferous formations are the Upper Devonian shales at Barnstaple (North Devonshire) and the Liassic in the Bristol district (Gloucestershire-Somersetshire). At Ashwick (Somersetshire) several barrels of petroleum entered a well after a slight earthquake in 1892.

Notwithstanding the general distribution of indications, oil was not discovered in commercial quantities at any point in Great Britain until about 1918,[4] when a Government well which was being drilled on a faulted dome at Hardstoft in Derbyshire, found natural gas and a few barrels of oil in rocks of Mississippian age at a depth of 3078 feet. Several wells in Derbyshire report good showings, so that England has an excellent chance of ultimately becoming a productive oil field.

Scotland

Petroleum has been reported in Scotland from the Orkney Mainland, from rocks of Carboniferous age at Dysart (Fife), Broxburn (Linlithgowshire) and Liberton (Edinburghshire). The greater part of the output of petroleum reported annually from Great Britain is derived from distillation of bituminous shales in Scotland. In 1920 some drilling was being done on domes at West Calder and Darcy, not far from Edinburgh, but no results had been obtained.

Wales

A little oil is distilled from the bituminous shales of Wales, and petroleum has been noted in the Permian series at Ruabon (Denbighshire).

[1] Shirley, Thomas: Philos. Trans., Vol. II, p. 482.
[2] Clayton, Rev. John: Ibid., Vol. XLI, p. 59.
[3] Dr. Plot: Ibid., Vol. XIV, p. 806.
[4] Veatch, A. C.: Petroleum resources of Great Britain, Min. and Met. Bull. No. 157, Sect. 3, Am. Inst. Min. and Met. Eng., February, 1920, 4 pp.

Belgium

In Belgium, traces of petroleum are found at a number of localities in the coal fields, but only in insignificant quantities in iron concretion and fossil cavities. Showings are also reported in probable Eocene strata at Bourlers, near Chimai, and in the Liassic shales of Jamoigne.

Holland

Practically no indications of petroleum have been noted in Holland, although an oily liquid is reported to be obtained from a well in chalk near Maestricht.

Sweden

In Sweden, natural gas is of somewhat widespread occurrence at a number of horizons, especially in Silurian, Liassic and Miocene rocks, but the volume is small. Petroleum indications are less common, but oil-filled fossil cavities have long been known in Silurian strata, and these led in 1867–1869 to the drilling of several holes on the flanks of Mt. Osmund of Delarne. Although the wells were carried to a depth of 900 feet, only traces of oil were found. Later drilling at Nullaberg and elsewhere has been equally unsuccessful.

Switzerland

Between the Alps and Jura Mountains lies a Tertiary Basin where oil seepages and well developed anticlines of frequently moderate dip have long been known. Some of the outcropping sands of Oligocene age are so rich that it might pay to mine them and distill the oil. Traces of petroleum exist in bituminous Miocene limestones near Mathod, Orbe, Chavornay, and Vaud, while small amounts of natural gas exist in Liassic rock-salt of Bex and at Montreux, Vaud. There are no commercial developments.

References

Heim, Arnold and Hartmann, Adolf: Untersuchungen über die petrolfuhrende Molasse der Schweiz, Beiträge zur Geologie der Schweiz, Geotechnische Serie, Vol. VI, 1919.

Bloesch, Edward: Petroleum investigations in Switzerland, Bull. Am. Assoc. Pet. Geologists, Vol. IV, No. 1, 1920, pp. 87–88.

Spain

In Spain, surface indications exist at a number of localities and small quantities of petroleum have been collected in tunnels at Huidbro, 30 miles north of Burgos. Although wells were drilled to a depth of over 1500 feet, mere traces of oil were encountered. A few gallons were obtained in Eocene shales in a mine shaft at Conil near Cadiz, but drilling was commercially unsuccessful, as was likewise a boring sunk in similar beds near Algar, 38 miles east-northeast of Cadiz. Other localities which have afforded indications, but none of which have yielded oil in commercial quantities, are Cueva de la Pez near Bayarque, Liguenza and Molina in Guadalajara, Soria Girona and San Lorenzo de la Muga and Pont de Molina lying west and north of Figueras. Asphalt rock is mined in northern Spain.

Portugal

Petroleum indications are apparently limited to certain calcitic amygdules in basalt at Sicario, near Cintra, and these are of absolutely no commercial importance.

Greece

In addition to oil springs described by the ancients as emerging from Tertiary deposits on the Island of Zante, petroleum has been reported in Miocene strata on the Island of Antipaxos, in springs from Cretaceous limestone at Dremisou Maurolithari, in the Parnassid and at Galazidi near Delphi, in seeps from lignitic Miocene strata near Lintzi (the ancient Cyllene) and from several localities on the Island of Milo. Earthquakes have sometimes been accompanied by discharges of oil, and petroleum and natural gas are known to rise in small quantities from beneath the sea. Near Florina, not far from Saloniki, French military engineers have recently obtained some oil by drilling.

Reference

Redwood, Sir Boverton: A Treatise on Petroleum, Vol. I, 1913, pp. 135–136.

Bulgaria

No petroleum is known in Bulgaria, but the existence of oil shale or bituminous schist has been known for years. The important deposits, according to a United States Commerce Report, dated April 27, 1920, are as follows:

1. Six miles north of the town of Bresnik, a bed 20 feet in thickness is said to contain 12 to 13 per cent of crude oil.
2. Graphitic deposits near Radomir, 5 to 7 miles from the Sofia-Batanovtzi-Radomir railroad line.
3. A bed of earthy bituminous shale, 9 to 90 feet in thickness, situated about one mile south of Popovtzi station on the Stara-Zagora-Tirnovo railroad line.
4. Deposits 30 feet thick west of Kasanlik near the village of Saltikovo. These are similar to those at Popovtzi.
5. Bituminous shales, 80 feet in thickness, near Sirbinovo, 9 miles west of the town of Gorna Djounaya.

Albania

Extensive asphalt beds exist in Pliocene conglomerate at Selenitza and Rompzi, on the Voyutza. Dioscorides described bitumen from Avlona (Apollonia) as floating on the Voyutza River (formerly the Aous).

Reference

Redwood, Sir Boverton: A Treatise on Petroleum, 3rd Ed., Vol. I, 1913, p. 135.

ASIA

General Discussion

Until the end of 1917 the Continent of Asia had produced less than 2 per cent of the world's petroleum, but in that year it produced about 3½ per cent, and Asia may in time surpass all other parts of the world in its production, though this is by no means certain. At any rate, the oil potentialities of Asia are of great importance. The countries at present producing petroleum are India, Persia, and Japan, with negligible amounts from China and Siberia. The occurrence of oil in these and other countries is described in the following pages.

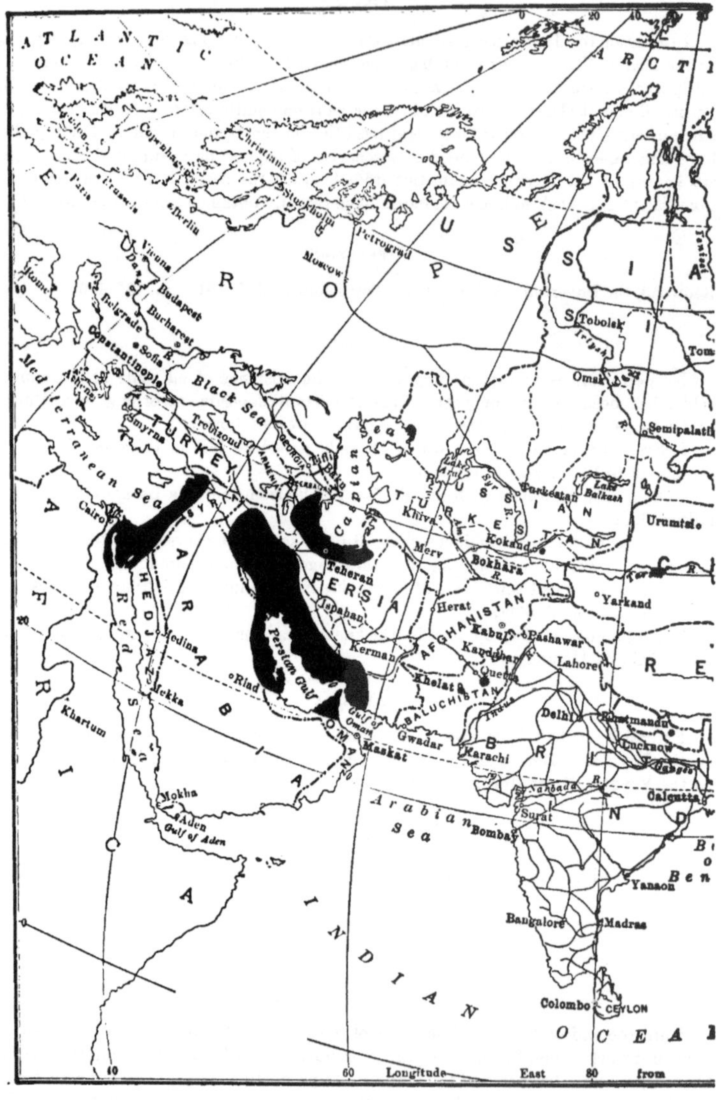

Fig. 60.—Principal oil fields

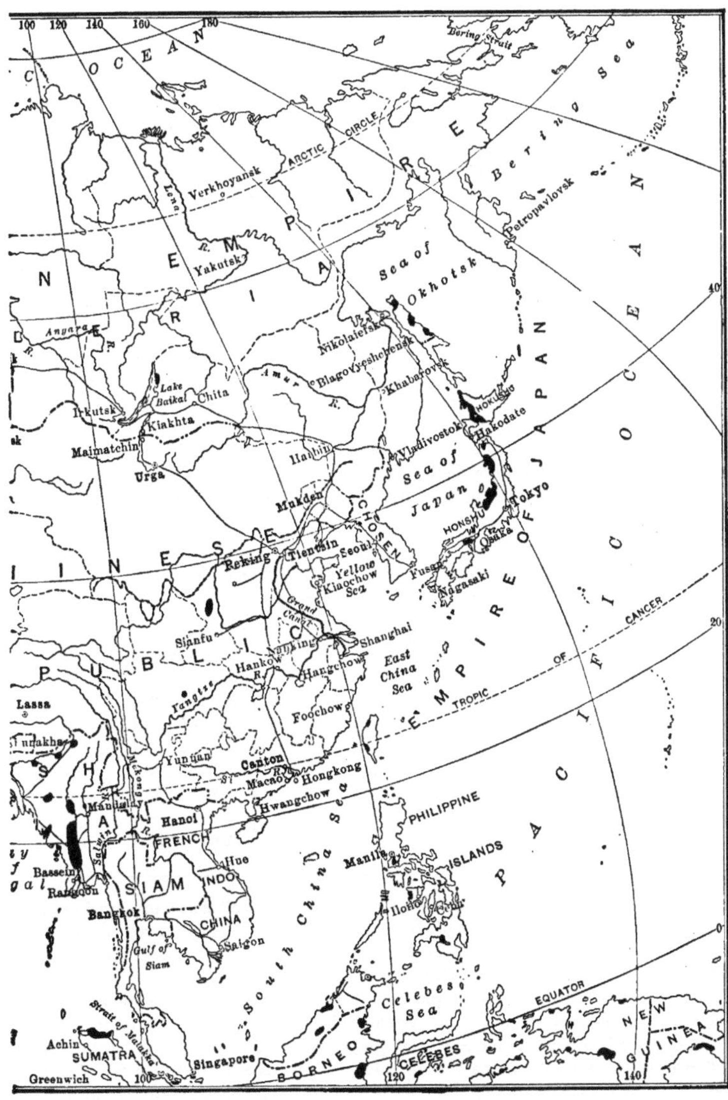

in Asia and neighboring islands.

India

The first mention of oil production in India appears to have been made in 1759 by Boetling in Baker's "Journal of an Embassy to the King of Burma," which states that "at this place (Raynangome, as he calls it) there are about two hundred families who are chiefly employed in getting earth-oil out of pits." In 1795 a Major Symes spent a few hours at Yenangyaung on the return journey from his embassy to the Court of Ava, being followed in 1797 by Cox, who wrote "Journal of a Voyage from Ranghong up the River Eraiwaddy to Amarapura, the present capital of the Burma Empire." This journal included "An Account of the Petroleum Wells in the Burmese Dominions." It is, however, only since 1889 that the oil business in Burma has been conducted in the modern fashion.

The Burma fields may be divided into two distinct groups; for besides the main field in Upper Burma, there is a small occurrence near the Aracan coast on the Islands of Ramri and Chaduba. The main district is situated on both sides of the Irrawaddy, about midway between Rangoon and Mandalay. It starts at Minbu, on the west bank of the river, just above the twentieth parallel of latitude, about 80 miles above the old military station at Thayetmyo, and apparently ends a few miles north of Yenangyat, about 70 miles farther up the river and a few miles below Pagan, the famous old capital of Burma. By far the most important part of the oil district lies on the east bank of the Irrawaddy, and is divided into the Yenangyaung field, 26 miles above Minbu, and the Singu field about 20 miles higher up. The relative importance of the principal fields in the main Burma district may be judged by their comparative outputs, which in 1916 were as follows:

Relative Importance of Burma Fields

Oil fields	Imperial gallons
Yenangyaung	199,152,938
Singu	77,005,880
Yenangyat	5,310,740
Minbu	2,043,542
Other sources	8,255,983
Total	291,769,083

At present the best known field is at Yenangyaung, but oil is also produced in insignificant quantities in the districts of Kyaukpyu, Akyab, and Thayetmyo. Some of the recent wells are as much as 2500 feet deep. Other fields are being developed, and are producing oil in increasing quantities.

Since Assam is the least known province in India in area and population, and since the whole of the eastern end of the Assam Valley, except where cleared for cultivation, is covered with extraordinarily thick jungle, the oil-bearing area in Assam is quite different in character from the Burma fields. The exposed strata consist of shales, clays and sandstones of varying degrees of hardness. The beds usually dip at a sharp angle, and the dips are unsymmetrical, some being perpendicular and some actually overturned. Petroleum is found in Assam in coal-bearing strata of Eocene age, exposed near the foot of the Naga hills southwest of the Bramapootra River.

Oil is found in the Punjab in Eocene strata which outcrop near Rawal Pindi north of Shahpur. The only locality worked to any extent is Gunda or Sudkal, about 23 miles west of Rawal Pindi, where oil wells were first dug in 1861. The principal well yielded at first only 5 gallons a day; when it was deepened the amount was increased, but it never yielded more than 50 gallons in one day. In 1889 the Punjab production was only 2873 imperial gallons. By 1917, however, the production had climbed to 619,517 imperial gallons.

Summary of Oil Occurrences in India

Province	Producing structure	Age	Production in 1917 (bbls.)	Fields
Burma	Anticlinal	Oligocene and Miocene	6,994,748	Yenangyaung Singu Yenangyat Minbu
Assam (including eastern Bengal)	Asymmetrical anticlines	Eocene	239,611	Makum field Dibgou
Punjab		Eocene	15,859	

References

Pascoe, E. H.: The oil fields of Burma, Mem. Geol. Survey, India, Vol. XL, Pt. I, 1912.

Maclean, Jack H. S.: In Petroleum World, April 17, 1917, from paper presented at Institute of Petroleum Technologists.

Mesopotamia

A great oil region is being developed in Mesopotamia, extending from Hit on the Euphrates to El Deir, a distance of 200 miles, northeast to Herboul near Zakhu, and east into Persia. The fields of Mesopotamia lie in the valleys of the Lower Euphrates, Tigris and Karun Rivers, north of latitude 30°, in the vilayets of Mosul, Bagdad and Busra, and extending north as far as Harbel in the Caza of Sakho, 60 miles north of Mosul, where Eocene outcrops exist. The fields have a general northwest-southeast trend paralleling Tertiary fields which lie farther east, the oil being found where the mountain ranges terminate near the ocean. In their geology the fields are anticlinal and domal. The surface indications consist of seepages of oil, salt and hydrogen sulphide from Miocene saline gypsiferous marls, limestones and sandstones, and there are additional oil-bearing strata of Cretaceous age. At the ruins of Natrae (or El Hadi) which lie west of Gayera, at Baba Gygar ("Father of Murmurs"), Guil at Kifri which lies at the foot of Neft Dagh ("Naphtha Mountain"), Jebel Oniki Iman, Zahrn and Mendals there are important indications, and at Tuz-Khurmath an exposed seepage can be seen for a mile along a stream bank.

The historic occurrence in the vilayet of Bagdad is in the form of circular pools which are situated south of Hit and emit bubbles of petroliferous gas; in the same locality oil issues from natural wells which deposit asphalt at the surface. An important locality for oil springs is Mendali, from which high grade oil has been refined and exported for years to Bagdad and India. Oil is also reported from Zacho, Saad and Doyet in the vilayet of Mosul, and at Darondich, Kizil-Rubati, Yakouba and El-Haim in the vilayet of Bagdad.

Bitumen from the "fountains of Is" (now Hit) was exploited by the ancient Babylonians for mortar, which still exists as an efficient binder (Herodotus, Book I, p. 179). Not only the Bible (Genesis IX, 3) but Pliny (Book II, 110) and Strabo (Book XVI, 15) refer to the abundant use of asphalt in Babylonia for mortar, embalming, ornamentation and tribute. In 311 B. C. Antigonus sent Hieronymus of Cardia with an army, to take the asphalt works of the Dead Sea from the Nabatei, but the invaders were repulsed. In the 17th or 18th Century an English traveler visited Hit and found many active oil springs.

References

Rawlinson's History of Herodotus, Vol. I, 1859, New York, p. 245.
Reich, E.: Atlas antiquus, text facing Map 18.
Launay: Géologie et richesse minerale de l'Asie, p. 444.
Redwood, Sir Boverton: A Treatise on Petroleum, 3rd Ed., Vol. I, 1913, pp. 134–135.
Dominion, Leon: Trans. Am. Inst. Min. Eng., Vol. LVI, p. 248–250.

Persia

The Persian fields have been grouped by Hunter [1] into those of (A) northern Persia, (B) western Persia and (C) southeastern Persia. The oil region of northern Persia lies between Lake Uremieh and the Caspian Sea, and is about 20 miles in breadth; it belongs to the Tertiary period and is traversed by asymmetrical folds. This field is presumably continuous geologically with that of Baku in southern Russia.

The fields of western and southern Persia are also anticlinal in structure and of Tertiary age, so far as known. A wide area of petroliferous territory is known to exist about the head of the Persian Gulf, extending northwards along the Turco-Persian frontier. The oil occurrences along the old Turkish frontier lie in the Province of Shurhistan (or Arabistan) and the field appears restricted to the valley of the Karun. The wells are situated at Shuster, and a small refinery has been built at Abudan on the Shat-el-Arab River 50 miles south of Basra. Oil has also been discovered at Ahraz, 140 miles inland from this place.

The producing properties of the Anglo-Persian or principal producing oil company are in the Maidan-i-Naphtun district about 50 miles northeast of Ahruz on Karun River and 140 miles north-northeast of Mohammerah, which lies at the Junction of the Shat-el-Arab and Karun Rivers, in the Province of Kermanshah. In 1920 Persia was credited with a production of 12,352,655 barrels of oil.

References

The Reclus: Géographie Universelle, Vol. IX, p. 389.
Ainsworth, W. F.: A personal narrative of the Euphrates Expedition, Vol. II, pp. 385 and 440.
Busk, H. G., and Mayo, B. T.: The geology of the Persian oil fields, Pet. Review, Oct. 26, 1918, pp. 271–272; Nov. 2, 1918, pp. 287–288; Nov. 9, 1918, pp. 307–308.
Chesney, F. R.: Narrative of the Euphrates Expedition, p. 78.
Dominion, Leon: Trans. Am. Inst. Min. Eng., Vol. LVI, 1916, pp. 250–254.
Hunter, C. M.: The oil fields of Persia, Min. and Met., No. 158, Sec. 11, Feb., 1920.

[1] Hunter, C. M.: The oil fields of Persia, Min. and Met., No. 158, Sec. 11, Feb. 1920.

JAPAN

Japan proper has five oil districts,[1] situated respectively in the Provinces of Echigo, Shinano, Aki, Totomi and the Island of Hokkaido; Formosa (Taiwan) has one district. Besides the fields situated in these districts, there appear to be valuable undeveloped fields in the Island of Sakhalin, part of which is owned by Japan.

The principal field thus far developed is the Echigo, situated on the northwestern coast of the Island of Nippon, Province of Echigo, about 200 miles northwest of Tokyo. Other localities where some oil has been produced on the same island are the extreme northern portion of the Province of Ugo, and a locality in the Province of Totomi, about 150 miles southwest of Tokyo.

On the Island of Hokkaido, or Yezo, oil has been found to a limited extent in the Provinces of Teshio and Ishikari near the western flank of the foothills of the great mountain chain running to the north. In 1903 and 1904 several wells were drilled in the Ishikari district and apparently some oil was found but later tests were discouraging. Indications are scattered over a large portion of this island.

Indications also exist on the Island of Formosa, and oil has been struck there at a depth of 810 feet. In northern Formosa the Byoritsu field has been developed.

Only four pools in the Echigo district are considerably productive at the present time, these being the Nishiyama, Higashiyama, Nittsu and Kubiki fields; in addition, the Ono and Uonuma pools are small producers.

The production in Echigo and the indications elsewhere are usually in Miocene or Pliocene rocks,[2] occurring on the flanks or along the crests of well-marked anticlines, sometimes faulted, and generally not over 2 or 3 miles long, although in a few cases an anticline can be traced for 10 or 15 miles. Both flanks of the anticline are generally steep, carrying the oil-bearing strata in a short distance to depths which are unreachable by the drill, or at which the strata are saturated with water. The oil-bearing formation is generally a loosely cemented sandstone 5 to 40 feet in thickness, having a bluish cast and perhaps not always siliceous. The depth of the wells ranges from 750 to 1800 feet. Probably 80 per cent of the production comes from drilled wells, the remainder being from dug wells or from shafts 200 to 500 feet deep. The average life of the wells is short.

The banks of Lake Suwa are said to contain a large amount of natural gas, though the present output comes from four wells. Since 1911 certain villages have been developing the gas business with good results.

The production of Japan, without Formosa, was 2,147,770 barrels in 1918. Formosa only produced 8101 barrels, giving a total Japanese production in that year of 2,155,871 barrels.

CHINA

The principal Province of China in which petroleum is known is Shensi, where small seepages have been operated by antiquated methods from time to time. The known seepage localities are situated in the districts of Yen-chang, Yen-chuan, Fu-chow, An-sai, I-chun, San-shui, Yen-an-fu, etc. None of the seepages is of great size. At Yen-chang in the hsien (or district) of the same name, four wells were drilled in 1904–1911 to depths of only 200 to 400 feet. The original well produced about 60 barrels per day, and in 1914 this had declined to 6 barrels, while yields of other wells were not over 1

[1] Clements, J. M.: Petroleum in Japan, Econ. Geol., Vol. XIII, No. 7, 1918, pp. 512–523.

[2] Iki, T.: Preliminary note on the geology of Echigo oil field, Mem. Geol. Survey, Japan, No. 2, 1910, p. 56.

to 4 barrels. A small refinery was built, the field was operated on this small scale, and kerosene was sold locally. In 1914, an American company, in association with the Chinese Government, commenced drilling operations, and several deep wells were sunk in the province, but the results were believed by the company to be discouraging, and this and other considerations caused the abandonment of the enterprise.

The oil is in sandstones and shales of the Shensi series of "Middle" Carboniferous age. In its structure, northern Shensi constitutes a great westward dipping monocline, in which folding is not prominent except near the edges of the basin. Such oil as is found appears to be at points of slight change in dip or moderate folds or terraces. The formations are rather devoid of water. The oil is similar to that of Pennsylvania.

In Szechuan Province, wells from 2500 to 3000 feet in depth have been drilled for centuries by most primitive and laborious methods and have produced small quantities of petroleum and large quantities of brine, accompanied by natural gas in limited quantities. The gas is used as fuel to evaporate the brine. Light grades of oil are obtained, and, in addition to being used in lamps, are also employed by the Chinese for medicinal purposes.

The presence of oil and gas in the salt wells and the use of gas as fuel in obtaining the salt makes it impossible to separate the two subjects in writing of them. In general the oil and gas occupy second place. Salt production is an ancient and highly developed industry, while oil and gas production are only of recent date and exist as by-product issues of the salt industry.

The largest and one of the oldest gas seepages is at Tzu-liu-ching, close to the mountain of the same name, while that of Chu-pai-ching has been operating day and night for forty years. In a few instances where a small well produces petroleum alone, the oil is conveyed to special reservoirs, but where oil is found mingled with brine the oil floats on top and is skimmed off. The gas appears to come from two horizons, one near the surface and the other at a depth of 2150 feet.

Palestine and Syria

Asphalt has been known on the western side of the Dead Sea for thousands of years, and was formerly exported to Egypt where it was used in embalming. Hence the word *mummy* is derived from the Coptic root "*mum*," signifying bitumen. Soft Cretaceous limestones extend from the southern end of the Dead Sea northward throughout the Jordan Valley, and emit oil, asphalt, and hydrogen sulphide gas, so that before the beginning of the historic period "The vale of Siddim was full of slime pits." (Genesis XIV, 10). Great asphalt deposits have been formed by the escape of petroleum through faults in the Jordan Valley, and oil issues from cliffs facing the Dead Sea north of the confluence of the Nahr Zerka Ma (river). The chief seepages [1] occur at Wadi Mahawat, Wadi Sebbeh and Nebi Musa on the Dead Sea, Hasbeya on the upper Jordan, Sahmur and Ain-et-Tineh on the Nehr Litany and another place called Ait-et-Tineh, which lies about 28 miles northeast of Damascus. Bitumen is reported by the same writer in Cretaceous sandstones on the Jebel el Dahr between the Jordan and the Litany. Petroleum is known at Alexandretta, bitumen is said to have been produced at Antioch and Aleppo, and asphalt deposits are reported near Latakia, in the villages of Kferie, Cassab, Chmeisse, Khorhe and Sonlas. Syrian localities from which petroleum is reported are the Ajiloon, Haroun and El Kork districts. In 1912 drilling was begun in the desert near Makarim on the Hedjaz Railroad, at the line of contact of preglacial lavas with Upper Cretaceous limestones.

[1] Redwood, Sir Boverton: A Treatise on Petroleum, 3rd Ed., Vol. I, 1913, p. 134; Dominion, Leon: Trans. Am. Inst. Min. Eng., Vol. LVI, 1916, p. 250.

Arabia

Indications of oil have oeen found in Eocene and Cretaceous strata at many points in Arabia. Redwood [1] mentions bitumen at a place called Wadi Ghorandel, 40 miles south of Suez, and he reports traces of oil at Wadi el Arabah, 30 miles farther south. Bitumen is known elsewhere in the Sinai Peninsula and near the Gulf of Akabah. On the surface of the Persian Gulf " between the islands of Kubbar and Garu and near Tarsi Island (28° N. 50° 10 E.)" films of oil have been found aft r earthquakes and storms. Bitumen is also reported " on Bahrein Island, and oil rises in the sea off Haulal Island (25° 40′ N. 52° 25′ E.) east of Bahrein Peninsula." Traces of oil are reported on the Gulf at Benaid el Oar near Koweit.

Russia in Asia

Oil was discovered five years ago in Ferghana, Russian Turkestan, in wells 500 to 1200 feet deep, as much as 480 barrels being reported in the first twenty-four hours from one of the wells. The structure is reported as anticlinal in nature. Oil is also said to exist near Lake Baikal, and great seepages occur on the Island of Sakhalin. The Ferghana region appears to be the only developed field in Asiatic Russia.

Siam

Petroleum is obtained in Muang Fang from pits dug by hand, and is said to occur elsewhere in Siam.

Armenia

In Turkish Armenia a bituminous spring emerges from Eocene limestone within the citadel of Van, and oil is eported at Parghiri The Cretaceous mountains in the Muzurdagh range on the upper Euphrates River contain asphalt, while the occurrence of maltha or rock tar at Samosta (or Someisat), 70 miles to the south, is mentioned by Pliny.

Reference

Redwood, Sir Boverton: A Treatise on Petroleum, 3rd Ed., Vol. I, 1913, p. 135.

Asia Minor

Our word chimera is derived from the Phoenician *chamirah* (burnt), which in turn came from the legend based on the " eternal fire " of the Chimera, which has been burning for thousands of years at Janartasch in Lycia, on the Gulf of Adalia. The burning gas issues from decomposed serpentine intrusives in Eocene rocks. On the island side of Samos an intermittent gas spring is reported, while traces of petroleum have been found near Smyrna and Cherokose Deli above Lake Isnik, south of the Sea of Marmora.

[1] Redwood, Sir Boverton: A Treatise on Petroleum, 3rd Ed., Vol. I, 1913, p. 134: Dominion, Leon: Trans. Am. Inst. Min. Eng., Vol. LVI, p. 250.

SOUTH AMERICA

General Discussion

In order to outline the distribution of petroleum in South America, it is desirable to classify the known and prospective districts of p oduction into groups similar to those in North America, of which the following may conveniently be used:

1. Pacific district.
2. Orinoco district.
3. Caribbean district.
4. Central Andean district.
5. South Atlantic district.
6. Western Argentina district.

1. The Pacific district is geographically a continuation of the California district of North America, which includes the fields and prospective fields between the mountains and the Pacific Ocean from Alaska to Tierra del Fuego. Occasionally some fields east of the extreme western range of mountains may conveniently be included in this group.

2. The Orinoco district includes the Orinoco Valley and delta, and in general the region between the Cordillera de Merida on the northwest and the Sierra de Pacàraima on the southeast.

3. The Carribbean district includes all fields east of the western Cordillera on the Caribbean coast and those north of the Cordillera de Merida and eastern Cordillera in Venezuela and Colombia.

4. The Central Andean district includes the fields occupying the plateaus and east Andean slopes in Peru and Bolivia, with possible extensions into Ecuador and Brazil.

5. The South Atlantic district contains the Comodoro Rivadavia fields and prospective fields on the Atlantic Coast of Argentine from the Rio de la Plata south to Tierra del Fuego.

6. The Western Argentine district includes the fields along the east border of the Andes and possibly some intermontane areas from Bolivia southward.

Since these districts merge in places and new districts may be discovered, this classification is subject to change, but it forms a convenient grouping for discussion of the geographical occurrence of oil in South America. Articles on the occurrence of oil in South America have been published by Arnold [1] and the writer.[2]

PERU

Peru holds the record for early production in South America, having asphalt roads or paths over two hundred years old. Before the petroleum fields of Russia or Pennsylvania were even suspected, the presence of oil in Peru was known. It was first mentioned by Acosta, who states that the captain of a ship detected the presence of land by an oily appearance on the water off Cape Blanco. Oil was used by the Incas for centuries, and in the time of Pizarro it was collected in trenches, boiled down and converted into pitch. Modern developments commenced in 1867, when a Mr. Prentice, a Pennsylvanian oil producer, visited Peru and subsequently drilled wells at Zorritos in the Province of Tumbez.

[1] Arnold, Ralph: Conservation of the oil and gas resources of the Americas, Econ. Geology, Vol. XI, 1916, pp. 203–222, 299–306.

[2] Clapp, F. G.: Review of present knowledge regarding the petroleum resources of South America, Trans. Am. Inst. Min. Eng., Vol. I, LVII, 1917, pp. 914–967.

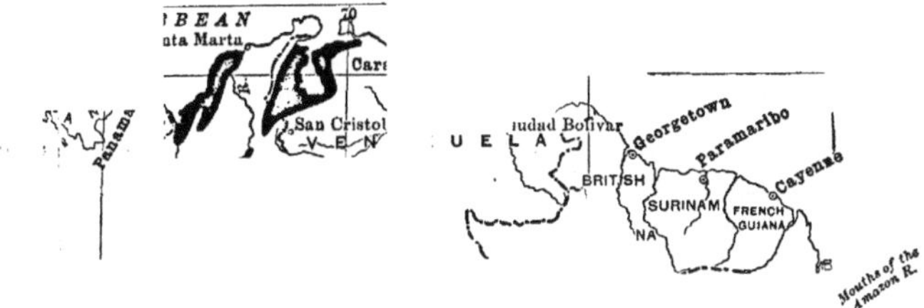

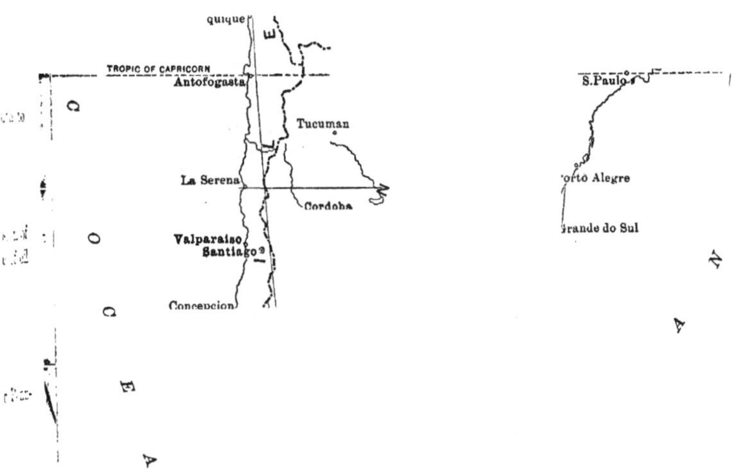

Fig. 61.—Occurrence and distribution of principal oil fields of South America.

The Peruvian fields fall into two groups, the Titicaca fields of the Andes and the fields of the Pacific Coast district, including the Zorritos, Lobitos and Negritos fields. The total oil area of Peru, according to Hunter, is over 5000 square miles, of which 100 square miles contain unproven oil possibilities, and 200 square miles are proven. At the end of 1917 there were more than 845 wells in the country, which produced 2,533,417 barrels in that year.

In addition to the districts enumerated above, traces of oil are found in the districts of Pirin, Chimbote, Jauja, Huancavelica, and Ica; in fact, oil is said to occur at frequent intervals from Tumbez to Lake Titicaca. Oil is reported also on Huallaga River, two days' journey from Yurimaguas. Traces of oil are reported south of Cerro de Pasco as far as Huancayo, Department of Junin, also in Condorocana Mountain of the Province of Angaraes, in the Calaveras Mountains of the Province of Çamana, and in Chumpi in the Province of Parinacochas. Not all of these reports have been confirmed.

A. *Central (Andean) District*

Oil is found at several points in the Andes, as in the Huallanca region, and in the Provinces of Jauja and Parinacochas, but the principal deposits are near Lake Titicaca.

Titicaca Field.—This field is in the district of Puno, 300 miles from the coast and 8 miles distant from Lake Titicaca, high in the Andes, and near the Bolivian frontier. The oil deposits are found in the Provinces of Canas, Lampa, Azangaro and Huancane. Work was begun in 1905, and by 1908 a company had sunk ten wells with an average production of 50 barrels per day. The field was abandoned after 1912, its production never having exceeded 76,103 barrels in any year.

B. *Pacific (Coastal) District*

The coastal belt extends 2000 miles southward from Ecuador to the frontier of Chile. It is now about 30 miles wide and occupies the Province of Tumbez and the northern part of Piura. The developed fields, which extend from the town of Tumbez, south of the Gulf of Guayaquil, for 180 miles to and beyond Paita, are bounded on the east by spurs of the Andes and on the west are believed to include the Islands of Lobos. The best description of the Pacific district of Peru is given by Deustua.[1]

Zorritos Field.—This field is the oldest and the most northerly in Peru. It is situated about 24 miles south of Tumbez, which in turn is south of the Gulf of Guayaquil. The producing territory extends along the coast for 4 miles, most of the wells being at the water's edge and some of them in the ocean. The greatest depth recorded is 3020 feet in Peroles ravine, but most of the wells are between 600 and 2000 feet deep.

Lobitos Field.—South of the Zorritos lies the Lobitos field in the Department of Piura, 12 miles north of Talara and first prospected in 1901. The proven area of this field is about 25 square miles, and it yields the second largest production in Peru. The deepest well in the field was sunk to a depth of 3435 feet without success. All the producing wells are over 2000 feet deep.

Negritos or Talara Field.—The southernmost and richest of the developed coastal areas is the Negritos field, the development of which was started in 1874. It is situated 40 miles north of Paita in the Hacienda La Mina Brea and Parinas. The area of this field is about 650 square miles. Negritos is the center of drilling operations.

Argentina

In the Province of Salta, Argentina, several wells were drilled many years ago to depths of 300 to 500 feet. At Cacheuta near Mendoza in the Province of Men-

[1] Deustua, R. A.: La Industria del Petroleo en el Peru durante, 1915, 2d Ed., Lima, 1916.

doza, five producing wells existed in 1889, and by 1893 the number had been increased to twelve. In northwestern Argentina, 40 miles from the Bolivian frontier, two holes were drilled prior to 1910 to depths of 300 to 500 feet. At Santa Catalina, some time previous to 1910, a well was drilled to a depth of 2300 feet, of which the last 1800 feet were said to be through "quicksand." In 1912 oil and gas were discovered in four wells on the coast of Patagonia.

The Argentina fields, like those of Peru, may be classified into (A) those of the Andean region, which belong to the western Argentina district; and (B) those on the coast, which are included in the South Atlantic district. According to Arnold,[1] petroleum is supposed to underlie 8000 square miles, of which 400 square miles have surface evidences; but the proven area does not exceed 2 square miles. According to Hermitte,[2] these oil districts can be subdivided into four fields, so that the known areas are classified as follows:

- A. Andean district.
  1. Cacheuta field.
  2. Mendoza-Neuquen field.
  3. Salta-Jujuy field.
- B. South Atlantic district.
  4. Comodoro Rivadavia field.

Oil has also been reported in the Province of Santa Fe, which might be called field No. 5, but this report has not been confirmed.

A. *Andean District of Argentina*

Since petroleum is known throughout a belt 150 miles wide in Bolivia, between 63° to 64° West longitude and 19° to 22° South latitude, little doubt exists that the zone of oil occurrence continues as far into Argentina as Tartogal and Aguaray.

Data regarding the *Cacheuta field* are not at hand.

Mendoza-Neuquen field.—This field lies 600 to 800 miles southwest of Buenos Aires, and may comprise 6000 square miles. It extends nearly 300 miles along the eastern Andean slope, from 40 miles north of Mendoza southward to latitude 42°. The area having oil evidences is estimated by Arnold not to exceed 300 square miles, and the proven area is less than 1 square mile.

In Mendoza and Neuquen Provinces, Arnold reports oil indications in the following localities:

San Rafael, Mendoza Province.
Cerro Auca Mahuida.
On the Rio Barrancas (boundary of Mendoza and Neuquen).
Curileuva.
Garrapatal (note repetition of name in Jujuy Province).
La Brea.
Vachenta.
La Carene.
"Plaza Huincul of Challaco" north of kilometer 81 on the railroad.
La Cortaderita (formerly Government land) in the south of Mendoza Province.

Salta-Jujuy field.—In the Andean region of northwestern Argentina is a roughly triangular area with sides approximately 100 miles in length, 800 to 900 miles northwest

[1] Arnold, Ralph: Conservation of the oil and gas resources of the Americas, Econ. Geology (1916) 11, 203–222, 299–326.

[2] Hermitte, E. M.: Proceedings of the Second Pan-American Congress (1916).

of Buenos Aires and adjoining the Bolivian frontier. The fields are practically continuous with those of Bolivia, and are distributed through 5000 square miles, of which about 250 square miles contain evidences of petroleum. However, less than 1 square mile has actually been tested, and the results have been indifferent. In Salta Province, oil springs have been known for years, and evidences were reported by Redwood in the following localities:

Yavi Chico, on the Bolivian frontier at Longitude 65° 30′ West.
Tejada, 60 miles south of Yavi Chico.
Abra de la Cruz, 29 miles east of Tejada.
Garrapatal, 21 miles east-northeast of Jujuy.
Laguna de la Brea, 42 miles farther in the same direction.
Cerro de Calilegua, 30 miles northwest of Laguna de la Brea.
Tartagal, on the frontier at Longitude 63° 44′ West.

The seepages in northern Argentina [1] lie along the faulted anticlines, paralleling the Andes, and the dips are frequently high. The petroliferous horizon in the north is believed to be of Cretaceous age. In the Salta-Jujuy district and in the Provinces of Mendoza and Neuquen the structure is rather complex, with faulting and high dips.

B. *South Atlantic District of Argentina*

This district includes the producing Comodoro Rivadavia field and other discoveries on the coast farther south. At a point 60 to 70 miles west and north of Comodoro Rivadavia, on Rio Chico, near Lake Conles Whapi, several large asphaltic seepages are reported.

Comodoro Rivadavia Field.—In 1907, oil was struck near the town of Comodoro Rivadavia, in the territory of Chubut, on the Gulf of St. George, about 850 miles southwest of Buenos Aires. Although the development has been comparatively local, the oil-bearing territory is believed to be of wide extent. A number of wells have been sunk, and in 1920 the output of the Comodoro Rivadavia field was 1,665,989 barrels, from 39 producing wells, this being all of the oil production of Argentina.

The productive strata of the Comodoro Rivadavia field are thought to be of Cretaceous age, and the structure is a mild dome, on a normal eastward dipping monocline.[2] The depth of wells in the field ranges from 1700 to 1900 feet.

Still farther to the south is the Gallegos-Punta Arena region, with many gas springs and some oil.[3]

Venezuela

The petroliferous localities of Venezuela, so far as can be learned, are classifiable into:

A. Caribbean district (mainly in the vicinity of Lake Maracaibo) where some big wells have been secured; and

[1] Herold, S. C.: The economic and geologic conditions pertaining to the occurrence of oil in the north Argentine–Bolivian field of South America, Trans. Am. Inst. Min. Eng., Vol. LXI, 1919, pp. 544–564.

Bonarelli, G.: Las Sierras Subandinas del Alto y Aguaragüe y los yacimientos petroliferos del distrito minero de Tartagal, Provincia de Salta; Republica Argentina, Anales del Ministerio de Agricultura, Sección Geologia, Mineralogia y Minera, Vol. VIII, 1913, No. 4; La Estructura Geológica y los yacimientos petroliferos del distrito minero de Oran, Provincia de Salta, Ministerio de Agricultura, Dirección, General de Minas, 1914, Num. 9, Ser. B.

[2] Herold, S. C.: Petroleum in the Argentine Republic, Min. and Met., Bull. 158, Sect. 12, Am. Inst. Min. and Met. Eng., Feb., 1920, 5 pp.

[3] Herold, S. C.: Ibid., p. 5.

B. Orinoco district known as the Pederneles field, situated in the Delta of the Orinoco River, at the place where one of its northern mouths empties into the Gulf of Paria. The field includes portions of the Islands of Capure, Pederneles, and Plata. It was first discovered owing to a deposit of asphalt half a mile long and 300 to 600 feet across on the northwest coast of the Island of Capure.

The oil in the Caribbean district is found on anticlines trending generally northeast-southwest on the flanks of a broad geo-syncline and having close relations with the Magdalena-Santander field of Colombia. The geologic formations of northwestern Venezuela are shown on page 152.

From 1914 to the present time drilling was carried on with good results in the Maracaibo district in the States of Zulia, Falcon, Trujilla and Lara. Initial productions up to 400 metric tons are reported from a number of wells, at depths of 400 to 2000 feet from the surface. The principal proven fields in the Maracaibo district are the Mene Grande field (70 miles from Maracaibo and 16 miles from San Lorenzo), La Sierrita field (in the district of Mara, State of Zulia) and the Colon field (district of Colon, State of Zulia). Many American and British companies have holdings in these and other prospective fields.

Oil production rose from 119,734 barrels in 1917 to 423,895 barrels in 1919. The oil has an asphalt base and ranges from 0.928 to 1.02 specific gravity. Gasoline content appears to be limited to 3 per cent and kerosene to 7 per cent.

References

Redfield, A. H.: Petroleum and asphalt in Venezuela, Eng. and Min. Jour., Vol. 111, No. 8, pp. 354–357, 1921.

Miller, J. S.: "The Lake Asphalt Industry," Chem. and Met. Eng., Vol. 22, pp. 749–754, April 21, 1920.

Ministerio de Fomento, Dirección de Minas: "Datos sobre la industrio petrolera de Venezuela," Boletin, Vol. 1, No. 1, pp. 1–14, October, 1920.

Wall, G. P.: "On the Geology of a Part of Venezuela and Trinidad," Quart. Jour. Geol. Soc., Vol. 16, pp. 460–470, London, 1860.

Colombia

It is only since 1915 that much interest has been evinced in the petroleum of Colombia though its existence was mentioned by Humboldt as far back as the 18th Century. Many oil springs have been found near the shores of the Gulf of Darien, on the Arboletes and San Juan Rivers. One is described as a pool of oil 50 feet in diameter, and one flows a noticeable amount of oil every day. Other seepages exist in many parts of the country. The most sensational oil strike was on the concession of the Tropical Oil Company, far up the Magdalena River, but many companies having concessions, are starting to drill or hold leases in the country.

The oil-bearing districts of Colombia are grouped here somewhat as proposed by Arnold, viz:

A. Caribbean district.
  1. Caribbean field.
    (*a*) Tubura pool.
    (*b*) Turbaco pool.
  2. Magdalena-Santander field.
  3. Tolima field.
B. Pacific district.

Generalized Geologic Section of Maracaibo Fields

System	Series	Formation	Character	Thickness (feet)	Productivity
Quaternary	Recent		Gravel and loam		
Tertiary	Maracaibo		Silt and gravel		
	?	"Upper Coroni" of Trinidad	Sandstone and clay		
	Miocene (?)	Cerro de Oro or "Lower Coroni" of Trinidad	Shale, thin sandstone and coal seams	2500	**Impregnated with oil**
	Oligocene		Black carbonaceous shale	3000	**Supposed by some geologists to be the "parent oil-rock"**
	Eocene		Shale of sandstone (generally absent)		
		UNCONFORMITY			
Cretaceous	Upper Cretaceous	Capacho	Limestone		Oil bearing
	Lower Cretaceous	Táchira and Barbacoas	Blue limestone	500	
		San Cristóbal	Dark bituminous limestones		
			Red, yellow and white sandstones		
Pre-Cretaceous rocks	Lagunillas		Conglomerate and breccias		
Undifferentiated Paleozoic			Metamorphosed shales		
Probably Pre-Cambrian			Crystalline gneisses, schists and igneous rocks		

A. *Caribbean District*

1. **Caribbean field.**—This field includes the petroliferous area which extends from Rio Hacha, on the west edge of the Guajiro Peninsula, southwestward along the Caribbean Coast to the Gulf of Darien and Gulf of Uraba, inland to include the Tubura and Turbaco pools, southward to a point 30 miies south of Chima on Sinu River and Monpos on San Jorga River, and up the Atrato River valley 90 miles from Punta Arenas. It occupies portions of the departments of Magdalena, Bolivar, and Cauca. This prospective field lies throughout 300 miles long and 50 miles wide, contains 15,000 square miles and, according to Arnold, has 300 square miles of probably productive territory, of which only 1 square mile has been proven by drilling. The Tubura pool is 20 miles east of Cartagena and quite close to Barranquilla. Several wells have been drilled, ranging from 700 to 3018 feet in depth, and one is reported to have yielded 7 or 8 barrels of oil. About 12 or 15 miles south of Cartagena in the Turbaco pool five wells were drilled, ranging from 500 to 2200 feet in depth.

2. **Magdalena-Santander field.**—The Magdalena-Santander field is probably more important, but at the same time more inaccessible than that of the Caribbean field. It includes the southern part of the Department of Magdalena, the Department of Santander, and the western edge of the Department of Boyaca, extending from Magdalena River to the eastern Cordilleras, and also occupies an area in the southeastern part of the Department of Bolivar. It includes a belt 200 miles long and 50 miles wide, with an area of 10,000 square miles, of which only 200 are believed to be productive, while only 1 square mile has been proven. In this field the Tropical Oil Company has brought in some highly productive wells. Close to the boundary of Venezuela and extending into that country concessions have been granted in a region prolific in seepages and of great promise.

3. **Tolima field.**—In the Tolima field are grouped the petroliferous occurrences in the upper Magdalena basin, in the Departments of Cuninamarca and Tolima, and on the edge of the San Martin and Casanare plains. While this district has an area of 7000 square miles, only 100 are likely to be productive, and none of it has yet been definitely proven.

B. *Pacific District of Colombia*

The Pacific district lies in the Department of Cauca, and, as defined by Arnold, includes a belt 60 or 70 miles long extending on the Pacific Coast from north of Buenaventura to Baudo River, reaching inland to Atrato River at Quibdo and as far south as Cali on Cauca River. The area of the Pacific District is given as 1800 square miles, of which the probable productive territory is believed to be only 18 square miles, none being yet proven.

BOLIVIA

Although not at all developed, Bolivia has extensive favorable areas for oil development, situated along the southeastern flank of the Andes, indications being quite frequent in a belt extending throughout a distance of 150 miles, as far as the Argentina boundary at Yacuiva. The zone is a diagonal one, traversing the eastern provinces of Santa Cruz, Sucre and Tarija, between the parallels of 63° to 64° West longitude and 19° to 22° South latitude. According to Rakusin,[1] there are at least three fields, viz., (1) Pienna, (2) Kuarazuti and (3) Lomas de Ipaguaciu. The petroliferous formations extend from near Santa Cruz in the center of Bolivia southward through Sauces to Piquirenda, Plata and Guarazuti in the Province of Tarija and thence into northern Argentina. Geological investigations in the area between the Incahuasi

[1] Rakusin, M. A.: In the Troudi of the Grosny branch of the Russian Technical Society, 1913.

and Aguaraygus ranges, south of Sucre, are also said to have shown the presence of a considerable area of prospective oil land.

Geological investigations of the eastern slope of the Bolivian Andes have confirmed the claim that an anticlinal belt exists along the entire range of these mountains from Ayacuiba to the Madre de Dios River, the possible length of the favorable structure being some 300 miles. Indications are also reported in the Beni district. In view of the cumulative evidence, it seems likely that Bolivia will some day be a great producer of petroleum.

The petroliferous formation belongs to the Salta series of probable Cretaceous age, while the structure consists of a series of prominent anticlines and great faults paralleling the Andes, with vertical dips in some places. Several seepages are reported to flow 2 or 3 quarts respectively per day; the best known being situated in the Departments of Tarija, Sucre and Santa Cruz.

Reference

Herold, S. C.: The economic and geologic conditions pertaining to the occurrence of oil in the North Argentine-Bolivian field of South America, Trans. Am. Inst. Min. Eng., Vol. LXI, 1919, pp. 544–564.

Ecuador

Petroleum has been known in Ecuador since 1700, when it was mentioned by Velasco in his "History of the Kingdom of Quito." Many shafts and holes had been sunk by hand on the Gulf of Guayaquil and between the Bay of Santa Elena and the Pacific Ocean. These were only 10 to 50 feet deep and about 10 feet square, producing 2 to 10 barrels per day from each pit, so that they have yielded a total of approximately 500 tons per month for many years. At Santa Paula, where sufficient oil is obtained for local use, about forty of these wells have been dug. A number of old dug wells also exist at Achagian, but these are no longer worked. No more serious attempt has been made at development.

Brazil

Brazil is commonly supposed to have no oil indications, and the reports of travelers into the far interior of that country are superficially not encouraging. Occasional reports, however, indicate bitumen at Morro do Taio, Santa Catharina, and at Piropora in São Paulo. On the Itahipe River north of Ilheos, on the Island of Joas Thania in Marahú River, about 80 miles southward of Bahia, and at Riachadoce and Camarajibe, in Alagóas, respectively 25 and 45 miles north of Maceio, bitumen is also reported. Petroleum is alleged to occur below São Paulo, at Jesus de Tremembe on the Parahyba del Norte, and oil shales exist in many places.

Disregarding the possible deposits of liquid petroleum, Brazil is known to have rich oil shale resources in different parts of the country. Williams states [1] that "such deposits are found in the Permian rocks of central and southern Maranhão; in the Tertiary and Cretaceous beds along the coast of Alagóas Sergipe, Bahia, and perhaps farther south in Espirito Santo; and in the Parahyba embayment north of Cape Frio and in the interior Tertiary basin of eastern São Paulo." Tests of oil shales in Maranhão showed 36.5 per cent of bitumen,[2] and those in Alagaós showed from 7 to 50 per cent [3] of volatile matter. In the vicinity of Marahú and southwards

[1] Williams, H. E.: Oil-shales and petroleum prospects in Brazil, Trans. Am. Inst. Min. and Met. Eng., St. Louis Meeting, Sept., 1920, 8 pp.

[2] Lisboa, M. A. R.: Permian geology of northern Brazil, Am. Jour. Sci., Vol. XLVII, 1914, p. 425.

[3] Redwood, Boverton and Topley, William: Report on the Riacho Doce and Camaragibe shale deposits on the coast of Brazil near Maceió, London, 1891.

along the coast in the state of Bahia, where a highly bituminous series of rocks exists,[1] including what is known as *turfa*, a soft elastic combustible bituminous material, which on distillation yields as high as 430 liters of crude oil per ton. Some of the richer shale beds of São Paulo hold as high as 13 per cent of crude oil, and extending through southern Brazil into Uruguay [2] is a persistent black shale of Permian age, having sometimes 7 per cent of crude oil.

REFERENCES

Branner, J. C.: Oil-bearing shales of the coast of Brazil, Trans. Am. Inst. Min. Eng., Vol. XXX, 1900, p. 537.

White, I. C.: Final report of the Brazilian Coal Commission, Rio de Janeiro, 1908; The coals of Brazil, Second Pan-American Congress, 1916.

White, I. C.: The coals of Brazil, Second Pan-American Congress, 1916.

Williams, H. E.: Oil-shales and petroleum prospects in Brazil, Trans. Am. Inst. Min. and Met. Eng., St. Louis Meeting, Sept., 1920, 8 pp.

Chile

Although surface evidences of petroleum are not abundant, some indications exist, viz.:

(1) In the Province of Tarapaca, south of Patillos, northern Chile;

(2) In the southern part of the republic, where an extensive area southward of the Maullin River is said to have indications of natural gas in Tertiary deposits;

(3) At Puerto Porvinir and Agua Fresca in the Magallanes Territory, where oil is reported.

Reports indicate that the country surrounding Punta Arena contains petroleum and asphalt.

British Guiana

There are persistent suspicions of oil in British Guiana, the most authentic being a quotation from a report by Chamberlain,[3] as follows:

> "After an extended investigation of the Waini River district in the autumn of last year, Mr. E. C. Buck, director of public works, reported that the oil indications were most favorable."

Nothing further is known of indications in British Guiana, beyond the fact that Redwood [4] mentions some " finds " of asphalt near the coast.

Dutch Guiana

A work of the 18th Century refers to the bitumen of Surinam, and Arnold [5] states that three petroliferous localities are known, viz.:

1. South side of Surinam River, 6 miles below Kabele Station and 97 miles by rail south of Paramaribo, where exposures of shale and sandstone yield small quantities of high-grade amber-colored oil.

[1] de Campos, Gonzaga: Reconhecimento geologico na Bacia do Rio Merahú, Bahia, São Paulo, 1902.

[2] de Oliveira, E. P.: Regiões carboniferas dos estados do sul, Servico Geologico, Rio de Janiero, 1918.

[3] Chamberlain, G. E.: Oil exploration in British Guiana, U. S. Commerce Reports (Feb. 23, 1916).

[4] Redwood, Sir Boverton: A Treatise on Petroleum, 3d Ed , Vol. I (1913), pp. 100–106 and 192–196.

[5] Arnold, Ralph: Conservation of the oil and gas resources of the Americas, Econ. Geology, Vol. XI, (1916), pp. 203–222 and 299–326.

2. An area on the Marowijne River, 100 miles above Albina, where seepages emerge from shale, the quality of the oil being excellent.

3. Small seepages between Surinam River and the railroad, about 48 miles above the head of deep-water navigation. The gravity of the oil is about 45° Bé. The seepages on the Marowijne River are found in shale, cut by serpentine dikes, but 5 miles from the nearest serpentine. In general, the geology of the country is believed to be unfavorable to considerable deposits of oil.

French Guiana

Arnold[1] mentions oil seepages southeast of the Marowijne River, the formations being continuous with those having seepages in Dutch Guiana, but he believes that the possibilities are insignificant.

Falkland Islands

An article on mineral deposits, in the Bulletin of the Imperial Institute in 1912, mentions a sample of dull black bitumen received from the Falkland Islands in 1908, and states that an accompanying letter referred to several similar outcrops in various parts of the islands. However, the geological conditions are not believed to be favorable for development of petroleum.

AFRICA

General Discussion

Africa is, at present, the continent of least importance as an oil producer and will probably continue to be so in the future. Nevertheless, the Egyptian fields have sprung into notice recently, and geologists are known to have made petroleum surveys on the Gold Coast, in the Congo, and elsewhere.

Egypt

Petroleum indications along the borders of the Red Sea have been known since ancient times, the Romans giving the name Mons Petrolius (Oil Mountain) to an elevation near which oil was found at Jebel Zeit near the mouth of the Gulf of Suez. In the more limited acceptation of the term, according to Hume,[2] the area of prospective Egyptian oil fields lies between 27° 10′ and 28° 10′ North latitude and between 33° and 33° 50′ East longitude, but having outlying prospective localities on the western and southwestern sea borders of Sinai. Pronounced oil indications have also been noted at Gemsah, 13 miles south of the first locality, but although wells were carried to a depth of over 2000 feet, only small showings were obtained. The rocks consist of Upper Miocene limestones, clays and gypsum, which form a belt extending from Ras El Gharib on the north to Abu Shaar on the south. Oil indications also occur in Upper Cretaceous sandstones and limestones along a ridge parallel to the coast and lying west of the Miocene series. The first producing field is that of Jemsa, where the strata are reported to be horizontal, but broken folds may exist there, and anticlinal and domal structure are present in other localities. The geological conditions are in general complex. Oil under heavy gas pressure was found in a boring at Zafaraña, while surface seeps have been noted on the Jebel Atakah, southwest of Suez. Oil indications are also reported at El Hamman (Mokattam) in the Eocene, in the hills 20 miles inland from Suakin, and on the upper Nile.

[1] Arnold, Ralph: Ibid.

[2] Hume, W. F.: Petroleum Times, Sept. 27, 1919, p. 296.

The production of oil in Egypt has increased from 9150 barrels in 1911 to 1,042,000 barrels in 1920. The principal fields are those of Gemsah and Hurgada.

References

Hume, W. F.: The oil fields region of Egypt, Petroleum Review, Jan. 12, Feb. 23, March 9, 16, 23, April 20, May 4, 11, 18, 1918, pp. 27–28, 125–126, 159–160, 171–172, 191–192, 255–256, 283–284, 301–302, 317–319.

Hume, W. F.: The oil fields of Egypt, A study of the geology, Petroleum Times, Sept. 20, 27, Oct. 4, 11, 18, 25 and Nov. 1, 1919, pp. 269–270, 295–296, 327–328, 351–352, 375–376, 399–400, 427–428.

Tunis

Seepages of heavy asphaltic oil are reported from supposed Cretaceous limestones and shales near Testour, about 70 kilometers southwest of the capital of Tunis.

Algeria

The presence of petroleum in Algeria [1] has been known since the time of the Romans, the principal occurrence being in the Department of Oran on the south flank of the Dahra range and at Dahra, Tilouanet and Bel-Hacel. In 1877 an attempt was made to collect the oil by tunneling, and in 1892 test wells were drilled without much result. In 1895, however, small quantities were obtained in wells less than 1000 feet deep in a series of alternating Miocene marls, clays, gypsum, and limestones in the same province, not over 50 barrels per day being obtained from any well. Otherwise, no material production has been developed in Algeria, although indications have also been reported from Cape Ivi, near La Stidia, from Port-aux-Poules near Arzeu in the valley of Oued-Ouarizane in the Province of Constantine.

Morocco

Oil indications are known in northern Morocco between El Araish and Fez, and some petroleum is said to have been obtained in 1912 by drilling. In 1920, petroleum was reported to have been discovered at Jebel Talfat (or Thelfet), 50 miles due west of Fez and near Petit Jean, and developed by makeshift means at a depth of less than 200 feet. The discovery was made through the efforts of Professor Gentil, of Sorbonne, who had been exploring that country for ten years.

West Africa

Oil rises from the sea bottom off the Cape Verde Islands. On the mainland the principal occurrences are in local Cretaceous areas along the western flanks of the crystalline highlands. Among the localities in which oil has been specifically reported are Portuguese Guiana, Island of St. Thomas, Camaroons, French Congo and Angola. Some drilling has been done, but has not resulted in commercial production.

Angola

Indications of asphalt and petroleum are reported near Dande, Libollo and Musserra in Angola, and some prospecting has been done by wells in the Alto Daude district, near Loanda and at Ambrizzette.

[1] Le Pétrole en Algérie: Journal du pétrole, Vol. XVII, No. 10, Oct., 1917, pp. 15–16.

Gold Coast

Liquid bitumen is reported at Bonyere, and some drilling for petroleum has been carried out near Half Assinie.

Southern Nigeria

Bitumen is reported at Ijebu, Errium Hill, Mafaoku, and the Errigu Valley in the Western Province, and some heavy petroleum has been found in wells.

South Africa

In a number of localities indications of petroleum are associated with igneous intrusions in bituminous Triassic beds, and gas has been reported to emerge from Upper Silurian and Devonian rocks in the Bokkeveldt (Ceres district), 90 miles northeast of Cape Town, and at Mossel Bay. Also in Orange River Colony indications are associated with igneous intrusions in bituminous shales. In Rhodesia showings are reported in the hills, 20 miles inland from Suakim. Oil showings are also reported 60 miles northwest of Potchefstroom, and are reported in lower Mesozoic shales in a belt extending across the Wakkerstroom, Piet Retief, and Ermelo districts, in association with or exuding from igneous intrusions.

No important oil probabilities are believed to exist in South Africa except those of oil shale deposits, which may be of considerable value. It has been said that 20 gallons of crude oil and 50 pounds of ammonium sulphate per ton can be obtained from much of this shale. The oil shales of commercial value are reported to be those of the Wakkerstroom, Gulek and some other districts mentioned above.

East Africa

Traces of petroleum are said to occur near the junction of the Imzingwani and Limpopo Rivers in Matabeleland, and Livingston reported seeps of a paraffin oil on the shores of Lake Nyanza. In Portuguese Nyassaland bitumen is reported on the coast near the Rovuma River. Indications of oil are reported in Somaliland, but nothing authentic is known of them.

Madagascar

On the west side of Madagascar oil is reported from Eocene strata, in Mesozoic rocks in the valley of the Ranobe and Mananubolo Rivers, and in the Jurassic coal fields of Ambavatoby on the northeast coast.

Showings of oil were reported to have been found in 1917 in a well at a depth of 450 feet in the Betsiriry Valley.

WEST INDIES

General Discussion

Several of the islands of the West Indies have oil indications. These have resulted in extended development only in Trinidad, although Cuba, Barbados and Santo Domingo have had some oil findings.

Trinidad

Various attempts were made to obtain oil in Trinidad from 1870 to 1900, but it was first obtained by drilling in 1901, and active drilling commenced in 1908.

Surface indications, especially seepages, are abundant and are confined to the southern part of the island, along several more or less clearly defined anticlines whose general trend is east and west. On the crest of one anticline is the famous asphalt or "Pitch Lake," situated at an altitude of 138 feet above the sea and covering an area of about 127 acres. The asphalt is at least 150 feet in thickness. Other indications of oil in Trinidad are manjak, deposited in fissures, and natural gas seepages. The oil is of Tertiary age, and in the Pitch Lake has escaped to the surface from the La Brea sand, which outcrops west of the lake and elsewhere in the island. The sand is overlain by a fine bluish clay which has been locally changed to porcellanite through combustion. Two main productive horizons are known, but that many sands may exist is indicated by the fact that the total thickness of the Tertiary strata is estimated at over 6000 feet.

The best paper on petroleum in Trinidad is by Macready.[1] The position of the petroleum deposits in the geologic column is shown by the table on page 160.

The most productive pools or fields are, according to Macready, as follows:

1. Brighton, or Pitch Lake pool.
2. Vessigny pool, 2 miles south of Pitch Lake.
3. Lot 1 pool, 3 miles south of Pitch Lake, on Lot 1 of Morne l'Enfer Forest Reserve.
4. Parry Lands pool, 3½ miles south of Pitch Lake.
5. Point Fortin pool, 6 miles southwest of Pitch Lake.
6. Fyzabad pool, 6 miles south-southeast of Pitch Lake and several miles southwest of Fyzabad Village.
7. Barracpore pool, 15 miles east of Pitch Lake and several miles south of San Fernando.
8. Tabaquite pool, 30 miles northeast of Pitch Lake and 4 miles southeast of Tabaquite Railroad Station.
9. Guayaguayare pool, in the extreme southeast corner of the island, 45 miles from Pitch Lake.

All of the fields are in the southern part of the island.

In 1918 the island contained 410 wells, the total production being 2,082,068 barrels in that year. Twelve companies were operating then, the main supplies coming from the Fyzabad and Barracpore pools. The best pool is that at Fyzabad. The greatest productive depth of Trinidad wells is less than 2000 feet, and many wells are only 500 to 1000 feet in depth. There has been no drilling below 3000 feet. Wel's have produced as high as 24,000 barrels in the first twenty-four hours, but they fall off rapidly. The heaviest oil in the island is of 12° Bé., while the best oil is sometimes as high as 46° Bé.

References

Cunningham-Craig, E. H.: Trinidad oil fields, Council Paper No. 12 of 1906, Port-of-Spain, 1906, 7 pp.

Council Paper No. 30 of 1906, Port-of-Spain, 1906, 14 pp.

Council Paper No. 131 of 1907, Port-of-Spain, 1907, 19 pp.

Cuba

There are numerous indications of petroleum in Cuba, with a range in gravity from that of naphtha to solid bitumen, but as yet there has been little commercial development. Deposits of asphalt exist in the Provinces of Pinar del Rio and Havana, and

[1] Macready, G. A.: Petroleum industry of Trinidad, Trans. Am. Inst. Min. and Met. Eng., St. Louis Meeting, Sept., 1920, 10 pp.

Producing and Petroliferous Formations of Trinidad

System	Series	Formation or member	Thickness (feet)	Oil indications or production
Quaternary	Pleistocene	Alluvium	40	Asphalt cones, **seepages and mud volcanoes**
		UNCONFORMITY		
		Llanos	100	Asphalt **seepages**
UNCONFORMITY				
Tertiary	Pliocene	(Bluish clay or porcelanite)	400	None
	LOCAL UNCONFORMITY			
	Oligocene or (and) Miocene	Morne l'Enfer (contains the La Brea "tar sand")	800-2500	Sometimes **saturated with** asphalt, **and may be productive near Pitch Lake**
	Eocene or Oligocene	Forest clay	500	None
		GREAT UNCONFORMITY		
		Stollmeyer sand	500	Most profitable oil **sand** thus far **discovered in Trinidad**
		Stollmeyer Cruse shale	600	Contains some **sand lenses** from **which a few wells** produce oil
		Cruse sand	40	Most persistent oil **and gas** sand in **Trinidad**
		Naperian clay	4000	Sometimes has **kerosene** odor or **irridescent film** on water at **outcrop**

shales highly charged with hydrocarbons are found scattered over nearly all of the provinces in the island. Oil was discovered in Cuba as long ago as 1887. Some oil was discovered in Havana Province in 1917, in the Bacuranao district, which in that year produced 19,167 barrels. While that field has not turned out to be a valuable one, testing of the island is still going on with good prospects for success in other localities, especially on a great anticlinal uplift which crosses the center of Havana Province from east to west, and in probably favorable domes farther north. Little real work seems to have been done in a systematic manner by experienced companies. One difficulty consists of complex unconformities owing to which it is difficult to make prediction for specific localities.

Reference

DeGolyer, E.: The geology of Cuban petroleum deposits, Bull. Am. Assoc. Pet. Geologists, Vol. II, 1918, pp. 133–167.

Porto Rico

At several points on this island exudations of petroleum are said to occur, possibly from the beds of bituminous lignite which are found in Tertiary beds at the southwest corner of the island. There are no known oil prospects of commercial importance.

Santo Domingo and Haiti

Small quantities of petroleum have been found on this island, principally in pits and along the dry beds of streams near the old town of Azua, in the southern part of Santo Domingo, where several wells have been drilled, and showings of oil and gas have been obtained.

Barbados

Oil indications are mainly confined to the Scotland district on the east side of the island, where the petroliferous rocks are Miocene sandstones and shales known locally as the Scotland beds. The most northerly occurrence is that on the Morgan-Lewis estate, about 1½ miles north of St. Andrews, where shallow wells have yielded a small quantity of petroleum, as they have also done on the Turner's Hall Wood estate, about 3 miles southwest. Tarry Gully, a short distance south of the latter place, is named from oil-saturated earth found there. Oil is also found in shallow wells in the Baxters district. On the Friendship and Groves estates, a short distance southwest, a large quantity of "manjak" or desiccated tar occurs about 4 feet from the surface. A little heavy oil has been obtained on Barrow Gully, about three-quarters of a mile farther south, while manjak and oil occur at Springfield, in the Lloyd wells on the coast, and inland at St. Joseph. Manjak is also found at Burnt Hill on Conset Bay, some distance south, outside the Scotland District, in similar shales. The rocks consist of thick-bedded sandstones, coarse grits, bituminous sandstones and shales, and dark gray and mottled clays. The strata are much disturbed, ranging from 13° to vertical, and are broken by faults.

OCEANIA

Dutch East Indies

The East Indies are among the richest fields of the world, as regards petroleum. Some large wells have been obtained in Borneo, Sumatra, Java, Ceram, and Papua, the relative importance of these islands, from the standpoint of production, being indicated by the order in which they have been named. The total production (including British Borneo) for 1918 aggregated 13,284,936 barrels.

In geology, the Islands of Java, Sumatra, Timor, Borneo, Celebes, Ceram and New Guinea (or Papua) are similar, and it is natural to expect similar occurrences of oil in them.

Borneo

The oil-fields of Borneo are situated in the Sultanate of Koetei and elsewhere on the east coast. Since drilling was begun, in 1897, several hundred wells have been sunk, and production has gradually increased. Oil is found mainly in Miocene sandstones along narrow anticlines on a great monocline, and is associated with heavy gas pressure. Most of the wells are gushers. At Sanga Sanga, in East Borneo, north of Balik Papan, ten oil sands have been proven, but not all have oil in commercial quantity. They are not supposed to be persistent horizons, but in the nature of lenses of little continuity. The Sanga Sanga anticline is some 30 miles long, with an undulating crest, and plunges at both ends. There are three grades of oil in this field—"heavy oil," "light oil" and "paraffin oil," ranging in specific gravity from 0.96 to 0.84.

Oil showings are numerous in the Tertiary coal-bearing series extending from the northwestern part of the Is'and of Borneo southwestward through the British possessions to Sarawak, and on the adjacent Islands of Labuan and Mengalon, but although some wells have been drilled, no developments of importance have resulted in those regions.

References

Jezler, H.: Das Oelfeld Sanga Sanga in Koetei, Mederal-Ost-Borneo, Zeitschr. für prakt. Geologie, Vol. XXIV, pp. 77–85, and 113–125, 1916.

Emmons, W. H., and Gruner, J. W.: The Sanga Sanga oil field of Borneo (being an abstract of the paper above listed), Eng. and Min. Jour., Vol. CXI, No. 10, 1921, pp. 431–432.

Sumatra

Many years ago oil was discovered in large quantities in Sumatra. This island produces a much larger quantity than the other two islands of the group, and the oil is of better quality than that of Borneo.

On the east side of Sumatra are the well-known fields at Telega Said in the Langkat district, in Atjeh district farther northwest, at Palembang in southern Sumatra, etc. The oil sands range from 150 to 2700 feet below the surface.

Java

Petroleum is widely scattered in Java but most of the production comes from the northeast corner of the island, from the Provinces of Semarang, Rembang, Soerabaja and the adjacent Island of Madoera. The principal production of the Province of Rembang comes from Tinawoen and Panolan, that of Soerabaja from the pools of Twalf Dessas, Lida Koelon and Made, all of which have produced for years. The most productive horizon is at the contact of the Middle and Upper Miocene strata. The depths of wells are from 400 to 2500 feet.

Papua or New Guinea

Parts of this country are exceedingly rich in oil. Some of the evidences of petroleum in New Guinea are seepages and mud volcanoes. On the north coast of Dutch New

Guinea oil is reported at the mouth of the Buti River; and 200 miles southwest it saturates coal-bearing rocks on Iwaka River, south of the Narsua Range in latitude 4° 21′ 30″ South, longitude 136° 52′ 30″ East, 40 miles from the sea.

Petroleum is reported near the Gira gold field and also on the Vailala River in the Gulf division. In addition to these occurrences, emanations of gas and seepages of oil are stated to have been met with over an area of about 900 square miles. The structure is ànticlinal, highly upturned and faulted. The oil-bearing beds are presumably Miocene in age.

References

Carne, J. E.: Bull. No. I of the Territory of Papua, Dept. of External Affairs, Melbourne, 1913, pp. 34–82.

Langford, W. G.: Idem, Bull. No. 4, 1918, pp. 7–16.

Chapman, Frederick, Idem, Bull. No. 5, 1918, pp. 3–17.

Wade, Arthur: Report on petroleum in Papua, Printed for the Commonwealth of Australia by Arthur J. Mullett, Government printer for the State of Victoria, 1914, 45 pp.

Timor

Indications of petroleum have been met with at several points in the Portuguese portion of the Island of Timor.

Philippine Islands

Petroleum and natural gas exist in Miocene shale and thin tuffaceous sand beds in the Philippines. The best known seepages of petroleum, some of which are accompanied by emanations of natural gas, are in four localities: (1) The southern part of Bondoc Peninsula, Tayabas Province, Luzon; (2) on the west coast of the island of Cebu; (3) in northwestern Leyte; and (4) in Cotabato Province, Mindanao. A little petroleum, accompanied by salt water, was encountered at a depth of 1200 feet in a well drilled in search of artesian water in Pangasinan Province, west-central Luzon. Other occurrences, few of which have been verified, have been reported.

The petroleum occurrences in Bondoc Peninusla, Luzon, are on the crests of non-symmetrical anticlines which are more or less faulted in places. Several shallow wells, 100 feet to 300 feet deep, have been drilled in that region without encountering more than slight seepages of oil. In Cebu, near Alegria, oil seeps from a steep anticline, and at Toledo two wells were drilled on monoclinal structure to depths of about 800 feet and 1100 feet respectively, but only a slight showing of oil was obtained. In Leyte the petroleum seeps, which are near the town of Villaba, are associated with noteworthy deposits of bituminous tuff and limestone, and smaller pockets of material resembling ozocerite. Folding and faulting have been recognized here, complicated by intrusions of igneous rocks. The bitumen-impregnated rock is quarried and shipped to the cities of Cebu and Manila for use in its natural state in street paving.

Natural gas has been found in a well 1772 feet deep near Janiuay, in the island of Panay, and it has been reported as issuing at places in the island of Mindoro and Bohol. In Mindoro it is said to have been piped and used for cooking purposes. Bubbles of gas accompany the seepages of petroleum at most of the places mentioned above.

Philippine petroleum has a paraffin base and much of it is of light gravity. The Islands have not yet been demonstrated to have oil in commercial quantities, but arrangements have been made by a strong American company to begin adequate drilling tests. Geologists have expressed themselves as not favorably impressed with chances of large production.

Exploration by the drill was begun in 1921, by the Standard Oil Co. of Cal.

References

Pratt, Wallace E.: Econ. Geol., Vol. XI, No. 3, 1916, pp. 246-265; Wallace E. Pratt and Warren D. Smith: Phil. Jour. Sci., Sec. A, Vol. VIII, 1913, No. 5, pp. 301–376; Wallace E. Pratt: Phil. Jour. Sci., Wallace E. Pratt, and Warren D. Smith: Mineral Resources of Philippine Islands for 1912, Bureau of Science, Div. of Mines, 1913, pp. 49–57.

Smith, Warren Du Pré: Petroleum in the Philippines, Min. and Met., Bull. 158, No. 6, Am. Inst. Min. and Met. Eng., Feb., 1920, 8 pp.

New Zealand

Small showings of oil have been found in borings near Brunner, 21 miles east of Greymouth, on the west coast of South Island, and indications have long been known to exist in North Island, at New Plymouth and on adjacent islands off the west coast, and in a belt from near East Cape Waiapu to the Okahuatin block, 30 miles west of Gisborne on Poverty Bay. Drilling began near New Plymouth in 1865, and later wells of small production were sunk there. Wells were also sunk at Poverty Bay and on Waiapu River, but no oil in commercial quantities has been found. The showings are in Miocene strata, but structural evidence appears to be lacking. The oil horizons near New Plymouth lie from 1000 to 5000 feet below the surface.

Reference

Carne, J. E.: Bull. No. I on the Territory of Papua, Dept. of External Affairs Melbourne, 1913, pp. 70–77.

New Caledonia or Loyalty Islands

Oil is said to occur on the northwest coast of New Caledonia.

Australia

New South Wales

There are rumors of oil springs in New South Wales, but none has yet been proved authentic. Tertiary lignites saturated with oil occur on the coast north of Cape Howe, at Twofold Bay and Bonda, and at Kiandra in the interior, while oil-bearing shales are extensively mined at various points along the borders of the coal basin. A small amount of oil is distilled from shale, amounting in 1917 to 31,661 long tons, but no commercial production exists other than that. Natural gas is said to have been struck in a hole 3100 feet deep drilled in Triassic rocks at Grafton in the northeastern part of the state.

References

Carne, J. E.: Mem. Geol. Survey N. S. Wales, Geology No. 3, 1903, p. 106.

Bulletin No. 1 on the Territory of Papua, Dept. of External Affairs, Melbourne, 1913, pp. 77–78.

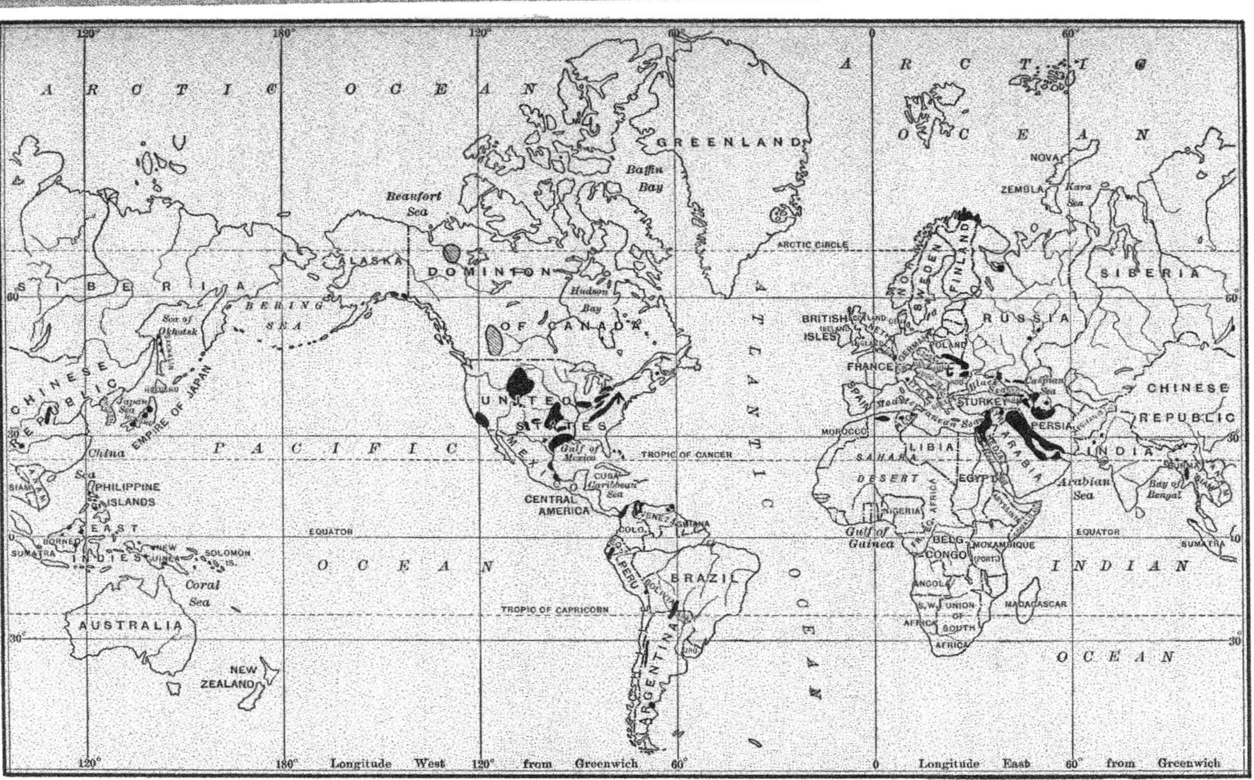

FIG. 62.—Map showing distribution of principal oil-fields throughout the world.

Queensland

Oil shales occur in the Tertiary basin of Dawson River and on the north flank of the McPherson range near the border of New South Wales; natural gas has also been reported in Triassic rocks. No petroleum is produced.

West Australia

Seepages are reported near the mouths of Warren and Connelly Rivers on the southwestern coast of West Australia. Oil showings are reported in the Permo-Carboniferous shales and sandstones near Fly Brook and Lake Jasper. No developments are known.

South Australia

Petroleum indications are reported in Miocene shales on Leighs Creek (latitude 31° South) and in the Gawler between Kapunda and Adelaide. Small quantities have also been reported on Salt Creek near Meningie and at Bordertown. Oil springs are reported near Yorktown.

At Robe, wells have been drilled to depths of 4000 to 5000 feet, and showings of oil have been reported, although none has ever been produced.

Victoria

Traces of solid bitumen exist at numerous points, and oil seepages are reported near Bridgewater, 100 miles northwest of Melbourne. No occurrences of economic value are known.

TASMANIA

Oil shales are found in the Carboniferous strata south of Table Cape in the northern part of the island, and in a belt extending from the Don Valley, past Mersey Run to Tamar estuary. Some oil is secured from the distillation of shales, but there is no production from wells.

FIELD METHODS IN PETROLEUM GEOLOGY

BY

FREDERIC H. LAHEE

GEOLOGICAL OBSERVATIONS

Geological field work for petroleum.—Geological field work for petroleum includes, on the one hand, the search for actual evidences of oil and the study of these evidences, and, on the other hand, the search for and study of rock structures which may control the distribution of oil. The geologist should be trained in both of these branches. He should be able to recognize evidences of oil, and, above all, he should be able to discriminate between those which are reliable and those which are deceptive or false. He should understand the methods by which geologic structures are discovered and mapped, and he should know how to investigate the bearing of such structures on underground oil storage and on oil production.

Surface indications of petroleum.—Indications of petroleum are of various kinds. The more important are oil seepages, asphalt deposits, gas seepages, outcropping petroliferous sands and shales, rainbow films on water, etc.

Oil seepages are simply springs of oil. Just as in the case of many water springs, the oil may rise through porous beds to the surface and escape along the outcrops of these beds, or it may come out along faults or fracture zones. It may vary in quality from very light grades to heavy, viscous kinds which are almost too thick to flow. Usually seepages are of the heavier dark-colored oils.

By the evaporation of the lighter constituents of oil, asphalt may accumulate near the point of escape or in the pores of the rock through which the oil has passed. In this way, rocks which were formerly porous may become clogged and so ultimately prevent the further escape of the oil.

Gas seepages are not easily detected unless they are below water. In this case, the gas comes bubbling up through the water. It is usually recognizable by its smell and its inflammability, but other natural gases—notably ordinary marsh gas—may be mistaken for it. Escaping gas may be collected for analysis by inverting a wide-mouthed bottle over the seepage for several minutes, and then carefully raising and corking it while held in the same inverted position.

In some localities petroliferous sandstones, shales or limestones are exposed. On their outcropping surfaces they may not appear to be oil-bearing, but if broken open they may yield a strong petroliferous odor. Such rocks, although black or dark gray inside, may weather light gray If the fractured surface gives no odor, the rock may be tested for petroleum by crushing a piece of it and putting the powder in a saucer containing chloroform. After a minute or two, the chloroform should be poured off into another dish and allowed to evaporate. If a black rim is left, it is an indication of the presence of petroleum in the rock.

Where gas or oil escapes through water, an iridescent film of oil may cover the surface of the water. To distinguish this from other similar "rainbow films," pass a stick across the water surface. If the film is of oil, it will unite again behind the stick, but if it consists of iron oxide, manganese oxide, or organic substances, it will break, and the detached pieces will not easily reunite.

In this connection we may refer to certain false ideas concerning evidences of petroleum. A very common fallacy is the assumption that the presence of oil underground is indicated by certain kinds of topography. Some men will pick out a depression surrounded by rocky ledges and will tell you that this offers the best chance for striking oil. Some will choose a broad, dome-like hill, and others will look for a small rise in a broad shallow valley. In most cases, this impression that one type of country "looks good" for oil, while another has poor prospects, is due to an unconscious comparison with some other producing field with which the observer is acquainted. Occasionally hills and valleys are pretty closely coincident with the underlying structural highs and lows, but usually there is no such close relation. It is generally unsafe to rely solely on topography as indicating whether or not a region has promise of oil.

Significance of surface indications of oil.—The importance of surface indications of oil is often very much overestimated. They may be evidence that *some* oil or gas exists in or near the localities where they are found, but they do not tell us whether such oil or gas is here in commercial quantities. As a matter of fact, seepages are excellent witness to the fact that the oil has been escaping, perhaps for long periods of time. Under these circumstances it is the business of the geologist to search, in the region where the seepages occur, for structures which may have caused the *retention* of oil. He will be less interested in those particular places where, on account of peculiar structural conditions, original accumulations of petroleum may have been depleted by seepage.

Object of geological mapping for petroleum.—The results of drilling in many of the producing fields of to-day have shown that one of the most important factors which govern the distribution of petroleum is geologic structure. In a majority of cases anticlinal folds are the controlling structures. Consequently, in regions which have been only partially or not at all developed, but where there may be a chance of obtaining oil, geologists usually endeavor to determine and map the structure of the rocks, with the object of ascertaining the position, form, size, and distribution of anticlines, or of any other types of structure which may have a bearing on the occurrence or extraction of oil.

Phases of geological mapping.—Geological field mapping may be said to have these three phases,—(1) the examination and study of the rock strata, (2) the location of points or outcrops where observations are made, and (3) the plotting of the field data on a map. All three of these phases are often carried along together, as we shall presently see. All three are important, and none should be slighted. Too frequently geologists are not sufficiently trained as engineers, the consequence being that men who may be proficient in the classification of rocks and in the interpretation of structures are sometimes inadequately prepared in the methods of locating or mapping what they may observe in the field. The results of the work of some good geologists have been useless because of carelessness in the locating of stations or outcrops.

Regions of low dip and regions of high dip.—Regions where geological work is done for petroleum may be classified under two heads, those in which the strata have prevailingly low dips and those in which the strata have prevailingly high dips. The discrimination between high and low dips is purely arbitrary. We may call dips high if they are more satisfactorily measured and recorded in degrees of inclination, and we may call dips low if they are best recorded as so many feet per mile. We may take 5° as the limiting angle. Dips of 5° and over may be referred to as high, and those which are under 5° (462 feet per mile) may be referred to as low.

We may look at this matter in a somewhat different way. If beds dip 5° or over, an error of 1° more or less in the determination of the dip will not be very serious, but if dips of less than 5° were to be measured by clinometer (page 178) in degrees, an error of 1° in the reading would be a large proportion of the whole dip. Obviously, then, the method to be adopted in measuring dips will depend upon whether they are high or low (page 177).

Outcrops and topography.—Without rock outcrops the geologist is helpless, as far as field mapping is concerned. In regions of strong topographic relief they are likely to be numerous and to stand out well, so that they can easily be examined, whereas, in regions of low relief, they are usually few, and those which are found are so badly weathered that they can not readily be interpreted.

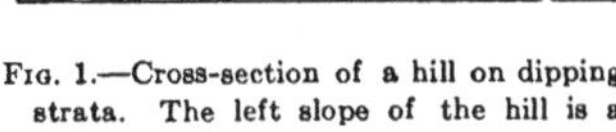

Fig. 1.—Cross-section of a hill on dipping strata. The left slope of the hill is a "dip slope."

In most places outcrops consist of relatively hard or resistant rocks, the softer, weaker varieties beng covered by soil. If, for example, the district is underlain by a series of alternating shales, sandstones and limestones, the sandstones, and particularly the limestones, are apt to be exposed, while the shales are nearly everywhere concealed.

Where the strata have very high dips, say between 45° and 90°, the resistant rocks may crop out as sharp ridges, sometimes like walls, often forming the backbones of long, more or less sinuous hills. If the dips are somewhat lower, the outcropping ledges may have the relations shown in Fig. 1, dipping into the hillside. One side of such a hill cuts across the bedding; the other, which is called a "dip slope," may be nearly coineident with the bedding. In the case of strata with still lower dips, the hard layers may outcrop in shelf-like benches, and these may sometimes be traced for long distances round the hillsides (Fig. 2). Ledges like those in Figs. 1 and 2 are called *scarps* or *escarpments.*

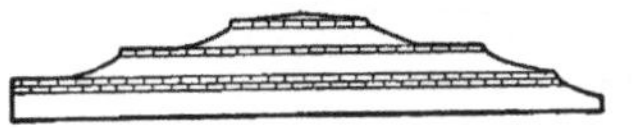

Fig. 2.—Cross-section of a hill on horizontal strata.

Top and bottom of beds.—In mapping geologic structures, it is customary to represent the form of the top or bottom surface of some chosen bed called the key bed. For this and other reasons the geologist must not only understand the probable nature of the top and bottom surfaces of strata of different kinds, from a theoretical standpoint, but he must also be careful to ascertain as closely as possible, in the field, the positions of the bounding surfaces of the outcropping beds.

In the processes of sedimentation, sands are generally laid down under more disturbed conditions than muds, clays, or other fine-grained materials. The surface of a sand deposit is characteristically irregular, being diversified by channels, bars, and other features due to current action. During sedimentation, a mud or clay deposit is likely to be more even. However, if mud-depositing conditions are changed to sand-depositing conditions, the stronger currents that bring the sand are likely to scour away some of the upper layers of the mud previously deposited, so that the upper surface of the mud, when it is finally buried by sand, will be decidedly irregular—often more so than the upper surface of the sand. Since the top of the mud, in this case, is identical with the bottom of the overlying sand, we may draw the conclusion that, where mapping must be done on sandstones, the upper surface of a sandstone layer is apt to be more even, or flatter, than its under surface. While this is not always true, it applies to many localities.

Thin limestone beds are generally fairly regular both above and below, and here the upper surface is commonly used for mapping or for measurements.

Stratigraphic intervals.—The perpendicular distance between the top surfaces, or the bottom surfaces, of any two strata in a formation is called the *stratigraphic interval* (Fig. 3). If the beds are horizontal, it is vertical; otherwise it is inclined. In places where the key bed, which has been selected for the mapping of a region, has been eroded away, or where it dips underground, the field mapping must be done on lower or higher beds, respectively, and the stratigraphic intervals between such beds and the key bed must be determined. Thus, in a region like that shown in Fig. 4, the geologist might be able to work on the bed *abcd*, which he selects as his key bed, in area *aefb*. In area *efgh* he must work on bed *efgh*. In this area he must find the stratigraphic interval, *gi*, and subtract it from his observations on bed *efgh* to make these observations apply to the underlying key bed. In area *abjk*, the key bed has been eroded. Here, then, he must work on bed *kjm*, and, to reduce his observations on this bed to the key bed, he must add the stratigraphic interval, *bm*.

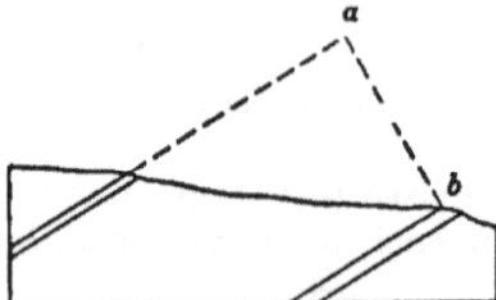

FIG. 3. — Cross-section of dipping strata, showing the stratigraphic interval, *ab*, between two beds. Observe that the distance, *ab*, is measured between corresponding parts of the two beds—in this case their upper surfaces.

The stratigraphic interval between any two given beds must never be assumed to be constant. In fact, variations in stratigraphic intervals are the rule rather than the exception. This merely amounts to saying that strata are not characteristically of constant thickness for any considerable distance. They thin and thicken in different directions. The geologist must always be on the watch for thinning and thickening, or lensing, of beds, and for the related variations in stratigraphic intervals. In mapping extensive areas, he may have occasion to use different intervals between the same beds in different places.

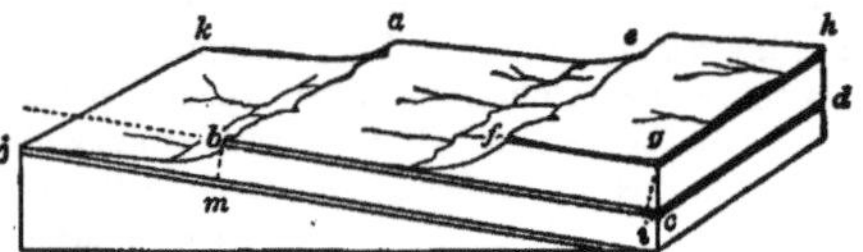

FIG. 4.—Block diagram of a region where field mapping must be done on several different beds.

Covering and settling of outcrops.—Difficulty is often experienced in finding the exact position of the contact surfaces bf an outcropping stratum. This may be due to (1) soil cover, (2) beveling of the outcrop by erosion, or (3) settling or slumping.

(1) In regions of moderate or low relief, the lower surfaces of outcropping beds are often soil-covered, while their upper surfaces may be partly exposed. Under these circumstances the top surface of the bed is used for mapping.

(2) Friable sandstones may be beveled or rounded off along their outcrops, so that the upper part of an exposure may be several feet below the original position of the top of the bed (Fig. 5). If such beds only are available for mapping, the bottom surface of a prominent sandstone bed may be used for the key horizon, provided it is exposed extensively enough. Otherwise, if the top surface is used, allowance must be made for the eroded portion.

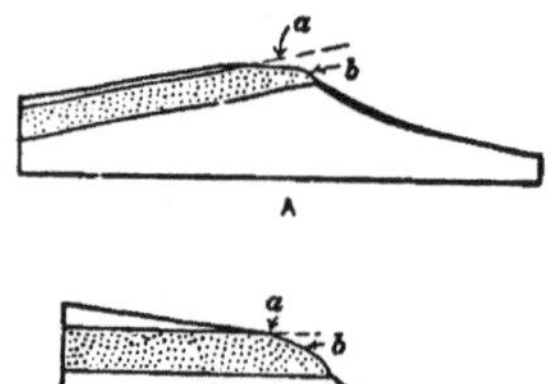

FIG. 5.—Cross-sections of an inclined sandstone bed (*A*), and of a horizontal sandstone bed (*B*), which have been beveled by erosion. The top, *a*, of the bed might be erroneously located near *b*.

(3) Settling of the more resistant strata on account of the downhill creep or the

removal of the weaker underlying strata is a very common phenomenon. Sometimes, when the settling is above soluble beds of salt, gypsum, or limestone, it may be localized in the form of sink holes. On hill slopes soil-covered shale or clay beds may slowly move downhill and thus let the overlying harder beds sag toward the valley (Fig. 6). This type of settling may amount to many feet, thus making very difficult the discrimination between true dip and the secondary dip due to the slumping. Even where the slumping is only a few feet, to estimate the original position of the exposed part of the bed requires considerable judgment.

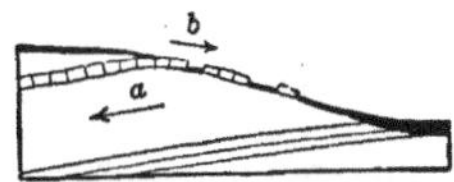

Fig. 6.—Cross-section of a limestone bed which has settled at its outcrop. On account of this settling the dip might be assumed to be toward the right (*b*), whereas it is actually toward the left (*a*).

Cross-bedding.—In working with sandstones the geologist has to be very careful not to mistake cross-bedding for true bedding. Cross-bedding is a feature caused by current action during the deposition of the beds, and it is always at an angle to the principal or true bedding of the series of strata in which it occurs. Usually it is on so small a scale—the cross laminæ being from a few inches to a few feet in length—that it is easily recognizable; but sometimes the cross-beds are so long that a single outcrop may expose only the cross-bedding, and not the true bedding. Under these circumstances the dip of the cross-bedding may be recorded for that of the true bedding and serious mistakes may be made in the interpretation of the structure.

Normal and abnormal dips.—In regions of moderate or slight folding there is usually an average dip of the formations in some general direction, a dip which may continue for many miles. For instance, in eastern and central Kansas and northern Oklahoma, the principal outcropping formations have an average westward to northwestward dip of a few feet per mile. Average dips of this kind are referred to as *regional* or *normal dips*. Any local dips which are in the same direction as the regional dip are called *normal*. In contradistinction, any local dips which are in any other direction are *abnormal*, and dips in a direction opposite from the normal dip are termed *reversals*. It is always important for the geologist to familiarize himself with the normal dip in a region, since abnormal local dips generally indicate the presence of folded structures in the strata.

Types of Geological Maps

Types of geological structure maps.[1]—Before discussing the methods of geological field surveying we may first consider the nature of the maps which are the result of such work. This will enable us better to understand the reasons for the various operations in the field work.

Most geological maps used in connection with the search for oil fall under the following heads: (1) structure contour maps; (2) formation maps; and (3) dip-arrow maps. These we shall consider in order.

Structure contour maps.—The structure contour map is by far the most convenient kind of geological map. It is a type very commonly used and should be thoroughly understood. Figs. 7 and 8 illustrate such a map. The curving lines are "contours." These contours are supposed to be drawn on the top or bottom surface of some definite stratum, the "key bed." This surface is the "key horizon."

Imagine that the key horizon is everywhere coincident with the earth's surface. If the sea should rise a certain amount its shore would be a sinuous line on this surface, bending in up the valleys and out around the hills. If the water level were to rise

[1] The term "map" is used here for the representation of features projected vertically on to a horizontal plane.

10 feet more, there would be a new shore-line 10 feet higher than the first one. With another 10-foot rise of the sea level, there would be a third shore-line; and so on. These imaginary shore lines are contours, and the vertical interval between them—here 10 feet—is the "contour interval." A contour map is a sketch of such contours on a selected key horizon as they would appear if viewed from above.

The contour interval can be any chosen quantity, but it should be the same on all parts of any particular map. Where the beds dip very gently a small contour interval is adopted, say 5 or 10 feet, or sometimes even 1 or 2 feet; but if the beds dip several hundred feet to the mile, a larger contour interval is employed. (See Figs. 7, 8.)

Each contour on a map passes through points of equal elevation on the key horizon. The usual method in preparing such a map is to determine the elevations of points scattered here and there on the key horizon, to plot these points, and then to sketch contours by interpolation. In Fig. 7, for instance, the contour *ab* is drawn through *a*, *c*, and *b*, halfway between *d* and *e*, one-fourth of the way from *f* to *g*; etc.

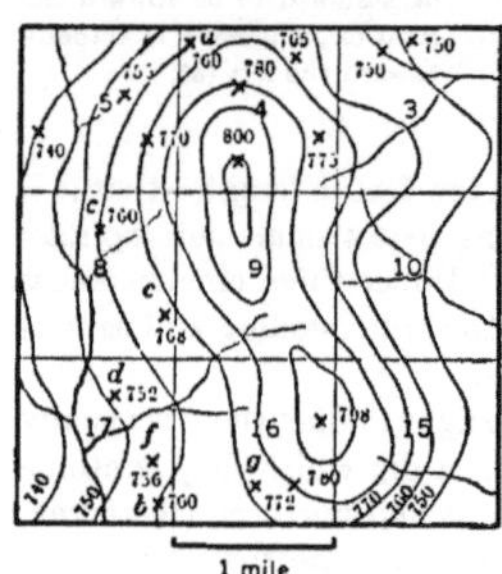

FIG. 7.—Structure contour map of an anticline of which the dips are low. The contour interval is 10 feet. Some of the stations where observations were made are indicated by crosses beside which are recorded the elevations of the key horizon.

If the key horizon were everywhere exposed, as we postulated above, the determination of elevations at selected stations would be comparatively easy. Usually, however, the bed which is finally contoured is exposed only locally, or it may be entirely covered, perhaps several hundred feet below the surface of the ground. In the former case the geologist may work on beds other than the key bed, but he must ascertain the stratigraphic interval between such other beds and the key horizon (see page 170).

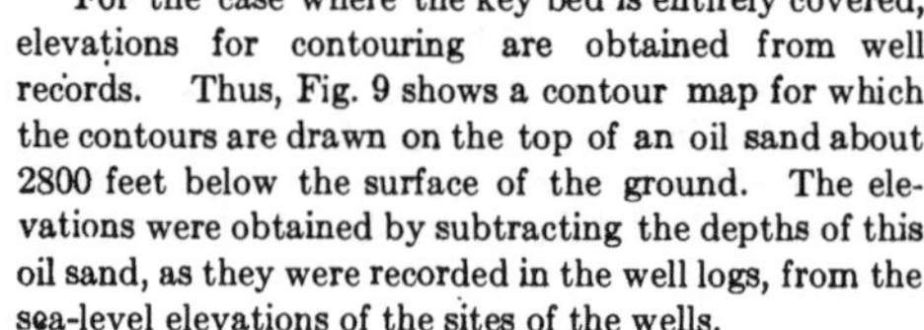

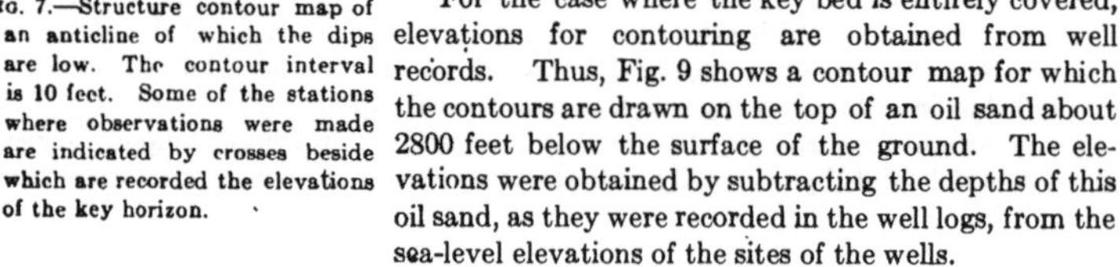

For the case where the key bed is entirely covered, elevations for contouring are obtained from well records. Thus, Fig. 9 shows a contour map for which the contours are drawn on the top of an oil sand about 2800 feet below the surface of the ground. The elevations were obtained by subtracting the depths of this oil sand, as they were recorded in the well logs, from the sea-level elevations of the sites of the wells.

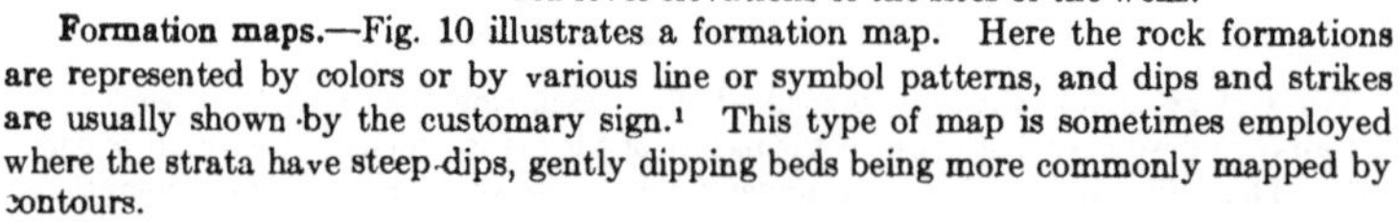

Formation maps.—Fig. 10 illustrates a formation map. Here the rock formations are represented by colors or by various line or symbol patterns, and dips and strikes are usually shown by the customary sign.[1] This type of map is sometimes employed where the strata have steep dips, gently dipping beds being more commonly mapped by contours.

Dip-arrow maps.—This is a very crude type of map which is used only when field work is rapid and the object is merely to indicate the approximate positions of structures which may deserve subsequent and more careful examination. Fig. 11 is an example. Frequently the arrows are for dip components and not for true dips. Thus, the arrows labeled *a* and *b* are really components of a southwestward dip. The amount of dip, or of the dip component, may be recorded beside the arrow or may be roughly shown by the length of the arrow. Maps of this kind may be employed to record the results of reconnaissance work (see page 174).

Mapping in General

Surface and subsurface mapping.—The type of contour mapping, mentioned above, in which the key bed is far below the ground surface, may be called *subsurface mapping*.

[1] The sign is like a T, the cross line being plotted parallel to the strike and the short line, corresponding to the upright of the T, being drawn to point *down* the dip.

It lies outside of the scope of the present chapter. It is a very important branch of petroleum geology, but it cannot be satisfactorily accomplished until enough wells have been drilled, and logs prepared, to enable the geologist to correlate the underground formations from well to well.

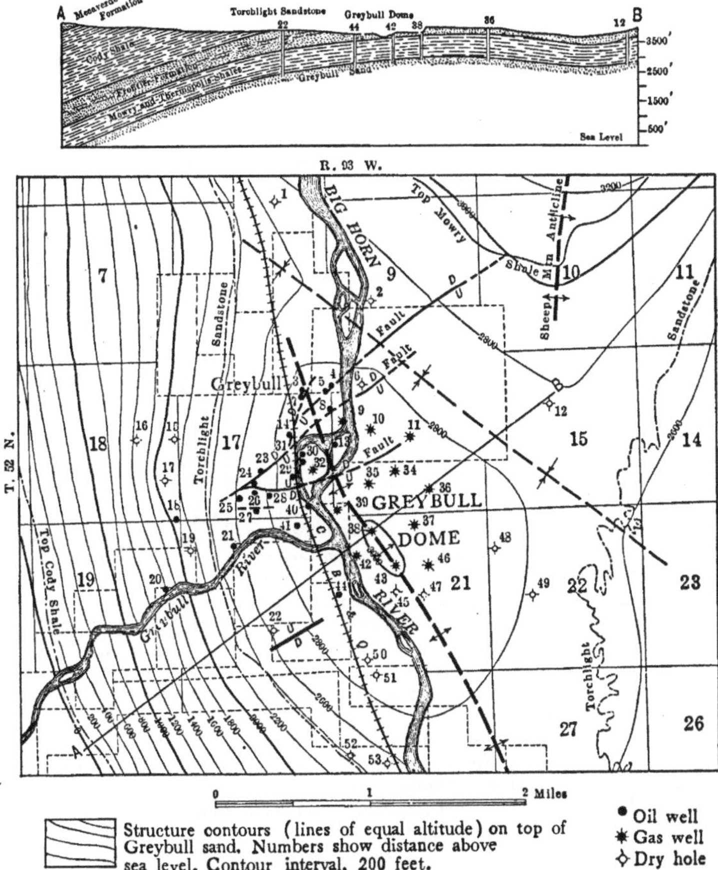

FIG. 8.—Structure contour map of the Greybull Dome, Wyoming. Dips are steep here, so that a 200-foot contour interval is used. (After D. F. Hewett and C. T. Lupton.)

Surface mapping, or, as it is generally termed, field mapping, may be done at any time. It may precede, accompany, or follow drilling operations. As a rule it precedes them, for the geologist, in this kind of mapping, is interested in locating structural features in order that he may recommend, or advise against, the purchase of acreage or the drilling of wells.

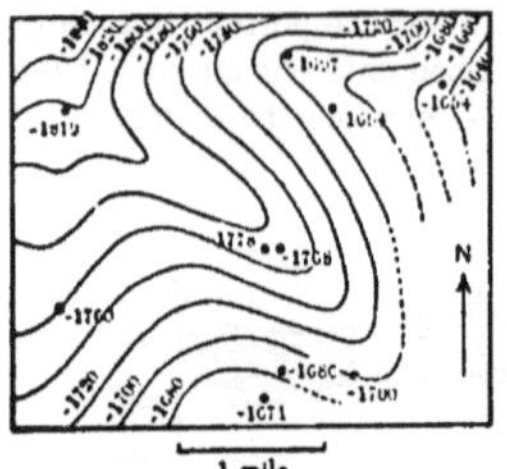

Fig. 9.—Contour map of a subsurface structure, as determined by studying and correlating well logs. The black dots mark the sites of the wells. The elevations are on a key horizon below sea level.

Relations of surface and subsurface structures.—In many cases there is a marked difference between the geologic structure in the outcropping strata and that in the beds several hundred feet below the surface. This may be true even where no unconformity exists between the surface beds and the deeply buried beds. For instance although anticlines and synclines at the surface may directly overlie anticlines and synclines in the subsurface the dips may change with depth (Fig. 12).

If a folded series of strata lies unconformably above an older series of strata, which was itself folded, previous to the deposition of the upper series, structures in the two groups may be widely divergent. As in the case above cited, a weak anticline may overlie a pronounced anticline, and the synclines may be similarly related (Fig. 13), as if renewed folding had occurred along the old lines of deformation. Or there may be no genetic relation between the surface and subsurface folds, as seems to be the case at Hewitt, in southern Oklahoma, where the producing subsurface anticline lies almost directly beneath a syncline in the surface beds (Fig. 14).

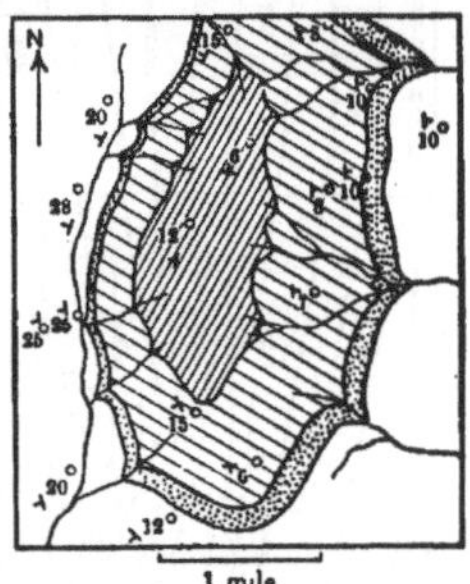

Fig. 10.—Formation map of an anticlinal dome. The oldest formation is in the center of the dome. Dips and strikes are shown by the customary T-shaped symbol, the amount of dip being recorded in degrees.

If the superficial and deeply buried structures in a region are so discordant, and the expected pay sands are in the lower series of beds, the value of the mapping of surface beds will naturally be questioned. We may answer that, where the discordance is merely in the *amount of dip* and not in the position of the folds, the surface mapping is often very useful. In proportion as the discordance in *position* of the surface and subsurface structures increases, the value of mapping the surface structure, where subsurface structure is wanted, decreases. In undrilled areas very little can be foretold as to the relations of surface and subsurface structures. As wells are drilled, more and more is learned. The geologist should always have in mind the possibility of discordances of this nature, either due to changing intervals within the same rock series or to actual unconformity between two series of strata.

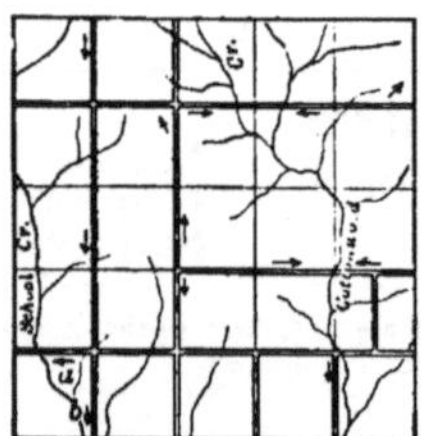

Fig. 11.—Example of a dip-arrow map made in a rapid reconnaissance of an area. Travel was by automobile along the roads (double lines).

Reconnaissance and detail mapping.—A somewhat elastic distinction is often made in petroleum geology between reconnaissance and detail mapping. The terms are relative. In reconnaissance work the geologist is supposed to survey an area as rapidly as he can, without overlooking any important structure. He explores new territory, always on the lookout for abnormal dips or for structures that may be favorable to oil accumulation. If he discovers any such, he reports them and perhaps roughly indicates them on a map, but he does not take the time to determine accurately the shape and size of the structures which he finds.

In detail mapping, on the other hand, the work is done more slowly and more thoroughly. The structure is studied and mapped to the best of the geologist's ability. This kind of work usually follows reconnaissance examinations. It is generally required when advice is wanted regarding leasing, abandoning leases, locating wells, etc.

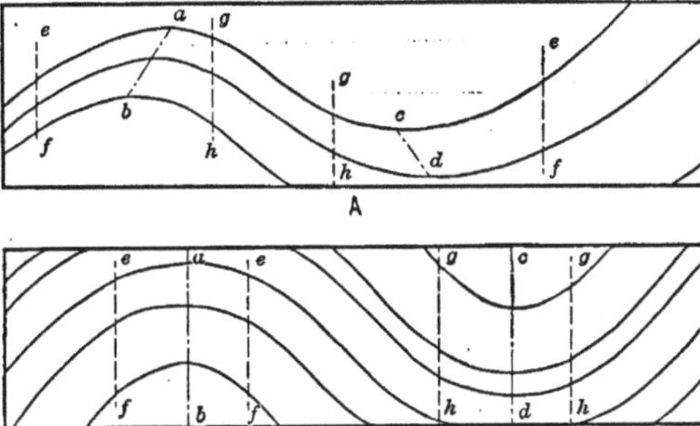

Fig. 12.—Cross-sections of folded strata, converging toward the left in A, and parallel in B. In A, anticlinal axes (ab) shift toward the convergence, and synclinal axes (cd) shift away from the convergence, with depth; and on anticlines dips become gentler toward the convergence (ef) and steeper away from the convergence, with increasing depth. In B there is no shift of axes, but, with increasing depth, dips become steeper on anticlines (ef) and gentler on synclines (gh).

Preparation for field work.—When an area is assigned to the geologist for investigation, he should secure all the available general information regarding its geology and all petroleum or drilling data, before he begins field mapping. This preliminary work may be divided into two parts, (1) that which includes studying the literature on the region and discussing conditions with other persons who have been in the same part of the country, and (2), where detail work is to be done, a brief preliminary reconnaissance for the purpose of obtaining a personal insight into existing conditions and determining the best method to be used in the subsequent mapping.

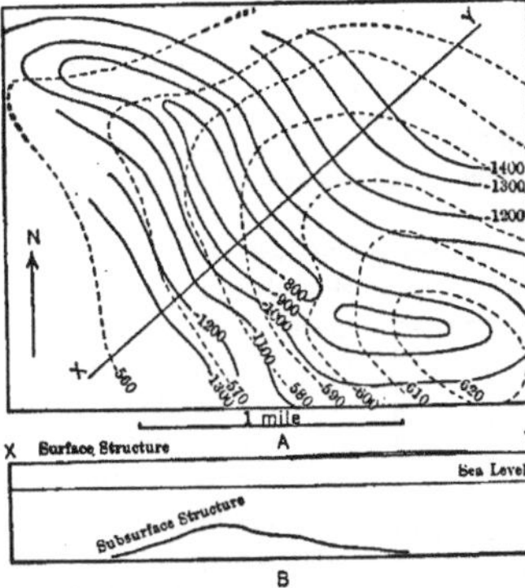

Fig. 13.—Structure contour map (A) of a region where a low surface anticline (dashed contour lines; contour interval 10 ft.) approximately coincides in position with a pronounced subsurface anticline (full contour lines; contour interval 100 ft.). The cross-section, B, is drawn where the surface anticline does not directly overlie the subsurface anticline. There is an unconformity between the upper gently folded series of beds and the lower strongly folded series. Both vertical and horizontal scales in B are the same as the scale of A.

In these studies one should endeavor to get information on the following points:

(1) The name and geologic age of each of the formations exposed in the district.

(2) The average or approximate thicknesses of these formations.

(3) The normal or average regional dip of the formations.

(4) The names and geologic ages of the formations that probably lie below ground in the area to be surveyed.

(5) Anything that can be learned concerning regional thinning or thickening of formations—its direction and amount.

(6) The nature of the rocks and exposures in the area.

(7) Whether folding is moderate, with low dips, or pronounced, with high dips.

(8) Whether the folds are irregular in form and distribution or have a marked elongation and trend.

(9) Whether any indications of oil or gas have ever been noticed in the region, either in seepages or in drilling operations.

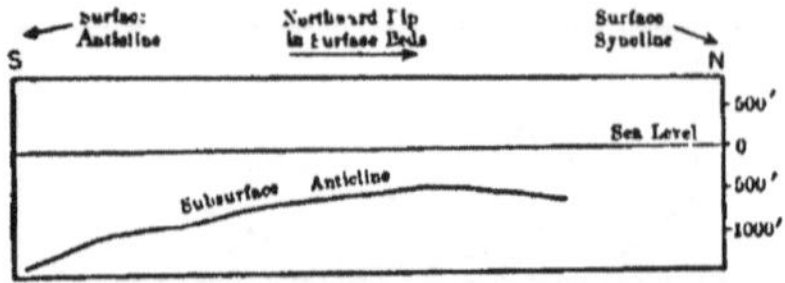

FIG. 14.—Outline cross-section of part of the Hewitt field in Carter County, Oklahoma. The crest of the subsurface anticline here lies nearly under the surface syncline. There is an unconformity between the gently folded surface beds and the strongly folded deeper beds. Length of section approximately 1¾ miles.

(10) At what approximate depths and in what geologic formations any possible oil sands lie.

(11) If possible, whether such oil sands lie conformably or unconformably below the formations exposed in the area.

Quantities measured in field mapping.—In geological field mapping four quantities are measured, namely, direction, horizontal distance, vertical distance or differences of elevation, and vertical angle. Not all four quantities are always necessary for any one map, but at least three are measured, and the fourth may be determinable from the others. For example, every point or station on the key horizon which is to be used for a contour map must have its position determined as regards its direction and distance from other similar points and s regards its relative elevation. When these three quantities have been found for every such point, the fourth quantity, the angle of dip of the key horizon, may be calculated.

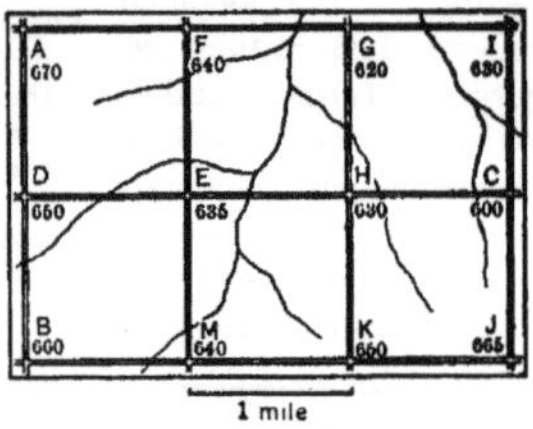

In the case of a formation map, the distribution of the several formations is sketched from scattered stations, the positions of which have been fixed in reference to direction and distance, and the angle of dip of the beds is recorded at these points. The elevations of the stations may or may not have been determined.

FIG. 15.—Outline map of roads and stations used in mapping by barometer. The numbers are elevations above sea level. *A*, *B*, and *C* were control stations.

On dip-arrow maps less care is taken in measuring the quantities, but the geologist should find the approximate positions (direction and horizontal distance) of localities where the dip arrows are plotted.

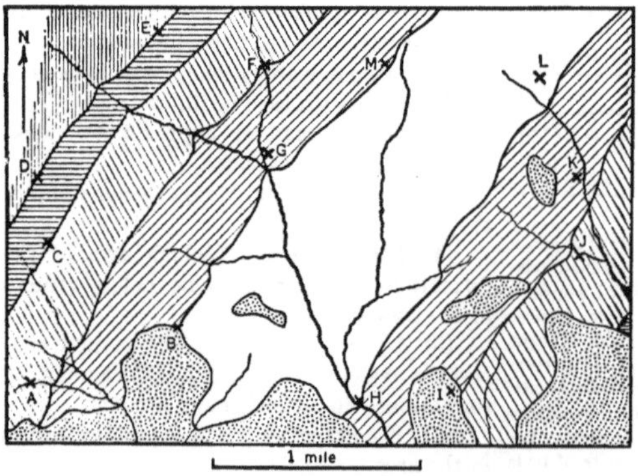

FIG. 16.—Formation map showing control stations (*A* to *M*).

Controls for mapping.—In any kind of geological field mapping, a system of stations, the positions of which have already been established so that they may serve as a guide in the location of subsequent stations, or in the plotting of geologic features, is called a *control.* Figs. 15 and 16 illustrate control systems. In Fig. 15, stations *A*, *B*, and *C* were first accurately located in the field and plotted on a map. Later, stations *D* to *M* were located in reference to the primary system which, therefore, may be said to have served as a control for the subsequent work. In Fig. 16 stations A to M on various rock outcrops were located and plotted. The distribution of the rock formations was then mapped in reference to these earlier control stations. Provided a system of control stations has been established with considerable care—which is generally supposed to be the case—it is of value in any kind of field mapping, in that it adds to the accuracy of the final results.

In many regions the geologist finds control stations already established. They may be United States Geological Survey triangulation stations and bench marks. Geodetic Survey monuments, section and township corners in sectionized land, etc. In other places he must make his own control in advance of the geological work. Sometimes there is no opportunity to prepare a control in advance, and all the mapping must be accomplished during the continued forward progress of the field party.

Base maps.—For most kinds of geological work, the field man finds it of great advantage to have some sort of a preliminary map of the country which he is to survey. This may be merely a plat showing the positions of a few established bench marks, a road plat, a sketch map of the drainage system, or, better, a topographic map. A map of this kind, previously prepared, upon which the geologist can plot his own stations, rock outcrops, etc., is a *base map.*

Any established control system should be plotted on such a base map, and if the scale of the base map happens to be different from that adopted for the geological field work, a copy should be made by reduction or enlargement to this adopted scale.

Traverses, "Walking Out" beds.—Where dips are high and rock exposures are isolated, the geologist may proceed, in his field work, from outcrop to outcrop, choosing his course as he goes, always aiming to " cover " the area assigned to him for investigation, that is, to leave no large gaps. If the outcrops are very numerous, he need not visit them all. Under all circumstances, he must map whatever route he pursues.

If dips are high and outcrops are continuous for long distances as roughly parallel ridges, the best plan is to trace and map some particular well-exposed stratum and from this to make an occasional traverse across the beds to one side and the other. If the ridge which was first traced disappears, another bed can usually be found and followed.

If strata, having a moderate or low dip, outcrop as hillside benches, the geologist should endeavor to follow or " walk out " a prominent bench. Here the main object is usually to plot the structure by contours on a key horizon (see page 171), and the most prominent bed, or the one most widely and best exposed, is often taken as the key. Trouble may be experienced in tracing these rock benches. The rock that causes the bench is not always exposed, even though the bench still persists as a flattening of the hill slope. In this case the bench should be followed as a guide to some locality farther on where the stratum may again be exposed. As a matter of fact, in regions of rather low slopes, benches are the principal features to be walked out, the rock itself being exposed only at rare intervals.

Principal methods employed in field mapping.—Four principal methods are commonly used by geologists in surface mapping for petroleum. Each has many variations which depend upon the nature of the country, the degree of accuracy required in the work, the individual preferences of the geologist, and a number of other factors. These four methods are, (1) the hand-level method, (2) the barometric method, (3) the planetable method, and (4) the method of compass and clinometer. The first three methods are best used in regions of low dips; the fourth, where dips are high. The writer believes that in regions where dips are prevailingly low no attempt should be made to measure them by clinometer, since errors may easily be made by mistaking cross-bedding and slumping for true bedding, and such errors are likely to be very large in proportion to the true dip.

In the following pages of this chapter will be described the four methods of field mapping and the instruments used in the performance of the work.

Instruments Used in Field Work

Compass and Clinometer.—The compass is an instrument for determining directions. A clinometer is used to measure vertical angles, such as the dip of strata or the slope of the ground. For geological purposes these two instruments are generally combined. The two types most commonly employed are the Gurley compass and the Brunton compass or "pocket transit."

The Gurley compass has a square metal base with beveled edges which are graduated on two sides in inches and on two sides in degrees of arc. It has a pair of folding open sights, a swinging (pendulum) clinometer, and rectangular spirit levels. Its advantages are that it can be used both as a sight alidade and as a compass and clinometer. Its disadvantages are that its edges are rather short for satisfactory alidade work and it is somewhat clumsy to carry in one's pocket. Its clinometer, while useful for dips, cannot always be used for slopes.

The Brunton compass consists of a compass, folding open sights, a mirror, rectangular spirit levels, and a spirit-level clinometer. In reading a direction toward some distant object, the geologist, while standing facing the object, holds the compass level, with the mirror nearest him and the open sight away from him. The sight stands vertical. The mirror is inclined at such an angle that he can see the distant object reflected in it. He then turns the compass a little, this way and that way, until, with the needle swinging free, the object as seen in the mirror is bisected by the center line of the mirror and the center line of the sight. In this position he reads the compass angle.

To measure the angle of a slope while looking up or down the slope, hold the plane of the Brunton compass vertical and parallel to the line of sight; set the open sight at about 90° to the plane of the compass; have the mirror about three-fourths closed, making an angle of about 45° with the compass; and hold the instrument so that the mirror is away from you. Looking through the open sight and the hole near the base of the mirror at a point up or down the slope, which is as high above the ground as your eye, bring this point, the cross-hair of the mirror, and the center line of the open sight into perfect alignment. Without changing the position of the instrument, turn the clinometer until the bubble in the longer level is centered, as seen in the mirror. Then read the angle of slope as indicated at the point on the inner dial of the compass opposite the zero on the vernier scale of the clinometer.

To read a vertical dip angle or a slope profile with the Brunton, open the sight and mirror wide, hold the compass with its plane in a vertical plane between the eye and the dipping bed or slope, and, looking across the upper edge of the instrument, make this edge apparently coincide with the edge of the dipping bed or slope. The plane of the compass should be perpendicular to the strike of the strata. While the instrument is in this position, rotate the clinometer arm on the back of the compass until the bubble in the longer level is centered. The angle of dip (or slope) will be indicated on the inner dial of the compass opposite the zero of the clinometer vernier.

When the geologist thinks he sees a slight dip or slope and wants to verify his opinion, he can set the clinometer vernier with its zero at the zero of the inner dial and then sight at the supposed dip across the top edge of the compa s, as described in the preceding paragraph. If the bubble does not move to the center, he knows that the supposed dip (or slope) exists. This method is often preferred when dips are so low that sighting at the object, adjusting the clinometer, and centering the bubble, all at the same time, are almost impossible without introducing errors as great as or greater than the supposed angle of dip or slope.

The advantages of the Brunton are that it is easily handled and carried in the pocket; it may be employed for measuring directions, dips, slope profiles, and slopes both upward and downward from the observer's position; and, by virtue of the mirror, fairly accurate alignments may be made without the necessity of a support. It has no straight fiducial edges like those of the Gurley, but when the geologist wants such edges, he generally prefers a small alidade which may or may not have a compass attached to it. For most petroleum work, a Brunton is used when a compass and clinometer are required.

Locke level.—A Locke level is a short straight metal tube with a level attached near one end, and a mirror inside which reflects the bubble of the level. The mirror covers only half the inside cross-section of the tube, so that the observer can sight through the instrument at a distant object and, at the same time, see the reflection of the bubble in the mirror. When the instrument is in use, the level should be uppermost. When the bubble is bisected by the cross-hair, a distant object which appears to be opposite the middle of the cross-hair is at the same level as the observer's eye, provided the instrument is in proper adjustment. The geologist should know the height of his eyes above the ground.

Abney level.—An Abney level differs from a Locke level mainly in that the former has a vertical graduated arc and movable vernier. With the vernier set at 0°, it can be used just as a Locke level. The Abney can also be used as a clinometer. The observer, sighting up or down a slope at an object which is as high above the ground as his eyes, turns the spirit level until the bubble is centered. The angle of the slope may then be read directly on the graduated arc.

Telescopic hand level.—Like the Locke level, the telescopic hand level consists of a straight tube with spirit level and inclined mirror. It is also provided with lenses and stadia hairs. It has a focusing device for the distant object at the end away from the observer, and a focusing device for the stadia hairs at the end held nearest his eye. The stadia hairs are usually spaced on the ratio of 1 vertical to 100 horizontal. In other words, an object 5 feet high and 500 feet away will appear to be just included between the two outer (stadia) cross-hairs. If the horizontal distance between the observer and the object is known, the height of the latter can be obtained; and, *vice versa*, if the height of the distant object is known, its approximate horizontal distance from the observer can be found. By the use of a graduated rod, held at the distant station by a rodman, fairly accurate measurements of horizontal distance can be made. Less exact measurements can be made if the observer estimates the heights of objects near the distant point.

Planetable.—A planetable is a drawing-board mounted on a tripod in such a way that it can be rotated, while the table is set up, without moving the tripod. Boards 15 by 15 inches, or 18 by 18 inches, are commonly preferred. The paper for the map is attached to the board by thumb-tacks or thumb-screws.

Rods.—For planetable work, a stadia rod is used in connection with the telescopic alidade. This is usually a jointed wooden rod, 12 to 15 feet long when open, and 2¾ inches wide, graduated in feet and tenths of feet. It is carried by a "rodman." When held up for reading, it should be vertical, and its graduated side should be placed directly toward the instrument man who is looking at it through his telescopic alidade.

In some kinds of mapping, a level rod and a telescopic hand level are used. The level rod may be from 5 to 15 feet long. It may be provided with a sliding disk or "target" which can be moved up or down to the level of the instrument man's eye.

Open-sight alidade.—An open-sight alidade is a flat metal strip, beveled and graduated on one edge, and with a folding open sight at each end. When these sights are upright, one can look through them and bring the cross-hair of one sight into alignment with a distant object and the middle of the opening of the other sight. This is done while the alidade rests on a planetable. When such an alignment has been made, a line drawn on paper along the beveled edge of the base will be parallel to the direction to the distant object.

The open-sight alidade is seldom used in petroleum field work. A few geologists employ it with a small planetable for rapid reconnaissance mapping.

Telescopic alidade.—The telescopic alidade used by geologists is commonly of the Gale type. This is an instrument with a short base, about 11 by 2½ inches, and a short telescope. The right edge of the base is beveled and graduated to fiftieths of an inch. On this side is a tangent screw for raising and lowering the telescope. On the left side of the base is a magnetic needle which should be kept locked while not in use. The edges of the base are parallel to the vertical plane which contains the line of sight of the telescope. With the alidade resting flat on the planetable, the latter can be leveled by means of the round spirit level on the alidade base. A detachable striding level is employed for leveling the telescope. A graduated arc, moving with the telescope, indicates, by reference to a vernier, the angle through which the telescope is rotated. The alidade may be equipped with a Beaman stadia arc.

Within the telescope are three horizontal cross-hairs and one vertical cross-hair. The outer horizontal hairs are the stadia hairs. As in the telescopic hand level, they are spaced so that the distance between them just covers 1 foot on an object 100 feet distant, or 2 feet on an object 200 feet distant, and so on. This is referred to as " the interval," or, in leveling operations, as a "step." The space between the middle cross-hair and either stadia hair is therefore a " half interval " or a " half step."

A Stebinger drum may be attached to the tangent screw. This instrument is a small metal cylinder which is graduated in 100 parts. One complete revolution of the drum raises or lowers the telescope one interval.

Stadia reduction tables.—When the telescopic alidade is used to make an observation on a station lower or higher than the instrument the solution of a right triangle is involved. The level line of sight from the observer toward the distant station is one leg of the triangle. It is the distance which is plotted on the map in accurate work. The vertical difference in elevation between the observer's position and the distant station is the other leg of the triangle, and the inclined line of sight to the distant object is the hypotenuse. The instrument man reads the length of the hypotenuse and the angle between the level line and the hypotenuse. By reference to a " stadia reduction table," he finds the length of the vertical leg of the triangle for the given angle and a distance of 100 feet. This factor he multiplies by the number of hundreds in the length of the hypotenuse, as he has determined it. In geological work it is not customary to reduce the length of an inclined "shot" (hypotenuse) to its horizontal equivalent.

Stadia slide rule.—A useful instrument for planetable work is the stadia slide rule, which may be substituted for stadia reduction tables. The stationary part of the rule is graduated above for horizontal distances and below for vertical distances. The sliding scale is in angular degrees. Knowing the horizontal distance to a station, *A*, and the angle of the line of sight to Sta. *A*, the vertical distance of *A* above the observer's station, *B*, can be determined as follows: Set " 0," marked on the side of the sliding scale nearest the fixed scale for horizontal distances, against the value of the horizontal distance obtained. The vertical distance, on the other fixed scale, will then be against the observed angle. Note that when the horizontal distance is greater than the vertical distance, one fixed scale is for horizontal distances, and when the reverse is true, the other fixed scale is for horizontal distances.

Scale for plotting.—A scale for plotting is a straightedge graduated in intervals representing feet, yards, meters, or paces, on whatever scale is adopted for mapping. For scales of 1 inch to 1000 feet or 2000 feet, the most satisfactory for planetable work, the plotting scale is best graduated in inches and tenths of inches.

Aneroid barometer.—An aneroid barometer is a case containing a metal vacuum-chamber so tightly closed that air cannot enter it. A mechanism inside the case causes a needle on the outside to rotate and so indicate differences of pressure exerted on the vacuum-chamber. Consequently, the aneroid barometer registers variations in atmospheric pressure such as are due to meterological conditions and differences of elevation above sea level.

The needle turns above a double circular scale, one series of graduations representing inches of mercury of barometric pressure, and the other, corresponding feet of elevation. On some aneroids one of these scales is movable, so that any point on it may be brought opposite the needle. By this means the instrument may be "set" at any desired elevation at a particular station. This, however, is a practice which should be avoided, since the size of the graduations on the scale of feet is not uniform around the circumference. In all barometric work the "0" of the scale in feet should be kept opposite the "31" of the scale in inches.

An aneroid barometer is always a delicate instrument. It should therefore be handled with care and should be carried in such a way that it is jolted as little as possible. It should be read while it rests flat in the open palm of the hand, and should not be grasped.

Aneroids are made for various ranges of altitude. They may register up to 3000, 5000, 6000, 10,000 feet, or more. One that registers to 5000 or 6000 feet is usually most satisfactory in regions where elevations above sea level are not over 4000 feet. A barometer should not be used at elevations near the limits of its range.

For success in barometric work, the aneroid must be reliable. It must be neither too sensitive nor too sluggish. It is often necessary to try many before a satisfactory one is found. When the instrument is gently tapped on the glass face with one's finger nails or with a pencil, the needle should return to the same point at which it stood before being disturbed, and it should return to this point after successive tappings, provided the instrument is kept in the same place. The best way to test a new aneroid is actually to use it in field work for several days. Its individual peculiarities may then be pretty definitely ascertained.

Barograph.—A barograph may be described as a self-recording aneroid barometer. Differences in atmospheric pressure cause the vacuum-box to expand or contract, as the case may be, and these movements are imparted to an arm which carries a pen. This pen plays upon a piece of ruled paper which is wrapped round a drum revolved at a constant rate by clockwork. By the movements of the pen, due to the expansion and contraction of the box, a curved line is drawn, recording the variations in atmospheric pressure.

Watch.—The geologist should be provided with a good watch. For barometric work, in which both time and barometric elevation are recorded at each station, it is indispensable. A wrist watch is most convenient.

Protractor.—A 4-inch or 5-inch protractor is highly desirable for laying off angles. It is used most frequently, perhaps, in plotting compass traverses.

Hand lens.—A hand lens is not so often used in petroleum work as it is in many other kinds of geological work, but it should nevertheless be included in the oil geologist's equipment.

Steel tape.—The steel tape may be required for measuring base lines for triangulation, or for making other direct measurements which must be fairly accurate.

Camera.—A camera is now considered an essential part of the field geologist's equipment. Some prefer one of vest-pocket size provided with a high-class lens which will give sufficiently good definition to permit enlargement. Others choose a camera taking photographs of picture post card size.

Odometer.—In some parts of the country a great deal of geological mapping is done by automobile. Under these circumstances the automobile must have a speedometer with odometer registering the distance traveled in tenths of miles. By keeping watch of the trip gauge, the geologist can estimate distances to less than .05 mile, which is close enough for some kinds of reconnaissance work.

Pace counter.—Pacing is one of the common methods of measuring distances. Some geologists use a pedometer which operates automatically, but this instrument is very unreliable. If its spring is too weak, it may register two steps for every step taken. The best plan is to hold an ordinary tally register or pace counter in one hand and press the lever for every other step taken. With only a little practice, this becomes so much a habit that one does it without any thought. Allowing a little over 5 feet to a pace (double step), this makes about 1000 paces equal 1 mile, a very convenient quantity to use in approximate measurements of distance.

Other equipment.—There is no need to go into details concerning such other articles as pencils, erasers and thumb-tacks, all of which belong in the equipment of the field geologist.

The Hand-Level Method

The hand-level method defined.—In the hand-level method of field mapping, the hand-level is the instrument used for obtaining differences of elevation. Directions are determined by compass, and horizontal distances by pacing, odometer, or other rough methods of measurement, or by reference to the stadia hairs in case the telescopic hand-level is employed.

Equipment.—For this kind of work the geologist needs a hand-level, either sight or telescopic, a compass, a tally register or other instrument for recording distances, a notebook, a base map if such is obtainable, pencils, camera, etc. For short sights a telescopic hand-level is more accurate than a Locke level. For either instrument, the relative error is similar for short or long sights, but the absolute error is greater for long distances.

Determination of strike.—To find the approximate strike of a dipping bed, stand on the bed at a point from which this same bed can be seen at some distance, either across a valley or along a dip slope. Sight through the hand-level and, holding it always horizontal, swing it slowly around until it is directed to a distant point on the bed at the elevation of the observer's eye above the bed. This will be the approximate strike. Find this direction by compass.

Determination of dip.—One cannot always make sure whether a hand-level " shot " is in the direction of the true dip or of a component of the dip. For instance, a series of strata may dip east and strike north and south. If a hand-level sight were due east or due west the inclination found would be approximately the true dip, but if the sight were, let us say, northeastward, the inclination of the bedding found in this direction would be merely a component of the dip. If several dip components are determined, both as regards their amount and direction, a rough value and direction for the true dip may be estimated.

To find the dip, or dip component, of an inclined bed, stand on the bed and choose some distant point on this bed as the station to be observed. Select some object, of which the height can be judged, near the distant point. Regarding this height as a unit of vertical measurement, a "step," see how many times it will be contained in the vertical distance between the distant object on the bed and the horizontal line of sight from the observer's position toward the distant point. The approximate dip (or dip component) *in this direction* will be the number of " steps " times the height of the distant object, minus the height of the observer's eye above the ground. The dip will, of course, be expressed in feet per mile, the distance to the object having been estimated or being known.

With a telescopic hand-level this process is made somewhat more accurate by the use of the stadia cross-hairs. Also, on account of the spacing of these hairs (see page 179), the distance to the object can be found if the height of the object is known. The direction to this object is found by compass.

Application of the method.—Obviously the hand-level method cannot be used for anything but the crudest type of geological mapping. It is the least precise of the three methods employed for low-dipping strata. Its principal application is found in making relatively short branch traverses or " side shots " from more accurately mapped traverses or stations. In its rapidity of execution, it is excellent for quick reconnaissance work where only rough results are expected or required. It can be used to advantage, however, only in regions where dip slopes are clearly recognizable as such, or where low or moderately dipping beds of varying hardness outcrop in distinct scarps or benches, which can be traced by the eye for considerable distances along hillsides or across valleys without danger of mistakes in correlation.

In some cases geologists do extensive areal mapping by hand-level. For instance, stepped traverses, or level traverses, may be run in such a way that they tie in on one another at certain stations. Errors in tying at these points are corrected by dividing the total error proportionally among the several stations of the traverse from the last tie point to the one in question.

The Barometric Method

The barometric method defined.—In the term, " barometric method," stress is laid on the fact that the aneroid barometer is used for determining the elevations of stations. The directions between the stations are usually obtained by compass. The distances between stations may be measured by pacing, by counting the paces of a horse, or the revolutions of a wheel, by odometer when an automobile is provided, or by timing when the approximate rate of progress is known.

Equipment.—For the barometric method of mapping, the geologist needs, in addition to the aneroid barometer, a compass (preferably a Brunton), a hand-level, a watch, a notebook, pencils, pocket lens, some suitable means for measuring distances, and a base map, if such is available.

After trying several varieties of notebook, the writer has found that the most satisfactory kind is one ruled in $\frac{1}{8}$-inch coordinate squares, with every fifth line printed red and the other lines printed blue or black. The most convenient size is about 5 by 8 inches.

Any reasonably accurate map, such as a county, state or property plat, will do for a base map. A good topographic map is to be preferred, but this is not essential. If no map can be obtained, the geologist can either have a simple base prepared by plane-table or transit in advance of his work, or he can make his own base map along with his geologic work. The last plan is often necessary in foreign lands where the party is constantly moving forward, generally at a fairly rapid rate.

Recording the notes.—The notes may be kept in tabulated form or partly in the form of a geologic profile section. When there is no particular necessity for plotting a graphic section, the tabulation method is to be preferred since it is quicker; but if the correlation of strata can be facilitated by representing the outcrops in section, then the profile method should by all means be followed.

Below is a plan for tabulating the field data:

Station	Time,	Barometer	Corrected Barometer	Direction from last station	Distance from last station	Remarks
	a.m.				Paces	
A1*	8.00	2460		No.		Top of 5 foot bed yellowish Fusilina ls. Call L1
A2	8.10	2500		N. 10 E.	110	Top 10-foot brown sandstone. Call S1
A3	8.22	2520		No.	140	S1
A4	8.30	2490		No.	55	L1
A5	8.35	2460		No.	60	Change of slope

The fourth column is left blank in the field. It is subsequently filled in with the corrected elevations for the different stations.

The profile method of recording notes is illustrated by Fig. 17. Let us assume that the geologist is starting at Sta. A1 at 7 a.m., and that his barometer indicates an elevation of 1000 feet for this spot. Knowing that he is on a high point in the area which he is to examine, he writes "1000 feet" against a line rather high on the

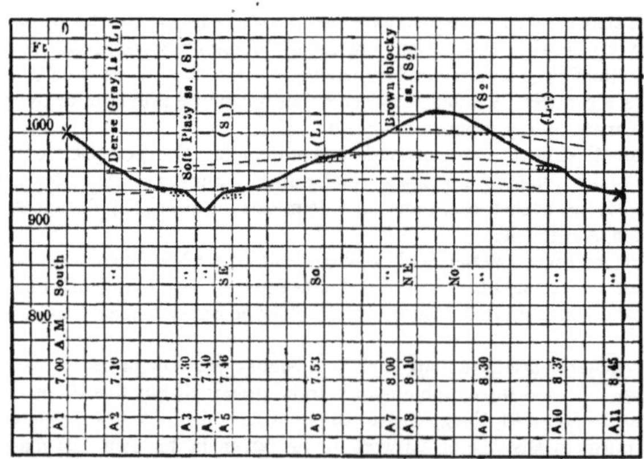

Fig. 17.—Illustration of the method of making profile section notes.

notebook page, thus leaving room for points of lower altitude. Holding the notebook with the length of the page right and left, and with a vertical scale of 1 inch = 100 feet, he labels the horizontal lines for 1100 feet, 900 feet, etc. The vertical lines are marked for distance, a scale of 2 inches = 1 mile, or $\frac{1}{5}$ inch = 100 paces being satisfactory.

* "A" represents the first day and "1" the first station on that day. "B" will be used for the second day, and so on.

At the intersection of the 1000-foot horizontal line and the "0" vertical line, he marks a point for Sta. 1 and, *along the vertical line* directly opposite the point for this station, he records the station name, "A1," the time, "7 a.m.," and the direction to the second station, here south. In barometric work it is usually desirable, if possible, to traverse north-south or east-west lines and not to zigzag this way and that. Outcrops off the course should be examined, but the geologist should return to his route and continue along it from where he turned aside; or, if it happens to be far to one side, he may make a right-angle traverse to it and then proceed parallel to his former course. The object in doing this is to facilitate the final interpretation of the profile sections. If the work must be rapid, time may be saved by going straight from outcrop to outcrop, provided they lie fairly near the line of progress.

To return to our illustration—let us suppose that the geologist goes 300 paces south and there, at 7.10 a.m., finds an outcrop of dense gray limestone with its top at an elevation of 960 feet. He records this by the customary symbol for limestone at the intersection of the horizontal line for 960 feet and the vertical line that represents 300 paces from Sta. A1. The other facts for this limestone outcrop, which he designates "A2," he writes along the vertical line that passes through the station point in the section (see Fig. 17).

He continues in this way, plotting stations and outcrops, sketching in the topographic profile, and recording all facts that may be useful in making out the geology of the region. Separate outcrops of the same bed, when known to be such, are connected by dashed lines.

When the profile section has been drawn to near the end of the page, it is stopped at a station, and this station is repeated at the beginning of the next page at a point corresponding to its position on the completed page. This makes possible bending back the new page to match the portion of the section on the lower page.

Fluctuations in atmospheric pressure.—If the weight of a vertical column of air were everywhere the same at the same elevation above sea level, and if it diminished uniformly from sea level upward, the geologist would have little trouble in determining elevations by the use of a barometer. However, atmospheric pressure is very inconstant. At any particular spot it fluctuates considerably and its variations are not the

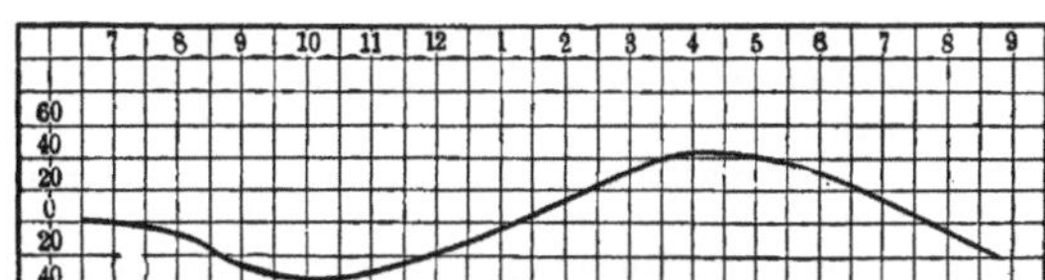

Fig. 18.—Curve of diurnal pressure variation.

same on different days, nor are they the same at different points of the same elevation on the same day. Under ordinary conditions, the barometer needle is falling [1] from some time before dawn until between 9 and 11 a.m., then rising until 4 or 5 p.m., and falling again until after dark. These changes are illustrated in Fig. 18, which may be called a curve of diurnal atmospheric pressure variation. If similar curves be plotted for different days, they will be found, with occasional exceptions, to be somewhat similar to Fig. 18, but not identical.

The regular curve of atmospheric pressure variation is often locally modified—some-

[1] Falling and rising of the barometer refer in this chapter to the movements of the needle as indicated in *feet*. When the needle shows successively lower elevations, expressed in feet, it indicates higher atmospheric pressures in inches of mercury.

times very much so—by weather changes, "twisters," thunderstorms, etc. A typical storm curve is shown in Fig. 19 at *s*.

The correction curve.—On account of the fluctuations in atmospheric pressure, barometric readings have to be corrected. This is accomplished by reference to an atmospheric pressure variation curve which is constructed to show how much should be added to, or subtracted from, the barometric readings to obtain the true elevation at each station. For brevity, we may refer to this as a "correction curve." Some geologists make a correction curve by averaging a large number of diurnal pressure variation curves plotted for different days. That this method may introduce considerable error may be seen if one compares an average curve of this sort with any curve for a single day. The only way to secure a satisfactory correction curve is to plot one for each separate day. The data used in plotting the correction curve may be obtained by various methods of checking, as explained below.

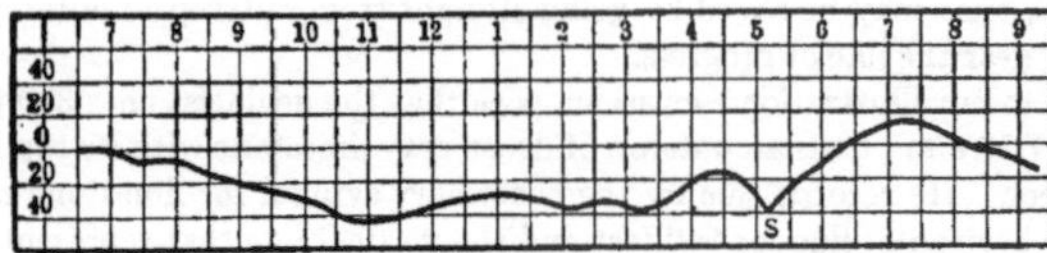

Fig. 19.—Irregular curve of diurnal pressure variation, showing a depression at *s* caused by a thunderstorm.

Methods of checking barometric readings.—1. Two or more readings at a station.—If the barometer is read more than once at the same place, any change in the position of the needle will indicate a change in atmospheric pressure during the time between readings. This exception should be noted, however, that if the instrument is read immediately after being brought to the station, the reading may not be correct within several feet, because two or three minutes are required for the needle to "settle." A reading just after a change in position of the barometer should be compared with a reading after it has remained undisturbed for some time. For satisfactory results at least fifteen minutes or half an hour should pass between readings at a single station.

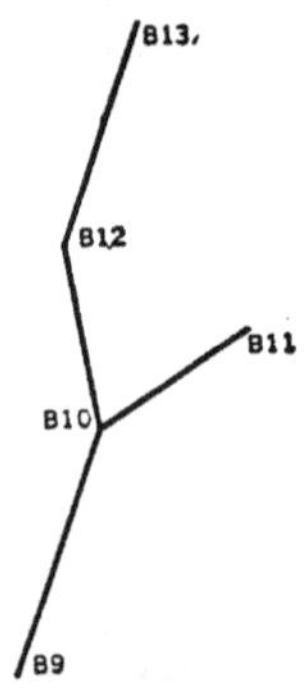

Fig. 20.—Map of a traverse. The geologist read his barometer at *B10* twice.

The geologist may go to a certain station, let us say B10, then continue to B11, and then return to B10 (Fig. 20); or he may proceed from B10 to B11 to B12 and so on, and later come back to B10, either by following a new course (Fig. 21), or by retracing his old course (Fig. 22). These three methods may be called checking by branch traverse, checking by closed or intersecting traverse, and checking by return traverse. These methods of repeating readings at revisited stations furnish the best means of discovering the fluctuations of the barometer.

2. Checking on established control stations.—Any point of which the elevation has been established, whether it is a station in a previous day's traverse made by the geologist, a Government bench mark, triangulation station, geodetic survey monument, or other fixed control point, may serve as a means for correcting barometric readings. The correction will, of course, be the quantity to be added to, or subtracted from, the barom tric reading to obtain the true elevation of the control point. Traverses should be planned to include stations of this description if possible, four or five times during a day.

3. **Checking by hand level.**—Both sight and telescopic hand-levels may be used to check the relative elevations of intervisible stations of approximately the same altitude. The geologist can level from his position to a point just left or to a point soon to be visited, and can compare the barometric elevations recorded at these two stations. When a hand-level check is made, it should be recorded in the notebook.

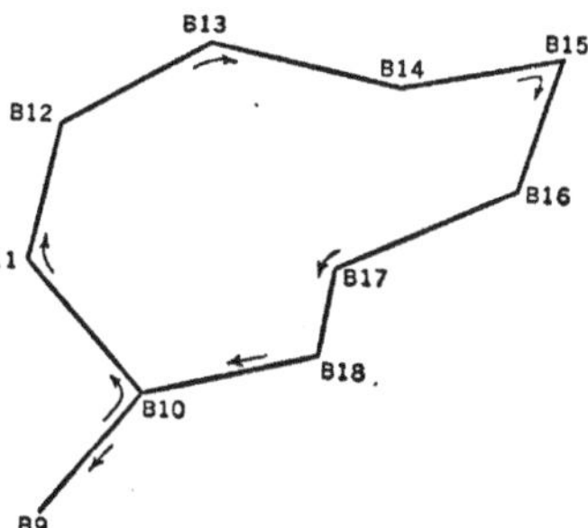

Fig. 21.—Map of a traverse. The geologist read his barometer at $B10$ twice.

4. **Checking by Eye.**—Always, during barometric work, the geologist should watch the topography and the structure, comparing by eye the slopes and dips traversed with the barometric elevations recorded. In this way he can usually detect sudden sharp rises and falls of the barometer, which might otherwise escape notice.

5. **Checking by reference to a second record.**—A not uncommon practice is to set a barograph somewhere near the middle of the area to be surveyed during the day, and

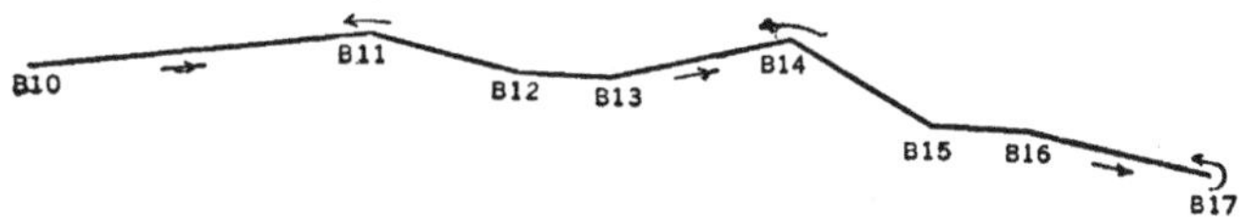

Fig. 22.—Map of a traverse where the geologist read his aneroid at each station, except Sta. $B17$, both going and coming.

to use its automatically plotted curve as a correction curve for the portable aneroid used by the geologist. Or an assistant in camp may keep a record of the readings of a second aneroid every fifteen minutes or every half hour through the day, and the curve of this camp aneroid may be employed, like the barographic record, to correct the field readings. Both of these schemes have the disadvantage that the stationary instrument may not catch sharp atmospheric pressure changes that may affect the field barometer, or *vice versa*, or if both instruments do record such changes they may not do so simultaneously. For this and other reasons, the writer believes that the other methods of checking, as above described, are to be preferred.

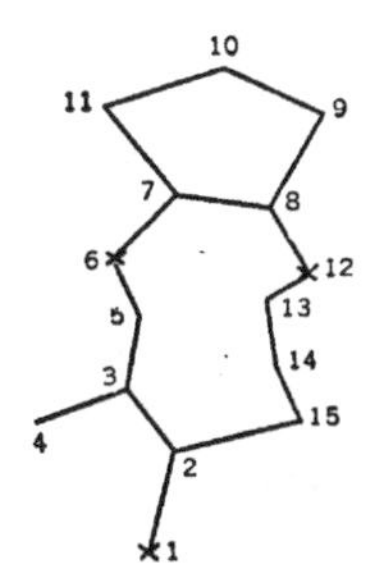

Fig. 23.—Map of a traverse, illustrating several methods of checking.

Plotting the correction curve.[1]—The following is a brief explanation of the plotting of the atmospheric pressure curve, or, as we have called it here, the *correction curve*. Let Fig. 23 represent a traverse in which several methods of checking have been employed, yielding the results listed in the table below:

[1] This description was taken from "The Barometric Method of Geologic Surveying for Petroleum Mapping," by Frederic H. Labee; Econ. Geology, XV, pp. 150–169, 1920.

A Station	*B* Time	*C* Barometer	*D* Corrected barometer	*E* Elevations previously established
	a.m.			
1	6.00	625	700	700
1	7.30	620	700	700
2	8.00	575	660	
3	8.20	580	665	
4	8.50	685	780	
3	9.15	565	665	
5	9.30	535	640	
6	9.40	525	630	630
7	10.00	540	650	
8	10.25	505	615	
9	10.45	520	630	
10	11.10	550	660	
11	11.45	595	700	
	p.m.			
7	12.15	550	650	
7	1.00	560	650	
8	1.15	535	615	
12	2.00	560	635	635
13	2.20	570	640	
14	2.35	590	655	
15	3.00	620	680	
2	3.30	605	660	
1	3.50	650	700	700
1	5.30	655	700	700
1	7.05	640	700	700

Camp is at Sta. 1 which happens to have its elevation already established as 700 feet above sea level. At 6.00 a.m. the barometric reading was 625 feet, or 75 feet too low. For plotting the curve, take a piece of coordinate-ruled paper and, holding it with its length right and left, label every other vertical line for the hours, beginning at 6 a.m. (see Fig. 24). Every space, right and left, will correspond to thirty minutes, and smaller intervals of time can be interpolated. Label a horizontal line near the middle of the page, " 0 "; label the next line, both above and below the 0 coordinate, " 20," the next " 40," and so on, as far as seems necessary. The horizontal coordinates below the "0" coordinate represent quantities, in feet, to be added to the barometric elevations to obtain corrected elevations, and those above the "0" line represent quantities to be subtracted from the barometric readings to obtain corrected elevations. Since the first reading, at 6 a.m., was 75 feet too low, 75 feet must be added for correction. Mark a cross (for an established station) at a point on the 6.00 a.m. vertical line one-fourth of a space up from the 80-foot horizontal line. At 7.30 a.m. the barometer read 80 feet too low. Mark the intersection of the 7.30 vertical line and the 80-foot horizontal line.

Between the two visits to Sta. 3, at 8.20 and 9.15 a.m., the barometer fell 15 feet. Somewhere well below the probable position of the curve to be drawn through the points just plotted, mark a point (m, Fig. 24) on a horizontal line at 8.20, and another point (n, Fig. 24) three-fourths of a space (=15 feet) below m on the 9.15 line. Draw a straight line between m and n. This line, mn, may be called a *guide line*. It is not necessarily parallel to the ultimate correction curve (cf. st and vw, two other guide lines), but merely indicates to the eye that points in the curve corresponding to the end of the guide line are related to one another exactly as they are in the guide line. This will become clearer presently.

At Sta. 6, at 9.40 a.m., the barometric reading was 105 feet too low. Mark a cross (again for an established elevation) where the 105 coordinate intersects the 9.40 coordinate.

At Sta. 7, readings were made at 10.00 a.m., and 12.15 and 1.00 p.m., 12.15 to 1.00 being the lunch period. Again, below the region of the curve, construct another guide line, opr, o being at the intersection of any horizontal line and the 10.00 a.m. vertical, p being 10 feet above o and on the 12.15 coordinate, and r being 10 feet above p and on

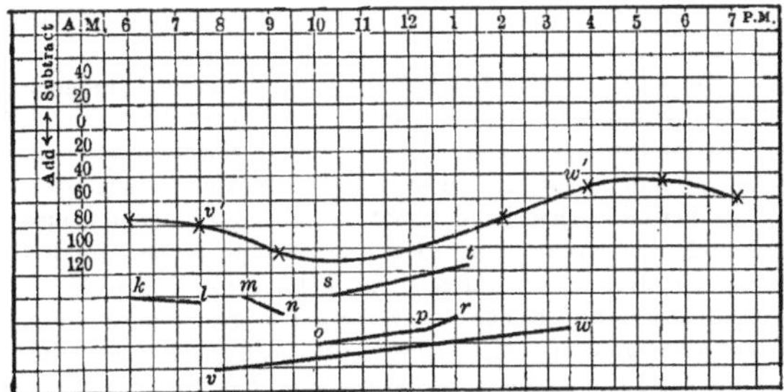

Fig. 24.—A correction curve for the barometric readings recorded on the traverse plotted in Fig. 23.

the 1.00 p.m. coordinate. The 10-foot intervals correspond to the rise of the barometer from 540 at 10.00 to 550 at 12.15, and from 550 at 12.15 to 560 at 1 o'clock.

Similarly, a guide line is plotted for Sta. 8, visited twice, and for Sta. 2, visited twice. Also, crosses are plotted at 2.00 p.m. and 75 feet for Sta. 12 (see columns C and E in the table), and at 3.50 and 50 feet, at 5.30 and 45 feet, and at 7.05 and 60 feet, these last three being readings at Sta. 1.

The correction curve is drawn through the plotted crosses, which indicate the actual corrections to be made, and in such a way that points in this curve are the same distances apart, both vertically and horizontally, as the end points of the several guide lines, vertically below. For example, v' is in the curve at 8.00 a.m. and w' is in the curve at 3.30 p.m., 30 feet higher than v'. These are exactly the relations of the end points of the guide line vw, but, to repeat what was stated above, the curve between v' and w' is not by any means necessarily parallel to the guide line vw.

Observe that two or more readings may be taken at the same station before starting in the morning and after returning in the afternoon. These additional readings are seldom made by geologists, but the writer has found them to be of considerable value in showing the slope and general tendency of the curve. Note also that the curve is drawn through the crosses, that is, through the points that indicate the corrections to be applied at stations previously established. The guide lines are obtained by checking. In this traverse (Fig. 23), checking was accomplished by reference to control stations at 1, 6, and 12, by closure at 2, 7, and 8, by branch traverse at 3, and by two readings with intervening wait at 7.

Use of the correction curve.—By means of the correction curve, constructed after the manner above described, any barometric readings made during the day may be corrected. To save time, only those stations should be corrected which are needed for making the structure contour map. If notes were kept by the graphic profile method, the corrected elevations may be written against the proper points in the profile section. If the notes were tabulated, the corrected elevations are to be written in a column for the purpose (see page 184).

To illustrate the way in which the corrections are made, let us take Sta. 14 (see table on page 188). The barometric elevation for Sta. 14 was observed to be 590 feet at 2.35 p.m. A point at 2.35 p.m. in the curve, Fig. 24, is just below the horizontal line that would represent "add 65 feet." Approximately 65 feet then, should be added to the recorded elevation, 590, thus giving the true elevation at 655 feet.

Where stratigraphic intervals must be added to or subtracted from the elevations of certain strata in order to obtain elevations on a chosen key bed (see page 170), this may be done at the same time as the correction of the barometric readings, and the *true elevation of the key horizon* may then be entered in the notebook, in the manner suggested above, ready to be copied on the map.

Frequency of checking for correction.—With a good aneroid barometer, the degree of accuracy of the results obtained by this method is largely dependent upon the frequeney of checking the readings and upon the spacing of the check stations. If checks are few, there is opportunity for missing evidence of unexpected atmospheric pressure changes. On the other hand, checks should not be too frequent, since too many readings retard the work and, rather than helping the process of correction, hinder it by introducing numerous unimportant but confusing small changes of 5 or 10 feet, such as may be due to lag, inexact reading of the barometer, etc. The error resulting from the smooth curve which disregards these frequent shifts is practically negligible. In the writer's experience, the aim should be to check the barometer in one way or another at intervals of half an hour to one hour. Intervals longer than two hours should certainly be avoided if possible.

Rate of mapping by the barometric method.—The rate at which barometric surveying can be accomplished is very variable and depends on such factors as the method of travel, the number and condition of roads, whether roads are very irregular and therefore need to be mapped in the absence of a base map or are laid out on section lines, whether the country is open or wooded, gently rolling or rugged, whether rocks are well exposed or not, etc. With conditions favorable, geologic structure can be mapped by barometer at the rate of 6 to 12 square miles per day, and with the relative error considerably less than unavoidable geologic inaccuracies. Work of this kind is properly classed as detail mapping, as contrasted with reconnaissance mapping.

By increasing the rate of travel, and taking readings and checking at longer distances, the limit of error is increased. The work may then be classed as reconnaissance. More than 100 square miles can thus be mapped in a day.

Applications of the method.—The barometric method may be used merely as an adjunct to planetable work to ascertain the elevations of side points which may not be visible from the main stations of a triangulation net or a stadia traverse (see page 197), or it may be used as the principal means of determining relative elevations of the stations in the area to be mapped.

In the former case, the geologist reads the elevation at a point, *A*, on the main traverse, and notes the time; then he goes to the side station, *B*, and records the elevation and the time; he then returns to *A* and again reads the elevation and time. The elevation of *B* is found with reference to that of *A*, making allowance for any rise or fall of the barometer during the time elapsed while the geologist is away from *A*.

As the principal means of ascertaining relative elevations, the barometric method is susceptible of considerable modification. A method of rapid reconnaissance, as used in China by Mr. M. L. Fuller and Mr. F. G. Clapp, has been described by Mr. Fuller in Economic Geology.[1] He summarizes the salient features of this method as follows:

" 1. A mounted guide or leader to head the party, select the route, and set a steady, even pace.

"2· A geologist, following immediately behind the leader, whose horse his own follows closely, making guidance unnecessary and leaving hands free for compass and barometric observations and note-taking.

"3· The taking of compass bearings to the *nearest* 5°, *while in motion.*

"4. The direct plotting of courses by *standard coordinates while in motion.*

"5. The determination of distances *by time while in motion.*

"6. The sketching of culture and topography *while in motion.*

"7 The determination and plotting of geology, usually in motion, but stopping when necessary.

"8· The inking in of map and notes, and the elaboration of the latter in the evenings."[2]

For a complete description of this method the reader is referred to the original article.

In the Mid-Continent field of North America, the present writer has made it a practice, in reconnaissance work, to drive by automobile along all the roads, establishing control stations at recognizable points, such as road corners, gates, windmills, etc., and watching for variations from the known normal dip of the region. Whenever such a variation was suspected, a side traverse was made, either on foot or by automobile, far enough and in a direction to prove or disprove the existence of abnormal dip. If abnormal dip was found, the area was reported for detail examination.

In the Red Beds country of southern Oklahoma, where roads are usually on the section lines, profile barometric traverses were run on the roads, in reconnaissance work, and any uncertain correlations were checked by foot traverses from road to road across the section.

Barometric work is ordinarily done by one man. When two or more men are assigned to the same area, the geologist with the best barometer should set the control, and all traverses should be tied in on his stations. The traverses of the two or more men should be planned to intersect or meet three or four times a day.

The Planetable Method

Definition.—In geological mapping by planetable, directions and horizontal distances are determined principally by the use of an alidade, and the map is plotted in the field.

Equipment.—The equipment for this method includes a planetable, a telescopic alidade, usually of the Gale type, a stadia rod, plotting scale, 3H or 4H and 7H or 8H pencils, eraser, several sheets of planetable paper, two notebooks, stadia slide rule or stadia tables, a dozen thumb-tacks, and a hand lens. The planetable, alidade, etc., are carried by an " instrument man." The rod and one of the notebooks are carried by a " rodman," who, in common practice, is the geologist himself. He should also have a Brunton compass and such other articles as he may need for geological work.

Paper and notebooks.—For planetable mapping, each of the two men is provided with a notebook. The geologist records the numbers and descriptions of the stations which he visits, and keeps full note of all geological information.

The instrument man's notebook has its pages divided into vertical columns for recording such data as may be included under these headings: station number; level reading

[1] "Quick Method of Reconnaissance Mapping." Econ. Geology, XIV, pp. 411–426. 1919.

[2] Op. cit., pp. 412–413.

(telescope level); vertical angle of line of sight; positive or negative angle; distance; difference of elevation; height of the instrument (H. I.); elevation of the distant station; and any other important matters.

The paper for the planetable should be hard and slightly tinted, with a surface that is a little rough. It should be fairly heavy and should be as free as possible from the tendency to warp or shrink.

Definitions of terms.—The *elevation of a point* is the vertical distance of the point above mean sea level. Sometimes, where the geologist cannot readily obtain the elevation above sea level, an arbitrary horizontal datum is chosen, and all the vertical readings in the work are referred to this.

The *height of the instrument*, or *H. I.*, is the vertical distance of the horizontal axis of the telescope of the alidade above mean sea level, or above any assumed horizontal datum.

Preparing for the work.— Before the actual field mapping is begun, the paper should be mounted on the planetable. It is customary to fasten several sheets on the board, one above another, so that there will be plenty for the work. These sheets should be ruled in coordinate squares on the scale adopted for the map. Common scales are 1 inch = 1000 feet, 1 inch = 2000 feet, 1 inch = 4000 feet, 1 inch = 1 mile, 2 inches = 1 mile, and 4 inches = 1 mile.

True north should be assumed to be parallel to one set of coordinates; the other set of lines will then run east and west. Through the intersection point of two of the coordinate lines, near the middle of the sheet, a line should be ruled parallel to the magnetic north-south line. This is the declination line. It may be found in several ways, as follows:

1. If the land is sectionized and roads follow the section lines, choose a stretch of road which can be seen for 2 or 3 miles. This condition is best realized where the road crosses a broad valley. Set up the planetable and roughly level it by eye. Then place the alidade near the middle of the board, with its beveled edge coinciding with a north-south coordinate, and, by use of the round spirit level on the alidade, carefully level the board. Clamp the angular motion of the board. Do not move the alidade. Center the vertical cross-hair on a point in the road 2 or 3 miles away, this point being in a position, across the road, corresponding to the position of the planetable. Release the magnetic needle of the alidade. Rotate the alidade slowly about a coordinate intersection point near the middle of the sheet (clockwise west of a line from northernmost Lake Superior to easternmost Georgia [1]) until the needle swings freely and points to magnetic north. Draw a line along the beveled edge of the alidade. This is approximately the declination line for the locality, and the angle between it and the true north-south coordinates is the angle of declination.

2. Where there are two intervisible Geological Survey monuments or other established stations, both of which are already plotted on the sheet, set up the planetable at one of these stations, carefully leveling it as described in the preceding case. Clamp the angular motion. Place the alidade with its beveled edge along the line connecting the two established station points, and rotate the table horizontally until the distant station is in line of sight of the alidade. Clamp the horizontal motion of the board. Release the magnetic needle. Rotate the alidade about a coordinate intersection point near the middle of the sheet until the needle points to magnetic north. Draw the declination line along the beveled edge of the alidade.

3. For the third method, one should know the constellations, Cassiopeia, Ursa Minor or "The Little Bear," and Ursa Major or "The Great Bear," and the stars, Delta Cassiopeia, Polaris or the Pole Star, Mizar, and the "pointers" in Ursa Major (see

[1] For information on the declination and the movement of isogonic lines at any locality in the United States, refer to the Isogonic Chart published by the U. S. Coast and Geodetic Survey.

Fig. 25). Find out the approximate time of night at which Delta Cassiopeia, Polaris, and Mizar will appear to be in a vertical line. Either Delta or Mizar may be above Polaris. A little while before this time, go out and carefully level the planetable, and place the alidade with its length exactly parallel to a north-south coordinate on the paper, and with its objective end supported on a rectangular block of wood of which the right and left edges are exactly perpendicular to the length of the alidade. Rotate the board until, by raising and lowering the telescope, Polaris is brought into the line of sight. Clamp the horizontal motion. Then raise and lower the telescope until the time when both Polaris and the "guide star," i.e., either Mizar or Delta, as the case may be, may be brought successively into the line of sight. The north-south coordinates on the paper are then lying due north and south. Place the alidade flat on the board, removing the block of wood, and, with the table steady, rotate the alidade about a coordinate intersection point near the center of the sheet until the released compass needle points to magnetic north. Draw the declination line along the beveled edge of the alidade.

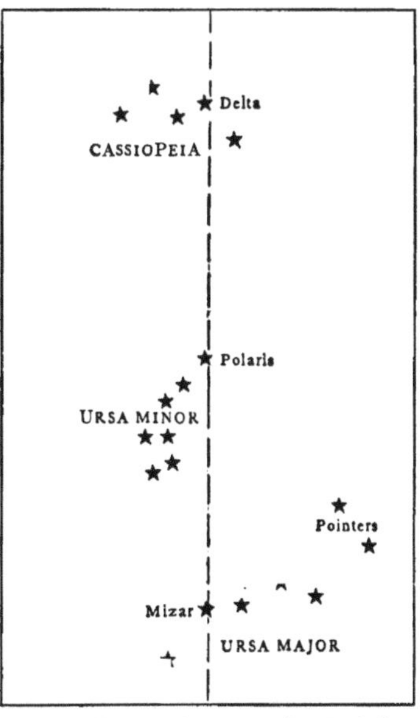

FIG. 25.—Constellations near the North Star (Polaris) used in finding the declination.

Setting up the planetable.—In ordinary planetable work, the table must be leveled, centered, and oriented at each station. These three operations together may be called "setting up" the planetable.

In petroleum mapping, leveling is usually done by means of the circular spirit level on the alidade, as explained in the foregoing article.

Centering the table means bringing the point for a station on the map over the corresponding point on the ground. Since, in petroleum work, stations are seldom occupied twice, centering is usually unnecessary. If by any chance it becomes necessary, it may be accomplished by eye without introducing any serious error.

Orienting is generally done as follows: Set the alidade on the planetable with its beveled edge exactly along the line for magnetic north (the declination line), and with its objective end toward the north. Then rotate the table in a horizontal plane until the compass needle points north.

Orientation may also be performed by backsight. Set up the table over the new station, *B*, with its north edge toward the north; lay the alidade with its beveled edge against the line drawn from the last station, *A*, to the one now occupied, *B*; and turn the board until *A* is brought into the line of sight ot the alidade.

Method of determining horizontal distances.—Horizontal distances are obtained by the Gale alidade as follows: Suppose that Sta. *A* is to be occupied and the distance from *A* to *B* is to be found. Set up the table at *A* and, with the beveled edge of the alidade cutting *a*, the point on the map for *A*, rotate the alidade until *B* is in the line of sight. Set one stadia hair on some foot mark on the rod (held vertically at *B*) and count the number of feet on the rod intercepted between the two stadia hairs. Multiply the intercept by 100. If the rod appears less than the distance between the stadia hairs, read the intercept on it between one stadia hair and the middle cross-hair. Multiply this intercept by 100×2. The result in each case will be the distance AB.

When distances are determined, they should be entered in the notebook in the column headed " Distance." For distances of more than a mile, correction should be made for refraction, curvature, etc. Explanations of these matters will be found in any good textbook on surveying methods.

Methods of determining elevations.—Several methods of determining elevations are described below:

1. Whenever possible, the elevation of a distant station, B, should be determined by a *direct level reading* on the rod as seen from A, the station occupied. Having oriented the table and the alidade, place the striding level on the telescope, and level the latter. If the middle cross-hair falls on the rod held at B, read the length of the rod between this cross-hair and the base of the rod. This interval is to be subtracted from the known elevation of the instrument (H.I.) for foresight readings, and added to the elevation of the distant station for backsight readings, in finding the unknown H.I.

2. The elevation of B may be calculated from the vertical angle between the horizontal line of sight through the telescope and the direct line of sight from A to B (see Fig. 26). Set the middle cross-hair on a convenient foot mark of the rod, preferably

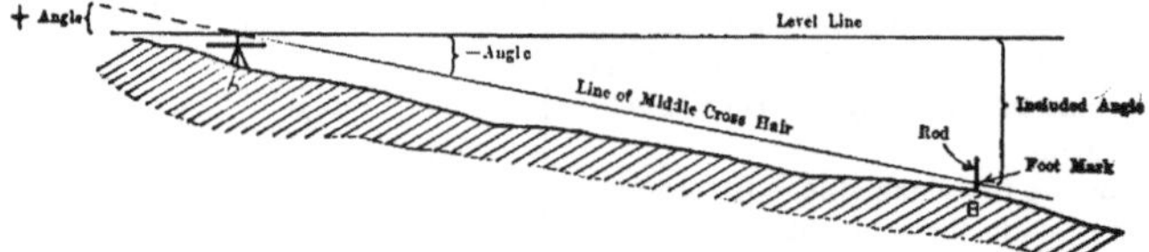

Fig. 26.—Section illustrating method of obtaining the elevation of a distant station.

the base. Record this foot mark in the notebook. Wave the rodman off the station. Read the angle of inclination of the telescope on the vertical arc, recording this under " Angle." Level the telescope, read the vernier again, and record this reading under " Level." Angles downward from the horizontal are called negative and angles upward from the horizontal are positive. Algebraically subtract the plus or minus angle from the *level* reading, thus obtaining the included angle (Fig. 26). By reference to a stadia conversion table or by using a stadia slide rule, one can find the difference of elevation corresponding to the angle for a distance of 100 feet. Multiply this figure by the number of hundreds of feet in the distance from A to B to get the difference of elevation between these stations. If the base of the rod was not used, the calculated difference of elevation must be corrected accordingly. Record the corrected quantity under " Difference of Elevation " and mark it plus (+) if B is above A and minus (−) if B is below A. Add this number algebraically to the H.I. at A to obtain the elevation of B.

3. The third method of finding the elevation of a point is called " stepping." Level the board, place the alidade, sight at the rod, and draw the alignment, ab. Determine the distance. Suppose that B is higher than A. Level the telescope by striding level. Select some object, such as a bush, fence-rail or rock, which is on the upper stadia hair. Move the bottom stadia hair to it. This is a " step." Repeat, keeping count of the " steps," until the lower stadia hair falls on the rod. Read the rod and signal the rodman off. The elevation of B will be the number of " steps " multiplied by the interval for the distance AB, less the rod reading. If N = the number of " steps," I = the vertical interval of the middle hair to the stadia hair for the distance AB, R = the rod reading, and HI = the height of the instrument above sea level, the elevation of B will be

$$HI+(N \times I)-R.$$

In "stepping" down, use the upper instead of the lower stadia hair. The formula then becomes

$$\text{Elevation of } B = HI - (N \times I) + R.$$

"Stepping" is not very accurate, and it should therefore be used only for side shots and not for control stations.

4 and 5. Two other methods of finding elevations are by the Beaman stadia arc and the Stebinger drum. For these the reader is referred to books on surveying.

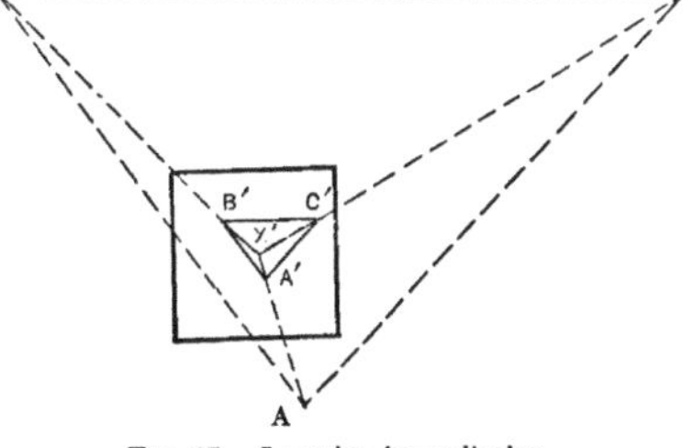

FIG. 27.—Location by radiation.

Methods of locating stations.—Stations may be located in four ways, namely, by radiation, traversing, intersection, and resection.

1. Fig. 27 illustrates location by radiation. A, B and C are points to be located. Level and orient the plane-table at any point, which we shall call X, from which the stations are visible. Mark a point, X', for X, on the map. With the beveled edge of the alidade against X', sight at the rod, held at A by the rodman, and draw a line from X' toward A. Determine the distance, XA (see page 193), and lay off this distance, on the scale of the map, from X' toward A. Mark A' on the map. Find the elevation of A (see page 194) and write this elevation beside A'.

In the same manner, determine and lay off $X'B'$ and $X'C'$.

2. The second method of location is traversing. Let A, in Fig. 28, be the station occupied, and let B, C and D be stations to be included in a traverse. Level and orient the table. Orientation is usually done by backsight from the station last occupied (see page 193). Let A' be the point for A on the map. Place the alidade with its beveled edge cutting A', and center the vertical cross-hair on the edge of the rod held at B. Draw a line from A' toward B. Find the distance, AB, and record it in the notebook under "Distance." Set the telescope for a plus or a minus vertical angle, wave the rodman off the station, and determine the elevation of B as explained under paragraph 2 on page 194. Record this in the column for "Elevation." Scale the distance $A'B'$ and record the elevation opposite B' on the map.

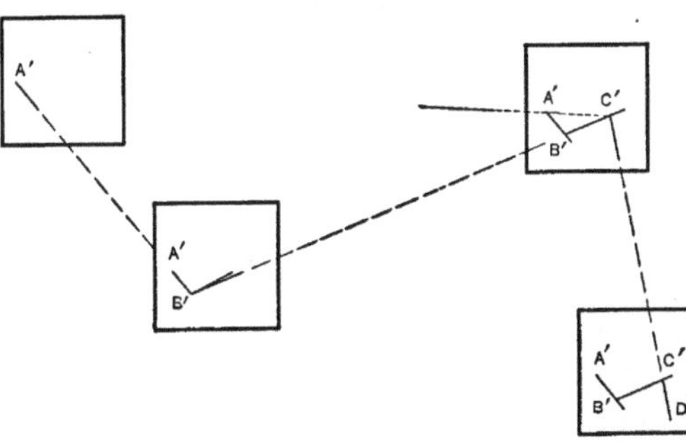

FIG. 28.—Location by traversing.

The rodman proceeds to C while the instrument man goes to B. At B, the alignment and distance to C and the elevation of C are found, as they were for B. In this way the advance is continued from station to station.

3. For the method of intersection, the careful measurement of a base line is necessary. This line should be chosen between two intervisible stations, where the land is as nearly level as possible. Let AB (Fig. 29) be this line. Set up the planetable at A, orienting by compass or by backsight on some established station. Sight at B and draw a line from A', the point plotted on the map for A, toward B. Sight and draw lines from A' toward other visible stations, C, D, E, etc. Read and record the vertical angle to these stations, including B, and calculate their elevations. Measure AB. Scale off $A'B'$ on the map. Set up the planetable at B, orienting by backsight on A. Sight at C, D, E, etc., and draw lines from B' toward these stations. The intersection points of these lines and the lines drawn from A' toward the corresponding stations should mark the positions of these stations on the map. The correctness of any locations made thus by intersection may be checked by sighting from a third station. The elevation of any station, as, for instance, C, may be checked by reading the vertical angle to this station from a second and a third station and calculating the elevation in each case.

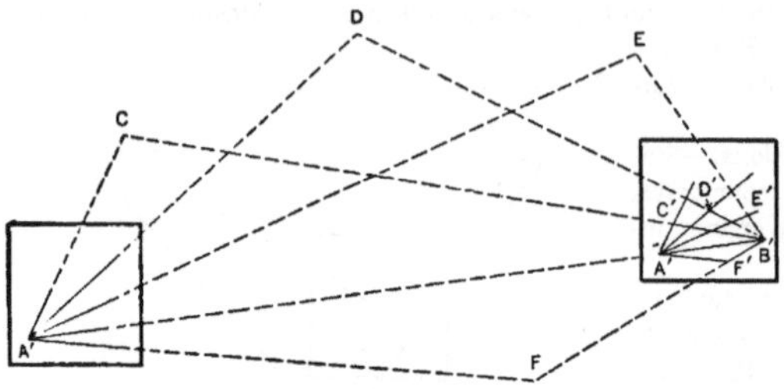

FIG. 29.—Location by intersection.

By proceeding from station to station and determining the locations of new points by two-line or three-line intersections from old points, a network of triangles is built up. This is called a triangulation net. To check the accuracy of the triangulation net, tie it on to a second measured base line at the far end of the net.

4. By resection, the geologist locates his position in reference to three or more established stations, all of which are visible from the station which he occupies. Let D be his position, and let A, B, and C (Fig. 30), be three stations already located and plotted on the map. Level and orient the table. Keeping the beveled edge of the alidade on A', the plotted point for A, rotate the alidade until A is in the line of sight. Draw a line from A' away from A. Take the vertical angle to A and record it. Repeat for B and C, always with the table in the same position. The lines from A', B' and C' will either intersect in a common point, which is the position of D', or they will cut one another so as to form a small triangle, called the *triangle of error*. The various methods of correcting for this triangle of error may be found described in books on surveying. They will not be explained here.

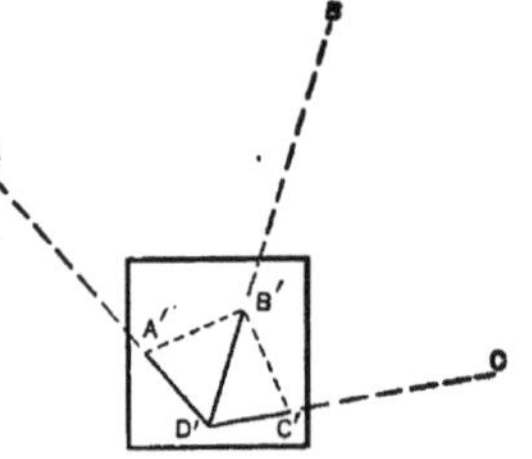

FIG. 30.—Location by resection.

A simple plan for avoiding the triangle of error is as follows: Level and orient the table at D, the geologist's position. Pin a piece of tracing paper on the board. Draw a north-south line on it. Mark a dot, D', near the middle of it for D. With the alidade edge cutting D', sight successively at A, B and C, and draw lines $D'A'$, $D'B'$, and $D'C'$. Observe and record the vertical angles to A, B, and C. Unfasten the tracing paper and slide it above the planetable map, with its north-south line correctly oriented, until the three lines, $D'A'$, $D'B'$ and $D'C'$, pass through A', B' and C', respectively, on the map. Then prick D' through on to the planetable sheet, and mark this point, D', for the plotted position of D.

From the recorded vertical angles measured from D to A, B and C, of which the elevations are known, the elevation of D can be calculated.

Resection, the reader will see, is really location of the new station by backsight on two or more of the stations already plotted.

Comparison of the methods of location.—Intersection and resection are together known as *triangulation*. Intersection is seldom used by the geologist except occasionally for the location of single stations. The method is too slow, for several readings must be made on every station in the net. In some kinds of open country, and in regions which have not been surveyed, a triangulation net may be made by engineers in advance of geological work. The engineering party surveys the land, builds monuments or sets flags at conspicuous triangulation points, and prepares an accurate map of the system of stations. Then the geologist goes out, provided with a copy of this map, and, whenever he wants to locate himself, he does so by resection from three or four visible stations of the triangulation net. He obtains distances from the maps by measuring the intercepts of intersecting lines, and elevations are calculated from vertical angle readings on the monuments or flags of the triangulation stations. An objection to this method is that the geologist, unless he has an assistant, is encumbered with a planetable and alidade. To lighten the equipment as much as possible, however, a small, light planetable and a Gurley compass (alidade) or other open-sight alidade may be used.

In traversing and radiation, the geologist is usually rodman. The instrument man maps his own course as a traverse and maps the side positions of the rodman by radiation. This method, combining traversing and radiation, is called *stadia traversing*. As compared with triangulation it has these advantages:

1. The plan of the course is more open to modification than in the case of the triangulation net. It is more readily applicable to the tracing of outcrops.
2. Ordinarily each station in stadia traversing needs to be visited only once.
3. Both distances and elevations are obtained from direct rod readings.

On account of these advantages, stadia traversing is the method almost universally practiced by petroleum geologists in planetable mapping.

Application of the planetable method.—The planetable method is capable of less variation than the barometer method. However, in reference to the stadia traverse only, some variation is possible both in the planning of the route and in the number of men in a party. As previously stated, the customary planetable party consists of an instrument man and a rodman, but sometimes there is a larger party. There may be an instrument man, a rodman, and a geologist, or an instrument man, two or three rodmen, and a geologist, who will then be chief of party. The geologist should make a rapid preliminary survey of the area in order to plan the best method of surveying it. This may take one or two days before the planetable work can begin. Then, with his party in the field, he ought to be able to keep instrument man and rodmen busy while he, himself, scouts ahead of the mapping, preparing for the advance of the party. At the same time he will have to supervise their work while it is in progress.

The Method of Compass and Clinometer

Definition of the method.—The three methods of field mapping, above described, are applied principally in regions where the dips are low and where they are determined by finding and plotting differences of elevation at different points on a given stratum, called the key horizon. Where the strata are inclined more steeply, their dips are more conveniently obtained by clinometer and their strikes by compass. In the name, *method of compass and clinometer*, then, emphasis is again placed on but one phase of the work, i.e., the determination of the attitude of the strata.

Equipment.—The most satisfactory compass for this method is the Brunton pocket transit with mirror and sights (page 178). The compass should be set for magnetic declination before any field work is begun.

In addition to the compass, with its attached clinometer, the geologist should be provided with a tally register, a notebook, a hand lens, pencils, etc., and, if possible, he should have a base map of the area to be surveyed. A simple hand-level may often prove to be useful. If he cannot procure a base map, he may have one prepared either in advance of, or along with, the geological work. In any case he will need some sort of a map on which to plot the outcrops, formations, etc., in their relations to established control points.

Traverses.—Traverses are generally run from outcrop to outcrop, the directions being obtained by compass and the distances, paced and recorded by tally register. Such traverses should be tied in on established control points every few miles. Errors of distance may then be divided proportionally between the several sections of the part of the traverse between two control stations. Where work must be rapid, the traverse should be planned to include not only such outcrops as are necessary for a rough determination of the structure. Time should not be taken to visit others. If, however, the work is to be more carefully done, all outcrops should be visited, dips should be determined and recorded at frequent intervals, and all formations should be mapped.

Recording observations.—Dips and strikes of strata are plotted by the customary T-shaped symbol (see page 172). Formation boundaries are sketched in on the map, and the several formations are represented by colors or special patterns. Geological information may be written on the map or in the notebook.

Variations of the method.—As in the case of the other three methods of instrumental work, the compass and clinometer method can be varied to meet conditions of speed, degree of precision required, topography, etc. Whatever plan is followed, the work should be tied to fixed control points. Variations in the accuracy of the results depend upon the system used in locating the stations. If a high degree of accuracy is demanded, the stations should be located by planetable and alidade, in respect to their horizontal and vertical positions. In this event, the geologist, as rodman, ascertains the dip and strike of the beds at each outcrop selected for a rod station.

If somewhat less precise results are wanted, the geologist may carry an aneroid, with which he will read the elevation of each station, these elevations being subsequently corrected by some approved method (see page 185). He will measure distances by pacing, horse paces, odometer, or in any other way that may best serve the purpose. The directions of his traverse he will obtain by compass.

Provided dips are pretty steep and the topographic relief is not very strong, the relative elevations of the stations may be neglected. Distances measured on the ground are not reduced to their horizontal equivalents. They are plotted and corrected be the proportional method noted on page 183. In this manner considerable time may be saved, and the errors introduced in the location of stations will not be very different from probable errors due to variations in the dip and strike of the strata. This method is useful for fairly rapid mapping in areas of high dips.

Choice of Method

Sources or error.—In all kinds of field work the geologist should ascertain the sources and limits of error which may affect his results. This is as important in geological surveying as it is in any other branch of engineering.

The sources of error may be classified under four headings, namely:

(1) Instrumental errors.
(2) Errors in method.
(3) Errors due to geologic conditions.
(4) Personal errors.

1. Instrumental errors are such as may be caused by defects in manufacture or by imperfect adjustments. Many instruments, which may have been in perfect condition when purchased, are likely to fall out of adjustment after use. Consequently, the geologist ought to know how to make simple adjustments himself. If he cannot put an instrument in proper condition, he should send it to a responsible firm for inspection. At least he should be able to test his instruments to see whether or not they are in adjustment.

2. Errors in method are due to inexperience, lack of training, or carelessness. We may also include here errors which may be due to taking dips on cross-bedding or slump instead of on true bedding, and other similar mistakes.

3. We may define as errors due to geologic conditions those indeterminable and variable quantities consequent upon such geologic factors as the number and extent of outcrops, the amount of settling or slumping (see page 170), the kind of weathering which affects the rocks, variations in the intervals between beds (see page 170), lensing of beds, and the degree of certainty with which the strata can be correlated. These are factors which, to a greater or less extent, may distinctly modify geologic structures as these are mapped from surface observations. The geologist should always have his mind open for the possibility of these errors, and should endeavor to reduce them by any reliable subsurface data that he can obtain.

4. Personal errors are due to those factors commonly classed under the name of the " personal equation." They are largely dependent upon the man himself.

Limits of error.—We cannot enter into a full discussion of the limits of error involved in the different kinds of geological field mapping. In general, the allowable limits of error are least in planetable work and greatest in hand-level and compass mapping. This amounts to saying that, provided the work is reasonably well done, planetable mapping is most accurate, barometer mapping less so, and mapping by hand-level and compass least accurate. Precise limits are rather hard to define, but ordinarily the error in planetable work should not be over 3 to 5 feet vertical and 200 feet horizontal, no matter what the length of the traverse may be, since these errors are largely compensating. These figures are for scales of 2 to 4 inches to the mile. The smaller the scale, the larger the actual limit of error may be. In barometric work, the permissible error in relative elevation should not be over half a contour interval. Errors in measurements of horizontal distance may vary according to the method employed. For pace and compass traverses it ought not to be over 3 per cent. Automobile traverses, where measurement is by odometer, may be between one-twentieth and one-tenth of a mile in error, if the odometer is properly regulated. The relative error in odometer measurements is naturally greater for short distances.

Choice of field method.—Several factors must always govern the choice of the method of field mapping to be used in any particular area. Such factors are:

(1) The general structural conditions.
(2) The vegetation.
(3) The topography.
(4) The roads and the method of travel.
(5) The rate at which the work must be done.
(6) The number of men available.
(7) The accuracy desired and the accuracy possible.

1. By general structural conditions is meant especially the kind of folding and the prevailing amount of dip—whether high or low. If the dips are generally low, any of the methods except the compass and clinometer method may be used. Under these circumstances, dips should not be read by clinometer, even if they are visible, because slump, settling, and cross-bedding may lead to serious errors.

2. The nature of the vegetation is important. If the country is largely covered by a thick growth of shrubs or trees, the planetable is at a disadvantage. Planetable work requires the intervisibility of stations. For a similar reason, the hand-level is not satisfactory. The barometric method and the method of compass and clinometer are to be preferred in heavily wooded districts.

3. The type of topography is not so important in governing the choice of method as is the vegetation. The planetable method gives most satisfaction in open country with considerable relief and conspicuous, easily traceable scarps. Under these conditions the barometer and hand-level methods are also successfully applied. Where the country is open and gently rolling, with few outcrops, either planetable and alidade or barometer and compass may be used, but the hand-level is less suitable.

4. The number, position, and condition of roads in an area to be surveyed help to control the choice of method. If there are many roads, an automobile is the best means of travel, and rapid work can be done by barometer, provided always that outcrops are sufficiently numerous and correlation is possible. If roads are few or very poor cross-country traverses may have to be made on foot or on horseback. The barometric method can be used in either case, but planetable work is best done on foot.

5. A fifth factor is the rate at which the work has to be done. Sometimes a quick judgment must be passed as to whether or not certain acreage is structurally favorable. The geologist must then use either the barometer or the hand-level, in low dipping beds, and endeavor to make a reconnaissance as rapidly as possible without overlooking any large or important structures. If more time is available, the work may be done less rapidly either by a more careful barometer survey or by planetable.

6. The number of men who may be used on an assignment is the next factor to be considered. One man, well grounded in the barometric method, can make a good map of an area as quickly as two men can map the same area by planetable. For this reason the barometric method may be preferred.

7. A most important point is the degree of precision attainable in a region, and the degree of precision desired in the mapping. We have already referred to the limits of error in geological mapping (see page 199). Granted that errors due to geologic conditions must always exist, the question arises as to what degree of accuracy is necessary or advisable in the field work. Some geologists hold that detail mapping should be as precise as possible, and they therefore use the planetable in nearly all their detail work. Others maintain that the precision of the field mapping should be comparable with—perhaps a little greater than—the limit of error in the geologic conditions. In other words, they think that time and money are often wasted in an attempt to make a very accurate surface geological map, where considerable errors are undoubtedly present on account of the underlying geologic conditions.

To a certain extent the writer is in sympathy with this idea. There is a satisfaction in having a nicely "tied" planetable map; but if reasonably accurate work can be done by barometer by one man, where two men would be required to map the area by planetable, and where the subsurface error would be just as large in either case, it seems that the barometer method may be justified for what is usually known as detail mapping.

In reconnaissance work less accuracy is demanded. Here the barometer can be used and the rate of travel can be increased; or the hand-level and compass may be employed.

OIL-FIELD DEVELOPMENT AND PETROLEUM PRODUCTION

BY

LOUIS C. SANDS [1]

The purpose of this chapter is not to furnish a compendium of all theoretical or even practical oil-field information, but to select and discuss only those methods and fundamental principles that have met with general approval and are responsive to modifications necessary to suit the practice of the operator and the requirements of the locality.

These discussions will, therefore, be limited to American methods and will necessarily be suggestive rather than exhaustive. This limitation does not imply criticism of other systems. The unbiased critic will readily concede that America's past and present preeminence in the petroleum industry may be attributed largely to the use of efficient machinery and tools and the application of correct principles and methods.

METHODS OR SYSTEMS OF DRILLING

The objective sought in oil-field drilling is to tap or penetrate a stratum that will produce petroleum or natural gas in commercial quantities, exclude all water from the producing sand and complete the work in the shortest possible time; but the obligation of the driller is fulfilled when he has drilled a hole in good order to a required depth, whether production is obtained or not.

Assuming that the drilling site or location for the well has been determined, the next problem involves consideration of the formations to be penetrated and selection of the proper drilling tools and machinery to be used. Geological factors constitute the determining influence.

Beginning with America's pioneer commercial producer, the Drake well, drilled near Titusville, Pennsylvania, in 1859, the cable-tool system (sometimes termed the "standard" or "percussion system") was universally employed until near the year 1900. About that time the "rotary" method first attracted serious attention. Many practical operators were then somewhat hostile or indifferent to the advent of this innovation and more or less skeptical regarding the merits claimed for it. It was not until several years later that the possibilities of the rotary system were recognized, but in recent years its evolution has been rapid. During the past ten years the rotary has grown steadily in favor and its adaptability and limitations are now quite well understood and perhaps definitely determined.

The rotary method is most economical and successful in drilling through sand, clays, soft shales and alluvial deposits, but, as rotary drilling must be conducted with the well

[1] Hearty acknowledgment is hereby made to Mr. W. H. Whittekin, who has aided me greatly in the preparation of these pages. He has been specially fitted for this work by sixteen years in practical engineering in oil country work, with only a lapse of two years in the American Army.

or hole filled with mud-laden water, this system is not so well adapated as the dry-hole, cable-tool method for "wildcatting" (prospecting), where the compilation of accurate records or logs of the formations penetrated is essential. Neither is the rotary system suitable for drilling through limestone and other hard rocks, most of which are readily shattered and penetrated by the heavy blows of cable-tools. Because of this broader adaptability, cable-tools are more generally used, but in soft or caving formations and heaving sands, the rotary method is conducive to greater progress and economy.

In a few restricted areas and for prospecting in remote regions, it has been found expedient to install a combination outfit. This comprises a complete cable-tool outfit and practically complete rotary equipment. The peculiar installation of this hybrid outfit facilitates rapid changing from one system to the other whenever changes in the formations so dictate.

Still another modification is the standard-tool drilling-in outfit. This consists of only sufficient cable-tools and machinery to drill into the sand and complete the well after it has been drilled to near the oil stratum with rotary tools. (A portable cable-tool outfit is frequently used for this work.) This method is commonly employed in the fields adjacent to Burkburnett, Texas, and at Panuco and Topila in Mexico. It must not be confused with the combination outfit, already mentioned, as these standard drilling-in tools are used, not in conjunction with the rotary, but after the latter has practically completed the well and is removed to a new location. For want of a better designation we may term this the "dual system." In regions to which this method is especially adapted it possesses certain advantages over any of the other systems. The initial cost of these two separate units (the regular rotary equipment and the light standard drilling-in tools) is considerably less than the cost of the complete combination outfit. Moreover, the two units of the combination outfit must operate alternately, whereas with the two separate outfits or dual system, now under discussion, each unit may be operated independently but simultaneously on different wells. It is quite generally conceded that cable-tools are best adapted to drilling-in, bailing, testing and finishing a well. Thus, this method gives the operator the benefit of speed and economy to be gained through drilling the well with rotary tools and the added security afforded through completing it with cable-tools. Furthermore, this light cable-tool machinery is then available for swabbing the well or for tubing and pumping it temporarily, should that be desired.

Though differing in design and construction, the several drilling outfits under discussion have many features in common. Regardless of the method or system employed, the power plant generally consists of a portable boiler capable of developing from 30 to 70 horse-power and a standard type of horizontal steam engine of from 20 to 50 horse-power. The lighting plant, usually a 1-K.W. steam turbo-generating outfit with suitable wiring and fixtures, is admirably adapted to all of the varied requirements of oil-field practice. The derricks for all systems are similar but differ in height, weight and design. Many of the small tools and accessories are identical for all outfits.

The evolution and development of oil-field machinery, tools and methods, while interesting and instructive, cannot be considered within the province of this undertaking and this chapter will accordingly be restricted to a discussion of modern methods. It may be pertinent to mention here that in a few fields electric motors, natural gas engines and crude-oil engines are being tried and considered for motive power in various phases of oil-field work, and while these are giving excellent service on pumping powers, they appear to be still in the experimental stage when applied to drilling operations.

(Separate, detailed descriptions of the leading systems are submitted in the following pages.)

THE STANDARD OR CABLE-TOOL SYSTEM

This discussion will be devoted to describing and explaining briefly the fundamental principles and general practices of cable-tool drilling.

It is regrettable that a uniform or more definite terminology has never been developed and adopted in oil-field parlance. Local influences and usages are potent and the absence of a universal nomenclature renders the compilation of specifications exceedingly difficult. The term "rig" is sometimes used in reference to a complete drilling outfit, including machinery, drilling tools and derrick. In other instances "rig" is employed to designate only the complete timber or steel foundation, the derrick (tower), band wheel, sand reel, calf wheel, bull wheels, and the customary housing or shelter for drilling machinery, and it is in this latter sense that the term "rig" will be herein employed.

Granting that the location for the well has been made and that the nature of the formations to be penetrated or the information or log to be obtained suggests the use of cable-tools, a rig of proper specifications with derrick of suitable height will be selected and built. Until recent years, perhaps on account of the abundance of timber available, only wooden rigs were commonly used, but the introduction of the steel derrick with wooden foundation about the year 1900, and the all-steel rig in 1907, marked the advent of a new and important factor in oildom. The steel rig has since grown in favor and has come into quite general use in certain regions.

Every distinct type of rig in use, regardless of material, size or design, possesses certain advantages that suggested its adoption. Rig requirements differ according to depths of wells and drilling methods in different fields and even in the same field, and any attempt to adopt a rig of standard type and size for universal use would prove futile. Wooden rigs of similar design are, however, in use throughout the world, the derrick having a height of 72, 74, 80, 82, or 84 feet and a base 20 or 22 feet square. In the United States the popular height is 72 feet in the Eastern fields and 80, 82, or 84 feet in the Mid-Continent and Western fields. Space will permit cuts and detailed specifications of only two standard rigs, and the 72 and 82-foot derricks with 20-foot base are accordingly selected, being popular in use and best suited to this work.

A rig-building crew consists of either four or five men, and the time required for this crew to build a complete rig is generally five to seven days, perhaps equal to twenty-five days for one man. A rig in use may be dismantled and rebuilt on another location in less time than is required to fit and erect a new one.

Rig Irons, Including Sand Reel and Woodwork

The lumber and rig timbers are delivered to the drilling location in the "rough," to be cut and fitted there by the rig-builders. Finished wooden cants, arms, braces and handles for construction of bull wheels, calf wheel, and band wheel are, however, furnished with the rig irons. A set of rig irons comprises the gudgeons, bearings, flanges, stirrup, main shaft, crank, wrist pin, and other steel and cast-iron appliances necessary to install and operate the bull wheels, calf wheel, band wheel, pitman and walking beam. The size of a set of rig irons is designated by the diameter of the band wheel shaft, sometimes termed the main shaft. This may be 4, 4½, 4¾, 5, 6 or 7½ inches in diameter. The 4½, 4¾ and 5-inch rig irons are suitable for moderate depths, perhaps not exceeding 2000 feet, the 6-inch size being best adapted to deeper drilling.

The introduction of the calf wheel, about 1898, marked the advent of a new and helpful factor facilitating the handling of heavy strings of casing. This improvement met with such general favor that it has been universally adopted in standard deep-

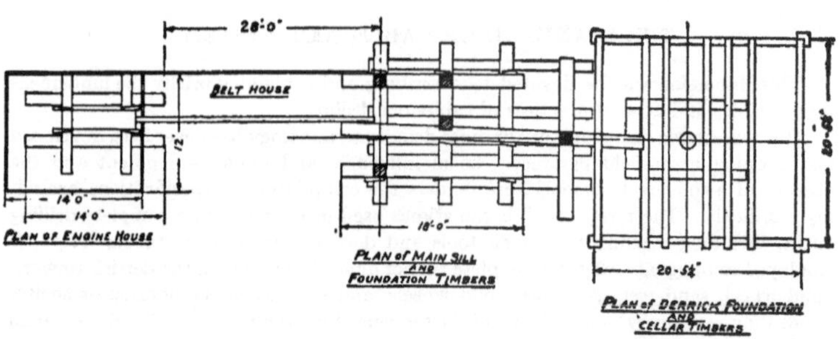

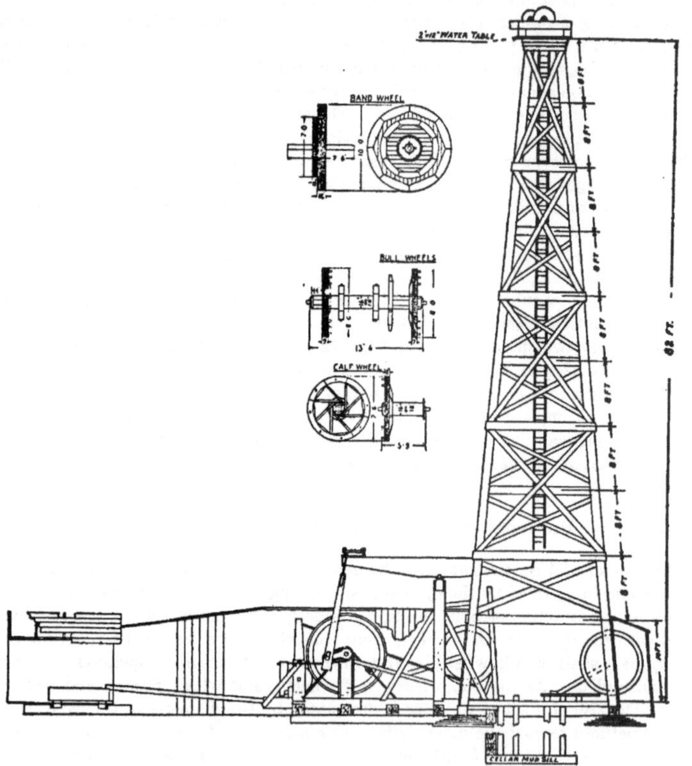

FIG. 1.—Standard rig construction plan. 82-foot derrick, 20-foot base.

Standard Rig Specification

Derrick 72 feet high, 20-foot Base

Pine Timbers			Dimension Material		
Pieces	Dimensions	Board ft.	Pieces	Dimensions	Board ft.
1	14″×24″×24′	672	36	2″×10″×20′	1,200
1	16″×16″×28′	597	30	2″×10″×18′	900
1	16″×16″×16′	341	90	2″×10″×16′	2,400
1	14″×16″×16′	299	18	2″×12″×18′	648
3	14″×14″×14′	686	90	2″×12″×16′	2,880
3	14″×14″×18′	882	16	2″× 8″×18′	384
1	14″×14″×20′	327	36	2″× 8″×16′	768
2	12″×12″×12′	288	24	2″× 6″×16′	384
1	12″×12″×18′	216	16	2″× 6″×14′	224
1	6″× 8″×22′	88	10	2″× 6″×20′	200
4	6″× 8″×16′	256	8	2″× 6″×18′	144
2	6″× 8″×14′	112	24	2″× 4″×16′	256
2	6″× 6″×14′	84	10	2″× 4″×12′	80
3	8″×10″×22′	440	12	1″× 6″×16′	96
6	8″× 8″×20′	640			
1	8″×18″×16′	192		Total.......	10,564
2	4″× 4″×16′	43			
4	3″×12″×16′	192			
	Total.......	6355			

Steel Wire Nails

100 pounds of 8d	100 pounds of 20d	100 pounds of 40d
100 pounds of 16d	200 pounds of 30d	50 pounds of 60d

Special for Cellar			Oak Timbers		
Pieces	Dimensions	Board ft.	Pieces	Dimensions	Board ft.
28	2″×10″×18′	840	1	6″×16″×16′	128
2	12″×12″×16′	384	1	6″×16″×12′	96
2	6″× 8″×18′	144	1	14″×16″×14′	261
			1	18″×18″×14′	378
	Total.......	1368	1	18″×18″× 7′	189
			1	8″×10″×10′	67
Boxing			1	8″×10″×12′	80
			2	12″×12″×10′	240
180	1″×12″×16′	2880	2	12″×12″×12′	288
70	1″×12″×14′	980	1	6″×14″×12′	84
70	1″×12″×12′	840	4	3″× 5″×16′	80
	Total.......	4700		Total.......	1891

Grand total, complete rig, 24,878 board feet.

Heavy Standard Rig Specification

82-foot Derrick, 20-foot Base

Pieces	Dimensions	Board ft.	Pieces	Dimensions	**Board ft.**
20	2″×12″×20′	800	1	14″×24″×24′	**672**
35	2″×10″×20′	1,167	1	16″×16″×28′	**597**
40	2″× 8″×20′	1,067	2	16″×16″×16′	**684**
20	2″×12″×18′	720	6	14″×14″×14′	**1372**
10	2″×10″×18′	300	2	14″×14″×16′	**522**
12	2″× 6″×20′	240	2	14″×14″×18′	**588**
18	2″× 6″×18′	324	2	12″×12″×12′	**288**
80	2″×10″×16′	2,134	2	6″× 8″×22′	**176**
20	2″× 8″×16′	427	3	6″× 8″×14′	**168**
24	2″× 4″×16′	256	8	6″× 8″×16′	**512**
12	2″× 4″×14′	112	4	8″× 8″×20′	**427**
40	1″× 6″×16′	320	9	8″×10″×20′	**1200**
70	2″×12″×16′	2,240	1	8″×20″×16′	**213**
56	2″× 6″×16′	896	2	12″×12″×18′	**432**
100	1″×12″×14′	1,400	1	14″×16″×12′	**224**
200	1″×12″×16′	3,200			
				Total.......	**8075**
	Total.......	15,584			

Steel Wire Nails

100 pounds of 8d — 100 pounds of 20d — 100 pounds of 40d

100 pounds of 16d — 200 pounds of 30d — 50 pounds of 60d

Oak Timbers

Pieces	Dimensions	Board ft.
1	18″×18″×14′	378
1	9″×10″×10′	75
2	12″×12″×10′	240
1	6″×14″×12′	84
5	3″× 5″×16′	100
2	6″×16″×16′	256
1	4″× 6″×16′	32
2	12″×12″×14′	336
1	18″×18″× 7′	189
	Total.......	1690

Legs.....2″×12″ Doublers..2″×10″

Braces...2″× 6″ Girts.....2″×12″

Extra for Sway Braces

Pieces	Dimensions	Board ft.
8	2″× 8″×16′	171
32	2″× 8″×20′	853
16	2″×12″×16′	512
10	2″×10″×20′	334
4	2″×12″×18′	144
	Total.......	2014

Grand total, for complete rig, 27,363 board feet.

well drilling. It is, therefore, included in these rig specifications, although it is not essential for shallow wells of less than 1500 feet in depth.

The rig and related appliances serve as a medium for transmitting power from the engine to the drilling bit and for convenient manipulation of the drilling equipment. A heavy canvas or rubber belt (usually 8-ply, 12 inches wide and 90 or 95 feet long) runs over the engine belt pulley and the band wheel, the latter encircling and operating the crankshaft (main shaft) which in turn rotates the calf wheel and the bull wheels and operates the pivoted walking beam. The sand reel, which is revolved by a direct friction drive on the band wheel, operates the sand line and bailer or sand pump. The old-style calf wheel is operated by an endless wire cable running (in grooved pulleys) from the main shaft and the bull wheels are operated from the band wheel in the same manner, but by means of one or two heavy, plain laid manila bull ropes. With the modern calf wheel a heavy sprocket chain replaces the old-style endless wire rope already mentioned. The drilling cable passes from the shaft or drum of the bull wheels over the crown pulley, on top of the derrick, and connects with the rope socket on the drilling tools, while the steel casing line passes from the calf wheel shaft over pulleys in the crown block and through a triple—or quadruple-sheave traveling pulley block, and moves the casing hook and elevators.

RIG IRON SPECIFICATIONS

(1) *Standard Rig Irons* (Figs. 1a–16), 4″, 4½″, 4¾″ or 5″ shaft:

1 Shaft and crank (shaft to be 5′ 6″ long for regular rig irons, but 6′ 6″ long if calf wheel attachment is desired), complete with collar and set screw.
1 Set (2) flanges, complete with 2 keys and 8 ¾″×10″ bolts.
1 Stirrup (2″ bar), complete with 3 ¾″×7½″ bolts.
1 Set center irons, complete (consisting of 1 saddle, 2 side irons, 2 side iron caps, 2 side iron cap bolts, 4 ¾″×28″ saddle bolts and 4 ¾″×18″ side iron bolts).
1 Set (2) bowl gudgeons, complete (2 gudgeons, 2 bands and 8 bolts ⅞″×18″).
1 Bull wheel brake iron, 3/16″×9″×28′, complete with lever and staple.
1 Set (2) pulleys (1 24″ sand line pulley and 1 30″ or 36″ crown (derrick) pulley).
1 Jack post box, complete (1 bearing with 4 ⅞″×10″ bolts, 1 cap with 4 cap bolts).
1 Open jack post bearing.
1 Wrist pin, complete with nut and washer, 1 brake iron, lever and staple.
Optional: 1 set bridle irons (consisting of 1 jack post stirrup, 1 jack post plate, 2 jack post eyebolts). 1 double-drum iron sand reel, complete. (Fig. 19.)
Extras, if calf wheel is to be used: 1 outer tug wheel, 1 calf wheel rim, 4 22″ casing line pulleys, 1 set calf wheel gudgeons, brake iron, lever and staple.

(2) *Imperial Ideal Rig Irons* (Figs. 21–51). 5″, 6″, or 7½″ shaft:

1 Shaft and crank, complete (with 2 flanges, 2 keys, 1 collar and set screw, 1 wrist pin with nut and washer, and 8 ⅞″×11″ bolts.
1 Stirrup (5 hole with 2½″ bar), complete with 5 bolts ⅞″×10″.
1 Set center irons, complete (1 saddle with 4 bolts ⅞″×34″, 2 side irons with side iron caps and cap bolts and 6 side iron bolts ⅞″×20″).
1 Set (2) bowl gudgeons, complete with 2 bands and 12 1″×20″ gudgeon bolts.
1 Set calf wheel gudgeons, complete (1 bowl gudgeon with band, 1 30″ flanged gudgeon with band, 12 1″×20″ gudgeon bolts.
1 Set closed jack post boxes, complete (consisting of 1 14″ crank side box and 1 12″ calf wheel side box, each complete with cap and cap bolts).

(Continued on page 211)

CALIFORNIA REGULAR RIG IRONS.

With Calf Wheel.

FIG. 1a.—Shaft, crank and flanges complete with outer tug wheel.

FIG. 2.—Shaft and crank.

FIG. 3.—Flanges and bolts.

FIG. 4.—Collar and set screw.

FIG. 5.—Wrist pin and nut.

FIG. 6.—Center irons.

FIG. 8.—Saddle.

FIG. 9.—Side iron.

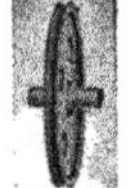

FIG. 10.—Sand line pulley.

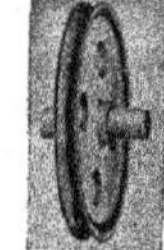

FIG. 11.—Derrick or spudding pulley.

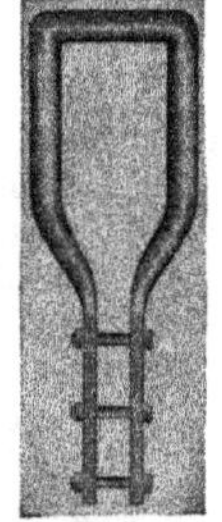

FIG. 7.—Round iron stirrup.

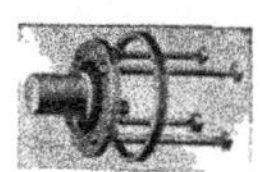

FIG. 12.—Bowl gudgeon with band and bolts.

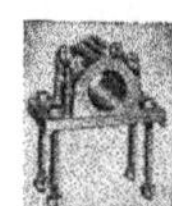

FIG. 13.—Jack post box.

FIG. 14.—Jack post stirrup.

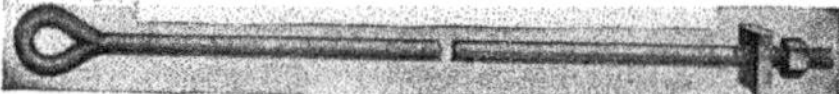

FIG. 15.—Jack post eye bolt.

FIG. 16.—Back brake hook bolt.

IRON SAND REELS.

FIG. 17.

NOTE.—Single drum iron sand reels are made in different sizes and with either cast-iron or pressed-steel flanges.

FIG. 19.—Beveled friction pulley.

FIG. 18.—Straight face pulley.

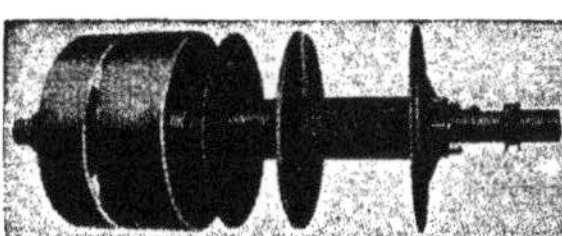

FIG. 20.—Straight face pulleys, double.

IMPERIAL IDEAL RIG IRONS.

FIG. 21.—Shaft complete with crank, flanges, clutch sprocket, boxes, etc.

FIG. 22.—Flanges and bolts.

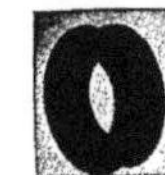

FIG. 23.—Collar and set screw.

FIG. 24.—Shaft and crank.

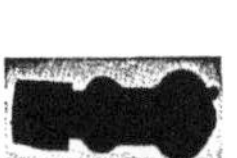

FIG. 25.—Wrist pin and nut.

FIG. 29.—Stirrup.

FIG. 26.—Center irons.

FIG. 27.—Saddle.

FIG. 28.—Side iron.

FIG. 30.—Jack post box.

IMPERIAL IDEAL RIG IRONS—*Continued.*

FIG. 31.—Open jack post bearing.

FIG. 32.—Jack post plate.

FIG. 33.—Anchor D. E. bolt.

FIG. 34.—Anchor eye bolt.

FIG. 35.—Turnbuckle rod.

FIG. 36.—Jack post D. E. bolt.

CALIFORNIA REGULAR RIG IRONS.

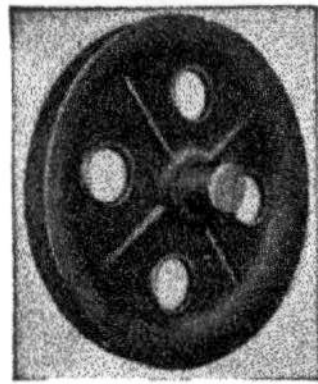

FIG. 37.—Derrick pulley.

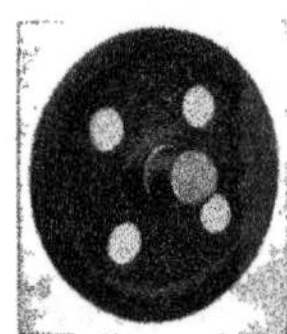

FIG. 38.—Sand line pulley.

FIG. 39.—Casing line pulley.

FIG. 40.—Bowl gudgeon.

FIG. 41.—Knuckle post D. E. rod.

FIG. 42.—Back brake D. E. bolt.

CALF WHEEL PARTS

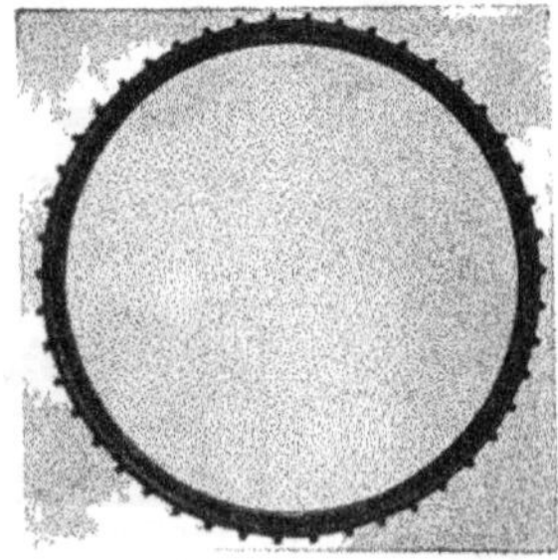

FIG. 43.—Sprocket tug rim.

FIG. 44.—Clutch.

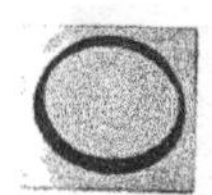

FIG. 46.—Gudgeon band.

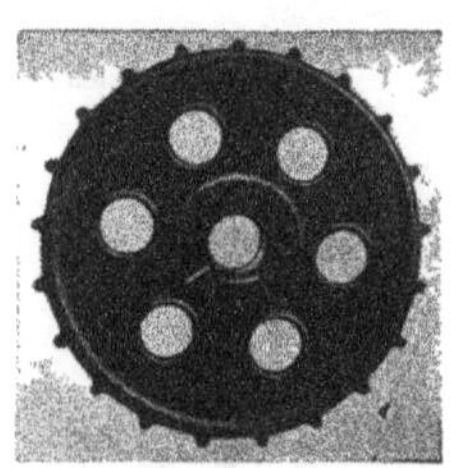

FIG. 45.—Clutch sprocket.

Calf Wheel Parts—*Continued*

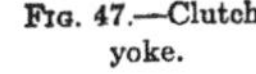

Fig. 47.—Clutch yoke.

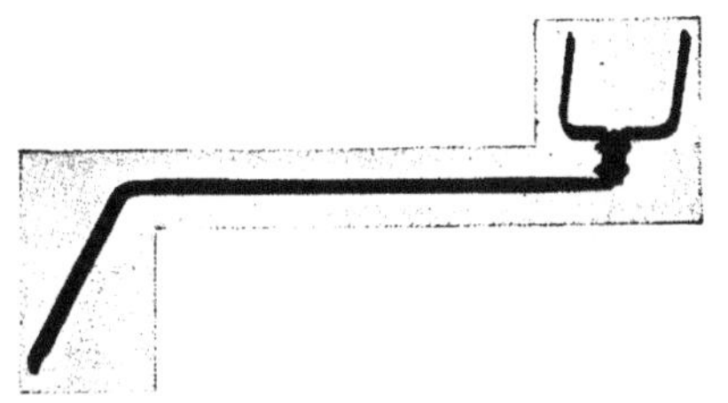

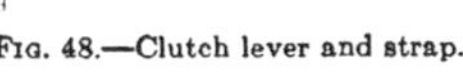

Fig. 48.—Clutch lever and strap.

Fig. 49.—Flanged gudgeon.

1 Set pulleys with $4\frac{1}{2}''$ cast-iron trunnions (1 30″ crown pulley, 1 24″ sand line pulley and 4 22″ casing line pulleys).
1 Clutch sprocket, 1 clutch with keys, 1 clutch yoke, 1 clutch lever and strap.
1 Calf wheel sprocket tug rim (84″ diameter of bolt hole centers).
1 Jack post plate (optional), 2 $1\frac{1}{2}''$X8′ 4″ jack post bolts (double end).
4 $1\frac{1}{4}''$×8′ 6″ turnbuckle rods, 2 $1\frac{1}{2}''$×4′ anchor eye bolts.
1 $1\frac{1}{2}''$×2′ 2″ anchor bolt (double end)
1 $\frac{7}{8}''$×8′ 10″ knuckle post double end rod, 2 $\frac{7}{8}''$×10′ 4″ back brake bolts.
55 feet of No. 1030 sprocket chain.
2 $\frac{3}{4}''$×12″×28′ bull wheel brake irons, complete with brake lever and staple (1 for bull wheels and 1 for calf wheel).
1 Double drum iron sand reel, complete (see Figs. 18–20).

72-Foot California Steel Drilling Rig

5-Inch Rig Irons—4-Pulley Crown Block—20-Foot Base

Specification and approximate weights in pounds

Derrick, straight-line crown block and ladder	12,500
Four (4) crown, sand line and casing pulleys	700
9-inch brake band, lever and staple	400
Bull wheel posts, braces and bearings	800
Calf wheel posts, braces and bearings	950
Derrick foundation	4,400
Machinery supports, braces and bearings	7,000
House framing	5,300
Walking beam W-1 and bearings	2,100
Center irons and stirrup	400
8-foot bull wheel—wooden tug rim	3,600
10-foot band wheel—wooden tug rim	3,000
5-inch ideal band wheel shaft, crank, wrist pins and keys	1,200
7-foot 6-inch calf wheel—sprocket rim	2,400
42-inch sprocket, clutch and keys	700
55-foot sprocket chain	500
7-inch brake band, lever and staple	300
38-inch California double-drum sand reel	1,600
Standard engine block	1,600
49 squares 26-gage corrugated sheet steel, painted	4,100
Total Complete Rig	53,550

Wood Work for Rig Iron Outfits

Cants or segments used in building the wheels and arms and handles of bull wheels. Cants, arms and handles are crated for shipment and convenience in handling, and vary in size and number to conform with the requirements of the various outfits.

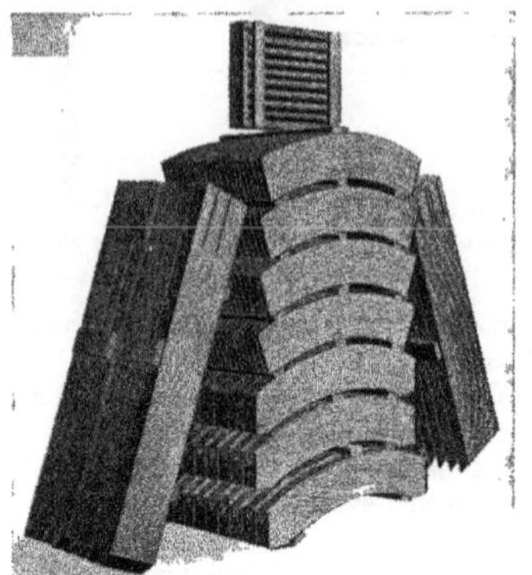

Fig. 50.

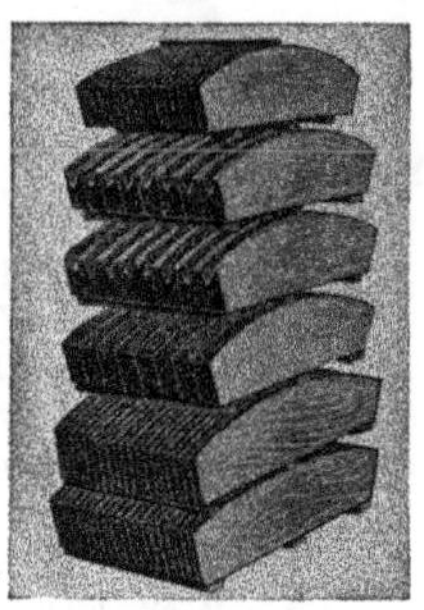

Fig. 51.

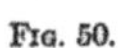

For 5 and 6-inch Calf Wheel Outfits (formerly known as the California) or for 5-inch, 6-inch and 7½-inch Imperial Ideal Rig Iron Outfits

		Weight, lbs.
Band wheel 10 feet diameter 12-inch face	56—1-inch pine cants	464
Tug pulley on band wheel 7 feet diameter	8—2½-inch grooved hemlock cants 8—2½-inch plain hemlock cants 16—1-inch pine cants	
Bull wheels 8 feet diameter	For tug side 8—2½-inch grooved hemlock cants 32—1-inch pine cants 8—10-inch oak arms 16—handles For brake side 8—2½-inch plain hemlock cants 40—1-inch pine cants 8—10-inch oak arms 16—handles	946
Calf wheel 7½ feet diameter	8—2½-inch plain hemlock cants 56—1-inch pine cants 8—2×12-inch oak arms 8—2×12-inch oak arm braces 16—handles	427

For double tug wheels, add—

For band wheel	8—2½-inch×7-foot grooved hemlock cants 8—1-inch×7-foot plain pine cants
For bull wheel, tug side	8—2½-inch grooved hemlock cants 8—1-inch pine cants
For bull wheel, brake side	16—1-inch pine cants

80-Foot Heavy California Steel Drilling Rig

6-Inch Rig Irons—6-Pulley Crown Block—20-Foot Base

Specification and approximate weights in pounds

Derrick, ladder and heavy crown block	17,600
Six (6) crown, sand line and casing pulleys	1,200
9-inch brake band, lever and staple	400
Bull wheel posts, braces and bearings	900
Calf wheel posts, braces and bearings	1,100
Derrick foundation	6,100
Machinery supports, braces and bearings	8,700
House framing	5,400
Walking beam W-1 and bearings	2,100
Center irons and stirrup	600
8-foot bull wheel—long shaft—wooden tug rim	3,800
11-foot band wheel—wooden tug rim	3,200
6-inch ideal band wheel shaft, crank and keys	1,400
7-foot 6-inch calf wheel—sprocket rim	2,400
42-inch sprocket, clutch, keys, etc	700
55-foot sprocket chain	500
7-inch brake band, lever and staple	300
42-inch California double-drum sand reel	2,000
Heavy engine block	1,800
49 squares 26-gage corrugated sheet steel, painted	4,100
Total Complete Rig	64,300

86-Foot Heavy California Steel Drilling Rig

6-Inch Rig Irons—6-Pulley Crown Block—24-Foot Base

Specification and approximate weights in pounds

Derrick, ladder and heavy crown block	21,000
Six (6) crown, sand line and casing pulleys	1,200
9-inch brake band, lever and staple	400
Bull wheel posts, braces and bearings	900
Calf wheel posts, braces and bearings	1,200
Derrick foundation	7,500
Machinery supports, braces and bearings	8,900
House framing	5,900
Walking beam W-7 and bearings	2,500
Center irons and stirrup	600
8-foot bull wheel—long shaft—wooden tug rim	3,800
11-foot band wheel—wooden tug rim	3,200
6-inch ideal band wheel shaft, crank and keys	1,400
7-foot 6-inch calf wheel—sprocket rim	2,400
42-inch sprocket, clutch, keys, etc	700
55-foot sprocket chain	500
7-inch brake band, lever and staple	300
42-inch California double-drum sand reel	2,000
Heavy engine block	1,800
51 squares 26-gage corrugated sheet steel, painted	4,300
Total Complete Rig	70,500

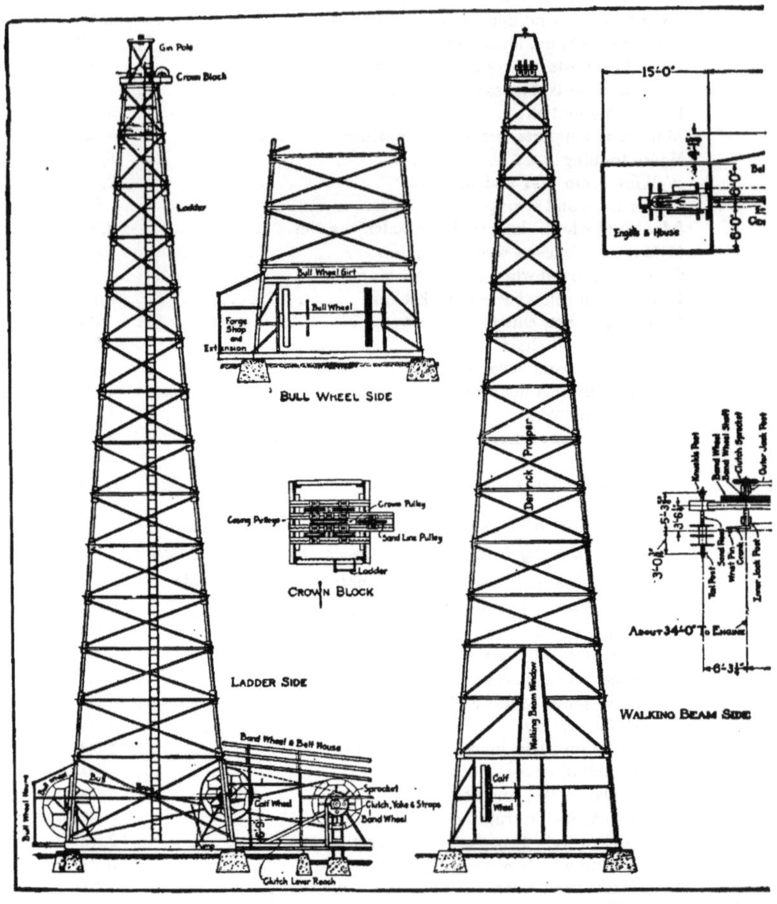

FIG. 52.—Dimensions

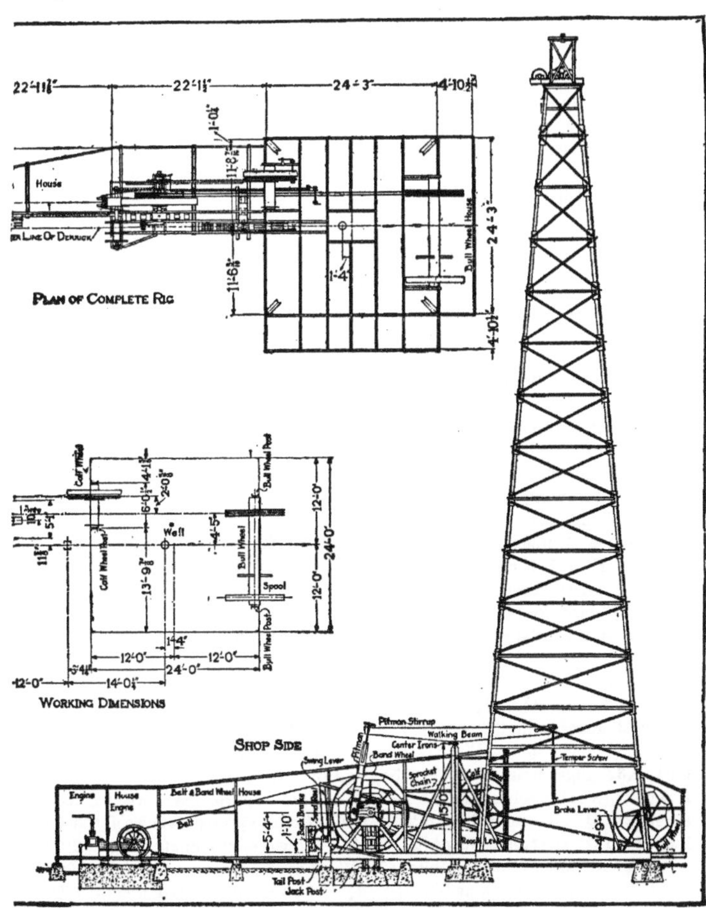

of steel derrick.

When the rig builders have finished the construction of the rig, the drilling crew comprising four men—two drillers and two tool dressers—appears on the scene to receive the drilling tools (as they arrive), and to begin "rigging-up." This operation embraces installation of the boiler, engine, tools and machinery and establishing a supply of fuel. and water for boiler and derrick use. All four men work together daytime, rigging-up. This requires from two to ten days' time. When actual drilling starts, one driller and one tool dresser work together by "tour" or shift, usually twelve hours, midnight to noon. The other driller and the other tool dresser operate the outfit during the second shift, noon to midnight. In certain fields two extra men are employed on each shift—one a fireman, who cares for the boiler and engine, the other a "derrick man" for common labor about the rig. For a hole beginning with 15½ inch and ending with 6⅝ or 5$\frac{3}{16}$ inch casing and to be drilled approximately 3000 feet in depth, the following standard cable-tool specification is submitted:

SPECIFICATION

FOR

Standard Drilling Outfit, for 2500 to 3000 Feet Depth

Machinery:

1 45 horse-power firebox, locomotive-type boiler, complete with stack, grate-bars and fittings.
1 Burner for oil or gas (if either fuel is used).
1 3″ flue cleaner, with joint ⅜″ pipe.
1 12″×12″ steam drilling engine complete with pulley, rimmed fly wheel, lubricator, pump and heater.
1 2″ iron body lubricator (tallow cup).
1 6″×4″×6″ Duplex boiler feed pump, 1 12″×90′ or 95′ 8-ply drilling belt.
1 Set 12″ E. H. belt clamps, 1 No. 16 belt punch (½″).
1 No. 1 grooved wheel, 1 150′ wire telegraph cord.
4 Joints ⅜″ pipe, for reverse rod.
1 1 KW. steam turbo-generating oufit, complete with wiring, etc.

Drilling Tools:

1 5½″×32′ auger stem, 2¾″×3¾″—7 pin, 4″×5″—7 box, 5″ squares.
1 5″×32′ auger stem, 2¾″×3¾″—7 pin, 3¼″×4¼″—7 box, 5″ squares.
1 4½″×32′ auger stem, 2¾″×3¾″—7 box and pin, 4″ squares.
1 3½″×32′ auger stem, 2″×3″—7 box and pin, 3¼ squares.
1 Set (2) 18″ all steel drilling bits, 4″×5″—7 pins, approximate weight, 4000 lbs.
1 Set (2) 15½″ all steel drilling bits, 4″×5″—7 pins, approximate weight, 3600 lbs.
1 Set (2) 12½″ all steel drilling bits, 4″×5″—7 pins, approximate weight, 2500 lbs.
1 Set (2) 10″ all steel drilling bits, 4″×5″—7 pins, approximate weight, 2000 lbs.
1 Set (2) 8¼″ all steel drilling bits 3¼″×4¼″—7 pins, approximate weight, 1400 lbs.
1 Set (2) 6⅝″ all steel drilling bits 2¾″×3¾″—7 pins, approximate weight, 900 lbs.
1 Set (2) 5$\frac{3}{16}$″ all steel drilling bits 2″×3″—7 pins, approximate weight, 500 lbs.
1 14″×16′ bailer, 1 11″×19′ bailer, 1 7″×25′ bailer, 1 5½″×30′ bailer.
1 4¼″×30′ bailer, 1 5½″ new era rope socket, 2¾″×3¾″—7 box.
1 5½″ swivel rope socket, 2¾″×3¾″—7 box.
1 4⅜″ swivel rope socket 2″×3″—7 box.
1 5½″ regular wire line socket, 2¾″×3¾″—7 box.
1 Set 5½″ drilling jars, 2¾″×3¾″—7 joints.
1 Set 4⅜″ drilling jars 2″×3″—7 joints.

1 18", 1 15½", 1 12½", 1 10", 1 8¼", 1 6⅝" and 1 5$\frac{3}{16}$ bit gage (complete set of 7 gages).
1 2¼"×5' 10" Mannington pattern temper screw, complete with elevating head and clamps for manila rope.
1 Set extra heavy wire line clamps, ⅞" liners.
1 Set 5" 400-lb. tool wrenches, with 5"×4" bushings.
1 Set 3¼" 175-lb. tool wrenches.
1 Ball bearing I-beam derrick crane, complete with 2-ton triplex hoist and No. 22 swivel wrench.
1 No. 2 new style jack and circle, complete, 1 No. 3 iron tool box.
1 Spudding shoe and clevis for manila rope, 1 spudding ring.
1 Iron slack tub, 1 300 lb. ram, 1 bit pulley with 8' chain.

Blacksmith Tools:
1 No. 4 steam blower, 1 350 lb. new style anvil, 2 14-lb. straight peen sledges with handles.
1 No. 71 ball peen hammer 1⅞ lbs., 1 pr. 24" blacksmith tongs.
1 Steel forge.

Cordage:
1 2¼"×1000 manila drilling cable, about 2000 lbs.
2 2½"×90' pure manila bull ropes, about 330 lbs.
1 ⅞"×3500' 6×19 California special steel drilling line.
1 $\frac{9}{16}$"×3500' 6×7 steel sand line.
1 ¾"×850' plow steel casing line.
1 1"×200' pure manila derrick line about 100 lbs.

Casing Tools:
1 Set (2) 15½" double gate, safety, detachable-link elevators with 2 2½"×44" steel links.
1 Set (2) 12½" double gate, safety, detachable-link elevators, less links.
1 Set (2) 10" double gate, safety, detachable-link elevators, less links.
1 Set (2) 8¼" double gate, safety, detachable-link elevators with 2 2¼"×36" steel links.
1 Set (2) 6⅝" double gate, safety, detachable-link elevators, less links.
1 Set (2) 5$\frac{3}{16}$" double gate, safety, detachable-link elevators, less links.
1 5½" double swivel casing hook.
1 Set (2) (front and rear) casing wagons, 6⅞" wire rope clips.
1 Rhinelander never-slip pipe-grip.
1 Set 12½" casing tongs, with bushings for 10", 8¼", 6⅝" and 5$\frac{3}{16}$" casing.
2 No. 16 and 1 No. 14 flat link chain tongs.

Pipe Cutting and Threading Outfit:
1 No. 2 combination vise.
1 Stock and dies to cut 1" to 2" pipe threads.
1 Stock and dies to cut ¼" to ¾" threads inclusive.
1 Square end pipe cutter to cut pipe ½" to 2".
1 Pipe cutter to cut pipe ⅛" to ¾".

Fittings:
6 1" standard brass globe valves.
2 1" standard horizontal brass check valves.
4 1" No. 1 BW. iron stop cocks, 2 2" No. 1 BW. iron stop cocks.

18 1" assorted nipples, 24 2" assorted nipples.
6 2" oil country flange unions, 6 1" mall. lip unions.
12 2" mall. ells, 12 1" mall. ells, 6 2" and 6 1" mall. street ells.
12 2" mall. tees, 24 1" mall. tees, 12 2" and 24 1" C. I. plugs.
6 2"×1" C. I. bushings, 6 2"×1" M. I. reducers.
400′ of 1" steel line pipe.
500′ of 2" steel line pipe.

Miscellaneous:

1 Swan's auger handle, with 1", 1¼", 1½" and 2" bits.
1 ⅞", 1 1" and 1 1⅛" ship auger, 1 4½-lb. single bit axe and handle.
1 14", 1 18", 1 24" and 1 36" Trimo pipe wrench.
1 18" and 1 15" combination wrench.
1 10" adj. S wrench, 1 6-lb. mattock and handle, 1 6-lb. pick and handle.
2 No. 2 L.H.R.P. shovels, 2 No. 2, S.H.R.P. shovels.
1 No. 2 hatchet, 1 No. 7 handsaw, 1 1¼ lb. cold chisel.
1 1¼-lb. punch and 1 1¼-lb. diamond point.
1 5-lb. splitting chisel and 1 5-lb. casing ripper.
1 14" H. R. Bast. file, 1 12" mill file, 1 6" taper file.
2 Wire thread brushes, 6 5" new style hay fork pulleys, 1 6" melting ladle.
50 Lbs. No. 4 babbitt metal and 50 lbs. high grade babbitt metal.
1 ½-pint gem oiler.
40 Gallons engine oil, 40 gallons cylinder oil.
50 Lbs. No. 3 cup grease and 50 lbs. beef tallow, 50 lbs. white waste, 2 lbs. Dixon's graphite.
2 12-quart galvanized iron pails, 1 derrick broom.
1 Pr. 10" combination pliers, 1 10" adjustable hack saw and 12 blades.
1 No. 14 steel square, 1 pr. 12 calipers, O. S., 1 25′ and 1 100′ metallic tape line.
1 2′ box-wood rule, 2 padlocks, 1 low down force pump, with 15′ suction hose.
1 20-lb. crow bar, 1 No. 2 Barrett lifting jack, 1 2-bbl. steel derrick tank.

Fishing Tools and Underreamers (Suggested):

1 Set 5½" 36"-stroke fishing jars, 2¾"×3¾"—7 joints.
1 6⅝" combination socket, 2¾"×3¾"—7 pin, 2 sets slips.
1 10" combination socket, 2¾"×3¾"—7 pin, 2 sets slips.
1 6⅝" latch jack, 2¾"×3¾"—7 pin.
1 10" latch jack, 2¾"×3¾"—7 pin.
1 6⅝" 3-prong grab, 2¾"×3¾"—7 pin.
1 8¼" Rogers patent grab, 2¾"×3¾"—7 pin.
1 6⅝" center rope spear, 2¾"×3¾"—7 pin.
1 10" center rope spear, 2¾"×3¾"—7 pin.
1 12½", 1 10", 1 8¼", and 1 6⅝" M. and F. steel die nipple (threaded to suit casing used).
1 5-3/16" underreamer, 2"×3"—7 pin.
1 6⅝" underreamer, 2¾"×3¾"—7 pin, square 4".
1 8¼" underreamer, 2¾"×3¾"—7 pin, square 4".
1 10" underreamer, 2¾"×3¾"—7 pin, square 4".
1 12½" underreamer, 3¼"×4¼"—7 pin, square 5".
Total weight of complete drilling outfit, approximately 65,000 lbs.

(By substituting a drilling cable and sand line each 5000 feet in length, this outfit would be suitable for drilling to a depth of 4000 to 4800 feet.)

CASING

When wells or shafts of any description are dug or sunk by hand, caving of the walls is prevented by the use of timbers, stone, brick or tile. In drilling, this tendency to cave is overcome through the medium of casing. A "string" of casing consists of a number

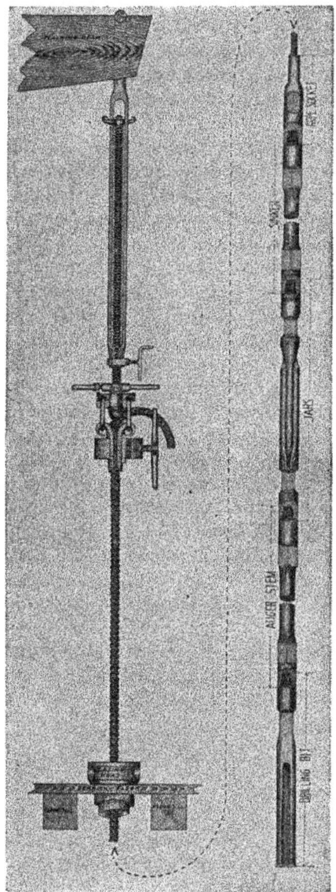

Fig. 53.—Temper screw with cable attached.

Fig. 54.—A "string" of drilling tools.

Fig. 55.—Hydraulic circulating system. Used in connection with cable tools.

of circular steel or iron tubes of a required diameter and with necessary couplings. Each tube, termed a "joint of casing," is about 20 feet in length, and when coupled together and set in the well, the string constitutes a strong, impervious metal cylinder or tube from the top of the hole downwards to the bottom or point of setting. A reinforcing steel casing "shoe" is ordinarily screwed on the bottom end of the first (lowest) joint of casing to protect it from damage and collapsing. When the rapid ingress of fluid, mud, sand or cavings into the drill-hole seriously interferes with the "free-fall" or drop of the tools, drilling progress is impeded and if this condition does not sufficiently respond to bailing or other corrective measures it becomes imperative to "set" a string of casing.

The diameter of a well at starting is much greater than at completion. Good judgment must be exercised in deciding upon the initial diameter of the well and due consideration given to the probable depth to be attained, the nature of the formations to be penetrated, the possible number of strings of casing to be used, and the resulting reduction in the diameter of each succeeding string; otherwise an impracticably small diameter may be prematurely reached and so necessitate the abandonment of the well without attaining the desired depth.

To avoid this contingency, in many fields casing is "carried," that is, inserted in the well but not "set." For example, a string of 15½-inch casing may be set on the bottom of an 18-inch hole, 500 feet deep, to exclude water or cavings. Drilling, with a 15½-inch bit, may then be resumed and a depth of possibly 1000 feet reached when more caving occurs. As little or no progress is possible and the tools are in constant danger of being buried, and perhaps lost, a string of 12½-inch casing is now inserted and "carried" on a spider, or casing hanger, but not set. This string may be lowered to the bottom but must be moved (lifted and lowered) frequently to prevent "freezing" (sticking) by the unstable walls. (At this juncture the hydraulic circulating system, illustrated in Fig. 55, is occasionally employed with success.) After cleaning out the accumulated cavings, drilling with a 12½-inch bit may now be resumed; but after drilling ahead a short distance, an underreamer must be substituted for the bit, and this newly-drilled 12½-inch hole is thus reamed out to 15½ inches to permit lowering the 12½-inch casing to the bottom again. This procedure is repeated as often as the external friction on the 12½-inch casing, and other deterrent factors will permit. The 12½-inch casing is then set, possibly, in this case at a depth of 1200 to 1600 feet. Regular drilling is thereafter continued with the 12½-inch bit until 10-inch casing is required, inserted, carried and underreamed, as with the previous string of 12½-inch. If necessary, this operation is repeated with each successive string of casing (10, 8¼, 6⅝ and $5\frac{3}{16}$-inch) and facilitates reaching the greatest possible depth with each string and, finally, with the well.

The size, weight and quantity of casing needed depend upon the depth to be attained, the nature of the formations to be drilled through, and the required diameter of the hole at completion. The following examples typify the regional differences in depth and the variations in casing requirements:

PENNSYLVANIA

Oil City field.

Average depth of wells, 1000 feet.
Casing required:
75 to 150 feet of 8¼" 24-lb.
300 to 400 feet of 5⅝" 10½-lb.

Washington field.

Average depth of wells, 3000 feet.
Casing required:
525 feet of 10" 35-lb.
1250 feet of 8¼" 24-lb.
1600 feet of 6⅝" 17-lb.
2600 feet of $5\frac{3}{16}$" 17-lb.

Kansas

Eldorado Field.

Average depth of wells, 2500 to 2700 feet.
Casing required:
- 180 feet of $15\frac{1}{2}''$ 70-lb.
- 950 feet of $12\frac{1}{2}''$ 50-lb.
- 1600 feet of 10″ 35-lb.
- 2100 feet of $8\frac{1}{4}''$ 28-lb.
- 2500 feet of $6\frac{5}{8}''$ 20-lb.

Oklahoma

Cushing Field.

Average depth of wells, 3200 feet.
Casing required:
- 400 to 600 feet of $15\frac{1}{2}''$ 70-lb.
- 900 to 1200 feet of $12\frac{1}{2}''$ 50-lb.
- 1600 to 1800 feet of 10″ 40-lb.
- 2000 to 2400 feet of $8\frac{1}{4}''$ 28-lb.
- 2500 to 3000 feet of $6\frac{5}{8}''$ 24-lb.
- 3000 to 3150 feet of $5\frac{3}{16}''$ 17-lb.

Texas

(Rotary Drilling)

Beaumont Field.

Depth of wells, 1500 to 4000 feet.
Casing required:
- 200 to 800 feet of 10″ 32-lb. casing.
- 1425 to 3925 feet of 6″ $19\frac{1}{2}$-lb. rotary drill pipe, set as casing.

Burkburnett Field.

Depth of wells, 1500 to 1800 feet.
Casing required:
- 1350 to 1750 feet of $6\frac{5}{8}''$ 20-lb.

California

(Taft Field (deep sand)

Average depth of wells, 3200 feet

(Drilled with either cable-tool, rotary or combination system)

Casing required if drilled with cable tools:
- 1500 feet of $12\frac{1}{2}''$ 40-lb. or 45-lb.
- 2000 feet of 10″ 40-lb.
- 2500 feet of $8\frac{1}{4}''$ 32-lb.
- 3200 feet of $6\frac{1}{4}''$ 24-lb.

Casing required if drilled with rotary:
- 1600 feet of $12\frac{1}{2}''$ 40-lb. or 45-lb.
- 3200 feet of $8\frac{1}{4}''$ 32-lb. or $6\frac{1}{4}''$ 24-lb.

Mexico

(Either Rotary or Cable Tools)

Panuco Field (Northern).

Average depth of wells 2000 feet.
Casing required if drilled with cable tools:
- 200 feet of $12\frac{1}{2}''$ 40–50-lb.
- 850 feet of 10″ 35 or 40-lb.
- 1600 feet of $8\frac{1}{4}''$ 28 or 32-lb.

Casing required if drilled with rotary:
- 200 feet of $12\frac{1}{2}''$ 40 or 50-lb.
- 1600 feet of $8\frac{1}{4}''$ 28 or 32-lb.

(Rotary drill pipe used, 2000 feet of 4″ 15-lb. or 6″ 19.55-lb.)

Southern Fields.

(Casiano, Tepetate, Chinampa, Amatlan, Cerro Azul, Potrero del Llano, Zacamixtle).
Depth, 2100–2600 feet.
Cable tools generally used.
Casing required:
- 100 feet of $15\frac{1}{2}''$ 70-lb.
- 800 feet of $12\frac{1}{2}''$ 50-lb.
- 1500–2000 feet of 10″ 40-lb.
- 2100–2500 feet of $8\frac{1}{4}''$ 32 or 36-lb.

Factors in Drilling

The several preparatory steps involved in drilling have now been briefly treated and it is proper at this point to consider separately a few of the more important factors that largely determine the measure of success to be achieved in any drilling venture.

(a) The **driller.**—The technique of drilling can be acquired only through practice and experience in actual drilling. There is no royal road. The tool-dresser, as assistant to the driller, may begin with a very superficial knowledge of the work and, under the sympathetic tutelage of an able driller, eventually become proficient and accordingly qualify as driller himself. As the duties of the drilling crew are exacting and laborious the members must be men of intelligence, energy and good physique. The successful driller possesses good judgment, bold initiative and considerable mechanical ability. He must be able not only to sense and avoid dangers, exercise patience and constant vigilance, but also solve numerous puzzling problems. Moreover, as the itinerant, expert tool-fisherman of bygone days is now extinct, his mantle and likewise his troubles, are the heritage of the modern driller.

As the tools operate at great depths below the surface, their action in drilling cannot be observed and must be determined or " sensed " by touch, through the action of the drilling cable, or by the varying motion of the engine. It is particularly in this sensing or technique that the superior knowledge of a competent driller becomes apparent and effective. The formations being drilled change frequently and suddenly in structure and hardness and to make due progress and avoid fishing jobs and bad hole, the proper tension on the drilling cable must be constantly maintained. The ability to accomplish this cannot be hastily acquired. Thus, the importance of the driller and the difficulties to be surmounted by him may be readily appreciated.

(b) The **power plant.**—Because of the difficulty and expense involved in securing fuel and fresh water in sufficient quantity and kind for boiler use, vain efforts have been made to displace steam with some cheaper motive power and, while some progress is claimed for other methods, only steam power is in general use for deep-well drilling. The steam plant is simple in construction and method of operation and delivers a steady flow of flexible power which is essential to successful drilling.

In the early days of drilling the power plant delivered perhaps 10 horse-power, but as the average drilling depth has steadily increased and heavier tools have been adopted, a proportionate increase in the power plant has necessarily followed. The popular steam drilling engine now develops 25 to 30 and the boiler 40 or 45 horse-power, with larger sizes occasionally preferred for certain regions and for other drilling systems. Where an abundance of natural gas or other cheap fuel is available or where wells are shallow and changes in location frequent, boiler installations are generally hastily improvised and crudely inefficient. But for deeper drilling and more permanent locations (such as pumping stations) more care and attention should be given to the selection and installation of boilers. Because of the frequent moving and the ease with which it may be installed the common locomotive-type, firebox boiler is commonly used throughout the United States, except for central power plants and in California where the tubular boiler meets with favor. These boilers, and the installation of the latter are illustrated in Figs. 57-8-9. The tubular boiler thus installed is more efficient as the heating surfaces are larger and the sand or earthen covering prevents the rapid radiation of heat. Both boiler and engine should be of sturdy construction to withstand hard usage and heavy strains and should also be responsive to sudden demands for the maximum power.

A standard type of horizontal single-cylinder, reversible steam engine is in common use everywhere in oildom (although a twin-cylinder engine of similar type is now undergoing experiment for rotary drilling). The engine is equipped with a belt pulley and a flywheel, or balance-wheel, with two or three detachable rims. These rims are not needed until a considerable depth has been reached, after which they are attached to the flywheel progressively, the additional weight contributing to uniform engine speed. The weight of the 30 horse-power (12 by 12-inch) steam engine, complete, is 4600 pounds, and the 45 horse-power firebox boiler, complete, weighs 10,900 pounds.

(*c*) The **rig.**—Geological factors influence or determine not only the selection of the drilling system, but also the features of the rig and characteristics of the drilling tools and machinery to be used, as these are all governed by the depth and nature of the formations to be drilled through. Detailed rig specifications have already been submitted and these remarks will refer only to rig construction and operation.

An expert rig-building crew should always be employed in rig construction, as accuracy and thoroughness are essential and have a direct bearing on the later success of the driller. If the latter is obliged frequently to suspend drilling to align, adjust or repair some part of the rig, drilling progress will be correspondingly retarded. At certain stages of drilling, as in white lime, caving formations or quicksand, immediate action and the maximum power may be urgently needed and if defects in the rig, machinery or equipment obstruct the continuity of drilling progress or the prompt application of remedial measures, serious results may follow. The use of doubtful materials or inefficient labor is false economy in connection with any drilling operation as the strains imposed are sudden and extreme and the hazards numerous.

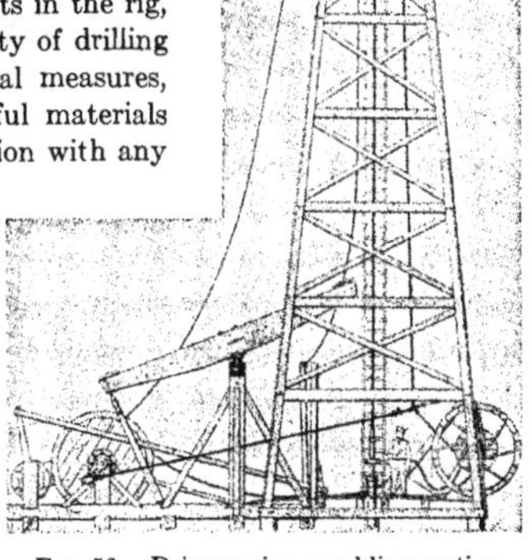

Fig. 56.—Driving pipe, spudding motion.

(*d*) **Drilling tools and their joints.**—As the elements composing a string of tools (rope socket, jars, auger stem and bit) are subjected to heavy blows and severe strains, the steel used in their manufacture must be specially adapted to withstand hard usage. Every tool used in drilling should, therefore, be of highest quality and most reputable manufacture. The dimensions of each tool should also be recorded by the driller, as a knowledge of these measurements will facilitate recovering it, should it later be lost in the well.

The work required of the rope socket and the auger stem is not destructive and, unless the steel of which they are made becomes crystallized, they rarely fracture. But as drilling bits and jars are subjected to hard blows and much greater strains, fractures and fishing jobs occasionally result. In ordinary drilling there is no movement within the jars as they remain completely extended, but if through caving or other cause the tools become stuck the tension on the drilling cable is slackened by " letting out " with the temper screw and sharp upward blows (" jars ") are delivered with the upper, free half of the jars, thus freeing the tools. This action of the jars obviates hard pulling and consequent strain on the drilling cable and rig.

Drilling bits of numerous sizes and designs are in common use, each being adapted to a certain requirement (Figs. 62–68). Formerly they were made of iron with a piece of tool steel welded on the bottom to form the cutting edge and withstand the impact. Later the all-steel bit met with favor and has been universally adopted. Spudding bits, used for setting the first string of casing, or " conductor," are generally of great width but short and thin, thus meeting the requirements of the soft, upper formations. The thin bit allows ample clearance in the drill-hole for mixing and emulsifying soft formations, but a heavy thick bit is better suited to drilling rock where less clearance but greater strength is required. For hard formations the bit is dressed chisel-shaped as a

OIL COUNTRY BOILERS

FIG. 57.—Locomotive type (fireboX) boiler. These boilers are of the style generally used for drilling wells. They are built for 125 pounds working pressure and are tested to 190 pounds hydrostatic pressure before shipment.

SPECIFICATIONS OF OIL COUNTRY BOILERS

Locomotive Type

Nominal horse-power	20	25	28	30	40	45	66
Heating surface ... sq. ft.	282	335	369	443	498	566	686
Length of boiler	13′ 10″	13′ 10″	14′ 10″	15′ 10″	15′ 10″	17′ 6″	19 ′4″
Height of boiler	7′ 6″	8′ 2″	8′ 2″	8′ 4″	8′ 11″	8′ 11″	9′ 4″
Diameter of boiler ... inches	36	40	40	42	44	48	48
Diameter of dome ... inches	28	30	30	32	36	36	36
Height of dome ... inches	33	36	36	36	40	40	40
Length of furnace ... inches	49	49	49	49	49	56	60
Height of furnace ... inches	36	40	40	43	44	44	48
Width of furnace ... inches	36	40	40	42	44	44	48
Number of hand-holes	8	14	14	14	8	14	8
Number of tubes (3-inch)	36	43	43	48	54	56	62
Length of tubes ... feet	8	8	9	10	10	11	12
Diameter of smoke stack ... inches	18	20	24	22	24	24	26
Length of smoke stack (coal or wood) .feet	25	25	25	30	32	32	35
Length of smoke stack (gas) ... feet	14	14	14	14	14	21	21
Steam opening in dome ... inches	3	3	3	3	3	3	4
Diameter of pop valves ... inches	2	2	2	2	2	2	2½
Weight boiler, only ... lbs.	5800	6800	7000	8100	8900	10,200	11,680
Weight complete ... lbs.	6400	7600	7800	9000	9800	10,900	13,000

cutting edge is required, but in soft formations a blunt or concave bit is desired so that the emulsification of detritus may synchronize with drilling progress. The wear on the bit varies with the changing formations. In shales a bit may drill upwards of 200 feet with one dressing, while in hard sand or gritty rocks dressing may be required after every "screw" (about 6 feet), or even after every foot drilled. In preparing a bit for dressing, it should be heated slowly and turned frequently so that the expansion may be even and harmless; otherwise uneven strains and consequent cracking or fractures may follow. Special care is also necessary in tempering a bit. If it is set in the slack tub when too hot and not again removed temporarily to " draw " the temper properly, brittleness and fractures will result. But if insufficiently heated when immersed, little or no temper will be imparted and the bit will " mushroom " or batter in a hard

FIG. 58.—Tubular boiler, California type. (With pressed steel brackets for setting.) 35, 40 and 45 H.P. boilers are furnished with pressed-steel loops and hanger hooks for setting.

structure. The driller inspects the bit at every opportunity and examines the cutting surface to discern when it is sufficiently worn out of gauge to require dressing. After a badly-worn bit has been removed, difficulty may be experienced in " reaching bottom" with a " fresh " (newly-dressed) one. A vertical, friction " wearing surface " develops at each cutting corner of the bit and causes it to stick in certain formations. For these

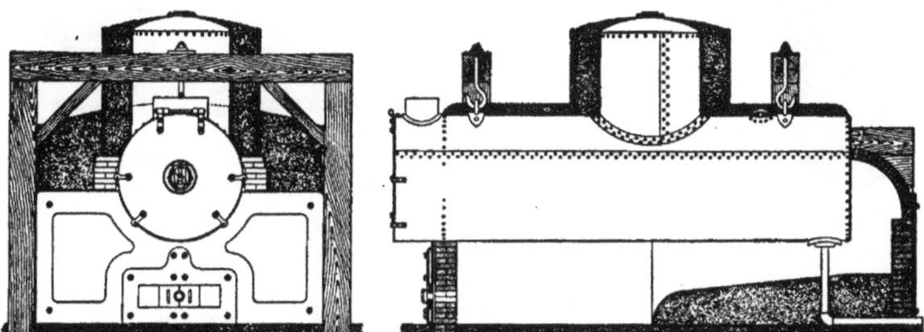

FIG 59 —Installation of tubular boiler.

reasons and in order that the drill-hole may later accommodate casing, if necessary, the bit should at all times be maintained approximately " to gage," or full size.

In oil-field parlance the term " joint " is commonly employed to designate that part of a drilling or fishing tool by means of which connection is made with another member of the string; but the complete joint consists of two parts, a threaded truncated cone (the pin) and a corresponding coupling (the box). For many years a straight joint, similar to a common pipe connection, was used but unsatisfactory service and lack of a shoulder for sustaining friction suggested an improvement, and the taper joint,

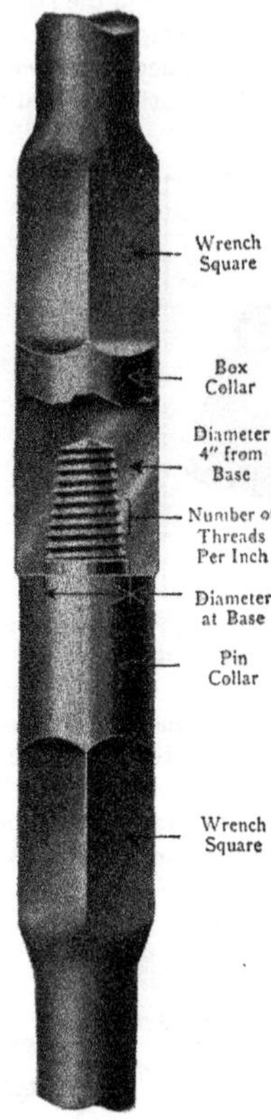
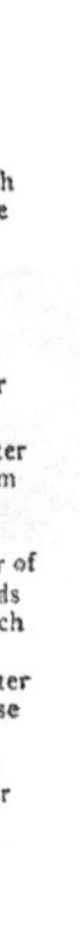

Fig. 60.

8-thread sharp

7-thread flat

Fig. 60a.—Joint and threads.

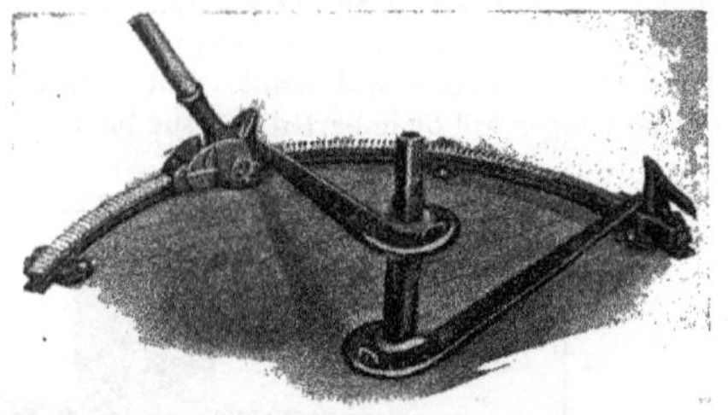

Fig. 61.—Jack and circle (showing method of setting up joints).

Tool wrenches

Two in a set, right and left hand.

Fig. 61A.—Left hand.
Form of wrench under 175 pounds.

Fig. 61B.—Right hand.
Form of wrench, 175 pounds and heavier.

Fig. 61C.—Left hand with liner or bushing.

Drilling Bits

All steel, annealed pin. Bits are usually sold in pairs or sets of two.

Fig. 62.—Regular form bits, in sizes 6¾ inches and smaller. From 4 to 8 feet in length.

Fig. 63.—Regular form bits, in sizes 8 inches and larger. From 4½ to 8 feet in length.

Fig. 64.—Mother Hubbard pattern.

Fig. 65.—California pattern.

Fig. 66.—Diagram.

When other than standard or stock bits are required, measurements conforming to the diagram, Fig. 66, should be given as follows: (*A*) Diameter of collar; (*B*) Length of collar; (*C*) Size of wrench square; (*D*) Length of water channel; (*E*) Width of water channel; (F) Width of blade; (*G*) Thickness of blade; (*H*) Thickness of bit in channel; (*I*) Length of bit. Also state size of pin, style of taper, number of threads to the inch, sharp or flat.

Derrick crane oufit

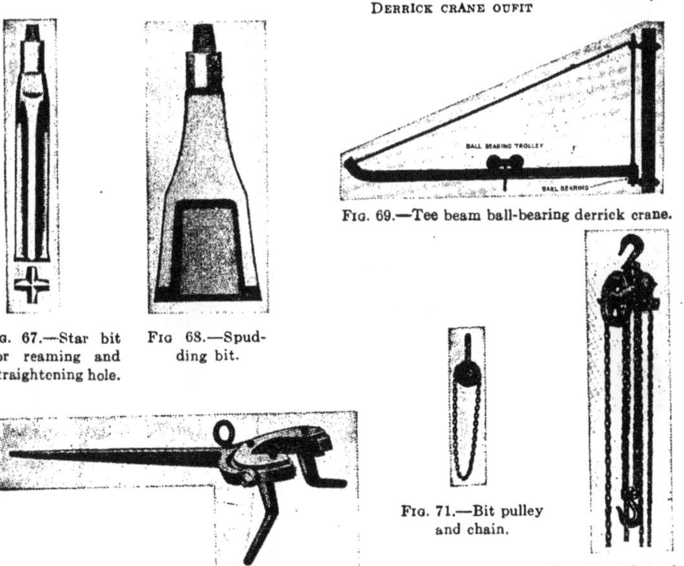

Fig. 67.—Star bit for reaming and straightening hole.

Fig. 68.—Spudding bit.

Fig. 69.—Tee beam ball-bearing derrick crane.

Fig. 70.—Chain hoist, duplex or triplex, 1 to 2 tons capacity.

Fig. 71.—Bit pulley and chain.

Fig. 72.—Swivel wrench for handling bits, jars, fishing tools, etc.

BAILING TOOLS

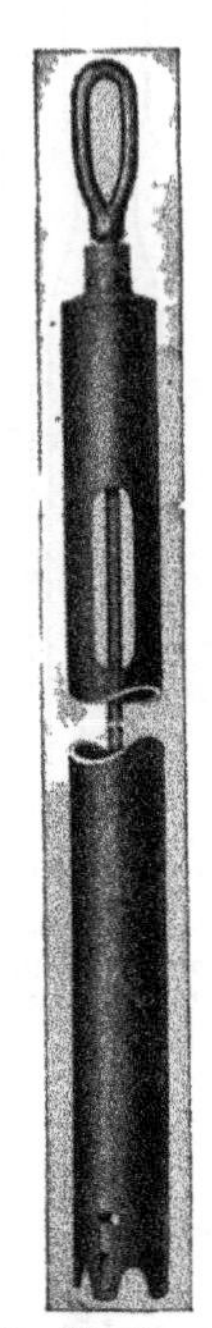

FIG. 73.—Bayonet joint sand pump.

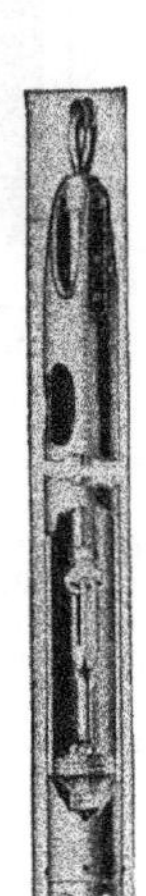

FIG. 75.—Sand pump.

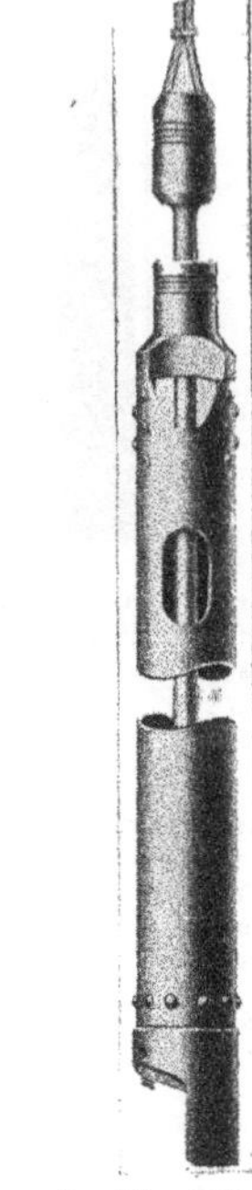

FIG. 76.—"Oilwell" sand pump.

FIG. 78.—Sectional flush joint bailer. Made in two or three sections. The middle section can be removed from a three-section, making a regular length bailer.

FIG. 74.—Bottom for bayonet joint sand pump.

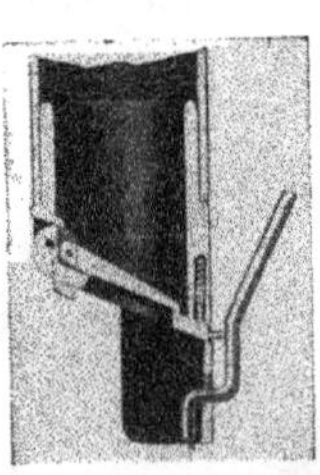

FIG. 77.—"Oilwell" sand pump bottom.

FIG. 79.—Texas bailer top.

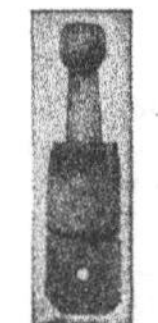

FIG. 80.—Texas bailer bottom.

which has since been universally adopted for standard tools, was accordingly introduced. Probably no factor in drilling is more important than the quality and condition of these joints.

The three dimensions of a joint designate (1) the diameter 4 inches from the base, (2) the diameter at the base and (3) the number of threads per inch, respectively; thus, a 3″×4″-7 box or pin is 3 inches in diameter at a point 4 inches from the base, 4 inches at the base and bears 7 threads per inch.

The I. & H. (Ireland & Hughes) 24° taper joint with seven flat threads per inch is now the recognized oil-field standard, but on joints smaller than 2″×3″, 8 threads sharp are still favored. The threads and shoulders of joints in use should be cleaned and inspected at every opportunity and when laid aside should be coated with grease or oil and protected against damage. The space remaining between the shoulders of a new joint, when screwed up by hand, should be about $\frac{1}{16}$ inch and it must be "broken-in" with care and patience. The most popular joints now in use are the $2\frac{3}{4}$″×$3\frac{3}{4}$″—7, the standard for drilling tools in many fields and practically everywhere for fishing tools, the 2″×3″—7, $2\frac{1}{4}$″×$3\frac{1}{4}$″—7, $2\frac{1}{2}$″×$3\frac{1}{2}$″—7, 3″×4″—7, $3\frac{1}{4}$″×$4\frac{1}{4}$″—7, $3\frac{1}{2}$″×$4\frac{1}{2}$″—7, and 4″×5″—7.

Fig. 81.—Hawser laid manila rope.

Fig. 82.—Plain laid manila rope.

Complete specifications covering these joints are shown in table on p. 230.

(*e*) **The cordage.**—This caption was formerly applied only to fiber rope—manila, sisal hemp, and cotton—but in recent years steel wire rope has also been included under the general term "cordage" and will be so treated herein. Manila rope is easily knotted, spliced and handled and is more elastic than wire cable. It is, therefore, used for general purposes about the rig and as it conduces to better progress at shallow depths, it is usually recommended for spudding and drilling the first 500 to 1000 feet although wire cable may be used exclusively in spudding and drilling the entire well. The standard manila drilling cable is hawser laid, $2\frac{1}{4}$ inches in diameter, and 1000 to 3500 feet in length. The standard bull rope is $2\frac{1}{2}$ inches by 90 feet, plain laid, while ropes for general purposes are from $\frac{1}{2}$ to $1\frac{1}{2}$ inches in diameter, plain laid, and cut to required lengths. (See Figs. 81-82.) In America the diameter, but in England the circumference, designates the size of manila rope.

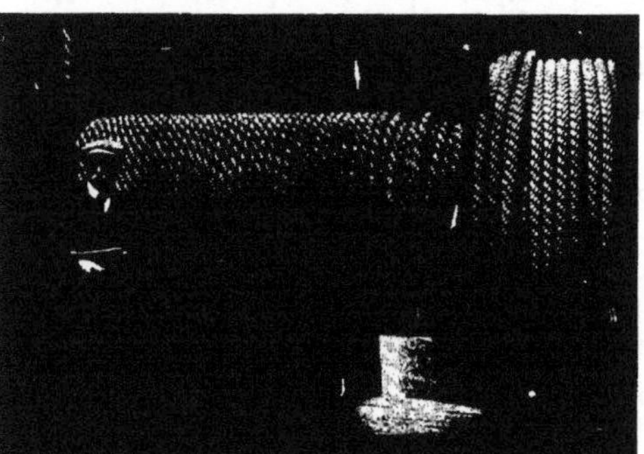

Fig. 83.—No. 1 Discovery Well of Calgary. Showing a 3500-foot manila cable coiled on bull-wheel shaft after drilling to a depth of 2500 feet.

Specifications of Tool Joints

Size casing, inches	Casing weight per foot, pounds	Casing inside diameter, inches	Size joint, inches	Size wrench, square inches	Box collar diameter, inches	Pin collar diameter, inches
3	**4.10**	3.01	1¼×1¾	1¾	2⅝	2½
3½	**5.10**	3.49	1⅞×2	2¼	3	2⅞
4	**6.20**	3.97	1⅞×2	2¼	3	2⅞
4¼	**6.75**	4.21	1⅞×2½	2½	3⅝	3½
4¼	**9.50**	4.09	1⅞×2½	2½	3⅝	3½
4½	**7.25**	4.46	1¾×2¾	2¾	3⅞	3¾
4½	**9.50**	4.36	1¾×2¾	2¾	3⅞	3¾
4¾	**8.00**	4.69	1¾×2¾	2¾	3⅞	3¾
4¾	**15.00**	4.40	1¾×2¾	2¾	3⅞	3¾
5	**8.56**	4.94	2 ×3	3¼	4⅜	4¼
5	**10.07**	4.88	2 ×3	3¼	4⅜	4¼
5	**13.07**	4.76	2 ×3	3¼	4⅜	4¼
5	**16.06**	4.64	1¾×2¾	2¾	3⅞	3¾
5 3/16	**9.00**	5.19	2 ×3	3¼	4⅜	4¼
5 3/16	**13.00**	5.04	2 ×3	3¼	4⅜	4¼
5 3/16	**17.00**	4.89	2 ×3	3¼	4⅜	4¼
5⅝	**10.50**	5.67	2¼×3¼	3½	4¾	4½
5⅝	**12.00**	5.62	2¼×3¼	3½	4¾	4½
5⅝	**14.00**	5.55	2¼×3¼	3½	4¾	4½
5⅝	**17.00**	5.45	2¼×3¼	3½	4¾	4½
6¼	**12.00**	6.28	2¾×3¾	4	5½	5¼
6¼	**13.00**	6.25	2¾×3¾	4	5½	5¼
6¼	**17.00**	6.13	2¾×3¾	4	5½	5¼
6¼	**20.00**	6.04	2½×3½	4	5⅜	5
6¼	**24.00**	5.92	2½×3½	4	5⅜	5
6¼	**26.00**	5.85	2½×3½	4	5¼	5
6⅝	**13.00**	6.65	2¾×3¾	4	5½	5¼
6⅝	**17.00**	6.53	2¾×3¾	4	5½	5¼
6⅝	**20.00**	6.45	2¾×3¾	4	5½	5¼
6⅝	**24.00**	6.33	2¾×3¾	4	5½	5¼
6⅝	**26.00**	6.27	2¾×3¾	4	5½	5¼
6⅝	**28.00**	6.21	2¾×3¾	4	5½	5¼
6⅝	**30.00**	6.15	2¾×3¾	4	5½	5¼
8¼	**17.50**	8.24	3¼×4¼	5	6½	6¼
			or			
			3½×4½	5	6¾	6½
8¼	**20.00**	8.19	3¼×4¼	5	6½	6¼
			or			
			3½×4½	5	6¾	6½
8¼	**24.00**	8.09	3¼×4¼	5	6½	6¼
			or			
			3½×4½	5	6¾	6½
8¼	**28.00**	8.00	3¼×4¼	5	6½	6¼
			or			
			3½×4½	5	6¾	6½
8¼	**32.00**	7.92	3¼×4¼	5	6½	6¼
8¼	**36.00**	7.82	3¼×4¼	5	6½	6¼
8¼	**38.00**	7.77	3¼×4¼	5	6½	6¼
8¼	**43.00**	7.65	3¼×4¼	5	6½	6¼
10 and larger	any weight	....	3¼×4¼	5	6½	6¼
			or			
			3½×4½	5	6¾	6½
			or			
			4 ×5	5½	7⅜	7
			or			
			4¼×6	6	8⅜	8¼

The chief virtue of a manila cable and its advantage over a wire line arise from its superior elasticity. Because of this the "lift" or vertical movement of the drilling tools is much greater than indicated by the corresponding movement of the walking beam and the maximum blow is thus delivered. The elasticity of a wire cable is less pronounced but gives this same effect, although in lesser degree, after reaching a depth of perhaps 1000 feet.

As manila rope is susceptible to abrasion, and as short kinks are destructive, a drilling cable should be carefully handled, must never be dragged en masse, and, whenever possible, should be protected from the parching effect of intense heat or sunlight. To insert the cable in a rope socket, first draw it through the socket 6 or 8 feet, then draw back to the proper position for inserting the "filler." This will correct or adjust any uneven or misplaced strands. Heavy pulling that taxes the capacity of a cable is also destructive.

Approximate Weight and Strength of Manila Rope

Diameter in inches	Circumference in inches	Weight, pounds per foot	Estimated working strength in pounds	Approximate breaking strength in pounds
$\frac{1}{2}$	$1\frac{1}{2}$	$\frac{3}{32}$	2,000	3,000
$\frac{5}{8}$	2	$\frac{1}{8}$	3,800	5,700
$\frac{3}{4}$	$2\frac{1}{4}$	$\frac{1}{5}$	4,500	6,750
$\frac{7}{8}$	$2\frac{3}{4}$	$\frac{3}{8}$	6,200	9,300
1	3	$\frac{7}{16}$	7,000	10,500
$1\frac{1}{4}$	4	$\frac{1}{2}$	12,000	18,000
$1\frac{1}{2}$	$4\frac{3}{4}$	$\frac{3}{4}$	16,500	25,000
2	$6\frac{1}{4}$	$1\frac{1}{2}$	30,000	45,000
$2\frac{1}{4}$	7	$1\frac{9}{10}$	37,000	55,500
$2\frac{1}{2}$	$7\frac{3}{4}$	$2\frac{1}{8}$	43,000	65,000

Wire drilling lines, sand lines and casing lines are now in general use for deep drilling. They are essential to progress in wells where water strata are frequently encountered. In such regions sufficient strings of casing cannot be set to exclude the water while drilling and because of the greater friction on a manila cable it is not so suitable as wire rope in water-logged wells. Wire rope is also less susceptible to damage and abrasion. Crucible cast-steel wire rope is generally used for sand lines and casing lines, because of its convenient size and lack of elasticity, but when a stronger, more flexible and more durable line is required, a "plow-steel" cable is recommended.

The strength of a wire rope was formerly assumed to equal the combined breaking tests of the individual wires but the breaking strength of a cable has since been correctly rated at from 80 to 95 per cent of the combined strength of the individual wires, the ratio depending upon the method of construction. The commonest steel cable for moderate requirements is composed of 6 strands of 7 wires each but for the more exacting requirements of deep drilling or heavy hoisting a cable consisting of 6 strands of 19 wires each is preferable, being more flexible and less likely to fracture. The sizes most commonly used are $\frac{1}{2}$, $\frac{9}{16}$ and $\frac{5}{8}$ inch for sand lines and $\frac{3}{4}$, $\frac{7}{8}$ and 1 inch for drilling and casing lines.

There is no accepted rule or formula for determining the proper size of wire cable for any drilling operation. Several considerations are involved but the judgment of the driller is usually the deciding factor. In general, ¾-inch cables are satisfactory for drilling to depths not exceeding 2000 feet, ⅞ inch for depths under 3500 feet, and 1 inch for greater depths. In a few instances 1¼-inch and 1½-inch steel cables have been used with unusually heavy tools but, as these over-size lines cannot be run over the standard-size shafts and pulleys furnished with regular oil-field equipment, they are not practical for general use.

The manufacturers of wire rope recommend that only drums and sheaves (pulleys) of the largest practicable diameter be used and that these be aligned in the best possible manner to avoid contact of the cable with any part of the sheave except the bottom of the groove. Proper attention to this alignment, to the condition of the sheaves, and to lubrication of the cable will be amply compensated by increased efficiency and better service. Reversed bending (bending the cable in one direction then in the opposite direction) should also be avoided wherever possible.

As wire rope is shipped from the mills on wooden spools or reels it must not be uncoiled like manila rope. In transferring a wire cable from the spool to the bull wheel shaft a spindle is run through the aperture or hub of the spool. This is elevated and suspended on blocks or trusses permitting the spool to revolve on this horizontal spindle (Figs. 84 and 85). One end of the cable is then passed inside the derrick, over the crown pulley, and down to the bull wheel shaft on which it is carefully wound by the power of the engine. This method may also be employed in transferring the casing line to the calf wheel shaft and the sand line to the sand reel drum.

Kinks.—The kinking tendency of wire rope can be easily avoided if proper care is taken. The accompanying illustrations show how kinks are placed in wire rope, as well as the damage resulting therefrom. It is, therefore, very important that wire rope be kept free from kinks, otherwise the life of the rope will be greatly shortened, even if a sudden breakage does not occur.

The life of a wire cable is materially shortened if subjected to kinks, "dog-legs" (short bends), or to "lifted" strands so frequently caused by "slipping" the drilling cable through the wire line clamps. The grooves of the sheaves should be smooth and slightly larger than the cable to avoid undue friction and binding. The power should be gradually applied as sudden jerks subject a cable to excessive stresses and cause rapid deterioration. Contact with drums, sheaves, casing and walls causes wear on the external wires and the constant bending produces friction on the internal wires of a cable. The need of a proper lubricant that will not only protect the outer wires against the action of water and acids but also reduce friction on the internal wires in thus apparent. The amount of service that any cable will render is directly dependent upon the care and treatment accorded it. These factors are so variable that no standard of efficiency is recognized. A ¾ or ⅞-inch by 2000 foot, 6 by 19 steel-wire cable may serve for drilling from one to four 1500-foot wells whereas a similar 1 inch by 5000-foot cable might fail before drilling 4000 feet.

The diameter of a wire rope is determined by measuring from the top of one strand to the top of the strand directly opposite (see Fig. 89).

Only wire rope of regular lay is used in oil-field drilling. This consists of individual wires laid to the right to form strands and the strands laid to the left to form left-lay cable (Fig. 90), or wires laid to the left and strands laid to the right to form right-lay cable (Fig. 91).

Some operators successfully use a unique combination embodying the virtues of both wire and manila cable. This consists of a regular wire drilling cable and a "cracker" (about 100 feet of 2¼ inch manila cable), the latter connecting the wire line with the drilling tools and furnishing the desired elasticity. This quality of manila rope and the advanatge of wire-line economy and adaptability are thus conveniently combined.

Care of Wire Rope.

Fig. 84.—This shows how wire rope should be taken from a reel.

Fig. 85.—The wrong way to take wire rope from a reel.

Fig. 86.—The start. A rope should never be allowed to take this position, but if the loop is thrown out *now*, the kink can be avoided.

Fig. 87.—The kink. The damage is done.

Fig. 88.—Pulled tighter. Showing very sharp stress upon wires.

Fig. 88*A*.—The damage. Note the broken wires.

Lay of Wire Rope

By the "lay" is meant the direction or twist of the wires and strands composing a rope.

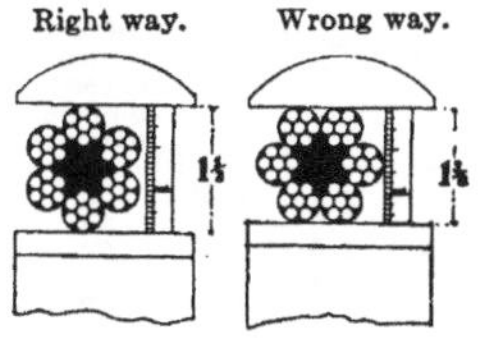

Fig. 89.—Measuring wire rope.

Fig. 90.—Left lay, regular lay hoisting rope. The regular lay of a wire rope is when the wires in each strand twist in the opposite directions from the strands themselves.

Fig. 91.—Right lay, regular lay, hoisting rope. A rope is "right lay," or "left lay," according to the direction in which the strands are laid, as shown.

Statistics covering standard steel hoisting rope commonly used in drilling are submitted in the following tables. It should be observed that while drums and sheaves of the diameters specified are recommended these sizes are not practicable for drilling purposes and sheaves and drums of about one-half these diameters are actually used. Note also that the factor of safety (breaking-strength to working-load) is approximately five to one.

Steel Transmission, Haulage, or Standing Rope

(Used for sand lines)

6 Strands, 7 Wires per Strand, 1 Hemp core

Diameter, inches	Approx. circum. inches	Approx. weight per foot, pounds	Approximate Breaking Strength, Tons (2000 lbs.)				Allowable Working Load, Tons (2000 lbs.)				Suggested Minimum Diameter of Drum or Sheave, Inches
			Crucible cast steel *	Extra strong crucible cast steel	Plow steel	Special plow steel	Crucible cast steel *	Extra strong crucible cast steel	Plow steel	Special plow steel	(Pulleys of about one-half these diameters are used in oil-field practice)
$\frac{3}{8}$	$1\frac{1}{8}$	0.22	4.6	5.25	5.9	6.5	0.92	1.05	1.2	1.3	32
$\frac{7}{16}$	$1\frac{1}{4}$	0.30	5.5	6.25	7.0	7.75	1.1	1.25	1.4	1.5	36
$\frac{1}{2}$	$1\frac{1}{2}$	0.39	7.7	8.85	10.0	11.0	1.5	1.8	2.0	2.2	42
$\frac{9}{16}$	$1\frac{3}{4}$	0.50	10.0	11.0	12.0	12.0	2.0	2.2	2.4	2.6	48
$\frac{5}{8}$	2	0.62	13.0	14.5	16.0	17.5	2.6	2.9	3.2	3.5	54

* Only the crucible cast-steel grade is commonly used for sand lines.

Standard Steel Hoisting Rope

(Used for drilling and casing lines)

6 Strands, 19 Wires per Strand, 1 Hemp Core

Diameter, inches	Approx. circum. inches	Approx. weight per foot, pounds	Approximate Breaking Strength, Tons (2000 bls.)				Allowable Working Load, Tons (2000 lbs.)				Suggested Minimum Diameter of Sheave, or Drum Inches
			Crucible cast steel *	Extra strong crucible cast steel	Plow steel	Special plow	Crucible cast steel *	Extra strong crucible cast steel	Plow steel	Special plow steel	(Pulleys of about one-half these diameters are used in oil-field practice)
$\frac{3}{8}$	$1\frac{1}{8}$	0.22	4.8	5.30	5.75	6.75	0.96	1.06	1.15	1.35	18
$\frac{7}{16}$	$1\frac{1}{4}$	0.30	6.5	7.25	8.0	9.4	1.30	1.45	1.6	1.9	21
$\frac{1}{2}$	$1\frac{1}{2}$	0.39	8.4	9.2	10.0	12.1	1.68	1.84	2.0	2.4	24
$\frac{9}{16}$	$1\frac{3}{4}$	0.50	10.0	11.2	12.3	14.5	2.0	2.24	2.4	2.9	27
$\frac{5}{8}$	2	0.62	12.5	14.0	15.5	19.0	2.5	2.80	3.1	3.8	30
$\frac{3}{4}$	$2\frac{1}{4}$	0.89	17.5	20.2	23.0	26.3	3.5	4.04	4.6	5.3	36
$\frac{7}{8}$	$2\frac{3}{4}$	1.20	23.0	26.0	29.0	35.0	4.6	5.20	5.8	7.0	42
1	3	1.58	30.0	34.0	38.0	45.0	6.0	6.80	7.6	9.0	48
$1\frac{1}{8}$	$3\frac{1}{2}$	2.00	38.0	43.0	47.0	56.0	7.6	8.6	9.4	11.0	54
$1\frac{1}{4}$	4	2.45	47.0	53.0	58.0	69.0	9.4	10.6	12.0	14.0	60
$1\frac{1}{2}$	$4\frac{3}{4}$	3.55	64.0	73.0	82.0	98.0	12.8	14.6	16.0	20.0	72

* Extra strong crucible cast steel is commonly used for drilling cables, and while all four grades are suitable for casing lines, the special plow steel casing line is recommended.

As the cost of cordage is one of the chief factors in the expense of drilling and as discarded wire rope has little or no salvage value drillers resort to the practical expedient of splicing wire cables to repair breaks and secure the maximum service. "Blind" splicing, in which the hemp core is removed to accommodate the ends of the strands, is the modern practice and meets drilling requirements.

Splicing Wire Rope

In oil-field practice the length of the "blind" splice, commonly used, is about 360 times the diameter of the rope; that is, 15 feet for a ½-inch sand line and 30 feet for a 1-inch drilling cable, although for unusually hard service longer splices may be expedient. The tools comprising the splicing outfit vary with the preferences of the drillers.

The following instructions apply to a 30-foot drilling cable splice but are responsive to modification for any required length.

1. Unlay the strands of both ends of the rope for a distance of 15 feet each. Then cut off the exposed hemp cores and interlock the strands as shown in Figs. 92–93.

2. Unlay any strand (*A*), and follow up by inserting a strand (1) of the other end. Proceed, making the twist of the inlaid strand conform exactly to the twist of the open groove until all but 30 inches of strand 1 is inserted. Then cut off strand *A* leaving 30 inches to match strand 1 (Fig. 94).

3. Proceed similarly with the opposite end, unlaying strand 6 and inserting strand *F* in the resulting groove, then back to strands *B* and 2, strands 5 and *E*, 3 and *C*, and finally with 4 and *D*, all of the strands now being laid with ends at 5-foot intervals (Figs. 95–96).

4. These overlapping 30-inch ends must now be tucked and secured without enlarging the diameter of the rope. With two rope clips and wrenches, or small ropes and short levers, placed opposite the points of intersection (Fig. 97), twist the rope in opposite directions to open the lay; then remove the hemp core for a short distance to the right and (by means of a marlin spike inserted under an adjacent strand and moved along as in Fig. 98) force the remainder of strand *A* into the space previously occupied by the hemp core cutting off the core where the strand ends and forcing the end of the strand and the end of the core, which should be abutting, back into the center of the rope. Repeat this operation to the left with strand 1, then shift the twisting apparatus to the next intersection and proceed in like manner with the other strands. If the spliced portion of the rope is to be run over sheaves of small diameter or to be subjected to excessive strains, the part of the strand that is tucked into the center of the rope should be wrapped with tarred twine or tape to increase the friction and keep the ends from working out. After the splice is completed the cable should be hammered lightly with a copper or wooden mallet at the points of tucking (Fig. 99).

(*f*) The **casing.**—The term "casing" is applied to steel or iron pipe used in a well to prevent caving of the walls or the ingress of water or both. Regular casing is made in sections or "joints" about 20 feet long and in nominal inside diameters of from 2 to 20 inches. The material is either iron or mild steel of approximately 27 tons tensile strength per square inch and the commonest type is lap-welded with screwed (threaded) ends and couplings. In appearance and construction casing is similar to line pipe except that casing couplings are longer and heavier and the ends of the joints are reamed internally to facilitate free passage of the drilling tools.

For shallow wells common line pipe is often successfully used but deep wells require heavy-weight casing with long couplings. In many wells the cost of the casing is the largest item of expense. Good judgment should, therefore, be used to select casing of proper size and weight and to insure starting the well with sufficient diameter to admit the several successive strings of casing that will probably be required. The initial diameter of a "wildcat" well is generally 18 inches or larger to permit the use of 15½ inch or larger casing for a "surface" (first) string, but in proven fields the probable number of strings necessary is known and the diameter desired for the completed well determines the initial diameter and, in general, the casing requirements.

Four kinds of tubular goods are used as casing—line pipe, drive pipe, regular screw casing, and riveted "stovepipe" casing.

Line pipe, because of its short couplings, is suitable only for short strings, although it is occasionally used at depths of 1500 to 2000 feet, but at great hazards and frequently

SPLICING WIRE ROPE (Illustrated).

FIG. 92.—First step, measuring.

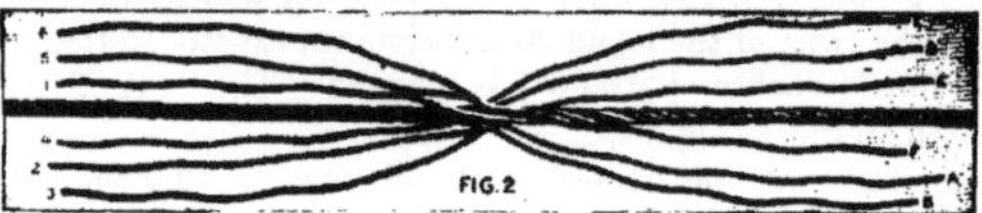

FIG. 93.—Second step, interlocking.

FIG. 94.—Laying first strand.

FIG. 95.—Laying of strands completed

FIG. 96.—Tucking of ends completed.

These illustrations show a splice 30 feet long. When a splice of a different length is to be made use proportionate dimensions.

with most unsatisfactory results. Because of weak couplings it should never be "carried" or used where underreaming is necessary.

Drive pipe is heavier than ordinary casing. The ends of the tubes are faced squarely without scarfing so that they form a strong butt joint in the 8-thread couplings. The drilling tools and drive clamps are employed to "drive" this pipe through heaving or unstable formations such as quicksand or gravel. (See Fig. 56.)

Regular casing is made in several grades, differing chiefly in the kinds of couplings furnished and in the number of threads per inch. South Penn casing is made in a wide variety of sizes and weights but is fitted with couplings of medium length and is, therefore, not suitable for strings exceeding perhaps 2500 feet in length. California casing, always furnished with long couplings and 10 threads per inch, is made only in the heavier weights especially adapted to deep drilling.

Riveted stovepipe casing of large diameter is commonly used in Rumania and Russia and occasionally in California where the loose formations and unstable walls necessitate starting the well with the largest convenient diameter. Under such conditions stovepipe casing of 12 to 36 inches in diameter is usually used. The joints are generally 24 inches long made of 8 to 12 U. S. gage steel sheets. These are rolled and riveted together at the edges, forming a tube with a longitudinal seam. A special steel casing shoe is riveted to the lowermost joint (or to a joint of drive pipe which is then attached to the bottom of the string of stovepipe casing) to form a cutting edge and reinforcing ring.

An ingenious type of "shoe" known as a "turnback" is sometimes made by constructing the starter joint of three thicknesses of steel sheets and allowing the innermost tube to extend 6 to 8 inches beyond the other two. This projection is then turned back over the outside tube and riveted there. This forms an improvised shoe and, while it is not so rigid and durable as the forged steel shoe, it may be more easily driven because of its smaller external diameter.

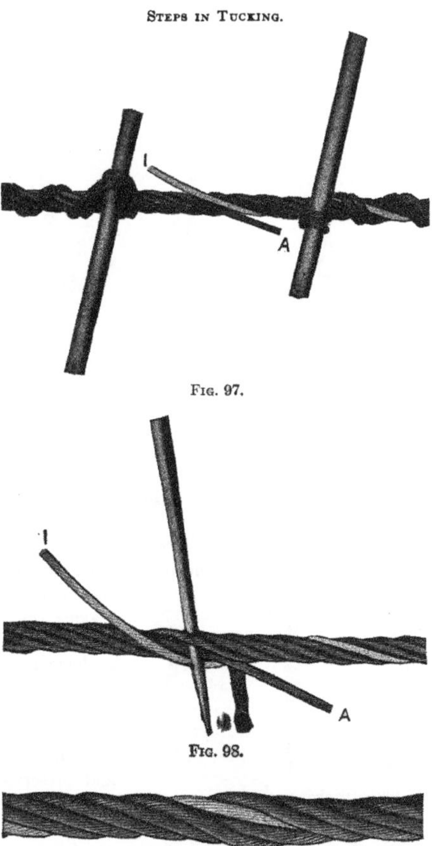

Fig. 97.

Fig. 98.

Fig. 99.—A view of the complete tuck before being hammered into shape.

Various methods of construction are utilized in the manufacture of riveted casing but the building-up (dove-tailing) process is probably most popular. By this method a double tube is built after riveting on the "starter" joint, the succeeding inside and outside joints being of equal length (usually 24 or 36 inches) and with square-cut ends, to form a butt joint. Thus, in starting the string the inner joint is made to extend exactly 12 inches above the outer joint. The next outer joint (24 inches long) then encircles and extends 12 inches above the inner joint, forming two continuous tubes. This arrangement is continued throughout the string, but for convenience in handling, several of these 2-foot sections may be assembled to form a joint 10 to 20 feet long, which is then joined to that already in the well and driven. Only the longitudinal seams are riveted, the sections being loosely held together by friction of the outer tubes against the inner ones. This friction is greatly increased by denting the outer joints with a common railroad pick. While riveted casing is less expensive than screw casing, it is obviously of limited adaptability. Being flush-jointed, it is easily driven through caving formations but, unlike screw casing which is water-tight and may be pulled with elevators, riveted casing can be moved upwards or withdrawn from the well only with a spear, if at all. During recent years the rotary has been generally

TOOLS FOR SPLICING WIRE ROPE SETS.

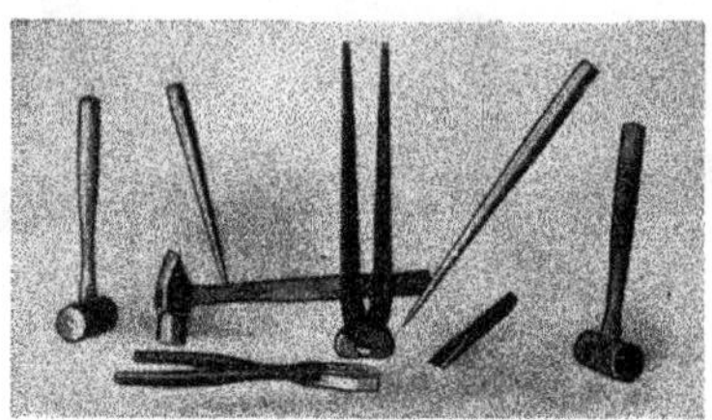

FIG. 100.—Regular style splicing outfit. A set consists of 1 pair nippers, 1 pair pliers, 2 copper mallets, 2 marlin spikes, 1 cold chisel, 1-2 pound C. P. hammer.

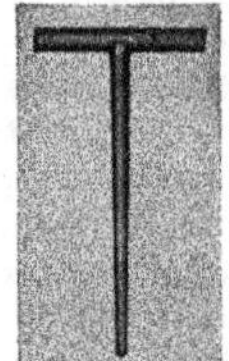

FIG. 101.—Special needle.

employed in California to drill through the soft upper formations and its action in plastering and stabilizing the walls of the bore hole renders the use of stovepipe casing unnecessary.

(Detailed information and casing specifications, including collapsing pressures, are submitted on pages 302–306.)

(*g*) **Formations and logs.**—The "*Log*" of a well is a record characterizing the formations penetrated. This delineation includes the color, nature, thickness and contents of every formation encountered. In certain regions the laws require that logs of all wells drilled be furnished for official records and as geological factors are important considerations in selecting the proper drilling machinery, tools and casing, and in deciding upon the method of procedure, public interest demands that all available information be accuractly recorded and preserved. This applies with equal force to the prospector in new territory and the exploiter in proven fields, although the guarded secrecy of the "wildcatter" is temporarily justified.

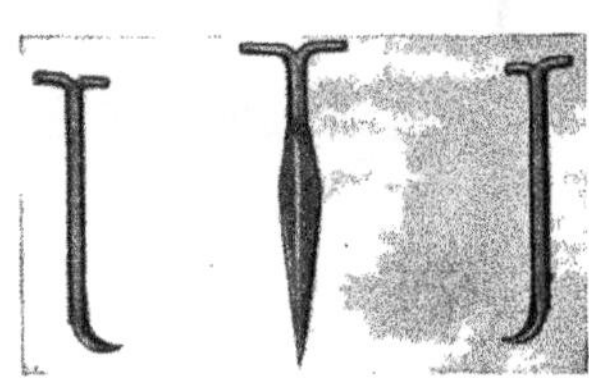

FIG. 102.—Oklahoma style splicing outfit.

Access to this accumulated knowledge will be of great value in subsequent drilling operations. Much time and money have been wasted and many wells (including the famous Dos Bocas and Lakeview gushers) have been lost chiefly because accurate logs were not available. Under such circumstances it is not possible to foresee and provide for all probable contingencies.

The data necessary for compilation of the log may be secured from the daily drilling reports. These reports are made by the driller, usually in triplicate. The original is sent to the main office, the duplicate being for the field superintendent while the triplicate is retained at the well for reference of the driller.

The required information is obtained by the driller from an examination of the drillings removed from the well by the bailer or sand pump. Sometimes particles of detritus cling to the bit when it is withdrawn and are of special significance. It must be noted that some formations when pulverized and mixed with water do not display their natural colors and for this reason a small portion of the cuttings should be dried and examined carefully whenever any change in formation is indicated.

It must be remembered that in describing clays, shales and rocks there is no established classification. Opinions and descriptions differ and what one driller may regard as light blue shale may be designated as dark gray by another. Likewise sandy shale may be termed "gritty," etc. Also when the formations change suddenly, or where the strata drilled are very thin, the slush or mud removed at a single bailing may be a composition of two or three distinct formations, the identity of which can be determined only by washing the drillings and minutely examining the larger fragments.

DAILY DRILLING REPORT

....................Oil Company

Well No.	Lease or Tract	Date			Tour (X)	
		Day	Month	Year		
						Morning
						Afternoon

Depth		Description, Contents and thickness of formations	Struck		
			Water	Oil	Gas
At beginning of Tour	At end of Tour				
Depth Drilled this Tour					

Casing

Strings Already Set								String Being Run			
Size		$15\frac{1}{2}''$	$12\frac{1}{2}''$	$10''$	$8\frac{1}{4}''$			Size	No. ft. this Tour		Total feet in well
No. feet									Inserted	Removed	

Remarks:

	Hours
Tool-dresser:	
Driller:	

Typical Well Records "Logs"

McClosky No. 1. T. & P. Coal Co., Eastland Co., Texas.

Contractor, Warren Wagner. Commenced July 8, 1917; well came in October 21, 1917.

Located 1 Mile SW. of Ranger, Texas.

	Top	Bottom
Soil	0	35
Yellow clay	35	45
Blue shale	45	175
Lime	175	215
Black slate	215	276
Water sand	276	295
Light slate	295	354
Lime	354	365
Blue shale	365	380
Lime	380	395
Blue shale	395	460
Lime	460	498
Blue shale	498	520
Lime	520	540
Blue shale	540	570
Lime	570	580
Hard sand	580	590
Blue shale	590	770
Lime	770	800
Light shale	800	1160
Sandy shale	1160	1180
Light shale	1180	1330
Sand	1330	1390
Light shale	1390	1615
Lime	1615	1635
Hard shale	1635	1660
Sand	1660	1680
Shale and lime	1680	1945
Water sand	1945	1967
Lime	1967	1990
White slate	1990	2470
Broken	2470	2566
Sand	2566	2572½
Sand and shales	2572½	2740
Black shales	2740	2920
Black lime	2920	2985
Black slate	2985	3035
Light shale	3035	3090
Black lime and black shale	3090	3240
Top, black lime	3240	....
Top, oil sand	3427	....
Total depth	....	3432

Casing:
Set 12½" at 16 ft.
10" at 1189 ft.
8¼" at 1869 ft.
6⅝" at 3204 ft.

Spudding and Drilling

There is no established method of procedure followed in rigging-up and, as the details of that undertaking are not pertinent here, we may now consider briefly the principles of cable-tool drilling and the characteristics of the tools.

Where long strings of casing are necessary (requiring a calf wheel, casing line, 37-inch–60-inch triple-sheave traveling pulley block, and heavy casing hook) it is customary to dig a "cellar" (6 to 10 feet square and 10 to 20 feet deep) under the derrick to facilitate handling the casing by means of elevators and a spider with slips (Figs. 112–113).

A dump box or trough under the derrick floor carries the detritus, as removed from the well by the sand pump or the bailer, to the sump-hole (a small reservoir dug near the derrick).

Until recent years the drilling crew handled the separate bits, jars, and fishing tools without mechanical aids but deeper drilling and the consequent advent of heavier tools necessitated the use of a derrick crane, trolley, chain hoist and swivel wrench (Fig. 69–72). The upright post to which the crane is attached stands in a corner at the right side of the derrick and the sweeping movement of the crane and attached appliances facilitates the moving of heavy tools and bits to and from the forge and about the derrick.

TYPICAL WELL RECORDS—"LOGS"—*Continued*

Dickson No. 1. Atlas Oil Co., NW¼ of SW½ Sect. 3-2ON-R15W, Caddo Co., La. Elev. 176′. Rigged up Nov. 1, 1913, commenced drilling Nov. 11, 1913; finished Dec. 23, 1913.

	Top	Bottom
Red clay	1	25
Water sand	25	45
Yellow clay	45	50
Blue sand	50	60
Clay	60	98
Rock	99	100
Gumbo and gravel	100	105
Rock	105	106
Gumbo	106	113
Rock	113	114
Gumbo	114	206
Shale	206	250
Rock	250	251
Shale	251	285
Gumbo	285	290
Shale	290	402
Shale	402	460
Gumbo	460	470
Shale	470	580
Rock	580	581
Shale	581	770
Gumbo	770	780
Gas rock	780	786
Rock	786	807
Gas rock	807	810
Rock	810	812
Shale	812	840
Rock	840	850
Gas rock	850	878
Gumbo	878	918
Shale	918	985
Gumbo	985	998
Shale	998	1060
Gumbo	1060	1070
Shale	1070	1072
Gumbo	1072	1146
Shale	1146	1340
Chalk rock	1340	1568
Gumbo	1568	1648
Shale	1648	1660
Rock	1660	1661
Shale	1661	1680
Gumbo	1680	1684
Sandy shale oil showing	1684	1700
Good showing oil and gas from 1680 to 1700; reamed and set 8″ casing at 1348′		
Made test; no result. Pulled 8″ casing and continued drilling		
Shale	1700	1780
Gumbo	1780	1785
Shale	1785	1800
Rock	1800	1805
Shale	1805	1806
Gumbo	1809	1886
Shale	1886	1920
Gumbo	1920	1930
Shale	1930	1974
Gumbo	1974	2050
Shale	2050	2124
Gumbo	2124	2150
Rock	2150	2157
Gumbo	2157	2160
Rock sand	2160	2165
Shale	2165	2170
Sand and shale	2170	2185
Soft rock	2185	2186
Oil sand	2186	2200
Set liner; well's initial production about 80 bbls.		

Casing record:
Set 10″ at 202 Gumbo
8″ at 1348 Chalk
6″ at 2030 Gumbo
4½ perf. 172.1″ blank 46′

Strata are penetrated by breaking, pulverizing and mixing with water the formations encountered. This mixing of the drillings with water forms a "mud" or emulsion that is readily removed from the hole by means of a circular tubular bailer or sand pump. (Figs. 73–80.) To drill through the various formations a drilling bit, auger stem, set of jars and a rope socket (constituting what is commonly termed a "string" of tools) are used. The tools comprising this "string" are screwed together by means

Typical Well Records—"Logs"—*Continued*

Daniel Baker No. 1, near Lone Pine. Amwell Twp., Washington Co., Pa., **Approx.** Elev. 1135′.

	Top	Bottom
Surface	....	16
Waynesburg coal	205	209
Sand	490	525
Pittsburgh coal	525	533
Sand	570	585
Red rock	790	810
Little Dunkard sand	938	953
Big Dunkard sand	1078	1118
Gas sand	1205	1240
Salt sand	1441	1564
1st Water (5 bailers)	1486	....
2d Water (hole full)	1531	....
Black slate	1564	1639
Lime	1639	1659
Black slate	1659	1664
Big lime	1664	1705
Big Injun sand	1705	1965
Squaw sand	2010	2050
Slate and shale	2112	2192
Lime	....	2212
Thirty-foot sand	2277	2292
Slate	2297	2328
Slate	2328	2385
Gantz sand	2431	2458
Broken slate	2458	2465
Fifty-foot sand	2465	2541
Sand	2567	2630
Red sand	2650	2670
Sand	2683	2700
Gordon sand	2705	2723
Sand	2784	2791
Sand	2817	2822
Slate	....	2865
Lime shells	2905	2923
Slate	2935	3000
Lime	3005	3026
Elizabeth sand	3031	3041
Lime	3080	....
Total depth	....	3107
Casing 10″	....	551
$8\frac{1}{4}$″	....	1445
$6\frac{5}{8}$″	....	1664
Plugged 10/24/03		

of two heavy steel wrenches and a powerful rachet jack and horizontal track (Fig. 61), and are suspended on a manila or wire drilling cable which, ordinarily, is connected to a pivoted walking beam by means of a temper screw (Figs. 52–54). In many fields the sinker illustrated is omitted. The power transmitted from the engine (by means of a belt, band wheel, main shaft, crank, wrist pin, and pitman) imparts to the walking beam a vertical, oscillatory motion which lifts the tools and allows them to drop back, striking the formation at the bottom of the hole with destructive force. This operation accomplishes the result sought, as above stated, and it is from this action that the term " percussion system " is derived.

The initial step in actual drilling, however, is termed " spudding," in which a jerk line and spudding shoe are used (instead of the pitman, walking beam, and temper screw) to transmit the desired motion from the wrist pin to the drilling cable and tools as illustrated (Fig. 56). It is obvious that the walking beam and temper screw cannot be used before reaching a depth exceeding the length of the tools, but due to greater resilience, which spudding affords at shallow depths, this operation is usually continued until a depth of from 100 to 300 feet is reached. Then the driller decides to " hitch-on " and use the walking beam and temper screw thenceforth, although the spudding apparatus, with drive head and clamps added, will be used in driving pipe or casing at any subsequent stage (Figs. 56, 107–108).

In drilling through dry formations it is necessary, after removing the drilling tools and bailing out the detritus every few feet, to run water into the drill-hole in sufficient quantity to convert the drillings into an emulsion. This " mud " must be bailed out before it becomes too consistent to permit the free-fall of the drilling tools. If drilling is thus continued too long, the tools become "mudded-up," fail to turn properly, and a "flat" or "crooked" hole results. This condition may also be caused by undue haste in drilling and improper emulsification of the detritus. It can usually be remedied by filling the flat or crooked portion with crushed rock, small pieces of cast iron or other durable material and then carefully redrilling with a thick, specially-dressed bit or a star bit (Fig. 67).

This method is known as "dry-hole" drilling and, particularly when using a manila drilling line, is conducive to much greater speed than can be attained with the drillhole full of water. The friction of water on a steel cable is less obstructive but finally becomes a serious impediment to progress in water-logged wells of great depth.

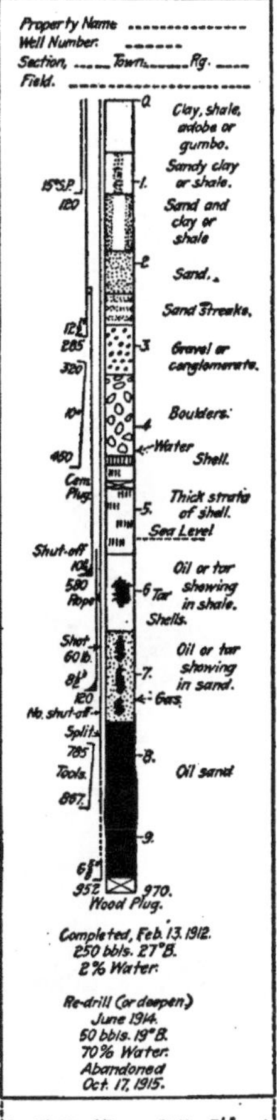

FIG. 103B.—Form of well log used by California State Mining Bureau.

To avoid a flat hole and to assist the bit in mixing the cuttings with water, it is necessary to "turn" the tools continually while spudding or until a depth of 100 to 200 feet has been reached, depending upon the nature of the formation. In hard strata at any depth and in soft formations at greater depths the action of the drilling cable and the glancing blows of the bit tend to impart this turning movement to the tools, but in soft formations at shallow depths an improvised "twister" is applied to the cable just above the derrick floor and manipulated by the driller to insure proper rotation of the bit.

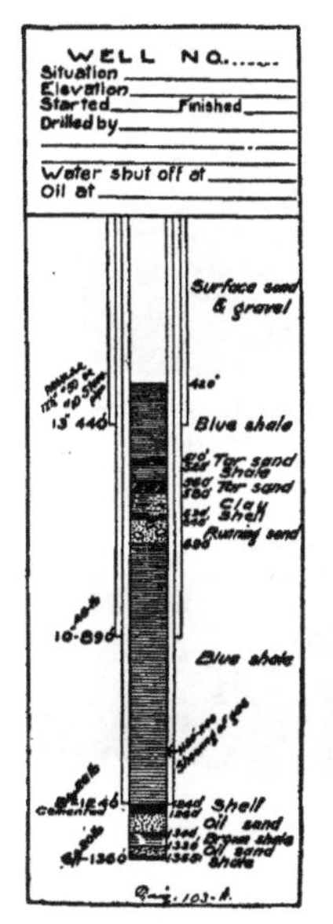

FIG. 103A.—Form of well log.

When the introduction of wire drilling cable was contemplated, hostile critics predicted that it would not impart sufficient rotary movement to the drilling tools and that flat holes and consequent trouble would result. But the use of wire rope in deep drilling is now universal and the development of the swivel rope socket, permitting the tools to rotate independently of the wire cable, has largely overcome the objection mentioned.

Drilling bits wear rapidly and require frequent dressing. When drilling operations are conducted near large camps, it is common practice to have this dressing done at the central blacksmith shop; otherwise the bits are dressed at the well by the crew with sledges and a "ram" or perhaps with a light portable steam derrick hammer. Either oil or natural gas may be used as fuel for heating bits, but soft coal is generally used in most regions. Rapid combustion

CASING APPARATUS AND SHOES

FIG. 104.—Traveling block.

FIG. 107.—Drive clamps.

FIG. 108.—Drive head.

FIG. 112.—Double gate, detachable-link casing elevator.

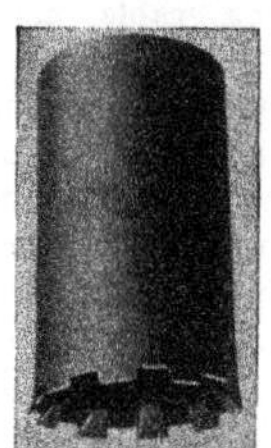

FIG. 109.—Baker casing shoe.

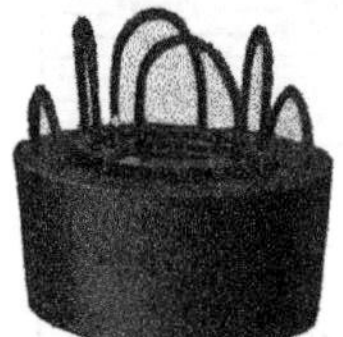

FIG. 113A.—Bushing and slips.

FIG. 105.—Casing hook.

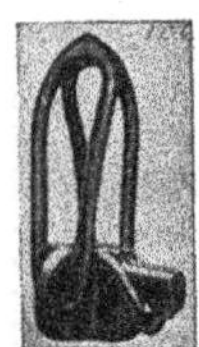

FIG. 106.—Common type of casing elevator.

FIG. 110.—California casing shoe.

FIG. 113.—Spider and slips.

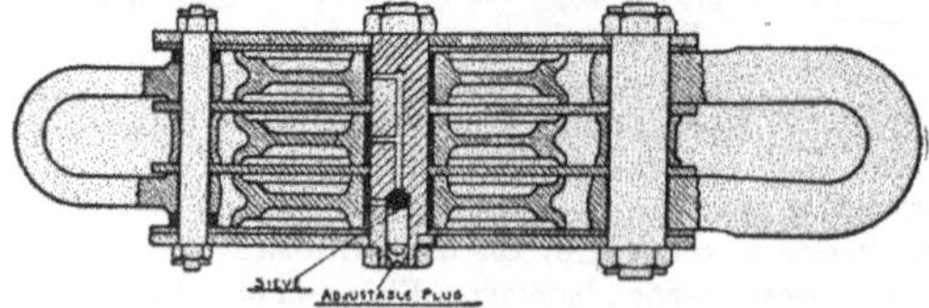

FIG. 111.—Steel pulley block showing method of lubricating the center pin.

of the coal is effected, and intense heat thus produced, by means of an air blast furnished by a "blower" (an enclosed rotary fan operated at high speed by a small, direct-connected steam turbine).

Dressing a large bit by hand (with sledge hammers) is a very laborious procedure and the intense heat radiating from the bit subjects the men, at close proximity, to much discomfort.

Assuming that completion of a well is contemplated with 6⅝-inch 24-pound casing at a depth of approximately 3000 feet, it is probable that an 18-inch hole will be drilled to

TYPES OF UNDERREAMERS

WILSON.

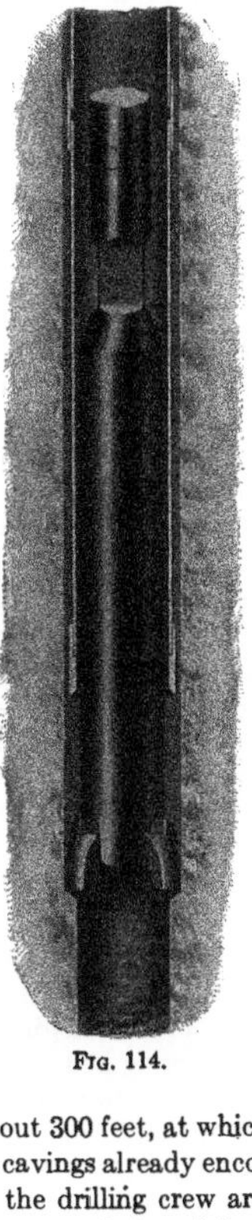

DOUBLE.

SWAN.

FIG. 114.

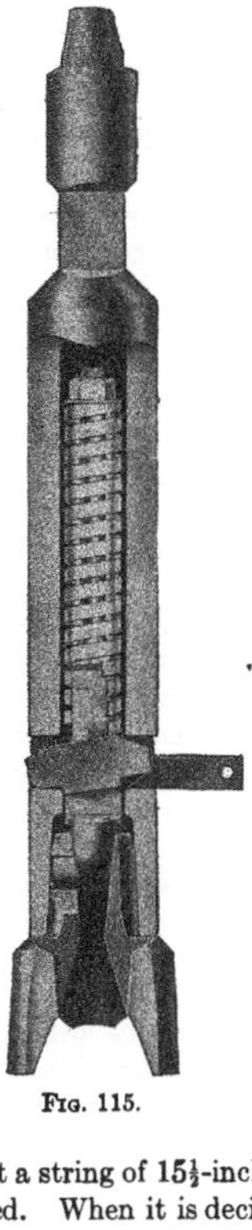

FIG. 115.

FIG. 116.—Showing cutters collapsed ready to enter casing.

FIG. 117.

about 300 feet, at which point a string of 15½-inch casing will be set to exclude any water or cavings already encountered. When it is decided to "run" casing, all four members of the drilling crew are present, working by daylight only. In many regions a casing crew, consisting of five to eight men but under the supervision of the drillers, makes a specialty of this work, relieving the drilling crew of its rigors and expediting the operation.

Casing requirements are anticipated when possible so that each string may be available at the well when needed. It is unloaded from the trucks or wagons at a convenient point near the rig and from there rolled to the floor of the derrick by hand. At that point the joints are individually engaged in a casing elevator, couplings upward, and by engine power (through the calf wheel, casing line, crown block, traveling block, casing hook, and elevator, in the order named) are hoisted to a vertical position in the center of the derrick. After the first joint, bearing the shoe, is inserted in the well and suspended on the spider, or, in some cases, on a second elevator, the succeeding joints are elevated in like manner and joined to the last preceding joint, then lowered into the well. Although this coupling operation is started by hand, heavy casing tongs, the jerk lines and engine power are used to complete it and insure a tight joint. If the joints in a long string of casing are improperly screwed together or if the threads or couplings are defective, a parted string and an expensive fishing job will inevitably result. Many wells have thus been ruined and abandoned.

When the 15½-inch casing has been set, the 18-inch bits will be laid aside and drilling continued with 15½-inch bits to perhaps 600 feet, at which point caving formations may again be encountered and drilling progress impeded. Water has a solvent action on certain strata and causes caving, but in other cases the hole may be purposely filled with water, the pressure of the column of fluid tending to stabilize the walls and overcome this crumbling and caving. If this expedient fails, it becomes necessary to insert the string of 12½-inch casing, which will be " run " to bottom and " carried " on the spider, as sufficient depth to warrant " setting " this string has not been reached. The 12½-inch bits are then substituted for the 15½-inch and drilling proceeds as before, except that the 12½-inch casing must be moved (shifted up and down) at frequent intervals to prevent " freezing." As the depth of the well and the length of the string of casing increase, the friction caused by the caving and unstable walls make it increasingly difficult to keep the casing free. Sometimes driving may loosen and free it, but great care must be exercised in driving casing as the ends of the joints do not butt in the coupling and the stress is thus borne entirely by the threads. Before and after driving either casing or drive pipe, the casing tongs, jerk line, and engine power should be applied to tighten the joints throughout the string.

If the walls of the 12½-inch hole do not crumble or cave badly a distance of from 20 to 200 feet, or even more, may be drilled without further interruptions, but if caving occurs the hole must be reamed out to approximately 15 inches so that the 12½-inch casing may be periodically lowered to the new depths progressively attained.

Underreaming is accomplished through the medium of an ingeniously designed bit (termed an " underreamer," because it reams under the casing), of which there are several types in use. When it is necessary to lower the casing to a new level, the drilling tools are detached from the rope socket and temporarily stood aside in the derrick. An underreaming string (consisting of the same rope socket, an auger stem about 4½-inches by 32 feet, a set of jars and an underreamer, in the order named or with the jars between the rope socket and the auger stem or between two sinker bars) is then substituted. The cutters of the underreamer (12½ inches in this instance) are " collapsed " so that the tool may enter the casing, but upon passing out at the bottom of the 12½-inch casing the cutters are automatically expanded by a powerful internal spring and when the drilling motion is resumed, the reaming proceeds, enlarging the hole to approximately 15 inches, a size that exceeds the external diameter of the 12½-inch couplings by about 1 inch and thus admits the free passage of the casing. (See Figs. 114–117.)

The underreamer cutters should not be permitted to strike the casing shoe (which occurs when the casing is allowed to follow too closely). If the reaming is to begin at the shoulder or bottom of the 15½-inch hole, as is necessary with certain types of underreamers, the casing must be raised a few feet to prevent damage to the shoe. In any event and with any type of underreamer, the casing must be elevated later in order that the " shoulder " may be reamed off and the casing lowered.

Underreaming progress is generally slower than drilling, as the constant external friction on the cutters prevents the free-swinging motion of the tools so essential to speed in drilling and also causes " key-seating " (vertical channeling of the walls of the hole) if the underreamer is not rotated properly. If the distance underreamed is 20 feet or more an equal amount of casing must be attached to the string, and lowered into the well as needed. The underreamer cutters require frequent attention and must be dressed (sharpened), tempered, and kept approximately to gage, the general principles governing the care of drilling bits being applicable.

When the 12½-inch hole has been reamed to bottom the underreaming string is set aside in the derrick for subsequent use. The regular drilling tools are again attached to the rope socket and drilling proceeds as before until underreaming or other procedure is necessary. The drilling and underreaming tools may thus be interchanged as often as conditions dictate until the 12½-inch casing is " landed " (set) at the desired depth, probably 800 feet in this instance.

Drilling will then proceed to the greatest possible depth with the 12½-inch bits. When this has been reached the 10-inch casing will be run, carried, and landed (with 10-inch bits and 10-inch underreamer) at a depth of about 1500 feet, or at less depth if necessary to shut off water at a higher level. Drilling in the 10-inch hole will be continued as far as possible, after which the 8¼-inch casing will be run (the 5½-inch auger stem and the 10-inch bits will be laid aside at this point and the 5-inch stem and 8¼-inch bits substituted), and if necessary, carried, underreamed, and finally landed at about 2400 feet, or sooner if necessary to exclude water. Drilling with the 8¼-inch bits will be further conducted until necessary to insert the 6⅝-inch casing when the drilling tools must again be changed, smaller tools (6⅝-inch bits and underreamer and perhaps a 4½-inch by 32-foot auger stem) are used thenceforth until the desired or greatest possible depth is reached, probably 2800–3200 feet.

At this point many factors and possible contingencies must be duly considered. It may have been impossible to make reasonable progress by employing the methods suggested or originally contemplated, and various slight modifications or even radical changes (including the Hydraulic Circulating System, Fig. 55, described later in this chapter) may have been necessary at one or more stages. Where water-bearing strata are frequently encountered, casing troubles will be numerous and difficult of solution and drilling progress correspondingly retarded. Because of the external pressure of boulders or of water at great depth, the casing may collapse and require swedging or the removal of the entire string to replace damaged joints. Casing " seats " may give way, necessitating deeper drilling to find a harder or more suitable formation for setting. A column of water or mud-laden fluid may be required to hold back troublesome gas and cementing (discussed later on) may be necessary at some point. Finally, unfavorable indications or insuperable difficulties at any depth may suggest or necessitate pulling the casing, abandoning the uncompleted well and the entire region, or merely moving the rig and outfit to a new location, 25 to 100 feet distant, to start anew.

If, at any stage of drilling, the probability of encountering oil or gas in volume is indicated, drilling is suspended until necessary precautions, including provisions for controlling the expected flow, are taken. This may suggest the setting or cementing of the casing, even if necessary to plug the bottom of the hole temporarily until casing can be set and cemented in a suitable formation at a higher level, after which drilling to the oil stratum may safely proceed. Many valuable wells have been damaged or ruined because precautionary measures were totally disregarded.

Where the formations are known it is customary to set and, if a gusher is expected, to cement the last string of casing in the cap rock or hardest stratum overlying the producing formation. A heavy gate valve or other high-pressure control is attached to this string of casing at or under the derrick floor and drilling is continued through this appliance until the well is completed or the pay stratum is penetrated, when the drilling tools are withdrawn or blown from the hole and the well is " shut-in " or the flow of oil is directed into storage tanks. If unusually high pressures are expected, as in the Mexican fields, the several strings of casing are clamped together at the top (by means of casing clamps) and securely anchored, in order that their combined weight and the external friction against them may be sufficient to overcome the ejecting pressure from below.

In California, gushers producing upwards of 10,000 barrels daily have been promptly and successfully capped without special preparations or appliances, but the famous Lakeview gusher (producing 40,000 barrels daily) was lost as was the much larger Dos Bocas gusher in Mexico, through lack of authentic logs and proper precautionary measures. But after tremendous losses in oil and at great expense two other Mexican gushers (Potrero del Llano No. 4, flowing perhaps 100,000 barrels and Cerro Azul No. 4, spouting more than 200,000 barrels daily) were successfully capped and saved, two achievements that rank high in the annals of practical oil-held engineering.

If a prolific oil sand is encountered and considerable gas is present, the well may produce naturally, that is, flow steadily or " by heads " (intermittently). If the oil sand is rich but hard and compact, a charge of nitroglycerine may be exploded to shatter it and facilitate the ingress of oil into the well after which it may be pumped to the surface. But if testing for deeper sands is desired drilling may be continued with the $6\frac{5}{8}$-inch bits, running, underreaming, and setting $5\frac{3}{16}$-inch casing, obtaining production or drilling to the greatest possible depth with $5\frac{3}{16}$-inch bits, and finally, pulling all casing that can be withdrawn and abandoning the well if a " dry hole " (non-producer) results.

Removal of drillings.—The movement of the tools in drilling mixes the cuttings with water (either present or purposely run into the drill-hole) and at frequent intervals —usually after drilling about 6 feet—the tools are withdrawn and the mud or emulsion is " bailed out " by means of a tubular bailer or sand pump (Figs. 73–80). The cuttings resulting from drilling most formations remain in solution for hours or even days, and readily enter and are trapped in a dart-bottom bailer, Fig. 78. White lime, gravel or sands that quickly precipitate may be removed by means of a suction sand pump, Figs. 73, 75 and 76.

The hydraulic circulating system (Fig. 55) may be adopted as a modification of the cable-tool system of drilling, but the fullest consideration should be given to all factors before installing this adjunct as its adaptability is limited and the extra expense is considerable. The principle embodies the addition of mud-flushing (a fundamental characteristic of rotary drilling) to regular cable-tool methods and its application has given gratifying results in stabilizing caving walls and smothering or controlling troublesome gas-bearing strata.

The additional equipment necessary to install the circulating system comprises one or two heavy slush pumps (the 10 by $5\frac{7}{8}$ by 12-inch size being suitable) with connections, discharge hose, and a circulating head. The latter consists of a special casing head with one or two lateral inlet openings and a hollow polished plunger which encircles and grips the wire drilling cable and oscillates vertically with it through a packed stuffing box, thus preventing the escape of the circulating fluid while drilling progresses. Two, three, or four sump holes are dug near the derrick (Fig. 118) and as the mud-laden fluid (water and plastic clay, thoroughly mixed) is pumped down inside the innermost string of casing, it must return outside this string to the surface, carrying the drillings with it and plastering the walls of the hole en route. The fluid emerges under the derrick floor and flows into the outermost sump hole, then through a circuitous settling ditch to the inner sumps. The cuttings are thus permitted to settle before reaching the innermost sump in which the clay and water are mixed and periodically agitated. It is from this sump that the slush pump secures the mud-laden fluid.

If water alone is used as the circulating medium its solvent action may induce more caving or it may disappear in porous strata, but if clay of suitable quality is added in proper proportion to meet the requirements, the walls of the hole will be thoroughly plastered and further caving prevented. A column of pure water exerts a pressure of 43 pounds per square inch for each 100 feet in height but with mud-laden fluid this pressure may be increased to 50 or 55 pounds for each 100 feet or to 1100 pounds at a

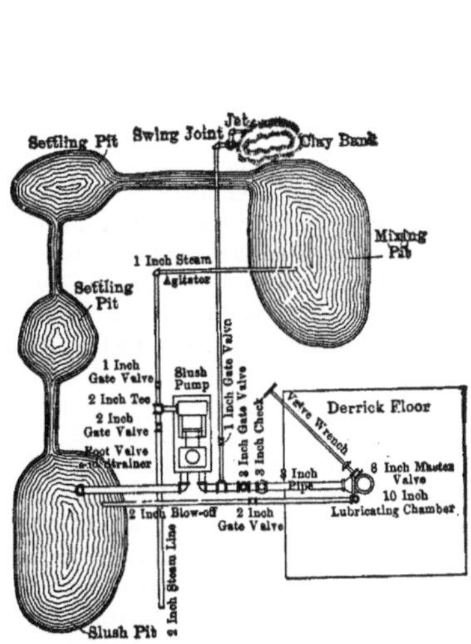

Fig. 118.—Ground plan of equipment used in handling mud-laden fluid (Bulletin 134, U. S. Bureau of Mines).

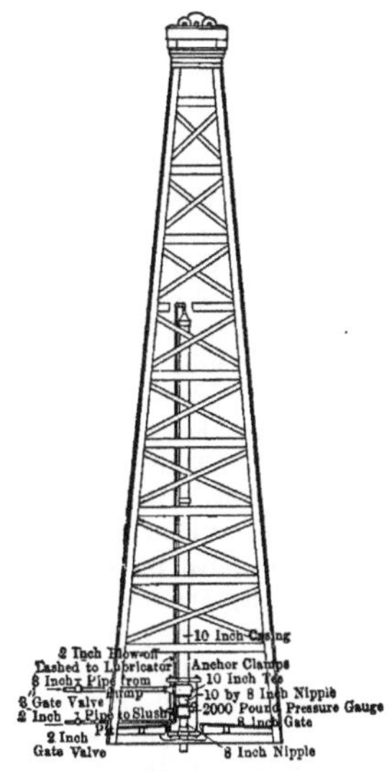

Fig. 119.—"Lubricator," a device for introducing mud-laden fluid against heavy natural gas pressure. (Bulletin 134, U. S. Bureau of Mines.)

depth of 2000 feet. This lateral pressure forces the mud into sands and porous structures, stabilizes caving formations and effectively shuts off water, gas or oil, and because of this, the circulating system should not be used in "wildcatting" nor in close proximity to known sands from which production is expected.

The addition of the circulating apparatus does not materially alter the regular drilling procedure already described, except that in order to maintain an effective circulation which will wash out drillings, the casing must be more frequently lowered to keep the shoe near the bottom of the hole. The interchange of drilling and underreaming tools must, therefore, occur at shorter intervals—from 10 to 30 feet generally being drilled at each "run."

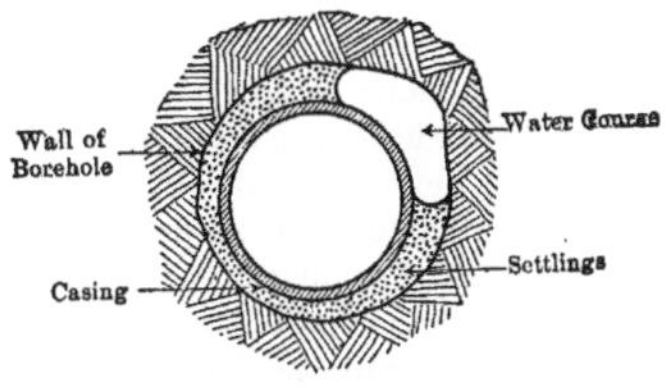

Fig. 120.—Cross-section of well. (Bulletin 134, U. S. Bureau of Mines.)

Due attention must be given to the consistency of the mud-laden fluid; otherwise it may become too thin and be absorbed by porous formations or too heavy and retard drilling progress by obstructing the free-fall of the tools.

The hydraulic circulating system may also be employed to free "frozen" casing, but in any operation involving this principle, the innermost casing must be completely encircled by the rising column of fluid in order that channeling (Fig. 120) and permanent "freezing" of the casing may be avoided.

The mud lubricator (Fig. 119) is sometimes installed to smother or control strong gas flows. A gate valve is attached to the casing and closed, then surmounted by a swedged nipple, tee, and one, two or three joints of casing of large diameter (8¼, 10 or 12½ inch) as shown in the figure. By means of another swedged nipple or reducer the top of the column is reduced and a 2-inch line connected to serve as a vent or blow-off, as illustrated. This vent or relief is opened and mud is pumped into the "lubricator" until it overflows at the end of the 2-inch line, showing that the lubricator is filled. The pump is then stopped the relief valve is closed, and the large gate valve on the casing is opened allowing the mud to gravitate to the bottom of the well. By sounding (tapping) on the lubricator the driller determines when it is empty. The casing gate valve is then closed, the relief gate valve opened to permit the gas to escape, and the lubricator is again filled, the operation being repeated as often as necessary to overcome the gas pressure. The lubricator may, however, be erected in a corner of the derrick and connection with the well made through a lateral opening in a casing head or casing head control thus permitting the mud-laden fluid to be run into the well while drilling progresses. The steam pump used may be installed to work automatically, stopping when the relief valve is closed and starting when it is opened again.

Portable drilling machines.—These are light, compact, modified cable-tool drilling outfits mounted on wheels (Fig. 121). Because of the greater initial cost of the standard rig and heavy cable-tool outfit and the expense involved in dismantling and moving it, the portable machine has come into common use for shallow-well drilling.

These portable outfits are made in many sizes and designs. The lighter ones, suitable for depths not exceeding 800 feet, are operated either by an internal-combustion engine or by a light steam plant mounted as an integral part of the machine, either with or without the tractor feature. These lighter outfits generally manipulate the drilling tools by the spudding principle and are often called "spudders." They weigh from 6000 to 20,000 pounds complete. The heavier machines, built in sizes to drill from 600 to 4000 feet in depth, generally adopt the pivoted walking-beam feature. They use a 45 to 60-foot mast (substitute for derrick), a 15 to 30 horse-power steam engine mounted with the machine, and a 20 to 30 horse-power firebox boiler, individually mounted. These machines, complete with boiler and tools, weigh from 25,000 to 85,000 pounds.

The portable drilling machine is a radical modification of the standard percussion outfit, much of the power and many of the conveniences of the latter being sacrificed to attain the desired compactness and portability. In many regions the portable outfit is conducive to greater economy in shallow drilling and while it is possible, under favorable conditions, to reach depths of from 2000 to 4000 feet with the heavier portable machinery, only standard cable-tool or rotary outfits are in general use for drilling wells exceeding 1000 to 1500 feet in depth.

The cost of drilling is dependent upon so many variable factors that averages, if they could be computed, would be of little or no value. The determining influences are numerous, but the following are the most important: the distance of the well from available sources of supplies; the depth to be reached; the character of the formations to be penetrated (these in turn determining the system to be employed and the kind and quantity of casing to be used); the number of water-bearing, gas-bearing, caving and other troublesome strata encountered and the cost of labor, fuel and water. A 1000-foot well may cost from $1500 to $15,000, and a 2500-foot well may cost from $20,000 to $150,000, accidents, fishing jobs, and consequent delays contributing largely to unusually high costs. The time required may vary from fifteen or twenty days on a 1500-foot well to a year or more on a 5000-foot one, although the time necessary to drill wells of from 2500 to 3200 feet in depth ordinarily varies from about thirty to perhaps one hundred days.

The lighting plant.—As drilling operations, regardless of the system employed, usually continue throughout the twenty-four hour day, artificial lighting is necessary at night, and the gas lights, lanterns and open derrick lamps formerly used have been largely displaced by a compact 1 or 1½ kilowatt, direct-connected, steam turbo-generating electric-lighting outfit. This plant is efficient, durable, simple in operation and eliminates the fire hazards peculiar to the other systems of lighting.

THE ROTARY SYSTEM

Boring and mud-flushing, the basic principles of rotary drilling, have long been known, but their combination, as now applied to deep-well drilling, is of comparatively recent occurrence. Authorities have never designated a definite date to mark the advent of the rotary in oil-field drilling and the facts concerning its introduction, experimental stage, and early development are somewhat obscure. It appears to have been used successfully in Texas, both at Corsicana and at Beaumont, about the year 1900. But it made very little impression outside of Texas and Louisiana until 1910, since when its evolution has been rapid. Its possibilities have been recognized, and its adoption is now almost universal in regions to which it is peculiarly adapted—Louisiana, Trinidad, and parts of Texas, Oklahoma, California, Mexico, Russia and South America.

The operation of the common rotary bit resembles boring with an auger, except that mud-laden water is pumped, under high pressure, downward within the rotary drill pipe and jetted out with great force through two small apertures in the rotary bit against the bottom of the hole. This fluid then returns outside the drill pipe to the surface, carrying the borings with it, and (subjected to the pounding and the wobbling motion of the rotary drill pipe) plasters and stabilizes the unlined walls of the bore-hole en route.

The objective sought in rotary drilling is the same as in cable-tool drilling, already explained, although the methods of procedure are radically different. Whereas the power plant of a cable-tool outfit transmits power through several intervening appliances to the temper screw and drilling cable, imparting to the drilling bit a vertical, oscillating motion, the similar power plant of a rotary outfit, by means of sprocket chains, draw works (a countershaft and drum substituted for band wheel, calf wheel, bull wheels and sand reel), geared turntable and drill pipe, imparts to the rotary bit a continuous rotary motion. Instead of temporarily suspending drilling to bail out the detritus, as with cable tools, the rotary cuttings are washed out while drilling proceeds.

The rotary method is conducive to great speed in drilling wells where the strata are not hard. Caving and heaving formations that baffle and delay the cable-tool driller are easily penetrated and sealed (mudded off) by this method. But the rotary is not so sensitive as cable tools to sudden changes in strata nor so responsive to the varying requirements of different conditions and regions and is, therefore, still subject to hostile criticism. As the rotary drillings are discolored and mixed with other materials and since much time must elapse before the borings can reach the surface and be examined by the driller, it is exceedingly difficult to determine their nature and the exact depth at which they were encountered. Drillers, who through long experience in one region, can accurately estimate the value of " showings " there, may be unable to determine the significance of similar or different indications in other fields. Slight showings that would be clearly manifested in dry-hole, cable-tool drilling may be held back and effectively mudded off, due to the presence of mud-laden fluid under high pressure, but rotary drillers are acquiring proficiency in detecting slight changes in formations and in evaluating any information obtained from the drillings or from the intonation of the engine and machinery.

It will be observed from the list of regional casing requirements, that wells drilled with rotary machinery require less casing than if drilled with cable-tools, the plastering of the walls being a substitute, to a certain extent, for casing. As the cost of casing is one of the chief factors in the expense of drilling, this advantage is frequently the deciding factor in choosing the system of drilling to be employed.

But, while conceding that the rotary method possesses many advantages in drilling soft formations, it is not so suitable as cable tools for prospecting, for drilling through hard rock, or for "bringing in" a producing well. The following statement by an eminent petroleum engineer is illuminating:

Fig. 121.—Portable drilling machine, heavy type.

"One of the several visits I have made to the Texas oil-fields was in October, 1901, and on that occasion I saw a well on the Hogg-Swayne tract of Spindle Top that had been drilled into the known oil horizon by the hydraulic rotary method, but made no showing of oil, notwithstanding that within a radius of 250 feet around it were a number of gushers, and the original pressure of Spindle Top wells had been reduced but little. After this well had been agitated and bailed for six days, it finally gushed and flowed a solid 6-inch stream of oil until shut in. If this had been a prospect well in a new district, where the oil horizon had not been definitely located, there evidently would have been a strong possibility of passing the oil stratum without revealing its productive capacity."

It should be noted that the foregoing incident occurred during the pioneer stage of rotary drilling when drillers were inexperienced and perhaps very inefficient compared with the expert rotary drillers of the present, and while it is true that a thin producing sand, exhibiting little or no gas, might be passed unnoticed by the rotary crew (or even by a cable-tool driller, if drilling with considerable water in the drill-hole) this disadvantage is far outweighed by the gain in speed and economy. The rotary is steadily growing in popularity in regions where the producing strata are known and the overlying formations contain little or no hard rock.

Factors in Rotary Drilling

The factors in rotary drilling comprise those mentioned in the discussion of cable-tool drilling with certain modifications, and in addition the hoist, rotary, grief stem, drill pipe, pumps, swivel, bits and the mud-laden fluid. The power plant used in rotary work is similar to that furnished with cable-tool outfits, except that a larger boiler (45 to 66 horse-power) is required, as the engine and one of the slush pumps must run simultaneously, requiring a greater volume of steam. In California electric motors are now being used on a few outfits to operate not only the draw works and rotary but also the geared or belt-driven slush pumps.

The rotary derrick may be 84, 96, 106, 112, 120 or 130 feet in height, with either a 20-, 22-, 24-, or 26-foot base. It is similar to the cable-tool derrick, except that additional height is added to permit standing the rotary drill pipe in the derrick in " stands " (sections) of three or four joints each. The three-joint stand is about 67 feet and the four-joint stand 90 feet in height. The rotary hoist supplants the band wheel, calf

FIG. 122.—Standard type of drilling engine equipped with sprocket for use on rotary outfit.

wheel, bull wheels, and sand reel used in cable-tool drilling. These and the rig irons are, therefore, not included with a rotary rig, but must be added later if it is desired to " standardize " and drill in or complete the well with cable-tools, as explained in the first section.

(*a*) **The hoist or draw works.**—(Figs. 123 and 124).—The hoist is placed between

FIG. 123.—Standard type of draw works—front view.

the engine and the rotary machine (turntable) to transmit power to the latter and, through the casing line to the traveling block, casing hook and elevators or swivel. It consists of a countershaft, drum shaft, and drum, with necessary sprockets, sprocket chains, three clutches and three upright wooden or steel posts acting as supports or framework. The first hoists used were single-brake, one-speed machines made largely of cast iron and weighing about 2500 pounds, but the modern two-speed types are made of steel fitted with double brake and weigh from 7000 to 12000 pounds.

(*b*) **The rotary.**—(Sometimes called the rotary machine or the "turntable.")—The rotary (Figs. 125, 126) is made almost entirely of cast steel. Its principal parts are a sturdy base and a geared turntable. The turntable rotates on roller bearings running in oil and by means of a central drive bushing imparts a rotary motion to the grief stem, drill pipe and bit. Grip rings were formerly attached to the turntable to grasp the drill pipe but often crushed or damaged it, resulting in the introduction of the fluted grief stem. This is a circular steel bar about 5½ inches in diameter and 30 feet long with a 2½-inch water course bored longitudinally through the center and with four V-shaped grooves cut lengthwise on the surface to accommodate, and make rigid contact with, the grip rings or with the central drive bushing. The 4-inch by 30-foot or 6-inch by 45-foot square grief stem is also commonly used now and is perhaps superior to any other. It is fitted with two couplings, to connect with the swivel above and with the drill pipe below. By means of the hoist the grief stem, passing through the rotary turntable,

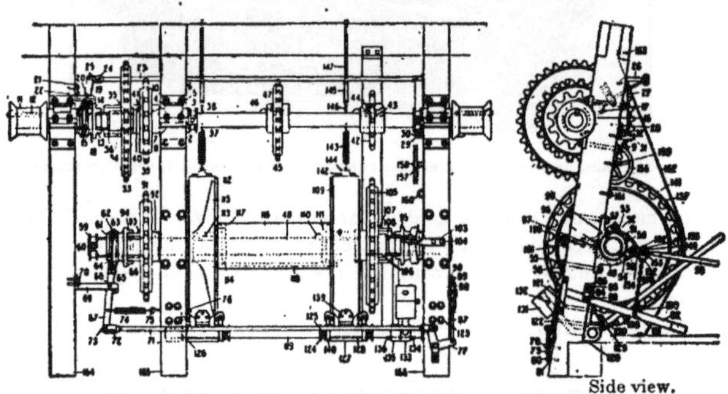

Fig. 124.—Hoist, or draw works. (*a*)

(*a*) For further explanations see "Oil Well Supply Co. Catalogue."

may be raised and lowered about 25 feet imparting the same movement to the swivel, drill pipe and bit without interfering with the rotary movement.

The size of a rotary is designated by the diameter of the circular opening in the turntable. Sizes vary from 10-inches to 30 inches, but there is an apparent tendency to standardize on the 20-inch size because of its adaptability to practically all requirements. The weights vary from 2000 to 7500 pounds, depending upon the depth to be drilled and the sizes of the drill pipe and casing to be used.

The grip-ring type of rotary is now practically obsolete, being displaced by the "flat-top" machine and the more modern double-turntable rotary (Figs. 125, 125-A, 125-B). The latter embodies many new and very advantageous features, especially in the handling of drill pipe or casing.

It is necessary to withdraw, examine and replace the rotary bit at frequent intervals, an operation which involves unscrewing the drill pipe at every third or fourth joint and standing these sections in the derrick. When this is desired the drive bushing is laid aside and serrated slips (as shown in Figs. 125–126) are substituted. The drill pipe is then elevated and suspended in these slips (seated in the turntable) the swivel and grief stem are set aside or inserted in the "rat-hole," and an elevator, Fig. 125, is used in connection with the casing hook and traveling block to lift the drill pipe, the uppermost joint of which protrudes above the turntable about 30 inches (including the coupling or rotary tool joint, to be described later). The entire string is lifted sufficiently to free the slips which are then removed but immediately replaced when the three- or four-joint stand has been raised clear of the turntable. The upper half of the tool-joint coupling extending just above the rotary is then grasped and rigidly held by a heavy pipe wrench ("breakout tongs," Fig. 133). The turntable is immediately rotated turning the suspended string of pipe, as in drilling, and this "breaks" (unscrews) the tool-joint coupling. The stand is then set aside in the derrick. The

Fig. 125.—A modern rotary, showing slips, break-out tongs, drill pipe, double gate elevator and draw works.

"derrick man," stationed on a platform at suitable height, disengages the elevator, after which it is lowered to the derrick floor to repeat the operation.

When the bit is reached and the examination or necessary change is made, the string of drill pipe is replaced in the well. The withdrawing procedure is reversed, except that with the older style rotaries it is necessary to "make up" (screw together) the joints by hand and to complete the tightening process by means of tongs and a manila jerk line running over one of the "catheads" (winch heads) on the countershaft of the draw works. The latest rotary is equipped with a double turntable and, while simple in construction and operation, it "makes up" both drill pipe and casing, thereby contributing greatly to speed and safety and relieving the drilling crew of these arduous tasks. With this improved machine the joints or couplings may be unscrewed without rotating the suspended drill pipe and bit, thus eliminating the danger of injuring the

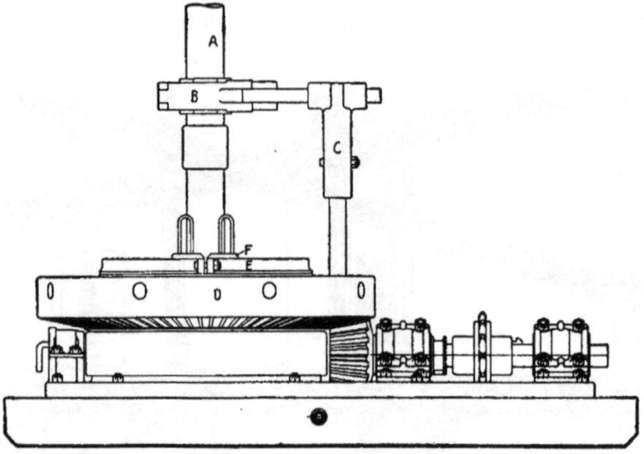

FIG. 125*A*.—The latest type of rotary (double turntable). Side view. Showing method of "making-up" or "breaking-out" casing or drill pipe. The inner table (*E*) and slips (F) remain stationary, holding the lower joint of casing, while the outer table (*D*), the post (*C*), and the tongs (*B*) rotate the upper joint (*A*) in either direction.

unlined walls of the bore hole and overcoming an acknowledged imperfection in other rotaries. By the addition of a simple cutter and stock-and-die attachment the improved rotary may also be used as a pipe machine to cut and thread drill pipe and casing (Fig. 125-B).

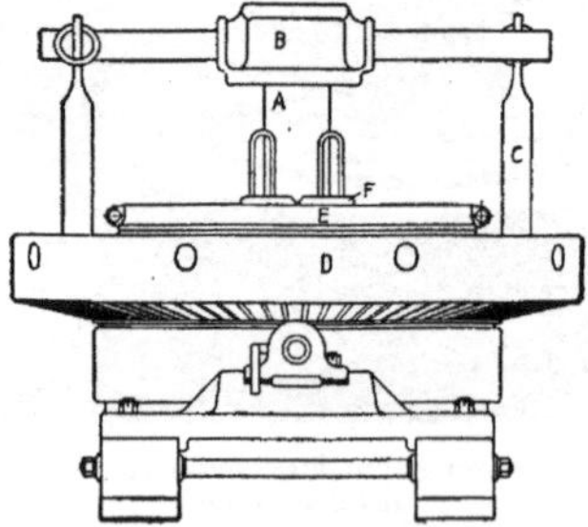

FIG. 125*B*.—End view. Showing position of pipe threading attachment (stock and dies, *B*) in threading rotary drill pipe (*A*), the outer table turning clockwise, parts *A*, *E* and F remaining stationary.

(c) **The drill pipe.**—Drill pipe bears about the same relation to a rotary outfit that drilling cable bears to a cable-tool outfit. Both are subject to great stresses and constant wear and are important factors in the expense of drilling. It is impossible to complete a well economically with weak or defective drill pipe and its condition is a matter of great importance and constant concern to the driller. The slipping of the grip rings on the old-style rotary, while occasionally damaging a joint of drill pipe and delaying drilling accordingly, acted as a safety appliance and danger signal when hard or unyielding formations were encountered, but the positive action of the rigid grief stem, while contributing to speed and economy, requires keener perception and greater vigilance on the part of the driller to avoid twist-offs and consequent fishing jobs.

Rotary drill pipe is made in three grades (regular or "special," upset, and seamless upset) of either wrought iron or mild steel and in diameters of 2½, 3, 4, 5 and 6 inches, the 4 and 6-inch sizes being most popular. The joints are from 20 to 22 feet long,

Fig. 126.—Interior view of combination rig in California, showing the rotary, grief stem, swivel and hose in the foreground and the spider, slips and draw works in the background.

including the coupling. The 4-inch commonly used weighs from 12½ to 15 pounds and the 6-inch from 19½ to 29 pounds per foot. Eight threads per inch are standard and the taper is ¾ inch diameter per foot of length for all sizes.

A break or twist-off may occur at any point and the frequent occurrence of fractures at the threads prompted manufacturers to introduce the upset end pipe (Fig. 135). This improvement makes the threaded portion as strong as the central part of the joints and has done much to minimize the danger of twist-offs. The greatest strain on the drill pipe obviously occurs just above the rotary bit and to strengthen and prevent wobbling at that point and to insure a better hole, drillers sometimes use one or two

REGULAR AND GIANT MUD PUMPS.

(10″×5¼″×12″) (12″×6¾″×14″)

FIG. 127.—A popular slush pump.

Size	Diameter of steam cylinder, inches	Diameter of water cylinder, inches	Stroke, inches	Gallons per revolution	Steam pipe, inches	Exhaust pipe, inches	Suction pipe, inches	Discharge pipe, inches	Discharge from air chamber, inches	Weight, lbs.
"Regular"..	10	5¼	12	5.6	2	2½	6	4	3	4500
"Giant"...	12	6¾	14	8.675	2½	3	8	6	5	6900

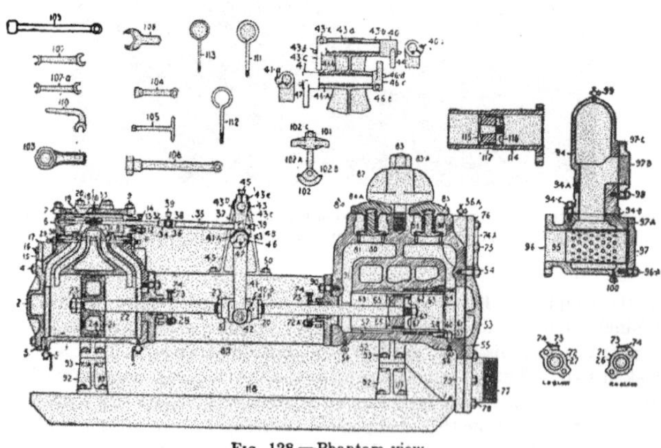

FIG. 128.—Phantom view.

joints of 6-inch drill pipe between the bit and the bottom of the regular string of 4-inch drill pipe.

Regular, or special, rotary drill pipe is made by the lap-weld process, leaving a welded, longitudinal seam which is sometimes opened by the torsional strain, causing the pipe to " corkscrew " and twist off. To overcome this danger in deep drilling and in hard formations, seamless upset drill pipe is now being manufactured, representing another milestone in the evolution of rotary drilling equipment.

Rotary tool joints (Figs. 136–140) are now in general use. These are special drill-pipe couplings similar in principle to the standard tool joint (Fig. 60) described on page 161, except that a circular water course is bored longitudinally through the center and the ends are finished to serve as a regular drill-pipe coupling. Rotary tool joints, because of the 24° taper and the coarser threads, can be " broken " (unscrewed) more quickly than common pipe couplings, and in addition, are much more durable. The box and pin may be fitted with either 4, 5, 6, or 7 flat threads or 4 acme, or improved acme, threads per inch. Some operators use one tool joint on each stand of pipe while others discard all regular couplings and use a tool joint on every joint of drill pipe, claiming that increased speed and ultimate economy justify the extra initial expenditure.

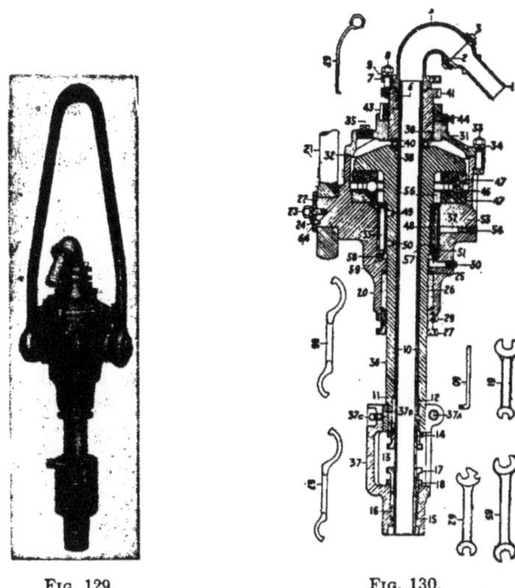

Fig. 129. Fig. 130.

The improved oilbath swivel. (a)

(a) For further explanations see " Oil Well Supply Co. Catalogue."

(*d*) **The slush pumps.**—These pumps (Figs. 127, 128) must deliver mud-laden fluid under varying pressures and are subject to extremely hard usage. They require considerable attention and, while it is not necessary to operate two pumps simultaneously, repairs and replacements of worn parts must be made at frequent intervals, resulting in costly delays if the outfit is equipped with only one pump. Slush pumps are made in two sizes, the regular 10 by 5⅞ by 12-inch being suitable for ordinary drilling requirements and the 12 by 6¾ by 14-inch " Giant " which is adapted to the most exacting service. Air chambers are recommended and customarily furnished and the pumps must be so constructed that worn-out parts may be quickly renewed.

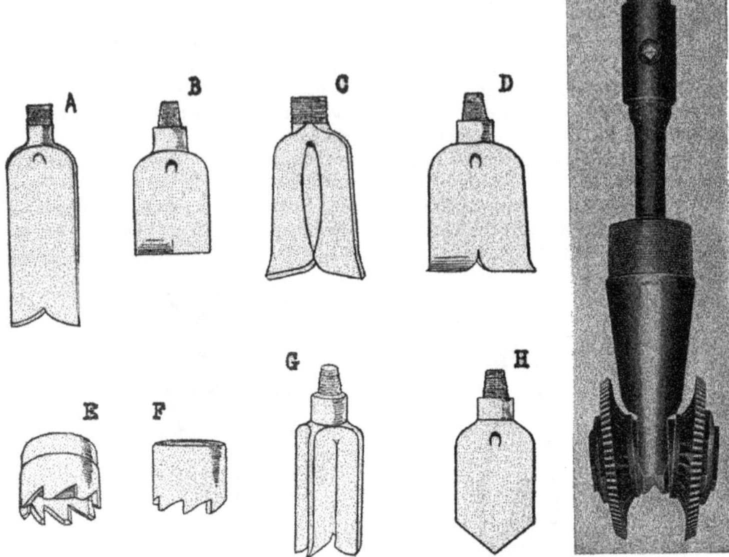

I, disk bit.

FIG. 131.—Types of rotary bits and serrated rotary shoes. *A*, *B*, *C* and *D* fish tail bits; *E* and F, rotary shoes; *G*, 4-wing bit; *H*, diamond point bit;

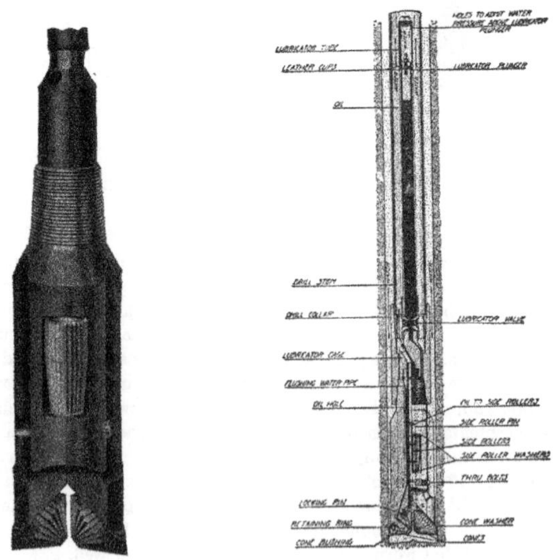

FIG. 132.—"SimpleX" type. FIG. 132*a*.—Sectional view of simpleX rock bit with lubricator.

(*e*) The rotary swivel.—The swivel used in rotary drilling is a sturdy mechanical device designed to permit the free turning movement of the grief stem, drill pipe and bit without imparting similar motion to the hose and suspending appliances above, and meanwhile to facilitate the introduction of mud-laden fluid, under pressure from the slush pump, into the grief stem and drill pipe. The swivel must carry the weight of the grief stem and drill pipe, entirely or in part, and is subjected to sudden stresses and constant wear. The older types of swivels were not free from binding, quickly wore out,

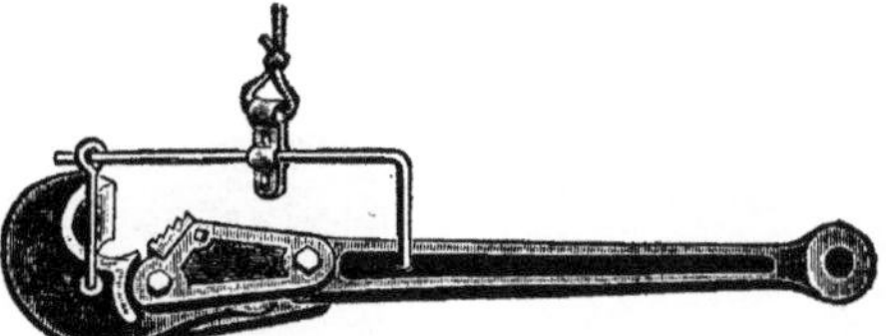

FIG. 133.—Break-out tongs.

required frequent lubrication, and much care and attention but the modern " oil-bath " swivel (Figs. 129, 130) fitted with ball bearings and roller bearings, has overcome these imperfections, thereby avoiding delays and relieving the drilling crew of much trouble and annoyance.

(*f*) The bits (Figs. 131 to 132-A).—Regular fish-tail bits (Fig. 131, C and D), are generally used in all soft formations, disc bits (I, Fig. 131) being occasionally used for

ROTARY DRILL PIPE COUPLINGS

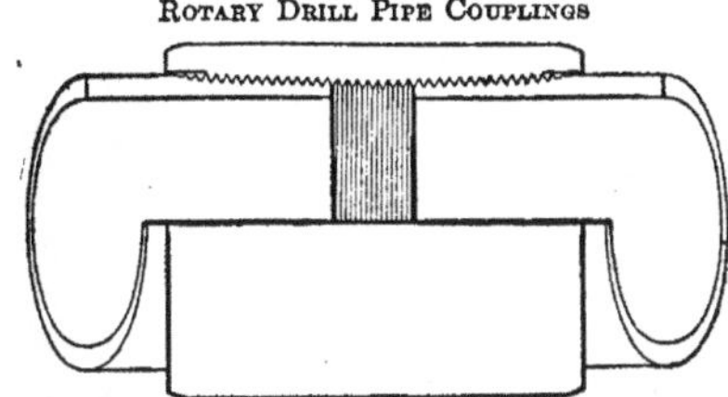

FIG. 134.—Typical section of special rotary pipe coupling and joint.

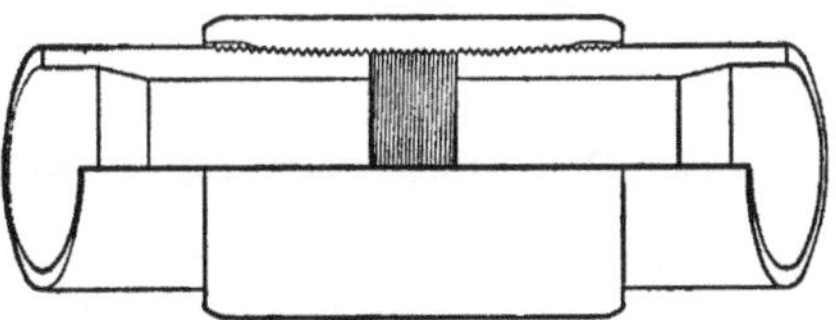

FIG. 135.—Typical section of special upset rotary pipe coupling and joint.

harder shales. The reaming cone rock bit (Fig. 132 and 132-A) is adapted to cutting through hard rocks and the diamond point bit is used in " side-tracking," fishing operations, and sometimes for " rat-holing " (reducing the size of the bore hole and " feeling ahead "). The four-wing bit is used to prove and, if necessary, to ream the hole and eliminate any " knots " or obstructions that might interfere with the free passage of casing.

Serrated rotary shoes (E and F, Fig. 131) may be run and rotated on either drill pipe or casing to remove projecting wall fragments, and in fishing to encircle and free "frozen" drill pipe. When worn the cones and discs are renewed on the rock bits and disc bits, respectively, but the other bits may be sharpened by the drilling crew.

All rotary bits contain water courses, generally two ⅞-inch circular holes bored vertically through the shank of the bit through which the mud-laden circulating fluid is jetted with great force against the bottom of the bore-hole, thereby dislodging the dis-

ROTARY TOOL JOINT

FIG. 136.—Box end.

FIG. 137.—Pin end.

FIG. 138. Box end showing pipe connection.

FIG. 139. Pin end showing pipe connection.

integrated material from the cutting edges of the bit. For drilling in "gumbo" or sticky formations the blades of the bits should be short (6 to 12 inches in length) in order that the fluid may be jetted against the strata with the greatest possible force, thus aiding in removing and pulverizing the cuttings, but in loose, porous or sandy formations, longer bits should be used as additional length causes them to stand erect in the drill-hole, and facilitates their recovery, if lost. For this same reason drillers prefer a long

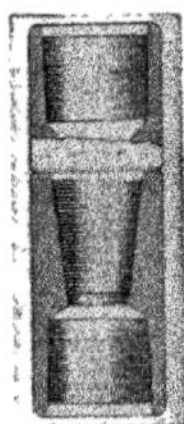

FIG. 140.—Complete joint.

FIG. 141.—Drill coupling.

"drill collar," or drill coupling (3 to 10 feet in length, Fig. 141), to connect the bit with the drill pipe.

The rock bit is now widely used and its effectiveness in cutting (milling) through hard structures has reduced the need of the combination outfit (cable-tools in combination with the rotary), except for prospecting in remote regions. This bit is equipped with a lubricator (Fig. 132-A) which is operated by the pressure of the circulating fluid on the plunger and the action of the revolving steel cones in cutting through hard rock generally results in very satisfactory progress.

ROTARY RIG SPECIFICATIONS

Timbers, Lumber and Nails

Light Derrick

88 Feet High, 22-foot Base

10 pieces 8″×10″×22′ for sills
8 pieces 2″×10″×18′ for starting legs
8 pieces 2″× 8″×20′ for starting legs and top legs
6 pieces 2″×10″×24′ for bottom doublers and V boards
45 pieces 2″×10″×16′ for legs and girts
60 pieces 2″× 8″×16′ for legs and doublers
8 pieces 2″×12″×20′ for first set girts and swivel boards
26 pieces 2″×12″×22′ for floor
9 pieces 2″× 6″×22′ for braces
8 pieces 2″× 6″×20′ for braces
8 pieces 2″× 6″×18′ for braces
16 pieces 2″× 6″×16′ for braces
8 pieces 2″× 6″×14′ for braces
16 pieces 2″× 6″×12′ for braces
8 pieces 1″× 6″×16′ for braces, etc.
10 pieces 2″× 4″×16′ for ladder
7 pieces 1″× 4″×16′ for ladder strips
8 pieces 2″×10″×20′ for derrick top
3 pieces 4″× 4″×14′ for gin poles

Total 7500 feet. 150 lbs. nails.

Estimated weight, 29,000 lbs.

Derrick

100 Feet High, 24-foot Base

6 pieces 10″×10″×26′ for bottom sills
8 pieces 8″×10″×26′ for top sills
8 pieces 2″×12″×24′ for foundation blocks
24 pieces 2″×12″×24′ for floor
4 pieces 2″×12″×24′ for extension for pumps
4 pieces 2″×12″×18′ for starting legs
22 pieces 2″×12″×16′ for legs
4 pieces 2″×12″×20′ for legs
24 pieces 2″×10″×16′ for legs
56 pieces 2″×10″×16′ for doublers
2 pieces 2″×10″×26′ for V board
4 pieces 2″×12″×24′ for girts
4 pieces 2″×12″×22′ for girts
4 pieces 2″×12″×20′ for girts
8 pieces 2″×12″×18′ for girts
20 pieces 2″×12″×16′ for girts
10 pieces 2″× 8″×24′ for braces
8 pieces 2″× 8″×22′ for braces
16 pieces 2″× 8″×20′ for braces
8 pieces 2″× 6″×18′ for braces
16 pieces 2″× 6″×16′ for braces
8 pieces 2″× 6″×14′ for braces
16 pieces 2″× 6″×12′ for braces
8 pieces 2″× 6″×10′ for braces
8 pieces 2″×10″×20′ for top
3 pieces 5″× 5″×16′ for gin pole
12 pieces 2″× 4″×16′ for ladder
10 pieces 1″× 4″×16′ for ladder
4 pieces 1″× 6″×16′ for ladder
3 pieces 3″×12″×24′ for head board to set up hoists
2 pieces 2″×12″×22′ for swivel boards
2 pieces 2″×12″×20′ for double boards
2 pieces 2″×12″×16′ for triple boards
2 pieces 2″×12″×12′ for quadruple boards
4 pieces 2″×10″×16′ for finger boards

Total, 11,780 feet
100 lbs. 30d nails
200 lbs. 20d nails
50 lbs. 10d nails

Estimated weight, 45,550 lbs.

Engine Foundation

2 pieces 16"×18"×14' mud sills
2 pieces 16"×16"×12' pony sills
2 pieces 8"×24"× 9' split block

4—¾"×24" machine bolts and washers
2—⅞"×11' d. c. bolts, washers and nuts
1 piece 1"×4"×2$\frac{10}{12}$' steel plate

Material used in constructing a 112-foot derrick, 24-foot base, for use with rotary outfit in Oklahoma.

Pine Timbers

Pcs.	Dimensions	Board ft.
4	8"×10"×24'	640
2	8"×10"×22'	293
4	8"× 8"×24'	512
1	8"× 8"×20'	108
1	10"×12"×12'	120
2	4"× 6"×14'	56
1	12"×12"×12'	144
1	16"×16"×28'	597
2	14"×14"×18'	588
1	14"×14"×16'	261
2	14"×14"×14'	457
	Total	3776

Girts

Pcs.	Dimensions	Board ft.	
6	2"×12"×18'	216	1st and 2d sets
4	2"×12"×22'	176	
10	2"×10"×20'	333	3d and 4th sets
10	2"×10"×18'	300	5th set
4	2"×10"×16"	107	6th set
22	2"×10"×16'	587	7th to 13th sets
4	2"×10"×22'	147	
	Total	1866	

For Boxing and Engine House

Pcs.	Dimensions	Board ft.
90	1"×12"×18'	1620
45	1"×12"×16'	720
6	2"× 6"×18'	108
	Total	2448

Braces

Pcs.	Dimensions	Board ft.	
10	2"×8"×24'	320	1st set
10	2"×8"×22'	293	2d set
18	2"×8"×20'	480	3d and 4th sets
10	2"×8"×18'	240	5th set
10	2"×6"×18'	180	6th set
18	2"×6"×16'	288	7th and 8th sets
10	2"×6"×14'	140	9th set
18	2"×6"×12'	216	10th and 11th sets
20	2"×6"×10'	200	12th set
	Total	2357	

Miscellaneous

Pcs.	Dimensions	Board ft.	
33	2"×10"×24'	1320	Floor-V-braces
90	2"×10"×16'	2400	Legs and braces
90	2"×12"×16'	2880	Legs and doublers
16	2"× 4"×16'	171	Ladder
12	1"× 6"×16'	96	Ladder
20	2"× 8"×18'	480	Corners
15	2"× 8"×16'	320	Corners
4	3"×12"×16'	192	Corners
	Total	7859	

300 pounds 16d wire nails
100 pounds 30d wire nails
25 pounds 8d wire nails
3 reels ⅜" guy wire (1500')

Grand total board feet...... 18,306 Total weight, approximately... 55,000 lbs.

SPECIFICATION

FOR

ROTARY DRILLING OUTFIT

(Suitable for drilling to 3000 feet)

(*a*) *Machinery, Pipe and Tools*:

1 66- or 2 45-horse-power drilling boilers, complete with stack and fittings.
2 Texas oil burners.
1 3″ flue cleaner with 1 joint of ⅜″ pipe.
1 12×12″ drilling engine, fitted with sprocket for No. 1030 or No. 1240 chain, 50″ fly wheel, lubricator, pump and heater, but less belt pulley.
1 Set engine brace sockets.
2 10″X5⅞″X12″ or 2 12″×6¾″×14″ slush pumps.
1 Set extras (replacements) for pumps.
1 6″×4″×6″ boiler feed pump.
1 20″ steel rotary machine, with spider and slips for 4″ and 6″ pipe and for 6⅝″ and 8¼″ casing, also with bushings for 4″ and 6″ square or fluted grief stems.
1 Extra heavy two-speed semi-steel hoist or draw works, complete with 4⅜″ line shaft, 5″ drum shaft, double brake and 3 wood posts.
1 4″ improved oilbath rotary swivel, complete with 4″ coupling and wrenches.
1 6″ improved oilbath rotary swivel complete with 6″ coupling and wrenches.
2 Pieces 2½″×33′ 6-ply wire-wound rotary hose, complete with couplings for swivel and stand pipe.
1 Piece 1″×25′ 4-ply derrick hose.
1 8″ or 10″ heavy structural steel I-beam crown block, fitted with five sheaves and closed bearings.
1 42″ or 48″ standard triple, or quadruple, bronze-bushed, self-lubricating traveling block.
1 ⅞″×1100′ 6×19 plow-steel casing line.
1 9/16″×3500′ 6×7 steel sand line.
6 ⅞″ wire-rope clips.
6 ⅝″ wire-rope clips.
1 6″ double-swivel spring casing hook   1 10-lb. sucker rod hook for cathead line.
1 4″ new style strapped C-hook.
1 10″ drop-link hartz steel snatch block.
1 4″×30′ square grief stem, with 4″ couplings.
1 6″ ×45′ square grief stem, with 6″ couplings.
1 5½″×30′ dart-bottom bailer.
1 Set 4″ double gate, safety, detachable-link elevators with 2¼″×36″ links.
1 Set 6″ double gate, safety, detachable-link elevators, less links.
50 4″×18″ rotary tool joints, 3½″×4½″—7 box and pin.
30 6″×18″ rotary tool joints, 5″×6″—4 box and pin.
3000′ of 4″ 15 lb. upset-end rotary drill pipe.
2000′ of 6″ 20 lb. upset-end rotary drill pipe.
1 No. 12 flat link chain tongs, 1 No. 13½ flat-link chain tongs.
1 No. 15 flat link chain tongs, 1 No. 16 flat-link chain tongs.
1 4″×60″ steel drilling coupling, 3½″×4½″—7 box.
1 6″×60″ steel drilling coupling, 5″×6″—4 box.
1 1¼″×275′ manila cathead line.

2 4"×6" hydraulic swedged nipples.
25′ of No. 1240 sprocket chain (for line shaft-drum).
65′ of No. 1030 sprocket chain (for engine and rotary drive).
4 6⅝" fish-tail rotary bits (3½"×4½"—7 shanks).
2 6⅝" diamond point rotary bits (3½"×4½"—7 shanks).
12 9" fish-tail rotary bits (3½"×4½"—7 shanks) ⅞" water holes.
12 11" fish-tail rotary bits (5"×6"—4 shanks) ⅞" water holes.
2 9" diamond point bits (3½"×4½"—7 shanks) ⅞" water holes.
1 9" special side-reaming rock bit complete with lubricator and 6"×4" taper coupling.
1 11" special side-reaming rock bit complete with lubricator and 6" straight coupling.
1 Breakout tongs with jaws for 4" drill pipe, 4" tool joints, 6" drill pipe, 6" tool joints and 6⅝" casing.
1 1 or 1½-K.W. turbo generator complete with wiring and lamps.

(b) *Fittings for making Boiler and Engine Connections.*

18 2" malleable tees, 18 2" malleable ells, 12 2" cast-iron plugs.
36 assorted 2" nipples, 10 2" standard globe valves.

10 2" ground-joint lip unions	2 1¼" No. D patent injectors
1 No. 1 grooved wheel	1 Roll wire telegraph cord
24 1" malleable-iron tees	24 1" malleable iron ells
12 1" cast-iron plugs	30 1" assorted nipples
12 1" standard globe valves	12 1" ground-joint lip unions
12 Assorted ½" nipples	6 2"×½" malleable-iron reducers
6 ½" malleable iron ells	500′ of 2" line pipe
100′ of black pipe	

(c) *Manifold Connections for Slush Pumps* (Regular).

2 6" iron-body foot valves with strainers	2 3"×8" nipples
2 Pieces 6"×10′ suction pipes	4 3" cast-iron tees
3 3" standard cast-iron flange unions	4 2½" cast-iron ells
2 3"×2" swedged nipples, about 8" long	2 6"×10" nipples
2 3"×2½" swedged nipples, about 8" long	2 3"×10" nipples
2 2" No. 2 flat-head iron stop cocks, B. W.	8 3"×4" nipples
2 2½" quick-opening gate valves	4 3" cast-iron ells
4 2½" oil-country flange unions	2 2½"×8" nipples
4 2½"×20′ pcs. line pipe (for stand pipe)	2 2½"X10" nipples
2 3" iron-body wedge gate valves	2 6" cast-iron ells
2 4"×3" cast-iron bushings	

(d) *Miscellaneous Items and Small Tools.*

10 lbs. $\frac{1}{16}$" red rubber sheet packing	2 Handsaws
24 Rubber packing rings for pistons of slush pumps	24 10" hacksaw blades
	1 Large size expansion bit
1 Low-down force pump with suction hose and connections	2 Round-point long-handle shovels
	2 6-lb. R. R. picks and handles
1 Set solid-center auger bits ¼" to 1", in eighths	100 lbs. white cotton waste
	2 Round-point short-handle shovels
1 Pipe cutter to cut ¼" to 1" pipe	2 4½-lb. axes with handles
1 Square-end pipe cutter to cut 1" to 2" pipe	1 Adjustable hacksaw
	1 10" ratchet brace
1 Square-end pipe cutter to cut 2½" to 4" pipe	2 No. 2 hatchets
	1 Pr. bit tongs
1 Stock and dies to thread ¼" to 1" pipe	1 24" pipe wrench

- 1 Ratchet stock and dies to thread 1" to 2" pipe
- 1 Ratchet stock and dies to thread $2\frac{1}{2}$" to 4" pipe
- 10 lbs. $\frac{3}{8}$" square flax packing
- 5 lbs. $\frac{1}{2}$" square hydraulic packing
- 1 15" combination wrench
- 2 1-pint oilers
- 2 Sledges with handles (1–8 lbs. and 1–12 lbs.).
- 2 No. 71 ball peen hammers and handles
- 2 Steel center punches
- 2 14" half-round bastard files
- 1 2' steel square
- 1 No. 4 steam blower
- 1 Blacksmith's fuller, 3 lbs.
- 2 20 lbs. crow bars
- 50 lbs. No. 1 Babbitt metal
- 50 lbs. graphite axle grease
- 1 Blacksmith's set hammer
- 1 Blacksmith's hot cutter
- 1 14" pipe wrench
- 1 36" pipe wrench
- 1 18" pipe wrench
- 1 10" combination wrench
- 10 lbs. $\frac{1}{16}$" fiber sheet packing
- 10 lbs. $\frac{1}{2}$" square rubber packing
- 1 No. 2 combination pipe vise
- 2 1-quart long spout oilers
- 1 Auger handle
- 4 1-lb. cold chisels
- 3 12" flat bastard files
- 2 6" slim taper files
- 1 75' metallic tape line
- 1 300-lb. cast steel anvil
- 1 Blacksmith's flatter, 3 lbs.
- 1 36" babbitt ladle
- 50 lbs. No. 4 babbitt metal
- 50 lbs. white lead
- 1 Pr. blacksmith's tongs
- 1 Blacksmith's cold cutter

Rotary Drilling Procedure

The rotary drilling crew, working twenty-four hours per day, consists of from eight to twelve men, or one driller and three, four or five helpers on each twelve-hour shift. The method of procedure to be followed in "rigging-up" is determined by the "head driller," the details of that operation not being relevant here. It includes the installation of the boiler, engine, draw works, rotary, pumps and crown block and the construction of the settling ditch and sumps (small reservoirs similar to those shown in Fig. 118). A slanting "rat-hole" is also bored and cased (by hand or by inclining the rotary machine) to a depth of about 25 feet, at one side of the derrick, the function of this being to accommodate the grief stem and facilitate packing, oiling and adjusting the swivel at the derrick floor while the drill pipe is being withdrawn from and re-run in the well.

In starting to drill unusual care must be exercised to maintain the drill stem in a vertical position, otherwise a "crooked" or slanting hole will result and necessitate moving the rig and starting anew. A quantity of suitable clay must also be secured and thoroughly mixed with water. This mud-laden fluid is then pumped by one of the slush pumps through the stand pipe, rotary hose, swivel, grief stem, and drill pipe to the cutting edge of the bit. At that point the fluid encounters the drillings and carries them, in suspension, upward outside the drill pipe to the settling ditch and outer sump hole. This fluid must be agitated in the central slush pit and new clay added at intervals to maintain the proper consistency and plaster the wall of the drill-hole. In California some operators guard this mixture carefully, moving what remains at a completed well to a new location for further use.

If it is necessary to suspend drilling, even temporarily, the drill pipe should be elevated at least 25 feet so that the drillings, in settling, will not "freeze" and stick the bit. As drilling progresses, the driller releases the brake on the drum of the draw works and allows the grief stem to slide through the drive bushing in the rotary turntable thus lowering the drill pipe and bit. When a distance of about 20 feet has been drilled, rotating is temporarily suspended: the drive bushing is then removed, the drill pipe is elevated until the coupling of the uppermost joint protrudes through the turntable, and serrated slips (similar to Fig. 113-A) are inserted, suspending the drill pipe in the rotary. The grief stem is then detached and a joint of drill pipe is added to the string between the protruding coupling and the grief stem. The slips are then removed, the string is lowered until the lower end of the grief stem passes through the turntable when the drive bushing is again inserted and drilling proceeds as before. After each change, or period of suspension, the bit is rotated, not dropped, to bottom as it is necessary to agitate and emulsify any drillings that may have settled in order to re-establish circulation.

In clays or other soft formations a bit may drill from 50 to 100 feet without changing and from 100 to 250 feet may be drilled during the twenty-four hour day. In hard, or gritty structures it may be impossible to drill more than a few feet per day or with each change of bit. In ordinary drilling the weight of the drill pipe and grief stem is supported in part by the swivel, the fish-tail bit not being forced, except at shallow depths, but in very hard strata it may be necessary to use all available weight to attain satisfactory speed with the rock bit.

If the depth of the well is not to exceed 2000 feet, it is probable that the hole will be started with a 9-inch fish-tail bit, setting $6\frac{5}{8}$ or $6\frac{1}{4}$-inch casing just above the oil stratum, perhaps at 1900 or 1950 feet, 4-inch drill pipe being suitable for drilling the entire distance. It may, however, be necessary to start the well with an 11-inch bit on 6-inch drill pipe, setting $8\frac{1}{4}$-inch casing at from 500 to 1000 feet, finishing as above with an 8-inch bit, 4-inch drill pipe and $6\frac{1}{4}$-inch casing. If the depth is to approximate or greatly exceed 3000 feet, 6-inch drill pipe will be used to a depth of probably 2000 feet, using a 15-inch bit to set a short surface string of $12\frac{1}{2}$-inch casing, or a 13-inch bit to set 10-inch surface casing, later inserting $8\frac{1}{4}$-inch casing, and finishing with 4-inch drill pipe and $6\frac{1}{4}$-inch casing.

It is often difficult to maintain circulation while passing through porous formations, as these rapidly absorb the circulating fluid. It may be necessary to add clay and pump great quantities of this heavier fluid into the hole before such strata can be effectively sealed, circulation re-established and drilling resumed.

If heavy gas pressures are encountered, a blow-out preventer is used. This is a heavy casing-head control fitted with special gates or disks which may be closed around the drill pipe or which completely close the top of the casing, if the drill pipe happens to be withdrawn. The pressure and discharge may then be controlled. The "lubricator" (Fig. 119) may also be used to insert heavy mud and thus "kill" or control high gas pressures.

When nearing the pay stratum many rotary drillers prefer to drill by daylight only, in order that the drillings may be more carefully examined. It is customary to set casing just above the producing formation, reduce the size of the hole and drill into the oil sand, setting a liner (short string of casing) later, if necessary. But if the oil stratum should inadvertently be penetrated before the casing has been set and all water excluded from overlying formations, it will be expensive, if not impossible, to make a successful producer of the well. To forestall this danger it is common practice to "rat-hole" ahead at intervals. For example, when approaching the oil sand the 9-inch bit may be removed and drilling continued with a 6-inch bit for perhaps 40 to 50 feet, thus leaving a shoulder or seat on which casing can be set, but if the oil sand is not encountered, a 9-inch bit will be used to ream this 6-inch hole to bottom, after which the 6-inch bit is again used, the interchange being repeated until the oil stratum is reached. The 9-inch hole will then be reamed down to form a seat for the casing in suitable formation between the last water-bearing stratum and the oil sand. The $6\frac{1}{4}$-inch or $6\frac{5}{8}$-inch casing may then be set and the well completed.

This usually involves consideration of many possible contingencies. In many regions where the strata are known the last string of casing is set on a shell or in other suitable formation immediately overlying the oil sand. The rotary outfit is then removed and the well is bailed, drilled-in, and completed with either a portable drilling machine or a light cable-tool drilling-in outfit. (This " dual " system is described on page 202.) This method (sometimes called " standardizing ") is steadily growing in favor as it avoids subjecting the oil sand to injurious contact with the mud-laden fluid employed in rotary drilling.

If cable-tools are not available the well may be completed with the rotary as already indicated, but if the oil-bearing stratum is porous, it may absorb quantities of clay and become mudded, and oil production will be correspondingly impeded. In regions subject to water troubles the final string of casing is generally cemented or a liner and a packer may be used (as described later) to exclude any water or cavings appearing after the casing is set. In any event, after the rotary has penetrated the oil sand to the required depth, clear water is pumped into the well through the drill pipe in order to wash out all mud and drillings, the circulation being continued until clear water returning to the surface indicates that the suspended clay and detritus have been thoroughly washed out. While washing proceeds, the drill pipe or tubing used must be moved frequently to keep it from " freezing." The water is then bailed or pumped out and if the sand is hard or free from heaving or caving, the driller may consider his work completed, as the well is then ready for production—flowing, swabbing, pumping, or perhaps shooting.

If the well is completed with cable-tools and shot, the drillers usually assist the " shooter " and afterwards clean out the disintegrated material. This cleaning out operation (drilling, pulverizing and bailing out the cavings,—light drilling tools and a sand pump being used) may require from a few hours to several days, after which the well is generally pumped.

If the oil-bearing formation consists of loose, heaving or caving sands, as in parts of California, Texas, Louisiana, Russia, Rumania, and Argentine, much additional work may be necessary to prepare the well for pumping. The grinding action of sand suspended in oil rapidly cuts and destroys pumping equipment and numerous devices have been designed to overcome this destructive action.

In Russia and Rumania wells of large diameters make bailing feasible, if the wells do not flow naturally, but in the American fields wells of smaller diameters render bailing impracticable and various types of sand screens, and perforated pipes and casing are used to separate the sand from the oil or to exclude cavings. These appliances are described on pages 295–299.

If it is decided to supplement or complete the work of the rotary with cable-tools, the rig must be " standardized." This involves the addition of rig irons, band wheel, sand reel, walking beam, and bull wheels. If casing is to be handled, a calf wheel will also be included, although this is not ordinarily required for the drilling-in operation.

The Combination System

While the effectiveness of the combination system is unquestioned and it is occasionally employed in California and in a few foreign fields, it has never met with universal approval and is, therefore, not extensively used. It has been largely supplanted by the " dual " system already described, and by the rotary. For deep-well prospecting and for drilling in certain regions, the combination system is still the most practicable. But its initial cost is high and the increased efficiency of the rotary and the success lately achieved with the rock bit (Fig. 132) in boring through hard strata have largely rendered the combination of cable-tools with the rotary unnecessary in most fields.

In California it is common practice to use both rotary and cable-tools on the same well. They may be embodied in the combination system or used separately—the rotary to start the well and mud-off the caving upper formations, after which surface casing is set and cable-tools used to complete the well.

When they are used in combination, a tall derrick is built, rotary style, but the rest of the rig is equipped with cable-tool devices—band wheel, sand reel, walking beam, calf wheel and bull wheels. The rotary hoist is then installed on the right (forge) side of the derrick, the two slush pumps being set at the opposite (ladder) side, and facilities are provided for the rapid interchange from one system to the other.

Specifications embracing all materials, tools and machinery comprising a cable-tool rig and outfit were submitted in the second section, and for a rotary rig and outfit in this section. The machinery and tools constituting a combination outfit comprise substantially everything required on the separate cable-tool and rotary outfits, including both engines but omitting all other duplications. One large boiler (45 to 66 horsepower) is generally sufficient.

Specifications for "standardizing" a rotary rig and for constructing a combination rig follow:

Material used for standardizing a 112-foot derrick when the rotary is to be replaced with cable tools for "drilling-in" ("dual system").

Pine Timbers			Oak Timbers		
Pieces	Dimensions	Board feet	Pieces	Dimensions	Board feet
1	14″×24″×24′	672	2	12″×12″×12′	288
1	14″×16″×16′	299	2	12″×12″×10′	240
1	16″×16″×16′	341	1	8″×10″×10′	67
1	12″×12″×18′	216	1	6″×13″×16′	104
2	12″×12″×12′	288	1	6″×13″×12′	78
2	14″×14″×14′	457	1	16″×16″×16′	341
1	6″× 8″×22′	88	1	16″×16″× 7′	150
2	6″× 8″×16′	128	1	14″×16″×14′	261
3	6″× 8″×14′	168	1	5″×12″×12′	60
3	6″× 6″×12′	108	5	3″× 5″×16′	100
1	6″×18″×18′	162			
				Total........	1689
	Total........	2927			

Pine or Hemlock			Boxing (Pine or Hemlock)		
Pieces	Dimensions	Board feet	Pieces	Dimensions	Board feet
3	2″×12″×24′	144	75	1″×12″×16′	1200
5	2″×12″×22′	220	60	1″×12″×14′	840
16	2″× 6″×16′	256	50	1″×12″×12′	600
10	2″× 4″×12′	80			
8	2″× 4″×16′	86		Total........	2640
	Total........	786			

Grand total, board feet.............................. 8,042

Wire Nails

100 pounds..........	16d	50 pounds.............	40d
150 pounds..........	30d	50 pounds.............	8d

1 set standard 4¾″ or 5″ rig irons, complete (with or without calf-wheel attachments).

Combination Rig

TIMBER AND LUMBER REQUIRED TO BUILD A COMPLETE COMBINATION RIG

Derrick 106 Feet High, Base 24 Feet Square

The sizes of timbers may be varied depending upon the relative strength of the material as compared with the strength of the timbers listed below.

Pieces	Size, inches	Length, feet	Name	Kind of wood
			For Main Foundation	
1	16 ×16	30	Main sill	Oregon pine
1	16 ×16	20	Sub sill	Oregon pine
1	16 ×16	12	Tail sill	Oregon pine
1	16 ×16	16	Samson post	Oregon pine
2	16 ×16	6	Jack posts	Oregon pine
2	16 ×16	6	Knuckle and tail posts	Oregon pine
1	16 ×16	5	Back brake	Oregon pine
1	6 × 8	12	Back brake support	Oregon pine
1	14 ×30	26	Walking beam	Oregon pine
2	6×6×6×16	14	Pitman and swing lever	Oregon pine
1	2 ×14	5	For ends of beam	Hardwood
1	6 × 6	30	Bunting pole	Oregon pine
32	2½×4—4×4	22-in.	Keys	Oregon pine
4	16 ×16	16	Mud sills	Redwood
1	14 ×14	14	Nose sill	Redwood
5	3 ×12	22	For under mud sills	Redwood
			For Derrick	
2	12 ×12	26	Side sills	Oregon pine
8	10 ×12	24	Sills	Oregon pine
24	3 ×12	6	Blocking for corners	Redwood
20	3 ×12	5	Blocking for corners	Redwood
16	3 ×12	4	Blocking for corners	Redwood
4	14 ×14	4	Blocking	Redwood
2	14 ×14	18	Casing sills	Redwood
2	8 ×10	20	Posts	Redwood
30	2 ×12	24	Derrick floor	Oregon pine
12	2 ×12	24	Girts	Oregon pine
16	2 ×12	22	Girts	Oregon pine
12	2 ×12	20	Girts	Oregon pine
12	2 ×12	18	Girts	Oregon pine
12	1½×12	18	Girts	Oregon pine
42	1½×12	16	Girts	Oregon pine
5	2 ×12	20	Top and water table	Oregon pine
4	2 ×12	20	Bottom legs	Oregon pine
4	2 ×10	26	Bottom legs	Oregon pine
29	2 ×10	16	Legs	Oregon pine
35	2 ×12	16	Legs	Oregon pine

COMBINATION RIG—*Continued*

Pieces	Size, inches	Length, feet	Name	Kind of wood
			FOR DERRICK—*Continued*	
12	2 ×12	24	Doublers	Oregon pine
12	2 ×12	18	Doublers	Oregon pine
29	2 ×12	16	Doublers	Oregon pine
16	2 × 4	16	Ladder	Oregon pine
60	$1\frac{1}{2}$× 6	16	Braces and ladder strip	Oregon pine
16	2 × 6	18	Braces	Oregon pine
16	2 × 6	20	Braces	Oregon pine
16	2 × 6	22	Braces	Oregon pine
16	2 × 6	24	Braces	Oregon pine
16	2 × 6	16	Sway bracing	Oregon pine
16	2 × 6	18	Sway bracing	Oregon pine
16	2 × 6	20	Sway bracing	Oregon pine
16	2 × 6	22	Sway bracing	Oregon pine
16	2 × 6	24	Sway bracing	Oregon pine
16	2 ×12	18	Sway bracing	Oregon pine
16	2 ×12	20	Sway bracing	Oregon pine
16	2 ×12	22	Sway bracing	Oregon pine
2	6 × 6	22	Roof rollers	Oregon pine
2	4 × 6	20	Roof stringers	Oregon pine
2	6 ×18	16	Crown block	Oregon pine
2	6 × 6	12	Crown block	Oak
4	12 ×12	$7\frac{1}{2}$	Bumpers for crown block	Oregon pine
50	1 ×10	14	Roof	Oregon pine
10	1 ×10	20	Roof	Oregon pine
50	1 ×10	16	Roof	Oregon pine
30	1 ×10	22	Pump shed siding	Oregon pine
25	1 ×10	24	Pump shed siding	Oregon pine
2	6 × 8	16	Samson post braces	Oregon pine
1	6 × 8	16	Headache post	Oregon pine
1	8 × 8	16	Crane post	Oregon pine
16	2 × 6	26	Sway bracing	Oregon pine
1	8 × 8	24	Rotary girt	Oregon pine
			FOR ENGINE HOUSE	
2	16 ×16	16	Mud sills	Redwood
2	3 ×12	16	Under mud sills	Redwood
2	14 ×14	14	Pony sills	Oregon pin
1	24 ×24	9	Engine block	Oregon pin
6	2 × 6	16	Frame	Oregon pin
6	2 × 4	16	Frame	Oregon pin
20	1 ×10	16	Flooring	Oregon pin
64	1 ×10	8	Siding	Oregon pin

COMBINATION RIG—*Continued*

Pieces	Size, inches	Length, feet	Name	Kind of wood
			FOR ENGINE HOUSE—*Continued*	
4	1 ×10	16	Gables	Oregon pine
4	1 ×10	12	Gables	Oregon pine
*40	1 ×10	10	Roofing	Oregon pine
*24	1 × 6	10	Battens	Oregon pine
			FOR BELT HOUSE	
2	2 ×10	26	Bottom plate	Oregon pine
1	2 ×10	22	Bottom plate	Oregon pine
5	2 ×10	26	Stringers	Oregon pine
2	2 ×10	20	Stringers and brace	Oregon pine
†80	1 ×10	10	Siding and roofing	Oregon pine
†30	1 ×10	12	Siding and roofing	Oregon pine
† 5	1 ×10	24	Siding and roofing	Oregon pine
†10	1 ×10	12	Siding and roofing	Oregon pine
† 5	1 ×10	16	Siding and roofing	Oregon pine
†40	1 ×10	8	Roofing	Oregon pine
†40	1 ×10	16	Roofing	Oregon pine
†20	1 × 6	16	Roofing battens	Oregon pine
†20	1 × 6	10	Roofing battens	Oregon pine
			FOR ROTARY ENGINE FOUNDATION	
2	16 ×16	16	Mud sills	Redwood
2	3 ×12	16	Under mud sills	Redwood
2	14 ×14	8	Pony sills	Oregon pine
1	24 ×24	14	Engine block	Oregon pine
			BULL, BAND AND CALF WHEELS	
			Same as shown in specification of 82-ft Rig	
			ADDITIONAL	
25	2 ×12	24	Sump boxes	Oregon pine
16	1 × 6	16	Sump boxes	Oregon pine
6	4 × 6	20	Stringers for walk and braces	Oregon pine
5	4 × 6	18	Stringers for walk and braces	Oregon pine
5	4 × 6	8	For walk	Oregon pine
14	2 ×12	20	For walk	Oregon pine
2	6 × 8	20	Casing rack	Oregon pine
			500 ft. extra of 1″×10″×16′ Oregon pine	

* If preferred, 18 sheets of 10-ft. 26-in. galvanized corrugated steel may be substituted for the above.

† If preferred, 10 sheets of 10-ft. and 10 sheets of 8-ft. galvanized corrugated steel may be substituted for the above.

COMBINATION RIG—*Continued*

TIMBER AND LUMBER REQUIRED TO BUILD A COMPLETE COMBINATION RIG

Derrick 122 Feet High, Base 24 Feet Square

The size of timbers may be varied depending upon the relative strength of the material as compared with the strength of the timbers listed below.

Use specifications for 106-ft. rig with the addition of the following:

Pieces	Size, inches	Length, feet	Name	Kind of wood
4	1½×12	16	Girts	Oregon pine
2	2 ×10	16	Legs	Oregon pine
4	2 ×12	16	Doublers	Oregon pine
16	1½× 6	16	Braces for derrick and ladder strip	Oregon pine

TIMBER AND LUMBER REQUIRED TO BUILD A COMPLETE COMBINATION RIG

Derrick 130 Feet High, Base 24 Feet Square

The sizes of timbers may be varied depending upon the relative strength of the material as compared with the strength of the timbers listed below.

Use specifications for 106-ft. rig with the addition of the following:

Pieces	Size, inches	Length, feet	Name	Kind of wood
8	1½×12	16	Girts	Oregon pine
6	2 ×10	16	Legs	Oregon pine
16	2 ×12	16	Doublers	Oregon pine
24	1½× 6	16	Braces for derrick and ladder strip	Oregon pine

(Above rigs to be equipped with 6-inch Imperial Ideal Rig Irons, listed on pages 207 and 211, also with 5- or 6-sheave Steel Crown Block, if desired.)

FISHING AND FISHING TOOLS

Drilling hazards are numerous and the causes of fishing jobs are so manifold and the resulting problems so varied in character and methods of solution that elaborate treatment of the subject would require volumes. Therefore, only a brief exposition of principles can be attempted here.

The commonest causes of fishing jobs are the breaking or unscrewing of taper joints, and defects in the cordage, jars, bits, auger stems, underreamers, casing, and other equipment; but undue haste, reckless handling of drilling tools, caving or inclined strata, the ingress of boulders, and collapsed or parted casing are also contributory factors. Hazards are inseparable from drilling and fishing operations, but frequent inspection of cables, joints, and drilling tools will aid in discovering defects or incipient fractures and accordingly minimize the dangers. The driller should record accurate measurements of every tool used in drilling or fishing as a knowledge of the dimensions of anything lost in the well will facilitate recovering it.

Fishing tools are constantly being improved and new ones introduced. Problems that perplexed the pioneer drillers are readily solved by modern tools and methods, and certain types of fishing jobs that formerly resulted in abandoned wells now cause only brief delays. Much depends upon the skill and patience of the driller. Simple problems (like recovering a loose bailer) may require less than an hour, while a lost bit or underreamer, inclining and buried in a caving hole, may cause weeks or months of fishing and final abandonment of the well. Very few fishing problems, however, are now impossible of solution. Many lost tools that cannot be recovered are drilled up or "side-tracked" (driven into or against the wall), and passed in drilling.

Fishing for lost cable tools.—In fishing, the regular drilling machinery and tools are used, except that the bit and the short-stroke ($4\frac{1}{2}$ to 12-inch stroke) drilling jars are laid

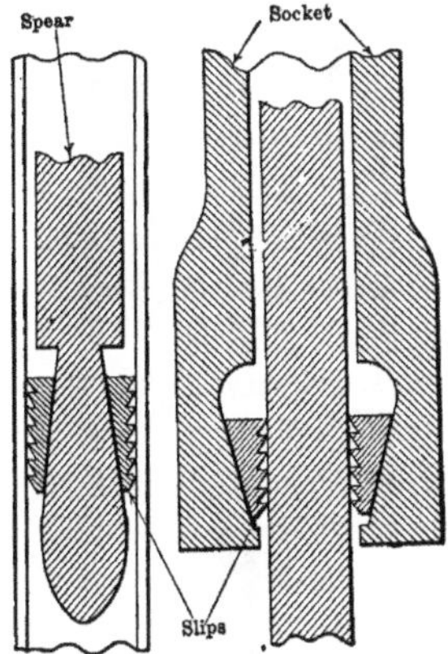

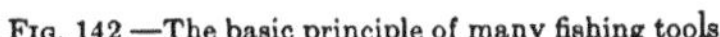

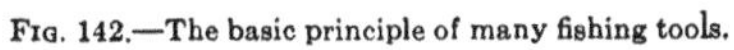

FIG. 142.—The basic principle of many fishing tools.

FIG. 143.—Impression block.

aside, the rope socket is attached directly to the auger stem, and fishing jars (24 to 48-inch stroke) are placed between the lower end of the auger stem and the fishing tool employed. Several fishing tools in general use are illustrated in Figs. 144–192. Each is designed to catch some tool or part that has been broken or lost in the drill-hole, and while thousands of different fishing tools have been improvised and hundreds have been patented and successfully used, the possibilities for improvement are practically unlimited and a fertile field is still open to the creative mechanic and the driller of inventive genius. Occasionally a fishing tool may be specially designed for a single operation, being remodeled or discarded after recovering the object sought.

The basic principle of many fishing tools involves an obliquely-sliding movement of hardened steel "slips" (milled wedges) on a vertical cone or within the concavity of a circular bowl (Fig. 142). This principle is embodied in spears for recovering tubing and casing and in the various types of combination sockets, slip sockets, jar sockets, and casing bowls and spiders (Figs. 144–153*a*).

Simpler types of fishing tools comprise horn sockets, corrugated friction sockets, rope grabs, rope spears, bit hooks, spuds, whipstocks, fluted swedges, rasps, bell sockets, rope knives, boot jacks, casing knives, and die nipples (Figs. 154–192).

FISHING TOOLS

FIG. 144.—Henderson casing spear. Side view—showing method of setting slips.

FIG. 145.—Henderson casing spear. Front view—showing slips released from casing hold and locked to insure free passage from well.

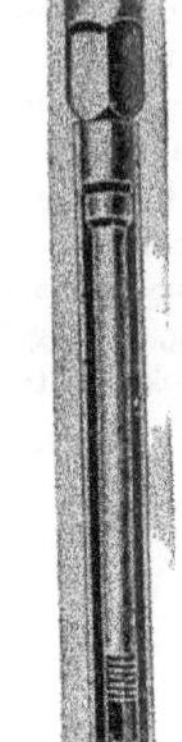

FIG. 146.-Tubing spear with socket.

FIG 147.—Side slot combination socket. To take hold of pin or neck of rope socket. Working parts. adjusted through the slots.

SLIP SOCKETS.

FIG. 148. Without bowl.

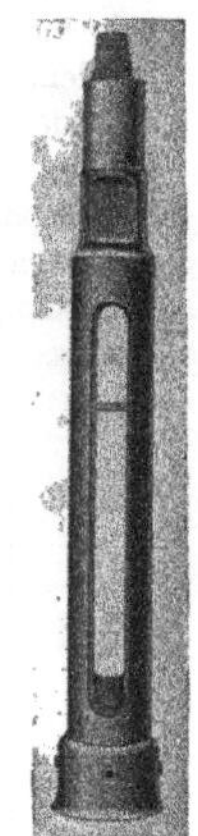

FIG. 149. With bowl.

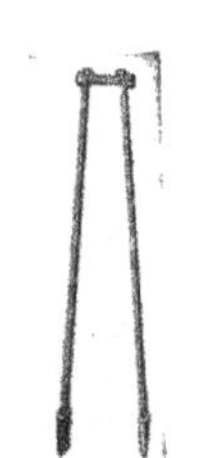

FIG. 150.—Slips and cross bar.

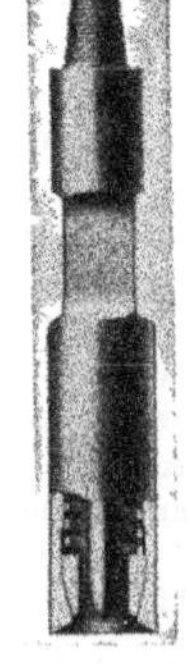

FIG. 151.—Combination socket with mandrel.

FIG. 152.—Bowl spear, bull dog pattern.

For taking inside hold of bowl or any hollow tool.

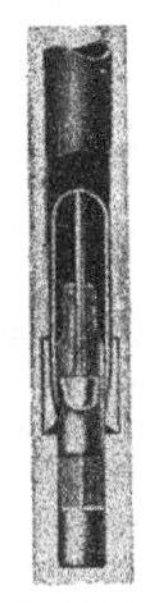

FIG. 153. Casing bowl.

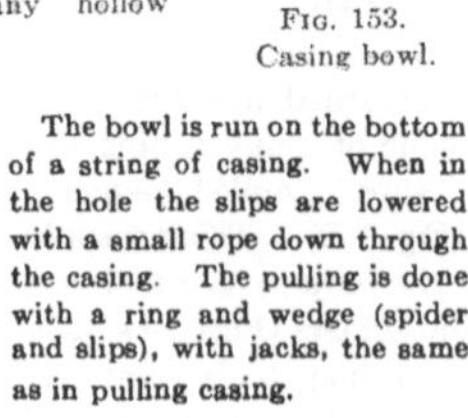

The bowl is run on the bottom of a string of casing. When in the hole the slips are lowered with a small rope down through the casing. The pulling is done with a ring and wedge (spider and slips), with jacks, the same as in pulling casing.

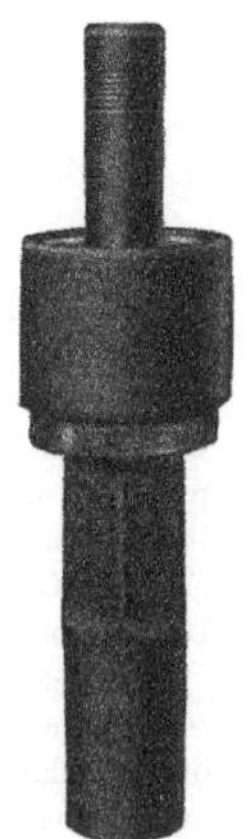

FIG. 153*A*.—Casing substitute.

The jars, essentially and universally used in fishing with cable tools, consist of two heavy, forged-steel links, interlocking as the links of a cable chain, but fitting together more snugly. Each end is equipped with the regular wrench square and taper joint. Their operation, in conjunction with the auger stem immediately above, enables the driller to secure a powerful upward stroke or " jar " and thus dislodge and recover the tool being sought. These jars, like all-steel bits and certain other tools, are made of a special steel, the nature of which the tool manufacturers regard as a trade secret.

Any drilling or fishing tool used within a well may be broken (or dropped) and lost, a fishing job inevitably resulting. Lost tools may rest in any one of three positions:

(*a*) Erect, and easily removed without jarring, as a bit, underreamer cutter, bailer, or entire string that is not covered or stuck by cavings, drillings or other impediments;

(*b*) The top of the object free and exposed to contact with fishing tool, but lower portion imbedded in cavings, sediment or held fast by other obstacles;

(*c*) Object (either erect or inclined) buried beneath heaving sand, cavings, boulders, or other obstructions.

If the top of the lost instrument extends within the casing, its recovery will probably be simple, as contact can readily be made with fishing tools, but if it is lost in the " open hole " below the casing, its recovery may be difficult or impossible.

If the lost tool extends within the casing, its " axis " will be vertical, or only slightly inclined, and a suitable socket can generally be forced over it without difficulty.

But if it is lost below the casing it may be inclining and any attempt to force a fishing tool over it will tend to batter and force it into the wall, making its recovery more perplexing. In such cases a bit hook, or wall hook (Fig. 164), may be used to right the object, permitting a socket to engage and remove it.

There is no definite or established method of procedure recognized in fishing operations. Different drillers may accomplish similar results by varying the application of accepted principles. Occasionally a bit unscrews in the hole and if the accident is discovered before the pin is battered or damaged by subsequent blows of the auger stem, the lost bit may be recovered by using either a horn socket (Fig. 154), a corrugated friction socket (Fig. 156), a combination socket fitted with pin slips (Fig. 147), a slip socket (Fig. 149), or a cherry picker (Fig. 191), all methods being equally effective. The plan to be followed is determined by the preference of the driller and the fishing tools available. Certain fundamental principles, however, are recognized by all drillers, although methods of applying them differ.

Fishing tools are generally run beneath the jars and auger stem on the regular drilling cable, although fishing may be conducted with a string of casing, drill pipe or heavy tubing through the medium of a casing connection (sometimes called a casing substitute, Fig. 153-A) or by directly attaching a casing bowl (Fig. 153).

Fishing with regular tools.—(Fishing tool, jars, auger stem and rope socket attached to drilling cable.) When the bailer, sand pump, or drilling tools are lost by the breaking of the sand line or drilling cable, recovery may readily be effected through the medium of rope grabs or a spear (Figs. 158–162), provided that the lost tools are not imbedded or stuck and that the attached fragment of broken cable is of sufficient length to become enmeshed in the barbs of the fishing tool. If the lost object is held fast by sediment or cavings, jarring will probably tear the cable completely free from it. In that event a latch jack (Fig. 161 or 169) may be used to engage the loop of a bailer or sand pump, and if jarring jerks the bail free, the tube (body) may then be retrieved with a casing spear (Fig. 144) or a bell socket (Fig. 175).

By means of a setting tool (shown in use, Fig. 144) the slips of the casing spear may be set at a point which will permit them to enter the tube (bailer, sand pump or lost casing) snugly, but upon being raised the milled " teeth " of the slips engage the interior walls of the object and take a firm hold. Then if consistent jarring upward fails to dislodge the imbedded tool, jarring downward will " trip " and release the spear, an internal spring forcing the slips back to the position shown in Fig. 145. The fishing tools may then be withdrawn and the bailer drilled up or sidetracked, an operation that is feasible though generally tedious.

In fishing for a lost string of tools, bit or underreamer, it is sometimes necessary to run an impression block (Fig. 143). This consists of a swedged nipple connected to the taper pin of a tool joint for attaching to the jars, or a funnel-shaped piece of sheet iron nailed to a wooden plug for inserting in the lower end of a bailer. Nails are then driven into the wooden block, as shown, and the concavity about them is filled with soap, gumbo or plastic clay, the heads of the nails serving to support this impressible substance in position. Contact with the lost tool is then made by lowering the impression block upon it, the resulting dent or mold indicating its position in the hole and suggesting the method and tools to be employed in retrieving it.

Friction sockets (Figs. 154–156) are tapered or corrugated tubes which, when driven over any loose object in the hole, usually grip it with sufficient force to insure its removal. But if the object is imbedded and held fast, its withdrawal may necessitate the application of great force. In that event either a combination socket (Fig. 147) or a slip socket (Figs. 148, 149), embodying the principle illustrated in Fig. 142, may be employed. The three or four slips in the combination socket completely encircle the object grasped, and as this socket is sturdily built, and its hold is strengthened by increasing the pull, the greatest force may be applied in pulling and jarring, if necessary. On account of its thick walls and heavy slips, this socket will accommodate only objects of small diameters, as the neck of a rope socket, the taper pin of a tool joint, the mandrel of a casing connection, or the coupling or thread of a joint of tubing.

The slip socket may also be used to recover similar tools. Because of its lighter construction, thinner walls and longer body, it will pass over larger objects and is generally used to catch the collar (instead of the neck of a rope socket or the pin of a bit) of a lost tool. The slips (Fig. 150) are narrower but susceptible to varied adjustments thereby admitting entrance to objects of different diameters. The slip socket thus possesses greater adaptability, as a set of medium-size slips in a slip socket of proper size may be adjusted to catch a wide range of diameters, while the slips in the combination socket are designed to catch only one object of a given diameter and are not so responsive to modified requirements and adjustments.

Fishing tools involving the sliding slip principle (Fig. 142) may be of either the " bulldog " or the trip pattern. If it is found impossible to dislodge the tool being sought, fishing tools of the trip pattern may be disengaged and withdrawn at will by lowering or jarring downward, after which other means of recovery may be tried. But tools of the bulldog type (such as slip sockets, combination sockets, casing bowls and certain kinds of spears) do not embody this releasing feature, and while their holds may occasionally be broken, by alternate jarring upward and downward, there is no assurance that this can be accomplished and because of this inflexibility a very complicated fishing job sometimes results.

The reins (links) of the jars are the most fragile part of a string of drilling tools and to avoid the danger of fracturing them, some operators prefer to drill without jars in formations known to be free from troublesome elements. If, under such conditions, the tools should stick and not respond to pulling on the drilling cable, a " bumper " may be used. This may consist of only one piece (Fig. 190), which is attached to the sand line (the lower portion of the bumper encircling the drilling cable) and lowered until it touches the top of the rope socket. Tension is then maintained on the drilling cable while the bumper is repeatedly raised (by the sand reel) from 10 to 30 feet and allowed to drop on the rope socket with great force. The tools usually spring free after a few blows, but if this expedient fails to dislodge them, it becomes necessary to cut the

FISHING TOOLS

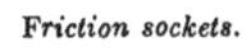

Friction sockets.

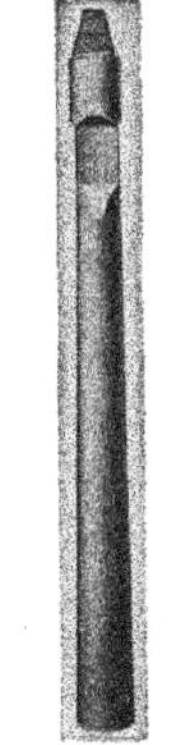

FIG. 154.—Horn socket. To pick up a loose tool in the well.

FIG. 155. Forged steel.

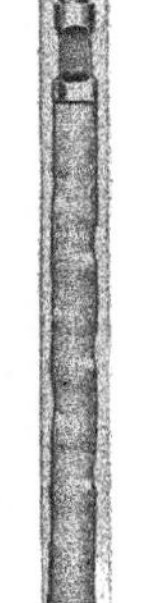

FIG. 156. Corrugated.

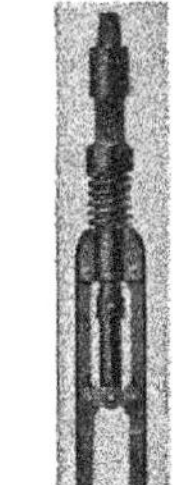

FIG. 157.—Rodger's grab.

Rope spears and grabs.

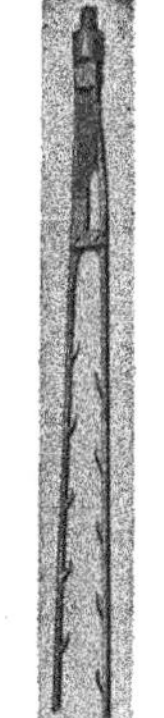

FIG. 158.—Rope grab. Hall patent.

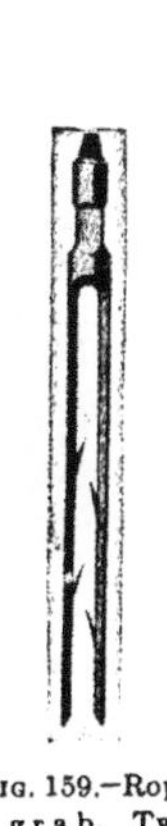

FIG. 159.—Rope grab. Two wing.

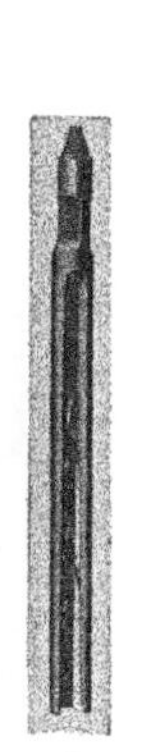

FIG. 160.—Rope grab. Three wing.

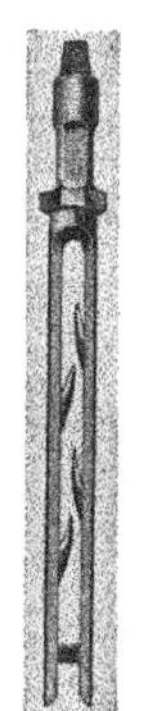

FIG. 161.—Two winged with latch jack for wire line.

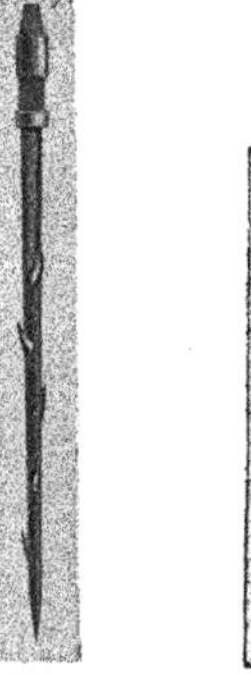

FIG. 162. Rope Spear.

FIG. 163.—Rope spear. Side hole.

Fishing Tools

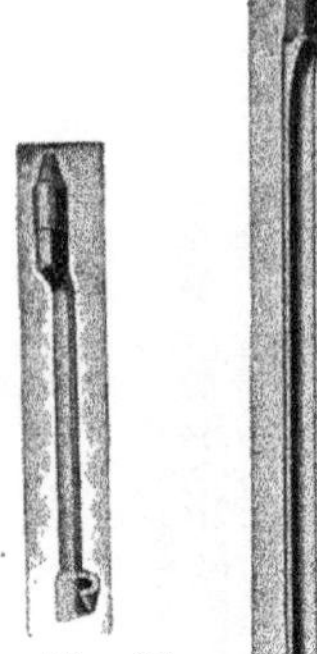

Fig. 164.—Bit hook. To center tools leaning to side of well and sometimes pick them up, catching under the collar of a tool.

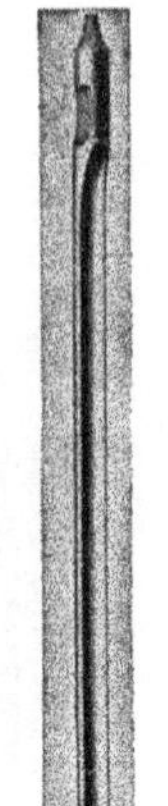

Fig. 165.—Spud. For spudding around and loosening a bit or reamer lost and fast in the well.

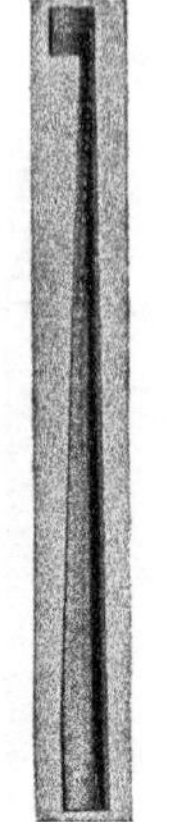

Fig. 166.—Whip stock. With top threaded to suit casing. Used in drilling past lost tools. There is a hole in the bottom of the whip stock which goes over the end of the lost tool. The whip stock guides the tools used in drilling past, into the side of the well. Length, 15 feet or longer.

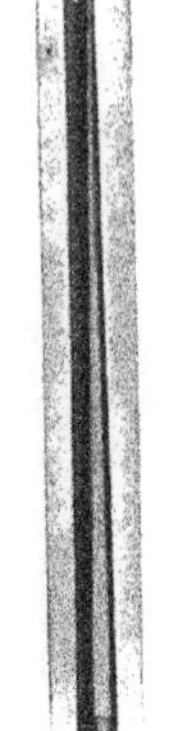

Fig. 167. Whip stock.

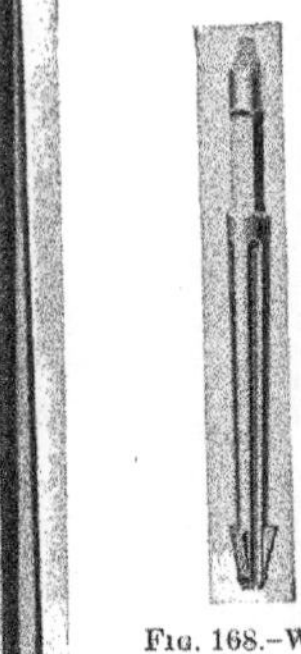

Fig. 168.—Whip stock grab. For removing whip stock from well.

Rasps

For rasping off battered or mushroom tops on lost tools, so that a fishing tool can take hold.

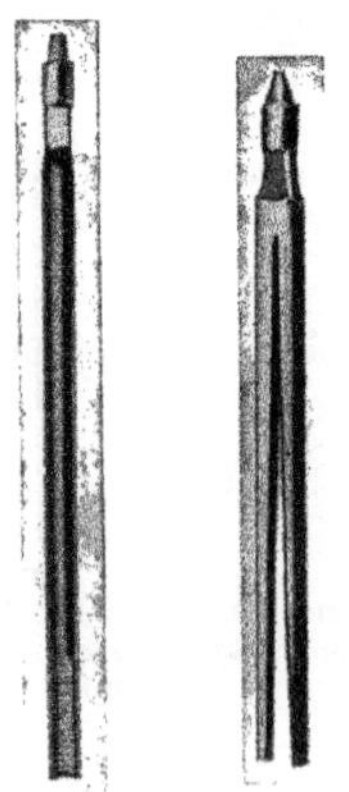

Fig. 170. Side rasp.

Fig. 171. Two-wing rasp.

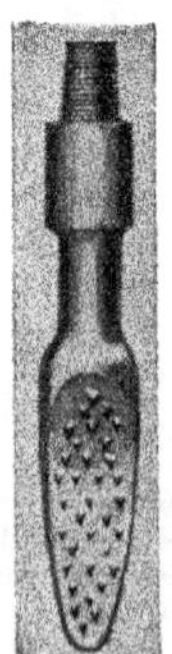

Fig. 172. Cow-foot rasp.

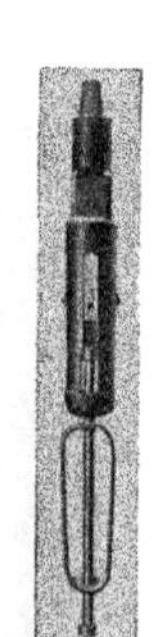

Fig. 173.—Perforator. For perforating casing in well. Can be successfully operated at any depth.

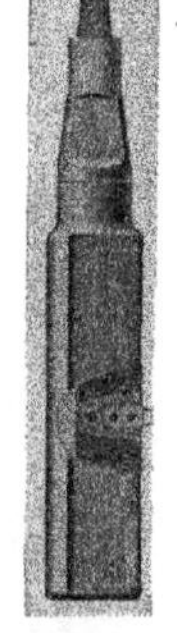

Fig. 174.—Turtle back perforator. For perforating casing in well. F

rope and fish the tools out with a combination socket or a slip socket. The bumper may also be similarly used as a jar knocker, the impact freeing the reins of the jars should they become wedged or locked while drilling. Another type of bumper consists of three pieces—knocker, jars and sinker (Fig. 192). This may be run in like manner producing a double blow or solely by the operation of the jars and sinker, with the knocker resting on the rope socket.

When the drilling cable breaks, pulls out, or "swivels-off" at the rope socket, the tools may sometimes be recovered, if loose, by driving a friction socket over them. The usual procedure, however, is to run a combination socket with slips of proper size to catch the neck of the rope socket (3 to 4½ inches in diameter), or a slip socket with slips

TOOLS FOR CUTTING DRILLING CABLE AT ROPE SOCKET NECK

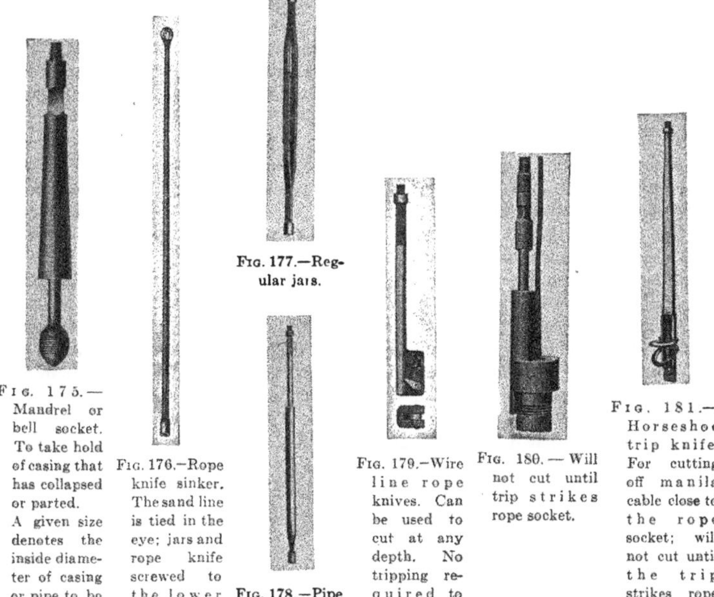

FIG. 175.—Mandrel or bell socket. To take hold of casing that has collapsed or parted. A given size denotes the inside diameter of casing or pipe to be caught.

FIG. 176.—Rope knife sinker. The sand line is tied in the eye; jars and rope knife screwed to the lower end.

FIG. 177.—Regular jars.

FIG. 178.—Pipe or swivel jars.

FIG. 179.—Wire line rope knives. Can be used to cut at any depth. No tripping required to operate.

FIG. 180.—Will not cut until trip strikes rope socket.

FIG. 181.—Horseshoe trip knife. For cutting off manila cable close to the rope socket; will not cut until the trip strikes rope socket.

set to grasp the body (collar) of the rope socket (4⅝ to 6¼ inches in diameter). Due to the length of the full string of tools, it is not possible for them to be greatly inclined in the hole, and their recovery is usually prompt and simple.

When the joint at any point in a string of tools unscrews, the lost portion can usually be fished out promptly by this same procedure, which may also be invoked when a box or pin breaks and the accident is discovered before the top of the lost piece is battered or driven into the wall.

Broken jars may sometimes be retrieved with a friction socket or a slip socket. If an irregular projection results from broken reins, jar sockets of various types (Fig. 189) may be used. Jars break in such a variety of ways that it is occasionally necessary to manufacture or improvise a special device to secure a firm hold, and effect recovery, the patience and ingenuity of the driller being severely taxed in such operations.

When an auger stem breaks, a slip socket is generally used to withdraw the lost portion. If the box of the stem should break, or the pin of a bit or of an underreamer be "jumped" (severed), the top of the lost tool may be battered or driven into the wall and left in a reclining position if the break is not immediately detected. Difficult fishing jobs result from such accidents. A bit hook (Fig. 164) may be run on either tools or tubing to pry the tool loose from the wall so that a friction socket or a slip socket may be forced over it. A spud (Fig. 165) or a round reamer (similar to Fig. 191) may be used to drill around and right it, scrapings from the wall meanwhile being tamped around the broken tool and assisting to sustain it in an erect position, after which it may be picked out with a suitable socket. If the top of the lost bit or underreamer has been battered, or "mushroomed" a rasp (Figs. 170–172) may be run to reduce (file) it to a size that will enter the proper fishing socket. A milling tool, or rotary rasp, may be run on 2-inch tubing if the simpler side rasps fail to accomplish the necessary reduction. Running the bumper or jamming a fishing tool upon the rope socket may batter and enlarge the neck to such an extent that a combination socket will not pass over it. This condition can usually be remedied through the manipulation of the "blind box" (drive-down socket, Fig. 188), the milled teeth of which act as cutters in rasping off the burred portion of a taper pin or rope socket neck.

Should a small portion of the bit break off, or the cutters of an underreamer or other small pieces of hard steel be lost in the hole, the special grab (Fig. 157) will generally effect recovery. By means of a setting tool the powerful spring is compressed and the heavy jaws are spread, then held apart by a small piece of wood. When the grab is lowered over the lost object, the piece of wood is dislodged and the spring forces the jaws together, firmly grasping the elusive piece of steel. The grab has also been successfully used in clearing wells that were "junked" through the agency of broken strings of tubing and other wreckage.

If the lost tools are imbedded or covered it will be necessary to clean out the cavings or obstructions before running the fishing tools. This may be accomplished either with drilling tools or by water-flushing already described. Finally, if all efforts fail to retrieve the lost object, it may perhaps be side-tracked by means of a wedge or whipstock (Figs. 166, 167), or by exploding a charge of dynamite or nitroglycerine alongside of it, thereby forcing it aside so that drilling tools and casing may pass by. A professional shooter should be employed for this work, as it is a delicate and dangerous operation, and frequently necessitates abandonment of the well, if not successful. The wedge or whipstock should, therefore, be tried first. If this is unsuccessful, it can be withdrawn with the whipstock grab (Fig. 168) after which the charge may be lowered and exploded opposite the lost tools.

It may again be stated that it is not possible to deliver a heavy upward blow with the jars used in drilling as the stroke is short and the weight of the rope socket (and of the short sinker bar, when used) above is not sufficient to deliver an upward blow powerful enough to dislodge the tools if firmly held by caving or heaving formations. Because of this limitation it is often necessary to cut the drilling cable and extricate the tools with a fishing string in which the auger stem is placed above the long-stroke fishing jars, thereby effecting dislodgment through the delivery of more powerful upward blows (jarring).

Many types of rope knives are in use, Figs. 176–181 illustrating the parts and principles. If the tools become immovable the rope knife is attached to the jars and sinker and lowered on the sand line until it rests on the neck of the rope socket as shown in Fig. 180. Upward or downward blows (depending on the type of knife) are then delivered with the jars and sinker until the drilling cable is severed, slight tension being maintained on it meanwhile. The drilling tools may then be dislodged and withdrawn through the agency of fishing tools, as previously indicated.

MISCELLANEOUS TOOLS

FIG. 182.—Fluted casing swedge.

FIG. 183.—Forged steel nipple. Male and female. The forged steel nipple has both ends filed and case-hardened.

FIG. 184.—Casing cutter.—To cut pipe or casing at any point in the well when the entire "string" cannot be pulled. Used on tubing. With rope socket neck so that a fishing tool may take hold if lost in the well.

FIG. 185.—One wheel casing cutter. Used in small diameter thick pipe.

FIG. 186.—California style casing cutter. Extra long. For cutting heavy pipe or casing at any point in the well. The extra length with rollers insures a perfect tread for cutters. It has a rope socket neck so that a fishing tool may take hold of it if lost in the well.

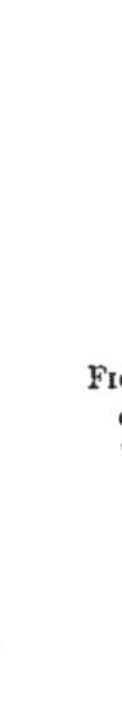

FIG. 187.—Wedge and jar for casing cutter. To expand the cutters in Figs. 184-186. The wedge is lowered through tubing by means of a small rope attached to eye in head.

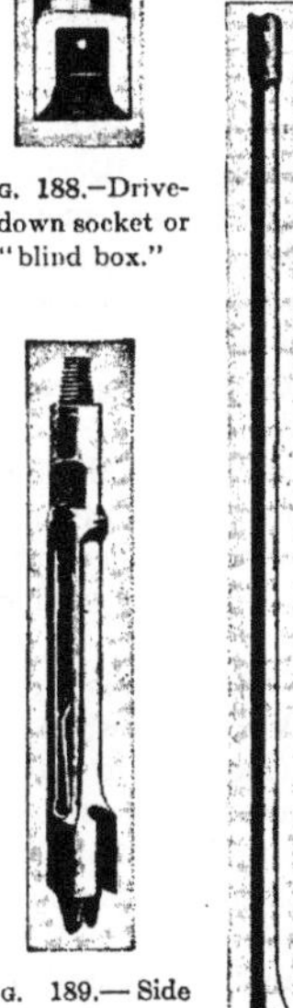

FIG. 188.—Drive-down socket or "blind box."

FIG. 189.—Side jar socket. To catch one rein of broken jars.

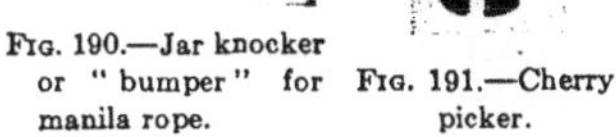

FIG. 190.—Jar knocker or "bumper" for manila rope.

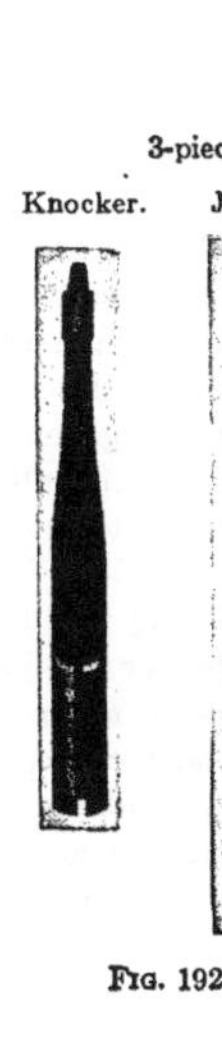

FIG. 191.—Cherry picker.

3-piece bumper.

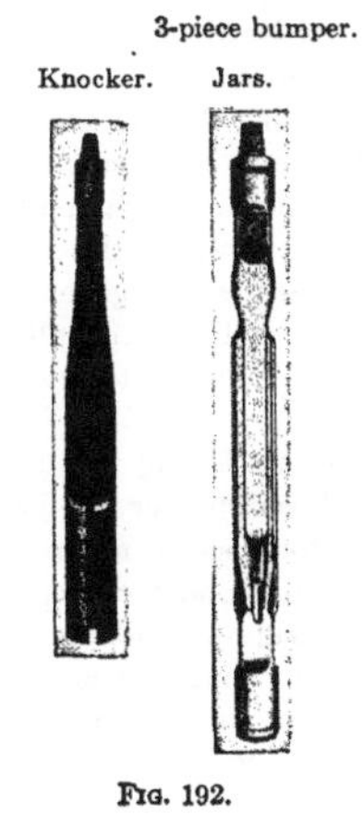

Knocker. Jars. Sinker.

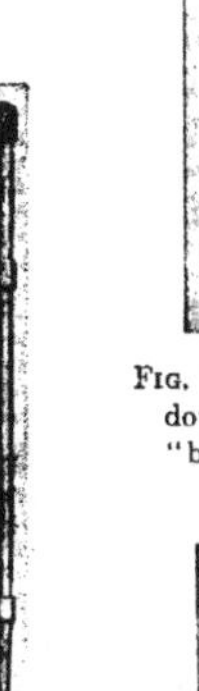

FIG. 192.

FIG. 193.

Fishing with casing, instead of a drilling cable, may sometimes be advantageous. If a string of imbedded tools resists jarring, or if other fishing tools are not available, a casing bowl (Fig. 153) may be used effectively. By lowering the slips on light rope or wire attached to the bottom of the bailer and manipulating the casing, a firm hold on the body of the rope socket may be secured. The bailer is then jerked free, breaking the rope and (if special slips are used) leaving the neck of the rope socket exposed. Thereafter, if the tools do not respond to the lifting pressure applied by jacks operating under a spider and slips (Fig. 113) at the surface, a regular string of fishing tools may be lowered within the casing, grasping the neck of the rope socket and jarring upward while a powerful pull is steadily sustained on the casing. Removal of the tools by this method, after securing the proper hold, is practically certain. After the tools have been loosened and lifted a reasonable distance, the drilling cable may be cut at the rope socket of the fishing string, to avoid the necessity of "stripping." The cable is then withdrawn from the well and the casing is pulled in the regular way, bringing along both strings of tools. If the security of the hold is doubted it may be considered unsafe to cut the drilling cable, in which case it must be unwound from the bull wheel shaft and passed through every joint or stand of casing removed until the tools arrive at the derrick floor. This process is called "stripping the cable" and is not only laborious but exceedingly tedious.

By means of a casing "connection," or casing substitute (Fig. 153-A), other fishing tools may be successfully run on casing or rotary drill pipe, but the operation is obviously much slower than when lowered on a drilling cable. The protruding mandrel of the casing substitute is similar to the neck of a rope socket and serves to make connection with a combination socket should it be desired to operate fishing tools for purposes of jarring within the casing, as above described.

Fishing for lost rotary tools.—In rotary drilling the only tools ordinarily used in the well are the drill pipe and bits, although slips and other small objects may be accidentally dropped in the hole and cause tedious fishing jobs. Because of this simplicity in drilling equipment, rotary fishing jobs and fishing tools are comparatively free from the complexities of cable-tool work. Notwithstanding this apparent advantage, abandonment of uncompleted wells because of insoluble fishing jobs occurs more frequently in rotary drilling than in cable-tool operations. In cable-tool work a string of drilling tools and several complete strings of fishing tools may be temporarily lost, one above the other, baffling complications frequently resulting. But as these tools often rest mainly or wholly within the casing, the exercise of patience and the application of proper remedial measures will generally overcome all obstacles and ultimately save the well. The unstable walls of an unlined hole and the irregular projections of the fractured portions of the tools often render the recovery of lost rotary equipment vastly more uncertain.

Most rotary fishing jobs are caused by "twist-offs" (broken drill pipe) although the bit, drill coupling or tool joints may either break or unscrew. It may be advisable, as in cable-tool fishing, to run an impression block (Fig. 143) to determine the position of anything lost and the proper fishing tool to be used in recovering it. This can be lowered on drill pipe, tubing, or with a sand line or cable. Many other tools designed for cable-tool fishing may also be attached to the box end of a rotary tool joint and successfully run on drill pipe in rotary fishing. Fishing for rotary tools lost at great depths is slow work compared with corresponding cable-tool procedure. Fishing tools may be lowered on a cable to a depth of 3000 feet, or more, and be withdrawn in a few minutes but hours would be required to make a similar run with a string of drill pipe, the manipulation of the pipe and joints being not only tedious but, with the ordinary rotary machine, but also very laborious.

When a twist-off occurs an overshot (Figs. 194–196) is usually lowered (on one or more joints of casing attached to the lower end of the string of drill pipe) and passed over the broken pipe until the upper ends of the steel springs (Fig. 194) become engaged under a coupling or tool joint, the casing and recovered drill pipe then being withdrawn together. If the pipe breaks or twists off without collapsing, a trip spear (Fig. 197) equipped with a guiding diamond point bit may be run on drill pipe, rotated, and

ROTARY FISHING TOOLS

Types of Overshots

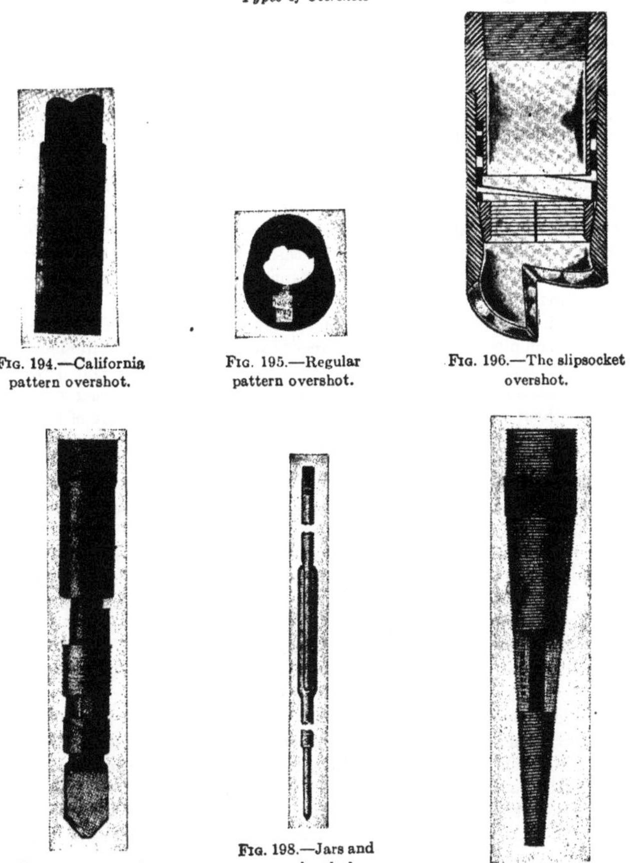

FIG. 194.—California pattern overshot.

FIG. 195.—Regular pattern overshot.

FIG. 196.—The slipsocket overshot.

FIG. 197.—Wash-down rotary trip spear.

FIG. 198.—Jars and mandrel for washdown trip spear.

FIG. 199.—Rotary tool joint grab.

forced into the broken pipe, when the teeth of the milled slips engage the interior walls facilitating removal of the lost pipe. But if this should prove immovable, the jars and mandrel (Fig. 198) may be lowered on a sand line within the drill pipe, "tripping" (releasing) the hold of the spear, after which it may be withdrawn and other expedients employed.

If the threads on a joint of drill pipe, rotary tool joint, or coupling give way or become "stripped," either a steel die nipple (Fig. 183) or a screw grab (Fig. 199) may be lowered on drill pipe and rotated until new threads are cut and a firm connection made after which the lost pipe may be withdrawn. A hook similar to Fig. 164 may be effectively used in righting an inclining object. The special grab (Fig. 157) and the three-prong grab (Fig. 160) have also been successfully employed in recovering slips and other small objects or fragments of bits lost in the hole. Means for maintaining circulation must be provided if fishing is to be conducted near the bottom of the hole; otherwise the object sought and perhaps the fishing tools may become imbedded and permanently held in sediment or cavings.

As a large part of a rotary hole remains "open" (uncased) in the regular course of drilling, disastrous caving may occur at any time. The purpose of the plastic clay in the circulating fluid is to plaster the walls and circumvent this danger, but notwithstanding all precautionary measures, the walls occasionally collapse and imbed and bind the bit and drill pipe so firmly that they can be liberated only by clearing away the cavings and releasing the accompanying friction. This can generally be accomplished by cutting (rotating) around the "frozen" drill pipe with a rotary shoe (Fig. 131-E) attached to a string of larger drill pipe or casing, meanwhile establishing circulation with mud-laden fluid. The fluid passes downward between the imbedded drill stem and the rotating string, then upward outside this pipe or casing.

Drill pipe fitted with left-hand threads is best adapted to fishing requirements. If a fishing tool run on pipe with right-hand threads cannot be disengaged from an imbedded string, a serious situation and a complicated problem may result. But if the spear or other fishing tool is run on pipe with left-hand threads, the fishing string may be turned to the left, without danger of parting, and in this manner one or more of the lost joints may be unscrewed and withdrawn at a time.

Should it be impossible to recover the lost pipe or bit, it may be feasible to drill by it with a diamond point bit in conjunction with a wedge or side-tracking apparatus similar to Fig. 167, casing being run as soon as the lost object is safely passed, thereby preventing its falling into the drill-hole again.

Fishing for casing.—The parting of casing may result from various causes. Contributing factors are defective materials, poor threads, careless handling, hard driving, excessive strains in pulling, the breaking of casing equipment (casing lines, elevators, hooks, etc.) and external friction caused by unstable walls. Because of the high cost of heavy casing there is sometimes a disposition to economize by substituting lighter than standard weights, or casing of inferior quality, thereby inviting disaster. Almost any grade of new or used casing or line pipe may serve for short surface strings but when casing is to be carried to great depths only heavy, high-grade, new casing with long couplings should be used. The weight of a 3000-foot string of 10-inch 45-pound casing (used in California) is 135,000 pounds and the result of suspending such loads on defective threads, couplings and equipment is generally an expensive fishing job.

In addition to this inherent weight, friction caused by binding walls (and amounting to several tons) must also be overcome placing a total strain of from 67½ to more than 100 tons on the casing equipment and couplings.

When a string of casing parts the separation commonly occurs at the coupling, due either to a fracture, the stripping of the threads, the collapsing of the end of a joint, the expanding of the coupling, or two or more of these factors in combination. If the lost portion stands concentrically in the hole, little difficulty will be experienced in making connection with it by means of a forged-steel die nipple lowered on the free portion of the casing. This tool is in effect a thread-cutting coupling, being case-hardened and fluted to accommodate the metal cuttings. If the top of the lost casing does not bear a coupling the die nipple will be run with the interior thread downward (as shown in

Fig. 183) and connection will be made by rotating and cutting new threads (or repairing damaged ones) on the end of the exposed joint, the die nipple being reversed should the coupling remain on the top of the lost casing. After making connection it is advisable to withdraw the entire string of casing and remove the die nipple and damaged joint but external friction may prevent this in which event the connection thus made, while perhaps not water-tight, will serve to exclude cavings and permit the resumption of drilling—a longer string of smaller casing being run later if necessary.

If the coupling or exposed thread at the top of the lost string should become badly damaged, it may be expedient to lower a casing cutter (Figs. 184–187) on tubing to cut off a portion (1 or 2 feet) so that the die nipple may start anew under favorable conditions, the detached piece being withdrawn on the cutter or by other means.

If the lost casing consists of only a few joints loosely held in the hole, recovery may be effected through the agency of a casing spear (Figs. 144, 145) run on regular fishing tools (the operation of this spear is explained in the discussion of cable-tool fishing), but a string of casing is ordinarily used in such fishing operations, the spear being connected to the bottom joint by means of a casing connection, or substitute (Fig. 153-A). The slips of the spear are set at the proper position to engage the interior wall of the lost casing and if it does not respond to pulling (with elevators, or with jacks and spider) a combination socket on regular fishing tools and drilling cable may be lowered within this casing, grasping the protruding mandrel of the casing substitute and delivering powerful upward blows (jars) to assist in retrieving the parted casing.

In the foregoing examples it was assumed that the walls of the hole, above the point at which the casing had parted, were stable and would safely permit the withdrawal of the casing above the fracture or point of separation. But this condition does not always obtain and if the free or upper portion of the casing is removed from the hole, caving may occur, imbedding and burying the lost casing and rendering its recovery exceedingly difficult, if not impossible. To circumvent this danger the two sections of the parted string may be united and subsequently withdrawn together. This joining operation (assuming that the parted ends can be forced approximately together) is accomplished by lowering on a string of fishing tools a casing spear which is then engaged in the casing just below the rupture. Jarring upward may then be employed to make certain that the parted ends are joined, or brought as closely together as possible. Concentric double slips (with interior milled teeth on the inner slips and exterior teeth on the outer slips) are then tied to the sand line and lowered, the inner slips grasping the rope socket and the outer ones engaging in the casing, above the break. The sand line is then jerked free from the slips and a rope knife is lowered to cut the drilling cable at the rope socket. This completes the operation of joining the two sections of casing and the entire string may now be withdrawn.

A string of casing may part at any point during the process of inserting, carrying, or pulling it—a fishing job following. If the falling portion passes through the larger casing or falls unobstructed in open hole, it will acquire great momentum and strike the bottom of the hole with terrific force, the impact causing one or more joints to corkscrew and collapse, or possibly telescope. A tedious fishing job results from such accidents as the crooked and collapsed joints can be withdrawn only with the greatest difficulty, if at all. If a long string of casing drops only a few feet, unimpeded, it may be damaged at many points, particularly near the lower end where the weight and strain are greatest. But in a caving hole, with considerable water and cavings present, the force of the fall may be broken by lateral friction caused by projections and " tight " places in the hole and the shock further absorbed by the emulsified detritus at the bottom. While much trouble has often resulted from falls of only a few feet, in other instances portions of strings of casing have dropped hundreds of feet without causing serious damage.

The external pressure caused by water or boulders pressing against the casing may cause it to dent or collapse. If a joint is badly dented or collapsed it must be replaced, but smaller dents may be repaired (straightened out) by running a fluted casing swedge (Fig. 182) and driving it through the damaged portion, the jars and auger stem being used as in regular fishing.

When several strings of casing (e.g. 15½-inch, 12½-inch, 10-inch and 8¼-inch) are inserted, one or more of these may be withdrawn for use in another well after the final string has been successfully set. In the above example the 10-inch and 12½-inch could perhaps be "pulled," entirely or in part. But water from the surface or from a high level must not be permitted to accumulate around a long string of casing, as the pressure of a column of water may cause the innermost string of casing to collapse, thus ruining the well or necessitating a heavy expenditure to repair the damage. When all known means of recovering parted casing have been tried, it may perhaps be side-tracked or passed in drilling after exploding a charge of dynamite at the top of the lost portion or by inserting a wedge or whipstock (Fig. 167) at that point. Failure to recover lost casing may, however, necessitate final abandonment of the well.

When a drilling venture results in failure it is customary to salvage everything of value. This involves "pulling" the casing, the operation and order of inserting the several strings being reversed. Due to external friction on the part of each string of casing which is exposed to caving or which makes contact with the wall of the hole, great difficulty may be experienced in getting the string started. If it does not respond to pulling with the elevators or jacks, a casing spear may be lowered and engaged in a joint near the bottom of the string, upward jarring then being employed to assist in liberating it, while the lifting apparatus at the rig maintains tension on the column. If this method does not succeed, the spear can be tripped and withdrawn, after which a casing cutter (Figs. 184–186) may be lowered on 2-inch tubing within the casing to a point at which the latter is believed to be free. The jars and pointed mandrel (Fig. 187) are then run on the sand line within the tubing to force the circular knives out against the casing, after which the column of tubing is rotated by the drilling crew. After the cutter has been rotated sufficiently to insure severance of the casing, a "pull" on the latter may be tried to ascertain whether the "frozen" portion has been severed from the free part of the string. If the casing will not move the cutter may be raised and another cut made at a higher point or the cutter may be withdrawn to permit the use of the spear as before. These operations may be repeated as often as necessary to eliminate the "frozen" portion of the casing, which is then abandoned in the well. Modern practice favors the use of a "collar buster" (somewhat similar to Fig. 174) or a casing splitter (Fig. 174-A). Either of these may be lowered on a regular string of fishing tools and a casing coupling at any desired depth may be split, after which the casing above that point, if not frozen, may be withdrawn. As these appliances are lowered on a drilling cable the severing operation is much faster and simpler than with a casing cutter, a feature that largely accounts for their growing popularity. A string of casing may also be parted by lowering dynamite on a rope and exploding it at the desired point by means of a detonating cap and a fuse or with an electric charge.

Fishing for tubing and sucker rods.—Hazards do not always end with the completion of the well. Accidents to producing wells are numerous and varied and while fishing jobs are of frequent occurrence their solution is generally simple. The commonest cause is the parting of the tubing or the rods but as the lost portions are not endangered by cavings, their recovery seldom involves complications, unless they have been dropped a considerable distance. In that event, the broken, crooked, collapsed and telescoped portions of a string of tubing may render fishing exceedingly tedious, if not difficult.

Lost tubing may be recovered with a combination socket or a spear (types represented by Figs. 146 and 151, to be run on regular fishing tools; Figs. 201 and 202, to be run on tubing), also with a friction socket (Fig. 156) or a grab, (Fig. 157), although the hold obtained by the latter two will not withstand heavy jarring. The bowl or shoe (Figs. 146 and 202) is designed to guide the spear and force the tubing to the center of the hole. This centering action facilitates engaging the milled slips in the lost tubing. The principle of these tools is illustrated in Fig. 142.

A very simple but successful tubing catcher (Fig. 203) has lately been introduced. This practically eliminates the hazards that heretofore accompanied the manipulation of strings of tubing. This appliance is placed at any convenient point in the column of

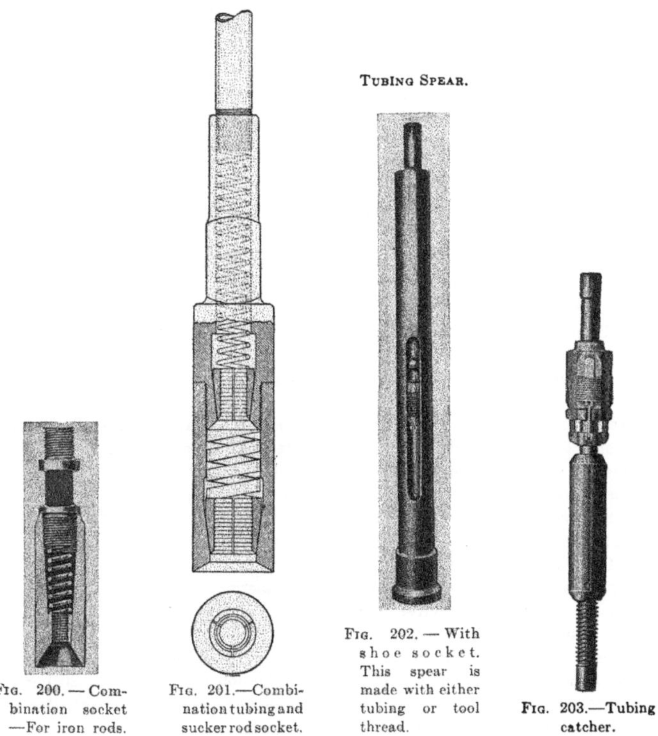

FIG. 200.—Combination socket—For iron rods.

FIG. 201.—Combination tubing and sucker rod socket.

FIG. 202.—With shoe socket. This spear is made with either tubing or tool thread.

FIG. 203.—Tubing catcher.

tubing, usually just above the working barrel, and by an ingenious application of the principle of inertia, the string of tubing, if it should break, is prevented from falling more than a few inches. This is accomplished through the action of three steel slips which, when the tubing parts, are automatically expanded and grasp the interior of the casing, thus arresting the fall of the tubing. The upper portion of the tubing (above the break) is then withdrawn and a spear used to recover the part below the break. When this part is lifted the slips on the tubing catcher are released, thereby permitting removal of this lower portion of the string without damage or interference.

If the sucker rods unscrew in the tubing they can usually be screwed together again simply by rotating that part of the string above the point of separation. In that event pumping may be resumed without withdrawing the rods; but when they break, a fishing job or the withdrawal of the tubing is the only remedial measure. Their recovery is generally a very simple operation, being effected by withdrawing the rods above the fracture and lowering on them a small combination socket (Fig. 200). If a suitable socket is unavailable, or if its operation is unsuccessful, it may be necessary to withdraw the tubing until the broken rod is reached.

Wooden rods with iron joints were formerly used exclusively, but both iron and steel rods have since been adopted in many fields. The wire-line pumping outfit is also quite extensively used, but as similar methods of fishing are applicable in recovering all pumping equipment, elaborate treatment of the subject is unnecessary.

While the improvement of fishing tools and the elaboration of fishing methods have probably kept pace with other phases of oil-field evolution, the application of electricity to fishing has made practically no headway. Efforts to utilize electromagnets in the recovery of lost tools have encountered many difficulties due to short-circuiting. Experiments are still being conducted, but because of the presence of water, metallic casing and other conductors, insulating problems have thus far proved insurmountable.

MISCELLANEOUS FACTORS IN OIL-FIELD DEVELOPMENT

1. The exclusion of water from oil sands (*a*) by the use of packers; (*b*) by cementing.
2. Shooting.
3. Perforating and screening.
4. Plugging and abandoning wells.
5. Tubular goods and pipe trade customs.

The exclusion of water from oil-bearing formations is a matter of the gravest importance, not only to the owner of a well in which water may appear but also to the holders of surrounding leases. Some oil-fields are comparatively free from water troubles, there being no water-bearing strata in close proximity to the producing sand. In other regions water may be frequently encountered in drilling and if it is admitted to the oil sand, permanent injury to the well and adjacent territory inevitably follows. A small quantity of water entering a well may readily be separated from high-gravity oil after reaching the tanks, but considerable difficulty is experienced in effecting its separation from heavy oil. Should a defective string of casing or casing seat admit 20 barrels of water daily into an oil well of large capacity the water would scarcely be perceptible, but coincident with the decline in the quantity of oil produced the water increases in percentage, if not also in volume. In a total production of 500 barrels, 20 barrels of water would constitute only 4 per cent but, the ingress remaining constant, this would eventually represent 40 per cent of a 50-barrel production.

Water entering at a leaky casing seat gradually cuts a channel which admits increasing quantities and, if this attains ample volume and reaches a sufficient level in the well, the oil will be " drowned " (pushed back laterally by reason of the greater specific gravity of water), perhaps permanently ruining the well and flooding a portion of the adjacent territory. The legislatures of several states have recognized this danger and have enacted stringent laws in an endeavor to cope with this serious situation.

In setting the last (innermost) string of casing, commonly called the water string, the drillers usually succeed in excluding from the oil sand all water encountered at higher levels. The achievement of this object is not difficult if an impervious formation suitable for a casing " seat " is found between the lowermost water-bearing stratum and the producing sand, as the weight of the casing, supplemented by driving if necessary, will press the casing shoe into this formation and effectively shut off the water, after which the well may be completed as already described. But if for any reason the water string cannot be successfully "landed," as above indicated, deviations from customary procedure must be adopted.

The exclusion of water by the cementing process, while involving considerable expense and delay, is probably the most satisfactory in regions where wells of large capacity are encountered and a permanent solution of the difficulty is desired. In other fields water is promptly, economically and effectively excluded by means of packers lowered on either casing or tubing. In the older fields of the United States the packer method is the one generally used, about thirty types of improved packers now being manufactured to meet the varied requirements.

(*a*) The **Packer method.**—The fundamental principle of all types of packers embodies the vertical compression and lateral expansion of a resilient substance (rubber, canvas or burlap) between casing or tubing and the wall of the hole, between two strings of casing, or between tubing and casing, the packer being run on the interior string and expanded to fill the annular space surrounding it. Rubber packers vary from 3 to 8 feet in length, the rubber sleeve generally being from 10 to 32 inches long. The standard canvas packer is about 8 feet in length with a canvas or burlap sleeve of 3 feet, but these packers are also made in special lengths of from 16 to 20 feet with 6 or 8 feet of canvas or burlap. The packing material (rubber or fabric) surrounds a sliding tubular sleeve at the bottom of which another sleeve of slightly larger diameter is suspended. When sufficient weight is superimposed the two sleeves partially telescope ("collapse") vertically compressing and laterally expanding the packing. The packer may be withdrawn integrally with the casing at any time, although the packing material, which is renewable, may be damaged or destroyed by this action.

After the water string of casing has apparently been successfully set and the well completed, a leak may develop at any point in the casing or at the seat. In a pumping well this trouble may be promptly remedied by withdrawing the tubing and running a casing tester (Fig. 215), determining the location of the leak, and setting a packer between that point and the oil sand. The effectiveness of a packer if set within the casing is ordinarily more certain and permanent than if set in open hole as the unlined wall may cave and permit the passage of water. For this reason packers are commonly set at some point above the casing shoe unless the ingress of water at or below the seat necessitates placing the packer below that point.

The adaptations of only a few of the commonest types of packers can be described here.

If the formation in which it is proposed to set casing is of doubtful strength or hardness a *bottom hole packer* (Fig. 204) may be used to give added security to the casing seat. The copper rivets which prevent collapsing of the packer while it is being lowered into the well on the bottom of the casing are sheared by the superimposed weight when the bottom of the hole is reached, thereby forcing the packer to telescope. The ends of the rubber packing sleeve are attached to the upper and lower portions of the packer. When telescoping reduces the distance between the shoulders of these two parts the rubber sleeve between them is compressed vertically and expanded laterally, thereby snugly filling a small vertical section of the annular space surrounding the packer and intercepting the downward flow of water between the casing and the wall of the hole. This illustrates the basic principle of all packers. As the interior diameter of the packer equals that of the casing on which it is run, drilling through it may proceed in the regular way.

The *casing anchor packer* (Fig. 205) may be inserted at any point in a string of casing, its operation being similar to that of the bottom hole packer, except that the bottom of the latter is fitted with a light setting shoe, whereas this packer is fitted with a casing

Fig. 204.—Bottom hole packer.

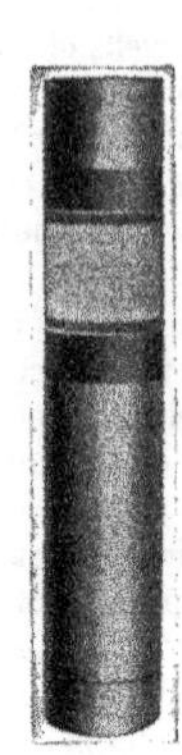

Fig. 205.—Casing anchor packer.

Fig. 206.—Texas canvas packer.

Fig. 207.—Disk anchor packer.

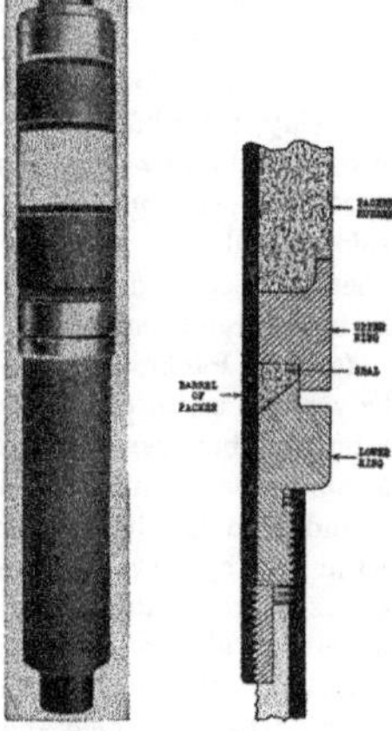

A. *B.*

Fig. 208.—Special gas anchor packer, with automatic packing ring.

The Duplex Screw-down Liner Packer

Fig. 209.—Bottom packer.

Fig. 210—Top packer.

Fig. 211.—Letting-in tool.

Fig. 211*A*.—Showing position of letting-in tool for setting lower packer.

Fig. 211*B*.—Showing position of letting-in tool for setting upper packer.

coupling at each end and a number of joints of casing may be attached and run below it. It is frequently used to pack between strings of casing or between casing and a hard wall and by attaching the required number of joints below it may be set (anchored) at any point in the hole. Should water from a high level break in near the oil sand this packer or a Texas Canvas packer (Fig. 206) may be run on a liner (short string of casing) and set below the leak. Another similar packer is run on the same string of casing and set just above the leak thus confining the water in the space between these two packers. While the string of casing bearing the two packers is being inserted a nipple with a left-hand thread should be placed on the second or third joint of casing above the upper packer. After both packers have been set (collapsed) by the weight of the column of casing above them and allowed to stand for a few hours the casing is turned to the right, unscrewing the left-hand thread, and all of the casing above that point is then withdrawn. The portion ("liner") left in the well, usually consisting of from three to twelve joints, thus substitutes for a complete string of casing, the lowermost joint having previously been perforated to admit the oil. The liner with the packers may later be withdrawn by means of a casing spear run on tools or casing.

A *special duplex screw-down liner packer* (Figs. 209-211-B) has been designed to set at any point (within casing or in open hole) and accomplish this same result without either the anchor or the superimposed weight. The upper and lower units of this double packer are connected by one or more joints of casing, depending upon the length required to segregate the fluid or the cave between the packers. After the proper amount of casing has been inserted between the two packers the letting-in tool attached to a string of tubing is set in the position shown in Fig. 211-A and the complete double packer lowered to the desired position, the lower packing rubber being below the leak or cave and the upper rubber resting above it. A small weight is then dropped through the tubing, breaking the disk in the lower packer and permitting the spring to expand. This forces the slips upward on the conical body and engages them in the surrounding wall or casing. The packer is then lowered slightly to expand the slips further and engage them more firmly, after which the tubing is turned to the right to screw down the upper part of this lower packer and thereby expand and set the rubber. Then, after turning the tubing one-fourth turn to the left and raising it about 2 feet (Fig. 211-B), the upper packer is likewise screwed down and set. The tubing and letting-in tool may then be withdrawn, leaving the two packers and intervening casing suspended in the well.

The *common bottom hole packer* (Fig. 204) and the *casing anchor packer* (Fig. 205) cannot be forced through tight places in the hole as the lateral friction on the portion below the rubber will retard the downward movement sufficiently to shear the copper rivets and cause premature setting. Steel rivets may be substituted, increasing the resistance to shearing, but if this trouble is anticipated, *a packer of the disk type* (Fig. 207) should be used. A metallic disk prevents telescoping while the packer is being manipulated on the casing and lowered to the proper position. The disk may then be dislodged by dropping a small weight upon it, thereby permitting the packer to function.

Because of the large amount of packing material carried the *canvas or burlap packer* (Fig. 206) is the most suitable type to set in open hole if the wall at the point of contact is likely to crumble or cave. This packer is collapsed when the rivets, just below the canvas, are sheared by the weight of the casing.

The *special gas anchor packer* (Fig. 208) is designed to be run on tubing and exclude water or cavings from the producing stratum. It is inserted in the string of tubing at a point that will set it just below the cave or leak, collapsing by the weight of the tubing above. Pumping may be conducted through this packer, a joint of tubing at the bottom of the string being perforated to admit the oil. The special rubber seal prevents leakage at the joint in the packer.

The *hook wall packer* (Fig. 212) represents a very popular type. It may be run on either casing or tubing (with tubing top Fig. 213), is easily raised or lowered in re-setting, and can be finally withdrawn. The V-shaped hook penetrates any cavings or sediment that becomes lodged on the shoulder and it may thus be re-engaged at will.

How To Set Hook Wall Packer

Lower packer into the well with hook engaged as shown in illustrations. When packer is 10 or 12 inches above the point at which it is desired to pack, turn the casing or tubing one-half turn to the left, which disengages the hook. Then lower until slips catch the wall of the well and support the weight of the casing or tubing.

Fig. 212.—Hook wall packer. Four friction springs on all sizes up to and including 6¼-inch hole. Eight friction springs on sizes for 8-inch hole and larger.

Fig. 213.—Hook wall packer. Pumping top.

Fig. 214.—Disk wall packer. Four friction springs on all sizes.

Fig. 215.—Casing tester.

Should it be necessary to reset the packer at a lower point, raise the casing or tubing until the packer swings clear, and turn the casing or tubing one half turn to the right, thereby re-engaging the hook. Lower packer to the desired point, turn to left, and set as before.

The packer can be removed from the well by simply pulling up.

How to Set Disk Wall Packer

Lower packer to within 10 or 12 inches above the point at which it is desired to pack and drop weight to break disk. This releases the spring which forces the slips upward on the taper cone.

Then lower until slips catch the wall of the well and hold the weight of the casing. A piece of pipe, 6 to 8 inches long, may be used for breaking the disk. After the disk is broken the packer can be raised but not lowered.

(*b*) **Cementing.**—In known regions where wells of large capacity or high pressure are anticipated or where the exclusion of water is difficult, it is customary to cement the "water" string of casing in the most suitable formation between the oil sand and the last water-bearing stratum encountered above it. The objective sought in cementing is to construct a water-tight barrier between the exterior of the lower portion of the casing and the wall of the well in order to exclude from the drill-hole all water previously encountered.

For reasons apparent, the plastic cement must be lowered to bottom within the casing, after which the column of casing is raised a few feet and by mechanical means the cement is then forced upward into the annular space between the casing and the surrounding wall and there allowed to solidify. Several distinct methods have been devised to accomplish this. In earlier practice a dump bailer was successfully used to lower the cement into shallow wells in which the water did not rise to a high level. The casing was then raised slightly and a wooden plug was quickly driven to bottom by the weight of a string of tools, drill pipe, or tubing. This forced the cement into the desired position as above indicated. By another method the cement was pumped through tubing to the bottom of the hole and forced upward behind the casing by a plug propelled by a column of water, a packer (Fig. 213) preventing the return of the cement between the tubing and the casing. Although the rapid "setting" of the cement renders these methods dangerous and they are not generally approved, extreme necessity may occasionally justify recourse to them.

Another method (the two-plug method) commonly employed in many parts of the United States during the past ten years, has been widely discussed in several publications and will be briefly described here. Any good grade of quick-setting (one to two hours) cement will serve. The crystallizing action of cement in the process of hardening is irrelevant to this work but a uniform mixture should be obtained with the least amount of fresh water possible to effect complete emulsification. With this method the mixing of the cement is usually done by hand but everything (cement, tools, vat and water) should be so arranged that the mixing operation may be completed in the briefest possible time. Cement of the proper grade for this work begins to solidify in from one to two hours after being mixed with water and for this reason the entire cementing operation should not consume more than about one and one-half hours. Cold water retards and hot water hastens the process of solidification. From 75 to 80 pounds of cement should be used for each cubic foot of space to be filled. The total amount required depends upon the diameter of the hole and the nature of the formation, and varies from 5 to 35 tons. The cement should be screened to eliminate all lumps.

Two wooden plugs (*A* and *B*, Fig. 216) are prepared, slightly smaller in diameter than the interior of the casing through which they are to pass and a circular piece of heavy belting to fit the casing is nailed to the bottom of each. The combined length of the two plugs should be about 5 feet and the casing should be raised about 4 feet to permit the egress of the cement. If the casing shoe is inadvertently raised too high it may lodge on the top plug in being lowered again after the cement has been forced out.

If the well is full of mud-laden fluid (still prescribed in some fields) this method will prevent contamination of the cement, a factor largely ignored by modern practice, although it must be possible to renew circulation immediately when the cement is mixed and ready to run. At this juncture the bottom plug is inserted in the casing followed by the cement and the top plug. Some operators then throw some empty sacks or a piece of burlap and some clay on the top plug. The slush pump is then connected with the casing and the cement and plugs are rapidly forced to the bottom of the hole by the fluid from the pump, as in *C*, Fig. 216. When the lower plug reaches bottom the cement will pass upward between the casing and the wall and when the plugs come together (*D*), the action of the pump will so indicate. The casing should then be rotated a few turns, (to distribute the cement now outside of the casing) and lowered to the bottom (*E*). The cement will permanently solidify in from twelve to fifteen days after which the well may be drilled in and completed in the regular way. Before inserting the plugs, however, a casing swedge (Fig. 182) or other tool should be run to ascertain that the casing is clear and free from dents or "blisters." Otherwise the plugs might lodge and a serious situation result from the premature setting of the cement at some point in the casing.

The two-plug method is still commonly used except in California and, while it clearly illustrates the principle of cementing, it is based largely on erroneous assumptions. It was formerly believed that the mixed cement would become diluted or contaminated if exposed to contact with liquid in the hole and that the mud-laden fluid must be retained in the well to prevent caving of the walls—two contentions that modern practice in California has disproved, as both are disregarded in the Perkins and the Scott patented methods.

In California it is now customary to wash the hole thoroughly with clear water before cementing. But during the washing process the innermost string

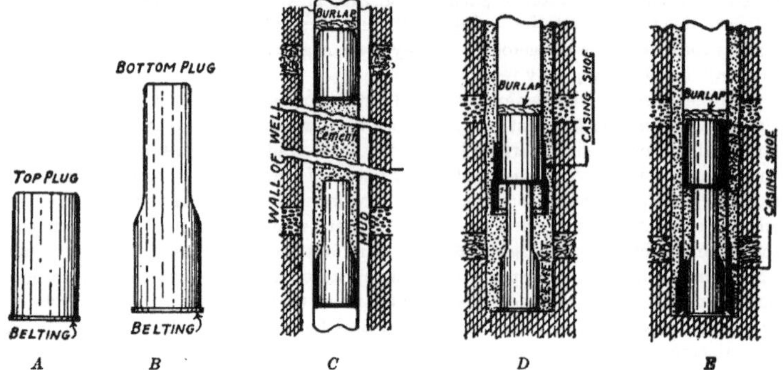

FIG. 216.—Illustrating the two-plug method of cementing.

of casing must be slightly raised and lowered constantly to prevent freezing. When the mud has been washed out some operators pump in about a ton of slaked lime to purify the water and aid the cement in setting. The cement (10 to 30 tons) is then pumped in and followed by a special plug (*the Perkins One-plug Method*). A few operators prefer to leave 10 or 15 feet of cement within the bottom of the casing, and to accomplish this a piece of timber about 4 inches by 4 inches by 10 feet is capped with a circular piece of heavy belting to fit the casing, and dropped in between the cement and the plug. This stops the plug at the desired distance from the bottom.

The Scott Method, simple and very successful, is similar except that the cement is mixed with a mounted mixer (instead of by hand) and no plugs are used. After the cement is run, pure water accurately measured is pumped into the well in sufficient quantity to fill the casing and force the cement outside, after which the casing is lowered to bottom, the cementing operation being complete. Two pumps, generally moved about on a truck, are used. The larger is similar to a regular rotary slush pump and the smaller one is a higher-pressure pump, as great power is often necessary to force the 20 or 30 tons of cement upward between the casing and the wall of the hole.

Upon completing any cementing operation, sufficient time must be allowed for the cement to solidify, after which the plugs or the portion of cement remaining within the the casing may be drilled out and the well completed in the regular way.

SHOOTING A WELL

When the stratum from which production is expected appears hard and compact, a heavy charge of nitroglycerine may be employed to rend and shatter it, a process termed "shooting." The explosive is poured into "shells" (circular tinned cans about 4½ by 30 to 60 inches) and lowered, by means of a rope and reel to the bottom of the well. Then it is detonated by the impact of a go-devil (cast-iron weight) dropped from the derrick floor, or by an electric charge. The uppermost shell contains a firing head fitted with a percussion cap which, when "touched-off," explodes the nitroglycerine. The explosive charge varies from 5 to 300 quarts and the rock or sand is fissured by the force of the explosion. From 100 to 600 feet of water is used as "tamping" on top of the explosive to produce the desired lateral-shattering effect. This water and quantities of detritus are usually blown from the well, forming a pocket or reservoir into which the oil, if any, flows through the newly-created fissures. Shooting may also be successfully employed to increase or restore the production of partially exhausted or even temporarily abandoned wells, but unless good judgment is used and great care exercised, shooting may ruin a well and considerable adjacent territory.

Where a water-bearing sand or formation lies immediately above, below, or in close proximity to the oil sand, a heavy explosion may produce crevices admitting the water and ruining the well. If, as in many fields, the oil sand lies just under an impervious cap-rock, any water previously encountered may be successfully excluded from the oil-bearing stratum by setting (and, if a gusher is expected, by cementing) the last string of casing in this hard cap-rock. This method of setting and cementing casing is commouly employed but for obvious reasons measurements should be accurate and a heavy shot must not be placed immediately under the casing "seat."

Neither will shooting prove beneficial where the oil sand is thin with an overlying formation of a crumbling nature, as the resulting cavings will clog the fissures, clutter the sand and impede or stop production. When it is desired to pull casing that is immovable or fast in the well it is often expedient to insert dynamite or nitroglycerine to "shoot-off," or part, this casing at any desired point, thereby facilitating removal of the upper, free portion of the string.

PERFORATING AND SCREENING

In many fields the water string of casing may be set at a point hundreds of feet above the oil sand, no caving or heaving strata being encountered thereafter. This is an ideal condition, as under such circumstances a well may be quickly completed and the production obtained without trouble from unstable formations. But in other regions—particularly in Louisiana, South Texas, California, Rumania and Russia—loose, caving or heaving sands require different treatment and the use of special appliances. The large diameters of the Rumanian and Russian wells facilitate bailing but in parts of America an extra string of casing (the "oil string") or a liner, with screened or perforated sections is used to overcome the unstable walls and at least partially exclude the loose sand.

(*a*) Perforated **pipe or casing.**—If loose oil sand is not cased it will quickly cave and perhaps heave and bridge solidly; thereby completely stopping production. To prevent this, one or more joints of perforated casing may be attached to the lower end of the oil string or liner, the number depending on the thickness of the oil sand or the depth to which it has been penetrated. The perforations may be shop-made with a drill press or an acetylene torch before running the casing or liner, or afterwards with a perforator (similar to Fig. 173) run on either a string of fishing tools or on tubing. The nature of the sand and the gravity of the oil determine the size and number of the perforations. These may comprise from three to six longitudinal rows of circular holes varying from $\frac{3}{8}$ to $\frac{7}{8}$ inch in diameter, or vertical slots ranging in size from $\frac{3}{8}$ by $\frac{5}{8}$ inch to $\frac{3}{4}$ by 2 inches. The shop method of perforating is preferable where conditions and requirements are known; otherwise a perforator of the type preferred by the operator or driller may be used to perforate any desired number of joints at any point in the string. If the perforator is lowered on tubing the work may be done more accurately but with less speed and at greater hazards than if run on tools. It is apparent that perforations materially weaken the casing and for that reason only high-grade tubular goods should be subjected to this operation.

A perforator may sometimes be employed to free "frozen" casing, the perforations permitting the sand (causing the external friction) to flow from behind to within the casing.

Joints of perforated casing in conjunction with packers (Fig. 205 or 206) may be lowered on a single string of casing to secure production from two or more oil sands even if intervening caves or water-bearing strata be present.

TYPES OF WELL STRAINERS

FIG. 217.—Keystone wire-wrapped screen.
This screen, made in any desired gage or mesh, consists of a piece of perforated pipe or casing and a screen covering of specially wrapped and soldered brass, galvanized or steel wire. The keystone shape of the wire is designed to prevent clogging.

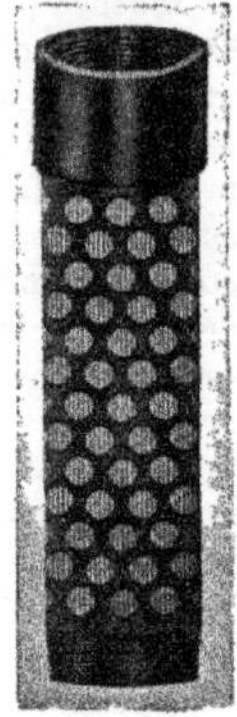

FIG. 218—Wireless strainer.
The wireless strainer consists of perforated pipe or casing with slotted steel or brass plugs or buttons, of the desired gage, countersunk and riveted in the perforations. These strainer-plugs are thus protected against injury from either external or internal friction.

(*b*) **Sand screens or strainers.**—As loose sand rapidly cuts and destroys valves and working barrels, various means of excluding it have beenadopted. The setting of screens or strainers (Figs. 217, 218) in the oil sand is common practice in parts of California, Louisiana and South Texas. The proper mesh to be used in any locality can be determined only by experimentation or after several wells have been completed, finer meshes being used where the oil is of light gravity; but the texture of the sand must also be considered. There is an apparent tendency, however, to use finer mesh than is necessary and this mistake frequently accounts for declining production.

If the oil sand does not crumble, cave or heave, a liner, a screen or a strainer will not be necessary. If it crumbles slowly a screen, a strainer or a joint of perforated casing may be run without difficulty either on a complete "oil" string or on a liner; but if it caves or heaves badly special care and apparatus may be necessary to "wash-in" and set a sand screen or a strainer properly. If the customary "water" string of casing has been set the screen or strainer may be inserted as shown in Fig. 219-H, representing the shoe of the water string of casing, set at any convenient distance above the oil sand. The screen is lowered on casing, drill pipe or large tubing connected with *E* by a coupling at *B*, the steel wash ring (*B*) being inserted within the coupling. The wash pipe (*C*) extends only from *A* to the wooden wash plug (*L*) in which it is seated. The diameter of the longitudinal bore through the wash plug is about equal to the interior diameter of the wash pipe (2 or 3 inches), the plug being seated on the back-pressure valve (*N*). Sufficient casing must be inserted between the screen (*J*) and the lead seal (*F*) to bring the latter within the water string of casing when the shoe (*P*) is resting on bottom. After this appliance has been assembled and lowered to bottom or until it is stopped by the loose oil sand, clear water is pumped down through the inner casing, drill pipe, or tubing (connected with *E* at *B*), and as it cannot pass the wash ring (*B*) it is forced through the wash pipe and back pressure valve, washing out the mud and loose sand as it returns to the surface, traveling upward outside the screen and between *D* and *E*. After the washing has been completed the inner casing or pipe is turned to the right, the left-hand thread being unscrewed at *G*, after which withdrawal of the inner casing will also remove parts *A*, *B*, *C* and *E*. The top of the lead seal is then expanded (by means of a swedge or a special expanding tool) to form a tight bond with the casing (*D*) and the well is ready for production.

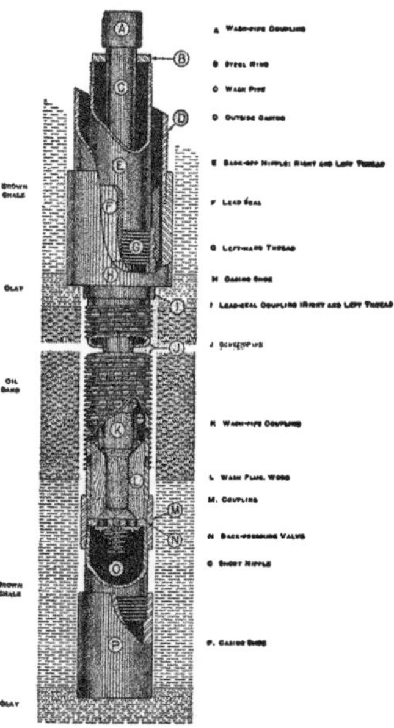

Fig. 219.—Method of setting screen pipe when it is desired to recover that part of the oil string above the screen. (From Technical Paper, 247, U. S. Bureau of Mines.)

If a water string has not been set or if, for any reason, it is desired to run and set the strainer on a full string of casing, the screen will be connected to the casing with a regular coupling and the wash ring (*B*) will be made of wood or soft metal. After washing is completed the wash pipe may be removed with a steel die nipple run on tubing or with a spear run on a fishing string or on tubing, the wash plug breaking when forced upward by the coupling *K*. The well is then ready to be bailed, swabbed or pumped.

Perforated pipe, sand screens and strainers may become partially or completely clogged with sand or clay, impeding or obstructing production accordingly. This condition usually responds to washing with hot oil pumped down through tubing. This washing may be necessary weekly, monthly or annually.

Plugging Abandoned Wells

When a " dry hole " or an exhausted well is abandoned it is customary to " pull " (withdraw) the casing. During the early period of oil-field development in America, the hazards incident to this operation were either unrecognized or ignored and in many localities the sands were penetrated and then carelessly flooded by water from higher levels when the casing was subsequently salvaged. Due to head pressure such water may travel considerable distances laterally through porous sands or overlying formations, finally reaching and injuring a producing area.

To prevent such disasters, laws have been passed in petroleum producing states to safeguard oil-bearing strata through the regulation and supervision of this abandonment procedure. As structures and conditions existing in the oil-fields of different states are dissimilar there is a corresponding lack of uniformity in the governing regulations. It is, therefore, incumbent upon every operator contemplating the abandonment of a well to familiarize himself with the legal requirements before pulling the casing. These laws may be changed frequently; for example, Kansas Senate Bill No. 638, introduced (1921) by the Committee on Gas and Oil, requires the plugging of all dry and abandoned oil or gas wells and prescribes the method to be followed in each case, a "County Plugger" being given authority to direct plugging operations.

In some regions the effective sealing of the oil sands against the ingress of water from above or below has been sufficient, but in others more elaborate procedure is required, depending upon the number and location of the water-bearing strata and the nature of the formations between them and the oil sand. In some regions only wooden plugs from 2 to 4 feet long are used, these fitting the hole snugly and being driven to a point between the oil sand and the nearest water-bearing stratum above. Absorption of water causes the wooden plug to swell, after which clay may be dropped upon it, thereby forming a firm bond with the wall of the hole, a permanent bridge and impediment to water finally resulting. A plug may be used in this manner to form a bridge in combination with cement instead of clay. Or cement alone may be used to seal the oil sand and the hole for a considerable distance above, the plastic cement being lowered either by a dump bailer through tubing or by the lubricator process (Fig. 119) if gas or water pressure from below is encountered. In sealing the sand by this process it may be necessary to maintain pump pressure on the well to prevent the ingress of gas or water through the cement until it has solidified. Lead plugs are also used in hard formations, being battered with tools and expanded to form a bond with the wall of the hole.

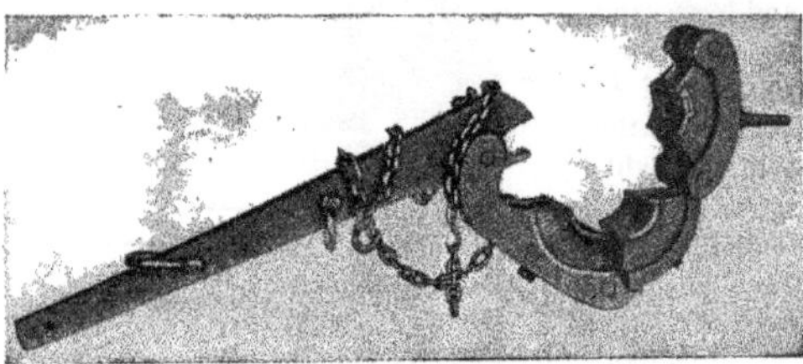

Fig. 220.—Dunn tongs. Shows 4-inch bushings in use with 10-inch tong.

TUBULAR GOODS FOR OIL COUNTRY USE

The term "tubular goods" generally covers all classes of pipe, casing and tubing used in drilling or operating oil or gas wells, and comprises the following distinct types; casing, tubing, drive pipe, line pipe and rotary drill pipe. Each type has been developed for a definite purpose and possesses individual characteristics, but all are distinct from standard pipe generally known to the engineering or commercial trade.

Casing.—Screw casing ranges in size from 3 to 15½ inches, these dimensions representing the nominal inside diameters. The table for casing shows numerous instances in which the nominal inside diameter remains constant while the weight increases. It will be observed, for example, that 6¼-inch casing weighing 24 pounds per foot must have a smaller internal diameter than 6¼-inch casing weighing 13 pounds per foot; but for trade purposes both weights are designated as 6¼-inch. The extra thickness of pipe or casing heavier than standard weight is added to the inside of the tube; the outside, notwithstanding variations in weight, remains the same to accommodate standard couplings. A proper description of casing, therefore, includes the nominal inside diameter, the weight per foot and the number of threads per inch. The usual taper of threads is $\frac{3}{8}$ inch diameter per foot of length, but in some of the heavier sizes the taper is $\frac{3}{4}$ inch. The taper of thread on the casing corresponds to a similar taper in the coupling, thus insuring a strong, impervious joint.

Drive pipe.—The joints on drive pipe differ from those of casing, the taper being only $\frac{3}{16}$ inch, and the sizes from 6 to 20 inches in diameter. The reason for this difference is that it is necessary for the ends of the tubes to meet within the coupling. While such joints may not be water-tight they present a continuous column for sustaining the blows incident to driving and relieve the threads from the impact. For sizes above 15 inches the outside diameter is the basis of measurement.

Tubing.—Oil-well tubing is used in a completed well to secure the production. It must withstand unusual strains due to pressure and vibration and is subjected to constant wear caused by the movement of sucker rods. Tubing is finished with the utmost care and is subjected to extreme tests; but because of the weakening effect of threading common tubing, quantities of exterior upset-end tubing are now being used in deep-well production.

Line pipe.—The line-pipe specifications cover sizes ranging from $\frac{1}{4}$ inch to 15 inches, but the smallest size in actual use is 2 inches inside diameter. Line pipe differs from standard pipe, always being lap-welded unless otherwise specified. The couplings are longer than those used on standard pipe and the ends of both couplings and tubes are always reamed. The combination of these features presents a smooth interior surface for the passage of fluids and it is sometimes used as tubing in shallow wells. But line pipe is not so carefully tested as tubing and is not recommended for that purpose.

(Stovepipe casing is described on page 237 and rotary drill pipe on page 256.)

Threads.—The standard thread for tubular goods has a vertical angle of 60° but is very slightly rounded at the top and bottom. The height or depth of a thread may be determined by dividing the decimal 0.8 by the number of threads per inch. The threaded ends of pipe or casing are slightly conical, the taper generally being $\frac{3}{4}$ inch diameter per foot of length (i.e. 1 in 32 to the axis of the pipe). Thus a threaded end 4 inches long is tapered $\frac{1}{4}$ inch.

Hydrostatic and collapsing pressures.—While the hydrostatic test (bursting-pressure) is important in relation to line pipe and tubing, it is a secondary consideration relative to casing and drive pipe. But when a long string of casing is set and subsequently surrounded by a column of water extending almost or quite to the surface, the external pressure (0.43 pound per square inch for each foot in height) may cause the casing to collapse at some point near the bottom after the fluid has been bailed or pumped from within. The collapsing pressure to which casing may be subjected must, therefore, be carefully considered. The formula from which the approximate collapsing pressure of mild steel pipe or casing may be determined follows:

$$\text{Collapsing pressure} = 86{,}670\left(\frac{\text{thickness of wall of tube, in inches}}{\text{outside diameter of tube, in inches}}\right) - 1386.$$

Thus, the approximate collapsing pressure in pounds per square inch of new $8\frac{1}{4}$-inch 32-pound California DBX Casing is $86{,}670\left(\frac{.352}{8.675}\right) - 1386 = 2131$. Allowing for a factor of safety of two, this would be equal to a column of water 2478 feet in height $\left(\frac{2131}{2 \times 0.43} = 2478.\right)$

In applying this formula to drive pipe and casing consideration must be given to the harmful, crystallizing effects of any previous driving or abnormal stresses and a larger factor of safety accordingly allowed. Manufacturers sometimes suggest a factor of from 4 to 10 in extreme cases. In the following tables the approximate collapsing pressure of California Casing is indicated without allowance for a factor of safety.

Casing

All weights and dimensions are nominal

Size	Diameters		Thickness	Weight per Foot		Threads per inch	Couplings		
	External	Internal		Plain ends	Threads and couplings		Diameter	Length	Weight
3	3 250	3.010	0.120	4.011	4.100	14	3.771	3½	2.612
3¼	3.500	3.250	.125	4.505	4.600	14	4.021	3½	2.799
3½	3.750	3.492	.129	4.988	5.100	14	4.271	3½	2.987
3¾	4.000	3.732	.134	5.532	5.650	14	4.521	3½	3.174
4	4.250	3.974	.138	6.060	6.200	14	4.771	3⅝	3.923
4¼	4.500	4.216	.142	6.609	6.750	14	5.021	3⅝	4.141
4¼	4.550	4.090	.205	9.403	9.500	14	5.021	3⅝	4.141
4½	4.750	4.460	.145	7.131	7.250	14	5.271	3⅝	4.360
4½	4.750	4.364	.193	9.393	9.500	14	5.271	3⅝	4.360
4¾	5.000	4.696	.152	7.870	8.000	14	5.521	3⅝	4.578
5	5.250	4.944	.153	8.328	8.500	14	5.828	4½	5.929
5	5.250	4.886	.182	9.851	10.000	14	5.828	4½	5.929
5	5.250	4.886	.182	9.851	10.000	11½	5.800	4½	5.742
5	5.250	4.768	.241	12.892	13.000	11½	5.800	4½	5.742
5	5.250	4.648	.301	15.909	16.000	11½	5.800	4½	5.742
5 3/16	5.500	5.192	.154	8.792	9.000	14	6.078	4½	6.200
5 3/16	5.500	5.044	.228	12.837	13.000	11½	6.050	4⅝	6.759
5 3/16	5.500	4.892	.304	16.870	17.000	11½	6.155	5½	8.849
5⅝	6.000	5.672	.164	10.222	10.500	14	6.664	4½	7.729
5⅝	6.000	5.620	.190	11.789	12.000	11½	6.636	4½	7.516
5⅝	6.000	5.552	.224	13.818	14.000	11½	6.636	4½	7.516
5⅝	6.000	5.450	.275	16.814	17.000	11½	6.636	4½	7.516
6¼	6 625	6.287	.169	11.652	12.000	14	7.308	4⅝	9.825
6¼	6.625	6.255	.185	12.724	13.000	14	7.308	4⅝	9.825
6¼	6.625	6.257	.184	12.657	13.000	11½	7.280	5½	10.630
6¼	6.625	6.135	.245	16.694	17.000	11½	7.280	5½	10.630
6¼	6.625	5.913	.356	23.835	24.000	11½	7.280	5½	10.630
6⅝	7.000	6.652	.174	12.685	13.000	14	7.692	4⅝	10.497
6⅝	7.000	6.538	.231	16.699	17.000	11½	7.664	4⅝	10.225
6⅝	7.000	6.538	.231	16.699	17.000	10	7.642	5½	11.133
6⅝	7.000	6.450	.275	19.751	20.000	10	7.699	6½	14.458
6⅝	7.000	6.334	.333	23.711	24.000	10	7.699	6½	14.458

Casing—Continued

All weights and dimensions are nominal

Size	Diameters		Thickness	Weight per Foot		Threads per inch	Couplings		
	External	Internal		Plain ends	Threads and couplings		Diameter	Length	Weight
7¼	7.625	7.263	0.181	14.390	14.750	14	8.317	4⅝	11.401
7⅝	8.000	7.628	.186	15.522	16.000	11½	8.788	5⅛	15.308
7⅝	8.000	7.528	.236	19.569	20.000	11½	8.788	5⅛	15.308
8¼	8.625	8.249	.188	16.940	17.500	11½	9.413	5⅛	16.461
8¼	8.625	8.191	.217	19.486	20.000	11½	9.413	5⅛	16.461
8¼	8.625	8.097	.264	23.574	24.000	11½	9.413	5⅛	16.461
8¼	8.625	8.097	.264	23.574	24.000	8	9.358	6⅛	18.577
8¼	8.625	8.003	.311	27.615	28.000	8	9.358	6⅛	18.577
8⅝	9.000	8.608	.196	18.429	19.000	11½	9.788	5⅛	17.153
9⅝	10.000	9.582	.209	21.855	22.750	11½	10.911	6⅛	26.136
10	10.750	10.192	.279	31.201	32.515	8	11.958	6⅝	39.772
10	10.750	10.146	.302	33.699	35.000	8	11.958	6⅝	39.772
10⅝	11.000	10.552	.224	25.780	26.750	11½	11.911	6⅛	28.536
11⅝	12.000	11.514	.243	30.512	31.500	11½	12.911	6⅛	31.051
12½	13.000	12.482	.259	35.243	36.500	11½	14.025	6⅛	37.499
12½	13.000	12.278	.361	48.730	50.000	8	14.085	7⅛	46.464
13½	14.000	13.448	.276	40.454	42.000	11½	15.139	6⅛	44.495
14½	15.000	14.418	.291	45.714	47.500	11½	16.263	6⅛	52.401
15½	16.000	15.396	.302	50.636	52.500	11½	17.263	6⅛	55.779

The permissible variation in weight is 5 per cent above and 5 per cent below.

Furnished with threads and couplings and in random lengths unless otherwise ordered.

Taper of threads is ⅜-inch diameter per foot length for all sizes, eXcept the 8¼-inch 24-pound, 8¼-inch 28-pound 8-thread, 10-inch, and 12½-inch 50-pound, which are ¾-inch taper.

Thickness of walls makes it impracticable to cut threads of coarser pitch than shown on table.

The weight per foot of casing with threads and couplings is based on a length of 20 feet, including the coupling, but shipping lengths of small sizes will usually average less than 20 feet.

All weights given in pounds. All dimensions given in inches.

On sizes made in more than one weight or thread, weight and number of threads desired must be specified.

California Casing

All weights and dimensions are nominal

Size	Diameters		Thickness	Weight per Foot		Threads per inch	Couplings		Weight	Approximate collapsing pressure
	External	Internal		Plain ends	Threads and couplings		Diameter	Length		
4¼	4.750	4.082	0.334	15.752	16.000	10	5.364	6⅝	9.963	4635
4¾	5.000	4.500	.250	12.682	12.850	10	5.491	6⅝	8.533	2950
4¾	5.000	4.408	.296	14.870	15.000	10	5.491	6⅝	8.533	3740
5⅝	6.000	5.352	.324	19.641	20.000	10	6.765	7¼	15.748	3300
6¼	6.625	6.049	.288	19.491	20.000	10	7.390	7⅝	18.559	2350
6¼	6.625	5.921	.352	23.582	24.000	10	7.390	7⅝	18.559	3200
6¼	6.625	5.855	.385	25.658	26.000	10	7.390	7⅝	18.559	3650
6¼	6.625	5.791	.417	27.648	28.000	10	7.390	7⅝	18.559	4050
6⅝	7.000	6.456	.272	19.544	20.000	10	7.698	7⅝	17.943	2000
6⅝	7.000	6.276	.362	25.663	26.000	10	7.698	7⅝	17.943	3100
6⅝	7.000	6.214	.393	27.731	28.000	10	7.698	7⅝	17.943	3500
6⅝	7.000	6.154	.423	29.712	30.000	10	7.698	7⅝	17.943	3850
7⅝	8.000	7.386	.307	25.223	26.000	10	8.888	8⅛	27.410	1850
8¼	8.625	8.017	.304	27.016	28.000	10	9.627	8⅛	33.096	1650
8¼	8.625	7.921	.352	31.101	32.000	10	9.627	8⅛	33.096	2150
8¼	8.625	7.825	.400	35.137	36.000	10	9.627	8⅛	33.096	2650
8¼	8.625	7.775	.425	37.220	38.000	10	9.627	8⅛	33.096	2850
8¼	8.625	7.651	.487	42.327	43.000	10	9.627	8⅛	33.096	3500
9⅝	10.000	9.384	.308	31.881	33.000	10	11.002	8⅛	38.162	1250
10	10.750	10.054	.348	38.661	40.000	10	11.866	8⅛	45.365	1400
10	10.750	9.960	.395	43.684	45.000	10	11.866	8⅛	45.365	1800
10	10.750	9.902	.424	46.760	48.000	10	11.866	8⅛	45.365	2000
10	10.750	9.784	.483	52.962	54.000	10	11.866	8⅛	45.365	2500
11⅝	12.000	11.384	.308	38.460	40.000	10	13.116	8⅛	50.445	800
12½	13.000	12.438	.281	38.171	40.000	10	14.116	8⅛	54.508	450
12½	13.000	12.360	.320	43.335	45.000	10	14.116	8⅛	54.508	750
12½	13.000	12 282	.359	48.467	50.000	10	14.116	8⅛	54.508	1050
12½	13.000	12.220	.390	52.523	54.000	10	14.116	8⅛	54.508	1200
13½	14.000	13.344	.328	47.894	50.000	10	15.151	9⅛	67.912	650
15½	16.000	15.198	.401	66.806	70.000	10	17.477	9⅛	98.140	800

Taper of threads is ¾-inch diameter per foot length for all sizes.

Drive Pipe

All weights and dimensions are nominal

Size	Diameters		Thickness	Weight per Foot		Threads per inch	Couplings		
	External	Internal		Plain ends	Threads and couplings		Diameter	Length	Weight
2	2 375	2.067	0.154	3.562	3.730	11½	2.923	3⅝	2.380
2½	2.875	2.469	.203	5.793	5.906	8	3.486	4⅛	3.748
3	3.500	3.068	.216	7.575	7.705	8	4.111	4⅛	4.493
3½	4.000	3.548	.226	9.109	9.294	8	4.723	4⅛	5.973
4	4.500	4 026	.237	10.790	10.995	8	5.223	4⅛	6.740
4½	5.000	4.506	.247	12.538	12.758	8	5.723	4⅛	7.439
5	5.563	5.047	.258	14.617	14.989	8	6.410	5⅛	11.871
6	6.625	6.065	.280	18.974	19.408	8	7.473	5⅛	13.956
7	7.625	7.023	.301	23.544	24.021	8	8.474	5⅛	15.955
8	8.625	8.071	.277	24.696	25.495	8	9.588	6⅛	24.343
8	8.625	7.981	.322	28.554	29.303	8	9.588	6⅛	24.343
8	8.625	7.917	.354	31.270	32.334	8	9.882	6⅛	31.320
9	9.625	8.941	.342	33.907	34 711	8	10.588	6⅛	27.035
10	10.750	10.192	.279	31.201	32.631	8	11.950	6⅝	40.108
10	10.750	10.136	.307	34.240	35.628	8	11.950	6⅝	40.108
10	10.750	10.020	.365	40.483	41.785	8	11.950	6⅝	40.108
11	11.750	11.000	.375	45.557	46.953	8	12.950	6⅝	43.664
12	12.750	12.090	.330	43.773	45 358	8	13.950	6⅝	47.220
12	12.750	12.000	.375	49.562	51.067	8	13.950	6⅝	47.220
13	14.000	13.250	.375	54.568	56.849	8	15.438	7⅛	66.024
14	15.000	14.250	.375	58.573	61.005	8	16.438	7⅛	70.533
15	16 000	15.250	.375	62.579	65.161	8	17.438	7⅛	75.043
17 O. D.	17.000	16.214	.393	69.704	73 000	8	18.675	7⅛	91.746
18 O. D.	18.000	17.182	.409	76.840	81.000	8	19 913	7⅛	109.669
20 O. D.	20.000	19.182	.409	85.577	90.000	8	21 913	7⅛	121.298

The permissible variation in weight is 5 per cent above and 5 per cent below.
Taper of threads is ¾ inch from 2 inches to 5 inches, and 3/16-inch from 6 inches to 20 inches.

Line Pipe

All weights and denominations are nominal

Size	Diameters		Thickness	Weight per Foot		Threads per inch	Couplings		
	External	Internal		Plain ends	Threads and couplings		Diameter	Length	Weight
⅛	0.405	0.269	0 068	0.244	0 246	27	0.582	1⅛	0.043
¼	.540	.364	.088	.424	.426	18	.724	1⅜	.069
⅜	.840	.493	.091	.567	.571	18	.898	1⅝	.126
½	.840	.622	.109	.850	.856	14	1.085	1⅞	.205
¾	1.050	.824	.113	1.130	1.138	14	1.316	2⅛	.316
1	1.315	1.049	.133	1 678	1.688	11½	1.575	2⅜	.445
1¼	1.660	1.380	.140	2.272	2.300	11½	2.054	2⅝	.974
1½	1.900	1.610	.145	2.717	2.748	11½	2.294	2⅞	1.103
2	2.375	2.067	.154	3.652	3.716	11½	2.841	3⅝	2.146
2½	2.875	2.469	.203	5.793	5.881	8	3.389	4⅛	3.387
3	3.500	3.068	.216	7.575	7.675	8	4.014	4⅛	4.076
3½	4.000	3.548	.226	9.109	9.261	8	4.628	4⅛	5.510
4	4.500	4.026	.237	10.790	10.980	8	5.233	4⅛	6.673
4½	5.000	4.506	.247	12.538	12.742	8	5.733	4⅛	7.379
5	5.563	5.047	.258	14.617	14.966	8	6.420	5⅛	11.730
6	6.625	6.065	.280	18.974	19.367	8	7.482	5⅛	13.869
7	7.625	7.023	.301	23.544	23.975	8	8.482	5⅛	15.883
8	8.625	8.071	.277	24.696	25.414	8	9.596	6⅛	24.130
8	8.625	7.981	.322	28.554	29.213	8	9.596	6⅛	24.130
9	9.625	8.941	.342	33.907	34.612	8	10.596	6⅛	26.838
10	10.750	10.192	.279	31.201	32 515	8	11.958	6⅝	39.772
10	10 750	10.136	.307	34.240	35.504	8	11.958	6⅝	39.772
10	10.750	10.020	.365	40.483	41 644	8	11.958	6⅝	39.772
11	11.750	11.000	.375	45.557	46 805	8	12.958	6⅝	43.326
12	12.750	12.090	.330	43.773	45 217	8	13.958	6⅝	46.898
12	12.750	12.000	.375	49.562	50.916	8	13.958	6⅝	46.898
13	14.000	13.250	.375	54.568	56.649	8	15.446	7⅛	65.506
14	15.000	14.250	.375	58.573	60 802	8	16.446	7⅛	70.031
15	16.000	15.250	.375	62.579	64 955	8	17.446	7⅛	74.555

Taper of threads is ¾-inch diameter per foot length for all sizes.

Oil-well Tubing

All weights and dimensions are nominal

Size	Diameters		Thickness	Weight per Foot		Threads per inch	Couplings		
	External	Internal		Plain ends	Threads and couplings		Diameter	Length	Weight
1¼	1.660	1.380	0.140	2.272	2.300	11½	2.054	2¾	0.974
1½	1.900	1.610	.145	2.717	2.748	11½	2.294	2¾	1.103
2	2.375	2.041	.167	3.938	4.000	11½	2.841	3⅛	2.146
2	2.375	1.995	.190	4.433	4.500	11½	2.841	3⅛	2.146
2½	2.875	2.469	.203	5.793	5.897	11½	3.449	4⅛	3.636
2½	2.875	2.441	.217	6.160	6.250	11½	3.449	4⅛	3.636
3	3.500	3.068	.216	7.575	7.694	11½	4.074	4⅛	4.366
3	3.500	3.018	.241	8.388	8.500	11½	4.074	4⅛	4.366
3	3.500	2.922	.289	9.910	10.000	11½	4.074	4⅛	4.366
3½	4.000	3.548	.226	9.109	9.261	8	4.628	4⅛	5.510
4	4.500	4.026	.237	10.790	10.980	8	5.233	4⅛	6.673
4	4.500	3.990	.255	11.561	11.750	8	5.233	4⅛	6.673

Taper of threads is ¾ inch diameter per foot length for all sizes.

Special Rotary Pipe

All weights and dimensions are nominal

Size	Diameters		Thickness	Weight per Foot		Threads per inch	Couplings		
	External	Internal		Plain ends	Threads and couplings		Diameter	Length	Weight
2½	2.875	2.323	0.276	7.661	7.830	8	3.603	5⅛	5.888
2½	2.875	2.143	.366	9.807	10.000	8	3.693	5⅛	7.316
4	4.500	3.958	.271	12.240	12.500	8	5.228	5⅛	8.901
4	4.500	3.826	.337	14.983	15.000	8	5.240	6⅛	11.720
4½	5.000	4.388	.306	15.340	15.500	8	5.604	5⅛	8.270
4½	5.000	4.290	.355	17.611	18.000	8	5.740	6⅛	12.950
5	5.563	4.955	.304	17.074	17.500	8	6.373	6⅛	14.620
5	5.563	4.813	.375	20.778	21.000	8	6.272	7⅛	16.442
6	6.625	5.937	.344	23.076	23.500	8	7.435	6⅛	17.254
6	6.625	5.761	.432	28.573	29.000	8	7.334	7⅛	19.451

Special Upset Rotary Pipe

All weights and dimensions are nominal

Size	Diameters		Thickness	Weight per Foot		Threads per inch	Couplings		
	External	Internal		Plain ends	Threads and couplings		Diameter	Length	Weight
2½	2.875	2.323	0.276	7.661	7.841	8	3.564	6⅛	6.743
2½	2.875	2.143	.366	9.807	10.000	8	3.678	6⅛	7.844
4	4.500	3.958	.271	12.240	12.632	8	5.256	7⅜	14.296
4	4.500	3.826	.337	14.983	15.323	8	5.256	7⅜	14.296
5	5.563	4.975	.294	16.544	17.000	8	6.303	8⅛	18.472
5	5.563	4.859	.352	19.590	20.000	8	6.303	8⅛	18.472
6	6.625	6.065	.280	18.974	19.551	8	7.350	8⅜	22.994
6	6.625	5.761	.432	28.573	28.948	8	7.350	8⅜	22.994

Taper of threads is ¾ inch diameter per foot length for all sizes.

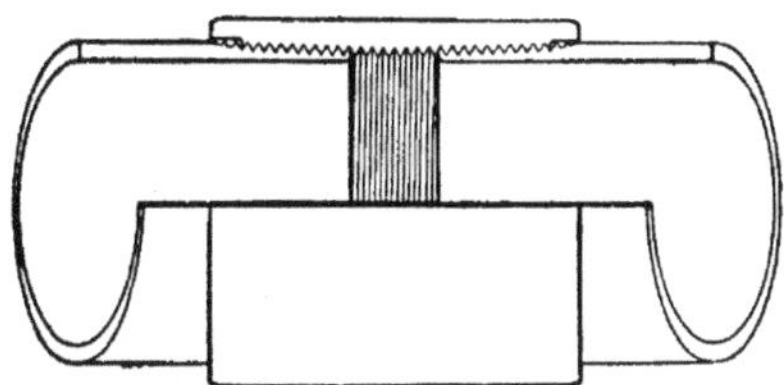

FIG. 221.—Typical section of line pipe coupling and joint.

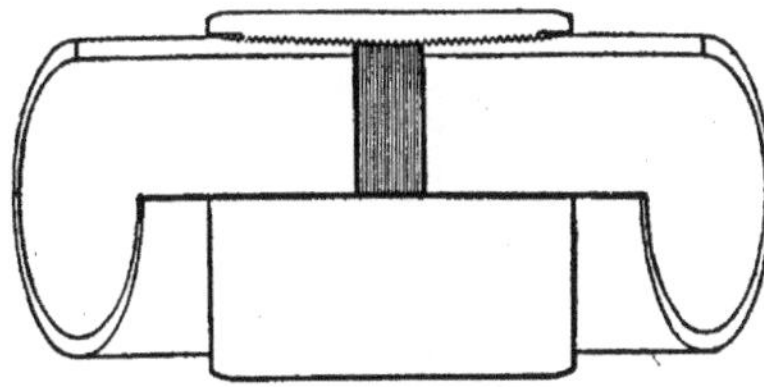

FIG. 222.—Typical section of oil-well tubing coupling and joint.

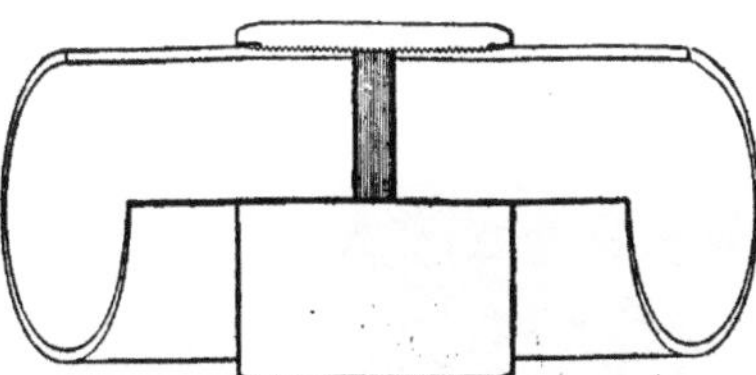

FIG. 223.—Typical section of standard Boston casing coupling and joint.

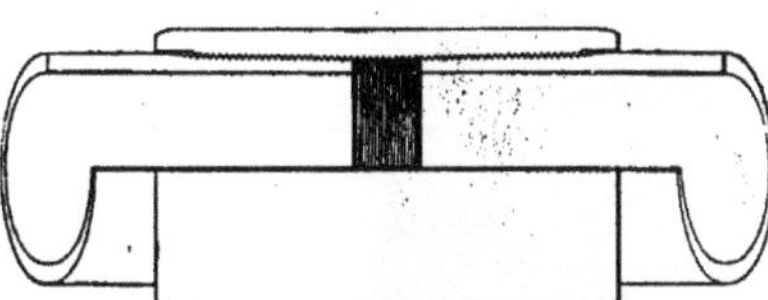

FIG. 224.—Typical section of California Diamond BX casing coupling and joint.

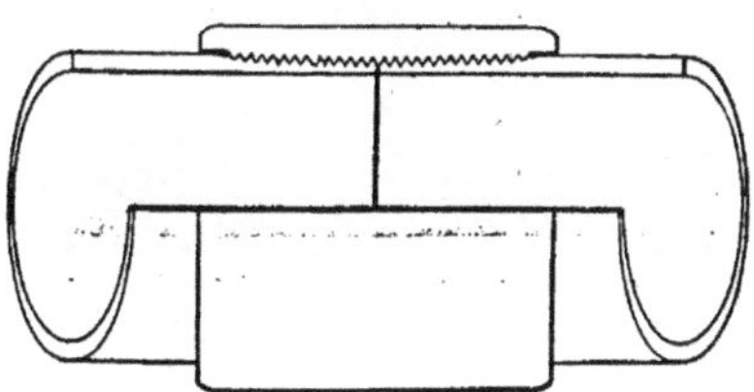

FIG. 225.—Typical section of drive-pipe coupling and joint.

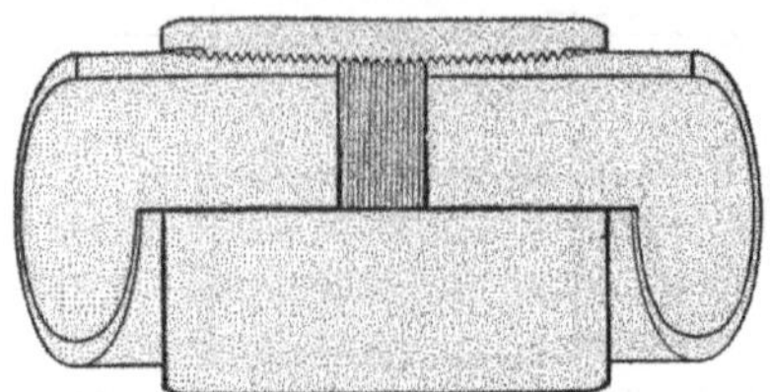

Fig. 226.—Typical section of special rotary pipe coupling and joint.

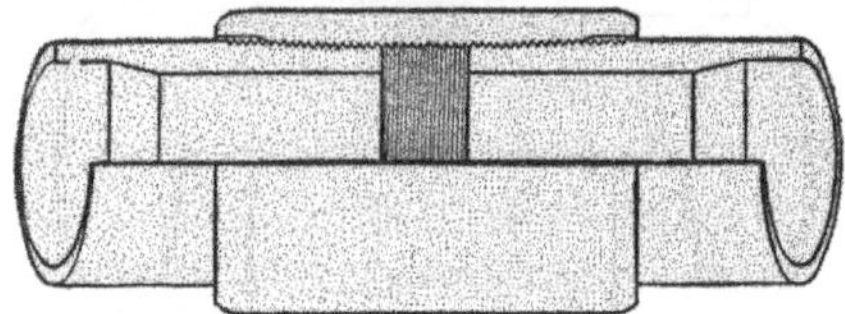

Fig. 227.—Typical section of special upset rotary pipe coupling and joint.

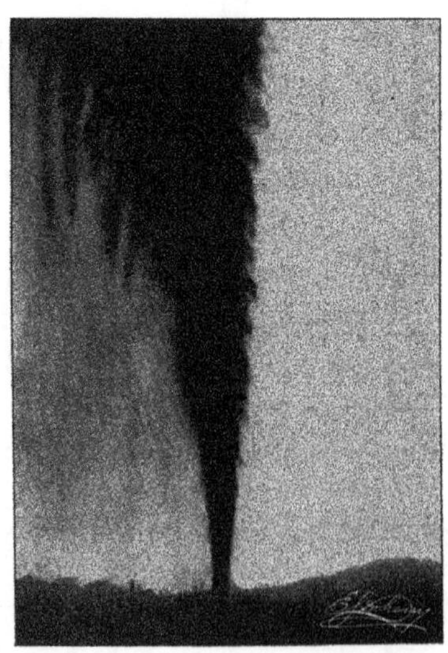

Cerro Azul No. 4, Mexican Petroleum Company, Mexico.

METHODS OF PRODUCING OIL

While the production of petroleum on a commercial basis in America began about 1860, oil springs and seepages appear to have been known to the ancient peoples of Eurasia and also to the Toltecs of Mexico and the Incas of South America, hundreds of years ago. In more recent times, but previous to the advent of drilling machinery, oil was bailed from hand-dug pits in Rumania and Asia Minor to supply the inhabitants of those regions. But only during the past fifty years has appreciable progress been made in discovering the derivatives and determining the uses of petroleum. Many valuable by-products are now derived from substances that were formerly wasted. Improved methods of drilling, producing, refining and marketing are constantly being introduced, all conducing to economy in operation and to the manufacture of superior petroleum products. Because of these improvements it is now profitable to operate wells producing not more than one-tenth of a barrel each per day.

There were on October 31, 1920, approximately 258,600 producing oil wells in the United States, distributed as follows:[1]

	Approximate number of producing wells	Approximate average production per well (barrels)
California	9,490	32.3
Colorado	70	4.1
Illinois	16,800	1.7
Indiana	2,400	1.1
Kansas	15,700	6.7
Kentucky	7,800	3.1
Louisiana	2,700	31.8
New York	14,040	0.2
Ohio	39,600	0.5
Oklahoma	50,700	6.0
Pennsylvania	67,700	0.3
Texas	11,100	27.0
West Virginia	19,500	1.1
Wyoming and Montana	1,000	55.9
Total	258,600	General average 5 bbls. each

After a producing well has been successfully completed, it may produce naturally (gush, or flow) or the use of mechanical appliances may be necessary to lift the oil to the surface, the commonest methods being bailing, swabbing, pumping and the air-lift method.

Wells of large initial capacity generally flow until the production has declined to perhaps 200 barrels per day, although some wells yielding only a few barrels flow and others capable of producing as much as 1000 barrels must be bailed, swabbed or pumped. The method to be employed in securing the production is determined largely by the gravity of the oil, the amount of water and gas accompanying it, the level to which it rises, the amount of sand it carries in suspension, and the diameter and depth of the well.

[1] U. S. Goelogical Survey.

Flowing wells.—When the oil in a prolific sand is accompanied by high-pressure gas in great volume or when it is followed by salt water under pressure, or both (as in Mexico), a gusher generally results. Gushers of varying capacities have been found in nearly all oil-fields, but particularly in Mexico, Russia, California, Texas, Louisiana,

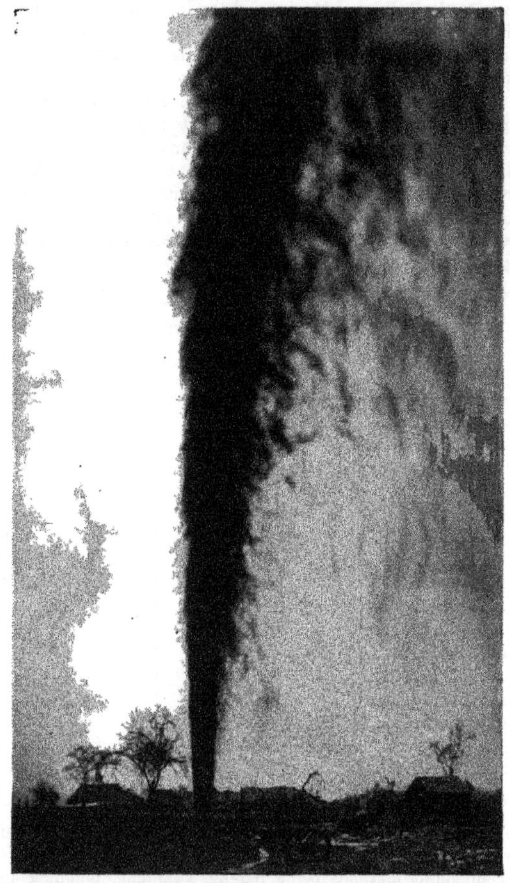

Fig. 228.—A Mexican gusher. While drilling was suspended this well began flowing unexpectedly and several weeks later, while gushing at a rate of more than 100,000 barrels daily, as shown here, was successfully capped (February, 1911), as illustrated in Fig. 229.

Trinidad and South America. Several Russian gushers developed between 1873 and 1890 were credited with an initial daily output of approximately 100,000 barrels each. Some Texas gushers (Spindle Top, 1900–1901 and West Columbia, 1920) produced from 20,000 to 30,000 barrels each; several California spouters developed during 1910–1911 produced upwards of 15,000 barrels, one (the Lakeview) being estimated at 40,000 barrels. But the world's largest gushers have been found in the state of Vera Cruz, Mexico. The first of these and possibly the largest, the Dos Bocas, was completed in 1908, imme-

Fig. 229.—The gusher in Fig. 228, is here shown under control, the capping operation being accomplished by assembling the hinged connections on the ground, then raising them upright and forcing the forged bell-nipple down over the top of the casing, the oil meanwhile being permitted to gush freely through the $8\frac{1}{4}$-inch standpipe. After the nuts at the top of the anchor bolts were tightened (making a firm joint between the bell-nipple and the casing), the gate valve on the standpipe was closed and the flow controlled by means of gate valves on the lead lines.

diately caught fire and resulted in a total loss. Casiano No. 7, completed in 1910, produced approximately eighty million barrels before final exhaustion in 1919; the initial output of Potrero del Llano No. 4 (1910–1918) was estimated in excess of 100,000 barrels daily, and the measured total was more than one hundred million barrels, a fountain of salt water in each instance succeeding the oil when exhausted. Cerro Azul No. 4 (completed in Feb. 1916 and still producing) gushed 261,000 barrels in twenty-four hours, a record. This exceeds the total daily production (in 1920) of the 168,000 oil wells in the United States east of the Mississippi River. Due to accidents or carelessness a flowing well may become ignited. When this occurs the intense heal prevents work at the surface near the well and in such cases the flames have been extinguished by tunneling and either tapping or collapsing the casing underground in order to stop or reduce the flow at the surface. If the tapping (cutting) operation is successful the casing may be collapsed above that point, by means of jacks, thereby diverting the flow of oil through the lateral lead line in the tunnel. If for any reason it is impossible to make a satisfactory underground connection with the casing, it may perhaps be collapsed sufficiently to diminish the flow, after which the flames may be more easily extinguished.

It is impossible to approach a burning well if the oil is gushing as shown in Fig. 228 and no progress whatever can be made by applying any known means of attack to the column of flame. The Dos Bocas gusher typically illustrated this condition, the flames finally dying out with the exhaustion of the oil. This was followed by a tremendous volume of hot salt water which (as proved by subsequent developments) ruined the oil-stratum for miles around. But in other cases, where gushing was intermittent or the flames were confined to the surface, success resulted from surrounding the well or burning area with a series of jets and thereby converging a great volume of high-pressure steam (or non-inflammable gas) upon the flames. The principle of this operation is to exclude the air and prevent combustion, a battery of forty-eight boilers once being used for this purpose in Mexico.

Great losses have resulted from such unfortunate occurrences and it is now customary (and in some places legally required) to take precautionary measures when approaching the oil sand. Various types of control casing heads (Fig. 232) have been devised to permit the continuation of drilling while affording a safe means of controlling the flow of oil should a gusher suddenly be encountered.

In Mexico the drilling tools, upon encountering the oil stratum, have been blown from the well with great force, tremendous volumes of oil or gas, or both in combination, immediately following. A high-pressure, fullway gate valve was formerly attached to the innermost (cemented) string of casing when nearing the oil horizon, this being closed and connections made with tanks after the oil had been encountered and the tools had been withdrawn or ejected from the hole. The several strings were clamped together at the top by means of casing clamps, but a more modern appliance, termed a safety "swinging nipple," has been designed to control the flow of Mexican gushers. By means of this mechanism, embodying the principle of the lead seal in conjunction with a heavy gate valve, the strings of casing (12½-inch, 10-inch and 8¼-inch) may be firmly sealed and bound together at the top and the oil or gas flow safely controlled.

Many wells capable of producing large quantities of oil do not flow because the volume or pressure of the accompanying gas is insufficient. Such wells are usually swabbed or pumped, but flowing (steadily or intermittently) may sometimes be induced by inserting a string of 2-inch tubing with a packer (Fig. 208 or 213) attached near the bottom. A small volume of gas that might readily break through a column of oil without materially raising the level in 6¼-inch casing would eject the oil through 2-inch tubing, the jetting action of the gas (confined to the smaller area) being sufficient to cause the well to flow. In other regions an "oil" string of casing or tubing may be used to "agitate" (stir) the loose sand, release greater quantities of gas and thereby induce flowing. The running of the bailer or sand pump sometimes produces this same result, spasmodic flows or "gushes" following.

Many types of "traps" or gas tanks (Fig. 245) have been designed to separate and save the gas emitted by flowing wells. The gas, under pressure from the well, enters with the oil at or near the top of the tank and is propelled downward through a pipe within, subsequently rising and escaping through another vent in the top of the tank. The oil is trapped at the bottom of the tank until it has accumulated in sufficient quantity to open the float valve, the operation of which keeps the fluid at a constant level above the oil exit. The oil thus acts as a seal and prevents the escape of gas at that point. If the pressure of the well is not sufficient to force the gas through a pipe line to a desired point, the aid of pumps or compressors may then be invoked. In some cases (e.g. the great Mexican gushers) the wells flow naturally for years or until the oil is completely exhausted, but elsewhere the period of active gushing is generally of brief duration—an hour, a week or a few months—the cessation of gushing then being followed by mechanical methods of production. Thus wells that flow for a period may afterward be pumped for many years.

Bailing.—Immense quantities of loose sand accompany the oil found in Rumania and parts of Russia and under such conditions bailing is the most satisfactory method of obtaining the oil after the wells stop flowing. Pumps (working barrels and valves) will not withstand the cutting effect of sand-laden oil, but the large diameters of the Russian and Rumanian wells render bailing feasible. While considerable loose sand is also encountered in many wells in California, Texas and Louisiana, bailing is uneconomical and impracticable on account of the greater depths and smaller diameters. Several types of screens and strainers (Figs. 217, 218) have been designed to exclude this sand and make pumping possible. The bailing process is, therefore, not well adapted to conditions in America.

Swabbing.—If the potential production of a well that does not flow is greater than the capacity of a regular pumping outfit, the maximum production may be obtained by the swabbing process. This is analogous to bailing but conducive to a much larger output, especially in wells of small diameter.

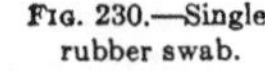

Fig. 230.—Single rubber swab.

Fig. 231.—Double rubber swab.

The swab (Figs. 230, 231) is lowered with a wire drilling cable on a regular string of drilling (or fishing) tools. The rubber (single or double) fits over a perforated mandrel made of 2 or 3-inch line pipe and may be expanded repeatedly, as required by wear, to fit the interior of the casing. This is accomplished by tightening the pipe coupling under the lower ring, thereby vertically compressing and laterally expanding the rubber. The weight of the column of oil being raised also contributes to this lateral expansion by pressing (through the perforations in the mandrel) against the rubber from within. The top of the mandrel screws into the hollow upper portion (shank) of the swab and the lower end is fitted with a vertical check valve. When the swab is lowered into the oil the check valve opens and permits the fluid to pass within the mandrel and emerge through the circular holes in the wrench square above the rubber. When the swab is raised the check valve closes and the oil above is trapped and lifted to the surface, a control casing head (Fig. 232) or a common casing head with an oil saver (Fig. 233) being attached to the top of the casing to divert the flow into the receiving tank.

As the swab rubber makes a tight fit within the casing, the latter must be free from dents and "blisters." Under favorable conditions from 200 to 2000 barrels daily may be swabbed from a well, the amount and the expense depending upon the depth of the oil level in the well, the diameter of the casing, and the power of the swabbing machinery. The work is generally performed by a regular drilling crew operating a steam plant, but either a gas engine, a crude oil engine or an electric motor may be used. A pumping outfit is usually installed when the production declines to "the point of diminishing returns," about 200 barrels per day.

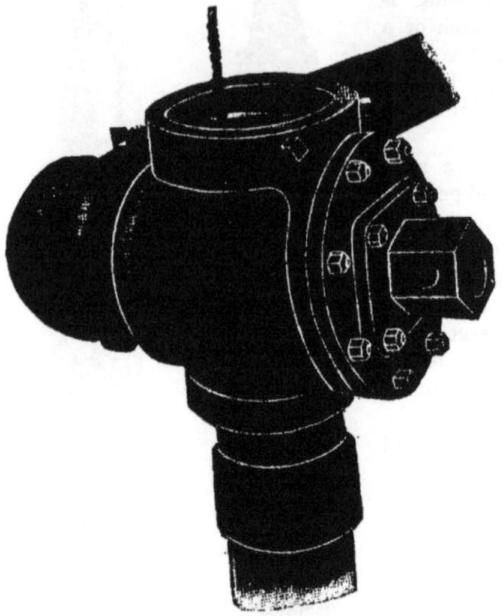

FIG. 232.—Control casing head. Shows how the head may be closed to deflect the flow of oil or gas into the flow line and prevent waste without withdrawing the tools.

If the swab is lowered too far within the fluid a load exceeding the capacity of the outfit may be trapped, stalling the engine or causing the check valve, swab or other equipment to fracture. To avoid such mishaps, a "swab load regulator" has lately been introduced. This is an appliance combining the principles of a vertical check valve and a pop safety valve. It may be set to trap and lift any desired amount (weight) of fluid, automatically releasing any excess if the swab should inadvertently be lowered too far. Some operators condemn the swabbing process, claiming that the vacuum thus produced occasionally aids or causes water to break in, either from above or below the oil sand, if a water-bearing stratum lies in close proximity to it, or if the innermost ("water") string of casing has been set in a formation of uncertain strength or quality.

The use of compressed air.—The principle of lifting or jetting water from a well by means of compressed air has long been known, but its advent as a factor in oil production several years ago has not been followed by the unqualified success and universal adoption predicted for it. This failure to attain a more general distribution may be ascribed largely to the rapid evolution of other methods of production, particularly that of multiple pumping. From the simplicity of the air-lift installation (only the compressor plant, pipe lines, and tubing being necessary, thus eliminating all surface equipment, sucker rods, working barrels, etc.) it would seem that great economy should result, but there are many drawbacks. One engineer states, "During the past twenty-five years I can safely say that in the designing or selection of pumping plants, it has been my experience that in no case was the air-lift ever introduced where it was found practicable to install a deep-well pump (working barrel)." While the foregoing perhaps represents an extremely unfavorable impression and notwithstanding the fact that most producing wells are pumped, certain advantages are claimed for the air-lift by its advocates, but a thorough investigation should be made before making a permanent installation of this kind. The irregular pumping or swabbing of adjacent wells may cause the fluid level in the well to change frequently and impair the efficiency of the air-lift, as this depends largely upon the maintenance of the proper ratio between the submergence and the lift. In general, the efficiency is augmented by an increase in the depth of the submersion or a decrease in the height of the lift.

An oil-field air-lift installation ordinarily consists of a two-stage direct-gas-engine-driven air compressor capable of delivering 300–500 cubic feet of air (under a pressure of from 300–700 pounds per minute, also the necessary line pipe laterals and well tubing, —the size and power of the plant being governed by the number, depth and capacity of the wells, the gravity and viscosity of the oil, the height to which it rises in the well, and the amount of water or sediment present with it. The basic principle involves delivering into the bottom of this submerged tubing compressed air forced down through an interior or exterior string of pipe or tubing, many modifications of the general principle being recognized. The rising column consists of oil intermingled with air bubbles, and this mixture, being lighter than the surrounding fluid, rises above the natural level. The friction caused by the rapidly rising air tends to move the oil in the same direction, either in solid column or in spray. If the level of the fluid in the well falls perceptibly, the air will break through the column intermittently, ejecting the oil in gushes or sprays. Under such conditions the efficiency of the plant is materially reduced. On the other hand, the maximum production of any well cannot be obtained (either by bailing, swabbing, pumping or the air-lift) if the oil is permitted to stand at a high level in the well, as the weight of the column of fluid thus retards and finally prevents the natural flow of oil from the sand into the bore hole.

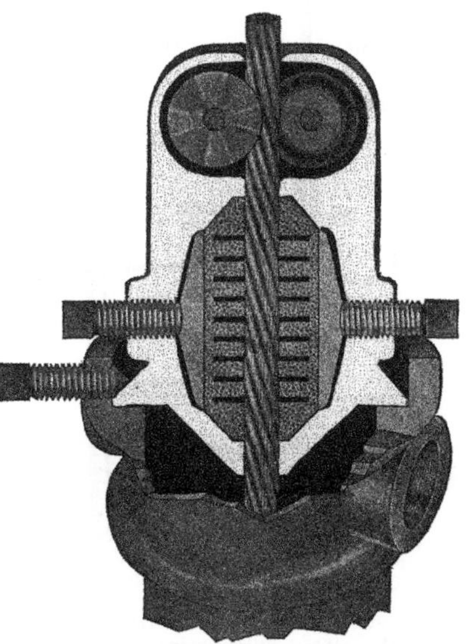

Fig. 233.—Roller oil saver (attached to common two-way casing head).

Several modifications of the air-lift, as applied to oil production and mining operations, have been described in the engineering publications, and an ingenious adaptation of it (the Smith-Dunn Process) is covered by the Mitchell and Dunn Patents Nos. 745,825 and 1,067,868. The apparatus may be simple or complicated, depending upon the conditions and requirements.

Air flooding (the "Air-Drive") represents another adaptation of compressed air to oil production. The owner of a large lease may install the air-drive (wherever lawful) in one or more of the centrally located wells, the object being to drive the oil laterally through the sand from the center of the lease outward so that it may be recovered more rapidly from the surrounding wells.

A compressor plant similar to that used for the air-lift is suitable. A string of tub-

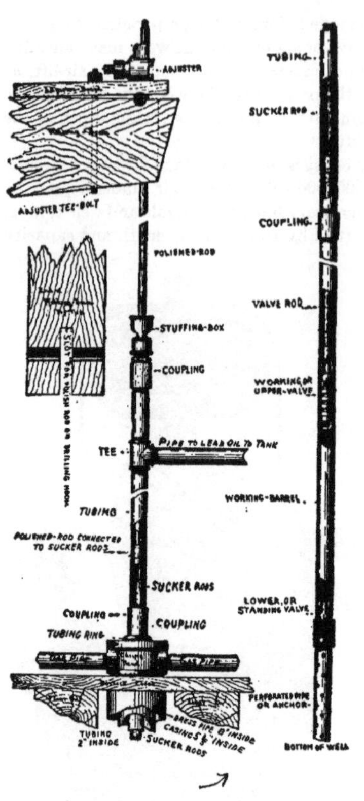

Fig. 234.—Parts comprising a deep-well pumping outfit (When walking beam and sucker rods are used).

Wire line pumping outfit.

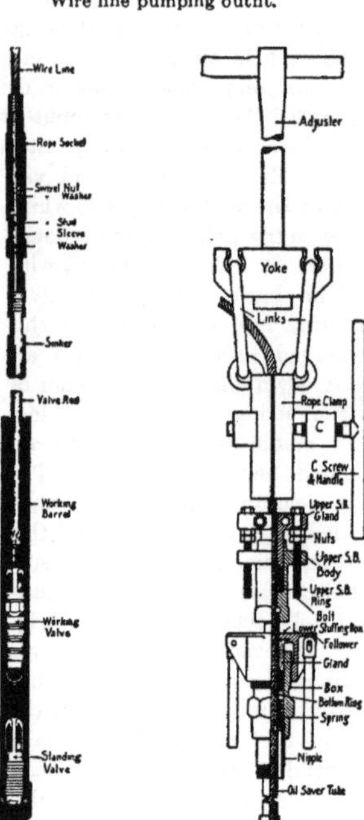

Fig. 235.—Pump.

Fig. 236.—Upper connections.

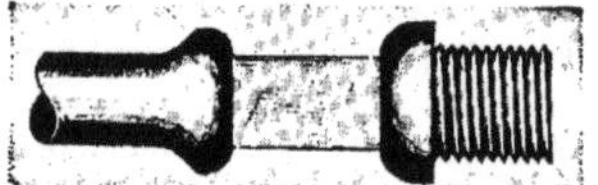

Fig. 234A.—Iron or steel sucker-rod joint.

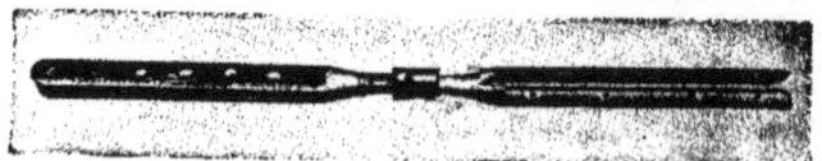

Fig. 234B.—Wooden sucker-rod joint.

ing with a packer (Fig. 208 or 213 with perforations in the tubing below it) attached to the lower end is inserted in the well, the packer being set at the top of the sand. The compressed air passes downward within the tubing and as the packer prevents its return to the surface it is forced into the oil sand, radiating laterally towards the surrounding wells and carrying or forcing the oil and gas along with it. After the air pressure has been maintained on the central well for a time an increased flow of gas, and later of oil also, will be observed in the other wells, the time required and the amount of increase depending upon the distance from the air inlet, the porosity of the sand, and the pressure and volume of air entering it. Results may not be apparent for several months, but after prolonged tests, and maintaining a pressure of from 60 to 150 pounds, oil production has been increased as much as 500 per cent. All of the sixteen wells on one 80-acre tract subjected to 125 pounds pressure showed improvement, the increase in the volume of gas and oil being in inverse ratio to the distance of the producing well from the air inlet. But the intermingling of the air diluted the gas, the percentage of air pervading it being 10½ per cent at 2200 feet distant, 25 per cent at 1500 feet and 81 per cent at 500 feet, thereby rendering it unfit for commercial use. The use of gas under pressure, instead of air, would, however, overcome this objection and much of the gas thus used could be recovered.

Fig. 237.—Unit pumping outfit. Equipped with either a gas engine, gasoline engine, crude-oil engine or an electric motor.

Pumping.— Because of the simplicity of the plant, the economy of operation and the adaptability of the principle, pumping is decidedly the most popular method of extracting petroleum from wells, being almost universal to the United States.

Fig. 238.—A common type of multiple power.

Practically all deep-well pumps are modifications of the common plunger-pump principle, a multiplicity of designs to meet the varied requirements having been developed. A steel or cast-iron cylinder (working barrel) from 5 to 10 feet in length is lowered on tubing (generally 2-inch, although the 2½, 3, 3½ and 4-inch sizes are occasionally used) to a point in the well near or within the oil sand, the position being determined by the height of the oil level and the amount of water, gas or loose sand present with the oil. A short piece of perforated pipe and one or more joints of tubing are generally attached to the lower end of the working barrel, the perforations serving to admit the oil into the pump and the tubing below it to act as an "anchor." The lower (standing) valve (Fig. 235) is generally dropped in after the tubing is completely inserted. This permits any loose scales or debris in the tubing to fall freely through the working barrel to the bottom of the well (instead of lodging on this valve) while the tubing is being run. The upper (working) valve is then attached to a valve rod and lowered on either steel, iron or wooden sucker rods (Fig. 234) or on a wire-line pumping outfit (Fig. 235). The string of sucker rods or wire line may then be attached, through intervening appliances, to a walking beam or to a pumping jack (Figs. 237 to 242) in multiple pumping.

Fig. 239.—Band-wheel power (double eccentric).

The commonest types of working barrels and valves are shown in Figs. 234, 235 and 243. The packing rings on the standing valve and the cups on the "traveling" (working) valve may be composed of leather, heavy canvas, or fiber and must fit the interior of the working barrel snugly, thus requiring frequent renewal. The working valve is withdrawn with the sucker rods, after which the standing valve may be recovered by means of a threaded tap lowered on the rods, or by withdrawing the tubing. Many types of plunger pumps (Fig. 243) are in use and are especially effective in handling sand-laden oil and resisting its cutting action. The steel or brass plunger fits the working barrel closely and thereby eliminates the need of cups on the working valve. A late type of deep-well pump, similar to Fig. 243 but fitted with a plunger bearing steel snap rings, is very efficient in pumping and unusually resistant to the cutting effect of sand. The valves and the plunger shown in Fig. 243 are hollow, permitting the free passage of the oil within, and the Garbutt rod serves to remove the standing valve when the sucker rods are withdrawn.

Each steel sucker rod ($\frac{9}{16}$, $\frac{5}{8}$, $\frac{3}{4}$, $\frac{7}{8}$ or 1 inch in diameter by 25 feet in length) bears a threaded pin at one end and a corresponding box at the other end (Fig. 234-A), and wooden sucker rods are similarly equipped with riveted screw joints (Fig. 234-B).

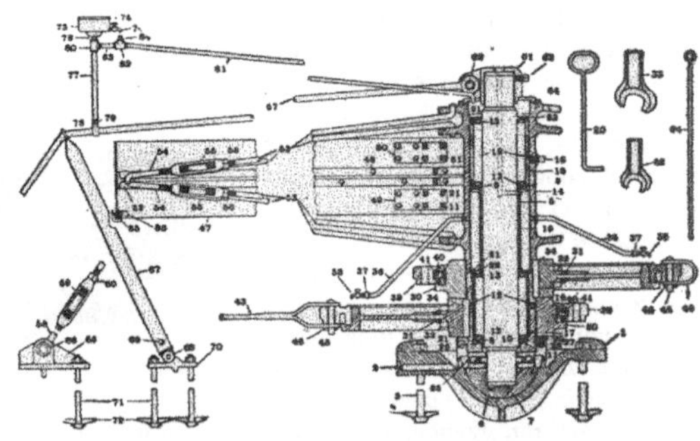

Fig. 240.—Band wheel power (sectional view).

A packed stuffing-box (shown in Figs. 234–236 and 237) through which the polished rod or oil-saver tube oscillates, caps the top of the tubing and diverts the oil into the receiving tank. The sucker rods (Fig. 234) or the wire-line outfit (Fig. 236) may be manipulated by means of a regular walking beam, a unit pumping power (Fig. 237), or by a pumping jack (Figs. 241, 242) operated by surface lines connected to a multiple power (Figs. 238–240). The surface lines may consist of iron rods or wire cable and usually glide back and forth through lubricated grooves in wooden posts or blocks set from 15 to 40 feet apart.

The mechanisms shown in Figs. 238, 239, called "powers" are standard types and serve to transmit power, in multiple pumping, from the engine or motor to the surface lines and pumping jacks. A belt runs from the engine or motor and passes over the pulley of the geared power (Fig. 238) or over the rim of the band-wheel power, and the horizontal revolutions of the eccentrics impart the necessary gliding movement to the surface lines which lead from the eccentric straps to the pumping jacks. By this system of multiple pumping from two to thirty wells may be successfully pumped with one central power plant, the number depending upon the size of the plant, the diameter of the tubing, the depth of the wells and the size of the area over which they are scattered. Some wells must be pumped more frequently or for longer periods than others and by

Fig. 241.—Jones & Hammond Pumping jack No. 2, lower connection.

Fig. 242.—No. 2, Oklahoma pattern pumping jack.

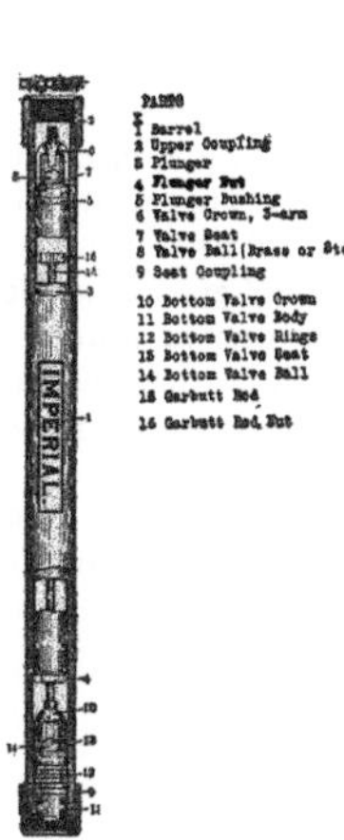

Fig. 243.—Working barrel (plunger type with Garbutt rod).

Fig. 244.—Gas pump.

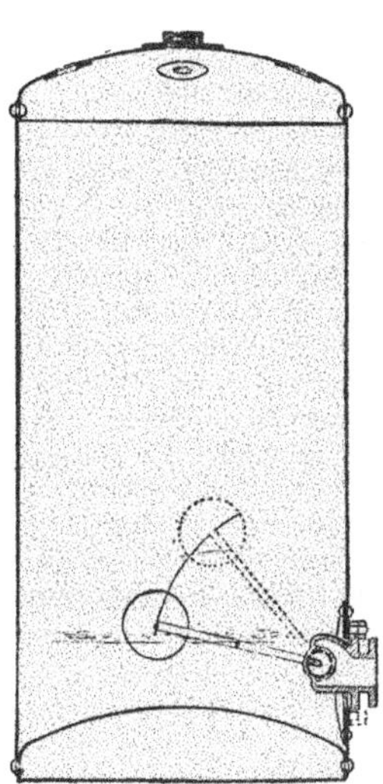

Fig. 245.—Automatic gas tank (trap) with A. B. C. rotary valve.

means of throw-off books the surface lines may be attached to or disconnected from the power at will, thereby facilitating the pumping of each well according to its individual needs. The unit pumping outfit (Fig. 237) also meets this requirement in a very economical manner.

If the well is to be pumped " on the beam " (walking beam) it is customary to leave the drilling derrick standing, at least temporarily, or to erect a special pumping derrick about 56 feet in height for use in " pulling " rods and tubing when necessary to repair or replace any part of the pump. But if the unit pumping outfit or the multiple pumping system is installed, a three-pole derrick, or a single gin pole may be permanently erected at the well-site, or a portable pulling machine may be used to withdraw and replace the rods or tubing whenever necessary.

Casing-head gasoline.—In any producing oil sand there is usually present a volume of gas containing a recoverable quantity of gasoline. In a pumping well this gas rises between the casing and the tubing and escapes through the lateral openings in the casing head (Fig. 233), or may be piped to any desired point. Until a few years ago it was customary to utilize this gas for fuel or to waste it on the lease, but the increased demand for gasoline in recent years has given great impetus to a new industry—the manufacture of gasoline from natural gas. In some fields the lease owner installs an " absorption " plant commensurate with the size of his lease; while in other instances larger central compressor plants are installed to acquire and treat the gas from surrounding leases, in some cases pumping it back to the producer for use as fuel after extracting the gasoline. This novel industry has experienced a very rapid expansion and has grown to large proportions in regions in which conditions are favorable, i.e., where the oil is light, the gas rich, and the leases intensively drilled, as in parts of Kansas, Oklahoma and Texas. (In the Spindletop, Texas field 1003 wells were drilled on 144 acres, perhaps a record.)

The Burkburnett field, where the leases are small, has also been closely drilled and, to augment production of both gas and oil, gas pumps (Fig. 244) have been installed on many leases. These are simple, vertical " suction " pumps, about 14 by 20 inches and are designed to be operated by a walking beam or other oscillating apparatus. The pump inlet is connected to the gas line leading to the casing head and its action places a vacuum on the oil sand thereby inducing a greater flow of oil and gas. The beneficial effects of this procedure may not become apparent until it has been in continuous operation for several days, or possibly weeks, depending upon the nature and thickness of the oil sand. The gas is then pumped to the gasoline plant where it is compressed and the gasoline condensed and extracted.

The various steps involved in drilling a well and securing the production by established American methods have been briefly described, and the oil, now in the receiving tank, is ready for transportation to the loading rack or direct to the refinery.

STATISTICS OF PETROLEUM AND NATURAL GAS PRODUCTION

BY

ANNE B. COONS[1]

As an industry of consequence, the production of petroleum has entered into its seventh decade. Production from drilled wells started in 1860 and the great industrial development which then began marked a sharp point of separation from the occasional production of medicinal oil, incidental to the salt industry, previously carried on. During the period of less than sixty-five years which has since elapsed, the world has produced more than 8,735,800,000 barrels of crude oil. The total amount of this necessary commodity produced by the United States during this time aggregates 5,430,000,000 barrels or 62 per cent of all the world has produced. Sporadic production of petroleum in other countries antedates the industry in the United States, but immediately upon the introduction of drilled wells the United States became and has remained the dominant factor of production and trade. The striking feature of the development of the industry in recent years has been the phenomenal rise of Mexico as a source of crude oil. Whether large production will continue long enough to make Mexico the dominant influence in the oil trade remains to be seen.

The statistics of production given in the following tables are taken from published reports of the United States Geological Survey, which through a long series of years has become an authoritative source of original information on this subject. This is a proper place also for recording the pioneer work of the Oil City Derrick, under the management of the late Patrick Boyle, in collecting statistics of the pipe-line runs of crude oil. This source has contributed statistics which have been found accurate in many tests and have closely checked with information collected by the Government itself. For many years the tables of the Oil City Derrick were the only reliable statistics of oil production available to the public. The Government work began by supplementing such published pipe-line runs by estimates of the amount of crude oil shipped direct to refineries and consumed as fuel at the wells, etc. Eventually these estimates have become carefully collected statistics compiled from returns filed by pipe-line and marketing companies. Finally the Geological Survey instituted an elaborate system of collecting annual statements from petroleum producers. These statements have not been used to replace the returns of the pipe-line and marketing companies but to complete those reports by the addition of the amount of oil consumed on the leases as fuel and the net changes in the storage held on the leases by the producers.

Because of limited space only general tables are given. Detailed statistics of production, stock, wells, etc., of the United States fields may be found in the annual reports of the United States Geological Survey, and in the reports of the various State Surveys.

Current information is published monthly by many Federal, State, and private organizations. Monthly statistics of production, stocks, and consumption of crude

[1] Assistant Statistician U. S. Geological Survey.

oil are issued by the United States Geological Survey; monthly statistics of production, stocks, and consumption of refinery products by the Bureau of Mines; monthly imports and exports by the Bureau of Foreign and Domestic Commerce; weekly estimates of production by the American Petroleum Institute. Monthly figures of production, stocks, and consumption of crude oil and wells for California are issued by the Standard Oil Co. (Calif.) and the Independent Oil Producers Agency; weekly reports of drilling operations in California by the California State Mining Bureau; monthly reviews of drilling operations in states east of California by the Oil City Derrick and the Oil and Gas Journal.

The following tables and graphs show the World's production since 1857. The tables are arranged in the order in which the countries began commercial production. The figures are from the United States Geological Survey, except where noted. The graphs are presented in two parts on different scales, in order that the annual variations of the smaller countries may be brought out for comparison.

In the United States oil statistics are expressed in barrels of 42 United States gallons. The equivalents of the various units of weight used for oil in foreign countries are given on page 761, Volume II. The conversion of the various units of weight into barrels, and the reverse conversion, is difficult. The information is too meager as to the specific gravity of the oils actually marketed. The following list shows the sources from which the petroleum statistics are obtained, and the specific gravity used for the petroleum product of each country.

Foreign Statistical Authorities and Average Specific Gravity

Country	Authority for statistics	Specific gravity used
United States	U. S. Geological Survey	0.8837
Mexico	Boletin del Petroleo	......
Russia	Petroleum Times; Oil News (London)	0.859
Dutch East Indies	Bureau of Mines, Dutch East Indies	0.8761
India	Geological Survey of India	0.8403
Rumania	Moniteur du Petrole Roumain	0.8766
Persia	American Consul-General at London	0.86
Poland (Galicia)	Legation of Poland	0.86
Peru	Sociedad de Ingenieros del Peru	0.8403
Japan	Oriental Economist Yearbook	0.8403
Trinidad	Trinidad Dept. Mines	0.8766
Egypt	American Consul-General at London	0.97
Argentina	Comodoro Rivadavia Company	0.9174
Venezuela	Boletin del Ministerio de fomento	0.959
Alsace	Bulletin Soc. de l'industrielle minerale	0.89
Canada	Canada Dept. Mines	0.8403
Germany	Private statistics through Consular Office, State Dept.	0.89
Italy	Economista d'Italia	0.876
Algeria	Algerian Bureau of Mines	0.98
England	H. M. Petroleum Executive	0.828

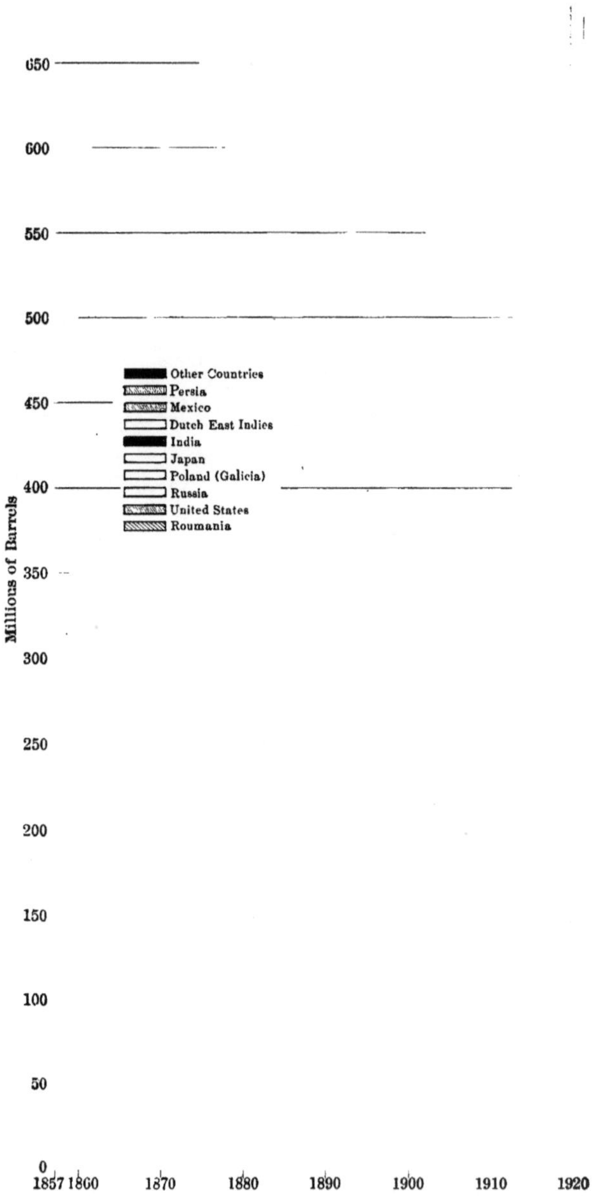

Millions of Barrels

650
600
550
500
450
400
350
300
250
200
150
100
50
0

Other Countries
Persia
Mexico
Dutch East Indies
India
Japan
Poland (Galicia)
Russia
United States
Roumania

1857 1860 1870 1880 1890 1900 1910 1920

FIG. 1.—World's production of petroleum, 1857–1920.

[*To face page 323*]

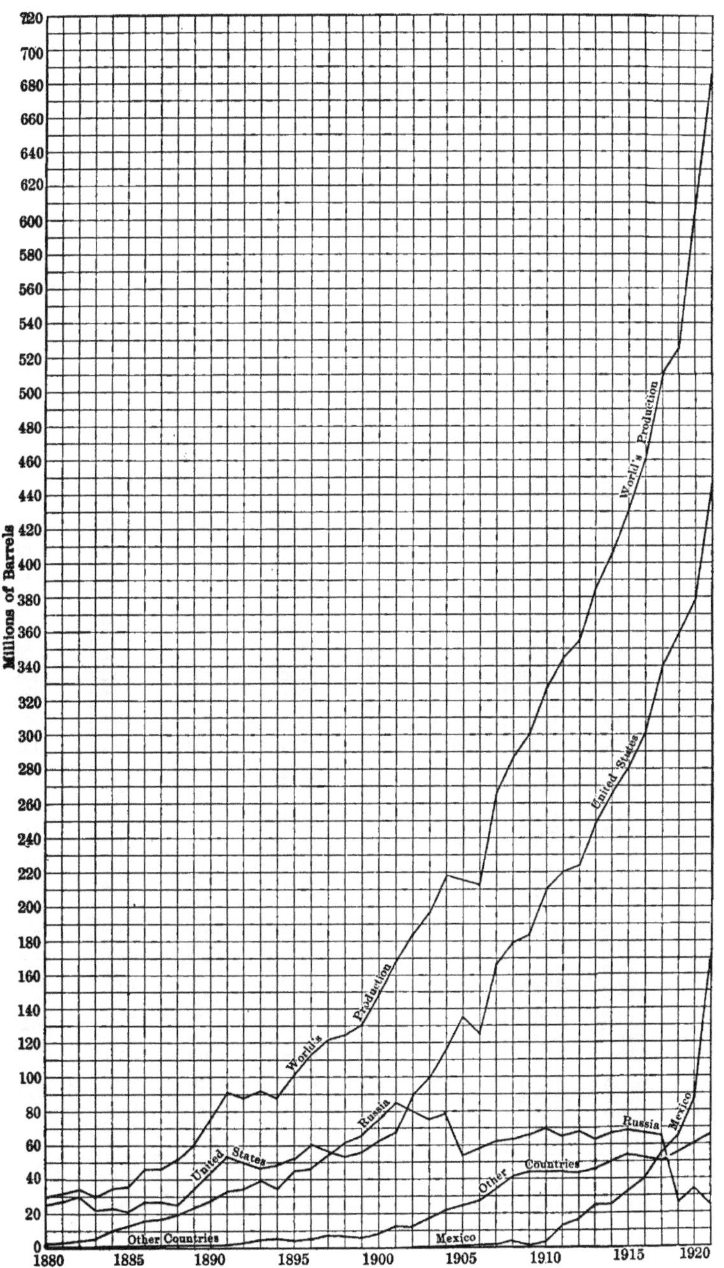

FIG. 2.—World's production of petroleum 1880–1920. Total, United States, Russia, Mexico and

World's Production of Petroleum

(In thousands of barrels

	Rumania	United States	Italy	Canada	Russia	Poland (Galicia)	Japan and Formosa	Germany
1857	2	......	....	......	......	......	......	......
1858	4	......	....	......	......	......	......	......
1859	4	2	....	......	......	......	......	......
1860	9	500	(a)	......	......	......	......	......
1861	17	2,114	(a)	......	......	......	......	......
1862	23	3,057	(a)	12	......	......	......	......
1863	28	2,611	(a)	83	41	......	......	......
1864	33	2,116	(a)	90	65	......	......	......
1865	39	2,498	2	110	67	......	......	......
1866	42	3,598	1	175	83	......	......	......
1867	51	3,347	1	190	120	......	......	......
1868	56	3,646	(a)	200	88	......	......	......
1869	59	4,215	(a)	220	202	......	......	......
1870	83	5,261	(a)	250	205	......	......	......
1871	90	5,205	(a)	270	165	......	......	......
1872	91	6,293	(a)	308	185	......	......	......
1873	104	9,894	(a)	365	475	......	......	......
1874	103	10,927	1	169	583	150	......	......
1875	108	8,788	1	220	697	159	4	......
1876	111	9,133	3	312	1,320	164	8	......
1877	108	13,350	3	312	1,801	170	10	......
1878	109	15,397	4	312	2,401	176	18	......
1879	110	19,914	3	575	2,761	215	23	......
1880	115	26,286	2	350	3,001	229	26	9
1881	122	27,661	1	275	3,602	286	17	29
1882	136	30,350	1	275	4,538	330	16	58
1883	139	23,450	2	250	6,002	365	20	27
1884	211	24,218	3	250	10,805	408	28	46
1885	194	21,859	2	250	13,925	465	29	41
1886	169	28,065	2	584	18,006	305	38	74
1887	182	28,283	1	526	18,368	344	29	74
1888	219	27,612	1	695	23,049	467	37	85
1889	298	35,163	1	705	24,610	515	53	68
1890	383	45,824	3	795	28,691	659	52	108
1891	488	54,293	8	755	34,573	631	53	109
1892	593	50,515	18	780	35,775	646	69	101
1893	536	48,431	19	798	40,457	693	106	99
1894	507	49,344	21	829	36,376	949	172	123
1895	575	52,892	26	726	46,140	1,453	142	121
1896	543	60,960	18	727	47,221	2,443	197	145
1897	571	60,476	14	710	54,399	2,226	218	166
1898	776	55,364	15	758	61,609	2,376	266	184
1899	1,426	57,071	16	809	65,955	2,313	536	192
1900	1,629	63,621	12	914	75,779	2,347	867	358
1901	1,678	69,389	16	756	85,168	3,252	1,111	314
1902	2,060	88,767	19	531	80,540	4,142	1,193	354
1903	2,763	100,461	18	487	75,591	5,235	1,210	446
1904	3,599	117,081	26	552	78,537	5,947	1,419	637
1905	4,421	134,718	44	634	54,960	5,765	1,473	561
1906	6,378	126,494	54	569	58,897	5,468	1,710	579
1907	8,118	166,095	60	789	61,851	8,456	2,001	757
1908	8,252	178,527	51	528	62,187	12,612	2,070	1,009
1909	9,327	183,171	42	420	65,970	14,933	1,890	1,019
1910	9,724	209,557	51	316	70,336	12,674	1,931	1,032
1911	11,108	220,449	75	291	66,184	10,519	1,659	1,017
1912	12,976	222,935	54	244	68,019	8,535	1,672	1,031
1913	13,555	248,446	47	228	62,834	7,818	1,942	(c) 1,000
1914	12,827	265,763	40	215	67,020	6,436	2,738	(c) 1,000
1915	12,030	281,104	44	215	68,548	6,964	3,118	(c) 1,000
1916	8,945	300,767	51	198	65,817	6,729	2,997	(c) 1,000
1917	3,721	335,316	41	214	63,072	6,451	2,882	(c) 1,000
1918	8,730	355,928	35	305	27,168	6,157	2,449	(c) 700
1919	6,618	(f) 377,719	35	241	31,752	6,079	2,172	234
1920	7,435	(f) 443,402	34	197	25,430	5,606	2,140	212
	165,462	5,429,693	1,042	24,864	1,904,021	171,263	42,810	17,120

(a) Less than one thousand. (b) Included with Germany previous to 1919. (c) Esti-

since 1857, by Years and Countries

of 42 U. S. gallons)

India	Dutch East Indies	Peru	Mexico	Argentina	Trinidad	Egypt	Other countries	Total
.....	.....	.....	.....	.....	......	......	......	2
.....	.....	.....	.....	.....	......	......	......	4
.....	.....	.....	.....	.....	......	......	......	6
.....	.....	.....	.....	.....	......	......	......	509
.....	.....	.....	.....	.....	......	......	......	2,131
								3,092
								2,763
								2,304
								2,716
								3,899
								3,709
								3,990
								4,696
								5,799
								5,730
								6,877
								10,838
								11,933
								9,977
								11,051
								15,754
								18,417
.....	.....	.....	.....	.....	......	......	......	23,601
.....	.....	.....	.....	.....	......	......	......	30,018
.....	.....	.....	.....	.....	......	......	......	31,993
.....	.....	.....	.....	.....	......	......	......	35,704
.....	.....	.....	.....	.....	......	......	......	30,255
.....	.....	.....	.....	.....	......	......	......	35,969
.....	.....	.....	.....	.....	......	......	......	36,765
.....	.....	.....	.....	.....	......	......	......	47,243
.....	.....	.....	.....	.....	......	......	......	47,807
.....	.....	.....	.....	.....	......	......	......	52,165
94	.....	.....	.....	.....	......	......	......	61,507
118	.....	.....	.....	.....	......	......	......	76,633
190	.....	.....	.....	.....	......	......	......	91,100
242	.....	.....	.....	.....	......	......	......	88,739
299	600	.....	.....	.....	......	......	......	92,038
327	688	.....	.....	.....	......	......	......	89,336
372	1,216	.....	.....	.....	......	......	......	103,663
430	1,427	48	.....	.....	......	......	......	114,159
546	2,552	71	.....	.....	......	......	......	121,949
542	2,964	71	.....	.....	......	......	......	124,925
941	1,796	89	.....	.....	......	......	......	131,144
1,078	2,253	275	.....	.....	......	......	......	149,132
1,431	4,014	275	10	.....	......	......	(c) 20	167,434
1,617	2,430	287	40	.....	......	......	(c) 26	182,006
2,510	5,770	278	75	.....	......	......	(c) 36	194,880
3,386	6,508	290	126	.....	......	......	(c) 40	218,149
4,137	7,850	373	251	.....	......	......	(c) 30	215,217
4,016	8,181	531	503	.....	......	......	(c) 30	213,410
4,344	9,983	751	1,005	(a)	......	......	(c) 30	264,241
5,047	10,284	945	3,933	12	(a)	......	(c) 30	285,486
6,677	11,042	1,411	2,714	18	57	......	(c) 20	298,711
6,138	11,031	1,258	3,634	21	143	......	(c) 20	327,865
6,451	12,173	1,465	12,553	13	285	9	(c) 20	344,271
7,117	10,845	1,752	16,558	47	437	210	(c) 20	352,451
7,930	(d) 11,309	2,071	25,696	131	504	88	(e) 1,778	385,377
7,410	(d) 11,730	1,837	26,236	275	643	683	(e) 2,776	407,629
8,203	(d) 12,299	2,579	32,911	516	750	206	(e) 3,438	433,925
8,491	(d) 13,166	2,593	40,546	797	929	396	(e) 4,271	457,693
8,078	(d) 13,704	2,577	55,292	1,145	1,602	939	(e) 6,914	502,949
8,188	(d) 13,266	2,527	63,828	1,242	2,082	1,900	(e) 8,503	503,009
8,733	(d) 16,010	2,627	87,073	1,341	1,841	1,517	(e) 9,380	553,373
(c) 7,500	(d) 18,545	2,816	163,540	1,666	2,083	1,042	(e) 13,206	694,854
122,583	223,634	29,797	536,524	7,225	11,356	6,990	50,588	8,744,972

Production of Countries included in "Other Countries"

Year	France (Alsace)	Persia	Venezuela	Cuba	Algeria	England
1913	(b)	1,758	.....	.....	......	......
1914	(b)	2,755	.....	.....	1	......
1915	(b)	3,424	.....	.....	4	......
1916	(b)	4,238	.....	.....	8	......
1917	(b)	6,766	120	19	9	......
1918	(b)	8,163	333	..	7	......
1919	334	8,613	425	..	6	1.9
1920	389	12,353	457	..	4	2.9

mated. (d) Includes British Borneo. (e) See details in insert. (f) Preliminary figures.

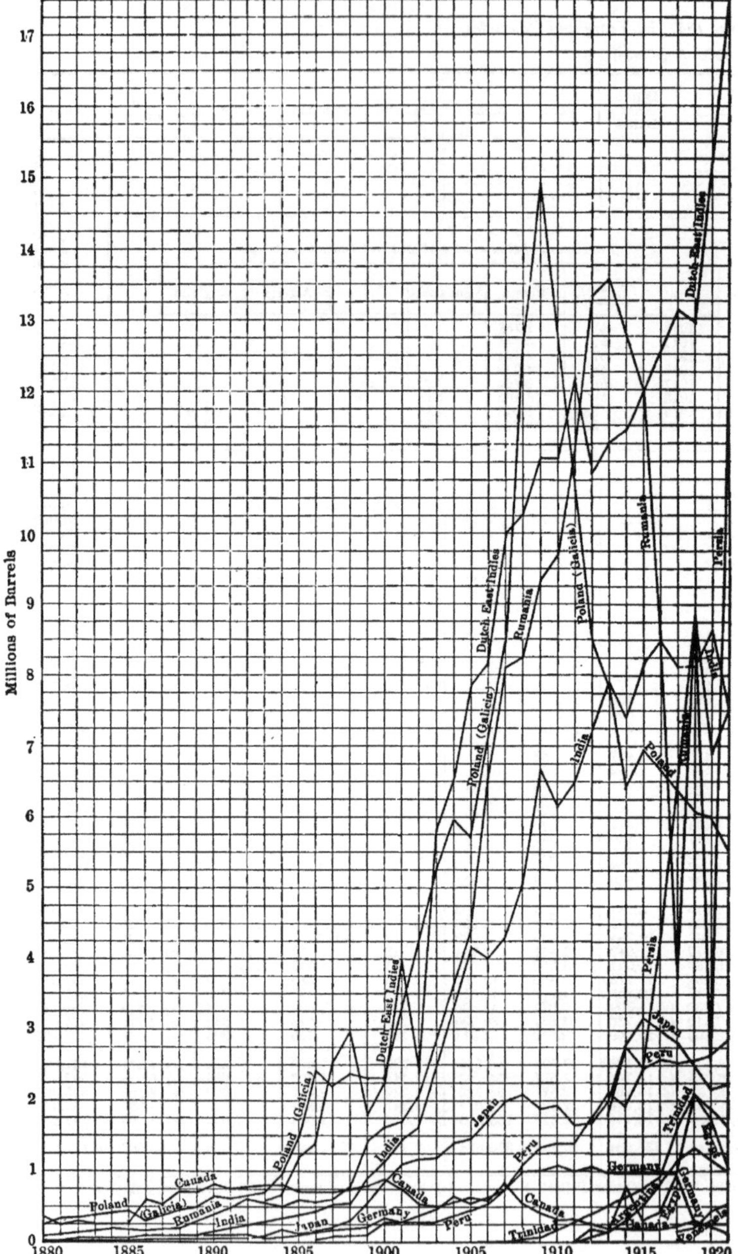
Millions of Barrels
0
1
2
3
4
5
6
7
8
9
10
11
12
13
14
15
16
17
1880
1885
1890
1895
1900
1905
1910
1915
1920
Poland
(Galicia)
Canada
Rumania
India
Japan
Germany
Peru
Trinidad
Poland (Galicia)
Dutch East Indies
India
Canada
Japan
Peru
Dutch East Indies
Rumania
Poland (Galicia)
India
Rumania
Persia
Poland
Japan
Peru
Germany
Argentina
Trinidad
Canada
Egypt

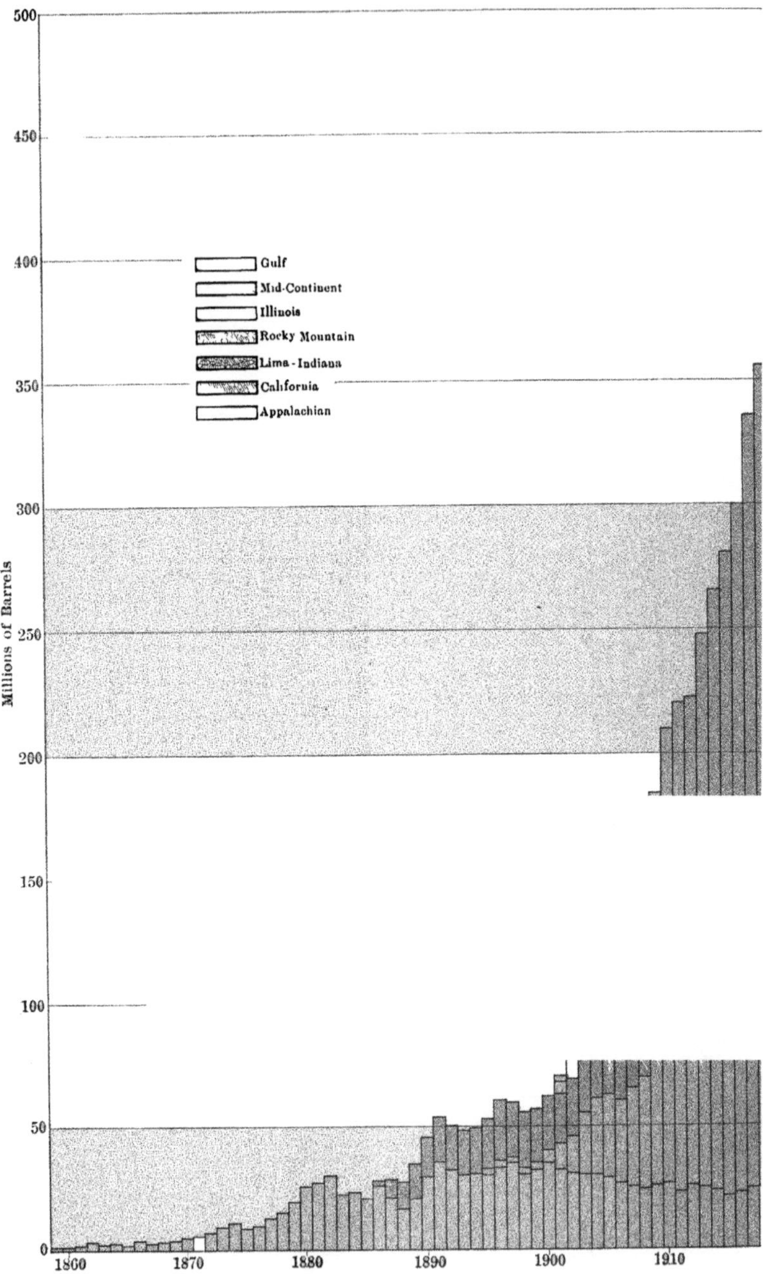

Fig. 4.—Production of petroleum in the United States from 1860 to 1920, by fields.

[*To face page 3*

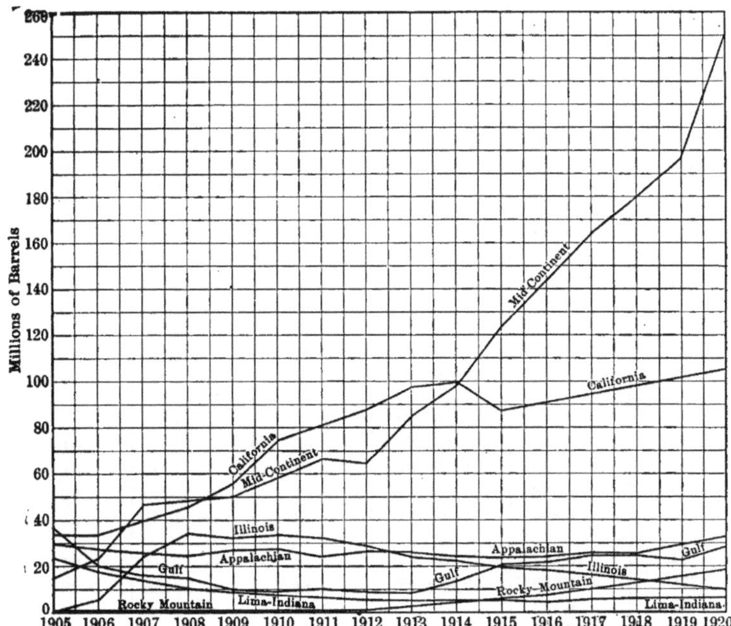

Fig. 5.—Production of petroleum in the United States by fields, 1905–1920.

The following tables showing the production and value of petroleum in the United States by fields are arranged geographically by fields and by states, in the order in which commercial production was begun.

The colored graph accompanying these tables does not show the quantity produced by any field before it began to form an appreciable percentage of the total production of the United States.

Quantity and Value of Petroleum Produced in the United States since 1859

(In thousands of barrels of 42 U. S. gallons)

Year	Quantity	Value	Average value per barrel	Percentage of yearly change of production	Percentage of World's production
1859	2	$ 32,000	$16.00	....	33
1860	500	4,800,000	9.60	+24,900.00	98
1861	2,114	1,036,000	.49	+322.76	99
1862	3,057	3,210,000	1.05	+ 44.62	99
1863	2,611	8,226,000	3.15	− 14.57	94
1864	2,116	20,897,000	9.87	− 18.96	91
1865	2,498	16,460,000	6.59	+ 18.03	92
1866	3,598	13,455,000	3.74	+ 44.04	92
1867	3,347	8,067,000	2.41	− 6.96	90
1868	3,646	13,217,000	3.63	+ 8.93	91
1869	4,215	23,730,000	5.63	+ 15.60	90
1870	5,261	20,504,000	3.90	+ 24.81	91
1871	5,205	22,591,000	4.34	− 1.06	91
1872	6,293	21,441,000	3.41	+ 20.90	92
1873	9,894	18,100,000	1.83	+ 57.21	91
1874	10,927	12,648,000	1.16	+ 10.44	92
1875	8,788	7,368,000	.84	− 19.58	88
1876	9,133	22,983,000	2.52	+ 3.92	83
1877	13,350	31,789,000	2.38	+ 46.17	85
1878	15,397	18,045,000	1.17	+ 15.33	84
1879	19,914	17,211,000	.86	+ 29.33	84
1880	26,286	24,601,000	.94	+ 32.00	88
1881	27,661	25,449,000	.92	+ 5.23	86
1882	30,350	23,631,000	.78	+ 9.72	85
1883	23,450	25,790,000	1.10	− 22.74	77
1884	24,218	20,596,000	.85	+ 3.29	67
1885	21,859	19,198,000	.88	− 9.74	59
1886	28,065	19,996,000	.71	+ 28.39	59
1887	28,283	18,877,000	.67	+ .78	59
1888	27,612	17,948,000	.65	− 2.37	53

Quantity and Value of Petroleum Produced in the United States since 1859—Continued

(In thousands of barrels of 42 U. S. gallons)

Year	Quantity	Value	Average value per barrel	Percentage of yearly change of production	Percentage of World's production
1889	35,164	$26,963,000	$0.77	+27.35	57
1890	45,824	35,364,000	.77	+30.32	60
1891	54,292	30,527,000	.56	+18.48	60
1892	50,514	25,906,000	.51	− 6.96	57
1893	48,431	28,951,000	.60	− 4.16	53
1894	49,344	35,522,000	.72	+ 1.92	55
1895	52,893	57,632,000	1.09	+ 7.19	51
1896	60,960	58,519,000	.96	+15.25	53
1897	60,476	40,874,000	.68	− .80	50
1898	55,364	44,194,000	.80	− 8.45	44
1899	57,071	64,604,000	1.13	+ 3.08	44
1900	63,621	75,989,000	1.19	+11.48	43
1901	69,389	66,417,000	.96	+ 9.07	41
1902	88,767	71,179,000	.80	+27.93	49
1903	100,461	94,694,000	.94	+13.17	52
1904	117,081	101,175,000	.86	+16.54	54
1905	134,717	84,157,000	.62	+15.06	63
1906	126,494	92,445,000	.73	− 6.10	59
1907	166,095	120,107,000	.72	+31.31	63
1908	178,527	129,079,000	.72	+ 7.48	63
1909	183,171	128,328,000	.70	+ 2.59	61
1910	209,557	127,900,000	.61	+14.41	64
1911	220,449	134,044,000	.61	+ 5.20	64
1912	222,935	164,213,000	.74	+ 1.13	63
1913	248,446	237,121,000	.95	+11.44	65
1914	265,763	214,125,000	.81	+ 6.97	65
1915	281,104	179,463,000	.64	+ 5.77	65
1916	300,767	330,900,000	1.10	+ 6.99	66
1917	335,316	522,635,000	1.56	+11.49	67
1918	355,928	703,944,000	1.98	+ 6.15	70
1919 (*a*)	377,719	775,000,000	2.05	+ 6.12	68
1920 (*a*)	443,402	1,360,000,000	3.07	+17.39	64
	5,429,693	6,683,867,000	1.23	....	62

a Preliminary figures—fuel consumed on leases and net change in producers storage not included.

Quantity and Value of Petroleum Produced

(In thousands of barrels

Year	Pennsylvania		West Virginia		Southeastern and Central Ohio		New
	Quantity	Value, in thousands	Quantity	Value, in thousands	Quantity	Value, in thousands	Quantity
1859	2	$ 32	....	......	....	......	....
1860	500	4,800	....	......	....	......	....
1861	2,114	1,036	....	......	....	......	....
1862	3,057	3,210					
1863	2,611	8,226					
1864	2,116	20,897					
1865	2,498	16,460					
1866	3,598	13,455					
1867	3,347	8,067					
1868	3,646	13,217					
1869	4,215	23,730					
1870	5,261	20,504					
1871	5,205	22,591					
1872	6,293	21,441					
1873	9,894	18,100					
1874	10,927	12,648	....	......	....	......	....
1875	8,788	7,368	....	......	....	......	....
1876	8,969	(a) 22,562	120	(a) $298	32	(a) $92	....
1877	13,135	(a) 31,280	172	(a) 413	30	(a) 63	....
1878	15,164	(a) 17,758	180	(a) 211	38	(a) 41	....
1879	19,685	(a) 16,985	180	(a) 154	29	(a) 26	....
1880	(b) 26,028	(ab)24,303	179	(a) 167	39	(a) 37	(c)
1881	(b) 27,376	(ab)25,049	151	(a) 136	34	(a) 33	(c)
1882	23,368	(a) 18,042	128	(a) 98	40	(a) 33	6,685
1883	19,125	(a) 20,890	126	(a) 137	48	(a) 53	4,003
1884	20,541	(a) 17,142	90	(a) 74	90	(a) 74	3,231
1885	18,118	(a) 15,524	91	(a) 77	662	(a) 566	2,658
1886	(b) 25,798	(ab)18,153	102	(a) 71	645	(a) 454	(c)
1887	(b) 22,356	(ab)15,909	145	(a) 104	372	(a) 264	(c)
1888	(b) 16,489	(b) 14,440	119	105	328	287	(c)
1889	19,591	21,407	544	654	318	351	1,897
1890	(b) 28,458	(b) 28,465	493	501	1,110	1,150	(c)

Year	Tennessee	
	Quantity	Value
1916	(a)	$ 1,000
1917	12	28,000
1918	8	22,000
1919	(b)	(b)
1920	13	(b)

(a) Less than one thousand.
(b) Not available.

(a) Separation by States estimated. (b) Includes

in the Appalachian Field since 1859

of 42 U. S. gallons)

York	Kentucky and Tennessee		Total		Average value per barrel	Percentage of total production of United States	Percentage of yearly change of production
Value, in thousands	Quantity	Value, in thousands	Quantity	Value, in thousands			
....	....	......	2	$32	$16.00	100.00	......
....	....	......	500	4,800	9.59	100.00	+24900.00
....	....	......	2,114	1,036	.49	100.00	+322.72
....	....	......	3,057	3,210	1.05	100.00	+ 44.62
....	....	......	2,611	8,226	3.15	100.00	− 14.57
....	....	......	2,116	20,897	8.06	100.00	− 18.96
....	....	......	2,498	16,460	6.59	100.00	+ 18.03
....	....	......	3,598	13,455	3.74	100.00	+ 44.04
....	....	......	3,347	8,067	2.41	100.00	− 6.96
....	....	......	3,646	13,217	3.63	100.00	+ 8.93
....	....	......	4,215	23,730	5.64	100.00	+ 15.60
....	....	......	5,261	20,504	3.86	100.00	+ 24.81
....	....	......	5,205	22,591	4.34	100.00	− 1.06
....	....	......	6,293	21,441	3.64	100.00	+ 20.90
....	....	......	9,894	18,100	1.83	100.00	+ 57.21
....	....	......	10,927	12,648	1.17	100.00	+ 10.44
....	....	......	8,788	7,368	1.35	100.00	− 19.58
....	....	......	9,121	22,953	2.56	99.87	+ 3.79
....	....	......	13,337	31,756	2.42	99.90	+ 46.23
....	....	......	15,382	18,010	1.19	99.90	+ 15.33
....	....	......	19,894	17,165	.86	99.90	+ 29.34
(c)	....	......	26,246	24,507	.95	99.85	+ 31.93
(c)	....	......	27,561	25,218	.92	99.64	+ 5.01
(a)$5,161	....	......	30,221	23,334	.77	99.58	+ 9.65
(a) 4,372	5	(a) $8	23,307	25,460	1.09	99.39	− 22.88
(a) 2,697	4	(a) 4	23,956	19,991	.83	98.92	+ 2.79
(a) 2,276	5	(a) 4	21,534	18,447	.86	98.51	− 10.11
(c)	5	(a) 4	26,550	18,682	.70	94.60	+ 23.29
(c)	5	(a) 3	22,878	16,280	.71	80.90	− 13.83
(c)	5	5	16,941	14,837	.88	61.36	− 25.95
(a) 2068	5	5	22,355	24,485	.94	63.57	+ 31.96
(c)	6	6	30,067	30,122	.87	65.61	+ 34.50

New York. (c) Included in Pennsylvania.

Quantity and Value of Petroleum Produced

(In thousands of barrels

Year	Pennsylvania		West Virginia		Southeastern and Central Ohio		New
	Quantity	Value, in thousands	Quantity	Value, in thousands	Quantity	Value, in thousands	Quantity
1891	31,424	$21,241	2,407	$1,613	424	$295	1,585
1892	27,149	15,303	3,810	2,120	1,193	683	1,273
1893	19,283	12,564	8,446	5,426	2,603	1,676	1,031
1894	18,077	15,343	8,578	7,222	3,184	2,675	942
1895	18,232	24,901	8,120	11,039	3,694	5,026	913
1896	19,379	22,982	10,020	11,830	3,366	3,970	1,205
1897	17,983	14,296	13,090	10,310	2,878	2,265	1,279
1898	14,743	13,608	13,615	12,427	2,148	1,960	1,205
1899	13,054	17,054	13,910	18,015	4,765	6,247	1,320
1900	13,258	18,088	16,196	21,923	5,478	7,418	1,301
1901	12,625	15,430	14,177	17,173	5,472	6,622	1,207
1902	12,064	15,266	13,513	17,041	5,137	6,473	1,120
1903	11,355	18,171	12,899	20,517	5,587	8,883	1,163
1904	11,126	18,222	12,645	20,584	5,527	8,995	1,113
1905	10,437	14,653	11,578	16,133	5,017	6,993	1,118
1906	10,257	16,597	10,121	16,170	4,906	7,839	1,243
1907	9,999	17,580	9,095	15,852	4,214	7,345	1,212
1908	9,425	16,881	9,523	16,912	4,110	7,317	1,160
1909	9,300	15,425	10,745	17,642	4,717	7,774	1,135
1910	8,794	11,909	11,753	15,723	4,822	6,470	1,054
1911	8,248	10,894	9,795	12,767	4,281	5,591	953
1912	7,838	12,806	12,129	19,928	5,013	8,177	874
1913	7,918	19,690	11,567	28,829	4,964	12,230	948
1914	8,170	15,574	9,680	18,468	4,809	8,937	939
1915	7,838	12,432	9,265	14,468	4,432	6,761	888
1916	7,592	19,150	8,731	21,914	4,609	11,245	874
1917	7,733	25,154	8,379	27,247	4,840	15,473	881
1918	7,408	29,606	7,867	31,652	4,942	18,042	808
(e) 1919	(f)	(f)	(f)	(f)	(f)	(f)	(f)
(e) 1920	7,454	(f)	8,173	(f)	5,285	(f)	906
	......	......	....	......	....	......	....

(d) Less than one thousand. (e) Preliminary figures—fuel consumed on

in the Appalachian Field since 1859—Continued

of 42 U. S. gallons)

York	Kentucky and Tennessee		Total		Average value per barrel	Percentage of total production of United States	Percentage of yearly change of production
Value, in thousands	Quantity	Value, in thousands	Quantity	Value, in thousands			
$1,062	9	$9	35,849	$24,220	$0.67	66.03	+ 19.23
708	7	17	33,432	18,831	.56	66.19	− 6.74
660	3	1	31,366	20,327	.64	64.76	− 6.18
790	2	(*d*)	30,783	26,030	.84	62.38	− 1.86
1,241	2	(*d*)	30,961	42,207	1.36	58.54	+ .58
1,421	2	1	33,972	40,204	1.18	55.73	+ 9.73
1,006	(*d*)	(*d*)	35,230	27,877	.79	58.25	+ 3.70
1,098	6	3	31,717	29,096	.91	57.29	− 9.97
1,709	18	17	33,069	43,042	1.29	57.94	+ 4.26
1,759	62	47	36,295	49,235	1.35	57.05	+ 9.76
1,460	137	111	33,618	40,796	1.21	48.45	− 7.38
1,530	185	141	32,019	40,451	1.24	36.07	− 4.76
1,849	554	486	31,558	49,906	1.59	31.41	− 1.44
1,812	998	985	31,409	50,598	1.63	26.83	− .47
1,558	1,217	943	29,367	40,280	1.39	21.80	− 6.50
1,996	1,214	1,032	27,741	43,634	1.60	21.93	− 5.54
2,128	822	862	25,342	43,767	1.75	15.26	− 8.65
2,071	728	707	24,946	43,888	1.78	13.97	− 1.57
1,878	639	518	26,536	43,237	1.63	14.49	+ 6.38
1,415	469	325	26,892	35,842	1.33	12.83	+ 1.33
1,249	472	329	23,749	30,830	1.30	10.77	− 11.37
1,402	485	425	26,339	42,818	1.63	11.81	+ 10.90
2,284	525	676	25,922	63,709	2.46	10.43	− 1.58
1,761	503	499	24,101	45,239	1.88	9.07	− 7.02
1,390	437	418	22,860	35,469	1.55	8.13	− 5.15
2,190	1,203	2,190	23,009	56,689	2.46	7.65	+ .55
2,850	3,100	7,062	24,933	77,786	3.12	7.44	+ 8.36
3,308	4,376	11,309	25,401	93,917	3.70	7.14	+ 1.88
(*f*)	(*f*)	(*f*)	29,232	110,000	3.76	7.74	+ 15.08
(*f*)	8,693	(*f*)	30,511	165,000	5.41	6.88	+ 4.38
....	....	......	1,281,580	2,025,954	1.58	23.60	......

leases and net changes in producers' storage, not included. (*f*) Not available.

Quantity and Value of Petroleum Produced in Lima, Ohio-Indiana Field since 1886

(In thousands of barrels of 42 U. S. gallons)

Year	Lima, Ohio		Indiana		Total		Average Value per barrel	Percentage of total production of United States	**Percentage of yearly change of production**
	Quantity	Value, in thousands	Quantity	Value, in thousands	Quantity	Value, in thousands			
1886	1,138	$ 444	....	.....	1,138	$ 444	$0.39	4.06	......
1887	4,650	953	....	.....	4,650	953	.21	16.44	+308.69
1888	9,683	1,452	....	.....	9,683	1,452	.15	35.07	+108.21
1889	12,153	1,823	34	$ 11	12,187	1,834	.15	34.66	+ 25.86
1890	15,014	4,505	64	32	15,078	4,537	.30	32.91	+ 23.73
1891	17,316	5,280	137	54	17,453	5,334	.31	32.15	+ 15.75
1892	15,170	5,555	698	260	15,868	5,815	.37	31.41	− 9.08
1893	13,647	6,448	2,335	1,015	15,982	7,499	.47	33.00	+ .75
1894	13,608	6,532	3,689	1,774	17,297	8,306	.48	35.05	+ 8.82
1895	15,851	11,373	4,386	2,811	20,237	14,184	.70	38.26	+ 17.00
1896	20,575	13,724	4,681	2,954	25,256	16,678	.66	41.43	+ 24.80
1897	18,683	8,968	4,122	1,880	22,805	10,848	.48	37.71	+ 9.70
1898	16,590	10,245	3,731	2,214	20,321	12,459	.61	36.71	− 1.89
1899	16,377	14,719	3,849	3,364	20,226	18,083	.89	35.44	− .47
1900	16,885	16,673	4,874	4,694	21,759	21,367	.98	34.20	+ 7.58
1901	16,176	13,911	5,757	4,823	21,933	18,734	.85	31.61	+ .80
1902	15,878	14,284	7,480	6,527	23,358	20,811	.89	26.31	+ 6.50
1903	14,894	17,351	9,186	10,474	24,080	27,825	1.15	23.97	+ 3.09
1904	13,350	14,735	11,339	12,236	24,689	26,971	1.09	21.09	− 2.53
1905	11,330	10,062	10,964	9,405	22,294	19,467	.87	16.55	− 9.70
1906	9,881	9,158	7,674	6,770	17,555	15,928	.91	13.88	− 21.26
1907	7,993	7,425	5,128	4,537	13,121	11,962	.91	7.90	− 25.26
1908	6,748	6,862	3,284	3,204	10,032	10,066	1.00	5.62	− 23.54
1909	5,915	5,451	2,296	1,998	8,211	7,449	.91	4.48	− 18.15
1910	5,094	4,182	2,160	1,568	7,254	5,750	.79	3.46	− 11.66
1911	4,536	3,888	1,695	1,229	6,231	5,117	.82	2.83	− 14.10
1912	3,956	3,909	970	886	4,926	4,795	.93	2.21	− 20.95
1913	3,817	5,309	956	1,279	4,773	6,588	1.38	1.93	− 3.10
1914	3,727	4,435	1,335	1,548	5,062	5,983	1.18	1.90	+ 6.07
1915	3,394	3,301	876	813	4,270	4,114	.96	1.52	− 15.66
1916	3,136	4,909	769	1,208	3,905	6,117	1.57	1.30	− 8.54
1917	2,911	5,632	759	1,470	3,670	7,102	1.94	1.09	− 6.01
1918	2,343	5,423	878	2,028	3,221	7,451	2.31	.90	− 12.25
1919 (*a*)	(*b*)	(*b*)	(*b*)	(*b*)	3,444	8,000	2.32	.91	+ 6.92
1920 (*a*)	2,127	(*b*)	932	(*b*)	3,059	11,000	3.60	.69	+ 11.17
	....	.....	....	.....	455,028	361,023	.79	.84	....

(*a*) Preliminary figures—fuel consumed on leases and net changes in producers' storage not included.
(*b*) Not available.

Quantity and Value of Petroleum Produced in Illinois since 1889

(In thousands of barrels of 42 U. S. gallons)

Year	Quantity	Value	Average Value per barrel	Percentage of total production of United States	Percentage of yearly change of production
1889	2	$ 5,000	$3.36	....	....
1890	1	3,000	3.33	....	− 38.36
1891	1	3,000	3.50	....	− 25.00
1892	(*a*)	2,000	3.50	....	− 22.81
1893	(*a*)	1,C00	3.50	....	− 23.22
1894	(*a*)	2,000	6.00	....	− 25.00
1895	(*a*)	1,000	6.00	. ..	− 33.33
1896	(*a*)	1,000	5.00	...	+ 25.00
1897	(*a*)	2,000	4.00	....	+ 100.00
1898	(*a*)	2,000	5.00	....	+ 28.00
1899	(*a*)	2,000	5.00	....	....
1900	(*a*)	1,000	5.00	....	− 44.44
1901	(*a*)	1,000	5.00	....	+ 25.00
1902	(*a*)	1,000	5.00	....	− 20.00
1903	....	......	....	....	− 100.00
1904	....	......	....	....	....
1905	181	116,000	.64	0.13	....
1906	4,397	3,274,000	.75	3.47	+2,328.18
1907	24,282	16,433,000	.68	14.62	+ 452.23
1908	33,686	22,650,000	.67	18.87	+ 38.73
1909	30,899	19,789,000	.64	16.87	− 8.28
1910	33,144	19,669,000	.59	15.82	+ 7.27
1911	31,317	19,734,000	.64	14.21	− 5.51
1912	28,602	24,333,000	.85	12.83	− 8.67
1913	23,894	30,972,000	1.30	9.62	− 16.45
1914	21,920	25,426,000	1.16	8.25	− 8.26
1915	19,042	18,656,000	.98	6.77	− 13.13
1916	17,714	29,237,000	1.65	5.89	− 6.97
1917	15,777	31,358,000	1.99	4.71	− 10.94
1918	13,366	31,230,000	2.34	3.76	− 15.28
1919(*b*)	12,436	31,000,000	2.49	3.29	− 6.96
1920(*b*)	10,772	40,000,000	3.71	2.42	− 13.38
	321,433	363,905,000	1.13	5.92	....

(*a*) Less than one thousand.
(*b*) Preliminary figures—fuel consumed on leases and net changes in producers' storage not included.

Quantity and Value of Petroleum Produced in Mid-Continent Field since 1889

(In thousands of barrels of 42 U. S. gallons)

Year	Kansas		Oklahoma		Northern and Central Texas		Northern Louisiana		Total		Average value per Barrel	Percentage of total production of United States	Percentage of yearly increase of production
	Quantity	Value, in thousands	Quantity	Value, in thousands	Quantity	Value, in thousands	Quantity	Value, in thousands	Quantity	Value, in thousands			
1889	(a)	$3	......	......	......	......	......	......	(a)	$ 3	$5.00	......	
1890	1	8	......	......	......	......	......	......	1	8	7.00	......	140.00
1891	1	10	(a)	(a)	......	......	......	......	1	10	6.96	......	19.18
1892	5	5	(a)	(a)	......	......	......	......	5	5	1.08	......	255.24
1893	18	18	(a)	(a)	......	......	......	......	18	18	1.00	0.04	254.53
1894	40	40	(a)	$ 1	......	......	......	......	40	41	1.02	.08	122.82
1895	45	27	(a)	(a)	......	......	......	......	45	27	.61	.08	10.81
1896	114	51	(a)	(a)	1	$ 1	......	......	115	52	.46	.19	158.93
1897	81	32	(a)	2	66	38	......	....	147	72	.49	.24	28.23
1898	72	36	......	......	545	270	......	......	617	306	.50	1.11	317.62
1899	70	53	......	......	668	470	......	......	738	523	.71	1.29	19.72
1900	74	69	7	5	836	872	......	......	917	946	1.03	1.44	24.25
1901	179	155	10	7	801	616	......	......	990	778	.79	1.43	7.90
1902	332	293	37	33	618	420	......	......	987	746	.76	1.12	− .30
1903	932	988	139	142	502	516	......	......	1,573	1,646	1.05	1.57	59.42
1904	4,251	4,122	1,367	1,326	569	412	......	......	6,187	5,860	.95	5.28	293.28
1905	(b)12,014	6,547	(c)	(c)	520	361	......	......	[illegible]	6,908	.55	9.30	102.60
1906	(b)21,719	9,615	(c)	(c)	1,118	741	3	$ 2	22,840	10,358	.45	18.05	82.23
1907	2,409	965	43,524	17,514	913	721	50	39	46,896	19,239	.41	28.23	1033
1908	1,801	746	45,799	17,695	724	479	500	214	48,824	19,134	.39	27.35	4.11
1909	1,264	491	47,859	17,429	682	394	1,029	549	50,834	18,863	.37	27.75	4.12
1910	1,129	445	52,029	19,923	969	505	[illegible]	2,291	59,217	23,164	.39	28.26	16.49
1911	1,279	608	56,070	26,452	2,251	1,214	896	3,654	66,596	31,928	.48	30.21	12.46
1912	1,593	1,095	51,427	34,672	5,275	4,113	7,178	5,420	65,473	45,300	.69	29.48	− 1.68
1913	2,375	2,249	63,579	59,582	9,185	[illegible]	9,782	9,812	84,921	80,768	.95	34.18	29.70
1914	3,104	2,433	73,632	57,254	9,451	7,779	11,808	11,206	97,995	78,672	.80	36.87	5140
1915	2,823	1,702	97,915	56,706	7,474	4,657	15,082	9,366	123,294	72,431	.59	43.86	25.82
1916	8,738	10,340	107,072	128,464	9,303	11,835	11,822	12,178	136,935	162,817	1.18	45.53	106
1917	36,536	67,120	107,508	181,647	10,900	19,953	8,562	14,076	163,506	282,796	1.73	48.76	19.40
1918	45,451	100,546	103,347	231,136	17,280	38,313	13,305	23,036	179,383	393,031	2.19	50.40	9.71
1919 (d)	(b) 115,897	(e)	(c)	(e)	67,419	(e)	13,575	(e)	196,891	447,000	2.27	52.12	9.76
1920 (d)	38,501	(e)	105,725	(e)	70,952	(e)	33,896	(e)	249,074	837,000	3.36	56.17	26.50
	....	......	......	......	......	......	......	......	1,617,594	2,540,000	1.57	29.79	......

(a) Less than one thousand. (b) Includes Oklahoma. (c) Included in Kansas. (d) Preliminary figures—fuel consumed on leases and net changes in producers' storage not included. (e) Not available.

Quantity and Value of Petroleum Produced in Gulf Field since 1889

(In thousands of barrels of 42 U. S. gallons)

Year	Coastal Texas		Coastal Louisiana		Total		Average value per barrel	Percentage of total production of United States	Percentage of yearly change of production
	Quantity	Value, in thousands	Quantity	Value, in thousands	Quantity	Value, in thousands			
1889	(a)	(a)	....	.....	(a)	(a)	$7.08	......	......
1890	(a)	(a)	....	.....	(a)	(a)	4.20	......	+ 12.50
1891	(a)	(a)	....	.....	(a)	(a)	4.20	......	......
1892	(a)	(a)	....	.....	(a)	(a)	5.00	......	− 16.67
1893	(a)	(a)	....	.....	(a)	(a)	4.20	......	+ 11.11
1894	(a)	(a)	....	.....	(a)	(a)	5.00	......	+ 20.00
1895	(a)	(a)	....	.....	(a)	(a)	5.00	......	− 16.67
1896	(a)	(a)	....	.....	(a)	(a)	5.00	......	......
1897	(a)	(a)	....	.....	(a)	(a)	5.00	......	......
1898	2	$ 7	....	.....	2	$ 7	5.00	......	+2800.00
1899	(a)	3	....	.....	(a)	3	5.00	......	− 63.45
1900	....	.....	....	.....	......	......	......	......	−100.00
1901	3,593	631	....	.....	3,593	631	.18	5.18	......
1902	17,466	3,578	548	$ 189	18,014	3,767	.21	20.29	+401.36
1903	17,454	7,002	918	416	18,372	7,418	.41	18.29	+ 1.98
1904	21,672	7,744	2,959	1,073	24,631	8,817	.36	21.03	+ 34.07
1905	27,616	7,191	8,910	1,601	36,526	8,792	.24	27.11	+ 48.29
1906	11,450	5,825	9,074	3,556	20,524	9,381	.46	16.23	− 43.81
1907	11,410	9,681	4,951	4,024	16,361	13,705	.84	9.85	− 20.29
1908	10,483	6,222	5,289	3,289	15,772	9,511	.60	8.83	− 3.60
1909	8,852	6,399	2,031	1,474	10,883	7,873	.72	5.94	− 30.00
1910	7,930	6,101	1,751	1,283	9,681	7,384	.76	4.62	− 11.05
1911	7,275	5,341	3,725	2,015	11,000	7,356	.67	4.99	+ 13.63
1912	6,460	4,740	2,085	1,604	8,545	6,344	.74	3.83	− 22.32
1913	5,825	5,550	2,717	2,444	8,542	7,994	.94	3.44	− .03
1914	10,617	7,164	2,501	1,681	13,118	8,845	.67	4.94	+ 53.56
1915	17,469	8,370	3,110	1,439	20,579	9,809	.48	7.32	+ 56.87
1916	18,342	13,925	3,426	2,492	21,768	16,417	.75	7.24	+ 5.78
1917	21,513	22,939	2,830	3,149	24,343	26,088	1.07	7.26	+ 11.83
1918	21,470	36,555	2,738	4,409	24,208	41,054	1.70	6.80	− .56
1919 (b)	(c)	(c)	(c)	(c)	20,568	23,000	1.12	5.45	− 15.51
1920 (b)	25,048	(c)	1,753	(c)	26,801	64,000	2.39	6.05	+ 30.35
	....	.....	....	.....	353,831	288,196	.81	6.52	......

(a) Less than one thousand.
(b) Preliminary figures—fuel consumed on leases and net changes in producers' storage not included.
(c) Not available.

Quantity and Value of Petroleum Produced in the Rocky Mountain Field since 1887
(In thousands of barrels of 42 U. S. gallons)

Year	Colorado		Wyoming		Montana		Total		Average value per barrel	Percentage of total production of United States	Percentage of yearly change of production
	Quantity	Value, in thousands	Quantity	Value, in thousands	Quantity	Value, in thousands	Quantity	Value, in thousands			
1887	76	$76	....	......	....	......	76	$ 76	$1.00	.27	....
1888	298	268	....	......	....	......	298	268	.90	1.07	+29.01
1889	317	280	....	......	....	......	317	280	.89	.90	+ 6.34
1890	369	310	....	......	....	......	369	310	.84	.80	+16.54
1891	665	559	....	......	....	......	665	559	.84	1.22	+80.42
1892	824	692	....	......	....	......	824	692	.84	1.63	+23.82
1893	595	498	....	......	....	......	595	498	.84	1.22	−27.86
1894	516	304	2	$ 16	....	......	518	320	.62	1.05	−12.83
1895	438	336	4	28	....	......	442	364	.82	.83	−14.75
1896	361	319	3	23	....	......	364	342	.94	.59	−17.51
1897	385	332	4	30	....	......	389	362	.93	.64	+ 6.66
1898	444	368	6	38	....	......	450	406	.90	.80	+15.77
1899	390	404	6	39	....	......	396	443	1.20	.69	−12.01
1900	317	324	6	38	....	......	323	362	1.12	.50	−18.44
1901	461	461	5	38	....	......	466	499	1.07	.66	+44.32
1902	397	484	6	44	....	......	403	528	1.31	.45	−13.47
1903	484	432	9	63	....	......	493	495	1.00	.43	+22.26
1904	502	578	11	81	....	......	513	659	1.28	.43	+ 4.14
1905	376	338	9	51	....	......	385	389	1.01	.28	−25.06
1906	328	263	7	49	....	......	335	312	.93	.26	−13.03
1907	332	273	9	21	....	......	341	294	.86	.20	+ 1.98
1908	379	346	18	28	....	......	397	374	.94	.22	+16.48
1909	311	318	20	34	....	......	331	352	1.07	.18	−16.74
1910	240	243	115	94	....	......	355	337	.95	.17	+ 7.35
1911	227	228	187	124	....	......	414	352	.85	.18	+16.44
1912	206	200	1,572	798	....	......	1,778	998	.56	.80	+329.95
1913	189	175	2,406	1,187	....	......	2,595	1,362	.53	1.04	+ 45.94
1914	223	201	3,560	1,679	....	......	3,783	1,880	.50	1.43	+ 45.77
1915	208	183	4,246	2,217	....	......	4,454	2,400	.54	1.59	+ 17.73
1916	197	217	6,234	5,644	45	$44	6,476	5,905	.91	2.15	+ 45.40
1917	121	128	8,979	11,048	99	146	9,199	11,322	1.23	2.74	+ 42.05
1918	143	188	12,596	18,160	70	126	12,809	18,474	1.44	3.60	+ 39.24
1919(a)	(b)	(b)	(b)	(b)	(b)	(b)	13,584	21,000	1.55	3.60	+ 6.05
1920(a)	110	(b)	17,071	(b)	336	(b)	17,517	48,000	2.74	3.95	+ 28.95
	....	......	....	......	....	......	82,654	121,214	1.47	1.52	

(a) Preliminary figures—fuel consumed on leases and net changes in producers' storage not included.
(b) Not available.

Quantity and Value of Petroleum Produced in California since 1876

(In thousands of barrels of 42 U. S. gallons)

Year	Quantity	Value	Average value per barrel	Percentage of total production of United States	Percentage of yearly change of production
1876	12	$ 30,000	$2.50	.13	....
1877	13	33,000	2.50	.10	+ 8.33
1878	15	35,000	2.31	.10	+ 17.13
1879	20	46,000	2.31	.10	+ 30.41
1880	40	94,000	2.31	.15	+104.21
1881	100	231,000	2.31	.36	+146.26
1882	129	297,000	2.31	.42	+ 28.81
1883	143	330,000	2.31	.61	+ 11.06
1884	262	605,000	2.31	1.08	+ 83.40
1885	325	751,000	2.31	1.49	+ 24.46
1886	377	870,000	2.31	1.34	+ 16.05
1887	679	1,568,000	2.31	2.39	+ 79.92
1888	690	1,391,000	2.01	2.50	+ 1.73
1889	303	356,000	1.17	.86	− 56.08
1890	308	384,000	1.25	.67	+ 1.37
1891	324	401,000	1.24	.59	+ 5.28
1892	385	561,000	1.46	.76	+ 18.99
1893	470	608,000	1.29	.97	+ 22.11
1894	706	823,000	1.17	1.43	+ 50.15
1895	1,208	849,000	.70	2.28	+ 71.18
1896	1,253	1,241,000	.99	2.05	+ 3.67
1897	1,904	1,713,000	.90	3.15	+ 51.93
1898	2,257	1,918,000	.85	4.08	+ 18.59
1899	2,642	2,508,000	.95	4.63	+ 15.67
1900	4,325	4,077,000	.94	6.80	+ 63.67
1901	8,787	4,975,000	.57	12.66	+103.17
1902	13,984	4,874,000	.35	15.75	+ 59.16
1903	24,382	7,399,000	.30	24.27	+ 74.36
1904	29,649	8,625,000	.28	25.33	+ 21.60
1905	33,427	8,202,000	.25	24.81	+ 12.74
1906	33,099	9,553,000	.29	26.17	− .98
1907	39,748	14,700,000	.37	23.93	+ 20.09
1908	44,855	23,433,000	.52	25.13	+ 12.87
1909	55,471	30,757,000	.55	30.29	+ 23.67
1910	73,010	35,749,000	.49	34.84	+ 31.62
1911	81,134	38,719,000	.48	36.80	+ 11.13
1912	87,273	39,625,000	.45	39.15	+ 7.57
1913	97,789	45,709,000	.47	39.36	+ 12.05
1914	99,776	48,066,000	.48	37.54	+ 2.03
1915	86,591	36,559,000	.42	30.81	− 13.21
1916	90,952	53,703,000	.59	30.24	+ 5.04
1917	93,878	86,162,000	.92	28.00	+ 3.22
1918	97,532	118,771,000	1.22	27.40	+ 3.89
1919 (*a*)	101,564	(*b*) 135,000,000	1.33	26.89	+ 4.13
1920 (*a*)	105,668	(*b*) 195,000,000	1.85	23.94	+ 4.04
	1,317,459	966,941,000	1.73	24.26	

(*a*) Average of figures reported by Standard Oil Co. (Calif.) and Independent Oil Producers' Agency
(*b*) Estimated.

Quantity and Value of Petroleum Produced in "Other" States since 1889

(In barrels of 42 U. S. gallons)

Year	Quantity		Value	Average value per barrel
1889	(a)	20	$ 40	$2.00
1890	(a)	278	556	2.00
1891	(a)	25	84	3.36
1892	(a)	10	40	4.00
1893	(a)	50	154	3.08
1894	(a)	8	40	5.00
1895	(a)	10	50	5.00
1896	(a)	43	185	4.30
1897	(a)	19	174	9.16
1898	(a)	10	105	10.50
1899	(a)	132	205	1.55
1900	(b)	1,602	1,177	.74
1901	(b)	2,335	2,600	1.11
1902	(b)	757	1,066	1.41
1903	(b)	3,000	4,650	1.55
1904	(b)	2,572	4,769	1.85
1905	(b)	3,100	3,320	1.07
1906	(b)	3,500	4,890	1.40
1907	(b)	4,000	6,500	1.63
1908	(b)	15,246	22,345	1.47
1909	(b)	5,750	7,830	1.36
1910	(b)	3,615	4,794	1.33
1911	(b)	7,995	7,995	1.00
1912		(c)	(c)	
1913	(d)	10,843	19,263	1.78
1914	(e)	7,792	14,291	1.83
1915	(e)	14,265	24,295	1.70
1916	(e)	7,705	14,410	1.87
1917	(f)	10,300	20,600	2.00
1918	(f)	7,943	15,986	2.01
1919		(g)	(g)	
1920		(g)	(g)	

(a) Missouri. (b) Missouri and Michigan. (c) Michigan, included in Lima, Indiana field.
(d) Missouri, Michigan, Alaska and New Mexico. (e) Missouri, Alaska and Michigan.
(f) Alaska and Michigan. (g) Not available.

As the production and value of petroleum in Ohio, Texas, and Louisiana are subdivided by districts, totals for these states are here shown.

Quantity and Value of Petroleum Produced in Ohio, Texas, and Louisiana
(In thousands of barrels of 42 U. S. gallons)

Year	Ohio		Texas		Louisiana	
	Quantity	Value	Quantity	Value	Quantity	Value
1876	32	$92,000				
1877	30	63,000				
1878	38	41,000				
1879	29	26,000				
1880	39	37,000				
1881	34	33,000				
1882	40	33,000				
1883	48	53,000				
1884	90	74,000				
1885	662	566,000				
1886	1,783	898,000				
1887	5,022	1,217,000				
1888	10,011	1,739,000				
1889	12,471	2,174,000	(*a*)	(*a*)		
1890	16,124	5,655,000	(*a*)	(*a*)		
1891	17,740	5,584,000	(*a*)	(*a*)		
1892	16,363	6,238,000	(*a*)	(*a*)		
1893	16,250	8,124,000	(*a*)	(*a*)		
1894	16,792	9,207,000	(*a*)	(*a*)		
1895	19,546	16,399,000	(*a*)	(*a*)		
1896	23,941	17,694,000	1	$ 1,000		
1897	21,561	11,233,000	66	38,000		
1898	18,738	12,205,000	546	277,000		
1899	21,142	20,966,000	669	473,000		
1900	22,363	24,091,000	836	872,000		
1901	21,648	20,533,000	4,394	1,247,000		
1902	21,015	20,757,000	18,084	3,998,000	549	$189,000
1903	20,481	26,234,000	17,956	7,518,000	918	416,000
1904	18,877	23,730,000	22,241	8,156,000	2,959	1,073,000
1905	16,347	17,055,000	28,136	7,552,000	8,910	1,601,000
1906	14,787	17,097,000	12,568	6,566,000	9,077	3,558,000
1907	12,207	14,770,000	12,323	10,402,000	5,000	4,063,000
1908	10,858	14,179,000	11,207	6,701,000	5,789	3,503,000
1909	10,632	13,225,000	9,534	6,793,000	3,060	2,023,000
1910	9,916	10,652,000	8,899	6,606,000	6,841	3,574,000
1911	8,817	9,479,000	9,526	6,555,000	10,721	5,669,000
1912	8,969	12,086,000	11,735	8,813,000	9,263	7,024,000
1913	8,782	17,539,000	15,010	14,675,000	12,499	12,256,000
1914	8,536	13,372,000	20,068	14,943,000	14,309	12,887,000
1915	7,826	10,062,000	24,943	13,027,000	18,192	10,805,000
1916	7,745	15,654,000	27,645	25,760,000	15,248	14,670,000
1917	7,750	21,105,000	32,413	42,892,000	11,392	17,225,000
1918	7,285	23,465,000	38,750	74,868,000	16,043	27,535,000
1919	(*b*)	(*b*)	(*b*)	(*b*)	(*b*)	(*b*)
1920 (*c*)	(*c*) 7,412	(*b*)	96,000	(*b*)	35,649	(*b*)

(*a*) Less than one thousand. (*b*) Not available. (*c*) Preliminary figures.

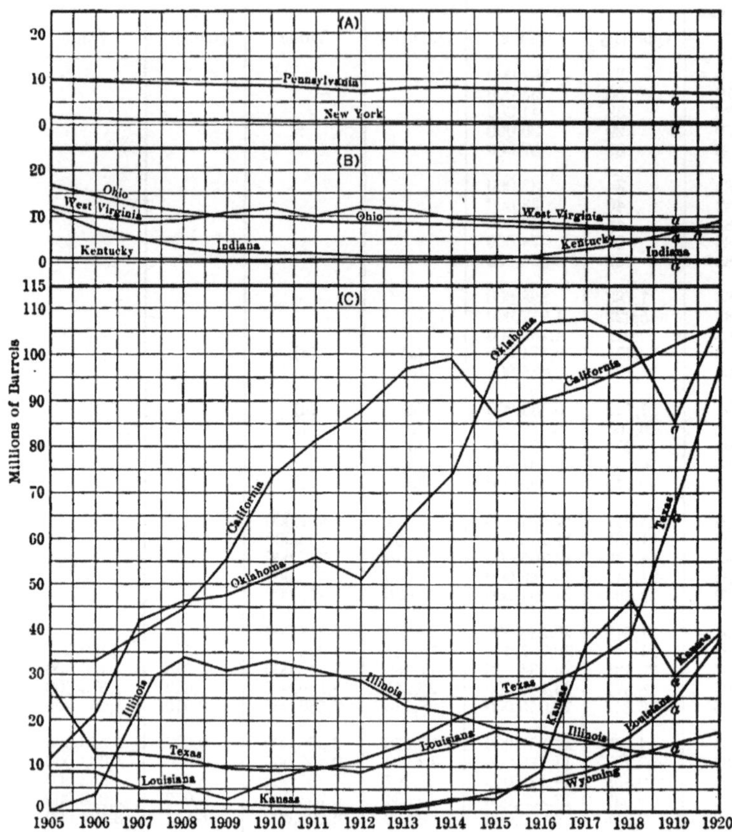

FIG. 6.—Production of petroleum in the United States, by states, 1905–1920. Tennessee, Colorado and Montana not shown because of necessary smallness of scale. Grouping of states for clearness of diagram only. (a) Estimated. Figures by states not available.

The amount of oil on hand is the most important barometer for changes in prices of the different crude oils. The following table shows the stocks held by pipe-line and marketing companies, December 31, 1906–1920. Separation is made by field sources.

Stocks of Domestic Petroleum Held by Pipe-line and Marketing Companies, December 31, 1906–1920, by Fields

(In thousands of barrels of 42 U. S. gallons)

Year	Appala-chian	Lima-Indiana	Illinois	Mid-Con-tinent	Gulf	Rocky Moun-tain	Cali-fornia	Total
1906	8,119	6,794	1,900	22,941	5,772	14	6,235	51,775
1907	6,537	6.701	15,848	38,605	2,535	1	6,055	76,282
1908	4,142	3,825	29,210	50,942	2,969	4	5,839	96,930
1909	5,939	4,011	32,344	53,542	2,839	12	18,000	116,687
1910	5,007	4,730	31,325	(*a*)54,179	(*a*)2,674	30	33,085	131,030
1911	4,735	3,195	24,064	(*a*)57,680	(*a*)3,294	25	44,240	137,233
1912	4,031	2,420	15,710	51,538	1,472	147	47,552	122,870
1913	4,619	1,746	8,243	57,392	2,059	442	48,302	122,803
1914	5,725	1,648	13,564	60,818	3,964	170	55,661	141,550
1915	5,742	2,919	11,328	91,731	7,022	425	44,588	163,755
1916	3,850	2,088	6,600	100,401	9,315	745	39,398	162,397
1917	3,822	1,906	3,506	99,427	8,385	515	28,427	146,042
1918	3,616	1,029	2,367	79,094	8,168	915	26,538	121,727
1919	3,757	1,323	4,194	76,712	12,575	164	(*b*)29,142	127,867
1920	3,673	1,286	3,495	91,696	12,311	299	(*b*)20,930	133,690

(*a*) Separation estimated.
(*b*) Includes producers' stocks.

NOTE.—Because of a re-classification of stocks since 1920 by the United States Geological Survey, figures for current months are not directly comparable. For comparison, add 17.8 millions to current figures.

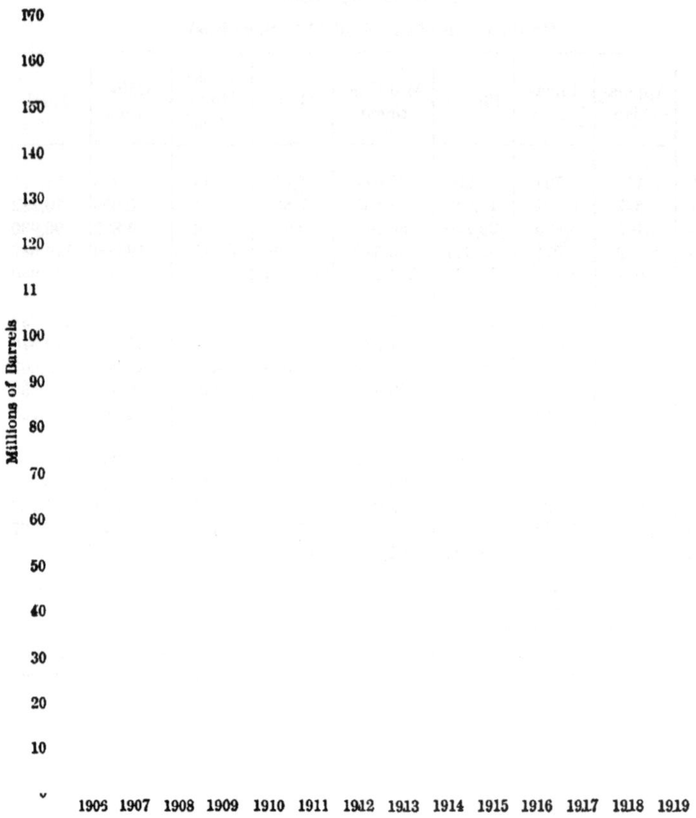

Millions of Barrels

170
160
150
140
130
120
11
100
90
80
70
60
50
40
30
20
10

1906 1907 1908 1909 1910 1911 1912 1913 1914 1915 1916 1917 1918 1919

FIG. 7.—Stocks of domestic petroleum held by pipe line and marketing companies. By fields, Dec. 31, 1906–1920 (a) Includes producers' stocks.

The following table shows the amount of domestic petroleum delivered by pipe-line and marketing companies, 1906–1920. The figures are obtained by adding net decrease or subtracting net increase of stocks during the year to the production of the year.

Deliveries of Domestic Petroleum by Pipe-line and Marketing Companies, 1906–1920, by Field Sources

(In thousands of barrels of 42 U. S. gallons)

Year	Appalachian	Lima-Indiana	Illinois	Mid-Continent	Gulf	Rocky Mt.	California	Other	Total
1906	23,164	23,397	2,497	13,149	37,295	341			
1907	26,924	13,214	10,334	31,232	19,598	354	39,923	4	141,588
1908	27,341	12,908	20,324	36,487	15,338	394	45,071	15	157,879
1909	24,739	8,025	27,765	48,234	11,013	322	43,310	6	163,414
1910	27,824	6,535	34,163	58,580	9,846	337	57,925	4	195,214
1911	24,022	7,766	38,578	63,095	10,380	418	69,979	8	214,246
1912	27,042	5,701	36,956	71,615	10,367	1,656	83,961	..	237,298
1913	25,333	5,447	31,361	79,067	7,955	2,300	97,039	11	248,513
1914	22,995	5,160	16,599	94,569	11,213	4,055	92,417	8	247,016
1915	22,843	2,999	21,278	92,381	17,521	4,199	97,664	14	258,899
1916	24,901	4,736	22,442	128,266	19,475	6,156	96,142	8	302,126
1917	24,961	3,852	18,817	164,480	25,273	9,429	104,849	10	351,671
1918	25,607	4,098	14,559	199,716	24,425	12,409	99,421	8	380,243
1919	29,091	3,150	10,609	199,273	16,161	14,335	98,960	..	371,579
1920	30,595	3,096	11,471	234,090	27,065	17,382	113,880	..	437,579

The effect of variation in production, consumption, and stocks upon the prices of crude oils is shown in the following graph. The figures upon which the curves are based may be found in the tables for the respective fields and represent the average value per barrel received for the crude at the wells in the different fields. This average value is obtained by dividing the total amount received by the total number of barrels. It should not be confused with average price, which is the average of the prices offered by the marketing companies.

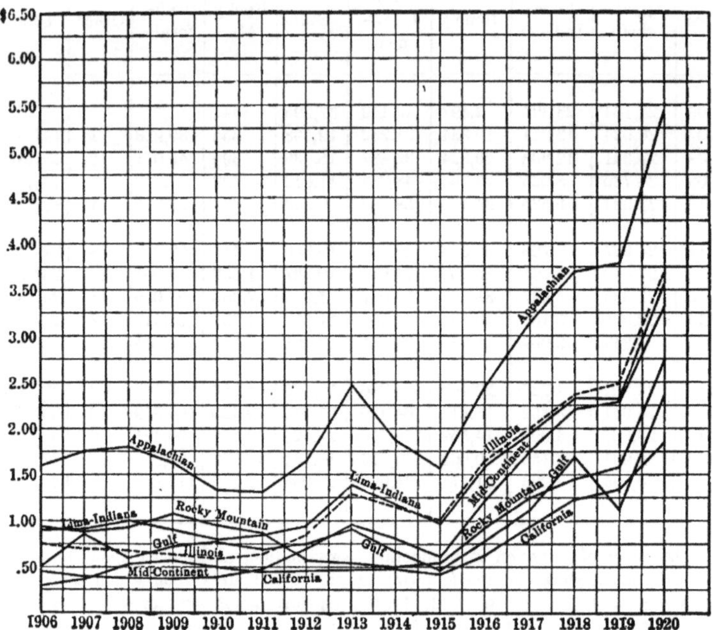

Fig. 8.—Average value per barrel at wells of petroleum by fields, 1906–1920.

The following table is a comparison of the quantity and value of the petroleum produced in the United States, 1909–1920, with the quantity and value of the crude oil imported and exported and the amount consumed during that time. The figures for imports and exports are taken from the records of the Bureau of Foreign and Domestic Commerce, and since that bureau did not separate mineral crude oil from its products prior to 1909, the comparison for previous years could not be made. The estimated consumption is obtained by adding net imports to production and adding the decrease or subtracting the increase in stocks. Since December, 1920, the United States Geological Survey has made a re-classification of stocks and in order to compare the figures of a current month with those of 1920 or any year previous, 17.841 millions should be added to the current stock figures.

Production, Imports, Exports, Consumption, and Stocks of Petroleum, 1909–1920

(In thousands of barrels of 42 U. S. gallons)

Year	Production		Imports		Exports		Estimated domestic consumption	(a)Stocks Dec. 31
	Quantity	Value, in thousands	Quantity	Value, in thousands	Quantity	Value, in thousands	Quantity	Quantity
1909	183,171	$128,328	70	$197	4,056	$ 6,028	167,089	116,687
1910	209,557	127,900	557	1,399	4,288	5,404	191,483	131,030
1911	220,449	134,044	1,710	2,411	4,806	6,165	211,150	137,233
1912	222,935	164,213	7,383	6,083	4,493	6,770	240,188	122,870
1913	248,446	237,121	17,809	12,947	4,630	8,448	261,692	122,803
1914	265,763	214,125	17,247	11,465	2,970	4,959	261,293	141,550
1915	281,104	179,463	18,140	10,389	3,768	4,283	273,271	163,755
1916	300,767	330,900	20,570	12,603	4,096	7,030	318,599	162,397
1917	335,316	522,635	30,163	16,400	4,098	7,668	377,736	146,042
1918	355,928	703,944	37,736	21,319	4,901	12,084	413,078	121,727
1919	(b)377,719	775,000	52,822	26,043	5,924	14,806	418,477	127,867
1920	(b)443,402	1,360,000	106,175	55,799	8,045	28,990	531,186	133,690

(a) Held by pipe-line and marketing companies. (b) Preliminary figures.

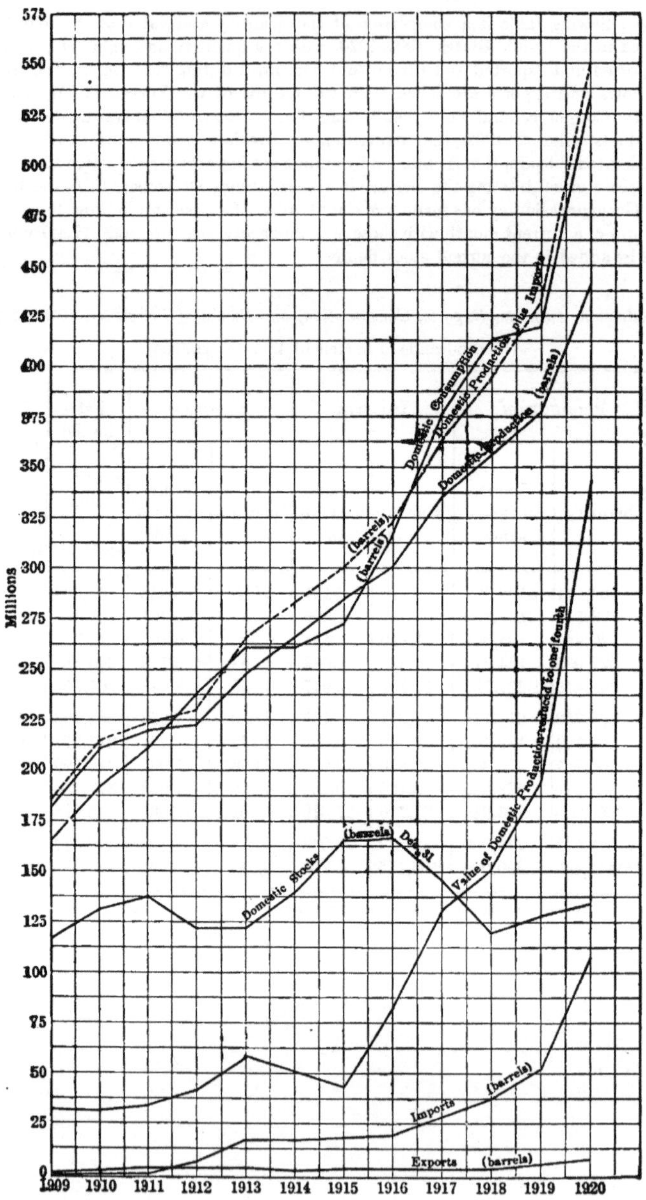

FIG. 9.—Domestic production, consumption and stocks 1909–1920. Exports and imports, 1909–1920.

The following figures are from the United States Geological Survey, those for wells completed being compiled from the Oil City Derrick and the Oil and Gas Journal and those for productive wells from the statements of oil producers.

Wells Completed and Productive in the United States, 1909–1920

Year	Wells Completed				Wells productive, December 31
	Oil	Gas	Dry	Total	
1909	13,875	1,048	3,404	18,327	147,118
1910	11,018	1,500	2,422	14,940	148,027
1911	9,818	1,580	2,363	13,761	153,288
1912	12,512	1,811	2,855	17,178	157,850
1913	19,101	2,207	4,282	25,920	169,440
1914	16,668	2,327	4,142	23,137	179,383
1915	9,154	2,022	2,981	14,157	181,219
1916	18,777	1,803	4,039	24,619	190,025
1917	16,595	1,966	4,851	23,407	197,149
1918	17,845	2,229	5,613	25,684	203,375
1919	21,288	2,077	5,706	33,957	(*a*)
1920	24,310	2,272	7,375	29,021	(*b*)258,600

(*a*) Not available. (*b*) October 31.

Producing Oil Wells in the United States, October 31, 1920

State	Aproximate number of producing oil wells	Approximate production per well per day (barrels)
California (*a*)	9,490	32.3
Colorado	70	4.1
Illinois	16,800	1.7
Indiana	2,400	1 1
Kansas	15,700	6.7
Kentucky	7,800	3.1
Louisiana:		
Northern	2,560	31.6
Coastal	140	34.6
Total Louisiana	2,700	31.8
New York	14,040	0.2
Ohio:		
Central and Eastern	18,500	0.8
Northwestern	21,100	0.3
Total Ohio	39,600	0.5
Oklahoma	50,700	6.0
Pennsylvania	67,700	0.3
Texas:		
Central and Northern	9,400	22.9
Coastal	1,700	49.7
Total Texas	11,100	27.0
West Virginia	19,500	1.1
Wyoming and Montana	1,000	55.9
Total	258 600	Average 4.9

(*a*) Average of wells reported by the Standard Oil Company and the Independent Producers' Agency.

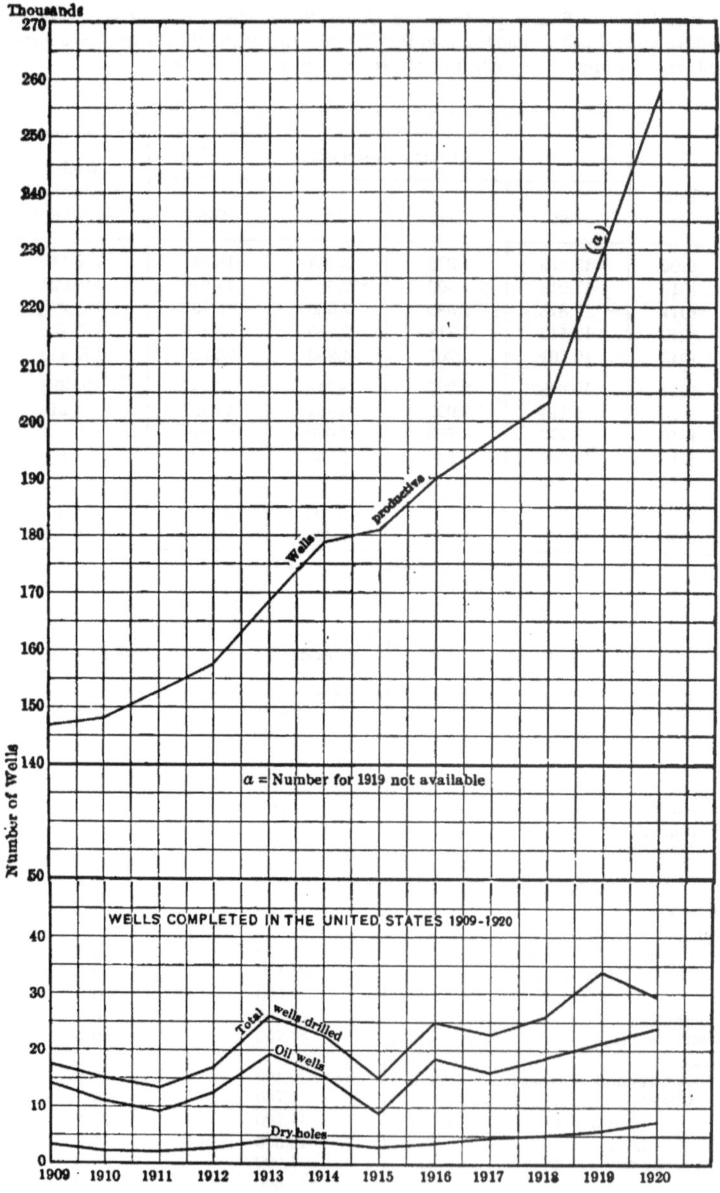

Fig. 10.—Wells productive in the United States, Dec. 31, 1909–1920.

Prior to 1911 statistics for gasoline from natural gas were not available. The following table is a general summary of the industry in the United States, 1911–1920, as reported by the United States Geological Survey:

Natural-gas Gasoline Produced in the United States, 1911–1920

Year	Number of operators	Plants		Gasoline Produced			Gas Used		
		Number	Daily capacity	Quantity	Value	Price per gallon	Estimated volume	Value (a)	Average yield of gasoline per thousand cubic feet.
			Gallons	Gallons		Cents	M. cu. ft.		Gallons
1911	132	176	37,100	7,425,839	$531,704	7.16	2,475,697	$176,961	3.00
1912	186	250	61,268	12,081,179	1,157,476	9.60	4,687,796	331,985	2.60
1913	232	341	152,415	24,060,817	2,458,443	10.22	9,889,442	566,224	2.43
1914	254	386	179,353	42,652,632	3,105,909	7.28	16,894,557	889,906	2.43
1915	287	414	232,336	65,364,665	5,150,823	7.88	24,064,391	1,202,555	2.57
1916	460	596	495,448	103,492,689	14,331,148	13.85	208,705,023	(b)14,609,351	.496
1917	750	886	902,385	217,884,104	40,188,956	18.45	429,287,797	(b)34,343,000	.508
1918	(c) 503	1004	(b)1022,072	282,535,550	50,363,535	17.83	449,108,661	(b)40,419,700	.63
1919	(c) 611	1191	......	351,535,026	64,196,763	18.26	480,403,963	(b)41,314,700	.73
1920	(c) 577	1151	......	383,311,817	71,534,118	18.70	495,883,700	(b)41,700,000	.772

(a) The value of the gas is based on sales to gasoline producers, not on sales for domestic or industrial purposes.

(b) Estimated.

(c) The number of operators in 1918, 1919 and 1920 is not comparable with that for earlier years as the method of listing has been changed.

The quantity and value of natural gas produced and consumed in the Unites States, 1910–1918, by states, are given in the following table. The figures are from the reports of the United States Geological Survey. The term "production" is used to designate only that portion of the natural gas produced during the year that was commercially utilized, and, for the total, is synonymous with "consumption." But as natural gas is freely transported from one state to another, the word "production" is used for the output of natural gas in a given state without regard to place of consumption and "consumption" is used for the amount utilized in a given state without regard to place of production. The word "value" means the market value of the natural gas at the point of ultimate consumption, not that of production.

Quantity and Value of Natural Gas

(In millions of

	1910		1911		1912		1913	
	Quantity	Value, in thousands	Quantity	Value, in thousands	Quantity	Value, in thousands	Quantity	**Value, in thousands**
West Virginia....	190,706	$23,817	206,891	$28,436	239,007	$33,324	245,454	**$34,165**
Oklahoma.......	50,430	3,491	67,276	6,732	73,799	7,407	75,018	**7,437**
Pennsylvania....	126,867	21,057	108,869	18,521	112,150	18,540	118,860	**21,696**
Ohio...........	48,232	8,627	49,450	9,367	56,210	11,891	50,612	**10,522**
California.......	6,764	477	6,390	801	9.355	1,135	11,035	**1,884**
Louisiana........	(*a*)8,111	(*a*) 957	(*b*)9,786	(*b*) 858	(*b*)4,493	(*b*)1,747	*b*26,653	(*b*)**2,120**
Kansas..........	59,380	7,755	38,799	4,855	28,069	4,265	22,885	**3,288**
Texas..........	(c)	(*c*)	5,503	1,015	7,470	1,405	12,160	**2,074**
New York.......	6,010	1,679	5,240	1,419	8,626	2,343	8,515	**2,426**
Arkansas........	(*d*)2,705	(*d*) 301	(*d*)2,294	(*d*) 296	(*d*)1,742	(*d*) 310	(*d*)1,106	(*d*) **269**
Illinois..........	6,723	614	6,762	688	5,603	617	4,767	**574**
Wyoming........	(*e*)	(*e*)	(*e*)	(*e*)	(*e*)	(*e*)	(*e*)	(*e*)
Kentucky.......	1,357	456	1,275	408	1,951	522	1,821	**510**
Tennessee.......	1	(*g*)	1	(*g*)	1	(*g*)	2	(*g*)
Indiana.........	5,760	1,473	4,365	1,192	3,618	1,014	2,921	**843**
Montana........	.....	.....	.....	.....	.....	.....	.....	.....
South Dakota....	43	32	26	17	(*h*) 54	(*h*) 31	(*h*) 66	(*h*) 31
Maryland.......	.....	.....	.....	.....	.....	.....	.....	.....
Utah...........	.....	.....	.....	.....	.....	.....	.....	.....
Washington......	.....	.....	.....	.....	.....	.....	.....	.....
Missouri.........	47	12	50	10	53	12	21	**7**
Colorado........	(*e*)	(*e*)	(*e*)	(*e*)	(*e*)	(*e*)	(*e*)	(*e*)
Alabama........	(*c*)	(*c*)	(*c*)	(*c*)	(*c*)	(*c*)	(*c*)	(*c*)
Oregon..........	.....	.....	.....	.....	.....	.....	.....	.....
Iowa...........	(*g*)	(*g*)	(*g*)	(*g*)	(*g*)	(*g*)	(*g*)	(*g*)
Michigan........	1	1	2	1	2	1	2	**1**
North Dakota....	18	7	14	6	(*k*)	(*k*)	(*k*)	(*k*)
	509,155	70,756	512,993	74,622	563,203	84,564	581,898	**87,847**

(*a*) Includes Alabama and Texas.
(*b*) Includes Alabama.
(*c*) Included in Louisiana.
(*d*) Includes Colorado and Wyoming.
(*e*) Included in Arkansas.
(*f*) Includes Colorado.

Produced in the United States, 1910–1918

cubic feet)

1914		1915		1916		1917		1918	
Quantity	Value, in thousands	Quantity	Value, in thousands	Quantity	Value, in thousands	Quantity	Value, in thousands	Quantity	Value, in thousands
236,489	$35,077	244,004	$36,424	299,319	$47,604	308,617	$57,389	265,161	$41,324
78,167	8,050	87,517	9,196	123,517	12,015	137,384	13,985	124,317	15,805
110,745	20,840	113,691	21,140	130,484	24,513	133,397	28,717	128,812	38,609
68,270	14,668	79,510	17,391	69,888	15,601	68,917	18,435	61,261	24,235
17,829	2,911	21,891	4,069	31,643	5,440	49,427	6,817	39,719	7,952
(b)26,775	(b)2,228	25,540	2,164	32,081	2,660	31,287	3,263	36,094	4,912
22,628	3,340	27,046	4,037	31,710	4,855	24,439	5,702	27,825	6,641
13,434	2,470	13,323	2,594	15,810	3,144	17,047	3,193	13,440	5,028
8,935	2,601	7,977	2,335	8,036	2,355	8,372	2,499	8,460	5,673
(d) 963	(d) 214	992	193	2,388	211	5,610	316	5,295	575
3,548	437	2,691	350	3,534	396	4,439	479	4,473	621
(e)	(e)	(f) 342	(f) 60	(f) 575	(f) 86	(f) 1,223	(f) 144	4,339	156
1,422	491	1,667	615	2,107	753	2,802	580	3,022	666
1	(g)	2	(g)	2	1	11	2	1,827	361
2,580	756	2,261	695	1,716	503	1,712	453	1,667	900
.....	.....	6	3	213	39	334	81	177	62
(h) 61	(h) 27	(i) 89	(i) 36	(i) 77	(i) 32	(i) 60	(i) 25	42	19
.....	.....	.....	.....	.....	.....	.....	.....		
.....	.....	.....	.....	.....	.....	.....	.....	26	3
.....	.....	.....	.....	.....	.....	.....	.....		
18	5	28	8	69	18	31	8	22	6
(e)	(e)	(j)	(j)	(j)	(j)	(j)	(j)	10	3
(c)	(c)	(k)	(k)	(k)	(k)	(k)	(k)	5	2
.....	.....	.....	.....	.....	.....	.....	.....	2	(g)
(g)	(g)	(g)	(g)	(g)	(g)	(g)	(g)	2	(g)
2	1	2	2	1	1	1	1	1	1
(k)	(k)	(k)	(k)	(k)	(k)	(k)	(k)	1	(g)
591,867	94,116	628,579	101,312	753,170	120,227	795,110	142,089	721,001	153,554

(g) Less than one thousand.
(h) Includes North Dakota.
(i) Includes North Dakota and Alabama.
(j) Included in Wyoming.
(k) Included in South Dakota.

Quantity and Value of Natural Gas

(In millions of

	1910		1911		1912		1913	
	Quantity	Value, in thousands	Quantity	Value, in thousands	Quantity	Value, in thousands	Quantity	Value, in thousands
West Virginia.	77,068	$5,618	80,869	$6,240	(a) 95,402	(a) $7,001	(a) 96,645	(a) $7,334
Oklahoma....	26,470	1,911	28,214	2,093	41,550	3,149	51,249	3,741
Pennsylvania.	168,876	23,935	159,104	23,940	173,656	26,486	177,463	28,710
Ohio.........	108,075	21,211	112,123	22,792	126,855	27,196	128,205	27,056
California....	2,764	477	6,390	801	9,355	1,135	11,035	1,884
Louisiana.....	(c) 8,111	(a) 957	(b) 9,786	(d) 858	(d) 14,493	(d) 1,747	(d) 26,653	(d) 2,120
Kansas.......	(f) 83,340	(f) 9,335	(f) 77,861	(f) 9,494	(f) 60,318	(f) 8,522	(f) 46,653	(f) 6,984
Texas........	(h)	(h)	5,503	1,015	7,470	1,405	12,160	2,074
New York....	14,195	3,964	14,894	4,277	16,928	4,867	16,739	4,889
Arkansas.....	(i) 2,705	(i) 301	(i) 2,294	(i) 296	(i) 1,742	(i) 310	(i) 1,106	(i) 269
Illinois.......	6,723	614	6,762	688	(j) 5,603	(j) 617	(j) 4,767	(j) 574
Wyoming.....	(k)	(k)	(k)	(k)	(k)	(k)	(k)	(k)
Kentucky....	4,958	908	4,735	902	5,103	1,071	5,911	1,225
Tennessee....	1	(m)	1	(m)	1	(m)	2	(m)
Indiana......	5,760	1,473	4,365	1,192	(n) 3,618	(n) 1,014	(n) 3,221	(n) 948
Montana.....	....	....	....	....	....	....	....	....
South Dakota.	43	32	26	17	54	(o) 31	(o) 66	(o) 31
Maryland....	....	....	....	....	(q)	(q)	(q)	(q)
Utah.........	....	....	....	....	....	....	....	....
Washington...	....	....	....	....	....	....	....	....
Missouri.....	47	12	50	10	53	12	21	7
Colorado.....	(k)	(k)	(k)	(k)	(k)	(k)	(k)	(k)
Alabama.....	(h)	(h)	(h)	(h)	(h)	(h)	(h)	(h)
Oregon.......	....	....	....	....	....	....	....	....
Iowa.........	(m)	(m)	(m)	(m)	(m)	(m)	(m)	(m)
Michigan.....	1	1	2	1	2	1	2	1
North Dakota.	18	7	14	6	(s)	(s)	(s)	(s)
	509,155	$70,756	512,993	$74,622	562,203	$84,564	581,898	$87,847

(a) Includes gas piped to Maryland.
(b) Includes gas piped to Missouri.
(c) Includes Texas and Alabama.
(d) Includes Alabama, also gas piped to Texas and Arkansas.
(e) Includes gas piped to Texas and Arkansas.
(f) Includes gas piped from Kansas and Oklahoma to Missouri.
(g) Includes gas piped to Missouri.
(h) Included in Louisiana.
(i) Includes Colorado and Wyoming.
(j) Includes gas piped to Indiana.

Consumed in the United States, 1910–1918

cubic feet)

1914		1915		1916		1917		1918	
Quantity	Value, in thousands	Quantity	Value, in thousands	Quantity	Value, in thousands	Quantity	Value, in thousands	Quantity	Value, in thousands
(*a*) 95,147	(*a*)$7,335	(*a*)94,977	(*a*)$7,451	(*a*)105,104	(*a*) $8,610	(*a*)115,488	(*a*)$10,559	108,674	$12,285
(*b*) 55,544	(*b*) 4,227	(*b*)65,691	(*b*) 5,058	93,704	7,062	122,178	10,901	106,661	12,724
164,834	28,440	176,367	30,088	201,461	35,016	202,260	40,774	177,140	44,777
138,389	29,937	146,725	31,901	169,480	37,394	165,782	44,743	143,585	43,694
17,829	2,911	21,891	4,069	31,643	5,440	49,427	6,817	39,719	7,952
(d) 26,775	(*d*) 2,228	(*e*)25,540	(*e*) 2,164	(*e*) 32,081	(*e*) 2,660	(*e*) 31,287	(*e*) 3,263	26,324	2,597
(*g*) 45,251	(*g*) 7,164	(*g*)48,871	(*g*) 8,174	(*g*) 60,564	(*g*) 9,732	(*g*) 37,963	(*g*) 8,464	33,912	7,064
13,434	4,670	13,324	2,594	15,810	3,144	17,902	3,433	20,283	5,519
18,402	5,510	18,776	5,676	20,594	6,201	22,467	6,913	20,341	6,779
(*i*) 963	(*i*) 214	992	193	3,348	287	6,438	307	12,229	1,971
(*j*) 3,548	(*j*) 437	(*j*) 2,691	(*j*) 350	(*j*) 3,534	(*j*) 396	(*j*) 4,439	(*j*) 479	4,473	621
(*k*)	(*k*)	(*l*) 342	(*l*) 60	(*l*) 575	(*l*) 86	(*l*) 1,229	(*l*) 144	4,339	156
7,226	1,787	7,746	1,942	9,888	2,332	12,053	3,114	12,200	3,094
1	(*m*)	2	(*m*)	2	1	11	2	1,827	361
(*n*) 4,443	(*n*) 1,423	(*n*) 4,519	(*n*) 1,543	5,022	1,746	5,766	1,971	4,517	1,511
....	......	6	3	213	39	334	81	177	62
(*o*) 61	(*o*) 27	(*p*) 89	(*p*) 36	(*p*) 77	(*p*) 32	(*p*) 60	(*p*) 25	42	19
(*q*)	(*q*)	(*q*)	(*q*)	(*q*)	(*q*)	(*q*)	(*q*)		3
....	......	......	......	....	....	....	....	} 26	
....	......	......	......	....	....	....	....		
18	5	28	8	69	18	31	8	4,511	2,359
(*k*)	(*k*)	(*r*)	(*r*)	(*r*)	(*r*)	(*r*)	(*r*)	10	3
(*h*)	(*h*)	(*s*)	(*s*)	(*s*)	(*s*)	(*s*)	(*s*)	5	2
....	......	......	......	....	....	....	....	2	(*m*)
(*m*)	(*m*)	(*m*)	(*m*)	(*m*)	(*m*)	(*m*)	(*m*)	2	(*m*)
2	1	2	2	1	1	1	1	1	1
(*s*)	(*s*)	(*s*)	(*s*)	(*s*)	(*s*)	(*s*)	(*s*)	1	(*m*)
591,867	$94,116	628,579	$101,312	753,170	$120,227	795,110	$142,089	721,001	$153,554

(*k*) Included in Arkansas.
(*l*) Includes Colorado.
(*m*) Less than one thousand.
(*n*) Includes gas piped to Chicago, Illinois.
(*o*) Includes North Dakota.
(p) Includes Alabama and North Dakota.
(*q*) Included in West Virginia.
(*r*) Included in Wyoming.
(*s*) Included in South Dakota.

Summary of Statistics of Petroleum for 1921—Production by States

The following preliminary figures to the U. S. Geological Survey, show the quantity of petroleum transported from producing properties. They do not include oil consumed on the leases. This item and net changes in producers' stocks at the beginning and end of the year are obtained by annual canvass and are included in the final-statistics of production.

(Thousands of barrels of 42 U. S. gallons)

State	Jan.	Feb.	March	April	May	June	July	August	Sept.	Oct.	Nov.	Dec.	Jan.-Dec., inc., 1921
California (*a*)	10,177	9,184	10,482	10,170	10,448	10,120	10,247	10,026	7,894	7,065	8,804	10,092	114,709
Oklahoma: Osage County	1,383	1,452	1,782	1,677	1,764	1,723	1,746	1,914	1,901	1,976	2,122	2,152	21,592
Remainder	7,106	6,485	7,837	7,855	8,218	7,970	8,380	8,253	7,814	7,653	7,252	7,563	92,386
Total Oklahoma	8,489	7,937	9,619	9,532	9,982	9,693	10,126	10,167	9,715	9,629	9,374	9,715	113,978
Texas: Central and North	6,419	5,774	6,721	6,178	6,228	5,842	5,523	5,265	4,710	4,846	5,908	7,478	70,892
Coastal	2,610	2,566	2,976	2,757	2,669	2,448	2,557	2,873	2,796	2,730	2,813	2,635	32,430
Total Texas	9,029	8,340	9,697	8,935	8,897	8,290	8,080	8,138	7,506	7,576	8,721	10,113	103,322
Kansas	2,368	2,478	3,086	3,300	3,505	3,472	3,418	3,332	3,005	2,825	2,726	2,717	36,232
Louisiana: Northern	2,531	2,172	2,105	1,959	2,391	2,130	1,998	1,789	1,825	1,820	1,757	2,316	24,793
Coastal	156	156	168	156	160	150	155	138	130	126	115	120	1,730
Total Louisiana	2,687	2,328	2,273	2,115	2,551	2,280	2,153	1,927	1,955	1,946	1,872	2,436	26,523
Wyoming: Salt Creek	804	868	915	1,174	1,185	1,180	669	1,001	826	959	1,049	1,503	12,133
Other districts	658	629	699	739	791	678	474.6	471	538	522.8	455.4	433	7,088.8
Total Wyoming	1,462	1,497	1,614	1,913	1,976	1,858	1,143.6	1,472	1,364	1,481.8	1,504.4	1,936	19,221.8
Arkansas (*b*)	...	...	10	300	550	880	1,400	1,900	1,400	1,300	1,250	1,200	10,190
Illinois	852	785	963	765	863	823	805	846	806	890	837	809	10,044
Kentucky	743.8	695	768.7	767.7	826.3	730	706.6	786	729.9	716.4	717.8	762.4	8,950.6
West Virginia	645	634	760	691	712	631	727	681	613	628	607	674	8,003
Pennsylvania	620	603	675	636	678	615	566	675	602	580	594	590	7,434
Ohio: Central and Eastern	433	427	469	443	466	464	403	444	417	405	409	414	5,194
Northwestern	186	175	193	179	191	191	173	183	172	166	157	173	2,139
Total Ohio	619	602	662	622	657	655	576	627	589	571	566	587	7,333
Montana	78	115	97	96	118	114	116	119	137	153	143	149	1,435
Indiana: Southwestern	71	70	77	76	81	80	77	78	73	70	70	68	891
Northeastern	24	22	24	26	26	25	22	24	21	22	16	20	272
Total Indiana	95	92	101	102	107	105	99	102	94	92	86	88	1,163
New York	84	66	86	84	104	78	79	86	88	77	68	80	980
Colorado	9	9	10	10	10	9	8.4	9	9	8.2	8.6	8	108.2
Tennessee	1.2	1	1.3	1.3	.7	1	1.4	1	1.1	.6	1.2	.6	12.4
	37,959	35,366	40,905	40,040	41,985	40,354	40,252	40,894	36,508	35,539	37,880	41,957	469,639

(*a*) Average of figures reported by Standard Oil Co. and Independent Oil Producers' Agency. (*b*) Estimated in part.

Production by Fields

(Thousands of barrels of 42 U. S. gallons)

Field			Jan.-Dec., inc., 1921 Estimated value at wells
Appalachian	673	2,451	$89,500,000
Lima-Indiana	207	193	5,400,000
Illinois and Southwestern Indiana	924	879	23,000,000
Mid-Continent	22,453	20,655	408,100,000
Gulf Coast	3,011	2,926	39,200,000
Rocky Mountain	1,600	1,510	26,100,000
California	0,026	7,894	162,000,000

Stocks of Petroleum on Last Day of Each Month in 1921

(Thousands of barrels of 42 U. S. gallons)

		Jan.	Feb.	March	April	May	June	July	August	Sept.	Oct.	Nov.	Dec.
A. Domestic Petroleum—													
East of California. Pipe-line and tank-farm stocks: (Source of oil by fields)													
Appalachian:													
N. Y., Pa., W. Va., Eastern and Central Ohio:	Gross	2,705	3,051	3,586	3,760	4,152	4,335	4,424	4,417	4,495	4,602	4,801	4,933
	Net	2,486	2,772	3,307	3,484	3,888	4,061	4,150	4,138	4,229	4,339	4,536	4,670
Kentucky:	Gross	1,024	1,389	1,487	1,447	1,677	1,870	1,867	2,107	2,140	1,973	2,064	2,264
	Net	962	1,326	1,424	1,385	1,614	1,807	1,804	2,044	2,053	1,881	1,971	2,168
Lima-Indiana:	Gross	1,452	1,407	1,519	1,555	1,536	1,575	1,621	1,686	1,548	1,536	1,501	1,386
	Net	1,040	1,010	1,121	1,156	1,121	1,156	1,201	1,264	1,110	1,158	1,136	1,014
Illinois-S.W. Indiana:	Gross	3,717	4,239	4,860	5,014	5,554	6,052	6,566	6,887	6,951	7,227	7,622	8,066
	Net	3,189	3,710	4,332	4,490	5,030	5,526	6,075	6,396	6,473	6,733	7,122	7,567
Mid-Continent:													
Oklahoma, Kansas, Central and North Texas:	Gross	78,532	80,322	84,135	87,445	90,938	94,436	98,518	99,713	100,107	99,732	99,293	102,012
	Net	68,651	70,153	73,910	77,369	81,086	84,319	88,427	89,591	90,140	90,194	89,317	91,983
North Louisiana and Arkansas:	Gross	8,129	8,145	8,070	8,065	8,558	9,379	10,096	10,298	10,163	9,766	9,963	10,912
	Net	7,535	7,686	7,652	7,606	8,090	8,902	9,485	9,765	9,650	9,199	9,493	10,418
Gulf Coast:	Gross	12,701	14,200	14,805	15,751	16,645	17,900	17,975	17,886	18,702	18,706	18,795	17,973
	Net	11,248	12,730	13,386	14,413	15,295	16,340	16,476	16,360	17,150	17,159	17,422	17,606
Rocky Mountain:	Gross	740	773	735	851	899	885	1,260	1,013	1,120	1,111	1,374	1,647
	Net	727	760	726	842	888	874	1,247	1,007	1,113	1,098	1,360	1,635
Total pipe-line and tank-farm stocks east of California	Gross	109,060	113,526	119,197	123,888	129,959	136,432	142,327	144,007	145,226	144,653	145,413	149,193
	Net	95,838	100,147	105,858	110,745	117,012	122,985	128,865	130,565	131,918	131,761	132,357	137,061
California: Gross pipe-line, tank-farm and producers' stocks: (a)	....	21,261	21,566	22,896	23,974	26,602	28,354	30,165	32,338	32,158	31,574	31,716	33,289
B. Mexican Petroleum Held in the U. S. by Importers:													
At Atlantic Coast stations:	Crude	3,501	4,245	3,957	3,671	3,363	2,694	2,426	1,654	2,914	4,186	5,561	5,712
	Topped	775	905	1,155	1,420	1,564	1,131	1,055	583	914	1,182	1,896	1,599
At Gulf Coast stations:	Crude	2,348	3,162	3,637	4,458	4,241	4,496	3,573	2,455	2,598	2,564	3,523	5,413
	Topped	533	859	678	748	882	1,243	1,123	581	859	1,340	1,437	816
At Pacific Coast stations:	Crude	......	......	......	......	150	145	145	14				
Total Mexican petroleum	....	7,157	9,171	9,427	10,297	10,200	9,709	8,322	5,287	7,285	9,272	12,417	13,540
Total of net pipe-line and tank-farm stocks east of California, gross pipe-line, tank-farm and producers' stocks in California, and stocks of Mexican petroleum held in the United States by importers		124,256	130,884	138,181	145,016	153,814	161,048	167,352	168,190	171,361	172,607	176,490	183,890

(a) Average of figures reported by the Standard Oil Co. and the Independent Oil Producers' Agency.

Imports and Exports of "Mineral Crude Oil" (a)

(Thousands of barrels of 42 U. S. gallons)

	Jan.	Feb.	March	April	May	June	July	August	Sept.	Oct.	Nov.	Dec.	Jan.-Dec., inc., 1921
Imports into U. S.:													
From Mexico	13,192	11,384	12,303	10,104	9,148	10,255	8,047	3,352	9,094	11,635	12,986	13,684	125,184
From other countries	1	......	......	......	......	......	......	......	45	......	8	69	123
	13,193	11,384	12,303	10,104	9,148	10,255	8,047	3,352	1,939	11,635	12,994	13,753	125,307
Ports of entry:													
Atlantic Coast	7,195	5,975	6,444	4,679	4,428	4,907	4,239	657	4,853	7,319	7,173	7,159	65,028
Gulf Coast	5,687	5,270	5,529	5,109	4,217	5,142	3,725	2,695	4,196	4,316	5,591	6,361	57,838
Pacific Coast	242	139	43	208	367	143	83	......	57	......	70	65	1,417
Northern Border	1	......	......	......	......	......	......	......	......	......	......	......	1
Mexican Border	68	......	287	108	136	63	......	......	33	......	160	168	1,023
	13,193	11,384	12,303	10,104	9,148	10,255	8,047	3,352	9,139	11,635	12,994	13,753	125,307
Exports from U. S.:													
Domestic crude oil:													
To Canada	580	552	651	676	716	454	437	748	728	577	602	446	7,167
To other countries	155	217	78	67	150	132	101	137	153	170	262	77	1,699
Foreign crude oil	8	25	21	5	8	......	......	......	......	......	5	2	74
	743	794	750	748	874	586	538	885	881	747	869	525	8,940
Excess of imports over exports	12,450	10,590	11,553	9,356	8,274	9,669	7,509	2,467	8,258	10,888	12,125	13,228	116,367

(a) Compiled from records of the Bureau of Foreign and Domestic Commerce; includes oil topped in Mexico.

Estimated Deliveries of Domestic Petroleum by Pipe-line and other Marketing Companies

(Production plus domestic stocks at beginning of month minus domestic stocks at end of month)

(Thousands of barrels of 42 U. S. gallons)

Source of oil by fields	Jan.	Feb.	March	April	May	June	July	August	Sept.	Oct.	Nov.	Dec.	Jan.-Dec., inc., 1921
Appalachian	2,506	1,776	2,127	2,485	2,154	2,153	2,387	2,445	2,351	2,469	2,110	2,190	27,163
Lima-Indiana	208	227	106	170	252	181	150	144	347	140	195	315	2,435
Illinois and Southwestern Indiana	699	334	418	683	404	407	333	603	802	700	518	432	6,333
Mid-Continent	19,311	16,708	17,818	17,856	18,455	17,972	17,774	21,009	20,221	20,817	21,598	19,835	229,374
Gulf Coast	2,681	1,240	2,488	1,886	1,947	1,553	2,576	3,127	2,136	2,847	2,665	2,571	27,717
Rocky Mountain	1,458	1,588	1,755	1,903	2,058	1,995	895	1,840	1,404	1,658	1,394	1,818	19,766
California	9,846	8,879	9,152	9,092	7,820	8,368	8,436	7,853	8,074	7,649	8,662	8,519	102,350
	36,709	30,752	33,864	34,075	33,090	32,629	32,561	37,021	35,335	36,280	37,142	35,680	415,138

Estimated Consumption (Deliveries to Consumers) of Domestic and Imported Petroleum

(Domestic production and excess of imports over exports plus decrease or minus increase of stocks)

(Thousands of barrels of 42 U. S. gallons)

TRANSPORTATION OF PETROLEUM PRODUCTS

BY

FORREST M. TOWL[1]

THE earliest form of petroleum transportation was the handling of the commodity in packages. The first commercial packages transported were probably bottles and contained petroleum under such names as "Seneca Oil," "American Medical Oil" or "Kier's Rock Oil." S. M. Kier, in his imitation bank-note advertisement under date of January 1, 1852, says, "The oil is pumped up with the salt water; flows into the cistern, floats on top; when a quantity accumulates, is drawn off into barrels, is bottled . . ." This gives an idea of the small scale on which the transportation of these products was begun.

Transportation methods developing as the industry increased can be divided into the following classes: packages, tank wagons and tank cars, boats and pipe-lines. These will be discussed in the order named.

Package transportation.—The transportaion of packages containing petroleum or its products is dangerous, on account of the inflammable nature of the substance, and should be conducted with great care and with consideration of the likelihood of damage to other products carried in the same vehicle.

Packages should conform to the regulations of the Interstate Commerce Commission, The American Railroad Association rules, the regulations of the Bureau of Explosives and the regulations of the carriers. Those intending to ship petroleum products should procure the pamphlet of the Bureau of Explosives, 30 Vesey Street, New York, N. Y., containing "Specifications for shipping containers." Inquiry should be made as to foreign regulations before arranging to prepare packages for shipment abroad. The manufacture of packages and the specifications for the same will be treated in another portion of the handbook.

Tank wagons are a feature of the marketing, rather than of the transportation of oil products and therefore will not be considered in this chapter on transportation.

RAILROAD TRANSPORTATION

History.—With the bringing in of the first oil wells, it became necessary to provide transportation for the crude oil, in bulk, from the wells to the refineries. The first tank car was then improvised, two large wooden tubs or vats being mounted on a railroad flat-car. An iron tank, having a capacity of about 3000 gallons, was later substituted for the tubs, and from this developed the modern all-steel tank car.

Petroleum products are also transported by railroad in containers and packages. In general, these are subject to regulations similar to those in reference to tank cars.

[1] Mr. E. C. Sicardi, vice-president of the Union Tank Car Company, is mainly responsible for the article on Railroad Transportation. In the discussion of the Application of Theory, and in the checking up of tables, formulas, etc., great assistance has been given by Mr. J. H. Peper, M.E.; Mr. Daniel C. Towl, C.E.; Mr. E. W. Linscott, M.E.; and Mr. T. C. Towl, C.E.

Petroleum Tank Cars

A tank car is a metal cylindrical tank mounted on underframe and trucks. Tank cars vary in capacity from 6000 to 13,000 gallons. The car is a container as well as a vehicle of transportation.

There is a manhole in the dome through which cars are loaded from an overhead standpipe, and there is also a discharge valve at the bottom of the tank immediately beneath the dome opening through which the contents are unloaded. In some cases, contents are unloaded by siphoning through the dome opening. (See Fig. 1.)

Types of cars.—(1) *Cars without heater pipes.*—These cars are used for shipment of such products as refined oil, gasoline, and the more fluid lubricating oils.

(2) *Heater-pipe cars.*—Heater pipes are applied to the interior of tanks for use in heating viscous crude petroleum and such products as fuel oils, waxes and asphalts

Fig. 1.—A modern tank car.

with various melting points and the viscous products at destination, so as to facilitate unloading.

(3) *Compartment cars.*—As the name indicates, the tank is divided into two or three compartments so as to admit of the shipping of two or three different grades of petroleum.

(4) *Insulated cars.*—The tanks of these cars are covered with an insulating material such as hair felt, cork or magnesia, with an additional sheathing of metal outside, so as to reduce evaporation to a minimum. The Bureau for the Safe Transportation of Explosives and Other Dangerous Articles by Freight and Express requires the use of insulated cars for shipment of casing-head or compression gasolines within prescribed vapor tension limits.

Interchangeability.—Generally speaking, a tank car used for one grade of petroleum is not suitable, without cleaning, for transporting another grade. While cars previously used for lighter-colored grades of petroleum may be used for darker-colored grades without cleaning, cars used for the darker grades cannot be used for the lighter-colored grades without cleaning. In many cases it is also necessary in order to avoid contamination, to clean between petroleum products of about the same color but of different characteristics. There are some twenty or more grades of petroleum between which cars cannot be interchanged without cleaning.

Construction.—Tank cars must conform to specifications prescribed and published by the American Railroad Association, Section III, Mechanical, which include U. S. Safety Appliance Regulations. Tank cars not conforming to these specifications are not accepted by railroads for transportation.

Tank ears are divided into various classes according to the time when they were built and the service required of them. To give an idea of the general requirements for such cars, a copy of the specifications for a tank of a class-III tank car (built after May 1, 1917) which cover tank cars for general service, is here given with a few omissions and slight changes.

Specifications for Class-III Tank Car

(Built after May 1, 1917)

Tank

1. **Bursting pressure.**—The calculated bursting pressure, based on the lowest tensile strength of the plate or casting and the efficiency of the seam or attachment, shall not be less than 300 pounds per square inch.

2. **Material.**—(*a*) For cars built after April 1, 1920, all plates for tank and dome shall be of steel complying with the American Society for Testing Materials specifications for boiler plate steel, flange quality.

The use of material other than steel will be permitted where desirable because of the nature of the product to be transported, provided that the tank shall be subject to the same requirement as to bursting pressure, and that the thicknesses of the plates shall be increased to correspond to those specified in the case of steel, based upon the lowest tensile strength of the material, and be approved by the American Railroad Association.

(*b*) Rivets shall comply with the American Railroad Association Specifications for rivet steel and rivets for passenger and freight equipment cars.

(*c*) Minimum thicknesses (thickness at rivet seam) of plates shall be as follows:

Diameter of tanks	Bottom sheet, inch	Shell sheet, inch	Dome sheet, inch	Tank heads, inch	Dome heads, inch
60 inches or under.......	$\frac{7}{16}$	$\frac{1}{4}$	$\frac{5}{16}$	$\frac{1}{2}$	$\frac{5}{16}$
Over 60 inches to 78 inches	$\frac{7}{16}$	$\frac{5}{16}$	$\frac{5}{16}$	$\frac{1}{2}$	$\frac{5}{16}$
Over 78 inches to 96 inches	$\frac{1}{2}$	$\frac{3}{8}$	$\frac{5}{16}$	$\frac{1}{2}$	$\frac{5}{16}$

For tanks without underframes the minimum thickness of the bottom sheet shall be $\frac{5}{8}$ inch.

For tanks built in rings the thickness specified for the bottom sheet shall apply to the entire cylindrical shell.

For car with underframe the minimum width of the bottom sheet of the tank shall be 60 inches, but in all cases the width shall be sufficient to bring the entire width of the longitudinal seam, including overlaps, above the cradle.

3. **Riveting.**—(*a*) All seams shall be double riveted, with the exception of the dome head seam, which may be single riveted.

Seams shall be riveted metal to metal, without the interposition of other material.

(*b*) For double riveting the size and pitch of rivets shall be as follows:

Thickness of plate, inch	Diameter of rivet, inch	Longitudinal pitch, inches	Back pitch, inches
$\frac{1}{4}$	$\frac{5}{8}$	$2\frac{1}{2}$ to $2\frac{3}{4}$	$1\frac{1}{2}$ to $1\frac{3}{4}$
$\frac{5}{16}$	$\frac{5}{8}$ / $\frac{3}{4}$	$2\frac{1}{2}$ to $2\frac{3}{4}$ / See Note	$1\frac{1}{2}$ to $1\frac{3}{4}$
$\frac{3}{8}$	$\frac{3}{4}$	$2\frac{3}{4}$ to 3	$1\frac{3}{4}$ to 2

NOTE.—For longitudinal seams with $\frac{3}{4}$-inch rivets, $2\frac{3}{4}$ inches longitudinal pitch is necessary to insure 70 per cent strength of seam.

The efficiency of the seam shall be not less than 70 per cent of the strength of the thinnest plate.

For single riveting the size and pitch of rivets shall be as follows:

Thickness of plate, inch	Diameter of rivet, inch	Pitch, inches
$\frac{5}{16}$	$\frac{5}{8}$	$1\frac{3}{4}$ to 2
$\frac{3}{8}$	$\frac{3}{4}$	$1\frac{7}{8}$ to $2\frac{1}{8}$

(*c*) The extreme caulking edge distance, measured from center line of rivet hole, shall not be less than one and one-half times the diameter of the hole, and not more than one and one-half times the diameter of the hole plus $\frac{1}{4}$ inch.

The angle of the caulking edges shall be between 60 and 70° with the flat surface of the plate.

4. **Caulking.**—Seams shall be caulked both inside and outside. The purpose of the inside caulking is to prevent access of contents of tank to the seams. Caulking may be done by the electric welding process.

Split caulking shall not be permitted.

5. **Tank heads.**—Tank heads shall be dished to a radius of 10 feet for pressure on concave side.

In the case of compartment cars each compartment shall have two heads, convex outward.

6. **Dome.**—(*a*) The tank shall have a dome with a minimum capacity, measured from the inside top of shell of tank to the top of dome, of not less than 2 per cent of the total capacity of the tank, that is, the shell and dome capacity combined.

(*b*) The dome head shall be of steel plate, or the dome head and ring in one casting may be of cast steel.

The dome ring and cover shall be of cast or pressed steel or of malleable iron.

If of steel plate, the dome head shall be dished to a radius of 8 feet.

The opening in the shell for the dome shall not exceed 22 by 28 inches.

(c) The dome cover shall be secured either by screw joint, or by bolting, or by yoke with center screw.

If a screwed dome cover is used, the depth of inside ring of cover shall not be less than 2½ inches.

The joint of the dome cover shall be made tight against vapor pressure, and when necessary to secure this a satisfactory gasket shall be used.

For cars to be used for the transportation of inflammable liquids with flash points below 20° F. the mechanical arrangement for closing the dome cover shall either be such as to make it practically impossible to remove the dome cover while the interior of the car is subjected to pressure, or suitable vents shall be provided that will be opened automatically by starting the operation of removing the dome cover. Other methods may be used if approved by the American Railroad Association.

7. Bottom outlet valve.—(*a*) If the tank is provided with a bottom outlet valve, the outlet-valve casting shall be so designed that breakage will not unseat the valve. The preferable construction is to have the outlet-valve casting scored to confine the breakage at the scoring.

The bottom of the main portion of the outlet-valve casting, or some fixed attachment thereto, shall have external " V " threads 5¼ inches in diameter, and a pitch of four threads to the inch. Additional attachments thereto, having threads of other dimensions, may be used.

Where a 6-inch bottom outlet valve is used, the bottom outlet-valve casting shall be designed to have a diameter of 8 inches over threads, and a pitch of four threads to the inch, in addition to connections as above.

Cars used for the transportation of acids or other corrosive substances, if fitted with bottom outlet-valve castings, need not have threads as above, but may be designed for the use of a bolted cover, to insure a tight joint. A suitable gasket shall be used when necessary.

Bottom outlet-valve castings when applied to tanks having center sills shall be of such length that the threaded end of the casting will project below the bottom face of the sills sufficiently to facilitate the application and removal of caps and other attachments.

All outlet-valve casting caps and attachments shall be secured to the car to prevent loss.

(*b*) Where the tank is anchored rigidly to the underframe there shall be a longitudinal clearance of not less than 3 inches on each side of the bottom outlet-valve casting, Where the anchorage used provides for a limited longitudinal movement of the tank, there shall be a clearance of not less than 3 inches on each side of the bottom outlet-valve casting when the tank has reached the limit of longitudinal movement provided for by such anchorage. There shall be a transverse clearance of not less than ½ inch on each side of the bottom outlet-valve casting.

(*c*) If a bottom-outlet-valve is used, it is preferred that the handle shall be within the tank or dome, but in the event that the rod is carried through the dome, leaking shall be prevented by packing and cap nut.

The specification of the other parts of the car is not given.

Safety regulations.—The American Railroad Association has established a bureau called " The Bureau for the Safe Transportation of Explosives and Other Dangerous Articles by Freight and Express." This bureau formulates and publishes rules and regulations designed to protect life and property by preventing the escape of gases or leakage of inflammable liquids. These rules and regulations have the endorsement of the Interstate Commerce Commission and can be amended only by recommendation of that body.

As crude petroleum and most of the products derived from it are dangerous to handle, such material should not be offered for shipment over the lines of any carrier without notifying the carrier as to what is being shipped, and finding out what regulations apply.

Tank cars used for the shipment of dangerous articles other than explosives must comply with the Master Car Builders' specifications, and a tank car that leaks or one that has any defect which would make leakage during transit probable or that has not been tested and stenciled in compliance with the Master Car Builders' specifications must not be used for the shipment of any inflammable liquid.

The tanks and their fittings must be examined by the shipper to see that they are in proper condition for loading. Tanks must be examined for evidence of previous leaks; safety and outlet valves, dome covers, and outlet-valve caps must be in proper condition before loading; tanks must be loaded with outlet-valve caps off; after loading, tanks must not show any dropping of liquid contents at the seams or rivets, and should such dropping appear cars must be properly repaired by caulking; outlet valves must not permit more than a dropping of the liquid with valve caps off, otherwise the valve must be reground and repaired. Dome covers, and valve caps provided with suitable gaskets, must be properly screwed in place before cars are tendered to the carrier.

Loaded tank cars tendered for shipment must be inspected by the carrier to see that they are not leaking; that the air and hand brakes, journal boxes, trucks, and safety appliances are in proper condition for service; and that the car has been tested within limits prescribed by the Master Car Builders' specifications.

Tests of all tank cars and their safety valves, as made in compliance with the Master Car Builders' specifications, must be certified by the person making the tests to the owner of the tank car and to the chief inspector, of the Bureau of Explosives, and this certification must show the initials and number of the tank car, the service for which it is suitable, the date of test, place of test, and by whom made.

All inflammable liquids must be shipped in packages complying with specifications that apply.

They may be shipped in tank cars complying with the Master Car Builders' specifications, provided the vapor tension of the inflammable liquid corresponding to a temperature of 100° F. does not exceed 10 pounds per square inch. A tank car must not be used for shipping inflammable liquids with flash point lower than 20° F., unless it has been tested with cold-water pressure of 60 pounds per square inch and stenciled as required by Master Car Builders' specifications, and is equipped with safety valves set to operate at 25 pounds per square inch, and with mechanical arrangement for closing dome cover.

Liquid condensates from natural gas or from easing-head gas of oil wells, made either by the compression or absorption process, alone or blended with other petroleum products, must be described as liquefied petroleum gas when the vapor pressure[1] at 100° F. (90° F. November 1 to March 1) exceeds 10 pounds per square inch.

When the liquid condensate alone or blended with other petroleum products has a vapor pressure not exceeding 10 pounds per square inch, it must be described and shipped as gasoline or casing-head gasoline.

Liquefied petroleum gas of vapor pressure exceeding 10 pounds per square inch and not exceeding 15 pounds per square inch, from April 1 to October 1 and 20 pounds per square inch from October 1 to April 1, must be shipped in metal drums or barrels which comply with Shipping Container Specification No. 5; or in special insulated tank cars approved for this service by the Master Car Builders' Association.

Liquefied petroleum gas of vapor pressure exceeding 15 or 20 pounds per square

[1] In measuring the vapor pressure the container may be vented momentarily at a temperature of 70° F. See also chapter on testing methods.

inch as provided herein, and not exceeding 25 pounds per square inch, must be shipped only in metal drums or barrels which comply with shipping container specification.

Liquefied petroleum gas of vapor pressure exceeding 25 pounds per square inch must be shipped in cylinders as prescribed for compressed gases.

When the liquid condensate, alone or blended with other petroleum products, has a vapor pressure not exceeding 10 pounds per square inch it must be described as gasoline or casing-head gasoline and must be shipped in metal drums or barrels complying with the specification; or in ordinary tank cars, 60 pounds test class, equipped with mechanical arrangement for closing of dome covers as specified in the Master Car Builders' specifications for tank cars.

Every tank car containing liquid condensates, either blended or unblended, including liquefied petroleum gas, as defined herein, must have safety valves set to operate at 25 pounds per square inch with a tolerance of 3 pounds above or below, and the mechanical arrangements for closing the dome covers of such cars must either be such as to make it practically impossible to remove the dome cover while the interior of the car is subjected to pressure; or suitable vents that will be opened automatically by starting the operation of removing the dome cover must be provided.

The shipper must attach securely and conspicuously to the dome and dome cover three special white dome placards measuring 4 by 10 inches, bearing the following wording:

10 Inches

4 Inches

CAUTION

AVOID ACCIDENTS

Do Not Remove This Dome Cover While Gas Pressure Exists in Tank

KEEP LIGHTED LANTERNS AWAY

One placard must be attached to each side of the dome and one placard must be attached to the dome cover. The presence of these special dome placards must be noted on the shipping order by the shipper and by the carrier on the billing accompanying the car. Placards must conform to samples furnished by the chief inspector of the bureau of explosives.

Packages containing inflammable liquids must not be entirely filled. Sufficient interior space must be left vacant to prevent leakage or distortion of containers, due to increase of temperature during transit. In all such packages this vacant space must not be less than 2 per cent of the total capacity of the container. In tank cars the vacant space must not be less than 2 per cent [1] of the total capacity of the tank, i.e.,

[1] An outage of 2 per cent is frequently insufficient for light petroleum products, owing to the fact that they expand more than heavier petroleum products when the temperature increases, and this rate of expansion varies with the specific gravity of the material. It is recommended that when tank

the shell and dome capacity, combined. If the dome of a tank car does not provide this 2 per cent, sufficient vacant space must be left in the shell of the tank to make up the difference.

Tank cars supplied by shippers.—Railroads do not furnish tank cars to shippers. They are owned or leased by the shipper and used by the shipper in the distribution of his products. Thus he is able to protect himself by keeping for his exclusive service, a sufficient number of tank cars suitable in every respect for his business; and in controlling his own equipment he is assured of a regular car supply. The tank car being a container, the shipper also safeguards the liquid commodity which he loads, against the contamination which would occur if he used cars that were indiscriminately used by other shippers. He knows what his car was last loaded with and he knows what it is in a condition to receive on its next loading. This subject was very carefully considered by the Interstate Commerce Commission in their investigation of the whole private car situation, and the private ownership and control of tank cars received their full concurrence in the decision handed down by them.[1]

Railroad operating rules.—The carriers make a mileage allowance to the shipper for furnishing the car and also move the car empty without charge at the time of movement, under an agreement with the car owner that the number of miles the carriers haul the car empty will be equalized with an equivalent number of loaded miles within a specified period. The excess empty mileage not so equalized is paid for by the car owner at current tariff rates. This is covered by rules in the carriers' published tariffs and filed with the Interstate Commerce Commission.

Tank cars delayed in loading and unloading after placement are subject to demurrage charges in accordance with the carriers' tariffs.

Method of handling by carriers.—The exact method of handling a private car from the time it is ordered placed for loading until returned to the owner is as follows:

The shipper gives the carrier an order to place his private car, which is then standing on his tracks in his yards, at the point at which the shipper wishes to load. It is then moved by the carrier's engine and placed for loading. The car is loaded by the shipper and a bill of lading is then taken out by the shipper; the shipper indicating the name of the consignee, destination and the route through to destination. The car is then taken by the carrier's engine to their classification yard, is put into a train and moved through to destination in accordance with the instructions of the shipper as to routing as provided by the bill of lading. When the car reaches its destination, it is delivered to the consignee and unloaded by him. The consignee then notifies the railroad agent that the car is empty and in accordance with instructions the consignee has

cars are loaded with gasoline or casing-head gasoline, the outage in the tank shall not be less than the following:

Temperature of product when loaded	Minimum Outage Required when Baumé Gravity is—		
	50–60°	60–70°	70–80°
	Per cent	Per cent	Per cent
0– 60° F.	3.2	3.5	4.1
61– 70° F.	2.5	2.8	3.3
71– 80° F.	2.0	2.1	2.4
81–100° F.	2.0	2.0	2.0

[1] See I. C. C. Docket 4906, "In The Matter of Private Cars."

received from the shipper, he issues written instructions to the railroad agent to forward the car home to the owner or to some other point via a given route. The car is then moved under these instructions on a memorandum or card waybill and when it reaches its destination is delivered back to the tracks of the owner or is again placed for loading as may be required.

Duties of the shipper.—The shipper must ascertain that the tank, fittings and containers are in proper condition before they are shipped. Tanks should be examined for evidence of previous leaks; safety and outlet valves, dome covers and outlet-valve caps must be in proper condition before loading. Tanks must be loaded with outlet-valve cap off. After loading, tanks must not show any dropping of liquid contents at seams or rivets. Should such leakage occur, the car should be properly repaired. Dome covers and valve caps should be provided with suitable gaskets and must be properly screwed in place before the cars are offered for shipment. Loaded tank cars tendered for shpiment must be inspected by the carrier to see that they are not leaking. The carrier should also see that all other appliances are in proper condition for service and that the car has been tested within the limits prescribed by the Master Car Builders' specifications. Crude oils and the various products derived from them vary in nature and are subject to requirements varying with the nature of the product.

Unloading tank cars.—In unloading tank cars the following rules should be observed:

The dome cover should be unscrewed by placing a bar between the dome-cover lug and the knob; the valve-rod handle in the dome should be moved back and forth a few times to ascertain if the valve is properly seated, and if seated, the valve cap should then be removed with a suitable wrench, having a pail to catch any liquid that may be in the valve nozzle.

The unloading connection should be securely attached to the valve nozzle, and the valve should then be raised by working the valve-rod handle. *The dome cover should be placed over the dome opening, resting on a piece of wood, to allow air to enter the tank. The dome cover should not be replaced while unloading, as this action may result in collapse of the tank.* After the tank is unloaded the valve should be seated, valve cap and dome cover replaced. "Inflammable" placards must not be removed.

When necessary to unload tank cars from the dome, or when necessary to transfer the contents of one tank car through the outlet valve into the dome of another tank car, care should be observed to see that all of the connections are tight and that the pipe or hose, when inserted into the open manhole for pumping or filling purposes, is surrounded by wet burlap to prevent the escape of vapors and to avoid igniting them.

When the "blowing" of safety valves of a car containing inflammable liquids is noted, any available means for cooling the car shell and contents, such as spraying with water, should be utilized; and if practicable the car should be moved to an isolated point, to minimize the fire risk. Covering the safety valves with wet cloth, wet blankets, or wet gunny sacks will decrease the danger of igniting vapors escaping from a "blowing" valve. The burning of these vapors at the safety valve is not liable to cause an explosion. The valves are designed to permit, in emergencies, the burning in this way of the entire contents of the car.

Inflammable placard.—A placard of diamond shape, printed on strong, thin, white paper for pasting on tank cars, and on a strong tag board for tacking to wooden cars or to wooden boards of suitable size attached for this purpose to metal box cars or tank cars, measuring $10\frac{3}{4}$ inches on each side, and bearing in red and black letters the following inscription, must be securely attached to each outside end and to each side door of a box or stock car containing one or more packages protected by the *red* or the *yellow* diamond label, and to each side and end of a tank car containing an inflammable liquid:

Note.—Cars containing cylinders of compressed noninflammable gases (green label) do not require placards.

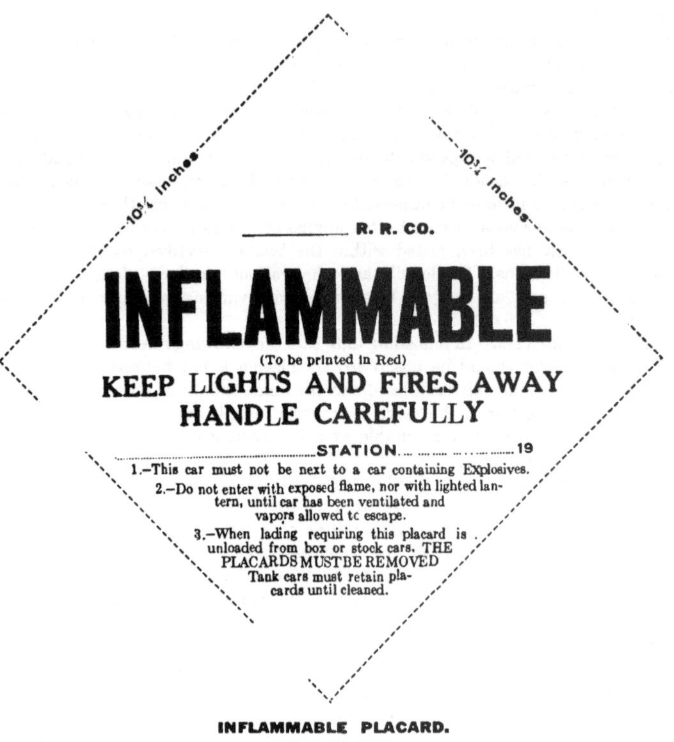

INFLAMMABLE PLACARD.
(REDUCED SIZE)

Placards must conform to standards, and samples will be furnished by the Chief Inspector of the Bureau of Explosives on request.

In case of a wreck.—In case of a wreck involving a car containing inflammable freight it should be assumed that packages are broken and that leakage has occurred which may cause fire if lighted lanterns or other flames are taken into or near these cars. As much of the train as possible should be moved to a place of safety. A car containing inflammable freight should be opened for ventilation and packages protected by red labels and cylinders of compressed gases should be removed to a safe place. Substances spilled from broken packages protected by yellow labels should also be carefully removed. Cylinders of compressed gases may be exploded if they are exposed to fire or struck a sharp blow, and the flying fragments are then dangerous. Inflammable liquids spilled from broken packages or tank cars should be well covered with dry earth before a lighted lantern, torch, or an engine is used in the vicinity.

Acids spilled in cars should be covered with dry earth and the car floor should be thoroughly swept.

Leaking tank cars.—Action in any particular case will depend upon existing conditions, and good judgment will be necessary to avoid disastrous fires on the one hand and the useless sacrifice of valuable property on the other.

Volatile (or combustible) liquids, such as gasoline, naphtha, etc., in large quantity and spread over a large surface will form vapors that will ignite at a considerable distance, depending on the kind and quantity of liquid and the direction and force of the wind. Many of the liquids, regarded as safe to carry under ordinary conditions and to transport in tank cars without the inflammable placard, should still be treated as dangerous in handling a wreck.

When oil cars are leaking all lights or fires near them that can possibly be dispensed with should be extinguished or removed. Incandescent electric lights or portable electric flash lights should be used when available. Whenever practicable the work of handling a wrecked oil car should be done during daylight.

Lanterns necessarily used for signaling should be kept on the side from which the wind is blowing and at as high an elevation as can be obtained. The vapors will go with the wind but not against it. The ash pan and firebox of a locomotive or steam derrick, especially on the side of a wrecked or leaking tank car toward which the wind is blowing, is a source of danger. Wrecks involving oil cars should in no case be approached with lighted pipes, cigars, or cigarettes.

Effort should be made to prevent the spread of oil over a large surface by collecting it in any available vessels or draining it into a hole or depression at a safe distance from the track. When necessary, trenches should be dug for this purpose.

It is not safe to drain inflammable oil in large quantities into a sewer, since vapors may thus be carried to distant points and there ignited. Care should be exercised also not to permit oil to drain into streams of water which may be used by irrigation plants or for watering stock. Dry earth spread over spilled oil will decrease the rate of evaporation and the danger. A stream of oil on the ground should be dammed and dry earth be thrown on the liquid as it collects.

Sudden shocks or jars that might produce sparks or friction should be avoided. When possible, jack the wrecked cars carefully into position after removing other cars and freight that might be injured by fire. Only as a last resort, to meet an emergency, should a wrecked tank car be moved by dragging, and when this is done all persons should be kept at a safe distance.

No unnecessary attempt should be made to transport a damaged tank car from which inflammable liquid is leaking. Safety in short movements may be secured by attaching a vessel under small leaks to prevent spread of inflammable liquid over tracks. Cover the tracks at intervals in the rear of a moving car with fresh earth to prevent fire overtaking the car. Keep engines away; also spectators who may be smoking. If wrecked or derailed, and not in a position to obstruct or endanger traffic, the car should have its leak stopped as far as possible and be left under guard until another tank car or sufficient vessels can be provided for the transfer of the liquid, which should be transferred by pumping when practicable.

Even a tank that is not leaking is liable to be ruptured by the use of slings, and slipping of chain slings may produce sparks. Saving the contents of the tank is not as important as the prevention of fire.

An empty or partially empty tank car, with or without placards, is very liable to contain explosive gases, and lights must be not brought near it.

If the fire can not be smothered by use of earth, steam, or wet blankets, effort should be concentrated on confining it and saving other property.

Should a leak occur by the breakage or displacement of the unloading valve and pipe at the bottom of the tank car, it may be stopped by removing the dome cap on the top of the tank and dropping the plunger into the plunger seat, as a shock sufficient to damage the outlet valve and pipes may also have unseated the plunger.

The dome cover should be unscrewed by placing a bar between the dome-cover lug and knob. The dome cover should not be hammered, and should not be unscrewed until the absence of vapor pressure in the tank is verified by lifting the safety valve.

To ascertain whether the valve is properly seated, the valve-rod handle in the dome should be moved back and forth a few times. The following drawing indicates the general plan covering the valve rod and the unloading or discharge valve.

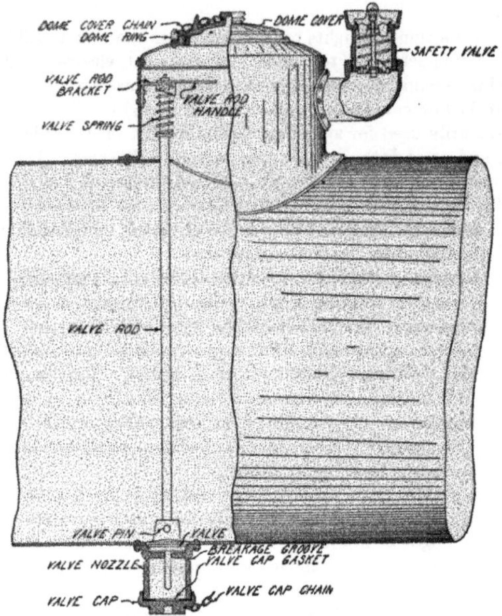

Fig. 2.—General arrangement of tank-car connections.

BOAT TRANSPORTATION

BY

N. J. PLUYMERT

Petroleum shipments by boat are either handled in packages or in bulk. See "Package transportation," page 359. In places where water transportation is available, the oil is often handled in barges. These barges vary from the small, simply built, wooden boats, often used in the oil-fields, to the more complicated steel barges approaching in size and construction to the tankers. As the tank steamer is the most prominent vehicle for bulk transportation, only this type will be considered here. It involves the same principles of construction and is liable to the same dangers as other bulk boats.

THE TANK STEAMER

History

Prior to 1886 all petroleum products were carried overseas by ordinary cargo vessels, in packages, cases or barrels, whereby much space was lost and heavy cost incurred. As early as 1870, ships were built in England for the conveyance of oil in bulk, but were never tried. Subsequently, some ordinary cargo vessels were fitted with tanks in the holds, but these were only partially successful. The first steamer which successfully carried petroleum in bulk was the *Gluckauf*, which was built in England in 1886.

Up to the year 1900, the world's tonnage of seagoing tank vessels reported by Lloyd's Register comprised 193 vessels with a total of 637,014 deadweight tons. From 1900 on, rapid strides were made in tank vessel building, and by 1914 the world's tonnage had grown to 441 vessels with a total of 2,343,877 deadweight tons. The greatest development in tank tonnage occurred during the World War, because of the tremendous demand for fuel oil for the navies and for refined products for the various war appliances. Lloyd's 1920 Register gives the total tank tonnage as 780 vessels with a total of 5,053,242 deadweight tons. On January 1, 1921, in the United States alone, 105 tank vessels with a total of 1,216,340 deadweight tons were being built; in other countries, 94 tank vessels with a total of 895,800 deadweight tons were in process of construction.

Up to the year 1908, all tankers were built on the transverse framing system, for which rules are prescribed by the ship classification societies. In that year the *Paul Paix* was built on the Isherwood system of longitudinal framing. The longitudinal system has now been universally adopted for the construction of tank vessels, and very few tankers are built on the transverse system.

General Characteristics

Vessels for carrying oil in bulk do not differ in general appearance and form of model from the conventional type of general cargo-carrying vessels. (For theoretical and practical naval architecture and marine engineering, see the list of reference books at the end of this section.)

Some tankers were built with machinery amidships, but it is now the universal practice to fit the machinery aft. This permits an unobstructed continuous oil-tight strueture for carrying oil. The hull is subdivided into small tanks by a center-line longitudinal bulkhead and closely spaced transverse bulkheads. The spacing of transverse bulkheads seldom exceeds the 30-foot limit prescribed by the rules of ship classification societies. At the end of the oil-carrying space, cofferdams are fitted, to separate the oil cargo from the machinery space, and the package freight space or the buoyancy space forward. The cofferdams consist of two transverse bulkheads spaced 3 to 4 feet apart. There is no fixed rule for the location of the pump room, although the general practice is to fit the pump room forward in the smaller vessels, and amidships in the larger vessels, or in some cases aft near the machinery space.

Much is to be said in favor of fitting the machinery aft. (See figure illustrating this subject.) The advantages of fitting the machinery aft are briefly as follows: (*a*) The elimination of the cofferdams at each end of the machinery space; (*b*) The elimination of the shaft tunnel, which is required to be completely oil-tight, isolated from machinery space, and fitted with trunks at each end for access from the weather deck; (*c*) The elimination of one pump room; (*d*) The elimination of required special structural stiffening in the way of the machinery and boiler space where the center-line bulkhead and expansion trunks have been terminated to accommodate the machinery.

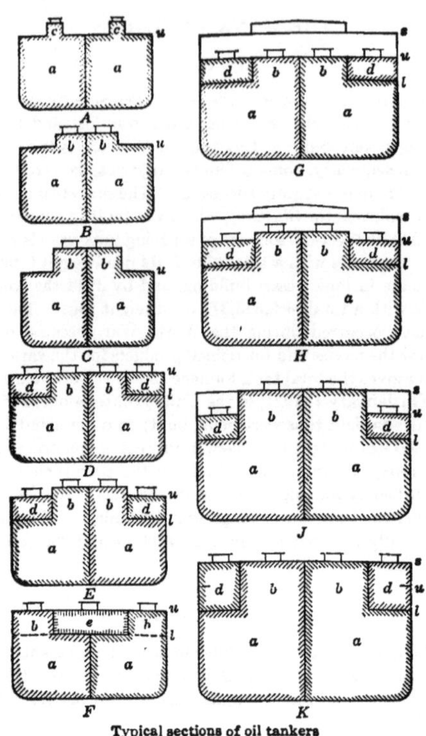

Typical sections of oil tankers

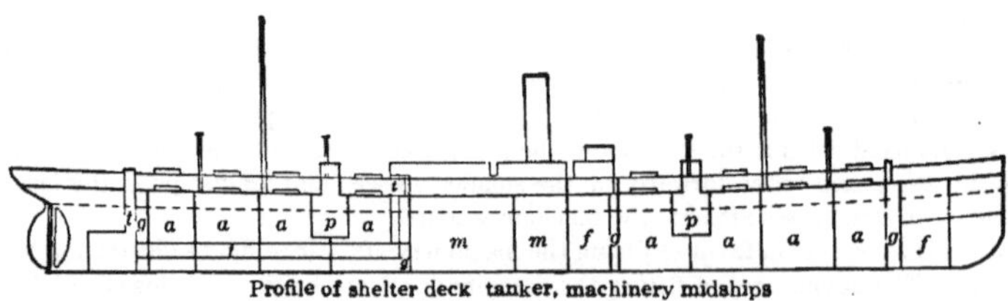

Profile of shelter deck tanker, machinery midships

Profile of shelter deck tanker, machinery aft.

Typical Cross-Sections of Tank Vessels

A. Single deck (*u*) with center-line bulkhead to deck and individual expansion trunks (*c*) over each tank.

B. Single deck (*u*) with center-line bulkhead to top of expansion trunk and continuous expansion trunk (*b*) above deck.

C. Single deck (*u*) with center-line bulkhead and continuous deep expansion trunk (*b*) above deck.

D. Two decks: upper deck (*u*); lower deck (*l*) with center-line bulkhead to upper deck; expansion trunk (*b*) between upper and lower deck; summer tanks (*d*) in the wing spaces; individual hatches on upper deck.

E. Same as *D* but with continuous expansion trunk carried up above upper deck.

F. Two decks, with center-line bulkhead to lower deck; continuous expansion trunks (*b*) at sides; space (*e*) for package freight. (This type is well-adapted for the conversion of general cargo vessels into tankers, where the existing lower deck need not be made oil-tight at the shipsides.)

G. Three decks: shelter deck (*s*); upper deck (*u*); lower deck (*l*), with center-line bulkhead to upper deck; continuous expansion trunks (*b*) between lower and upper deck; summer tanks (*d*) in the wing spaces; shelter deck fitted as strength member.

H. Same as *G*, but with continuous expansion trunks carried up above upper deck.

J. Three decks: shelter deck (*s*); upper deck (*a*); lower deck (*l*), with center-line bulkhead to shelter deck; continuous expansion trunks (*u*) carried from lower to shelter deck (two deck-expansion trunks); summer tanks (*d*) in wing spaces between upper and lower decks.

K. Three decks: shelter deck (*s*); upper deck (*u*) omitted; lower deck (*l*) with center-line bulkhead and continuous expansion trunks (*b*); and summer tanks (*d*) extending from lower deck to shelter deck.

Index

a. Cargo tanks
b. Expansion trunks
c. Expansion trunks (individual)
d. Summer tanks in wing spaces
e. Summer tanks in center space
f. Fuel-oil bunkers
g. Cofferdams
m. Machinery and boiler space
p. Pump rooms
t. Shaft alley and access trunks to same

Other types.—Jack patent tankers—vessels of ordinary cargo type fitted with large cylindrical tanks in the cargo holds, a convenient method for conversion.

Transverse Framing System

The type of transverse framing used in the construction of tank vessels is that known as the "web frame system." Web frames are widely spaced, with closely spaced intermediate transverse frames of lighter construction. Transverse and longitudinal bulkhead construction is on the same plan, with the web frames spaced in line with the web frames of the vessel's hull or keelsons or deck girders, and closely spaced intermediate stiffeners of lighter construction. This construction forms a rigid horizontal and vertical transverse as well as longitudinal web construction.

Longitudinal Framing System (Isherwood System)

In this system, similar web frame construction is used, but the intermediate framing is placed longitudinally and horizontally, thereby greatly increasing the longitudinal strength of the structure.

Particulars of Typical Tank Steamers

Length	Breadth	Depth	Type	Draft to L.S.F.*	Dis-place. Tons	Dead-wgt. Tons	Gross Ton-nage	Net Ton-nage	WEIGHTS IN TONS		
									Hull	Wood and out-fit	Machinery
330′	46′ 6″	27′	2-Dk., P. B.&.F.†	22′ 6″	7,615	5,010	3,663	2223	1790	412	403
360′	46′ 3″	27′ 4″	"	20′ 6¼″	7,690	4,970	3,737	2397	1925	340	455
370′	52′	30′	"	22′ 6″	10,065	6,910	4,711	2869	2270	455	430
419′ 5″	54′ 2″	31′ 6″	"	25′ 0″	12,950	8,800	6,563	4089	2870	550	730
440′	58′ 3″	40′ 9″	2Dk., & Shel. Dk.	27′ 11½″	16,250	10,100	6,631	4233	4880	540	730
468′ 6″	62′ 6″	39′ 6″	"	27′ ¼″	18,332	12,620	7,794	4801	4172	538	902
512′	63′ 3″	42′ 0″	"	28′ 3¾″	20,871	11,976	9,196	5776	6000	790	925
525′	68′ 6″	41′ 9″	"	28′ 4¼″	23,615	17,280	10,073	5903	5055	580	700
527′	66′ 6″	42′ 0″	"	27′ 11″	22,220	15,690	11,929	7584	5000	620	910
530′	66′ 6″	42′ ¼″	"	29′ 9″	23,970	17,110	12,074	7560	5420	530	910

* L. S. F., Lloyd's Summer freeboard.
† P. B. F., Poop, Bridge and Forecastle.

Tonnage and Capacity of Tankers

Load displacement, or displacement tonnage, expressed in tons of 2240 pounds, is equal to the weight of the vessel when loaded to the "load water line." According to the well-known law of physics, the weight of a vessel is equal to the weight of the volume of water displaced by it when afloat. This displacement is calculated from a "lines plan."

The load water line, referred to above, is the line of maximum immersion to which a ship may be loaded under the established "load line rules" of the principal maritime nations. Load displacement may be subdivided as follows: the weight of the vessel is usually referred to as "vessel light ready for sea." The term "deadweight" comprises the weight of the cargo, fuel, stores, water, crew, effects, etc.

Vessel **light ready for sea,** for purposes of design and comparison, is determined by:

(*a*) Hull steel.

(*b*) Machinery and boilers. (Boilers filled to working level and steam up, condenser and circulating system filled with circulating water.)

(*c*) Wood and outfit.

The subdivision of weights is given in detail under particulars of typical tank steamers.

Deadweight.—To determine the deadweight of a tanker, it is therefore necessary to calculate its load displacement, deducting therefrom the weight of the vessel "light ready for sea," which also can be determined by calculation from the actual measurements of the draft forward and aft of the vessel in "light ready for sea" condition. The difference will be the total deadweight. When the total deadweight has been determined in this way, it will be necessary to establish the amount of fuel, water, provisions, stores, etc., that the vessel requires for the intended voyage, in order to arrive at the amount of cargo the vessel can carry on the particular voyage. The total deadweight is also given in the list of particulars of typical tank steamers. The following table will be useful in the conversion of cargo deadweight tons into cubic feet, gallons and barrels for oils of varying specific gravity.

Gross register tonnage is the total cubic capacity of the ship, including all deck erections, with certain exemptions, expressed in units of 100 cubic feet per ton. It is calculated by the rules and regulations of the several maritime nations. (See Department of Commerce, Bureau of Navigation, booklet on measurement of vessels.)

Pluymert Conversion Table *

Specific gravity	Degrees Baumé	Weight U. S. gallons	Weight Cubic feet	Cubic feet per ton	Gallons per ton	Barrels of 42 gallons per ton	Barrels of 50 gallons per ton
1.0000	10	8.33	62.355	35.9	268.9	6.43	5.38
.9859	12	8.21	61.475	36.5	272.8	6.50	5.46
.9722	14	8.10	60.621	36.9	276.6	6.59	5.53
.9589	16	7.99	59.792	37.5	280.3	6.69	5.61
.9459	18	7.88	58.981	38.1	284.2	6.77	5.69
.9333	20	7.78	58.195	38.5	287.9	6.86	5.76
.9210	22	7.67	57.428	39.0	292.0	6.96	5.84
.9090	24	7.57	56.680	39.5	295.7	7.06	5.92
.8974	26	7.48	55.957	40.1	299.4	7.14	5.99
.8860	28	7.38	55.149	40.6	303.5	7.24	6.07
.8750	30	7.29	54.560	41.1	307.2	7.32	6.15
.8642	32	7.21	53.887	41.6	310.9	7.40	6.21
.8537	34	7.12	53.232	42.1	314.8	7.49	6.29
.8434	36	7.03	52.590	43.0	318.6	7.59	6.37
.8333	38	6.94	51.960	43.1	322.5	7.69	6.46
.8235	40	6.86	51.349	43.6	326.3	7.78	6.53
.8140	42	6.78	50.756	44.1	330.1	7.87	6.61
.8046	44	6.70	50.171	44.6	334.0	7.96	6.69
.7955	46	6.63	49.603	45.2	337.8	8.05	6.78
.7865	48	6.55	49.042	45.7	341.7	8.14	6.84
.7778	50	6.48	48.500	46.2	345.5	8.23	6.91
.7692	52	6.41	47.964	46.7	349.4	8.32	6.99
.7609	54	6.34	47.446	47.2	353.2	8.41	7.07
.7527	56	6.27	46.935	47.7	357.0	8.51	7.15
.7447	58	6.20	46.436	48.2	360.8	8.60	7.23
.7368	60	6.14	45.943	48.8	364.7	8.69	7.30
.7292	62	6.07	45.469	49.3	368.5	8.79	7.26
.7216	64	6.01	44.995	49.8	372.4	8.87	7.35

*The above table is based on the formula $\frac{140}{130° \text{ Bé}} = \text{Sp. Gr.}$

For each 10° F. above 60° F. add 0.7° Bé.
For each 10° F. below 60° F. subtract 0.7° Bé.
1 ton = 2,240 lbs.

Underdeck register tonnage is the cubic capacity of the vessel below the tonnage deck, expressed in tons of 100 cubic feet, and measured according to the established rules.

Net register tonnage is found by deducting from the gross register tonnage, certain deductible space, such as crew spaces, spaces for machinery and boilers, navigation spaces, space for storage of sails, etc., and is measured according to the established rules.

Panama and Suez **Canal** tonnages differ slightly from the register tonnages and are determined in practically the same way as the register tonnages.

Ship Classification Rules

Rules covering the scantlings and details of construction of oil tankers are published by the principal ship classification societies. Among the most important of these societies are the American Bureau of Shipping, British Lloyd's Register of Shipping, and Bureau Veritas. The principal features of these rules are given below:

Length of **oil compartments.**—Oil compartments shall not exceed 24 to 28 feet in length.

Expansion trunks.—Expansion trunks, shall be fitted over the oil compartments to allow the expansion or contraction of the oil due to temperature changes. They must be so arranged that the surface of the oil cargo will not fall below the sides of the trunk when the vessel is rolling or pitching. Plans must be submitted for special consideration when the breadth of the expansion trunks exceeds 60 per cent of the beam of the vessel, or when the height of the expansion trunks exceeds 8 feet above the top of the oil compartment.

Cofferdams.—Cofferdams shall be fitted at fore and aft ends of the oil cargo space. When propelling machinery is fitted amidships the cofferdams shall also be fitted at each end of the machinery space, not less than two frame spaces in length.

Poop and bridge erection.—Tankers with propelling machinery aft shall have the poop fitted to cover the machinery space. With the propelling machinery amidships the bridge erection shall be of a sufficient length to overlap the ends of the center-line bulkhead in the oil compartments.

Testing.—Before the vessel is launched or while in drydock the oil compartments shall be filled separately with water and tested under pressure with head of water 8 feet above the highest point of the expansion trunks. The cofferdams shall be tested by being filled with water to the top of the hatch.

Workmanship.—Workmanship must be of the highest character because of the penetrating character of oil and the stresses to which tank vessels are subjected. These are largely due to the fact that the cargo is carried directly on the vessel's outside plating. Special care must be exercised to provide against local stresses at the ends of the oil spaces, superstructures, etc., and also for the necessary compensation for the close spacing of rivets throughout the structure.

Riveting.—The riveting shall be of the most efficient character, and the points of the rivets shall be left full or convex. Special care shall be observed in punching and countersinking holes and in fitting the various parts. Where the holes are unfair they shall be reamed and larger rivets fitted. All oil-tight joints shall have the surface of the plates fitted close to each other and as far as is practicable shall be caulked without the use of canvas, etc.

Spacing of **rivets** shall be as follows:

> 3½ diameters C. to C. in butts of flat and vertical keel plates; butts of outside plating and deck stringer plates; seams and butts of tank deck plating and middle-line and transverse bulkheads; sides and top of expansion trunk and tunnel plating.

4 diameters C. to C. in quadruple riveted overlapped butts of outside plating and deck stringer plates.

5 diameters C. to C. in flat keel angles; boundary bars of bulkheads; tank deck stringer angles and angles at upper and lower corners of expansion trunks; bracket knee plate attachments to floor plates; side stringers and bulkhead stiffeners.

6 diameters C. to C. in frames, reversed frames, floors, keelsons; face angles on web frames and side stringers; bulkhead stiffeners; web plates and girders; expansion trunk stiffeners; tunnel stiffeners and deck plating; and top of expansion trunk to beams.

Attachment angles.—The breadths of the flanges of attachment angles are to be as follows:

For ⅝-inch and ¾-inch rivets....	3 inches
For ⅞-inch rivets	3½ inches
For 1-inch rivets	4 inches

and of the mean thickness of the plates they attach.

In oil-tight work all angles shall be welded at the corners and neither mitred nor overlapped. Where boundary bars of oil-tight bulkheads consist of double angles, both flanges of each bar shall be caulked.

Where engines are situated amidships, a "shaft tunnel" shall be fitted through after the oil compartment but entirely separated from the engine-room by the cofferdam, with a trunkway at each end leading from the upper deck to the tunnel. The trunkway at the forward end shall be abaft the cofferdam.

Oil-pumps.—Oil-pumps shall be kept entirely separate from ballast pumps.

Water ballast pipes.—Water ballast pipes shall not pass through oil compartments and vice versa. (A separate ballast pump shall be fitted forward of the oil compartments for forward ballast tanks).

Oil pump rooms.—Oil pump rooms shall be enclosed by water-tight bulkheads and shall have no direct communication with the machinery space.

Ventilation.—Efficient means shall be provided for clearing dangerous gases from the tanks by injecting steam or by artificial ventilation. Ventilators shall be fitted to double bottom compartments, to pump rooms, to cofferdams, between the decks at each end of the tunnel and to all other enclosed spaces, to permit the free escape of gases. All vent openings above the deck shall be protected by wire gauze.

Fire-extinguishing system.—Steam fire-extinguishing lines shall be fitted along the deck to convey the steam from the boilers to the oil compartments with a main stop valve on the main line. This apparatus shall be properly housed to protect its operator from heat and smoke. The valves on all branches to compartments shall at all times be left open. The fire-extinguishing agency known as foamite may be substituted for steam. Installation shall be subject to the approval of the Board of Supervising Inspectors, U. S. Steamboat Inspection Service.

Electrical equipment.—The following are the special requirements for vessels carrying oil having flash point less than 150° F.:

The pressure of supply shall not exceed 110 volts, whether the current be direct or alternating.

Every outgoing circuit from the main switchboard and every branch and subcircuit shall be provided with a double-pole linked switch.

The cases of all joint boxes, section and distribution boards shall be wholly of metal, and all cables shall enter them through water-tight glands.

All cables shall be lead-covered or lead-covered and armored.

Distribution shall be effected wholly on the two-wire two-conductor system with separate conductors, both insulated, for the respective poles. No part of the system shall be grounded.

Dangerous spaces.—No lamp, fitting or appliance of any kind, and no wiring may be fitted in, or enter, any of the following dangerous spaces: oil-holds; cofferdams.

Pump rooms may be lighted by lamps wired wholly outside the space and separated from the interior by an air-tight stout glass bowl.

Lamps in pump rooms, between decks and in spaces immediately adjoining oil-holds shall be contained in gas-tight fittings, the wiring being enclosed in gas-tight tubing. The switches controlling them shall be wholly outside these spaces. These switches shall in all cases be double-pole.

No portable lamps, other than self-contained battery-fed lamps of the type approved for use in fiery mines, may be used in dangerous spaces. All fittings therein shall be gas-tight.

Vessels fitted for burning oil fuel shall be approved for oil fuel flash point above 150° F. for the use of high-flash-point oil only, and for "low-flash oil fuel" for the use of oil with low flash point. The proposed arrangement of a vessel approved for low-flash oil fuel shall be submitted for approval.

Reference Books

The Naval Constructor, G. Simpson.
Know Your Own Ship, T. Walton.
The Design and Construction of Ships, J. H. Biles.
Naval Architecture, T. H. Watson.
Textbook of Theoretical Naval Architecture, E. L. Atwood.
Practical Shipbuilding, A. C. Holmes.
Steel Ships, T. Walton.
Oil Tank Steamers, H. J. White.
Ship Form Resistance and Screw Propulsion, G. S. Baker.
Speed and Power of Ships, D. W. Taylor.
Steamship Coefficients, Speeds and Powers, C. F. A. Fyfe.
Resistance of Ships and Screw Propulsion, D. W. Taylor.
Ship Stability and Trim, P. A. Hillhouse.
Manual of Marine Engineering, A. E. Seaton.
Marine Steam Engine, R. Sennett and H. J. Oram.
Marine Engines and Boilers, Dr. G. Bauer.
Design of Marine Engines and Auxiliaries, E. M. Bragg.
Internal Combustion Engines, R. C. Carpenter and H. Diederichs.
Marine Engineer's Handbook, F. W. Sterling.
Pocketbook of Marine Engineering Rules and Tables, A. E. Seaton and H. M. Rounthwaite.
Transactions of Institute of Naval Architects, England.
Transactions of American Society of Naval Architects and Marine Engineers.
Transactions of Institute of Marine Engineers, England.
Journal of American Society of Naval Engineers.
Rules for the Construction of Ships, American Bureau of Shipping.
Rules for the Construction of Ships, British Lloyd's Register of Shipping.
Rules for the Construction of Ships, Bureau Veritas.

PIPE-LINE TRANSPORTATION

The first successful pipe-line for the transportation of crude petroleum was completed in August, 1865, by Samuel Van Sickle. It was 4 miles long, from Miller's Farm to Pithole, Pa., and was constructed of 2-inch screw pipe. There were two pumping stations.

Under date of April 16, 1876, General Herman Haupt, one of the leading civil engineers of that time, made a report to Henry Harley on building a line to the Atlantic seaboard. This line was not built, but later experience showed results agreeing very closely with General Haupt's report.

A line of 4-inch pipe about 60 miles in length was built to Pittsburgh about 1875; this was followed in 1880 by a 5-inch line, 103 miles long, from Hillards, Butler County, Pa., to Cleveland, Ohio, and by lines to the Atlantic seaboard in 1881.

Practically 1,300,000 barrels of crude oil per day are transported by pipe-line. Some of it is moved only a few miles, while some may be carried more than 1500 miles. About 150,000 barrels per day are taken to the Atlantic seaboard through the connecting system of pipe-lines. A large quantity is transported to Baton Rouge and the Gulf ports through long lines. Probably the 1,300,000 barrels, or about 200,000 tons daily, move, on the average, more than 500 miles. This would make 100,000,000 ton miles per day. If this oil were moving over a railroad 500 miles long, in tank cars carrying 200 barrels each, it would require 6550 tank cars to be delivered each day, which would fill 50 miles of track. Fifty miles is probably more than the average daily travel of a tank car. A tank car has to be hauled back to its starting point to be reloaded; the principal advantage of a pipe-line is that only the freight moves.

The pipe-line system in the United States has been built up a little at a time, commencing in 1865, when the first lines were built to convey the oil from the wells in northern Pennsylvania to boats on the Allegheny River, followed later by the building of a line to Pittsburgh, then lines to Cleveland, Buffalo, Philadelphia, Baltimore and New York harbors. As the field in western Pennsylvania began to decline, search for oil resulted in extending the lines to the Ohio-Indiana field. Later they were extended to the Oklahoma fields.

Pipe-lines were made common carriers by a law passed in 1906, which the Supreme Court of the United States declared constitutional in 1914. No pipe-line has ever been built except for the purpose of delivering oil at a point where a refinery already existed or a demand for oil was to be established. The funds for building the lines have usually been supplied either by a company which expected to use the oil or by a company which had oil to sell. The common service has been entirely at the production end of the line. Much has been printed on pipe-lines as common carriers and their alleged unfair methods and unfair rates; but it is a fact that a pipe-line is only built to supply an existing or anticipated demand in which it can probably render better and cheaper service than the other available methods of transportation. A pipe-line may be built instead of certain tank car or boat transportation projects. Which method will best meet the requirements is a question to be decided in each case.

In general, the pipe-line rates in the United States are about one-half the railroad tank-car rates, but boat transportation, where it can be used, may offer a lower rate than either tank car or pipe-line. Fortunately for the development of the pipe-line system in this country, the oil has not, in general, been found near navigable streams or the coast.

During the construction of the Panama Canal a company built a line across the Isthmus to supply fuel for the construction of the canal. This line was moved from place to place during the work on the canal and permission to construct it was only given on the condition that on the completion of the canal, the line would be removed.

The government probably did not insist that this line should be removed. Most of it was removed, as the transportation which was needed after the completion of the canal did not warrant the maintaining of a pipe-line. Facilities were installed at each end of the canal for loading and unloading oil from tankers, so that the oil could later be delivered to boats for fuel.

In Mexico, lines have been built from the Panuco and Topila fields to transport the oil to Tampico, and from the field in the southern part of Ozulama and the northern part of Tuxpan to Tampico and the Gulf Coast. The general location of these lines is shown on the map of the oil-fields of Mexico. Some shorter lines were also built to take the oil from points near the Gulf Coast to load on boats, and a line was built from Puerto Mexico to the refinery at Minititlan.

In South America there are a few short lines. There are several transporting oil in Europe and Asia, but at no place has the pipe-line system of transportation been developed as it has in the United States.

During the late war (1918), an 8-inch pipe-line about 35 miles long was built across Scotland from a point near Glasgow to Grangemouth on the Firth of Forth, so that when tank steamers delivered fuel oil for the Allied navies at Old Kilpatrick near Glasgow, it could be conveyed across to the Firth of Forth where it was required by the navies. There was already a canal across Scotland but it was found that oil could not be transported over it in great enough quantities to supply the needs of the Allied navies. Tanker transportation to the Firth of Forth was very dangerous and it was decided to build this line as a war measure. The work of building the stations and ditching the line was done by the British Admiralty and the pipe was purchased by them in the United States. The line was laid by a United States naval unit and was completed only a short time before the armistice was signed. It was used after the war.

Flow of Oil through Pipes

In investigating the flow of a fluid through a pipe, it is important to consider the condition which exists when the fluid is at rest, what changes take place as the column gradually comes into motion, and finally, the condition of steady flow.

For convenience, assume a level line with a reservoir at each end. When the fluid is at rest the pressure at all points along the line will be the same. If the pressure at A is increased, the fluid will start to move toward B. As adhesion between the pipe and the fluid is usually greater than the cohesion between the particles of the fluid, a film of the fluid next to the pipe remains practically stationary, resulting in a fluid moving through a cylinder of the fluid. Consider the whole area of the pipe as filled with thin concentric cylinders of the fluid. The inner cylinders are moving through cylinders which are in motion, so that the velocity is greater as the center is approached. If the force causing the flow gradually increases, the velocity at the center will increase while the outer cylinder will remain nearly stationary. The pressure at any cross-section of the pipe can be resolved into two forces, one static and the other dynamic, depending on the velocity. The sum of the static and the dynamic heads at any cross-section being constant (Bernoulli's theorem), the static head, the one tending to produce flow, is greater at the edge of the pipe than at the center.

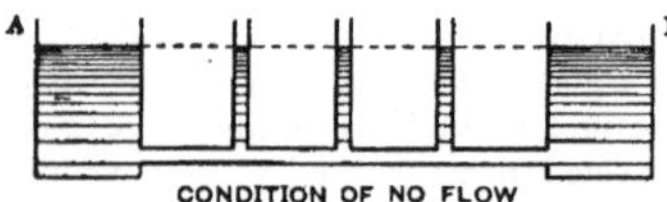

Fig. 3.—Condition of no flow in a pipe-line.

Disregarding gravity and static pressure there are four forces acting on each particle of the thin cylinder; one on the outside tending to retard, one on the inside tending to accelerate, one tending to push the particle forward and force it toward the center, and also inertia, tending to keep it moving in the same direction.

When these forces become sufficient to overcome the cohesion of the particles, a cross flow will be established and there will no longer be cylinders moving inside of cylinders, but an indeterminable mixing of the fluid in the pipe. The point at which this change of condition takes place is called the critical velocity. The flow in cylinders is called the sinuous or straight-line flow, and the flow at the higher velocities is called eddying or turbulent flow.

A theory has been developed that the coefficient of friction varies as a function of the diameter multiplied by the velocity and by the density of the fluid and divided by the so-called absolute viscosity, or $\frac{Vd\sigma}{\mu}$ when expressed in any consistent system of units.

Theory of the flow of a liquid through a long line of pipe.—The force necessary to produce a flow in a long line of pipe is equal to the resistance. Given a level pipe of uniform cross-section and surface, a uniform liquid without change in temperature, and a steady flow, the force becomes equal to what is usually called "frictional resistance." This resistance has been found to be practically independent of the pressure under which the operation takes place and is usually indicated by a coefficient

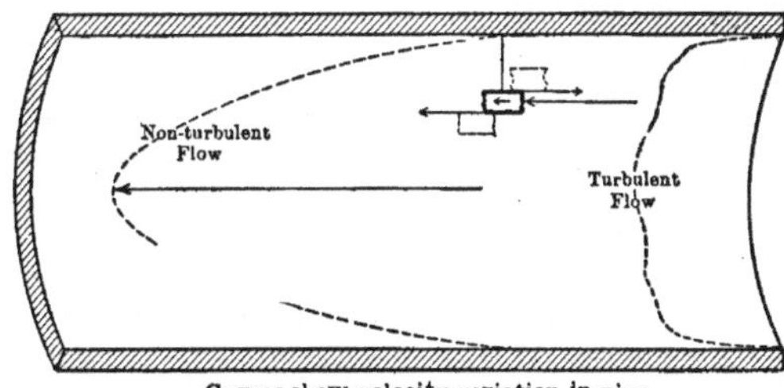

Curves show velocity variation in pipe

FIG. 4.

of friction. Many experiments have been made to determine the coefficient of friction for water, and reference is made to the many textbooks on hydraulics for discussion of what is often called "hydraulic flow."

Early in the present century there arose a demand for handling viscous oils by pipe-line. Study of the conditions and many experiments have resulted in a distinct advance in the method of solving these problems. The following discussion by W. F. Durand, Professor of Mechanical Engineering at Stanford University, states in a comprehensive manner the present development of the theory.

PROFESSOR DURAND'S STATEMENT

The purpose of the present paper is to call attention to the fundamental problem presented by the flow of viscous liquids in pipes and to note the extent to which existing information may be made to furnish the data necessary for a reasonably satisfactory treatment of actual problems, especially as they arise in connection with the design and operation of oil pipe-lines.

What are the characteristics or factors which will be determinative in any given problem of pipe-line flow? It should be noted that for simplicity the line is assumed to be horizontal. The changes necessary to allow for differences in gravity level are readily made in accordance with familiar hydraulic principles.

It seems clear that the following is a complete census of all characteristics or factors which can be considered determinative in a problem of this character.

Viscosity of liquid (μ) (absolute units).
Density of liquid (σ) (pounds per cubic foot).

These two characteristics define the liquid in a mechanical sense.

Diameter of pipe (d) (feet).
Length of pipe (l) (feet).
Roughness of pipe surface.

These three characteristics define the pipe in a mechanical sense. **For roughness** we have no method of direct measure and no symbol.

Velocity of flow (v) (feet per second).

This is a characteristic of operation. To these we must add

Gravity (g).

This determines the field of gravitational force within which the phenomena of pipe-line flow have their being. Viewed as a resultant of the interaction of the preceding seven factors we shall then have:

Loss of head (h) (feet).

This is, in effect, a measure of the pressure difference required between the two ends of the line in order to insure continued steady motion, or flow, under the conditions assumed.

Fundamental Relation of Factors

If then we are agreed that the first seven of the above factors are all that are involved in the determination of the loss of head, and if we further agree to neglect variation in roughness, or otherwise if we agree that, for the moment, we are dealing with a standard or uniform quality of surface, then an application of the theory of dimensions [1] gives the following relation:

$$h = \text{some function of} \left(\frac{dv\sigma}{\mu}\right) \text{times} \left(\frac{lv^2}{d^2g}\right).$$

The significance of this equation is far-reaching and it will repay the most careful study on the part of all interested in the subject. The treatment outlined in the following numbered sections will serve to indicate the more direct and simple applications to the phenomena of the flow of viscous liquids in pipes.

(1) On the basis assumed, viz., that we are dealing with a standard or uniform quality of pipe surface and that all other quantities and factors entering into the problem are represented in the equation, then it follows that the relation between these various quantities is of the form given in equation (1).

The derivation in detail of this relation from basic principles is beside the purpose of the present paper and it is furthermore unnecessary, since this has been provided long since by other writers and is well known in the literature of the subject.

(2) In order to show the relation of this equation to more familiar forms let us write it again alongside the d'Arcy equation, familiar to all students of hydraulics, thus:

$$h = \phi\left(\frac{dv\sigma}{\mu}\right) \times \frac{l}{d}\frac{v^2}{2g}. \qquad (1)$$

$$h = f\frac{l}{d}\frac{v^2}{2g}. \qquad (2)$$

[1] See among other references, Buckingham, Trans. A. S. M. E., Vol. 37, p. 263.

Comparing these two equations it appears that in this form (1) is the same as (2) if we put

$$f=\phi\left(\frac{dv\sigma}{\mu}\right).$$

One advantage of the expression $\frac{dv\sigma}{\mu}$ is that when the individual values of d, v, σ, μ are expressed in any consistent absolute system of units (foot, pound-mass, second; or centimeter, gram, second) the value of the expression remains the same independent of the system of units employed. This is commonly expressed by the statement that $\frac{dv\sigma}{\mu}$ is a dimensionless expression.

That is, we have in (1) simply an extension or development of (2) to the point of defining the coefficient f as a function of the expression $\frac{dv\sigma}{\mu}$. This means that f depends on these four quantities and furthermore that it depends on them in such a way that for any one value of the expression $\frac{dv\sigma}{\mu}$ there is one and only one definite and specific value of f, regardless of the individual values of d, v, σ, μ, so long as the expression $\frac{dv\sigma}{\mu}$ has the given value.

This point is of such great importance that emphasis through illustration will be desirable. Thus, suppose some particular combination of values of d, v, σ, μ gives a value of $\frac{dv\sigma}{\mu}=6000$. Then corresponding to this value 6000 as argument will correspond a value of $f=$ say .036. Then no matter what other combination of individual values d, v, σ, μ, we may have, as long as the value of $\frac{dv\sigma}{\mu}=6000$, so long shall we have the value $f=.036$. Stated otherwise, the value $f=.036$ is applicable to all possible combinations of individual values d, v, σ, μ, as long as their combination in the form $\frac{dv\sigma}{\mu}=6000$.

In particular it should be noted that these factors may belong to the same or to widely different liquids. It is only necessary to know the values of the defining characteristics μ and σ. Thus a viscous oil flowing with a high velocity in a large pipe may give the same value of $\frac{dv\sigma}{\mu}$ as water flowing with a low velocity in a small pipe. In such case, then, the value of f should be the same. This does not mean of course, the same values of h, since the effect of the other factors (specifically v^2) will be to give a much greater loss of head h for the viscous oil than for the water.

Again, the general laws for the flow of gases and vapors are entirely similar to those for liquids. Hence the same general relation holds for gases and vapors as well as for liquids, or generally for all fluid substances. It is therefore only necessary to know the value of the defining characteristics μ and σ in order to assimilate together in one expression the laws of flow for liquids, gases and vapors, and thus generally for all flowing substances.

Thus again and broadly, no matter what the substance, so long as we know its viscosity and density; no matter what the size of conduit so long as we know its size; no matter what the velocity so long as we know its measure, we may find a value of

$\frac{dv\sigma}{\mu}$ and to this value should correspond one and only one definite value of the coefficient f.

(3) From this it immediately follows that if, through a series of actual measurements, experimental values of f are obtained for a graded series of values of $\frac{dv\sigma}{\mu}$, and if such experiments are extended over the entire working range of values of $\frac{dv\sigma}{\mu}$ and at reasonably close intervals, then we may, by the usual graphical process, plot such data and draw in a fair curve through or among such experimental spots.

Such curve should then give a graphical expression for the general relation between f and the argument $\frac{dv\sigma}{\mu}$.

(4) Such a curve once in hand and accepted as reliable, we should then have no further need for experimental determination of values of f—always remembering the assumption of a standard quality of pipe surface. No matter what individual values of d, v, σ, μ the particular case may present, we should have only to find the resultant value of $\frac{dv\sigma}{\mu}$, go to this point on our axis of abscissæ and read off the corresponding value of f; or otherwise, in case the data were presented in tabular form, enter the table with the value of $\frac{dv\sigma}{\mu}$ as argument and draw therefrom the value of f (either directly or by interpolation) in the usual manner with the use of tables and tabular data.

The value of f once known, we may then proceed immediately to find h by the familiar d'Arcy equation as in (2).

(5) We come next to the practical question, does such information exist? Are we now, on the basis of experimental work, in a position to prepare a chart or table setting forth the general relation between f and the argument $\frac{dv\sigma}{\mu}$—always for a standard quality of pipe surface?

The answer is largely in the affirmative. The chief limitations arise from the lack of any standard for quality of surface. We have no scale for roughness nor do we have any direct basis for the measurement of this characteristic as in the case of pipe diameter or velocity of flow.

Large numbers of experiments, however, have been carried out on the flow of various substances such as air, gas, water, oil, etc., in tubes and pipes designated as " smooth " or " new and clean." In so far as the quality of surface involved in these experiments may be assumed to have been fairly uniform, such experimental data should fall in line for the provision of a standard series of values of f as indicated above. We cannot define or specify roughness but we can define, at least in a commercial sense, a smooth surface as one without roughness in degree sensible with regard to its influence on the values of f. If, therefore, we should find by plotting results from a large number of such experiments on different substances and with widely varying values of d, v, σ, μ, that the resultant values of f plotted on an abscissa $\frac{dv\sigma}{\mu}$ all fall on or near a smooth continuous curve, a curve clearly representing a law or relation between physical quantities, we should then be justified in assuming that the data thus in hand do represent in effect the results from a series of pipes or tubes or conduits with a sensibly uniform quality of surface. If slight variations one way or the other are noted, if slightly different values are found for the same value of $\frac{dv\sigma}{\mu}$, or if differences between values nearly the same are somewhat too great to be reconciled with the obvious general trend of the curve, then we may naturally conclude that we have in evidence the influence of slight variations in the quality of the surface. If again we should find certain values which lie consistently higher than the general trend of values, while themselves following the same general slope or form of graphical law, we shall naturally conclude, especially if individual groups of such values each belong to one particular pipe or conduit, that we have in evidence the influence of a degree of roughness in a given pipe distinctly greater than the general average for "smooth" or new pipe values, and thus giving by the extent of increase of values above those for smooth or new pipe, a form of measure of the roughness as compared with what we may term "commercially smooth" pipe.

Actual examination of the data in hand reveals all of these characteristics. There are sufficient data relating to pipe which may be rated as commercially smooth to give a good indication of the general run of values for f for such pipe, and extending over a wide range of values of the argument $\frac{dv_{\sigma}}{\mu}$.

Suppose a smooth curve drawn through and among such values, giving in this manner the best indication to be drawn from existing data, of the values of f on the argument $\frac{dv\sigma}{\mu}$ for what may be termed commercially smooth pipe. We shall denote values of f given by such a curve as standard or normal values.

There are also to be found, lying above the curve of standard values, large numbers of spots which obviously belong to cases of rough pipe and with degrees of roughness, as indicated by the values of f, ranging from barely appreciable indications up to values 50 and 75 per cent and even more, in excess of standard values.

(6) There is one exception to the statement that clear indications are found for a general curve of values of f plotted on values of $\frac{dv\sigma}{\mu}$. This is found in the vicinity of values of the argument $\frac{dv\sigma}{\mu}$ close to 2000 or between 2000 and 2500.

The general character of the curve of f is shown in Chart I.

From very small values of the argument up to values approaching 2000 we find a clearly indicated branch of the curve *A B*. For values of the argument close to 2000 or slightly above, the tests are much scattered and give no definite indication of a consistent law. Soon, however, as we pass to larger values of the argument, we find again clear indication of a branch *C D*. The point *C* of this branch lies notably above *B*, implying a sudden and irregular jump in f close about the 2000 value of the abscissa. From *C* the curve gradually and smoothly declines for increasing values of the abscissa, approaching more and more nearly to parallelism with the axis of $\frac{dv\sigma}{\mu}$ as larger and larger values of the latter are reached.

Since the actual range of working values of $\frac{dv_{\sigma}}{\mu}$ may extend from lower limits of 10 and below to upper ranges of 300,000 and more, it is found convenient to use a logarithmic scale for $\frac{dv\sigma}{\mu}$ and for f instead of a natural scale. Chart I is shown to a scale of this character.

(7) Turning again to the two branches *A B* and *C D*, the first of these corresponds

to the so-called stream line mode of flow, the latter to turbulent flow. It is well known in hydromechanics that a liquid flowing between boundary surfaces, as in the case of a pipe, may exhibit two modes of flow.

In the first the liquid particles follow smooth paths generally parallel to the line of the pipe and without whirls, eddies or contorted motion. Such mode of motion is known as stream-line flow.

In the second mode the paths of the liquid particles are confused, comprising eddies, whirls and generally contorted, twisted, turbulent motion superimposed on the general movement of translation along the pipe. Such mode of motion is known as turbulent flow, and the general condition characterized by such motion is known as "turbulence."

Many experiments have been made showing the breaking up of stream-line flow into turbulent flow. In such experiments, for example, a colored liquid has been introduced in a thin thread into a stream flowing in a glass tube. As the velocity is gradually increased, a point is found where the coherent clearly defined colored thread suddenly breaks up and disappears as such, the color being generally diffused throughout the mass of turbulent flowing liquid. It is furthermore established by experiment that the combination of conditions which determine this point of abrupt transition from one mode of flow to another is found in the argument $\frac{dv_\sigma}{\mu}$, and that the value which determines such critical point is close to 2000. The appearance of this zone of uncertainty and confusion in the distribution of the values of f on a chart such as that of Chart I is therefore simply another indication of the existence of this double mode of liquid movement and a location of the transition point at or near the numerical value 2000 for the argument $\frac{dv\sigma}{\mu}$.

(8) Turning again to the branch $A\ B$, it appears from mathematical discussion, the details of which we shall not here repeat, that this curve is a rectangular hyperbola of which the equation is

$$xf = 64, \qquad (3)$$

or

$$f = \frac{64}{x}, \qquad (4)$$

where x = the argument $\frac{dv_\sigma}{\mu}$ and f = the friction coefficient as above.

It appears furthermore, from the theory of stream line flow, that the characteristics of such flow, and the resistance opposed thereto, are only in slight or perhaps negligible degree dependent on the relative roughness or smoothness of the pipe. If this is true it follows that experimental values of f lying within the zone of stream line movement, that is with values of $\frac{dv\sigma}{\mu}$ definitely less than 2000, should lie on or near the curve given by equation (3) above, regardless of the condition of the pipe surface.

Examination of experimental results shows that there is a marked tendency to realize this condition. Tests indicating experimental values, whether for commercially smooth or for rough pipe, are found to fall very near the hyperbola defined by equation (3), and in general with no wider departures one way and the other than may naturally result from experimental errors.

In one respect, however, there is indication of an influence due to roughness of pipe surface. This regards the point of transition from stream line to turbulent flow. There is some evidence that with rough pipe the break over from one to the other occurs at lower values of $\frac{dv\sigma}{\mu}$ than the 2000 value which marks the vicinity for commercially smooth pipe. There is need, however, for further experimental investigation regarding these matters.

However, with this reservation it appears that for all values of $\frac{dv\sigma}{\mu}$, well below 2000, whether with commercially smooth or with pipe relatively rough in varying degrees, the value of f should be given directly from Equation (4).

(9) Taking next the branch $C\ D$, for values of $\frac{dv\sigma}{\mu}$ definitely above 2000, there is no algebraic expression for this branch and the mode of procedure for such values will include the following steps:

(*a*) The numerical value of $\frac{dv\sigma}{\mu}$ is found.

(*b*) With this value as abscissa (or argument) take from the curve (or table according to the form of the data) the corresponding value of f. This will apply to commercially smooth pipe.

(*c*) If the pipe surface is assumed to be rough rather than commercially smooth, a correction is then made by increasing f, according to judgment, to allow for the quality of surface, as assumed.

The following table (page 388) gives values for f over the usual range for the branch $C\ D$ and for qualities of pipe which may be taken as "commercially smooth."

(10) This general method for determining the value of f involves then the following:

For values of $\frac{dv\sigma}{\mu}$ considerably less than 2000, a definite value is given by Equation (4), applicable to all qualities of pipe surface.

For values of $\frac{dv\sigma}{\mu}$ considerably greater than 2000, a definite value is given from the curve (or table) for the case of assumed commercially smooth pipe, leaving a correction for roughness of surface to be determined by judgment.

Regarding roughness and its effect on friction head, it is evident, since we are unable to directly measure or numerically evaluate this characteristic, that any allowance for its influence on the friction head must be made as an act of judgment based on the general observed effects of roughness in varying degrees. This must always necessarily be the case so long as we are unable to numerically evaluate roughness. The most that any method of treatment for the problem can hope to accomplish is to provide a definite procedure leading to a definite result, in so far as the four variables, d, v, σ, μ are concerned, and then to leave to judgment, applied to the particular case, the correction (if any) due to quality of pipe surface and involving an attempt to compare the actual or assumed degree of roughness with that constituting a "commercially smooth" surface and an estimate of relative increase in the value of f for the rough as compared with the smooth.

The procedure outlined in the present paper seems to realize this end. By the means indicated, definite values of f may be found for "commercially smooth" pipe, and correction (if required), based on observation and experience as it may accumulate, may be made for any assumed degree of roughness of surface.

(11) With regard to the determination of a definite value for f, there is, however, one exception. Close about the value 2000 for the argument $\frac{dv\sigma}{\mu}$ there is, as we have seen, confusion and uncertainty. For a value of f falling in this neighborhood there is, therefore, great uncertainty, and the only safe value of f will be one corresponding to the beginning of the branch $C\ D$ or a value about .044. All of this, however, corresponds with the facts. Actual conditions here are found to include a wide range

TABLE 1

	Abscissa	f	Abscissa	f
Sinuous flow	100	.6400	1,200	.0533
	200	.3200	1,400	.0457
	400	.1600	1,600	.0400
	600	.1067	1,800	.0356
	800	.0800	2,000	.0320
	1,000	.0640	2,400	.0267
Turbulent flow	2,500	.0442	30,000	.0238
	3,000	.0426	35,000	.0228
	3,500	.0412	40,000	.0219
	4,000	.0400	45,000	.0213
	4,500	.0390	50,000	.0208
	5,000	.0382	60,000	.0200
	6,000	.0364	70,000	.0195
	7,000	.0350	80,000	.0190
	8,000	.0340	90,000	.0185
	9,000	.0330	100,000	.0180
	10,000	.0320	150,000	.0168
	12,000	.0304	200,000	.0158
	14,000	.0292	250,000	.0150
	16,000	.0280	300,000	.0144
	18,000	.0271	350,000	.0140
	20,000	.0264	400,000	.0137
	25,000	.0249	450,000	.0134

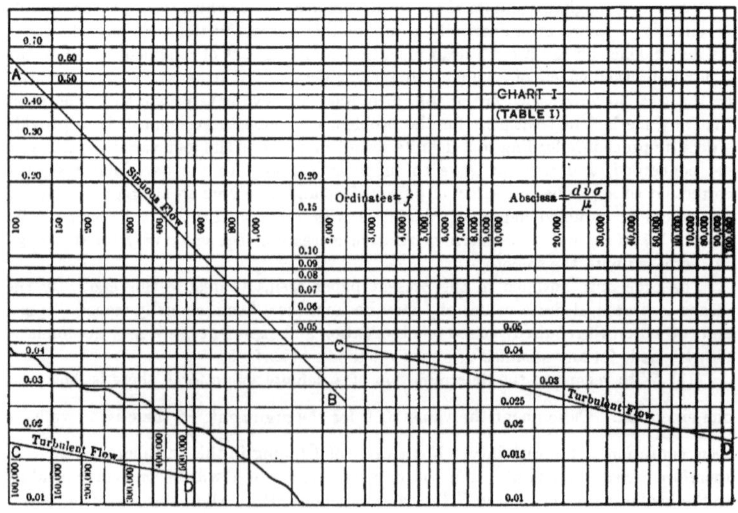

FIG. 5.—Chart to accompany Table I.

of uncertainty, and estimates, to be safe, must be made to cover the extreme values.

It results further that this value of $\frac{dv\sigma}{\mu}$ should be avoided as an operating condition. By a suitable lowering of the velocity the value may be reduced sufficiently to bring the value of f over to or near the lower end of the branch AB. This is, in effect, the most favorable point for operation, so far as pipe-line resistance is concerned, and may indicate that the conditions of operation should be so chosen as to bring the value of $\frac{dv\sigma}{\mu}$ down into the vicinity of 1500 to 1800 or by a safe margin below the critical value of 2000 to 2500.

(12) It now remains to note the units, measurements and numerical values involved in determining the four quantities, d, v, σ, μ.

The fundamental units are taken as the foot, pound-mass and second.

(1) The diameter d is therefore to be measured or expressed with the foot as the unit.

(2) The velocity v is to be expressed with the foot and second as units or numerically in feet per second.

(3) Density σ is to be expressed in pounds per cubic foot. We here meet the condition, however, that the densities of liquids, and in particular of petroleum oils, are commonly expressed in terms of degrees on the Baumé scale. We have also to remember that density will vary with temperature and hence, if the assumed working temperature is different from that used for the determination of the Baumé density, a suitable correction must be made.

In United States practice, density is usually indicated in degrees Baumé on a Tagliabue hydrometer constructed on a modulus of 141.5. To convert degrees Baumé to weight in pounds per cubic foot at 60° F., use the following formula:

$$\sigma=\frac{62.4\times141.5}{131.5+\text{Bé}^{\circ}}, \qquad (5)$$

gravity assumed to have been taken at 60° F. (This is not in agreement with the U. S. Bureau of Standards, which uses a modulus of 140.)

To estimate the density of a petroleum oil at any assumed working temperature, we have then as follows:

$$\sigma=\sigma_1-\left(\frac{121.2-\sigma_1}{2713}\right)(T-60), \qquad (6)$$

where σ_1 = density in pounds per cubic foot for oil at 60° and T = working temperature (Fahrenheit).

Equation (6) is simply an empirical relation between density and temperature which is found to be approximately correct for California oils.

(4) Viscosity μ is to be expressed in absolute English units; foot, poundal and second. In common practice, however, viscosity is more usually expressed in terms of t, a time of flow, on an arbitrary or comparative scale, and varying according to the particular instrument employed. The three types of instrument most commonly met with are the Saybolt, Engler and Redwood.

The following equations may be used to convert the time t, representing the reading on these instruments, into values of μ in the units desired.

$$\text{Saybolt,}\quad \frac{\mu}{\sigma} = .00000237t - \frac{.00194}{t}. \qquad (7)$$

$$\text{Engler,}\quad \frac{\mu}{\sigma} = .00000158t - \frac{.00403}{t}. \qquad (8)$$

$$\text{Redwood,}\quad \frac{\mu}{\sigma} = .00000280t - \frac{.00185}{t}, \qquad (9)$$

where μ = viscosity in absolute English units (foot, poundal, second);
σ = density in pounds per cubic foot;
t = time on instrument (Saybolt, Engler or Redwood) (seconds).

(The above formulæ should not be used to construct a formula for conversion from one standard to another.)

If then the viscosity under the assumed working temperatures is known or assumed, in terms of Saybolt, Engler or Redwood units, we have only to substitute the value t in the suitable formula, and find the value of $\frac{\mu}{\sigma}$ and hence of its reciprocal $\frac{\sigma}{\mu}$. The latter may then be directly used in finding the value of the argument $\frac{dv_\sigma}{\mu}$ or otherwise the values of μ and σ may be individually employed if desired.

It often happens, however, that values of the viscosity may be known for certain standard or reference temperatures, but not for the temperatures assumed as representing the conditions of operation. To meet this situation there is needed some relation between viscosity and temperature. Such relation, however, is apparently exceedingly complex and does not admit of simple algebraic expression.

Broadly, viscosity varies in inverse direction with the temperature. Thus, starting with a high temperature and assuming a gradual cooling it is found that the increase in viscosity is at first slow, the rate of increase rising rapidly as the temperature drops.

Professor W. R. Eckart of Stanford University has pointed out that the relation between viscosity μ and temperature t when plotted on double logarithmic scales shows a close approximation to a straight line, at least over the working range of temperatures for which the liquid may be said to retain its identity. At high temperatures, the more volatile constituents will begin to vaporize and at very low temperatures certain constituents may begin to solidify and separate out, thus in either case changing the character of the liquid itself. Between these limits of temperatures, however, Professor Eckart has shown by a large number of cases that within the limits of observational error the straight line relation may, for all practical purposes, be assumed to hold.

It thus results that if the value of μ for a given oil is known for two temperatures the logarithmic plot may be drawn as a straight line between these points and extended over the working range of temperatures, thus giving a direct and practical form of relation between viscosity and temperature and specifically, the value of μ for any specified temperature within the working range.

Among various numerical data regarding viscosity the following may be cited:

A. C. McLaughlin,[1] gives diagrams showing the relation between temperature F. and absolute viscosity for a series of American oils from which the following may be drawn:

[1] Jour. A. S. M. E., 1914. p. 263.

TABLE 2

Temperature F.	Viscosity *
60°	.060 to .300
80°	.040 to .250
100°	.020 to .160
120°	.016 to .080
140°	.013 to .045

* The units involved in these values are the foot, poundal, second.

The densities of the oils to which these values refer range from 56.8 to 60.0 (Gravity Baumé 24 to 18 approximately).

From tests made by Cooper on sixty samples of California petroleums, R. P. McLaughlin [1] gives values for the relation between viscosity and gravity from which the following tabular values are derived.

The very rapid increase of viscosity with density will be noted for 60° beginning about density 57 and for 185° beginning about density 60.

TABLE 3

Gravity Baumé	Density at 60° F., pounds per cubic foot	Viscosity* at 60° F.	Viscosity* at 185° F.
36	52.6	.0040	.00076
34	53.3	.0050	.00090
32	53.9	.0070	.00120
30	54.6	.0103	.00160
28	55.3	.0145	.00206
26	56.0	.0206	.00260
24	56.7	.0280	.00320
22	57.5	.0370	.00380
20	58.2	.0760	.00440
18	59.0	.2640	.00650
16	59.8		.01120
14	60.7		.02450
12	61.5		.14800

* The units involved in these values are the foot, poundal, second.

From a number of tests made on California oils, Dyer [2] gives for oils of four different gravities values of the viscosity for varying temperatures from which the following tabular values are derived:

[1] Jour., A. S. M. E., 1914, p. 264.
[2] Jour., A. S. M. E., 1914, p. 259.

TABLE 4

Gravity, Baumé....	18.2	16.6	15	12.1
Density, pounds per cubic foot.......	58.95	59.59	60.26	61.48
Temperature F.	Viscosity			
50°	.7360			
60°	....	1.0530		
75°	.2450	.4970	1.5670	
100°	.1220	.2230	.3440	
110°	....	....	....	1.6610
125°	.0670	.0990	.1500	.6140
150°	.0365	.0490	.0750	.2110
175°	.0242	.0245	.0310	.1020
200°	.0216	.0220	.0220	.0510

The above oils are noted to have contained about 2 per cent of water.

Again viscosity is frequently met with expressed in terms of metric units. Here again two units may be encountered—the absolute unit based on the centimeter, dyne, second, or the centipoise which is .01 absolute unit.

To convert these various measures into one another we have the following:

Viscosity (foot, poundal, second) = viscosity.
(centimeter, dyne, second) ÷ 14.88.
Viscosity (foot, poundal, second) = viscosity.
(centipoise) ÷ 1488.

It is, of course, always desirable to have values of μ and σ derived directly by test from the oil in question and at the working temperatures assumed. In default of this precise information, the best practicable approach must be made by a comparison with known data as in Tables 1, 2, 3, and through formulæ and relations such as those indicated in the latter part of the present paper.

With reasonably accurate values of μ and σ in any given case and a quality of pipe surface not too far removed from what has been termed commercially smooth, the procedure herein indicated may be followed with confidence regarding the resulting values of f and h. With old and roughened pipe surface, due allowance must be made by judgment and in the light of such information as may be available regarding the values of f applicable in such cases.

In any case the procedure herein indicated may be urged as undoubtedly more reliable and more widely applicable than any empirical combination of density or viscosity or both which may be given in a d'Arcy or other hydraulic formula, but which does not rest on the basic principles of rational mechanics, as does the method presented herewith.

Application of Theory

To Obtain Viscosity Coefficient

A value for the coefficient μ may be obtained by following the method outlined by Professor Durand. While this value may not agree with that obtained by the use of a viscosimeter, it should be close enough to use when the absolute viscosity cannot be obtained. For example, under a given set of conditions, $f=.028$, $d=.6667$, $v=3$ feet per sec., and $\sigma=60$ pounds per cubic foot; from Table 1 the abscissa corresponding to $f=.028$ is 16,000. Substituting these values in the expression, abscissa $=\frac{vd\sigma}{\mu}$ gives

$$16{,}000=\frac{3\times .6667\times 60}{\mu},$$

$$\mu=.0075$$

For oil flowing through a single pipe of uniform diameter at a constant rate of weight, the velocity v will decrease as the temperature decreases, and the density σ will increase, $v\sigma$ being then constant. Also, diameter d will decrease with the temperature. This has a tendency to increase the velocity v, but, for the extreme variation in a line handling hot oil, this variation in d will not be more than $\frac{1}{200}$ inch for a 12-inch pipe.

For a steady flow, by weight, in any system of single pipe of uniform diameter which may be under consideration, $vd\sigma$ may be assumed as constant; f then becomes a function of $\frac{1}{\mu}$. That is, the variation in f is then dependent only on the viscosity.

Now, if there is a test on a line showing the temperatures and the pressures existing at various points when there is a steady flow by weight, it is possible to find f and the corresponding μ for the various points and the various temperatures. The μ for the various temperatures having been ascertained, the corresponding f for various conditions of v, d, and σ can be obtained.

To Flow of Oil through Pipes

To apply the theory as given in Professor Durand's statement, the expression $\frac{dv\sigma}{\mu}$ can conveniently be considered by dividing it into dv and $\frac{\sigma}{\mu}$. If the value of dv is approximately known, use such value as a multiplier of $\frac{\sigma}{\mu}$. If not known, assume some convenient value of dv, say from the following:

Table 5

	Inches	Inches	Inches	Inches	Inches	Inches
D	2	3	4	6	8	12
d	$\frac{1}{6}$	$\frac{1}{4}$	$\frac{1}{3}$	$\frac{1}{2}$	$\frac{2}{3}$	1
v	6	4	3	4	3	3
dv	1	1	1	2	2	3

The value of $\frac{\sigma}{\mu}$ can be obtained from the time expressed in seconds as shown by a Saybolt Standard Viscosimeter from Table 6 which is computed by Formula (7).

TABLE 6

Saybolt seconds	$\frac{\sigma}{\mu}$	Saybolt seconds	$\frac{\sigma}{\mu}$	Saybolt seconds	$\frac{\sigma}{\mu}$
40	21,600	100	4596	400	1060
50	12,550	125	3562	500	847
60	9,100	150	2919	750	564
70	7,236	175	2477	1000	422
80	6,050	200	2154	1500	281
90	5,216	300	1419	2000	211

The approximate value of $\frac{dv\sigma}{\mu}$ having been obtained, the coefficient of friction f can be obtained from Table 1. (In general, except for extreme cases and small pipes, it will be found that where the Saybolt Viscosity S is less than 150, a turbulent or eddying flow will be indicated.)

FOR A PIPE OF UNIT (1-FOOT) DIAMETER

If we assume a coefficient of friction $=0.024$ and use $2g=64.4$ then

$$h=f\frac{l}{d}\frac{V^2}{2g}, \qquad (2)$$

becomes by transposition and substitution

$$V=51.8\sqrt{\frac{hd}{l}}.$$

$$\text{Volume (cubic feet per second)} =\frac{V\pi d^2}{4}.$$

This is in the form of

$$\text{Volume} =K\sqrt{\frac{hd^2}{l}},$$

or when $d=1$ or when taken care of in the constant, this can be written

$$\text{Volume} =K\sqrt{\frac{h}{l}}, \qquad (10)$$

for head and length in feet, or

$$\text{Volume} =K\sqrt{\frac{h}{L}}, \qquad (11)$$

when length is in miles.

Constants K for various diameters, quantities and times are given in Table 7.

Caution.—In order to secure reliable results, it is necessary that the proper coefficient f should be used. It is recommended that where possible the coefficient, as obtained

TABLE 7

STANDARD LINE PIPE

Nominal size		Actual dimensions				Cubical contents per foot			Cubical contents per mile		
d feet	D inches	D	Area, inches	$D^{5/2}$	$d^{5/2}$	U. S. gallons	Barrels, 42 gallons	Barrels, 50 gallons	U. S. gallons	Barrels, 42 gallons	Barrels, 50 gallons
1	12	12.000	113.09	498.83	1.00000	5.876	0.139886	0.1175	31,025	738.6	620.5
$\frac{5}{6}$	10	10.020	78.85	317.81	0.63712	4.097	0.09752	0.08193	21,630	515.0	432.6
$\frac{2}{3}$	8	7.981	50.03	179.94	0.36073	2.600	0.06188	0.05198	13,723	326.7	274.5
$\frac{1}{2}$	6	6.065	28.89	90.59	0.18160	1.501	0.03573	0.03001	7,924	188.7	158.5
$\frac{5}{12}$	5	5.047	20.00	57.22	0.11472	1.039	0.02473	0 02078	5,4 6	130.6	109.7
$\frac{1}{3}$	4	4.026	12.73	32.52	0.06520	.661	0.01574	0.01323	3,492	83.1	69.8
$\frac{1}{4}$	3	3.068	7.39	16.49	0.03305	.384	0.00914	0.00767	2,027	48.3	40.5
$\frac{1}{6}$	2	2.067	3.35	6.14	0.01231	.174	0.00414	0.00348	919	21.9	18.4

CONSTANT K FOR VOLUME WHEN $f=0.024$

Nominal size inches	Volume $=K\sqrt{\frac{h}{l}}$ or $=K(h/l)^{1/2}$							Volume $=K\sqrt{\frac{h}{L}}$		
	Cubic feet			U. S. gallons		Barrels per hour		Per hour	Barrels per day	
	Per sec.	Per min.	Per hour	Per min.	Per hour	42 gallons	50 gallons	42 gallons	42 gallons	50 gallons
12	40.68	2440.8	146,464	18,260	1,095,630	26,086	21,913	359	8616	7237
10	25.92	1555.0	93,300	11,630	698,000	16,620	13,960	228.7	5486	4611
8	14.673	880.7	52,830	6,587	395,200	9,410	7,904	129.5	3108	2611
6	7.388	443.2	26,600	3,316	199,000	4,737	3,979	65.2	1565	1314
5	4.667	280.0	16,800	2,095	125,700	2,992	2,514	41.2	988	830
4	2.653	159.1	9,562	1,190	71,440	1,700.6	1,428	23.4	562	472
3	1.344	80.7	4,842	603	36,210	862.2	724	11.86	285	239
2	.501	30.0	1,803	225	13,488	321.1	270	4.42	106.1	89.1

by the method described in Professor Durand's article, be checked with the actual coefficient obtained under working conditions. A coefficient obtained under ordinary working conditions (such as are used in the transportation of oil as a business) may safely be used for small variations of velocity, say from 2 to 10 feet per second where eddying flow is undoubtedly indicated. While it does not appear to be definitely determined that each factor in the expression $\frac{dv_\sigma}{\mu}$ affects the function in proportion to the first power of the factor, experience indicates that these results can be relied upon. The roughness of the pipe, the presence of fittings and bends, and the pulsating effect of a pump probably have some bearing on the velocity at which the eddying flow begins.

TABLE 8

Showing factor by which constant K, Table 7 is to be multiplied when f is not 0.024

$$\text{Factor} = \sqrt{\frac{0.024}{f}}$$

f	Factor	f	Factor	f	Factor
0.020	1.096	0.030	0.894	0.040	0.775
.021	1.069	.031	.880	.05	.693
.022	1.044	.032	.866	.06	.632
.023	1.021	.033	.853	.07	.585
.024	1.000	.034	.840	.08	.548
.025	.980	.035	.828	.09	.516
.026	.961	.036	.817	.1	.490
.027	.943	.037	.805	.2	.346
.028	.926	.038	.795	.3	.283
.029	.910	.039	.784	.4	.245

PIPE-LINE COMPUTATIONS

It is convenient in making all pipe-line computations to use in formulæ, the head in feet rather than pounds per square inch or any other pressure unit. (See Table 11.)

In order to estimate how much oil will flow through a line, it is necessary to have the proper coefficient f, the size of pipe, the length of the line, the elevation of the initial and terminal points, the intermediate high points as some of them may be controlling points, also the low points, as some of them may limit the pressure to be carried on the line. A typical profile with grade line and controlling points of a single uniform diameter section of a system, is here shown.

A, F, C, B is the profile to scale.

A is the initial point or station; C is the controlling point; B, the delivery or terminal station. If the available head h is set off on the profile above A at the same scale as the profile, and a straight line is drawn from this point D to the terminal station, the straight line will represent a tentative grade line which, provided no intermediate points of the profile extend above this line, will determine the flow. If points extend above such line, one of them will control. The real grade line will be the line of least grade which strikes any point between D and B. On the above profile, this controlling point is C. This grade line means a grade at which the required amount of oil would flow if it flowed directly from a reservoir or tank located at D. For the problem, the available head becomes the difference in the elevation of D and C or $D\,H$, and the controlling length is the length from A to C. The elevations beyond C are such that the whole head is not required to produce the flow. This condition is very common as the location of fuel and water supply often require that the pumping stations be located in valleys or along railroads. The vertical distance between the grade line and the profile represents the head on the pipe at the various points. The head $G\,B$ is only in existence if the flow at B is throttled enough to make the head

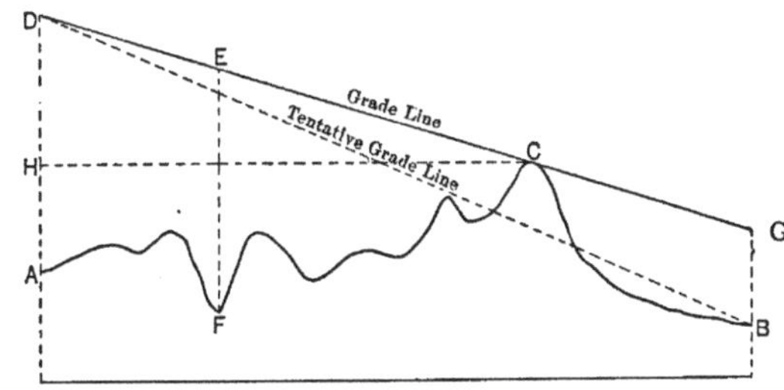

Fig. 6.—Profile showing controlling points on a single uniform diameter pipe system.

equal to $G\,B$, otherwise the pipe will not run full. This head $G\,B$ is often utilized as "intake pressure" which is described later. The flow capacity can be simply computed by Tables 7 and 8.

Often the system consists of more than a single line and also of various sizes of pipe. In such cases, it is convenient to use, in making computations, what is known as "equivalent length." That is a length of some standard size of pipe which will, when used in computations, give a computed carrying capacity of the same amount as the system under consideration. Neglecting the slight change in f due to velocity, the amount that will flow through 4 miles of double line will be the same as would flow through 1 mile of single line under like conditions. In the following discussion the change in the coefficient f due to velocity is neglected. The equivalent length of 4 miles of double line would be 1 mile of single line. To obtain the factor for reducing any size or combination of sizes of pipe to equivalent length of any size selected as a standard, add together the $\frac{5}{2}$ powers taken from Table 7 and divide by the $\frac{5}{2}$ power of the size selected for the standard, and square the result.

Example.—What would be the equivalent length, in terms of 8-inch pipe, of a line constructed of 16 miles of double 8-inch line and 24 miles of 8-inch and 12-inch laid parallel?

16 miles of double 8-inch is equivalent to 4 miles of single 8-inch.

From Table 7, $12^{5/2} = 498.83$

$8^{5/2} = 179.94$ [1]

678.77

$(678.77 \div 179.94)^2 = 14.23$

$24 \div 14.23 = 1.687$

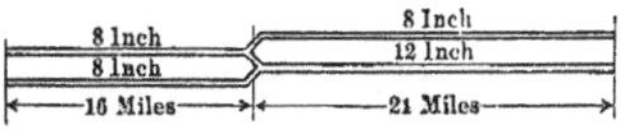

Fig. 7.—Diagram showing manner of connecting.

16 miles double 8-inch equivalent to............ 4 miles of single 8-inch.
24 miles 8-inch and 12-inch parallel, equivalent to 1.687 miles of single 8-inch.

Equivalent length of combination.............. 5.687 miles of single 8-inch.

[1] 8 inch actual size used.

That is, a line of 8-inch pipe 5.687 miles long would show the same computed flow capacity as the 40-mile combination.

The following table shows the equivalent length of different sizes of pipe:

TABLE 9

Diameters D = (actual pipe size)

2-inch	3-inch	4-inch	5-inch	6-inch	8-inch	10-inch	12-inch
1.	7.22	28.05					
	1.	3.89	12.04	30.32			
		1.	3.10	7.76	30.61		
			1.	2.51	9.88	30.85	
				1.	3.94	12.30	30.33
					1.	3.12	7.68
						1.	2.46
							1.

This fiction of equivalent length is used to facilitate the making of computations and can be used for comparing flow capacities only if f is the same in each case.

The following profile and grade line shows method of treating problem.

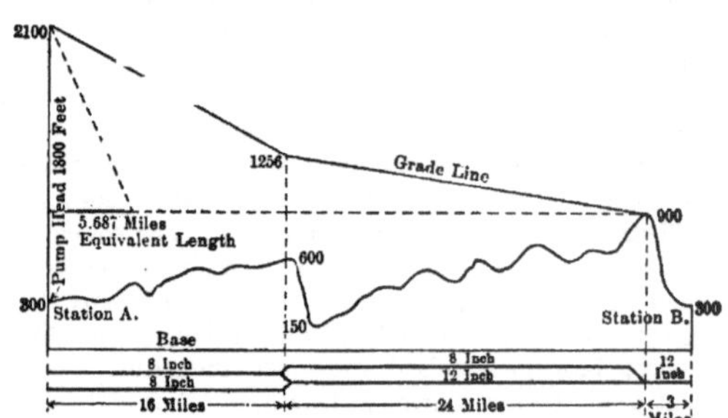

FIG. 8.—Profile and grade line of problem given below.

Problem—Fluid Oil $f = 0.024$.

16 miles double 8-inch line.
24 miles 8-inch and 12-inch parallel.
3 miles of 12-inch beyond controlling point.
Elevation initial station 300 feet.
Elevation connections 600 feet.
Elevation of a low point 2 miles beyond connections 150 feet.
Elevation controlling point 900 feet.
Head due to pump pressure 1800 feet.
Available head 1200 feet.

The equivalent length to controlling point is 5.687 miles equivalent 8-inch, $h \div L = 1200 \div 5.687 = 211$ feet of head per mile of equivalent length available for overcoming pipe friction.

$$(211)^{1/2} = 14.525.$$

Then from Table 7:

$$\text{Bbls. (42 gal.) per day} = 3108 \times (h \div L)^{1/2}.$$
$$\text{Bbls. (42 gal.) per day} = 3108 \times 14.525 = 45{,}144.$$

The elevation above the initial station of the grade line at the connections would be 1800 feet minus the friction resistance of 16 miles of double 8-inch line (4 miles of

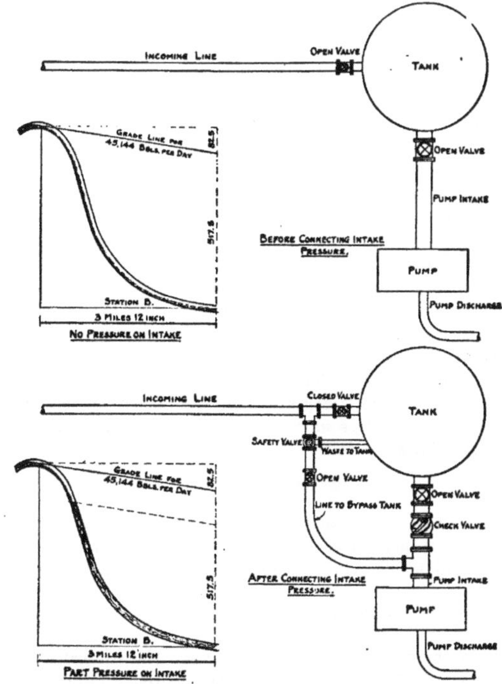

Fig. 9.—Diagram showing connections before and after installing "intake pressure."

single), or 211 feet×4 = 844 feet. 1800 feet minus 844 feet = 956 feet above station or 1256 feet above the base. The pressure at the connections would be 1256 feet minus 600 feet or 656 feet. The equivalent length, in terms of 8-inch of the 3 miles of 12-inch beyond the controlling point would be (see Table 9) 3÷7.68 or 0.391 mile. The friction head per mile of equivalent length in the above problem is 211 and the flow capacity would require 211×0.391 foot to overcome the friction, or 82.5 feet. This leaves 517.5 feet to be wasted or utilized.

Intake pressure can often be employed to utilize the head which would otherwise be lost where the oil flows into a station under conditions like those shown in the above example. If the station at B were to pump the oil on again at 1800 feet head, by making use of the available intake pressure, it would be possible to prevent most of this loss. In practice, at least 450 feet could easily be made available and would result in a saving of 25 per cent of the power necessary to do the pumping at station B. The following sketch shows a method of connecting up to make this saving possible and also shows an enlarged profile of the last 3 miles of the example above. (Fig. 10.)

The fluid in the intake line increases when the pump at A runs faster than the one at B. This results in an increase in the pressure on the intake which tends to reduce the load on pump B, therefore causing it to speed up. If the pump at A runs slower than the one at B, the intake pressure is diminished, thus causing B to slow down. The intake pressure tends to act as a pump governor. In practice it is found that all but about 50 feet of this head can thus be utilized.

In 1894 a test was made at Cameron, Mills, N. Y. The pump, when running with 900 pounds pressure, required 39,000 pounds of coal per day; after connecting up the intake pressure system, the pump running at same speed and pressure and with 265 pounds on the intake, required only 27,000 pounds of fuel and the pump ran better with the intake pressure. This test showed that the saving was more than proportionate to the work done. The reason for this was undoubtedly that the boilers were, in that case, more economical under the reduced load.

In building a line across country, if it were possible to locate each station so that the work on the pumps would be the same, the same number of stations would be required as would be needed if the grade were a straight line from start to controlling point, independent of the contour of the intervening country except as a limit to pressure on the pipe. It is necessary to have a profile in order properly to locate the stations. It is often found desirable to locate stations at points where the amount which can be pumped through each section with the same pressure is different. When this condition exists or when it becomes desirable to increase the amount which can be conveyed, recourse is often had to looping the lines.

"*Looping lines*" consists in laying a parallel line for part of the distance and connecting it in at both ends. Take a simple case of a level 8-inch line 40 miles long having a capacity of 20,000 barrels per day.

$$20{,}000 = K(h/L)^{\frac{1}{2}}. \qquad (11)$$

If it is desired to increase this capacity to 30,000 barrels per day, K and h remaining constant, the only factor changed would be L. As it is not possible to change the actual length, recourse is had to changing the *equivalent length* of the line between stations; that is,

$$30{,}000 : 20{,}000 :: 40^{\frac{1}{2}} : x^{\frac{1}{2}}.$$

This gives $x = 17.78$ the required equivalent length. The equivalent length must be shortened $40 - 17.78$ or 22.22 miles. As each mile of loop shortens the equivalent length $\frac{3}{4}$ mile, it is evident that $\frac{4}{3}$ of 22.22, or 29.63 miles of loop will be required. The flow capacity of the line would be the same no matter where this loop was placed, provided it did not extend beyond the controlling point or change the grade line so that some other point became the controlling point. The profile and grade lines below show the condition of pressure along the line.

It is usual to place the loop at the low pressure end of the line as it gives a lower pressure all along the system. Sometimes, on account of changing the controlling point or some local condition, it becomes advisable to locate the loop at some other point.

Loops in more than one line or in a composite line can be computed in the same manner. The following formula is for computing the amount of loop required from the difference in the equivalent lengths:

Let N = number of lines to which loop is to be added;

E_1 = equivalent length of existing lines;

E_2 = required equivalent length.

$$\text{Loop required} = (E_1 - E_2)\left(\frac{N^2(N+1)^2}{2N+1}\right).$$

The value of the multiplier $\frac{N^2(N-1)^2}{2N-1}$ for several cases is given in the table below.

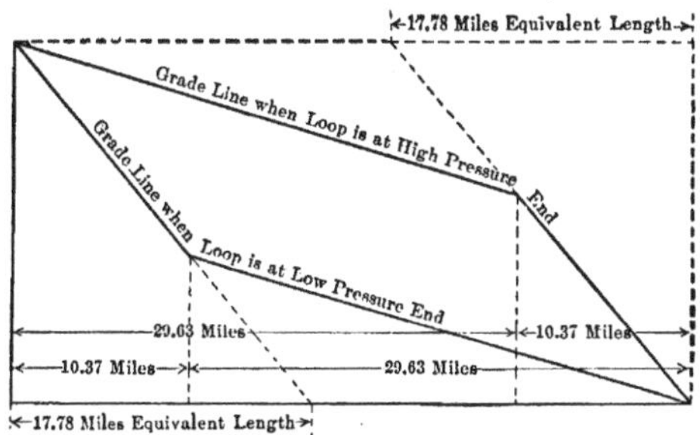

FIG. 10.—Profile and grade lines graphically showing pressure along the line.

TABLE 10

N	1	2	3	4	5	6
Multiplier....	1.33	7.20	20.57	44.44	81.81	135.5

Pressure in pounds per square inch can be converted to head in feet from formula (5).

$$\sigma = \frac{62.4 \times 141.5}{131.5 + B\acute{e}^{\circ}}. \qquad \frac{144}{\sigma} = \text{head}.$$

$$\text{Head per pound} = 0.016309 \times (131.5 + B\acute{e}^{\circ}).$$

The following Table 11 shows head in feet for 1 pound pressure per square inch, also specific gravity and pounds per U. S. gallon corresponding to Baumé degrees for liquids lighter than water. All are computed from modulus 141.5 Those marked (*) are from Saybolt's tables.

Charts can be used to great advantage in solving the problems which frequently occur in pipe-line work. Each pipe-line has its own particular problems and many ingenious charts have been produced. A flow capacity chart is here shown on logarithmic paper to indicate how simply some of these labor saving charts can be made.

The chart below is constructed from Formula (11) and Table 7 by taking two points at different rates per mile and drawing a straight line through them. Take the 8-inch flow line as a sample; if $h=1$ and $L=1$ then volume = 129.5; then taking $h=100$ and $L=1$ the volume becomes 1295. A straight line through these points shows on the chart all flow capacities from 1 foot per mile to 1000 feet per mile. It must not be forgotten that $f=0.024$ and the result as shown by the chart will have to be changed

TABLE 11

Baumé degrees	Head in feet	* Specific gravity	* Pounds per U. S. gallon	Baumé degrees	Head in feet	* Specific gravity	* Pounds per U. S. gallon
10	2.308	1.000	8.331	40	2.797	.8251	6.874
11	2.324	0.9930	8.273	41	2.813	.8203	6.834
12	2.340	.9861	8.215	42	2.830	.8156	6.795
13	2.356	.9792	8.158	43	2.846	.8109	6.756
14	2.373	.9725	8.102	44	2.862	.8063	6.717
15	2.390	.9659	8.047	45	2.878	.8017	6.679
16	2.406	.9593	7.992	46	2.895	.7972	6.641
17	2.422	.9529	7.939	47	2.911	.7927	6.604
18	2.438	.9465	7.885	48	2.927	.7883	6.567
19	2.454	.9402	7.833	49	2.944	.7839	6.531
20	2.471	.9340	7.781	50	2.960	.7796	6.495
21	2.487	.9279	7.730	51	2.976	.7753	6.459
22	2.503	.9218	7.680	52	2.993	.7711	6.424
23	2.520	.9159	7.630	53	3.009	.7669	6.389
24	2.536	.9100	7.581	54	3.025	.7628	6.355
25	2.552	.9042	7.533	55	3.042	.7587	6.321
26	2.569	.8984	7.485	56	3.058	.7547	6.287
27	2.585	.8927	7.437	57	3.074	.7507	6.254
28	2.601	.8871	7.390	58	3.091	.7467	6.221
29	2.618	.8816	7.345	59	3.107	.7428	6.188
30	2.634	.8762	7.300	60	3.123	.7389	6.156
31	2.650	.8708	7.255	65	3.205	.7201	5.999
32	2.667	.8654	7.210	70	3.286	.7022	5.850
33	2.683	.8602	7.166	75	3.368	.6852	5.708
34	2.699	.8550	7.123	80	3.449	.6690	5.573
35	2.716	.8498	7.080	85	3.531	.6536	5.445
36	2.732	.8448	7.038	90	3.612	.6388	5.321
37	2.748	.8398	6.996				
38	2.764	.8348	6.955				
39	2.781	.8299	6.914				

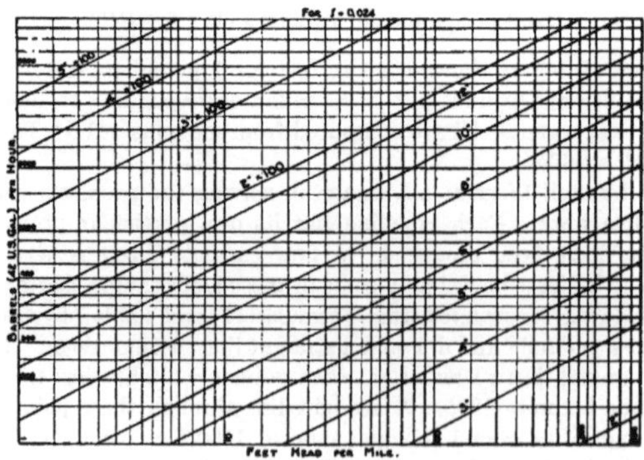

FIG. 11.—Chart showing flow capacity.

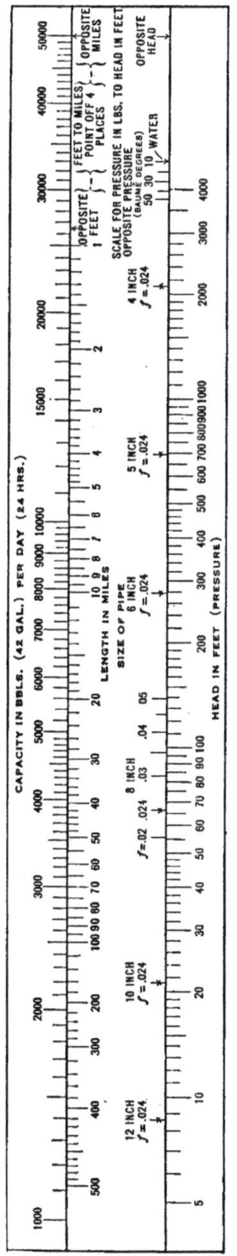

Fig. 12.—Slide rule for computing flow capacity.

accordingly, provided unusually high or low velocities are indicated. In practice, both high and low velocities of flow are avoided; the high velocities, because of the expense of pumping and cost of pumping plant; the low velocities, because of the cost of the pipe system. Generally a velocity of from 3 to 6 feet per second is found to give satisfactory results. A velocity of 3 feet per second corresponds to nearly 50 miles per day or a little over 2 miles per hour.

The slide rule is very convenient in computing flow capacities. A special slide rule designed for this purpose is here shown as a suggestion. It enables one to perform these computations quickly and is of great assistance to the pipe-line engineer. The rule shown is a slight modification of the one copyrighted by Forrest M. Towl, C. E., in 1889.

While the more viscous oils obey the same laws as the more fluid ones, the computation of the flow capacity is more complicated. The effect of temperature on the more viscous oils is very marked and a slight change in temperature often makes a great difference in the viscosity. On account of the increase in flow capacity at higher temperatures, the more viscous oils when transported by pipe-lines, are usually pumped "hot," that is, at a temperature above the normal for the locality where the operation is taking place. This method was first used on a long-distance line by the Pacific Coast Oil Company (now Standard Oil Company of California). An 8-inch pipe-line was commenced in 1902 from Waite station, near Bakersfield, California, to Point Richmond on San Francisco Bay. Operation was begun in July, 1903. This line was about 280 miles long and originally intended to convey 10,000 barrels (42 gal.) per day. The first set of stations were located from 23 to 30 miles apart. Under normal conditions, with a pump pressure of 600 pounds, it was only possible to transport about 900 barrels per day of 15° to 16° Baumé oil through this line. The stations were designed to heat the oil with the exhaust steam from the pumps.

The heaters were of the return tubular type. The oil passed through the lower set of 2-inch tubes, 16 feet long, to the rear chamber and then returned through the upper set. The tubes were surrounded by the exhaust steam, the whole arrangement being similar to a surface condenser. The pumps were of the high-pressure, compound, duplex type and furnished more than enough heat to increase the temperature of the oil from a normal of about 60° F. to 180°.

It was found that it took too long to " warm up " these sections of the line, and intermediate pumping stations were built. These stations not only increased the available head per mile but also reheated the oil. (Probably more than 85 per cent of the heat of the required steam remained in the exhaust after the pumping operation. The heating of the oil alone would probably have taken 85 per cent of the fuel necessary to run the pumping station.)

It is difficult to ascertain the specific heat of the oil as the material itself is a very poor conductor of heat and the particles of the more viscous oils do not circulate freely to carry the heat to other particles. This increased the length of time it took the oil, pipe and surrounding material in a 28-mile section to warm up. At 900 barrels

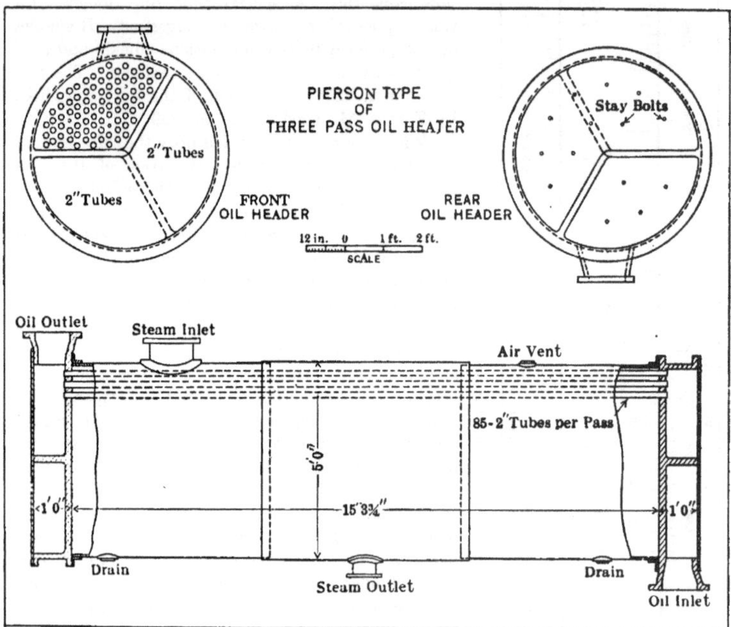

FIG. 13.—A recent type of three-pass oil heater. This type was used on the Scotland pipe-line mentioned on page 380.

per day, less than 3 miles of the cubic capacity of the 8-inch line containing cold oil was displaced by hot oil in twenty-four hours. It was found that the top of the line would warm up long before any change was noted in the temperature of the under side. It took over three weeks to " warm up " one of the original sections of the line in the summer of 1903, and if the ground had been damp and the normal temperature lower, it would have taken much longer. With the reduction of the distance between stations to about 14 miles, it was found that the line could be " warmed up " in from thirty-six to forty-eight hours. After one of the first stations was finally heated to a good working condition, it was found that the flow capacity was increased to about 18,000 barrels per day.

A pumping record is given (Fig. 14) for the second 14-mile section of the Bakersfield line and shows three rates of flow and the corresponding pressures and temperatures. The data are incomplete and are only given to show that each rate of flow presents different conditions. Each section of the line as well as each kind of oil is a distinct problem. It is entirely reasonable to expect that there may be conditions where

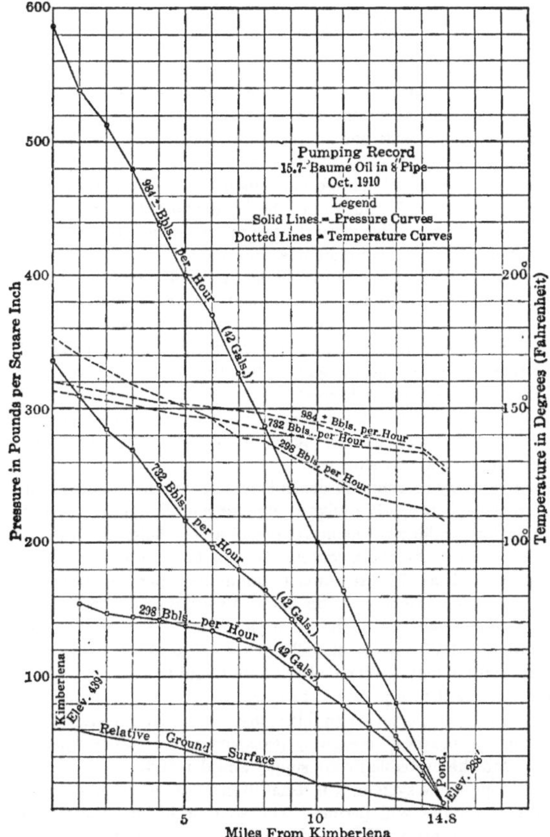

Fig. 14.—A pumping record for the second section of the Bakersfield pipe-line, showing three rates of flow, and the corresponding pressures and temperatures.

it will require more pressure to pump a small amount of oil than to pump twice that amount through the same line. In making tests on these long lines handling hot oil, it is necessary to see that the condition of steady flow is continued until all parts of the plant have become normal for that rate. This may take days or even weeks.

Pumping a mixture of oil and water was tried in several places. Oil and water will not mix. The cohesion of the oil is greater than that of water and when the two fluids are slightly agitated together, the oil forms globules in the water. When in this state the mixture will flow or can be pumped through a pipe-line with much less friction than if the oil alone were moved. The continued agitation of the pump and the moving of the combination of material through the line soon results in a more intimate mixture of the oil and the water. When the globules become very small, the oil and water form an emulsion which is often more viscous than oil at normal temperature. These emulsions are very difficult to separate and dangerous to handle if the oil is used for either fuel or refining purposes. Some oils are more easily emulsified than others. In some cases it is almost impossible to separate the oil and the water. The Pacific Coast Oil Company engineers experimented with this method in 1901 before they decided to adopt the "hot oil" system. The Associated Pipe-Line Company later built a line nearly parallel to the Bakersfield line and adopted the system of oil and water mixture for reducing the friction. This line adopted the "rifled pipe," a description of which can be found in The Engineering Record, Vol. 57, No. 21. The inventor of the system claimed that the centrifugal motion caused by the rifling would carry the water to the outside of the pipe. An analysis of the forces shows that no advantage could be expected from the rifling. After several years of experience the plan was abandoned and the method of pumping the oil hot was adopted.

Heating the oil to **high temperatures** under sufficient pressure to keep it in liquid form changes the nature of the oil. Experiments by the Pacific Coast Oil Company in 1902 showed that this treatment when applied to the Bakersfield oil, greatly reduced the viscosity and also favorably changed the oil for refining purposes. The contemplated use of this system on the Bakersfield line was abandoned as it was found that the treated product emulsified with great rapidity when it came in contact with water and the resulting emulsion was much harder to pump than the unheated oil. It was found almost impossible to separate the oil from the water except by a process of distillation carried on at a very slow rate.

The **power necessary** to **pump** oil depends on the volume handled and the pressure required. Two systems properly designed, one for handling a given volume of viscous oil and the other for handling the same volume of a very fluid oil, will require the same power, but the one for the more viscous oil will have shorter or larger pipes. It requires about two indicated horse-power to pump 1000 barrels (42 gallons) of oil per day for each 100 pounds pressure. This allows 15 per cent for loss between the steam cylinder of the engine and the discharge of the pump. The work done on the fluid is 1.7 horse-power per 1000 barrels (42 gallons) per day for each 100 pounds pressure.

Surveys and right of way.—When oil has been found in apparently paying quantity, the question of getting it to a point where it will be used is one of the most important to be solved. This transportation problem involves a careful study of the quantity available, its quality, nearness to existing transportation facilities, location and extent of the demand, and the probable future supply and demand.

In designing a pipe-line system, it is usual to keep as nearly in a straight line across country as the conditions will permit. The contour of the country, difficult river crossings, swamps, distance which the pipe must be transported, road conditions, and the nature of the country through which the line is to pass, all have to be considered, together with the kind of soil or rock through which the ditch must be dug. These conditions as well as probable difficulty in procuring a right of way, often make it advisable to deviate from a direct line. If after due consideration has been given to these phases, it is decided that as a business proposition, transportation by pipe-line is a feasible plan, then a preliminary survey is made. The engineers lay out a route and make a map showing the ownership of the properties, the location of buildings, fences, streams and other physical features, a profile of the elevations along the line, and the proposed location of the necessary pumping stations. In making surveys, all measurements are taken as surface measurements in order to show as accurately as possible the amount of pipe needed. This is important in rough country. A copy of the map is furnished to the right-of-way men for use in securing the necessary consents to the laying of the line. These men then go upon the ground and explain the matter to the landowners and take such papers as may be necessary, using some standard form as far as possible. The taking of these papers differs only slightly from any other business involving buying, selling and dealing with different individuals. In many cases, the person securing the papers is the one who after construction work is completed, again deals with the landowner in settling for the damages that may have been done. Care should therefore be taken to have the papers drawn up in as clear a form as possible and to leave the landowner in a friendly mood. Many forms of grants have been and are in use, some giving the right to lay but one line, some more than one. They include the right to erect a telegraph or telephone line if necessary. The form considered the best is one which in the granting clause, gives the right to construct from time to time, and maintain and operate one or more lines of pipe with such accompanying telegraph or telephone lines as the grantee may deem necessary for the conduct of its business, with free ingress and egress to construct, operate, maintain and from time to time, alter, repair and remove the same. This form also provides that in case more than one line is laid under the grant, the landowner shall receive a fixed compensation for the right of way for each additional line so laid and that all lines under the grant shall be constructed on a strip of ground not exceeding 10 feet in width. The grantee agrees to bury the line so as not to interfere with the ordinary cultivation of the soil and to pay any damages which may arise to crops and fences, such damages, if not mutually agreed upon, to be determined by arbitration, each side selecting an arbitrator and they a third, the award of any two to be binding. In cases where it is necessary to resort to arbitration, a paper is signed by the grantor and grantee naming their arbitrators and giving a copy of the grant under which they are appointed. The grantor specifically reserves the right to use and enjoy the premises fully except for the purposes granted, and the right of way cannot be fenced. In most cases it is hard to locate a line exactly on any individual property a few months after the work has been completed.

After the grants have been secured, " requirements " are made out for the construction men. These requirements set out all particular information as to the location or work which may be in the original grants. In laying a long line, a company usually encounters a few landowners with whom agreements cannot be made. Resort is then had to condemnation, pipe-line companies having the right to condemn property for pipe-line purposes in most of the states. This method is, however, of quite infrequent use when the vast number of properties is considered. All grants of right of way are executed in strict accordance with the laws of the state where the land is located and with as much care as would be exercised in the case of actual purchase of the land. Existing mortgages are usually ignored, however, as experience has shown that the cost of securing releases from the same would not warrant the trouble; the instances in which the grant has been cut off by foreclosure of the mortgage are very few and may practically be ignored. The grants are recorded in the office of the proper party in the county where the land is located, so that subsequent purchasers of the land may have legal notice of the same. Sites for building the pumping stations are secured either by leasing or purchasing the necessary property.

Preparing right of way.—Preparing the secured right of way " for the laying of the line " is the next step. This is done by the " clearing gang " or the " right-of-way gang " as it is sometimes called. In some localities very little of this kind of work has to be done, while in other places it may be necessary to cut a path through the woods and build roads not only along the right of way but even to it from the nearest transportation facilities. It is important that the pipe should be shipped to a point from which it can be conveniently and expeditiously distributed along the right of way. The pipe is usually shipped in open cars or by boat. Care should be taken to see that the thread end of the joints is properly protected and that the pipe is stowed so that it will not be damaged during transportation. The loading, stowing and unloading in and from a steamer is a job for an expert, and is usually done by special stevedores. Great care must be taken to avoid a shifting of the cargo. When the pipe is shipped by car it is properly loaded at the mill. If the load has been transferred during shipment, it should be carefully inspected before unloading. The unloading from the cars where power is not available is generally done by placing skids at the side of the car and allowing the pipe to roll down and along the surface to the pipe pile. A gin pole can often be used to facilitate the unloading. Where it is necessary to check the momentum of the pipe, a rope-break is often employed. This break consists of a rope fastened to the top of the car near the skid. When the pipe is about ready to start down, the rope is passed *over* and around the pipe a couple of turns. The friction causes the pipe to roll slowly. The speed can be regulated by the number of turns of the rope around the pipe or by tension on the free end. In the manual handling of the pipe, it is rolled, and where necessary, lifted by means of a "handspike" about 4 feet long placed in the end of the joint.

Stringing the pipe.—The "*stringing*" or distributing of the pipe along the proposed route is very important. The pipe should be placed so that the joints will be located where they are to be screwed together. Each joint screws about 2½ inches into the collar; that is about 1 per cent of the length. The pipe when joined together, is like a big stiff cable and has to be bent to conform to the surface. An intelligent "stringing" foreman soon becomes accustomed to making these allowances so that it is seldom necessary to carry a joint of pipe more than a step or two. This can easily be done when the joint is placed in position for screwing into the line. Many methods of distribution are used, local conditions indicating what is best.

The joints of pipe are usually screwed together before the ditch is dug but sometimes the conditions are best met by digging the ditch first or even screwing the pipe together after it is placed at the bottom of the ditch. If the ditch is dug before the line is laid, it is in the way and delays the work of screwing the sections into the line. There is also trouble to be anticipated from the possible caving in of the ditch and from washing out, should rains occur. The best results are usually obtained by screwing the joints together and ditching afterward.

Most of the lines in the United States have been built with screw-joint pipe. A few lines have been built with welded joints. There are usually from 250 to 260 joints in a mile of line pipe. Leaky joints are an exception and do not occur where the line is properly laid and where care is taken to clean and inspect the threads of the pipe and the collars before screwing together. With carefully inspected pipe, there should not be more than one defective piece in ten thousand.

The only possible arguments in favor of a welded line would be that it was better, or that it was cheaper than a screw-joint line. Neither of these claims has been substantiated up to the present time (1921). A screw-joint line may leak high-pressure gas or air and show no perceptible leak when handling a liquid under a higher pressure. The screw-joint line is the only kind that has been thoroughly tested with satisfactory results for oil lines, and other kinds will therefore not be considered in this section.[1]

Laying the line, that is, screwing the joints together in place, is done by hand or by pipe-laying machines. The following list of tools for a gang laying pipe by hand is copied from that furnished the U. S. Navy pipe-line unit which built the line across Scotland in 1918. It was designed to supply two gangs laying 8-inch pipe:

[1] For lines for gas see the section on Gas Transportation by Mr. T. R. Weymouth.

14 pairs 8-inch tongs
2 pairs 8-inch chain tongs
8 8-inch jack boards
8 8-inch pipe jacks
4 8-inch carrying irons
8 8-inch spike bars
4 8-inch swabs
4 sucker rods for swabs
4 8-inch caliper tongs
12 8-inch growler boards
8 whitewash brushes
12 wire brushes
12 14-inch half-round files
12 15-inch monkey-wrenches
12 hand chisels
12 machinists' hammers
12 pairs hand leathers
3600 tong keys
60 tong wedges
1 8-inch stock and die
1 8-inch pipe cutter
6 8-inch bull plugs
1 set extra dies for stock

The stock and dies should be for the regular line-pipe thread known as "Brigg's Standard Gage as made by Pratt & Whitney Company." To the above list should be added sufficient rope or chain for "rolling" the pipe, and material for making handspikes, if the same cannot be obtained on the job.

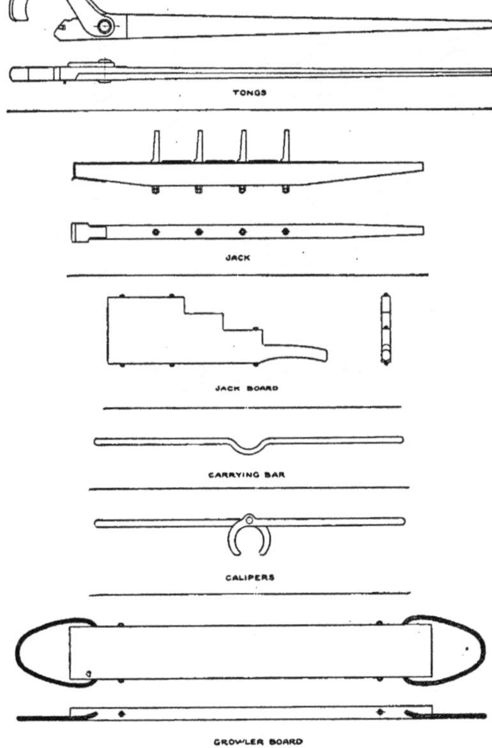

FIG. 15.—Tools used in laying a pipe-line.

Pipe tongs are long-handled tools used in screwing the pipe together. The steel tong key bites into the pipe when the handles are pressed together. One or two pairs of tongs are used to " back up " the last joint and prevent its turning. About four or six pairs are usually used to screw the joint into the finished line.

The pipe jacks are used to support the pipe while it is being turned. The jack boards are a brace for the jack, and the growler boards are used as a foundation on which the jack and jack board rest.

In putting a line together the first thing to be done is to see that any dirt in the joint is removed. This is accomplished by running a " swab " through the pipe. This swab consists of disks fastened on a rod. The rod is usually of the wooden " sucker rod " type, which is the jointed rod used in pumping wells.

FIG. 16.—Hand pipe-laying tools in position for use.

After the pipe is cleaned, the thread in the collar and at the end of the joint is carefully cleaned and then oiled. Sometimes compounds are used which are intended to help make the joint tight, but clean threads properly lubricated and screwed together do not need a compound to make them tight. In laying the pipe, the joint is usually screwed into the collar of the last joint laid, as experience has shown that this method facilitates the operation. Care should be taken in stringing to see that the collar end is " ahead," that is, in the direction toward which the laying gang will be working. To avoid having to put connections into a line, two gangs often start at a point and work in opposite directions. The stringer should know in advance if this is contemplated. The operation of screwing a joint into position is as follows:

The last joint is lowered to a convenient point and the jacks are moved ahead to the approximate position they will occupy while the new joint is screwed in. The joint is lifted into place and a man, the " stabber," standing at the end of the joint with a handspike, moves the joint until it is straight with the last joint laid. Word is then given to roll the joint. As soon as the thread catches, the stabber indicates that it is properly started. The rolling is done first by hand and then by ropes or chains which are wound around the pipe three or four turns. One man pulls back on one end to make the rope or chain grip the pipe, while several others pull, causing the joint to rotate in the required direction. Regular line pipe, say 8-inch size, has about

Fig. 17.—River clamp for strengthening joint and adding weight.

twenty taper threads and about one revolution can be gained each pull without the men moving from their tracks. When the friction becomes so great that this method cannot be used, the tongs are placed on the line. They are at first operated in two sets, the one screwing while the other recovers to get ready for the next stroke. At the last, when the friction becomes greater, all of the tongs work together. As many men as can conveniently be used are employed on each pair of tongs. In order that all may work together, they are kept in unison by a man with a hammer who beats time on the collar. It is the business of this man to signal by the hammer for the changes, and to watch and see when the joint has been sufficiently screwed up. The threads of line pipe are very accurately cut and the last thread disappears into the collar when the joint is properly made.

Fig. 18—U-bolt clamps used for repairing lines.

Pipe-laying machines.—One of the first pipe-laying machines used on the oil lines in the United States was a steam-operated, coal-burning machine built by the National Transit Company in about 1890. This was used on the line of the Indiana Pipe Line Company in 1891. It had a head which clamped on the collar of the pipe and tightened as the head started to revolve. This wrench head, which automatically gripped the pipe, was on the end of a telescopic shaft and operated as a universal joint. The shaft also had a joint, so that the head could be placed at almost any point within a few feet from where the traction engine stopped. The whole apparatus was fitted on an ordinary agricultural traction engine. This machine was abandoned and it was not until the internal combustion engine came into use that other machines were made. In about 1905, the National Transit Company built a machine which was similar to the one described above, except that the power was furnished by an internal combustion engine. This machine was used experimentally on the Prairie Oil & Gas Company line and on the line built by the Ohio Oil Company. Later, the development of this type of machine was taken up by the Buckeye Traction Ditcher Company of Findlay, Ohio, and was perfected. This latter type of machine has been used among others, by the Illinois Pipe Line Company in Wyoming, Illinois, Indiana, and Ohio; by the Gulf Pipe Line Company in Texas, and by The Prairie Pipe Line

Fig. 19.—Stuffing-box clamps.

Company in the Mid-Continent field. This machine screws up the joints at the rate of from 11 to 19 revolutions per minute. The general construction of the machine is shown in Fig. 20.

In the California and Mid-Continent fields, a pipe-laying machine has been

Fig. 20.—Buckeye pipe-laying machine.

developed by the California Pipe Line Machine Company now of Kansas City, Mo. This machine consists of an internal combustion engine of the automobile type mounted on a frame which travels on the pipe as a track. A crane hung in front of the machine is used to take up the weight of the pipe and assist in placing the joint in position. With this machine it is easy to line up the joints of pipe. The method of operation of this machine is to take the pipe and place it in position by the crane, after which the tongs or clutches are closed on the pipe. The hold-back or back-up tong is attached to the frame of the machine which rests solidly on the ground while the pipe is being screwed up. The pipe is supported ahead and back of the machine. The hold-back clamp and the screwing clamp or head are very near to each other, so that the strain

Fig. 21.—California pipe-laying machine shown in working position.

Fig. 22.—New model California 8- and 10-inch machine.

of screwing is only applied to a short section of the pipe. As soon as the joint has been screwed up the spuds or legs supporting the machine are withdrawn with the growler board leaving the weight of the machine on the line; the machine then travels under its own power to the head of the next joint and the operation is repeated. The machine is used to supply the weight necessary to make bends in the pipe. This, of course, requires all the bends to be made either as over bends or sag bends. After bending, the joint is rotated to the required position. Fig. 22 shows general construction of this machine.

It is plain that by the use of pipe-laying machines, lines can be laid with less than half the number of men necessary when they are laid with hand tongs, and that the pipe can be laid more rapidly. Care must be taken to see that the machine does not screw up the pipe too rapidly, as the collars may heat and damage the threads. Recent developments in the threading of pipe indicate that this trouble can be avoided by using a new type of electroplated thread which prevents galling. With a pipe-line machine of any type it is necessary in some places to lay a few joints by hand. In a farming country where wire fences are frequent, objection is often made to the use of a machine or any apparatus which requires the cutting of the fences. As a pipe-line machine will not pass through the ordinary wire fence without cutting, it is necessary to move it around. This process is slow and where this requirement is insisted on, the lines are usually laid by hand. It is claimed that the pipe-laying machines screw up the pipe faster than can be done by hand. In case of trouble, like cross-threading or dirt which has not been removed from the threads, the pipe machine does not usually slow up enough to give warning of the trouble. When laying by hand, the additional force required is noticed, and the joint can be removed and the trouble rectified. There are about twenty threads on the ends of a joint of 8-inch pipe. This requires more than twenty revolutions. An experienced gang working in favorable country where but few bends have to be made in the pipe, will lay about twenty-five joints per hour. A gang working nine hours per day in the San Joaquin Valley in California, in the summer of 1902, laid 28 miles of pipe in thirty-four days. This is similar country to that in which the California machine lays at the rate of about 1 mile a day. In rough country requiring a number of bends, road crossings, stream crossings, and other more difficult construction, the work is much slower.

Bends are usually made on the ground, the jacks being used to support the pipe where the bend is to be made, while the weight of the men or the machine does the bending. Shop bends are sometimes used around stations or as a starting point from which to lay pipe in both directions. Bends of open-hearth steel, wrought iron or soft Bessemer pipe can be made cold by the gang on the ground in less time than it would take to screw in a bent joint. Hot bends are sometimes made in places where it is difficult to apply the force necessary to bend the pipe cold. The heat is applied either by a kerosene torch with a shield to keep the flame around the pipe, or by building a wood fire at the point where the bend is to be made. A number of pipe-bending machines have been devised, but most of them are too heavy to carry and do not save time when a line is being built across country.

In starting a line, a bull plug is usually screwed into a collar which is too tight to be removed without tongs and a fair-sized gang. A similar plug is screwed into a collar at the close of each day's work to prevent small animals from entering and to prevent malicious filling of the line. No light material should be put into the line for safe keeping before the plug is screwed into place, as during the night the air in the pipe will cool sufficiently to cause a strong inrush of air when the bull plug is removed the next morning.

The pipe, when in the hot sun during the day, is expanded and will contract during the night. The elasticity of the steel is sufficient to take care of the ordinary expansion and contraction if the pipe is held firmly in place. When the line is to be used for pumping hot oils, proper consideration should be given to the strains on the pipe and the surrounding material. Expansion joints are not needed except in a few special cases for facilitating the removing of fittings or in taking care of the strains on the pumping machinery. It has been found where expansion joints have been put in along the line that they seldom, if ever, indicate any movement. The laying or screwing together of the joints must be properly done in order to insure a good line, but a well-laid line can be greatly damaged if improperly handled after it is put together.

"The ditching, lowering, and filling" gangs are responsible for the work after the joints are properly connected. The ditch may be dug by hand, or in some places by a ditching machine. The work on a hand-made ditch requires no extensive description. The important points are that the ditch should be dug in the proper place, that it should be uniform, and that the sides should be as firm and regular as the soil will permit.

Ditching machines can be used where the soil is firm and comparatively free from rock or large boulders. The modern "trench excavators" or ditching machines will handle boulders if they are not too large for the buckets. This kind of work delays the machine and reduces the cost advantage when compared to hand ditching. Typical of this kind of trenching machines are those made by the Austin Machinery Corporation, Railway Exchange Building, Chicago, Ill., and by the Buckeye Traction Ditcher Company of Findlay, Ohio. A very simple machine was invented and built by one of the early pipe-line engineers, the late John Page, C.E., for ditching in firm, rock-free soil in northern Ohio. It consists of a large, heavy cylinder with sharp steel disks extending about 8 inches on each side. These disks were slightly dished outward so that when the cylinder was rolled over the surface the disks cut into the soil and wedged it in between the disks. The dirt was carried up and over as the wheel revolved and so removed by a plane-like knife which extended out between the disks. The earth was scooped out and fell into an inclined trough which was shifted so as to leave the material on either side of the ditch. The apparatus was guided by wheels and drawn by horses. By operating the machine three times over the ground, several miles of ditch 2 feet deep can be excavated in a day.

Ditch location.—The ditch must be located in the proper place so that the pipe can be properly put into it. The ditching gang or machine should be able to make the same speed per day as the tong gang or pipe-screwing machine. No set rule can be made as to the proper location of the ditch in order that the pipe will properly fit into it. Experience under the conditions actually existing is the only proper guide. Where the pipe is exposed to the sun it will heat and expand. The expansion will force the pipe in any direction in which the strain becomes sufficient to overcome resistance. The contraction at night will not necessarily return the pipe to its original position. A pipe-line rounding the point of a hill may slide down during the day and not return during the night. Sometimes this expansion accumulates a sufficient force to pull the line apart. A line buried in the summer may pull apart in the winter or may strain the connections enough to cause a leak. It is considered the best practice in the summer or when the conditions cause the line to expand during the day, to put it in the ditch early in the morning before the line has been heated by the sun.

The line is "prepared for lowering" by placing it upon skids over the ditch. To remove these skids the pipe is raised by means of levers, windlasses, or chain blocks, usually in connection with a heavy horse or a tripod.

Back filling is usually done by a light, flat wood scraper pulled by horses, by a road machine, or by hand with shovels and tamping bars. Care must be used in hilly or rolling country to prevent the line from being washed out.

Road Crossings are usually put in by the tong gang at the time the pipe is put together. The filling should be thoroughly tamped and the top should be of the same material as the road bed or of some material which will not settle, if such is procurable. Sometimes it is possible to bore under the road with an auger made of a piece of pipe a little larger than that used for the line. This auger is made by cutting and bending the end of the pipe in such a way that, when rotated by the tongs, it will cut away the earth so that it will fall inside the pipe. The auger must be forced ahead by some means such as levers, screw-jacks, or blocks.

Railroad crossings are made in the same manner, usually under the direction of the section foreman. It is sometimes necessary to tunnel or excavate the crossing, protecting the road-bed while the work is in progress. Often it is possible to use some existing opening to facilitate the crossing. It is easier to cross a railroad in a fill than in a cut.

Bridging of railroads, roads, or streams is sometimes resorted to. The pipe may be used as part of the bridge. It is generally better to have the line buried.

Stream crossings are put in to meet the various conditions that are encountered.

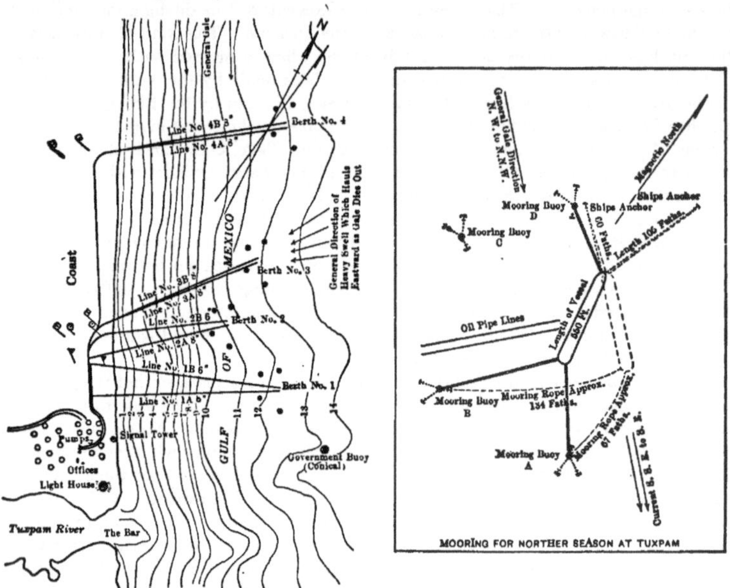

Fig. 23.—Boat anchorage off Tuxpam, Gulf of Mexico.

Where the bottom is hard and the lines are liable to be disturbed by ice or other floating material which may jam and partly dam the stream, the line must be put in a ditch or trench. The line must also be trenched where there is danger that large anchors may pull up or break the line. In many streams like the Mississippi River and its tributaries, the bottom is gradually shifting down stream. In such cases, the line is laid with a bow up the stream. As the bottom shifts, the line will be buried deeper where the movement down stream is slower. If the line is laid straight across such streams, the shifting bottom will probably soon pull the line apart. The local U. S. Engineers have data as to the shifting of the bottom. They should be consulted as to this and also as to the best place for crossing rivers. War department permits must be secured for crossing all navigable streams and for building lines in navigable waters.

Sea lines are built for loading and unloading tankers at points where harbors are not available, like the coast of the Gulf of Mexico off Tuxpam, see Fig. 23. These lines are either pulled out from the shore or laid out with barges. A 12-inch line of 50-pound line pipe will float in fresh water. Sometimes floats can be used to support the smaller-sized lines until they are in the proper position.

Main line fittings are usually of the flanged type. Flange unions, or right and left fittings, are used to join two ends in the field, where necessary. These fittings as well as others needed for the construction and repair work, are shown in the chapter on well-drilling methods.

Pumping stations or power plants of the pipe-line systems vary in size from small gathering stations having a capacity of only a few barrels per day to the large trunk line stations having a capacity of over 100,000 barrels per day. While other sources of power are occasionally used, the principal sources are explosive type engines, using either oil or natural gas for fuel, and steam-driven engines, using any available fuel. The work of designing the stations is in general similar to that of laying out any other power plants with the few modifications incident to the oil business. The danger of fire is always present when handling a very inflammable material like oil, and this must constantly be considered. The principal features of a station are tankage, gate houses, pumping plant, telegraph (or some other system of communication), fire protection, and water system.

Fig. 24.—Gate house.

Tanks for petroleum are used in all branches of the business and as they are not peculiarly a pipe-line accessory, will not be described in this section. Details as to construction of tanks and reservoirs are given in a careful study of the subject made by Mr. C. P. Bowie.[1] The amount of tankage necessary at a station depends on the operating conditions. Where a station is both receiving and pumping oil, it is necessary to have at least three tanks if an accurate " check " of the amount of oil being handled is required. If only two tanks are used the pumps must be shut down when changing tanks. This check is only required at occasional points and at many stations only one tank is used.

Gate houses are only built where more than one line is in use These buildings contain most of the gates or valves which are used to switch or turn the stream from one point to another. The general arrangement in these gate houses is a system of "manifold" castings crossing each other at different grades at right angles with a

[1] (Bulletin 155—Petroleum Technology 41), Oil-storage Tanks and Reservoirs, published by the Bureau of Mines, Washington, D. C.

valve between each pair of outlets. See Fig. 24. If, for example, any one of three pumps is to be arranged so that it can pump through any one of four lines, the gate house would require 4 three-branch manifolds and 3 four-branch manifolds, and 3 times 4 or 12 gate valves. These manifolds permit placing the valves in a small space and also enable those in charge of the plant to operate with less chance of making an error than if the valves were scattered all over the property. With the lines distinctly

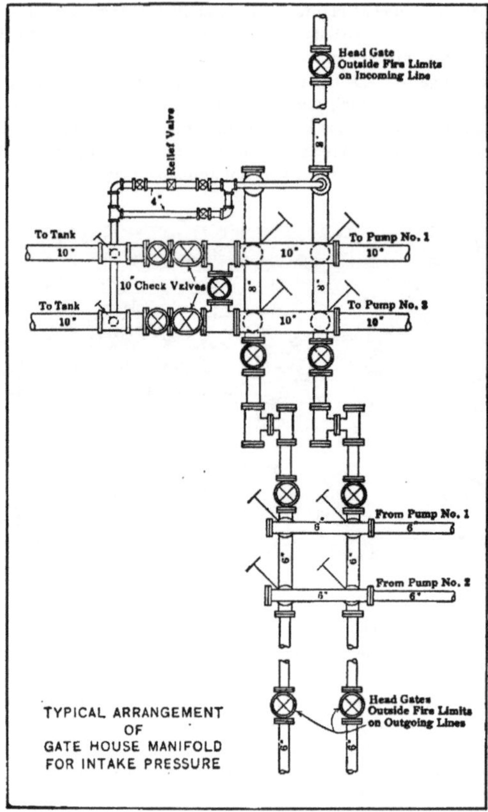

Fig. 25.

marked and valves which indicate whether they are opened or closed, there is very little excuse for a mistake in handling the stream of oil passing through the gate house.

The pumping plant can be divided into two parts, the pump and the power supply. The pump is usually of the outside packed plunger type except where pumps are used to draw in the oil from the field or from distant sources. These suction pumps are usually of the piston type. A plunger pump is not a good suction pump, so it is desirable, where possible, to have the oil flow freely to the pump, or in other words, to have the supply for the pump high enough above the suction valves to overcome the friction of the lines and to lift the valves. The most common source of power for pumping is steam. All kinds of boilers are used and engines of various types from the high-

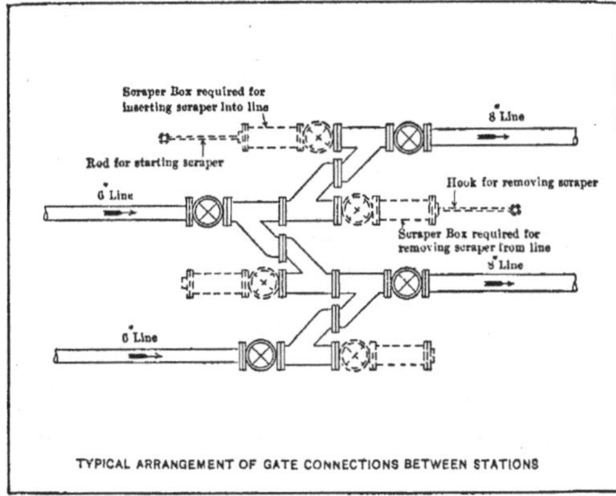

Fig. 26.

pressure, direct-acting, requiring about 150 pounds of steam per horse-power per hour, to the triple expansion condensing engine which only requires about 15 pounds of steam per horse-power per hour. The economy depends on the steam pressure carried in the

Fig. 27.—Horizontal cross compound pumping engine (lagged).

boilers as well as the mechanical work and design of the engines. Explosive type engines in general are more economical in their use of fuel than the steam type. This type of engine is treated in the chapter on explosive engines.

Fig. 23.—Steam end of a vertical triple-expansion pumping engine.

Fig. 29.—Steam end of a horizontal triple-expansion pumping engine.

A well-designed explosive type engine with direct-connected pump of more than 50 horse-power will require about one-half pound of oil fuel per horse-power hour, while a well-designed direct-connected steam pumping engine of the compound or triple type, running condensing, will require about nine-tenths of a pound of oil fuel per horse-power hour. The advantages of a steam plant are that it can use other fuels, is generally easier to understand and more flexible as to speed, and that steam is often of great assistance in fighting oil fires. An electrically operated plant can be used to advantage where the current is obtained from water or some other low-cost source, but it is not feasible to convert power that can be used directly into electric power on account of the loss due to such conversion. The difference in the indicated horse-power of a steam-driven pump and the horse-power delivered by the pump is generally less than

Fig. 30.—Valve deck of horizontal triple-expansion pumping engine.

10 per cent, while the loss due to change to electricity and its re-conversion into power is several times that amount. Electrically driven pumps can often be used to advantage where the service is intermittent or where a large generator of an economical type can be used to supply a number of pumps in different locations. Pumps for viscous oils require a larger area of opening through the valves than pumps for the more fluid oils. The type of valve, kind of packing, and material for plungers, valves, and gaskets, vary with the kind of oil to be pumped and whether the oil is to be pumped hot or cold. The accompanying illustrations, page 422, Fig. 31, show pumps which have been developed for various services and manufactured by the National Transit Company of Oil City, Pa. Almost all of the large pump manufacturing companies have had considerable experience in building pumping engines for oil. The following list is not intended to be complete but simply to furnish the names of a few companies operating

in different localities, so that the engineer may consult with their special engineering force and find out what kinds and sizes of pumping engines they make which would be suitable for the contemplated purpose. In writing for information, all of the avail-

Fig. 31.—Pumps developed for various services. 1. Horizontal power pump. 2. Small fuel oil pump. 3. Jerker power pump. 4. Compound duplex pump. 5. Single duplex pump. 6. Single-acting vertical power pump with herring-bone gear.

able information should be furnished or it may be impossible to recommend a satisfactory pump for the purpose.

Allis-Chalmers Company,	Milwaukee, Wis.
Dean Brothers,	Indianapolis, Ind.
Dow Pump & Diesel Engine Co.,	Alameda, Cal.
Goulds Manufacturing Company,	Seneca Falls, N. Y.
National Transit Pump & Machine Company,	Oil City, Pa.
The Worthington Pump & Machinery Co.,	New York, N. Y.
Wilson-Snyder Company,	Pittsburgh, Pa.

The pumping stations should have some automatic device for shutting down the station in case of a break in the line or the accidental closing of a valve. This usually takes the form of a device for shutting off the power in case there is a sudden rise or drop in line pressure or increase in the speed of the pump. The general details of the National Transit Company device for shutting down a flywheel pump when speed increases is shown in Fig. 32. It is not usually necessary or even desirable to have regulators on pipe-line pumps, as the service is a good regulator. Sometimes on account of an irregular source of power such apparatus is needed. Pressure alarms and recording gages are to be recommended, as they often give notice or show facts which avoid or account for trouble.

Communication between **stations** and points on lines is necessary and should be as quick and dependable as possible. This is usually obtained by building a telegraph or telephone line along the pipe-line, as the regular commercial lines cannot give the quick response necessary to the proper and continuous operation of pipe-lines.

Fig. 32.—Automatic shutdown for flywheel pump.

Fire protection involves proper location and construction of tanks and buildings as well as means for extinguishing the fire after it has started and of preventing it from spreading to other tanks or structures. The two most important causes of fire are lightning and open flames or matches. Occasional fires have resulted from other causes, such as hot bearings and accidental electrical contact. Mr. C. P. Bowie [1] estimates that 80 per cent of the oil fires are due to lightning and most of the rest are caused by carelessness. The most common fire extinguishing agent is water, but this must be used with caution when dealing with oil fires, as a dash of water often results in a semi-explosion which permits a large surface of burning oil to come in contact with the oxygen of the air and results in intense heat which may extend the area of the fire. Oil floats on water and may be carried to points which would otherwise not be reached. Water as an agent for fighting oil fires is a source of danger and is not to be recommended except where it can be used to cool off the outside of a tank or structure which is on fire and by reducing the temperature help in getting the fire under control. The use of steam for smothering a fire is usually effective where the roof of the tank is not blown

[1] Bureau of Mines, Bulletin 170, Petroleum Technology, 48.

off by the explosion or where the fire is in an enclosed space. Chemical solutions producing a lasting foamy or frothy blanket are also used and when properly applied are very successful in putting out oil fires.[1] Wet blankets may sometimes be used to stop up openings and so cause the fire to smother itself by preventing the admission of air. Carbon tetrachloride, the basis of "Pyrene" and other similar extinguishers, is a valuable agent for extinguishing small fires. Sand, dust, and various chemical powders are also used for this purpose. Fire banks and walls are used to prevent the spread of fire from one structure to another. One bank will not usually stop the stream of burning oil which results from the "boiling over" or splitting of a burning tank. When a tank boils over, the stream of burning oil moves with sufficient velocity to carry it over a bank several feet high. This first bank will break up the stream and slow its velocity so that a second bank will probably stop the stream. Water should not be allowed to accumulate inside of fire banks. It decreases the capacity of the enclosed space and increases the intensity of the fire when the burning stream of oil strikes the water.

The water supply needed for a station depends on the kind of station to be built and, as it is similar to the supply necessary for any similar power plant, it will not be discussed in detail. A steam plant requires much more water than an internal combustion plant. This is often a controlling factor in deciding what source of power to use.

Pipe-Line Operation

Pipe-line transportation starts when the oil commences to move from the producer's tank. Consider that the tank has been previously measured, numbered, registered,

Fig. 33.—Station thief.

and tables showing its capacity by inches prepared. The "gauger" goes to the tank, inspects the oil and examines the seals on the valves or stop-cocks to see that they have not been opened. The inspection consists in finding out whether there is any water or foreign matter in the tank. This is ascertained by means of outlets near the bottom. If they show water, this is drawn off before the tank is gaged. The condition of the oil at the bottom of the tank is also determined by a "thief." Figs. 33 and 34. If the oil is in condition to be run, the gauger measures the height of the oil in the tank with a

[1] The Foamite Firefoam Company of 200 Fifth Avenue, New York, is one of the principal companies engaged in furnishing this system of extinguishing fires.

graduated gage pole or a tape line having a weight with a sharp-pointed end, records this measurement and then opens the valve or stop-cock and allows the oil to enter the line. When the oil has run into the line, the outlet of the tank is closed and sealed and the measurement again recorded.

If the oil in the tank is fresh from the wells, it often contains gas. As soon as the body of oil is agitated, the gas bubbles will begin to escape. (The action is similar to that noted in a glass of aërated water which has been left standing until the bubbles stopped rising and then agitated again with the result that more bubbles rise.) This is the

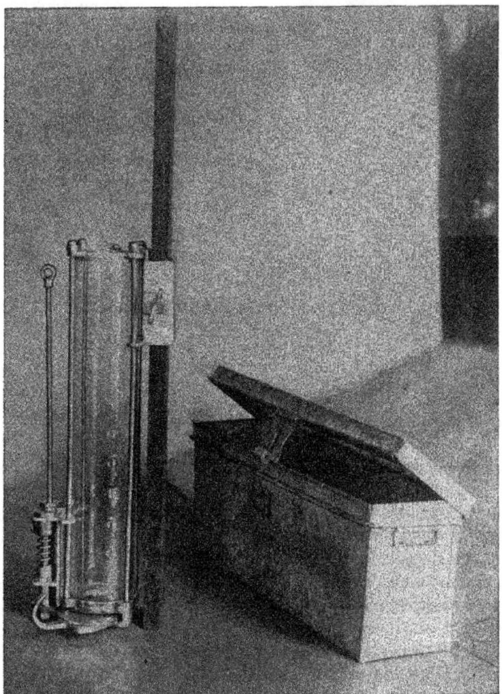

FIG. 34.—Gager's thief.

beginning of the inevitable loss in transportation. During its progress the light products of the oil continue to escape, whenever open to the air. Other transportation losses result from the settling-out of water and other foreign substances. The oil runs into the first of the field lines and then into the first of the pipe-line tanks. There is almost always an appreciable loss between the producers' tanks and the first of the pipe-line tanks. As the lines are not always full of oil (since the oil runs by gravity and there are places where the grade of the line is such that the oil flows without filling the pipe), it is impossible to check up these losses accurately each day, but a record must be kept to account as nearly as possible for the oil in the system. These checks are continued from tank to tank throughout the system and until the oil is finally delivered. To obtain the more accurate checks on the oil when the lines are in operation, it is necessary to have at a station at least three tanks, one in which the oil is being received, one from which the oil is being pumped, and a third to use when making a change of tanks. The temporarily idle tank is gaged before it is put in use. When a change is made, it is substituted for the receiving or the delivering tank and that tank is then gaged as soon as the change is made. As it is not possible to gage all tanks in use at the same time, where accuracy is required, it is often necessary to make an allowance for the rate at which the oil is coming in or going out. The custom in the United States is to keep track of the oil by days. The day usually starts at 7 A.M., at which hour it is usually light enough to read the gage-rod or tape accurately. Hourly gages are taken to check up between stations so that any loss may be detected or error shown and conditions remedied. Floats with a tape line extending down the outside of the tanks are often used, but unless this form of measuring apparatus is accurately made and practically frictionless, it is not always reliable. To prevent sway of the tape by the wind, it is usual to enclose the tape in a box or metal tube. If this kind of apparatus is used, fine wire gauze should be fitted in the box around the tape to prevent accidental explosion which might result from careless or ignorant persons trying to read the tape at night by the light of a match or lantern.

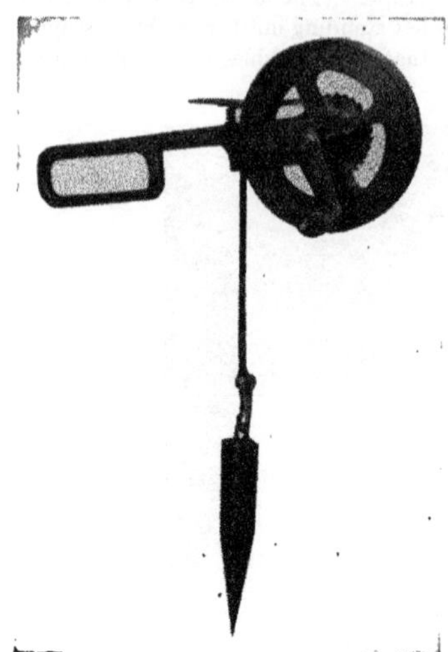

FIG. 35.—Gage-line reel with bob.

In accounting for the oil in the pipe-lines and tanks, it is evident that each day cannot be exactly checked. It is impossible to measure accurately the oil in the lines and tanks, as the lines are not always entirely filled with oil, variations being due to temperature and actual leakages. The difference is taken up in an "over and short" account. In other respects, oil accounts are handled in the same way as financial accounts. In the gathering field, where the inevitable losses are greatest, it is customary to make an allowance in oil to cover the unavoidable losses. Experience has shown that the usual loss in handling the volatile oils of Pennsylvania is more than the 2 per cent which custom has established in the eastern part of the United States. Many of the newer fields have adopted 3 per cent as their estimate of these losses. To avoid complications and a multitude of computations, it is the general practice to make the tables for the producers' tanks 2 per cent or 3 per cent as the case may be, less than their actual capacity. This is understood by all men in the industry and a note to this effect should be printed on such tank tables. A profile of the country enables the engineer to locate points which will not be filled with oil when the line is in operation. If such a line is in steady operation, the amount in the line will be nearly constant, but when the rate of operation changes, the amount will vary and may show in the "over and short" account. The gravity and field lines often have but little pressure and if the grade of such lines is not laid out with care, they may become "air bound." It is usually better to take care of such conditions when the line is laid, as air vents have not proved satisfactory. In lines under the heavier operating pressures in common

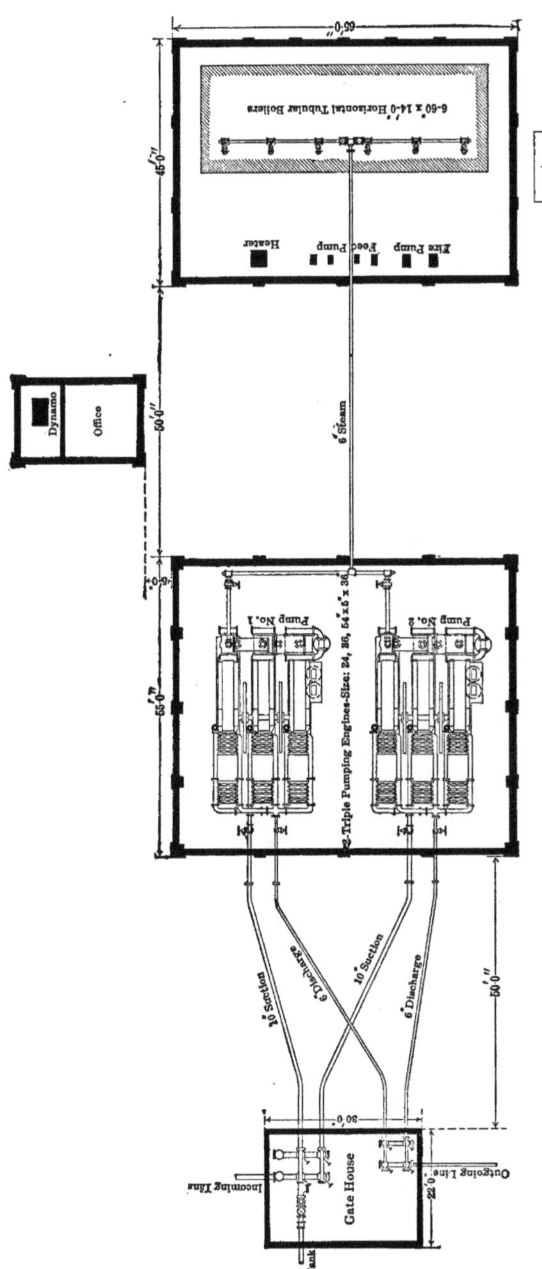

Fig. 36.—General arrangement of pumping station.

use, it is not necessary either to put in traps or pay much attention to the changes of grade. The air at the high points will collect and be compressed into a relatively small space. The oil flowing over the summit made by the bottom of the pipe will run in a small stream through the compressed air or gas and soon absorb this compressed gaseous material, thus removing the obstruction. In new lines, the air under high pressure will leak through joints that are practically oil-tight. A profile of the line also enables the engineers to estimate the amount of loss in case of the accidental breaking of the line.

The pumping stations take the oil from the tanks and pump it forward to other tanks or stations. As the stations are several miles apart, it is necessary to have some system of communication between stations; this is usually by telephone or telegraph. Each method has its advantages and disadvantages. In many cases a telegraphone is used in connection with the telegraph. The telegraphone is an instrument of the telephone order which can be used over the telegraph wire without interfering with the use of the wire for telegraphic purposes. Before pumping is begun, it is usual for the person in charge of a pumping station to examine the gate valves at the plant and see that the right valves are open and that all other connections with the system are closed. He then advises the station that is to receive the oil that he is ready to commence pumping, but does not start until the receiving station advises that the line is clear.

Oil pipe-lines are common carriers under the laws of the United States, and as such are operated under the rules and regulations laid down by the Interstate Commerce Commission. This commission has regulations to govern the construction and filing of freight tariffs and classifications, recording and reporting of extensions and improvements, etc., and the destruction of records. They also issue a classification of investment in pipe-lines, pipe-line operating revenues and pipe-line operating expenses. These and all other regulations and requirements can be obtained from the Interstate Commerce Commission at Washington, D. C. Pipe-line tariffs are issued under the rules of the Commission. The pipe-line service is divided for convenience, into "gathering" and "trunk-line" service. This division is arbitrary as to where the gathering service ends and the trunk-line service begins. The Commission requires that separate accounts be kept. This is accomplished by considering part of the system as trunk line and the remainder as gathering line. When part of the plant of one system is used for the service assigned to the other, a charge is made for the use of the part of the plant or the service rendered. The State regulations as to handling the oil vary, as do also the regulations of the carriers. The following regulations are taken from one of the West Virginia tariffs for gathering oil. These regulations are intended to meet the West Virginia law and the local requirements.

Gathering Oil Regulations in West Virginia

"The company will receive crude petroleum oil of a gravity exceeding 35° Baumé (at 60° F.) for storage, or for transportation to established delivery points on its own lines on the following conditions:

"First: It will deduct for evaporation and waste 2 per cent of the amount of oil run.

"Second: The oil is to be steamed or otherwise heated and settled before running, whenever the representative of the pipe-line company deems it necessary. When oil has been heated and is run at a temperature above normal, it will deduct $\frac{1}{20}$ of 1 per cent for each degree, Fahrenheit, above normal temperature.

"Third: All oil will be accepted for transportation only on condition that delivery may be made to consignees out of a common stock of oil of the same general kind and quality as that received from the shipper and that the oil shall be subject to such changes in gravity or quality while in transit, as may result from the mixture of said oil with other oil in the pipe-lines or tanks of this company.

"Fourth: It will charge storage for the month, at the close of each month, on all oil in production accounts, which was in its custody at the beginning of the preceding month and still remains in such accounts.

"Fifth: It will charge storage, beginning ten days after oil has been transferred from production accounts, until tender of shipment is accepted; but if shipment be thereafter canceled or changed, so that the period on which no storage is charged is extended, then storage will be charged as if the original tender had never been made.

"Sixth: Whenever the total charge for storage on any batch of oil shall amount to less than 5 cents for any one month, no charge will be made for such storage in such month.

"Seventh: Payment of accumulated charges may be demanded before accepting tender of shipment.

"Eighth: Payment of transportation charges may be demanded in advance.

"Ninth: When steam or other power for pumping oil is furnished by a producer, he will be paid quarterly by check at the rate of 2 cents per barrel of oil pumped.

"Tenth: Without additional charge, patrons may use the private telegraph lines of the company for messages incident to the business, but the company shall not be liable for delivery of messages away from its offices, and it shall not be liable for errors or delay in transmission or for interruption of the service."

The trunk-line part of the service is represented by another form of tariff containing clauses referring especially to such service. In such a tariff is generally shown a regulation in reference to the minimum amount of any one kind of oil which can be handled in a lot or "batch" without undue mixing of the oils passing through the lines. These amounts vary according to conditions from 1000 to 100,000 barrels. The following regulations are also from a tariff originating in West Virginia:

Sample of Regulations.—"This company receives crude petroleum oil for interstate transportation to established delivery stations on its own lines, and lines of connecting pipe-line companies, on the following conditions:

"First: The oil will be received for interstate transportation when the shipper or consignee has provided the necessary facilities for receiving the oil as it arrives at destination.

"Second: When the oil shipped hereunder in conjunction with other oil of the same kind and quality, received at ————, ————, from any other points and consigned to the same destination, shall, in the aggregate, amount to at least ———— at ————, then such aggregate amount will be forwarded to destination.

"Third: All oil will be accepted for transportation only on condition that deliveries may be made to connecting carriers and consignees out of a common stock of oil of the same general kind and quality as that received from the shipper and that the oil shall be subject to such changes in gravity or quality, while in transit, as may result from the mixture of said oil with other oil in the pipe-lines or tanks of this or the connecting company or companies.

"Fourth: The shipper or consignee shall pay the transportation and all other lawful charges accruing on oil tendered for shipment and if required, shall pay the same either in advance or before delivery. Oil accepted for transportation shall be subject to a lien for all such charges.

"Fifth: No carrier of any of the oil herein described shall be liable for any loss thereof or damage thereto or delay caused by the act of God, the public enemy, quarantine, the authority of law, strikes, riots, or the act or default of the shipper or owner, or from any other cause not due to the negligence of these carriers; in such cases the shipper shall stand the loss in the same proportion as the amount of his tender bears to the whole amount of the consignment of which it is a part, and shall be entitled to receive only such portion of his tender as is left after deducting his due proportion of the loss, as above; but this exemption shall not apply to loss by fire (whether originating from lightning or any other cause), while the oil is in the custody of this company or of any other participating carrier if insured against loss by fire.

"Sixth: Except where property is damaged in transit by carelessness or negligence of the carrier, claims for loss or damage must be made in writing to the initial or delivering carrier within five months after delivery of the property, or in case of failure to make delivery, then within five months after a reasonable time for delivery has elapsed.

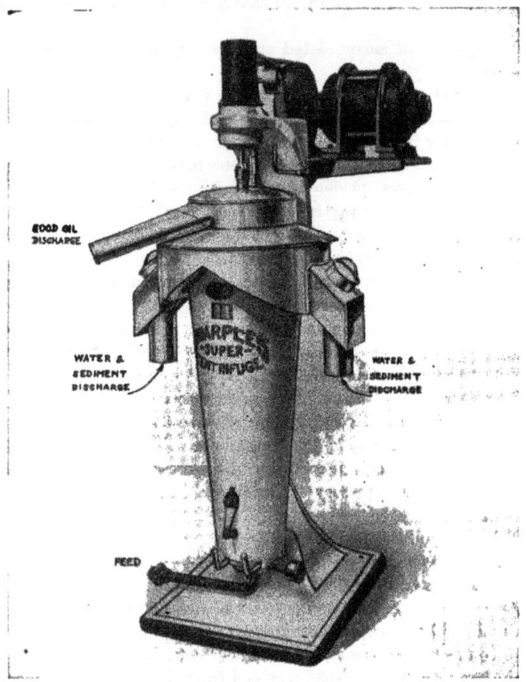

FIG. 37.—Sharples machine for separating the "B.S."

Suits for recovery of claims for loss or damage, notice of which is not required, and which are not made in writing to the carrier within five months as above specified, shall be instituted only within two years and one day after the delivery of the property, or in case of failure to make delivery, then within two years and one day after a reasonable time for delivery has elapsed. No claims not in suit will be paid after the lapse of two years and one day as above, unless made in writing to the carrier within five months as above specified.

"Seventh: A deduction of 1 per cent will be made from the amount of each tender when the oil is received, to cover losses due to shrinkage and evaporation incident to the transportation of oil by pipe-lines." A tariff in which only one carrier company is interested is known as a "local tariff." If there is more than one company joining in the movement, the tariff is known as a "Joint Tariff." Local requirements, special privileges, charges, and any other matters which may change the nature of the service are covered by tariffs. A clause calling the attention of the shipper to the possibility of such regulations generally appears on the tariff in about the following form:

"Shipments transported under this tariff are entitled to such privileges and subject to such charges as are or shall be published by this company or any of the carriers

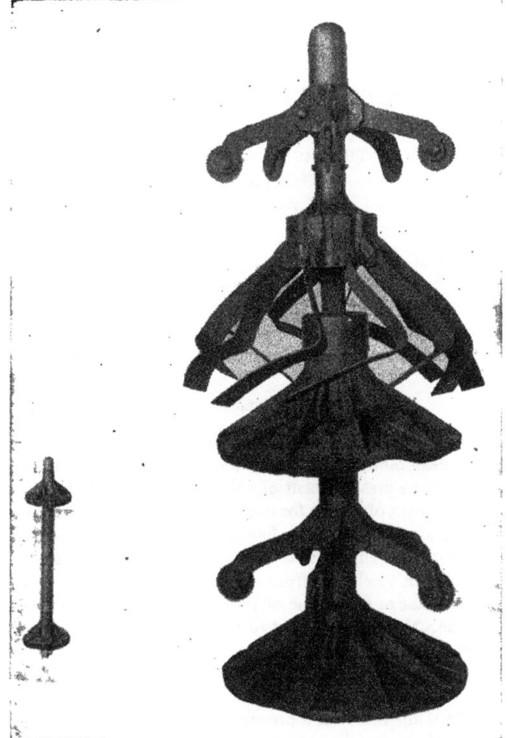

FIG. 38.—Pipe-line scraper for removing dirt, B. S. and paraffin from inside of pipe while line is in operation.

party to this tariff and such as are lawfully in effect on date of shipment and lawfully on file with the Interstate Commerce Commission as to interstate traffic, providing for reconsignment, storage, transit privileges, or any other privileges, charges or rules which in any way increase or decrease the amounts to be paid on any shipment transported under this tariff, or which increase or decrease the value of the service to the shipper."

"Emulsions" of oil, water and solid matter settle to the bottom of the tanks and often collect on the sides of the pipes. Paraffin wax often enters into this bad sediment, which is known in the trade as "B.S." The recovery of the valuable part of this "B.S." is a slow and difficult process. The material can be run through a still operating at a slow speed or treated to separate the good part from the bad. Part of the good can be separated by heating the mass and allowing the water and solid matter to settle out by gravity. Heating and using a centrifugal separator will remove most of the foreign matter from the oil. The Sharples separator designed for this purpose is shown in Fig. 37. The sediment and paraffin which collect on the inside surface of the pipe are removed by a scraper or "go devil" which is pumped through the pipes. Fig. 38.

TRANSPORTATION OF NATURAL GAS

BY

THOMAS R. WEYMOUTH

AND

FORREST M. TOWL

The transportation of natural gas and its measurement are closely allied with the transportation of the liquid hydrocarbons. Laws which govern the flow of gas through orifices and lines are well established, and when it is known what a line will do with one kind of gas under certain conditions, it is easy to find out what it will do with any other gas. The problem then becomes one of coefficients and constants.

Meters

When gas having a commercial value is to be transported, it becomes necessary to measure it with a considerable degree of accuracy and, as the problem of the flow of gas must be based on some measurement of volume or weight of the gas, the first thing required is the establishment of a basis for measurement and an apparatus for measuring. The basis usually employed is the cubic foot at atmospheric pressure and at a stated temperature, although many engineers use 1 pound of air as the unit. By Mariotte's Law it is a simple problem to change from one basis to the other.

On account of the change of volume of the gas, for differences in temperature and pressure, the actual accurate measurement of the gas becomes a very difficult problem. It is usually considered that a gas meter is accurate if it registers within 2 per cent of the standard. The commercial gas meter has been perfected so that when it is in good condition and working under normal speed it can be relied on to give results within that amount, provided the temperature and pressure remain practically constant.

The measurement of gas, particularly at high pressures, presents many difficulties. The following types of meters are at present in use:

The displacement, or regular, type of meter,
The orifice;
The proportional, which is a combination of the first and second;
The anemometer;
The dynamic;
The electric.

Each of these forms of apparatus has its special advantages and limitations when employed in measuring gases at high pressure. For measuring large volumes at high pressure, the proportional meter, the orifice, the dynamic, and the electric seem to be the only ones available.

The Orifice

In a number of meters the orifice is used to measure the gas. The meter is usually calibrated by gas or air flowing through the orifice into a gasometer under a constant difference in head. After the orifice has been calibrated, one or more orifices are placed in the line and the pressure is noted on each side of the opening. This requires constant watching and readings, or dependence on some type of recording meter in order to compute the amount of gas flowing. The St. Johns meter of this type uses a variable orifice and records on a chart the position of a plug in an opening. There is no attempt in this meter to make corrections for variations in pressure. The charges are averaged by a planimeter. This plan of measurement is used largely by the New York Steam Heating Company.

The Proportional Meter

In the proportional meter, it is necessary to make corrections on account of change of pressure. This requires either an observer to note the readings of the meter and the pressure or a recording apparatus to show the readings and pressures simultaneously. Such an apparatus is manufactured by several of the companies, but it is difficult to make the computations from the charts.

The Dynamic Meter

The principal representative of the dynamic class is the Pitot tube. The General Electric Company makes a recording Pitot tube which is automatically corrected for variation in pressure. The general practice in measuring gases with the Pitot tube is to take readings at stated intervals and make computations from these readings.

The Electric Meter

The electric meter is a recent development in gas engineering resulting from work done by Professor Carl C. Thomas of the University of Wisconsin. It is based on the principle that to make a given increase in the temperature of a given weight of a gas requires the addition of a corresponding amount of heat. The heat is supplied electrically and the amount of energy required is measured.

The calibration of the proportional meter and the Pitot tube at Buffalo, N. Y., was undertaken because the ordinary displacement meter was too large and not strong enough to answer the purpose when it is required to measure large volumes at high pressures. The coefficients for the orifice meters had not then been accurately obtained. A proportional meter was selected for the purpose and taken to a 60-foot gas holder in Buffalo to be calibrated. At the same time, it was decided to test the Pitot tube with the idea of obtaining coefficients to see how accurately it would measure the gas. During the test it was found that the proportional meter did very well when everything was working all right, but it was very liable to sudden changes in its rate. This led to more extensive experiments with the Pitot tube as a possibly more accurate and reliable meter. The method of calibrating the Pitot tube at the Buffalo gas holder is here given in order to establish a basis for the measurement of gas through the lines.

The tube selected was made from a piece of 4-inch pipe bored out to 4⅛ inches internal diameter. A diagram showing the way this tube was connected at the gas holder is shown in Fig. 1, which also shows the form of tip used in the test. It was found that the pressure gages ordinarily used were not reliable, and a mercury column was connected to the static side of the Pitot tube in order that the pressures might be accurately obtained. The temperature of the flowing gas was obtained by means of thermometers placed about 1 foot on the downstream side of the Pitot tube nozzle. The temperature of the gas in the holder was obtained by placing thermometers in the gas holder and observing the temperature at three different points. All of the tests were made early in the morning, late in the afternoon, or during cloudy weather, so that the variation of temperature might be as small as possible. The duration of each test was five minutes. The pressure of the flowing gas and the difference in water level as shown by the U-tube connected to the Pitot tube was observed at the beginning and close of the tests and at minute intervals during the tests. The average of the amounts observed was taken as the average condition under which the tube was operating. In instances where the variation was considerable, the average of the square roots of the

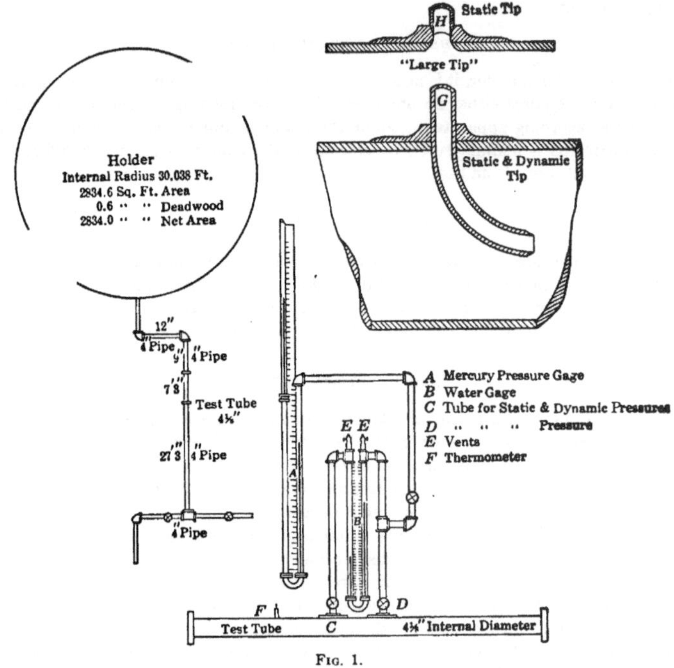

Fig. 1.

pressures was taken, but this was found not to change the results perceptibly. The amount of gas flowing into the holder was obtained by marking the holder at the sound of a whistle, two marks being made at points diametrically opposite each other. The data of this test are given below.

The dimensions of the holders were as follows: external circumference 188.88 feet; allowances for thickness of iron and lap of plates 0.012 foot; internal radius 30.038 feet; area in square feet 2834.6; deadwood, braces, etc., 0.6 foot.

A cubic foot of gas was to be measured under an absolute pressure of 14.65 pounds, that being 4 ounces above average barometric pressure at Buffalo.

The simplified formula used for computing results is $V = C\sqrt{H(Br+P)}$, in which V represents the number of cubic feet of gas per hour; C a constant depending on the various conditions of the tube, gas, etc.; H the difference in water level in inches between the two sides of the Pitot tube; Br the barometric reading in pounds; and P the gage pressure of the flowing gas in pounds per square inch.

The following table shows the result of 13 observations made to standardize this tube.

Tabulation Showing Data Used in Computing Coefficient for Standard Tube.

Test	Difference in water level in inches, as shown by Pitot tube	Absolute pressure in pounds per square inch		Lift in holder in feet	Temperature ° F.		Coefficient for 50° storage and 40° flowing temperature and a pressure of 14.65 pounds per square inch
		Flowing gas	Stored gas		Flowing gas	Stored gas	
1	5.0258	39.63	14.43	2.56500	36	48	6086
2	9.6792	34.54	14.44	2.30500	37	45	6096
3	14.7293	29.00	14.44	3.74750	38.75	46	6112
4	17.8025	20.98	14.50	3.47375	40	49.5	6051
5	11.9200	23.09	14.50	3.03750	42	50.8	6160.4
6	7.1100	27.87	14.50	2.58250	43	51.5	6172.5
7	4.8605	31.83	14.50	2.26000	43.5	52	6110.5
8	21.9950	26.03	14.53	4.36250	43.5	53.6	6124
9	16.5555	27.94	14.53	3.93050	43	52.6	6147.5
10	12.1690	32.45	14.54	3.60650	43	51.8	6118
11	19.2200	31.38	14.54	4.47750	36.5	48.2	6148
12	14.3625	32.66	14.54	3.98150	40.5	49.6	6209
13	9.5700	33.98	14.54	3.30100	41	51	6169

Specific gravity of gas 0.64 based upon air as 1.

The above observations indicated that the coefficient for the Standard tube was practically independent of the varying velocity of the gas. In order to ascertain if this was so, a comparison was made with the Standard tube and another tube similarly constructed. In one series of tests the gas was allowed to pass through the Standard tube at high pressure and, by means of a throttle valve between the two tubes, passed through the other tube at a considerably lower pressure. This gave a greater velocity and higher readings on the second tube. The places of the two tubes were changed, and the gas was allowed to pass at a higher pressure and the same reading through the tube that was being tested, and then throttled before passing through the Standard tube. By means of this comparison, it was found that the coefficient of the Standard tube was practically a constant.

This test established the coefficient of the Standard Pitot tube which was used in calibrating most of the other tubes then used for measuring natural gas. Since that time the tubes have been calibrated at gasometers and by comparison with a large number of displacement meters, and also by comparison with standardized orifices until it is believed that the measurement of gas by the Pitot tube, under favorable conditions, gives commercially accurate results.

A test has recently been made comparing the Pitot tube with the Thomas Electric meter. It shows that the two methods for measuring give practically the same result.

Buffalo Line Test

After the Pitot tube was standardized at Buffalo, a number of tests were made on various natural gas lines to ascertain the proper coefficient to be used in making computations as to the capacity of lines.

In 1901 there were two lines supplying Buffalo with natural gas from the Pennsylvania field. In addition to this supply, some gas was obtained from Canada and a

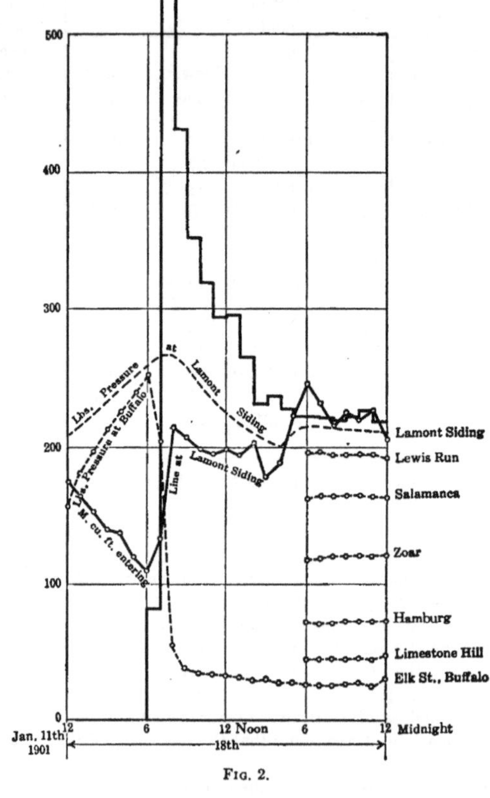

Fig. 2.

little from wells near the city. One of the lines, built of 8-inch pipe, had just been repaired and tested and was known to be practically gas-tight. It was decided to use this line, about 90 miles long, in making the test, the other lines being used to supply the irregularities of the daily demand for gas at Buffalo. The 8-inch line was continuous north of Lamont Siding, Pa., and all local consumption was shut off so that it was taken from another line. Pressure gages were carefully calibrated and put in at Lamont Siding, at Elk Street, Buffalo, and at five intermediate points. Observers were placed at these points, and readings of the pressure were taken every fifteen minutes

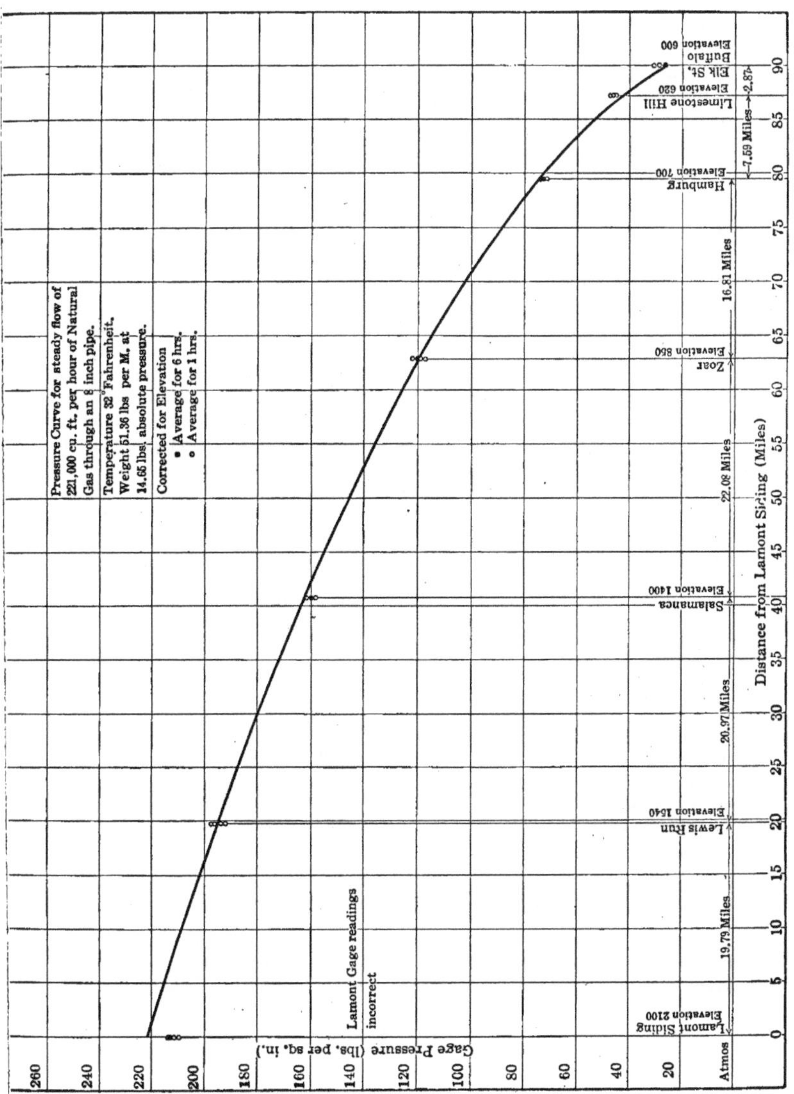
Pressure Curve for steady flow of 221,000 cu. ft. per hour of Natural Gas through an 8 inch pipe.
Temperature 32° Fahrenheit.
Weight 51.36 lbs per M. at 14.65 lbs. absolute pressure.
Corrected for Elevation
▪ Average for 6 hrs.
○ Average for 1 hrs.
Lamont Gage readings incorrect
Gage Pressure (lbs. per sq. in.)
260
240
220
200
180
160
140
120
100
80
60
40
20
Atmos
Lamont Siding
Elevation 2100
Lewis Run
Elevation 1640
Salamanca
Elevation 1400
Zoar
Elevation 850
Hamburg
Elevation 700
Limestone Hill
Elevation 620
Elk St.
Buffalo
Elevation 600
19.79 Miles
20.97 Miles
22.08 Miles
16.81 Miles
7.59 Miles
2.87
0
5
10
15
20
25
30
35
40
45
50
55
60
65
70
75
80
85
90
Distance from Lamont Siding (Miles)

Fig. 3.

night and day for a period of one week before the line showed a satisfactory steady flow of gas. The gas delivered at the city of Buffalo was measured by the Pitot tube, and the amount of gas flowing into the line at Lamont Siding was computed from the known capacity of the line and difference in pressure. From 7:00 to 11:00 P.M. on January 18th, an even steady flow was obtained, as is shown by the pressures at Lamont Siding, Buffalo, and intermediate points (Fig. 2). The Pitot tube at that time showed that the delivery at Buffalo was at the rate of 221,000 cubic feet per hour. The temperature was 32° F. The weight of the gas per thousand was 51.36 pounds, measured at an absolute pressure of 14.65 pounds. After the test the gages were again returned and compared with a mercury column. It was found that the gage located at Lamont Siding did not read the same as when it was sent out, the readings being 16 pounds too low. As there was no way to ascertain the condition this gage was in during the test, the readings at Lamont Siding are only shown to indicate that the pressure at this point was steady. All of the other gages were found to be correct.

Fig. 3 shows graphically the result of this test. The curved line is put in from formula (5), the readings at Lewis Run and at Buffalo being used as the determining points of the curve. The black dot shows the average pressure at the various points for the six hours. The small circles show as many of the individual hours as can be put on at the scale of the drawing. Taking the average for the six hours, the pressure at Limestone Hill showed 1.89 pounds higher than the curve. The pressure at Hamburg was 2.05 pounds below the curve. The pressure at Zoar was 0.10 pound higher than the curve. The pressure at Salamanca was 1.85 pounds lower than the curve.

When it is considered that pressure gages are read with difficulty as close as these amounts, this test may be said to confirm the theory as to the flow of gas through lines, and will give as accurate coefficients as any test on record.

The average data of the test from which the curve is plotted are as follows:

$P_1 = 210$ pounds per square inch absolute;
$P_2 = 41$ pounds per square inch absolute;
$D = 7.981$ inches (regular 8-inch pipe);
$v_1 = 12.5$ feet per second;
$y_1 = 0.7362$ pound per cubic foot:
$L = 70.32$ miles.

Substituting these values in Formula (7) page 439, the value of the coefficient f is found to be .003645.

A number of attempts have been made to ascertain the coefficients which should be used for other sizes of pipe and different velocities. These coefficients vary but little from the one above given.

Pipe-Line Formula Development

The head lost by fluid friction in a pipe-line, when the flow is turbulent, has been found by experiment to vary directly as the length of line and the square of the velocity and inversely as the diameter of line. This may be expressed by the relationship,

$$h = 4f\frac{l}{d}\frac{u^2}{2g}, \qquad \text{(1)}$$

where h is the friction head, expressed in height of a homogeneous column of the fluid, l is the length and d the diameter of line, u is the mean velocity of flow, g the acceleration due to gravity, and f is the coefficient of friction, all expressed in any consistent system of units.

The value of f depends upon the condition of the interior surface of the pipe and also upon the diameter, as will be shown later.

Applying Bernoulli's theorem with a consideration of friction to gas flow where the drop in pressure is small, the density may be considered sensibly constant and the velocity, therefore, sensibly the same at all points along the pipe. If z represents the elevation or head above the datum plane, p the absolute static pressure and w the specific weight, and subscripts 1 and 2 respectively refer to the inlet and outlet ends of the pipe, then

$$\frac{u_2^2}{2g}+\frac{p_2}{w_2}+z_2=\frac{u_1^2}{2g}+\frac{p_1}{w_1}+z_1-4f\frac{l}{d}\frac{u^2}{2g}, \quad . \quad . \quad . \quad . \quad . \quad . \quad (2)$$

where u represents the weighted mean velocity. Hence,

$$\frac{u_1^2-u_2^2}{2g}+\frac{p_1}{w_1}-\frac{p_2}{w_2}+z_1-z_2=4f\frac{l}{d}\frac{u^2}{2g}, \quad . \quad . \quad . \quad . \quad . \quad . \quad . \quad (3)$$

But, since the velocity and density are assumed to be practically constant at all points along the pipe, $u_1^2-u_2^2=0$ and $w_1=w_2=w$ (the mean specific weight). The pipe may, without great error, be considered to be horizontal, so that $z_1-z_2=0$. Hence,

$$\frac{p_1-p_2}{w}=4f\frac{l}{d}\frac{u^2}{2g}. \quad . \quad . \quad . \quad . \quad . \quad . \quad . \quad . \quad . \quad . \quad . \quad (4)$$

That is, the difference in head will be considered as due entirely to the frictional loss.

Flow of gas at high pressures and with large pressure drop.—In long pipe-lines transmitting gas, it has been found by experience that the temperature is practically constant and the flow therefore isothermal. In such lines, however, the pressure drop is large, producing a corresponding diminution in density and increase in velocity. Considering a short length of pipe, δl, however, Formula (4) may be applied, in which case p_1-p_2 becomes $-\delta p$ and (4) becomes

$$-\frac{\delta p}{w}=4\frac{f}{d}\frac{u^2}{2g}\delta l. \quad . \quad . \quad . \quad . \quad . \quad . \quad . \quad . \quad . \quad . \quad . \quad (5)$$

For isothermal flow, $w=p\frac{w_1}{p_1}$, and with pipe of constant cross-section $uw=u_1w_1$ or

$$u^2=u_1^2\frac{w_1^2}{w^2}=u_1^2\frac{w_1^2p_1^2}{p^2w_1^2}=u_1^2\frac{p_1^2}{p^2}.$$

Hence, by substitution and reduction, (5) becomes

$$-p\,\delta p=4\frac{f}{d}\frac{u_1^2}{2g}w_1p_1\,\delta l. \quad . \quad . \quad . \quad . \quad . \quad . \quad . \quad . \quad . \quad . \quad (6)$$

Integrating between pressure limits p_1 and p_2 and length limits o and l,

$$\tfrac{1}{2}(p_1^2-p_2^2)=4\frac{f}{d}\frac{u_1^2}{2g}w_1p_1l, \quad . \quad . \quad . \quad . \quad . \quad . \quad . \quad . \quad . \quad . \quad (7)$$

$$u_1^2=\frac{gd(p_1^2-p_2^2)}{4fw_1p_1l}, \quad . \quad . \quad . \quad . \quad . \quad . \quad . \quad . \quad . \quad . \quad (8)$$

or since $w_1=p_1\div RT$.

$$u_1^2=\frac{gdRT(p_1^2-p_2^2)}{4flp_1^2}. \quad . \quad . \quad . \quad . \quad . \quad . \quad . \quad . \quad . \quad (9)$$

In Equation (9), T is absolute temperature of the flowing gas and R is the Clapeyron constant, or so-called gas constant in the equation of a perfect gas, $pv=RT$. For equal volumes of two gases, the corresponding values of R are proportional to the respective masses of gas, or for equal masses R is inversely proportional to the densities of the two gases. Therefore, if R is the constant for any gas and R_a that for air,

$$R=R_a\frac{w_a}{w} \quad \text{or} \quad R=\frac{R_a}{G},$$

where G is the specific gravity of the gas referred to air. Substituting in Equation (9) and extracting the square root, the initial velocity is

$$u_1=\sqrt{\frac{gdR_aT(p_1^2-p_2^2)}{4fGlp_1^2}}. \qquad (10)$$

Equation (10) holds true for any system of units, and upon the foot-pound-second system u will be expressed in feet per second. When pressures are taken as pounds per square inch absolute, represented by P_1 and P_2, volumes in cubic feet, temperatures in Fahrenheit degrees (absolute), diameter of line (D) in inches, and length of line (L) in miles, $R_a=53.33$, $g=32.17$, $d=D\div12$, $l=5280\ L$ and u_1 will remain feet per second.

If Q = flow of gas in cubic feet per hour at standard temperature, T_0 degrees F., absolute and standard pressure P_0 pounds per square inch absolute, and since $Q=\frac{P_1T_0}{P_0T}$ times the flow as expressed in terms of P_1 and T

$$Q=3600\frac{P_1T_0}{P_0T}\frac{\pi d^2}{4}u_1=3600\frac{\pi D^2}{4\times144}\frac{P_1T_0}{P_0T}\sqrt{\frac{32.17}{4}\frac{D}{12}\frac{53.33}{5280L}\frac{T(P_1^2-P_2^2)}{fGP_1^2}}.$$

$$Q=1.616\frac{T_0}{P_0}\sqrt{\frac{D^5(P_1^2-P_2^2)}{GTfL}}. \qquad (11)$$

From a study of a great number of experiments on the flow of air and gas by various investigators (see Weymouth, Trans. A. S. M. E., Vol. 34, 1912, p. 197), it appears that the coefficient of friction, f, has the average value of $0.008\div D^{1/3}$, where D is in inches, and the pipes are of wrought iron or steel, similar to those used in natural gas practice.[1] This value of f introduced into equation (11) makes the flow, in cubic feet per hour.

$$Q=18.06\frac{T_0}{P_0}\sqrt{\frac{D^{5\frac{1}{3}}(P_1^2-P_2^2)}{GTL}}. \qquad (12)$$

This is the general formula for the flow of gas in long lines with large pressure drop.

For any particular gas and assumed average flowing temperature, equation (12) may be written

$$Q=K\sqrt{\frac{D^{5\frac{1}{3}}(P_1^2-P_2^2)}{L}}, \qquad (13)$$

where

$$K=18.06\frac{T_0}{P_0\sqrt{GT}}. \qquad (14)$$

[1] This coefficient of friction f is one-fourth of the coefficient f used in Professor Durand's statement, and the empirical formula used assumes that f for a given gas is only dependent upon d. This is practically in line with Professor Durand's statement, but makes no allowance for change in velocity and weight. The late Mr. F. H. Oliphant gave a coefficient in which f was a function of $d^{5\frac{1}{2}}+\frac{d^3}{30}$.

The formula expressed by Equation (13) is the one generally used in practice for natural gas problems by evaluating K for the conditions obtaining for the particular case in hand. Usually T_0 is taken as $60+460 = 520°$ F. absolute, P_0 as 4 ounces + 14.4 pounds = 14.65 pounds per square inch absolute, and T as $40+460=500°$ F. absolute. If G is not known exactly it may be taken as 0.65 for the average natural gas problem, whence

$$K=18.06\frac{520}{14.65\sqrt{0.65\times500}}=35.56, \text{ for hourly flow rates,} \quad \ldots \quad (15)$$

or

$$K=853.5 \text{ for twenty-four-hour rates.} \quad \ldots \quad (15a)$$

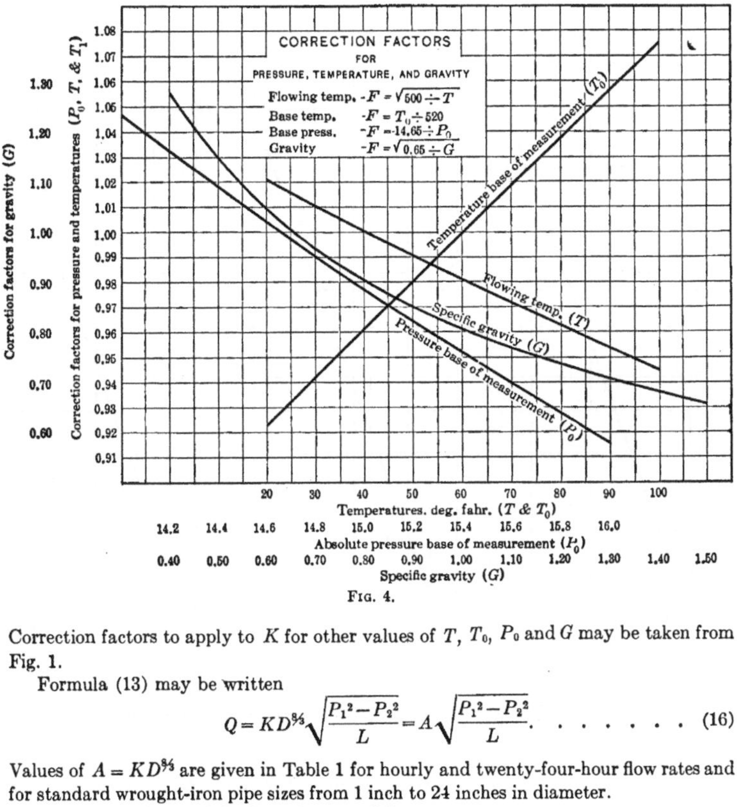

Fig. 4.

Correction factors to apply to K for other values of T, T_0, P_0 and G may be taken from Fig. 1.

Formula (13) may be written

$$Q=KD^{8/3}\sqrt{\frac{P_1^2-P_2^2}{L}}=A\sqrt{\frac{P_1^2-P_2^2}{L}}. \quad \ldots \quad (16)$$

Values of $A=KD^{8/3}$ are given in Table 1 for hourly and twenty-four-hour flow rates and for standard wrought-iron pipe sizes from 1 inch to 24 inches in diameter.

Other useful forms of the formula are as follows:

$$P_1=\sqrt{P_2^2+L\left(\frac{Q}{A}\right)^2}, \quad \ldots \quad (17)$$

$$P_2=\sqrt{P_1^2-L\left(\frac{Q}{A}\right)^2}, \quad \ldots \quad (18)$$

$$L=\left(\frac{A}{Q}\right)^2(P_1^2-P_2^2), \quad \text{(19)}$$

$$A=Q\sqrt{\frac{L}{P_1^2-P_2^2}}. \quad \text{(20)}$$

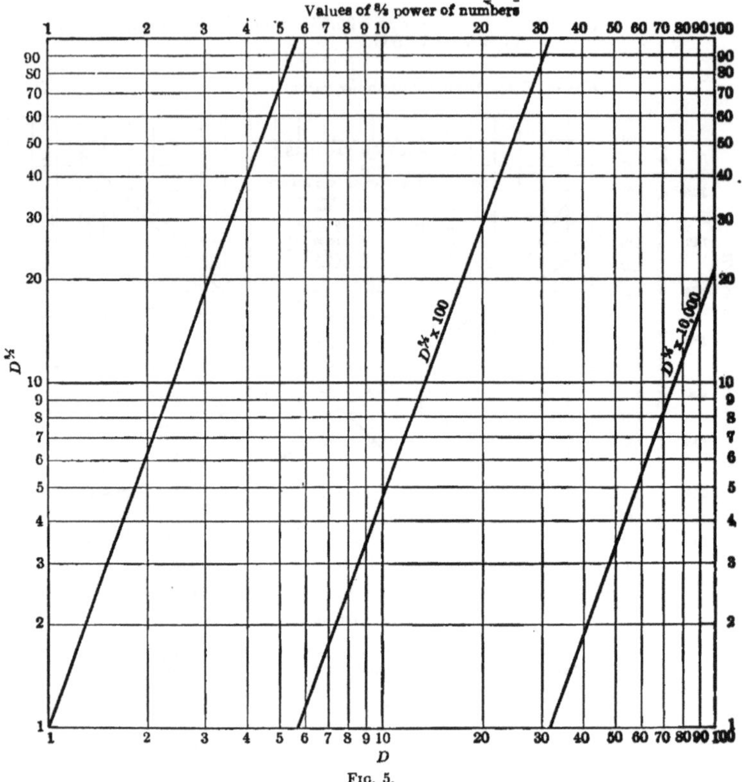

Fig. 5.

For values of A corresponding to pressure, temperature and gravity bases, other than those on which Table 1 is based, multiply the values in Table 1 by the correction factors obtained from Fig. 4.

Values of $D^{8/3}$ for standard line pipe are given in Table 1, and for any value of D from 1 inch to 75 inches in Fig. 5.

Problems with Single Line of Several Diameters of Pipe

Formula (13), as given, or in a transposed form, may be used for all problems relating to flow of gas in a single line of uniform diameter throughout its length. If a line is composed of several lengths, L_1, L_2, . . . L_n, of diameters D_1, D_2, . . . D_n, each of these lengths must be transformed into equivalent lengths, L_1', L_2', . . . L_n', of one chosen diameter, D_0, by means of the relationship

$$L_1' = L_1\left(\frac{D_0}{D_1}\right)^{5\frac{1}{3}}, \qquad L_2' = L_2\left(\frac{D_0}{D_2}\right)^{5\frac{1}{3}}, \text{ etc.,} \qquad (21)$$

whence

$$L_0 = L_1' + L_2' + \ldots + L_n' = D_0^{5\frac{1}{3}}\Sigma\frac{L}{D^{5\frac{1}{3}}},$$

or

$$\frac{L_0}{D_0^{5\frac{1}{3}}} = \Sigma\frac{L}{D^{5\frac{1}{3}}},$$

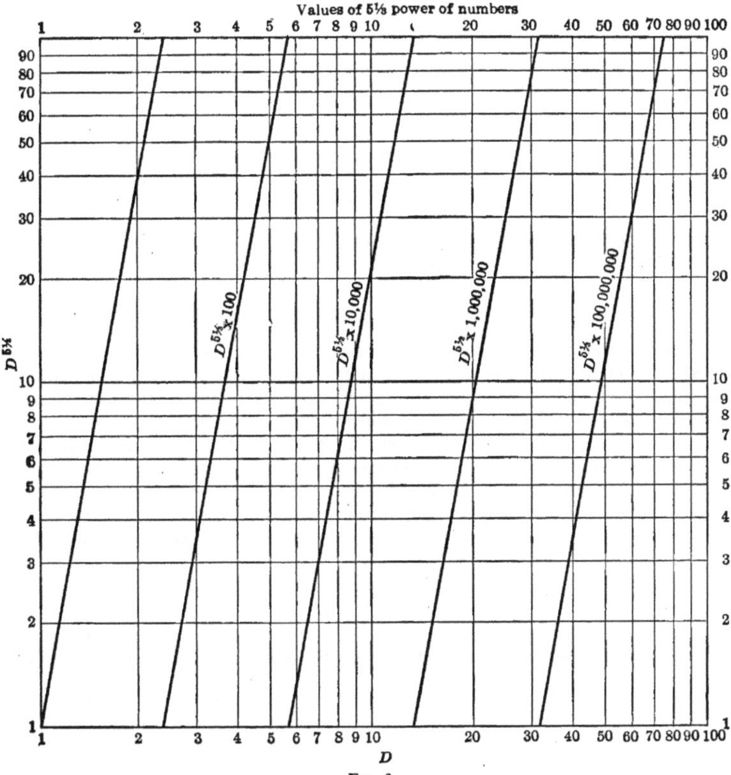

FIG. 6.

and (13) then becomes

$$Q = K\sqrt{\Sigma\frac{D^{5\frac{1}{3}}}{L}(P_1^2 - P_2^2)} = K\sqrt{\frac{D_0^{5\frac{1}{3}}(P_1^2 - P_2^2)}{L_0}}. \qquad (22)$$

Values of $D^{5\frac{1}{3}}$ for standard line pipe are given in Table 2. Values of $D^{5\frac{1}{3}}$ for any value of D from 1 inch to 75 inches may also be obtained from Fig. 6.

Problems with Looped System of Various Dimensions

In order to apply the formula to a looped system consisting of several pipes, side by side, of different diameters and lengths, it is necessary to determine the equivalent single line of length L_0 and diameter D_0. This is done by first transforming the diameters of the various individual pipes forming the loop into equivalent diameters of pipe having a common length, L_0, by means of the relationship

$$D_1 = D\left(\frac{L_0}{L}\right)^{3/16}, \quad \text{or} \quad D_1^{8/3} = D^{8/3}\left(\frac{L_0}{L}\right)^{1/2}, \quad \ldots\ldots \quad (23)$$

wherein D and L are actual diameter and length of one of the pipes, L_0 is the equivalent common length, and D_1 the corresponding equivalent diameter. The flow through the whole system is then equal to the sum of the flows through the individual pipes, or

$$Q \propto \sqrt{\frac{D_1^{5\frac{1}{3}}}{L_0}} + \sqrt{\frac{D_2^{5\frac{1}{3}}}{L_0}} + \ldots + \sqrt{\frac{D_n^{5\frac{1}{3}}}{L_0}} \propto \frac{\Sigma D^{8/3}}{\sqrt{L_0}}.$$

In other words, the diameter of a single line of length L_0 equivalent to the looped system is

$$D_0 = (\Sigma D^{8/3})^{3/8}.$$

Problems with the looped system may then be worked by Formula (13) with the equivalents D_0 and L_0, as if this system consisted of a single line of these dimensions.

Problems with Complex Line System in General

If a line consists of several sections composed variously of single lines and looped pipes, each of these sections may be transformed into equivalent lengths of single line of some chosen uniform diameter, and these lengths added together to obtain a single line equivalent to the entire system, to which Formula (13) may be applied.

In many cases line problems arise wherein gas is admitted to or withdrawn from a given system, and therefore all parts of the system do not convey the same quantity of gas. In such cases, where the quantities passing through the various sections bear a constant relationship to each other for different rates of flow, the lengths of the sections may be transformed to equivalent lengths corresponding to a single uniform rate of flow for all sections, by the relationship

$$L_0 = L\left(\frac{Q}{Q_0}\right)^2, \quad \ldots\ldots\ldots \quad (24)$$

wherein L and L_0 are the actual and equivalent lengths respectively and $\frac{Q}{Q_0}$ is the ratio of actual to equivalent rates of flow. The equivalent lengths thus determined may then be combined to form a line equivalent to the existing line, as long as the proportionality of flow rates through the various sections remains as assumed.

If it is desired to make use of equivalent diameters in this case, it may be done by the relationship

$$D_0 = D\left(\frac{Q_0}{Q}\right)^{3/8}. \quad \ldots\ldots\ldots \quad (25)$$

Values of ratios to the $5\frac{1}{3}$ and $\frac{3}{8}$ powers can be obtained from Figs. 5 and 6 respectively.

Effect of Position of Pipes of Different Diameter

With a given quantity of gas flowing through a line consisting of various sections having different diameters, the total pressure drop is independent of the order in which these sections occur. That is, if a line is composed of 25 miles of 8-inch pipe followed by 25 miles of 12-inch pipe, the total pressure drop will be the same as if the gas passed first through the 12-inch pipe and then through the 8-inch. The pressure distribution throughout the system will not be the same in the two cases, however. For example, with initial and terminal pressures of 200 and 25 pounds gage respectively, the pressure at the junction of the 8- and 12-inch pipes in the first case will be 63 pounds, and in the second case 189 pounds.

While the total pressure drop will be the same in either case, other considerations render it desirable to use the smaller pipes at the intake or high-pressure end of the system. The small pipes are better able to withstand the high pressures with the same weight of metal, and tighter joints can be made with them, thus keeping leakage losses at a minimum. Furthermore, the pressure distribution is such that the average pressure on the system is not as great as if the large pipes were at the initial end, thus further minimizing the leakage loss. Another distinct advantage in having the large pipes near the outlet end is that in the process of " packing " the system, that is, building up the pressure by introducing gas into the line at a rate greater than that at which it is being withdrawn, the surplus gas, or storage supply, is held nearer the point of consumption for use during peak loads.

Tests of pipe-line flows.—In 1901, Forrest M. Towl made a test of a pipe-line 70.32 miles long, with an internal diameter of 7.981 inches. (Page 436.) The specific gravity (G) of the gas was 0.64, the flowing temperature was 32° F., or $T=492°$ F., absolute. The temperature basis of measurement was $T_0=50+460=510°$ F. absolute and the pressure base, $P_0=14.65$ pounds per square inch absolute. The gas was measured at the outlet end by standardized Pitot tubes which gave results accurate to 1 per cent, and the line had been examined and repaired and was proved by test to be practically gas-tight. Pressure gages were installed at six points on the line at the distances noted in Table 3, which also gives the average pressures which were observed over a period of four hours during which the flow was steady. The initial and terminal pressures during this period were 210 and 41 pounds absolute, respectively, and the observed rate of flow was 221,000 cubic feet per hour. When the above data are inserted in the general Formula (12) the flow rate becomes 221,400 cubic feet per hour, or a difference of less than 0.2 of 1 per cent of the actual flow as measured.

This test also corroborates the relationship between pressures and length, as given in the general formula. With steady flow and uniform diameter, $P_1^2-P^2=LK$, where P is the pressure at any distance L from the intake end. With K determined from the initial and terminal average pressures in Table 3, the pressures P for the intermediate observation points on the line have been computed and are also shown in Table 3; and it will be seen that these computed values check with the observed values within the limits of precision to be expected with spring gages used in making the observations.

E. O. Hickstein (American Gas Engineering Journal, Feb. 2, 1918, p. 102) has made many tests of lines from 2 inches to 18 inches in diameter, with pressures ranging from a low vacuum to over 300-pound pressure, measuring the gas either by meter or by compressor displacement and found Formula (12) accurate in all cases. In the same publication he gives a table of squares of absolute pressures based on an atmospheric pressure of 14.4 pounds per square inch, which is reproduced here as Table 4. This table greatly facilitates computations of pipe-line flows.

Flow of Gas at Low Pressures and with Small Pressure Drop

Equation (12) may be written

$$Q=18.06\frac{T_0D^{8/3}}{P_0}\sqrt{\frac{(P_1+P_2)(P_1-P_2)}{GTL}}. \quad \ldots \ldots \quad (26)$$

But P_1+P_2 is twice the average pressure in the line, and for small drop in pressure may, without appreciable error, be considered equal to $2P_0$ for gas measured at atmospheric pressure, which is usually the case in designing low-pressure distributing systems. Hence,

$$Q=18.06\sqrt{2}\frac{\sqrt{P_0}T_0D^{8/3}}{P_0}\sqrt{\frac{P_1-P_2}{GTL}}$$

$$=25.54\frac{T_0D^{8/3}}{\sqrt{P_0}}\sqrt{\frac{P_1-P_2}{GTL}}. \quad \ldots \ldots \quad (27)$$

In low-pressure computations, however, it is preferable to express the pressure drop in inches of water head, and the length in feet. The average flowing temperature may also be taken as 40° F., and the standard temperature and pressure bases of measurement as 60° F. and 14.65 pounds per square inch respectively. Therefore, if H is the pressure drop in inches of water, $P_1-P_2=0.03608\ H$, $L=l\div5280$, $T=40+460=500$, $T_0=60+460=520$ and $P_0=14.65$.

Inserting these values in (27)

$$Q=2141D^{8/3}\sqrt{\frac{H}{Gl}}. \quad \ldots \ldots \quad (28)$$

Assuming a specific gravity (G) of 0.65, the working formula becomes

$$Q=2656D^{8/3}\sqrt{\frac{H}{l}} \text{ cubic feet per hour.} \quad \ldots \ldots \quad (29)$$

Table 5 gives values of $2656D^{8/3}$ for standard pipe sizes from ¾ inch to 12 inch based upon $P_0=14.65$, $T_0=520$, $T=500$ and $G=0.65$. Values for other bases of measurement can be obtained by multiplying the values in Table 5 by correction factors to be obtained from Fig. 4.

Pipe Line Storage Capacity

Under ordinary conditions of natural gas service the peak demands during a cold day may call for a delivery rate two or more times the average daily rate and possibly many times the minimum rate. In spite of this, however, it is not necessary to use a gas holder to equalize the load, inasmuch as the pipe-line itself performs a similar service because of the compressibility of the gas. When the demand drops off, the supply at the intake end is maintained, resulting in a building up of the pressure all along the line, due to the surplus gas forced into it. This surplus constitutes the storage capacity of the line.

The quantity of gas contained in a line with uniform flow, expressed in cubic feet at a base pressure of P_0 and the temperature existing in the line, is

$$V=19.20\frac{D^2L}{P_0}\left[P_1+P_2-\frac{P_1P_2}{P_1+P_2}\right], \quad \ldots \ldots \quad (30)$$

where D = diameter of pipe in inches, L = length of line in miles, P_1 and P_2 are the initial and terminal pressures respectively, in pounds per square inch absolute. (See Weymouth, Trans. A. S. M. E., Vol. 34, 1912, p. 203.)

The storage capacity is obtained by computing V from Equation (30) for the "packed" line condition, wherein P_1 may be taken as the maximum pressure for which the line is designed and P_2 is derived therefrom by means of Formula (16), on the assumption of minimum daily rate of flow; then deducting from this volume the value of V obtained by taking P_2 equal to the lowest delivery pressure consistent with satisfactory service and deriving P_1 therefrom on the assumption of maximum daily rate of flow.

In a complex line of various loops and sizes of pipe, it is necessary to compute the pressures prevailing at the extremities of each individual length of line of a given diameter, and from these to obtain the storage capacity of such line, the sum of the several capacities being the total storage capacity of the system.

Compression of Natural Gas

Theoretical power required.—If gas could be compressed isothermally, the theoretical work required, as derived from the laws of perfect gases, and with no clearance in the compressor, would be

$$W = p_1 v_1 \log_e \frac{p_2}{p_1}, \quad \ldots \ldots \ldots \quad (31)$$

where W = foot-pounds of work per pound of gas compressed;
p_1 = absolute suction pressure, pounds per square foot;
p_2 = absolute discharge pressure, pounds per square foot;
v_1 = volume of 1 pound of gas at pressure p_1 cubic foot.

Let p_0 = absolute pressure, pounds per square foot on which the measurement of the gas is based;
T_0 = absolute temperature basis of measurement, degrees Fahrenheit;
v_0 = volume of 1 pound of gas at p_0 and T_0;
T_1 = absolute temperature at which the gas is compressed.

Then

$$\frac{p_1 v_1}{T_1} = \frac{p_0 v_0}{T_0},$$

and

$$p_1 v_1 = p_0 v_0 \frac{T_1}{T_0},$$

hence

$$W = p_0 v_0 \frac{T_1}{T_0} \log_e \frac{p_2}{p_1} \text{ foot-pounds.} \quad \ldots \ldots \quad (32)$$

At p_0 and T_0 1,000,000 cubic feet of gas will weigh $\frac{1,000,000}{v_0}$ pounds, and if 1,000,000 cubic feet are compressed in one day of twenty-four hours (1440 minutes) there will be $\frac{1,000,000}{1440 v_0}$ pounds compressed per minute. Hence it will require.

$$\frac{1,000,000}{1440 v_0} \frac{p_0 v_0 T_1}{T_0} \log_e \frac{p_2}{p_1} \text{ foot-pounds per minute,}$$

or

$$\frac{1,000,000}{1440 \times 33,000} \frac{T_1}{T_0} p_0 \log_e \frac{p_2}{p_1} \text{ H.P.}$$

If P_0 = the base pressure in pounds per square inch absolute, and P_2 and P_1 are absolute pressures per square inch corresponding to p_2 and p_1.

$$p_0 = 144P_0,$$

and the theoretical isothermal horse-power required per million cubic feet of gas per twenty-four hours is

$$\text{H.P.}_{\cdot t} = \frac{1{,}000{,}000 \times 2.3026 \times 144}{1440 \times 33{,}000} P_0 \frac{T_1}{T_0} \log_{10} \frac{P_2}{P_0} = 6.978 P_0 \frac{T_1}{T_0} \log \frac{P_2}{P_1}. \quad . \quad . \quad (33)$$

If compression were performed strictly according to the adiabatic law, where n is the exponent of v in the adiabatic equation pv^n = constant, the work, neglecting clearance, required for 1 pound of gas would be, theoretically,

$$W = \frac{n}{n-1} p_1 v_1 \left[\left(\frac{p_2}{p_1} \right)^{\frac{n-1}{n}} - 1 \right]. \quad . \quad . \quad . \quad . \quad . \quad . \quad . \quad . \quad (34)$$

But, since

$$p_1 v_1 = p_0 v_0 \frac{T_1}{T_0},$$

$$W = \frac{n}{n-1} p_0 v_0 \frac{T_1}{T_0} \left[\left(\frac{p_2}{p_1} \right)^{\frac{n-1}{n}} - 1 \right]. \quad . \quad . \quad . \quad . \quad . \quad . \quad . \quad (35)$$

In order to compress 1,000,000 cubic feet per day, it would require,

$$\text{H.P.}_{\cdot a} = \frac{1{,}000{,}000}{1440 \times 33{,}000 \times v_0} \frac{n}{n-1} 144 P_0 v_0 \frac{T_1}{T_0} \left[\left(\frac{P_2}{P_1} \right)^{\frac{n-1}{n}} - 1 \right]$$

$$= 3.03 P_0 \frac{T_1}{T_0} \frac{n}{n-1} \left[\left(\frac{P_2}{P_1} \right)^{\frac{n-1}{n}} - 1 \right] \text{H.P.} \quad . \quad . \quad . \quad . \quad . \quad . \quad . \quad . \quad (36)$$

Taking as measurement bases, $P_0 = 14.65$ pounds per square inch and $T_0 = 60 + 460 = 520°$ and assuming $T_1 = T_0 = 520°$, and $n = 1.266$ for natural gas, formula (33) for isothermal compression becomes,

$$\text{H.P.}_{\cdot t} = 102.23 \log \frac{P_2}{P_1}, \quad . \quad . \quad . \quad . \quad . \quad . \quad . \quad . \quad . \quad . \quad . \quad (37)$$

and Formula (36) for adiabatic compression, becomes

$$\text{H.P.}_{\cdot a} = 211.27 \left[\left(\frac{P_2}{P_1} \right)^{0.21} - 1 \right]. \quad . \quad . \quad . \quad . \quad . \quad . \quad . \quad . \quad (38)$$

Equations (37) and (38) give the theoretical H.P. per 1,000,000 cubic feet per day, for isothermal and adiabatic compression respectively, on the assumption that the compressors have no clearance and that there are no losses of any kind. Obviously these conditions are impossible of realization in practice, and experience shows that while one would expect the true formula to be somewhere between those for adiabatic and isothermal conditions, on account of jacket cooling, in reality this is not the case, on account of the effects of mechanical clearance and slippage losses.

Theoretical mean effective pressure.—For isothermal compression, the mean effective pressure in pounds per square inch, as derived from Equation (31), is

$$\text{m.e.p.}_{\cdot t} = 2.3026 P_1 \log \frac{P_2}{P_1}. \quad . \quad . \quad . \quad . \quad . \quad . \quad . \quad . \quad . \quad . \quad (39)$$

For adiabatic compression, derived from Equation (34), it is

$$\text{m.e.p.}_a=\frac{n}{n-1}P_1\left[\left(\frac{P_2}{P_1}\right)^{\frac{n-1}{n}}-1\right], \quad (40)$$

which, for natural gas, when $n=1.266$, is

$$\text{m.e.p.}_a=4.76P_1\left[\left(\frac{P_2}{P_1}\right)^{0.21}-1\right]. \quad (41)$$

If P_2 in Equations (39) and (41) be regarded as constant and equal to 100 per cent, and P_1 be allowed to vary from 0 to 100, and the results plotted with values of m.e.p.

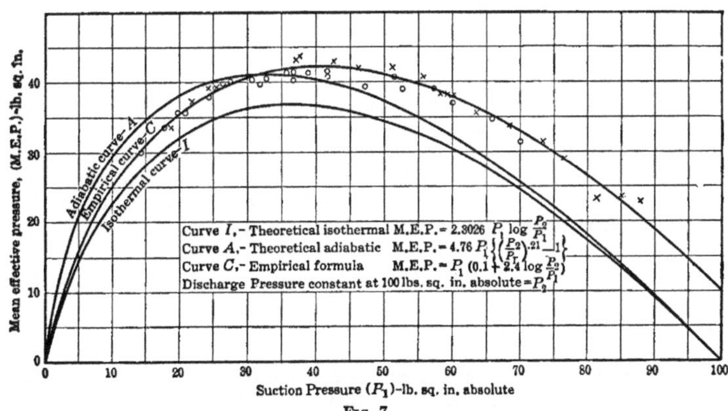

Fig. 7.

as ordinates and values of P_1 as abscissae, the resulting curves will be shown as I and A respectively, of Fig. 7. The maximum or peak load point, for the isothermal curve I occurs at

$$P_1=0.3679P_2 \quad \text{or} \quad \frac{P_2}{P_1}=r=e=2.718.$$

For the adiabatic curve A, it occurs at

$$P_1=0.3254P_2 \quad \text{or} \quad \frac{P_2}{P_1}=r=n^{\frac{n}{n-1}}=3.073.$$

In the above equations, r is the compression ratio.

By referring to Equations (37) and (38), it is seen that the power required to pump a given volume of gas is dependent solely upon the ratio of discharge to suction pressures, irrespective of the actual value of either of them, whereas Equations (39) and (41) show the mean effective pressure to depend upon the absolute value of the suction pressure, as well as upon the number of compressions. They further show that as the suction increases from zero to the value of the discharge pressure, the latter remaining fixed, the mean effective pressure increases, theoretically, from zero to a maximum value, and then decreases to a theoretical zero when the suction is equal to the discharge. As a matter of fact, however, the indicated mean effective pressure in the compressor cylinder will not fall to zero with suction and discharge pressures equal, because of the wire-drawing of the dense gas through the valves, resulting in a diminution dur'ng the admission stroke, of the pressure in the cylinder below that in the suction line, and in an accumulation, during the discharge stroke, of the pressure above that in the discharge line. Consequently, with suction and discharge pressures equal, power will always be expended.

With $P_1=0$, that is, when no gas is admitted to the compressor, the indicator card will show the compression and re-expansion lines to be practically superimposed if the valves are tight, and consequently the condition of zero work with zero suction pressure is very nearly attained in practice.

Actual mean effective pressure.—It has been the custom of most engineers to consider that the actual mean effective pressure of compression follows the adiabatic law, and they have therefore used Equations (39) and (41), with constants depending upon assumed values of inefficiency, or lost work, according to the judgment and experience of the engineer. Repeated tests have shown, however, that the work curve has characteristics more nearly approaching those of the isothermal than those of the adiabatic, and can be represented satisfactorily by a simple empirical equation. The actual curve begins at zero, for zero suction, and rises between the isothermal and adiabatic curves until a point is reached, at about 3.2 compressions, where it crosses the adiabatic and thereafter continues above it, as the compression ratio decreases.

In Fig. 7 is given a plot of the observations of two independent tests made under different conditions (Trans. A. S. M. E., Vol. 32, 1912, p. 210) showing actual mean effective pressures as compared with theoretical isothermal and adiabatic curves. The curve (*c*) through the observed points is plotted from the empirical equation

$$\text{m.e.p.}=P_1\left(0.1+2.4\log\frac{P_2}{P_1}\right). \qquad (42)$$

The maximum value of m.e.p. according to this equation, occurs at $P_2 \div P_1=2.47$ compressions, or $P_1=0.405P_2$.

The **volumetric efficiency of a compressor** is the ratio of the quantity of gas actually delivered into the discharge line to that computed from piston displacement and suction pressure and temperature. There are a number of sources of loss in volumetric efficiency, the most important of which are

(*a*) Mechanical clearance.
(*b*) Wire-drawing of the gas passing through suction valves and passages.
(*c*) Heating of the inlet gas while entering the cylinder.
(*d*) Valve and piston leakage.

It will thus be seen that the volumetric efficiency depends not only upon the diameter of compressor, length of stroke, design of compressor and valves, and on speed, but also upon the number of compressions through which the gas is pumped. For purposes of preliminary computation with the modern reciprocating compressor, the volumetric efficiency may be expressed by the empirical relationship

$$E_v=0.97-0.0306r, \qquad (43)$$

where $r=\dfrac{P_2}{P_1}$=number of compressions. This equation corresponds to a slippage loss of 6 per cent and an average clearance of 3 per cent, and gives a fair representation of the usual efficiencies encountered in practice.

Actual power required to compress natural gas.—If l is the length of stroke of a compressor, in feet, a the piston area, square feet, and N the number of strokes per minute, the quantity of gas pumped in one day of 1440 minutes would be, in millions of cubic feet, based on P_0 and T_0,

$$Q=\frac{1440}{1{,}000{,}000}\,\frac{a}{144}\,l\,N\,\frac{T_0}{T_1}\,\frac{P_1}{P_0}\,E_v. \qquad (44)$$

The horse-power required on the compressor piston will be

$$\text{H.P.}=\frac{a\times l\times N\times \text{m.e.p.}}{33{,}000}.$$

The horse-power per 1,000,000 cu. ft. per day will thus be equal to

$$\text{H.P.}_{\cdot m}=\frac{\text{H.P.}}{Q}=\frac{100}{33}\,\frac{P_0}{P_1}\,\frac{T_1}{T_0}\,\frac{\text{m.e.p.}}{E_v}. \qquad (45)$$

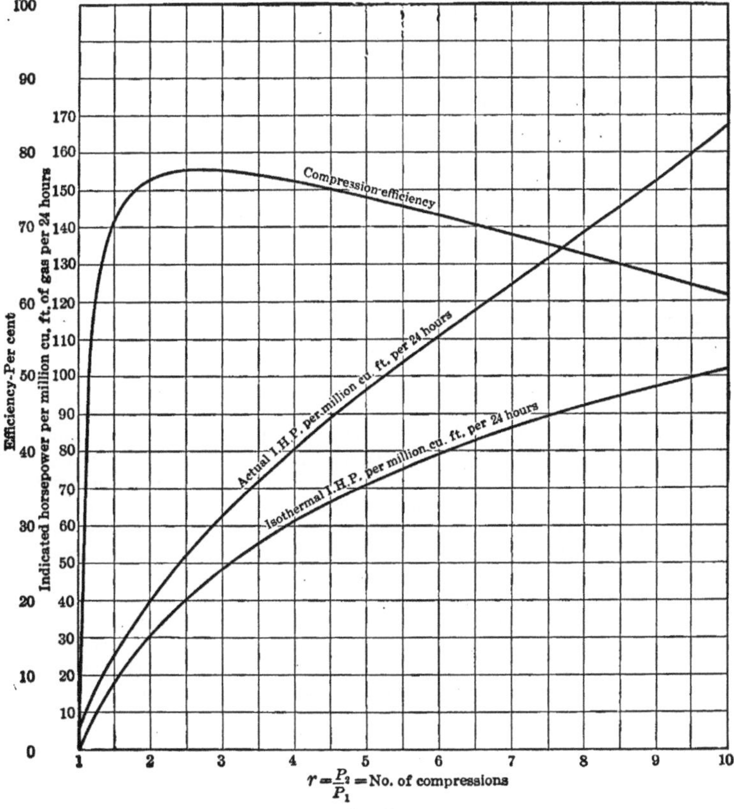

FIG. 8.

Substituting the value of m.e.p. as given in Equation (42) and the value of E_v from Equation (43),

$$\text{H.P.}_{\cdot m}=\frac{100}{33}\,\frac{P_0T_1}{T_0P_1}\,\frac{P_1\left(0.1+2.4\log\frac{P_2}{P_1}\right)}{0.97-0.0306\frac{P_2}{P_1}}, \qquad (46)$$

or, if

$$\frac{P_2}{P_1}=r,$$

$$H.P._m=3.03P_0\frac{T_1}{T_0}\frac{0.1+2.4\log r}{0.97-0.0306r}. \qquad (47)$$

For $T_1=T_0$ and $P_0=14.65$,

$$H.P._m=\frac{4.44+106.6\log r}{0.97-0.0306r}. \qquad (48)$$

Equation (48) is plotted in Fig. 8.

Compression efficiency.—The isothermal horse-power Formula (33) gives the least possible power required to compress gas at the rate of 1,000,000 cu. ft. per day, and is therefore used as a measure of the efficiency of compression. In other words, "compression efficiency" is the figure obtained by dividing the horse-power as determined from Equation (33) by that actually required on the compressor piston, as given by equation (47), or

$$E_c=\frac{H.P._i}{H.P._m}=2.303\frac{E_v\log r}{0.1+2.4\log r}=2.3\frac{(0.97-0.0306r)\log r}{0.1+2.4\log r}. \qquad (48)$$

Formula (48) is shown plotted in Fig. 8.

The formulæ and curves given in this section are for the indicated horse-power required on the compressor piston. Therefore, in order to determine the amount necessary in the power cylinders, the values taken from the curves must be increased according to the mechanical efficiency of the machine. This efficiency will range from 75 to 80 per cent in direct-connected gas-engine-driven compressors of large size.

Single-stage compression is advisable for ranges up to seven or eight compressions for the reason that the benefit that would be derived from two-stage operation is overbalanced by the excess first cost of the units, combined with the greater cost of operating, installing and housing them. As the compression range increases above this point, however, two-stage operation becomes necessary from an operating standpoint, and its advantage from an economic standpoint becomes more and more pronounced.

In compressing natural gas through two or more stages, it is found to be impracticable to drive the compressors of all stages by the same power unit, on account of the variable pressure conditions which unbalance the loads of the different stages except at one particular ratio of compression. For this reason it is the usual practice to have the compressors of each stage driven by their own individual power units.

It is seldom, if ever, advisable to pump through more than two stages in natural-gas transmission, for the reason that the pressure is limited by the strength of the pipe-lines to a value that would not warrant the added complication and cost of multi-stage units higher than two.

PIPE-LINES

The transporting of gas requires a pipe-line which shall be practically "air-tight." It is more difficult to make a line hold gas under pressure than to make it hold a liquid. Trouble has been experienced in almost all lines built for high pressure on account of the leaking of the gas at the couplings. The first high-pressure lines were laid with bell-and-spigot joints, caulked with lead. Such lines might have been tight when first laid, but expansion and contraction soon caused them to leak large amounts. The next lines used were of wrought iron or steel pipe, with screw joints such as are usual in low-pressure lines in buildings. These joints held much better than the bell-and-spigot joints, but there was still enough leakage under high pressure to make it desirable to have a more perfect joint. Actual tests showed that some of the joints on 8-inch and 10-inch lines, which would probably be satisfactory for fluids, leaked from 20 to 50 cubic feet of gas per hour. As there are about 250 joints per mile it is evident that the loss on long high-pressure lines would be considerable. Rubber stuffing-box clamps were at first used, and from these were evolved the Dresser coupling, the Hammond coupling and several others of a similar type. These couplings are similar to the ordinary stuffing-box and depend on rubber or on an elastic, rubber-like ring to keep the joints tight. The Dresser coupling is shown in Fig. 9 and the Hammond coupling in Fig. 10.

Lines of pipe can be built in almost any kind of country, but in some places it is necessary to take extra precautions to keep the line from acting as a Bourdon tube and expanding until the ends are pulled out of the couplings. To avoid this trouble, it is usual in such places as river crossings to use screw-joint couplings and to place a special rubber joint stuffing-box fitting outside the collars. The packing manufacturers have developed a special elastic composition which is not acted on by the gas or oil which it

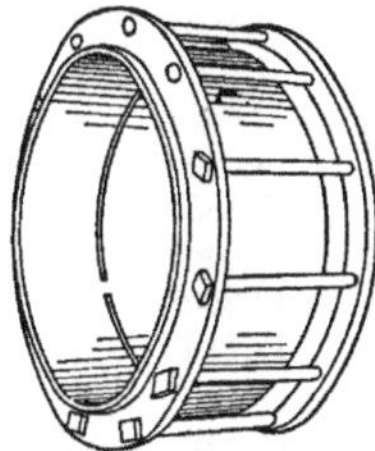

Fig. 9.

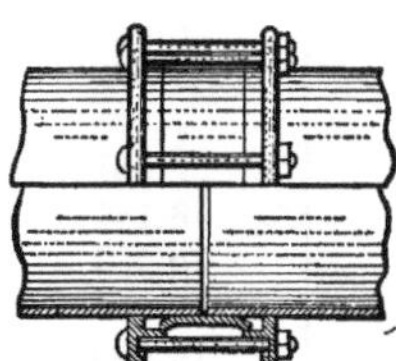

Fig. 10.

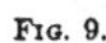

may contain, and which does not lose its elasticity by oxidation. A type of these couplings is also used to prevent the line from acting as an electrical conductor for stray currents.

Electrolysis of Pipe-lines

Oil and gas pipe-lines, as well as other underground structures, are liable to be seriously damaged by stray electric currents. Attention is called to the "Report of the American Committee on Electrolysis, 1921," and to the investigations which are being conducted by a sub-committee in connection with the United States Bureau of Standards and the United States Department of Agriculture. "A pipe line laid with every joint an insulating joint has a comparatively high resistance, and no substantial current can flow on such a line." Insulating joints improperly applied may be a source of danger instead of a remedy. Other remedies for reducing the damage liable to be done by stray currents are also available but must be properly applied or they may increase rather than reduce the damage.

TABLE 1

Values of D; A (one hour) $= 35.56D$; and A (24 hours) $= 853.5D$; based upon $= 14.65$, $T = 500$, $G = 0.65$

Nominal pipe size	Actual internal diameter-inches $= D$	$D^{8/3}$	$A = 35.56D^{8/3}$	$A = 853.5D^{8/3}$
1	1.049	1.136	40.40	969.6
1¼	1.380	2.361	83.96	2,015
1½	1.610	3.561	126.6	3,039
2	2.067	6.827	242.8	5,827
2½	2.469	11.14	396.1	9,508
3	3.068	19.87	706.6	16,960
3½	3.548	29.28	1,041	24,990
4	4.026	41.02	1,459	35,010
4½	4.506	55.39	1,970	47,280
5	5.047	74.95	2,665	63,970
6	6.065	122.3	4,349	104,400
7	7.023	180.9	6,433	154,400
8	7.981	254.4	9,046	217,100
9	8.941	344.4	12,250	293,900
10	10.136	481.2	17,110	410,700
11	11.000	598.5	21,280	510,800
12 i. d.	12.000	754.8	26,840	644,200
12 o. d.	11.250	635.4	22,590	542,300
14	13.250	983.1	34,960	839,100
16	15.250	1,430	50,850	1,221,000
18	17.250	1,987	70,660	1,696,000
20	19.250	2,662	94,660	2,272,000
22	21.250	3,464	123,200	2,957,000
24	23.250	4,403	156,600	3,758,000

TABLE 2

Values of D for standard line pipe

Nominal pipe size	Actual inside diameter, inches	$D^{5/3}$	Nominal pipe size	Actual inside diameter, inches	$D^{5/3}$
1	1.049	1.291	8	7.981	64,710
1¼	1.380	5.572	9	8.941	118,600
1½	1.610	12.68	10	10.136	231,500
2	2.067	46.61	11	11.000	358,200
2½	2.469	124.0	12	12.000	569,700
3	3.068	395.0	12	11.250	403,800
3½	3.548	857.5	14	13.250	966,500
4	4.026	1,683	16	15.250	2,045,000
4½	4.506	3,068	18	17.250	3,947,000
5	5.047	5,617	20	19.250	7,084,000
6	6.065	14,970	22	21.250	12,000,000
7	7.023	32,720	24	23.250	19,390,000

TABLE 3

Pressures on 8-inch Pipe-line with Steady Flow

Station No.	Distance from initial station, miles	Observed absolute pressure	Computed absolute pressure
1	0.00	209.4	209.4
2	20.97	179.0	177.0
3	43.05	134.9	134.3
4	59.86	87.1	89.1
5	67.45	57.3	58.2
6	70.32	40.8	40.8

TABLE 4

Values of $2656D^{8/3}$ for standard line pipe and for hourly flow rates, based on 14.65 pounds square inch absolute, 60° F. measurement temperature, 40° F. flowing temperature, 0.65 specific gravity.

Nominal pipe size	Actual internal diameter, inches = D	$D^{8/3}$	$2656D^{8/3}$
1/8	.269	0.03015	80.10
1/4	.364	0.06755	179.4
3/8	.493	0.1517	402.9
1/2	.622	0.2819	748.9
3/4	.824	0.5968	1,585
1	1.049	1.136	3,018
1 1/4	1.380	2.361	6,270
1 1/2	1.610	3.561	9,459
2	2.067	6.827	18,140
2 1/2	2.469	11.14	29,580
3	3.068	19.87	52,790
3 1/2	3.548	29.28	77,790
4	4.026	41.02	109,000
4 1/2	4.506	55.39	147,100
5	5.047	74.95	199,100
6	6.065	122.3	325,000
7	7.023	180.9	480,500
8	7.981	254.4	675,700
9	8.941	344.4	914,800
10	10.136	481.2	1,278,000
11	11.000	598.5	1,590,000
12 i. d.	12.000	754.8	2,005,000
12 o. d.	11.250	635.4	1,688,000

TABLE 5

Squares of Absolute Pressures. B=14.4 pounds

	0	1	2	3	4	5	6	7	8	9	
0	207.36	237.16	268.96	302.76	338.56	376.36	416.16	457.96	501.76	547.56	0
10	595.36	645.16	696.96	750.76	806.56	864.36	924.16	985.96	1,049.8	1,115.6	10
20	1,183.40	1,253.2	1,325.0	1,398.8	1,474.6	1,552.4	1,632 2	1,714.0	1,797.8	1,883.6	20
30	1,971.4	2,061.2	2,153.0	2,246.8	2,342.6	2,440.4	2,540.2	2,642.0	2,745.8	2,851.6	30
40	2,959.4	3,069.2	3,181.0	3,294.8	3,410.6	3,528.4	3,648 2	3,770.0	3,893.8	4,019.6	40
50	4,147.4	4,277.2	4,409.0	4,542.8	4,678.6	4,816.4	4,956.2	5,098.0	5,241.8	5,387.6	50
60	5,535.4	5,685.2	5,837.0	5,990.8	6,146.6	6,304.4	6,464.2	6,626.0	6,789.8	6,955.6	60
70	7,123.4	7,293.2	7,465.0	7,638.8	7,814.6	7,992.4	8,172.2	8,354.0	8,537.8	8,723.6	70
80	8,911.4	9,101.2	9,293.0	9,486.8	9,682.6	9,880.4	10,080	10,282	10,486	10,692	80
90	10,899	11,109	11,321	11,535	11,751	11,968	12,188	12,410	12,634	12,860	90
100	13,087	13,317	13,549	13,783	14,019	14,256	14,496	14,738	14,982	15,228	100
110	15,475	15,725	15,977	16,231	16,487	16,744	17,004	17,266	17,530	17,796	110
120	18,063	18,333	18,605	18,879	19,155	19,432	19,712	19,994	20,278	20,564	120
130	20,851	21,141	21,433	21,727	22,023	22,320	22,620	22,922	23,226	23,532	130
140	23,839	24,149	24,461	24,775	25,091	25,408	25,728	26,050	26.374	26,700	140
150	27,027	27,357	27,689	28,023	28,359	28,696	29,036	29,378	29,722	30,068	150
160	30,415	30,765	31,117	31,471	31,827	32,184	32,544	32,906	33,270	33,636	160
170	34,003	34,373	34,745	35,119	35,495	35,872	36,252	36,634	37,018	37,404	170
180	37,791	38,181	38,573	38,967	39,363	39,760	40,160	40,562	40,966	41,372	180
190	41,779	42,189	42,601	43,015	43,431	43,848	44,268	44,690	45,114	45,540	190
200	45,967	46,397	46,829	47,263	47,699	48,136	48,576	49,018	49,462	49,908	200
210	50,355	50,805	51,257	51,711	52,167	52,624	53,084	53,546	54,010	54,476	210
220	54,943	55,413	55,885	56,359	56,835	57,312	57,792	58,274	58,758	59,244	220
230	59,731	60,221	60,713	61,207	61,703	62,200	62,700	63,202	63,706	64,212	230
240	64,719	65,229	65,741	66,255	66,771	67,288	67,808	68,330	68,854	69,380	240
250	69,907	70,437	70,969	71,503	72,039	72,576	73,116	73,658	74,202	74,748	250
260	75,295	75,845	76,397	76,951	77,507	78,064	78,624	79,186	79,750	80,316	260
270	80,883	81,453	82,025	82,599	83,175	83,752	84,332	84,914	85,498	86,084	270
280	86,671	87,261	87,853	88,447	89,043	89,640	90,240	90,842	91,446	92,052	280
290	92,659	93,269	93,881	94,495	95,111	95,728	96,348	96,970	97,594	98,220	290
300	98,847	99,477	100,109	100,743	101,379	102,016	102,656	103,298	103,942	104,588	300
310	105,235	105,885	106,537	107,191	107,847	108,504	109,164	109,826	110,490	111,156	310
320	111,823	112,493	113,165	113,839	114,515	115,192	115,872	116,554	117,238	117,924	320
330	118,611	119,301	119,993	120,687	121,383	122,080	122,780	123,482	124,186	124,892	330
340	125,599	126,309	127,021	127,735	128,451	129,168	129,888	130,610	131,334	132,060	340
350	132,787	133,517	134,249	134,983	135,719	136,456	137,196	137,938	138,682	139,428	350
360	140,175	140,925	141,677	142,431	143,187	143,944	144,704	145,466	146,230	146,996	360
370	147,763	148,533	149,305	150,079	150,855	151,632	152,412	153,194	153,978	154,764	370
380	155,551	156,341	157,133	157,927	158,723	159,520	160,320	161,122	161,926	162,732	380
390	163,539	164,349	165,161	165,975	166,791	167,608	168,428	169,250	170,074	170,900	390
	0	1	2	3	4	5	6	7	8	9	

CHARACTERISTICS OF PETROLEUM

BY

DAVID T. DAY

Petroleum is the name commonly given to the liquid oils found in the earth. It is usually a homogeneous intersolution of many oils which are compounds of carbon and hydrogen. These hydrocarbons, as they constitute petroleum, usually carry many foreign bodies dissolved with them, or in other cases emulsified with them. These foreign bodies include water, clay, sand, asphalt, resinous bodies, and many compounds of carbon and hydrogen with oxygen, sulphur and nitrogen, or with any one of them. Occasionally other elements are found in chemical combination with "organic" substances in petroleum, but usually such elements are merely present in some mineral compound which is emulsified with the petroleum itself.

Inasmuch as hydrocarbons are the essential constituents of petroleum, a list of them is given (pages 460 to 523), which includes nearly all that have been identified, whether or not they have actually been isolated from petroleum. Others, doubtless, are already known, and still others will be isolated, while some included in this list will eventually prove to be mixtures of other hydrocarbons.

The statement is continually made, by those who have not studied petroleum very exhaustively, that it is made up of many hundreds, or even thousands, of different individual oils, while only about four hundred and forty-five hydrocarbons are tabulated, and more than half of these have never been found in petroleum. The discrepancy is the fault of the popular writer rather than of the table. If additional hydrocarbons are found in some variety of petroleum they will certainly make up only an infinitesimal part of the oil and will be of no industrial importance. It is a matter of great significance, however, that many more substances are *obtained from* petroleum than are shown in this table. Most of these are only partial separations of the constituents of petroleum, prepared to suit some special industrial use, often with an unjustified hope of filling some special need better than the oil which has previously been used. By this means petroleum derivatives have been given a great and needless complexity.

Another reason for the large number of hydrocarbons spoken of in connection with petroleum is a result of a peculiar characteristic of the hydrocarbons themselves—of all of them to a certain extent, but of certain oils in particular. This peculiarity is the ease with which one hydrocarbon may be transformed into another through such an external agency as heat. Doubtless other forms of energy, light rays, electricity, etc., may be equally potent, but their effects are less known.

The following is a familiar example: If any clear, water-white oil, such as kerosene, or even gasoline, be partly distilled, no matter how gently, the residue will promptly darken in color, and the products obtained will vary according to the temperature at which the distillation has taken place. If the oil be distilled at atmospheric pressure the products obtained will differ from those obtained by distilling at the lower temperatures at which distillation may be conducted in a vacuum. Even where the heating of a hydrocarbon mixture is so controlled as to avoid entirely the "cracking" of the constituents into hydrocarbons of greatly different boiling points, there is every reason to believe that many important changes in the structure of the hydrocarbons, isomeric changes, etc., are constantly taking place. It may be generally stated that few hydrocarbons can long remain entirely unchanged at their boiling points.

In a handbook of the petroleum industry all the methods of producing, transporting, storing, separating and refining various products may be regarded as dependent upon the fundamental properties of the various mixtures of hydrocarbons which are called petroleums. It is convenient to give a concise statement of these fundamentals in one place, where they may be readily referred to for all purposes. Various physical properties of crude oils, such as color, odor, specific gravity, coefficient of expansion, and viscosity, are necessarily considered in detail in other chapters to which they are more pertinent, and are therefore not described here.

The ease with which the constitution of a hydrocarbon may be affected by heat was shown by the author for the hydrocarbon ethylene, which, on being heated in a tube under merely atmospheric pressure to 350° C., gradually changed with the formation of higher members of the same series.

The significance of the action of heat in complicating the composition of an oil mixture has been felt in the industry for many years. It was first regarded as a disadvantage because it broke up oils and produced others less desirable on account of their odor, etc. More recently, it has been regarded as a great advantage because it allows the production of oils, such as gasoline, for which there happens to be a great industrial demand.

It is also desirable to note in this connection that the action of heat has been used chiefly to produce oils of lower vaporizing temperature from those of higher boiling points. On the other hand, experience with the action of heat upon the gas ethylene reveals the fact, which may be of equally great significance in the future, that it is possible to reverse the usual "cracking" effect of heat and to produce more complex hydrocarbons from those of low boiling points.

Elements occasionally found in petroleum.—Oxygen, sulphur and nitrogen are usually found in petroleum though in very small proportions. Frequently they are of such industrial significance that they deserve special consideration. Other elements are occasionally found, but are of merely scientific interest. Oxygen, sulphur and nitrogen may be considered as occurring in petroleum in compounds which are, in general, not oily in their nature. They are simply compounds of carbon, of one sort or another, which through some accidental circumstance have become dissolved in the crude oil. An important exception may be noted in the case of organic sulphur compounds. These are themselves oils, and cannot easily be separated as extraneous bodies from petroleum. It is generally necessary to break them up and extract the elementary sulphur by combination with some metal, such as copper.

Oxygen.—Oxygen reacts only very slightly on the lighter members of the paraffin hydrocarbons. Members of certain other series, notably terpenes, absorb oxygen readily, with the formation of organic acids and resins. Furthermore, the terpenes are able to transfer part of the oxygen to other substances which ordinarily are unaffected by it. Most hydrocarbons are acted upon slightly by oxygen when they are heated; light also has a marked influence upon the reaction. Moreover, when a direct electric current is passed through an emulsion of oil and caustic soda, soaps are formed quite readily.

Part of the oxygen present in petroleum oils must be ascribed to oxidation by the air, while part of it may represent incomplete reduction of the organic matter from which the petroleum may have originated. What proportion of the total oxygen may be ascribed to the latter cause is entirely an open question. Study of the character of the oxygen compounds may ultimately lead to a method of settling this question. The compounds which result from slight oxidation of petroleum are, in general, alcohols and fatty or naphthenic acids, and the esters of both. This has been made evident by the work of many authorities, partly by analtyic examination of natural oils, and especially by direct partial oxidation of oils by the action of air, oxygen or ozone. Mr. T. T. Gray has been especially successful in oxidizing 90 per cent of a paraffin wax to oxidized compounds, by exposing the molten material in a shallow pan to ultra-violet light while air in fine bubbles is jetted through it. The products are chiefly fatty acids, together with smaller proportions of alcohols and esters. The examination of the oxygen compounds in natural oils shows similar substances, although the nature of the oxidized compounds in asphalts is very obscure. Phenols have been found in many crude oils, but not in interesting quantities; in fact, they are usually the merest traces.

Oxygen usually occurs in petroleum as an element of the asphalt, which, dissolved, in the other oils, makes up a large percentage of the petroleums found in Mexico and California, as well as of the petroleum of the Gulf Coast of Texas and Louisiana. Occasionally asphalt is found in Oklahoma and Illinois oils, and to a very slight extent in some other petroleums of the United States. It is an important ingredient of Russian petroleum. Various oils found in other foreign countries contain asphalt, as will be shown in the tables of analyses of crude petroleum which follow on a later page of this chapter.

THE HYDROCARBONS IN PETROLEUM

In compiling the following table of hydrocarbons the effort has been made to include all that have been found in natural petroleums and the more important hydrocarbons prepared synthetically, as well as those formed by destructive distillation of other hydrocarbons (cracking), or during the distillation of coal or other carbonaceous material. Such artificially prepared hydrocarbons are liable at any time to become economic products of some heat treatment of petroleum.

Where the physical properties of a given hydrocarbon have been obtained by several competent authorities they have all been given, with references to the sources of information, so that the user may select the most accurate data. Frequently where several specific gravities are given for the same hydrocarbon, this has been measured at differing temperatures, thus giving a clue to the rate of expansion. Similarly, boiling points have been quoted for different pressures.

It is not to be understood that data from all investigations have been presented. Obviously crude approximations published many years ago have been replaced by more accurate data. Much material has also been omitted where the results could not possibly be brought even into approximate harmony with standard authorities. Much material has been taken from the Handbook of Chemistry and Physics and from Van Nostrand's Chemical Annual, and these sources are especially acknowledged. The great number of places left blank from lack of reasonably accurate data, and the many cases of conflicting figures, bear eloquent testimony to the need of much more work in establishing the fundamental facts concerning many oils. But those who recognize the enormous labor required in obtaining hydrocarbons sufficiently pure for determining their physical and chemical characteristics will find these tables a monument to such men as Mabery, Engler, Landolt, Warren and Storer, and a small army of other contributors, whose patience and skill through many years have been necessary for the results thus far compiled, and are an inspiration to coming investigators.

THE HYDROCARBC

Together with the Princ

CHAIN HYDROCARBONS

Name	Description	Specific gravity	Index of refraction	
Methane (Marsh gas)	CH_4	0.4150 (−164°)		−184
		*.5547 (air = 1) (15.6° C.)		
		....		
Ethane	C_2H_6	.4460 (0°)*		−172.1†
		1.0490 (air = 1)		−172
..		1.0360 (15.6° C.)		
Propane	C_3H_8	0.5360 (at M.P.)		
		1.5580 (ir = 1)*		†−187.8
Butane (Diethyl)	$CH_3 \cdot CH_2 \cdot CH_2 \cdot CH_3$	0.6000 (at M.P.)		Hexagonal tals
..		2.0460 (air = 1)		−135
Isobutane (Trimethylmethane)	$(CH_3)_2{>}CH \cdot CH_3$	0.6030 (0°) 2.046 (air = 1)		
Pentane (normal)	$CH_3 \cdot (CH_2)_3 \cdot CH_3$	0.6454 (4/0)		
..		0.6263 (17°)		
.		.6475 (0°)		
.		.6250 (25/25)		−130.8
Isopentane Ethyl isopropane	$C_2H_5—CH—(CH_3)_2$	.6380 (14°)		
		.6393 (0°)		
..		.6394 (4/0)		
	$(CH_3)_2CH \cdot CH_2 \cdot CH_3$	....		
	$CH_3—CH(C_2H_5)—CH_3$	0.6393 (0°)		
		.6385 (14)		
		....		
		....		
(Tetramethyl methane)	$CH_3—C(CH_3)_2—CH_3$	....		−20
Hexane (normal)	$CH_3 \cdot (CH_2)_4 \cdot CH_3$	.6771 (0/4)		Freezes to m clinic cr clinic cry
		.6775 (0) .6606 (20)		
.. ..		.67697 (0/4)		

IN PETROLEUM

Hydrocarbons from Other Sources

PARAFFIN SERIES, C_nH_{2n+2}

Boiling point ° C. Pressure, m.m.	Occurrence	Authority
−164 at 760	Petroleum generally and "dry" gas	Meyer and Jacobson, Lehrbuch der Organ. Chemie, Ed. 2, I-1, 164, (1906–7)
−164		Landolt and Börnstein, Tabellen, **81**, (1894)
†−160		*Burrell, Seibert and Oberfell, Bureau of Mines Bull. 88, Petroleum Technology 20, 25 (1918)
		†Holleman, Organic Chemistry, Ed. A. J. Walker, 41 (1910)
− 84.1 at 749†	Petroleum generally	Richter, Organ. Chemie, 53 (1888)
		Meyer and Jacobson, Lehrbuch der Organ. Chemie, Ed. 2, I-1, 164, (1906–7)
−86		Handbook Chem. and Phys., 178 (1920–21)
		Landolt and Börnstein, Tabellen
− 37 at 760	Do.	Meyer and Jacobson, Lehrbuch der Organ. Chemie, Ed. 2, I-1, 164, (1906–7)
*−38-39		*Handbook Chem. and Phys., 228 (1920–21)
		†Van Nostrand's Chem. Ann., 352 (1918)
	Do.	Warren, Mem. Am. Acad., 2d ser., 121, **9**, 135 (1367)
0.3 at 760		Van Nostrand's Chem. Ann., 268 (1918)
− 11		Handbook Chem. and Phys., 199 (1920–21)
36 3	Do.	Chavanne and Simon, Compt. rend., **1'8**, 1324–6 (1919)
37 at 760	Russia	Markownikoff, Ber., **30**, 975 (1897)
36 3 at 760		Thorpe and Jones, Jour. Chem. Soc., **63**, 290 (1893)
36.3 at 760		Mabery and Young, Jour. Frank. Inst., **162**, 57–70, 81–128
27.95 at 760	Petroleum; also produced synthetically	Young and Thomas, Jour. Chem. Soc., **71**, 445 (1897)
28		Chavanne and Simon, Compt. rend., **1 8**, 1324–6 (1919)
30.4		Thorpe and Jones, Jour. Chem. Soc., **63**, 290 (1893)
29–32 at 763.4	Produced synthetically	Thorpe and Jones, Jour. Chem. Soc., **63**, 275 (1893)
30		Richter, Organ. Chemie, 11th Aufl., I, 90 (1909)
30.2	Pennsylvania and New Brunswick	Warren, Mem. Am. Acad., **9**, 135, 166 (1867)
29–30	Ohio and Pennsylvania	Mabery and Hudson, Proc. Am. Acad., **32**, 101–118 (1896)
	Rumania	Engler and Höfer, Das Erdöl, **1**, 242
	Caucasus	
68.95 at 711	Appalachian Oil	Mabery, Proc. Am. Acad., **31**, 33, 1894
68.5–69 at 756		Kymer, Jour. prakt Chemie, **64**, 126
68.95 at 760	American petroleum	Young, Jour. Chem. Soc., **73**, 906 (1898)
69.0		Bacon and Hamor, Am. Pet. Ind., **1**, 3

Name	Description	Specific gravity	Index of refraction	
Hexane (normal)				
" "				
" "				
		.6603 (20)		
		.660 (20)		−94
		.6603 (20/4)		−94
		.6630 (17)	1.3780	
		.663 (17)		
		{ .6769 } (0/4)		
		{ .6771 }		
		.6769		
		.6603 (20/4)		
		.6645 (14.8)	1.3780	
Caproylhydride	C_6H_{14}			
Is heXane	C_6H_{14}	.676		
		.658 (4/15)		
(Methyl diethyl methane)	*$CH_3 \cdot CH<(CH_2 \cdot CH_3)(CH_2 \cdot CH_3)$	.677 (20°)		
"		.6762 (0		
		.6760 (0)		
	$(CH_3 \cdot CH_2)_2 \cdot CH \cdot CH_3$	.6765 (20/4)		
Di-isopropyl	$(CH_3)(CH_3)>CH \cdot CH<(CH_3)(CH_3)$	{ .6701† .66485* .668 } (17.5)		
"	$(CH_3)_2 \cdot CH \cdot CH \cdot (CH_3)_2$	.668 (17)		
Dimethyl propyl methane	$(CH_3)_2—CH—CH_2—CH$			
Do.		.6599 (20/4)		
Do.		.6766 (0)		
Dimethyl iso-propyl methane	$(CH_3)_2—CH—(CH)—(CH_3)_2$	.666 (20)		
Propyl dimethyl methane Propyl iso-propyl	$CH_3—CH_2—CH_2—CH<(CH_3)(CH_3)$	....		
Isohexane (Trimethyl ethyl methane)	$CH_3—C(CH_3)(CH_2—CH_3)—CH_3$	.6646 (0/0)		
		.664		
		.6730 (0.4)		
		.6488 (20)		
2-3 Dimethyl butane	$CH_3CH(CH_3)—CH(CH_3)—CH_3$	.6488 (0/0)		

PARAFFIN SERIES—*Continued*

Boiling point, ° C. Pressure, m.m.	Occurrence	Authority
68*	Canada†	Pelouze and Cabours, Compt. rend., **56**, 505†, 513*, (1863)
68.5	Lima, Ohio and Can.	Mabery, Proc. Am. Acad., **31**, 61 (1894)
68–70	Pennsylvania	Morgan, Chem. News, **30**, 292
69		Meyer and Jacobson, Ed. 2, I-1, 164 (1906–7)
69	Rumania	Handbook Chem. and Phys., 190 (1920–21)
69.05 at 760		Van Nostrand's Chem. Ann., 308 (1918)
	Sumatra	Do.
71.5		Richter, Organ. Chemie, Ed. 5, 54 (1888)
71.5	Produced synthetically	Schorlemmer, Lieb. Ann., **161**, 275 (1872)
69 at 760		Thomas and Young, Chem. News, **72**, 277 (1895)
68.5–69 at 743	Russia	Markownikoff, Chem. Ztg., **21**, 352 (1890)
68		Landolt and Jahn, Zeitschr. Phys. Chem., **10**, 303 (1892)
67.5–68		Warren and Storer, Mem. Am. Acad., **9**, 177–208 (1867)
61.3	Peru	Warren, Proc. Am. Acad., **27**, 61–74 (1891); 92 (1898)
61.7–62.4		Chavanne and Simon, Compt. rend., **168**, 1324–6 (1919)
64		Richter, Organ. Chemie, Ed. 5, 54 (1888)
61.2	Lima, Ohio and Can.	Mabery, Proc. Am. Acad., **31**, 25 (1889–90)
62		Thorpe and Jones, Jour. Chem. Soc., **63**, 290 (1893)
64		Van Nostrand's Chem. Ann., 308 (1918)
{ 58	Baku petroleum; produced synthetically	
57–59		Schorlemmer, Lieb. Ann., **144**, 186*, 185†, (1867)
58 }		Schorlemmer, Lieb. Ann., **144**, 185, 186 (1867)
58–59	Russia	Markownikoff, Jour. prakt. Chemie, **59**, 564 (1899)
58	Sumatra	Van Nostrand's Chem. Ann., 308 (1918)
		Van Nostrand's Chem. Ann., 318 (1918)
62		Meyer and Jacobson, Lehrbuch der Organ. Chem., Ed. 2, I-1, 164, (1906–7)
58		Meyer and Jacobson, Lehrbuch der Organ. Chem., Ed. 2, I-1, 164, (1906–7)
62	Ohio Canada	Richter, Organ. Chemie, Ed. 5, 54 (1888)
48.5 at 739	American petroleum; also produced synthetically	Markownikoff, Ber., **32**, 1449 (1899)
61	Pennsylvania and New Brunswick	Redwood, A Treatise on Petroleum, Ed. 3, **1**, 242 (1913)
61 at 711		Mabery, Jour. Frank. Inst., **162**, 57–70, 81–128
		Young, Do.
49.6		Meyer and Jacobson, Lehrbuch der Organ. Chem., Ed. 2, I-1, 164, (1906–7)
49.6–49.7 at 760	Produced synthetically	Markownikoff, Ber., **32**, 1447 (1899)

CHAIN HYDROCARBONS

Name	Description	Specific gravity	Index of refraction	
Ethyl iso-butane	$CH_3-CH_2-CH\begin{smallmatrix}C_2H_5\\CH_3\end{smallmatrix}$	.6762 (0)		
Do.				
Heptane (normal)	$CH_3-(CH_2)_5-CH_3$	.7304 (0)	1.4025 (16–17)	
" "		....		
		.709 (17.5)		
		.7006 (0)		
		.70186† (0/4)		
		.68288 (20)	1.3895 (20)	
		....		
		....		
" "		.73315 (18/4)		−97.1
		.6879 (4/15)		
Dimethy. diethyl methane	$(CH_3)_2-C-(C_2H_5)_2$	.67207 (25/4)		
Triethyl methane	$HC(C_2H_5)_3$	.6940 (29/4)		
Methyl ethyl propyl methane	$C_2H_5\cdot CH(CH_3)C_3H_7$	.7806 (17)		
2, Methyl hexane	$\begin{smallmatrix}CH_3\\CH_3\end{smallmatrix}CH-CH_2-CH_2-CH_2-CH_3$	.7116		
Ethyl isoamyl Dimethyl pentane		.7067 (0/4)		
Tetramethyl propane 2, 4-Dimethyl pentane	$\begin{smallmatrix}CH_3\\CH_3\end{smallmatrix}CH-CH_2-CH\begin{smallmatrix}CH_3\\CH_3\end{smallmatrix}$	.6971 (0/0) .6805 (0/20)	1.3825	
		.7022 (0/0) .6879 (0/22)	1.38477(22)	
(Trimethyl propyl methane)	$CH_3-C(CH_3)(CH_2-CH_2-CH_3)-CH_3$	.6910 (0/0) .6743 (20/0)		
Methyl ethyl iso-propyl methane or	$(CH_3)_2CH-CH\begin{smallmatrix}CH_3\\C_2H_5\end{smallmatrix}$			
Dimethyl diethyl methane	$(CH_3)_2=C=(C_2H_5)_2$	.709 (16)		
Octane (normal)	$CH_3\cdot(CH_2)_6\cdot CH_3$	.7374 (0)		
" "		.7068 (15/15)	1.3963 (n 23)	
		.7074 (15.1)	1.4007 (15.1)	
		.732 (12.1)		
		.7032 (17)		
		.706		−56.6
		.7188 (20/20)		
		.7134		
		....		
		.7056 (20)		
		.7063 (4/15)		
" "		.7516 (0)		
Iso-octane		.703		
"		.7190		
"		....		

PARAFFIN SERIES—*Continued*

Boiling point, ° C. Pressure, m.m.	Occurrence	Authority
	Penna.	Warren, Proc. Am. Acad., **31**, 55 (1894)
	***Rumania**	Skita, Ber., **41**, 2940; *Saligny, Chem., I, 60 (1900)
	Russia	Engler and Höfer, Das Erdöl, **1**, 90 (1913)
	Sumatra	
97.5–99	**Pennsylvania**	Schorlemmer, Lieb. Ann., **161**, 218, 280 (1872)
98–99	**Italy**	Schorlemmer, Jour. Chem. Soc., **15**, 423 (1863)
98.**4** at **760**	**Galicia**	Richter Organ. Chem. Ed. 2, **1**, 90 (1909)
98.**4***	**American petroleum**	Young, Jour. Chem. Soc., **73**, 906,* 922† (1898)
98.**25–98**.**45**	**Rangoon**	Young, Jour Chem. Soc., **73**, 918 (1898)
97.**33** at 760		Kremers, E., and Kremers, R., Jour. Am. Pharm. Assoc., **9**, 860–4 (1920)
98.5–99.5	**American petroleum**	Beilstein and Kurbatoff, Ber., **13**, 2028 (1880)
96–**97**	**Ohio**	Mabery, Proc. Am. Acad., **31**, 30 (1895)
98.4	**Ontario**	Van Nostrand's Chem. Ann., 306 (1918)
98–98.**3**		Chavanne and Simon, Compt. rend., **168**, 1324–6 (1919)
	Alsace	Van Nostrand's Chem. Ann., 306 (1918)
95–98		Van Nostrand's Chem. Ann., 308 (1918)
		Do.
91.5–**92.5**	Rumania	Costăchescu, Ann. Sci. Université de Jassy, **6**, 294–301 (1910)
89.9–**90.4**		Francis and Young, Jour. Chem. Soc., **73**, 921 (1898)
83–84	Russia	Chonin, Chem. Centralbl., II, 813 (1905)
83–84	Russia	Konowaloff, Chem. Centralbl., I, 330 (1906)
.5–79	**Russia**	Markownikoff, Ber., **33**, 1906–1908 (1900)
	Produced synthetically	
.**5–90**	**Coal distillate**	Schorlemmer, Jour. prakt. Chemie, **8**, 220 (1874)
.**5**	**Pennsylvania**	Warren, Proc. Am. Acad., **27**, 78–79 (1891–2)
124.**7** at **760**	**Sumatra**	Clarke, Jour. Am. Chem. Soc., **33**, 520 (1911)
125.**5**	Germany	Landolt and Jahn, Zeitschr. physikal. Chemie, **10**, 303
121 at **779**	America	Lemoine, Bull. Soc. Chim., Paris, **41**, 163 (1884)
123–**125**	Produced synthetically	Schorlemmer, Philos Trans., **162**, 122 (1872)
125.**5**	Italy	Handbook Phys. and Chem., 222 (1920–21)
125 at **760**		Mabery, Proc. Am. Acad., **31**, 33 (1894)
124–**125**	Ohio	Mabery and Hudson, Proc. Am. Acad., **32**, 101–118 (1896)
121–**123**	**India**	Warren and Storer, Mem. Am. Acad., **9**, 208 (1867)
125	**Java**	Kijner, Jour. prakt. Chemie, (2), **64**, 127
125.**8**		Chavanne and Simon, Compt. rend., **168**, 1324–6 (1919)
127.**6**		Warren, Proc. Am Acad., **27**, 78 to 79 (1891 2)
118		Redwood, A Treatise on Petroleum, Ed. 3, I, 242 (1913)
119.**5** at **760**	Canada and Ohio	Mabery and Hudson, Proc. Am. Acad., **32**, 101–118 (1896)
116–**118**	American petroleum	Pelouze and Cahours, Compt. rend., **56**, 511–512

Name	Description	Specific gravity
Iso-octane		.719 (17.5)
2-Methyl heptane	$CH_3-CH(CH_2)_4-CH_3$ CH_3	.7035
		.7019
		.712 (11°)
3-Methyl heptane	$CH_3-CH_2-CH-(CH_2)_3-CH_3$ CH_3	.7167
4-Methyl heptane	$CH_3-(CH_2)_2-CH-(CH_2)_2-CH_3$ CH_3	.7217
2-, 3-Dimethyl hexane	$CH_3-CH-CH-(CH_2)_2-CH_3$ CH_3 CH_3	.7246 (15/15)
2-, 4-Dimethyl hexane	$CH_3-CH-CH_2-CH-CH_2-CH_3$ CH_3 CH_3	.7083 (15/15)
2-, 5-Dimethyl hexane	$CH_3-CH-(CH_2)_2-CH-CH_3$ CH_3 CH_3	.6991 (15/15)
3-, 4-Dime hyl hexane	$CH_3-CH_2-CH-CH-CH_2-CH_3$ CH_3 CH_3	.7270 (15/15)
Di-iso-butyl Tetramethyl butane	CH_3, CH_3 > $CH-CH_2-CH_2-CH$ < CH_3, CH_3	0.711 (0/4)
3-Ethyl hexane	$CH_3-CH_2-CH-(CH_2)_2-CH_3$ C_2H_5	.7175 (15/15)
2-Methyl 3-ethyl pentane	$CH_3-CH-CH-CH_2-CH_3$ CH_3 C_2H_5	.7084 (15/15)
2-, 2-, 3-Trimethyl pentane	CH_3 $CH_3-C-CH-CH_2-CH_3$ CH_3 CH_3	.7219 (15/15)
Nonane (normal)	$CH_3\cdot(CH_2)_7\cdot CH_3$	.7555
.. ”		
		.7124
		.741 (15°)
		....
		....
		.718 (20)
		.7177 (20)
4-Methyl octane	$CH_3-(CH_2)_2-CH-(CH_2)_3-CH_3$ CH_3	.7320 (15/15)
2-, 4-Dimethyl heptane	$CH_3-CH-CH_2-CH-(CH_2)_2-CH_3$ CH_3 CH_3	.7206 (15/15)

PARAFFIN SERIES—*Continued*

Boiling point, ° C. Pressure, m.m.	Occurrence	Authority
119–120		Schorlemmer, Jour. Chem. Soc., **15**, 425 (1863)
116.0 at 761		Clarke, Jour. Am. Chem. Soc., **33**, 520 (1911)
		Eijkman, Chem. Centralbl., II, 1209 (1907)
118–120		Hofman, Ber., **18**, 13 (1885)
117.6 at 760		Clarke, Jour. Am Chem. Soc., **33**, 520 (1911)
118.0 at 760		Clarke, Jour. Am. Soc. Chem., **33**, 520 (1911)
113.7 at 758		Do.
110.0 at 763		Do.
108.3 at 760		Do.
116.5 at 759		Do.
108.53	Italy Sumatra	Engler and Höfer, Das Erdöl, **1**, 245
118.9 at 766		Clarke and Riegal, Jour. Am. Chem. Soc., **34**, 674 (1912)
114 at 760		Clarke, Jour. Am. Chem. Soc., **34**, 674 (1912)
110.5		Clarke and Jones, Jour. Am. Chem. Soc., **34**, 170 (1912)
150.8	Pennsylvania petroleum	Warren, Proc. Am. Acad., **27**, 80 (1891–2)
151 at 760	Ohio	Mabery, Proc. Am. Acad., **31**, 33, 1894
147.5–148.5	Galicia	Lachowicz, Lieb. Ann. **220**, 168–206 (1883)
136–138	Canada	Felouze and Cabours, Compt. rend., **56**, 505–13
138	Germany	Engler and Grodnitzky, v. d. v. z. Bef. d. Gewerbefl., 662 (1887)
135–139	Produced synthetically	Do Dipl. Karlsruhe, 29 (1884)
149.7	Russia	Handbook Chem. and Phys., 222 (1920–21)
149.5 at 760	India	Krafft, Ber., **15**, 1692 (1882)
141.8 at 771		Clarke, Jour. Am. Chem. Soc., **34**, 680 (1912)
132.9 at 752		Clarke, Jour. Am. Chem. Soc., **34**, 54 (1912)

Name	Description	Specific gravity	Index of refraction	
2-, 5-Dimethyl heptane	CH_3—CH—$(CH_2)_2$—CH—(CH_2)—CH_3 (with CH_3 branches on both CH)	.7190 (15/15)	:.4020 (n 25/D)	
2-, 6-Dimethyl heptane	CH_3—CH$(CH_2)_3$—CH—CH_3 (with CH_3 branches on both CH)	.7122 (15/15)		
a. Nonane		.742 (12°)		
b. Nonane	C_9H_{20}	.734 (12°)		
		.743 (0°)		
Decane (normal)	$CH_3(CH_2)_8CH_3$	.7456 (0°)		− 32
" "		.7278 (14.9)	1.4108 (14.9)	
" "		.7467 (20/20)		
2-, 6-Dimethyl octane		.7322 (20)	1.4105 (D)	
(Di-isoamyl)	$(CH_3)_2$—CH—CH_2—CH_2—CH_2—CH_2—$CH(CH_3)_2$			
"		.7629 (20)		
"		.757		
		.7479 (20/20)		
		....		
"		.7187 (21)		
Amyl butylcaproyle	$C_{10}H_{22}$	....		
" "		.753 (15)		
" "		....		
Undecane (normal)	$CH_3(CH_2)_9CH_3$	.7411 (20°)		−26.5
" "		.7581 (20/20)		
		.7622 (20)		
" "		.7581 (20)		−25.6
Dodecane (normal)	$(CH_3)(CH_2)_{10}$—CH_3	.7730		−12.0
" "		....		
" "		.7676 (20/20)		
Di-isohexyl	CH_3, C_4H_9 } CH—CH { CH_3, C_4H_9	....		
"		....		
"		....		
		.7511 (20)		
"		.768 (20)		−12
Tridecan	$C_{13}H_{28}$	.775		− 6.2
"		.792		
"		....		
		.7571 (20°)		−6.2
		....		
		.7834 (20/20)	1.451	
"		....		
		....		
Tetradecane (normal)	$C_{14}H_{30}$	....		
" "		.7814 (20/20)	1.436	
" "		....		
" "		.7751 (at M. P.)		+4.5
" "		.7697 (20)	1.4307	
" "		.7645 (20)		4.5
		.765 (20)		5.5

PARAFFIN SERIES—*Continued*

Boiling point, ° C. Pressure, m.m.	Occurrence	Authority
135–135.9 at 760		Clarke, Jour. Am. Chem. Soc., **34**, 54 (1912)
132		Clarke, Latham, Jour. Am. Chem. Soc., **34**, 54 (1912)
135–137	America	Lemoine, Bull. Soc. Chim., Paris, **41**, 164 (1884)
129.5–131.5 at 751	America	Do.
173 at 760		Krafft, Ber., **15**, 1695 (1882)
173		Landolt and Jahn, Zeitschr. physikal. Chemie, **10**, 303 (1892)
173–174 at 760	Pennsylvania and Ohio, Canada	Mabery, Proc. Am. Acad., **32**, 121–176 (1897)
61–63 at 25	Formosan lemon-grass oil	Kafuku, Jour. Chem. Ind., Tokyo, **20**, 825–33 (1917)
162–163 (760)	Ohio	Mabery and Dunn, Am. Chem. Jour., **18**, 215–36
161	Pennsylvania and Ohio	Dingler's polytech. Jour., **232**, 363 (1879)
	Germany	Ahrens and Riemer, Zeitschr. angew. Chemie, **20**, part 2, 1557 (1907)
163–164 at 760		Mabery, Proc. Am. Acad., **32**, 121–176 (1896–7)
160–162	American petroleum	Pelouze and Cahours, Compt. rend., **56**, 511
152–153	Galician petroleum	Lachowicz, Lieb. Ann., **220**, 195 (1883)
158	Boghead coal	Williams, C. Greville, Philos. Trans., **147**, 450 (1857)
151–160	America	Lemoine, Bull. Soc. Chim., Paris, **41**, 165 (1884)
155		Williams, C. Greville, Philos. Trans., **147**, 450 (1857)
194.5 at 760		Krafft, Ber., **15**, 1697 (1882)
196–197 at 760	Pennsylvania petroleum	Mabery, Proc. Am. Acad., **32**, 121–176 (1896–7)
195 760	Berea grit	Mabery and Dunn, Am. Chem. Jour., **18**, 215–236
194.5		Van Nostrand's Chem. Ann., 374
214		Richter, Organ. Chemie, Ed. 5, 55 (1888)
214		Mabery and Goldstein, Am. Chem. Jour., **28**, 69 (1912)
214–216 at 760	Pennsylvania	Mabery, Proc. Am. Acad., **32**, 121–176 (1896–7)
196 200	Canada	Felouze and Cahours, Compt. rend., **57**, 62–63
202		Schorlemmer, Chem. News, **23**, 253 (1871)
185–190	Produced synthetically	Delacre, Bull. Soc. Chim., Series (4), **9**, 1024 (1911)
201–202		Krafft, Ber., **15**, 1697 (1882)
214.5		Handbook Chem. and Phys., 176 (1920–1)
234		Richter, Organ. Chemie, Ed. 5, 55 (1888)
216		Redwood, A Treatise on Petroleum, Ed. 3, I, 242 (1913)
226		Mabery, Proc. Am. Acad., **37**, 565, 595 (1902)
234 at 760		Krafft, Ber., **15**, 1697 (1882)
226		Mabery and Goldstein, Am. Chem. Jour., **28**, 69 (1902
226 at 760		Mabery (Richardson, Jour. Frank. Inst., **162**, 57–70, 81-128)
216–218	Canada	Felouze and Cabours, Compt. rend., **57**, 63–64
225		Mabery, Jour. Soc. Chem. Ind., **19**, 508 (1900)
242		Mabery and Goldstein Am. Chem. Jour., **28**, 67 (1902)
236–238 at 760		Mabery (Richardson, Jour. Frank. Inst., **162**, 57–70, 81–128)
236–240	Canada	Felouze and Cabours, Compt. rend., **57**, 64–65
252.5		Richter, Organ. Chemie, Ed 5, 5.5 (1888)
126 at 12		Ubbelohde and Agthe, Dipl. Karlsruhe (1912)
252.5 at 760		Krafft, Ber., **15**, 1695 (1882)
252.5		Handbook Chem. and Phys., 238 (1920–1921)

Name	Description	Specific gravity	Index of refraction	
Pentadecane	$(CH_3(CH_2)_{13}CH_3)$	.7896 (20/20)	1.4413	
"		.7758 (at M. P.)		10
"		.769 (20)		10
		....		
"		.7689 (20)		+10
Hexadecane	$C_{16}H_{33}$	.7911 (20/20)	1.4413	
"	"	.7754 (at M. P.)		+18
"	"	.7803 (20)	1.4362 (23)	
"	"	.8000 (20/20°)	1.4435	10
Heptadecane	$C_{17}H_{36}$	.7767 (at M.P.).		+22.5
"				
"				
		{ .794 (0/4)	1.4376 (20/D)	
		{ .780 (20/4)		
Octadecane	$C_{18}H_{38}$	.8017 (20/20°)	1.440	20
"		.7768 (28)		+28
"		.7793 (28)	1.4390 (27)	
"				
Nonadecane	$C_{19}H_{40}$	.8122 (20/20)	1.4522	33–34
"		.7774 (at M. P.)		32
"		....		33–34
		.7774 (at M. P.)		32
"		....		
Eicosane	$C_{20}H_{42}$	.7779 (at M. P.)		36.7
"		....		
Heneicosane	$C_{21}H_{44}$	.8230 (20)		40–41
"		.7783		40.4
"		....		
Docosane	$CH_3(CH_2)_{20}CH_3$	.7796 (15°)		44
"		.7782 (at M. P.)		44.4
"		.7782 (44/4)		44.4
"		....		
Tricosane	$C_{23}H_{48}$	....		
"		.7900 (60°)		45
"		.7836 (60)		48.0
		.7814 (70)		
		.7807 (80)		
" (normal)	$CH_3(CH_2)_{21}CH_3$; Glittering leaflets	.7799 (48)		47.7
Tetracosane	$C_{24}H_{50}$	.7902 (60)		48
" (normal)	"	.7786 (at M. P.)		51.1
" "	"	....		
Pentacosane	$C_{25}H_{52}$	.7911 (60)		53–54
"		.7881 (70)		
"		.7870 (80)		
"		.7854 (90)		
Hexacosane	$C_{26}H_{54}$	.7977 (60)		58
"		.7938 (60)		55–56
"		.7918 (70)		
		.7893 (80)		
"		.7879 (90)		
Myricyl paraffin	$C_{27}H_{56}$	.7796 (at M. P.)		
Heptacosane	"	.7796 (at M. P.)		59.5
Octacosane	$C_{28}H_{58}$	.7945 (70)		60
Nonacosane	$C_{29}H_{60}$	.7797 (at M. P.)		62–63

PARAFFIN SERIES—*Continued*

Boiling point, ° C. Pressure, m.m.	Occurrence	Authority
256–257 at 760		Mabery (Richardson, Jour. Frank. Inst., **162**, 57–70, 82–128)
270.5 at 760		Richter, Organ. Chemie, Ed. 5, 54 (1888)
270.5		Handbook Chem. and Phys., 224 (1920–21)
255–260	Canada	Pelouze and Cabours, Compt. rend., **57**, 65 (1863)
270.5 at 760		Krafft, Ber., **15**, 1691 (1882)
274–275 at 760		Mabery, Jour. Frank. Inst., **162**, 57–70, 81–128
287.5 at 760		Richter, Organ. Chemie, Ed. 5, 54 (1888)
151 at 12		Ubbelohde and Agthe, Dipl. Karlsruhe (1912)
288–289 at 760	Pennsylvania	Mabery, Proc. Am. Acad., **37**, 565–595 (1902)
{ 303 at 760 { 81 at 0	Produced synthetically	Krafft, Ber., **15**, 1697 (1882)
170 at 15		Allen, Commercial Organic Analysis, **2**, 2
161–162 at 15		Willstätter, Schuppli and Mayer, Ann., **418**, 121–47 (1919)
300–301 at 760	Pennsylvania	Mabery, Proc. Am. Acad., **37**, 565–595 (1902)
317		Krafft, Ber., **15**, 1697 (1882)
174 at 12		Ubbelohde and Agthe, Dipl. Karlsruhe (1912)
181 at 15		Allen, Commercial Organic Analysis, **2**, 2
210–212 at 50	Pennsylvania	Mabery, Proc. Am. Acad., **37**, 565, 595 (1902)
		Krafft, Ber., **15**, 1711 (1882)
316		Mabery, Jour. Soc. Chem. Ind., **19**, 508 (1900)
330 at 760		Richter, Organ. Chemie, Ed. 5, 54 (1888)
193 at 15		Allen, Commercial Organic Analysis, **2**, 2
205 at 15		Richter, Organ. Chemie, Ed. 5, 54 (1888)
300.2	Pennsylvania	Young, Jour. Chem. and Phys., **3**, 248
230–231 at 50		Mabery, Proc. Am. Acad., **37**, 565, 595 (1902)
215 at 15	Pennsylvania	Richter, Organ. Chemie, Ed. 5, 54 (1888)
313.1		Young, Jour. Chim. phys., **3**, 248
240–242 at 50	Pennsylvania	Mabery, Proc. Am. Acad., **37**, 565, 595 (1902)
224.5 at 15		Richter, Organ. Chemie, Ed. 5, 54
317.4		Van Nostrand's Chem. Ann., 290 (1918)
316.7		Young, Jour. Chim. phys., **3**, 248
320.2		Do.
258–261 at 50		Mabery, Proc. Am. Acad., **37**, 565, 595 (1902)
		Mabery, Proc. Am. Acad., **40**, 349 (1904)
		Mabery, Am. Chem. Jour., **33**, 251 (1905)
		Do.
320.7		Van Nostrand's Chem. Ann., 370
272–274 at 50		Mabery, Proc. Am. Acad., **37**, 565–595 (1902)
243 at 15		Richter, Organ. Chemie, Ed. 5, 54 (1888)
323.6		Young, Jour. chim. phys., **3**, 248
280–282 at 50		Mabery, Proc. Am. Acad., **37**, 565–595 (1902)
		Mabery, Proc. Am. Acad., **40**, 349 (1904)
		Mabery, Jour. Am. Chem. Soc., **33**, 251 (1905)
		Do.
292–294 at 50		Mabery, Proc. Am. Acad., **37**, 565–595 (1902)
		Mabery, Proc. Am. Acad., **40**, 349 (1904)
		Mabery, Jour. Am. Chem. Soc., **33**, 251 (1905)
		Do.
		Do.
322.2		Young, Jour. chim. phys., **3**, 248
270 at 5		Richter, Organ. Chemie, Ed. 5, 54 (1888)
310–312 at 50		Mabery, Proc. Am. Acad., **37**, 565–595 (1902)
346–348 at 40		Mabery, Am. Chem. Jour., **33**, 289 (1905)

CHAIN HYDROCARBONS

Name	Description	Specific gravity	Index of refraction	
Ceryl paraffin	$C_{30}H_{62}$	....		
Triacontane	··	....		65.2–65.5
Hentriacontane	$C_{31}H_{64}$	.7997 (70°)		66
··		.7799 (at M. P.)		68.4
··		.7808		68.1
··		....		
Dotriacontane	$C_{32}H_{66}$	.8005 (75°)		67–68
··		.7798 (at M. P.)		69.5
··	··	.7810 (at M. P.)		70
··	··	....		
Tritriacontane	$C_{33}H_{68}$	.7801 (at M. P.)		71.8
Tetratriacontane	$C_{34}H_{70}$	.8009 (80°)		71–72
	··	.7806 (at M. P.)		About 73
	··	....		73.2
Pentatriacontane	$C_{35}H_{72}$	.8052 (80)		76
··		.7813 (at M. P.)		About 74
··	··	.781 (at M. P.)		74.7
Hexatriacontane	$C_{36}H_{74}$	.7819 (at M. P.)		About 76
Octatriacontane	$C_{38}H_{78}$	....		79.4
Hexacontane	$C_{60}H_{122}$	....		102
(Dimyricyl)		....		

CHAIN HYDROCARBONS

Name	Description	Specific gravity	Index of refraction	
Ethylene	Gas	.6095 (liquid)		−169
··	$CH_2 : CH_2$	.978 (air = 1)		−169
··		.5650 (liquid)		−169
Propylene	Colorless gas; $CH_3 \cdot CH : CH_2$	1.498 (air = 1)		
··	·· ··	....		−180
Butylene	Gas; $C_2H_5 \cdot CH : CH_2$	....		
(Ethyl ethylene)		....		
··				
Isobutylene	$(CH_3)_2C : CH_2$	....		
Dimethyl ethylene	$CH_3 \cdot CH : CH \cdot CH_3$	....		
Pseudobutylene		.635		
Amylene	$CH_3 \cdot CH_2 \cdot CH_2 \cdot CH : CH_2$	....		
Propyl ethylene (normal)		....		
Isopropyl ethylene	$(CH_3)_2 \cdot CH \cdot CH : CH_2$	.648 (0)		
Trimethyl ethylene	$(CH_3)_2C : CH \cdot CH_3$	.(85 (0)		
·· ··		.666 (15/15)		
Methyl ethyl ethylene	$CH_3 \cdot CH : CH \cdot C_2H_5$	....		
·· ·· ··		....		
·· ·· ··	$CH_3(C_2H_5)C : CH_2$	.670 (0)		

PARAFFIN SERIES—*Concluded*

Boiling point, ° C. Pressure, m.m.	Occurrence	Authority
370	Pennsylvania petroleum	Redwood, Petroleum and Its Products, **1**, 242
	Produced synthetically	Gascard, Compt. rend., **153**, 1485 (1911)
328–330 at 50		Mabery, Am. Chem. Jour., **33**, 282 (1905)
193.5 at 0		Krafft, Ber., **40**, 4783 (1907)
302 at 15		Richter, Organ. Chemie, Ed. **4**, 54 (1888)
342		Young, Jour. chim. phys., **3**, 248
342–345 at 50		Mabery, Proc. Am. Acad., **40**, 349–361 (1904)
201 0		Krafft, Ber., **40**, (4), 4783 (1907)
310 at 15		Richter, Organ. Chemie, Ed. 5, 54 (1888)
344		Young, Jour. chim. phys., **3**, 248
208 at 0		Krafft, Ber., **40** (4), 4783 (1907)
366–368 at 50		Mabery, Am. Chem. Jour., **33**, 285 (1905)
215 0		Krafft, Ber., **40** (4), 4783 (1907)
	Produced synthetically	Gascard, Compt. rend., **153**, 1486 (1911)
380–384 at 50		Mabery, Am. Chem. Jour., **33**, 285 (1905)
222 at 0		Krafft, Ber., **40** (4), 4783 (1907)
331 at 15		Richter, Organ. Chemie, Ed. 5, 54 (1888)
350		Young, Jour. chim. phys., **3**, 248
230 at 0		Krafft, Ber., **40** (4), 4783 (1907)
		Allen, Commercial Organic Analysis, **2**, 2
	Produced synthetically	Richter, Organ. Chemie, Ed. 5, 54 (1888)
374.3		Young, Jour. chim. phys., **3**, 248

OLEFINE SERIES, C_nH_{2n}

Boiling point, ° C. Pressure, m.m.	Occurrence	Authority
−102.7 at 757		Höfer, Das Erdöl, Ed. 2, 65
−102.7		Handbook Chem. and Phys., 182 (1920–21)
−102.5		Van Nostrand's Chem. Ann., 300
− 50.2		Handbook Chem. and Phys., 230 (1920–21)
− 47.8		Van Nostrand's Chem. Ann., 354
1.0		Höfer, Das Erdöl, Ed. 2, 65; Bacon and Hamor, Am. Pet. Ind., **1**, 4
1.5–2.5		Handbook Chem. and Phys., 158 (1920–21). *Also* Van Nostrand's Chem. Ann., 268
−6		Engle and Höfer, Das Erdöl, **1**, 268
1		Do.
		Willstätter and Bruce, Ber., **40**, 3979
37–41		Warren and Storer, Mem. Am. Acad., **9**, 177–208 (1865)
39–40		Engler and Höfer, Das Erdöl, **1**, 268
20–21		Do.
36		Do.
37–42		Van Nostrand's Chem. Ann., 258
37.1		Handbook Chem. and Phys., 150 (1920–21)
36 at 741		Engler and Höfer, Das Erdöl, **1**, 268
36.5		Handbook Chem. and Phys., 150 (1920–21)
31–32		Engler and Höfer, Das Erdöl, **1**, 268

Name	Description	Specific gravity	Index of refraction	
Hexylene (normal)	$C_4H_9 \cdot CH : CH_2$	....		
" "		.6886 (15.2)	1.3995 (15.2)	
		.6825 (20.4)	1.3939 (20)	
		.683 (20)		−98.5
		.6825 (20/4)		8.5
Tetramethyl ethylene	$(CH_3)_2C : C(CH_3)_2$	.7006 (19)		
Dimethyl ethyl ethylene	$(CH_3)_2C : CH \cdot C_2H_5$	.687 (19)		
Heptylene	$C_5H_{11} \cdot CH : CH_2$; Colorless			
		.703 (19 5)		
		.7122 (16)		
Oenanthylene				
Trimethyl butene	$(CH_3)_3C \cdot C(CH_3) : CH_2$	....		
Dimethyl pentene		....		
2, (2, 3)	$(CH_3)_2C : C_2H_3(C_2H_5)$	.7185 (21)		
2, (2, 4)	$(CH_3)_2 \cdot C : CH \cdot CH(CH_3)_2$	.6985 (14)		
Octylene	$C_6H_{13} \cdot CH : CH_2$	.722 (17)		
		.72559 (16)	1.4157	
Di-isobutylene	$(CH_3)_2C{=}CH{-}C(CH_3)_3$	....		
Nonylene	C_9H_{18}	....		
Diamylene	$(CH_3)_2C{=}CH{-}CH_2{-}C(CH_3)_3$	....		
"	C_9H_{18}	.7992 (29°/4)	1.4370	
Decylene	$C_{10}H_{20}$			
" (normal)		.7512 (15°)		
"		.7721 (17°)	1.4385 (17)	
" "	$CH_3 \cdot (CH_2)_7 \cdot CH : CH_2$ colorless	.763 (0°)		
Rutylene	$C_{10}H_{20}$	.823 (0°)		
"		{ .7703 (0°)		
		.7598 (15°)		
Undecylene	$C_{11}H_{22}$	....		
"	$C_9H_{19} \cdot CH : CH_2$	....		
"		.773 (20°)		
Margarylene	$C_{11}H_{22}$	.8398 (0°)		
"	"	{ .7822 (0°)		
		.7721 (15)		
"	"	.7729 (20°)		
Dodecylene	$C_{10}H_{21} \cdot CH : CH_2$	.795 (at M. P.)		−31
"		....		
"		.785 (20)°		−31.5
Laurylene	$C_{12}H_{24}$	.865 } .845 } (0°)		
"		.7905 (0°)		
		.7804 (15°)		
		.7851 (20°)		
		....	1.4649 (20°)	
		.8327 (20°)		

OLEFINE SERIES, C_nH_{2n}—*Continued*

Boiling point, ° C. Pressure, m.m.	Occurrence	Authority
69.0		Höfer, Das Erdöl, Ed. 2, 65
68	Canada	Mabery and Quayle, Proc. Am. Acad., **41**, 110 (1905)
		Landolt and Jahn, Zeitschr. physikal. Chemie, **10**, 303 (1892)
68		Engler and Höfer, Das Erdöl, **1**, 288
68–70		Handbook Chem. and Phys., 190 (1920–21)
67.7		Van Nostrand's Chem. Ann., 286 (1918)
72		Landolt and Jahn, Zeitschr. physikal. Chemie, **10**, 303 (1892)
65–67 at 757		Van Nostrand's Chem. Ann., 286 (1918)
		Mabery and Quayle, Proc. Am. Acad., **41**, 110 (1905)
95		Engler and Höfer, Das Erdöl, **1**, 268
95		Schorlemmer, Chem. Centralbl., 1041 (1863)
90–94		Warren and Storer, Mem. Am. Acad., **9**, 177–208
78–80		Van Nostrand's Chem. Ann., 372 (1918)
75–80		Van Nostrand's Chem. Ann., 286 (1918)
83–84		Do.
		Mabery and Quayle, Proc. Am. Acad., **41**, 110 (1905)
122–123		Engler and Höfer, Das Erdöl, **1**, 268
		Landolt and Jahn, Zeitschr. physikal. Chemie, **10**, 303 (1892)
146.5–147.5		Butleroff, Chem. Centralbl., 2, (1877)
141.0		Höfer, Das Erdöl, Ed. 2, 65
		Mabery and Quayle, Proc. Am. Acad., **41**, 110 (1905)
....		Butleroff, Chem. Centralbl., 19 (1877)
145.7	Jennings, Louisiana	Coates and Best, Jour. Am. Chem. Soc., **28**, 387
175.9	Pennsylvania petroleum	Warren, Mem. Am. Acad., ser. 2, **9**, 167 (1867)
172		Engler and Höfer, Das Erdöl, **1**, 268
		Landolt and Jahn, Zeitschr. physikal. Chemie, **10**, 303 (1892)
172		Handbook Chem. and Phys., 166 (1920–21)
175.8 760		Warren and Storer, Mem. Am. Acad., ser. 2, **9**, 208 (1867)
174.9		Warren, C. M., Proc. Am. Acad., **27**, 61–83 (1891–2)
195.8	Pennsylvania petroleum	Warren, Mem. Am. Acad., ser. 2, **9**, 167 (1867)
84 at 18		Engler and Höfer, Das Erdöl, **1**, 268
About 195		Handbook, Chem. and Phys., 248 (1920–21)
195.9 760		Warren and Storer, Mem. Am. Acad., **9**, 208 (1867)
		Warren and Storer, Mem. Am. Acad., **27**, 82 (1891–92)
195.8		Warren, C. M., Proc. Am. Acad., **27**, 61–83 (1892)
196–197 at 760	Canada	Mabery, Proc. Am. Acad., **32**, 119–176 (1897)
96 below 15		Engler and Höfer, Das Erdöl, **1**, 268
216.2	Pennsylvania petroleum	Warren,Mem. Am. Acad., ser. 2, **9**, 167 (1876)
213–215		Handbook, Chem. and Phys., 176 (1920–21)
208.3–214.6 760		Warren and Storer, Mem. Am. Acad., **9**, 211
216.2		Warren, C. M., Proc. Am. Acad., **27**, 61–83 (1892)
212–214 at 745	Canada	Mabery, Proc. Am. Acad., **32**, 119–176 (1897)
216	California	Mabery, Hudson and Sieplein, Proc. Am. Acad., **36**, 255–295 (1901); Am. Chem. Jour., **25**, 253–297 (1901)
	Japan	Mabery and Takano, Proc. Am. Acad., **36**, 295–304 (1901)

CHAIN HYDROCARBONS

Name	Description	Specific gravity	Index of refraction	
Laurylene		.7970	1.435 (20°)	
Tridecylene	$C_{13}H_{26}$	....		
"		.7979 (20°)	1.4440 (20°)	
"		.8134 (10°)	1.4745 (20°)	
		.8055 (20°)		
		....	1.4400	
		.8087 (20°)		
"		.845 (15°)		
Cocinylene	$C_{13}H_{26}$	.8445 (0°)		
Tetradecylene (normal)	$C_{12}H_{25}\cdot CH : CH_2$	.794 (at M. P.)		−12
"		.8129 (20°)	1.4437 (20°)	
"		.8099 (20°)	1.449 (20°)	
		.8154 (20°)	1.4423 (20°)	
		.7745 (15°/4)		−12
	$C_{15}H_{30}$	.8204 (20°)	1.4480 (20°)	
		.8192 (20°)	1.452 (20°	
		.8171 (20°)		
Hexadecylene	$C_{14}H_{29}\cdot CH : CH_2$	.792 (at M. P.)		
		.7849		
Cetene	$C_{16}H_{32}$	....		
"	"	.8254 (20°)	1.451 (20)	
Heptadecylene	$C_{17}H_{34}$	.790 (20/4)		
Octadecylene	$C_{16}H_{33}\cdot CH : CH_2$	.791 (at M. P.)		18
Eicosylene	$C_{20}H_{40}$	....		
		.8181 (24°)		
....	$C_{22}H_{44}$			
....	$C_{23}H_{46}$	....		
....	$C_{24}H_{48}$	....		
....	$C_{26}H_{52}$	....		
Cerotene	$C_{27}H_{54}$	....		58
Melene	$C_{30}H_{60}$			62
	Colorless solid	.8900 (15)		62

OLEFINE SERIES—*Concluded*

Boiling point, ° C. Pressure, m.m.	Occurrence	Authority
	Ohio	Mabery and Palm, Proc. Am. Acad., **40**, 323–334 (1904)
232	Argentina	Engler and Otten, Dingler's polytech. Jour., **268**, 380 (1888)
228–230 at 760	Canada	Mabery, Proc. Am. Acad., **40**, 334–340 (1904)
230–232 at 760	California	Mabery, Hudson and Sieplein, Proc. Am. Acad., **36**, 295–304 (1901); Am. Chem. Jour., **25**, 253–297
223–225 at 30	Trenton limestone, Ohio	Mabery and Palm, Proc. Am. Acad., **40**, 323–334 (1904)
123–125 at 30	Do.	Mabery and Palm, Proc. Am. Acad., **40**, 323–334 (1904)
228–230	Canada	Mabery, Proc. Am. Acad., **40**, 334–340 (1904)
233		Handbook, Chem. and Phys., 244 (1920–21)
232.7	India (Rangoon)	Warren and Storer, Mem. Am. Acad., **9**, 208 (1867)
127 below 15		Engler and Höfer, Das Erdöl, **1**, 268
138–140 at 30	Ohio	Mabery and Palm, Proc. Am. Acad., **40**, 323–334 (1904)
141–143 at 50	Canada	Mabery, Proc. Am. Acad., **40**, 334–340 (1904)
144–146 at 50	California	Mabery, Hudson and Sieplein, Proc. Am. Acad., **36**, 255–295 (1901); Am. Chem. Jour., **25**, 253–297 (1901)
240–246		Van Nostrand's Chem. Ann., 362 (1918)
252–253	Caucasia	Markownikoff and Ogloblin, Chem. Centralbl., 754 (1882)
152–154 at 30	Trenton limestone, Ohio	Mabery and Palm, Proc. Am. Acad., **40**, 323–334 (1904)
159–160 at 50	Canadian petroleum	Mabery, Proc. Am. Acad., **40**, 334–340 (1904)
160–162 at 50	California	Mabery, Hudson and Sieplein, Proc. Am. Acad., **36**, 255–295 (1901); Am. Chem. Jour., **25**, 253–297
155 below 15		Engler and Höfer, Das Erdöl, **1**, 268
		Eijkman, Chem. Cenralbl., 2, 1209 (1907)
275.0		Höfer, Das Erdöl, Ed. 2, **40**, 65
164–168 at 30	Trenton limestone, Ohio	Mabery and Palm, Proc. Am. Acad., **40**, 223–234 (1904)
{ 288–291 at 719 153–155.5 at 10 }		Willstätter, Schuppli, and Mayer, Ann. **418**, 121–147 (1919)
179 below 15		Engler and Höfer, Das Erdöl, **1**, 268
395.0		Höfer, Das Erdöl, Ed. 2, 65
314–315		Engler and Höfer, Das Erdöl, **1**, 268
240–242 at 50	Pennsylvania petroleum	Mabery, Proc. Am. Acad., **37**, 565–595 (1902)
258–260 at 50	Do.	Mabery Proc. Am. Acad. **37**, 565–595 (1902)
272–274 at 50	Do.	Do.
280–282	Do.	Do.
		{ Höfer, Das Erdöl, Ed. 2, 65 Engler and Höfer, Das Erdöl, **1**, 268 }
		{ Höfer, Das Erdöl, Ed. 2, 65 Engler and Höfer, Das Erdöl, **1**, 268 }
370–380		Handbook Chem. and Phys., 204 (1920–21)

CHAIN HYDROCARBONS

Name	Description	Specific gravity	Index of refraction	
Methyl allylene	Colorless liquid $CH_2=C=CH-CH_3$	....		
Divinyl, Erythrene, 1-3 Butadiene, Pyrrolylene	$CH_2=CH-CH=CH_2$	....		
Piperylene α-Methyl butadiene	$CH_2=CH-CH=CH-CH_3$	....		
Isoprene	$CH_2=CH-C(CH_3)=CH_2$	.6804 (21/21)	1.4107	
β-Methyl butadiene		.6910 (0)		
Di-iso-propenyl β, γ-Dimethyl butadiene	$CH_2=C(CH_3)-C(CH_2)=CH_2$	....		
Diallyl-1-5-hexadiene	Crystallizable tablets from alcohol and			
"	ether	.6880 (20/4)		
	$CH_2=CH-CH_2-CH_2-CH=CH_2$	....		
1-1-3-Trimethyl butadiëne	$(CH_3)_2C=CH-C(CH_3)=CH_2$	....		
		.7232 (20/5)		
2, 5-Dimethyl 1, 5-hexadiëne	$CH_2=C(CH_3)-CH_2-CH_2-C(CH_3)=CH_2$	....		
Conylene 1, 4-octadiëne	$CH_2=CH-CH_2-CH=CH-CH_2-CH_2-CH_3$	....		
1, 1, 5-Trimethyl = 1, 5-hexadiene	$(CH_3)_2C=CH-CH_2-CH_2-C(CH_3)=CH_2$	....		
Tri-olefine, β-tetra methyl menthene	$(CH_3)_2-C=CH-C(=CH_2)-CH=C-(CH_3)_2$	....		
Acetylene	$CH\equiv CH$	.906 (a = 1) (15)		−81
"		.91 (a = 1) (15) .613 −(80°)		−81.5
Allylene (normal) (Methylacetylene, propine)	$CH\equiv C-CH_3$	....		−110
Crotonylene (Dimethyl acetylene)	$CH_3-C\equiv C-CH_3$	....		
Ethyl acetylene	$CH\equiv C\cdot C_2H_5$	....		−130
Methyl ethyl acetylene (Valerylene)	$CH_3-C\equiv C-C_2H_5$	....		
Propylacetylene (normal)	$CH\equiv C-CH_2-CH_2-CH_3$	....		
Isopropyl acetylene	$CH\equiv C-CH(CH_3)_2$	.685 (0)		
Butyl acetylene	$C_4H_9\cdot C \vdots CH$	....		
Methyl propyl acetylene	$CH_3\cdot C \vdots C\cdot C_3H_7$	.7377 (13/4)		
Heptine	$CH_3(CH_2)_4C \vdots CH$	.7384 (12.6/4)		
Hexyl acetylene	$CH_3(CH_2)_5C \vdots CH$	.7701 (0)		
Methyl amyl acetylene (normal)	$CH_3(CH_2)_4C \vdots C\cdot CH_3$	....		
Methyl decyl acetylene	$CH_3-C \vdots C(CH_2)_9CH_3$	.8016 (20/4)		30
Hexadecyl acetylene	$C_{17}H_{34}$	.8335	1.4545 (20)	
" "	$HC \vdots C\cdot(CH_2)_{15}CH_3$	.7983 (26)		26
Olefinacetylenes	$CH\equiv C-CH=CH_2$			
Valylene	$CH_2 : C\cdot(CH_3)\cdot C \vdots CH$	....		
Diallylene	$CH\equiv C-CH=CH(CH_3)$	.8579 (18)		

CHAIN HYDROCARBONS, DI-OLEFINES, C_nH_{2n-2}

point, ° C. re, m.m.	Occurrence	Authority
8–19		Van Nostrand's Chem. Ann., 326 (1918)
−5		Engler and Höfer, Das Erdöl, **1**, 338
1		Van Nostrand's Chem. Ann., 290 (1918)
		Engler and Höfer, Das Erdöl, **1**, 338
3.5	Japanese light camphor oil	Van Nostrand's Chem. Ann., 318 (1918)
5.8		Handbook Chem. and Phys., 200 (1920–21)
1		Engler and Höfer, Das Erdöl, **1**, 338
59.5		Van Nostrand's Chem. Ann., 278
59		Engler and Höfer, Das Erdöl, **1**, 338
93		Do.
91–93 at 745	Russia	Markownikoff, Ber., **30**, 976
37		Engler and Höfer, Das Erdöl, **1**, 338
26		Do.
41		Do.
55–57		Engler and Höfer, Das Erdöl, **1**, 351

85 at 760	Produced synthetically	Handbook Chem. and Phys., 144 (1920–21)
		Van Nostrand's Chem. Ann., 252 (1918)
83.6		
23.5		Handbook Chem. and Phys., 146 (1920–21); Van Nostrand's Chem. Ann., 256
27		Engler and Höfer, Das Erdöl, **1**, 337
18.3 760		Handbook Chem. and Phys., 178 (1920–21)
55.5–55.6		Handbook Chem. and Phys., 212 (1919)
48–49		Handbook Chem. and Phys., 230 (1920–21)
28–29		Handbook Chem. and Phys., 200 (1920–21)
70.5–72.0		Van Nostrand's Chem. Ann., 268 (1918)
83.4		Van Nostrand's Chem. Ann., 330 (1918)
99		Van Nostrand's Chem. Ann., 308 (1918)
131–132		Do.
133–134		Van Nostrand's Chem. Ann., 326 (1918)
184 at (15)		Van Nostrand's Chem. Ann., 330 (1918)
77–179 at 30	Trenton limestone, Ohio	Mabery and Palm, Proc. Am. Acad., **40**, 323–334 (1904)
80 at (15)		Van Nostrand's Chem. Ann., 308 (1918)
		Engler and Höfer, Das Erdöl, **1**, 344
50		Van Nostrand's Chem. Ann., 374 (1918)
70		Van Nostrand's Chem. Ann., 278 (1918)

CHAIN HYDROCARBONS

Name	Description	Specific gravity		Index of refraction	
Dipropargyl	CH ⋮ C·CH_2—CH_2—C ⋮ CH	.8049	(20/4)		−6
"		.805	(15)		−6
Dimethyl diacetylene	CH_3—C≡C—C≡C—CH_3				64
Ethyl phenyl acetylene	C_6H_5—C ⋮ C—C_2H_5	.923	(21)		

CYCLIC HYDROCARBONS

Name	Description	Specific gravity		Index of refraction	
Benzol (benzene)	CH=CH—CH |　　　|| CH=CH—CH	*.8785	(20/4)	1.5005 (20°)	*5.4
		.8820	(15)		
		.8987	(0)		
		....			
		.88041	(20)	1.5016(20°)	
		.8790	(20)		5.4
		.8839	(14/4)		
		.8850	(15.5)	1.5022 (20/D)	
		.8867	(13/4)		
		.880	(20/4)		
Toluol (toluene) (Methyl benzene)	C_6H_5—CH_3				Solidifies dy-like from rhombi tals a tained
		....			−94.2
		....			
		.870	(14.7)	1.4992	
		.8824	(0)		
		.872	(15)		
		.866	(20)		
		.8614	(25/4)		−94.5
		.8625	(25/25)		−93.2
		.8711	(4)		
		.8684	(16/4)	1.4962 (20/D)	
"		.8660	(20/4)		
Xylol (Xylene) (ortho) (Dimethyl benzene)	C_6H_4<$CH_3(1)$, $CH_3(2)$	.8852	(14.1)	1.5082 (14.1)	
"		.8810	(20)		−28
		.8811	(20/4)		−27.1
		.8760	(25/25)		−29

DIACETYLENES

	Occurrence	Authority
5.4		Van Nostrand's Chem. Ann., 290 (1918)
5		Handbook Chem. and Phys., 176 (1920-21)
0		Engler and Höfer, Das Erdöl, **1**, 351
1-203		Van Nostrand's Chem. Ann., 298 (1918)

AROMATIC OR BENZOL HYDROCARBONS, C_nH_{2n-6}

ing point, ° C. essure m.m.	Occurrence	Authority
0.4 at 760	Italy	‡Meyer and Jacobson, **2**, Part 102
ezes to rhombic	Galicia	*Landolt and Börnstein, Phys. chem. Tab., Ed. 3, 673
risms	Germany	†Kramer and Böttcher, Ber., **20**, 606 (1887)
	Sumatra	
	Java	
	Canada	Schorlemmer, Zeitschr. physikal. Chem., **8**, 242 (1865)
	Borneo	
0.2	Pennsylvania	Young, Jour. Am. Chem. Soc., **73**, 906 (1898)
0.4	Rumania	Ubbelohde and Agthe, Dipl. Karlsruhe (1912)
i0.36	California	Handbook Chem. and Phys., 154 (1920-21)
i0.3	Russia	Markownikoff, Lieb. Ann., **301**, 176 (1898)
i0-1	Japan	Nakagawa and Kawai, Jour. Chem. Ind. (Japan), **23**, 453-485 (1920)
i0		Von Auwers, Ann., **419**, 91-120 (1919)
ı.5-110		Zelinski, Ber., **28**, 1022
ı0	Galicia	Ladenburg and Krügel, Ber., **32**, 1821
ı0.8	Pennsylvania	Young, Jour, Am. Chem. Soc., **73**, 906 (1898)
.0	Galicia	Lachowicz, Lieb. Ann., **220**, 199 (1883)
ı0.3	Rumania	Landolt and Jahn, Zeitschr. physikal. Chemie, **10**, 303 (1892)
ı0.3	Coal tar naphtha	Warren, Mem. Am. Acad., **9**, 143-149 (1867)
ı1		Handbook Chem. and Phys., 240 (1920-21)
ı0.70		
ı0-111	Russia	Van Nostrand's Chem. Ann., 366 (1920-21)
ı0-111	Japan	Nakagawa and Kawai, Jour. Chem. Ind. (Japan), **23**, 453-485 (1920)
	Sumatra	
)9-110	Borneo	
	Java	Von Auwers, Ann., **419**, 92-120 (1919)
ı2		Landolt and Jahn, Zeitschr. physikal. Chemie, **10**, 303 (1892); Engler and Hö er, Das Erdö, **1**, 351-114
ı2		Handbook Chem. and Phys., 250 (1920-21)
ı2.6		Van Nostrand's Chem. Ann., 376
ı2.3		Do.

Name	Description	Specific gravity	In refr
Orthoxylol		.8798 (17.9/4)	1
		.8780 (20/4)	
Metaxylol	$C_6H_4\langle {CH_3(1) \atop CH_3(3)}$	.8660 (20)	
"		.8666 (17.1/4)	1
		.8667 (20/4)	
		.8686 (14.85/4)	1
Paraxylol	$C_6H_4\langle {CH_3(1) \atop CH_3(4)}$		
"	Monoclinic crystals		
		.8800 (0)	1.4
		.8688 (15.7)	1.4
		.8627 (17.2/4)	1.4
		.8659 (16.1/4)	1.4
		.8630 (20/4)	
Ethyl benzol	$C_6H_5(CH_2—CH_3)$	.8746 (14.5)	1.4
(Ethyl benzene)		.8708 (14.5/4)	1.4
		.8660 (20/4)	
(Phenyl ethane)		.8740 (14)	
" "		.8678 (20/4)	
" "			
		.8697 (20/4)	
		.8678 (20/4)	
" "			
Ethylene benzol	$C_6H_5—CH{=}CH_2$	.9250 (0)	
Styrol	C_8H_{10}	.9121 (13/4)	
Meta-styrol	$(C_8H_{10})x$	1.0541	
Styrol (cinnamene)			
Mesitylene	$C_6H_3 \begin{cases} CH_3(1) \\ CH_3(3) \\ CH_3(5) \end{cases}$	.8649 (14.6)	1.4
Trimethyl benzol			
"		.8634 (20/4)	
		....	
		.8646 (17.4/4)	1.4
		.8620 (20/4)	
Pseudocumol	$C_6H_3 \begin{cases} CH_3(1) \\ CH_3(2) \\ CH_3(4) \end{cases}$	.8950 (0)	1.5
"		.8790 (20)	
Hemellithol	$C_6H_3(CH_3)_3$		
(Hemimellithene)		.8949 (19.55/4)	1.5
Propyl benzol (normal)	$C_6H_5(C_3H_7)$	.8616 (20/4)	

AROMATIC OR BENZOL HYDROCARBONS—*Continued*

oiling point, ° C. Pressure m.m.	Occurrence	Authority
142–142.5		Von Auwers, Lieb. Ann., **419**, 92–120 (1919)
	Russia	
139.2	Rumania	Handbook Chem. and Phys., 250 (1920–21)
	Galicia	
135–136	Canada	Von Auwers, Lieb. Ann., **419**, 92–120 (1919)
	East Indies	
	California	
137.5	Ohio	Do.
136	Galicia	Lachowicz, Lieb. Ann., **220**, 188 (1883)
138	Java	Handbook Chem. and Phys., 250 (1920–21)
138	Canada	Engler and Höfer, Das Erdöl, **1**, 361
139	California	Landolt and Jahn, Zeitschr. physikal. Chemie, **10**, 303 (1892)
135–136	Ohio	Von Auwers, Lieb. Ann., **419**, 92–120 (1919)
136–137		Do.
134		Landolt and Jahn, Zeitschr. physikal. Chemie, **10**, 303 (1892)
135–136		Von Auwers, Lieb. Ann., **419**, 92–120 (1919)
136.5		Handbook Chem. and Phys., 178 (1920–21)
		Van Nostrand's Chem. Ann., 292 (1918)
135–136		Ladenburg and Krügel, Ber., **32**, 1415 (1899)
6.4–136.5 at 767	Produced synthetically	Richards and Shipley, Am. Chem. Soc., **41**, 2008 (1919)
6.3–136.4	Produced ynthetically	Do.
31–132		Kursanoff, Ber., **32**, 2972
46		Handbook Chem. and Phys., 164 (1920–21)
46 (cor.) at 759		Van Nostrand's Chem. Ann., 358 (1918)
		Van Nostrand's Chem. Ann., 324 (1918)
	Russia	
64.5	Rumania	Landolt and Jahn, Zeitschr. physikal. Chemie, **10**, 303 (1892)
	Penna	
64.8 at 760	Ohio	Van Nostrand's Chem. Ann., 324 (1918)
64	Galicia	Ladenburg and Krügel, Ber., **32** (1899)
65–166		Von Auwers, Lieb. Ann., **419**, 92–120 (1919)
9.8	Germany and Pennsylvania	Engler and Höfer, Das Erdöl, **1**, 363
9.8	Ohio	Handbook Chem. and Phys., **230** (1920–21)
7		Engler and Höfer, Das Erdöl, **1**, 351
75–176 at 744		Von Auwers, Lieb. Ann., **419**, 92–120 (1919)
– at 765		Van Nostrand's Chem. Ann., 352 (1918)

CYCLIC HYDROCARBONS

Name	Description	Specific gravity	Index of refraction	
Propyl benzol (normal)		.8681 (12.25/4)	1.4920 (20/D)	
		.8620 (20/4)		
" " "				
" " "		.8659 (15.7)	1.4942 (15.7)	
" " "				
Isopropyl benzol	$C_6H_5—CH(CH_3)_2$	.8792 (0)		
		.8675 (15)		
" "		.8662 (16.8/4)	1.4930 (20/D)	
		.8640 (20/4)		
Cumol		.8633 (15.1)	1.4947 (15.1)	
(Cumene)		.8620 (20)		
Allyl benzol	C_9H_{10}	.914 (15)		
1, 2-Hydrindene	$C_6H_4 : C_2H_4 : CH_2$	.9570 (15)		
Ethyl methyl benzol Ethyl toluol	$C_6H_4<^{C_2H_5}_{CH_3}$	.8640 (20/4		
Phenyl acetylene	$C_6H_5—C{\equiv}CH$	.9370 (12)		
Rhinyl allylene	$C_6H_5—C{\equiv}C—CH_3$			
Ethyl toluol (ortho)				
Methyl ethyl benzol (ortho)	$C_2H_5—C_6H_4—CH_3$	.8730 (15)		
Methyl ethyl benzol (ortho)		.8841 (15.7/4)	1.5042 (20/D)	
		.8810 (20/4)		
Ethyl toluol (meta)		....		
Methyl ethyl benzol (meta)		.8690 (20)		
Methyl ethyl benzol (meta)		.8690 (17.9/4)	1.4975 (20/D)	
		.8670 (20/4)		
Ethyl toluol (para)		.8650 (21)		
Methyl ethyl benzol (para)		.8601 (22.3/4)	1.4943 (20/D)	
"		.8620 (20/4)		
		.8687 (13.1/4)	1.4971 (20/D)	
Tetramethyl benzol (Durol)				
"	C_6H_2 $CH_3(1)$, $CH_3(2)$, $CH_3(4)$, $CH_3(5)$	.8380 (81.3/4)		79
.	Monoclinic leaflets	.838 (81)		79
		....		79–80
1, 2, 3, 4-Tetramethyl benzol		.9044 (16/4)	1.5185 (20/D)	
		.901 (20/4)		
Isodurol	C_6H_2 $CH_3(1)$, $CH_3(2)$, $CH_3(3)$, $CH_3(5)$	·		
		.896 (0)		
		.8961 (0/4)		Low
1, 2, 3, 4-Prehnitol		.882 (9)		
"		.8816 (9)		– 4

AROMATIC OR BENZOL HYDROCARBONS—*Continued*

Boiling point, ° C. Pressure, m.m	Occurrence	Authority
158–159		Von Auwers, Lieb. Ann., **419**, 92–120 (1919)
1		Engler and Höfer, Das Erdöl, **1**, 351, 114
158.5		Landolt and Jahn, Zeitschr. physikal. Chemie, **10**, 303 (1892)
1	Cuminic acid	Warren, Mem. Am. Acad., **9**, 149–155 (1867)
152.8–153.4		Von Auwers, Lieb. Ann., **419**, 92–120 (1919)
153		Landolt and Jahn, Zeitschr. physikal. Chemie, **10**, 303 (1892)
152.5–153		Handbook Chem. and Phys., 166 (1920–1921)
176–177		Handbook Chem. and Phys., 146 (1920–21); Van Nostrand's Chem. Ann., 254
176		Handbook Chem. and Phys., 190 (1920–21)
162	California	Mabery and Hudson, Proc. Am. Acad., **36**, 259 (1901)
139–142		Van Nostrand's Chem. Ann., 346 (1918)
185		Do.
158–159		Van Nostrand's Chem. Ann., 300 (1918)
164.8–165		Von Auw rs, Lieb. Ann., **419**, 92-120 (1919)
158–159		Van Nostrand's Chem. Ann., 300 (1918)
158–159		Handbook Chem. and Phys.. 208 (1920–21)
161.5–162.5		Von Auwers, Lieb. Ann., **419**, 92–120 (1919)
		Van Nostrand's Chem. Ann., 300 (1918)
161–162		Von Auwers, Lieb. Ann., **419**, 92–120 (1919)
161–162		Do.
179–198	Russia	Markownikoff, Lieb. Ann., **234**, 100 (1886)
	Canada	
	Penna	
195	Ohio	Engler and Höfer, Das Erdöl, **1**, 364
	California	
About 190		Handbook Chem. and Phys., 238 (1920–21)
193–195		Van Nostrand's Chem. Ann., 292 (1918)
203–204		Von Auwers, Lieb. Ann., **419**, 92–120 (1919)
195–197	Canada	Engler and Höfer, Das Erdöl, **1**, 364
195-197	Penna	Handbook Chem. and Phys., 200 (1920–21)
	Ohio	Van Nostrand's Chem. Ann., 318 (1918)
		Engler and Höfer, Das Erdöl, **1**, 351
		Van Nostrand's Chem. Ann., 352 (1918)

Name	Description	Specific gravity	Index of refraction	°C
2-, 3-, 5-Dimethyl ethyl benzol	$C_2H_5—C_6H_3(CH_3)_2$	.861 (20)*		−20†
1-, 3-, 4-Dimethyl ethyl benzol	"	.878 (20)		
1- Methyl-3, 5-di-iso-propyl benzol	$CH_3—C{=}CH—C(C_3H_7){=}CH—C(C_3H_7){=}CH$ (ring)	.8608 (20)	1.4950 (20/D)	58.66
Diethyl benzol	$C_6H_4(C_2H_5)_2$	.8645 (18.2/4)		
" " (ortho)		.866 (18)		
		.8662 (18/4)		−20
" " (meta)		.860 (20)		
		.8602 (20/4)		−20
" " (para)		.862 (18)		
		.8675 (14/4		−20
" " (para)		.8678 (16.3/4)	1.4973 (20/D)	
		.865 (20/4)		
Methyl propyl benzol (ortho)		.8770 (15.75/4)	1.4995 (20/D)	
		.8740 (20/4)		
Methyl propyl benzol (meta)		.8648 (17/4)	1.4951 (20/D)	
		.8620 (20/4)		
Methyl propyl benzol (para)		.8642 (15.4/4)	1.4954 (20/D)	
1-, 2-, 5-Trimethyl, -4-ethyl benzol		.8866 (15.85/4)	1.5086 (20/D)	
		.883 (20/4)		
1-, 3-, 5-Trimethyl, -2-ethyl benzol		.8885 (16.35/4)	1.5111 (20/D)	
		.886 (20/4)		
Methyl isopropyl benzol (meta)	$CH_3—C_6H_4—CH(CH_3)_2$	.862 (20)		
Cymol (ortho) (Cymene)	$CH_3—C_6H_4—CH_2—CH_2—CH_3$	.8748 (20/20)		
		.8619 (13.7)	1.5111	
		.8697 (0)		
		.8592 (14)		
		.8789 (16.15/4)	1.5003 (20/D)	
		.876 (20/4)		
Cymole		.8530 (15)		
4-Cumol		.8794 (15.3/4)	1.5046 (20/D)	
"		.8760 (20/4)		
Cymol (meta) Iso	$C_6H_4(CH_3)(C_3H_7)$	.862 (20)		
" "		.862 (20)		−25
" "				−75.1
Cymol (meta)		.8628 (17.05/4)	1.4925 (20/D)	
		.860 (20/4)		

AROMATIC OR BENZOL HYDROCARBONS—*Continued*

	Occurrence	Authority
5*		*Handbook Chem and Phys., 172 (1920–21); †Van Nostrand's Chem. Ann., 352 (1918)
3.4		Handbook Chem. and Phys., 172 (1920–21)
5–218		Schorger, Jour. Am. Chem. Soc., **39**, 2671–79 (1917)
3–182	Russia	Engler and Höfer, Das Erdöl, **1**, 364
5		Handbook Chem. and Phys., 170 (1920–21)
1–184.5		Van Nostrand's Chem. Ann., 282
1–182		Handbook Chem. and Phys., 170 (1920–21)
1–182		Van Nostrand's Chem. Ann., 282 (1918)
–183		Handbook Chem. and Phys., 170 (1920–21)
–183		Van Nostrand's Chem. Ann., 352
		Von Auwers, Lieb. Ann., **419**, 92–120 (1919)
		Von Auwers, Leib. Ann., **419**, 92–120 (1919)
1.5–182.5		Do.
183		Do.
		Do.
2 at 753		Do.
176		Handbook Chem. and Phys., 210 (1920–21)
–182		Van Nostrand's Chem. Ann., 278 (1918)
		Landolt and Jahn, Zeitschr. physikal. Chemie., **10**, 303 (1892)
5	Cuminic acid	Warren, Mem. Am. Acad., **9**, (135–177)
176		Von Auwers, Lieb. Ann., **419**, 92–120 (1919)
9	Coal tar naphtha	Warren, Mem. Am. Acad., **9**, 148 (1867)
7–169.2		Von Auwers, Lieb. Ann., **419**, 92–120 (1919)
176	Canada	Engler and Höfer, Das Erdöl, **1** 364
–176	Ohio	Van Nostrand's Chem. Ann., 278 (1918)
	Penna	Ladenburg and Krügel, Ber., **32**, 1821 (1899)
		Von Auwers, Lieb. Ann., **419**, 92–120 (1919)

CYCLIC HYDROCARBONS

Name	Description	Specific gravity		Index of refraction	M
Cymol (para)		.856	(20)	1.4926 (13.7)	−73.5
" "		.8551	(20/4)		−73.5
		.853	(25/25)		−73.5
		.8631	(15/4)	1.4925 (24/D)	
		.859	(20/4)	1.4908 (20)	
" "		.8575	(20/15)		−45.29
Isopropyl toluene (para)	$CH_3—C_6H_4—CH : (CH_3)_2(1, 4)$	.860	(16)		−73.5
		.8619	(13.7)	1.5111	
		.8697	(0)		
		.8592	(14)		
Butyl benzol	$C_6H_5(CH_2)_3·CH_3$	.8620	(20/0)		
Isobutyl benzol	$C_6H_5—C_4H_9$	.87163	(17.5)	1.4957 (17.5)	
" "		.873	(15)		
" "		.8596	(20/0)		
Pentamethyl benzol	$C_6H(CH_3)_5$	....			53
Diethyl toluol	C_6H_3 $C_2H_5(1)$, CH_3 (3), $C_2H_5(5)$	....			
" "		.8790	(20)		
Amyl benzol (normal) (Phenyl pentane)	$C_6H_5—C_5H_{11}$	.8602	(22)		
		.8600	(22)		
Iso-amyl benzol	$C_6H_5—C_5H_{11}$	.8350	(18)		
" "		.8850	(18)		
" "		.8870	(14/4)		
Hexamethyl benzol	$C_6(CH_3)_6$				164
Triethyl benzol symmetrical (1-, 3-, 5-)	$C_6H_3(C_2H_5)_3$	.8640	(17)		
		.8636	(17/4)		
Monoethyl trimethyl benzol	$C_6H_2·C_2H_5·(CH_3)_3$				
Tetraethyl benzol (symmetrical) (1,2,4,5)	$C_6H_2(C_2H_5)_4$	.8880	(15)		13
Pentaethyl benzol	$C_6H(C_2H_5)_5$	.8090	(19)		
" "		.8963	(20/4)		−20
Hexaethyl benzol	$C_6(C_2H_5)_6$	.8310	(20)		129
" "	Large monoclinic prisms	.8305	(20/4)		129
Dihydrobenzol	CH—CH₂—CH | | CH—CH —CH₂	.8421	(20/20)	1.4755 (n 2/D)	
(1-, 2-) Dihydrobenzol	Colorless liquid	.8480	(20)		
"		.8478	(20/4)		
(1-4-) Dihydrobenzol	Colorless liquid	.8470	(20)		
"		.84710	(20/4)		

AROMATIC OR BENZOL HYDROCARBONS—*Continued*

Boiling point, ° C. Pressure, m.m.	Occurrence	Authority
175	Russia	Engler and Höfer, Das Erdöl, **1**, 364
	Galicia	
176.5	Japan	Van Nostrand's Chem. Ann., 278 (1918)
174–176	Canada	Do.
175–176	Penna	Von Auwers, Lieb. Ann., **419**, 92–120 (1919)
	Ohio	
176–176.6	Sulphite turpentine	Schorger, Jour. Am. Chem. Soc., **39**, 2671–79 (1917)
175–176.5		Handbook Chem. and Phys , 148 (1919)
175		Landolt and Jahn, Zeitschr. physikal. Chemie, **10**, 303 (1892)
179.5	Cuminic acid	Warren, Mem. Am. Acad., **9**, 149–155 (1867)
183–185		Van Nostrand's Chem. Ann., 268 (1918)
167		Landolt and Jahn, Zeitschr. physikal. Chemie, **10**, 303 (1892)
171–171.5		Handbook Chem. and Phys., 198 (1920–21)
170–170.5		Van Nostrand's Chem. Ann., 316 (1918)
230		Handbook Chem. and Phys., 224 (1920–21)
199–205	Russia	Markownikoff, Lieb. Ann., **234**, 108 (1886)
199–200		Handbook Chem. and Phys., 170 (1920–21)
201 at 743		Van Nostrand's Chem. Ann., 258 (1918)
		Handbook Chem. and Phys , 132–133 (1919)
187.6–196.6	Russia	Markownikoff, Lieb. Ann., **234**, 100 (1886)
193–201		Handbook Chem. and Phys., 196 (1920–21)
201–202 at 760		Van Nostrand's Chem. Ann., 316
264		Engler and Höfer, Das Erdöl, **1**, 351
214–218		Handbook Chem. and Phys., 244 (1920–21)
217		Van Nostrand's Chem. Ann., 370
212–214		Jannasch and Wagner, Chem. Centralbl., 815 (1895)
250		Handbook Chem. and Phys., 238 (1920–21)
277		Handbook Chem. and Phys., 224 (1920–21)
277 (cor.)		Van Nostrand's Chem. Ann., 344 (1918)
298		Handbook Chem. and Phys., 188 (1920–21)
288		Van Nostrand's Chem. Ann., 308
80–80.5	Synthetic	Harris, Ber., **45**, 811 (1912)
65–70	American petroleum	Lefebvre, Compt. rend., **67**, 1352
82–85		Handbook Chem. and Phys., 170 (1920–21)
		Van Nostrand's Chem. Ann., 282 (1918)
85–86		Handbook Chem. and Phys., 170 (1920–21)
85.5		Van Nostrand's Chem. Ann., 282 (1918)

Name	Description	Specific gravity	Index of refraction
Cyclopropane (Trimethylene)	$H_2C\langle CH_2 \mid CH_2$	....	
Methyl cyclopropane Methyl trimethylene	CH_3—CH / \ CH_2—CH_2	{ .6912 (−20/0) .6760 (−8/0) }	
"		.6912 (20/0)	
Dimethyl trimethylene 1-, 1-Dimethyl trimethylene	$(CH_3)_2C$—CH_2 \ | CH_2	.6604 (20°/4)	
1, 2- "		.6806	1.3763
1-, 2-Dimethyl cyclopropane	$CH_3\cdot CH$ | $CH_3\cdot CH$ $\rangle CH_2$	.6755 (22.2°/4)	
1-1-2-Trimethyl dimethyllene	CH_3, CH_3 $\rangle C\langle$ $CH\cdot CH_3$ | CH_2		
		.6822 (19.5°/4)	1.3848 (19.5°
		.6946	1.3945 (18°
		.6888 (15.3°/4)	1.3870 (14.5°
Trimethyl-1, 2-, 3-cyclopropane	$CH_3\cdot CH$ $\langle$ $CH\cdot CH_3$ | $CH\cdot CH_3$	.6921 (22°/4)	1.3942
		.7560 (15°)	
Cyclobutane	CH_2—CH_2 | | CH_2—CH_2	.7038 (0°/4)	1.3752
Cyclobutene	C_4H_6	.7330 (0°/4)	
Methyl cyclobutane	CH_3—CH—CH_2 | |	.7510 (20.5°/20.5)	
(Methyl tetramethylene)	CH_2—CH_2	....	

AROMATIC OR BENZOL HYDROCARBONS—*Concluded*

Boiling point, ° C. Pressure, m.m	Occurrence	Authority
110–110.5 770		Van Nostrand's Chem. Ann., 282 (1918)
134–135		Do.
132–134		Do.
134–135		Do.
130.5 10.5		Prunier and David, Compt. rend., **87**, 991 (1878)

NAPHTHENE SERIES

Boiling point, ° C. Pressure, m.m.	Occurrence	
−35 at 749	Produced synthetically	Ladenburg and Krüger, Ber., **32**, 1821 (1899)
4–5	Produced synthetically	Engler and Hofer, Das Erdöl, 1, 298
4–5′		Demjanoff, Ber., **28**, 22
21		Gustavson and Popper, Jour. prakt. Chemie, **58**, 458, 459, Ser (1898)
32–33		Zelinsky and Ujedinoff, Chem. Centralbl., **1**, 216 (1912)
32.6–33.2 at 761	Produced synthetically	Östling, Jour. Chem. Soc., **101**, 468 (1912)
77–80		Zelinsky and Zelikoff, Ber., **34**, 2857 (1901)
56–57 at 750	Produced synthetically	Zelinsky and Zelikoff, Ber., **34**, 2863 (1901)
65–76 at 755		Do
at 768	Produced synthetically	Östling, Jour. Chem. Soc., **101**, 457 (1912)
65–66 at 748	Produced synthetically	Zelinsky and Zelikoff, Ber., **34**, 2863 (1901)
78–80		Zelinsky, Ber., **28**, 1024 (1895)
11–12	Produced synthetically	Willstätter and Bruce, Ber., **40**, 3979–3988 (1907)
62 at 13		Willstätter and Bruce, Ber., **40**, 3979 (1907)
39–42	Produced synthetically	Perkin and Calman, Jour. Chem. Soc. (Transactions), **53**, 201 (1
50–51	Balachany Russian oil	Markownikoff, Ber., **20**, 975 (1897)
39–42		Van Nostrand's Chem. Ann., 332 (1918)

CYCLIC HYDROCARBONS

Name	Description	Specific gravity	Index of refraction	
Ethyl cyclobutane (Ethyl tetramethylene)	$C_2H_5\cdot CH—CH_2$ $CH_2—CH_2$	.7540 (10°/4)	1.4080 (19.5°)	
		.7483 (20/5)	1.4094 (20°)	
Diethyl cyclobutane				
Cyclobutyl diethyl methane	C_2H_5 $CH_2—CH—CH$ $CH_2—CH_2\cdot C_2H_5$	.7807 (18/4)		
		.7945 (19/0)	1.4334	
		.7881 (20/0)	1.4298 (20)	
Methyl propyl tetramethylene		.7770 (0°)		
Do.		.7640 (0°)	1.4190	
Cyclopentane	$CH_2—CH_2$ $>CH_2$ $CH_2—CH_2$	.7506 (20.5°/4)		
Pentamethylene		.7000 (0/4)		
..		.7635	1.4086	−80
Methyl cyclopentane	H $CH_2—C—CH_2$ $>CH_2$ $HC_2—CH_2$	.7489 (20°)	1.4101(20°)	
Methyl pentamethylene		.7683 (0°)		Liquid at
		.7501 (20°)	1.4104(18°)	
		.7414 (21°/4)	1.4088(21°)	
		.753 (4/15)		
		.7502 (18/4)	1.4099 (n 18)	
		.7660 (0°/4)		
		.7758 (20°/0)		
		.7879 (0°/0)		
Dimethyl-1, 1-cyclopentane	$CH_2—CH_2$ $>C<$ CH_3, CH_3 $CH_2—CH_2$	.7547 (20°/0)	1.4131	
		.7547 (20°/0)	1.4131(20°)	
		.7581 (18.5°/0)	1.4150 (18.5°)	
		.742 (20/20)		
Dimethyl-1, 2-cyclopentane	$CH_2—CH_2$ $>CH\cdot CH_3$ $CH_2—CH$ $\cdot CH_3$	.7510 (18.5°/0)	1.4160 (18.5°)	
		.7534 (20/0)	1.4126(20°)	
		.7581	1.4150	
Dimethyl-1, 3-cyclopentane	CH_3 $CH—CH_2$ $H_3C<$ $CH_2—CH\cdot CH_3$	.7410 (24/4)		
		.7543 (20°)	1.4130(20°)	
		.7497 (18/4)	1.4110(18°)	
		.7410 (24°/4)	1.4066(24°)	
		.7543 (20/4)	1.4130	

NAPHTHENE SERIES—*Continued*

Boiling point, ° C. Pressure, m.m.	Occurrence	Authority
72.2–72.5 760		Zelinsky and Guth, Ber., **41**, 2433 (1908)
72–72.2		Do.
124–125	Produced synthetically	Wedekind, Miller, Ber., **44**, 3285 (1912)
143–144	Produced synthetically	Kishner and Amosoff, Jour. Russ. Phys. Chem. Gesell., **37**, 517 (1905)
151–154	Produced synthetically	Kishner and Amosoff, Chem Centralbl., **76**, pt. 3, 816 (1905)
151–152 at 755 } 148–149 at 767 }		Do.
117–120 } 115–120 }	Produced synthetically	Wreden, Lieb. Ann., **187**, 156 (1877)
115–118		Wallach, Ber., **25**, 923 (1892)
50.25–50.75		Aschan, Chemie der Alicyclischen Verbindungen
50 at 760	Pennsylvania	Young and Richardson, Jour. Frank. Inst., **47**, 50–70
49	Russia	Willstätter and Bruce, Ber., **40**, 3981 (1907)
	Rumania	
49.5–49.6	America	Young (S.), Jour. Chem. Soc., **73**, 906 (1898)
72–73 at 752	Produced synthetically	Kijner, Jour. prakt. Chemie, **56**, 364 (1897)
70–72 at 746	Russia	Markownikoff, Lieb. Ann., **307**, 337 (1899)
71.5–72.5	Rumania	Zelinsky, Ber., **30**, 387 (1897)
72–72.2	Produced synthetically	Zelinsky, Ber., **44**, 2, 2781 (1911)
72	Italy	Chavanne and Simon, Compt. rend., **168**, 1324–26 (1919)
72–74		Markownikoff, Chem. Centralbl., **1**, 1211–1212 (1899)
72–71.5 at 760		Young, S., Jour. Chem Soc., **73**, 913 (1898)
72–72.2 760		Zelinsky, Ber., **35**, 2686 (1902)
72 at 754		Markownikoff, Chem. Centralbl., 1211 (1899)
87.8–87.9	Produced synthetically	Kishner, Jour. Russ. Phys. Chem. Gesell., **40**, 944 (1908)
88 at 762		Kishner, Chem. Centralbl., **79**, 1860 (1908)
92–93 at 758		
95–98	Caucasian naphtha; also produced synthetically	Markownikoff, Jour. prakt. Chemie, **45**, 569 (1892)
92.7–93	Produced synthetically	Kishner, Jour. Russ. Phys. Chem. Gesell., **40**, 944 (1908)
92–93		Kishner, Jour. Russ. Phys. Chem. Gesell., **40**, 944 (1908)
		Kishner, Chem. Centralbl., **79**, 1859 (1908)
91–95 760		{ Aschan, Chemie der Alicyclischen Verbindungen, 472–73 Zelinsky and Rudsky, Ber., **35**, 2676 (1902)
93–96 at 743	Produced synthetically	Zelinsky and Rudsky, Ber., **29**, 405 (1896)
90.5–91		Zelinsky, Ber., **35**, 2678–2679 (1902)
	Rumania	Zelinsky, Ber., **30**, 976 (1897)
94 at 760		Zelinsky, Ber., **30**, 1540 (1897)
90.5–98	Balachany Russian oil	Markownikoff, Ber., **30**, 976 (1897)

CYCLIC HYDROCARBONS

Name	Description	Specific gravity	Index of refraction	
Methyl-1 ethyl-2-cyclopentane	$CH_2—CH_2$ / $CH_2—CH$ >$CH·CH_3$; C_2H_5 on CH			
Methyl-1 ethyl-3-cyclopentane	$CH_2—CH_2$ / $CH—CH_2$ >$CH·CH_3$; C_2H_5 on CH	.7669 (16°/4)	1.4214(16)	
		.7661 (26°/9)	1.4218(26°)	
Cyclohexane	C_6H_{12}			
Hexamethylene		.7764 (20/4)	1.4258	2
Hexanaphthene		.7722 (0/4)	1.4245	
Hexahydrobenzol		.7903		
		{ .7828 (14.5°/15)	}	4.7
		.7968 (0°/4) }		
		.7934	1.4266	6.4
		.7808 (18.7)		
		.7790 (20°)		6.4
Hexahydrotoluene	$CH_2—CH_2—CH·CH_3$ / $CH_2—CH_2—CH_2$	.7690 (20°)		
		.7720		
Methyl hexamethylene		.7697 (20°/0)		
(Heptanaphthene)		.7737		
"		.7859 (0°/0)		−147.5
(Methyl cyclohexane)		.7700 (15/15)		
" "		.7859		
" "		.7800 (4/15)		
		.7647 (18°/4)	1.4205 (18°)*	
" "				
" "		†.7647 (18°/4)	1.4243(19°)	
		.7662 (18.5°)	1.4170 (18.5°)	
" "				
1-, 1-Dimethylhexamethylene	CH_3, CH_3 >C—$C(H_2)$—$C(H_2)$ / C(H_2)—C(H_2)—CH_2	.75667 (0°/4)		
		.7728 (18)	1.4236 (18)	
1-, 2-Dimethyl cyclohexane	CH_3 / $H_2C—H_2C—CH$ / $H_2C—CH_2—CH$ / CH_3	.7835		
		.7681 (20°)		
		.7835		
1-, 3-Hexahydrometaxylol	$CH_2—CH_2—CH_2$ / $CH_3CH—CH_2—CH·CH_3$	.7770 (0/0)		
Hexahydrometaxylol		.7835		
		.7640 (19/19)		
		.7585 (17°)	1.4186(17°)	
		.7707 (21/4)		

NAPHTHENE SERIES—*Continued*

Boiling point, ° C. Pressure, m.m.	Occurrence	Authority
124	Produced syntheti ally	Mar.hall and Perkin, Jr., Jour. Chem. Soc., **57**, 241 (1890)
120 5-121 at 750		Zelinsky, Ber., **35**, 2686 (1902)
120		Do.
81-82	Produced synthetically	Zelinsky, Ber., **34**, 2799-2803 (1901)
80 55-80 65	America	Fortey, E. C., Jour. Chem. Soc., **73**, 932, 949 (1898)
80.8 at 760	Galicia	Do.
80.85	Do.	Young and Fortey, Jour. Chem. Soc., **75**, 873-83 (1899)
81		Willstätter and Bruce, Ber., **40**, 981-3982 (1907)
81 at 755	Caucasian petroleum	Sabatier and Senderens, Compt. rend., **132**, 1255
80.8		Handbook Chem. and Phys., 199 (1920-21)
101-102		Handbook Chem. and Phys , 190 (1920-21)
97	Russia, Bibi Eibat	Markownikoff, Ber., **18**, 1819 (1880)
100.2 at 751	Hanover an (German) naphtha	Markownikoff, Lieb. Ann., **341**, 118 (1901)
100-102	Caucasia	
100.1	Produced synthetically	Sabatier and Senderens, Compt. rend, **132**, 1255 (1908)
100.4		
101-102 at 744		Zelinsky and Generosoff, Ber., **29**, 731-733 (1896)
101-102†		Zelinsky, Ber., **30**, 1532,† 1537* (1897)
102-104 at 760	Produced synthetically	Knoevenagel and Lübben, Lieb. Ann., **297**, 159 (1897)
102	Galicia	Poni, Chem. Centralbl., **2**, 452 (1900)
102 at 760	American petroleum	Young, Jour. Chem. Soc., **73**, 916-918 (1898)
		Zelinsky and Lepeschkin, Chem. Centralbl., **1**, 34 (1902)
	Russia	Pautynskij, Jour. Russ. Phys. Chem. Gesell., **18**, II, 256-258 (1887)
		Do.
		Pautynskij, Chem. Centralbl., 237 (1887)
	Japan	Markownikoff, J. Russk. Ph. Kh. Obsch., **15**, 237 (1883); **24**, 141 (1892)
	California	
122-124	Russian petroleum	Do.
120-123	Colophony	Do.
117-118		Wischen, Naphthene, Braunschweig, 58 (1901)
119.4 at 751		Van Nostrand's Chem. Ann., 308 (1918)

CYCLIC HYDROCARBONS

Name	Description	Specific gravity	Index of refraction
Dimethyl-1, 1-cyclohexane	CH_3 CH_3 \/ C—CH_2—CH_2 | | CH_2—CH_2—CH_2		
Hexahydro-orthoxylol	CH_3 CH_3 | | CH —CH —CH_2 | | CH_2—CH_2—CH_2	.7835	
		.7681 (20°/0)	
Dimethyl cyclohexane (ortho)		.7980 (4/15)	
		.8008	
Dimethyl metacyclohexane	CH_3 | CH_2—CH—CH_2 | | CH_2—CH_2—CH | (CH_3)	.7874	
		.7687 (20°/4)	1.4234(20°)
		.7741 (15/0)	1.4270
Hexahydro-isoxylol		.7770	
1-, 3-Dimethyl hexamethylene		.7687 (20°)	1.4234(20°)
"		.7710 (20/D)	1.4256 (20/D)
		.7706 (0°)	1.4186(17°)
		.7788 (19°/0)	
		.7736 (18/4)	1.4270 (18/4)
		.7822 (20°)	
Do.		.7750 (4/15)	
1-, 4-Dimethyl hexamethylene		.7830 (4/15)	
Trimethyl-1, 1, 3-cyclohexane (Dimethylheptane)	CH_3 CH_3 \/ C—CH_2—CH—CH_3 | \\ CH_2—CH_2—CH_2	.7848 (15°/4)	1.4324
		.7120 (20°)	
		.7736 (18°/4)	
Nononaphthene (Hexahydro-pseudocumol, 1-, 2-, 4-)	C_9H_{18} CH_3 | CH—CH_2—CH_2 | | CH—CH_2—CH | | CH_3 CH_3	{ .7808 (0°) { .7664 (20°/20)	
Do.		.7647 (20/0)	
		.7662 (15)	
Do. (para)		.7690 (20/4)	1.4244(20°)
		.7866 (0°/4)	
	CH_3 CH_3 | | CH —CH —CH_2 | | CH_2—CH_2—CH—CH_3	.8052 (0°/4)	
		.7807 (18°/4)	
Hexahydroparaxylol	CH_2—CH—CH_2—CH_2 | | CH_2—CH_2—CH—CH_3	.7690 (20°)	

NAPHTHENE SERIES—*Continued*

Boiling point, ° C. Pressure, m.m.	Occurrence	Authority
17 at 743	Produced synthetically	Crossley and Renouf, Proc. Chem. Soc., **20**, 243 (1904)
		Aschan, Chemie der Alicyclischen Verbindungen
.6–129		Chavanne and Simon, Compt. rend., **168**, 1324–6 (1919)
		Sabatier and Senderens, Compt. rend., **32**, 1255 (1851)
		Sabatier and Senderens, Compt. rend., **132**, 1566 (1901)
		Zelinsky, Ber., **30**, 1539 (1897)
		Balbiano and Angeloni, Chem. Centralbl., **2**, 955 (1904)
118		Beilstein and Kurbàtow, Ber., **18**, 1819 (1883)
120		Ubbelohde and MalaXa, Dipl. Karlsruhe (1910)
118–120 at (20)		Tausz and Peter, Zent. Bakt. u. Parasit, II, Chem. abs. **49**, 4 7, 554 (1919)
117–118		Aschan, Ber., **24**, 2718–2719 (1891)
119.5 at 751		Zelinsky, Ber., **28**, 781 (1895)
120 at 760		Knoevenagel, Lieb. Ann., **297**, 167 (1877)
119–120		Skita, Ber., **41**, 2942 (1908)
121.2–121.8		Chavanne and Simon, Compt. rend., **168**, 1324–6 (1919)
122.7–123		Do.
137–138		Knoevenagel and Fisher, Lieb. Ann , **297**, 202 (1897)
132–134		Skita, Ber., **41**, 2941 (1908)
118 (cor.)		Markownikoff and Ogloblin, Jour. Russ. Phys. Chem. Gesell., **15**, 237, 307 (1883)
135–136	Baku, Russia	Markownikoff and Ogloblin, Ber., **16**, 1873 (1883)
	Produced synthetically	Konowaloff, Chem. Zeitung, 113, 145, (1890)
135–137	Russia; also produced synthetically*	Zelinsky and Reformatsky, Ber., **29**, 215, 216,* (1896)
137–139	California	Ahrens and Mozdzenski, Chem. Centralbl., **2**, 402 (1908)
	Japan	
		Sabatier and Senderens, Compt. rend., **132**, 1254 (1901)
	Produced synthetically	Do.
144		Zelinsky and Reformatzky, Ber., **29**, 214–216 (1896)
		Do.
120–121.0		Handbook Chem. and Phys., 190 (1920–21)

CYCLIC HYDROCARBONS

Name	Description	Specific gravity	Index of refraction
Ethyl cyclohexane (Ethyl naphthene)	$CH_2-CH_2-CH_2$ | | $>C_9H_{18}$ $CH-CH_2-CH_2$ | C_2H_5	.8025 (0/4)	
Ethyl hexamethylene		.7989 (4)	1.4325 (20.4)
Trimethyl-1, 2-, 4-cyclohexane		.8025	
		.7772 (20/0)	
Trimethyl-1, 2-, 5-hexamethylene			
"			
1-, 3-, 4-Trimethyl cyclohexane		.8052 (0/4)	
		.7890 (20/4)	1.4330 (20/4)
1-, 3-, 5-Trimethyl cyclohexane (Hexahydromesitylene)	CH_3 | $CH-CH_2-CH-CH_3$ | | $CH_2-CH-CH$ | CH_3	.7884	
"		.7857 (4)	1.4262 (22)
"			
Methyl-1-ethyl-2-cyclohexane	CH_3 C_2H_5 | | $CH-CH-CH_2$ | | $CH_2-CH_2-CH_2$		
Methyl-1-ethyl-3-cyclohexane	CH_3 | $CH-CH_2-CH-C_2H_5$ | | $CH_2-CH_2-CH_2$	.7989 (20/4)	
Do.	Optically active	.7896 (17/4)	1.4353 (17)
Methyl-1-ethyl-4-cyclohexane	CH_3 · $CH-CH_2-CH_2$ | | $CH_2-CH_2-CH_2-C_2H_5$	.8041 (0/4)	
Diethyl-1-3-cyclohexane (Mentonaphthene)	C_2H_5 | $CH-CH_2-CH-C_2H_5$ | | $CH_2-CH_2-CH_2$	.7957 (20/4)	1.4388 (20)
Propyl cyclohexane	$C_3H_7-C_6H_{11}$	.8091 (0/4)	
Propyl hexamethylene		.7671 (20/0)	
Hexahydro-4-cumol		.7808	
Hexahydrocumol	$C_6H_{11}\cdot C_3H_7$	.7870 (20)	
Hexahydropropyl benzol		.7811	
Menthane	$(CH_3)_2CH\cdot C_6H_{10}\cdot CH_3$	.7949 (20)	
" (ortho)	"	.8135 (21/0)	
" (meta)	"	.7965 (24/0)	
(para)		.8028 (25/0)	

NAPHTHENE SERIES—*Continued*

Boiling point, ° C. Pressure, m.m.	Occurrence	Authority
130		Sabatier and Senderens, Ann. Chem. and Phys., **4** (1905)
130	Rumania	Ubbelohde and Agthe, Dipl. Karlsruhe (1912)
130		Sabatier and Senderens, Compt. rend., **132**, 1255
132–133 at 755	Produced synthetically	Kursanoff, Ber., **32**, 2973 (1899)
142–144	Russia	Zelinsky and Reformatzky, Ber., **29**, 214–216 (1896)
140–142	Rumania	Poni, Chem. Centralbl., **2**, 1370 (1902)
140–142	Rumania	Poni, Chem. Centralbl., **2**, 1370 (1902)
143–144		Knonwaloff, Jour. Russ. Phys. Chem. Gesell., **19**, 255 (1887)
139–140		Tausz and Peter, Zent. Bakt u. Parasit, II, abs. **49**, 497–554 (1919)
137–139	*Baku; also produced synthetically	Sabatier and Senderens, Compt. rend , **132**, 1254 (1901); Ann. Chem Phys , **4**, 366,* (1905)
134–136		Ubbelohde and Agthe, Dipl. Karlsruhe, 1912
135–138		Handbook Chem. and Phys., 190 (1920–21)
135–138		Markownikoff, J. Russk. Phys. Kh. Obsch , **15**, 237 (1883); **24**, 1 (1892)
150–152	Produced synthetically	Perkin, Jr., and Kipping, Jour. Chem. Soc , **57**, 25 (1890)
149–150	Produced synthetically	Zelinsky and Oberländer, Jour. Russ. Phys. Chem. Gesell., **31**, 496 (18
148–149	Produced synthetically	Zelinsky, Ber., **35**, 2681 (1902)
150	Produced synthetically	Sabatier and Senderens, Compt. rend., **132**, 1254 (1901)
169–171		Zelinsky and Rudewitsch, Chem. Centralbl., **2**, 86 (1895)
153–154	Produced synthetically	Sabatier and Senderens, Ann. Chemie and Phys., **4**, 367 (1905)
147–149.5		Van Nostrand's Chem. Ann., 352 (1918)
135–136	Russian petroleum	Markownikoff, J. Russk. Ph. Kh. Obsch , **15**, 237 (1883); **21**. 141 (18
147–150		Handbook Chem. and Phys., 190 (1920–21)
140–142		Markownikoff, J. Russk. Ph. Kh. Obsch., **15**, 237 (1883); **24**, 141 (189
163–170 at (20)		Ipatieff, Ber., **43**, 3551 (1910)
171		Van Nostrand's Chem. Ann , 324 (1918)
166–167		Do.
167–168		Do.

Name	Description	Specific gravity	Index of refraction
Menthane (para) Methy -isopropyl-4-cyclo-heXane	CH_3 | $CH—CH_2—CH_2$ | | $CH_2—CH_2—CH—C_3H_7$	.8130 (0/4)	
Decanaphthene	$C_{10}H_{20}$	.8043 (0)	
..		.8072 (0)	
		.7808 (17.40/40)	
		{ .8073 (15/15) .8140 }	
		.7970 (15/15)	
		.8114 (0/0)	
		.8060 (0/0)	
		.7970	
		.7936 (0)	
		.7950	
(1-, 3-Dimethyl-5-ethyl cycloheXane)	CH_3 CH_3 | | $CH—CH_2—CH$ | | $CH_2—CH—CH_2$ | C_2H_5	.8076 (0)	
		.7929 (20)	
Hexahydrocymene (para)	$CH_3—C_6H_{10}—C_3H_7$	.7960 (15)	
Undecanaphthene	$C_{11}H_{22}$	.8119 (0)	
Isobutyl cycloheXane	$H_2C—CH_2—CH—CH_2—CH—CH_3$ | | | $H_2C—CH_2—CH_2$ CH_3	.8042 (21)	1.4472 (15.6)
	$C_{10}H_{20}$	.7810	
Dimethyl 1-3-5-isobutyl cycloheXane	CH_3 CH_3 \/ $C—CH_2—CH—C_4H_9$ | | $CH_2—CH_2—CH_2$	.8227 (0)	
Methyl-1-ethyl-3-isopropyl-4-cycloheXane	CH_3 | $CH—CH_2—CH—C_2H_5$ | | $CH_2—CH_2—CH—C_3H_7$	.8146 (20/0) .8120 (18.4/40)	
Dodecanaphthene	$C_{12}H_{24}$	.8055 (14) .8050	
Tetradecanaphthene	$C_{14}H_{28}$	.8215 (18.6/40)	
	..	.8390 (0)	
	..	.8321 (0)	
Cycloheptane Suberane (Heptamethylene)	$CH_2—CH_2—CH_2$ | $>CH_2$ $CH_2—CH_2—CH_2$	.8160 (15°) .8253 (0°) .8152 (20°) .8930 (20°) .8252	

103–104

NAPHTHENE SERIES—*Continued*

Boiling point, ° C. Pressure, m.m.	Occurrence	Authority
169–170	Produced synthetically	Sabatier and Senderens, Compt. rend., **1**, 1254 (1901)
150–152	Balachany, Bibi Eibat — Russian petroleum	Starodubski, Chem. Zeitung, 808 (1890); Jour. Russ. Phys. Ch Gesell., 64 (1890)
160–162	Petroleum	Markownikoff, J. Russk. Ph. Kh. Obsch., **15**, 237 (1883); **24**, 141 (18
168–170	Russia	Do.
168.5–170	Terpene hydrate	Markownikoff, J. Russk. Ph. Kh. Obsch., **15**, 237 (1883); **24**, 141 (18
167–169	Camphor (Starodulski)	Do.
162–167	Tetrahydroterpene (Orloff)	Do.
166–169		Ipatieff, Ber., **43**, 3552 (1910)
165–175*	Russia	Subkow, Chem. Centralbl., **2**, 857*, (1893)
160–162	Russia	Markownikoff and Ogloblin, Ber., **16**, 1873–1879 (1883)
168.5 at 752		
162–164	Caucasus	Subkow, Chem Centralbl., **2**, 857 (1893)
168.5–170	Caucasus	Markownikoff, Chem. Centralbl., **1**, 176 (1899)
171–173		Handbook Chem. and Phys., 190 (1920–21)
180–185		Markownikoff and Ogloblin, Ber., **16**, 1873–1879 (1883)
157		Ubbelohde and Agthe, Dipl. Karlsruhe (1912)
156–160	Germany	Kraemer and Böttcher, Ber., **20**, 597–598 (1887)
193–195		Am. Chem. Phys., Series 8, **4**, 367 (1904)
7–208 at 736	Produced synthetically	Kursanoff, Jour. Russ. Phys. Chem. Gesell., **33**, 289 (1908)
		Markownikoff, J. Russk. Ph. Kh. Obsch., **15**, 237 (1883); **24**, 141 (18
6.5–197	Balachany, Russia	Markownikoff and Ogloblin, Ber., **16**, 1873–1879 (1883)
	Germany	Kraemer and Bottcher, Ber., **20**, 597 (1887)
240–241		Markownikoff, J. Russk. Ph Kh. Obsch., **15**, 237 (1883); **24**, 141 (18
240–241 (cor.)	Balachany, Russia	Markownikoff and Ogloblin, Ber., **16**, 1873–1879 (1883); **15**, 734 (18
230–232	Russia	Schützenberger and Ionine, Compt. rend., **91**, 823–25 (1880)
	Produced synthetically	Markownikoff, Jour. prakt. Chemie, **49**, 426 (1894)
117–117.5 at 743	Russia	Do.
63–163.5		Jakub, Chem. Centralbl., I, 568 (1903)
	Produced synthetically	Markownikoff, Lieb. Ann., **327**, 65 (1903)
17–117.5		Willstätter and Bruce, Ber., **40**, 3981–3982 (1907)

Name	Description	Specific gravity	Index of refraction	
Suberane (Heptamethylene)		.8094		
"		.7791		
		.8274 (0°/4)		
"		.8108 (20/4)		
Dimethyl 1-, 2-cyclo-heptane	CH_3 CH_3 CH —CH —CH_2 > CH_2 CH_2—CH_2—CH_2			
Ethyl cycloheptane	C_2H_5 CH —CH_2—CH_2 > CH_2 CH_2—CH_2—CH_2	.8152 (20/0)		
Cyclo-octane	CH_2—CH_2—CH_2—CH_2	.8390 (20/4)	1.4586 (20)	
"		.8410		
"	CH_2—CH_2—CH_2—CH_2	.8390		
		.8500 (0/4)	1.4577 (16)	11.5
(Octomethylene)		.8350 (20/4)		9.5
		.8390 (20/4)	1.4586 (20)	14.2–14.
Dimethyl cyclo-octadiene	CH_3·CH·CH : CH·CH_2 CH_3·CH·CH : CH·CH_2	.8623 (13)	1.4903 (13)	
Cyclononane	C_9H_{18}	.7850	1.4328	
Nonanaphthene	CH_2—CH_2—CH_2—CH_2 > CH_2 CH_2—CH_2—CH_2—CH_2	.7652 (20)		
		.7733 (16)	1.4328 (16)	
Cyclo nonomethylene		.7733 (16/4)	1.4328 (16)	
		.7850 (0/4)		
Pentadecanaphthene	$C_{15}H_{30}$	.8210 (18.8/40)		
Dihexanaphthene	$C_{12}H_{22}$	.8265 (20)*		
		.8294 (12)		
"				
	Optically inactive	.8553 (20)		
		.8479 (27)		
		.8551 (22)	1.4662 (25)	
		.8511 (20/4)	1.4640 (25)	
"	$C_{16}H_{32}$			41–42
Methyl dihexanaphthene	$C_{13}H_{24}$	.8649 (22)	1.4692 (25)	
	"	.8629 (22/4)	1.4692	
	"	.8679 (25)	1.4666 (25)	
	"	.8557 (25)	1.4691 (25)	
	"	.8621 (20)	1.4681 (25)	
	"	.8765		
	$C_{14}H_{26}$	.871 (20)	1.4713	
	"	.8746 (20)	1.473	
	"	.8785 (29)		
" "	$C_{15}H_{28}$	.8788 (20)	1.4746	
		.8915	1.484	
	"	.8498 (21.5)	1.4703 (23)	

NAPHTHENE SERIES—*Continued*

Boiling point, ° C. Pressure, m.m.	Occurrence	Authority
117 at 743		Van Nostrand's Chem. Ann., 306 (1918)
98–101	Produced synthetically	Markownikoff, Compt. rend., **110**, 467 (1890)
118		*Willstätter and Kametaka, Ber., **41**, 1483 (1908)
153–154*	Produced synthetically	Kipping and Perkin, Chem. News, **63**, 122; Chem. Centralbl., I, 656 (1891)
150–152		Do.
163–163.5	Produced synthetically	Kipping and Jakub, Jour. Russ. Phys. Chem. Gesell., **34**, 916 (1903); Chem. Centralbl., I, 568 (1903)
150		Willstätter and Wasser, Ber., **44**, 3428–3444 (1911)
145–147 at 720		Willstätter and Veraguth, Ber., **40**, 959 (1907)
147.3–148.3		
145.3–148		Willstätter and Wasser, Ber., **43**, 1181 (1910)
149.6–150.6		Willstätter and Veraguth, Ber., **40**, 968 (1907)
60–71 at 15		Doehner, Ber., **35**, 3136 (1902)
170–172		Willstätter and Bruce, Ber., **40**, 3981 –3982 (1902)
135–136	Russia	Tichwinsky, Petrol., **7**, 935 (1912)
170–172		Zelinsky, Ber., **40**, 3277 (1907)
170–172	Produced synthetically	Zelinsky, Ber., **40**, 3277 (1907); Engler and Höfer, Das Erdöl, **1**, 303
		Ahrens and Mozdzenski, Chem. Centralbl., II, 402 (1908)
246–248		Markownikoff and Ogloblin, Ber., **15**, 734 (1882); **16**, 1873 (1883)
*246–248 (cor.)	Russia	Markownikoff and Ogloblin, Ber., **16**, 1873–1879 (1883)*
		Markownikoff and Ogloblin, Ber., **15**, 734 (1882)
234–236 at 760	Russia	Markownikoff, Chem. Centralbl, I, 1064 (1899); Ber., **32**, 1441 (1899)
130 at 135	Lucas Well, Spindle Top, Texas	Mabery, Jour. Am. Chem. Soc., **23**, 264 (1901)
110–115 at 60 }	Anse la Butte, La.	Coates and Best, Jour. Am. Chem. Soc., **25**, 1157 (1904)
205–210 at 760 }		
145–150 at 80	Welsh, La.	Coates and Best, Jour. Am. Chem. Soc., **27**, 1319 (1905)
215–217 at 760	Jennings, La.	Coates, Jour. Am. Chem. Soc. **28**, 387 (1906)
283–285		Zelinsky, Ber., **40**, 3278 (1907)
150–155 at 80	Jennings, La.	Coates and Best, Jour. Am. Chem. Soc., **27**, 1319 (1905)
235–238 at 760		Coates, Jour. Am. Chem. Soc., **28**, 37 (1906)
165–170 at 100	Welsh, La.	Coates and Best, Jour. Am. Chem. Soc., **27**, 1319 (1905)
140–145 at 33	Bayou Bouillon, La.	Coates and Best, Jour. Am. Chem. Soc., **27**, 1320 (1905)
150–155 at 60	California	Mabery, Proc. Am. Acad., **40**, 323–362, 340–346 (1904)
151.5		Eijkman, Chem. Centralbl., II, 989 (1903)
125–130 at 25	Heavy oil, Texas	Mabery and Buck, Jour. Am. Chem. Soc., **22**, 553 (1900)
155 at 160	Lucas Well, Spindle Top, Texas	Mabery, Jour. Am. Chem. Soc., **23**, 264 (1900)
160–165 at 60	Anse la Butte, La.	Coates and Best, Jour. Am. Chem. Soc., **25**, 1158 (1904)
140–145 at 25	Heavy oil, Texas	Mabery and Buck, Jour. Am. Chem. Soc., **22**, 553 (1900)
190–195	Lucas Well, Texas	Mabery, Jour. Am. Chem. Soc., **23**, 266 (1901)
186–188 at 50	Produced synthetically	Nastjukiow, Chem. Zeitung, 1220 (1912); Eighth Inter. Cong. Ap. Chem., **10**, 201 (1912)

CYCLIC HYDROCARBONS,

Name	Description	Specific gravity		Index of refraction	Melti °
Dehydrobetulene	$C_{15}H_{22}$	.9186	(23)	1.5052 (23/D)	
Disuberyl $CH_2—CH_2—CH_2$ / $CH_2—CH_2—CH_2$ >CH·CH< $CH_2—CH_2—CH_2$ / $CH_2—CH_2—CH_2$		.9069	(20)		
	$C_{16}H_{30}$	.8915	(20)	1.484	
	$C_{18}H_{36}$				
	$C_{19}H_{38}$	.8208	(20)	1.415 (20)	
	$C_{21}H_{42}$	.8424	(20/20)		
	$C_{22}H_{44}$	{ .8262	(20/20)	} 1.454	
		.8296	(20)		
	$C_{23}H_{46}$	.8569	(20)	1.4714 (20)	
	$C_{24}H_{48}$	.8598	(20)	1.4726	
	$C_{26}H_{52}$	.8580		1.4725	
	C_9H_{16}; Optically inactive	.7992	(24/4)	1.4370	
	$C_{10}H_{18}$; Optically inactive	.8146	(22/4)	1.4460 (25)	
	$C_{10}H_{18}$				155
	"				
	$C_{16}H_{30}$	.8808	(20)	1.470	
		.8894	(20)	1.4672	
	$C_{17}H_{32}$	.8966	(20)	1.4721	
		.8736	(28)	1.4760 (25)	
	$C_{19}H_{36}$	.8364	(20)	1.4614	
		.8935	(15/15)		
		.9020	(20)	1.4928	
	$C_{21}H_{40}$	.8417	(20)		
		.8546	(20)	1.465	
		.905	(15/15)		
	$C_{22}H_{42}$	.8614	(20)	1.469	
		.908	(15/15)		
	$C_{23}H_{44}$	.8639	(20)		
	$C_{20}H_{46}$	.9130	(15/15)		
		.8639	(20)	1.4715	
	$C_{25}H_{48}$	.8116	(20)		
	$C_{27}H_{52}$; Liquid at −10	.8688	(20/20)	1.4722	
	$C_{28}H_{54}$; Liquid at −10	.8694	(20/20)	1.4800	
		.8212			
	$C_{29}H_{56}$	.8235	(70)		
	$C_{30}H_{58}$	.8288	(75)		
	$C_{31}H_{60}$	.8312	(75)		
	$C_{34}H_{66}$	.8336	(75)		
	$C_{35}H_{68}$	.9160	(15/15)		
Disuberyl	$C_{16}H_{28}$	.8801	(22)	1.4805 (25)	
		.8871	(29)	1.4828 (25)	
	$C_{17}H_{30}$	.8919	(20)	1.4778	80
		.9009	(29)		88

NAPHTHENE SERIES—*Continued*

Boiling point, ° C.	Pressure, m.m.	Occurrence	Authority
112–114 at	9	Birch bud oil	Semmler, Jonas and Richter, Ber., **51**, 417–24 (1918)
290–291			Jakub, Chem. Centralbl., I, 568 (1903)
		Spindle Top, Texas	Richardson, Jour. Frank. Inst., **162**, 117 (1906)
271.5–274.5		Grozny, Russia	Markownikoff, Ber., **32**, 1441 (1899)
272–274			Markownikoff, Jour. prakt. Chem., **59**, 556 (1899)
		Pennsylvania	Mabery, Proc. Am. Acad., **37**, 565–595 (1902)
210–212 at	(50)	Pennsylvania	Do.
230–232 at	(50)	Pennsylvania	Do.
240–242 at	(50)	Pennsylvania	Do.
258–260 at	(50)	Pennsylvania	Do.
272–274 at	(50)	Pennsylvania	Do.
280–282 at	(50)	Pennsylvania petroleum	Do.
145–147 at	760	·	Engler and Höfer, Das Erdöl, **1**, 341
168–170 at	760		Do.
		Jennings, La.	Coates, Jour. Am. Chem. Soc., **28**, 387 (1906)
160–162			Semmler, Ber., **33**, 777 (1900)
169–175			Ipatieff, Ber., **43**, 3552 (1910)
175–180 at	60	California petroleum	Mabery, Proc. Am. Acad., **40**, 323–362 (1904)
160–165 at	25	Texas	Mabery and Buck, Jour. Am. Chem. Soc., **22**, 553 (1900)
190–195		Spindle Top, Texas	Markownikoff, Jour. prakt. Chemie, **59**, 556 (1899)
175–180 at	25	Heavy oil, Texas	Mabery and Buck, Jour. Am. Chem. Soc., **22**, 553 (1900)
175–180 at	33	Welsh, La.	Coates and Best, Jour. Am. Chem. Soc., **27**, 1319 (1905)
198–202 at	30	Trenton limestone, Ohio	Mabery and Palm, Proc. Am. Acad., **40**, 323–334 (1904)
		Grozny, Russia	Charitschkoff, Chem. Centralbl., I, 409 (1904)
195–200 at	25	Heavy oil, Texas	Mabery and Buck, Jour. Am. Chem. Soc., **22**, 553 (1900)
213–217 at	30	Trenton limestone, Ohio	Mabery and Palm, Proc. Am. Acad., **40**, 323–334 (1904)
213–217 at	30	Ohio	Mabery and Palm, Proc. Am. Acad., **40**, 323 (1904)
		Grozny, Russia	Charitschkoff, Chem. Centralbl., I, 409 (1904)
224–227 at	30	Trenton limestone, Ohio	Mabery and Palm, Proc. Am. Acad., **40**, 323–334 (1904)
		Grozny, Russia	Charitschkoff, Chem. Centralbl., I, 409 (1904)
237–240 at	30	Trenton limestone, Ohio	Mabery and Palm, Proc. Am. Acad., **40**, 323–334 (1904)
		Grozny, Russia	Charitschkoff, Chem. Centralbl., I, 409 (1904)
237–240 at	30	Ohio	Mabery and Palm, Proc. Am. Acad., **40**, 323 (1904)
272–274 at	50	Pennsylvania	Mabery, Proc. Am. Acad., **40**, 349 (1904)
290–294 at	50	Pennsylvania	Mabery, Proc. Am. Acad., **37**, 565–595 (1902)
310–312 at	50	Pennsylvania	Do.
316–318 at	50		Mabery, Proc. Am. Acad., **40**, 349 (1904)
328–330 at	50	Pennsylvania	Do.
342–344 at	50	Pennsylvania	Do.
366–368 at	50	Pennsylvania	Do.
380–384 at	50	Pennsylvania	Do.
		Grozny, Russia	Charitschkoff, Chem. Centralbl., I, 409 (1904)
200–205		Jennings, La.	Coates and Best, Jour. Am. Chem. Soc., **27**, 1317 (1905)
170–175		Bayou Bouillon, La.	Coates and Best, Jour. Am. Chem. Soc., **27**, 1318 (1905)
190–195		California	Mabery, Proc. Am. Acad., **40**, 340 (1904)
210–215		Anse la Butte, La.	Coates and Best, Jour. Am. Chem. Soc., **25**, 1158 (1903)

CYCLIC HYDROCARBONS

Name	Description	Specific gravity	Index of refraction	Melting p ° C.
		.8966 (27)	1.4883 (25)	33
	$C_{18}H_{32}$	.8996 (20)	1.4840 (20)	60
		.9006 (27)	1.4916 (25)	33
	$C_{19}H_{34}$	.9104 (28)	1.4972 (25)	33
	$C_{21}H_{38}$	.9163 (20)	1.4979	25
	$C_{23}H_{42}$	.8842 (20)	1.4797	−10
	$C_{24}H_{44}$	.8864 (20)	1.4802	30
		.9299 (20)		60
	$C_{25}H_{46}$	.8912 (20)	1.4810	30
		.9410 (20)	1.5152	
	$C_{27}H_{46}$	.9451 (20)	1.5146	
	$C_{29}H_{50}$	.9978 (20)		
Diphenyl	Colorless tablets	1.165		70.5
·	$C_6H_5—C_6H_5$	.9845 (82)		
Phenyl toluol (ortho)	$C_6H_5 \cdot C_6H_4CH_3$			
" " (meta)	"	1.031 (0)		
" " (para)	"	1.015 (27)		−2–3
Diphenyl methane	Colorless rhombic needles	1.001 (26)		26–27
	$(C_6H_5)_2CH_2$	1.0056 (25/25)		26.5
Dibenzyl	Colorle s monoclinic crystals	.995 (15)		52
	$C_6H_5 \cdot CH_2 \cdot CH_2 \cdot C_6H_5$	.9752 (50/50)		51.8
Ditolyl (m. m.)	$CH_3 \cdot C_6H_4 \cdot C_6H_4 \cdot CH_3$	....		
" (o. o.)		....		
" (o. m.)		.9993 (10/4)		
"				
(p. p.)	Monoclinic prisms from ether	.9172 (121)		121
"				
				103–105
	$C_{14}H_{14}$			
Fluorene	(C_6H_4) : CH_2; colorless leaflets			113–116
Diphenyl ethane	$CH_3 \cdot CH(C_6H_5)_2$	1.0033 (20/0)		
Phenyl ditoyl methane	$C_6H_5—CH(C_6H_4—CH_3)_2$			55–56
	Small prisms			
Piceneikosihydride	$C_{22}H_{34}$			
Dicaryophyllane	$C_{30}H_{50}$			144–145

NAPHTHENE SERIES—*Concluded*

Boiling point, ° C. Pressure, m.m.	Occurrence	Authority
190–195	Bayou Bouillon, La.	Coates and Best, Jour. Am. Chem. Soc., **27**, 1318 (1905)
210–215	California	Mabery, Proc. Am. Acad., **40**, 340 (1904)
200–205	Bayou Bouillon, La.	Coates and Best, Jour. Am. Chem. Soc., **27**, 1318 (1905)
225–230	Bayou Bouillon, La.	Do.
215–220	Heavy oil, Texas	Mabery and Buck, Jour. Am. Chem. Soc , **22**, 553 (1900)
253–255	Trenton limestone, Ohio	Mabery and Palm, Proc. Am. Acad., **40**, 323 (1904)
263–265	Do.	Mabery and Palm, Proc. Am. Acad., **40**, 323 (1904)
250–255	California	Mabery, Proc. Am. Acad., **40**, 340 (1904)
275–278	Trenton limestone, Ohio	Mabery and Palm, Proc. Am. Acad., **40**, 323 (1904)
270–275	Heavy oil, Texas	Mabery and Buck, Jour. Am. Chem. Soc., **22**, 553 (1900)
310–315	California petroleum	Mabery, Proc. Am. Acad., **40**, 340–346 (1904)
340–345	California petroleum	Do.
254.6		Handbook Chem. and Phys., 176 (1920–1921)
254.93 (cor.)		Van Nostrand's Chem. Ann., 288 (1918)
258–260		Van Nostrand's Chem. Ann., 348; Handbook Chem. and Phys , 226 (1920–21)
272–277		Do.
263–267		Do.
261–262		Handbook Chem. and Phys., 156 (1920–21)
264.7		Van Nostrand's Chem. Ann , 290 (1918)
284		Handbook Chem. and Phys., 168 (1920–21)
		Van Nostrand's Chem. Ann., 278 (1918)
280–281		Handbook Chem. and Phys., 176 (1920–21)
272		Do.
288		Van Nostrand's Chem. Ann., 290 (1918)
		Handbook Chem and Phys., 176 (1920–21)
		Van Nostrand's Chem. Ann., 290 (1918)
121	Synthetic	Handbook Chem. and Phys., 184 (1920–21)
	Synthetic	Ipatieff Ber., **41**, 997 (1908)
295		Handbook Chem. and Phys., 184 (1920–21)
286		Van Nostrand's Chem. Ann., 290 (1918)
		Van Nostrand's Chem. Ann., 346 (1918)
over 360		Liebermann and Spiegel, Ber., **22**, I, 781 (1889)
225–230 at (13)		Wallach, Terpene and Camphor, 576, Leipzig (1909)

DICYCLIC HYDROCARBONS

Name	Description	Specific gravity	Index of refraction	
Naphthalene	CH=CH—C—CH=CH | || | CH=CH—C—CH=CH			80
	Colorless monoclinic crystals	1.152 (15)		80
"		1.0070 (25/25)		80.05
Methyl naphthalene α	Colorless liquid	1.0010 (19)		−22
" "	$C_{10}H_7 \cdot CH_3$	1.0005 (19)		−22
" " β	Colorless monoclinic crystals	....		32.5
	$C_{10}H_7 \cdot CH_3$	....		32.5
Dimethyl naphthalene β	$(CH_3)_2 \cdot C_{10}H_6$	1.008		−20
1-4-Dimethyl naphthalene	$C_{12}H_{12}$	1.1803 (14/4)		−18
Ethyl naphthalene α	$C_{10}H_7—C_2H_5$	1.064 (15/15)		
				−14
" " β	$C_{10}H_7—C_2H_5$	1.008 (0)		−19
Propyl naphthalene				
Pseudopropyl naphthalene	$C_{10}H_7—C_3H_7$			
Naphthalene tetrahydride	$C_{10}H_8$	{ .9840 (0) .9660 (20) }	1.5402 (20)	
1-, 4-Dihydronaphthalene	$C_{10}H_8 \cdot H_2$			15–15.5
"		1.0031 (10/4)		−8 −7
"	$C_{10}H_{10}$	....		25
Tetrahydronaphthalene ε (p and o)	$C_{10}H_{12}$	.9810 (13)		
Tetrahydronaphthalene (a)	"	.9340 (23/0)		
	CH——CH—CH——CH | | | | CH—CH₂—CH CH—CH₂—CH	.9756 (35/4)	1.505 (35)	
Tetrahydronaphthalene		.8020 (0/0)		
	$C_{11}H_{12}$	....		
Acenaphthene	$C_{10}H_6(CH_2)_2$; Needles	1.0687 (15)		95
Naphthalene α-phenyl	$C_{10}H_7C_6H_5$	....		
	Colorless leaflets	....		102–102.
Tetrahydronaphthalene		.9530 (17)		
Dihydrodimethyl naphthalene	$C_{12}H_{16}$	.9306 (17.1)		

NAPHTHALENES

	Occurrence	Authority
18	Germany (?) California (?)	Engler and Höfer, Das Erdol, **1**, 370
18		Handbook Chem. and Phys., 212 (1920–21)
17.96 (cor).		Van Nostrand's Chem. Ann., 334 (1918)
40-242		Handbook Chem and Phys., 210 (1920–21)
40–242		Van Nostrand's Chem. Ann., 330 (1918)
42		Handbook Chem. and Phys., 210 (1920–21)
41.2		Van Nostrand's Chem. Ann., 330 (1918)
64–266		Van Nostrand's Chem. Ann., 286 (1918)
65–266		Jones and Wooten, Jour. Chem. Soc., **91**, 1149 (1907)
62–264		Van Nostrand's Chem. Ann., 286 (1918)
58 (slightly decomposes		Handbook Chem. and Phys., 180 (1920–21) Van Nostrand's Chem. Ann., 298
51		Handbook Chem. and Phys., 180 (1920–21)
bove 250	Russian oil	Markownikoff, Lieb. Ann., **234**, 110 (1886)
06		LerouX, Compt. rend., **139**, 673 (1904)
12		Handbook Chem. and Phys., 170 (1920–21)
12		Van Nostrand's Chem. Ann., 282 (1918)
94.5 at 17		Do.
05		Handbook Chem. and Phys., 298 (1920–21)
08–212		Van Nostrand's Chem. Ann., 362 (1918)
		Kraemer and Spilker, Ber., **29**, 558 (1896)
53–158		Markownikoff, J. Russk. Ph. Kh. Obsch., **15**, 237 (1883); **24**, 141 (1892)
50–255	Rumania { Possibly decomposition product	Edeleanu and Gans. Osterr. Chem. Technic. Zeitung, 156, 163 (1908)
77.5		Handbook Chem. and Phys., 142 (1920–21)
24–325		Handbook Chem. and Phys., 226 (1920–21)
45		Do.
41–246	Russian oil	Markownikoff and Ogloblin, Chem. Centralbl., 754 (1882)
38 at 770		Eijkman, Chem. Centralbl., II, 989 (1903)

CYCLIC HYDROCARBONS

Name	Description	Specific gravity	Index of refraction
Santene	C_9H_{14}	.8698	
Terpene (Monocyclic)	CH—CH₂ CH₃ ‖ | | CH₃—C CH—C | | ‖ CH₂—CH₂ CH₂	.8618	
	$C_{10}H_{16}$	.8507 (20/20)	
Citrene	"	{ .8446 { .8470	{ 1.4659 { 1.4700
"		.8595 (20)	
e-Limonene	"	.8460	1.4746
d- or *l*-Limonene	$C_{10}H_{16}$	.8530 (10)	
Hesperidene		.8441 (20/20)	
Dipentene	Optically inactive	.8450	1.47644
		.8500 (15)	
		.8390 }	1.467 }
		.8460 }	1.471 }
Dihydromyrcene		.7795 (18)	1.4485 (18/D)
Terpinolene	CH—CH₂ CH₃ ‖ | | CH₃—C C═C | | | CH₂—CH₂ CH₃	....	
Sylvestrene	$C_{10}H_{16}$	.8510 (16)	1.4780
		.8460	1.4746
"	"	.84924 (16/4)	
Terpinene	$C_{10}H_{16}$	.8650 (20)	
"		.8647 (20)	
"		.8550	
A-Terpinene	CH—CH CH₃ ‖ ‖ | CH₃—C C—CH | | | CH₂—CH₂ CH₃	{ .8460 { .8420	1.4789 1.4719
B-Terpinene	CH₂—CH CH₃ | ‖ | CH₂═C C—CH | | | CH₂—CH₂ CH₃	.8380	1.4754
A-Phellandrene	CH—CH₂ CH₃ ‖ | ╱H | CH₃—C C—CH | | | CH═CH CH₃	.8465 .8440 .8410 (22)	1.4880 (19) 1.4320 (19) 1.4760

MONOCYCLIC TERPENES

Boiling point, ° C. Pressure, m.m.	Occurrence	Authority
140		Van Nostrand's Chem. Ann., 358 (1918)
173–177		Markownikoff, Chem. Centralbl., I, 176 (1899)
169–171	California; oil of Rubieva multifida	Nelson, Jour. Am. Chem. Soc., **42**, 1286 (1920)
172–179		Brühl, Ber., **21**, 149–150 (1881)
176		Ipatieff, Ber., **43**, 3547 (1908)
175–177	P ne-needle oil	Wallach, Terpene and Camphor, 292–295, Leipzig, (1909)
176.5		Handbook Chem. and Phys., 202 (1920–21)
177.5 at 759		Van Nostrand's Chem. Ann., 322
178–180	Swedish and Russian turpentine. From plants, limonene, pinene, camphene	Wallach, Terpene and Camphor, 303, Leipzig (1909)
170–180		Ipatieff, Ber., **43**, 3551 (1908)
181–182		Van Nostrand's Chem. Ann., 288 (1918)
172–175		Brühl, Ber., **21**, 148 (1881)
70–71 at 25	Formosan lemon-grass oil	Kafuku, Jour. Chem. Ind., Tokyo, **20**, 825–33 (1917)
About 185	From plants; synthetic	Wallach, Terpene and Camphor, 311, Leipzig (1909)
183–185		Handbook Chem. and Phys , 238 (1920–21)
185–190		Wallach, Lieb. Ann., **230**, 254 (1885)
176–177	Plants; Swedish and Russian turpentine; synthetic	Wallach, Terpene and Camphor, 503, Leipzig (1909)
		Brühl, Ber., **21**, 154(1881)
176–177		Van Nostrand's Chem. Ann., 360 (1918)
179–182		Handbook Chem. and Phys., 236 (1920–21)
179–182		Van Nostrand's Chem. Ann., 360
179–182	Synthetic	Wallach, Lieb. Ann., **230**, 254–260 (1885)
179–181 174–179	Ceylon; oil of cardamon; plants	Wallach, Terpene and Camphor, 482, Leipzig (1909)
73–174	Plants	Wallach, Terpene and Camphor, 486, Leipzig, (1909)
65 at 12	Plants	Wallach, Terpene and Camphor, 499, Leipzig (1909)
61 at 11	Dry distillation of Elemi resin	Do.
75–176	Synthetic	Do.

CYCLIC HYDROCARBONS

Name	Description	Specific gravity	Index of refraction	
B-Phellandrene	CH=CH CH₃ / CH₂=C CH—CH / CH₂—CH₂ CH₃	.8520	1.4788 (20)	
Isoamyl-*a*-dihydro-phellandrene		.8679 (20)	1.49478 (20/D)	
Isoamyl menthane		.8250 (22)	1.45562 (22/D)	
Cyclene	$C_{10}H_{16}$	.8268		
Laurene and methene	"	.8690 to .8805	1.4597 to 1.4614	
Pinene	CH₃ / HC=C—CH / CH₂ / CH₃—C / H₂C—CH—CH₂	....		
		.8635		
		.8630-.8730	1.4620-1.4650	
		.8560		
		.8546		
Terpentine (pinene)	Oil; $C_{10}H_{16}$	.8587 (20/4)		
a-Pinene	$C_{10}H_{16}$	.8590 (20)		
"	"	.8580 (20)	1.4655 (21)	
		.8540 (25)		
"	"	.8587 (20/4)	1.4652 (20)	
b-Pinene	"	.8573 (20/4)	1.4641 (20)	
"	"	.8660	1.4724 (22)	
l-Pinene				
Hydropinene	$C_{10}H_{18}$	.8567 (20/4)	1.4605 (20)	
Camphene	CH₃ / CH₂—C—CH / CH₃—C—CH₃ / CH₂—CH—CH	....		48-48.5
		.8223		
"		....		48-49
		.8500	1.4581 (48)	49
		....		50
		....		47
	Feather needles	....		47
	$C_{10}H_{16}$	....		49.5-50
" (deXtro or laevo rotary)	Feather needles; from alcohol	....		51-52
		.8446 (50/4)		46.7
Terpene (bicyclic)		.8628 (20/4)		
Borneene	$C_{10}H_{16}$	....		
Bornylene	CH₂—CH—CH / CH₃—C—CH₃ / CH₂—C—CH / CH₃	....		97.5-98
		....		65-66

MONOCYCLIC TERPENES—*Continued*

Boiling point, ° C. Pressure, m.m.	Occurrence	Authority
57 at 11		Wallach, Terpene and Camphor, 501, Leipzig (1909)
130–132 at 11	Gum ammoniac oil	Semmler, Jonas and Roenisch, Ber., **50**, 1823–37 (1917); **also** Jour. Chem. Soc., **114**, I, 118–9 (1918)
131–133 at 14	Gum ammoniac oil	Do.
152.5		Eijkman, Chem. Centralbl., II, 1616 (1907)
173 to 175		Brühl, Ber., **21**, 157 (1888)
159–161		Wallach, Lieb. Ann., **230**, 272 (1888)
	Turpentine	Eijkman, Chem. Centralbl., II, 1209 (1907)
155–160		Brühl, Ber., **21**, 155 (1888)
155–156		Ba'hiano and Paolini, Ber., **35**, 2997 (1902)
155–156		Ipatieff, Ber., **43**, 3551 (1910)
156; **50** at 15		Van Nostrand's Chem. Ann., 360 (1918)
156		Handbook Chem. and Phys., 228 (1920–21)
155–156	Swedish and Russian turpentine	Wallach, Terpene and Camphor, 261, Leipzig (1909)
155–155.5 at 760		Zelinsky, Ber., **44**, 2783 (1911)
158.5–159.5	American turpentine	Wallach, Terpene and Camphor, 261, Leipzig (1909)
162–163	French turpentine	
159–161		Wallach, Lieb. Ann., **230**, 272 (1888)
167.5–168 at 748		Zelinsky, Ber., **44**, 2784 (1911)
158–159		Wagner and Brykner, Ber., **33**, 2122 (1900)
		Eijkman, Chem. Centralbl., II, 1209 (1907)
160–161	**Synthetic**	Wallach, Lieb. Ann., **230**, 234 (1888)
161	**From Bornylchloride**	Wallach, Terpene and Camphor, 567, Leipzig (**1909**)
160–161	**From Bornylamine**	Wallach, Terpene and Camphor, 568, Leipzig (**1909**)
156–157		Brühl, Ber., **21**, 158 (1888)
157		Handbook Chem. and Phys., 158 (1920–21)
		Van Nostrand's Chem. Ann., 270 (1918)
159		Handbook Chem. and Phys., 158 (1920–21)
159		Van Nostrand's Chem. Ann., 270 (1918)
163–165	Finnish turpentine	Aschan, Technikern, 3 (1918)
176–180	Synthetic	Wallach, Lieb. Ann., **230**, 237 (188)
149–150 at 750	Synthetic	Wagner and Brykner, Ber., **33**, 2123, 2124 (1900)
153		Godlenski and Wagner, Chem. Centralbl., I, 1055 (1897)

CYCLIC HYDROCARBONS.

Name	Description	Specific gravity	Index of refraction	
Terpene isomers	$CH_2-CH-C=CH_2$ CH_2 $CH_2-CH-C(CH_3)_2$	.794		
		....		
Fenchene	$C_{10}H_{16}$	.8690	1.4724 (19)	
		.8667 (18)		
Sabinene	$C_{10}H_{16}$	.8420	1.4678 (20)	
	"	.8400 (20)		
	"	.8420 (20)		
Sesquiterpene cardinene	$C_{15}H_{24}$	.9180 (20)	1.5065	
Sesquiterpene	"	.9187 (20/4)		
"	"	{ .9255	1.491 }	
		.9041	1.504	
Do. Caryophyllene	"	.9085	1.5009 (15)	
Do. Clovene	"	.9390	1.5009 (23)	
Do. Gaujene	"	.9100	1.5011 (20)	
Do. Tetrahydrocardinene	"	.8720	1.4740 (18)	
Sesquiterpene (tricyclic)	"	{ .9408 (15)	1.5031 (20/D)	
		.9398 (20)		
Do. Dicaryophyllene	$C_{30}H_{48}$			144–145
Tetrahydroferulene		.8400 (20)	1.4581 (20/D)	

ANTHRACENES

Name	Description	Specific gravity	Index of refraction	
Anthracene	$C_{14}H_{10}$	1.147		216.5
"	Colorless, leafless, monoclinic			
	$C_6H_4 : (CH)_2 : C_6H_4$	1.147		216.55 cor.
Iso-anthracene	Pearly leaflets; $C_{14}H_{10}$	....		133.5–134.5
α-Methyl anthracene	Colorless leaflets; $C_{14}H_9 \cdot CH_3$	1.063 (100)		100
		....		
β- " "	Colorless scales	....		
		....		280
2-, 3-Dimethyl anthracene	Colorless leaflets; $C_{14}H_8(CH_3)_2$	....		364
2-, 4-Dimethyl anthracene	Needles from alcohol; $C_{14}H_8(CH_3)_2$	....		
Ethyl anthracene	Leaflets from alcohol; $C_2H_5 \cdot C_{14}H_9$			

MONOCYCLIC TERPENES—*Concluded*

Boiling point, ° C. Pressure, m.m.	Occurrence	Authority
150	Sy thetic	Engler and Höfer, Das Erdöl, **1**, 346
		Eijkman, Chem. Centralbl., II, 1209 (1907)
146 at 750		Van Nostrand's Chem. Ann., 266 (1918)
155–158	Synthetic, from fenchel oil and thuja oil	Wallach, Terpene and Camphor, 549, Leipzig (1909)
153		Engler and Höfer, Das Erdöl, **1**, 346
155–156		Van Nostrand s Chem. Ann., 302 (1918)
163–165	From sad-tree oil, Ceylon; Cardamon oil, oil of Majgran	Wallach, Terpene and Camphor, 507, Leipzig (1909)
162–166		Handbook Chem. and Phys., 234 (1920–21)
165		Van Nostrand's Chem. Ann., 356 (1918)

TERPENES

Boiling point, ° C. Pressure, m.m.	Occurrence	Authority
274–275	Plants, synthetic	Wallach, Terpene and Camphor, 574 (1909)
260–263	Finnish turpentine	Aschan, Finska Kem. Medd., 3 (1918)
250–260		Brühl, Ber., **21**, 148 (1888)
258–260	Carnation and copaiba oils	Wallach, Terpene and Camphor, 574 (1909)
254–256	Patchouli camphor	Wallach, Terpene and Camphor, 575 (1909)
124–128 at 13		Do.
230–257		Wallach, Terpene and Camphor, 576 (1909)
92–93 at 2.5	Pine resin from Wenchow, China	Shinosaki and Ono, Jour. Chem. Ind., Tokyo, **23**, 45–56 (1920)
225–230 at 13		Wallach, Terpene and Camphor, 576 (1909)
118–120 at 10	Gum ammoniac oil	Semmler, Jonas and Roenisch, Ber., **50**, 1823–37 (1917); also Jour. Chem. Soc., **114**, I, 118–9

ANTHRACENES

Boiling point, ° C. Pressure, m.m.	Occurrence	Authority
351	Russian pet. residue; (Nobel Bros.)	Handbook Chem. and Phys., 150 (1920–21)
{ 360	American pet. residue	} Van Nostrand's Chem. Ann., 260 (1918)
{ 103–104 at (0)		} Maher, Proc. Am. Acad., **31**, 8 (1894)
		Van Nostrand's Chem. Ann , 316 (1918)
340		Prunier and David, Compt. rend., **87**, 991 (1878)
199–200		Handbook Chem. and Phys., 206 (1920–21)
199–200		Do.
	Petrocene	Sadtler and McCarter, Jour. Am. Chem. Soc., **1**, 114 (1879)
518–520		Handbook Chem. and Phys., 210 (1919)
246		Handbook Chem. and Phys., 172 (1920–21)
71		Do.
60–61		Van Nostrand's Chem. Ann., 292 (1918)

AHTHRACENES

Name	Description	Specific gravity		Index of refraction	
Trimethyl anthracene					
1 : 2 : 4	$(CH_3)_3C_{14}H_7$				243
1 : 3 : 6	"				222
1 : 4 : 6	Fluorescent leaflets; $(CH_3)_3C_{14}H_7$				227
Dihydro-anthracene	$C_6H_4 : (CH_2)_2 : C_6H_4$				108.5
Hexahydro-anthracene	Colorless leaflets				63
Phenanthrene	$C_{14}H_{10}$	1.063	(100)		100
"					145
Methyl phenanthrene					90–95
Octahydrophenanthrene	C_6H_8—CH | || C_6H_8—CH				
Retene	Leaflets; $C_{18}H_{18}$	1.1300	(15)		98.5
	" from alcohol; $C_{18}H_{18}$	1.1300	(15)		98.5
Leucacene	Voluminous and silky.				250
Rhodacene	Dark-violet crystal mass with greenish metallic luster				338–340
Chalcacene	BordeauX red, flat needles or tables with bronze luster				358–360
Stilbene	Colorless tablets				124
Sym-diphenyl ethylene	$C_6H_5 \cdot CH : CH \cdot C_6H_5$				
" "	Monoclinic	.9707	(119)		124–125
	$C_{14}H_{12}$				105–106
					135–137
Tolane	Monoclinic; $C_6H_5C \vdots C \cdot C_6H_5$				60
Fluoranthene	$C_{15}H_{10}$; Colorless monoclinic				109–110
Diphenyl diacetylene	Needles from dilute alcohol				96
Pyrene	$C_6H_5 \cdot C \vdots C \cdot C \vdots C \cdot C_6H_5$; monoclinic tablets				148–149
"	$C_{16}H_{10}$				148.9
Diphenyl benzol (para)	Colorless leaflets; $C_6H_5 \cdot C_6H_4 \cdot C_6H_5$				205
Diphenyl fulvene	CH : CH\\ /C_6H_5 | >C : C< CH : CH/ \\C_6H_5				82
Methyl phenyl fulvene	CH : CH\\ /CH_3 | >C : C< CH : CH/ \\C_6H_6				
Triphenyl methane	$(C_6H_5)_3CH$	1.0570	(95/95)		92
	Colorless leaflets	1.0568	(95/95)		
Diphenyl-*m*-tolyl methane	$(C_6H_5)_2$—CH·CH—CH_3	1.0700	(16)		60.5–61.5
	Irregular prisms				
	Monoclinic leaflets				
Triphenyl ethane	$C_6H_5 \cdot CH_2 \cdot CH(C_6H_5)_2$	....			54
Phenyl ditolyl methane	$C_6H_5 : CH(C_6H_4 \cdot CH_3)_2$; Needles	....			55–56
Chrysene	$C_{18}H_{12}$	....			250
"	"	....			
Phenyl anthracene	C_6H_5—$C_{14}H_9$; Leaflets from alcohol	....			152–153
xx-Dinaphthyl	$C_{10}H_7$—$C_{10}H_7$; Colorless tablets				154
"					160.5
bb-Dinaphthyl	Colorless prisms from alcohol				187
	$C_{10}H_7$—$C_{10}H_7$				
a-Dinaphthyl methane	$(C_{10}H_7)_2CH_2$; Prisms from alcohol				109
b-Dinaphthyl methane	$(C_{10}H_7)_2CH_2$; Fine needles				92
Picene	Colorless leaflets				345
"	$C_{22}H_{14}$				280
"					364

ANTHRACENES

Boiling point, ° C. Pressure, m.m.	Occurrence	Authority
		Van Nostrand's Chem. Ann., 370 (1918)
		Do.
		Van Nostrand's Chem. Ann., 372 (1918)
313		Handbook Chem. and Phys., 170 (1920–21)
290		Handbook Chem. and Phys., 190 (1920–21)
340		Handbook Chem. and Phys., 224 (1920–21)
		Krämer and Spilker, Ber., **33**, 2267 (1900)
299 at 766 167.5 at 13		Schroeter, Brennstoff, Chem., **1**, 39–41 (1920)
390	From petrocene	Handbook Chem. and Phys., 234 (1920–21)
135 at 0°		Van Nostrand's Chem. Ann., 356 (1918)
		Dziewonski, with Podgörska, Lemberger and Suszka, Ber., **53**, 2173–92 (1920)
		Do.
		Do.
306–307		Handbook Chem. and Phys., 236 (1920–21)
306–307		Van Nostrand's Chem. Ann., 358 (1918)
		Ipatieff, Ber., **41**, 999 (1908)
275–300		Van Nostrand's Chem. Ann., 366 (1918)
217 at 30		Van Nostrand's Chem. Ann., 302 (1918)
		Van Nostrand's Chem. Ann., 290 (1918)
		Handbook Chem. and Phys., 232 (1920–21)
Above 360		Van Nostrand's Chem. Ann., 354 (1918)
383		Handbook Chem. and Phys., 176 (1920–21)
		Thiele, Ber., **33**, 672 (1900)
130.5 at 10.5		Prunier and David, Compt. rend., **87**, 991 (1878)
358–359		Handbook Chem. and Phys., 230 (1919)
		Van Nostrand's Chem. Ann., 374 (1918)
354 at 706		Van Nostrand's Chem. Ann., 290 (1918)
348–349 760		Van Nostrand's Chem. Ann., 372 (1918)
		Handbook Chem. and Phys., 226 (1920–21)
448 at 760	From petrocene	Van Nostrand's Chem. Ann., 274 (1918)
436		Graebe and Walter, Ber., **14**, 177 (1881)
417		Van Nostrand's Chem. Ann., 346 (1918)
About 360		Handbook Chem. and Phys., 174 (1920–21)
		Van Nostrand's Chem. Ann., 288 (1918)
		Handbook Chem. and Phys., 174 (1920–21)
Above 360		Van Nostrand's Chem. Ann., 288 (1918)
		Do.
520	California residuum	Graebe and Walter, Ber., **14**, 177 (1881)
	Petrocene	Sadtler and McCarter, Am. Chem. Jour., **1**, 114 (1879)
518–520		Handbook Chem. and Phys , 228 (2910–21)

Name	Description	Specific gravity	Index of refraction	Melting point, ° C.
Crackene	$C_{24}H_{18}$	....		308
(Probably homologous with picene)	(Decomposing on dis. at 500° C.)			
(Probably same as benzerythrene)				
(Probably identical with "crackene")	$C_{24}H_{18}$	....		280–285
1-, 3-, 5-Triphenyl benzol	Rhombic tablets from ether	1.2060 (15)		169–170
" " "	$C_6H_3(C_6H_5)_3$	1.2055		
Tetraphenyl ethane	Colorless rhombic needles from chloro-	1.1820 (15)		209
(symmetrical)	form; $(C_6H_5)_2CH \cdot CH(C_6H_5)_2$	....		211
Tetraphenyl ethylene	Colorless monoclinic			221
" "	$(C_6H_5)_2C : C(C_6H_5)_2$	....		223.5–224.5
		1.0960 (15)		About 120
Dianthryl	Leaflets; $C_{28}H_{18}$	....		300
Carbopetrocene	$C_{24}H_8$			268
Petrocene	$C_{32}H_{22}$			Above 300
				300
				300
Fluoranthrene	Colorless monoclinic crystals	....		109–110
"	$C_{15}H_{10}$	....		109–110
"	"	....		250.5
		....		221.5

CYCLIC HYDROCARBONS,

Name	Description	Specific gravity	Index of refraction	° C.
Dimethyl fulvene	$\begin{matrix} CH{=}CH \\ \vert \\ CH{=}CH \end{matrix} \!\!>C{=}C<\!\! \begin{matrix} CH_3 \\ CH_3 \end{matrix}$	.8858 (17/4)		

Boiling point, ° C. Pressure, m.m.	Occurrence	Authority
	Galicia	Klaudy and Fink, Monatsb. Chem., **21**, 118–136 (1900)
	Japan residuum	Divers and Nakamura, Jour. Chem. Soc., **47**, 925 (1885)
		Handbook Chem. and Phys., 248 (1920–21)
Dist.		Van Nostrand's Chem. Ann., 372 (1918)
279–283		Handbook Chem. and Phys., 240 (1920–21)
379–383		Van Nostrand's Chem. Ann., 364 (1919)
415–425		Handbook Chem. and Phys., 240 (1920–21)
415–425		Van Nostrand's Chem. Ann., 364 (1919)
	From petrocene	Prunier and David, Compt. rend., **87**, 991 (1879); Ann. Chim. Phys., Series 5, **16**, 28 (1879)
		Van Nostrand's Chem. Ann., 264 (1918)
	From petrocene	Prunier and David, Compt. rend., **87**, 991 (1879); Ann. Chim. Phys., Series 5, **16**, 28 (1879)
	From Pennsylvania residuum	Hemilian, Ber., **9**, 1604 (1876)
	American petroleum	Hemilian, Chem. Centralbl., 2 (1877)
		Zaloziecki and Gans, Chem. Zeitung, 536 (1900)
217		Van Nostrand's Chem. Ann., 302 (1918)
		Handbook Chem. and Phys., 184 (1920–21)
	From Russian and Galician petroleum residues	Zaloziecki and Gans, Chem. Zeitung, 535, 553 (1900)
	In Russian and Galician résidues	Do.

DIOLEFINES

Boiling point, ° C. Pressure, m.m.	Occurrence	Authority
153–154 717 46 11		J. Thiele, Ber., **33**, 671–672 (1900)

Name	Description	Specific gravity		Index of refraction	
	Colorless crystals				
Cyclopentadiëne	HC=CH╲ │ CH₂ → $HC=CH$ > CH_2 HC=CH╱	1.012	(17/5)		32.9
		.8047	(19/4)		
		.8150	(15)		
Dicyclo-pentadiëne	$(C_5H_6)_2$				
Dicyclo-octadiëne	C_8H_{12}	....			114
b-Cyclo-octadiëne	"	.8870			
a- "	"	.8840			
Cyclo-octadiëne	$CH_2 \cdot CH : CH \cdot CH_2$	.8890	(0/4)		
"	· ·	.8564	(20.7)		
	$CH_2 \cdot CH : CH \cdot CH_2$				
Dicyclo-octene	C_8H_{12}	.9097	(0/4)		
Cyclo-octene	C_8H_{14}	.8550	(20/4)	1.4360–1.4739	14.5
	"	.7984	(20/4)	1.4461	
	"	.7747	(24/4)	1.4260 (25)	
	" optically inactive	.7992	(24/4)	1.4370	
		.7669	(16/4)	1.4216 (16)	
Dicyclo-octane	C_8H_{14}	.8775	(0/4)	1.4615 (20)	
		.8604	(20/0)		
	Yellow liquid				
Dicyclododecatriëne	$CH_2—CH=CH—CH_2$ │ │ $CH—CH=CH—CH$ │ │ $CH_2—CH=CH—CH_2$	.9764	(20.7)		

SATURATED DICYC

Name	Description	Specific gravity		Index of refraction
Dicyclononane	$CH_2—CH_2—CH—CH_2$ │ │ │ │ │ CH_2 │ │ │ $CH_2—CH_2—CH—CH_2$	.8759	(23)	1.4667
	$C_4H_6=C(C_2H_5)_2$	.8092	(20)	1.4510 (20)
Decahydronaphthalene (Dicyclo-o, 4-, 4-decane)	$H_2C—CH_2—CH—CH_2—CH_2$ │ │ │ $CH_2—CH_2—CH—CH_2—CH_2$			1.4675
		.8930	(0)	
		.8770	(20)	
Dihydropinene (Pinane) (Trimethyl 2, 7-, 7 bi-cyclo-1-, 1-heptane)	$CH—CH—CH_3$ CH_3 ╲ │╲ │ > C CH │ CH_3 ╱ │╱ │ $CH—CH_2$	.8620	(0/4)	

DIOLEFINES

Boiling point, ° C.	Pressure, m.m.	Occurrence	
			Roscoe, Jour. Chem. Soc., **47**, 669–671 (1885)
42.5			Van Nostrand's Chem. Ann., 278 (1918)
41	760	From coal tar	Kramer and Spilker, Ber., **29**, 544–560 (1896)
		Polymerization product from cyclopentadiëne	Engler and Höfer, Das Erdöl, **1**, 346
			Willstätter and Veraguth, Ber., **40**, 964 (1907)
143–144			Do.
			Do.
135–150			Willstätter and Veraguth, Ber., **38**, 1979 (1905)
50–52	17		Doebner, Ber., **35**, 2134 (1902)
137.5–139			Willstätter and Veraguth, Ber., **40**, 966 (1921)
145–150			Willstätter and Wasser, Ber., **43**, 1178–1180 (1910)
124–125.5	760		Zelinsky and Goutta, Bull Soc. Chem., **4**, 1562
120–125			Engler and Höfer, Das Erdöl, **1**, 341
120.5		Jennings, La.	Coates, Jour. Am. Chem. Soc., **28**, 384 (1906)
120.5–121			Zelinsky, Ber., **35**, 2679 (1902)
139.5–140.5			Willstätter and Kametaka, Ber., **41**, 1485 (1908)
92–95			Doebner, Ber., **35**, 2135 (1902)

HYDROCARBONS

Boiling point, ° C.	Pressure, m.m.	Occurrence	
163–164		Synthetic	Eijkman, Chem. Centralbl., II, 1989 (1903)
			Kishner and Amosow, Chem. Centralbl., II, 816–817 (1905)
			Leroux, Compt. rend., **139**, 673 (1904)
187–188		Synthetic	Ipatieff, Ber., **40**, 1287 (1907)
189–191			
166		Synthetic	Sabatier and Senderens, Compt. rend., **132**, 1254 (1901)

SATURATED DICYCLIC

Name	Description	Specific gravity		Index of refraction	Melting point, ° C.
Methyl-1-bicyclo-1-3-3-nonane	CH_3 | CH_2—C—CH_2 | | | CH_2 CH_2 CH_2 | | | CH_2—CH—CH_2	.8416	(20/4)	1.4929	
Camphane (Trimethyl-1-7-7-bicyclo-1-2-2-heptane)	CH_3 | CH_2—C—CH_2 | | | | —C—CH_3 | | CH_3 | | CH_2—CH—CH_2 C_8H_{14}< CH_2 | CH_2				152 153–154

SATURATED POLYCYCLIC

Name	Description	Specific gravity		Index of refraction	Melting point, ° C.
Fichtelite (Perhydrorecene)	$(C_4H_7)n$				46
Do.	$C_{12}H_{22}$				45
Do.					39
Do.	$C_{18}H_{32}$				46
Retendodecahydride	$C_{18}H_{30}$				
α-Tricyclodecane	$C_{10}H_{16}$	.9492 .9128	 (79)	 1.4726	 77
b-Tricyclodecane	CH_2—CH—CH—CH_2 / | | \\ CH_2 | | CH_2 \\ | | / CH_2—CH—CH—CH_2	.9021 .9049	(80) (60)	1.4931	9
Perhydroacenaphthene	CH_2——CH_2 | | CH_2—CH —CH—CH —CH_2 | | | CH_2—CH_2—CH—CH_2—CH_2				
Perhydrofluorene	CH_2—CH_2—CH—CH—CH_2—CH_2 | | | | CH_2—CH_2—CH CH CH_2—CH_2 \\ / CH_2				
Perhydroanthracene	CH_2—CH_2—CH—CH_2—CH—CH_2—CH_2 | | | | CH_2—CH_2—CH—CH_2—CH—CH_2—CH_2				88–89
Perhydrophenanthrene	CH_2—CH_2—CH—CH—CH_2—CH_2 | CH_2—CH_2—CH CH—CH_2—CH_2 | | CH_2—CH_2	.9330	(20)		−3
Perhydropicene	$C_{22}H_{36}$				175

HYDROCARBONS

Boiling point, ° C. Pressure, m.m.	Occurrence	Authority
176–178 751	Synthetic	Rabe, Ber., **37**, 1674 (1904)
157–158		Aschan, Ber., **33**, 1008 (1900)
160–162	Synthetic	Aschan, Lieb. Ann., **316**, 238 (1901)

HYDROCARBONS

Above 320	Pine needles	Clarke, Lathan, Lieb. Ann., **103**, 236
Above 350		Hell, Fehling's Wörterb., III, 251
		Schrötter, Fehling's Wörterb., III, 252
355.2 at 719	In peat bogs	Aschan, Chemie der Alicyclischen Verbindungen, 1107
336	Synthetic (from re-cene)	Liebermann and Spiegel, Ber., **22**, 780 (1889)
		Eijkman, Chem. Centralbl., II, 1209 (1907)
	Synthetic	Engler and Höfer, Das Erdöl, **1**, 305
	Synthetic	Do
235–236	Synthetic	Liebermann and Spiegel, Ber., **22**, 781 (1889)
230 ;		Do
266–276	Synthetic	Ipatieff, Ber., **41**, 998–999 (1908)
	Synthetic	Liebermann and Spiegel, Ber., **22**, 779 (1889)
360	Synthetic	Liebermann and Spiegel, Ber., **22**, 781 (1889)

The subject of asphalt is treated in a chapter devoted especially to that subject. It should be pointed out here, however, that asphalt differs from petroleum oil by containing oxygen in addition to carbon and hydrogen. In many cases this oxygen is partially or even entirely replaced by the allied element, sulphur. In either case the customary treatment of asphalt in petroleum is to remove it, as such, by distilling off the other oils and obtaining the asphalt as a large proportion of the residuum. It then receives various kinds of treatment to suit it for the special uses which are described in the chapter on Asphalt. This is true of asphalts, whether they merely contain oxygen or whether they are sulphur asphalts. It is only within the last few years that successful results have been accomplished in the breaking up of asphalts into light petroleum products by cracking processes. By this means light oils, useful for gasoline and other products, are obtained. They are far more valuable than asphalt. A complex residue is left composed partly of undecomposed asphalt and partly of comminuted coke, in varying proportions.

Oxygen compounds.—The table below gives the total percentage of oxygen found in typical crude petroleums. Some of the data may include sulphur:

Oxygen in Petroleum *

Locality	Per cent
Pennsylvania	
Oil Creek	3.2
Allegheny	1.4
West Virginia	
Guthry Well, Rogers Gulch	3.6
White Oak	3.2
Burning Springs	1.6
Ohio	1.4
Canada	
Bothwell	2.3
Petrolia	2.0
East Galicia	5.7
West Galicia	2.1
Rumania	4.9
Burma (Rangoon)	3.5
China (Foo-choo-foo)	3.6
Java (Tjibodas-Fanggah)	2.4
France, Pechelbronn, Alsace	4.5

* The figure for oxygen may also include some sulphur and nitrogen.

Resins.—The occurrence of resins as important ingredients of some petroleums has only recently been recognized, although such resins have been casually identified in petroleum for many years. During 1919 and 1920 the writer was led to a consideration of the resins in California and other oils. He found that the oil shales of Elko, Nevada, contain a waxy material which was at first considered to be entirely natural paraffin wax or ozokerite. Investigation showed that this waxy substance is easily divided, by solution and precipitation from hot alcohol, into true paraffin wax and a white material easily identified as resinous in its character. The resin crystallizes from petroleum ether in fine needles which melt between 160° and 165° F. and are easily soluble by boiling with caustic soda. This forms a soap solution from which an acid substance is precipitated by acids. The organic matter in the California oil shales was also studied and it was found that a similar resin could be isolated in the same way. In this case the resin proved to be a very sticky, viscous liquid, deep red in color. On being heated in a glass flask, this Monterey-California shale resin broke up into water and a succession of oils which showed the usual fractions. It was noted that very little water was driven out from this material until considerably above the boiling point of water. It has long been known that the petroleum produced by the ordinary well-drilling methods in the Casmalia oil-fields is always very heavy and frequently sinks in water. Examination showed that in some cases one-third of the entire oil consisted of resinous matter very similar to the resin extracted from the nearby Monterey shale, if not identical with it. It is well known that this petroleum yields its water with great difficulty. Frequently the oil of this region will contain 25 per cent of water. It is manifest from the amount of resinous material contained in it that much of this water is chemically combined in the form of resins dissolved in the oil. It is altogether probable that this resinous material will be worthy of industrial consideration.

Sulphur.—Sulphur occurs, at least in traces, in practically all petroleums, but there is a trade distinction between those in which the sulphur is negligible, such as the standard Pennsylvania petroleum, and those which require special refining processes for eliminating the sulphur, such as the oils of Mexico and coastal Texas. Other important examples of the latter class are the petroleums of northwestern Ohio and Indiana and of Ontario, Canada. The following table has been modified from that given in Engler and Höfer's treatise.[1]

Sulphur Content of American Petroleum

Locality	Sulphur, per cent	Authority
Pennsylvania		
General	0.0445	Kissling, Chem. Zeitung, p. 371, 1896
Washington District	0.0468	" "
Warren District	0.0343	" "
Residuum	0.60	Richardson, C., Jour. Frankl. Inst., **162**, p. 68, 1906
Ohio		
General	0.70	Mabery, C. F., Am Chem. Jour., **18**, p. 57
County		
Auglaize, St. Mary's Township	0.61	Mabery, C. F., Proc. Am. Acad., **21**, pp. 12–14
Hancock, Liberty Township	0.71	
Heilstone Oil Co	0.63	
Langemade Well	0.68	
Peerless Rfg. Co	0.68	
Wood, Montgomery Township	0.37	
Ohio Oil Co	0.56	
Peerless Rfg. Co	0.61	
Liberty Township	0.76	
Sandusky, Woodville Township	0.49	
Wells, Nottingham Township	0.56	
Allen, Lima	0.81	
Miscellaneous		
Welker	0.48	

[1] Vol. I, p. 468.

Sulphur Content of American Petroleum—Continued

Locality	Sulphur, per cent	Authority
Macksburg field................	0.025	Lord, N. W. U. S. Geol. Survey, Ann. Rept., p. 624, 1886–1887
Trenton limestone, N.W. Ohio...	0.535	
Indiana...........................	0.3619	Kissling, Chem. Zeitung, p. 371, 1896
West Virginia.....................	0.0456	" "
Texas		
Texas oil (0.9500 sp. gr.).........	0.94	Mabery, C. F., Jour. Am. Chem. Soc., **22,** 1900
Beaumont......................	1.96	Coates, C. E., and Best, A., Jou . Chem. Soc., **25,** p. 1154, 1903
Beaumont......................	1.75	Richardson, C., Jour. Frankl. Inst., **162,** p. 113
Beaumont......................	2.16	Mabery, C. F., Am. Chem. Soc., **23,** p. 265
Higgins.......................	2.4	Palm, O. H., ref. cited by A. F. Lucas, Cong. Int. Petrol., Bukarest, p. 353, 1907
Lucas Well....................	2.04	ditto
Louisiana		
Welsh..........................	0.32	Coates, C. E., and Best, A., op. cit.
Jennings......................	0.39	ditto
Breaux Bridge..................	0.26	ditto
California		
Fresno County..................	0.21	Mabery, C. F., Am. Chem. Jour., **19,** p. 802, 1897
Santa Barbara County		
Summerland.................	0.84	Mabery, C. F., Proc. Am. Acad. A. Sci., **40,** p. 341, 1904
Los Angeles County		
Puente Hills.................	0.80	Mabery, C. F. Am. Chem. Jour., **25,** pp. 253–285, 1901
Ventura County		
Bardsdale....................	1.5	ditto
Torrey Wells.................	0.49	ditto
Sespe........................	0.38	ditto
Adams Canyon.................	0.92–0.87	ditto
Canada		
Petrolia District................	0.98	Mabery, C. F., Am. Chem. Jour., **18,** p. 56, 1896
Petrolia District................	0.99	ditto
Petrolia District................	1.06	ditto
Oil Springs....................	0.60	ditto
Gaspé Peninsula................	0.09–0.20	Redwood, Sir B., Petroleum and its Products, **1,** p. 63

Sulphur in Canadian Distillates

Fraction:	115°–150°,	150°–200°,	200°–250°,	250°–300°,	300°–350°,	Residuum.
Sulphur, per cent:	0.28	0.42	0.50	0.51	0.86	0.70

Fraction:	To 100°,	100°–150°,	150°–200°,	200°–250°,	250°–300°,	300°–350°,	Residuum.
Sulphur, per cent:	0.25	0.45	0.47	0.75	0.78	0.81	0.83

Authority: Mabery, C. F., Am. Chem. Jour., **18**, p. 59, 1896.

Sulphur occurs in petroleum in three or more easily differentiated conditions. (*a*) Elementary sulphur: Crude petroleums will dissolve sulphur in considerable percentages when hot, and at a temperature of 70° F. at least 4 per cent will remain in solution. Sulphur is found in this elementary condition in many petroleums, especially in Mexico. (*b*) Hydrogen sulphide: This is also recognized in petroleum in various localities, especially in the northern oil-fields of Mexico and even more markedly in certain petroleums of the Gulf Coast of the United States. (c) Sulphur-bearing asphalts: These have been referred to above and are very common in Mexican oils. (*d*) Organic compounds of sulphur: Reference is here made to such compounds of sulphur as those found in the oils of northwestern Ohio and Indiana. These oils contain such fluid compounds as mercaptans consisting of hydrogen, carbon and sulphur. The removal of the sulphur from these compounds is accomplished with great difficulty, and requires such special processes as that of Frasch, in which the sulphur is removed by very intimate mixture of the oil with a copper compound, such as copper oxide. The details of these methods of treating sulphur oils will be referred to in the chapter on Refining.

Hydrogen sulphide is found in many crude oils when they are taken directly from the wells, but it rapidly disappears when the oil is exposed to the air. The quantity of hydrogen sulphide evolved with the oil of the Texas Coastal Plain has frequently been so great as seriously to " gas " the workmen, and it has been necessary to expel it by blowing steam through the oil tanks. Such hydrogen sulphide must be carefully distinguished from that noticed in oil distillation where hydrogen sulphide is frequently produced by the action on the oil itself of sulphur in organic sulphur compounds contained in the oil.

Thiophanes ($C_nH_{2n}S$).—Thiophanes are sulphur compounds which can be detected by such a delicate test as the isatine reaction; they have been detected in many oils. These compounds form a considerable part of the organic sulphur compounds which render difficult the refining of Canadian oils, as shown by the classic work of Mabery and of Mabery and Smith.[1] These authors have found the following members of this series as shown on page 528.

Mercaptans have been identified in Baku petroleum, according to Kwjatkowsky.[2]

Many other sulphur compounds have been actually identified in petroleum products, or their presence therein has been indicated, but it is questionable whether they are actual constituents of crude petroleum or are products of decomposition in the course of distillations. Much has been written concerning the possible origin of the sulphur in petroleum, but such theoretical considerations have no place in a book of this scope. The same may be said of the theory that the percentage of sulphur increases with the depth at which the oil is found and with the gravity of the oil. Neither of these theories is sufficiently accurate for discussion. It is, in general, true that the percentage of sulphur is greater in heavy distillates than in lighter ones.

[1] The publications by these authors on this subject are many. Consult Proc. Am. Acad., **41**, 89, 1905, and Am. Chem. Jour., **35**, 404, 1906.

[2] Kwjatkowsky, Anleitung z. Verarb. d. Naphta, 12, 1904.

Thiophanes in Canadian Petroleum

Name	Boiling Point, °C.		Specific gravity, (20°)	Index of refraction	Remarks
	At atmospheric pressure	(In vacuo)			
Hexylthiophane ($C_6H_{12}S$)	125-130	55–57	......	......	Not obtained pure
Hexylthiophane-sulphone ($C_6H_{12}SO_2$)	......	......	......	......	Very thick, sweet-smelling oil
Heptylthiophane ($C_7H_{14}S$)	158–160	74–76	0.8878	1.468	
Heptylthiophane-sulphone ($C_7H_{14}SO_2$)	......	......	1.1138	......	Very thick, heavy oil
Octylthiophane ($C_8H_{16}S$)	167–169	81–83	0.8929	1.4860	
Octylthiophane-sulphone ($C_8H_{16}SO_2$)	......	......	1.1142		
Iso-octylthiophane ($C_8H_{16}S$)	183–185	94–96	0.8937		
Nonylthiophane ($C_9H_{18}S$)	193–195	106–108	0.8997	1.4746	
Nonylthiophane-sulphone ($C_9H_{18}SO_2$)	......	......	1.1161		
Decylthiophane ($C_{10}H_{20}S$)	207–209	114–116	0.9074	1.4766	No sulphur could be isolated by oxidation
Undecylthiophane ($C_{11}H_{22}S$)	......	128–130	0.9147	1.480	
Undecylthiophane-sulphone ($C_{11}H_{22}SO_2$)	......	......	1.1126	......	Pleasant, sweet odor
Dodecylthiophane-sulphone ($C_{12}H_{24}SO_2$)	......	......	1.1372	......	From the fraction boiling, in vacuum, at 142°–144°, C. the sulphur compound was obtained by oxidation
Quartdecylthiophane ($C_{14}H_{28}S$)	266–268	168–170	0.9208	1.4892	
Sexdecylthiophane ($C_{16}H_{32}S$)	283–285	184–186	0.9222	1.4903	
Octodecylthiophane ($C_{18}H_{36}S$)	290–295	198–202	0.9235	1.4903	

Alkyl Sulphides Found in Ohio Oil

Name	Formula	Boiling Point	
		At atmospheric pressure, °C.	Pressure at 100 m.m. °C.
Methyl sulphide	$(CH_3)(CH_3)S$	100–120	Under 50
Ethyl sulphide	$(C_2H_5)(C_2H_5)S$	135–140	88–92
Ethylpropyl sulphide	$(C_2H_5)(C_3H_7)S$	....	110–112
Normal propyl sulphide	$(C_3H_7)(C_3H_7)S$	....	127–132
Ethyl pentyl sulphide	....	156–160	95–100
Isobutyl sulphide	....	170–176	110–115
Normal butyl sulphide	....	180–185	117–125
Butyl pentyl sulphide	....	185–190	135–140
Pentyl sulphide	....	205–210	150–155
Hexyl sulphide	....	225–235	160–170

Sulphur, like oxygen, reacts very slightly on the lighter paraffin hydrocarbons, even when heated. Engler and Spanier have shown [1] that hexane is not attacked by sulphur at the boiling point, and that there is only a perceptible evolution of hydrogen sulphides at 210° C. It is well known that paraffin wax heated with sulphur yields hydrogen sulphide abundantly, at first even violently. This reaction has frequently been used as a method of obtaining hydrogen sulphide. Acetylene is acted upon by sulphur, with the formation of hydrogen-sulphide, carbon disulphide and thiophanes.[2]

Markownikoff and Spady [3] have been able to remove hydrogen from octonaphthene by means of sulphur with the formation of metaxylol, etc. Charitschkoff [4] has been able to remove hydrogen from hydrocarbons, step by step, leaving coke only.

Nitrogen.—The following table gives the percentage of nitrogen found in various petroleums. The forms of combination of this nitrogen have not been completely determined. Studies by Peckham, Mabery, Engler and many others have shown the presence of nitrogen bases having the characteristic odor of the pyridine series, and several authorities have isolated pyridine-like bases which have then been quite carefully examined. The writer has also detected similar bases in oil extracted by solvents from Utah oil shales. The occurrence of other nitrogen compounds in petroleum is evident, for none of the investigators has accounted for the total nitrogen in various oils by the amount of pyridine bases extracted, and Mabery has shown that the total nitrogen in California oil would be equivalent to 10 to 25 per cent of pyridine bases of the average composition $C_{10}H_{13}N$. The investigations of Engler and MacGarvey[5] make it probable that hydrogenated pyridine bases are also contained in German petroleum.

Nitrogen Content of American Petroleum

Locality	Nitrogen, per cent	Authority
West Virginia		
Cumberland County	0.54	See note
Ohio		
County		
Auglaize, St. Mary's Township	0.068	Mabery, C. F., Proc. Am. Acad., **21**, 12–14
Hancock, Liberty Township	0.047	ditto
Heilstone Oil Co	0.023	ditto
Langemade Well	0.13	ditto
Peerless Rfg. Co	0.35	ditto
Wood, Montgomery Township	0.054	ditto
Liberty Township	0.056	ditto
Peerless Rfg. Co	0.08	ditto
Ohio Oil Co	0.021	ditto

1 Karlsruhe Dissertation, 30, 1910.
2 Capelle, Bull. Soc. Chim. (4) 3, 4, 150.
3 Ber., **20**, 1850–1887.
4 Chem. Zeitung, 731, 1903.
5 MacGarvey, Fred., Karlsruhe Dissertation, 16, 1896.

Nitrogen Content of American Petroleum—Continued

Locality	Nitrogen, per cent	Authority
Sandusky, Woodville Township	0.049	Mabery, C. F., Proc. Am. Acad., **21**, 12–41
Wells (Ind.), Nottingham Township	0.060	ditto
Allen, Lima	0.024	ditto
Belmont, Barnesville	0.26	ditto
Trumbull District (Mecca)	0.230	See note (Page 531)
California		
County		
Santa Barbara, Summerland	1.25	Mabery, C. F., Proc. Am. Acad., **40**, 340, 1904
Los Angeles, Puente Hills	0.564–0.587* 1.18–1.22†	Mabery, C. F., Am. Chem. Jour., **25**, 253–285, 1901
Ventura, Bardsdale	0.50* 1.25†	ditto
Torrey	0.38* 1.15†	ditto
Sespe	1.25†	ditto
Adams Canyon	0.58* 1.46†	ditto
Sulphur Mountain	1.1095	Peckham, S. F., Am. Jour. Sci., **3**, XLVIII, 250, 1894
Pico Spring	1.0165	ditto
Canada Loga	1.0855	ditto
Ojai Ranch	0.5645	ditto
Fresno	0.10–0.12	Mabery, C. F., Am. ChJom. ur., **19**, 802, 1897
Texas		
Jefferson County, Lucas Well	More than 1 per cent	Mabery, C. F., Proc. Am. Acad., **23**, 265, 1901
Beaumont field	0.92 oxygen and nitrogen	Richardson, Jour. Frankl. Inst., **162**, 113
Canada		
Province of Ontario		
Petrolia Oil-field	0.16	Mabery, C. F., Proc. Am. Acad., **31**, 43–58
Oil Spring Oil-field	0.18	ditto

* Nitrogen determined by Kjeldahl method.
† Per cent nitrogen by volume, absolute method.

NOTE.—In his tables, Engler attributes the references on "the nitrogen content of oils" from Cumberland, W. Va. and Mecca, Ohio, to Redwood, Petroleum and its Products, 1, 225, 1906. Redwood in turn refers to Peckham, S. F., Am. Jour. Sci. 3, 280, 1884.

This reference does not give the two determinations referred to, but gives the additional references which follow and indicates that the Cumberland and Mecca references are cited in the appendix to the second volume on Geology, Geological Survey of California by S. F. Peckham.

It should be pointed out that the oils of California which have recently been shown to contain resins are also unusually rich in nitrogen. The resins themselves contain nitrogen in sufficient quantity to account for all of the nitrogen found in these oils, and it is altogether probable that such resinous bodies are mainly responsible for the nitrogen in California oils. In other countries nitrogen occurs in petroleum in amounts which have no serious significance in the petroleum industry. In spite of considerable study by various authorities, comparatively little has been learned as to the nature of the nitrogen compounds in petroleum.

Water.—Pure petroleums, free from oxidized compounds, such as asphalt, resins, fatty acids, etc., generally separate sharply from water, even when the latter is very impure. It is therefore not possible to consider water as in any way a constituent of petroleum. It is well, however, to draw a sharp distinction between water which, more or less emulsified, is present, as such, in petroleum and water which is present as a constituent of some organic compound, such as a resin. The importance of the distinction rests upon the difficulty of separation when the water is in combination—even distillation often producing disastrous foaming.

In general, it may be stated that sufficient attention has not been given by the refinery chemists to the sulphur, nitrogen and oxygen compounds which occur in crude petroleums, and that the investigation of these compounds will not only lighten the refiners' burden, but eventually, especially in the case of resins, will lead to securing by-products which are more valuable when separated than they are when broken up by distillation.

PHYSICAL PROPERTIES

Specific heat.—There is strikingly slight variation in specific heat between different oils. The average specific heat is about 0.4978. In other words, about half the total quantity of heat is required to raise the temperature of most oils through a given number of degrees as would be required for water—a very important matter in distillation. This is shown in the following table:

Specific Heat of Various Aliphatic Hydrocarbons

PARAFFIN HYDROCARBONS

Hydrocarbon	Specific heat
C_6H_{14}	0.5272
C_7H_{16}	0.5074
C_8H_{18}	0.5052
C_9H_{20}	0.5034
$C_{10}H_{22}$	0.5021
$C_{11}H_{24}$	0.5013
$C_{12}H_{26}$	0.4997
$C_{13}H_{28}$	0.4986
$C_{14}H_{30}$	0.4973
$C_{15}H_{32}$	0.4966
$C_{16}H_{34}$	0.4957

ETHYLENE HYDROCARBONS

Hydrocarbon	Specific heat
C_6H_{12}	0.5062
C_7H_{14}	0.4879
C_8H_{16}	0.4863
C_9H_{18}	0.4851
$C_{11}H_{22}$	0.4819
$C_{15}H_{30}$	0.4708
$C_{16}H_{32}$	0.4723
$C_{18}H_{36}$	0.4723
$C_{20}H_{40}$	0.4706
$C_{23}H_{46}$	0.4612
$C_{24}H_{48}$	0.4586

Optical activity.—Under this caption the rotatory power of oils is usually all that is considered. This is useful for the identification of oils, and for the detection of such impurities as resin oils. Nearly all crude petroleums rotate the plane of polarization to the right. No petroleums have been found which rotate to the left. The actual power of a crude petroleum to rotate the plane of polarization of a light ray is frequently due, to a great extent, to some impurity, rather than to the hydrocarbons themselves, and the presence of such an impurity is thereby indicated.

Index of refraction.—The index of refraction of various hydrocarbons increases with the specific gravity, and, for the same number of carbon atoms, decreases with the proportion of hydrogen, as shown in the table on page 533.

The index of refraction has been determined for various petroleums by Francis and Bennett:[1]

Refractive Indices of Petroleum

Petroleum	Gravity	Color	Refractive indices at 20° C.
Cabin Creek, W. Va.	48.0	Yellow	1.4468
Montana*	46.4	Dark green	1.4350
Riverton, Wyo.	46.2	White	1.4410
Bull Bayou, La.*	40.8	Green	1.4549
Corning, Ohio *	39.9	Greenish-brown	1.4550
Cushing, Okla.*	39.4	Dark green	1.4540
Homer, La.*	38.9	Very dark green	1.4430
Osage, Okla.*	36.6	Dark green	1.4650
Bixby, Okla.*	35.7	Dark green	1.4485

The crudes marked * were too dark to be tested alone, and the refractive index was obtained by testing a blend of the crude and kerosene, then calculating the refractive index of the crude.

The study by these authors shows that the refractive indices of products from the same crude increase as the specific gravity increases. Oils of lower boiling point have lower refractive index. Unsaturated hydrocarbons appear to refract light to a greater extent than saturated ones. The refractive index of oils increases with the viscosity. Removal of color compounds from lubricants decreases the refractive index. The presence of wax in lubricants lowers the refractive index.

[1] Petroleum Magazine, May, 1921, p. 134.

Index of Refraction of Various Hydrocarbons

Number of C atoms	Paraffin Series				Olefines		
	Name	Boiling point C°	Specific gravity	Index of refraction	Name	Specific gravity	Index of refraction
C_6	Hexane	68	0.6645 (14.8°)	1.3780 (14.8°)	Hexylene†	0.6886 (15.2°)	1.3995 (15.2°)
C_7	Heptane * ‡	92–94	0.70855 (15°) 0.730353 (0°)	1.40362 (16–17°)	Heptylene		
C_8	Octane	125.5	0.70743 (15.1°)	1.4007 (15.1°)	Octylene†	0.72559 (16°)	1.4157
C_9	Nonane * ‡	136–138	0.7404 (15°) 0.76236 (0°)	1.42073 (16–17°)	Nonylene		
C_{10}	Decane †	173	0.72784 (14.9°)	1.4108 (14.9°)	Decylene †	0.77207 (17°)	1.4385 (17°)
C_{14}	Tetradecane	126 (12 m.m.)§ 122.5 (11 m.m.)‖	0.7697 (20°)§ 0.7645 (20°)‖	1.4307 (23°)§			
C_{16}	Hexadecane	151 (12 m.m.)§ 151 (11 m.m.)‖	0.7803 (20°)§ 0.7742 (20°)‖	1.4362 (23°)§			
C_{18}	Octodecane	174 (12 m.m.)§ 174.5 (11 m.m.)‖	0.7793 (28°)§ 0.7768 (28°)‖	1.4390 (27°)§			

* Isolated from petroleum.
† Landolt and Jahn, Zeitschr. physikal Chemie, **10**, 303, 1892.
‡ Bartoli and Stracciati, Ann. Chim. Phys., **7**, 382, 1886.
§ Ubbelohde and Agthe, Diplomarbeit, Karlsruhe, 1912.
‖ Krafft, Ber., **15**, 1697, 1882.

Capillarity.—This property may be defined as the tendency of the liquid in a capillary tube to stand at a higher level than the body of liquid into which the capillary is dipped. It varies sufficiently in different oils to affect their rise through a lampwick, and is affected by the specific gravity of the oil. The rate of rise in a capillary tube is affected by the viscosity of the oil, and hence has been made a measure of relative viscosity in light oils, as shown in the chapter on Oil-testing Methods. Capillarity determines the ability of an oil to flow up from a lamp reservoir, through a wick to the flame. In general, the force involved in capillary movement is so much greater than the force of gravity that it causes movements of oils in the earth that are opposed to the normal effects of gravity or of hydrostatic head. These movements have not been carefully studied. Again, capillary force, varying inversely as the size of the capillary, tends to hold oil in rocks with unusually fine pores, even against rather high gas pressures. Capillary action is the chief force depended upon to move oils through porous media, when certain ingredients are to be held back by selective adsorption and thus separated, as in the filtration of lubricating oils.

Adsorption.—The individual oils which make up crude petroleum differ greatly from each other in regard to the force with which they adhere to the surfaces of other substances. Thus, some constituents of petroleum will adhere to the surface of a bottle or other container more than other constituents. Hence there is a tendency for such constituents to concentrate to a minute extent in the surface of the oil. Certain substances affect the oils more than others. If the surface of a solid container is greatly increased the concentrating effect is also increased. Thus, some petroleums contain a small proportion of asphaltic or gummy oils usually dark in color, and therefore undesirable. If such oils are poured into a very porous substance like a dry clay or charcoal, this substance will offer such a very large surface to the oil that all of the gummy substance may concentrate on the surfaces of the clay, although the amount at any one point of the surface is exceedingly minute. If the oil is carried by capillary movement through a body of such porous clay, all of the undesirable substance is gradually concentrated upon the clay surfaces, and the oil in passing through, is freed from the undesirable material. This application of "adsorption" is used to a great extent in "filtering" lubricating oils, vaseline, etc., and will be referred to in the chapter on oil refining. Another method of utilizing the same property of oils is somewhat simpler in theory and in practice, and may be applied to oils that are not very viscous. This is the "clay wash" method. It is most frequently used in brightening lamp oils or gasoline, when treatment with sulphuric acid and subsequently with water has left the oil slightly turbid. Only a very small amount of clay is added to the oil, which is then thoroughly agitated so as to bring the clay into contact with all parts of the oil and allow the substances with relatively greater affinity for the clay to concentrate on its surface. The difference between the capacities of the various constituents to concentrate on clay surfaces is so very great that it is possible to remove practically all of certain constituents from solution (or colloidal suspension) in the other oils of the mixture.

In general, the oils poorest in hydrogen show the greatest adsorptive capacity. Among oils of the same series (such as the oils of the paraffin series) those of greater viscosity are somewhat more easily removed. In general, very insufficient study has been given to pure substances in this connection.

By empirical tests, great differences have been found in the adsorptive capacities of different porous bodies, such as various clays; and lately the "activating" of charcoal is said to have increased its selective capacity to as much as twenty-five times that of such standard clays as Florida fuller's earth. What is needed, however, is quantitative study of the pure hydrocarbons of different series, to show their relative adsorptive capacities. A table published by Gilpin and Cram is given on the next page. It shows the fractionation of certain oils and is a summary of results of work suggested by the writer of this chapter. It gives an indication of what may be expected from such an investigation if it should be extended to the examination of pure substances, and should cover such question as the relative adsorption of pure heptane compared with that of benzol or the members of other series, such as the acetylene series, and further with various asphalts and resins.

Fractionation of Crude Petroleum in Single Tubes *

	1		2		3	
Time required, in hours	23.5		17.5		17.5	
Distance from top of tube to oil when opened, in centimeters	31		28		28	
	Specific gravity	Cubic centimeters	Specific gravity	Cubic centimeters	Specific gravity	Cubic centimeters
A. 8 centimeters at top	0.796	42	0.8012	30	0.8022	18
B. Next 8 centimeters	.808	45	.804	37	.803	35
C. Next 18 centimeters	.8125	75	.807	47	.8075	66
	.8137	24	.809	22	.810	25
D. Next 30 centimeters	.815	130	.8125	148	.812	140
E. Next 35 centimeters	.818	170	.8185	190	.8175	145
F. Rest	.8205	125	.823	100	.821	105
		611		574		534

* Gillpin, J. Elliott, and Cram, Marshall P., The Fractionation of Crude Petroleum by Capillary Diffusion, U. S. Geol. Survey, Bull. 365.

Radio activity.—Many petroleums show radio activity, especially when they have just been taken from the ground. Radio activity diminishes when the oil is stored, and finally disappears. No radium compound has yet been found in oils, and the rate at which radio activity disappears seems to indicate that it is caused by radium emanation, absorbed during passage through the earth.

Solubility.—Generally speaking, petroleums in the liquid form are soluble in each other. The semi-solid and solid varieties are also soluble, to a limited extent, in the liquid varieties, the proportion increasing rapidly with the temperature. The least soluble are asphaltic compounds in light oils of the paraffin series. This low solubility has been used as a method of separating certain asphalts from oils. The solid varieties of paraffin wax will under certain conditions, crystallize from heavy distillates when slowly cooled to low temperatures; this is the usual method for the commercial separation of of paraffin wax, as shown in the chapter on refining.

Petroleums are soluble in ether, benzol and its homologues, carbon tetrachloride, carbon bisulphide, alcohol, methanol, etc. Of these the alcohols are the poorest solvents. The following table, showing the solubility of paraffin wax in various media, will serve as an example.

Solubility of Various Solid

(Figures Mean Grams of Substance

Formula	Alcohol	Ether
Naphthalene, $C_{10}H_8$	5.29 (15°) absolute 30 (78°) absolute	very soluble
Phenanthrene, $(C_6H_4CH_2)_2$	2.62 (15°) absolute 10.08 (78°) absolute	very soluble
Anthracene, $C_6H_4 : (CH_2)_2 : C_6H_4$	0.59 (15°) 0.076 (15°) absolute 0.83 (78°)	1.17 (15°)
Pyrene	1.37 (15°) 3.08 (78°)	very soluble
Chrysene	0.097 (16°) 0.097 (15°) absolute 0.17 (78°)	very slightly soluble
Diphenyl, $C_6H_5C_6H_5$	9.98	soluble
Acenaphthene, $C_{10}H_6(CH_2)_2$	3.2 (20°) soluble in hot	very soluble
Fluorene, $(C_6H_4)_2 : CH_2$	slightly soluble	very soluble
Fluoranthene, $C_{15}H_{10}$	slightly soluble in cold	very soluble
Retene, $C_{18}H_{18}$	3	soluble
Dinaphthyl, $(C_{10}H_7)_2$	moderately soluble	moderately soluble
Stilbene, $C_6H_5 \cdot CH : CH \cdot C_6H_5$	0.088 (17°) absolute	7.88 (14°)
Phenyl naphthalene, B $C_{10}H_7C_6H_5$	very soluble	very soluble
Triphenyl methane, $(C_6H_5)_3CH$	slightly soluble	very soluble
Methyl anthracene B, $C_6H_4 : (CH_2)_2 : C_6H_3—CH_3$	slightly soluble	slightly soluble
Benzerthyrene, $C_6H_3(C_6H_5)_3$	slightly soluble	
Picene, $C_{22}H_{14}$	slightly soluble	slightly soluble
Diphenyl methane, $C_{13}H_{12}$	very soluble	very soluble

Hydrocarbons in Different Media

Dissolved in 100 Grams of Solvent)

Toluol	Other solvents	Authority
31.94 (15°) 30 (100°)	soluble in benzene (Allen, p. 275)	G. von Bechi, Allen, p. 274
0.92 (15°) 12.94 (100°)	soluble in hot benzene (Allen, p. 275)	G. von Bechi, Allen, p. 274
16.54 (15°) very soluble (100°)	soluble in hot benzene (Allen, p. 275)	G. von Bechi, Allen, p. 274
0.24 (15°) 5.39 (100°)	soluble in hot benzene (Allen, p. 275)	G. von Bechi, Allen, p. 274
....	....	G. von Bechi, Allen, p. 274
....	....	Chem. Ann., p. 288, 4th issue, 1918
....	very soluble in benzene	
....	soluble in CS_2	Chem. Ann., p. 302, 4th issue, 1918
....	....	Chem. Ann., p. 302, 4th issue, 1918
....	....	Chem. Ann., p. 356, 4th issue, 1918
....	very soluble in benzene	Chem. Ann., p. 288, 4th issue, 1918
....	....	Chem. Ann., p. 358, 4th issue, 1918
....	very soluble in benzene	Chem. Ann., p. 348, 4th issue, 1918
....	soluble in hot benzene	Chem. Ann., p. 374, 4th issue, 1918
....	soluble in benzene	Chem. Ann., p. 326, 4th issue, 1918
....	soluble in hot benzene	
....	soluble in benzene, soluble in CS_2	Allen, p. 271
....	soluble in hot benzene	Allen, p. 271

Solubility of Paraffin Wax *

Melting Point 64° to 65° C., Specific Gravity 0.9170, Temperature 20° C.

Solvent	Amount of paraffin dissolved by 100 grams of solvent. Grams	Amount required to dissolve 1 gram of ozokerite paraffin. Grams
Carbon bisulphide	12.99	7.6
Benzine (initial boiling point 75° C., sp. gr. 0.7233)	11.73	8.5
Turpentine (boiling range 158°–166° C., sp. gr. 0.857)	6.06	16.1
Cumol (boiling point 160° C., sp. gr. 0.867)	4.28	23.4
Cumol (fract. 150–160° C., sp. gr. 0.847)	3.99	25.0
Xylol (boiling range 135–143° C., sp. gr. 0.866)	3.95	25.1
Xylol (fract. 136–138° C., sp. gr. 0.864)	4.39	22.7
Toluol (boiling point 108–110° C., sp. gr. 0.866)	3.83	26.1
Toluol (fract. 108.5–109.5° C., sp. gr. 0.866)	3.92	25.5
Chloroform	2.42	41.3
Benzol	1.99	50.3
Ethyl ether	1.95	50.8
Isobutyl alcohol (sp. gr. 0.804)	0.285	352.9
Ethyl acetate	0.238	419.0
Ethyl alcohol (99.5° Tralles)	0.219	453.6
Amyl alcohol (boiling point 127°–129° C., sp. gr. 0.813)	0.202	495.3
Methyl alcohol (fract. boiling point 65.5–66.5°, sp. gr. 0.798)	0.071	1447.5

* Pawlewski and Filemonowicz, Ber., **21**, Part 2, pp. 2973–2976, 1888.

ANALYSES OF PETROLEUM

A comparison of the petroleums from the principal producing fields of the world is given in the following tables. The data include the chief features of interest to the refiner. They include the gravity in Baumé degrees, the color, the chief substances of interest, aside from the oil itself, such as sulphur, asphalt and paraffin wax, and the products yielded by distillation. These are shown by two methods. The older method gives the percentage yield of gasoline and kerosene, while a more detailed statement is given for the important oils, showing the temperatures at which each 10 per cent is given off and the gravity of that cut. Unfortunately, this information is not available for all petroleums. Enough examples are given of such petroleums, however, to show the usual characteristics of the important oils.

Several years ago the writer published a table similar to this. It showed the yield of the principal petroleums in gasoline and kerosene as those substances were then defined. Gasoline then included only the cut up to 150° C. or 302° F. To-day this range extends to 437° F. The yield of many oils is not known for that range, so that 200° C., or 392° F., has been taken as the limit for gasoline, except where otherwise stated. The safe upper limit for kerosene varies with the nature of the crude oil, and frequently extends to 300° C., but 275° C. or 527° F. is here adopted for comparative purposes.

Analyses of Petroleum

Source	Physical Properties *						Distillation					Remarks
	Baumé gravity	Color	Sulphur	Paraffin	Asphalt	Water	Begins to boil, ° F.	to 392° F.		392° F. to 527° F.		
			Per cent	Per cent	Per cent	Per cent		c.c.	Baumé gravity	c.c.	Baumé gravity	
Alaska												
Katalla	39.1	....	trace	....	....	....	....	24	54.9	51	40.6	
Oil Bay (seepage)	16.5	....	0.098	....	....	....	446	....	....	13.2	29.5	
Cold Bay (seepage)	16.6	....	0.116	....	....	....	437	....	....	13.3	29.6	
Arkansas												
Union County												
[1]Eldorado	27.0	dark brown	0.724	......	......	......	....	....	....	....	....	
California												
Coastal and Southern												
Alameda County	37.7	dark brown	....	5.4	0.9	....	347	....	....	†48.0	39.2	* To 302° F. † 302° to 527° F.
Contra Costa County	14.8	black	....	....	21.0	....	....	*3.7	....	†54.5	....	
Los Angeles County												
[2]Bardsdale	30.4	....	0.83	......	......	......	....	....	....	....	....	

Percentage by Volume

	10		20		30		40		50		60		70		80		90		95	
	° F.	Gravity	° F.	Gravity	° F.	Gravity	° F.	Gravity	° F.	Gravity	° F.	Gravity	° F.	Gravity	° F.	Gravity	° F.	Gravity	° F.	Gravity
1	400	52.2	460	46.4	522	41.4	586	37.7	626	34.7	656	33.1	666	32.0	676	30.4	710	26.5		
2	273	58.8	425	42.2																

* And components other than oil.

ANALYSES OF PETROLEUM—*Continued*

Source	Physical Properties						Distillation					Remarks
	Baumé gravity	Color	Sulphur	Paraffin	Asphalt	Water	Begins to boil, ° F.	to 392° F.		392° F. to 527° F.		
			Per cent	Per cent	Per cent	Per cent		c.c.	Baumé gravity	c.c.	Baumé gravity	
CALIFORNIA—*Continued*												
Coastal and Southern												
Los Angeles County												
Los Angeles City	16.9	....	....	....	....	....	....	trace	....	†20.0	35.8	* To 302° F.
[1] Montebello	24.2	....	....	....	....	0.9	....	..	....	....	....	† 302° to 527° F.
Newhall	17.2	....	....	....	2.59	....	....	....	....	†21.9	36.9	
	42.7	....	....	....	....	....	....	*51.0	43.8	†43.0	38.0	
Puente	22.3	....	....	....	....	....	....	trace	....	†36.0	47.4	
	40.3	....	....	....	....	....	....	*27.0	58.6	†23.4	37.5	
Whittier	21.9	....	....	....	20.6	....	....	6.9	50.4	32.7	33.9	
Santa Barbara County												
Casmalia	19.9	....	2.08	....	22.0	....	....	*6.9	54.3	†30.8	34.5	
Lompoc	16.2	....	4.43	....	20.6	7.0	....	*5.2	....	†31.5	36.3	
Santa Maria	16.9	....	....	....	41.9	....	....	*1.0	....	†24.0	39.5	
	27.6	....	1.56	....	12.0	....	....	*25.9	57.6	†30.6	35.3	
Summerland	14.9	....	....	....	....	....	....	0	....	†28.4	....	
Santa Clara County	18.6	....	....	....	37.0	....	....	*6.0	55.0	†19.5	36.0	

PERCENTAGE BY VOLUME

	10		20		30		40		50		60		70		80		90		95	
	° F.	Gravity	° F.	Gravity	° F.	Gravity	° F.	Gravity	° F.	Gravity	° F.	Gravity	° F.	Gravity	° F.	Gravity	° F.	Gravity	° F.	Gravity
1	295	49.0	435	36.0																

Analyses of Petroleum—*Continued*

Source	Physical Properties						Distillation					Remarks
	Baumé gravity	Color	Sulphur	Paraffin	Asphalt	Water	Begins to boil, ° F.	to 392° F.		392° F. to 527° F.		
			Per cent	Per cent	Per cent	Per cent		c.c.	Baumé gravity	c.c.	Baumé gravity	
California—*Continued*												
Coastal and Southern—Continued												
1 Ventura County	52.2	....	....	....	....	trace	....	....	....	....	....	
Santa Paula	26.0	....	....	....	16.0	....	....	*10	56.0	†27.0	38.8	* To 302° F. † To 302° to 527° F.
San Joaquin Valley												
Fresno County												
Coalinga	25.6	black	0.56	....	heavy	....	....	8.14	....	17.32	....	
Kern County												
2 Belridge	24.6	....	0.63	....	....	0.4	....	....	....	....	....	
Kern River	15.2	black	0.71	....	heavy	......	....	....	23.54	....	....	
3 Lost Hills	32.8	....	0.48	....	....	0.1	....	....	....	....	....	
4 McKittrick	32.4	....	....	....	....	0.3	....	....	....	....	....	

Percentage by Volume

	10		20		30		40		50		60		70		80		90		95	
	° F.	Gravity	° F.	Gravity	° F.	Gravity	° F.	Gravity	° F.	Gravity	° F.	Gravity	° F.	Gravity	° F.	Gravity	° F.	Gravity	° F.	Gravity
1	159	77.5	227	60.6	278	55.0	338	49.0												
2	206	66.4	305	49.8	402	38.7														
3	188	70.6	260	54.8	329	47.6	415	40.3												
4	187	69.0	265	54.5	329	48.0	412	41.4												

ANALYSES OF PETROLEUM—*Continued*

Source	Physical Properties						Distillation					Remarks
	Baumé gravity	Color	Sulphur	Paraffin	Asphalt	Water	Begins to boil, ° F.	to 392° F.		392° F. to 527° F.		
			Per cent	Per cent	Per cent	Per cent		c.c.	Baumé gravity	c.c.	Baumé gravity	
CALIFORNIA—*Continued*												
San Joaquin Valley—Continued												
[1] Midway	15.4 26.4	....	0.83	....	....	17.0	....	....	....	....	....	
[2] Sunset	26.3	....	0.68	......	......	......	....	....	....	....	....	
COLORADO												
Boulder County												
Boulder	38.6	....	....	....	....	....	....	*16.0	60.0	†40.0	45.0	
Fremont and Pueblo Counties												
Florence	29.1	....	0.17	....	....	....	190	8.9	54.7	14.5	43.3	
Mesa County												
De Beque	37.7	yellow	....	19.6	....	....	293	*1.0	....	†42.0	64.9	* To 302° F. † 302° to 527° F.
Rio Blanco County												
Rangely Field	40.9	....	0.06	....	....	....	77	34.6	57.2	19.3	42.8	

PERCENTAGE BY VOLUME

	10		20		30		40		50		60		70		80		90		95	
	° F.	Gravity	° F.	Gravity	° F.	Gravity	° F.	Gravity	° F.	Gravity	° F.	Gravity	° F.	Gravity	° F.	Gravity	° F.	Gravity	° F.	Gravity
[1]	411	39.2																		
	229	56.8	294	47.7	378	39.6	457	33.7												
[2]	220	59.2	311	47.2	428	37.3														

ANALYSES OF PETROLEUM—*Continued*

Source	Physical Properties						Distillation					Remarks
	Baumé gravity	Color	Sulphur	Paraffin	Asphalt	Water	Begins to boil, ° F.	to 392° F.		392° F. to 527° F.		
			Per cent	Per cent	Per cent	Per cent		c.c.	Baumé gravity	c.c.	Baumé gravity	
ILLINOIS												
Clark County												
Casey Township	38.7	dark green	....	....	....	....	185	* 8.0	63.9	†33.0	48.8	
Johnson Township	27.8	olive green	0.13	....	....	....	190	*4.4	57.9	†33.2	42.4	U. S. Geol. Survey, Min. Res., Part II (1907)
Parker Township	30.4	olive-green	0.73	....	....	....	....	*8.6	58.9	†34.8	42.2	U. S. Geol. Survey, Min. Res., Part II (1907)
Crawford County												
[illegible]ry [illegible]hip	23.0	brown	0.39	....	....	....	176	*1.0	....	†24.5	36.8	* To 302° F.
[illegible]ng Township	30.9	olive-green	0.10	....	....	....	140	*3.5	56.9	†36.8	43.7	† 302 to 527° F.
Robinson Pool	35.5	medium green	0.16	....	....	....	176	*8.0	64.9	†39.0	46.3	
Cumberland County												
Union Township	36.9	light green	0.20	....	....	....	140	*18.0	64.1	†38.0	41.9	
Lawrence County												
Bridgeport Township	38.9	dark green	....	4.31	....	trace	163	*12.0	63.6	†35.0	47.8	
Petty T[illegible]wnship	35.2	da·k green	....	1.96	....	trace	220	*6.0	57.1	†40.0	46.3	
Macoupin County	27.0	black	....	....	....	2.74	311	....	....	†21.0	40.1	
Marion County	33.6	dark green	....	4.21	0.53	....	203	*6.5	59.4	†34.0	49.2	

ANALYSES OF PETROLEUM—*Continued*

Source	Physical Properties						Distillation					Remarks
	Baumé gravity	Color	Sulphur	Paraffin	Asphalt	Water	Begins to boil, ° F.	to 392° F.		392° F. to 527° F.		
			Per cent	Per cent	Per cent	Per cent		c.c.	Baumé gravity	c.c.	Baumé gravity	
ILLINOIS—*Continued*												
Montgomery County												
Litchfield............	21.6	....	....	....	....	....	220	*1.3	....	†16.9	34.3	* To 302° F. † 302 to 527° F.
Randolph County												
Sparta...........	36.6	....	....	....	....	....	....	*14.0	62.0	†37.0	45.6	
INDIANA												
[illegible]is [illegible]ty												
Birdseye.......	34.7	. .	....	. .	. ..	. .	....	*17.4	....	†26.9	....	U. S. Geol. Survey, Min. Res., Part II (1907)
Grant County.....	34.1	. .	0.83	. .	....	. .	....	*7.2	64.7	†32.6	45.8	
[illegible]er [illegible]ty....	20.9	. .	1.26	. .	2.9	. .	248	....	....	....	....	
[illegible]a district.....	35.5	. .	0.48	. .	....	. .	81	26.0	55.9	19.2	41.4	
Vigo [illegible]ty												
Terre Haute....	29.3	. .	0.72	. .	. ..	. ..	....	....	....	39.6	39.7	
KANSAS												
Allen County												
Chanute.......	31.9	dark green	....	4.25	1.23	....	228	*5.0	60.4	†36.0	45.2	
Humboldt.....	27.7	dark green	....	3.93	2.33	....	253	*1.0	....	†29.0	41.7	
Moran....	29.2	black	....	1.21	2.63	....	203	*8.0	62.5	†29.0	42.2	

ANALYSES OF PETROLEUM—*Continued*

Source	Physical Properties						Distillation					Remarks
	Baumé gravity	Color	Sulphur	Paraffin	Asphalt	Water	Begins to boil, ° F.	to 392° F.		392° F. to 527° F.		
			Per cent	Per cent	Per cent	Per cent		c.c.	Baumé gravity	c.c.	Baumé gravity	
KANSAS—*Continued*												
Butler [illegible]ty												
[1] Augusta Pool	33.1	dark green	0.15	3.0	....	....	174	....	....	....	....	
[2] [illegible]do Pool	37.5	dark green	0.17	3.8	....	....	152	....	....	....	....	
[3] Smock Pool	39.3	dark green	0.086	....	....	....	126	....	....	....	....	
[illegible] [illegible]ty	34.2	black	....	3.81	0.35	....	188	*7.0	63.3	†3.25	46.5	* To 302° F.
[illegible]rd [illegible]	28.0	black	0.59	0.49	heavy	....	....	2.06	....	11.55		† 302° to 527° F.
Elk [illegible]ty												
Longton Pool	32.1	black	....	7.19	0.89	....	208	*12.0	64.4	27.5	46.7	
Franklin County												
Rantoul Pool	33.6	black	....	3.45	2.29	....	168	*11.5	66.9	†29.5	46.1	
[4] Greenwood [illegible]ty	41.3	dark green	0.160	....	....	0.15	100	....	....	....	....	

PERCENTAGE BY VOLUME

	10		20		30		40		50		60		70		80		90		95	
	° F.	Gravity	° F.	Gravity	° F.	Gravity	° F.	Gravity	° F.	Gravity	° F.	Gravity	° F.	Gravity	° F.	Gravity	° F.	Gravity	° F.	Gravity
1	306	60.3	360	54.0	424	47.8	486	41.7	550	36.8	614	33.1	672	30.0	682	28.5				
2	262	69.2	320	59.5	384	52.9	452	46.3	516	41.3	580	37.1	640	33.7	680	31.8				
3	230	71.5	290	61.3	362	53.8	444	46.8	524	41.0	608	36.4	674	33.1	690	31.6				
4	224	76.8	290	63.0	350	54.5	428	47.7	508	41.8	592	37.0	648	34.1	676	33.0	690	31.8		

ANALYSES OF PETROLEUM—*Continued*

Source	Physical Properties						Distillation					Remarks
	Baumé gravity	Color	Sulphur	Paraffin	Asphalt	Water	Begins to boil, ° F.	to 392° F.		392° F. to 527° F.		
			Per cent	Per cent	Per cent	Per cent		c.c.	Baumé gravity	c.c.	Baumé gravity	
KANSAS—*Continued*												
Miami County												
Paola Pool	34.5	black	....	7.44	2.94	....	176	*10.0	64.4	†33.0	45.8	* To 302° F. † 302° to 527° F.
Montgomery County												
Bolton Pool	34.8	black	....	6.31	0.55	....	228	*7.0	60.4	†36.5	45.4	
Cherryvale	30.4	....	....	....	....	....	....	16.3	....	53.9	....	
Coffeyville	28.7	dark green	....	5.31	0.17	....	343	....	....	†31.0	43.2	
Wayside	31.0	black	....	4.66	0.61	....	167	*9.5	67.1	†28.0	43.2	
Neosho County												
Erie Pool	30.2	black	....	4.78	3.20	....	275	*1.0	....	†34.0	45.0	
Thayer	34.9	....	....	....	....	....	....	*16.7	62.3	†39.2	43.0	
Wilson County												
Neodesha	37.7	....	....	....	....	....	....	*19.1	64.4	†38.1	43.2	
KENTUCKY												
Allen County	39.4	brown	....	3.65	2.10	trace	....	*12.5	59.9	†41.0	41.9	
Barren County												
Oskamp Pool	36.0	green	....	....	....	....	....	*18.1	....	†46.3	....	Dept. of Geology and Forestry, Series V, Vol I, No. 3, p. 333
Bath County												
Olympia Pool	34.1	green	0.23	....	....	....	194	11.2	54.9	24.5	44.1	
Ragland Pool	25.2	black	0.31	....	....	0.5	....	12.6	52.5	16.0	39.5	
Cumberland County	39.6	dark green	....	....	....	....	132	*22.0	58.5	‡38.0	44.3	‡ 302° to 554° F.

ANALYSES OF PETROLEUM—*Continued*

Source	Physical Properties							Distillation				Remarks
	Baumé gravity	Color	Sulphur	Paraffin	Asphalt	Water	Begins to boil, ° F.	to 392° F.		392° F. to 527° F.		
			Per cent	Per cent	Per cent	Per cent		c.c.	Baumé gravity	c.c.	Baumé gravity	
KENTUCKY—*Continued*)												
Floyd County	29.6	green	....	....	....	....	....	0	....	†32.8	....	Dept. of Geology and Forestry, Series V, Vol. I, No. 3, p. 333
Lawrence County	37.7	....	0.21	2.13	....	....	75	33.1	58.2	18.4	41.1	
Lee and Estill Counties												
Big Sinking Pool	35.9	....	0.14	....	....	....	86	31.2	56.4	17.3	40.9	U. S. Geol. Survey Bull. 661D
Irvine	30.5	dark brown	....	....	....	....	....	*7.4	51.1	†32.0	38.4	
Logan County	37.5	dark green	....	....	....	....	124	*20.2	56.5	‡33.2	44.2	* To 302° F.
Menifee County	33.9	dark green	....	....	....	....	140	*17.0	59.2	‡25.5	44.5	† 302° to 527° F. ‡ 302° to 554° F.
Morgan County	43.0	dark green	....	5.40	....	....	203	*13.0	62.5	†35.0	50.4	
Warren County												
Sunnyside	36.2	....	....	....	....	....	152	*19.0	....	†37.0	....	
Wayne County												
Cooper Pool	41.2	black	....	2.65	0.80	....	95	*25.0	65.8	†29.0	43.7	U. S. Geol. Survey Bull. 579
Johnson Fork	36.5	brown	....	3.01	2.66	....	137	*13.5	64.4	†32.0	45.4	U. S. Geol. Survey Bull. 579
Oil Valley Pool	41.7	....	....	3.34	1.78	....	122	*20.0	66.6	†36.0	45.4	U. S. Geol. Survey Bull. 579
Parmleysville	37.7	dark green	....	5.09	....	trace	168	*13.0	65.2	†36.0	46.1	U. S. Geol. Survey Bull. 579
Parnell Pool	43.2	light green	....	2.47	....	....	109	*27.0	68.8	†33.0	44.7	U. S. Geol. Survey Bull. 579
Rocky Branch	25.2	black	....	5.49	....	trace	338	0	....	†26.0	41.1	U. S. Geol. Survey Bull. 579

ANALYSES OF PETROLEUM—*Continued*

Source	Physical Properties							Distillation				Remarks
	Baumé gravity	Color	Sulphur	Paraffin	Asphalt	Water	Begins to boil, ° F.	to 392° F.		392° F. to 527° F.		
			Per cent	Per cent	Per cent	Per cent		c.c.	Baumé gravity	c.c.	Baumé gravity	
KENTUCKY—*Continued*												
Sinking Pool	41.0	dark green	....	3.73	0.56	....	149	*22.0	62.5	†36.0	44.1	U. S. Geol. Survey Bull. 579
Turkey Rock	41.5	dark green	....	2.31	0.36	....	140	*23.0	64.9	†36.0	46.3	U. S. Geol. Survey Bull. 579
Winchester County	38.6	dark green	....	....	....	....	110	*22	59.2	‡32.2	43.3	* To 302° F. † 302° to 527° F. ‡ 302° to 554° F.
Wolfe County												
Campton Pool	36.3	....	0.23	....	....	....	79	30.8	57.4	16.7	40.5	
Wolfe and Estill Counties												
Cow Creek Pool	31.7	....	0.13	....	....	....	145	19.7	51.6	16.7	39.1	
Wolfe, Lee, and Jackson Counties												
Ross Creek Pool	37.1	....	0.12	....	....	....	77	35.9	58.2	14.6	39.5	
[1] Somerset	37.7	dark brown	0.30	....	....	....	132	....	....	....	....	
LOUISIANA												
[2] Louisiana light crude	40.0	dark red	0.101	....	....	....	234	....	....	....	....	

PERCENTAGE BY VOLUME

	10		20		30		40		50		60		70		80		90		95	
	° F.	Gravity	° F.	Gravity	° F.	Gravity	° F.	Gravity	° F.	Gravity	° F.	Gravity	° F.	Gravity	° F.	Gravity	° F.	Gravity	° F.	Gravity
1	239	68.8	308	57.3	392	49.9	482	43.1	572	38.5										
2	348	54.2	390	50.8	422	48.4	456	46.1	488	44.0	530	41.9	594	39.1	662	35.3	696	32.1		

ANALYSES OF PETROLEUM—*Continued*

Source	Physical Properties							Distillation				Remarks
	Baumé gravity	Color	Sulphur	Paraffin	Asphalt	Water	Begins to boil, ° F.	to 392° F.		392° F. to 527° F.		
			Per cent	Per cent	Per cent	Per cent		c.c.	Baumé gravity	c.c.	Baumé gravity	
LOUISIANA—*Continued*												
1 Caddo Parish	25.1	dark green	0.200	....	....	....	296	....	....	....	....	
2 Caddo Parish	41.3	dark green	0.076	....	....	....	136	....	....	....	....	
3 Caddo Parish	46.1	....	0.105	....	....	0.2	....	....	....	....	....	
Calcasieu Parish Edgerly	18.7	black	....	....	....	1.25						
4 [illegible]	25.0	almost black	0.31	....	....	....	224	....	....	....	....	
5 Vinton	18.9	....	0.304	....	....	....	420	....	....	....	....	
6 Claiborne Parish	37.8	green	0.400	....	....	....	107	....	....	....	....	

PERCENTAGE BY VOLUME

	10		20		30		40		50		60		70		80		90		95	
	° F.	Gravity	° F.	Gravity	° F.	Gravity	° F.	Gravity	° F.	Gravity	° F.	Gravity	° F.	Gravity	° F.	Gravity	° F.	Gravity	° F.	Gravity
1	464	37.0	514	31.7	570	29.3	628	28.0	666	28.3	680	29.6	690	30.6	704	30.3				
2	262	63.9	322	58.6	380	55.4	421	51.6	483	46.5	546	43.2	628	38.6	670	36.0				
3	272	65.0	318	58.9	364	55.2	406	51.5	446	48.5	488	46.0	530	43.6	594	40.7	682	37.1		
4	370	49.6	482	38.9	548	31.5	604	27.9	648	26.0	646	26.8	666	27.7	678	27.5				
5	521	30.6	568	26.5	612	24.0	638	21.9	662	21.5	660	22.6	644	24.1	620	26.1	688	24.3		
6	250	73.0	326	61.8	400	53.5	484	46.0	568	39.7	648	35.7	680	32.8	690	32.3				

ANALYSES OF PETROLEUM—*Continued*

Source	Physical Properties: Baumé gravity	Color	Sulphur, Per cent	Paraffin, Per cent	Asphalt, Per cent	Water, Per cent	Begins to boil, ° F.	Distillation: to 392° F., c.c.	to 392° F., Baumé gravity	392° F. to 527° F., c.c.	392° F. to 527° F., Baumé gravity	Remarks
LOUISIANA—*Continued*												
i [illegible]e Parish												
[1] Homer	36.9	dark brown	0.780	....	....	....	154	....	....	....	....	
Concordia Parish												
[2] Bull Bayou	39.8	....	0.137	....	....	....	238	....	....	....	....	
De Soto Parish												
[3] Naborton	42.7	dark green	0.101	....	....	....	156	....	....	....	....	
Jefferson Davis Parish												
Jennings	24.0	....	0.39	....	5.0	....	200	....	....	†41.0	....	
Welsh	19.3	olive-green	none	....	....	....	91	*1.05	38.5	†30.07	34.4	* To 302° F. † 302° to 527° F.
[4] Red River Parish	40.7	green	0.108	....	....	....	215	....	....	....	....	
St. Martin Parish												
Anse La Butte	19.1	....	0.20	....	9.0	....	464	....	....	16.0	....	

PERCENTAGE BY VOLUME

	10, ° F.	10, Gravity	20, ° F.	20, Gravity	30, ° F.	30, Gravity	40, ° F.	40, Gravity	50, ° F.	50, Gravity	60, ° F.	60, Gravity	70, ° F.	70, Gravity	80, ° F.	80, Gravity	90, ° F.	90, Gravity	95, ° F.	95, Gravity
1	248	72.3	338	60.6	413	52.8	500	45.1	586	39.9										
2	354	54.2	392	50.2	430	48.0	465	45.7	500	43.4	538	41.1	590	38.6	660	35.6	674	33.8		
3	284	63.4	340	57.3	386	53.5	420	49.4	464	46.9	514	43.8	588	40.3	682	35.3				
4	325	55.6	375	51.5	420	48.8	455	46.6	487	44.5	525	41.9	577	39.6	645	36.1				

ANALYSES OF PETROLEUM—*Continued*

Source	Physical Properties							Distillation				Remarks
	Baumé gravity	Color	Sulphur	Paraffin	Asphalt	Water	Begins to boil, ° F.	to 392° F.		392° F. to 527° F.		
			Per cent	Per cent	Per cent	Per cent		c.c.	Baumé gravity	c.c.	Baumé gravity	
Michigan												
Saginaw County	43.6	light	....	0.78	0	0	98	*15.0	63.1	†52.0	48.25	* To 302° F. † 302° to 527° F.
St. Clair County												
Port Huron	38.0	dark green	....	....	....	....	....	*15.0	....	†55.0	....	
Missouri												
Cass County												
Belton	....	....	....	8	....	....	....	*10.0	....	†19.	....	
Montana												
Fergus and Garfield Counties												
Cat Creek Canyon	47.7	....	0.27	....	....	....	....	54.2	....	30.4	....	Oil and Gas Jour., **19**, Mar. 18, 1921
Musselshell County												
Winnett Field	49.3	....	0.36	....	....	....	100	63.2	57.4	25.5	42.0	
New Mexico												
Eddy County	26.4	black	....	0.56	....	....	422	....	....	30.0	36.8	
New York												
Allegany County	39.1	....	0.10	....	....	....	....	30.0	57.2	17.5	44.6	
Ohio												
Allen County												
Lima	35.5	....	0.48	....	....	....	81	....	55.9	....	41.4	
North Lima	37.7	....	0.55	....	....	....	82	31.9	56.9	19.2	41.8	
South Lima	36.3	....	0.55	....	....	....	86	27.0	54.7	20.0	41.1	

ANALYSES OF PETROLEUM—*Continued*

Source	Physical Properties							Distillation				Remarks
	Baumé gravity	Color	Sulphur	Paraffin	Asphalt	Water	Begins to boil, ° F.	to 392° F.		392° F. to 527° F.		
			Per cent	Per cent	Per cent	Per cent		c.c.	Baumé gravity	c.c.	Baumé gravity	
OHIO—*Continued*												
Fairfield County												
Bremen Pool	48.4	medium green	....	8.33	....	....	154	*15.0	69.1	†40.0	52.0	* To 302° F. † 302° to 527° F.
a [illegible]lle Pool	44.0	dark green	....	5.36	....	....	204	*10.0	64.7	†43.0	50.6	
Hancock County	35.3	dark green	....	....	....	....	....	*9.75	62.3	†37.1	47.8	
Knox County												
Bladensburg Pool	35.3	dark green	....	4.17	....	much	167	*14.0	64.4	†26.0	45.6	
Me[illegible] County												
Bethel [illegible]p	50.9	dark amber	....	2.91	....	....	....	....	....	....	....	
[illegible]n Pool	45.7	medium amber	....	5.65	....	....	194	*15.0	64.1	†40.0	50.1	
Decker Pool	46.0	light green	....	2.82	....	much	170	*18.0	65.2	†30	49.9	
Graysville Pool	49.9	amber	....	3.35	....	....	158	*25.0	69.4	†38.0	51.5	
Griffith Pool	46.4	light amber	....	3.56	....	....	167	*18.5	63.6	†42.0	47.4	
Jerusalem Pool	37.2	dark amber	....	5.65	....	trace	212	*5.0	49.2	†30.0	40.9	
Olive Township	39.5	dark amber	....	11.24	....	....	152	*12.0	62.0	†33.0	46.5	

ANALYSES OF PETROLEUM—*Continued*

Source	Physical Properties							Distillation				Remarks
	Baumé gravity	Color	Sulphur	Paraffin	Asphalt	Water	Begins to boil, ° F.	to 392° F.		392° F. to 527° F.		
			Per cent	Per cent	Per cent	Per cent		c.c.	Baumé gravity	c.c.	Baumé gravity	
OHIO—*Continued*												
Morgan County												
Milner Pool	44.0	dark green	....	5.36	....	....	165	*14.5	66.3	†38.5	49.2	* To 302° F. † 302° to 527° F.
Noble County												
Macksburg Field	41.7	dark green	....	6.15	....	....	239	*5.0	58.4	†43.0	49.4	U. S Geol. Survey, Mineral Resources, II, 1919
Belle Valley Pool	39.9	dark green	....	5.44	....	....	134	*13.0	63.1	†28.0	57.6	
Perry County												
Santoy Pool	39.9	dark green	....	6.16	....	trace	140	*18.0	63.6	†27.0	45.4	
Crooksville Pool	44.7	light amber	....	6.63	....	....	208	*7.0	64.1	†45.0	51.1	
New Straitsville Pool	46.7	....	....	8.30	....	....	176	*17.5	66.0	†33.0	50.6	
Vinton County	45.9	medium green	....	5.62	....	....	215	*4.0	66.0	†51.0	52.5	
Washington County												
Germantown Pool	44.5	medium green	....	6.43	....	....	149	*16.0	65.2	†35.0	49.7	
Fifteen Pool	48.0	dark amber	....	7.25	....	....	140	*20.0	70.8	†37.0	51.5	
Macksburg	42.5	light amber	....	....	....	....	....	*15.4	....	†45.8	....	

ANALYSES OF PETROLEUM—*Continued*

Source	Physical Properties							Distillation				Remarks
	Baumé gravity	Color	Sulphur	Paraffin	Asphalt	Water	Begins to boil, ° F.	to 392° F.		392° F. to 527° F.		
			Per cent	Per cent	Per cent	Per cent		c.c.	Baumé gravity	c.c.	Baumé gravity	
[illegible] Carter [illegible]ty Healdton	31.8	....	0.66	0.67	0.71	....	....	12.0	62.4	35.0	42.0	U. S. Geol. Survey Bull. 621, p. 26
[1] [illegible]	31.4	dark green	0.565	....	....	much	....	....	....	....	....	
[2] Hewitt	34.5	black	0.434	....	....	....	185	....	....	....	....	
[illegible] [illegible]ty Brd [illegible]ek	32.3	dark green	....	7.30	0.28	....	248	*2.0	....	†37.5	43.4	* To 302° F. † 302° to 527° F.
[3] Gl[illegible]n Pool	39.3	dark green	0.15	....	....	....	120	....	....	....	....	
Skiatook	35.1	dark green	....	7.35	0.23	....	203	*6.0	60.7	†37.0	47.4	

PERCENTAGE BY VOLUME

	10		20		30		40		50		60		70		80		90		95	
	° F.	Gravity	° F.	Gravity	° F.	Gravity	° F.	Gravity	° F.	Gravity	° F.	Gravity	° F.	Gravity	° F.	Gravity	° F.	Gravity	° F.	Gravity
1	302	59.4	386	50.6	452	44.5	522	39.8	590	36.4	640	33.8	644	32.4	662	32.0	686	31.4		
2	306	60.4	381	53.7	476	46.1	557	39.3	620	35.6	636	34.0	631	34.2	642	33.3				
3	232	70.1	292	59.4	356	52.9	428	47.1	504	41.7	576	37.1	640	34.0	670	32.7				

Analyses of Petroleum—*Continued*

Source	Physical Properties: Baumé gravity	Color	Sulphur Per cent	Paraffin Per cent	Asphalt Per cent	Water Per cent	Begins to boil ° F.	Distillation: to 392°F. c.c.	to 392°F. Baumé gravity	392° F. to 527° F. c.c.	392° F. to 527° F. Baumé gravity	Remarks
Oklahoma—*Continued*												
[1] Garfield County	46.6	dark green	0.04	....	....	....	136	....	....	....	....	
Kay County [illegible]	34.5	....	....	....	....	....	....	9.1	....	43.7	....	
Kiowa County [illegible] Pool	35.1	black	....	5.56	1.30	....	239	*3.5	59.4	†45.5	49.0	* To 302° F. † 302° to 527° F.
Marshall County [illegible]	47.5	light green	....	7.41	....	....	149	*22.0	66.9	†38.0	49.9	
Muskogee County [illegible] Pool	38.1	green	....	7.64	0	....	206	*11.0	61.0	†36.0	45.8	
[illegible] County [illegible]	35.7	dark green	....	4.51	0.75	....	176	*14.0	58.9	†33.0	44.3	
Delaware Pool	34.8	light green	....	4.18	1.69	....	149	*16.5	58.4	†35.0	40.9	
Shal dw Sand	34.0	black	....	3.04	0.12	....	212	*7.0	59.7	†39.0	44.3	

Percentage by Volume

	10 ° F	10 Gravity	20 ° F.	20 Gravity	30 ° F.	30 Gravity	40 ° F.	40 Gravity	50 ° F.	50 Gravity	60 ° F.	60 Gravity	70 ° F.	70 Gravity	80 ° F.	80 Gravity	90 ° F.	90 Gravity	95 ° F.	95 Gravity
1	208	71.6	238	66.0	270	61.2	306	56.7	368	52.5	444	47.5	534	42.0	630	37.1				

ANALYSES OF PETROLEUM—*Continued*

Source	Physical Properties: Baumé gravity	Color	Sulphur, Per cent	Paraffin, Per cent	Asphalt, Per cent	Water, Per cent	Begins to boil, ° F.	Distillation: to 392° F., c.c.	to 392° F., Baumé gravity	392° F. to 527° F., c.c.	392° F. to 527° F., Baumé gravity	Remarks
[illegible]ld [illegible]e [illegible]ty Baldhill Pool	34.1	dark green	....	3.43	0.15	....	230	*5.5	56.4	†36.0	44.7	* To 302° F. † 302° to 527° F.
[1] Beggs Field	37.1	green	0.139	....	....	....	152	....	....	....	....	
[illegible]s Pool	35.5	light green	....	11.9	0.	....	233	*3.0	....	†34.0	46.7	
[2] Osage Co[illegible]nty	34.1	dark green	0.078	....	....	....	200	....	....	....	....	
[illegible]le Pool	33.8	dark green	....	7.90	1.12	....	235	*3.5	56.4	†44.0	48.1	
[illegible]rt [illegible]ld	35.5	....	0.19	....	....	....	97	28.1	55.7	19.1	39.3	
[illegible]e [illegible]ty Cleveland Pool	34.4	dark green	....	6.62	0.30	....	242	*4.5	52.5	†39.0	43.7	
Payne [illegible]ty [3] [illegible]g Field	43.4	dark green	0.167	....	....	....	90	....	....	....	....	

PERCENTAGE BY VOLUME

	10 ° F.	10 Gravity	20 ° F.	20 Gravity	30 ° F.	30 Gravity	40 ° F.	40 Gravity	50 ° F.	50 Gravity	60 ° F.	60 Gravity	70 ° F.	70 Gravity	80 ° F.	80 Gravity	90 ° F.	90 Gravity	95 ° F.	95 Gravity
1	266	67.4	320	56.7	402	49.7	475	43.3	555	38.4	634	34.7	692	32.0	698	31.8				
2	310	53.9	404	47.1	478	41.9	536	38.4	604	35.5	666	33.3	708	31.7	708	31.1				
3	211	77.0	2[illegible]9	65.6	330	56.9	393	50.2	470	44.7	550	40.1	641	35.6	709	32.5				

ANALYSES OF PETROLEUM—*Continued*

Source	Physical Properties							Distillation				Remarks
	Baumé gravity	Color	Sulphur	Paraffin	Asphalt	Water	Begins to boil, ° F.	to 392° F.		392° F. to 527° F.		
			Per cent	Per cent	Per cent	Per cent		c.c.	Baumé gravity	c.c.	Baumé gravity	
[illegible]d												
Rogers [illegible]ty												
Alluwe [illegible]l	36.4	dark green	....	6.14	0.55	....	152	*10.0	66.3	†33.0	46.9	* To 302° F. † 302° to 527° F.
[illegible]a Bol	34.4	dark brown	....	4.16	2.19	....	206	*6.0	64.4	†35.0	46.7	
1 Collinsville Pool	32.2	dark green	....	3.46	....	....	216	....	....	....	....	
Sem[illegible]nle [illegible]ty												
[illegible]a Pool	28.3	black	....	6.28	0.90	....	262	*1.5	....	†30.0	39.4	
2 [illegible]s [illegible]y	32.1	....	0.618	....	....	0.1	124	....	....	....	....	
3 [illegible]	34.5	....	0.329	....	....	0.50	160	....	....	....	....	
4 [illegible]e	37.8	....	0.426	....	....	0.1	130	....	....	....	....	
5 Duncan	24.9	....	0.900	....	....	6.75	....	....	....	....	....	

PERCENTAGE BY VOLUME

	10		20		30		40		50		60		70		80		90		95	
	° F.	Gravity	° F.	Gravity	° F.	Gravity	° F.	Gravity	° F.	Gravity	° F.	Gravity	° F.	Gravity	° F.	Gravity	° F.	Gravity	° F.	Gravity
1	356	52.4	420	45.8	482	41.3	546	38.2	604	35.0	654	33.2	684	31.9						
2	264	72.9	346	58.1	460	48.4	550	40.4	624	33.9	638	31.6	630	31.3	632	30.2	510	31.4		
3	274	69.2	344	57.8	412	50.7	488	45.1	566	40.0	620	36.1	640	34.4	660	32.4	674	31.0		
4	234	73.5	292	62.0	356	55.6	434	48.5	520	42.7	608	37.4	644	33.9	660	33.0	684	31.5		
5	310	65.8	364	52.4	500	44.3	596	35.8	628	31.3	646	29.3	672	27.3	684	26.3	704	22.6		

ANALYSES OF PETROLEUM—*Continued*

Source	Physical Properties							Distillation				Remarks
	Baumé gravity	Color	Sulphur	Paraffin	Asphalt	Water	Begins to boil, ° F.	to 392° F.		392° F. to 527° F.		
			Per cent	Per cent	Per cent	Per cent		c.c.	Baumé gravity	c.c.	Baumé gravity	
OKLAHOMA—*Continued*												
Tulsa County												
Red Fork Pool	37.3	green	....	2.60	0	....	190	*15.0	64.7	†36.0	46.3	* To 302° F. † 302° to 527° F.
Washington County												
Bartlesville Pool	34.3	dark green	....	3.75	0.23	....	217	*8.0	59.9	†37.0	43.0	
Webber Pool	37.3	dark green	....	3.01	0.94	....	158	*13.0	64.1	†33.0	45.0	
PENNSYLVANIA												
Northern												
] [illegible]n [illegible]	40.1	....	0.10	....	....	....	77	32.5	59.4	17.8	44.6	
[illegible]r [illegible]	32.2	....	0.09	....	....	....	104	9.0	39.9	15.1	36.3	
[illegible]o [illegible]ty	40.9	....	0.10	....	....	....	82	29.6	57.7	17.3	43.7	
Venango [illegible]ty	38.3	....	0.08	....	....	....	86	24.4	54.0	16.4	42.6	
Southern												
[illegible]y [illegible]ty	45.0	....	0.08	....	....	....	82	37.8	61.0	20.2	46.8	
Greene [illegible]	41.8	....	0.08	....	....	....	75	29.0	57.9	18.7	44.5	
[illegible]n County	43.9	....	0.05	....	....	....	90	....	59.4	....	45.4	
Central												
Perry [illegible]ty	47.2	....	....	....	....	....	....	*17.0	72.0	†43.0	53.0	
TENNESSEE												
Scott County												
Glenmary	36.2	dark brown	0.17	....	....	trace	113	21.4	62.0	32.6	41.6	Resources of Tennessee, 8, Jan. 18, pp. 211–219

ANALYSES OF PETROLEUM—*Continued*

Source	Physical Properties							Distillation				Remarks
	Baumé gravity	Color	Sulphur	Paraffin	Asphalt	Water	Begins to boil, ° F.	to 392° F.		392° F. to 527° F.		
			Per cent	Per cent	Per cent	Per cent		c.c.	Baumé gravity	c.c.	Baumé gravity	
TEXAS												
Northern and Central												
Brown County												
1 Brownwood	26.3	brownish green	....	....	....	1.50	506	....	....	....	....	
Clay County												
2 Petrolia	37.4	....	0.427	....	....	0.05	140	....	....	....	....	
Coleman County												
3 Coleman	36.4	light green	0.141	....	....	....	223	....	....	....	....	
Eastland County												
4 Eastland	35.2	medium	....	....	....	....	260	....	....	....	....	
5 [illegible]	34.7	green	0.200	....	....	0.20	184	....	....	....	....	
Jack [illegible]	22.0	blackish brown	....	....	....	....	278	*.88	....	†17.53	36.8	* To 302° F. † 302° to 527° F.

PERCENTAGE BY VOLUME

	10		20		30		40		50		60		70		80		90		95	
	° F.	Gravity	° F.	Gravity	° F.	Gravity	° F.	Gravity	° F.	Gravity	° F.	Gravity	° F.	Gravity	° F.	Gravity	° F.	Gravity	° F.	Gravity
1	568	35.0	595	33.6	625	32.8	643	32.4	655	32.3	666	32.6	673	33.0	670	33.1	660	33.2		
2	264	70.6	324	59.4	390	52.9	460	46.6	534	41.6	608	37.6	634	34.5	658	33.5	670	31.0		
3	332	54.0	376	50.6	430	46.8	489	42.8	540	39.4	616	36.0	679	33.0	668	31.5	691	31.6		
4	338	54.9	386	51.9	454	46.2	510	41.6	574	38.5	634	35.2	658	33.9	670	32.9	690	31.4		
5	328	54.2	389	50.1	450	45.1	514	41.0	570	37.8	628	35.0	650	32.8	684	31.6	692	31.2		

ANALYSES OF PETROLEUM—*Continued*

Source	Physical Properties							Distillation				Remarks
	Baumé gravity	Color	Sulphur	Paraffin	Asphalt	Water	Begins to boil, ° F.	to 392° F.		392° F. to 527° F.		
			Per cent	Per cent	Per cent	Per cent		c.c.	Baumé gravity	c.c.	Baumé gravity	
TEXAS—*Continued*												
Northern and Central—Continued												
Limestone County												
[1] Mexia	36.1	....	0.326	....	....	....	180	....	....	....	....	
McLennan County												
[2] South Bosque	39.4	....	0.200	....	....	0.10	224	....	....	....	....	
Waco	36.0	....	....	....	....	....	....	1.39	....	40.03	....	
Marion County												
Caddo Pool	43.6	brown	0.119	7.02	....	0.75	212	*6.0	61.7	†50.5	53.2	U. S. Geol. Survey, Mineral Resources, II, 1913
Nacogdoches County												
[3] Cushing	37.2	....	0.333	....	....	7	146	....	....	....	....	
Navarro County												
Corsicana	39.0	dark brown	....	....	....	....	334	....	....	†19.5	33.3	* To 302° F. † 302 to 527° F.

PERCENTAGE BY VOLUME

	10		20		30		40		50		60		70		80		90		95	
	° F.	Gravity	° F.	Gravity	° F.	Gravity	° F.	Gravity	° F.	Gravity	° F.	Gravity	° F.	Gravity	° F.	Gravity	° F.	Gravity	° F.	Gravity
1	378	54.3	432	48.8	496	44.7	530	42.4	570	40.1	608	37.9	666	35.4	660	33.0	668	32.8		
2	330	64.5	414	55.1	476	49.7	522	44.7	576	41.7	628	39.2	676	37.5	677	37.1	676	36.6		
3	236	70.1	300	59.7	360	53.5	436	48.6	514	41.4	580	37.3	666	33.5	700	30.8	712	29.7		

Analyses of Petroleum—*Continued*

Source	Physical Properties: Baumé gravity	Color	Sulphur, Per cent	Paraffin, Per cent	Asphalt, Per cent	Water, Per cent	Begins to boil, ° F.	Distillation: to 392° F., c.c.	to 392° F., Baumé gravity	392° F. to 527° F., c.c.	392° F. to 527° F., Baumé gravity	Remarks
Texas—*Continued*												
Northern and Central—Continued												
Navarro County												
[1] Corsicana	27.5	....	....	....	....	....	370	....	....	....	....	
Powell	23.5	black	....	....	....	....	280	*1.0	....	†46.0	46.5	* To 302° F. † 302° to 527° F.
[2] Palo Pinto County	38.5	green	0.104	....	....		114					
Panola County												
[3] Carthage	45.9	dark green	0.125	...	....	0.55	169	....	....	....	....	
Shackelford County												
[4] Moran	32.7	black	0.032	....	....	....	264	....	....	....	....	
[5] Stephens County	38.2	....	0.265	....	....	.3	150	....	....	....	....	
[6] Brown Farm	38.6	transparent red	0.264	....	....	....	166	....	....	....	....	

Percentage by Volume

	10 ° F.	10 Gravity	20 ° F.	20 Gravity	30 ° F.	30 Gravity	40 ° F.	40 Gravity	50 ° F.	50 Gravity	60 ° F.	60 Gravity	70 ° F.	70 Gravity	80 ° F.	80 Gravity	90 ° F.	90 Gravity	95 ° F.	95 Gravity
1	465	40.3	518	36.9	565	34.6	602	32.6	645	31.2	652	31.0	660	31.4	650	31.6	635	32.8		
2	268	64.6	336	55.1	398	49.1	470	44.0	546	39.8	610	36.0	668	33.8	676	33.5				
3	326	65.1	368	58.4	412	54.0	446	50.4	492	47.8	526	45.2	578	42.9	636	40.5	710	37.4		
4	362	54.1	426	49.1	488	45.1	550	41.1	606	37.7	644	35.3	656	34.7	674	33.2				
5	240	68.9	310	60.4	385	52.8	475	46.3	560	40.7	645	35.9	690	32.9	703	31.0	715	29.9		
6	284	62.7	336	54.7	386	49.5	460	44.3	512	40.1	584	36.6	644	33.3	658	32.4	692	31.5		

ANALYSES OF PETROLEUM—*Continued*

Source	Physical Properties							Distillation				Remarks
	Baumé gravity	Color	Sulphur	Paraffin	Asphalt	Water	Begins to boil, ° F.	to 392° F.		392° F. to 527° F.		
			Per cent	Per cent	Per cent	Per cent		c.c.	Baumé gravity	c.c.	Baumé gravity	
TEXAS—*Continued*												
Northern and Central—Continued												
Stephens County—*Continued*												
1 Brown Farm	37.9	....	0.194	....	....	0.05	148	....	....	....	....	
2 Ranger	38.6	light green	0.077	....	....	....	148	....	....	....	....	
3 Ranger	38.9	dark green	0.100	....	....	....	140	....	....	....	....	
4 Ranger	38.2	red	0.120	....	....	....	162	....	....	....	....	
Travis County												
Waters Park	14.2	black	....	....	....	....	....	....	....	†58.1	23.1	† 302° to 527° F.
Wichita County												
5 Burkburnett	38.8	dark green	0.300	....	....	....	130	....	....	....	....	

PERCENTAGE BY VOLUME

	10		20		30		40		50		60		70		80		90		95	
	° F.	Gravity	° F.	Gravity	° F.	Gravity	° F.	Gravity	° F.	Gravity	° F.	Gravity	° F.	Gravity	° F.	Gravity	° F.	Gravity	° F.	Gravity
1	262	64.9	318	56.7	384	50.5	456	44.6	530	39.8	606	35.9	666	33.1	680	31.7	704	31.0		
2	230	68.6	316	58.5	394	52.0	474	46.3	550	41.8	630	36.7	683	33.4	700	32.3				
3	250	65.4	306	57.5	366	51.7	440	46.4	510	41.4	586	37.1	662	33.5	700	31.5	706	30.9	706	30.7
4	276	64.4	326	55.4	386	49.6	460	44.9	520	40.5	594	36.8	652	34.1	672	32.7	680	32.9		
5	226	72.1	278	63.4	326	57.0	372	52.7	430	46.7	494	42.2	558	38.3	628	35.2				

ANALYSES OF PETROLEUM—*Continued*

Source	Physical Properties: Baumé gravity	Color	Sulphur, Per cent	Paraffin, Per cent	Asphalt, Per cent	Water, Per cent	Begins to boil, ° F.	Distillation: to 392° F., c.c.	to 392° F., Baumé gravity	392° F. to 527° F., c.c.	392° F. to 527° F., Baumé gravity	Remarks
TEXAS—*Continued*												
Northern and Central—Continued												
Wichita County—*Continued*												
1 Burkburnett	38.9	....	0.375	....	....	trace	138	....	....	....	....	* To 302° F.
2 Electra	41.0	dark green	0.650	3.47	....	....	102	....	....	....	....	† 302° to 527° F.
3 Electra	31.1	black	0.570	2.80	....	....	190	....	....	....	....	
4 Electra	39.3	....	0.308	....	....	....	142	....	....	....	....	
Williamson County	39.1	red	....	....	....	....	....	*20	....	†35	....	University of Texas Bull. 66, pp. 7–78
5 Thrall	38.7	....	0.127	....	....	....	198	....	....	....	....	
Young County												
6 Graham Farm	38.1	....	0.204	....	....	....	154	....	....	....	....	
7 McCusky Farm	37.3	....	0.205	....	....	0.1	158	....	....	....	....	

PERCENTAGE BY VOLUME

	10 ° F.	10 Gravity	20 ° F.	20 Gravity	30 ° F.	30 Gravity	40 ° F.	40 Gravity	50 ° F.	50 Gravity	60 ° F.	60 Gravity	70 ° F.	70 Gravity	80 ° F.	80 Gravity	90 ° F.	90 Gravity	95 ° F.	95 Gravity
1	222	72.9	284	62.5	348	54.5	430	47.8	514	41.8	596	37.0	670	33.3	710	30.1	730	28.8		
2	212	76.1	272	63.6	317	56.7	412	48.7	492	42.7	560	38.3	638	34.6	650	33.1				
3	338	57.8	400	51.1	472	46.1	538	42.2	592	38.4	610	37.3								
4	242	70.1	300	60.3	360	54.0	428	47.3	506	41.6	574	37.6	644	34.0	686	31.9	710	30.1		
5	372	60.9	356	53.9	412	49.2	482	45.0	546	41.2	610	37.9	686	35.1	700	33.8	716	32.8		
6	268	64.8	322	56.4	382	51.2	452	45.8	520	40.0	592	36.2	642	33.3	606	32.9	662	32.4		
7	278	63.6	328	55.6	392	49.7	454	44.4	528	39.6	602	35.7	660	32.6	680	31.8	704	30.4		

ANALYSES OF PETROLEUM—*Continued*

Source	Physical Properties							Distillation				Remarks
	Baumé gravity	Color	Sulphur	Paraffin	Asphalt	Water	Begins to boil, ° F.	to 392° F.		392° F. to 527° F.		
			Per cent	Per cent	Per cent	Per cent		c.c.	Baumé gravity	c.c.	Baumé gravity	
TEXAS—*Continued*												
[illegible]												
Hardin [illegible]ty												
[illegible]n	27.7	....	....	....	6.0	....	....	*6.5	....	†20.4	....	* To 302° F.
Saratoga	17.8	black	....	....	0.74	trace	313	....	....	†18.0	28.1	† 302° to 527° F.
1 Sourlake	22.9	dark green	0.445	....	....	....	424	....	....	....	....	
2 [illegible]ke	17.8	black	0.447	....	....	....	454	....	....	....	....	
Harris County												
3 [illegible]se [illegible]k	23.2	dark green	0.117	....	....	....	420	....	....	....	....	
4 [illegible]le	25.7	green	0.154	....	....	....	210	....	....	....	....	
5 Humble	19.2	dark green	0.167	....	....	....	436	....	....	....	....	

PERCENTAGE BY VOLUME

	10		20		30		40		50		60		70		80		90		95	
	° F.	Gravity	° F.	Gravity	° F.	Gravity	° F.	Gravity	° F.	Gravity	° F.	Gravity	° F.	Gravity	° F.	Gravity	° F.	Gravity	° F.	Gravity
1	502	32.4	544	29.6	584	27.4	624	25.7	666	24.1	704	23.3	728	22.5	736	22.7				
2	546	28.7	594	25.0	640	22.5	674	21.0	704	20.4	716	20.3	716	21.7	714	22.5				
3	464	32.5	510	30.2	548	28.5	586	26.9	630	25.5	684	24.2	700	23.8	704	24.6				
4	380	44.9	492	36.4	542	30.5	588	26.1	654	23.3	664	23.0	677	25.0	677	25.5				
5	536	29.2	574	26.6	616	24.3	638	23.6	670	22.9	680	24.7	684	25.7	704	24.3				

Analyses of Petroleum—Continued

Source	Physical Properties: Baumé gravity	Color	Sulphur, Per cent	Paraffin, Per cent	Asphalt, Per cent	Water, Per cent	Begins to boil, ° F.	Distillation: to 392° F., c.c.	to 392° F., Baumé gravity	392° F. to 527° F., c.c.	392° F. to 527° F., Baumé gravity	Remarks
TEXAS—*Continued*												
Coastal—Continued												
Jefferson County												
Beaumont	23.6	dark green	....	....	....	....	....	*3.12	....	†12.5	....	* To 302° F. † 302° to 527° F.
Spindle Top	21.4	dark green	1.046	....	....	....	275	* .5	....	†37.0	32.6	
Liberty County												
[1] Hull	36.0	dark green	0.113	....	....	....	109	....	....	....	....	
[2] Hull	21.3	....	0.320	....	....	....	340	....	....	....	....	
[3] Hull Creek	21.2	dark green	0.301	....	....	....	186	....	....	....	....	
Southern												
Bexar County												
San Antonio	29.1	reddish-brown	2.02	....	....	....	....	*4.73	57.9	†36.9	42.6	

Percentage by Volume

	10 ° F.	10 Gravity	20 ° F.	20 Gravity	30 ° F.	30 Gravity	40 ° F.	40 Gravity	50 ° F.	50 Gravity	60 ° F.	60 Gravity	70 ° F.	70 Gravity	80 ° F.	80 Gravity	90 ° F.	90 Gravity	95 ° F.	95 Gravity
1	224	67.2	251	55.9	293	51.9	340	47.7	415	41.7	508	33.3	590	26.5	650	25.3				
2	494	33.7	540	29.0	576	26.5	620	24.4	662	22.9	670	22.3	690	25.1	710	24.7	664	26.6	---	
3	432	37.5	502	32.6	544	27.2	592	24.7	642	21.8	666	22.4	674	23.3	680	23.8				

ANALYSES OF PETROLEUM—*Continued*

Source	Physical Properties							Distillation				Remarks
	Baumé gravity	Color	Sulphur	Paraffin	Asphalt	Water	Begins to boil, ° F.	to 392° F.		392° F. to 527° F.		
			Per cent	Per cent	Per cent	Per cent		c.c.	Baumé gravity	c.c.	Baumé gravity	
TEXAS—*Continued*												
Southern—Continued												
Brazoria County												
1 Damon Mound	22.8	dark green	0.301	....	....	....	302	....	....	....	....	
2 West Columbia	19.8	dark green	0.171	....	....	....	415	....	....	....	....	
3 West Columbia	24.3	green	0.030	....	....	....	325	....	....	....	....	
4 West Columbia	23.5	verydark brown	0.260	....	....	....	432	....	....	....	....	
5 West Columbia	22.4	dark green	0.176	....	....	....	386	....	....	....	....	

PERCENTAGE BY VOLUME

	10		20		30		40		50		60		70		80		90		95	
	° F.	Gravity	° F.	Gravity	° F.	Gravity	° F.	Gravity	° F.	Gravity	° F.	Gravity	° F.	Gravity	° F.	Gravity	° F.	Gravity	° F.	Gravity
1	468	34.2	504	29.1	536	27.1	571	25.3	613	23.5	649	22.6	660	23.8	676	25.3				
2	505	29.4	548	26.6	590	24.6	632	22.7	674	21.2	696	20.5	635	25.5	715	24.1				
3	465	33.5	503	29.5	533	27.6	565	25.8	590	23.9	630	23.7	680	23.5	708	24.3				
4	507	30.9	542	28.8	569	27.3	605	26.3												
5	484	35.5	526	31.0	564	28.0	598	26.6	654	24.9	658	24.3	708	24.5	708	25.2	700	25.6		

ANALYSES OF PETROLEUM—*Continued*

Source	Physical Properties: Baumé gravity	Color	Sulphur, Per cent	Paraffin, Per cent	Asphalt, Per cent	Water, Per cent	Begins to boil, ° F.	Distillation: to 392° F., c.c.	to 392° F., Baumé gravity	392° F. to 527° F., c.c.	392° F. to 527° F., Baumé gravity	Remarks
TEXAS—*Continued*												
Southern—Continued												
Medina County												
Dunlay	25.0	reddish-brown	2.09	....	....	....	341	*6.1	56.4	†29.9	38.0	* To 302° F. † 302° to 527° F.
[1] Pecos County	25.3	....	0.672	....	....	....	290	....	....	....	....	
[2] Pecos County	33.5	....	0.523	....	....	trace	206	....	....	....	....	
Reeves County												
Ross Pool	24.2	dark green	....	....	....	....	341	....	....	†23.0	35.8	U.S. Geol. Survey, Mineral Resources, II, 1909
Toyah	23.4	dark brown	1.00	....	....	....	....	*16.0	59.1	†24.6	35.4	
UTAH												
San Juan County												
Goodridge	39.4	bl ck	0.26	6.09	0.80	trace	70	12.0	63.2	36.0	46.3	
Uinta County												
Whiskey Run	17.2	brown	....	....	....	trace	250	....	....	28.0	23.2	

PERCENTAGE BY VOLUME

	10		20		30		40		50		60		70		80		90		95	
	° F.	Gravity	° F.	Gravity	° F.	Gravity	° F.	Gravity	° F.	Gravity	° F.	Gravity	° F.	Gravity	° F.	Gravity	° F.	Gravity	° F.	Gravity
1	448	44.2	524	36.1	586	32.1	620	29.5	695	28.4	715	28.3	730	28.8	745	28.7	758	28.1		
2	298	56.3	350	49.9	412	44.8	492	39.4	568	35.2	628	32.5	656	32.0	670	31.6	696	31.0		

ANALYSES OF PETROLEUM—*Continued*

Source	Physical Properties						Begins to boil, ° F.	Distillation				Remarks
	Baumé gravity	Color	Sulphur	Paraffin	Asphalt	Water		to 392° F.		392° F. to 527° F.		
			Per cent	Per cent	Per cent	Per cent		c.c.	Baumé gravity	c.c.	Baumé gravity	
UTAH—*Continued*												
Uintah County												
Uintah River	22.5	black	0.45	29.4	5.9	....	60	2.1	....	19.5	48.6	
WEST VIRGINIA												
Northern												
Doddridge County												
Eagle Mills	46.3	dark green	....	5.10	....	....	127	*16.0	68.0	†40.0	49.7	U. S. Geol. Survey, Mineral Resources, II, 1909
Sullivan Pool	47.8	dark amber	....	6.02	....	....	131	*19.0	66.0	†40.0	49.2	
Morgansville Pool	44.7	medium green	....	5.19	....	....	136	*16.0	65.5	†38.0	48.1	* To 302° F. † 302 to 527° F.
Hancock County	48.8	....	....	....	....	....	77	45.9	65.0	16.5	46.3	
Harrison County												
Shinnston Pool	45.5	medium amber	....	9.73	....	....	161	*14.0	67.1	†40.0	50.1	
Lewis County												
Gantz sand	40.0	black	....	7.10	....	....	176	*9.5	62.8	†36.0	48.1	
Monongalia County												
[1] Morgantown	43.5	green	0.14	....	....	....	162	....	....	....	....	

PERCENTAGE BY VOLUME

	10		20		30		40		50		60		70		80		90		95	
	° F.	Gravity	° F.	Gravity	° F.	Gravity	° F.	Gravity	° F.	Gravity	° F.	Gravity	° F.	Gravity	° F.	Gravity	° F.	Gravity	° F.	Gravity
1	257	68.6	32[illegible]	59.7	401	54.0	478	49.1	550	44.1										

ANALYSES OF PETROLEUM—*Continued*

Source	Physical Properties: Baumé gravity	Color	Sulphur	Paraffin	Asphalt	Water	Begins to boil, ° F.	Distillation: to 392° F.		392° F. to 527° F.		Remarks
			Per cent	Per cent	Per cent	Per cent		c.c.	Baumé gravity	c.c.	Baumé gravity	
WEST VIRGINIA—*Continued*												
Northern—Continued												
Pleasants County												
Jefferson Township	41.3	light green	....	7.72	....	....	284	*.7	....	†48.0	49.7	* To 302° F. † 302° to 527° F.
McKim Township	42.1	dark green	....	6.71	....	....	246	*4.0	57.6	†49.0	49.2	
Arvilla Pool	47.6	medium green	....	3.18	....	....	253	*2.0	....	†68.0	537	
Lytton Pool	51.2	dark amber	....	4.87	....	....	154	*24.5	68.3	†37.0	50.6	
Tyler County	43.3	medium amber	....	6.11	....	....	158	*14.0	65.5	†38.0	48.5	
Central												
Kanawha County												
Cabin Creek	45.7	....	0.19	....	....	....	77	40.5	61.8	21.0	45.7	
Kelly Creek	45.2	....	0.11	....	....	....	73	39.6	61.5	18.4	45.4	
Blue Creek	42.8	....	0.11	....	....	....	81	40.2	62.6	18.0	45.7	
Western and Central												
Ritchie County												
Grant Township	27.4	brown	....	4.08	....	....	438	....	....	11.0	34.3	
Murphy Township	44.9	dark green	....	4.22	....	trace	194	*16.0	65.2	†39.0	50.1	
Clay Township	45.[illegible]	light green	....	3.61	....	trace	174	*11.0	68.8	†44.0	51.8	
Union Township	48.6	light amber	....	7.44	....	....	159	*20.0	66.0	†39.0	50.4	

ANALYSES OF PETROLEUM—*Continued*

Source	Physical Properties							Distillation				Remarks
	Baumé gravity	Color	Sulphur	Paraffin	Asphalt	Water	Begins to boil, ° F.	to 392° F.		392° F. to 527° F.		
			Per cent	Per cent	Per cent	Per cent		c.c.	Baumé gravity	c.c.	Baumé gravity	
WEST VIRGINIA—*Continued*												
Western and Central—Continued												
Wood County												
Williams Township	42.0	dark green	. .	5.89	. .	trace	206	*10.0	64.7	†39.0	50.4	* To 302° F. † 302° to 527° F.
Union Township	39.7	dark green	. .	5.61	. .	trace	267	*3.0	....	†44.0	50.4	
Walker Township	30.0	dark green	. .	....	. .	much	329	....	....	†16.0	37.6	
WYOMING												
Northern												
Big Horn County												
Greybull Field	44.3	....	0.08	....	....	....	81	38.6	59.7	17.8	43.7	
Torchlight Dome	39.5	dark wine	....	....	....	....	75	26.0	58.0	34.5	42.7	U. S. Geol. Survey Bull. 621, pp. 157–190
Park County												
Elk Basin Field	39.3	....	0.14	....	....	....	79	40.5	57.2	17.4	38.9	
Elk Basin Field	45.4	....	....	....	....	....	....	....	....	10	39.9	
Weston County												
New Castle	36.7	....	0.115	....	....	....	118	31.6	55.7	17.4	39.5	
Osage Range	37.3	....	0.29	....	....	....	75	34.8	57.7	15.8	39.7	

ANALYSES OF PETROLEUM—*Continued*

Source	Physical Properties							Distillation				Remarks
	Baumé gravity	Color	Sulphur	Paraffin	Asphalt	Water	Begins to boil, °F.	to 392° F.		392° F. to 527° F.		
			Per cent	Per cent	Per cent	Per cent		c.c.	Baumé gravity	c.c.	Baumé gravity	
WYOMING—*Continued*												
Northern—Continued												
Weston County—*Continued*												
[1] Osage Range	41.0	....	....	....	....	....	100	....	56.0	....	....	
Upton-Thornton	40.7	light olive-green	0.14	....	....	....	75	*25.6	53.2	†31.0	35.4	* To 392° F. † 392° to 527° F.
Central												
Converse County												
Big Muddy Field	32.2	....	0.17	....	....	0.7	136	22.2	53.7	15.7	38.3	
Big Muddy Field	34.4	....	....	....	....	....	123	....	....	15.0	42.0	
Fremont County												
Dallas	23.2	....	2.42	....	....	1.0	77	12.8	51.3	14.5	39.3	
Lander	23.3	....	2.62	....	....	trace	129	11.0	55.4	16.0	39.7	
Maverick	21.8	dark brown	2.46	....	....	....	219	8.6	53.0	14.7	39.9	
Pilot Butte	35.1	....	0.22	....	....	....	140	24.0	53.0	19.7	41.8	
Popo Agie	25.5	black	0.66	3 to 5	....	....	....	2 to 5	....	30to40		

PERCENTAGE BY VOLUME

	10		20		30		40		50		60		70		80		90		95	
	°F.	Gravity	°F.	Gravity	°F.	Gravity	°F.	Gravity	°F.	Gravity	°F.	Gravity	°F.	Gravity	°F.	Gravity	°F.	Gravity	°F.	Gravity
1	220	.....	278	.....	350															

ANALYSES OF PETROLEUM—*Continued*

Source	Physical Properties: Baumé gravity	Color	Sulphur, Per cent	Paraffin, Per cent	Asphalt, Per cent	Water, Per cent	Begins to boil, ° F.	Distillation: to 392° F., c.c.	to 392° F., Baumé gravity	392° F. to 527° F., c.c.	392° F. to 527° F., Baumé gravity	Remarks
WYOMING—*Continued*												
Central—Continued												
Fremont County—*Continued*												
Plunkett	35.5	....	0.55	....	....	....	189	21.0	49.7	22.6	39.3	
Shoshone	35.6	dark green	....	7.6	....	....	190	....	....	37.0	41.3	U. S. Geol. Survey Bull. 656
Hot Springs County												
Grass Creek	43.1	....	0.14	....	....	....	....	42.6	58.9	20.0	40.7	
Gas Creek	43.7	....	....	....	....	....	102	....	....	6.0	41.3	
Hamilton Dome	25.0	....	2.09	....	....	....	82	....	57.7	....	39.5	
Warm Spring	11.8	....	2.61	....	....	31.6	180	....	49.9	....	38.1	
Natrona County												
Pine Mtn.	16.9	....	0.51	....	....	7.0	....	....	....	....	....	
Salt Creek	36.5	....	0.18	6.93	2.77	7.0	77	29.3	56.7	15.7	39.9	
Shannon Field	24.0	green	0.20	....	....	....	282	3.1	37.1	11.1	31.5	
Niobrara County												
Mule Ck.	31.8	....	....	....	....	....	185	8.88	56.0	19.18	42.7	
Mule Ck.	31.5	....	0.14	....	....	....	212	11.7	52.3	17.4	40.5	
[illegible] Ck.	43.5	black	0.043	....	....	....	117	33.5	55.7	16.2	42.4	
Lance Creek	36.4	....	....	....	....	0.7	....	....	....	23.0	42.5	
Southern												
Carbon County												
Ferris Field	36.3	....	0.19	....	....	....	75	31.1	57.4	13.4	40.1	
Lost Soldier	30.0	....	0.11	....	....	....	180	16.7	44.1	18.5	33.9	
Rock Creek	36.1	....	0.27	....	....	....	77	31.4	58.2	14.2	39.5	

ANALYSES OF PETROLEUM—*Continued*

Source	Physical Properties: Baumé gravity	Color	Sulphur, Per cent	Paraffin, Per cent	Asphalt, Per cent	Water, Per cent	Begins to boil, ° F.	Distillation: to 302° F., c.c.	to 302° F., Baumé gravity	302° F. to 527° F., c.c.	302° F. to 527° F., Baumé gravity	Remarks
WYOMING—*Continued*												
Southern—Continued												
[1] Southern Wyoming	39.7	dark green	0.107	....	....	....	135	....	....	....	....	
[2] Southern Wyoming	31.5	dark green	0.125	....	....	....	180	....	....	....	....	
[3] Southern Wyoming	23.7	black	0.164	....	....	....	220	....	....	....	....	
Southwestern												
Sweetwater County												
Rock Springs	....	....	....	....	....	....	....	*21.3	65.0	†39.7	44.0	U. S. Geol. Survey Bull. 702, p. 103
Uinta County												* To 302° F.
Spring Valley	44.0	....	0.03	6.2	....	....	....	*21.3	65.0	†39.7	44.0	† 302° to 527° F.

PERCENTAGE BY VOLUME

	10 ° F.	10 Gravity	20 ° F.	20 Gravity	30 ° F.	30 Gravity	40 ° F.	40 Gravity	50 ° F.	50 Gravity	60 ° F.	60 Gravity	70 ° F.	70 Gravity	80 ° F.	80 Gravity	90 ° F.	90 Gravity	95 ° F.	95 Gravity
1	230	70.7	280	62.1	340	55.3	440	48.3	535	41.4	615	37.3	680	34.6	715	32.5				
2	335	54.9	430	40.9	510	35.9	565	33.9	615	33.2	600	32.5	705	31.2	730	31.7				
3	424	51.5	504	42.3	572	36.5	632	32.1	668	29.2	684	26.9	694	25.7	702	24.8				

ANALYSES OF PETROLEUM—*Continued*

Source	Physical Properties							Distillation				Remarks
	Baumé gravity	Color	Sulphur	Paraffin	Asphalt	Water	Begins to boil, ° F.	to 392° F.		392° F. to 527° F.		
			Per cent	Per cent	Per cent	Per cent		c.c.	Baumé gravity	c.c.	Baumé gravity	
MEXICO												
[illegible]as	28.6	....	....	....	....	1.29	....	16.5	37.8	28.3	24.5	
Guerrero	33.5	dark brown	....	....	....	....	....	*1.3	61.7	†35.9	40.3	* To 302° F. † 302° to 527° F.
[illegible]o	....	....	....	....	....	....	....	*9.8	32.4	†39.5	22.8	
San Luis Potosi												
[illegible]o	12.4	......	......	......	......	......	....	....	....	....	....	
San [illegible]	48.4	reddish brown	....	....	....	....	132	*40.2	62.5	†35.0	39.2	Boletin de Petrolio, Vol. 6, p. 354
[illegible]o [illegible]t	38.6	......	......	......	......	......	....	....	....	....	....	
Vera Cruz												
[illegible]o	23.3	black	2.7	....	....	....	....	....	....	....	....	Boletin de [illegible], Vol. 5, p. 245
Furbero	26.4	coffee-colored	2.46	....	20	....	....	*15.0	52.5	†36.0	31.2	Boletin de Petrolio, Vol. 5, pp. 244–245
Tanhuijo	23.0	black	1.69	....	....	6.66	....	3.3	21.5	11.7	43.4	Boletin de [illegible], Vol. 5, pp. 20–36
Tecuanapa	33.5	brown	1.00	....	4	....	....	*13.0	61.7	†36.0	34.3	Boletin de [illegible], Vol. 6, p. 33
[illegible]a	34.5	....	0.9	....	....	....	....	*10.0	....	†40.0	....	Boletin de Petrolio, Vol. 2, p. 442
[illegible]a	48.6	....	trace	....	....	....	311	....	....	†98.0	49.0	
Tuxpam	13.1	black	1.12	....	34.4	....	....	*10.6	32.2	†52.6	22.1	Boletin de Petrolio, Vol. 4, pp. 320–323
Chinampa	20.5	black	3.28	....	....	....	118	*8.7	55.4	†53.0	....	Boletin de Petrolio, Vol. 5, pp. 246–247

Analyses of Petroleum—*Continued*

Source	Physical Properties: Baumé gravity	Color	Sulphur, Per cent	Paraffin, Per cent	Asphalt, Per cent	Water, Per cent	Begins to boil, ° F.	Distillation: to 302° F., c.c.	to 302° F., Baumé gravity	302° F. to 527° F., c.c.	302° F. to 527° F., Baumé gravity	Remarks
Mexico—*Continued*												
Heavy Mexican	12.8	....	3.6	0.50	....	....	....	2.9	48.0	5.9		
Heavy Mexican	12.9	black	4.9	....	....	....	260	....	....	....	....	10 per cent cut, 500° F.
[1] Heavy Crude	12.5	black	4.17	....	....	....	....	6.0	49.9	6.0	37.7	
Light Mexican	20.3	....	2.6	1.62	....	much	....	4.4	49.0	16.0		
Light Mexican	19.2	black	3.19	....	....	....	158	....	....	....	....	10 per cent cut, 350° F., gr. 61.0 Bé., 20 per cent. 518° F., gr. 45.8 Bé.
[2] Light Crude	20.5	black	3.06	....	....	....	....	15.0	53.0	5.75	42.5	
Cuba												* To 302° F.
Havana	29.0	....	....	....	....	....	....	*13.0	....	†35.0		† 302° to 527° F.
Bacauranao	26.5	....	....	....	....	....	210	25.0	56.0	20.0	40.5	
Santiago	28.0	black	....	....	....	much	....	*13.0	60.7	†31.0	41.5	Boletin de Minas, 1, 1916
Pinar del Rio Candelaria	16	....	....	....	....	1.43	....	0	....	21.6	....	
Trinidad	14 to 42	....	....	....	....	....	....	*41.0	....	†37.0	....	Nat. Pet. News, Vol. 8, p. 92, July, 1916
Canada												
Alberta		....	....	....	....	....	....	....	....	....	....	
Ft. McKay	14.5	....	....	....	....	....	....	*2	26.9	†70	....	Canada Dept. of Mines, Mines Branch, 291

Percentage by Volume

10		20		30		40		50		60		70		80		90		95	
° F.	Gravity	° F.	Gravity	° F.	Gravity	° F.	Gravity	° F.	Gravity	° F.	Gravity	° F.	Gravity	° F.	Gravity	° F.	Gravity	° F.	Gravity
[1] 463	40.4	570	30.5	615	26.9	605	30.2	615	28.1	630	27.7	520	30.1						
[2] 363	56.7	491	44.2	571	36.4	646	30.6	645	29.6	651	27.1	656	24.9						

ANALYSES OF PETROLEUM—*Continued*

Source	Physical Properties: Baumé gravity	Color	Sulphur, Per cent	Paraffin, Per cent	Asphalt, Per cent	Water, Per cent	Begins to boil, ° F.	Distillation: to 392° F., c.c.	to 392° F., Baumé gravity	392° F. to 527° F., c.c.	392° F. to 527° F., Baumé gravity	Remarks
CANADA—*Continued*												
South Kootenai	....	....	....	....	....	....	....	*5.7	52.0	†56.3	41.0	Canada Dept. of Mines, Mines Branch, 291
Mackenzie												
Camp Creek												
Crude	36.0	light	....	....	....	....	....	*22.5	....	†38.5	....	Pet. Times, Vol. 5, 121, p. 490
Seepage	24.7	greenish black										
New Brunswick	36.8	....	....	....	....	....	....	14.0	65.2	37.6	46.1	Fifty-eighth Annual Report of the Crown Land Dept. of New Brunswick
Newfoundland	43.2	dark brown	....	....	....	....	....	12.7	....	55.2	....	Engler and Höfer, Das Erdöl
Ontario												
Kent Region												
Both [illegible]ll	37.0	....	....	....	....	....	....	19.6	....	32.0	....	Engler and Höfer, Das Erdöl
[illegible]x County												
Oil Springs	35.8	....	....	....	....	....	....	7.0	61.0	33.0	52.5	Engler and Höfer, Das Erdöl
Petrolia	....	....	....	....	....	....	....	12.5	60.4	35.8	40.7	Canadian Mining Institute, pp. 371–398, 1916
Petrolia	32.4	....	....	....	....	....	....	11.0	52.5	14.0	40.3	Engler and Höfer, Das Erdöl
[illegible]ec												
[illegible]pé	17 to 35	light amber	.09 to .20	....	....	....	....	....	....	....	....	Canada Dept. of Mines, Mines Branch, 291
ARGENTINA												
[illegible]lt												* To 302° F.
Comodoro												† 302° to 527° F.
Rivadavia	21.6	....	0.24	....	....	1.0	....	*3.3	....	†15.2	....	

ANALYSES OF PETROLEUM—*Continued*

Source	Physical Properties							Distillation				Remarks
	Baumé gravity	Color	Sulphur	Paraffin	Asphalt	Water	Begins to boil, ° F.	to 392° F.		392° F. to 527° F.		
			Per cent	Per cent	Per cent	Per cent		c.c.	Baumé gravity	c.c.	Baumé gravity	
ARGENTINA—*Continued*												
Mendoza	....	....	....	....	....	....	....	5.0	59.1	30.0	41.9	Pet. Age, Vol. 4, pp. 33-34, Aug., 1917
Punta Arenas	....	reddish-brown	....	....	....	....	....	....	....	28.0	....	Chemical Abstracts, Vol. 12, 1, p. 93
BOLIVIA												
Yacuiba	25.9	....	....	none	....	....	338	2.0	....	29.0	....	Engler and Höfer, Das Erdöl
BRAZIL	49.4	....	....	....	....	....	....	*9.0	44.5	†34.0	22.5	Roches Petroliferas do Brazil, Boletin No. 1
COLOMBIA												
Arboletes	15.3	reddish-brown	....	....	....	....	....	0	....	†25.0	....	Oil and Gas Jour., Vol. 19, p. 84, Sept. 10, 1920
Magdalena River	17.6	....	....	....	....	....	....	....	....	....	....	Engler and Höfer, Das Erdöl
Pinar del Rio	....	red	....	....	....	....	....	*6.0	....	†24.0	....	Oil and Gas Jour., Vol. 19, p. 84, Sept. 10, 1920
Sinu River	33.1	....	....	3	....	....	....	2.9	....	31.0	....	Viscosity 0.98
ECUADOR	22.5	dark greenish-brown	0.63	....	....	....	....	....	....	36.9	37.3	* To 302° F. † 302° to 527° F.
Ecuador	20.9	....	....	....	....	....	....	2.1	....	15.6	....	
Santa Elena	29.2	....	....	....	....	....	....	*2.8	....	†48.0	....	U. S. Dept. of Commerce, No. 165, pp. 186-188 July, 1916

ANALYSES OF PETROLEUM—*Continued*

Source	Physical Properties							Distillation				Remarks
	Baumé gravity	Color	Sulphur	Paraffin	Asphalt	Water	Begins to boil, ° F.	to 392° F.		392° F. to 527° F.		
			Per cent	Per cent	Per cent	Per cent		c.c.	Baumé gravity	c.c.	Baumé gravity	
PERU												
Peruvian Crude	35.7	dark green	0.60	....	heavy	....	....	....	18.7	....	28.6	* To 302° F. † 302° to 527° F.
Peruvian Crude	33.0	....	....	....	....	....	....	....	15.4	....	20.3	
Lima	16.7	dark brown	0.186	....	heavy	....	....	....	....	....	....	
Lobitos Pool	31 to 35	....	....	....	....	....	....	....	....	....	....	Oil and Gas Jour., Vol. 14, March 19, 1916
Northern	....	....	....	....	....	....	....	15to25	....	24.0	....	Oil and Gas Jour., Vol. 14, March 19, 1916
Puno Area	....	....	....	....	....	....	....	none	....	7 to 20	....	Oil and Gas Jour., Vol. 14, March 19, 1916
Zorritos	41.7	green	....	....	....	....	....	22.0	....	44.0	....	Engler and Höfer, Das Erdöl
VENEZUELA	20.8	....	....	....	....	....	....	3.0	....	7.0	....	Chemical Abstracts, Vol. 15, No. 9, p. 1395
Pesquero	22.5	very dark	....	....	....	....	....	....	....	30.0	....	Engler and Höfer, Das Erdöl
FRANCE												
Alsace	22.1	....	....	....	....	....	....	5.0	....	20.0	....	Revue Générale de l'électricité, Vol. 6, Part 1, pp. 246–247
Durrenbach	26.0	black	....	....	....	....	....	....	....	....	....	Nat. Pet. News, April 13, 1921
Pechelbronn	30.1	....	....	7.05	1.51	....	....	*1.3	....	†24.0	....	Bulletin de la Société pour l'encouragement de l'industrie nationale

ANALYSES OF PETROLEUM—*Continued*

Source	Physical Properties							Distillation				Remarks
	Baumé gravity	Color	Sulphur	Paraffin	Asphalt	Water	Begins to boil, °F.	to 392° F.		392° F. to 527° F.		
			Per cent	Per cent	Per cent	Per cent		c.c.	Baumé gravity	c.c.	Baumé gravity	
FRANCE—*Continued*												
Alsace—*Continued*												
Pechelbronn	....	....	....	3	....	....	....	5.5	....	21	....	Revue Générale de l'électricité, Vol. 6, Part 1, pp. 246–247
GERMANY												
Hanover												
Horst	30.6	....	....	....	....	....	....	1.9	....	35.1		
Wietsendorf	27.6	dark brown	....	4.54	....	....	....	10.0	....	25.0	....	Engler and Höfer, Das Erdöl
Wietsendorf	19.7	dark brown	....	....	....	....	....	0.5	....	12.0	....	Engler and Höfer, Das Erdöl
Königsee	34.7	....	....	2.0	....	....	....	13.3	....	29.4	....	Engler and Höfer, Das Erdöl
Odesse	34.9	....	....	1.1	....	....	....	12.8	63.1	31.4	47.2	Engler and Höfer, Das Erdöl
Oelheim	25.7	....	....	....	....	....	338	....	....	32.0	....	Engler and Höfer, Das Erdöl
Oelheim	28.1	....	....	....	....	....	....	0.7	56.6	11.0	43.9	Engler and Höfer, Das Erdöl
Oelheim	24.0	....	....	....	....	....	194	0.8	....	22.7	39.2	Engler and Höfer, Das Erdöl
GREAT BRITAIN												
England												
Hardstoft	39.0	dark brown	....	....	....	....	....	....	....	....	....	Chemical Abstract, p. 2591, Oct. 20, 1919
Scotland												
Shale Oil	38.6	green	....	1.25	....	....	....	10.2	....	34.1	....	Proc. Roy. Soc. Edinburgh, Vol. 36, pp. 44–86, 1916
Dunnett Mine Shale Oil	31.6	....	....	....	....	....	....	4.5	....	36.1	....	do.

ANALYSES OF PETROLEUM—*Continued*

Source	Physical Properties							Distillation				Remarks
	Baumé gravity	Color	Sulphur	Paraffin	Asphalt	Water	Begins to boil, ° F.	to 392° F.		392° F. to 527° F.		
			Per cent	Per cent	Per cent	Per cent		c.c.	Baumé gravity	c.c.	Baumé gravity	
GREAT BRITAIN—*Continued*												
Scotland—*Continued*												
Sandhole Pit	38.6	....	....	....	....	....	....	10.2	....	34.1	....	Proc. Roy. Soc. Edinburgh, Vol. 36, pp. 44–86, 1916
ITALY												
Parma												
Borgo San Donino	....	....	....	....	....	....	193	15.8	....	10.8	....	Engler and Höfer, Das Erdöl
Marzolaro	24.7	black	....	....	....	....	....	7.7	....	17.9	....	Engler and Höfer, Das Erdöl
Neviano de Rossi	51.8	bright red	....	....	....	....	....	18.9	....	69.5	....	Engler and Höfer, Das Erdöl
N[illegible]ano de [illegible]ssi	47.8	straw-yellow	....	....	....	....	....	52.0	53.9	38.0	....	Chemical Abstracts, Vol. 11, No. 23, p. 3424
Pavia	....	pale pink	....	....	....	....	....	....	....	52.0	28.7	Chemical Abstracts, Vol. 13, No. 2, p. 183
Retorbido	28.3	dark	....	....	....	....	....	7.5	....	91	....	Engler and Höfer, Das Erdöl
Rivanazzano	23.0	....	....	....	....	....	....	none	....	35to40	....	Pet. World (London), Vol. 13, p. 275, 1916
Piacenza												
Montechiaro	50.4	....	....	....	....	....	194	36.0	....	58.9	....	Engler and Höfer, Das Erdöl
Terra di Lavoro	14.3	black	....	....	....	....	....	0.34	....	27.8	....	Engler and Höfer, Das Erdöl
POLAND												
Galicia	33.6	dark green	....	....	....	....	....	12.5	....	37.5	....	
Galicia	35.7	....	....	....	....	....	....	8.8	....	37.4	....	* To 302° F.
Galicia	32.8	....	....	....	....	....	....	35.0	....	21.0	....	Chemical Abstracts, Vol. 14,
Tustanowice	29.8	....	....	8.37	6	....	....	*1.3	....	27.8	....	No. 22, p. 3523

Analyses of Petroleum—*Continued*

Source	Physical Properties							Distillation				Remarks
	Baumé gravity	Color	Sulphur	Paraffin	Asphalt	Water	Begins to boil, ° F.	to 392° F.		392° F. to 527° F.		
			Per cent	Per cent	Per cent	Per cent		c.c.	Baumé gravity	c.c.	Baumé gravity	
Rumania												
Campina	32.1	black	....	....	....	....	....	11.0	77.1	0.15	49.0	Pet. Review, p. 247, Sept. 23, 1916
Russia												
Azerbaijan												
Baku	30 to 32	....	....	....	....	....	....	....	....	25	....	Pet. World (Los Angeles), Vol. 5, p. 7, March, 1920
Balakhani	30.4	....	....	....	....	....	....	6.3	....	32.5		
Bibi-Eibat	33.0	....	....	....	....	....	....	10.5	....	40.0	....	Redwood, Sir B., A Treatise on Petroleum
Grozni	28.4	....	....	....	....	....	....	20.0	....	20.0	....	
Ilsky	34.1	....	....	....	....	....	....	20.0	....	40.0	....	
Ilsky	18.9	....	....	....	....	....	....	1.0	....	9.0	....	
Surakhani	49.5	....	....	....	....	....	....	48.9	....	43.9	....	
Spain												
Cadiz (near Villa Martin)	45.6	light amber	0	2.52	....	....	87	27.5	59.8	53.0	43.1	
India												
Assam	33.5	black	....	much	....	....	....	....	67.1	....	28.1	Chemical Abstracts, Vol. 11, No. 23, p. 3425
Burma												
Lower Burma	32	reddish-brown	....	3.7	....	....	158	39.0	....	18.7	....	Engler and Höfer, Das Erdöl
Upper Burma												
Khodoung	32	reddish-brown	....	....	....	....	275	11.0	....	30.0	....	Engler and Höfer, Das Erdöl

ANALYSES OF PETROLEUM—*Continued*

Source	Physical Properties							Distillation				Remarks
	Baumé gravity	Color	Sulphur	Paraffin	Asphalt	Water	Begins to boil, ° F.	to 392° F.		392° F. to 527° F.		
			Per cent	Per cent	Per cent	Per cent		c.c.	Baumé gravity	c.c.	Baumé gravity	
INDIA—*Continued*												
Burma—*Continued*												
Minba	30	brown	....	....	....	....	....	....	....	2.0	....	Engler and Höfer, Das Erdöl
Twingon	32	reddish-brown	....	....	....	....	266	13.0	....	31.0	....	Engler and Höfer, Das Erdöl
Yenangyat	30	greenish-brown	....	....	....	....	266	33.0	....	33.0	....	Engler and Höfer, Das Erdöl
JAPAN												
Amaze	39.9	....	....	1.5	....	....	122	25.0	....	53.6	....	Engler and Höfer, Das Erdöl
Hi.	32.4	....	....	....	....	....	95	27.8	....	38.3	....	Engler and Höfer, Das Erdöl
Hitatany	26.4	....	....	none	....	....	122	12.0	....	24.0	....	Engler and Höfer, Das Erdöl
Ushi	18.4	....	....	....	....	....	....	....	....	25.0	....	Engler and Höfer, Das Erdöl
Katsura	29.6	....	....	....	....	....	113	20.2	....	38.8	....	Engler and Höfer, Das Erdöl
Kusodzu	21.0	....	....	....	....	....	....	....	....	25.0	....	Engler and Höfer, Das Erdöl
Miyagawa	26.0	....	....	....	....	....	149	15.0	....	36.8	....	Engler and Höfer, Das Erdöl
PERSIA	41.7	brown	0.4	....	....	....	....	9.4	....	57.6	46.7	Redwood, Sir B., Petroleum and Its Products, Vol. I, p. 39
TURKESTAN												
Ferghana												
Leakon	....	. .	. .	....	. .	. .	....	0.5	64.4	31.0	45.6	Engler and Höfer, Das Erdöl
Mailly Say	28.1	. .	. .	....	. .	. .	....	10.0	67.7	22.0	47.4	Engler and Höfer, Das Erdöl
Tschimeon	32.7	. .	. .	8	. .	. .	....	14.5	....	37.4	....	Engler and Höfer, Das Erdöl
URALSK	32.6	green	....	....	....	....	....	0.36	....	23.6	....	Chemical Abstracts, p. 821, 1916

ANALYSES OF PETROLEUM—*Continued*

Source	Physical Properties							Distillation				Remarks
	Baumé gravity	Color	Sulphur	Paraffin	Asphalt	Water	Begins to boil, ° F.	to 392° F.		392° F. to 527° F.		
			Per cent	Per cent	Per cent			c.c.	Baumé gravity	c.c.	Baumé gravity	
Egypt												
Hurghada	35.2	very dark olive brown	....	1.55	....	....	....	....	....	....	....	
Hurghada		brownish-black	2.43	7.73	10.57	....	....	7.7	61.2	14.2	40.5	Pet. Times, Vol. 5, No. 115, pp. 327–328
Abu Durba	15.0	....	2.00	....	8.20	0.2	....	1.3	49.4	13.8	....	
Zeitia	22.3	black	1.00	2.10	2.10	....	....	....	....	....	....	
Gemsa	39.2	dark greenish brown	0.65	4.94	0.63	none	....	27.1	64.6	32.4	41.0	Pet. Times 5, 328–362.
Morocco	25.5	dark brown	....	....	....	....	194	20.0	....	52.0	....	Chemical Abstracts, Vol. 13, No. 5, p. 513
Borneo												
Light-asphalt crude	28.3	....	....	....	....	....	....	*26.8	....	†70.9	....	* To 302° F.
Paraffin-wax crude	28.9	....	....	....	....	....	....	*20.4	....	†74.0	....	† 302° to 527° F.
British North Borneo	15.0	....	....	....	....	....	201	25.0	....	32.0	....	
Koetei	33.0	....	....	....	....	....	....	17.4	....	46.0	....	
Koetei	31.8	....	....	....	....	....	....	4.6	....	43.3	....	
Sanga Sanga	32.4	....	....	none	....	....	....	18.5	51.9	51.9	....	Engler and Höfer, Das Erdöl
Sanga Sanga	15.9	....	....	....	....	....	....	2.4(*a*)	44.1	33.8(*b*)	26.2	(*a*) = to 302° F.
Sambodja	15.5	....	....	....	....	....	....	2.2(*a*)	42.6	37.4(*b*)	26.0	(*b*) = 302° to 527° F.
Tarakan	17 to 18	....	0.3	....	....	....	....	0.8	34.3	10.1	23.1	Jour. Inst. Pet. Tech., Vol. 7, July 1921, pp. 209-233.
Miri	21 to 27	brown	0.4	....	....	....	138	6.9	41.9	7.1	23.3	

ANALYSES OF PETROLEUM—*Continued*

Source	Physical Properties							Distillation				Remarks
	Baumé gravity	Color	Sulphur	Paraffin	Asphalt	Water	Begins to boil, ° F.	t° 392° F.		392° F. to 527° F.		
		Per cent	Per cent	Per cent	Per cent			c.c.	Baumé gravity	c.c.	Baumé gravity	
JAVA												
Berbeck	31.4	reddish-brown	....	....	....	....	185	*20.5	55.4	†44.8	35.8	Engler and Höfer, Das Erdöl
Cheribon	47.6	....	....	....	....	....	....	36.0	....	49.0	....	Engler and Höfer, Das Erdöl
Gorgor	17.5	light green	....	....	....	....	320	....	....	42.3	27.8	Engler and Höfer, Das Erdöl
Kutei	34.7	greenand reddish-brown	....	6.43	....	....	149	*27.7	56.6	†43.2	37.4	Engler and Höfer, Das Erdöl
Rembang	27.6	....	....	....	....	....	....	23.0	....	74.0	....	Engler and Höfer, Das Erdöl
Roengkoet	13.8	dark brown	....	....	....	....	....	....	....	25.4	27.3	Engler and Höfer, Das Erdöl
Surabaya	18.1	....	....	....	....	....	....	....	....	60.0	....	Engler and Höfer, Das Erdöl
NEW [illegible]A												
Papua												
Upoia Field	45.8	....	....	....	....	....	....	34.0	65.8	8.0	47.8	Jour. Chem. Ind., Vol. 38, No. 16, p. 319 T
Vailala Field	45.8	light brownish-yellow	....	....	....	....	....	31.0	68.0	4.2	47.6	Jour. Chem. Ind., Vol. 38, No. 16, p. 319 T
NEW ZEALAND												
Gisborne	32.0	green	....	8.88	....	some	210	....	....	43.0	61.8	
New Plymouth	36.4	dark	....	....	....	....	240	*20.0	53.2	†40.0	40.7	Henry, J. H., Oil-fields of New Zeland
Taranaki												
Motu Rua	....	....	....	21	....	0.15	....	*0.85	....	†30.0	....	New Zealand Official Year-book, 1916

ANALYSES OF PETROLEUM—*Continued*

Source	Physical Properties							Distillation				Remarks
	Baumé gravity	Color	Sulphur	Paraffin	Asphalt	Water	Begins to boil, ° F.	to 392° F.		392° F. to 527° F.		
			Per cent	Per cent	Per cent	Per cent		c.c.	Baumé gravity	c.c.	Baumé gravity	
NEW ZEALAND—*Continued*												
Taranaki	34.8	brown	....	14.7	....	some	206	*10.0	49.4	†50.0	39.0	
PHILIPPINES												
Cebu	28	brown	....	....	....	....	....	6.2	53.7	42.3	38.2	Econ. Geology, Vol. 11, pp. 246–265, April-May, 1916
Algeria (Seepage)	....	dark brown	....	....	....	....	....	*17.2	....	†30.5		* To 302° F. † 302° to 527° F.
Toledo Well	28.2	dark brown	....	....	....	....	....	*6.2	53.8	†42.3	38.2	* To 302° F. † 302° to 527° F.
Leyte	21.2	....	....	8.1	....	....	91	....	....	....	....	
Leyte	33.0	brown	....	8.14	....	....	....	*5.4	....	†33.7	....	Econ. Geology, Vol. 11, pp. 246–265, April-May, 1916
Mindanao												
Pidatan	20.6	....	1.56	....	....	trace	....	none	....	45.0	....	Oil and Gas Jour., Vol. 19, p. 78, May 6, 1921
Tayabas	38 to 39	wine red	none	8.1	....	....	....	*39.0	51.8	44.5	34.7	Econ. Geology, Vol. 11, pp. 246–265, April-May, 1916
East Coast	38.3	claret	....	....	....	....	100	18.0	51.9	58.0	38.6	
Bahya Well	38.2	light brown to wine	....	....	....	....	....	*39.0	51.9	†44.5	34.7	
SUMATRA	51.6	....	....	....	....	....	....	37.2	....	52.0	....	Engler and Höfer, Das Erdöl
East Sumatra												
Langkat No. 1	51.5	light brown	....	....	....	....	....	37.2	....	52.0	....	

ANALYSES OF PETROLEUM—*Continued*

Source	Physical Properties							Distillation				Remarks
	Baumé gravity	Color	Sulphur	Paraffin	Asphalt	Water	Begins to boil, ° F.	to 392° F.		392° F. to 527° F.		
			Per cent	Per cent	Per cent	Per cent		c.c.	Baumé gravity	c.c.	Baumé gravity	
SUMATRA—*Continued*												
Langkat No. 2	49.9	light brown	....	....	....	....	....	26.0	....	63.4	....	
Langkat No. 3	33.3	reddish-brown	....	....	....	....	....	17.5	....	32.8	....	
North Sumatra												
Perlak Field	10.0	light	....	0.56	....	....	134	56.2	....	34.0	....	Engler and Höfer, Das Erdöl
South [illegible]												
[illegible]Bat District	....	dark brown	....	0.22	....	....	167	36.0	....	19.0	....	Engler and Höfer, Das Erdöl
Bandjar Sari	15.0	....	....	....	....	....	....	42.8	....	11.5	....	Engler and Höfer, Das Erdöl
Kampong Minjak	15.0	dark brown	....	....	....	....	114	50.7	....	37.5	....	Engler and Höfer, Das Erdöl
Karang Ringgen	15.0	light green	....	0.4	....	....	....	56.4	....	39.0	....	Engler and Höfer, Das Erdöl
Palembang	18.0	grey	....	0.27	....	....	....	59.3	....	35.5	....	Engler and Höfer, Das Erdöl
Palembang	38.0	....	....	....	....	....	....	19.7	....	46.9	....	Engler and Höfer, Das Erdöl

PETROLEUM TESTING METHODS

BY

T. G. DELBRIDGE[1]

The importance of an adequate knowledge of testing methods as applied to petroleum and its products is increasing rapidly as the development of civilization becomes more and more dependent on these materials. Nor is interest in the subject confined to any one branch of the industry. Crude petroleum, as it comes from the ground, is sold and purchased on the basis of tests. The refiner conducts judicious separations and purifications only as progressive improvement is denoted by other tests. The marketer of petroleum products of every kind is in most cases entirely dependent upon tests to show quality. Finally the intelligent purchaser and consumer wisely seeks, by means of tests, assurance that he receives what he has purchased and that not once only but continuously.

The function of a petroleum testing laboratory is to furnish to producer, refiner, marketer and purchaser of petroleum products, that information which will serve as a sure language for straightforward, clear and clean business. It is imperative that this language be not only mutually intelligible but entirely unmistakable. Unfortunately at the present time there is not one universal language in petroleum testing but a veritable Babel with its confusion of tongues. Certainly there is plenty of information available to those who are able to distinguish it from misinformation. With so many varied interests both within and without the petroleum industry it is not surprising that there should be different and conflicting opinions as to proper testing methods. However, the average individual who has need of petroleum testing methods is interested, not in the doubts and bickerings of over-zealous chemists and physicists, but only in securing certain reasonably accurate and reproducible measurements of qualities he believes to be desirable. With this thought in mind, the purpose of this chapter is to describe as concisely as is consistent with clearness, all of the fundamental and some of the more important special methods used in testing petroleum and its products.

While some progress in standardization of methods has been made by such eminently commendable organizations as the American Society for Testing Materials, the National Petroleum Association, the Bureau of Mines and the Bureau of Standards, such progress is not very encouraging in view of the large number of test methods not yet given the authority of any such body. Such unstandardized methods are being used, however, in petroleum laboratories throughout the land and not very successfully on account of lack of uniformity in details. The subject of petroleum testing methods appears, therefore, to require a somewhat bold and fearless treatment. Consequently the author has not hesitated to describe methods as he believes they should be used. Accepted standard methods are described, but where these are incomplete or indefinite, the issue has been met by giving straightforward directions which have given satisfac-

[1] Chief Chemist, The Atlantic Refining Company, Philadelphia, Pa.

Table 1

Test Methods for Petroleum Products

A = always used; F = frequently; R = rarely

Method	Crude petroleum	Gasoline and naphtha	Illuminating oil	Gas oil and distillate fuel	Residual fuel	Spindle and engine oil	Cylinder oil and black oil	Road oil	Asphaltic flux oil	Solid asphalt	Paraffin wax	Petrolatum
1. Gravity, liquids, hydrometer	A	A	A	A	A	A	A	A	A			
2. Gravity, liquids, weighing	F	F	F	F	F	F	F	F	F			
3. Gravity, solids, hydrometer										F	F	F
4. Gravity, solids, weighing										F	F	F
5. Flash, Cleveland open cup	R		R	R	R	A	A	F	F	F	R	R
6. Flash, Tag. closed tester	R	F	F	R								
7. Flash, Elliott	R		F									
8. Flash, Pensky-Martens	R			F	F	F	F		R			
9. Flash, Abel	R	F	F									
10. Viscosity, Saybolt universal	F		R	F	F	A	A	F	F			F
11. Viscosity, Saybolt furol	R				A		R	F	F			
12. Viscosity, Saybolt thermo		F	A									
13. Viscosity, Redwood	F		R	F	F	F	F	F				R
14. Viscosity, Engler	F			F	F	F	F	F	F	R		R
15. Cloud	F		F	F		F	R					
16. Pour	R		R	F	F	F	A					
17. Color, Saybolt chromometer		A	A			R						
18. Color, N.P.A.				R		F	F					R
19. Color, Lovibond			R	R		F	F				F	R
20. Water by distillation	A			R	F							
21. B. S. by weighing	F			F	F							
22. B. S. and water, centrifuge	A			F	F							
23. End Point distillation		A	F									
24. Modified Assay distillation	A	F	F	F				R				
25. M. P. Saybolt, wax											F	
26. M. P. Solidification, wax											F	
27. M. P. American, wax											R	
28. M. P. Saybolt, petrolatum												F
29. M. P. ring and ball									F	A		
30. M. P. cube										F		
31. Corrosion and gumming		F										
32. "Doctor"		F	F									
33. Acid heat		F										
34. Floc			F									
35. Lamp burning		R	A									
36. R. E. emulsification				R		F	R					R
37. Conradson steam test						F	R					
38. Herschel demulsibility						F						
39. Carbon residue					R	F	F					
40. Heat test			R	R		F						
41. Corrosion of metals	R	R	R	R	R	F	F				F	F
42. Sulphur, lamp		R	A	F								
43. Sulphur, bomb	A	R	R	F	A	R	R		R	R		
44. Acidity		R	R			F	F					F
45. Saponification						R	F					
46. Gasoline precipitation							F					
47. Tar test	R			R		R	R					
48. Solubility, CS_2; gasoline								F	F	F		
49. Evaporation						R	R	F	F	F		
50. Penetration								R	F	A		
51. Asphalt content	R							F	F			
52. Ductility									F	F		
53. Float test								R	F	R		
54. Fixed carbon, ash	R			R	R	R	R	R	R	F		
55. Oil and moisture, wax											F	
56. Vapor pressure		R										

tion in some of the oldest and most reliable testing laboratories in the petroleum industry. Mistakes of judgment may have crept in but it is suggested that the methods be accepted by the petroleum industry in the same spirit in which they are offered, namely that of impartial selective presentation, reliable but not dictatorial. It is to be hoped that suggestions and criticisms will be made with the utmost frankness by all who, like the author, are sincerely desirous of wiping out differences and securing uniformity. Such an objective as uniform methods of petroleum testing can be attained only by cooperation. In the author's opinion, every individual and corporation using these methods should be actively associated with the work of the American organization best fitted by prestige and personnel to secure uniformity, namely the American Society for Testing Materials, whose Committee D-2 on Petroleum Products and Lubricants is unquestionably deserving of whole-hearted support.

The subject of petroleum testing methods does not lend itself readily to any logically systematic treatment. No single basis for classification appears to be satisfactory. Hence the order in which the various methods are presented is rather arbitrary, though in general the commoner tests are given first. While each method has been numbered to facilitate reference, such number has no significance other than for purpose of designation. Since different petroleum products are examined by different test methods, the relations between the various products and tests is shown in Table 1, an attempt being made to distinguish between tests used always (A), frequently (F) and rarely (R).

Essentials of Proper Testing

The testing of petroleum products is merely a specialized application of suitable physical, chemical and mechanical operations. The testing methods given do not require the services of a chemist or physicist but may be and usually are performed by operators who lack the advantages of scientific training. Such operators are nearly always the most successful and dependable. Nevertheless, petroleum testing is laboratory work and as such must be conducted with due regard for the general principles covering good laboratory practice. The following suggestions are of prime importance to every operator continuously or occasionally engaged in testing petroleum products. The value of such suggestions, however, depends upon their being put into practice not only in letter but in spirit.

Neatness.—This subject has been worn almost threadbare in countless laboratory manuals; in petroleum laboratories, at least, it requires further emphasis. Few operators fail to reflect mentally the physical condition of their work-tables and apparatus. "Sloppy conditions mean sloppy results." While this statement is rather inelegant, it is highly expressive and absolutely true. The habit of neatness is not hard to acquire and, once acquired, forms a most valuable asset. In a petroleum laboratory as in everyday life, "cleanliness is next to godliness," but with this difference, that in the laboratory, cleanliness should come first.

Apparatus.—No good workman is unmindful of the condition of his tools nor of their suitability for his purpose. Nor should a petroleum testing operator neglect similar attention to the apparatus used. No amount of knowledge of directions nor of skill in application can compensate for a defective, inaccurate or wrongly chosen instrument. For example, correct results are not to be expected from a viscosimeter whose capillary has been distorted, nor from an inaccurate stop-watch, nor from a Baumé (Tag) hydrometer when the gravity reading required is Baumé (B. of S.). The apparatus used in petroleum testing consists chiefly of scientific instruments which are delicate and require delicate treatment. Three principles should always govern: First, new instruments should be of suitable type, quality and accuracy, the last being carefully determined by a competent agency, preferably the Bureau of Standards. Second, such new apparatus should be kept new by careful handling while in use, by adequate protection when not in use, and by the avoidance of unskilled tampering at all times. In this connection, it seems wise to caution against injudicious cleaning of extremely delicate and important parts, such as the capillary of a viscosimeter. Third, all instruments should be checked frequently and periodically for the purpose of revealing possible changes, whose occurrence is often unexplainable. Adjustments, if necessary, should be made only by properly qualified experts.

Precision.—Possession or lack of this characteristic differentiates a first-class from an average operator. To a considerable degree the present status of petroleum testing is due to a tendency toward approximations. The proper aim should be to attain the highest accuracy possible under the necessary limitations of apparatus and methods used. Quality of work done rather than speed in the doing is preferable. Precision, like neatness, is a habit rather easily formed and its assiduous cultivation should never be neglected. Speed is the result of experience and practice but without accuracy it becomes futile.

Records.—The value of the results of a properly conducted test depends frequently upon the written record. Failure to recognize this fact is often the cause of repetitions involving a waste of time and is always dangerous, particularly where immediate use is not made of such results. Every observation should be recorded when and as made. Both form and substance of the record should be concise but complete, and should be capable of only one interpretation. Systematic methods should be employed for recording observed readings and calculations and the latter should always be checked. Tabular forms are preferable where possible; here again the advantages of neatness and precision are apparent. The practice of carrying results "in the head" can be approved only when such results have also been reduced to a written record intelligible without recourse to memory.

Sampling.—The origin of disagreements and disputes is far too often found in differences in samples. Perfection as to method, apparatus and execution is useless if the sample tested is not representative of the material on which the test is desired. While it is not always possible for a laboratory to take its own sample, such procedure is always preferable. The real problem in sampling is to obtain from a large bulk a small portion, such as a gallon or quart or pound, representative of the entire lot of material and therefore suitable for testing. The method for obtaining such a sample varies according to the nature and quantity of material to be sampled and also according to the container. The majority of petroleum products are liquids at ordinary temperatures or are stored at temperatures high enough to keep them liquid. For such products, the containers are usually vertical cylinders (storage and working tanks), horizontal cylinders (stills and tank cars) or comparatively small packages (barrels and cans).

Large tanks used for petroleum products often contain a layer of water of varying thickness on the bottom. Determination of the thickness of this layer by thieving and gaging should precede the taking of a test sample of the petroleum product. This factor being known, the actual sampling may be performed by either of two methods.

For most purposes it is satisfactory to use a sample container of suitable size with a single rather small opening at the top. Aside from the objection of fragility, glass bottles possess many advantages, being easily cleaned and inspected for cleanliness and not being subject to leaks or corrosion. This container is placed in a suitable cage, weighted at the bottom and connected at the top with a light chain long enough to suit the height of the tank to be sampled. The *open* container is then lowered through an opening in the top of the tank until it touches the surface of the liquid. With due consideration of the size of the sample container and of the depth of liquid in the tank, the container is lowered at a uniform rate of speed until its bottom reaches the bottom of the tank (or the top of the water layer when this is present) and is then immediately drawn up again at the same uniform rate as before, this rate being that which leaves the container nearly but not quite full when it leaves the liquid. Obviously this requires dexterity, skill and above all practice. Where this procedure is not considered sufficiently accurate, a number of samples may be taken by the same method and mixed together in a larger sample container.

For very careful sampling or where variations in quality of different layers are in question, the same sampling apparatus is used with the addition of a cork or stopper attached to a stout cord by means of which the stopper may be removed after the container has been lowered into the liquid. The sample container is closed by the stopper, inserted not too tightly, and is then lowered until its mouth is at the surface of the liquid. By means of any suitable measuring device, the container is then lowered to a suitable fixed depth below the surface and the stopper is pulled out by a sharp jerk on the cord; after a time sufficient to insure complete filling, the container is rapidly removed from the tank. In this way samples may be obtained at any desired depths. Such samples may be examined separately or may be combined into one average sample as the occasion may require. In the case of vertical tanks of approximately uniform cross-section or diameter at all heights, the samples should be taken at equal vertical distances apart and should be blended in equal quantities to obtain the composite sample. Other tanks, such as tank cars, should be sampled and the samples blended with due regard for the variations in cross-sectional area at different vertical heights. The method of sampling at different levels is rarely necessary except when an examination of the individual samples is desired.

A third method of sampling petroleum products in bulk is available when such products are flowing through pipe-lines, as for example, during loading or unloading. In this case the sample is drawn from a pet-cock on the line. It is best taken by opening the pet-cock slightly and allowing a small stream to run into the sample container throughout the duration of the pumping. If this is not desirable, small samples of exactly the same size should be drawn from the pet-cock at equal intervals of time and emptied into the container for the composite sample, but in no case should the number of such samples be less than 25. Both continuous and intermittent line samplings presuppose that liquid is being delivered through the pipe-line at a fairly uniform rate.

Sampling of liquid products in barrels or cans should be preceded by thorough mixing either by rolling or shaking according to the package. Special care is required when oils containing wax or fatty oil have been stored at low temperatures. In such cases separation of a semi-solid layer at the bottom is frequent and this layer must be thoroughly mixed with the rest of the oil. Good results are usually obtained by allowing the package to stand for fifteen minutes in an inverted position before attempting to mix. After suitable thorough mixing, the sample should be withdrawn by means of a thief.[1] If a barrel thief is not available, a 4-oz. oil sample bottle fastened

[1] The term "thief" means a device which permits taking a sample from a definite predetermined location in the body of the material to be sampled. In general oil-thieves belong to one of two classes: (a) those giving a sample all of which is taken from only one level. (b) those permitting removal of a vertical column of oil in the exact condition in which it exists in the tank or package.

To the first class belong closed containers which are lowered empty and filled only when the desired level is reached. Such containers are usually opened by means of a valve operated either by a string from above or automatically by a rod extending downward so that the weight of the thief opens the valve when the rod touches bottom.

The second class includes containers of considerable length as compared with their cross-section and which are lowered into the desired position with both ends open. The vertical column of liquid thus included is confined by closing the container by any suitable means. The ordinary barrel thief is a glass or metal tube of suitable length and diameter, narrowed at each end. The open thief is lowered

to a rod or stiff wire may be used in the manner described in connection with tank sampling. Where a large number of packages of the same material are to be sampled, it is ufficient to sample every tenth package. In every case of package sampling, the quantity of sample removed from each package should be the same and not less than 0.1 per cent of the contents.

Sampling solid or semi-solid products [1] is a difficult operation and descriptions of methods are satisfactory only when accompanied by demonstration. The purpose of all such methods is to secure representative samples.

Composite samples taken by any of the methods given should be thoroughly mixed before being reduced to a size suitable for laboratory work. The same general precaution should be observed with such samples as are received by the laboratory. Any portion of a laboratory sample removed for any test must be thoroughly representative of the sample itself just as this sample must in turn be representative of the bulk or packages from which it was taken. A case requiring specific mention is that of liquids such as petroleum products containing wax, fatty oils, or other materials which tend to separate from solution when exposed to low temperatures. In all such cases the sample should be warmed with occasional agitation until it becomes homogeneous. All laboratory samples should be protected from contamination in any manner and should be very plainly marked for purposes of identification. Such markings are very conveniently made in ink on printed tags, bearing the date, kind of material and source on one side and a list of tests suitable for that class of material on the other. Such tags with the test results recorded on them may be filed for reference. Different materials will require different tags and the use of different colors is often desirable. Three such tags are shown in Fig. 1.

Special precautions should be observed to prevent the evaporation of volatile liquids like gasoline; failure to do so is a frequent source of errors and discrepancies. Extremely volatile liquids like 86 gasoline and "casing-head" should be kept in a refrigerated compartment until there is no further use for the sample.

Thermometers.—During the past few years, there has been a growing tendency on the part of those interested in petroleum testing to seek to report the actual, true temperatures being measured. Formerly all the thermometers used in connection with petroleum were of the type in common use by chemists and physicists, namely an instrument so graduated that the true temperature was indicated only when the entire body of mercury, including the thread, was uniformly at that temperature. This type of thermometer is known as a "total immersion" thermometer. Since in the practical use of a thermometer a considerable length of the mercury thread is approximately at room temperature, the reading on the thermometer scale is not a true temperature but must be corrected by a so-called "stem correction." While true temperature may be very accurately calculated in this way, it has been found much simpler and almost as accurate to use thermometers which will give true temperature readings directly upon the scale when used under the conditions for which the instrument was made. This means that the manufacturer of the instrument has applied the stem correction to suit certain prescribed conditions before graduating the scale. Such thermometers are known as "partial immersion" thermometers. The term "bulb immersion" though often used is wrong, for the reason that all "partial immersion" thermometers are scaled with at least part of the mercury thread immersed. "Partial immersion" thermometers are daily gaining in popularity and it is highly important

slowly through the bung hole till it touches the opposite side of the barrel. The filled thief is then withdrawn, the oil being retained either by a valve at the lower end or by closing the upper end tightly.

For additional information see A. R. Elliott, Oil and Gas, Jour. 18, No. 51, p. 66; May 21, 1920.

[1] See "Methods of Analysis of the Coal Tar Industry," J. M. Weiss, p. 2; also Jour. Ind. Chem. Eng., 10, 732 (1918).

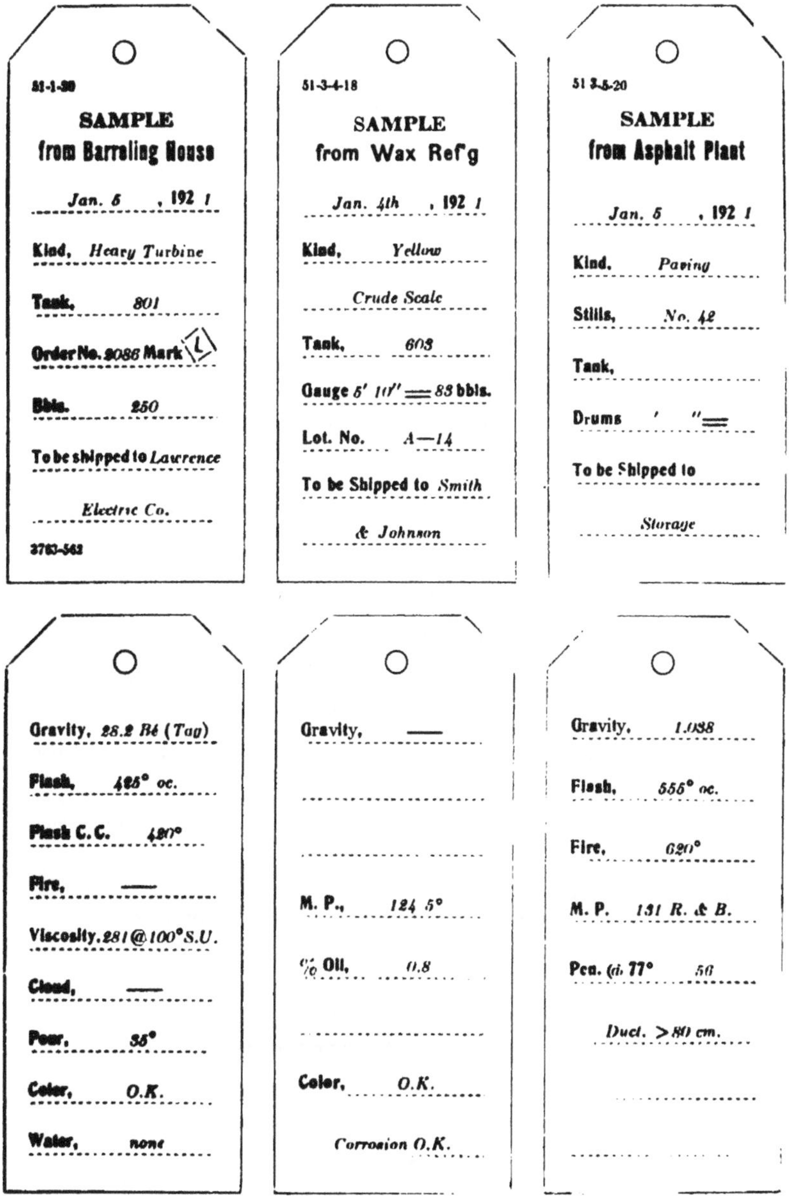

51-1-20

SAMPLE
from Barreling House

Jan. 5 , 192 1

Kind, Heavy Turbine

Tank, 801

Order No. 2086 Mark [L]

Bbls. 250

To be shipped to Lawrence Electric Co.

3763-563

51-3-4-18

SAMPLE
from Wax Ref'g

Jan. 4th , 192 1

Kind, Yellow Crude Scale

Tank, 603

Gauge 5' 10" = 83 bbls.

Lot. No. A—14

To be Shipped to Smith & Johnson

51 3-5-20

SAMPLE
from Asphalt Plant

Jan. 5 , 192 1

Kind, Paving

Stills, No. 42

Tank,

Drums ' " =

To be Shipped to Storage

Gravity, 28.2 Bé (Tag)

Flash, 425° oc.

Flash C. C. 420°

Fire, —

Viscosity, 281 @ 100° S.U.

Cloud, —

Pour, 35°

Color, O.K.

Water, none

Gravity, —

M. P., 124.5°

% Oil, 0.8

Color, O.K.

Corrosion O.K.

Gravity, 1.038

Flash, 555° oc.

Fire, 620°

M. P. 131 R. & B.

Pen. @ 77° 50

Duct. >80 cm.

Fig. 1 —Tags for samples.

that their proper use be understood. Every reliable "partial immersion" thermometer is marked with a line to indicate the depth to which it was immersed when the scale was fixed. Obviously such a thermometer will not give correct readings in use if immersed to a depth appreciably different from the immersion mark for which it was scaled. This precaution should never be disregarded nor should there ever be any confusion as to whether the thermometer in use is a "total" or "partial" immersion.

When partial immersion thermometers are specified and only total immersion are available, a so-called stem correction must be applied to the thermometer reading in order to obtain the true temperature. This correction for emergent stem should be added to the observed reading and may be calculated for mercury in glass, engraved thermometers by the following formula:

$$\text{Stem correction} = (0.000089° \text{ F.}) \times N(T° - t°),$$

in which N = number of degrees F. emergent from bath;
T = indicated temperature of the bath;
t = average temperature, F. of the emergent steam as measured by one or more small auxiliary thermometers.

Example.—A correctly graduated total immersion thermometer indicated 415° F. when immersed to the 60° mark, the average temperature of the emergent stem being 110° F. Then $N = 515 - 60 = 355°$ F.

$$\text{Stem correction} = (0.000089) \times 355(415 - 110) = 9.6° \text{ F.}$$

Corrected temperature 424.6° F.

GRAVITY

This is the most frequently made test in a petroleum laboratory. It is easily taken with a fair degree of accuracy and in spite of the fact that its importance from a specification standpoint is rapidly diminishing, yet it will always be used as a means of identification and particularly for the purpose of volume calculations. Gravity as applied to petroleum products may mean one of three kinds, all of which are in use.

(*a*) Specific gravity is the ratio,[1] expressed as a decimal, between the weight of any volume of the substance at 60° F. (in vacuo) to the weight of the same volume of pure water at 60° F. (in vacuo). It is always used for solid petroleum products and frequently for liquids. Practically all of the latter are lighter than water and therefore show specific gravities less than unity.

(*b*) Baumé gravity, Bureau of Standards (modulus 140) is expressed in degrees on the Baumé scale for liquids lighter than water. It is based fundamentally on true specific gravity as defined above, according to the following mathematical relations:

$$\text{Degrees Baumé (B. of S.)} = \frac{140}{\text{Sp. gr. } 60°/60° \text{ F.}} - 130,$$

or,

$$\text{Specific gravity, } 60°/60° \text{ F.} = \frac{140}{130 + \text{degrees Baumé (B. of S.)}}.$$

[1] It is customary in scientific writing to express this ratio with an added abbreviated definition of the conditions under which this ratio is determined. Thus the expression "Specific gravity, 77/60° F." (read as specific gravity at 77 over 60° F.) means the ratio of the weight of any volume of the substance at 77° F. to the weight of the same volume of pure water at 60° F. In other words, the numerator of the fraction is the temperature at which gravity is reported and the denominator is the temperature at which water is assumed to have a specific gravity of 1.0000. Further information may be found in Bureau of Standards, Circular No. 19, Standard Density and Volumetric Tables.

(*c*) Baumé gravity, Tagliabue (modulus 141.5) is merely a variation from the standard Baumé scale for liquids lighter than water. It is based on the following relation to true specific gravity:

$$\text{Degrees Baumé (Tag.)} = \frac{141.5}{\text{Sp. gr. } 60°/60° \text{ F.}} - 131.5,$$

or,

$$\text{Specific gravity } 60°/60° \text{ F.} = \frac{141.5}{131.5 + \text{Degrees Baumé (Tag.)}}.$$

Gravity on liquid petroleum products is usually expressed in degrees Baumé according to one of these scales. The numerical values range from about 100° to 10°, the latter being equal to the gravity of water on both these Baumé scales. Occasionally such gravities as 9.5° and 9.0° are given, indicating that the products are heavier than water. This practice cannot be condemned too severely. Gravity on products heavier than water should always be given in terms of specific gravity.

It is extremely deplorable that two Baumé scales are in use in the petroleum industry. The relative merits [1] of these have been the subject of much discussion attended by some bitterness. Impartial consideration of the matter from the standpoint of historical development would appear to justify the Bureau of Standards scale and the modulus 140. However, since a large proportion of the petroleum industry is using the Tagliabue scale and since this scale at the present time has the endorsement of such authoritative bodies as the New York Produce Exchange and the National Petroleum Association, it is absolutely necessary always to specify which scale is meant. Hence a Baumé gravity report or specification should never read, for example, 70.6° Bé., but always 70.6° Bé. (B. of S.) or 70.6° Bé. (Tag.). It is to be hoped that the influence of the comparatively young but powerful American Petroleum Institute may be exerted to eliminate any modulus other than that of the Bureau of Standards, namely 140. Arguments against such action are rapidly losing force as gravity becomes less and less important from a specification standpoint. The use of the Bureau of Standards scale is therefore strongly recommended as conforming with European practice and also with American practice outside the petroleum industry. As a compromise between the two Baumé scales, the use of the specific gravity scale has been urged. Such a solution is not likely to be adopted on account of the established position of the Baumé scale and its great convenience. The Baumé scale is both accurate and scientific.[2]

The differences between the Baumé (B. of S.) and Baumé (Tag.) vary from zero at 10° to 0.9° at 100°, the Tagliabue readings being the higher. A comparison between specific gravity and the corresponding Baumé readings according to each modulus is shown in the following table:

Inspection of this table indicates a simple working rule for converting Baumé (B. of S.) to Baumé (Tag.), namely: Add to the Bureau of Standards reading one-tenth degree for every 10° by which that reading exceeds 10°. Expressed mathematically the relations are very nearly as follows:

$$\text{Baumé (Tag.)} = (1.0107 \times \text{Baumé (B. of S.)} - 0.109,$$

or,

$$\text{Baumé (B. of S.)} = \frac{\text{Baumé (Tag.)} + 0.109}{1.0107}.$$

Except for very accurate work, all gravity determinations on liquid petroleum products are made by means of a hydrometer, the instrument used being graduated to

[1] See Bureau of Standards, Circular No. 59, United States Standard Hydrometer Scales.

[2] An excellent article on the subject by G. H. Taber, Jour. Ind. and Eng. Chem. 12, 593 (1920), should be read by every user of the Baumé scale.

TABLE 2

Comparison of the Three Gravity Scales for Petroleum Products

Specific gravity, 60°/60° F. in vacuo	Baumé gravity, Bur. of Stand. modulus 140	Baumé gravity, Tagliabue, modulus 141.5
1.0000	10.0°	10.0°
0.9333	20.0°	20.1°
0.8750	30.0°	30.2°
0.8235	40.0°	40.3°
0.7778	50.0°	50.4°
0.7368	60.0°	60.5°
0.7000	70.0°	70.6°
0.6667	80.0°	80.7°
0.6364	90.0°	90.8°+
0.6087	100.0°	100.9°+

indicate one of the three kinds of gravity described, specific gravity, Baumé (B. of S.) or Baumé (Tag.). Obviously the accuracy of a determination depends primarily upon the accuracy of the hydrometer, which should always be known. In general petroleum hydrometers of any type should not be in error by more than one of the smallest scale subdivisions. Specific gravity hydrometers should be only those made expressly for petroleum and reading true specific gravity at 60°/60° F. as stated. Baumé hydrometers of each type should conform to the respective mathematical formulæ given. Since gravity varies with temperature, a temperature measurement is part of a gravity determination. For this reason, combination hydrometer-thermometers are sometimes used. Separate hydrometers and thermometers are preferable, the combination instruments being unsatisfactory both as to economy and continued accuracy after long use. While petroleum gravities are always reported at 60° F., it is not necessary to make the determination at this temperature provided proper temperature corrections are made.

DETERMINATION OF GRAVITY OF LIQUIDS BY HYDROMETER METHOD

(Test Method No. 1)

This method is applicable to all petroleum products liquid at 60° F. When properly used it gives results whose error is not likely to exceed 0.2° Bé. or 0.001 specific gravity.

Apparatus.—See Figs. 2, 3, and 4.

1. Hydrometer of type, range and accuracy desired. It must be so constructed that it will float in an exactly vertical position. It must be perfectly clean and dry before each test.

2. Thermometer, scaled for total immersion, graduated in single degrees, error at any point on the scale not to exceed 0.5° F. The floating type with paper scale is not as satisfactory as that with the engraved stem.

3. Gravity cylinder of glass or metal (see Fig. 4), of length suitable to the hydrometer and of inside diameter at least ¾ of an inch greater than the largest outside diameter of the hydrometer. A convenient stock size of cylinder is 2 inches inside diameter and 14 inches high. This cylinder must stand exactly vertical.

4. Water or oil bath, when gravity determinations are made under conditions such that the temperature of the liquid changes more than 0.5° F. in one minute. This bath may be of any suitable type or dimensions but must be maintained uniformly at the desired temperature. For determinations at or near room temperature, the bath is seldom required.

While not essential, the following definite specifications for hydrometers and gravity thermometers are recommended as having given entire satisfaction in some of the largest petroleum testing laboratories.

Gravity Thermometer Specifications

Type: Glass engraved.
Total length: 305 mm.
Glass; Stem: Plain front, enamel back, suitable thermometer tubing.
Diameter 7 to 8 mm.
Bulb: Corning normal, Jena 16, III, or equally suitable thermometric glass. Extra heavy wall.
Length 25 mm. maximum.
Diameter less than stem.
Actuating liquid: Mercury.
Range: −20° to +220° F.
Immersion: Total.
Range distance from bottom of bulb: 45 to 55 mm.
Range distance from top of tube: 28 to 38 mm.
Ice point: Yes.
Contraction chamber: No.
Expansion chamber: No.
Filled: Nitrogen gas.
Top finish: Heavy glass ring.
Graduating: All lines, figures and letters clear cut and distinct.
Scale graduated in 2° divisions.
Scale numbered every 20°.
Special markings: Serial number and manufacturer's name or trade mark engraved on the tube.
Accuracy: Error at any point on scale shall not exceed one-half smallest scale division.
Points to be tested for certification: 32°, 92°, 152°, 212° F.

General specifications for both Baumé and specific gravity hydrometers.

Type: Plain, without thermometer, long sloping lines.
Ballast: Mercury.
Total length: 324–336 mm.
Ballast bulb length: 30–38 mm.
Float bulb length: 108–121 mm.
Spindle length: 171–184 mm.
Scale length: 140–152 mm.
Float bulb diameter: 22–24 mm.
Spindle diameter: 5–7 mm.
Standardization: at 60° F. according to specific gravity 60°/60° F. in vacuo or the proper mathematical formula for the Baumé type desired.
Accuracy: Error at any point not to exceed one smallest scale division.

Distinctive specifications.

	Specific gravity	Baumé
Set of eight hydrometers:	1.000–0.930	10–21°
	0.940–0.870	19–31°
	0.880–0.820	29–41°
	0.830–0.770	39–51°
	0.780–0.730	49–61°
	0.740–0.690	59–71°
	0.700–0.660	69–81°
	0.670–0.630	79–91°
Hydrometer scale graduated in..........	0.0005 divisions	0.1° divisions
Hydrometer scale numbered in..........	0.01 divisions	5° divisions
Points to be tested for certifications.......	two extremes and center of scale range	Multiples of 5°

Procedure.—Make sure that the sample is uniform and homogeneous. The practice of shaking in the container should not be followed unless time is available to permit complete escape of suspended air. Mixing by stirring is preferable. Glass bottles are far superior to metal containers.

Pour into the gravity cylinder (3) a quantity of oil sufficient to float the hydrometer at least ½ inch off the bottom, avoiding introduction of air bubbles by letting the oil run down the inside of the cylinder wall. Introduce the thermometer (2) and stir gently in a combined rotary and up and down motion for five seconds, being careful not to mix air with the oil. Take the temperature reading. Stir again as before and again take the temperature reading. Repeat this procedure until there is no readable temperature difference between successive operations. Record the thermometer reading as "Observed temperature." Whether or not the bath (4) is necessary will be apparent at this stage of the determination. Volatile liquids like 86° gasoline should be kept at low temperature to avoid change of gravity by evaporation during the determination.

Having obtained and recorded the "Observed temperature," quickly remove the thermometer and immediately introduce the proper hydrometer (1) immersing it with the utmost care until it just floats and then depressing it in such a manner that its line of immersion differs by not more than two nor less than one of the smallest scale subdivisions from its line of immersion when at rest in its natural floating position. If this is not accomplished on the first trial, remove the hydrometer, wipe the stem thoroughly, again measure the "Observed temperature" and introduce the hydrometer in the manner prescribed. Make sure that both hydrometer and oil are free from air bubbles and are at rest and that the hydrometer is floating freely in a vertical position without touching the cylinder wall. Quickly but carefully take the hydrometer reading in the following manner:[1]

The eye should be placed slightly below the plane of the surface of the oil (Fig. 2) and then raised slowly until this surface, first seen as an ellipse, becomes a straight line (Fig. 3). The reading of the hydrometer should be taken where this line crosses the scale of the instrument.

In case the oil is not sufficiently clear to allow the reading to be made as described, it is necessary to read from above the oil surface and to estimate as accurately as possible

[1] Bureau of Standards, Circular No. 57, United States Standard Tables for Petroleum Oils, page 5.

the line to which the oil rises on the hydrometer stem. It should be remembered, however, that the instrument is calibrated to give correct indications when read at the surface of the liquid. It will be necessary, therefore, to correct the reading at the top of the meniscus by an amount equal to the height of this meniscus. The amount of this correction may be determined with sufficient accuracy for most purposes by taking a few readings with the same hydrometer in a clear oil and noting the differences on the hydrometer scale between the top of the meniscus and the level of the oil surface.

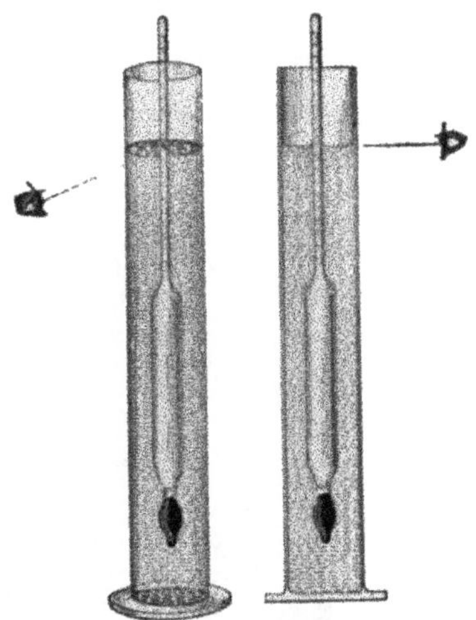

Fig. 2 and Fig. 3.—Method of reading hydrometer.

A specific gravity hydrometer will read too low and a Baumé hydrometer too high when read at the top of the meniscus. The correction for meniscus height should therefore be added to a specific gravity reading and subtracted from a Baumé reading. The magnitude of the correction will obviously depend upon the length and value of the subdivisions of the hydrometer scale and must be determined in each case for the particular hydrometer in use. It is impossible to read the hydrometer as accurately with opaque oils as with transparent oils because the height of the meniscus varies slightly in different oils. With Baumé hydrometers like those for which specifications have been given, it is customary to subtract 0.1° Bé. from the reading at the top of the meniscus. Accurate gravity measurements on opaque oils are usually made by weighing, as in Test Method No. 2.

Record the hydrometer reading, obtained in the manner described, as "Observed gravity." While the directions for obtaining this value have necessarily been given in detail with explanations, the actual operations involved are quite simple and must be carried on with speed as well as with precision, so that the hydrometer reading is taken while the oil is still at the "Observed temperature." The entire operation between reading "Observed temperature" and "Observed gravity" should require not more than 25 seconds and usually considerably less.

In laboratories where a large number of gravity determinations are made and particularly on dark-colored oils, the apparatus [1] shown in Fig. 4 has been found to be extremely convenient. In this apparatus the gravity cylinders (3) are No. 16 B.W. gage brass tubes, 1½ inch outside diameter and 14 inches long. These are supported by draining lines to which they are connected by the stop-cocks (7). The overflow trough (5) soldered to the cylinders makes the assembly rigid. The overflow drain line (8) as well as the cylinder drain lines discharge into a suitable receptacle beneath the work-table. The overflow trough is made wide enough to hold hydrometers (1) and thermometer (2) when these are not in use. The two end cylinders are equipped with pet-cocks (9) and a removable bent tube (10) through which the sample may be recovered when it is required for further testing. Ordinarily the sample is discarded after a gravity determination by opening the stop-cock (7) whereupon the cylinder empties itself.

[1] Devised by H. M. Nichols, Superintendent, The Atlantic Refining Company, Philadelphia, Pa.

In using this apparatus, a cylinder which has drained thoroughly is selected. The stop-cock is closed and the cylinder completely filled with the oil to be tested. The temperature is taken in the usual manner as described and the hydrometer immersed in such manner that in coming to rest it causes oil to overflow. The hydrometer reading is then made with the cylinder full of oil. Under these conditions no meniscus is observed and very accurate readings can be obtained even on the blackest oils. The advantages of this apparatus fully justify its endorsement by all operators who have used it.

Correction of observed gravity for temperature.—Since gravity on petroleum products is always reported at 60° F. regardless of the temperature at which measurement

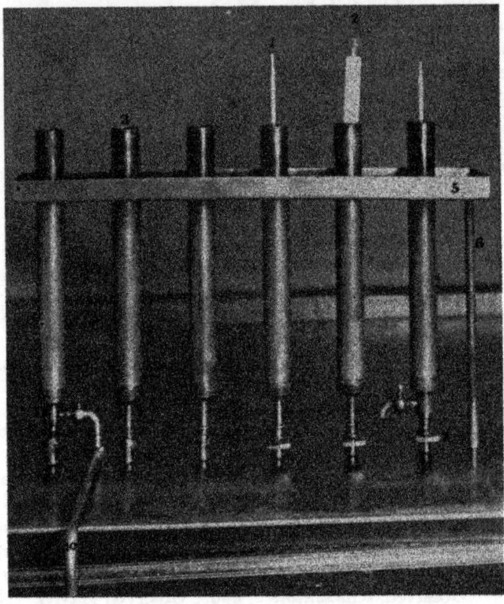

FIG. 4.—Nichols gravity apparatus.

is made, it is necessary to apply a correction for temperature to the "Observed gravity" whenever the "Observed temperature" is other than 60° F. The amount of this correction depends on the gravity of the oil, on the temperature and on the kind of gravity determined (Specific or Baumé). Fortunately this correction is independent of the nature or source of the oil, provided this oil is completely liquid at 60° F.[1]

For convenience, tables have been prepared by means of which an "Observed gravity" at an "Observed temperature" may be converted into a corrected gravity at 60° F. For specific gravity and for Baumé (B. of S.) gravity, the Bureau of Standards[2] has issued excellent tables which should be available in every petroleum laboratory

[1] Bureau of Standards, Technologic Paper No. 77, Density and Thermal Expansion of American Petroleum Oils, p 19.

[2] Bureau of Standards, Circular No. 57, United States Standard Tables for Petroleum Oils. Obtainable from the Superintendent of Documents, Government Printing Office, Washington, D. C. Price 15 cents.

in the original form and are therefore not reprinted here. The Bureau of Standards temperature correction table for Baumé (B. of S.) may also be used without perceptible error for Baumé (Tag.) corrections.

In plant-control work it is frequently necessary to measure Baumé gravity at temperatures higher than those given by the Bureau of Standards. Table No. 3 (pages 602-617, this volume) gives temperature corrections for Baumé gravity (Tag.) over a temperature range of 21–160° F. where necessary. This table has been calculated [1] partly from Bureau of Standards Circular 57 by exterpolating where necessary and partly from other measurements. At some points this table will be found to differ by 0.1° Bé. from Circular 57, Bureau of Standards. Except for very accurate work, it may be used for either Baumé (Tag.) or Baumé (B. of S.).

The method of obtaining corrected gravity at 60° F. from the various tables is as follows:

Specific gravity.—Turn to Bureau of Standards, Circular 57, Petroleum Oil Table 1. Locate the column having at its top the "Observed specific gravity" found. The figure in this column, which is horizontally opposite the "Observed temperature" found, is the corrected specific gravity at 60° F. Interpolation for temperatures or gravity or both may be necessary.

Examples.

Observed specific gravity	Observed temperature	Corrected specific gravity at 60° F.
0.854	88° F.	0.864
0.706	32°	0.692
0.791	75°	0.797

Baumé gravity.—Use Petroleum Oil Table 2, Bureau of Standards Circular 57 or Table No. 3, this volume, (pages 602-617) in a manner similar to that given for specific gravity.

Examples.

Observed Baumé gravity	Observed temperature	Corrected Baumé gravity at 60° F.
24.0	78° F.	22.9
72.0	40°	74.9
72.0	77°	69.9
34.8	96°	32.3
22.0	147°	16.2
22.2	160°	15.6

In laboratories making many Baumé gravity determinations, the Atlantic gravity-temperature corrector shown in Fig. 5 will facilitate this work both as to speed and particularly as to eliminating careless line-reading. In this device, Table No. 3 (pages 602-617) printed as a chart on one sheet, is mounted on a cylinder which can be rotated around a horizontal axis by the knobs at each end. The outside casing has a

[1] By G. M. Atherholt, Engineer, The Atlantic Refining Company, Philadelphia, Pa.

TABLE 3

Observed Degrees Baumé (Tag) at Observed Temperature

Observed temp., ° F.	10	11	12	13	14	15	16	17	18	19	20	21	22	23	24
	CORRECTED DEGREES BAUMÉ (TAG) AT 60° F.														
21	11.9	13.0	14.0	15.0	16.1	17.1	18.1	19.2	20.2	21.2	22.3	23.3	24.3	25.4	26.4
22	11.9	12.9	14.0	15.0	16.0	17.1	18.1	19.1	20.1	21.2	22.2	23.2	24.3	25.3	26.3
23	11.8	12.9	13.9	14.9	16.0	17.0	18.0	19.1	20.1	21.1	22.1	23.2	24.2	25.2	26.3
24	11.8	12.8	13.9	14.9	15.9	17.0	18.0	19.0	20.0	21.1	22.1	23.1	24.2	25.2	26.2
25	11.7	12.8	13.8	14.8	15.9	16.9	17.9	18.9	20.0	21.0	22.0	23.1	24.1	25.1	26.2
26	11.7	12.7	13.7	14.8	15.8	16.9	17.9	18.9	19.9	20.9	22.0	23.0	24.0	25.1	26.1
27	11.6	12.7	13.7	14.7	15.8	16.8	17.8	18.8	19.8	20.9	21.9	22.9	24.0	25.0	26.0
28	11.6	12.6	13.6	14.7	15.7	16.8	17.8	18.8	19.8	20.8	21.9	22.9	23.9	24.9	26.0
29	11.5	12.6	13.6	14.6	15.6	16.7	17.7	18.7	19.7	20.8	21.8	22.8	23.9	24.9	25.9
30	11.5	12.5	13.5	14.6	15.6	16.7	17.6	18.7	19.7	20.7	21.7	22.8	23.8	24.8	25.9
31	11.4	12.5	13.5	14.5	15.5	16.6	17.6	18.6	19.6	20.7	21.7	22.7	23.7	24.8	25.8
32	11.4	12.4	13.4	14.4	15.5	16.5	17.5	18.6	19.6	20.6	21.6	22.7	23.7	24.7	25.7
33	11.3	12.4	13.4	14.4	15.4	16.4	17.5	18.5	19.5	20.5	21.6	22.6	23.6	24.6	25.7
34	11.3	12.3	13.3	14.3	15.4	16.4	17.4	18.4	19.4	20.5	21.5	22.5	23.6	24.6	25.6
35	11.2	12.3	13.3	14.3	15.3	16.3	17.4	18.4	19.4	20.4	21.4	22.5	23.5	24.5	25.5
36	11.2	12.2	13.2	14.2	15.3	16.3	17.3	18.3	19.3	20.3	21.4	22.4	23.4	24.4	25.5
37	11.1	12.2	13.2	14.2	15.2	16.2	17.2	18.3	19.3	20.3	21.3	22.3	23.4	24.4	25.4
38	11.1	12.1	13.1	14.1	15.2	16.2	17.2	18.2	19.2	20.2	21.3	22.3	23.3	24.3	25.3
39	11.0	12.1	13.1	14.1	15.1	16.1	17.1	18.2	19.2	20.2	21.2	22.2	23.2	24.3	25.3
40	11.0	12.0	13.0	14.0	15.0	16.1	17.1	18.1	19.1	20.1	21.1	22.2	23.2	24.2	25.2
41	10.9	12.0	13.0	14.0	15.0	16.0	17.0	18.0	19.1	20.1	21.1	22.1	23.1	24.1	25.2
42	10.9	11.9	12.9	13.9	14.9	16.0	17.0	18.0	19.0	20.0	21.0	22.0	23.1	24.1	25.1
43	10.8	11.9	12.9	13.9	14.9	15.9	16.9	17.9	18.9	20.0	21.0	22.0	23.0	24.0	25.0
44	10.8	11.8	12.8	13.8	14.8	15.9	16.9	17.9	18.9	19.9	20.9	21.9	22.9	24.0	25.0
45	10.7	11.8	12.8	13.8	14.8	15.8	16.8	17.8	18.8	19.8	20.9	21.9	22.9	23.9	24.9
46	10.7	11.7	12.7	13.7	14.7	15.7	16.8	17.8	18.8	19.8	20.8	21.8	22.8	23.8	24.9
47	10.6	11.7	12.7	13.7	14.7	15.7	16.7	17.7	18.7	19.7	20.7	21.8	22.8	23.8	24.8
48	10.6	11.6	12.6	13.6	14.6	15.6	16.7	17.7	18.7	19.7	20.7	21.7	22.7	23.7	24.7
49	10.5	11.6	12.6	13.6	14.6	15.6	16.6	17.6	18.6	19.6	20.6	21.6	22.7	23.7	24.7
50	10.5	11.5	12.5	13.5	14.5	15.5	16.5	17.6	18.6	19.6	20.6	21.6	22.6	23.6	24.6
51	10.4	11.4	12.5	13.5	14.5	15.5	16.5	17.5	18.5	19.5	20.5	21.5	22.5	23.5	24.6
52	10.4	11.4	12.4	13.4	14.4	15.4	16.4	17.4	18.4	19.4	20.5	21.5	22.5	23.5	24.5
53	10.3	11.3	12.4	13.4	14.4	15.4	16.4	17.4	18.4	19.4	20.4	21.4	22.4	23.4	24.4
54	10.3	11.3	12.3	13.3	14.3	15.3	16.3	17.3	18.3	19.3	20.3	21.3	22.4	23.4	24.4
55	10.2	11.2	12.3	13.3	14.3	15.3	16.3	17.3	18.3	19.3	20.3	21.3	22.3	23.3	24.3
56	10.2	11.2	12.2	13.2	14.2	15.2	16.2	17.2	18.2	19.2	20.2	21.2	22.2	23.2	24.2
57	10.1	11.1	12.2	13.2	14.2	15.2	16.2	17.2	18.2	19.2	20.2	21.2	22.2	23.2	24.2
58	10.1	11.1	12.1	13.1	14.1	15.1	16.1	17.1	18.1	19.1	20.1	21.1	22.1	23.1	24.1
59	10.0	11.0	12.1	13.1	14.1	15.1	16.1	17.1	18.1	19.1	20.1	21.1	22.1	23.1	24.1
60	10.0	11.0	12.0	13.0	14.0	15.0	16.0	17.0	18.0	19.0	20.0	21.0	22.0	23.0	24.0
61	10.0	11.0	11.9	12.9	13.9	14.9	15.9	16.9	17.9	18.9	19.9	20.9	21.9	22.9	23.9
62	.....	10.9	11.9	12.9	13.9	14.9	15.9	16.9	17.9	18.9	19.9	20.9	21.9	22.9	23.9
63	.....	10.9	11.8	12.8	13.8	14.8	15.8	16.8	17.8	18.8	19.8	20.8	21.8	22.8	23.8
64	.....	10.8	11.8	12.8	13.8	14.8	15.8	16.8	17.8	18.8	19.8	20.8	21.8	22.8	23.8
65	.....	10.8	11.7	12.7	13.7	14.7	15.7	16.7	17.7	18.7	19.7	20.7	21.7	22.7	23.7
66	.....	10.7	11.7	12.7	13.7	14.7	15.7	16.7	17.7	18.7	19.7	20.7	21.6	22.6	23.6
67	.....	10.7	11.6	12.6	13.6	14.6	15.6	16.6	17.6	18.6	19.6	20.6	21.6	22.6	23.6
68	.....	10.6	11.6	12.6	13.6	14.6	15.6	16.6	17.5	18.5	19.5	20.5	21.5	22.5	23.5
69	.....	10.6	11.5	12.5	13.5	14.5	15.5	16.5	17.5	18.5	19.5	20.5	21.5	22.5	23.5
70	.....	10.5	11.5	12.5	13.5	14.5	15.5	16.5	17.4	18.4	19.4	20.4	21.4	22.4	23.4

TABLE 3—*Continued*

Observed Degrees Baumé (Tag) at Observed Temperature

Observed temp., °F.	10	11	12	13	14	15	16	17	18	19	20	21	22	23	24
	CORRECTED DEGREES BAUMÉ (TAG) AT 60° F.														
71	.....	10.5	11.4	12.4	13.4	14.4	15.4	16.4	17.4	18.4	19.4	20.4	21.4	22.4	23.3
72	.....	10.4	11.4	12.4	13.4	14.4	15.4	16.4	17.3	18.3	19.3	20.3	21.3	22.3	23.3
73	.....	10.4	11.3	12.3	13.3	14.3	15.3	16.3	17.3	18.3	19.3	20.3	21.2	22.2	23.2
74	.....	10.3	11.3	12.3	13.3	14.3	15.3	16.2	17.2	18.2	19.2	20.2	21.2	22.2	23.2
75	.....	10.3	11.2	12.2	13.2	14.2	15.2	16.2	17.1	18.2	19.2	20.1	21.1	22.1	23.1
76	.....	10.2	11.2	12.2	13.2	14.2	15.2	16.1	17.1	18.1	19.1	20.1	21.1	22.1	23.0
77	.....	10.2	11.1	12.1	13.1	14.1	15.1	16.1	17.0	18.1	19.0	20.0	21.0	22.0	23.0
78	.....	10.1	11.1	12.1	13.1	14.1	15.0	16.0	17.0	18.0	19.0	20.0	21.0	21.9	22.9
79	.....	10.1	11.0	12.0	13.0	14.0	15.0	16.0	16.9	17.9	18.9	19.9	20.9	21.9	22.9
80	.....	10.0	11.0	12.0	13.0	14.0	14.9	15.9	16.9	17.9	18.9	19.9	20.8	21.8	22.8
81	.....	10.0	10.9	11.9	12.9	13.9	14.9	15.9	16.8	17.8	18.8	19.8	20.8	21.8	22.7
82	.....	.....	10.9	11.9	12.9	13.8	14.8	15.8	16.8	17.8	18.8	19.7	20.7	21.7	22.7
83	.....	.....	10.8	11.8	12.8	13.8	14.8	15.8	16.7	17.7	18.7	19.7	20.7	21.6	22.6
84	.....	.....	10.8	11.8	12.8	13.7	14.7	15.7	16.7	17.7	18.6	19.6	20.6	21.6	22.6
85	.....	.....	10.7	11.7	12.7	13.7	14.7	15.6	16.6	17.6	18.6	19.6	20.5	21.5	22.5
86	.....	.....	10.7	11.7	12.7	13.6	14.6	15.6	16.6	17.5	18.5	19.5	20.5	21.5	22.5
87	.....	.....	10.6	11.6	12.6	13.6	14.6	15.5	16.5	17.5	18.5	19.5	20.4	21.4	22.4
88	.....	.....	10.6	11.6	12.6	13.5	14.5	15.5	16.5	17.4	18.4	19.4	20.4	21.4	22.3
89	.....	.....	10.5	11.5	12.5	13.5	14.5	15.4	16.4	17.4	18.4	19.4	20.3	21.3	22.3
90	.....	.....	10.5	11.5	12.5	13.4	14.4	15.4	16.4	17.3	18.3	19.3	20.3	21.2	22.2
91	.....	.....	10.5	11.4	12.4	13.4	14.3	15.3	16.3	17.3	18.3	19.2	20.2	21.2	22.2
92	.....	.....	10.4	11.4	12.4	13.3	14.3	15.3	16.3	17.2	18.2	19.2	20.2	21.1	22.1
93	.....	.....	10.4	11.3	12.3	13.3	14.2	15.2	16.2	17.2	18.1	19.1	20.1	21.1	22.0
94	.....	.....	10.3	11.3	12.3	13.2	14.2	15.2	16.1	17.1	18.1	19.1	20.0	21.0	22.0
95	.....	.....	10.3	11.2	12.2	13.2	14.1	15.1	16.1	17.1	18.0	19.0	20.0	20.9	21.9
96	.....	.....	10.2	11.2	12.2	13.1	14.1	15.1	16.0	17.0	18.0	18.9	19.9	20.9	21.9
97	.....	.....	10.2	11.1	12.1	13.1	14.0	15.0	16.0	17.0	17.9	18.9	19.9	20.8	21.8
98	.....	.....	10.1	11.1	12.1	13.0	14.0	15.0	15.9	16.9	17.9	18.8	19.8	20.8	21.7
99	.....	.....	10.1	11.0	12.0	13.0	13.9	14.9	15.9	16.8	17.8	18.8	19.8	20.7	21.7
100	.....	.....	10.0	11.0	12.0	12.9	13.9	14.9	15.8	16.8	17.7	18.7	19.7	20.7	21.6
101	.....	.....	10.0	10.9	11.9	12.9	13.8	14.8	15.8	16.7	17.7	18.7	19.6	20.6	21.6
102	.....	.....	.....	10.9	11.8	12.8	13.8	14.7	15.7	16.7	17.6	18.6	19.6	20.5	21.5
103	.....	.....	.....	10.8	11.8	12.8	13.7	14.7	15.7	16.6	17.6	18.5	19.5	20.5	21.5
104	.....	.....	.....	10.8	11.7	12.7	13.7	14.6	15.6	16.6	17.5	18.5	19.5	20.4	21.4
105	.....	.....	.....	10.7	11.7	12.7	13.6	14.6	15.6	16.5	17.5	18.4	19.4	20.4	21.4
106	.....	.....	.....	10.7	11.6	12.6	13.6	14.5	15.5	16.4	17.4	18.4	19.4	20.3	21.3
107	.....	.....	.....	10.6	11.6	12.6	13.5	14.5	15.5	16.4	17.4	18.3	19.3	20.3	21.2
108	.....	.....	.....	10.6	11.5	12.5	13.5	14.4	15.4	16.3	17.3	18.3	19.3	20.2	21.2
109	.....	.....	.....	10.5	11.5	12.5	13.4	14.4	15.4	16.3	17.3	18.2	19.2	20.2	21.1
110	.....	.....	.....	10.5	11.4	12.4	13.4	14.3	15.3	16.2	17.2	18.2	19.1	20.1	21.1
111	.....	.....	.....	10.5	11.4	12.4	13.3	14.3	15.2	16.2	17.2	18.1	19.1	20.1	21.0
112	.....	.....	.....	10.4	11.3	12.3	13.3	14.2	15.2	16.1	17.1	18.1	19.0	20.0	20.9
113	.....	.....	.....	10.4	11.3	12.3	13.2	14.2	15.1	16.1	17.1	18.0	19.0	19.9	20.9
114	.....	.....	.....	10.3	11.2	12.2	13.2	14.1	15.1	16.0	17.0	17.9	18.9	19.8	20.8
115	.....	.....	.....	10.3	11.2	12.2	13.1	14.1	15.0	16.0	16.9	17.9	18.8	19.8	20.7
116	.....	.....	.....	10.2	11.1	12.1	13.1	14.0	15.0	15.9	16.9	17.8	18.8	19.8	20.7
117	.....	.....	.....	10.2	11.1	12.1	13.0	14.0	14.9	15.9	16.8	17.8	18.7	19.7	20.6
118	.....	.....	.....	10.1	11.0	12.0	13.0	13.9	14.9	15.8	16.8	17.7	18.7	19.6	20.6
119	.....	.....	.....	10.1	11.0	12.0	12.9	13.9	14.8	15.8	16.7	17.7	18.6	19.6	20.5
120	.....	.....	.....	10.0	10.9	11.9	12.9	13.8	14.8	15.7	16.7	17.6	18.6	19.5	20.5

TABLE 3—*Continued*

Observed Degrees Baumé (Tag) at Observed Temperature

Observed temp. ° F.	10	11	12	13	14	15	16	17	18	19	20	21	22	23	24
	Corrected Degrees Baumé (Tag) at 60° F.														
121	....	....	....	....	10.9	11.9	12.8	13.8	14.7	15.7	16.6	17.6	18.5	19.5	20.4
122	....	....	....	....	10.8	11.8	12.8	13.7	14.6	15.6	16.6	17.5	18.5	19.4	20.4
123	....	....	....	....	10.8	11.8	12.7	13.6	14.6	15.6	16.5	17.5	18.4	19.4	20.3
124	....	....	....	....	10.7	11.7	12.7	13.6	14.5	15.5	16.5	17.4	18.4	19.3	20.3
125	....	....	....	....	10.7	11.7	12.6	13.5	14.5	15.5	16.4	17.4	18.3	19.3	20.2
126	....	....	....	....	10.6	11.6	12.6	13.5	14.4	15.4	16.4	17.3	18.2	19.2	20.2
127	....	....	....	....	10.6	11.6	12.5	13.4	14.4	15.4	16.3	17.3	18.2	19.1	20.1
128	....	....	....	....	10.5	11.5	12.5	13.4	14.3	15.3	16.2	17.2	18.1	19.1	20.0
129	....	....	....	....	10.5	11.5	12.4	13.3	14.3	15.3	16.2	17.1	18.1	19.0	20.0
130	...	....	....	....	10.5	11.4	12.4	13.3	14.2	15.2	16.1	17.1	18.0	19.0	19.9
131	....	....	....	....	10.4	11.4	12.3	13.2	14.2	15.1	16.1	17.0	18.0	18.9	19.9
132	....	....	....	....	10.4	11.3	12.3	13.2	14.1	15.1	16.0	17.0	17.9	18.8	19.8
133	....	....	....	....	10.3	11.3	12.2	13.1	14.1	15.0	16.0	16.9	17.8	18.8	19.7
134	....	....	....	....	10.3	11.2	12.2	13.1	14.0	15.0	15.9	16.9	17.8	18.7	19.7
135	....	....	....	....	10.2	11.2	12.1	13.0	14.0	14.9	15.9	16.8	17.7	18.7	19.6
136	....	....	....	....	10.2	11.1	12.1	13.0	13.9	14.9	15.8	16.7	17.7	18.6	19.6
137	....	....	....	....	10.1	11.1	12.0	12.9	13.9	14.8	15.7	16.7	17.6	18.6	19.5
138	.. ..	....	....	....	10.1	11.0	12.0	12.9	13.8	14.8	15.7	16.6	17.5	18.5	19.5
139	....	....	....	....	10.0	11.0	11.9	12.8	13.8	14.7	15.6	16.6	17.5	18.5	19.4
140	....	....	....	....	10.0	10.9	11.9	12.8	13.7	14.7	15.6	16.5	17.5	18.4	19.4
141	....	....	....	....	....	10.9	11.8	12.7	13.7	14.6	15.5	16.5	17.4	18.4	19.3
142	....	....	....	....	....	10.8	11.8	12.7	13.6	14.6	15.5	16.4	17.4	18.3	19.2
143	....	....	....	....	....	10.8	11.7	12.6	13.6	14.5	15.4	16.4	17.3	18.3	19.2
144	....	....	....	....	....	10.7	11.7	12.6	13.5	14.5	15.4	16.3	17.3	18.2	19.1
145	....	....	....	....	....	10.7	11.6	12.5	13.5	14.4	15.3	16.3	17.2	18.1	19.1
146	....	....	....	....	....	10.6	11.6	12.5	13.4	14.4	15.3	16.2	17.2	18.1	19.0
147	....	....	....	....	....	10.6	11.5	12.4	13.4	14.3	15.2	16.2	17.1	18.0	18.9
148	....	....	....	....	....	10.5	11.5	12.4	13.3	14.3	15.2	16.1	17.0	18.0	18.9
149	....	....	....	....	....	10.5	11.4	12.3	13.3	14.2	15.1	16.0	17.0	17.9	18.8
150	....	....	....	....	....	10.4	11.4	12.3	13.2	14.2	15.1	16.0	16.9	17.8	18.8
151	....	....	....	....	....	10.4	11.3	12.2	13.2	14.1	15.0	15.9	16.9	17.8	18.7
152	....	....	....	....	....	10.3	11.3	12.2	13.1	14.0	14.9	15.9	16.8	17.7	18.6
153	....	....	....	....	....	10.3	11.2	12.1	13.1	14.0	14.9	15.8	16.7	17.7	18.6
154	....	....	....	....	....	10.2	11.2	12.1	13.0	13.9	14.8	15.8	16.7	17.6	18.5
155	....	....	....	....	....	10.2	11.1	12.0	13.0	13.9	14.8	15.7	16.6	17.6	18.5
156	....	....	....	....	....	10.1	11.1	12.0	12.9	13.8	14.7	15.6	16.6	17.5	18.4
157	....	....	....	....	....	10.1	11.0	11.9	12.8	13.8	14.7	15.6	16.5	17.5	18.4
158	....	....	....	....		10.0	10.9	11.9	12.8	13.7	14.6	15.5	16.5	17.4	18.3
159	....	....	....	....	....	10.0	10.9	11.8	12.7	13.7	14.6	15.5	16.4	17.4	18.3
160	....	....	....	....	....	....	10.8	11.8	12.7	13.6	14.5	15.4	16.4	17.3	18.2

TABLE 3—*Continued*

Observed Degrees Baumé (Tag) at Observed Temperature

Observed temp., °F.	25	26	27	28	29	30	31	32	33	34	35	36	37	38	39
	Corrected Degrees Baumé (Tag) at 60° F.														
21	27.4	28.5	29.6	30.6	31.6	32.7	33.7	34.8	35.8	36.9	37.9	39.0	40.0	41.0	42.1
22	27.4	28.4	29.5	30.5	31.6	32.6	33.7	34.7	35.8	36.8	37.9	38.9	40.0	40.9	42.0
23	27.3	28.4	29.4	30.4	31.5	32.5	33.6	34.6	35.7	36.7	37.8	38.8	39.9	40.8	42.0
24	27.3	28.3	29.3	30.4	31.4	32.4	33.5	34.6	35.6	36.7	37.7	38.7	39.8	40.8	41.9
25	27.2	28.2	29.3	30.3	31.3	32.4	33.4	34.5	35.5	36.6	37.6	38.7	39.7	40.7	41.8
26	27.1	28.2	29.2	30.2	31.3	32.3	33.4	34.4	35.4	36.5	37.6	38.6	39.6	40.6	41.7
27	27.1	28.1	29.1	30.2	31.2	32.2	33.3	34.3	35.4	36.4	37.5	38.5	39.6	40.5	41.7
28	27.0	28.0	29.1	30.1	31.1	32.2	33.2	34.3	35.3	36.3	37.4	38.4	39.5	40.5	41.6
29	26.9	28.0	29.0	30.0	31.1	32.1	33.1	34.2	35.2	36.3	37.3	38.3	39.4	40.4	41.5
30	26.9	27.9	28.9	30.0	31.0	32.0	33.1	34.1	35.1	36.2	37.2	38.3	39.3	40.3	41.4
31	26.8	27.9	28.9	29.9	30.9	32.0	33.0	34.0	35.1	36.1	37.2	38.2	39.2	40.3	41.3
32	26.8	27.8	28.8	29.8	30.9	31.9	32.9	34.0	35.0	36.0	37.1	38.1	39.1	40.2	41.2
33	26.7	27.7	28.7	29.8	30.8	31.8	32.9	33.9	34.9	36.0	37.0	38.0	39.1	40.1	41.1
34	26.6	27.7	28.7	29.7	30.7	31.8	32.8	33.8	34.9	35.9	36.9	38.0	39.0	40.0	41.0
35	26.6	27.6	28.6	29.6	30.7	31.7	32.7	33.8	34.8	35.8	36.9	37.9	38.9	40.0	40.9
36	26.5	27.5	28.6	29.6	30.6	31.6	32.7	33.7	34.7	35.8	36.8	37.8	38.8	39.9	40.9
37	26.4	27.4	28.5	29.5	30.5	31.6	32.6	33.6	34.6	35.7	36.7	37.7	38.8	39.8	40.8
38	26.4	27.4	28.4	29.4	30.5	31.5	32.5	33.6	34.6	35.6	36.6	37.7	38.7	39.7	40.7
39	26.3	27.3	28.3	29.4	30.4	31.4	32.4	33.5	34.5	35.5	36.6	37.6	38.6	39.6	40.6
40	26.2	27.3	28.3	29.3	30.3	31.3	32.4	33.4	34.4	35.4	36.5	37.5	38.5	39.6	40.6
41	26.2	27.2	28.2	29.2	30.3	31.3	32.3	33.3	34.3	35.4	36.4	37.4	38.4	39.5	40.5
42	26.1	27.1	28.2	29.2	30.2	31.2	32.2	33.3	34.3	35.3	36.3	37.3	38.4	39.4	40.4
43	26.1	27.1	28.1	29.1	30.1	31.1	32.2	33.2	34.2	35.2	36.2	37.3	38.3	39.3	40.3
44	26.0	27.0	28.0	29.0	30.1	31.1	32.1	33.1	34.1	35.2	36.2	37.2	38.2	39.2	40.3
45	25.9	26.9	28.0	29.0	30.0	31.0	32.0	33.0	34.1	35.1	36.1	37.1	38.1	39.2	40.2
46	25.9	26.9	27.9	28.9	29.9	30.9	32.0	33.0	34.0	35.0	36.0	37.0	38.1	39.1	40.1
47	25.8	26.8	27.8	28.8	29.9	30.9	31.9	32.9	33.9	34.9	36.0	37.0	38.0	39.0	40.0
48	25.7	26.8	27.8	28.8	29.8	30.8	31.8	32.8	33.9	34.9	35.9	36.9	37.9	38.9	39.9
49	25.7	26.7	27.7	28.7	29.7	30.7	31.7	32.8	33.8	34.8	35.8	36.8	37.8	38.9	39.9
50	25.6	26.6	27.6	28.6	29.7	30.7	31.7	32.7	33.7	34.7	35.7	36.7	37.8	38.8	39.8
51	25.6	26.6	27.6	28.6	29.6	30.6	31.6	32.6	33.6	34.6	35.7	36.7	37.7	38.7	39.7
52	25.5	26.5	27.5	28.5	29.5	30.5	31.5	32.6	33.6	34.6	35.6	36.6	37.6	38.6	39.6
53	25.4	26.4	27.4	28.5	29.5	30.5	31.5	32.5	33.5	34.5	35.5	36.5	37.5	38.5	39.6
54	25.4	26.4	27.4	28.4	29.4	30.4	31.4	32.4	33.4	34.4	35.4	36.4	37.4	38.5	39.5
55	25.3	26.3	27.3	28.3	29.3	30.3	31.3	32.3	33.4	34.4	35.4	36.4	37.4	38.4	39.4
56	25.2	26.2	27.3	28.3	29.3	30.3	31.3	32.3	33.3	34.3	35.3	36.3	37.3	38.3	39.3
57	25.2	26.2	27.2	28.2	29.2	30.2	31.2	32.2	33.2	34.2	35.2	36.2	37.2	38.2	39.2
58	25.1	26.1	27.1	28.1	29.1	30.1	31.1	32.1	33.1	34.1	35.1	36.1	37.2	38.2	39.2
59	25.1	26.1	27.1	28.1	29.1	30.1	31.1	32.1	33.1	34.1	35.1	36.1	37.1	38.1	39.1
60	25.0	26.0	27.0	28.0	29.0	30.0	31.0	32.0	33.0	34.0	35.0	36.0	37.0	38.0	39.0
61	24.9	25.9	26.9	27.9	28.9	29.9	30.9	31.9	32.9	33.9	34.9	35.9	36.9	37.9	38.9
62	24.9	25.9	26.9	27.9	28.9	29.9	30.9	31.9	32.9	33.9	34.9	35.9	36.8	37.9	38.8
63	24.8	25.8	26.8	27.8	28.8	29.8	30.8	31.8	32.8	33.8	34.8	35.8	36.8	37.8	38.8
64	24.8	25.7	26.7	27.7	28.7	29.7	30.7	31.7	32.7	33.7	34.7	35.7	36.7	37.7	38.7
65	24.7	25.7	26.7	27.7	28.7	29.7	30.7	31.7	32.6	33.6	34.6	35.6	36.6	37.6	38.6
66	24.7	25.6	26.6	27.6	28.6	29.6	30.6	31.6	32.6	33.6	34.6	35.6	36.5	37.5	38.5
67	24.6	25.6	26.6	27.5	28.5	29.5	30.5	31.5	32.5	33.5	34.5	35.5	36.5	37.5	38.5
68	24.5	25.5	26.5	27.5	28.5	29.5	30.5	31.5	32.4	33.4	34.4	35.4	36.4	37.4	38.4
69	24.5	25.4	26.4	27.4	28.4	29.4	30.4	31.4	32.4	33.4	34.4	35.3	36.3	37.3	38.3
70	24.4	25.4	26.4	27.4	28.4	29.3	30.3	31.3	32.3	33.3	34.3	35.3	36.3	37.2	38.2

TABLE 3—*Continued*

Observed Degrees Baumé (Tag) at Observed Temperature

Observed temp., ° F.	25	26	27	28	29	30	31	32	33	34	35	36	37	38	39
	CORRECTED DEGREES BAUMÉ (TAG) AT 60° F.														
71	24.3	25.3	26.3	27.3	28.3	29.3	30.3	31.2	32.2	33.2	34.2	35.2	36.2	37.2	38.2
72	24.3	25.3	26.2	27.2	28.2	29.2	30.2	31.2	32.2	33.1	34.1	35.1	36.1	37.1	38.1
73	24.2	25.2	26.2	27.2	28.2	29.1	30.1	31.1	32.1	33.1	34.1	35.1	36.0	37.0	38.0
74	24.1	25.1	26.1	27.1	28.1	29.1	30.1	31.0	32.0	33.0	34.0	35.0	36.0	36.9	37.9
75	24.1	25.1	26.1	27.0	28.0	29.0	30.0	31.0	32.0	32.9	33.9	34.9	35.9	36.9	37.8
76	24.0	25.0	26.0	27.0	28.0	28.9	29.9	30.9	31.9	32.9	33.8	34.8	35.8	36.8	37.8
77	24.0	24.9	25.9	26.9	27.9	28.9	29.9	30.8	31.8	32.8	33.8	34.8	35.7	36.7	37.7
78	23.9	24.9	25.9	26.8	27.8	28.8	29.8	30.8	31.7	32.7	33.7	34.7	35.7	36.6	37.6
79	23.8	24.8	25.8	26.8	27.8	28.7	29.7	30.7	31.7	32.7	33.6	34.6	35.6	36.6	37.5
80	23.8	24.8	25.7	26.7	27.7	28.7	29.7	30.6	31.6	32.6	33.6	34.5	35.5	36.5	37.5
81	23.7	24.7	25.7	26.7	27.6	28.6	29.6	30.6	31.5	32.5	33.5	34.5	35.5	36.4	37.4
82	23.6	24.6	25.6	26.6	27.6	28.5	29.5	30.5	31.5	32.5	33.4	34.4	35.4	36.4	37.3
83	23.6	24.6	25.5	26.5	27.5	28.5	29.5	30.5	31.4	32.4	33.4	34.3	35.3	36.3	37.3
84	23.5	24.5	25.5	26.5	27.5	28.4	29.4	30.4	31.4	32.3	33.3	34.3	35.2	36.2	37.2
85	23.5	24.5	25.4	26.4	27.4	28.4	29.3	30.3	31.3	32.3	33.2	34.2	35.2	36.1	37.1
86	23.4	24.4	25.4	26.4	27.3	28.3	29.3	30.2	31.2	32.2	33.2	34.1	35.1	36.1	37.0
87	23.4	24.3	25.3	26.3	27.3	28.2	29.2	30.2	31.1	32.1	33.1	34.0	35.0	36.0	36.9
88	23.3	24.3	25.3	26.2	27.2	28.2	29.1	30.1	31.1	32.0	33.0	34.0	34.9	35.9	36.9
89	23.2	24.2	25.2	26.2	27.1	28.1	29.1	30.0	31.0	32.0	32.9	33.9	34.9	35.8	36.8
90	23.2	24.2	25.1	26.1	27.1	28.0	29.0	30.0	30.9	31.9	32.9	33.8	34.8	35.8	36.7
91	23.1	24.1	25.1	26.0	27.0	28.0	28.9	29.9	30.9	31.8	32.8	33.8	34.7	35.7	36.6
92	23.1	24.0	25.0	26.0	26.9	27.9	28.9	29.8	30.8	31.8	32.7	33.7	34.6	35.6	36.6
93	23.0	24.0	24.9	25.9	26.9	27.8	28.8	29.8	30.7	31.7	32.7	33.6	34.6	35.5	36.5
94	22.9	23.9	24.9	25.8	26.8	27.8	28.7	29.7	30.7	31.6	32.6	33.5	34.5	35.5	36.5
95	22.9	23.8	24.8	25.8	26.8	27.7	28.7	29.6	30.6	31.5	32.5	33.5	34.5	35.4	36.4
96	22.8	23.8	24.7	25.7	26.7	27.6	28.6	29.6	30.5	31.5	32.5	33.4	34.4	35.3	36.3
97	22.8	23.7	24.7	25.6	26.6	27.6	28.5	29.5	30.5	31.4	32.4	33.4	34.3	35.3	36.2
98	22.7	23.7	24.6	25.6	26.6	27.5	28.5	29.5	30.4	31.4	32.3	33.3	34.2	35.2	36.1
99	22.6	23.6	24.6	25.5	26.5	27.5	28.4	29.4	30.4	31.3	32.3	33.2	34.2	35.1	36.1
100	22.6	23.5	24.5	25.5	26.4	27.4	28.4	29.3	30.3	31.2	32.2	33.1	34.1	35.0	36.0
101	22.5	23.5	24.5	25.4	26.4	27.4	28.3	29.3	30.2	31.2	32.1	33.1	34.0	35.0	35.9
102	22.5	23.4	24.4	25.4	26.3	27.3	28.2	29.2	30.1	31.1	32.1	33.0	33.9	34.9	35.8
103	22.4	23.4	24.3	25.3	26.3	27.2	28.2	29.1	30.1	31.0	32.0	32.9	33.9	34.8	35.8
104	22.4	23.3	24.3	25.2	26.2	27.2	28.1	29.1	30.0	31.0	31.9	32.9	33.8	34.7	35.7
105	22.3	23.3	24.2	25.2	26.1	27.1	28.0	29.0	29.9	30.9	31.8	32.8	33.7	34.7	35.6
106	22.3	23.2	24.2	25.1	26.1	27.0	28.0	28.9	29.9	30.8	31.7	32.7	33.6	34.6	35.5
107	22.2	23.1	24.1	25.0	26.0	27.0	27.9	28.9	29.8	30.7	31.7	32.6	33.6	34.5	35.5
108	22.1	23.1	24.0	25.0	25.9	26.9	27.8	28.8	29.7	30.7	31.6	32.6	33.5	34.5	35.4
109	22.1	23.0	24.0	24.9	25.9	26.8	27.8	28.7	29.7	30.6	31.5	32.5	33.5	34.4	35.4
110	22.0	23.0	23.9	24.9	25.8	26.8	27.7	28.7	29.6	30.5	31.5	32.5	33.4	34.3	35.3
111	22.0	22.9	23.8	24.8	25.7	26.7	27.6	28.6	29.5	30.5	31.4	32.4	33.4	34.3	35.2
112	21.9	22.8	23.8	24.7	25.7	26.6	27.6	28.5	29.5	30.4	31.4	32.3	33.3	34.2	35.1
113	21.8	22.8	23.7	24.7	25.6	26.6	27.5	28.5	29.4	30.4	31.3	32.3	33.2	34.1	35.1
114	21.8	22.7	23.7	24.6	25.5	26.5	27.5	28.4	29.4	30.3	31.2	32.2	33.1	34.1	35.0
115	21.7	22.6	23.6	24.5	25.5	26.5	27.4	28.4	29.3	30.2	31.2	32.1	33.0	34.0	34.9
116	21.7	22.6	23.5	24.5	25.4	26.4	27.4	28.3	29.2	30.2	31.1	32.0	33.0	33.9	34.8
117	21.6	22.5	23.5	24.4	25.4	26.3	27.3	28.2	29.2	30.1	31.0	32.0	32.9	33.8	34.8
118	21.6	22.5	23.4	24.4	25.3	26.3	27.2	28.2	29.1	30.0	31.0	31.9	32.8	33.8	34.7
119	21.5	22.4	23.4	24.3	25.3	26.2	27.2	28.1	29.0	30.0	30.9	31.8	32.8	33.7	34.6
120	21.5	22.4	23.3	24.3	25.2	26.2	27.1	28.0	29.0	29.9	30.8	31.8	32.7	33.6	34.5

TABLE 3—*Continued*

Observed Degrees Baumé (Tag) at Observed Temperature

Observed temp., °F.	25	26	27	28	29	30	31	32	33	34	35	36	37	38	39
	Corrected Degrees Baumé (Tag) at 60° F.														
121	21.4	22.3	23.3	24.2	25.1	26.1	26.0	28.0	.9	29.8	30.7	31.7	32.6	33.6	34.5
122	21.3	22.3	23.2	24.1	25.1	26.0	29 27.0	27.9	29 28.8	29 29.8	29 30.7	31.6	32.5	33.5	34.4
123	21.3	22.2	23.1	24.1	25.0	26.0	26.9	27.8	28.8	.7	30.6	31.5	32.5	33.5	34.4
124	21.2	22.2	23.1	24.0	25.0	25.9	.8	27.7	.7	.6	.5	31.5	32.4	33.4	34.3
125	21.1	22.1	23.0	24.0	24.9	25.8	26.8	27.7	28.6	29.5	30.5	31.4	32.4	33.3	34.2
126	21.1	22.0	23.0	23.9	24.8	25.8	26.7	27.6	28.6	29.5	30.4	31.4	32.3	33.2	34.1
127	21.0	22.0	22.9	23.8	24.8	25.7	26.6	27.6	28.5	29.4	30.4	31.3	32.2	33.2	34.1
128	21.0	21.9	22.8	23.8	24.7	25.6	26.6	27.5	28.5	29.4	30.3	31.2	32.2	33.1	34.0
129	20.9	21.8	22.8	23.7	24.6	25.6	26.5	27.5	28.4	29.3	30.3	31.2	32.1	33.0	33.9
130	20.8	21.8	22.7	23.6	24.6	25.5	26.5	27.4	28.3	29.3	30.2	31.1	32.0	32.9	33.8
131	20.8	21.7	22.6	23.6	24.5	25.5	26.4	27.4	28.3	29.2	30.1	31.0	31.9	32.9	33.8
132	20.7	21.7	22.6	23.5	24.5	25.4	26.4	27.3	28.2	29.1	29 30.0	31.0	31.9	32.8	33.7
133	20.7	21.6	22.5	23.5	24.4	25.4	26.3	27.2	28.1	29.1	30.0	30.9	31.8	32.7	33.6
134	20.6	21.5	22.5	23.4	24.4	25.3	26.2	27.2	28.1	29.0	.9	30.8	31.7	32.6	33.6
135	20.6	21.5	22.4	23.4	24.3	25.2	26.2	27.1	28.0	28.9	29.8	30.7	31.7	32.6	33.5
136	20.5	21.4	22.4	23.3	24.3	25.2	26.1	27.0	27.9	28.9	29.8	30.7	31.6	32.5	33.5
137	20.5	21.4	22.3	23.3	24.2	25.1	26.0	26.9	27.9	28.8	29.7	30.6	31.6	32.5	33.4
138	20.4	21.3	22.3	23.2	24.1	25.0	26.0	26.9	27.8	28.7	29.6	30.6	31.5	32.4	33.3
139	20.4	21.3	22.2	23.1	24.1	25.0	25.9	26.8	27.7	28.7	29.6	30.5	31.5	32.4	33.3
140	20.3	21.2	22.2	23.1	24.0	24.9	25.8	26.7	27.6	28.6	29.5	30.5	31.4	32.3	33.2
141	20.3	21.2	22.1	23.0	23.9	24.8	25.8	26.7	27.6	28.5	29.5	30.4	31.3	32.2	33.1
142	20.2	21.1	22.0	29 .9	23.9	24.8	25.7	26.6	27.5	28.5	29.4	30.4	31.2	32.1	33.0
143	20.1	21.0	22.0	.9	23.8	24.7	25.6	26.6	27.5	28.4	29.4	30.3	31.2	32.1	33.0
144	20.1	21.0	21.9	22.8	23.7	24.6	25.6	26.5	27.4	28.4	29.3	30.2	31.1	32.0	32.9
145	20.0	20.9	21.8	22.8	23.7	24.6	25.5	26.5	27.4	28.3	29.2	30.1	31.0	31.9	32.8
146	20.0	20.9	21.8	22.7	23.6	24.5	25.5	26.4	27.3	28.2	29.2	30.0	30.9	31.9	32.7
147	19.9	20.8	21.7	22.6	23.6	24.5	25.4	26.4	27.3	28.2	29.1	30.0	30.9	31.8	32.7
148	19.8	20.7	21.7	22.6	23.5	24.4	25.4	26.3	27.2	28.1	29.0	29.9	30.8	31.7	32.6
149	19.8	20.7	21.6	22.5	23.5	24.4	25.3	26.2	27.1	28.0	28.9	29.8	30.7	31.6	32.5
150	19.7	20.6	21.6	22.5	23.4	24.3	25.3	26.2	27.1	28.0	28.9	29.8	30.7	31.6	32.5
151	19.7	20.6	21.5	22.4	23.4	24.3	25.2	26.1	26.0	27.9	28.8	29.7	30.6	31.5	32.4
152	19.6	20.5	21.5	22.4	23.3	24.2	25.1	29 26.0	29 26.9	27.8	28.7	29.6	30.5	31.5	32.4
153	19.5	20.5	21.4	22.3	23.2	24.1	25.1	26.0	26.9	27.7	28.7	29.6	30.5	31.4	32.3
154	19.5	20.4	21.4	22.3	23.2	24.1	25.0	.9	.8	27.7	28.6	29.5	30.4	31.4	32.3
155	19.4	20.4	21.3	22.2	23.1	24.0	24.9	25.8	26.7	27.6	28.5	29.5	30.4	31.3	32.2
156	19.4	20.3	21.3	22.2	23.0	23.9	24.9	25.8	26.7	27.6	28.5	29.4	30.3	31.2	32.1
157	19.3	20.3	21.2	22.1	23.0	23.9	24.8	25.7	26.6	27.5	28.4	29.4	30.3	31.1	32.0
158	19.3	20.2	21.1	22.0	22.9	23.8	24.7	25.6	26.5	27.5	28.4	29.3	30.2	31.1	32.0
159	19.2	20.2	21.0	22.0	22.9	23.8	24.7	25.6	26.5	27.4	28.3	29.2	30.1	31.0	31.9
160	19.2	20.1	21.0	21.9	22.8	23.7	24.6	25.5	26.4	27.4	28.3	29.2	30.0	30.9	31.8

TABLE 3—*Continued*

Observed Degrees *Baumé (Tag) at Observed Temperature*

Observed temp., °F.	40	41	42	43	44	45	46	47	48	49	50	51	52	53	54
	Corrected Degrees Baumé (Tag) at 60° F.														
21	43.2	44.2	45.3	46.4	47.4	48.5	49.6	50.7	51.8	52.9	54.0	55.1	56.1	57.2	58.3
22	43.1	44.2	45.2	46.3	47.3	48.4	49.5	50.6	51.7	52.8	53.9	55.0	56.0	57.1	58.2
23	43.0	44.1	45.1	46.2	47.3	48.3	49.4	50.5	51.6	52.7	53.8	54.8	55.9	57.0	58.1
24	43.0	44.0	45.1	46.1	47.2	48.2	49.3	50.4	51.5	52.6	53.7	54.7	55.8	56.9	58.0
25	42.9	43.9	45.0	46.0	47.1	48.1	49.2	50.3	51.4	52.5	53.6	54.6	55.7	56.8	57.9
26	42.8	43.8	44.9	46.0	47.0	48.1	49.1	50.2	51.3	52.4	53.5	54.5	55.6	56.7	57.8
27	42.7	43.8	44.8	45.9	46.9	48.0	49.0	50.1	51.2	52.3	53.3	54.4	55.4	56.6	57.7
28	42.6	43.7	44.7	45.8	46.8	47.9	48.9	50.0	51.1	52.2	53.2	54.3	55.3	56.4	57.5
29	42.5	43.6	44.6	45.7	46.7	47.8	48.8	49.9	51.0	52.1	53.1	54.2	55.2	56.3	57.4
30	42.4	43.5	44.5	45.6	46.7	47.7	48.8	49.8	50.9	52.0	53.0	54.1	55.1	56.2	57.3
31	42.3	43.4	44.4	45.5	46.6	47.6	48.7	49.7	50.8	51.9	52.9	54.0	55.0	56.1	57.2
32	42.3	43.3	44.3	45.4	46.5	47.5	48.6	49.6	50.7	51.8	52.8	53.9	54.9	56.0	57.1
33	42.2	43.2	44.3	45.3	46.4	47.4	48.5	49.5	50.6	51.7	52.7	53.8	54.8	55.9	56.9
34	42.1	43.1	44.2	45.2	46.3	47.3	48.4	49.4	50.5	51.6	52.6	53.7	54.7	55.8	56.8
35	42.0	43.1	44.1	45.1	46.2	47.2	48.3	49.3	50.4	51.5	52.5	53.6	54.6	55.7	56.7
36	41.9	43.0	44.0	45.1	46.1	47.1	48.2	49.2	50.3	51.3	52.4	53.4	54.5	55.6	56.6
37	41.9	42.9	43.9	45.0	46.0	47.0	48.1	49.1	50.2	51.2	52.3	53.3	54.4	55.4	56.5
38	41.8	42.8	43.8	44.9	45.9	47.0	48.0	49.0	50.1	51.1	52.2	53.2	54.3	55.3	56.4
39	41.7	42.7	43.8	44.8	45.8	46.9	47.9	48.9	50.0	51.0	52.1	53.1	54.2	55.2	56.3
40	41.6	42.7	43.7	44.7	45.7	46.8	47.8	48.8	49.9	50.9	52.0	53.0	54.1	55.1	56.2
41	41.5	42.6	43.6	44.6	45.7	46.7	47.7	48.8	49.8	50.8	51.9	52.9	54.0	55.0	56.0
42	41.4	42.5	43.5	44.5	45.6	46.6	47.6	48.7	49.7	50.7	51.8	52.8	53.9	54.9	55.9
43	41.3	42.4	43.4	44.4	45.5	46.5	47.5	48.6	49.6	50.6	51.7	52.7	53.8	54.8	55.8
44	41.2	42.3	43.3	44.3	45.4	46.4	47.4	48.5	49.5	50.5	51.6	52.6	53.6	54.7	55.7
45	41.2	42.2	43.2	44.3	45.3	46.3	47.3	48.4	49.4	50.4	51.5	52.5	53.5	54.6	55.6
46	41.1	42.1	43.2	44.2	45.2	46.2	47.3	48.3	49.3	50.3	51.4	52.4	53.4	54.4	55.5
47	41.0	42.1	43.1	44.1	45.1	46.1	47.2	48.2	49.2	50.2	51.3	52.3	53.3	54.3	55.4
48	40.9	42.0	43.0	44.0	45.0	46.1	47.1	48.1	49.1	50.1	51.2	52.2	53.2	54.2	55.3
49	40.9	41.9	42.9	43.9	44.9	46.0	47.0	48.0	49.0	50.0	51.1	52.1	53.1	54.1	55.2
50	40.8	41.8	42.8	43.8	44.9	45.9	46.9	47.9	48.9	49.9	51.0	52.0	53.0	54.0	55.1
51	40.7	41.7	42.8	43.8	44.8	45.8	46.8	47.8	48.8	49.9	50.9	51.9	52.9	53.9	55.0
52	40.6	41.7	42.7	43.7	44.7	45.7	46.7	47.7	48.7	49.8	50.8	51.8	52.8	53.8	54.8
53	40.5	41.6	42.6	43.6	44.6	45.6	46.6	47.6	48.6	49.7	50.7	51.7	52.7	53.7	54.7
54	40.5	41.5	42.5	43.5	44.5	45.5	46.5	47.5	48.6	49.6	50.6	51.6	52.6	53.6	54.6
55	40.4	41.4	42.4	43.4	44.4	45.4	46.4	47.4	48.5	49.5	50.5	51.5	52.5	53.5	54.5
56	40.3	41.3	42.3	43.3	44.3	45.3	46.4	47.4	48.4	49.4	50.4	51.4	52.4	53.4	54.4
57	40.2	41.2	42.2	43.2	44.3	45.3	46.3	47.3	48.3	49.3	50.3	51.3	52.3	53.3	54.3
58	40.2	41.2	42.2	43.2	44.2	45.2	46.2	47.2	48.2	49.2	50.2	51.2	52.2	53.2	54.2
59	40.1	41.1	42.1	43.1	44.1	45.1	46.1	47.1	48.1	49.1	50.1	51.1	52.1	53.1	54.1
60	40.0	41.0	42.0	43.0	44.0	45.0	46.0	47.0	48.0	49.0	50.0	51.0	52.0	53.0	54.0
61	39.9	40.9	41.9	42.9	43.9	44.9	45.9	46.9	47.9	48.9	49.9	50.9	51.9	52.9	53.9
62	39.8	40.8	41.8	42.8	43.8	44.8	45.8	46.8	47.8	48.8	49.8	50.8	51.8	52.8	53.8
63	39.8	40.8	41.8	42.8	43.7	44.7	45.7	46.7	47.7	48.7	49.7	50.7	51.7	52.7	53.7
64	39.7	40.7	41.7	42.7	43.7	44.6	45.6	46.6	47.6	48.6	49.6	50.6	51.6	52.6	53.6
65	39.6	40.6	41.6	42.6	43.6	44.6	45.6	46.6	47.5	48.5	49.5	50.5	51.5	52.5	53.5
66	39.5	40.5	41.5	42.5	43.5	44.5	45.5	46.5	47.5	48.4	49.4	50.4	51.4	52.4	53.4
67	39.5	40.4	41.4	42.4	43.4	44.4	45.4	46.4	47.4	48.3	49.3	50.3	51.3	52.3	53.3
68	39.4	40.4	41.4	42.3	43.3	44.3	45.3	46.3	47.3	48.3	49.2	50.2	51.2	52.2	53.2
69	39.3	40.3	41.3	42.3	43.2	44.2	45.2	46.2	47.2	48.2	49.1	50.1	51.1	52.1	53.1
70	39.2	40.2	41.2	42.2	43.2	44.1	45.1	46.1	47.1	48.1	49.1	50.0	51.0	52.0	53.0

TABLE 3—*Continued*

Observed Degrees Baumé (Tag) at Observed Temperature

Observed temp., °F.	40	41	42	43	44	45	46	47	48	49	50	51	52	53	54
	CORRECTED DEGREES BAUMÉ (TAG) AT 60° F.														
71	39.1	40.1	41.1	42.1	43.1	44.1	45.0	46.0	47.0	48.0	49.0	49.9	50.9	51.9	52.9
72	39.1	40.0	41.0	42.0	43.0	44.0	45.0	45.9	46.9	47.9	48.9	49.8	50.8	51.8	52.8
73	39.0	40.0	41.0	41.9	42.9	43.9	44.9	45.8	46.8	47.8	48.8	49.8	50.7	51.7	52.7
74	38.9	39.9	40.9	41.8	42.8	43.8	44.8	45.8	46.7	47.7	48.7	49.7	50.6	51.6	52.6
75	38.8	39.8	40.8	41.8	42.7	43.7	44.7	45.7	46.6	47.6	48.6	49.6	50.5	51.5	52.5
76	38.7	39.7	40.7	41.7	42.7	43.6	44.6	45.6	46.6	47.5	48.5	49.5	50.5	51.4	52.4
77	38.7	39.6	40.7	41.6	42.6	43.5	44.5	45.5	46.5	47.5	48.4	49.4	50.4	51.3	52.3
78	38.6	39.6	40.6	41.5	42.5	43.5	44.5	45.4	46.4	47.4	48.3	49.3	50.3	51.2	52.2
79	38.5	39.5	40.5	41.5	42.4	43.4	44.4	45.3	46.3	47.3	48.2	49.2	50.2	51.1	52.1
80	38.5	39.4	40.4	41.4	42.3	43.3	44.3	45.2	46.2	47.2	48.2	49.1	50.1	51.0	52.0
81	38.4	39.4	40.3	41.3	42.3	43.2	44.2	45.2	46.1	47.1	48.1	49.0	50.0	50.9	51.9
82	38.3	39.3	40.2	41.3	42.2	43.1	44.1	45.1	46.0	47.0	48.0	48.9	49.9	50.8	51.8
83	38.2	39.2	40.2	41.2	42.1	43.1	44.0	45.0	45.9	46.9	47.9	48.8	49.8	50.7	51.7
84	38.1	39.1	40.1	41.1	42.0	43.0	43.9	44.9	45.9	46.8	47.8	48.7	49.7	50.6	51.6
85	38.1	39.0	40.0	41.0	41.9	42.9	43.9	44.8	45.8	46.7	47.7	48.6	49.6	50.5	51.5
86	38.0	39.0	39.9	40.9	41.8	42.8	43.8	44.7	45.7	46.6	47.6	48.5	49.5	50.5	51.4
87	37.9	38.9	39.8	40.9	41.8	42.7	43.7	44.6	45.6	46.5	47.5	48.5	49.4	50.4	51.3
88	37.8	38.8	39.8	40.8	41.7	42.6	43.6	44.5	45.5	46.5	47.4	48.4	49.3	50.3	51.2
89	37.8	38.7	39.7	40.7	41.6	42.5	43.5	44.5	45.5	46.4	47.4	48.3	49.3	50.2	51.1
90	37.7	38.6	39.6	40.6	41.5	42.5	43.5	44.4	45.4	46.3	47.3	48.2	49.2	50.1	51.0
91	37.6	38.6	39.5	40.5	41.5	42.4	43.4	44.3	45.3	46.2	47.2	48.1	49.1	50.0	50.9
92	37.5	38.5	39.5	40.4	41.4	42.3	43.3	44.2	45.2	46.1	47.1	48.0	49.0	49.9	50.8
93	37.5	38.4	39.4	40.4	41.3	42.3	43.2	44.2	45.1	46.0	47.0	47.9	48.9	49.8	50.7
94	37.4	38.4	39.3	40.3	41.2	42.2	43.1	44.1	45.0	45.9	46.9	47.8	48.8	49.7	50.6
95	37.3	38.3	39.2	40.2	41.1	42.1	43.0	44.0	44.9	45.9	46.8	47.7	48.7	49.6	50.5
96	37.2	38.2	39.2	40.1	41.1	42.0	42.9	43.9	44.8	45.8	46.7	47.6	48.6	49.5	50.5
97	37.2	38.1	39.1	40.0	41.0	41.9	42.8	43.8	44.7	45.7	46.6	47.5	48.5	49.5	50.4
98	37.1	38.0	39.0	39.9	40.9	41.8	42.8	43.7	44.6	45.6	46.5	47.5	48.5	49.4	50.3
99	37.0	38.0	38.9	39.9	40.8	41.7	42.7	43.6	44.6	45.5	46.5	47.4	48.4	49.3	50.2
100	36.9	37.9	38.8	39.8	40.7	41.7	42.6	43.5	44.5	45.5	46.4	47.3	48.3	49.2	50.1
101	36.9	37.8	38.8	39.7	40.6	41.6	42.5	43.5	44.5	45.4	46.3	47.2	48.2	49.1	50.0
102	36.8	37.7	38.7	39.6	40.6	41.5	42.5	43.4	44.4	45.3	46.2	47.1	48.1	49.0	49.9
103	36.7	37.7	38.6	39.5	40.5	41.5	42.4	43.3	44.3	45.2	46.1	47.1	48.0	48.9	49.8
104	36.6	37.6	38.5	39.5	40.4	41.4	42.3	43.3	44.2	45.1	46.0	47.0	47.9	48.8	49.7
105	36.6	37.5	38.5	39.4	40.4	41.3	42.2	43.2	44.1	45.0	45.9	46.9	47.8	48.7	49.6
106	36.5	37.5	38.4	39.4	40.3	41.2	42.1	43.1	44.0	44.9	45.8	46.8	47.7	48.6	49.5
107	36.5	37.4	38.3	39.3	40.2	41.1	42.1	43.0	43.9	44.8	45.7	46.7	47.6	48.5	49.5
108	36.4	37.3	38.3	39.2	40.1	41.1	42.0	42.9	43.8	44.7	45.7	46.6	47.5	48.5	49.4
109	36.3	37.2	38.2	39.1	40.0	41.0	41.9	42.8	43.7	44.7	45.6	46.5	47.5	48.4	49.3
110	36.2	37.2	38.1	39.0	40.0	40.9	41.8	42.7	43.6	44.6	45.5	46.5	47.4	48.3	49.2
111	36.1	37.1	38.0	38.9	39.9	40.8	41.7	42.6	43.6	44.5	45.5	46.4	47.3	48.2	49.1
112	36.1	37.0	37.9	38.9	39.8	40.7	41.6	42.6	43.5	44.5	45.4	46.3	47.2	48.1	49.0
113	36.0	36.9	37.9	38.8	39.7	40.6	41.6	42.5	43.5	44.4	45.3	46.2	47.1	48.0	48.9
114	35.9	36.8	37.8	38.7	39.6	40.6	41.5	42.5	43.4	44.3	45.2	46.1	47.0	47.9	48.8
115	35.8	36.8	37.7	38.6	39.5	40.5	41.5	42.4	43.3	44.2	45.1	46.0	46.9	47.8	48.7
116	35.8	36.7	37.6	38.6	39.5	40.5	41.4	42.3	43.2	44.1	45.0	45.9	46.8	47.7	48.6
117	35.7	36.6	37.5	38.5	39.4	40.4	41.3	42.2	43.1	44.0	44.9	45.8	46.7	47.6	48.5
118	35.6	36.5	37.5	38.5	39.4	40.3	41.2	42.1	43.0	43.9	44.8	45.7	46.6	47.5	48.5
119	35.6	36.5	37.4	38.4	39.3	40.2	41.1	42.0	42.9	43.8	44.7	45.6	46.5	47.5	48.4
120	35.5	36.4	37.4	38.3	39.2	40.1	41.0	41.9	42.9	43.8	44.6	45.6	46.5	47.4	48.3

TABLE 3—*Continued*

Observed Degrees Baumé (Tag) at Observed Temperature

Observed temp., ° F.	40	41	42	43	44	45	46	47	48	49	50	51	52	53	54
	Corrected Degrees Baumé (Tag) at 60° F.														
121	35.	36.4	37.3	38.2	39.1	40.0	41.0	41.9	42.8	43.7	44.6	45.5	46.4	47.3	48.2
122	35.	36.3	37.2	38.1	39.0	40.0	40.9	41.8	42.7	43.6	44.5	45.5	46.4	47.3	48.1
123	35.	36.2	37.1	38.0	39.0	39.9	40.8	41.7	42.6	43.5	44.5	45.4	46.3	47.2	48.
124	35.	36.1	37.0	38.0	38.9	39.8	40.7	41.6	42.5	43.5	44.4	45.3	46.2	47.1	47.9
125	35.5	36.1	37.0	37.9	38.8	39.7	40.6	41.5	42.5	43.4	44.3	45.2	46.1	47.0	47.8
126	35.1	36.0	36.9	37.8	38.7	39.6	40.[illegible]	41.5	42.	43.3	44.2	45.1	46.0	46.9	47.7
127	35.0	35.9	36.8	37.7	38.6	39.6	40.5	41.4	42.	43.2	44.1	45.0	45.9	46.8	47.7
128	34.9	35.8	36.7	37.6	38.6	39.5	40.[illegible]	41.4	42.	43.1	44.0	44.9	45.8	46.7	47.6
129	34.8	35.7	36.6	37.6	38.5	39.5	40.4	41.3	42.4	43.1	43.9	44.8	45.7	46.6	47.5
130	34.8	35.7	36.6	37.5	38.5	39.4	40.3	41.2	42.1	43.0	43.8	44.7	45.6	46.5	47.5
131	34.7	35.6	36.5	37.5	38.4	39.3	40.2	41.1	42.0	42.9	43.8	44.6	45.	46.5	47.4
132	.6	35.5	36.[illegible]	37.4	38.3	39.2	40.1	41.0	41.9	42.8	43.7	44.6	45.	46.4	47.3
133	.6	35.5	36.9	37.3	38.2	39.1	40.0	40.9	41.8	42.7	43.6	44.5	45.	46.3	47.2
134	.5	35.4	36.4	37.2	38.2	39.1	39.9	40.8	41.7	42.6	43.5	44.5	45.	46.2	47.1
135	34.5	35.4	36.3	37.2	38.1	39.0	39.9	40.7	41.6	42.5	43.5	44.4	45.5	46.1	47.0
136	34.4	35.3	36.2	37.1	38.0	38.9	39.8	40.7	41.	42.5	43.4	44.	45.2	46.0	46.9
137	34.3	35.2	36.1	37.0	37.9	38.8	39.7	40.6	41.	42.4	43.3	44.	45.1	45.9	46.8
138	34.2	35.1	36.0	36.9	37.8	38.7	39.6	40.5	41.	42.4	43.2	44.3	45.0	45.8	46.7
139	34.2	35.1	36.0	36.8	37.7	38.6	39.5	40.5	41.	42.3	43.2	44.	44.9	45.7	46.6
140	34.1	35.0	35.9	36.8	37.7	38.6	39.5	40.4	41.6	42.2	43.1	43.9	44.8	45.6	46.5
141	34.0	34.9	35.8	36.7	37.6	38.5	39.4	40.4	41.2	42.1	43.0	43.8	44.7	45.	46.5
142	33.0	34.8	35.7	36.6	37.5	38.5	39.4	40.3	41.1	42.0	42.9	43.8	44.6	45.	46.4
143	33.8	34.8	35.6	36.5	37.5	38.4	39.3	40.2	41.1	41.9	42.8	43.7	44.6	45.	46.4
144	33.8	34.7	35.6	36.5	37.4	38.3	39.2	40.1	41.0	41.8	42.7	43.6	44.5	45.	46.3
145	33.7	34.6	35.5	36.4	37.4	38.3	39.1	40.0	40.9	41.7	42.6	43.5	44.5	45.5	46.2
146	33.6	34.	35.5	36.4	37.3	38.2	39.0	39.9	40.8	41.7	42.6	43.5	44.4	45.2	46.1
147	33.6	34.	35.4	36.3	37.2	38.1	39.0	39.8	40.7	41.6	42.5	43.4	44.3	45.1	46.0
148	33.5	34.	35.4	36.2	37.1	38.0	38.9	39.7	40.6	41.5	42.5	43.3	44.2	45.0	45.9
149	33.5	34.	35.3	36.2	37.0	37.9	38.8	39.7	40.6	41.5	42.4	43.2	44.1	44.9	45.8
150	33.4	34.6	35.2	36.1	36.9	37.8	38.7	39.6	40.5	41.4	42.3	43.2	44.0	44.8	45.7
151	33.4	34.2	35.1	36.0	36.9	37.8	38.6	39.5	40.5	41.4	42.2	43.1	43.9	44.7	45.6
152	33.3	34.2	35.1	35.9	36.8	37.7	38.6	39.5	40.4	41.3	42.1	43.0	43.8	44.7	45.5
153	33.2	34.1	35.0	35.8	36.7	37.6	38.5	39.4	40.3	41.2	42.0	42.9	43.7	44.6	45.5
154	33.1	34.0	34.9	35.8	36.6	37.5	38.5	39.4	40.2	41.1	41.9	42.8	43.6	44.5	45.4
155	33.1	33.9	34.8	35.7	36.6	37.5	38.4	39.3	40.2	41.0	41.9	42.7	43.5	44.5	45.4
156	33.0	33.8	34.7	35.6	36.5	37.4	38.3	39.2	40.1	40.9	41.8	42.	43.5	44.4	45.3
157	32.9	33.8	34.7	35.5	36.5	37.4	38.3	39.1	40.	40.8	41.7	42.	43.4	44.4	45.2
158	32.8	33.7	34.6	35.5	36.4	37.3	38.2	39.0	39.9	40.7	41.6	42.6	43.4	44.3	45.1
159	32.8	33.6	34.5	35.4	36.4	37.2	38.1	38.9	39.8	40.7	41.5	42.	43.3	44.2	45.0
160	32.7	33.5	34.5	35.4	36.3	37.1	38.0	38.9	39.0	40.6	41.4	42.5	43.2	44.1	44.9

TABLE 3—*Continued*

Observed Degrees Baumé (Tag) at Observed Temperature

Observed temp., °F.	55	56	57	58	59	60	61	62	63	64	65	66	67	68	69
	Corrected Degrees Baumé (Tag) at 60° F.														
21	59.4	60.5	61.6	62.7	63.8	64.9	65.9	67.0	68.1	69.2	70.2	71.3	72.4	73.5	74.5
22	59.3	60.3	61.5	62.6	63.7	64.7	65.8	66.9	67.9	69.0	70.1	71.2	72.2	73.3	74.4
23	59.2	60.2	61.3	62.4	63.5	64.6	65.7	66.7	67.8	68.9	70.0	71.1	72.1	73.2	74.3
24	59.1	60.1	61.2	62.3	63.3	64.4	65.6	66.6	67.7	68.8	69.8	70.9	72.0	73.1	74.1
25	58.9	60.0	61.1	62.2	63.2	64.3	65.4	66.4	67.6	68.6	69.7	70.8	71.9	72.9	74.0
26	58.8	59.9	61.0	62.0	63.1	64.2	65.2	66.3	67.4	68.5	69.6	70.7	71.7	72.8	73.9
27	58.7	59.8	60.9	61.9	63.0	64.1	65.1	66.2	67.2	68.3	69.5	70.5	71.6	72.6	73.7
28	58.6	59.7	60.7	61.8	62.9	63.9	65.0	66.1	67.1	68.2	69.3	70.3	71.4	72.5	73.6
29	58.4	59.6	60.6	61.7	62.8	63.8	64.9	65.9	67.0	68.1	69.1	70.2	71.2	72.3	73.4
30	58.3	59.4	60.4	61.6	62.6	63.7	64.8	65.8	66.9	67.9	69.0	70.1	71.1	72.1	73.2
31	58.2	59.3	60.3	61.4	62.5	63.6	64.6	65.7	66.7	67.8	68.9	69.9	71.0	72.0	73.1
32	58.1	59.2	60.2	61.3	62.3	63.4	64.5	65.6	66.6	67.7	68.7	69.8	70.8	71.9	73.0
33	58.0	59.1	60.1	61.2	62.2	63.3	64.3	65.4	66.4	67.5	68.6	69.7	70.7	71.8	72.8
34	57.9	58.9	60.0	61.1	62.1	63.2	64.2	65.3	66.3	67.4	68.4	69.5	70.6	71.6	72.7
35	57.8	58.8	59.9	60.9	62.0	63.0	64.1	65.1	66.2	67.2	68.3	69.3	70.4	71.5	72.5
36	57.7	58.7	59.8	60.8	61.9	62.9	64.0	65.0	66.1	67.1	68.2	69.2	70.3	71.3	72.4
37	57.6	58.6	59.7	60.7	61.8	62.8	63.8	64.9	65.9	67.0	68.0	69.1	70.1	71.2	72.2
38	57.4	58.5	59.5	60.6	61.6	62.7	63.7	64.8	65.8	66.8	67.9	68.9	70.0	71.0	72.1
39	57.3	58.3	59.4	60.4	61.5	62.6	63.6	64.6	65.7	66.7	67.8	68.8	69.9	70.9	71.9
40	57.2	58.2	59.3	60.3	61.4	62.4	63.4	64.5	65.6	66.6	67.6	68.7	69.7	70.7	71.8
41	57.1	58.1	59.2	60.2	61.2	62.3	63.3	64.4	65.4	66.4	67.5	68.6	69.6	70.6	71.7
42	57.0	58.0	59.1	60.1	61.1	62.2	63.2	64.2	65.3	66.3	67.3	68.4	69.4	70.4	71.5
43	56.9	57.9	58.9	60.0	61.0	62.0	63.1	64.1	65.1	66.2	67.2	68.2	69.3	70.3	71.4
44	56.8	57.8	58.8	59.9	60.9	61.9	63.0	64.0	65.0	66.0	67.1	68.1	69.2	70.1	71.2
45	56.6	57.7	58.7	59.7	60.8	61.8	62.8	63.9	64.9	65.9	67.0	68.0	69.0	70.0	71.1
46	56.5	57.6	58.6	59.6	60.7	61.7	62.7	63.7	64.8	65.8	66.8	67.9	68.9	69.8	70.9
47	56.4	57.4	58.5	59.5	60.5	61.6	62.6	63.6	64.6	65.7	66.7	67.7	68.8	69.7	70.8
48	56.3	57.3	58.3	59.4	60.4	61.4	62.5	63.5	64.5	65.5	66.6	67.6	68.6	69.6	70.7
49	56.2	57.2	58.2	59.3	60.3	61.3	62.3	63.3	64.4	65.4	66.4	67.4	68.5	69.5	70.5
50	56.1	57.1	58.1	59.1	60.2	61.2	62.2	63.2	64.2	65.3	66.3	67.3	68.3	69.3	70.4
51	56.0	57.0	58.0	59.0	60.0	61.1	62.1	63.1	64.1	65.1	66.2	67.2	68.2	69.2	70.2
52	55.9	56.9	57.9	58.9	59.9	61.0	62.0	63.0	64.0	65.0	66.0	67.0	68.1	69.0	70.1
53	55.8	56.9	57.8	58.8	59.8	60.8	61.8	62.9	63.9	64.9	65.9	66.9	67.9	68.9	70.0
54	55.7	56.7	57.7	58.7	59.7	60.7	61.7	62.7	63.8	64.8	65.8	66.8	67.8	68.8	69.8
55	55.5	56.6	57.6	58.6	59.6	60.6	61.6	62.6	63.6	64.6	65.6	66.7	67.7	68.7	69.7
56	55.4	56.4	57.4	58.5	59.5	60.5	61.5	62.5	63.5	64.5	65.5	66.5	67.5	68.6	69.5
57	55.3	56.3	57.3	58.3	59.3	60.4	61.4	62.4	63.4	64.4	65.4	66.4	67.4	68.4	69.4
58	55.2	56.2	57.2	58.2	59.2	60.2	61.2	62.2	63.2	64.3	65.3	66.3	67.3	68.3	69.3
59	55.1	56.1	57.1	58.1	59.1	60.1	61.1	62.1	63.1	64.1	65.1	66.1	67.1	68.1	69.1
60	55.0	56.0	57.0	58.0	59.0	60.0	61.0	62.0	63.0	64.0	65.0	66.0	67.0	68.0	69.0
61	54.9	55.9	56.9	57.9	58.9	59.9	60.9	61.9	62.9	63.9	64.9	65.9	66.9	67.9	68.9
62	54.8	55.8	56.8	57.8	58.8	59.8	60.8	61.8	62.8	63.7	64.7	65.7	66.7	67.8	68.7
63	54.7	55.7	56.7	57.7	58.7	59.6	60.7	61.6	62.6	63.6	64.6	65.6	66.6	67.0	68.6
64	54.6	55.6	56.6	57.5	58.5	59.5	60.5	61.5	62.5	63.5	64.5	65.5	66.5	67.5	68.5
65	54.5	55.5	56.5	57.4	58.4	59.4	60.4	61.4	62.4	63.4	64.4	65.4	66.3	67.3	68.3
66	54.4	55.4	56.3	57.3	58.3	59.3	60.3	61.3	62.3	63.3	64.2	65.2	66.2	67.2	68.2
67	54.3	55.3	56.2	57.2	58.2	59.2	60.2	61.2	62.1	63.1	64.1	65.1	66.1	67.1	68.1
68	54.2	55.1	56.1	57.1	58.1	59.1	60.1	61.0	62.0	63.0	64.0	65.0	66.0	66.9	67.9
69	54.1	55.0	56.0	57.0	58.0	59.0	59.9	60.9	61.9	62.9	63.9	64.8	65.8	66.8	67.8
70	54.0	54.9	55.9	56.9	57.9	58.8	59.8	60.8	61.8	62.8	63.7	64.7	65.7	66.7	67.7

TABLE 3—*Continued*

Observed Degrees Baumé (Tag) at Observed Temperature

Observed temp., ° F.	55	56	57	58	59	60	61	62	63	64	65	66	67	68	69
	Corrected Degrees Baumé (Tag) at 60° F.														
71	53.8	54.8	55.8	56.8	57.8	58.7	59.7	60.7	61.7	62.6	63.6	64.6	65.6	66.5	**67.5**
72	53.7	54.7	55.7	56.7	57.6	58.6	59.6	60.6	61.5	62.5	63.5	64.5	65.5	66.4	**67.4**
73	53.6	54.6	55.6	56.6	57.5	58.5	59.5	60.5	61.4	62.4	63.4	64.4	65.3	66.3	**67.3**
74	53.5	54.4	55.5	56.5	57.4	58.4	59.4	60.3	61.3	62.3	63.3	64.2	65.2	66.2	**67.1**
75	53.4	54.3	55.4	56.4	57.3	58.3	59.3	60.2	61.2	62.2	63.1	64.1	65.1	66.0	**67.0**
76	53.3	54.2	55.3	56.2	57.2	58.2	59.1	60.1	61.1	62.0	63.0	64.0	64.9	65.9	**66.9**
77	53.2	54.1	55.2	56.1	57.1	58.1	59.0	60.0	60.9	61.9	62.9	63.8	64.8	65.8	**66.7**
78	53.1	54.0	55.1	56.0	57.0	57.9	58.9	59.9	60.8	61.8	62.8	63.7	64.7	65.6	**66.6**
79	53.0	53.9	55.0	55.9	56.9	57.8	58.8	59.7	60.7	61.7	62.6	63.6	64.5	65.5	**66.5**
80	52.9	53.8	54.9	55.8	56.8	57.7	58.7	59.6	60.6	61.5	62.5	63.5	64.4	65.4	**66.4**
81	52.8	53.7	54.7	55.7	56.6	57.6	58.6	59.5	60.5	61.4	62.4	63.4	64.3	65.3	**66.2**
82	52.7	53.6	54.6	55.6	56.5	57.5	58.5	59.4	60.4	61.3	62.3	63.2	64.2	65.2	**66.1**
83	52.6	53.5	54.5	55.5	56.4	57.4	58.4	59.3	60.3	61.2	62.2	63.1	64.1	65.0	**66.0**
84	52.5	53.5	54.4	55.4	56.3	57.3	58.2	59.2	60.1	61.1	62.0	63.0	63.9	64.9	**65.8**
85	52.5	53.4	54.3	55.3	56.2	57.2	58.1	59.1	60.0	61.0	61.9	62.9	63.8	64.8	**65.7**
86	52.4	53.3	54.1	55.2	56.1	57.1	58.0	59.0	59.9	60.8	61.8	62.7	63.7	64.6	**65.6**
87	52.3	53.2	54.0	55.1	56.0	57.0	57.9	58.8	59.8	60.7	61.7	62.6	63.6	64.5	**65.5**
88	52.2	53.1	53.9	55.0	55.9	56.8	57.8	58.7	59.7	60.6	61.6	62.5	63.5	64.4	**65.4**
89	52.1	53.0	53.9	54.9	55.8	56.7	57.7	58.6	59.6	60.5	61.5	62.4	63.4	64.3	**65.2**
90	52.0	52.9	53.8	54.8	55.7	56.6	57.6	58.5	59.5	60.4	61.3	62.3	63.2	64.2	**65.1**
91	51.9	52.8	53.7	54.7	55.6	56.5	57.5	58.4	59.4	60.3	61.2	62.2	63.1	64.0	**65.0**
92	51.8	52.7	53.6	54.6	55.5	56.4	57.4	58.3	59.2	60.2	61.1	62.0	63.0	63.9	**64.8**
93	51.7	52.6	53.5	54.5	55.4	56.3	57.3	58.2	59.1	60.1	61.0	61.9	62.8	63.8	**64.7**
94	51.6	52.5	53.4	54.4	55.3	56.2	57.2	58.1	59.0	59.9	60.9	61.8	62.7	63.6	**64.6**
95	51.5	52.4	53.3	54.3	55.2	56.1	57.0	58.0	58.9	59.8	60.7	61.7	62.6	63.5	**64.5**
96	51.4	52.3	53.2	54.2	55.1	56.0	56.9	57.8	58.8	59.7	60.6	61.6	62.5	63.4	**64.4**
97	51.3	52.2	53.1	54.1	55.0	55.9	56.8	57.7	58.7	59.6	60.5	61.5	62.4	63.3	**64.2**
98	51.2	52.1	53.0	54.0	54.9	55.8	56.7	57.6	58.6	59.5	60.4	61.4	62.3	63.2	**64.1**
99	51.1	52.0	52.9	53.9	54.8	55.7	56.6	57.5	58.5	59.4	60.3	61.2	62.1	63.1	**64.0**
100	51.0	51.9	52.8	53.8	54.7	55.6	56.5	57.4	58.4	59.3	60.2	61.1	62.0	62.9	**63.8**
101	50.9	51.8	52.7	53.7	54.6	55.5	56.4	57.3	58.3	59.2	60.1	61.0	61.9	62.8	**63.7**
102	50.8	51.7	52.6	53.6	54.5	55.4	56.3	57.2	58.1	59.0	59.9	60.9	61.8	62.7	**63.6**
103	50.7	51.6	52.5	53.5	54.4	55.3	56.2	57.1	58.0	58.9	59.8	60.7	61.6	62.6	**63.5**
104	50.6	51.5	52.5	53.4	54.3	55.2	56.1	57.0	57.9	58.8	59.7	60.6	61.5	62.5	**63.4**
105	50.5	51.5	52.4	53.3	54.2	55.1	56.0	56.9	57.8	58.7	59.6	60.5	61.4	62.4	**63.3**
106	50.5	51.4	52.3	53.2	54.1	55.0	55.9	56.8	57.7	58.6	59.5	60.4	61.3	62.3	**63.1**
107	50.4	51.3	52.2	53.1	54.0	54.9	55.8	56.7	57.6	58.5	59.4	60.3	61.2	62.1	**63.0**
108	50.3	51.2	52.1	53.0	53.9	54.8	55.7	56.6	57.5	58.4	59.3	60.2	61.1	62.0	**62.9**
109	50.2	51.1	52.0	52.9	53.8	54.7	55.6	56.5	57.4	58.3	59.2	60.1	61.0	61.9	**62.8**
110	50.1	51.0	51.9	52.8	53.7	54.6	55.5	56.4	57.3	58.2	59.1	60.0	60.8	61.7	**62.6**
111	50.0	50.9	51.8	52.7	53.6	54.5	55.4	56.3	57.2	58.1	58.9	59.8	60.7	61.6	**62.5**
112	49.9	50.8	51.7	52.6	53.5	54.4	55.3	56.2	57.1	57.9	58.8	59.7	60.6	61.5	**62.4**
113	49.8	50.7	51.6	52.5	53.4	54.3	55.2	56.1	57.0	57.8	58.7	59.6	60.5	61.4	**62.3**
114	49.7	50.6	51.5	52.4	53.3	54.2	55.1	56.0	56.9	57.7	58.6	59.5	60.4	61.3	**62.2**
115	49.6	50.5	51.5	52.3	53.2	54.1	55.0	55.9	56.8	57.6	58.5	59.4	60.3	61.2	**62.1**
116	49.5	50.5	51.4	52.2	53.1	54.0	54.9	55.8	56.7	57.5	58.4	59.3	60.2	61.1	**62.0**
117	49.5	50.4	51.3	52.1	53.0	53.9	54.8	55.7	56.6	57.4	58.3	59.2	60.1	60.9	**61.8**
118	49.4	50.3	51.2	52.0	52.9	53.8	54.7	55.6	56.5	57.3	58.2	59.1	60.0	60.8	**61.7**
119	49.3	50.2	51.1	51.9	52.8	53.7	54.6	55.5	56.4	57.2	58.1	59.0	59.8	60.7	**61.6**
120	49.2	50.1	51.0	51.8	52.7	53.6	54.5	55.4	56.3	57.1	58.0	58.8	59.7	60.6	**61.5**

TABLE 3—*Continued*

Observed Degrees Baumé (Tag) at Observed Temperature

Observed temp., °F.	55	56	57	58	59	60	61	62	63	64	65	66	67	68	69
	CORRECTED DEGREES BAUMÉ (TAG) AT 60° F.														
121	49.1	50.0	50.9	51.7	52.6	53.5	54.4	55.3	56.1	57.0	57.8	58.7	59.6	60.5	61.4
122	49.0	49.9	50.8	51.6	52.5	53.4	54.3	55.2	56.0	56.9	57.7	58.6	59.5	60.4	61.3
123	48.9	49.8	50.7	51.5	52.5	53.3	54.2	55.1	55.9	56.8	57.6	58.5	59.4	60.3	61.2
124	48.8	49.7	50.6	51.5	52.4	53.2	54.1	55.0	55.8	56.7	57.5	58.4	59.3	60.2	61.0
125	48.7	49.6	50.5	51.4	52.3	53.1	54.0	54.9	55.7	56.6	57.5	58.3	59.2	60.1	60.9
126	48.6	49.5	50.5	51.3	52.2	....	....	....	....	....	....	....	....	....	....
127	48.6	49.5	50.4	51.2	52.1	....	....	....	....	....	....	....	....	....	....
128	48.5	49.4	50.3	51.1	52.0	....	....	....	....	....	....	....	....	....	....
129	48.5	49.3	50.2	51.0	51.9	....	....	....	....	....	....	....	....	....	....
130	48.4	49.2	50.1	50.9	51.8	....	....	....	....	....	....	....	....	....	....
131	48.3	.1	.0	.8	51.7	....	....	....	....	....	....	....	....	....	....
132	48.2	.0	.9	.7	51.6	....	....	....	....	....	....	....	....	....	....
133	48.1	.9	.8	.6	51.5	....	....	....	....	....	....	....	....	....	....
134	48.0	49.8	50.7	50.5	51.5	....	....	....	....	....	....	....	....	....	....
135	47.9	48.7	49.6	50.5	51.4	....	....	....	....	....	....	....	....	....	....
136	47.8	.6	.5	.4	51.3	....	....	....	....	....	....	....	....	....	....
137	47.7	.5	.5	.3	51.2	....	....	....	....	....	....	....	....	....	....
138	47.6	.5	.4	.2	51.1	....	....	....	....	....	....	....	....	....	....
139	47.5	48.4	49.3	50.1	51.0	....	....	....	....	....	....	....	....	....	....
140	47.5	48.4	49.2	50.0	50.9	....	....	....	....	....	....	....	....	....	....
141	47.4	.3	.1	.9	50.8	....	....	....	....	....	....	....	....	....	....
142	47.3	.2	.0	.8	50.7	....	....	....	....	....	....	....	....	....	....
143	47.2	.1	.9	.7	50.6	....	....	....	....	....	....	....	....	....	....
144	47.1	48.0	49.8	49.6	50.5	....	....	....	....	....	....	....	....	....	....
145	47.0	48.9	48.7	49.0	50.5	....	....	....	....	....	....	....	....	....	....
146	46.9	47.8	48.6	49.5	50.4	....	....	....	....	....	....	....	....	....	....
147	46.8	47.7	48.5	49.5	50.3	....	....	....	....	....	....	....	....	....	....
148	46.7	47.6	48.5	49.4	50.2	....	....	....	....	....	....	....	....	....	....
149	46.6	47.5	48.4	49.3	50.1	....	....	....	....	....	....	....	....	....	....
150	46.5	47.5	48.4	49.2	50.0	....	....	....	....	....	....	....	....	....	....
151	46.5	47.4	48.3	49.1	49.9	....	....	....	....	....	....	....	....	....	....
152	46.4	47.3	48.2	49.0	49.8	....	....	....	....	....	....	....	....	....	....
153	46.4	47.2	48.1	48.9	49.7	....	....	....	....	....	....	....	....	....	....
154	46.3	47.1	48.0	48.8	49.6	....	....	....	....	....	....	....	....	....	....
155	46.2	47.0	47.9	48.7	49.5	....	....	....	....	....	....	....	....	....	....
156	46.1	46.9	47.8	48.6	49.5	....	....	....	....	....	....	....	....	....	....
157	46.0	46.8	47.7	48.6	49.4	....	....	....	....	....	....	....	....	....	....
158	45.9	46.7	47.6	48.5	49.4	....	....	....	....	....	....	....	....	....	....
159	45.8	46.6	47.5	48.5	49.3	....	....	....	....	....	....	....	....	....	....
160	45.7	46.5	47.5	48.4	49.2	....	....	....	....	....	....	....	....	....	....

TABLE 3—*Continued*

Observed Degrees Baumé (Tag) at Observed Temperature

Observed temp., °F.	70	71	72	73	74	75	76	77	78	79	80	81	82	83	84
	CORRECTED DEGREES BAUMÉ (TAG) AT 60° F.														
21	75.7	76.8	77.9	78.9	80.0	81.1	82.2	83.3	84.4	85.5	86.6	87.7	88.8	89.9	91.0
22	75.6	76.6	77.7	78.8	79.9	81.0	82.0	83.1	84.2	85.3	86.4	87.5	88.6	89.7	90.8
23	75.4	76.4	77.6	78.7	79.7	80.8	81.9	83.0	84.1	85.1	86.2	87.3	88.4	89.5	90.6
24	75.2	76.3	77.4	78.5	79.6	80.7	81.7	82.8	83.9	85.0	86.1	87.1	88.2	89.3	90.4
25	75.1	76.1	77.2	78.3	79.4	80.5	81.6	82.6	83.7	84.8	85.9	87.0	88.1	89.1	90.2
26	74.9	76.0	77.1	78.1	79.2	80.3	81.4	82.5	83.6	84.7	85.7	86.8	87.9	89.0	90.0
27	74.8	75.9	76.9	78.0	79.1	80.1	81.2	82.3	83.4	84.5	85.6	86.6	87.7	88.8	89.9
28	74.6	75.7	76.8	77.8	78.9	80.0	81.0	82.1	83.2	84.3	85.4	86.4	87.5	88.6	89.7
29	74.4	75.6	76.6	77.7	78.8	79.8	80.9	82.0	83.0	84.1	85.2	86.2	87.3	88.4	89.5
30	74.3	75.4	76.4	77.5	78.6	79.7	80.7	81.8	82.9	83.9	85.0	86.1	87.1	88.2	89.3
31	74.2	75.2	76.3	77.3	78.5	79.5	80.6	81.6	82.7	83.8	84.8	85.9	87.0	88.0	89.1
32	74.0	75.1	76.1	77.2	78.3	79.3	80.4	81.5	82.5	83.6	84.7	85.7	86.8	87.9	88.9
33	73.9	74.9	76.0	77.0	78.1	79.2	80.2	81.3	82.3	83.4	84.5	85.5	86.6	87.7	88.8
34	73.7	74.8	75.8	76.9	78.0	79.0	80.1	81.1	82.2	83.2	84.3	85.3	86.4	87.5	88.6
35	73.6	74.7	75.7	76.7	77.8	78.9	79.9	81.0	82.0	83.1	84.1	85.1	86.2	87.3	88.4
36	73.4	74.5	75.5	76.6	77.7	78.7	79.7	80.8	81.9	82.9	84.0	85.0	86.1	87.1	88.2
37	73.3	74.3	75.4	76.5	77.5	78.5	79.6	80.7	81.7	82.7	83.8	84.9	85.9	87.0	88.0
38	73.1	74.2	75.2	76.3	77.3	78.4	79.4	80.5	81.5	82.6	83.6	84.7	85.7	86.8	87.8
39	73.0	74.0	75.1	76.1	77.2	78.2	79.2	80.3	81.3	82.4	83.5	84.5	85.6	86.6	87.7
40	72.8	73.9	74.9	76.0	77.0	78.1	79.1	80.1	81.2	82.2	83.3	84.3	85.4	86.4	87.5
41	72.7	73.7	74.8	75.8	76.9	77.9	78.9	80.0	81.0	82.1	83.1	84.2	85.2	86.2	87.3
42	72.5	73.6	74.6	75.7	76.7	77.8	78.8	79.8	80.9	81.9	82.9	84.0	85.0	86.1	87.1
43	72.4	73.5	74.5	75.5	76.6	77.6	78.6	79.7	80.7	81.7	82.8	83.8	84.9	85.9	86.9
44	72.2	73.3	74.3	75.4	76.4	77.5	78.5	79.5	80.6	81.6	82.6	83.7	84.7	85.7	86.8
45	72.1	73.1	74.2	75.2	76.2	77.3	78.3	79.3	80.4	81.4	82.5	83.5	84.5	85.6	86.6
46	71.9	73.0	74.0	75.1	76.1	77.1	78.1	79.2	80.2	81.2	82.3	83.3	84.3	85.4	86.4
47	71.8	72.9	73.9	74.9	75.9	77.0	78.0	79.0	80.0	81.1	82.1	83.1	84.2	85.2	86.2
48	71.6	72.7	73.7	74.8	75.8	76.8	77.8	78.9	79.9	80.9	81.9	83.0	84.0	85.0	86.1
49	71.5	72.6	73.6	74.6	75.6	76.7	77.7	78.7	79.7	80.8	81.8	82.8	83.8	84.9	85.9
50	71.4	72.4	73.5	74.5	75.5	76.5	77.5	78.6	79.6	80.6	81.6	82.6	83.7	84.7	85.7
51	71.3	72.3	73.3	74.3	75.3	76.3	77.4	78.4	79.4	80.4	81.5	82.5	83.5	84.5	85.5
52	71.1	72.1	73.1	74.2	75.2	76.2	77.2	78.2	79.2	80.3	81.3	82.3	83.3	84.3	85.4
53	71.0	72.0	73.0	74.0	75.0	76.1	77.1	78.1	79.1	80.1	81.1	82.1	83.2	84.2	85.2
54	70.8	71.8	72.9	73.9	74.9	75.9	76.9	77.9	78.9	80.0	81.0	82.0	83.0	84.0	85.0
55	70.7	71.7	72.7	73.7	74.7	75.8	76.8	77.8	78.8	79.8	80.8	81.8	82.8	83.8	84.9
56	70.6	71.6	72.6	73.6	74.6	75.6	76.6	77.6	78.6	79.6	80.6	81.7	82.7	83.7	84.7
57	70.4	71.4	72.4	73.4	74.4	75.5	76.5	77.5	78.5	79.5	80.5	81.5	82.5	83.5	84.5
58	70.3	71.3	72.3	73.3	74.3	75.3	76.3	77.3	78.3	79.3	80.3	81.3	82.3	83.3	84.3
59	70.1	71.1	72.1	73.1	74.1	75.1	76.1	77.2	78.2	79.2	80.2	81.2	82.2	83.2	84.2
60	70.0	71.0	72.0	73.0	74.0	75.0	76.0	77.0	78.0	79.0	80.0	81.0	82.0	83.0	84.0
61	69.9	70.9	71.9	72.9	73.9	74.9	75.9	76.9	77.8	78.8	79.8	80.8	81.8	82.8	83.8
62	69.7	70.7	71.7	72.7	73.7	74.7	75.7	76.7	77.7	78.7	79.7	80.7	81.7	82.7	83.7
63	69.6	70.6	71.6	72.6	73.6	74.6	75.5	76.5	77.5	78.5	79.5	80.5	81.5	82.5	83.5
64	69.5	70.4	71.4	72.4	73.4	74.4	75.4	76.4	77.4	78.4	79.4	80.4	81.4	82.3	83.3
65	69.3	70.3	71.3	72.3	73.3	74.3	75.3	76.2	77.2	78.2	79.2	80.2	81.2	82.2	83.2
66	69.2	70.2	71.2	72.1	73.1	74.1	75.1	76.1	77.1	78.1	79.1	80.1	81.0	82.0	83.0
67	69.0	70.0	71.0	72.0	73.0	74.0	75.0	75.9	76.9	77.9	78.9	79.9	80.9	81.8	82.8
68	68.9	69.9	70.9	71.9	72.8	73.8	74.8	75.8	76.8	77.8	78.7	79.7	80.7	81.7	82.7
69	68.8	69.8	70.7	71.7	72.7	73.7	74.7	75.6	76.6	77.6	78.6	79.6	80.5	81.5	82.5
70	68.6	69.6	70.6	71.6	72.5	73.5	74.5	75.5	76.5	77.5	78.4	79.4	80.4	81.4	82.3

TABLE 3—*Continued*

Observed Degrees Baumé (Tag) at Observed Temperature

Observed temp., °F.	70	71	72	73	74	75	76	77	78	79	80	81	82	83	84
	CORRECTED DEGREES BAUMÉ (TAG) AT 60° F.														
71	68.5	69.5	70.5	71.4	72.4	73.4	74.4	75.4	76.3	77.3	78.3	79.3	80.2	81.2	82.2
72	68.4	69.4	70.3	71.3	72.3	73.3	74.2	75.2	76.2	77.2	78.1	79.1	80.1	81.0	82.0
73	68.2	69.2	70.2	71.2	72.1	73.1	74.1	75.0	76.0	77.0	78.0	78.9	79.9	80.9	81.9
74	68.1	69.1	70.0	71.0	72.0	73.0	73.9	74.9	75.9	76.8	77.8	78.8	79.8	80.7	81.7
75	68.0	68.9	69.9	70.9	71.8	72.8	73.8	74.8	75.7	76.7	77.6	78.6	79.6	80.5	81.5
76	67.8	68.8	69.8	70.7	71.7	72.7	73.6	74.6	75.6	76.5	77.5	78.5	79.4	80.4	81.4
77	67.7	68.7	69.6	70.6	71.6	72.5	73.5	74.5	75.4	76.4	77.4	78.3	79.3	80.3	81.2
78	67.6	68.5	69.5	70.5	71.4	72.4	73.4	74.3	75.3	76.3	77.2	78.2	79.1	80.1	81.1
79	67.5	68.4	69.4	70.3	71.3	72.3	73.2	74.2	75.1	76.1	77.1	78.0	79.0	79.9	80.9
80	67.3	68.3	69.2	70.2	71.2	72.1	73.1	74.0	75.0	75.9	76.9	77.9	78.8	79.8	80.7
81	67.2	68.2	69.1	70.1	71.0	72.0	72.9	73.9	74.8	75.8	76.7	77.7	78.7	79.6	80.6
82	67.1	68.0	69.0	69.9	70.9	71.8	72.8	73.7	74.7	75.6	76.6	77.5	78.5	79.5	80.4
83	66.9	67.9	68.8	69.8	70.7	71.7	72.6	73.6	74.5	75.5	76.5	77.4	78.4	79.3	80.3
84	66.8	67.7	68.7	69.6	70.6	71.5	72.5	73.5	74.4	75.4	76.3	77.3	78.2	79.2	80.1
85	66.7	67.6	68.5	69.5	70.5	71.4	72.4	73.3	74.3	75.2	76.1	77.1	78.1	79.0	80.0
86	66.5	67.5	68.4	69.4	70.3	71.3	72.2	73.2	74.1	75.1	76.0	77.0	77.9	78.8	79.8
87	66.4	67.4	68.3	69.3	70.2	71.2	72.1	73.0	74.0	74.9	75.9	76.8	77.7	78.7	79.6
88	66.3	67.3	68.2	69.1	70.1	71.0	71.9	72.9	73.8	74.8	75.7	76.6	77.6	78.5	79.5
89	66.2	67.1	68.1	69.0	69.9	70.9	71.8	72.7	73 7	74.6	75.5	76.5	77.5	78.4	79.3
90	66.0	67.0	67.9	68.8	69.8	70.7	71.6	72.6	73.5	74.5	75.4	76.4	77.3	78.2	79.2
91	65.9	66.8	67.8	68.7	69.6	70.6	71.5	72.5	73.5	74 4	75.3	76.2	77.2	78.1	79.0
92	65.8	66.7	67.6	68.6	69.5	70.5	71.4	72.3	73.3	74.2	75.1	76.1	77.0	77.9	78.9
93	65.6	66.6	67.5	68.5	69.4	70.3	71.3	72.2	73.1	74.0	75.0	75.9	76.8	77.8	78.7
94	65.5	66.5	67.4	68.4	69.3	70.2	71.1	72.0	73.0	73.9	74.8	75.7	76.7	77.6	78.5
95	65.4	66.4	67.3	68.2	69.1	70.1	71.0	71.9	72.8	73.7	74.7	75.6	76.5	77.5	78.4
96	65.3	66.2	67.2	68.1	69.0	69.9	70.8	71.7	72.7	73.6	74.5	75.5	76.4	77.4	78.3
97	65.2	66.1	67.0	67.9	68.8	69.8	70.7	71.6	72.5	73.5	74.4	75.4	76.3	77.2	78.1
98	65.0	65.9	66.9	67.8	68.7	69.6	70.5	71.5	72.4	73.4	74.3	75.2	76.1	77.0	77.9
99	64.9	65.8	66.7	67.7	68.6	69.5	70.4	71.4	72.3	73.2	74.1	75.0	76.0	76.9	77.8
100	64.8	65.7	66.6	67.5	68.5	69.4	70.3	71.2	72.1	73.1	74.0	74.9	75.8	76.7	77.6
101	64.6	65.5	66.5	67.4	68.4	69.3	70.2	71.1	72.0	72.9	73.9	74.7	75.6	76.5	77.5
102	64.5	65.4	66.4	67.3	68.2	69.1	70.0	70.9	71.8	72.8	73.7	74.6	75.5	76.4	77.4
103	64.4	65.3	66.3	67.2	68.1	69.0	69.9	70.8	71.7	72.6	73.5	74.5	75.4	76.3	77.2
104	64.3	65.2	66.1	67.0	67.9	68.8	69.7	70.6	71.6	72.5	73.4	74.3	75.2	76.2	77.1
105	64.2	65.1	66.0	66.9	67.8	68.7	69.6	70.5	71.5	72.4	73.3	74.2	75.1	76.0	76.9
106	64.0	65.0	65.9	66.8	67.7	68.5	69.5	70.4	71.3	72.2	73.1	74.0	74.9	75.8	76.7
107	63.9	64.8	65.7	66.6	67.5	68.4	69.4	70.3	71.2	72.1	73.0	73 9	74.8	75.7	76.6
108	63.8	64.7	65.6	66.5	67.4	68.3	69.3	70.1	71.1	71.9	72.8	73 7	74.6	75.5	76.5
109	63.6	64.5	65.5	66.4	67.3	68.2	69.1	70.0	70.9	71.8	72.7	73.6	74.5	75.4	76.3
110	63.5	64.4	65.4	66.3	67.2	68.1	69.0	69.9	70.8	71.6	72.5	73.5	74.4	75.3	76 2
111	....	....	....	....	....	....	....	....	....	....	....	....	....	....	....
112	....	....	....	....	....	....	....	....	....	....	....	....	....	....	....
113	....	....	....	....	....	....	....	....	....	....	....	....	....	....	....
114	....	....	....	....	....	....	....	....	....	....	....	....	....	....	....
115	....	....	....	....	....	....	....	....	....	....	....	....	....	....	....
116	....	....	....	....	....	....	....	....	....	....	....	....	....	....	....
117	....	....	....	....	....	....	....	....	....	....	....	....	....	....	....
118	....	....	....	....	....	....	....	....	....	....	....	....	....	....	....
119	....	....	....	....	....	....	....	....	....	....	....	....	....	....	....
120	....	....	....	....	....	....	....	....	....	....	....	....	....	....	....

TABLE 3—*Continued*

Observed Degrees Baumé (Tag) at Observed Temperature

Observed temp., ° F.	85	86	87	88	89	90	91	92	93	94	95	96	97	98	99
	CORRECTED DEGREES BAUMÉ (TAG) AT 60° F.														
21	92.1	93.2	94.3	95.4	96.5	97.6	98.7	99.8	....	....	....	....	....	....	....
22	91.9	93.0	94.1	95.2	96.3	97.4	98.5	99.6	....	....	....	....	....	....	....
23	91.7	92.8	93.9	95.0	96.1	97.2	98.3	99.4	....	....	....	....	....	....	....
24	91.5	92.6	93.7	94.8	95.9	97.0	98.1	99.2	....	....	....	....	....	....	....
25	91.3	92.4	93.5	94.6	95.7	96.8	97.9	99.0	....	....	....	....	....	....	....
26	91.1	92.2	93.3	94.4	95.5	96.6	97.7	98.8	99.9	....	....	....	....	....	....
27	90.9	92.0	93.1	94.2	95.3	96.4	97.5	98.6	99.7	....	....	....	....	....	....
28	90.8	91.8	92.9	94.0	95.1	96.2	97.2	98.3	99.5	....	....	....	....	....	....
29	90.6	91.7	92.7	93.8	94.9	96.0	97.0	98.1	99.2	....	....	....	....	....	....
30	90.4	91.5	92.5	93.6	94.7	95.8	96.8	97.9	99.0	....	....	....	....	....	....
31	90.2	91.3	92.3	93.4	94.5	95.6	96.7	97.8	98.8	99.9	....	....	....	....	....
32	90.0	91.1	92.1	93.2	94.3	95.4	96.4	97.5	98.6	99.7	....	....	....	....	....
33	89.8	90.9	91.9	93.0	94.1	95.2	96.2	97.3	98.3	99.4	....	....	....	....	....
34	89.7	90.7	91.8	92.8	93.9	95.0	96.0	97.1	98.1	99 2	....	....	....	....	....
35	89.5	90.5	91.6	92.6	93.7	94.8	95.8	96.9	98.0	99.0	....	....	....	....	....
36	89.3	90.3	91.4	92.4	93.5	94.6	95.6	96.7	97.8	98.8	99.9	....	....	....	....
37	89.1	90.1	91.2	92.2	93.3	94.4	95.4	96.5	97.6	98.6	99.7	....	....	....	....
38	88.9	89.9	91.0	92.0	93.1	94.2	95.2	96.3	97.3	98.4	99.4	....	....	....	....
39	88.7	89.8	90.8	91.9	92.9	94.0	95.0	96.1	97.1	98.2	99.2	....	....	....	....
40	88.5	89.6	90.6	91.7	92.7	93.8	94.8	95.9	96.9	98.0	99.0	....	....	....	....
41	88.3	89.4	90.4	91.5	92.5	93.6	94.6	95.7	96.7	97.8	98.8	99.9	....	....	....
42	88.2	89.2	90.2	91.3	92.3	93.4	94.4	95.5	96.5	97.6	98.6	99.7	....	....	....
43	88.0	89.0	90.1	91.1	92.1	93.2	94.2	95.3	96.3	97.4	98.4	99.5	....	....	....
44	87.8	88.8	89.9	90.9	92.0	93.0	94.0	95.1	96.1	97.2	98.2	99.2	....	....	....
45	87.6	88.7	89.7	90.7	91.8	92.8	93.9	94.9	95.9	97.0	98.0	99.0	....	....	....
46	87.5	88.5	89.5	90.6	91.6	92.6	93.7	94.7	95.7	96.8	97.8	98.8	99.8	....	....
47	87.3	88.3	89.3	90.4	91.4	92.4	93.5	94.5	95.5	96.6	97.6	98.6	99.6	....	....
48	87.1	88.1	89.1	90.2	91.2	92.2	93.3	94.3	95.3	96.4	97.4	98.4	99.4	....	....
49	86.9	87.9	89.0	90.0	91.0	92.0	93.1	94.1	95.1	96.2	97.2	98.2	99.2	....	....
50	86.7	87.8	88.8	89.8	90.8	91.9	92.9	93.9	94.9	96.0	97.0	98.0	99.0	....	....
51	86.6	87.6	88.6	89.6	90.7	91.7	92.7	93.7	94.7	95.8	96.8	97.8	98.8	99.8	....
52	86.4	87.4	88.4	89.5	90.5	91.5	92.5	93.5	94.5	95.6	96.6	97.6	98.6	99.6	....
53	86.2	87.2	88.2	89.3	90.3	91.3	92.3	93.3	94.3	95.4	96.4	97.4	98.4	99.4	....
54	86.0	87.1	88.1	89.1	90.1	91.1	92.1	93.1	94.1	95.2	96.2	97.2	98.2	99.2	....
55	85.9	86.9	87.9	88.9	89.9	90.9	91.9	92.9	94.0	95.0	96.0	97.0	98.0	99.0	....
56	85.7	86.7	87.7	88.7	89.7	90.7	91.7	92.8	93.8	94.8	95.8	96.8	97.8	98.8	99.8
57	85.5	86.5	87.5	88.5	89.5	90.5	91.6	92.6	93.6	94.6	95.6	96.6	97.6	98.6	99.6
58	85.3	86.3	87.4	88.4	89.4	90.4	91.4	92.4	93.4	94.4	95.4	96.4	97.4	98.4	99.4
59	85.2	86.2	87.2	88.2	89.2	90.2	91.2	92.2	93.2	94.2	95.2	96.2	97.2	98.2	99.2
60	85.0	86.0	87.0	88.0	89.0	90.0	91.0	92.0	93.0	94.0	95.0	96.0	97.0	98.0	99.0
61	84.8	85.8	86.8	87.8	88.8	89.8	90.8	91.8	92.8	93.8	94.8	95.8	96.8	97.8	98.8
62	84.7	85.7	86.7	87.6	88.6	89.6	90.6	91.6	92.6	93.6	94.6	95.6	96.6	97.6	98.6
63	84.5	85.5	86.5	87.5	88.5	89.5	90.5	91.4	92.4	93.4	94.4	95.4	96.4	97.4	98.4
64	84.3	85.3	86.3	87.3	88.3	89.3	90.3	91.3	92.2	93.2	94.2	95.2	96.2	97.2	98.2
65	84.2	85.1	86.1	87.1	88.1	89.1	90.1	91.1	92.1	93.0	94.0	95.0	96.0	97.0	98.0
66	84.0	85.0	86.0	86.9	87.9	88.9	89.9	90.9	91.9	92.9	93.8	94.8	95.8	96.8	97.8
67	83.8	84.8	85.8	86.8	87.8	88.7	89.7	90.7	91.7	92.7	93.6	94.6	95.6	96.6	97.6
68	83.6	84.6	85.6	86.6	87.6	88.5	89.5	90.5	91.5	92.5	93.4	94.4	95.4	96.4	97.4
69	83.5	84.5	85.5	86.4	87.4	88.4	89.4	90.3	91.3	92.3	93.3	94.2	95.2	96.2	97.2
70	83.3	84.3	85.3	86.3	87.2	88.2	89.2	90.2	91.1	92.1	93.1	94.0	95.0	96.0	97.0

TABLE 3—*Continued*

Observed Degrees *Baumé (Tag) at Observed Temperature*

Observed temp., °F.	85	86	87	88	89	90	91	92	93	94	95	96	97	98	99
	CORRECTED DEGREES BAUMÉ (TAG) AT 60° F.														
71	83.2	84.1	85.1	86.1	87.1	88.0	89.0	90.0	90.9	91.9	92.9	93.9	94.8	95.8	96.8
72	83.0	84.0	84.9	85.9	86.9	87.8	88.8	89.8	90.8	91.7	92.7	93.7	94.6	95.6	96.6
73	82.8	83.8	84.8	85.7	86.7	87.7	88.6	89.6	90.6	91.5	92.5	93.5	94.4	95.4	96.4
74	82.7	83.6	84.6	85.6	86.5	87.5	88.5	89.4	90.4	91.4	92.3	93.3	94.3	95.2	96.2
75	82.5	83.5	84.4	85.4	86.4	87.3	88.3	89.3	90.2	91.2	92.2	93.1	94.1	95.0	96.0
76	82.3	83.3	84.2	85.2	86.2	87.2	88.1	89.1	90.0	91.0	92.0	92.9	93.9	94.8	95.8
77	82.2	83.1	84.1	85.1	86.0	87.0	87.9	88.9	89.8	90.8	91.8	92.7	93.7	94.6	95.6
78	82.0	83.0	83.9	84.9	85.8	86.8	87.8	88.7	89.7	90.6	91.6	92.5	93.5	94.5	95.4
79	81.8	82.8	83.8	84.7	85.7	86.6	87.6	88.5	89.5	90.5	91.4	92.4	93.3	94.3	95.2
80	81.7	82.6	83.6	84.5	85.5	86.5	87.4	88.4	89.3	90.3	91.2	92.2	93.1	94.1	95.0
81	81.5	82.5	83.5	84.4	85.4	86.3	87.2	88.2	89.1	90.1	91.0	92.0	92.9	93.9	94.8
82	81.4	82.3	83.3	84.2	85.2	86.1	87.0	88.0	89.0	89.9	90.9	91.8	92.7	93.7	94.6
83	81.2	82.2	83.1	84.1	85.0	85.9	86.9	87.8	88.8	89.7	90.7	91.6	92.5	93.5	94.5
84	81.1	82.0	82.9	83.9	84.8	85.8	86.7	87.6	88.6	89.5	90.5	91.5	92.4	93.3	94.3
85	80.9	81.8	82.8	83.7	84.6	85.6	86.5	87.5	88.5	89.4	90.3	91.3	92.2	93.1	94.1
86	80.7	81.7	82.6	83.5	84.5	85.5	86.4	87.3	88.3	89.2	90.1	91.1	92.0	92.9	93.9
87	80.5	81.5	82.5	83.4	84.4	85.3	86.2	87.2	88.1	89.0	90.0	90.9	91.8	92.7	93.7
88	80.4	81.4	82.3	83.3	84.2	85.1	86.0	87.0	87.9	88.8	89.8	90.7	91.6	92.5	93.5
89	80.3	81.2	82.1	83.1	84.0	84.9	85.9	86.8	87.7	88.6	89.6	90.5	91.4	92.4	93.3
90	80.1	81.0	82.0	82.9	83.8	84.8	85.7	86.6	87.5	88.5	89.4	90.3	91.3	92.2	93.1
91	80.0	80.9	81.8	82.7	83.7	84.6	85.5	86.5	87.4	88.4	89.3	90.2	91.1	92.0	92.9
92	79.8	80.7	81.6	82.6	83.5	84.4	85.4	86.3	87.2	88.2	89.1	90.0	90.9	91.8	92.7
93	79.6	80.5	81.5	82.5	83.4	84.3	85.2	86.1	87.1	88.0	88.9	89.8	90.7	91.6	92.5
94	79.5	80.4	81.4	82.3	83.2	84.1	85.0	86.0	86.9	87.8	88.7	89.6	90.5	91.5	92.4
95	79.4	80.3	81.2	82.1	83.0	83.9	84.9	85.8	86.7	87.6	88.5	89.5	90.4	91.3	92.2
96	79.2	80.1	81.0	81.9	82.9	83.8	84.7	85.6	86.5	87.5	88.4	89.3	90.2	91.1	92.0
97	79.0	79.9	80.9	81.8	82.7	83.6	84.5	85.5	86.4	87.3	88.2	89.1	90.0	90.9	91.8
98	78.9	79.8	80.7	81.6	82.5	83.5	84.4	85.3	86.2	87.1	88.0	88.9	89.8	90.7	91.6
99	78.7	79.6	80.5	81.5	82.4	83.3	84.2	85.2	86.0	86.9	87.8	88.7	89.6	90.5	91.5
100	78.5	79.5	80.4	81.3	82.2	83.2	84.0	85.0	85.8	86.7	87.7	88.5	89.5	90.4	91.3
101	78.4	79.4	80.3	81.2	82.1	83.0	83.9	84.8	85.7	86.6	87.5	88.4	89.3	90.2	91.1
102	78.3	79.2	80.1	81.0	81.9	82.8	83.7	84.6	85.5	86.5	87.4	88.3	89.2	90.1	90.9
103	78.1	79.0	79.9	80.8	81.7	82.6	83.5	84.5	85.4	86.3	87.2	88.1	89.0	89.9	90.7
104	77.9	78.9	79.8	80.6	81.5	82.5	83.4	84.3	85.2	86.1	87.0	87.9	88.8	89.7	90.5
105	77.8	78.7	79.6	80.5	81.4	82.4	83.3	84.1	85.0	85.9	86.8	87.7	88.6	89.5	90.4
106	77.6	78.5	79.5	80.4	81.3	82.2	83.1	84.0	84.9	85.7	86.6	87.5	88.5	89.3	90.3
107	77.5	78.4	79.4	80.3	81.1	82.0	82.9	83.8	84.7	85.6	86.5	87.4	88.3	89.2	90.1
108	77.4	78.3	79.2	80.1	81.0	81.8	82.7	83.6	84.5	85.5	86.4	87.2	88.1	89.0	89.9
109	77.2	78.1	79.0	79.9	80.8	81.6	82.6	83.5	84.4	85.3	86.2	87.1	87.9	88.8	89.7
110	77.1	77.9	78.9	79.7	80.6	81.5	82.4	83.4	84.2	85.1	86.0	86.9	87.7	88.6	89.5
111	....	....	....	....	....	....	....	....	....	....	....	....	....	....	....
112	....	....	....	....	....	....	....	....	....	....	....	....	....	....	....
113	....	....	....	....	....	....	....	....	....	....	....	....	....	....	....
114	....	....	....	....	....	....	....	....	....	....	....	....	....	....	....
115	....	....	....	....	....	....	....	....	....	....	....	....	....	....	....
116	....	....	....	....	....	....	....	....	....	....	....	....	....	....	....
117	....	....	....	....	....	....	....	....	....	....	....	....	....	....	....
118	....	....	....	....	....	....	....	....	....	....	....	....	....	....	....
119	....	....	....	....	....	....	....	....	....	....	....	....	....	....	....
120	....	....	....	....	....	....	....	....	....	....	....	....	....	....	....

longitudinal slot just wide enough to show one horizontal row of figures on the chart. Just above this slot is a paper scale on which are printed figures denoting "Observed Baumé gravities," both slot and scale being covered with glass for protection. "Observed temperatures" are printed in red at each side of the chart. In using the corrector, the cylinder should be turned until the proper "Observed temperature" appears in the slot. This brings into sight, throughout the length of the slot, a corrected Baumé gravity at 60° F. directly below each "Observed Baumé gravity" read on the upper scale. Obviously such a device practically eliminates the possibility of reading in the wrong line or wrong column as is so frequently done with the ordinary tables, even by experienced operators. The figures on the chart indicating tenths of degrees Baumé, as well as figures for temperature, are printed in red, all other figures in black. A suitable base on the corrector makes it possible to set the apparatus on the work-table or to fasten it on a wall as may be more convenient.

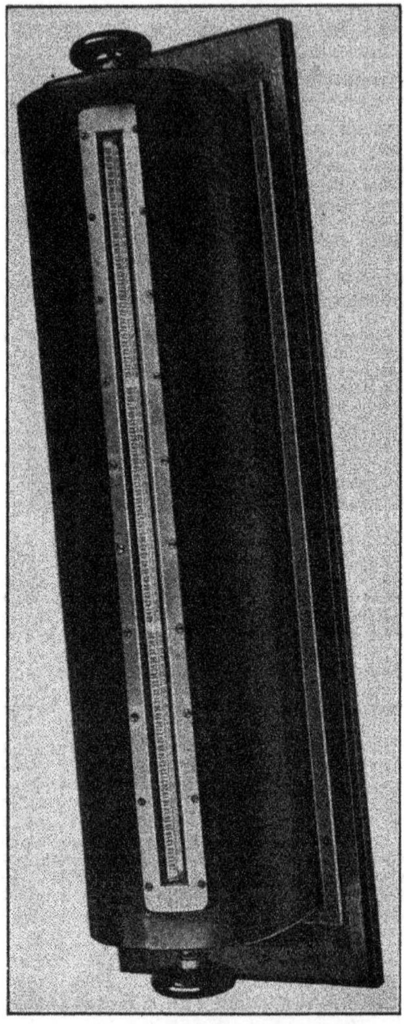

FIG. 5.—Atlantic gravity-temperature o rrect r.

DETERMINATION OF GRAVITY OF LIQUIDS BY THE WEIGHING METHOD

(Test Method No. 2)

This method is more accurate than Test Method No. 1 but requires a little more time and skill. It is applicable to all petroleum products which are not solid at 60° F. and is independent of any temperature correction. When properly used it gives results which are not likely to be in error by more than 0.02° Bé. or 0.0001 specific gravity. The use of Test Methods No. 1 and No. 2 on the same oil affords an excellent check on the accuracy of hydrometers.

Apparatus.—1. Measuring flask, 250 or 500 c.c. capacity, marked at a constriction on the neck as shown in Fig. 6.

2. Balance sensitive to 5 milligrams.

3. Water-bath for constant temperature at 60° F. This should not be too small. Where many determinations are to be made, this bath is conveniently made in trough form, with air agitation throughout its length. Lead rings of suitable size and weight should be used to keep the flasks immersed in an upright position. A bath with glass sides which has proved entirely satisfactory is shown in Fig. 6 (at the right-hand side).

This bath is made in trough form with a glass front and open top. An air line (*A*) with frequent $\frac{1}{32}$-inch perforations extends throughout the length of the bottom of the trough and the air bubbling through the water keeps the temperature uniform. Evaporation of water by this air assists in keeping the bath cool. A brine line (*B*) makes several turns (*E*) along the back of the trough when artificial cooling is needed. A thermostat (*C*) controls the flow of cold brine through an electrically operated valve. A constant level water-overflow and water-inlet line (*F*) make it possible to maintain the desired temperature wherever the feed water is sufficiently cold. Each flask is

Fig. 6.—Constant temperature bath.

weighted with lead rings (D) as shown and the corked mouth is further protected by a metal cap (*G*).

Procedure.—Weigh the clean, dry flask (1) with an accuracy of 0.01 gram and record as "Weight of flask." Make all subsequent weighings with the same degree of accuracy.

Fill the flask with pure distilled water at about 60° F. to slightly above the mark and immerse the flask so filled in the bath (3) for one hour, keeping the bath temperature at 60° F., plus or minus 0.1° F. With a pipette draw off sufficient water to bring the lowest part of the meniscus level with the filling mark on the constricted neck. Remove the flask from the bath and dry the outside thoroughly with a soft towel. Weigh at once and record as "Weight of flask plus water." Calculate by subtraction and record "Weight of water."

Empty the flask and dry the inside thoroughly by warming to about 190° F. and passing air in through a glass tube for five minutes after all visible moisture has disappeared. Cool the flask to room temperature and fill with the liquid to be tested just as in the case of the water-filling. Immerse the flask in the bath as before, again keeping the bath temperature at 60° F. plus or minus 0.1° F. After one hour, mark or measure the height of the liquid in the neck of the flask in any suitable manner. Leave the flask in the bath at 60° F. until this height shows no perceptible change during half an hour. This indicates that the oil has reached a constant, uniform temperature of 60° F. Then draw off sufficient oil to bring its surface (lowest part of meniscus) level with the mark on the constricted neck. Remove the flask from the bath, dry thoroughly with a soft towel and weigh as before. Record as " Weight of flask plus oil." Calculate by subtraction and record " Weight of oil."

Calculate the specific gravity of the oil at 60° F. by substituting the determined values in the following formula which is accurate to the fourth decimal place of specific gravity.

$$\text{Specific gravity } 60°/60° \text{ F.} = \left[\frac{\text{Weight of oil}}{\text{Weight of water}} \times 0.9988\right] + 0.0012.$$

The following examples show the method of recording results and calculating the specific gravity at 60° F.:

	Oil No. 1	Oil No. 2
Data:		
Weight of flask plus water	635.98 g.	635.99 g.
Weight of flask	137.06 g.	137.06 g.
Weight of water	498.92 g.	498.93 g.
Weight of flask plus oil	560.32 g.	595.15 g.
Weight of flask	137.06 g.	137.06 g.
Weight of oil	423.26 g.	458.09 g.

Calculations; substituting in formula:

$$\text{Oil No. 1.} \quad \left[\frac{423.26}{498.92} \times 0.9988\right] + 0.0012 = 0.8485 \text{ specific gravity at } 60° \text{ F.}$$

$$\text{Oil No. 2.} \quad \left[\frac{458.09}{498.93} \times 0.9988\right] + 0.0012 = 0.9183 \text{ specific gravity at } 60° \text{ F.}$$

For routine work by this method, it is convenient to use flasks on which " Weight of flask " and " Weight of water " have been determined very accurately. Each gravity determination thereafter requires only one weighing, namely that to obtain " Weight of flask plus oil." The calculation is then very simple, being as follows:

Specific gravity at 60° = (Weight of oil $\times K$) + 0.0012 where K is a constant for the flask used and is equal to 0.9988 ÷ " Weight of water." Thus the value of K for the flask used in the examples given is 0.002002.

If desired, Baumé gravity, either (B. of S.) or (Tag.), may be calculated from the specific gravity by use of the formulæ given on page 595 or may be obtained without calculation by use of tables.[1]

Where Test Method No. 2 is used for volatile petroleum products, the flask should be closed with a cork or better still a glass stopper ground to fit. The stopper should, of course, be considered as part of the flask and weighed as such.

[1] Bureau of Standards Circular 57, Table 5; C. J. Tagliabue, Manual for Coal Oil Inspectors, 14th edition.

Determination of Gravity of Solids by Hydrometer Method

(Test Method No. 3)

This method is applicable to petroleum products which are solid or semi-solid at 60° F. It is approximate only and its accuracy depends largely on the preparation of the sample. When properly used, it gives results that are not likely to be in error by more than 0.003 specific gravity or 0.4° Bé.

Apparatus and solutions.—1. Gravity cylinder of glass as in Test Method No. 1.

2. Hydrometer of suitable type as in Test Method No. 1.

3. Thermometer as in Test Method No. 1.

4. Water bath to be held at temperature of 60° F. within 1° F.

5. Denatured alcohol and a water solution of sodium chloride (common salt) as required.

Procedure.—Prepare a portion of the material to be tested, so that it will have the shape of a sphere, cylinder or cube whose smallest dimension is not less than ½ inch. This sample must be free from air bubbles. In most cases the material can be heated slightly above its melting point and poured into a mold, which should be made of brass and well amalgamated with mercury. Harder asphalts may be cut to shape with a hot knife.

Immerse the test sample in water at 60° F. for one hour. With some materials it is necessary to leave the test sample in the mold. In such cases immerse both mold and sample, removing the sample from the mold at the end of one hour. While the test sample is being brought to 60° F., the gravity cylinder about half full of water should also be left in the water bath. The temperature of the denatured alcohol and sodium chloride solutions should be approximately 60° F.

At the expiration of the hour, transfer the test sample to the gravity cylinder containing water at 60° F. Adjust the gravity of the liquid in the cylinder by adding denatured alcohol if the material floats or sodium chloride solution if it sinks, until the test sample neither floats nor sinks. Remove all air bubbles from the surface of the test sample by jarring or, if possible, with a soft brush. Make sure that the liquid in the cylinder is thoroughly mixed and is at 60° F. When all these conditions are secured, remove the test sample and immediately take the gravity of the liquid in the cylinder by means of the hydrometer as in Test Method No. 1. Since this liquid is at 60° F., no temperature correction is necessary. Record the gravity of this liquid as the gravity of the material at 60° F. Occasionally gravity on asphaltic materials from petroleum is required at 77° F. (25° C.). By using a temperature of 77° F. instead of 60° F. throughout the above method, specific gravity at 77°/60° F. is obtained. This figure should be multiplied by 1.002, if specific gravity at 77°/77° F. is desired.

Where many samples are to be tested by the above method, it is advisable to have a series of solutions covering the gravity range involved, at convenient intervals of gravity. Observation of the behavior of the test sample when successively immersed in two or three of these solutions (at 60° F.) will usually indicate the gravity with sufficient accuracy.

Determination of Gravity of Solids by Weighing

(Test Method No. 4)

This method is applicable to all petroleum products which are solid or semi-solid at 60° F. but which can be melted without decomposition at higher temperatures. It is more satisfactory than Test Method No. 3 and should always be employed when accuracy is required. When properly used it gives results which are not in error by more than 0.0003 specific gravity or 0.04 Bé.

Apparatus.—Same as in Test Method No. 2.

Procedure.—Weigh the clean, dry flask with an accuracy of 0.01 gram and record as " Weight of flask."

Obtain the " Weight of flask plus water " in the manner described in Test Method No. 2. Calculate by subtraction and record " Weight of water."

Empty the flask and dry the inside thoroughly by warming it to about 190° F. and passing air in through a glass tube for five minutes after all visible moisture has disappeared. Melt the material to be tested, being careful to prevent overheating and loss by evaporation. Pour the melted material into the warm, dry flask, using a funnel with a stem of large diameter and taking precautions to prevent the inclusion of air bubbles. About nine-tenths of the flask should be filled and none of the material should be allowed to contaminate the neck of the flask above the filling mark, either during filling or during the removal of the funnel. Immerse the flask in the water bath at 60° F. for one hour, remove, dry the outside thoroughly and weigh. Record as " Weight of flask plus material." Calculate by subtraction and record " Weight of material."

Fill the flask with pure distilled water at about 60° F. to slightly above the mark and immerse the flask so filled in the water bath at 60° F. for two hours, and bring the water level to the mark, observing all the precautions given in Test Method No. 2. When this level has become constant, remove the flask, dry the outside thoroughly and weigh. Record as " Weight of flask plus material plus added water." Calculate by subtraction and record " Weight of added water."

Calculate the specific gravity of the material being tested, by substituting the determined values in the following formula:

Specific gravity 60°/60° F.

$$= \left[\frac{\text{Weight of material}}{\text{Weight of water} - \text{Weight of added water}} \times 0.9988\right] + 0.0012.$$

The following example shows the method of recording results and calculating the specific gravity at 60° F.

Data:

Weight of flask plus water	635.98 g.
Weight of flask	137.06 g.
Weight of water (by difference)	498.92 g.
Weight of flask plus material	588.42 g.
Weight of flask	137 06 g.
Weight of material (by difference)	451.36 g.
Weight of flask plus material plus added water	626.27 g.
Weight of flask plus material	588.42 g.
Weight of added water (by difference)	37.85 g.

Calculations.—Substituting in formula:

$$\left[\frac{451.36}{498.92-37.85} \times 0.9988\right] + 0.0012 = 0.9790 \text{ specific gravity at } 60° \text{ F.}$$

Except where very accurate results are desired, an approximate formula for calculation may be taken as:

$$\text{Specific gravity at } 60° = \frac{\text{Weight of material}}{\text{Weight of water} - \text{Weight of added water}}.$$

The specific gravity calculated by this formula from the data given is 0.9789, showing a difference of 0.0001. When the specific gravity of the material is less than 0.9, the more accurate formula should be used.

Flash and Fire Points

These characteristics of petroleum products have been the subject of study since the earliest days of the industry. The introduction of kerosene into the world's economic life was quite naturally attended by more or less serious fires and explosions. Consequently, for the protection of the consumer, there arose an immediate necessity for some means of grading and controlling petroleum products according to their inflammability. As the result, many "testers" or instruments for measuring inflammability have been devised, and legislation on the subject is general though not uniform. Since "flash" and "fire" on the same material vary over a considerable range when determined on different instruments, it becomes necessary in every case, both in reports and in specifications, to state clearly the kind of instrument used. Except where legal requirements prevent, the selections made in the following pages will be found to be in accord with usual American practice.

Definitions.—Flash test or flash point is the lowest temperature at which, under definite specified conditions, a petroleum product vaporizes rapidly enough to form above its surface an air-vapor mixture which gives a flash or mild explosion when ignited by a small flame.

Fire test or burning point is the lowest temperature at which, under definite specified conditions, a petroleum product vaporizes rapidly enough to form above its surface an air-vapor mixture which burns continuously for at least five seconds when ignited by a small flame.

These definitions must not be misunderstood. The expression "under definite specified conditions" is a vital part of the definition and must not be disregarded. While a given oil may have a "flash, open cup" of 300° F. under the definite specified conditions of the test, there is no assurance that the same oil may not flash at a lower temperature under other conditions, such as those under which the material is actually used. Hence it is extremely important that there be general recognition of the fact that "flash test" and "fire test" are merely empirical qualities determined by laboratory methods.

The purpose for which these two tests were originated was to determine the fire hazard of petroleum products. From this standpoint fire test has little or no value since it may be as much as 100° F. higher than the flash test. On the other hand flash test, though the most important single factor influencing fire hazard, is not the only one.[1] At the present time, flash and fire tests are also of some importance as means of identification of the source of lubricating oils and particularly as an indication though not a measure of volatility.

While the list of flash "testers" is long, there are comparatively few of these in common use. There is a growing tendency in the United States to obtain better uniformity by eliminating all except a few standard and reliable types. The instruments not described in this section are purposely omitted, although some of them, such as the Foster and the Tagliabue Open Tester, are still the official testers in certain states. The selections made conform with the present views of a large majority of American authorities on the subject.

Petroleum products flashing below 135° F., and therefore tested to obtain primarily an index of fire hazard, should always be tested by means of a "closed tester." In all of the open-cup testers the most volatile and inflammable vapors escape but not in

[1] Bureau of Mines, Technical Paper 49, Allen and Crossfield (1913), p. 5.

sufficient quantity to be readily ignited. This defect of open cups is increased by drafts and air currents such as the breath of the operator. Hence open-cup flash tests are usually higher than those with a closed cup, the difference being as high as 50° F. in some cases.

The choice of flash testers, while influenced to some extent by legal requirements, should be along the following lines:

A. Petroleum products whose flash is 135° F. or less should be tested preferably by the " Tag Closed Tester"; or by the Abel or the Elliott instruments.

B. With the exception of fuel oils, all petroleum products flashing between 135° and 175° F. should be tested as in (*A*). Fuel oils having a flash in this range may be tested by the " Tag Closed Tester " but the Pensky-Martens Closed Cup is preferred.

C. Fuel oils flashing above 175° F. should be tested by the Pensky-Martens Closed Cup.

D. All petroleum products flashing above 175° F. except fuel oil should be tested by the Cleveland Open Cup except when a closed-cup flash is required, in which case the Pensky-Martens should be used.

E. All petroleum products of viscosity higher than 400 seconds Saybolt Universal at 212° F. require stirring during flash determination. For such materials the Pensky-Martens Cup should always be used. The practice of using the Elliott Cup with modifications for asphalts cannot be too severely condemned.

Comparisons of results of flash determinations by different instruments are dangerously inaccurate if used for purposes of conversion. This warning should be kept in mind in connection with Table 4.

TABLE 4

Comparison of Flash Tests by Different Instruments

Material	Abel	Tag Closed Tester	Elliott	Pensky-Martens	Cleveland Open Cup	Luchaire * (French)	Fire test
Naphtha	86	92	92	95	100	100	110
Naphtha	94	103	98	105	115	122	130
Water white kerosene	127	130	128	135	140	138	160
Petrolite	142	139	140	150	155	144	175
Gas oil	....	.....	.....	195	200	198	220
300 oil	....	.....	.....	255	265	266	305
Straw oil	....	.....	.....	315	325	322	385
Ice-machine oil	....	.....	.....	400	385	402	460
Engine oil	....	.....	.....	430	440	430	515
Cylinder oil	....	.....	.....	505	525	507	610
Heavy cylinder oil	....	.....	.....	510	560	510	635

* The Luchaire flash points have been calculated to Fahrenheit.

Fig. 7 shows a view of various flash cups, including some not described in this section.

General considerations in making flash and fire tests.—It is essential that the size and shape of the cup used agree with those given, within the tolerances allowed. Every direction given for operation is important and variations are not permissible. Tests should be made in a room, or better still in a compartment, which is free from air currents and which is dark enough to permit observation of the first flash.

FIG. 7.—Flash testers for petroleum products.

1) Luchaire 4) Fl... 7) Pensky-Martens

Thermometers.—Since true temperatures are desired, the thermometers used for flash determination should be scaled for suitable partial immersion. In case such thermometers are not available, correction [1] for emergent stem must be made in each case, and such corrections must be taken into account in applying the test flame.

Samples.[2]—In the case of materials flashing below 175° F. special precautions should be observed to ensure that the sample to be tested is truly representative and this sample must be protected from evaporation loss in a tightly closed container, preferably in a cool place. Under no circumstances should the material used for flash test be that on which some other test has been made.

Flash determinations cannot be made accurately, even if at all, on samples containing free water. Water, if present, must always be removed, but never by heating since this is likely to raise the flash point. In the case of thin oils, a few hours' standing at room temperature will permit water to settle out and the clear oil on top may be used for the test. Oils which still retain water should be dried with fused calcium chloride.[3] Removal of water by centrifuging or filtering through dry filter paper at room temperature is permissible only in case the flash point is not lower than 300° F.

For all accurate work it is advisable that a rough preliminary test be made to ascertain the approximate flash point when this is not known. Such preliminary tests may be made by heating at two or three times the specified rate and applying the test flame at 5° F. intervals. Their usefulness in facilitating strict adherence to instructions should be more generally recognized and will be apparent in the test methods described.

Heating.—The accuracy of all flash determinations depends to a large degree on constant maintenance of the uniform specified rate of heating. Guess-work should not be tolerated. An essential part of flash-testing apparatus is a stop-watch or other suitable time meter, such as a metronome. While ability to maintain a constant uniform rate of heating must be acquired by practice, the difficulties may be largely eliminated by selecting a suitable burner [4] and by equipping this burner with means for delicate control and adjustment. Where the gas pressure is not fairly constant, a pressure regulator should be provided. In most cases it is sufficient to use the combination of an ordinary gas stop-cock on the line and a needle valve at the burner itself. With the needle valve open, the flame should be made somewhat higher than is necessary by means of the stop-cock and then adjusted by needle valve control. The burner described under Test Method No. 23, Assay Distillation, will be found to be extremely convenient. The device for regulating the size of the test flame should permit delicate adjustment.

As a final recommendation, wherever possible, burners should be connected to the gas supply through permanent lines and fittings. Where flexible connections are necessary, metallic tubing is safer than rubber. Rubber connections, if used, should be as short as possible and should be subjected to frequent inspection and replacement. There are few laboratories where serious damage to person and property has not followed failure to observe these precautions.

Determination of Flash and Fire, Cleveland Open Cup Method

(Test Method No. 5)

This method may be applied to all petroleum products having an open-cup flash higher than 175° F. with the exception of fuel oils, which should be tested by the Pen-

[1] See p. 594.

[2] The ordinary flash testers require considerable oil for a test. For a method of taking f[illegible] fire points on a 20 c.c. sample, see Osmond and Abrams, Atlantic Lubricator, 4, No. 6, Feb. 1921.

[3] An electrical method for removing water from emulsions is described in Bureau of Mine[illegible]nical Paper 49, p. 16 (1913).

[4] Electric heaters are not satisfactory in any of their present forms.

sky-Martens Cup. When properly used it gives results which are not in error by more than 5° F.

Apparatus.—See Figs. 7 and 8.

1. Cleveland Open Cup made of brass according to the following specifications:

	Dimensions, inches	Tolerances, inches
Inside diameter	$2\frac{1}{2}$	$\frac{1}{32}$
Outside diameter	$2\frac{11}{16}$	$\frac{1}{32}$
Inside height	$1\frac{5}{16}$	$\frac{1}{32}$
Thickness of bottom	$\frac{1}{8}$	$\frac{1}{64}$
Distance from top of cup to filling mark scribed on inside	$\frac{3}{8}$	$\frac{1}{64}$
Outside diameter of flange around outside of cup at top	$3\frac{3}{4}$	$\frac{1}{4}$
Thickness of flange	$\frac{3}{16}$	$\frac{1}{16}$

2. Plate of suitable metal, preferably brass. It may be circular or square but must be $\frac{1}{4}$ inch thick and 6 inches wide, with a plane depression in the center, $\frac{1}{32}$ inch deep and of diameter just sufficient to fit the cup.

3. Sheet of hard asbestos board, $\frac{1}{4}$ inch thick, of the same size and shape as the metal plate and with a circular hole in the center to conform with the depression in the plate.

4. Thermometer, "Partial Immersion" to conform to the following specifications:

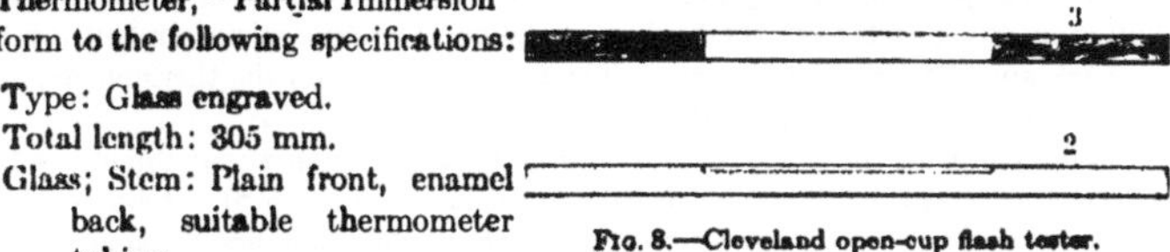

Fig. 8.—Cleveland open-cup flash tester.

Type: Glass engraved.
Total length: 305 mm.
Glass; Stem: Plain front, enamel back, suitable thermometer tubing.
Diameter 6 to 7 mm.
Bulb: Corning normal, Jena 16, III, or equally suitable thermometric glass.
Length 13 mm. maximum.
Diameter not greater than stem.
Actuating liquid: Mercury.
Range: Plus 20° to plus 760° Fahrenheit.
Immersion: 1 inch line around tube at immersion point and words "1 inch immersion" engraved on tube.
Range distance from bottom of bulb: 40 to 50 mm.
Range distance from top of tube: 30 to 45 mm.
Ice point: No.
Contraction chamber: No.
Expansion chamber: No.
Filled: Nitrogen gas.
Top finish: Red glass ring.

Graduating: All lines, figures and letters clear cut and distinct.
Scale graduated in 5° divisions.
Scale numbered every 20°.

Special markings: Open flash.
Serial number and manufacturer's name or trade mark engraved on the tube.

Accuracy: Error at any point on scale shall not exceed one-half smallest scale division up to 700° F.

Test for permanency of range: After being subjected to a temperature of 700° F. for twenty-four hours the accuracy shall be within the limit specified.

Points to be tested for certification: 32°, 212°, 400°, 700°.

5. Ring stand or other similar device suitable for support of the plate and flash cup and for suspension of the thermometer.

6. Gas burner, (or suitable lamp) of sufficient heating power to raise the temperature of the oil at the specified rate. The flame must be capable of very delicate control and adjustment.

7. Shield to protect heating flame from drafts. This shield must not extend above the level of the asbestos board but with this limitation may be of any size or shape.

8. Test-flame burner of any suitable type but so constructed that the tip is $\frac{1}{16}$ inch diameter at the end with an orifice $\frac{1}{32}$ inch diameter. If the test-flame burner is mounted so that it will swing on a pivot, the radius of this swing must be not less than 6 inches.

9. Stop-watch or other suitable time measuring device.

Procedure.—Clean the flash cup thoroughly so that no trace of oil from previous determinations remains on it. This is best accomplished by means of a clean dry cloth. Insert the cup into the hole in the asbestos board and then into the depression in the metal plate. Place the metal plate exactly level on the selected support, pushing the asbestos board down until it touches the plate. Suspend the flash thermometer in a vertical position by any suitable device so that the bottom of the thermometer bulb is exactly $\frac{1}{4}$ inch from the bottom of the cup and approximately halfway between the back and the center of the cup. The immersion line on the thermometer will then be $\frac{1}{16}$ inch below the rim of the cup.

Fill the cup with the material to be tested in such a manner that the top of the menisens is exactly at the filling line (scribed on the inside of the cup) at room temperature,[1] making sure that no oil is spilled on the cup above the filling line or on any part of the apparatus outside. Failure to observe this direction is a frequent cause of incorrect results.

Light the test-flame burner and adjust until it carries a bead of flame not more than $\frac{5}{32}$ and not less than $\frac{1}{8}$ inch in diameter. A bead of some light-colored material (such as ivory) and of the specified diameter may be mounted in a convenient position on the cup or stand so that the size of the test flame may be determined by comparison.

Place the heating burner, lighted but turned low, under the center of the cup and protect it from drafts by means of the shield. Regulate the heating flame very carefully so that the temperature read on the flash thermometer increases not less than 9° nor more than 11° F. each minute. This direction means heating at the uniform rate of about 10° F. per minute (not 5° for one minute and 15° for the next). The execution of this direction requires practice and cannot be accomplished without the use of a stop-watch or other suitable time meter.

[1] In the case of materials not liquid at room temperature, the sample should be heated to the lowest temperature at which it will flow. The cup should be filled to the mark and allowed to cool to room temperature. Then a second filling should be made to bring the level of the material to the filling mark as above.

Apply the test flame to the oil at each successive thermometer reading which is a multiple of 5° F. observing the following instructions:

a. Have the test flame of the right size.

b. Move it across the center of the cup in a straight line at a right angle to the diameter of the cup passing through the thermometer.

c. Keep the lower side of the test flame exactly level with the upper edge of the cup.

d. Make the sweep across the cup in one second.

e. Avoid breathing in the direction of the cup.

When the approximate flash is known, the application of the test flame need not be made until the temperature is 50° F. below the expected flash point. It is to be noted that with a proper rate of heating as specified, the test flame will be applied every half minute.

Note and record as "Flash, Cleveland Open Cup," the temperature read on the specified thermometer when a flash appears at any point on the surface of the oil. This flash must not be confused with a bluish halo which sometimes surrounds the test flame prior to the true flash. While some authorities claim that a flash originating at the edge of the cup is not a true flash, this statement is not true with the apparatus here described. Consequently a distinct burst or jump of flame between test flame and the surface of the oil at any point is to be considered as a flash. Obviously from the manner in which taken, every flash point by this method will be a multiple of 5° F.

Fire point.—After the flash point has been reached, without interruption continue heating at the same rate as before and applying the test flame in exactly the same manner at each thermometer reading which is a multiple of 5° F. Note and record as "Fire, Cleveland Open Cup" the temperature when the oil ignites and continues to burn for five seconds. The flame should then be extinguished, not by blowing but by smothering either with a sheet of metal or with the cover furnished with some instruments.

Determination of Flash by Tag Closed Tester[1]

(Test Method No. 6)

This method is suitable for all petroleum products flashing below 175° F. with the exception of fuel oils, which should be tested by the Pensky-Martens Cup. When properly used it gives results which are not in error by more than 1° F. As described, this is the Standard Test of the American Society for Testing Materials as adopted in 1919 for "Flash Point of Volatile Paint Thinners." One important change has been made, namely, the introduction of a more accurate and complete specification for the thermometer, which specification has the approval of the maker of the instrument.

Apparatus.—1. Tag Closed Tester as shown in Fig. 9 to conform with the following dimensions within the limits of tolerance given:

Dimensions	Normal	Tolerances
Depth of water surface below top of cup	$1\frac{3}{32}$ inches	$\frac{1}{64}$ inch
Depth of oil surface below top of cup	$1\frac{5}{32}$ inches	$\frac{1}{64}$ inch
Depth of top of bulb of oil thermometer when in place, below top of cup	$1\frac{5}{16}$ inches	$\frac{1}{32}$ inch
Inside diameter of oil cup at top	$2\frac{1}{8}$ inches	0.005 inch
Weight of oil cup	68 g.	1 g.
Diameter of bead on top of cover	$\frac{5}{32}$ inch	$\frac{1}{32}$ inch

[1] The "Tag Closed Tester" must not be confused with the Tagliabue Pyrometer or any other closed tester.

The plane of underside of cover to be between the top and bottom of the burner tip when the latter is fully depressed.

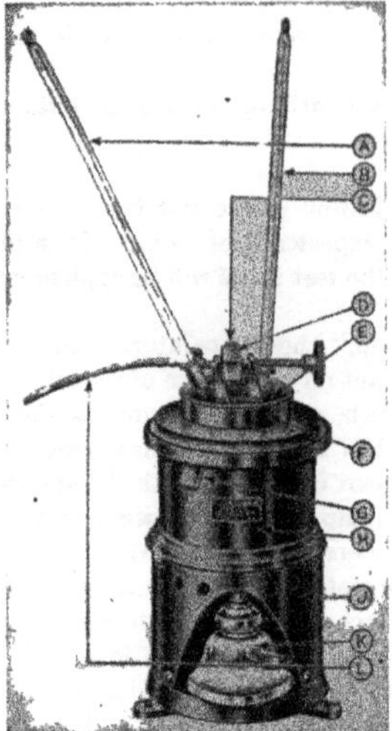

FIG. 9—Tag closed tester.

(*A*) Thermometer, indicating the temperature of the oil; (*B*) Thermometer, indicating the temperature of the water bath; (*C*) A miniature oil well to supply the test flame when gas is not available, mounted on the axle about which the test-flame burner is rotated, which axle is hollow and provided with connection on one end for gas hose and provided also with needle valve for controlling gas supply, when gas is available, the gas passing through the empty oil well; (*D*) Gas or oil tip for test flame; (*E*) Cover for oil cup, provided with three openings, which are in turn covered by a movable slide operated by a knurled hand knob, which also operates the test-flame burner in unison with the movable slide, so that by turning this knob, the test flame is lowered into the middle opening in the cover, at the same time that this opening is uncovered by the movement of the slide; (*F*) Oil cup (which cannot be seen in the illustration), standardized size, weight and shape, fitting into the top of the water bath; (*G*) Overflow spout; (*H*) Water bath, of copper, fitting into the top of the body, and provided with an overflow spout and openings in its top, to receive the oil cup and water-bath thermometer; (*J*) Body, of metal, attached to substantial cast metal base provided with three feet; (*K*) Alcohol lamp for heating water bath; (*L*) Gas hose.

1. (*a*) Thermometer to conform to the following specifications:

Type: Glass engraved.
Total length: 275 mm.
Glass; Stem: Plain front, enamel back, suitable thermometer tubing.
Diameter 6 to 7 mm.
Bulb: Corning normal, Jena 16 III, or equally suitable thermometric glass.
Length 9 to 13 mm.
Diameter not greater than stem.
Actuating liquid: Mercury.
Range: +20° to +230° Fahrenheit.
Immersion: 2¼-inches. Words "2¼ inch immersion" engraved on tube.
Range distance from bottom of bulb: 75 to 90 mm.
Range distance from top of tube: 25 to 40 mm.
Ice point: No.
Contraction chamber: No.
Expansion chamber: Yes.
Filled: Nitrogen gas.
Top finish: Glass ring.
Graduating. All lines, figures and letters clear cut and distinct.
Scale graduated in 1° divisions.
Scale numbered every 10°.
Special markings: P. M. and Tag.
Serial number and manufacturer's name or trade mark engraved on the tube.
Accuracy: Error at any point on scale shall not exceed one-half smallest scale division.
Test for permanency of range; After being subjected to the temperature of boiling water for twenty-four hours the accuracy shall be within the limit specified.
Points to be tested for certification 32°, 100°, 150°, 212°.

Procedure.—2. (*a*) If gas is available, connect a ¼-inch rubber tube to the corrugated gas connection on the oil cup cover. If no gas is available, unscrew the test flame burner-tip from the oil chamber on the cover, and insert a wick of cotton cord in the burner-tip and replace it. Put a small quantity of cotton waste in the oil chamber, and insert a small quantity of signal, sperm or lard oil in the chamber, light the wick and adjust the flame, so that it is exactly the size of the small white bead mounted on the top of the tester.

(*b*) Perform the test in a dim light so as to see the flash plainly.

(*c*) Surround the tester on three sides with an inclosure to keep away draughts.

A shield about 18 inches square and 2 feet high, open in front, is satisfactory, but any safe precaution against all possible room draughts is acceptable. Tests made in a laboratory hood or near ventilators will give unreliable results.

(*d*) See that the tester sets firm and level.

(*e*) For accuracy, the flash-point thermometers which are especially designed for the instrument should be used, as the position of the bulb of the thermometer in the oil cup is essential.

3. Put the water-bath thermometer in place, and place a receptacle under the overflow spout to catch the overflow. Fill the water-bath with water at such a temperature that, when testing is started, the temperature of the water-bath will be at least 20° F. (11° C.) below the probable flash point of the oil to be tested.

4. Put the oil cup in place in the water-bath. Measure 50 c.c. of the oil to be tested in a pipette or a graduate, and place in the oil cup. The temperature of the oil shall be at least 20° F. (11° C.) below its probable flash point when testing is started. Destroy any bubbles on the surface of the oil. Put on the cover, with flash-point thermometer in place and gas tube attached. Light the pilot light on the cover and adjust the flame to the size of the small white bead on the cover.

5. Light and place the heating lamp,[1] filled with alcohol, in the base of the tester and see that it is centrally located. Adjust the flame of the alcohol lamp so that the temperature of the oil in the cup rises at the rate of about 1.8° F. (1° C.) per minute, not faster than 2° F. (1.1° C.) nor slower than 1.6° F. (0.9° C.) per minute.

6. (*a*) Record the barometric pressure which, in the absence of a laboratory instrument, may be obtained from the nearest Weather Bureau Station.

(*b*) Record the temperature of the oil sample at start.

(*c*) When the temperature of the oil reaches 9° F. (5° C.) below the probable flash point of the oil, turn the knob on the cover so as to introduce the test flame into the cup, and turn it promptly back again. Do not let it snap back. The time consumed in turning the knob down and back should be about one full second, or the time required to pronounce distinctly the words " one-thousand-and-one."

(*d*) Record the time of making the first introduction of the test flame.

(*e*) Record the temperature of the oil sample at the time of first test.

(*f*) Repeat the application of the test flame at every 1° F. (0.5° C.) rise in temperature of the oil until there is a flash of the oil within the cup.

Do not be misled by an enlargement of the test flame or halo around it when it enters the cup, or by slight flickering of the flame; the true flash consumes the gas in the top of the cup and causes a very slight explosion.

(*g*) Record the time at which the flash point is reached.

(*h*) Record the flash point.

(*i*) If the rise in temperature of the oil, from the " time of making the first introduction of the test flame " to the " time at which the flash point is reached " was faster than 2° F. (1.1° C.) or slower than 1.6° F. (0.9° C.) per minute, the test should be

[1] Where gas is available, the substitution of a small low-form gas burner for the alcohol lamp of Section 5 is more convenient and has no effect on the results obtained.

questioned, and the alcohol heating lamp adjusted so as to correct the rate of heating. It will be found that the wick of this lamp can be so accurately adjusted as to give a uniform rate of rise in temperature within the above limits and that it will remain so adjusted.

7. (*a*) It is not necessary to turn off the test flame with the small regulating valve on the cover; leave it adjusted to give the proper size of flame.

(*b*) Having completed the preliminary test, remove the heating lamp, lift up the oil-cup cover, and wipe off thermometer bulb. Lift out the oil cup, empty and carefully wipe it. Throw away all oil samples after once used in making a test.

(c) Pour cold water into the water-bath, allowing it to overflow into a receptacle until the temperature of the water in the bath is lowered to 15° F. (8° C.) below the flash point of the oil, as shown by the previous test.

With cold water of nearly constant temperature, it will be found that a uniform amount will be required to reduce the temperature of the water-bath to the required point.

(*d*) Place the oil cup back in the bath and measure into it a 50 c.c. charge of fresh oil. Destroy any bubbles on the surface of the oil, put on the cover with its thermometer, put in the heating lamp, record the temperature of the oil, and proceed to repeat the test as described above in Sections 4 to 6, inclusive. Introduce the test flame for first time at a temperature of 10° F. (5.5° C.) below the flash point obtained on the previous test.

8. If two or more determinations agree within 1° F. (0.5° C.), the average of these results, corrected for barometric pressure, shall be considered the flash point. If two determinations do not check within 1° F. (0.5° C.), a third determination shall be made and, if the maximum variation of the three tests is not greater than 2° F. (1° C.), their average, after correcting for barometric pressure, shall be considered the flash point.

9. Correction for barometric pressure shall only be made in cases of dispute or when the barometer reading varies more than ½ inch (13 mm.) from the standard pressure of 29.92 inches (760 mm.). When the barometer reading is below this standard pressure, add to the thermometer reading 1.6° F. (0.9° C.) for each inch (25 mm.) of barometer difference to obtain the true flash point. When the barometer reading is above the standard pressure, deduct 1.6° F. (0.9° C.) for each inch (25 mm.) of barometer difference to obtain the true flash point.

Determination of Flash by the Elliott Tester

(Test Method No. 7)

This method is suitable for testing illuminating oils having flash points between 84° and 160° F. When properly used it gives results which are not likely to be in error by more than 2° F. The Elliott Tester is frequently known as the New York State Tester on account of its adoption in 1882 by the New York State Board of Health. It is open to the serious objection that about 300 c.c. of oil are required for the test. This tester has sometimes been used for flash on asphalts and lubricating oils by filling the bath with high-flash lubricating oil. This procedure should never be followed. All petroleum products flashing above 175° F. should be tested by the Cleveland Open Cup or if a closed-cup flash is desired, by the Pensky-Martens.

Apparatus.—1. Elliott Tester.

2. Test-flame burner, to give a bead of flame ⅛ inch in diameter.

3. Stop-watch or other timing device.

The Elliott Tester (Fig. 10) consists of a cylindrical stand, a water-bath and an oil cup, all made of copper. The water-bath is 4½ inches high and 4 inches inside diameter, with an opening in the top 2⅞ inches in diameter. The oil cup consists of a flat-bottom cylinder, 2⅞ inches inside diameter and 3⅛ inches high, above which is a vapor chamber 3¼ inches inside diameter and 1 inch high. Above the vapor chamber is a small flanged rim 3¾ inches diameter to hold the glass cover. This cover has two openings, a circular one at one side in which the thermometer is held by means of a cork and a U-shaped one at the opposite edge through which the test flame is inserted.

Fig. 10.—Elliott or N. Y. State closed tester.

The thermometer should be graduated in 1° F. divisions from 80° to 220° F. and accurate to 0.5° F.

Procedure.—Make sure that the instrument is level. Remove the oil cup and fill the water-bath with water at 77° F. up to the mark on the inside. Replace the oil cup and pour in enough of the oil to be tested to fill it to a line ⅛ inch below the flange joining the cup to the vapor chamber above. Make sure that no oil is allowed to get on the flange or cup wall above. Remove all air bubbles with a piece of dry paper. Place the glass cover on the cup and insert the thermometer so that the level of the oil comes exactly to the beginning of the capillary.

Light the test burner and adjust until the flame is exactly ⅛ inch in diameter.

Light the burner under the water-bath and heat so that the temperature of the oil rises at a uniform rate of 2° F. per minute. When this temperature reaches 84° F., start trying for flash as follows: Hold the test burner at an angle of 45° and insert the test flame through the side opening in the cover to a point half-way between the cover and the surface of the oil. The test flame should be introduced with a steady uniform motion, the time for introduction and withdrawal being one second. Continue to try for flash at each successive temperature reading which is a multiple of 2° F. Note and record as "Flash, Elliott," the temperature read on the thermometer when a slight bluish flame flashes over the surface of the oil. The temperature should be read just before introducing the test flame and not afterward. The test flame must not come in contact with the oil.

In the official description of the Elliott Method the following directions are given: "Testing for flash should be repeated at every 2° rise of the temperature until the temperature has reached 95° when the lamp should be removed and testing should be made for each degree of temperature until 100° is reached. After this, the lamp may be replaced, if necessary, and the testings continued for each 2° F." It will be found that the official directions give no more accurate results than those obtained by the Test Method No. 7.

Directions have also been given for taking fire point with the Elliott Cup, but these are not sufficiently satisfactory to warrant either description or use in practice.

Determination of Flash by the Pensky-Martens Tester

(Test Method No. 8)

This method is the proper one for all fuel oils, for lubricating oils where a closed-cup flash is desired and for asphalts and other very viscous products, all of which require stirring during flash determination.

Non-viscous liquids flashing below 175° F. may be tested on the Tag Closed Tester. When properly used the Pensky-Martens Cup gives results which are not likely to be in error by more than 5° F.

Apparatus.—1. Standard Pensky-Martens Closed Tester.

2. Stop-watch or metronome.

The Pensky-Martens Tester (Fig. 11) consists of a cast-iron air-bath with an annular chamber exposed to the flame and a brass jacket, which serves to check radiation. The jacket is separated from the iron casting by a considerable space at the sides and by a distance of ¼ inch at the top. The oil cup rests on the brass jacket and does not

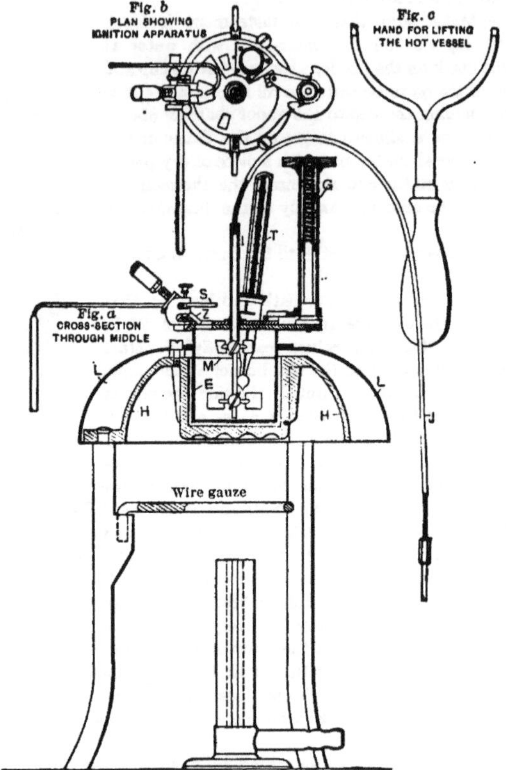

FIG. 11.—Pensky-Martens flash tester.

touch the cast-iron air-bath. The whole assembly is rigidly fastened to a supporting tripod.

The oil cup is a flat-bottom cylinder made of brass with a flange a little below the top, about ½ inch wide, with devices for locating the position of the cover on the cup and of the cup in the support. A filling mark is scribed on the inside of the cup. The cup may be of slightly larger inside diameter above the filling mark. The outside of the cup above the flange may be tapered but the wall thickness at the upper edge must be not less than 0.04 inch.

The essential dimensions of the oil cup are as follows:

Cup

	Minimum, inches	Normal, inches	Maximum, inches
Inside diameter below filling mark	1.95	2.00	2.05
Difference, inside and outside diameters below filling mark	0.12	0.12	0.13
Inside height	2.15	2.20	2.25
Thickness, bottom	0.07	0.09	0.12
Distance from rim to filling mark	0.85	0.86	0.88
Distance lower surface of flange to bottom of cup	1.78	1.80	1.81

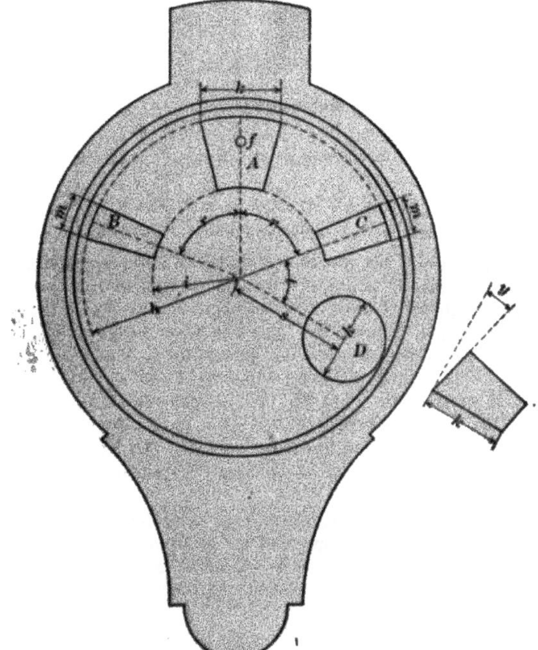

FIG. 12.—Openings in cover of the Pensky-Martens cup.

The oil-cup cover should have a rim projecting downward almost to the flange of the cup and fitting the outside of the cup closely. A stirring device consisting of a vertical steel shaft carrying two double-bladed propellers should be so mounted in the cover, that it is centered in the oil cup when the cover is in proper position. A proper locating device insures that the cover is always in the same position. The blades of both propellers should be approximately $\frac{5}{16}$ inch wide and should be set at an angle of 45°. The smaller (upper) propeller should have an over-all diameter of about $\frac{3}{4}$ inch and the plane of its center should be 0.4 inch below the level of the rim of the cup. The larger (lower) propeller should have an over-all diameter of about $1\frac{1}{4}$ inches and the plane of its center should be 2.0 inches below the level of the rim of the cup. The stirrer shaft should be connected outside the cover with a suitable device such as a flexible shaft by means of which it may be rotated.

The cover should have four openings as shown in Fig. 12. Openings B and C have equal areas such that their sum is not less than 75 per cent and not more than 100 per cent of the area of opening A. Opening D is for a thermometer collar $\frac{1}{2}$ inch inside diameter and set at an angle of not less than 10° nor more than 15° from vertical.

Essential dimensions for cover openings are as follows:

Dimensions	Maximum	Minimum
h, distance from center to outside of openings	0.97 in.	0.94 in.
i, distance from center to inside of openings	0.56 in.	0.53 in.
k, inside diameter of D and outside width of A	0.54 in.	0.50 in.
m, outside width of openings B and C	0.22 in.	0.18 in.
s, distance between centers of cover and opening D	0.8 in.	0.7 in.
Angle r	70°	65°
Angle t	60°	50°
Angle y	15°	10°

The openings A, B and C are opened and closed by a shutter capable of rotation through the necessary angle by means of a suitable mechanism terminating in a knob made of some non-conductor of heat. The test-flame burner should have a tip with an opening exactly $\frac{1}{32}$ inch in diameter. This test-flame burner should be automatically depressed by the mechanism operating the shutter so that when the shutter is wide open, the end of the test-flame burner is just level with the under surface of the cover, at a position indicated by (f) in Fig. 12. A small pilot-flame burner should be provided close to the test-flame burner position when the shutter is closed. This provides for automatic lighting of the test-flame burner. The burner for heating the cup during a test should be capable of maintaining the specified rate of heating at all temperatures.

The thermometers for this cup should correspond to the following specifications:

1. Low-range, P. M. and Tag thermometer as described under Test Method No. 6. This thermometer should be used for all oils having a flash below 230° F.

2. High-range, Pensky-Martens thermometer to conform to the following specifications:

Type: Glass engraved.
Total length: 275 mm.
Glass; Stem: Plain front, enamel back, suitable thermometer tubing.
Diameter 6 to 7 mm.
Bulb: Corning normal, Jena 16, III, or equally suitable thermometric glass.
Length 9 to 13 mm.
Diameter less than stem.
Actuating liquid: Mercury.

Range: 200° to 700° F.

Immersion: 2¼ inches line around tube at immersion point and words "2¼ inch immersion" engraved on tube.

Range distance from bottom of bulb: 75 to 90 mm.

Range distance from top of tube: 25 to 40 mm.

Ice point: No.

Contraction chamber: No.

Expansion chamber: Yes.

Filled: Nitrogen gas.

Top finish: Glass ring.

Graduating: All lines, figures and letters clear cut and distinct.
Scale graduated in 5° F. divisions.
Scale numbered every 50° F.

Special markings: P. M. High,
Serial number and manufacturer's name or trade mark engraved on the tube.

Accuracy: Error at any point on scale shall not exceed one-half smallest scale division.

Test for permanency of range: After being subjected to a temperature of 680° F. for 24 hours the accuracy shall be within the limit specified.

Points to be tested for certification: 212°, 450°, 700° F.

The P. M. high thermometer should be used for all oils having a flash of 230° F. or higher.

Procedure.—Make sure that all parts of the apparatus are thoroughly clean and free from oil. If gasoline is used for cleaning, it should be of such quality that it will evaporate quickly and completely.

Fill the cup with the oil to be tested, up to the level indicated by the filling mark. This filling should be done at room temperature in the case of liquids. With solid materials, the procedure should be that described in Test Method No. 5.

Place the cup in the bath and the cover on the cup, making sure that the locating devices are properly engaged and that the cup is exactly level. Insert the proper thermometer, making sure that the bottom of the bulb is 1¾ inches below the level of the rim of the cup. In the case of solid materials, this operation is deferred until the material has softened sufficiently.

Light the pilot and test-flame burners, adjusting the latter so that it carries a bead of flame exactly $\frac{5}{32}$ inch in diameter. Heat the bath so that the temperature of the oil is raised at a uniform rate of not less than 9° nor more than 11° F. per minute. In the case of oils whose flash is approximately known, the rate of heating may be 20° F. per minute up to a temperature 50° below this flash point. The 10° rate is essential during the rest of the heating. The stirrer should be rotated at the rate of one to two revolutions per second between trials for flash, beginning about 50° F. below the expected flash point.

Try for flash at each temperature reading which is a multiple of 5° F., stopping he stirring during this operation. Where the approximate flash is known, testing need not be started until the temperature is about 50° F. below the expected flash point. This trial for flash should be made by turning the knob controlling shutter and test-flame burner so that the test flame is lowered in one-half second, left in its lowest position for one second and raised to its high position quickly.

Note and record as "Flash, Pensky-Martens" the temperature when the first flash occurs in the interior of the cup. This true flash must not be confused with a bluish halo which appears around the test flame at temperatures slightly below the flash point. The appearance of this halo furnishes a valuable indication that the flash point is being approached.

Fire point.—As soon as the flash point has been reached, remove the flash thermometer from the cover and remove the cover from the cup. Suspend the flash thermometer in a vertical position so that the bottom of the bulb is $\frac{1}{4}$ inch from the bottom of the cup and over a point halfway between the center and the back of the cup. These operations should require not more than one minute and under this condition the heating burner should be left in place; otherwise the heating burner should be removed while the changes are being made.

With the apparatus thus arranged as an open cup, continue the heating at the specified uniform rate of 9° to 11° F. per minute. At each successive temperature reading which is a multiple of 5° F. try for fire point, observing the following instructions.

a. Use a test burner carrying a bead of flame not more than $\frac{5}{32}$ inch nor less than $\frac{1}{8}$ inch in diameter.

b. Move this flame across the center of the cup in a straight line at a right angle to the diameter of the cup passing through the flash thermometer.

c. Keep the lower side of the test flame exactly level with the upper edge of the cup.

d. Make the sweep across the cup in one second.

e. Avoid breathing in the direction of the cup.

Note and record as "Fire point, Pensky-Martens" the temperature at the time of the first flame application which causes the oil to ignite and burn continuously at least five seconds. The fire points determined by either Pensky-Martens or Cleveland Open Cup should differ by not more than 5° F.

Determination of Flash by The Abel Tester

(Test Method No. 9)

This method is intended for testing illuminating oils having an Abel flash not higher than 160° F. When properly used it gives results which are not in error by more than 1° F. The Abel Tester is the official English instrument and affords a stringent flash inspection. The Abel-Pensky Tester (Fig. 14) is a German modification of the Abel.

Apparatus.—1. Standard Abel Flash Tester. (Fig. 13.)

2. Stop-watch or metronome.

The oil cup is a cylindrical, flat-bottom vessel with projecting rim made of brass or gun-metal (17 B. W. gage), tinned inside. A bracket consisting of a short stout piece of wire bent upward and terminating in a point, is fixed to the inside of the cup to serve as a filling gage. The cup is provided with a close-fitting, over-lapping cover made of brass (22 B. W. gage) which carries the oil thermometer, the opening slide and the test burner. This test burner is suspended upon two supports at the sides by means of trunnions upon which it may be made to tilt up and down. The test burner is commonly made for the use of gas, though the original instrument carried a test burner using rapeseed or similar oil. A bead made of some suitable white material is mounted on the cover as a measure for the proper size of test flame to be used. The cover is provided with three rectangular holes, a large one in the center and two smaller ones close to the sides and opposite each other. These three holes may be closed and uncovered by means of a slide moving in grooves and having perforations corresponding to those on the lid. Movement of this slide automatically tilts the test burner in such a way as to bring the tip of this burner just below the surface of the lid when the holes in the lid are wide open., The essential dimensions of the instrument are as follows:

	Inches
Inside diameter of oil cup	2.0
Inside height of oil cup	2.2
Width of projecting rim	0.5
Distance of projecting rim from top of cup	$\frac{3}{8}$
Distance of projecting rim from bottom of cup	$1\frac{7}{8}$
Distance from filling gage point to bottom of cup, inside	1.5
Small holes in cover	0.2×0.3
Large hole in cover	0.4×0.5
Diameter of white bead	$\frac{5}{32}$
Diameter of inner bath cylinder	3
Diameter of outer bath cylinder	$5\frac{1}{2}$
Height of outer bath cylinder	$5\frac{3}{4}$
Distance between inside bath cylinder and the oil cup	$\frac{1}{2}$

FIG. 13.—Abel tester. FIG. 14.—Abel-Pensky tester.

The oil thermometer should be scaled for $2\frac{1}{4}$-inch immersion and graduated in 1° F. divisions from 55° to 150° F. (range on American thermometers is usually 40° to 165° F.). This thermometer should be mounted in a brass collar so that when in position, the center of the thermometer bulb is 1 inch above the center of the bottom of the cup. The bath thermometer may be of the same type but should have a range not less than 90° to 190° F.

Procedure.—The following directions differ somewhat from the official English method but give the same results and permit use of the Abel Tester over a wider temperature range to meet present-day needs.

Make the test in a room or compartment partially darkened and free from drafts. Use a fresh sample for each test. Make sure that the instrument is level and that oil cup, cover and oil thermometer are entirely free from oil.

Fill the water-bath through the funnel, until water begins to come out of the overflow tube. The temperature of this water should be about 80° F. when the oil being tested has an Abel flash of 60° F. or higher. For oils of flash lower than 60° F. the water-bath should be filled with water whose temperature is about 20° F. higher than the flash of the oil. Make sure that the air-bath in which the oil cup sets is clean and dry.

Cool the oil cup and the sample to be tested to 60° F. or to about 20° F. below the approximate flash point if this is below 80° F. Place the cup in the air-bath. Fill with the sample until the surface of the oil is exactly level with the point of the gage fixed in the cup, being careful that no oil is splashed on the sides of the cup above the filling level. Remove any foam, if present, by means of a strip of filter paper. Close the oil cup with the cover carrying the oil thermometer properly inserted. Make sure that the cover is pressed down as far as it will go, and that the apertures are closed.

Light the test flame and adjust its size to duplicate that of the white bead on the cover of the cup within $\frac{1}{64}$ inch.

Heat the water-bath with a small gas flame, so adjusted that the temperature of the oil in the cup is raised at a uniform rate of 1° F. per minute. Such adjustment will require occasional changes and involves experience with the apparatus.

When the thermometer in the oil cup indicates about 10° F. below the approximate flash point, start trying for flash at each rise of 1° F. as follows: Draw the slide open, taking one second for the operation. Leave the apertures open for one second and close quickly. Be very careful not to breathe toward the apparatus during these operations. Note temperature of the oil when the oil vapors flash or burst into flame with a slight explosion. The true flash must not be confused with a bluish halo around the test flame which may appear several degrees below the flash point. The appearance of this halo furnishes a valuable warning that the temperature is approaching the flash point.

Repeat the test on a fresh sample of oil, observing all the precautions given. If the difference between the two tests is not more than 1° F., record the average of the two, as "Abel flash." If the difference between the two tests is more than 1° F., make a third test, and if the maximum variation between the three tests is not more than 2° F., take the average of the three. In case the first test is known to be faulty it should be regarded only as an approximate test and should not be considered.

Viscosity

Viscosity is that property of a liquid which resists any force tending to produce flow. It should, therefore, be measured in units of force applied under definite conditions. The petroleum industry, however, uses the term "viscosity" in a rather loose sense and usually to indicate the time required for a definite volume of oil to flow through an orifice of definite size and shape. So firmly established and generally understood is this use of the word, that even scientific investigators in petroleum work are forced to characterize the correctly measured property as "Absolute Viscosity."[1]

Hence no apology nor explanation is required when an oil is said to have a viscosity of 200 seconds Saybolt Universal at 100° F.

Viscosity of petroleum products is measured most satisfactorily by observing the rate of flow through a capillary tube. For this reason no methods will be described other than those involving standard capillary tube viscosimeters. The determination is usually made for one or more of the following purposes:

[1] "Absolute viscosity" is that force which will move one square centimeter of plane surface with a speed of one centimeter per second relative to another parallel plane surface from which it is separated by a layer of the liquid one centimeter thick. Absolute viscosity is usually expressed in poises, the poise being equal to one dyne second per square centimeter. Water at 68° F. has an absolute viscosity of almost exactly 0.01 Poise. "Kinematic Viscosity" is the ratio of absolute viscosity to the density of the oil at the temperature of the viscosity measurement. See W. H. Herschel, Bureau of Standards Technologic Paper 100 (1917).

a. On lubricating oil as a partial indication of frictional resistance.

b. On all oils as a measure of quantity which can be delivered through pipe-lines under definite conditions.

c. On all oils as a verification of quality purchased or sold.

d. On illuminating oils, as an indication of feeding qualities through wicks.

Since the viscosity of all petroleum products decreases with increasing temperature, determinations must be made at a definite temperature and this temperature must always be given as an integral part of a reported viscosity. The viscosity (time of outflow) on a given sample will also vary according to the type of instrument used, even at the same temperature. For example, the same oil at the same temperature will have a viscosity of 145 seconds Saybolt Universal and 115 seconds Redwood. Hence the viscosimeter used as well as the temperature must be reported. Furthermore the viscosimeter must be distinguished clearly by its full name. Thus " 75 Saybolt " may mean either " Saybolt Universal " or " Saybolt Furol." Considerable confusion and even litigation has resulted from failure to specify clearly the instrument to be used. This opportunity is taken to point out that the Saybolt "A" Viscosimeter or " 70 Meter " is no longer manufactured by its designer and is not a standard meter. Its use cannot be condemned too severely and should never be permitted. When it is desirable to measure viscosity at 70° F., such measurements should be taken on one of the viscosimeters described in this chapter.

Comparison between viscosities on different meters.—It is frequently desirable to be able to convert viscosity as determined on a given sample with one viscosimeter to the corresponding viscosity on a different meter. Such a conversion can be made by calculation regardless of the nature of the oil but only when both viscosities are at the same temperature.

There is no fixed relation between viscosity readings taken on a given sample with the same meter at different temperatures, for the reason that different oils may show different rates of viscosity change with temperature change. Therefore, it is not possible to convert viscosity at one temperature on one meter to a viscosity at a different temperature, either on the same or on a different meter. Rough estimate may be made but for most purposes it is necessary to make a measurement on the given meter at the temperature of the required meter reading and then make the conversion.

The approximate relations between viscosity according to different viscosimeter at *the same temperature* are shown in Table 5. The conversion cannot be more accurate than the instruments themselves and variations of 5 per cent between used viscosimeters of the same make are common. For this reason Table 5 must be used with the knowledge that it applies only to meters conforming to the equations given for kinematic viscosity. The values chosen are those which appear to represent the average of instruments used by the most reliable petroleum laboratories of the United States.[1]

Accuracy of viscosity determinations.—New viscosimeters are liable to show variations of 2 per cent even when properly used. Old viscosimeters may vary considerably more than this unless they have been used with intelligent care to avoid corrosion, wear or other damage. The commoner sources of error in the actual measurement of viscosity are:

a, Outlet tube (capillary) not perfectly clean.
b, Suspended particles in oil.
c, Viscosimeter not level.
d, Oil tube not filled to proper level.
e, Stop-watch or receiving flask not accurate.
f, Receiving flask not perfectly clean and dry.

[1] Compare Bureau of Standards Technologic Paper 112. W. H. Herschel (1918).

TABLE 5

Approximate Viscosity Conversion Table

Saybolt universal time, seconds	Saybolt furol time, seconds	Redwood time, seconds	Engler degrees time, 51.3	Kinematic viscosity, poises	MULTIPLIERS FOR CONVERTING SAYBOLT UNIVERSAL SECONDS TO		
					Saybolt furol secs.	Redwood seconds	Engler degrees
32	....	....	1.05	0.0100	....	....	0.0328
34	....	....	1.11	.0174	....	....	.0325
36	....	31	1.16	.0245	....	0.856	.0322
38	....	32	1.21	.0313	....	.853	.0318
40	....	34	1.22	.0378	....	.850	.0316
45	....	38	1.40	.0531	....	.840	.0310
50	....	42	1.53	.0675	....	.834	.0306
55	....	46	1.66	.0811	....	.829	.0301
60	....	50	1.79	.0942	....	.825	.0298
65	....	53	1.92	.1068	....	.822	.0295
70	....	57	2.05	.1192	....	.819	.0293
80	....	65	2.32	.1431	....	.815	.0289
90	....	73	2.58	.1663	....	.811	.0287
100	....	81	2.85	.1890	....	.808	.0285
120	....	97	3.38	.2334	....	.805	.0282
140	....	112	3.92	.2770	....	.803	.0280
160	....	128	4.46	.3200	....	.801	.0279
180	....	144	5.00	.3626	....	.800	.0278
200	....	160	5.55	.4050	....	.800	.0277
225	....	180	6.22	.4577	....	.799	.0277
250	30 ?	200	6.91	.5103	0.111	.798	.0276
275	28	219	7.59	.5628	.110	.798	.0276
300	33	239	8.27	.6150	.108	.798	.0276
325	35	259	8.96	.6672	.108	.798	.0276
350	38	279	9.64	.7194	.107	.797	.0275
375	40	299	10.32	.7714	.107	.797	.0275
400	43	319	11.01	.8235	.106	.797	.0275
450	47	359	12.38	.9275	.105	.797	.0275
500	52	398	13.75	1.0314	.104	.797	.0275
550	57	438	15.11	1.1352	.104	.797	.0275
600	62	478	16.49	1.2390	.103	.797	.0275
700	72	558	19.23	1.4464	.103	.796	.0275
800	82	637	21.97	1.6538	.103	.796	.0275
900	92	717	24.71	1.8610	.102	.796	.0275
1000	102	796	27.46	2.0682	.102	.796	.0275

Saybolt universal kinematic viscosity $= 0.00207t - \frac{1.8}{t}$.

Saybolt furol: Kinematic viscosity $= .0204t - \frac{1.6}{t}$.

Redwood: Kinematic viscosity $= .0026t - \frac{1.715}{t}$.

Engler: Kinematic viscosity $= 0.00147t - \frac{3.74}{t}$.

g, Stop-watch and oil flow not started at exactly the same time.
h, Oil stream allowed to form foam in the receiving flask.
i, Temperature of oil not correct.
j, Stop-watch not stopped at the proper time.
k, Incorrect mental calculation from stop-watch reading.
l, Use of the same instrument over a wide temperature range.

Duplicate determinations of viscosity for the same oil on the same meter should not differ by more than 1 per cent even when taken by different operators.

Temperatures of measurement.—While viscosity on petroleum products may be measured at any desired temperatures, it is customary to use the following:

Saybolt Universal: 100, 130 and 210° F.; also 68, 70, 122, 140 and 158° F.
Saybolt Furol: 122° F.; also 70 and 100° F.
Redwood: 70, 100, 140 and 212° F.
Redwood Admiralty: 32° F.
Engler: 68, 77, 122 and 212° F.; also 302° F.
Saybolt Thermo: 60° F.

Thermometers.—All thermometers used for viscosity should be of the partial immersion type to suit the instrument. They should be graduated so that temperature readings may be made with an accuracy of 0.2° F.

Fig. 15 shows several types of viscosimeters set up ready for operation. The Barbey Ixometer shown on the right is the official French instrument.

FIG. 15.—Viscosimeters for petroleum products.

1. Saybolt Universal, electric heat.
2. Saybolt Universal, steam heat.
3. Saybolt Furol.
4. Redwood.
5. Redwood Admiralty.
6. Engler.
7. Barbey Ixometer.

Determination of Viscosity by The Saybolt Universal Viscosimeter

(Test Method No 10)

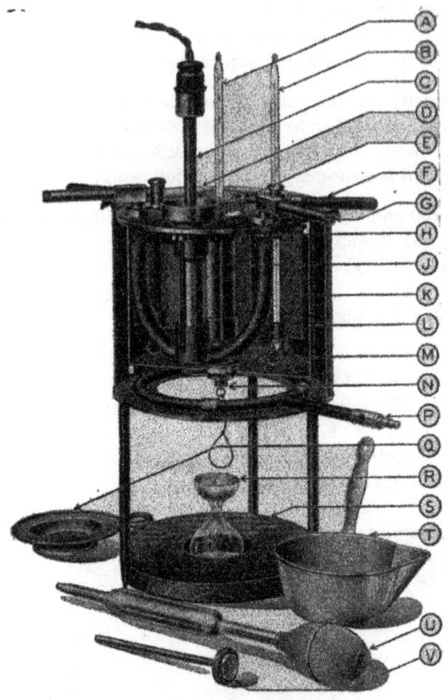

FIG. 16.—Saybolt universal viscosimeter.

(*A*) Oil tube thermometer; (*B*) Bath thermometer; (*C*) Electric heater; (*D*) Turn-table cover; (*E*) Overflow cup; (*F*) Turn-table handles; (*G*) Steam inlet or outlet; (*H*) Steam U-tube; (*J*) Oil tube; (*K*) Stirring paddles; (*L*) Bath vessel; (*M*) Electric-heater receptacle; (*N*) Outlet cork-stopper; (*P*) Gas burner; (*Q*) Strainer; (*R*) Receiving flask; (*S*) Base block; (*T*) Straining cup; (*U*) Pipette; (*V*) Tube-cleaning plunger.

The method given below is that adopted in 1920 by the American Society for Testing Materials as the "Standard Test for Viscosity of Lubricants" under serial designation D 47–20. It may be applied to other petroleum products which are entirely liquid at the temperature of measurement, with the exception that materials having a viscosity less than 32 seconds, Saybolt Universal at 100° F., should be tested on the Saybolt Thermo-Viscosimeter as in Test Method 12.

While the A. S. T. M. Standard Method prescribes only three temperatures, others may be used; but in such cases the oil should be maintained exactly at the true temperature reported. Attention is called to the fact that the practice of reporting Saybolt Universal Viscosity at 212° F. is obsolete. Since the oil is maintained at 210° F., the viscosity should be so reported. Viscosity at 70° F. on lubricating oils, when required, should always be taken on the Saybolt Universal and so reported.

The thermometers used in this method should conform to the following specifications:

Saybolt Viscosity Thermometer Specifications

Type: Glass engraved.
Total length: 225 mm.
Glass; Stem: Plain front, enamel back, suitable thermometer tubing.
Diameter 6 to 7 mm.
Bulb: Corning normal, Jena 16, III, or equally suitable thermometric glass.
Length 32 mm. maximum.
Diameter less than stem.
Actuating liquid: Mercury.
Range: +98° to +108° F. on A; 128 to 138° F. on B; 208 to 218° F. on C.

Immersion: 4½ inches, stem enlarged to a diameter of 9.5 mm. at point 4½ inches from bottom of bulb. Words "4½ inch immersion" engraved on tube.

Range distance from bottom of bulb; 130 mm. to 140 mm.

Range distance from top of tube; 22 to 26 mm.

Ice point: No.

Contraction chamber: Yes. Long, narrow.

Expansion chamber: Yes.

Filled: Nitrogen gas.

Top finish: Glass ring.

Graduating: All lines, figures and letters clear cut and distinct.
Scale graduated in ⅕° divisions.
Scale numbered every 2° F.

Special markings: Serial number and manufacturer's name or trade mark engraved on the tube.

Accuracy: Error at any point on scale shall not exceed one smallest scale division.

Points to be tested for certification: 100° F. on A; 130° F. on B; 210° F. on C.

A. S. T. M. Standard Method D 47–20

Viscosity.—1. Viscosity shall be determined by means of the Saybolt Standard Viscosimeter.

Apparatus.—2. (*a*) The Saybolt Standard Viscosimeter (see Fig. 17) is made entirely of metal. The standard oil tube *A* is fitted at the top with an overflow cup *B*, and the tube is surrounded by a bath. At the bottom of the standard oil tube is a small outlet tube through which the oil to be tested flows into a receiving flask, Fig. 18, whose capacity to a mark on its neck is 60 (0.15) c.c. The lower end of the outlet tube is enclosed by a larger tube, which when stoppered by a cork, *C*, acts as a closed air chamber and prevents the flow of oil through the outlet tube until the cork is removed and the test started. A looped string may be attached to

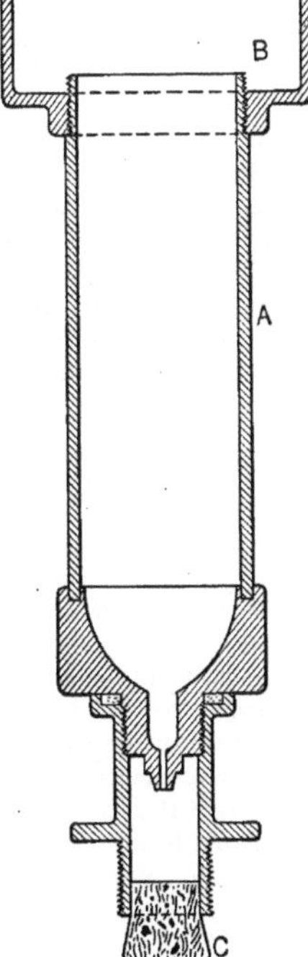

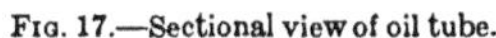
Fig. 17.—Sectional view of oil tube.

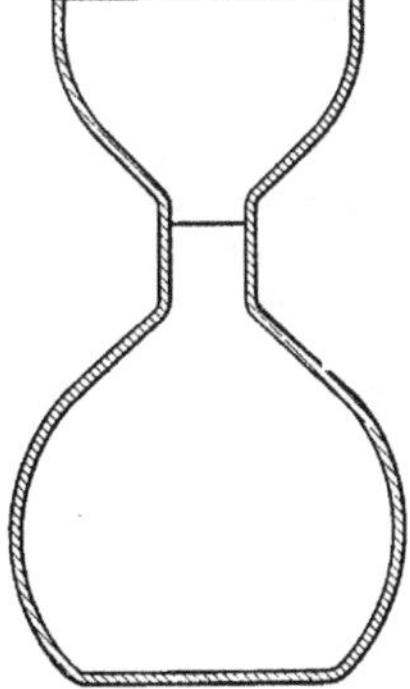
Fig. 18.—Receiving flask.

Saybolt universal viscosimeter.

the lower end of the cork as an aid to its rapid removal. The temperatures in the standard oil tube and in the bath are shown by thermometers. The bath may be heated by any suitable means. The standard oil tube shall be thoroughly cleaned, and all oil entering the standard oil tube shall be strained through a 60-mesh wire strainer. A stop-watch shall be used for taking the time of flow of the oil and a pipette shall be used for draining the overflow cup of the standard oil tube.

(*b*) The standard oil tube, which may be standardized by the U. S. Bureau of Standards, Washington, D. C., shall conform to the following dimensions:

Dimensions	Minimum, cm.	Normal, cm.	Maximum, cm.
Inside diameter of outlet tube	0.1750	0.1765	0.1780
Length of outlet tube	1.215	1.225	1.235
Height of overflow rim above bottom of outlet tube	12.40	12.50	12.60
Diameter of container of standard oil tube	2.955	2.975	2.995
Outer diameter of outlet tube at lower end	0.28	0.30	0.32

Method.—3. Viscosity shall be determined at 100° F. (37.8° C.), 130 F. (54.49° C.), or 210° F. (98.9° C.). The bath shall be held constant within 0.25° F. (0.14° C.) at such a temperature as will maintain the desired temperature in the standard oil tube. For viscosity determinations at 100 and 130° F., oil or water may be used as the bath liquid. For viscosity determinations at 210° F., oil shall be used as the bath liquid. Viscosity determinations shall be made in a room free from draughts, and from rapid changes in temperature. All oil introduced into the standard oil tube, either for cleaning or for test, shall first be passed through the strainer.

To make the test, heat the oil to the necessary temperature and clean out the standard oil tube. Pour some of the oil to be tested through the cleaned tube. Insert the cork stopper into the lower end of the air chamber at the bottom of the standard oil tube, sufficiently to prevent the escape of air, but not to touch the small outlet tube of the standard oil tube.

Heat the oil to be tested, outside the viscosimeter, to slightly below the temperature at which the viscosity is to be determined and pour it into the standard oil tube until it ceases to overflow into the overflow cup. By means of the oil tube thermometer keep the oil in the standard oil tube well stirred and also stir well the oil in the bath. It is extremely important that the temperature of the bath be maintained constant during the entire time consumed in making the test.

When the temperature of the bath and of the oil in the standard oil tube are constant and the oil in the standard oil tube is at the desired temperature, withdraw the oil tube thermometer; quickly remove the surplus oil from the overflow cup by means of a pipette so that the level of the oil in the overflow cup is below the level of the oil in the tube proper; place the 60-c.c. flask, Fig. 18, in position so that the stream of oil from the outlet tube will strike the neck of the flask so as to avoid foam. Snap the cork from its position, and at the same instant start the stop-watch. Stir the liquid in the bath during the run and carefully maintain it at the previously determined proper temperature. Stop the watch when the bottom of the meniscus of the oil reaches the mark on the neck of the receiving flask.

The time in seconds for the delivery of 60 c.c. of oil is the Saybolt Universal viscosity of the oil at the temperature at which the test was made.

Additional Notes.—The oil tube should be in an exactly vertical position. The bath liquid should be stirred frequently or, better still, a small motor-driven stirrer may be inserted into the bath. For careful work the bath should be held within 0.1° F. of the temperature necessary to maintain the desired temperature in the oil tube. The room temperature should not be below 70° F.

Cheek determinations on the same viscosimeter should not differ by more than 1 per cent. Determinations on different meters of the same type are likely to vary considerably more than this.[1]

A convenient arrangement of Saybolt Viscosimeter for determinations below 150° F. is shown in Fig. 19. The water-bath is set in a slightly larger bath *A* with a layer of suitable heat insulator between the two. The assembly is mounted on a tripod with two legs offset as shown, thus giving more room. Each water-bath is stirred constantly by a vertical flat-blade stirrer *B* driven from the line shaft *C* and electric motor *D*. Heat is applied by means of carbon filament incandescent lamps *E* immersed in the

Fig. 19.—Saybolt Universal Viscosimeter for routine work at 100° F.

bath as shown and controlled by the five-point rheostats *F*. A thin film of oil left on the surface of the water helps to retard loss of heat. Bath thermometers *G* are clamped in the bath. A vacuum line *H* facilitates drawing off the excess of oil. In place of stop-watches, time is measured by means of second-counters[2] *I* operated electrically from a master clock and controlled by switches *J*. A small electric lamp *K* is convenient for observing the end of a determination. Holders *L* furnish safe individual receptaeles for the oil thermometers when not in use.

[1] Measurement of dimensions alone is not sufficient to standardize a capillary viscosimeter. The subject of real standardization by measuring outflow time is now being investigated by Dr. Herschel of the Bureau of Standards, with the cooperation of the A. S. T. M.

[2] These second-counters are considerably more satisfactory than stop-watches both as to accuracy and cost of maintenance. The time is given directly in seconds, which eliminates a frequent source of error either from failure of watch to register minutes or of the operator to convert minutes and seconds into the correct number of seconds. A single master clock operates forty counters in one installation. While the actuating mechanism shown above is a solenoid, subsequent experience shows that an electromagnet operated by a relay is preferable.

Determination of Viscosity by The Saybolt Furol Viscosimeter

(Test Method No. 11)

The Saybolt Furol[1] Viscosimeter was devised in 1920 for testing viscous petroleum products such as fuel and road oils. Tests on such materials by the Saybolt Universal Viscosimeter require too much time for routine work. The Saybolt Furol is identical with the Saybolt Universal in construction except for the size of the capillary which is made large enough to give an outflow time only about one-tenth of that required for the Saybolt Universal. This ratio is not a constant but varies somewhat with the viscosity of the material.[2]

The Saybolt Furol Vi cosimeter will be found extremely convenient for taking the viscosity of very viscous lubricating oils at temperatures below 210° F. The choice of temperature of measurement should be made from one of those commonly used. When fuel and road oils are to be tested, the preferred temperature is 122° F. (50° C.).

The following instructions are those recommended by a joint committee of producers and consumers of fuel oils:

1. Viscosity of fuel oils and other oils of similar viscosity shall be determined by the Saybolt Furol Viscosimeter.

2. The apparatus and method of operating shall be the same as for the Saybolt Universal Viscosimeter (A.S.T.M. Standard D 47–20) except for temperature. All dimensions of the viscosimeter tube shall be the same except the diameter of the outlet tube, which shall be as follows:

	Minimum, cm.	Normal, cm.	Maximum, cm.
Inside diameter of outlet tube	0.313	0.315	0.317
Outside diameter at lower end	0.40	0.43	0.46

3. Viscosity shall be determined at 122° F. (50° C.) and shall be expressed as "—seconds, Saybolt Furol" according to the time in seconds required for the outflow of 60 c.c. of oil.

4. Oil showing a time of less than 25 seconds Saybolt Furol at 122° F. shall be tested on the Saybolt Universal at 122° F. Oils showing a time less than 32 seconds Saybolt Universal at 122° F., shall be measured in the Saybolt Universal at a temperature of 100° F. This specification does not apply to fuels having a viscosity of less than 32 seconds Saybolt Universal at 100° F., such products not being considered as fuel oils.

Determination of Viscosity by the Saybolt Thermo Viscosimeter

(Test Method No. 12)

This method has proved particularly suitable for lamp oils and may be used for naphtha. The instrument is considerably more satisfactory and reliable for routine work than the Ubbelohde viscosimeter[3] although no data are given as to tolerances in

[1] This name was devised by the A. S. T. M. and approved by Mr. G. M. Saybolt. It signifies "Fuel and Road Oil."

[2] (See Table 5).

[3] An authoritative estimate of the Ubbelohde is given by W. H. Herschel in Bureau of Standards Technologic Paper, No. 125, Viscosity of Gasoline.

its dimensions. Viscosities of less than 32 seconds Saybolt Universal have very little meaning. Petroleum products showing an outflow time of less than 32 seconds Saybolt Universal should be tested by the Saybolt Thermo Viscosimeter, and this instrument is suitable for materials having Saybolt Universal Viscosity as high as 42 seconds at 100° F. A Saybolt Thermo viscosity reading is ten times the number of seconds

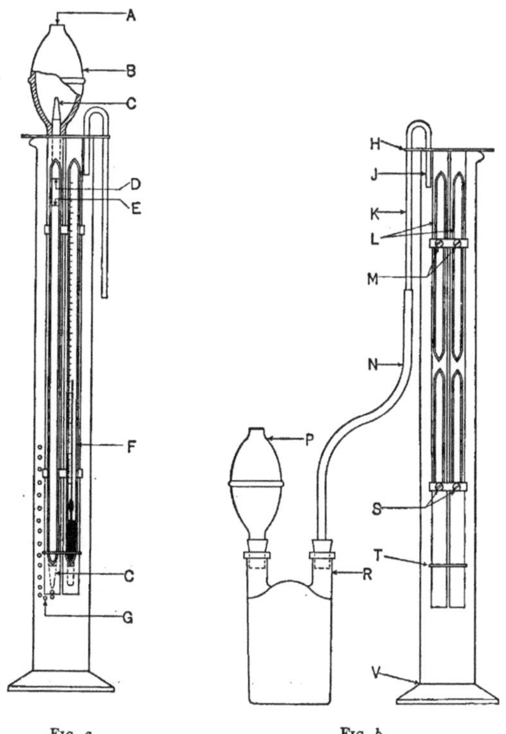

Fig. a. Fig. b.

Fig. 20.—Saybolt thermo viscosimeter.

Fig. a.—Front view of twin tubes with thermometer and capillary tube in place.
Fig. b.—Back view of twin tubes with thermometer and capillary tube removed.

(*A*) Air hole; (*B*) Compression bulb; (*C*) Capillary tube; (*D*) Head level upper mark; (*E*) Lower stop mark; (*F*) Thermometer; (*G*) Air bubbles; (*H*) Circular top hanger-plate; (*J*) Short leg of siphon; (*K*) Long leg of siphon; (*L*) Twin tubes; (*M*) Binding screws; (*N*) Rubber tubing; (*P*) Vacuum bulb; (*R*) Woulff bottle; (*S*) Binding screws; (*T*) Stirring disk; (*V*) Stand glass.

required for an oil level to rise through a glass capillary of standard size. The results of two properly conducted tests should differ by not more than 0.4 second or 4 Thermo units.

The following figures indicate the relation between viscosities at 100° F. on the Saybolt Universal and at 60° F. on the Saybolt Thermo.

Material	Saybolt universal at 100° F.	Saybolt thermo at 60° F.
Naphtha	30.0 seconds	215 units
Water white	32.6 seconds	340 units
Mineral seal	41.0 seconds	1260 units

Apparatus.—The Saybolt Thermo Viscosimeter[1] consists of a long glass capillary suitably mounted in a metal support beside an accurate thermometer. The illustration (Fig. 20) shows all the essential parts with reference letters.

Procedure.—1. The temperature of the room in which the tests are to be made must not be below 60° F. nor higher than 90° F., although the tests may be conducted at any temperature within this range.

2. After a sufficient period of time has elapsed to allow the samples and instrument to acquire a temperature approximately equal to that of the room in which the tests are to be made, pour perfectly clean or strained oil into the stand-glass (*V*). However, leave enough room for displacement of twin-tubes (*L*) when immersed in the oil.

3. Hang the twin-tubes (*L*) in the stand-glass (*V*) of oil and leave them there until a uniform temperature throughout the oil has been attained. If necessary, move the twin tubes (*L*) up and down to accelerate the desired result.

4. When the temperature is constant, attach rubber tubing (*N*) of the Woulff bottle (*R*) to the long leg (*K*) of the siphon and draw off the surplus oil until the oil level reaches the upper mark (*D*) on the long capillary tube. The oil level and the upper mark must always coincide when making a test, in order to preserve the same constant head-level for all tests.

5. Attach bulb (*B*) to top of capillary tube (*C*) as shown, so that it will firmly rest on hanger-plate (*H*). With the forefinger closing the air-hole (*A*) in compression bulb (*B*) on the capillary tube, gently and slowly press the bulb with thumb and middle finger until the capillary tube (*C*) is entirely emptied, an indication of which is a steady stream of air bubbles, (indicated by (*G*) ascending through the oil in stand-glass (*V*).

6. Hold a stop-watch in left hand and, while the air bubbles are still steadily ascending, gently slip the forefinger up and off the air-hole in bulb (*B*) and simultaneously start the watch.

7. The oil will now ascend the capillary tube (C), which it must do without any break in the thread-like column of oil. Watch carefully its near approach to the lower (stop) mark (*E*) on the capillary tube. The instant that the top of this oil column in the capillary tube reaches mark (*E*), stop the watch and mark down the reading (one-tenth of a second to be counted as a unit). Also make note of the temperature indication shown on thermometer (*F*). Repeat the foregoing procedure three or more times to insure against personal errors.

8. For example, the results may be, say at 80° temperature:

First trial	32.4 seconds
Second trial	32.4 seconds
Third trial	32.2 seconds
Fourth trial	32.4 seconds

Correct reading at 80° temperature equals 32.4 and, reading as a whole number, it is 324. Refer to Table 3 and in the column headed by "80" find the

[1] Devised by G. M. Saybolt and manufactured under his supervision and standardization.

TABLE 6.

Saybolt Thermo Viscosimeter Temperature Corrections.

	Working Temperatures										Net viscosity at
	70	**69**	**68**	**67**	**66**	**65**	**64**	**63**	**62**	**61**	**60**
INDICATIONS	210	212	214	216	218	220	222	224	226	228	**230**
	215	217	219	221	223	225	227	229	231	233	**235**
	219	221	223	225	227	229	231	233	235	237	**240**
	224	226	228	230	232	234	236	238	240	242	**245**
	228	230	232	234	236	238	240	242	244	247	**250**
	232	234	236	238	240	242	244	246	249	252	**255**
	237	239	241	243	245	247	249	251	254	257	**260**
	241	243	245	247	249	251	253	256	259	262	**265**
	246	248	250	252	254	256	258	261	264	267	**270**
	250	252	254	256	258	260	263	266	269	272	**275**
	253	255	257	259	262	265	268	271	274	277	**280**
	256	258	261	264	267	270	273	276	279	282	**285**
	260	263	266	269	272	275	278	281	284	287	**290**
	265	268	271	274	277	280	283	286	289	292	**295**
	270	273	276	279	282	285	288	291	294	297	**300**
	275	278	281	284	287	290	293	296	299	302	**305**
	280	283	286	289	292	295	298	301	304	307	**310**
	285	288	291	294	297	300	303	306	309	312	**315**
	289	292	295	298	301	304	307	310	313	316	**320**
	293	296	299	302	305	308	311	314	317	321	**325**
	297	300	303	306	209	312	315	318	322	326	**330**
	301	304	307	310	313	316	319	323	327	331	**335**
	305	308	311	314	317	320	324	328	332	336	**340**
	309	312	315	318	321	325	329	333	337	341	**345**
	313	316	319	322	326	330	334	338	342	346	**350**
	317	320	323	327	331	335	339	343	347	351	**355**
	321	324	328	332	336	340	344	348	352	356	**360**
	325	329	333	337	341	345	349	353	357	361	**365**
	329	333	337	341	345	349	353	357	361	365	**370**
	335	339	343	347	351	354	358	362	366	370	**375**
	339	343	347	351	355	359	363	367	371	375	**380**
	343	347	351	355	359	363	367	371	375	380	**385**
	347	351	355	359	363	367	371	375	380	385	**390**
	351	355	359	363	367	371	375	380	385	390	**395**
	355	359	363	367	371	375	380	385	390	395	**400**
	360	364	368	372	376	380	385	390	395	400	**405**
	364	368	372	376	380	385	390	395	400	405	**410**
	369	373	377	381	385	390	395	400	405	410	**415**
	373	377	381	385	390	395	400	405	410	415	**420**
	378	382	386	390	395	400	405	410	415	420	**425**
	382	386	390	395	400	405	410	415	420	425	**430**
	387	391	395	400	405	410	415	420	425	430	**435**
	392	396	400	405	410	415	420	425	430	435	**440**
	396	400	405	410	415	420	425	430	435	440	**445**
	401	405	410	415	420	425	430	435	440	445	**450**
	405	410	415	420	425	430	435	440	445	450	**455**
	410	415	420	425	430	435	440	445	450	455	**460**
	414	419	424	429	434	439	444	449	454	459	**465**
	419	424	429	434	439	444	449	454	459	464	**470**
	423	428	433	438	443	448	453	458	463	469	**475**
	428	433	438	443	448	453	458	463	468	474	**480**
	432	437	442	447	452	457	462	467	473	479	**485**
	437	442	447	452	457	462	467	472	478	484	**490**
	442	447	452	457	462	467	472	477	483	489	**495**
	446	451	456	461	466	471	476	482	488	494	**500**
	451	456	461	466	471	476	481	487	493	499	**505**
	456	461	466	471	476	481	486	492	498	504	**510**
	460	465	470	475	480	485	491	497	503	509	**515**
	465	470	475	480	485	490	496	502	508	514	**520**
	470	475	480	485	490	495	501	507	513	519	**525**
	475	480	485	490	495	500	506	512	518	524	**530**
	479	484	489	494	499	505	511	517	523	529	**535**
	483	488	493	498	504	510	516	522	528	534	**540**
	488	493	498	503	509	515	521	527	533	539	**545**
	491	496	502	508	514	520	526	532	538	544	**550**
	495	501	507	513	519	525	531	537	543	549	**555**
	500	506	512	518	524	530	536	542	548	554	**560**

TABLE 6.—*Continued.*

	Working Temperatures										Net viscosity at 60
INDICATIONS	80	79	78	77	76	75	74	73	72	71	
	200	201	202	203	204	205	206	207	208	209	**230**
	205	206	207	208	209	210	211	212	213	214	**235**
	208	209	210	211	212	213	214	215	216	217	**240**
	211	212	213	214	215	216	217	218	220	222	**245**
	213	214	215	216	217	218	220	222	224	226	**250**
	216	217	218	219	220	222	224	226	228	230	**255**
	218	219	221	223	225	227	229	231	233	235	**260**
	221	223	225	227	229	231	233	235	237	239	**265**
	226	228	230	232	234	236	238	240	242	244	**270**
	230	232	234	236	238	240	242	244	246	248	**275**
	233	235	237	239	241	243	245	247	249	251	**280**
	236	238	240	242	244	246	248	250	252	254	**285**
	239	241	243	245	247	249	251	253	255	257	**290**
	242	244	246	248	250	252	254	256	259	262	**295**
	246	248	250	252	254	256	258	261	264	267	**300**
	250	252	254	256	258	260	263	266	269	272	**305**
	254	256	258	260	262	265	268	271	274	277	**310**
	257	259	261	264	267	270	273	276	279	282	**315**
	261	263	265	268	271	274	277	280	283	286	**320**
	265	267	269	272	275	278	281	284	287	290	**325**
	269	271	273	276	279	282	285	288	291	294	**330**
	273	275	277	280	283	286	289	292	295	298	**335**
	277	279	281	284	287	290	293	296	299	302	**340**
	281	283	285	288	291	294	297	300	303	306	**345**
	285	287	289	292	295	298	301	304	307	310	**350**
	289	291	293	296	299	302	305	308	311	314	**355**
	293	295	297	300	303	306	309	312	315	318	**360**
	296	298	301	304	307	310	313	316	319	322	**365**
	299	302	305	308	311	314	317	320	323	326	**370**
	305	308	311	314	317	320	323	326	329	332	**375**
	309	312	315	318	321	324	327	330	333	336	**380**
	313	316	319	322	325	328	331	334	337	340	**385**
	317	320	323	326	329	332	335	338	341	344	**390**
	320	323	326	329	332	335	338	341	344	347	**395**
	325	328	331	334	337	340	343	346	349	352	**400**
	330	333	336	339	342	345	348	351	354	357	**405**
	334	337	340	343	346	349	352	355	358	361	**410**
	338	341	344	347	351	354	357	360	363	366	**415**
	342	345	348	351	354	357	360	363	366	369	**420**
	346	349	352	355	358	361	364	367	370	374	**425**
	349	352	355	358	361	364	367	370	374	378	**430**
	352	355	358	361	364	367	371	375	379	383	**435**
	356	359	362	365	368	372	376	380	384	388	**440**
	359	362	365	368	372	376	381	384	388	392	**445**
	363	366	369	373	377	380	385	389	393	397	**450**
	368	371	374	377	381	385	389	393	397	401	**455**
	371	374	378	382	386	390	394	398	402	406	**460**
	374	378	382	386	390	394	398	402	406	410	**465**
	379	383	387	391	395	399	403	407	411	415	**470**
	382	386	390	394	398	402	406	410	414	419	**475**
	387	391	395	399	403	407	411	415	419	423	**480**
	391	395	399	403	407	411	415	419	423	427	**485**
	395	399	403	407	411	415	419	423	427	432	**490**
	399	403	407	411	415	419	423	427	432	437	**495**
	402	406	410	414	418	422	426	431	436	441	**500**
	406	410	414	418	422	426	431	436	441	446	**505**
	411	415	419	423	427	431	436	441	446	451	**510**
	414	418	422	426	430	435	440	445	450	455	**515**
	417	421	425	430	435	440	445	450	455	460	**520**
	420	425	430	435	440	445	450	455	460	465	**525**
	425	430	435	440	445	450	455	460	465	470	**530**
	429	434	439	444	449	454	459	464	469	474	**535**
	433	438	443	448	453	458	463	468	473	478	**540**
	438	443	448	453	458	463	468	473	478	483	**545**
	441	446	451	456	461	466	471	476	481	486	**550**
	445	450	455	460	465	470	475	480	485	490	**555**
	448	453	458	463	468	473	478	483	488	494	**560**

TABLE 6.—*Continued.*

INDICATIONS

Working Temperatures										Net Viscosity at
90	**89**	**88**	**87**	**86**	**85**	**84**	**83**	**82**	**81**	**60**
192	192	192	193	194	195	196	197	198	199	**230**
196	196	197	198	199	200	201	202	203	204	**235**
198	199	200	201	202	203	204	205	206	207	**240**
201	202	203	204	205	206	207	208	209	210	**245**
203	204	205	206	207	208	209	210	211	212	**250**
206	207	208	209	210	211	212	213	214	215	**255**
208	209	210	211	212	213	214	215	216	217	**260**
211	212	213	214	215	216	217	218	219	220	**265**
214	215	216	217	218	219	220	221	222	224	**270**
217	218	219	220	221	222	223	224	226	228	**275**
220	221	222	223	224	225	226	227	229	231	**280**
223	224	225	226	227	228	229	230	232	234	**285**
226	227	228	229	230	231	232	233	235	237	**290**
229	230	231	232	233	234	235	236	238	240	**295**
233	234	235	236	237	238	239	240	242	244	**300**
237	238	239	240	241	242	243	244	246	248	**305**
240	241	242	243	244	245	246	248	250	252	**310**
243	244	245	246	247	248	249	251	253	255	**315**
247	248	249	250	251	252	253	255	257	259	**320**
250	251	252	253	254	255	257	259	261	263	**325**
253	254	255	256	257	259	261	263	265	267	**330**
256	257	258	259	261	263	265	267	269	271	**335**
260	261	262	263	265	267	269	271	273	275	**340**
263	264	265	267	269	271	273	275	277	279	**345**
266	267	269	271	273	275	277	279	281	283	**350**
270	271	273	275	277	279	281	283	285	287	**355**
273	275	277	279	281	283	285	287	289	291	**360**
276	278	280	282	284	286	288	290	292	294	**365**
279	281	283	285	287	289	291	293	295	297	**370**
284	286	288	290	292	294	296	298	300	302	**375**
287	289	291	293	295	297	299	301	303	306	**380**
290	292	294	296	298	300	302	304	307	310	**385**
293	295	297	299	301	303	305	308	311	314	**390**
294	296	298	300	302	305	308	311	314	317	**395**
299	301	303	305	307	310	313	316	319	322	**400**
303	305	307	309	312	315	318	321	324	327	**405**
305	307	310	313	316	319	322	325	328	331	**410**
308	311	314	317	320	323	326	329	332	335	**415**
312	315	318	321	324	327	330	333	336	339	**420**
316	319	322	325	328	331	334	337	340	343	**425**
319	322	325	328	331	334	337	340	343	346	**430**
322	325	328	331	334	337	340	343	346	349	**435**
326	329	332	335	338	341	344	347	350	353	**440**
329	332	335	338	341	344	347	350	353	356	**445**
333	336	339	342	345	348	351	354	357	360	**450**
338	341	344	347	350	353	356	359	362	365	**455**
341	344	347	350	353	356	359	362	365	368	**460**
344	347	350	353	356	359	362	365	368	371	**465**
347	350	353	356	359	362	365	368	371	375	**470**
349	352	355	358	361	364	367	370	374	378	**475**
353	356	359	362	365	368	371	375	379	383	**480**
356	359	362	365	368	371	375	379	383	387	**485**
359	362	365	368	371	375	379	383	387	391	**490**
362	365	368	371	375	379	383	387	391	395	**495**
365	368	371	374	378	382	386	390	394	398	**500**
368	371	374	378	382	386	390	394	398	402	**505**
371	375	379	383	387	391	395	399	403	407	**510**
374	378	382	386	390	394	398	402	406	410	**515**
377	381	385	389	393	397	401	405	409	413	**520**
380	384	388	392	396	400	404	408	412	416	**525**
384	388	392	396	400	404	408	412	416	420	**530**
387	391	395	399	403	407	411	415	419	424	**535**
389	393	397	401	405	409	413	418	423	428	**540**
392	396	400	404	408	413	418	423	428	433	**545**
395	399	403	407	411	416	421	426	431	436	**550**
398	402	406	410	415	420	425	430	435	440	**555**
401	405	409	413	418	423	428	433	438	443	**560**

number nearest to 324, namely 325. The figure in the column headed by " 60 " horizontally opposite this reading is the viscosity of the oil at 60° F. in Saybolt Thermo units, namely 400.

9. Where the actual result obtained at any temperature above 60° is exactly midway between two table readings, use the larger of the two, thus:

A similar oil to above example may have been tested at 66° temperature and given 369. The table shows 367 and 371. Select 371 and the viscosity will be 400 at 60° temperature.

Precautions.—The adding and removal of pressure on the bulb (*B*) must be so gentle as to prevent the possibility of the column of oil in capillary tube (*C*) being broken. Should this occur, force the oil column down slowly and start anew.

Have the pressure bulb (*B*) on the end of the capillary tube (*C*) down far enough so that it will touch the hanger-plate (*H*).

Care must be taken that no oil gets into the bulb (*B*).

Should any particles of rubber be found on top of capillary (*C*), they should be carefully cleaned off, (with gasoline), if necessary.

When not in use, the instrument should be hung in the empty stand-glass (*V*) and covered for protection against dust.

The uppermost (head-level) mark (*D*) must be exactly on a level with the oil surface in stand-glass (*V*) after the siphon has acted.

Oils having a viscosity greater than 560 Saybolt Thermo at 60° F. should be measured at 60° F. as the temperature corrections above this value are rather large. It is preferable to work as near 60° F. as possible on all oils. For plant-control work a temperature of 77° F. will be found much more convenient and is likely to be adopted as standard in the near future.

Determination of Viscosity by the Redwood Viscosimeter

(Test Method No. 13)

The Redwood Viscosimeter is the official English instrument. It may be used for the same products as the Saybolt Universal but the standard temperatures are 70, 140, and 200° F. Viscosities on fuel oils for English requirements are frequently measured at 32° F. on a modified form known as the Redwood Admiralty Viscosimeter. When the term Redwood is used alone, the original form is meant.

Apparatus.—The Redwood Viscosimeter (Fig. 21) consists of an oil tube *C* mounted in a bath *J* with a stirrer *L*, the assembly being supported on a tripod furnished with leveling screws. A heating tube *K* projects downward at an angle from the lower part of the bath. The oil-outlet tube *D* is made of agate and may be closed by a small brass ball fastened to the rod *E*. The stirring apparatus consists of four light metal vanes *L* fastened to a cylinder fitting loosely around the oil tube. This cylinder has a broad curved flange at the top to prevent any liquid from the bath being splashed into the oil tube. The bath thermometer *T′* and the stirrer handle *H* are mounted on opposite sides of this flange. The oil thermometer *T* is supported by a spring clamp as shown. The pointed gage *F* denotes the proper filling level. The receiving flask has a narrow neck and is marked to contain 50 c.c.

Procedure.—Make sure that the interior of the oil tube and of the capillary outlet tube are clean, dry and free from any loose particles which might obstruct the free flow of the oil. Level the instrument. Fill the bath nearly full, using water for temperatures up to 190° F. and high-flash petroleum oil for higher temperatures. Adjust the temperature of the bath liquid to that temperature which will maintain the desired temperature in the oil tube throughout the determination. This bath temperature must be determined experimentally and should be maintained by adding small portions of warm or cold water as needed or by the use of a flame under the heating tube if necessary.

Close the orifice by means of the ball valve. Clamp the oil thermometer in a vertical position at a height such that the top of the bulb will be slightly below the oil level at the end of the determination. Strain the oil sample through 60-mesh wire gauze and heat (or cool) it to the temperature at which the viscosity is to be measured. Fill the oil tube to slightly above the point of the filling gage.

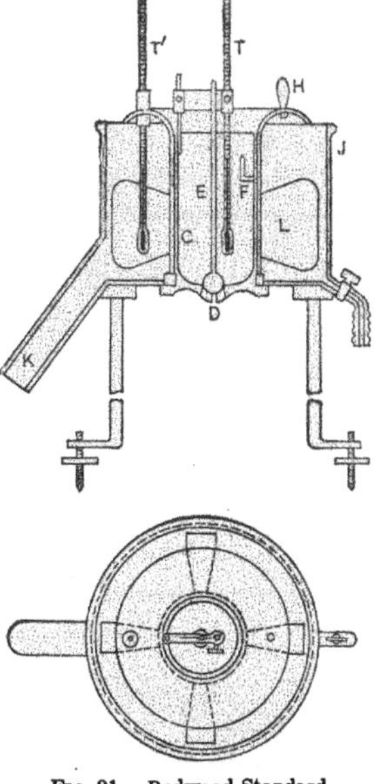

Fig. 21.—Redwood Standard Viscosimeter.

Stir the bath and regulate its temperature until the oil in the oil tube reaches and continues to hold the desired temperature. Adjust the height of the oil in the oil tube by removing enough so that the point of the filling gage is just at the surface. Place the clean, dry, 50 c.c. receiving flask directly below the capillary orifice. Raise the ball valve and at the same instant start the stop-watch. Support the ball valve at the side of the oil tube by hooking it over the oil-thermometer clamp. Stir the bath frequently and maintain both bath and oil sample at the respective required temperatures. Neither of these should vary more than 0.2° F. Stop the watch when the bottom of the meniscus reaches the mark on the neck of the receiving flask. Close the capillary with the ball valve. Note and record the time in seconds as the "Redwood Viscosity at —° F." according to the temperature of the oil during the test.

Redwood admiralty viscosimeter.—(See Fig. 15). Fuel oils to be tested on this instrument are held for six hours at 32° F., thoroughly stirred and immediately transferred to the viscosimeter, which should be maintained at 32° F., preferably in a refrigerated compartment. The rest of the determination is made as with the Standard Redwood and the viscosity in seconds is reported as "Redwood Admiralty Viscosity at 32° F."

The construction of the Redwood Admiralty viscosimeter is such that it gives viscosities approximately one-tenth of those obtained on the same oil at the same temperature by the Standard Redwood.

Determination of Viscosity by the Engler Viscosimeter

(Test Method No. 14)

The Engler viscosimeter is the official German instrument. It has been used in the United States for fuel oils and asphaltic materials. Fuel oils too viscous for the Saybolt Universal should be measured on the Saybolt Furol. Where viscosity determinations are required at temperatures higher than 210° F., as on asphaltic fluxes, the Engler is usually used. Engler viscosities are reported in Engler degrees or numbers obtained by dividing the outflow time in seconds for 200 c.c. of the material being tested by the outflow time in seconds for 200 c.c. of water at 68° F. (20° C.). The latter figure is known as the water factor and should lie between 50 and 52 seconds. The chief objections to the use of the Engler as a viscosimeter for routine work are the large amount of material required (240 c.c.) and the relatively small bath, which makes temperature control rather difficult. Determinations are usually made at one or more of these temperatures; 68, 77, 122, 212, and sometimes 300° F.

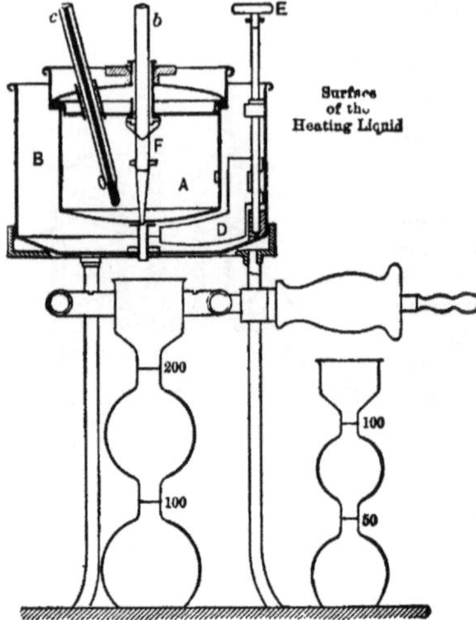

Fig. 22.—Engler viscosimeter.

Apparatus.—The standard Engler viscosimeter in use at the present time is shown in Fig. 22. It consists of an oil cup (A) rigidly supported in a bath (B) which is supported on a tripod of suitable height. A stirrer (D) operated by the knob (E) assists in maintaining the bath at the desired temperature. The oil cup is provided with a hollow cover to prevent loss of heat. There are two openings in the cover, one for the oil thermometer (c) and one for the wooden plug (b) which is tapered to a point for closing the outlet capillary. A stop (F) on this plug fixes the position of the plug during outflow. Equally spaced around the inside of the oil cup are three gage points to indicate the proper filling level. A thermometer (not shown) is supported in the bath by a spring clamp. A ring gas burner is used for heating when necessary. The inside of the oil cup is gilded and the capillary jet is made of or lined with platinum.

The essential dimensions of the instrument are as follows:

Inside diameter of oil cup	106±1.0 mm.
Height of vessel below gage points	25±1.0 mm.
Length of capillary jet	20±0.1 mm.
Inside diameter of capillary jet, upper end	2.9±0.02 mm.
Inside diameter of capillary jet, lower end	2.8±0.02 mm.
Length of jet projecting from lower part of bath	3.0±0.3 mm.
Outside diameter of jet at lower end	4.5±0.2 mm.
Exact volume of distilled water to fill oil cup to gage points at 68° F.	240 c.c.

The thermometers should be scaled for suitable partial immersion to conform with both American and German practice. The receiving flask should be accurately graduated to contain 200 c.c.

Procedure determination of the water factor.—Wash the inside of the oil cup, firs with ethyl ether or petroleum ether, then repeatedly with alcohol and finally with pure distilled water. Use a new wooden plug or one which has never been contaminated with oil, washing it first with alcohol and then with distilled water. Make sure that the oil cup is level and that the room temperature is within 2° of 68° F. Fill the bath with water at 68° F. (20° C.) to the level of the oil-cup cover. Insert the wooden plug and fill the oil cup to slightly above the gage points with pure distilled water at 68° F. Manipulate the plug so that the capillary jet is entirely filled with water and so that the bottom of the jet carries a suspended drop of water. Place

the excess of water from the oil cup with a pipette until the three gage points are exactly in the surface of the water. Place the clean dry 200 c.c. receiving flask below the jet, place the cover in position and wait a few seconds to let the liquid surface come to rest. Raise the plug as far as possible and at exactly the same time, start the stop-watch. Hold the bath temperature at exactly 68° F., stirring occasionally. Stop the watch at the instant the lower line of the meniscus reaches the 200 c.c. mark on the neck of the receiving flask. Repeat the determination exactly as before until six consecutive tests give results varying by not more than 0.5 second. Take the average of these six tests expressed to 0.1 second as the water factor of the instrument. Any instrument whose factor is not between 50 and 52 seconds is not a standard Engler. The water factor of an Engler viscosimeter should be redetermined every six months, or oftener if the meter is used constantly.

Determination of Engler viscosity.—Have the oil cup and wooden plug perfectly clean and dry if previously used for water. If used for oil on the previous determination, the oil cup may be cleaned by rinsing several times with small quantities of the oil to be tested. Bring the bath temperature to about the temperature at which the determination is to be made. When this is above room temperature, the required bath temperature must be one or two degrees higher than that of the oil. For viscosities at temperatures up to 190° F., water may be used, for higher temperatures a high-flash petroleum oil. Heat the oil to be tested to the desired temperature and pour it into the oil cup through a 60-mesh wire gauze strainer. As soon as the capillary jet is filled insert the wooden plug and complete the filling of the oil cup to the level indicated by the gage points.

Place the cover in position and determine the bath temperature necessary to keep the oil temperature constant at the desired point. As soon as oil temperature has remained constant for three minutes, remove the cover and make sure that the oil level is correct. Replace cover and with a glass rod remove any oil drop that may be suspended from the lower end of the capillary jet. Set the clean, dry 200 c.c. receiving flask in position below the capillary jet. Lift the wooden plug as far as it will go and at exactly the same instant start the stop-watch. Keep the temperature of the oil constant within 0.1° F. throughout the time of outflow, stirring the bath frequently. Stop the watch when the lower line of the meniscus reaches the 200 c.c. mark on the neck of the receiving flask. Divide the number of seconds for this outflow time by the water factor for the instrument and record as "Engler Viscosity at —° F." Thus an outflow time of 199 seconds at 77° F. on an Engler viscosimeter whose water factor is 51.3 seconds, is calculated to 199 ÷ 51.3, or 3.88 Engler Viscosity at 77° F.

Duplicate determinations on the same instrument should agree within 1 per cent. Differences of 3 per cent between determinations by different instruments are not uncommon.

Where great accuracy is not desired, the time required for a determination may be shortened by allowing only 100 c.c. to run out. The flask shown in Fig. 22 carries a graduation for this purpose. In this case the outflow time for 100 c.c. should be multiplied by the factor 2.35 to obtain the approximate time for outflow of 200 c.c. In every case, however, the viscosity should be expressed as an Engler number as defined. The importance of proper determination of the water factor is obvious since it enters into the calculation of every Engler viscosity.

COLD TESTS

Petroleum oils, when cooled, become more viscous and will eventually become solid if the temperature is sufficiently lowered. The temperature at which a petroleum oil solidifies or becomes too viscous to flow out of containers is obviously of great practical importance. With many petroleum products, particularly cylinder oils, viscous fuel oils and asphaltic materials, the temperature of solidification depends to a large degree on the time during which the oil is exposed to this temperature. It frequently happens, therefore, that in practice oils will be found to be solid at temperatures considerably above the various kinds of cold test reported by the laboratory. For this reason, cold tests of any kind are to be interpreted with much caution. Where the purpose of a cold test is to determine whether an oil will remain liquid at a specific temperature, an average sample of the oil should be maintained at this temperature for six hours without disturbance of any kind and then examined for fluidity at this temperature, either visually or, better still, by the use of the Saybolt Furol Viscosimeter (Test Method No. 11).

The term " cold test " is used commonly in the petroleum industry as a generic expression applying to any test of the sort. In other words, cloud and pour are " cold tests " taken in a definite manner. Strictly used, the term " cold test " is applied to a specific test which has been adopted by the American Society for Testing Materials. It is highly desirable that confusion be eliminated by using the terms cloud point and pour point whenever these are determined.

The methods for determining cloud and pour points, described in this section, are based on those adopted by the American Society for Testing Materials in 1918,[1] but represent revisions made to conform with the practice of large refiners of petroleum and to furnish additional information on points not clearly covered in the A. S. T. M. methods. The A. S. T. M. method for " cold test for steam cylinder and black oils " gives results much too low to serve as a basis for practical valuation of oils. It has proved entirely unsatisfactory and is therefore not described.

Determination of Cloud Point

(Test Method No. 15)

Cloud point of a petroleum oil is that temperature at which paraffin wax or other solid substances begin to crystallize out or separate from solution, when the oil is chilled under certain definite specified conditions. The test for cloud point should be used only for oils which are transparent in layers 1½ inches thick. The permissible difference between duplicate determinations of cloud point on the same sample is 2° F.

Apparatus.—(See Fig. 23.)

1. Cold-test jar of clear glass, cylindrical form, flat bottom, approximately 1¼ inches inside diameter and 4½–5 inches high. An ordinary 4 oz. oil sample bottle may be used if the cold-test jar is not available.

2. Cold-test thermometer, engraved-stem type, scaled for 4¼ inches immersion, range −20° to 140° F., graduated in 2° scale divisions, numbered every 20° and accurate within 2° throughout the range. Plain front, glass ring at top and expansion chamber to hold 212° F. Bulb, cylinder form, diameter $\frac{9}{32}$ to $\frac{5}{16}$ inch, length ⅜ inch. Tube diameter, $\frac{5}{16}$ inch, over-all length 8¾ inches, distance from bottom of bulb to −20° mark 4⅜–4⅝ inches, distance from top of tube to 140° mark ¾–1⅛ inch.

2. Cloud and Pour Point Thermometer[2] to conform to the following specifications:

Type: Glass engraved.
Total length: 222 mm.
Glass; Stem: Plain front, enamel back, suitable thermometer tubing.
Diameter 7 to 8 mm.

[1] A. S. T. M. Standards, 1918, 617.

[2] For temperatures below −36° F. a similar thermometer actuated by suitable non-freezing liquid should be used.

Bulb: Corning normal, Jena 16, III, or equally suitable thermometric glass.
Length 9.5 mm. maximum.
Diameter not greater than stem.

Actuating liquid: Mercury.

Range: Minus 36° to plus 120° F.

Immersion: $4\frac{1}{4}$ inches line around tube at immersion point and words "$4\frac{1}{4}$ inch immersion" engraved on tube.

Range distance from bottom of bulb: 120 to 130 mm.

Range distance from top of tube: 19 to 25 mm.

Ice point: No.

Contraction chamber: No.

Expansion chamber: Yes. To hold 212° F.

Filled: Nitrogen gas.

Top finish: Plain.

Graduating: All lines, figures and letters clear cut and distinct.
Scale graduated in 2° divisions.
Scale numbered every 20° F.

Special markings: Cloud and Pour.
Serial number and manufacturer's name or trade mark engraved on the tube.

Accuracy: Error at any point on scale shall not exceed one smallest scale division.

Points to be tested for certification: Minus 28°, +32°, +92° F.

3. Cork to fit cold-test jar, bored centrally to take cold-test thermometer.

4. Jacket of glass or metal, water-tight, cylindrical form, flat bottom, about $4\frac{1}{2}$ inches deep, inside diameter $\frac{1}{2}$ inch greater than outside diameter of the cold-test jar.

5. Disk of cork or felt, $\frac{1}{4}$ inch thick and of same diameter as inside of jacket.

6. Ring gasket, about $\frac{3}{16}$ inch thick, to fit snugly around the outside of the cold-test jar and loosely inside the jacket. This gasket may be made of cork, felt or other suitable material, elastic enough to cling to the cold-test jar and hard enough to hold its shape. The purpose of the ring gasket is to prevent the cold-test jar from touching the jacket.

7. Cooling bath of type suitable for obtaining the required temperatures. Size and shape of the bath are optional, but a support, suitable for holding the jacket firmly in a vertical position, is essential. The required bath temperatures may be maintained by refrigeration if available, otherwise by suitable freezing mixtures, those commonly used being as follows:

For temperatures down to 35° F. Ice and water.
For temperatures down to − 5° F. Crushed ice and sodium chloride.
For temperatures down to −25° F. Crushed ice and calcium chloride.
For temperatures down to −70° F. Solid carbon dioxide and acetone.

The last-named mixture is made as follows: In a covered metal beaker, chill a suitable amount of acetone to 10° F. or lower, by means of an ice-salt mixture. Invert a cylinder of liquid carbon dioxide and draw off carefully into a chamois-skin bag the desired amount of carbon dioxide, which through rapid evaporation will quickly become solid. Then add to the chilled acetone enough of the solid carbon dioxide to give the desired temperature.

Procedure.—The oil to be tested shall be brought to a temperature at least 25° F. above the approximate cloud point. Moisture, if present, shall be removed by any suitable method, as by filtration through filter paper, until the oil is perfectly clear, but such filtration shall be made at a temperature at least 25° F. above the approximate cloud point.

The clear oil shall be poured into the cold-test jar (1) to a height of not less than 1 inch nor more than 1¼ inches. The cold-test jar may be marked to indicate the proper level.

The cold-test jar shall be tightly closed by the cork (3), carrying the cold-test thermometer (2) in a vertical position in the center of the jar with the thermometer bulb resting on the bottom of the jar.

The disk (5) shall be placed in the bottom of the jacket (4) and the cold-test jar, with the ring gasket (6) 1 inch above the bottom, shall be inserted into the jacket. Disk, gasket and inside of jacket shall be clean and dry.

The temperature of the cooling bath (7) shall be adjusted so that it is below the cloud point of the oil by not less than 15° F. nor more than 30° F. and this temperature shall be maintained throughout the test. The jacket, containing the cold-test jar, shall be supported firmly in a vertical position in the cooling bath so that not more than 1 inch of the jacket projects out of the cooling medium.

At each cold-test thermometer reading which is a multiple of 2° F., the cold-test jar shall be removed from the jacket quickly but without disturbing the oil, inspected for cloud and replaced in the jacket. This complete operation shall require not more than three seconds.

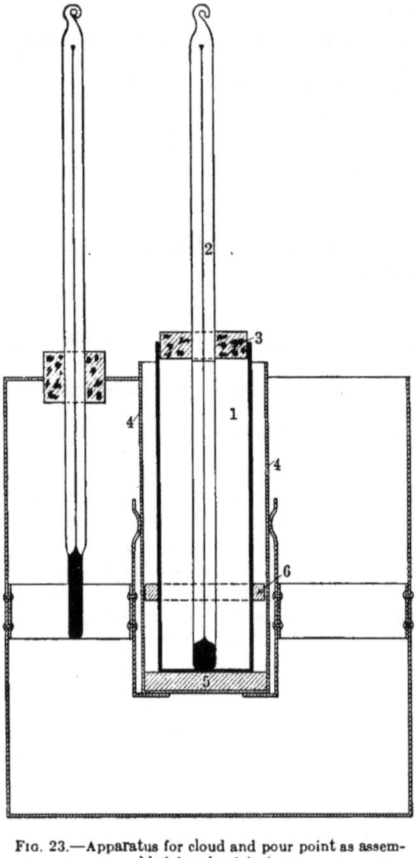

Fig. 23.—Apparatus for cloud and pour point as assembled for cloud test.

When the bottom of the oil has become opaque to a height of not less than ⅛ inch nor more than $\frac{3}{16}$ inch, the reading of the cold-test thermometer, corrected for error if necessary, shall be recorded as the cloud point. The required height of cloud is approximately at the middle of the thermometer bulb. The cold-test jar may be marked to indicate the proper level.

Determination of Pour Point

(Test Method No. 16)

Pour point of a petroleum oil is the lowest temperature at which this oil will pour or flow when it is chilled without disturbance under certain definite specified conditions.

The test for pour point should be used for all oils which are not transparent in layers 1½ inches thick and may be used for oils to which the test for cloud point is suited. The permissible difference between duplicate determinations of pour point on the same sample is 5° F.

Apparatus.—Same as for cloud point, Test Method No. 15.

Procedure.—Oils having a viscosity greater than 600 seconds, Saybolt Universal at 100° F. shall be allowed to stand in the cold-test jar at a temperature of 60°–85° F. for at least five hours prior to making the test for pour point. A viscous oil which has been stored in a warm place is liable to show an abnormally low, fictitious pour point unless this precaution is observed. Oils having a viscosity not greater than 600 seconds, Saybolt Universal at 100° F. may be tested without such preliminary standing.

After preliminary standing if necessary, the oil to be tested shall be brought to a temperature of 90° F. or to a temperature 15° F. higher than its pour point, if this pour point is above 75° F. and shall be poured into the cold-test jar (1) to a height of not less than 2 inches nor more than 2¼ inches. The jar may be marked to indicate the proper level.

The cold-test jar shall be tightly closed by the cork (3) carrying the cold-test thermometer (2) in a vertical position in the center of the jar with the thermometer bulb immersed so that the beginning of the capillary shall be ⅛ inch below the surface of the oil.

The disk (5) shall be placed in the bottom of the jacket (4) and the cold-test jar, with the ring gasket (6) 1 inch above the bottom shall be inserted into the jacket. Disk, gasket and inside of jacket shall be clean and dry.

The temperature of the cooling bath (7) shall be adjusted so that it is below the pour point of the oil by not less than 15° F. nor more than 30° F. and this temperature shall be maintained throughout the test. The jacket, containing the cold-test jar, shall be supported firmly in a vertical position in the cooling bath so that not more than 1 inch of the jacket projects out of the cooling medium.

At each cold-test thermometer reading which is a multiple of 5° F., the cold-test jar shall be removed from the jacket carefully and shall be tilted just sufficiently to ascertain whether the oil around the thermometer remains liquid. As long as the oil around the thermometer flows when the jar is tilted slightly, the cold-test jar shall be replaced in the jacket. The complete operation of removal and replacement shall require not more than three seconds. As soon as the oil around the thermometer does not flow when the jar is tilted slightly, the cold-test jar shall be held in a horizontal position for exactly five seconds and observed carefully. If the oil around the thermometer shows any movement under these conditions, the cold-test jar shall be immediately replaced in the jacket and the same procedure shall be repeated at the next temperature reading 5° F. lower. As soon as a temperature is reached at which the oil around the thermometer shows no movement when the cold-test jar is held in a horizontal position for exactly five seconds, the test shall be stopped.

The lowest reading of the cold-test thermometer, corrected for error if necessary, at which the oil around the thermometer shows any movement when the cold-test jar is held in a horizontal position for exactly five seconds, shall be recorded as the pour point.

NOTE.—It is evident that the pour point is 5° F. higher than the temperature at which the oil becomes entirely solid. This latter temperature is sometimes reported as the "solid point." In some laboratories, an oil which pours very freely at one 5° F. reading but is solid at the next, is said to have a pour point 2° F. higher than the solid point.

Color

By reflected light.—The color of petroleum products as examined by reflected light is frequently noted in the case of cylinder oil, crude petroleum and petrolatum. There is no standard of any sort nor is any attempt made to describe the color by anything more definite than general terms such as green, olive-green, bluish-green, etc. A comparison of color on samples by reflected light is best made by placing small drops of each close together on a smooth black surface. This procedure will frequently reveal a decided bloom in oils which appear to have none when examined in the ordinary sample bottle.

By transmitted light.—The determination of color of petroleum products by transmitted light is often required, but except in the case of nearly colorless materials cannot be made with much accuracy at the present time. The importance of color is usually overestimated, but since it is a property which is easily observed in a superficial way, it continues to be an item in specifications and the occasion for many a disagreement and dispute. The proper basis for the construction and enforcement of a color specification should be a full knowledge that color measurements on petroleum products are approximations only.

Many colorimeters have been proposed but nearly all are unsatisfactory, chiefly on account of the difficulty in securing permanent and reliable standards for comparison. Petroleum oils are not suitable color standards because they change color with age and particularly when exposed to light. Solutions of pure chemical substances, such as potassium bichromate or iodine in definite concentration, seldom match the color shade or tint of the petroleum product. The development of colored celluloid on a scientific basis during the last few years appears to hold considerable promise. Colored glass has long been used but except in a few cases it is neither uniform nor reliably accurate. For example, the glasses furnished with the Lovibond Tintometer vary considerably and cannot be used in the same additive manner as analytical weights.

Among the colorimeters used in the United States, the only ones which give approximately reproducible readings on petroleum products are the Saybolt Chromometer for nearly colorless oils, the colorimeter of the Union Petroleum Company [1] for lubricating oils, and the Lovibond Tintometer. All these instruments are devices to permit simultaneous visual observation of the color of light transmitted by a layer of oil and of that transmitted by a color standard. The Saybolt Chromometer possesses the distinct advantage of bringing the light from both sources into the field of vision as the two halves of a circle. The Hess-Ives Tint Photometer has not proved satisfactory for petroleum products in spite of the joint efforts of the inventors and one of the large petroleum laboratories. The American Society for Testing Materials has never adopted a standard colorimeter but is studying the subject with the purpose of developing one which will be as reliable for lubricating oil as the Saybolt Chromometer is for kerosene.

General precautions.—Samples must be perfectly clear when examined. Cloudiness due to separation of paraffin may be removed by warming the oil, but the necessary temperature must be maintained during the examination. Cloudiness due to moisture may be removed by filtration through a dry filter paper with repetition if necessary. Attention is called to the fact that even a slight haze or the presence of suspended air in the sample is sufficient to render color measurement useless.

Where an instrument is used to take color, with the exception of the color standards, the glass through which light passes must be colorless optical glass. Window glass is not colorless and must not be used to replace broken parts as for example in Lovibond cells. Where 4-oz. oil sample bottles are used they should be colorless and clear.

Source of light for color measurement is of great importance. Most text books

[1] Adopted by the National Petroleum Association.

TABLE 7

Comparison of Color Readings

Saybolt Chromometer			Lovibond Tintometer *			Union Colorimeter		Potassium Bichromate (mg. per liter)			Common descriptive term
No. of disks	Height of oil Inch	Color shade	Cell Inch	Series	Value	U. P. Co. letter	N. P. A. No.	*a*	*b*	*c*	
1	20	+25	12	510	.32					2.0	
1	18	+24								2.9	
1	16	+23								3.8	
1	14	+22								4.7	
1	12	+21	12	510	2.00			4.8	5.1	5.6	Water white
1	10.75	+20								6.5	
1	9.50	+19								7.5	
1	8.25	+18								8.5	
1	7.25	+17								9.5	
1	6.25	+16	12	510 200	4.0 1.02			12.0		10.5	Prime white
2	10.50	+15								12.5	
2	9.75	+14	12	510 200	4.0 1.1					13.5	
2	9.00	+13								14.5	
1	8.25	+12	12	510 200	5.0 1.1					16.5	
2	7.75	+11								17.5	
2	7.25	+10								18.6	
2	6.75	+9							19.0	19.7	
2	6.50	+8	12	510 200	6.0 1.1					20.8	
2	6.25	+7								21.9	
1	6.00	+6	12	510 200	7.0 1.00					23.0	
2	5.75	+5								24.1	
2	5.50	+4									
2	5.25	+3	12	510 200	9.0 1.2					25.3	
2	5.0	+2								26.6	
2	4.75	+1								28.0	Standard white
2	4.50	0							30.9		
2	4.25	−1									
2	4.00	−2	12	510 200	14.0 1.1						
2	3.75	−3									
2	3.625	−4	12	510 200	15.0 1.1						
2	3.50	−5									
2	3.375	−6									
2	3.25	−7									
2	3.125	−8									
2	3.00	−9									
2	2.875	−10									
2	2.75	−11	1	500	1.4–1.9	G	1				Lily white
2	2.625	−12									
2	2.500	−13									
2	2.375	−14									
2	2.250	−15									
2	2.125	−16	12	510 200	27.0 1.1						
2	1.875	−18									
2	1.625	−20									
2	1.375	−22									
2	1.125	−24	1	500	5–7	H	1½				Cream white
2	.875	−26									
2	.625	−28	12	510 200	27.0 25.5						
2	.375	−30									
2	.125	−32	1	500	20–23	I	2				Extra pale
						J					Ex. lemon pale
			1	500	50–63	K	3				Lemon pale
						L					Ex. orange pale
			1	500	110–125	M	4				Orange pale
						N					Pale
			1	500	220–250	O	5				Light red
			1	500	300–340	P	6				Dark red
						Q					Claret red

* 200 Series = Red; 500 Series = Amber; 510 Series = Yellow; 1180 Series = Blue.
a. N. P. A. Standard; *b.* Geo. M. Saybolt; *c.* 1% H_2SO_4—C. K. Francis.

recommend a "good northern exposure." While this might be obtained occasionally in open places, it is seldom available in petroleum-testing laboratories. Daylight is not suitable on account of its variable quality and especially because many color measurements must be made at night in plant-control work. The daylight electric lamps developed primarily for use in matching color of textiles are entirely satisfactory as the source of light for petroleum color measurements and furnish a uniform, reliable, artificial daylight, at all hours of day or night. The best results are obtained by taking colors in a room or compartment where light from all other sources is excluded. The intensity of the light should be kept as low as possible as any glare renders the eye less sensitive. Care should also be taken to avoid eye-fatigue due to observing a color field for too long a time without rest. Where both eyes are equally sensitive to color sensation, it is wise to use them alternately. Most operators, however, get the best results by using the same eye and closing it for a few seconds between observations.

While color measurements of petroleum products cannot be made with much accuracy, there are certain general approximate relations between the results by various methods. A rough comparison is given in Table 7.

Determination of Color by Saybolt Chromometer

(Test Method No. 17)

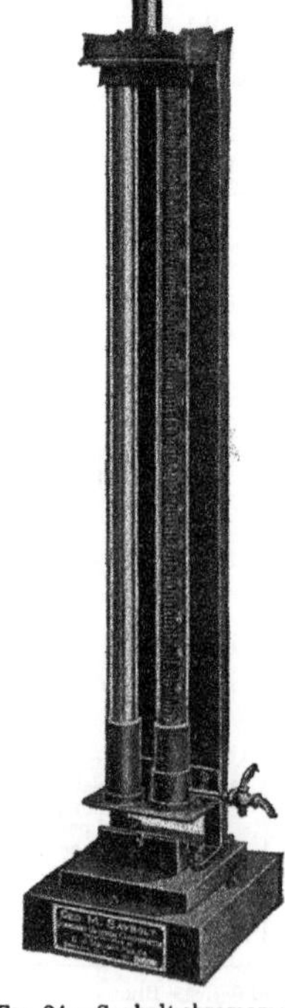

FIG. 24.—Saybolt chromometer.

This instrument is almost universally used in the United States for taking color on illuminating oils and naphthas. It has been officially adopted by the National Petroleum Association [1] and by the U. S. Committee on Standardization of Petroleum Specifications.[2] It may be used with confidence for all oils light enough to permit reading and is accurate within one color shade.

Apparatus.—The Saybolt Chromometer (Fig. 24) consists of two glass tubes, 20 inches long and about $\frac{5}{8}$ inch inside diameter. One of these, permanently closed at the bottom with a colorless, optical glass disk, is graduated to show height of oil column and is provided with a pet-cock. The other tube is open at each end with an arrangement for holding the standard color disks in proper position. Below the two tubes is a mirror for reflecting light up through them and above is an optical prism and eyepiece assembly which provides a circular field of vision, the two halves being illuminated by light transmitted by the sample and by the color standard respectively. A daylight lamp should be the source of light.

Procedure.—Set the instrument up before the daylight lamp which should not be bright enough to cause a glare. Exclude all light from other sources. Clean the oil tube by rinsing with some of the oil to be tested and allow to drain. Place two standard color disks in position. Close the pet-cock and fill the oil tube to the height of 10.5 inches. Observe the color through the eyepiece. If the oil at this height is unmistakably lighter than the two-disk standard, the measurement should be made with only one disk, otherwise with two. Having the determined number of standard color disks in position, fill the tube, if

[1] Sept. 22d, 1920.

[2] Bulletin No. 5, Dec. 20th, 1920, issued by the Bureau of Mines.

necessary, until the oil is decidedly darker than the standard. By means of the pet-cock, draw off oil slowly until the oil is slightly darker than the standard. Then draw down to the first reached height corresponding to a standard color shade shown in Table 7. If the color of the oil, observed through the eyepiece is still darker than the standard, draw down to the next height equal to a color shade and examine again. Continue this procedure until oil and standard colors match or show questionable difference. Lower the column of oil one shade more and if the oil is unmistakably lighter than the standard, record the previous color shade as the Saybolt Chromometer color.

Examples.

	No. 1	No. 2
Oil darker	Using one disk, at 16 inches	Using two disks, at 4.5 inches
Oil darker	Using one disk, at 14 inches	Using two disks, at 4.25 inches
Oil questionable	Using one disk, at 12 inches	Using two disks, at 4.0 inches
Oil lighter	Using one disk, at 10.75 inches	Using two disks, at 3.75 inches
Color is	plus 21	minus 2

Precautions.—Color-standard glasses should be kept clean and protected from light when not in use.

Eyepiece and prisms should be kept free from oil, dust and other contamination. Adjustments of the optical assembly should not be attempted.

Oil should be absolutely clear and free from suspended air. Color reading should always be made by having an excess of oil in the tube and drawing off so that the color match is obtained as the oil column grows lighter.

Determination of Color by the N. P. A. Colorimeter

(Test Method No. 18)

The National Petroleum Association in 1915 adopted color standards for lubricating oils as fixed by the Union Petroleum Company Colorimeter. This method, while not capable of fine distinction, is conveniently simple and, provided the color standards are correct, is sufficiently accurate for all commercial purposes.

FIG. 25.—Union Petroleum Company colorimeter.

Apparatus.—The Union Colorimeter (Fig. 25) is a covered box, blackened inside, with an eyepiece at one end and two rectangular openings at the other, both closed by a frosted glass plate. In the cover of the box at the end opposite the eyepiece are two felt-lined openings to fit closely around 4-oz. oil sample bottles. A slot is provided for the introduction of the color standards.

The color standards are permanent combinations of Lovibond glasses furnished with the instrument. The values and designations are given in Table 7.

Procedure.—Use a daylight lamp of suitable intensity. Secure two clear, colorless 4-oz. sample bottles, 1.5 inch outside diameter and of normal wall thickness. Fill one with pure distilled water and place in the opening next to the color standard slot. Fill the other 4-oz. bottle with the perfectly clear oil to be examined and place this in the other opening. Hold the instrument so that both sample and standard are equally illuminated and insert the color standards until the nearest possible match is obtained. In case the color of the oil lies between two standards, estimate which one is the closest match and report the N. P. A. number, adding " plus " if the sample is darker or " minus " if the sample is lighter than the standard. Thus an oil slightly darker than the Union O-Light Red standard would be reported as " Color, N. P. A. No. 5 plus."

In the case of filtered cylinder oils, these should be mixed with Water White Gasoline in the proportion of 15 volumes of cylinder oil and 85 volumes of gasoline. This mixture should then be examined as described, using and reporting the proper cylinder oil standard, namely one of the following:

A. Extra light filtered.
D. Light filtered.
E. Medium filtered.

The Union Colorimeter will also be found extremely convenient for comparing the colors of oils with standard color oil samples, both being placed in the instrument in 4-oz. bottles of the same correct size.

Determination of Color by Lovibond Tintometer

(Test Method No. 19)

While not entirely satisfactory, this instrument is used to some extent in the United States. It possesses several disadvantages. The field of vision is not optically suitable.

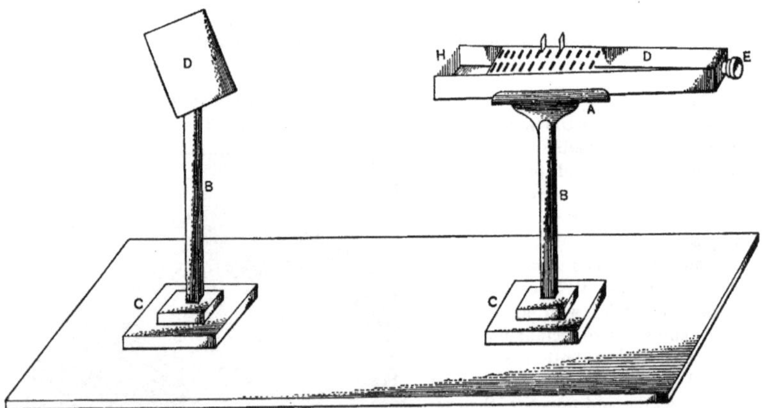

FIG. 26.—Lovibond Tintometer.

There is no standard practice as to length of cell to be used. The Lovibond glasses are hard to procure and are not accurate. A number of glasses, often from different series, are required to match a lubricating oil color and the summation of the value has very little meaning. On the other hand, under uniformly definite conditions, the Lovibond Tintometer is fairly suited to plant-control work. It is particularly valuable for examination of the color of paraffin wax. Its accuracy depends largely upon that of the standard glasses used.

Apparatus.—The Lovibond Tintometer (Fig. 26) consists of a covered box or trough divided longitudinally by a partition terminating in a thin vertical edge opposite an eyepiece at one end. The two channels, thus formed, diverge slightly from the eyepiece end. Both channels contain slots through which the standard color glass slips are inserted. The oil containers are metal cells with glass ends, these cells varying in length from ¼ inch to 18 inches according to the color of the material to be examined. A white reflector is mounted so that it may be set at any desired angle.

The Lovibond glasses are thin slips of tinted glass, each one numbered with color value in arbitrary Lovibond units on the principle of analytical weights. As used with petroleum products the standard glasses belong to four series as follows:

No. 510	Yellow.
No. 500	Amber
No. 200	Red.
No. 1180	Blue.

Procedure.—Set the Tintometer up in a compartment so that a daylight lamp gives the only illumination. If the reflector is used, the lamp should be directly above it but in most cases it is preferable to cover the daylight lamp with an evenly frosted optical glass plate and let the light enter the Tintometer directly.

Select a cell of suitable length. While there is no definite standard, it is convenient to use the following:

	Inches
Illuminating oils	18 or 12
Paraffin wax	6
Pale lubricating oils	2 to 6
Red lubricating oils	1 to 2
Filtered cylinder oils	½
Unfiltered cylinder oils	¼

Fill the cell with oil to be tested and place it on one side of the Tintometer so that the cell walls do not interfere with the field of vision on that side. Match the color as closely as possible, using yellow glasses (510 series) for refined oils and amber (500 series) for waxes and other oils. Use as few glasses as possible. Thus for a color of 15.2, the slips should be three in number, namely 10, 5 and 0.2. The same nominal color value could be made up from ten or more glasses but the shade would not be the same. In case it is not possible to match the color of the oil with the amber (500 series) glasses, add red (200 series) or blue (1180 series) or both, removing ambers as required. In any case use the smallest possible number of glasses.

Record the Lovibond color, giving both length of cell, and sum of glasses of each series separately. Thus, an engine oil examined showed:

Lovibond Color, 2 inch cell.

500 series Amber	50.0
200 series Red	60.0
1180 series Blue	0.5

The color should not be reported as the sum total of all glasses when these belong to different series, for the reason that such a total does not show the actual color of the oil and might apply to several oils of entirely different color. The necessity of giving the cell length is obvious. It is to be noted that colors measured on the same oil in cells of different length are not proportional to the length of the cells.

When the Lovibond is used for taking color of paraffin wax, this should be melted at a temperature not higher than 160° F. and filtered through dry filter paper to remove moisture and adventitious suspended matter. Even with these precautions, the color of re-melted wax is usually slightly darker than the original.

Water and B. S.

While crude petroleum is always necessarily examined quantitatively for water and B. S. (Bottom Sediment), the test methods described in this section are also frequently applied to semi-refined or very viscous petroleum products such as fuel oil, black oil, cylinder stocks and various materials only partly processed. The importance of obtaining a truly representative sample cannot be emphasized too strongly. Whenever possible, the sample should be taken under laboratory supervision.

Water suspended in petroleum products is frequently estimated by centrifuging after dilution with suitable thinning liquids and occasionally by gravity settling. Such methods are suitable only as the basis for a guess. When an accurate determination of water is required, the procedure described under Test Method No. 20 should always be followed.

B. S. (Bottom Sediment) is a semi-solid heterogeneous material of varying consistency which slowly settles out of certain petroleum products, conspicuously crude petroleum. In addition to emulsions of oil and water, it may contain (*a*) sand and similar mineral matter, (*b*) paraffin or petrolatum, (*c*) carbonaceous or insoluble asphaltic matter, (*d*) iron rust, scale and the like, (*e*) inorganic salts, either as solids or in solution in water. Certain facts relative to B. S. should be considered in connection with the selection of the proper method for its determination; (*a*) paraffin and petrolatum are not impurities and should not be included in sediment, (*b*) allowance for B. S. is made on the basis of volume and reports should be made accordingly, (*c*) for all exact work, B. S. must be differentiated from water even though both are undesirable impurities. For ordinary purposes, B. S. may be determined by Test Method No. 22 with sufficient accuracy and with considerably less trouble than by Test Method No. 21.

Determination of Water in Petroleum Products by Distillation

(Test Method No. 20)

This method is suitable for crude petroleum, fuel oil and other petroleum products whose water content requires accurate determination.

Sample.—The sample shall be thoroughly representative of the material to be tested, and the portion of the sample used for the test shall be thoroughly representative of the sample itself. Deviation from this rule shall not be permitted. The difficulties in obtaining proper representative samples are unusually great so that the importance of the sampling cannot be emphasized too strongly.

Apparatus.—*A*. The preferred form shall be that of Dean and Stark [1] as shown in Fig. 27, and shall include the following:

1. Distillation flask, 500 c.c. capacity, of well annealed glass, round bottom.
2. Reflux condenser, water-cooled, glass-tube type.
3. Special graduated distilling tube receiver, of well annealed glass, graduated upward from 0 to 10 c.c. in 0.1 c.c. divisions, accurate to 0.05 c.c.
4. Gas burner or suitable electric heater.

B. An optional form may be used and shall include the following:

1. Distillation flask, 500 c.c. capacity.
2. Condenser, water-cooled, glass-tube type, free from traps retarding complete draining.
3. Receiver of glass, capacity about 125 c.c., standard B. S. tube (Fig. 28) or its equivalent.
4. Gas burner or suitable electric heater.
5. Thermometer, engraved-stem type, to indicate 400° F.

[1] J. Ind. Eng. Chem. 12 (1920) 486.

Diluent.—The diluent used in this method shall be gasoline, free from water. When subjected to distillation, it shall show 5 per cent off at a temperature not above 212° F. (100° C.) nor below 194° F. (90° C.). It shall show 90 per cent off at a temperature not above 410° F. (210° C.). Motor gasoline may be allowed to evaporate until it meets these requirements.

Procedure.—1. Exactly 100 c.c. of the oil to be tested shall be measured in an accurate 100 c.c. graduated cylinder at room temperature and poured into the distillation flask (1). The oil adhering to the walls of the 100 c.c. graduated cylinder shall be transferred to the distillation flask by rinsing with two successive 50 c.c. portions of gasoline the cylinder being allowed to drain each time. The oil and gasoline in the distillation flask shall be thoroughly mixed by swirling the flask with proper care to avoid any loss of material. A boiling stone such as a piece of unglazed porcelain may be introduced for the purpose of preventing bumping during the subsequent distillation.

2. The apparatus shall be assembled in the following manner:

If the preferred form of apparatus is used, the distillation flask shall be connected with the special graduated distilling tube receiver (3) and this receiver with the reflux condenser (2) by means of tight-fitting corks as shown in Fig. 27.

If the optional form of apparatus is used, the vapor outlet tube of the distillation flask shall be inserted into the condenser (2), a tight connection being made by means of a cork. The receiver (3) shall be supported in a suitable position at the other end of the condenser without the use of a cork or other connection. Proper precautions shall be taken to prevent the introduction, into the receiver, of water from any source other than the subsequent distillation process.

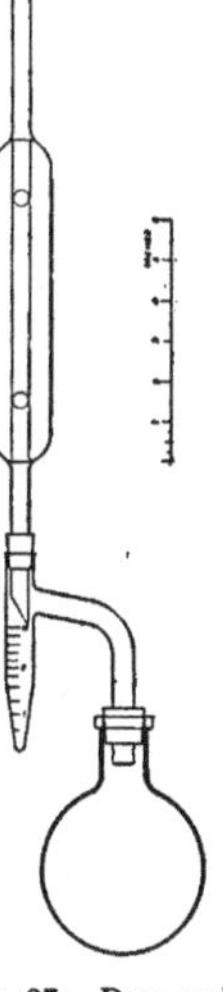

Fig. 27.—Dean and Stark apparatus.

The mouth of the distillation flask shall be closed by a tight-fitting cork supporting the thermometer (5) in such a position that the top of the thermometer bulb is level with the bottom of the vapor outlet tube, where the latter is joined to the neck of the flask.

3. Either form of apparatus being assembled as prescribed, heat shall be applied to the distillation flask and shall be so regulated that the condensed distillate falls from the end of the condenser at the rate of 2 to 4 drops per second.

4. With the preferred form of apparatus, distillation shall be continued at the specified rate until no water is visible on any part of the apparatus except at the bottom of the distilling-tube receiver. This operation usually requires less than one hour. A persistent ring of condensed water in the condenser tube shall be removed by increasing the rate of distillation for a few minutes.

With the optional form of apparatus, distillation shall be continued at the specified rate until the thermometer indicates a vapor temperature of 400° F. and until all condensed water has disappeared from the walls of the distillation flask and of the condenser.

5. With either form of apparatus, the volume of condensed water, measured in the graduated receiver at room temperature. shall be recorded as " — per cent Water, Distillation Method."

Accuracy.—With proper care and attention to details, duplicate determinations of water by this method should not differ from each other by more than 0.2 c.c. of water, provided the graduated receiver is accurate and readable to this degree.

Note.—When the sample to be tested contains more than 10 per cent of water, the volume of material used shall be decreased to that which will yield somewhat less than 10 c.c. of water. Otherwise the procedure shall be conducted as prescribed.

Gravimetric Determination of B. S.

(Test Method No. 21)

This method requires some practice on the part of the operator, chiefly in the preparation of the crucible. It is not intended for routine work except where an accurate measurement of the amount of solid suspended matter is desired. It is particularly valuable for determining the percentage of insoluble material in very dark lubricating oils.

The procedure is as follows: Prepare and ignite a Gooch crucible with an asbestos mat as described in Test Method No. 48.[1] Weigh accurately, in a 200 c.c. Erlenmeyer flask, about 10 grams of the oil to be tested. Add 90 c.c. of either pure or commercial benzol, previously filtered through paper if not perfectly clear. Close the flask tightly with a softened cork and shake vigorously until the contents are thoroughly mixed. Allow the mixture to stand at room temperature for one hour. Filter the solution through the weighed Gooch crucible, using the same apparatus and precautions described under Test Method No. 48. Transfer the precipitate from the flask to the crucible using fresh benzol and wash with small portions of fresh benzol until the filtrate is colorless. Dry the crucible at 212–220° F. for one hour, cool in a desiccator and weigh again. The increase in weight is due to the solid insoluble matter and should be calculated to percentage by weight of the oil used. The result should be reported as " B. S., Gravimetric Method." The results obtained by this method are nearly always lower than those obtained by Test Method No. 22, because B. S. measured by volume always contains some liquid, either water or oil or both.

The mineral matter in the B. S., may be determined by igniting the crucible containing the precipitate, cooling in a desiccator and weighing. The loss of weight in this operation is organic matter and the remaining increase of weight over the original weight of the crucible is the mineral matter.

Determination of B. S. and Estimation of Water by Centrifuge

(Test Method No. 22)

This is the routine method to be followed in control work. It involves the use of " 90 per cent benzol " as a diluent in place of the carbon bisulphide gasoline mixture often prescribed. While benzol is approved by the U. S. Committee on Standardization of Petroleum Specifications,[2] its selection has been criticised by some authorities. The reasons for using benzol are therefore briefly summarized:

1. 90 per cent benzol is cheaper than carbon bisulphide, less offensive as to odor and less inflammable.

2. Although the carbon bisulphide-gasoline mixture gives a somewhat lower B. S. on some asphaltic petroleum products there is no evidence that the results obtained with benzol are incorrect.

3. The specific gravity of the usual carbon bisulphide-gasoline mixture is so near that of water, even at 100° F., that it cannot be expected to yield a satisfactory separation of solid matter whose gravity is approximately the same as that of the mixture.

Scope.—This method may be used for crude mineral oils and fuel oils. A centrifuge method for " Water and Sediment " is not entirely satisfactory b cause the amount of water obtained is nearly always lower than the ac ual water content. Nevertheless, on account of the wide use of the centrifuge for this purpose, it is desirable that the method of making the determination be standardized as far as possible. It must be clearly understood that the reading of the centrifuge tube includes both the sedi-

[1] Page 715.

[2] Bulletin No. 5, December 29,1920.

ment and the precipitated water. Accurate determination of water content if desired, should be made by the distillation method for water.

Sample.—The sample shall be thoroughly representative of the material in question and the portion used for the test shall be thoroughly representative of the sample itself. Deviation from this rule shall not be permitted. The difficulties in obtaining proper representative samples for this determination are unusually great, so that the importance of sampling cannot be too strongly emphasized.

Apparatus.—1. Centrifuge.—The centrifuge shall be capable of whirling at the required speed at least two 100-c.c. centrifuge tubes filled with water. It shall be of sound design and rugged construction so that it may be operated without danger. The tube carriers shall be so designed that the glass centrifuge tubes may be cushioned with water, rubber or other suitable material. The tube holders shall be surrounded during operation by a suitable metal shield or case, strong enough to eliminate danger if any breakage occurs.

The preferred form of centrifuge shall have a diameter of swing (tip to tip of whirling tubes) of 15 to 17 inches and a speed of at least 1500 R.P.M. or equivalent. If the available centrifuge has a diameter of swing varying from these limits, it shall be run at the proper speed to give the same centrifugal force at the tips of the tubes as that obtained with the preferred form of centrifuge. The proper speed shall be calculated from the following formula in which D represents diameter of swing (tip to tip of whirling tubes) of the centrifuge used.

$$\text{R.P.M.} = 1500 \times \sqrt{\frac{16}{D}}.$$

2. Centrifuge tubes.—These tubes shall be made of suitable glass and thoroughly annealed. The total capacity shall be about 125 c.c. and the mouth shall be suitably constricted for closing with a cork. The graduations shall be clear and distinct, reading upward from the bottom of the tube as follows:

Range	Scale divisions	Limit of error	Numbered
0–3 c.c.	0.1 c.c.	0.05 c.c.	1, 2, 3 c.c.
3–5 c.c.	0.5 c.c.	0.2 c.c.	4, 5 c.c.
5–10 c.c.	1.0 c.c.	0.5 c.c.	6, 8, 10 c.c.
10–25 c.c.	5.0 c.c.	1.0 c.c.	15, 20, 25 c.c.
50–100 c.c.	50.0 c.c.	1.0 c.c.	50, 100 c.c.

The shape is optional provided it does not conflict with the other requirements. Satisfactory types are shown in Fig. 28.

3. Water or oil bath of depth sufficient for immersing the centrifuge tubes in a vertical position to the 100 c.c. mark. Means shall be provided for heating this bath to 100° F.

Procedure.—1. Exactly 50 c.c. of 90 per cent Benzol shall be measured into each of two centrifuge tubes (2) and exactly 50 c.c. of the oil to be tested shall then be added to each. The centrifuge tubes shall be tightly stoppered and shall be shaken vigorously until the contents are thoroughly mixed. The temperature of the bath (3) shall be maintained at 100° F. and the centrifuge tubes shall be immersed therein to the 100-c.c. mark for ten minutes.

2. The two centrifuge tubes shall then be placed in the centrifuge (1) on opposite sides and shall be whirled at a rate of 1400–1500 R.P.M. or equivalent, for ten minutes.

The combined volume of water and sediment at the bottom of each tube shall be read and recorded, estimating to 0.1 c.c. if necessary. The centrifuge tubes shall then be replaced in the centrifuge, again whirled for ten minutes as before and removed for reading the volume of water and sediment as before. This operation shall be repeated until the combined volume of water and sediment in each tube remains constant for three consecutive readings. In general, not more than four whirlings are required.

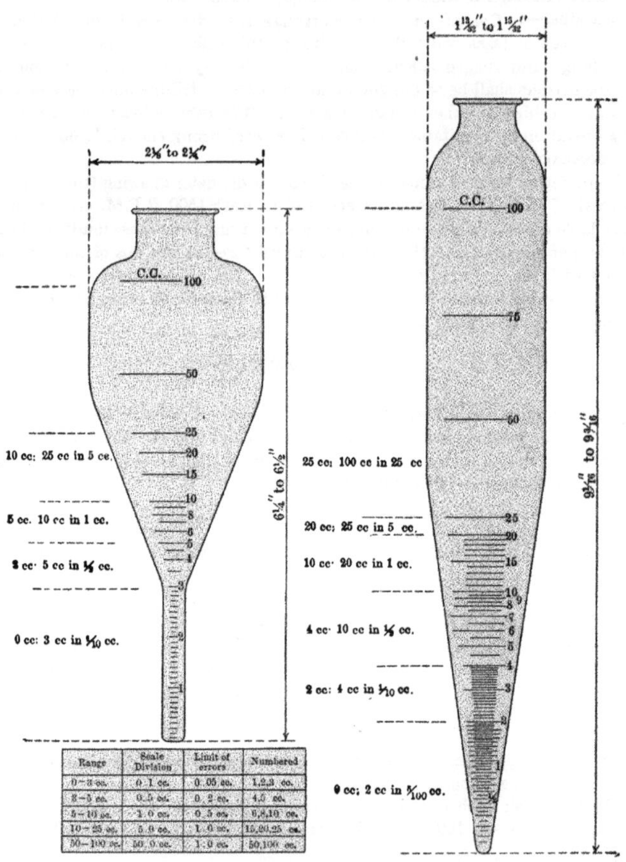

Range	Scale Division	Limit of errors	Numbered
0–3 cc.	0.1 cc.	0.05 cc.	1,2,3 cc.
3–5 cc.	0.5 cc.	0.2 cc.	4,5 cc.
5–10 cc.	1.0 cc.	0.5 cc.	6,8,10 cc.
10–25 cc.	5.0 cc.	1.0 cc.	15,20,25 cc.
50–100 cc.	50.0 cc.	1.0 cc.	50,100 cc.

Fig. 28.—Standard centrifuge B. S. tubes.

3. The combined total volume of water and sediment shall be read on each tube, estimating to 0.1 c.c. if necessary. The sum of the two readings shall be recorded as per cent Water and Sediment, Centrifuge Method.

Accuracy.—With care and proper attention to details, duplicate determinations of Water and Sediment by this method should not differ by more than 0.2 c.c., provided the centrifuge tubes are accurate and readable to this degree.

END-POINT DISTILLATION OF GASOLINE, NAPHTHA, AND KEROSENE

(Test Method No. 23)

This method is a modification of that of the Bureau of Mines[1] and of the American Society for Testing Materials Standard Method D-28-17.[2] The form given here is that recommended by a joint committee of producers and consumers and will probably be adopted by the various organizations interested in the subject. The results obtained will agree closely with those obtained by the U. S. Government method[3] published before the present revision of the method.

Apparatus.—1. Standard Engler 100-c.c. distilling flask.

2. Condenser.
3. Shield.
4. Ring support and hard asbestos boards.
5. Gas burner or electric heater.
6. Distillation thermometer.
7. Long-fiber absorbent cotton.
8. Corks.
9. Blotting paper.
10. 100-c.c. graduate.

Flask.—The standard 100-c.c. Engler flask is shown in Fig. 29, the dimensions and allowable tolerances being as follows:

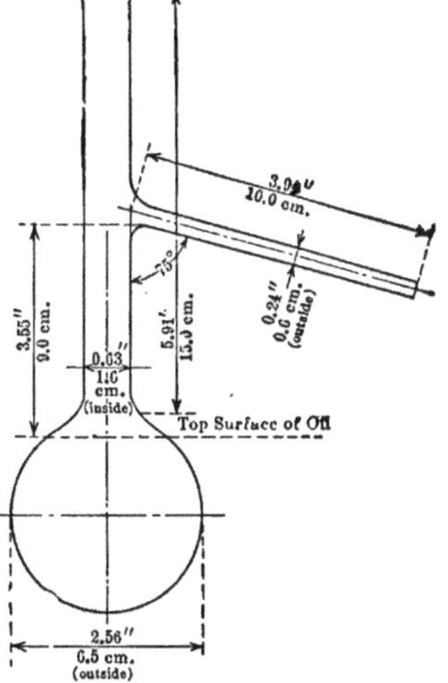

FIG. 29.—Engler distilling flask.

	Cm.	Inches	Tolerances (cm.)
Diameter of bulb (outside)	6.5	2.56	0.2
Diameter of neck (inside)	1.6	0.63	0.1
Length of neck	15.0	5.91	0.4
Length of vapor tube	10.0	3.94	0.3
Diameter of vapor tube (outside)	0.6	0.24	0.05
(inside)	0.4	0.16	0.05
Thickness of vapor tube wall	0.1	0.04	0.05

The position of the vapor tube shall be 9 cm. (3.55 inches) ±3 mm. above the surface of the liquid when the flask contains its charge of 100 c.c. The tube is approximately in the middle of the neck and set at a 75° angle (Tolerance ±3°) with the vertical.

Condenser.—The condenser (Fig. 30) consists of a $\frac{9}{16}$ inch OD number 20 Stubbs Gage seamless brass tube 22 inches long. It is set at an angle of 75° from the perpendicular and is surrounded with a cooling bath 15 inches long, approximately 4 inches wide by 6 inches high. The lower end of the condenser tube is cut off at an acute angle,

[1] Technical Paper 214, E. W. Dean (1919).

[2] A. S. T. M. 1918 Standards, 606.

[3] Bulletin No. 5, U. S. Committee on Standardization of Petroleum Specifications, pp. 5–7 (Dec., 1920).

and curved downward for a length of 3 inches and slightly backward so as to insure contact with the wall of the graduate at a point 1 to 1¼ inches below the top of the graduate when it is in position to receive the distillate.

Shield.—The shield (Fig. 30) is made of approximately 22-gage sheet metal and is 19 inches high, 11 inches long and 8 inches wide, with a door on one narrow side, with two openings, 1 inch in diameter, equally spaced, in each of the two narrow sides and with a slot cut in one side for the vapor tube. The centers of these four openings are 8½ inches below the top of the shield. There are also three ½-inch holes in each of the four sides with their centers 1 inch above the base of the shield.

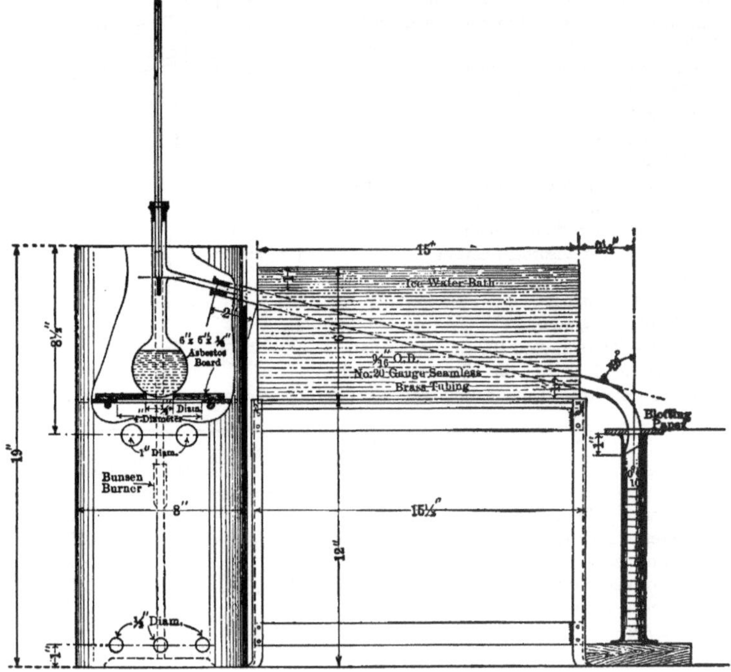

Fig. 30.—Distilling apparatus assembly.

Ring support and hard asbestos boards.—The ring support is of the ordinary laboratory type, 4 inches in diameter, and is supported on a stand inside the shield. There are two hard asbestos boards: (1) 6 by 6 by ¼ inch, with a hole 1¼ inches[1] in diameter in its center, the sides of which shall be perpendicular to the surface: (2) asbestos board to fit tightly inside the shield, with an opening 4 inches in diameter concentric with the ring support. These are arranged as follows: The asbestos board (2) is placed on the ring and the smaller asbestos board (1) on top so that it may be moved in accordance with the directions for placing the distilling flask. Direct heat is applied to the flask only through the 1¼ inch opening in the asbestos board (1).

[1] When distilling petroleum products having an end-point above 470° F., the hole in the asbestos board shall be 1½ inches in diameter.

Gas Burner or Electric Heater

1. Gas burner.—The burner is so constructed that sufficient heat can be obtained to distill the product at the uniform rate specified below. The flame should never be so large that it spreads over a circle of diameter greater than 3½ inches on the under surface of the asbestos board. A sensitive regulating valve is a necessary adjunct, as it gives complete control of heating.

2. Electric heater.—The electric heater, which may be used in place of the gas flame, shall be capable of bringing over the first drop within the time specified below when started cold, and of continuing the distillation at the uniform rate. The electric heater shall be fitted with an asbestos board top ⅛ to ¼ inch thick, having a hole 1¼ inches in diameter in the center. When an electric heater is employed the portion of the shield above the asbestos board shall be the same as with the gas burner.

Thermometers.—The thermometer shall comply with one of the following specifications:

Low-Distillation Thermometer Specifications

Type: Glass engraved.
Total length: 381 mm.
Glass; Stem: Plain front, enamel back, suitable thermometer tubing, Diameter 6 to 7 mm.
Bulb: Corning normal, Jena 16, III, or equally suitable thermometric glass.
Length 10 to 15 mm.
Diameter 5 to 6 mm.
Actuating liquid: Mercury.
Range: 30° F. or 0° C., to 580° F., or 300° C.
Immersion: Total.
Range distance from bottom of bulb: 100 to 110 mm.
Range distance from top of tube: 35 to 45 mm.
Ice point: No.
Contraction chamber: No.
Expansion chamber: No.
Filled: Nitrogen gas.
Top finish: Glass ring.
Graduating: All lines, figures and letters clear-cut and distinct.
Scale graduated in 2° F., or 1° C.
Scale numbered every 20° F., or 10° C.
Special markings: Low distillation,
Serial number and manufacturer's name or trade mark engraved on the tube.
Accuracy: Error at any point on scale shall not exceed one-half smallest scale division.
Test for permanency of range: After being subjected to a temperature of 560° F., or 290° C., for twenty-four hours the accuracy shall be within the limit specified.
Points to be tested for certification: 32°, 212°, 400°, 570° F., or 0°, 100°, 200°, 300° C.

High Distillation Thermometer Specifications

Type: Glass engraved.
Total length: 381 mm.
Glass; Stem: Plain front, enamel back, suitable thermometer tubing. Diameter 6 to 7 mm.

Bulb: Corning normal, Jena 16, III, or equally suitable thermometric glass.
Length 10 to 15 mm.
Diameter 5 to 6 mm.

Actuating liquid: Mercury.
Range: 30° F., or 0° C., to 760° F., or 400° C.
Immersion: Total.
Range distance from bottom of bulb: 25 to 35 mm.
Range distance from top of tube: 30 to 45 mm.
Ice point: No.
Contraction chamber: No.
Expansion chamber: No.
Filled: Nitrogen gas.
Top finish: Glass ring.
Graduating: All lines, figures and letters clear-cut and distinct.
Scale graduated in 2° F., or 1° C. divisions.
Scale numbered every 20° F., or 10° C.
Special markings: High Distillation.
Serial number and manufacturer's name or trade mark engraved on the tube.
Accuracy: Error at any point on scale shall not exceed one smallest scale division up to 700° F., or 370° C.
Test for permanency of range: After being subjected to a temperature of 700° F., or 370° C., for twenty-four hours the accuracy shall be within limit specified.
Points to be tested for certification: 32°, 212°, 400°, 700° F. or 0°, 100°, 200° 370° C.

Graduate.—The graduate shall be of the cylindrical type, of uniform diameter, with a pressed or molded base and a lipped top. The cylinder shall be graduated to contain 100 c.c., and the graduated portion shall be not less than 7 inches nor more than 8 inches long. It shall be graduated in single cubic centimeters and each fifth mark shall be distinguished by a longer line. It shall be numbered from the bottom up at intervals of 10 c.c. The distance from the 100 c.c. mark to the rim shall be not less than 1¼ inches nor more than 1¾ inches. The graduations shall not be in error by more than 1 c.c. at any point on the scale.

Procedure.—1. Fill the condenser bath with cracked ice [1] and add enough water to cover the condenser tube. The temperature shall be maintained between 32° F., and 40° F.

2. Swab the condenser tube to remove any liquid remaining from the previous test. A piece of clean, soft cloth attached to a cord or copper wire may be used for this purpose.

3. Cover the bulb of the distillation thermometer uniformly with long-fiber absorbent cotton weighing not less than 3 nor more than 5 milligrams. A fresh portion of clean cotton shall be used for each distillation.

4. Measure 100 c.c. of the product at 55° to 65° F., in the 100 c.c. graduated cylinder and transfer directly to the Engler flask. Do not permit any of the liquid to flow into the vapor tube.

5. Fit the thermometer, provided with a cork, tightly into the flask so that it will be in the middle of the neck and so that the lower end of the capillary tube is on a level

[1] Any other convenient cooling medium may be used.

with the inside of the bottom of the vapor outlet tube at its junction with the neck of the flask.

6. Place the charged flask in the $1\frac{1}{4}$-inch opening in the 6 by 6-inch asbestos board (1) with the vapor outlet tube inserted into the condenser tube. A tight connection is made by means of a cork through which the vapor tube passes. The position of the flask shall be adjusted so that the vapor tube extends into the condenser tube at least 1 inch and not more than 2 inches.

7. Without drying the graduated cylinder used in measuring the charge, place it at the outlet of the condenser tube in such a position that the condenser tube shall extend into the graduate at least 1 inch but not below the 100 c.c. mark. Unless the room temperature is between 55° F. and 65° F., the receiving graduate shall be immersed up to the 100 c.c. mark in a transparent bath maintained between these temperatures. Cover the top of the graduate during the distillation closely with a piece of blotting paper or its equivalent, cut so as to fit the condenser tube tightly.

Distillation.—When everything is in readiness, heat shall be applied and so regulated that the first drop of condensate falls from the condenser in not less than five nor more than ten minutes. When the first drop falls from the end of the condenser the reading of the distillation thermometer shall be recorded as the *initial boiling point.* The receiving cylinder shall then be moved so that the end of the condenser tube shall touch the side of the cylinder. The heat shall then be so regulated that the distillation will proceed at a uniform rate of not less than 4 nor more than 5 c.c. per minute. The reading of the distillation thermometer shall be recorded when the level of the distillate reaches each 10 c.c. mark on the graduate.

After the 90 per cent point has been recorded, the heat may be increased because of the presence of the heavy ends which have high boiling points. However, no further increase of heat should be applied after this adjustment. The 4 to 5 c.c. rate can rarely be maintained from the 90 per cent point to the end of the distillation, but in no case should the period between the 90 per cent and the end-point be more than five minutes.

The heating should be continued until the mercury reaches a maximum and starts to fall consistently. The highest temperature observed on the distillation thermometer shall be recorded as the *maximum temperature* or end-point. Usually this point will be reached after the bottom of the flask has become dry.

The total volume of the distillate collected in the receiving graduate shall be recorded as the *recovery*.

Pour the cooled residue from the flask into a small cylinder graduated in $\frac{1}{10}$ c.c. Measure when cool and record this volume as *residue.*

Calculate the difference between 100 c.c., and the sum of the recovery and the residue, and then record this as *distillation loss.*

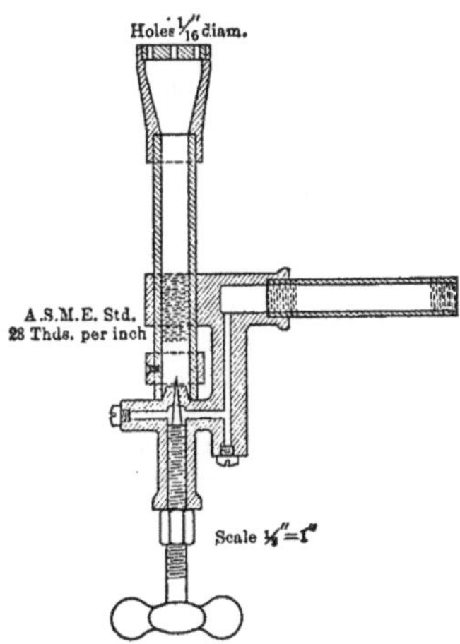

Fig. 31.—The Atlantic Refining Company Burner.

Accuracy.—With proper care and attention to detail, duplicate results obtained for initial boiling point and maximum temperature, respectively, should not differ from each other by more than 6°.

Note.—The small gas burner shown in Fig. 31, is extremely well adapted for use in gasoline distillations [1] and also in flash- and fire-point determinations. It consists of a special bronze casting with a very sensitive needle valve and a screw adjustment for regulating the air-gas mixture.

Modified Assay Distillation

(Test Method No. 24)

Crude petroleum, gas oils and other products containing constitutents boiling above 572° F. cannot be distilled to dryness as in Test Method No. 23 without decomposition. Furthermore, the volumes of distillate obtained from a 100 c.c. charge are too small to permit further examination. The object of an assay distillation is to obtain information as to the quality as well as the approximate yields of the distillates. The distillation of materials boiling above 572° F. must be conducted in a vacuum or with a lively flow of steam through the oil. Such steam or vacuum distillation with small quantities requires great skill and suitable apparatus [2] and will not be described here. Complete data as to yields, quality and necessary refining operations can be obtained only by distillation of not less than 10 gallons in miniature semi-plant scale apparatus. The following method is satisfactory for ascertaining the yields and quality of gasoline and illuminating oils from crude petroleum or its distillates.

Apparatus.—1. Metal [3] still of about 1300 c.c. capacity, preferably of the vertical cylinder type, about 4 inches inside diameter. The vapor line should be a vertical section of pipe or metal tubing, about 1 inch inside diameter and about 3½ inches long. A convenient form is a section of 1 inch steel pipe with a 1 by ½ by ½ inch tee at the top. The side opening of the tee should be connected to the condenser by a short section of ½ inch pipe and the top opening should be fitted with a stuffing box suitable for bolding the distillation thermometer. The lid of the still should be equipped with a vertical thermometer well, extending into the still to about ½ inch from the bottom. Unless the still is seamless all joints should be welded and if the lid is removable, a soft asbestos gasket should be used. The still should be held firmly in position on a strong rigid support.

2. Condenser, preferably of metal, straight inclined tube type, ½ inch inside diameter and about 40 inches long.

3. Two 720° F. distillation thermometers.

4. Efficient gas burner.

5. Graduated cylinders, 100 c.c. capacity.

Procedure.—Charge the still with 1000 c.c. of the oil to be examined, using a long-stem funnel inserted into the vapor line to below the lid of the still. Connect the still to the condenser and put the two thermometers in position, one in the well and the other in the center of the vapor line with the top of the thermometer bulb ¼ inch below the bottom of the side outlet. Do not tighten the stuffing box around this thermometer too tightly at first; use very soft asbestos packing. Supply cold water to the condenser and place a 100 c.c. graduated cylinder in position to receive the distillate.

Apply heat gradually, watching the well thermometer closely. If the oil contains water, serious foaming is liable to occur at a still temperature of 200–250° F. unless distillation begins first. As the still temperature rises, distillation will begin. The vapor temperature (reading of the thermometer in the vapor tube) at the instant the first drop falls from the end of the condenser is the approximate over-point of the material

[1] E. W. Dean, Jour. Ind. Eng. Chem., **10** (1918), 823.

[2] See Bureau of Mines, Bulletin **19** and **25**; also Tech. Paper **74**.

[3] Glass stills should never be used for distilling more than 200 c.c. of petroleum. Failure to obey this rule has been the occasion of many serious accidents.

and should be recorded. The rate of distillation should then be held at 10 c.c. per minute as closely as possible.

The method used in making cuts may be varied according to the information desired and according to the nature of the material under examination. In any case the vapor temperature should be recorded as each 5 per cent (50 c.c.) of distillate is collected in the receiver. Two methods in common use are (*a*) change the receiver for each 100 c.c. (10 per cent cuts) and (*b*) make three cuts corresponding to gasoline, kerosene and heavy burning oil distillates. The temperatures selected for these three cuts will vary according to individual preference but the following indicate a satisfactory choice:

Over to 392° F. Gasoline distillate.
392 to 527° F. Kerosene distillate.
527 to 572° F. Heavy burning oil distillate or light gas oil.

The distillation should be stopped at 572° F. vapor temperature unless it is desired to crack the residuum, in which case distillation may be carried further, but the temperature of the condenser should be raised to a point high enough to prevent any paraffin wax from chilling in the condenser tube.

The size of cuts obtained by either method is sufficient to permit fairly thorough examination. Among the more important tests which give valuable information are gravity, end-point distillation, color, odor and sulphur determination.

The residuum may also be examined for gravity, viscosity, pour, color by reflected light and carbon residue. An intelligent estimate of the value of this residuum for purposes other than fuel, cannot be made without further distillation which is not considered satisfactory with so small a quantity of material.

Melting Point

Determination of Melting Point by Saybolt Wax Method

(Test Method No. 25)

This method is the one in common use for melting point of paraffin wax and has been adopted by the U. S. Committee on Standardization of Petroleum Specifications.[1] It gives results which differ by not more than 0.25° F. from those obtained by the so-called "English Method"[2] or by the Solidification Method (Test Method No. 26). It is really a "freezing-point" method but is too well known under the name given to permit change.

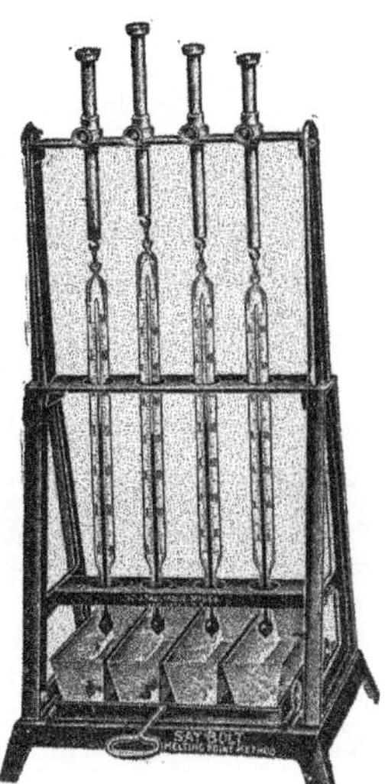

Fig. 32.—Saybolt improved wax tester.

Apparatus.—The Saybolt Improved Wax Tester (Fig. 32) consists of a stand supporting thermometers and a movable car with metal containers for wax. The car is fitted with four flanged wheels rolling over a double track in a linear horizontal reciprocating motion.

The amplitude of travel is about 6 inches. The car is recessed on top to hold firmly in position four sheet-metal boxes or "boats," having the following dimensions:

[1] Bulletin No. 5, pp. 33–34, Dec. 29, 1920.
[2] See Redwood, Treatise on Petroleum, 3rd Edition, Vol. II, p. 308. Also Test Method No. 27.

Length at top	$7\frac{3}{4}$ inches
Length at bottom	$6\frac{1}{2}$ inches
Width at top	$1\frac{9}{16}$ inches
Width at bottom	$\frac{7}{8}$ inch
Depth	$1\frac{5}{8}$ inches

The thermometer supports are adjustable as to height, with set screws for fastening. The thermometers should have a range of 80–150° F., graduated in $\frac{1}{4}$° divisions, numbered every 5°. The engraved-stem type of thermometer is preferable. If this is used, the bulb should be approximately $\frac{1}{4}$ inch diameter and not over $\frac{7}{16}$ inch long. Where the floating type thermometer is used, the bulb should be spherical, approximately $\frac{9}{16}$ inch diameter. In either case the thermometer should be scaled for $\frac{3}{4}$ inch partial immersion. The accuracy should be $\frac{1}{4}$°. An expansion bulb at the top is desirable. Over-all length is 10–12 inches.

Procedure.—Place the car and metal cells in proper position and make sure that the car rolls smoothly through its full amplitude of travel. Four determinations are made at a time if desired.

Adjust the thermometer height so that the bottom of the bulb is $\frac{1}{4}$ to $\frac{3}{8}$ inch above the bottom of the metal box. Heat the wax to be examined to a temperature about 10° F. above the expected melting point, using a clean glass or metal beaker. Pour the melted wax into the "boat" to a height of $1\frac{1}{8}$ inches corresponding to the immersion scale line. Start moving the car and boats back and forth in a smooth, uniform motion, making 15 to 20 complete oscillations per minute. Note and record the temperature every thirty seconds. As soon as three successive readings are the same, this temperature is the "Melting Point, Saybolt Wax Method."

Fig 33.—Nichols wax melting-point recorder.

Precautions.—Wax must not be overheated. The use of a steam or hot water bath is preferred and the sample should be stirred during melting and removed from the bath as soon as it is completely melted.

Where several samples are tested at once, they should be started at times and temperatures such that careful attention can be given to each one as it nears the melting point.

As the melting point on each sample is obtained, the thermometer should be removed to avoid breakage as the wax becomes hard. The chilled sample may be removed by turning the box upside down and tapping it gently against the top of the work-table.

Room temperature should be about 40 to 50° F. below the expected melting point.

Optional Procedure.—In case the Saybolt apparatus is not available, or in case the amount of material is too small, good results may be obtained by stirring the melted wax slowly with the wax thermometer described. The wax container may be a small beaker, evaporating dish or even a small crucible where very little sample is available. In the latter case, the test should be carried out at a temperature only about 25° F. below the expected melting point. The thermometer should not touch the sides of the container and should be kept immersed to the scale mark as closely as possible.

A modification [1] of the Saybolt Wax Melting Point Tester which has given good results in routine work is shown in Fig. 33. The metal boat (*A*) is supported on a plate sliding in guides on a platform (*B*). This platform can be raised or lowered by the eccentric cams (*C*) operated by a milled wheel (*D*). The sliding plate carrying the boat is moved back and forth by the crank (*E*) operated through a worm drive by a small electric motor. A sensitive recording thermometer (*F*) carries a chart graduated in ½° divisions from 100–150° F. This chart has four sectors each representing fifteen minutes of time, the maximum time required for any melting point. The lever (*G*) permits starting and stopping the clock mechanism. The thermometer bulb is of large size to permit great sensitiveness and the necessary correction after three years constant use is less than ½° F.

The procedure with this apparatus is as follows: The melted wax at a temperature not higher than 140° F. is poured into the boat to the proper level and the boat is placed on the support and raised by the cams to the position of maximum height. The slide is oscillated by starting the motor and the clock mechanism is started at the same time. There is obtained in this way a graphic record of the temperature-time cooling curve and from this the actual melting point may be read directly. Each chart is good for four determinations and may be filed as the original record after marking with proper identification. The only attention necessary after starting is to lower the boat before the wax becomes hard enough to put a strain on the thermometer bulb.

Determination of Melting Point of Wax by Solidification Method

(Test Method No. 26)

This method is more accurate than Test Method No. 25 on account of the better control of rate of cooling. As here described, it is a slight modification of the method used at the Bureau of Standards.[2] A method similar to this will probably be adopted by a joint committee of producers and consumers now considering the subject.

Definition.—The melting point[3] of paraffin wax is the temperature at which melted paraffin wax, when allowed to cool under definite specified conditions, first shows a minimum rate of temperature change.

Apparatus (Fig. 34).—1. *Wax container.*—Test-tube of standard form, 25 mm. (1 inch) outside diameter and 100 mm. (4 inches) long. It may be marked with a filling line, 2 inches above the bottom. This test-tube shall be closed by a tightly fitting cork having two openings, one at the center for the melting-point thermometer and the other for a stirrer at one side of the center. The opening for the stirrer may be lined with glass or metal tubing to act as a guide for the stirrer.

2. *Air-bath.*—Suitable water-tight cylinder, 2 inches inside diameter and 4½ inches deep. This air-bath shall be provided with a tightly fitting cork having a central opening for holding the test-tube (1) firmly in a vertical position in the center of the air-bath.

[1] Devised by H. M. Nichols. The Atlantic Refining Company, Philadelphia, Pa.

[2] Scientific Paper No. 340, Wilhelm and Finkelstein. A Standardized Method for the Determination of Solidification Points. (1919)

[3] The so-called "American Melting Point" is arbitrarily taken as 3° F. higher than the A. S. T. M. Wax Melting Point.

3. *Water-bath.*—Suitable cylinder, 5¼ inches inside diameter and 6 inches deep. This water-bath shall be provided with a suitable cover and with the guides and fasteners necessary to hold the air-bath (2) firmly in a vertical central position such that the sides and bottom of the air-bath shall be surrounded by a layer of water 1½ inches thick. The water-bath cover shall have a slot for introduction of a suitable stirrer and shall have an opening for the bath thermometer so that the latter may be suspended in a vertical position ¾ inch from the outside wall of the water-bath. Air-bath, water-bath and water-bath cover may be conveniently made of metal in one assembly as shown in Fig. 34.

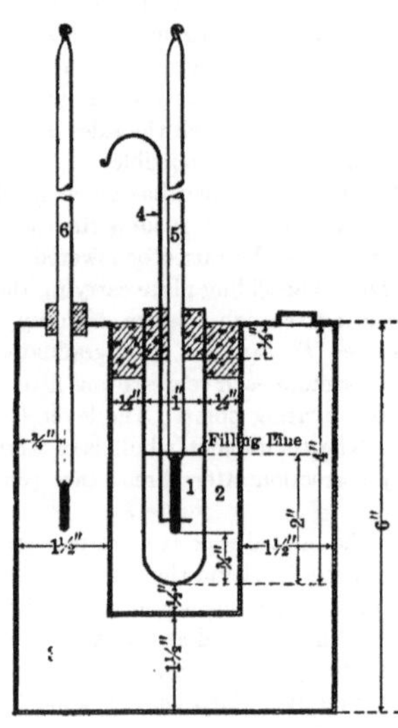

Fig. 34.—Wax melting-point apparatus.

4. *Stirrer in test-tube* (1).—Brass or copper wire, ⅛ inch diameter and about 12 inches long. A circular loop, ½ inch diameter, shall be formed at one end of this wire in such a manner that the loop lies in a horizontal plane when the rest of the wire is in a vertical position. The stirrer thus formed shall be passed through the proper opening in the test-tube cork and the upper end may then be bent into a shape convenient for holding.

5. Paraffin wax melting-point thermometer to conform to the following specifications:

Type: Glass engraved.
Total length: 368 mm.
Glass; Stem: Plain front, enamel back, suitable thermometer tubing.
Diameter: 6 to 7 mm.
Bulb: Corning normal, Jena 16, III or equally suitable thermometric glass.
Length: 28 mm. maximum.
Diameter: not greater than stem.
Actuating liquid: Mercury.
Range: 80° to 160° F.
Immersion: 3.125 inches. The words "3.125 inches immersion" shall be engraved on tube.
Range distance from bottom of bulb: 105 to 115 mm.
Range distance from top of tube: 25 to 40 mm.
Ice point: No.
Contraction chamber: Yes. Top to be not more than 41 mm. from bottom of bulb.
Expansion chamber: Yes. To hold 212° F.
Filled: Nitrogen gas.
Top finish: Plain.
Graduating: All lines, figures and letters clean-cut and distinct.
Scale graduated in 0.2° F.
Scale numbered every 2° F.
Special markings: Pffe. M. P.
Serial number and manufacturer's name or trade mark engraved on the tube.

Accuracy: Error at any point on scale shall not exceed one smallest scale division.

Points to be tested for certification: 80°, 100°, 120°, 140°, 160° F.

5. Bath thermometer of any suitable type, accurate to 2° F.

Procedure.—An average sample of the wax to be tested shall be melted in a suitable container in a water-bath whose temperature shall be not more than 35° F. above the approximate melting point of the wax sample. Direct heat, such as a flame or hot plate, shall not be used and the wax sample shall not be held in the melted condition any longer than necessary.

The test-tube (1) shall be filled with melted wax to a height of 2 inches. The test-tube cork, carrying the stirrer (4) and the melting-point thermometer (5) with the 3.125-inch immersion line at the under surface of the cork, shall be inserted into the test-tube for a distance of ½ inch. The lower end of the thermometer bulb shall then be ⅜ inch from the bottom of the test-tube.

The air-bath (2) being in its proper position in the water-bath (3), the latter shall be filled to within ½ inch of the top with water at a temperature 15–20° F. below the approximate melting point of the wax sample.

The test-tube containing the melted wax, with wax stirrer and thermometer in place, shall be inserted into the air-bath in a central vertical position so that the bottom of the test-tube is ½ inch from the bottom of the air-bath. The temperature of the water-bath shall be adjusted with stirring if necessary, so that it shall be lower than the temperature of the wax sample, by not more than 30° F. and not less than 25° F., when the wax sample has cooled to a temperature 10° F. above its approximate melting point.

When these conditions have been obtained, temperature adjustment and stirring of the water-bath shall be discontinued. The wax shall be stirred continuously during the remainder of the test, the stirring loop being moved up and down throughout the entire length of the test-tube in a steady motion at the rate of 20 complete cycles per minute. The melting-point thermometer reading, estimated to 0.1° F., shall be observed and recorded every 30 seconds. The temperature of the wax will fall gradually at first, will then become almost constant and will then again fall gradually.

The melting-point thermometer reading, estimated to 0.1° F., shall be observed and recorded every 30 seconds, for at least three minutes after the temperature again begins to fall after remaining almost constant. The record of temperature readings shall then be inspected and the average of the first four readings that lie within a range of 0.2° F. shall be considered as the uncorrected melting point. This temperature shall be corrected if necessary for error in the thermometer scale and the corrected temperature shall be reported as " Wax Melting Point."

Accuracy.—Duplicate determinations on the same sample should not differ by more than 0.2° F.

Determination of Melting Point of Wax, American Method

(Test Method No. 27)

This method is decidedly unsatisfactory and is given only because most wax sales in the United States are on the basis of melting point, " American Method." The most reliable method of obtaining the " American " melting point and the one used by practically all American laboratories is to add 3° F. to the melting point obtained by Test Methods 25 or 26. While there is considerable disagreement as to just how an " American " melting point is actually determined, the following directions are the most authoritative.

Apparatus.—Hemispherical cup $3\frac{3}{4}$ inches in diameter.

Wax-test thermometer graduated in $\frac{1}{4}$° F. divisions, with spherical bulb about $\frac{9}{16}$ inch in diameter.

Procedure.—Heat the wax to a temperature about 15° F. above the expected melting point. Fill the cup three-quarters full with this melted wax. Suspend the thermometer in the center of the cup so that the bulb is three-quarters immersed. Allow the wax to cool without any disturbance. Note and record as "American Melting Point" the temperature when a film of solidified wax extends from the thermometer bulb to any point at the edge of the cup. The test should be performed in a good light, thrown on the surface of the wax. Differences of $\frac{3}{4}$° F. between tests on the same sample are not unusual.

"English Method."—The directions for this test are as follows: A test-tube about 1 inch in diameter is filled with about 2 inches of melted wax. The wax is allowed to cool slowly while constantly stirred with a small thermometer. The temperature at which the thermometer remains stationary for a short time is the English melting (setting) point. It is obvious that these directions are extremely indefinite and "English Melting Point" will be more accurately determined by Test Methods 25 or 26. American Melting Point is 3° F. higher than English.

Determination of Melting Point by Saybolt Petrolatum Method

(Test Method No. 28)

This method has been used for many years by large producers of petrolatum. When carried out as described, determinations on the same sample by different operators should agree within $\frac{1}{2}$° F.

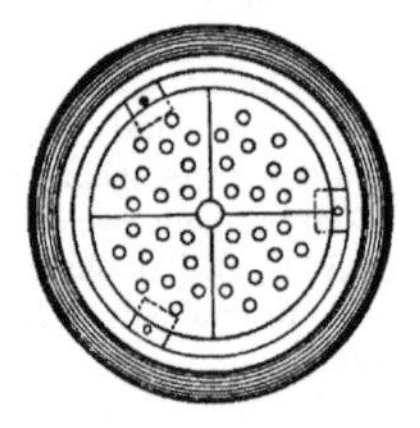

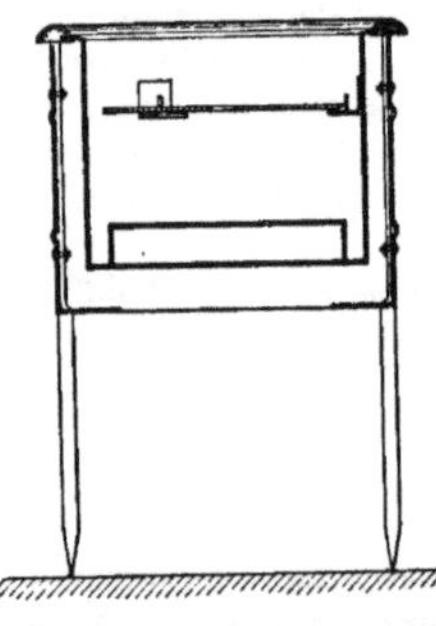

Fig. 35.—Saybolt petrolatum melting-point apparatus.

Apparatus.—(See Fig. 35). 1. Water-bath made of metal, preferably copper, $7\frac{1}{4}$ inches inside diameter and $5\frac{5}{8}$ inches deep. This bath is equipped with hangers and about 1 inch above the bottom is a 6 inch diameter metal disk to assure more uniform distribution of heat. The bath is supported in a jacket on a suitable stand.

2. Circular brass plate, 6 inches in diameter, made of two parts: (*a*) brass disk, 6 inches in diameter and $\frac{1}{8}$ inch in thickness (No. 8 B. & S. gage), marked into four sectors, each sector containing ten $\frac{1}{4}$ inch holes drilled through the plate perpendicular to its surface; (*b*) a brass disk, 6 inches in diameter and about $\frac{1}{16}$ inch thick, sweated on the other disk (*a*) so that each hole becomes a flat-bottom well. At the center of the plate is a hole, threaded to receive a short piece of $\frac{3}{8}$ inch pipe to serve as a handle.

3. Melting-point sticks consisting of smooth pine cylinders, well varnished, 0.102 inch in diameter (No. 10 B. & S. gage) and $1\frac{1}{4}$ inches long.

4. Melting-point thermometer, range 80° to 150° F., graduated in $\frac{1}{4}$° F. divisions and accurate to $\frac{1}{4}$° F. The thermometer used for Test Method No. 25 is entirely suitable, the difference in immersion being negligible.

Procedure.—Clean the well holes in the plate thoroughly by washing with gasoline and drying. Cool the plate to 70°–80° F. and lay it, with well holes at top, on a flat level surface. Melt the petrolatum to be tested and fill to overflowing each of the ten wells in one of the four sectors. Four samples may be tested at one time, using a different sector for each sample. Allow the petrolatum to solidify in air at room temperature. With a thin spatula, cut off the excess petrolatum flush with surface of the plate.

Place a clean, dry melting-point stick in a vertical position in the center of each well, pushing it down as far as it will go but without disturbing the petrolatum more than is necessary.

Fill the bath with water at about 70° F. so that the water level will be 1¾ inches above the upper surface of the plate, when this is in position. Place the plate in the bath using the ⅜ inch pipe as handle and taking great care not to disturb the sticks by jarring or otherwise. Remove the handle, and in the center of the hole thus left empty, suspend the melting-point thermometer at such a height that the center of the bulb is level with the center of the plate.

Heat the bath at a uniform rate of not more than 1° F. nor less than ½° F. per minute. A small gas flame affords a convenient source of heat. When the temperature becomes high enough, the petrolatum will soften and the wooden sticks will rise to the surface. Disregard the first two sticks which rise but note and record the temperatures at which the next six sticks rise. The average of these six temperatures is the "Melting Point, Saybolt Petrolatum Method." If care is taken in the details of the test, all these six sticks will rise within ¼° of the same temperature. The temperatures at which the first, second, ninth and tenth sticks rise, are disregarded to eliminate from consideration those sticks which rise abnormally early or late.

The sticks should not be allowed to remain in the water after a test is finished but should be removed and wiped with a clean, dry, soft cloth. Where several samples are tested at once their melting points should be far enough apart to permit careful observation in each case. The uniform rate of heating should be maintained without interruption regardless of the number of samples being tested. The melting-point sticks should be revarnished whenever this becomes necessary but the diameter should be kept the same as specified.

Determination of Melting Point by Ring and Ball Method

(Test Method No. 29)

This method is suitable for bituminous products and particularly for petroleum asphalts which are solid at 60° F. It is the standard method adopted in 1919 by the American Society for Testing Materials under the serial designation D 36–19.

1. The softening of bituminous materials generally takes place at no definite moment or temperature. As the temperature rises, they gradually and imperceptibly change from a brittle or exceedingly thick and slow flowing material to a softer and less viscous liquid. For this reason the determination of the softening point must be made by a fixed, arbitrary and closely defined method if the results obtained are to be comparable.

Apparatus.—2. The apparatus shall consist of the following:

(*a*) *Ring.*—A brass ring ⅝ inch in inside diameter and ¼ inch deep; thickness of wall $\frac{3}{32}$ inch; permissible variation on inside diameter and thickness of ring, 0.01 inch. This ring shall be attached in a convenient manner to a No. 15 B. & S. gage brass wire (diameter 0.0703 in.). See Fig. 36.

(*b*) *Ball.*—A steel ball ⅜ inch in diameter weighing between 3.45 and 3.55 grams.

(c) *Container*.—A glass vessel, capable of being heated, not less than 4.33 inches in diameter by 5.51 inches deep. (A 600 c.c. beaker, Griffin low form, meets this requirement.)

(*d*) *Thermometer.*—A thermometer which shall conform to the following specifications. (See note at end of method.)

Total length	370–400 mm.
Diameter	6.5–7.5 mm.
Bulb length	Not over 14 mm.
Bulb diameter	4.5–5.5 mm.

The scale shall be engraved upon the stem of the thermometer, shall be clear-cut and distinct, and shall run from 0 to 80° C. (32 to 176° F.) in $\frac{1}{5}$° C. divisions. It shall commence not less than 7.5 cm. above the bottom of the bulb. The thermometer shall

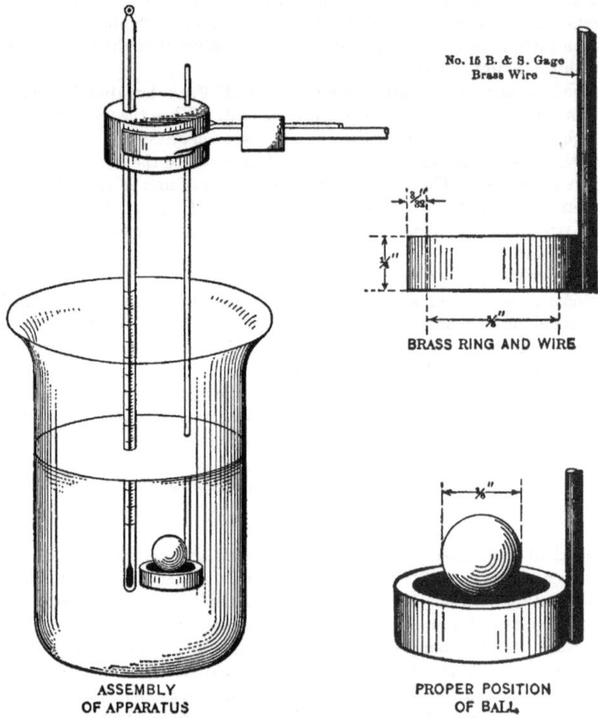

FIG. 36.—Apparatus for ring and ball method.

be furnished with an expansion chamber at the top and have a ring for attaching tags. It shall be made of a suitable quality of glass and be so annealed as not to change its readings under conditions of use. It shall be correct to 0.25° C. (0.45° F.) as determined by comparison at full immersion with a similar thermometer calibrated at full immersion by the United States Bureau of Standards.

Preparation of sample.—3. The sample shall be melted and stirred thoroughly, avoiding incorporating air bubbles in the mass, and then poured into the ring so as to leave an excess on cooling. The ring, while being filled, should rest on a brass plate which has been amalgamated to prevent the bituminous material from adhering to it. After cooling, the excesss material shall be cut off cleanly with a slightly heated knife.

Testing.—(A) Bituminous materials having softening points 90° C. (194° F.) or below.

Assembling.—4. Assemble the apparatus as shown in Fig. 36. Fill the glass vessel to a depth of substantially 3.25 inches with freshly boiled, distilled water at 5° C. (41° F.). Place the ball in the center of the upper surface of the bitumen in the ring and suspend it in the water so that the lower surface of the filled ring is exactly 1 inch above the bottom of the glass vessel and its upper surface is 2 inches below the surface of the water. Allow it to remain in the water for fifteen minutes before applying heat. Suspend the thermometer so that the bottom of the bulb is level with the bottom of the ring and within ¼ inch, but not touching, the ring.

Heating.—5. Apply the heat in such a manner that the temperature of the water is raised 5° C. (9° F.) each minute.

Softening point.—6. The temperature recorded by the thermometer at the instant the bituminous material touches the bottom of the glass vessel shall be reported as the softening point.

Permissible variation in rise of temperature.—7. The rate of rise of temperature shall be uniform and shall not be averaged over the period of the test. The maximum permissible variation for any minute period after the first three shall be 0.5° C. (0.9° F.) All tests in which the rate of rise in temperature exceeds these limits shall be rejected.

(B) Bituminous materials having softening points above 90° C. (194° F.)

Modifications for hard materials.—8. Use the same method as given under (A), except that glycerin shall be used instead of water.

Accuracy.—9. The limit of accuracy of the test is ±0.5° C. (0.9° F.).

Precautions.—10. The use of freshly boiled distilled water is essential, as otherwise air bubbles may form on the specimen and affect the accuracy of the results. Rigid adherence to the prescribed rate of heating is absolutely essential in order to secure accuracy of results.

A sheet of paper placed on the bottom of the glass vessel and conveniently weighted will prevent the bituminous material from sticking to the glass vessel, thereby saving considerable time and trouble in cleaning.

NOTE.—The A. S. T. M. thermometer is not suited to the melting-point range covered by the method. Results may be obtained with an accuracy of 2–5° F. on products melting as high as 300° F. An engraved-stem thermometer with range of 30°–400° F., graduated in 2° divisions and scaled for total immersion should be used for materials melting above 176° F. Otherwise the directions should be followed as given.

Determination of Melting Point by the Cube Method

(Test Method No. 30)

This method is often used for melting point of petroleum asphalts intended for road building and for waterproofing. It is not as satisfactory as Test Method No. 29, but gives results which should not vary more than 5° F. on the same sample. A modified form of this method adopted by the American Society for Testing Materials,[1] involves heating the sample suspended in water instead of air. In this case, the application of the method is limited to materials melting below 171° F.

Apparatus.—(See Fig. 37.)

a. Liquid bath. 800 c.c. glass beaker, low form.

b. Air bath. 400 c.c. glass beaker, high form without lip.

c. Ring stand and suitable supporting ring and clamp.

d. Beaker cover of metal or asbestos.

[1] A. S. T. M. Standard Method D 61–20, Softening Point of Tar Products.

e. An L-shaped right-angle hook made of No. 12 B. & S. gage copper wire. The foot to be 1 inch long.

f. Thermometer, engraved-stem type, range 30°–400° F., graduated in 2° F. divisions. To be scaled for total immersion and accurate to 2° F. throughout the range.

g. Wire gauze on tripod or other suitable support.

h. Brass mold, well amalgamated [1] to form a ½ inch cube. A convenient form is shown in Fig. 37.

Procedure.—Place the well-amalgamated mold on an amalgamated brass plate and fill to overflowing with asphalt melted at the lowest possible temperature. Overheating at this point is liable to raise the melting point of the sample through volatilization. Allow the mold and sample to cool to room temperature and with a hot knife cut off the excess asphalt flush with top of mold.

Assemble the apparatus as shown in Fig. 37. Adjust the temperature of the liquid bath (*a*) to 60° F., using water in the case of materials with melting points below 170° F. and cottonseed oil or a pale high-flash petroleum oil in the case of materials of higher melting points.

Remove the ½ inch cube of asphalt from the mold, place it on the wire hook as shown in Fig. 37 and suspend the hook so that the lowest edge of the cube is exactly 1 inch above the bottom of the inside beaker. Support the thermometer so that the bottom of the bulb is level with the lowest edge of the cube and so that the distance between bulb and cube is about ¼ inch.

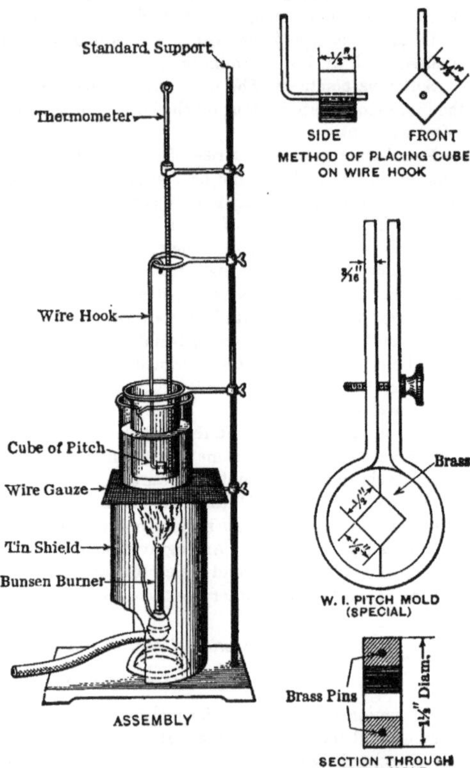

Fig. 37.—Cube method melting-point apparatus.

Heat the liquid in the outside beaker so that the temperature in the air-bath (inside beaker) increases at a uniform rate of not less than 9° F. and not more than 10° F. per minute. Note and record as "Melting Point, Cube Method" the reading of the thermometer at the instant the asphalt first touches a piece of paper laid on the bottom of the inside beaker.

[1] Brass may be amalgamated by cleaning with fine emery, washing with dilute mercuric nitrate solution and then rubbing mercury into the surface. The excess of mercury should be wiped off with a dry cloth.

Corrosion and Gumming Test of Gasoline and Naphtha

(Test Method No. 31)

This test was devised [1] for determining the purity of aviation gasoline. It has proved itself useful in the examination of motor gasoline and deserves a high recommendation.

Apparatus.—1. Freshly polished hemispherical spun copper cup about 9 cm. diameter, 4 cm. high, approximate capacity 150 c.c., approximate weight 60 grams. Edge to be plain without bead.

2. Steam-bath.

3. Air-bath.

4. Balance sensitive to 0.01 gram.

Procedure.—Weigh the "corrosion cup" to the nearest 0.01 gram and measure into it 100 c.c. of gasoline to be tested. Place the cup on the steam-bath taking necessary precautions to avoid fire. Evaporate to dryness on the steam-bath in the case of materials having a distillation end-point below 374° F. Materials of higher end-point should be evaporated first on the steam-bath and then in an air-bath at a temperature about 150° F. below this end-point. The increase in weight, after the cup has come to constant weight, is reported in grams per 100 c.c. of gasoline. If the gasoline contains sulphur in objectionable form, the cup will be colored gray or even black. A bluish green iridescence on the copper surface indicates a negligible amount of sulphur. Gasoline containing undesirable gum-forming compounds will yield a tarry residue whose weight in grams per 100 c.c. of gasoline is a measure of the gum-forming tendency.

For routine work, it is customary to conduct this test without weighing the cup. In this case it is convenient to evaporate first on the steam-bath and then with the cup set directly on an electric hot plate until evaporation is complete. Visual observation of the condition of the cup and residue will furnish valuable information as to the quality of the material under test.

"Doctor Test" for Gasoline, Naphtha, and Illuminating Oil

(Test Method No. 32)

This test is used for the purpose of detecting undesirable sulphur compounds that decompose imparting a yellow color and offensive odor to the petroleum products mentioned, especially when these are stored for some time. It gives no indication of the amount of sulphur present. In general, the "doctor test" is a reliable method of determining whether low-boiling petroleum products are "sour" and whether they will develop this quality on standing. The test has been officially adopted by the National Petroleum Association and by the U. S. Committee on Standardization of Petroleum Specifications.

Reagents.—"Doctor solution" is an aqueous solution of sodium plumbite and sodium hydroxide, made up as follows:—Dissolve about 125 grams of sodium hydroxide in one liter of distilled water. Add 70 grams of litharge (PbO) and shake vigorously for thirty minutes, or allow to stand for twenty-four hours with occasional vigorous shaking. Allow to settle and decant the clear liquid, filtering if necessary through a mat of asbestos. The solution should be kept in a glass bottle tightly stoppered with a cork (not with a glass or rubber stopper).

Sulphur.—Pure flowers of sulphur are entirely satisfactory. There has been some discussion as to the relative merits of "flowers," U. S. P. precipitated sulphur and pow-

[1] By the late F. C. Robinson, Chief Chemist, The Atlantic Refining Company. See Bureau of Mines Technical Paper 214, p. 25 (1919).

dered roll sulphur. In general it may be said that the delicacy of the reaction depends on the degree of fineness of the sulphur, but the ordinary flowers of sulphur may be used without hesitation.

Procedure.—Measure approximately one volume of doctor solution and two volumes of the material to be tested, into a clean 4 oz. sample bottle or a thick-walled test-tube. Cork tightly and shake vigorously for about fifteen seconds. Add a small pinch of flowers of sulphur, replace the cork and shake vigorously for about fifteen seconds. The quantity of sulphur used should be such that practically all of it will float on the surface separating the two layers of liquid.

Interpretation.—If the upper liquid layer is discolored or if the film of sulphur is appreciably darkened, the test is reported as "Doctor, Positive" indicating that the product under examination is "sour." If the upper liquid layer remains unchanged in color and if the film of sulphur remains yellow or only slightly discolored with gray or flecked with black, the test is reported as "Doctor, Negative" indicating that the product under examination is "sweet." The degree of sourness is sometimes indicated by descriptive terms such as "very sour," "slightly sour," "practically sweet," but such expressions have little value except for plant-control work where there is a definite understanding between laboratory and refinery as to the exact meaning.

Acid Heat Test

(Test Method No. 33)

Various modifications of this test have been used as a means of estimating the degree of unsaturation of gasoline and naphtha. There is no justification for the use of such a test for the following reasons: (*a*) many unsaturated hydrocarbons like benzol show little or no acid heat; (*b*) saturated compounds like alcohols show a high acid heat; (*c*) many excellent motor fuels and solvents derived from petroleum show a high acid heat. If the Corrosion and Gumming Test (Method No. 31) on a motor fuel or solvent is satisfactory, no test for degree of unsaturation should condemn the material.

The inaccuracies of the various methods for acid heat have been studied by Dean and Hill.[1] The form of test described is that adopted by the U. S. Committee on Standardization of Petroleum Specifications[2] and is given for the sole reason that it still forms one item in the specification for aviation gasoline. It is to be noted that an acid heat test gives no indication as to the suitability of a motor fuel.

Apparatus and reagent.—500 c.c. glass-stoppered bottle, weight about 525 grams.

250 c.c. graduated cylinder (or other measure suitable for 150 c.c.).

50 c.c. graduated cylinder.

Thermometer graduated in 1° F. scale divisions.

Commercial sulphuric acid containing 92.5 to 93.2 per cent H_2SO_4.

Procedure.—Have bottle and cylinders perfectly clean and dry. Bring the sulphuric acid, the material to be tested and all the apparatus to room temperature. Pour 150 c.c. of the sample into the 500 c.c. bottle and measure its temperature. Add 30 c.c. of the sulphuric acid, measured in the 50 c.c. cylinder. Immediately insert the ground-glass stopper and shake vigorously for exactly two minutes. The bottle should be held in such a manner that it is not warmed by the heat of the hands. It is convenient to wrap several thicknesses of cloth around the neck of the bottle and hold by the finger tips. Allow to stand for exactly one minute after shaking. Remove stopper cautiously and immediately take the temperature of the mixture. Note and record the increase in temperature as "Acid Heat, —° F."

[1] Bureau of Mines, Technical Paper 181 (1917).
[2] Bulletin No. 5 (1920).

Floc Test for Illuminating Oils

(Test Method No. 34)

This is a qualitative test applied to illuminating oils for the detection of substances rendered insoluble by heat. Such substances are undesirable in illuminating oils because they tend to clog wicks and to produce charred crusts. The methods described are those adopted by the U. S. Committee on Standardization of Petroleum Specifications [1] and by the National Petroleum Association.[2]

A. Floc test for all burning oils except 300° mineral seal oil. Take a hemispherical iron dish, and place a small layer of sand in the bottom. Take a 500 c.c. Florence or Erlenmeyer flask and into it put 300 c.c. of the oil (after filtering if it contains suspended matter). Suspend a thermometer in the oil by means of a cork slotted on the side. Place flask containing the oil in the sand-bath, and heat the bath so that the oil shall have reached a temperature of 240° F. at the end of one hour. Hold oil at temperature of not less than 240° F. nor more than 250° F. for six hours. The oil may become discolored but there should be no suspended matter formed in the oil. The flask should be given a slight rotary motion, and if there is a trace of " floc " it can be seen to rise from the bottom.

B. Floc test for 300° mineral seal oil. Take a 500 c.c. Florence or Erlenmeyer flask and into it put 300 c.c. of oil (after filtering if it contains suspended matter). Oil to be heated at the rate of 10° F. per minute to a temperature of 450° F. and held at that temperature for fifteen minutes. The oil shall show no floc or precipitate at that temperature or one hour after cooling.

Lamp Burning Tests

While physical and chemical examination of the oil itself may furnish valuable indications as to the quality of a lamp oil, the ultimate measure for suitability of such products is an actual burning test in a lamp of the proper type. It should be plainly understood that experience and judgment are required both for conducting and especially for interpreting lamp-burning tests. For this reason, the subject will be treated in such a manner that operators not familiar with lamp tests, may acquire such experience intelligently.[3]

Fig. 38.—A. Magowan Lamp, No. 1 Sunhinge Burner; B. Saybolt Lamp, No. 2 Sunhinge Burner; C. Duplex Lamp, No. 3 Dual Burner.

Lamps.—The type of lamp to be used will depend partly on the purpose for which the oil is to be used and partly on

[1] Bulletin No. 5, Dec. 29, 1920. [2] Sept. 22, 1920.

[3] The author desires to acknowledge the capable assistance of Mr. J. C. Henderson, Lamp Expert, The Atlantic Refining Co., whose thirty-six years' experience in such tests has been most valuable.

individual preference. Illustrations of standard lamps are shown in Figs. 38, 39 and 40, while Table 8 gives data relative to each. This table should be consulted for specific application to particular lamps of the following general description of test.

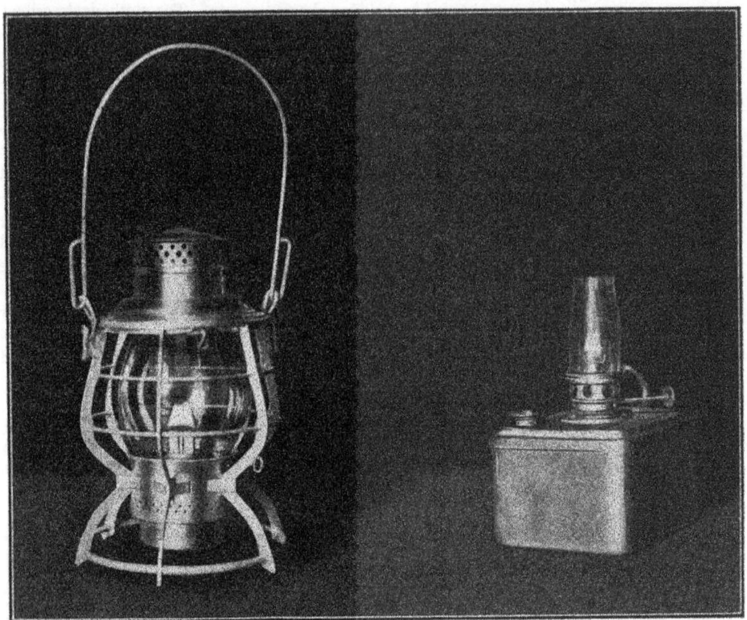

Fig. 39.—R. S. A. signal hand lantern. Fig. 40.—R. S. A. Semaphore lamp.

Lamp Test for Illuminating Oil

(Test Method No. 35)

Lamp.—The lamp selected should be of standard type and particular attention should be paid to the mechanical condition of the burner. The wick tube must not be bent or distorted and the top of the wick tube must be at the specified distance from the bottom of the fount (inside). The small spur gears engaging the wick must turn smoothly and freely when the wick is in place. Both fount and burner must be absolutely clean before each test.

Wick.—The wick should be of standard type for the burner used. New wicks must be used for referee tests, though for routine work the same wick may be used for several tests on oils of the same quality. In any case before each test, the wick whether new or old, must be thoroughly washed with redistilled ether and suspended in the air until it becomes odorless.

Chimney.—The chimney should be of the standard type furnished for the burner used. It should be held firmly but not tightly in the burner. Cracked or imperfect chimneys must not be used. Before each test, the chimney should be cleaned thoroughly with very soapy water and a soft brush, rinsed thoroughly with hot water and dried carefully with a clean, soft, absorbent towel. The chimney so cleaned must show no stains or grease marks.

Lamp room.—Except in special cases where outdoor burning is required, lamp tests must be conducted in a well-ventilated room, free from drafts at all times during the test. The average chemical laboratory is not a suitable place for lamp tests on account of the fumes usually present. In case a separate lamp room is not available, a compartment, ventilated without draft but with *fresh* air, is entirely satisfactory. If the lamp used has a glass fount, it should be protected from direct sunlight.

TABLE 8

Comparison of Details in Lamp Burning Tests

Lamp	Magowan	Saybolt	Duplex	R. S. A. signal hand lantern	R. S. A. Semaphore
Burner	No. 1 Sunhinge	No. 2 Sunhinge	No. 3 Dual	No. 1 Adlake	No. 3 Dressel
Shape of wick	Ribbon	Ribbon	Ribbon	Ribbon	Cylinder
Width of wick	$\frac{5}{8}''$	1″	$1\frac{1}{8}''$	1″	$\frac{3}{16}$–$\frac{1}{4}''$
Fount made of	Glass	Metal	Metal	Metal	Metal
Fount capacity	525–625	900 c.c.	900 c.c.	200 c.c.	31 oz. (900 c.c.)
Distance, top of wick tube to bottom of fount	$5\frac{1}{2}''$	$5\frac{1}{2}''$	$6\frac{1}{2}$–7″	4–$4\frac{1}{2}''$	4–$4\frac{1}{2}''$
Volume used	500–600 c.c.	800 c.c.	20 oz. (600 c.c.)	190 c.c.	25 oz. (750 c.c.)
Height of flame	1″	1″	1″	1″	$\frac{3}{4}''$
Duration of test	8–72 hrs.	8–72 hrs.	Dryness	18–24 hrs.	120–200 hrs.
Oil used	Kerosene	Kerosene	300 deg. Mineral Seal	Signal Oil	Long time burning

Procedure.—Select the proper lamp, burner and chimney, all perfectly clean. With clean fingers, stretch and manipulate the wick, cleaned as described, until it is soft and pliable. Insert the wick into the top of the burner tube until it can be engaged, turn down and stretch it again until it can be raised and lowered smoothly and freely. This stretching of the wick requires judgment but is essential, first because of slight variation in size of wicks and wick tubes; second, because lamp oil of good quality will burn poorly if the wick is either too tight or too loose.

Turn the wick down until about $\frac{1}{4}$ inch projects above the top of the wick tube and trim straight across the top of the tube, using very sharp scissors and drawing backward slightly during the operation, so that an even cut is obtained. Raise the wick about $\frac{1}{32}$ inch and trim off any strands projecting from the corners.

Weigh the lamp including burner, wick and chimney and pour into the fount the proper quantity of the oil to be tested. Weigh the lamp again; the increase in weight in grams divided by the specific gravity of the oil gives the volume of the oil in c.c. Allow to stand until the top of the wick becomes saturated.

Light the wick, being careful not to disturb the trim. Put the chimney in place. Adjust the height of flame to one-half the required height and allow to burn for ten minutes. The "height" of the flame is the distance from the top of the wick to the top of the flame. Then turn wick up to about three-quarters required flame height and allow to burn for five minutes, after which time the flame should be brought to the required height. All adjustments of flame height should be made by raising the wick slightly more than necessary and then turning *down;* after a few seconds the spur gears engaging the wick should be turned up just enough to relieve any tension on the wick but without changing the height of the flame. The flame should be symmetrical with a fairly flat or slightly rounded top as shown in Fig. 38. Sharp-pointed flames are usually due to faulty trimming of the wick.

During the first hour after the flame has been brought to the required height, further adjustments are permissible if necessary. Adjustments after this time should not be made unless specifically allowed by contract. The duration of a lamp test may be for a definite number of hours, for burning a definite quantity of oil or for burning to dryness. In the last case, a drop of flame height during the last ten minutes burning should

Fig. 41.—Proper method of extinguishing.

be disregarded. When one filling with oil is not sufficient for the duration of the test, the fount should be refilled to the original level, the required volume of oil being carefully measured. Unless otherwise specified, such refilling should be made when 75–80 per cent of the oil has been consumed.

When a lamp test is completed, the following rules should be observed in extinguishing the flame: (1) Turn wick down until flame is about $\frac{3}{8}$ inch high. (2) Place hand about 3 inches above the chimney with palm inclined at an angle of about 45° as shown in Fig. 41. (3) With lips slightly below the hand, extinguish the flame with one short quick puff. (4) Never blow directly down the chimney. These rules may be followed profitably by all users of kerosene lamps.

Record and interpretation.—The record of a lamp burning test should include the following items:

Kind of oil used and source of sample.
Physical and chemical characteristics of sample.
Type of lamp, burner and chimney used.

Volume of oil used for the test, in c.c. or fluid ounces.
Volume of oil consumed during test, in c.c. or fluid ounces.
Duration of the test, in hours.
Height of " char " (wick above top of wick tube), in fraction of an inch.
Amount and appearance of wick crust, in descriptive terms.
Drop of flame during test, usually reported " no drop " or " drop after —hrs."
Condition of chimney, in descriptive terms.

Wick crusts are usually reported as good, fair, bad or very bad, and " toadstools " or mushroom-like growths require special mention. In general, crusts may be light-colored, indicating presence of alkali soaps in the oil or they may be hard, coke-like deposits, indicating the presence or formation of resinous products. The ideal crust is soft and black, indicating absence of soaps and tar-forming material, but a crust of this type is not essential to a satisfactory lamp oil.

Drop of flame during test may indicate too high a viscosity for the increased " lift " or distance from top of wick tube to surface of the oil. More often it indicates the presence of impurities which are absorbed by the wick fibers with the resultant clogging and diminished feed. Condition of chimney is reported as excellent, good, fair or bad. Black soot on a chimney denotes either improper testing, drafts or very bad oil. A reddish deposit indicates high sulphur content. In some cases, careful examination will reveal tiny beads of condensate having a marked sour taste. A slight stain or cloud is to be expected with practically all oils, particularly in tests of many hours' duration.

It is evident that lamp-burning tests cannot be made and interpreted without knowledge and experience, but it is equally true that to the expert they furnish the only reliable means for the ultimate estimation of quality in illuminating oils.

Emulsification Tests

Lubricating oil used in circulating systems often comes in contact with water and under these conditions some oils form more or less stable emulsions which interfere with lubrication and particularly with filtering systems. For this reason a number of methods have been proposed for the purpose of obtaining, in the laboratory, information as to the emulsifying tendency of oils under service conditions. No single method has been generally adopted but the subject is now being studied by Committee D-2 of the American Society for Testing Materials and the three methods described below are those receiving favorable consideration. The problem is a difficult one but appears to be nearing a solution. General information on the subject may be obtained from an excellent article by W. H. Herschel.[1]

While it is not proper at the present time to condemn or recommend any of the following methods, certain observations may be helpful to the chemist or engineer who is attempting to make tests for emulsification tendency.

The Conradson Steam Test is capable of giving valuable information, but the observations and interpretation of results require experience and judgment, especially in the case of oils of doubtful quality. A satisfactory record of a Conradson Steam Test is somewhat too long for convenience and different operators are quite likely to vary both as to notation and interpretation.

The Herschel Demulsibility test possesses the decided advantage of giving a result consisting of a single numerical value. The chief objection to the Herschel Demulsibility test is the fact that it is not applicable to viscous oils without introduction of modifications of the standard procedure. Such oils are frequently reported as having a perfect demulsibility while as a matter of fact the emulsion was never formed. There is little excuse for such a report because Herschel in the paper cited, calls particular attention to the

[1] Bureau of Standards, Technologic Paper No. 86 (1917).

necessity of modifying the procedure in such cases. In the opinion of the author, the Demulsibility Test is liable to give misleading results in the case of oils having viscosities greater than 150 seconds Saybolt Universal at 130° F. Oils of lower viscosity may be graded with confidence by the Demulsibility Test.

The R. E. Test [1] combines the desirable features of both Conradson and Herschel tests with the additional advantage that it is applicable, without modification, to all lubricating oils regardless of viscosity. It is worthy of note that there is a rather definite mathematical relation between the R. E. and Demulsibility values on any given oil suitable for testing by the Herschel method. The R. E. Test has received considerable favorable comment from oil chemists and engineers, and is recommended as giving a reliable indication of the resistance of lubricating oil to emulsification under actual working conditions.

In connection with all emulsification tests it must be remembered that viscosity plays an important part in the time required for oil to separate from emulsion. For this reason, an emulsification test should be taken as an indication of degree of refining only when the viscosity is taken into account. A crude kerosene distillate will separate from emulsion more rapidly than a well-refined cylinder oil.

The R. E. Test

(Test Method No. 36)

This method [2] of test is based on the emulsifying action of steam injected into the

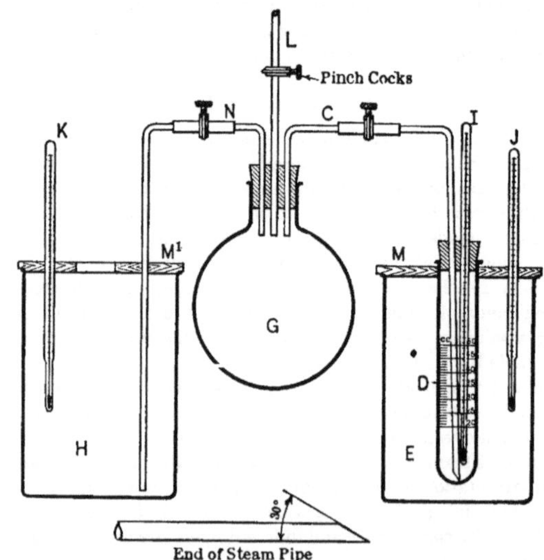

Fig. 42.—R. E. test with steam generator.

oil. The emulsion, during formation, is maintained at a definite temperature under conditions which give emulsions similar in appearance to those encountered in practice.

[1] "Resistance to Emulsification."

[2] The R. E. (Resistance to Emulsification) Test was devised in connection with the work of Committee D-2, A. S. T. M. and was first published in the A. S. T. M. Committee Reports (1920). See also The Atlantic Lubricator, Vol. IV, No. 3 (Aug. 1920), Abrams, Osmond and Delbridge.

The emulsion, thus formed is allowed to separate under definite, uniform conditions and the R. E. value of the oil is calculated in simple manner from the ratio of c.c. of separated oil to time of separation.

The apparatus is simple and the operations involved require but little experience. Twenty c.c. of oil are sufficient for the test and the total time for a test varies from six minutes for the best oils to twenty-six minutes for the poorest. Results obtained are reproducible and appear to give an accurate indication of the resistance of lubricating oils to emulsification in practice. When the R. E. value is used as a basis of specifications, it is to be remembered that the test is a severe one and that the viscosity as well as the degree of refinement is involved in the numerical figure. The magnitude of R. E. values is based on the fact that the definite R. E. value of 100 has been assigned to oils which separate from emulsion completely in one minute.

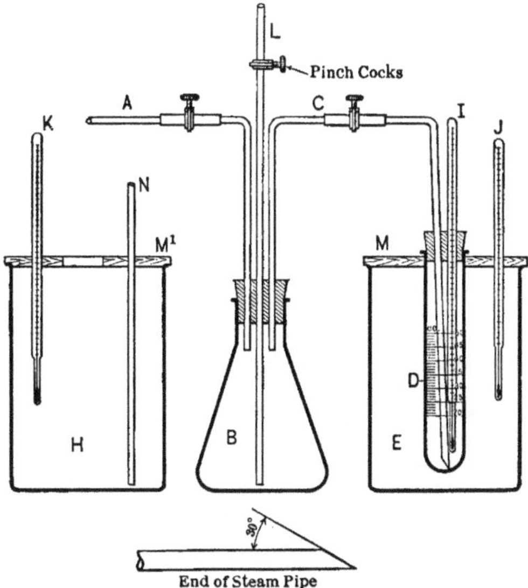

FIG. 43.—R. E. test with steam supply.

Apparatus.—Figs. 42 and 43.—1. Steam generator, of any suitable type, such as a liter boiling flask, *G* in Fig. 42. If steam from a line *A* is used, it must be passed through a suitable separator, such as a flask, *B* in Fig. 43. In either case a blow-off outlet tube, *L*, controlled with a screw pinch cock must be provided.

2. R. E. tube, *D*, glass test-tube, 25 by 200 mm., graduated from 20 to 50 c.c. in cubic centimeters.

3. Two water-baths, *H* and *E*, capacity 3 to 4 liters. They must be transparent and may be 4000 c.c. glass beakers, battery jars or gallon bottles with top cut off.

4. Two lids, *M* and *M′* of any suitable material, to fit baths and serve as supports for R. E. Tube and thermometers.

5. Thermometers, *I*, *J* and *K*, of engraved-stem type, range at least 60 to 212° F., graduated in 1° scale divisions and accurate to 1° F.

6. Steam delivery tube, of glass approximately 2.5 mm. inside diameter and cut off diagonally at an angle of 30° from axis of tube as shown.

7. Corks, pinch cocks, glass tubing and rubber connections as shown.

8. Stop-watch or equivalent.

Procedure.—Fill separating bath *H* with water and raise temperature slowly to 200° F., either by steam through line *N* or by an immersion electric heater. Do not attempt to heat with a burner or hot plate.

Fill emulsion bath *E* with water, using 3000 c.c.±60 c.c. and adjust temperature to 67° F. Pour exactly 20 c.c. of the oil to be tested into the clean, dry R. E. tube at room temperature. Place the R. E. tube in the lid *M* of bath *E*.

Place the cork (not a rubber stopper) carrying the thermometer *I* in position so that the bottom of the thermometer bulb is 1 inch above the bottom of the R. E. tube. The hole in the cork for the steam delivery tube must be slightly larger than the outside diameter of this tube, the clearance being the vent for the R. E. tube. Do not introduce the steam delivery tube into the R. E. tube but connect it with the steam outlet *C* and allow it to hang at an angle until needed.

With pinch cocks *L* and *C* open, generate steam from distilled water in *G*, Fig. 42 (or introduce steam through line *A* into separator *B*, Fig. 43) and steam out the delivery tube, partly closing pinch cock *L* if necessary. When the steam discharged is dry, open pinch cock *L*. Quickly introduce the steam delivery tube into the R. E. tube through the loosely fitting hole in the cork so that the tip of the delivery tube rests on the center of the bottom of the R. E. tube. Maintain the temperature of the emulsion, as indicated by the thermometer *I*, at 190–195° F. by regulating pinch cock *L*, pinch cock *C* being wide open. This will require a steam supply sufficient to insure generous discharge through the line *L*.

Continue steaming, holding the temperature of the emulsion at 190–195° F., until the total volume of oil and water from condensation is 40±3 c.c. This condition will be reached when the apparent volume of the steam-agitated mixture is approximately 52 c.c. This figure may vary slightly and should be determined by each operator for the conditions under which he is working. The time of steaming should be 4.0 to 6.5 minutes; a shorter time indicates wet steam or incomplete steaming out of the delivery tubes before starting the test.

When the required volume is obtained, quickly remove the steam delivery tube and at the same instant start the stop-watch. Transfer the R. E. tube immediately to the separating bath *H* whose temperature has previously been brought to 200° F. Remove the cork and thermometer, allowing the latter to drain for a few seconds. The temperature of the separating bath *H* during the remainder of the test must not be higher than 203° F. nor lower than 200° F.

Observe the progress of separation through the transparent walls of the bath, examining every thirty seconds until 20 c.c. of oil have separated; or, if less than 20 c.c. of oil have separated at the end of twenty minutes, read, to the nearest c.c. mark, the volume of separated oil observed at that time.

The R. E. value of the oil under examination may be taken from Table 9, or it may be calculated by substituting in the following formula:

$$\text{R. E. value} = \frac{\text{number c.c. separated oil} \times 5}{\text{number of minutes required}}.$$

Examples.—If 20 c.c. separated in 28 seconds, $\text{R. E.} = \frac{20 \times 5}{0.5} = 200.$

If 20 c.c. separated in 2.5 minutes, $\text{R. E.} = \frac{20 \times 5}{2.5} = 40.$

If 18.5 c.c. separated in 20 minutes, $\text{R. E.} = \frac{18.5 \times 5}{20} = 4.6.$

Accuracy.—Duplicate determinations on the same sample should not show variation greater than one minute in time required for 20 c.c. oil separation or greater than 4 c.c. in cases where less than 20 c.c. oil separation has taken place at the end of twenty minutes.

Interpretation.—While the R. E. test is recommended to those who desire to set their own specification limits, such limits should not be made so high as to exclude suitable products. For example, while the minimum R. E. value for a turbine oil might properly be specified as 25, such a limit is too high for an ordinary engine oil which may be considered satisfactory if the R. E. value is 10 or even lower. It is also to be noted that an R. E. value is a reciprocal function of time, thus R. E. values 100 and 67 denote a time difference of only 0.5 second. Numerical differences between R. E. values become more significant as the R. E. value decreases, as shown in Table 9.

TABLE 9

R. E. Values

Time, Minutes	Separated Oil, c.c.	R. E. Value	Time, Minutes	Separated Oil, c.c.	R. E. value
0.5	20.0	200.0	11.0	20.0	9.1
1.0	20.0	100.0	12.0	20.0	8.3
1.5	20.0	66.6	13.0	20.0	7.7
2.0	20.0	50.0	14.0	20.0	7.1
2.5	20.0	40.0	15.0	20.0	6.6
3.0	20.0	33.3	16.0	20.0	6.2
3.5	20.0	28.6	17.0	20.0	5.9
4.0	20.0	25.0	18.0	20.0	5.6
4.5	20.0	22.2	19.0	20.0	5.3
5.0	20.0	20.0	20.0	20.0	5.0
5.5	20.0	18.2	20.0	18.0	4.5
6.0	20.0	16.6	20.0	16.0	4.0
6.5	20.0	15.4	20.0	14.0	3.5
7.0	20.0	14.3	20.0	12.0	3.0
7.5	20.0	13.3	20.0	10.0	2.5
8.0	20.0	12.5	20.0	8.0	2.0
8.5	20.0	11.8	20.0	6.0	1.5
9.0	20.0	11.1	20.0	4.0	1.0
9.5	20.0	10.5	20.0	2.0	0.5
10.0	20.0	10.0	20.0	0.0	0.0

THE CONRADSON STEAM TEST

(Test Method No. 37)

The emulsifying tendency of lubricating oils has been roughly estimated for many years by noting the results obtained by passing steam through the oil. The method here described is due to Dr. P. H. Conradson.[1]

Apparatus.—(See Fig. 44.) The apparatus consists of a 4-pint copper retort, provided with a delivery tube, which is joined to a metal or glass pipe having an inside diameter of about $\frac{5}{16}$ inch and about 15 inches long from the elbow. The lower end of this pipe is cut off diagonally to prevent thumping.

[1] Proceedings A. S. T. M., XVI (1916), Part II, p. 274.

The glass cylinders are graduated to 250 c.c. They have an inside diameter of about $1\frac{7}{16}$ inches and a length of about $9\frac{1}{2}$ inches from the bottom to the 250 c.c. mark. They are $11\frac{1}{2}$ to 12 inches in over-all length, and are made of thin glass, with a flat bottom.

In place of a copper retort for the generation of steam, a glass flask or any other suitable source of steam supply may be used; likewise, ordinary 250 c.c. graduated glass cylinders, of dimensions given above, may be used where emulsion tests are required only occasionally.

Method of testing.—The cylinder is filled with distilled water up to the 20 c.c. mark, then 100 c.c. of the oil to be tested are added. To churn the mixture, steam at ordinary pressure is conducted through this oil-water mixture for ten minutes. The amount

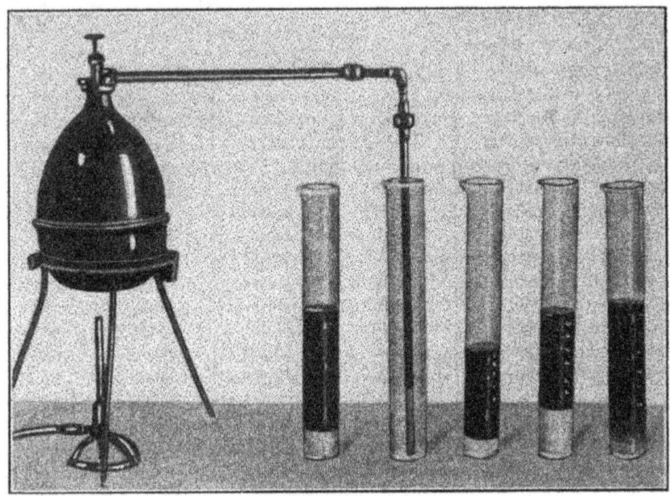

Fig. 44.—Conradson steam test.

of steam passed through is regulated in such a way as to prevent the mixture from splashing over the top of the cylinder, but the rate may be as rapid as is practical. This is easily regulated by the height of the gas flame.

The churning is begun from the time the temperature of the mixture has reached 200° F., or when the steam as such passes off the mixture. It usually takes from one to one and a half minutes to reach this temperature, depending somewhat on the body or viscosity of the oil. However, even churning with steam for fifteen minutes does not seem to make any difference in the results.

When the churning is completed, the cylinder is immersed for one hour in a water-bath, kept at a temperature of 130° F. During this time the cylinder and its contents are momentarily inspected at intervals to note the behavior of the oil mixture. At the expiration of one hour the cylinder is removed from the water-bath and the contents are examined for the following:

1. The number of cubic centimeters of separated clear or turbid water;
2. The number of cubic centimeters of separated emulsified layer;

3. The number of cubic centimeters of separated clear or turbid oil above the emulsified layer; and.

4. The percentage of water or moisture in the separated oil above the emulsified layer.

Interpretation.—Dr. Conradson makes the following statements [1]:

The number of cubic centimeters and condition of the emulsified layer is an indication of the emulsion-forming property or quality of the oil.

The number of cubic centimeters of clear or turbid oil above the emulsified layer, less the percentage of water or moisture contained in the oil, is the percentage of demulsibility of the oil.

The condition of the separated water or watery liquid under the emulsified layer, if any, gives an indication also of the behavior of the oil in actual service.

The Herschel Demulsibility Test

(Test Method No. 38)

This method is taken in slightly modified form from an excellent paper [2] by Dr. W. H. Herschel. Careful study of the original communication is strongly recommended to those interested in the subject of emulsification tests and their relation to lubrication. Particular attention is called to the following excerpts from the paper cited:

" When the test for demulsibility is applied to other oils than the high-speed engine and turbine oils for which it is primarily intended, difficulties may be encountered in obtaining readings."

" With some oils, under certain conditions, part of the water does not enter into the emulsion, and drops of oil are found adhering to the sides of the cylinder after the paddle is withdrawn. When this phenomenon occurs, the rate of settling, as indicated by the test when made in the usual way, is apt to be too high. In extreme cases the demulsibility might be called 1200 when the oil would emulsify readily under other conditions. The adhering drops are an important warning which should not be disregarded. This phenomenon is most apt to occur with very viscous or compounded oils, such as are used for gas or marine engines or steam-engine cylinders."

The method described in this section is suitable for lubricating oils having a viscosity not higher than 150 seconds, Saybolt Universal at 130° F. Modifications must be introduced for oils of higher viscosity than this, if no emulsion is formed or if part of the water does not enter into the emulsion, either with or without the " adhering drops " mentioned above. Failure to observe these limitations has been the cause of considerable unjust criticism of the method.

Apparatus.—A general view of one form of the apparatus is shown in Fig. 45. The essential items are:

1. Commercial 100 c.c. cylinder, inside diameter, 25.9 to 27.4 mm., reasonably accurately graduated from bottom up in 1 c.c. divisions.

2. Flat metal paddle, carried on a vertical shaft, rotated by an electric motor with means for controlling and measuring speed of rotation. The paddle dimensions should be 89 by 20 by 1.5 mm. ($3\frac{1}{2}$ by $\frac{13}{16}$ by $\frac{1}{16}$ inches).

3. Water-bath, preferably of metal, with a false bottom to facilitate circulation and support the cylinder.

4. Gas burner or other suitable heater to maintain the water-bath at 130° F. and thermometer to indicate this temperature.

Procedure.—Make sure that the 100 c.c. graduated cylinder (1) and the paddle (2) are perfectly clean. Fill the water-bath (3) with water and heat to 130° F. which

[1] For a more detailed explanation the original article should be consulted, loc. cit.

[2] Bureau of Standards Technologic Paper No. 86 (1917).

temperature should be maintained within 1° F. throughout the test. Pour 20 c.c. of the oil to be tested and 40 c.c. of distilled water into the graduated cylinder, making sure that the oil used for the test is thoroughlv representative of the sample.

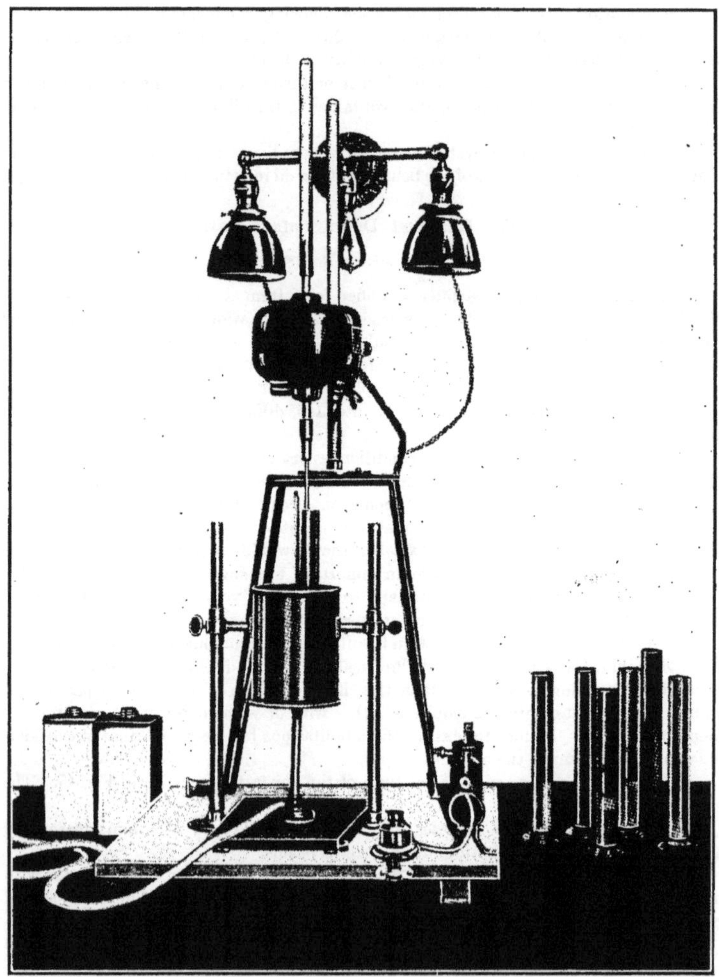

Fig. 45.—Herschel demulsibility apparatus.

Place the cylinder in the bath and lower the paddle or raise the bath until the paddle is completely submerged. Allow cylinder and contents to warm up to about 130° F. which will require about eight minutes. Then start the paddle and stir for five minutes at a speed of 1500 R.P.M. At the end of this time, stop the paddle and start a stop-

TABLE 10

Rate of Settling or Herschel Demulsibility

Elapsed time after stirring, minutes	Reading at Upper Surface of Emulsion, c.c.																			
	59	58	57	56	55	54	53	52	51	50	49	48	47	46	45	44	43	42	41	40
1	60	120	180	240	300	360	420	480	540	600	660	720	780	840	900	960	1020	1080	1140	1200
2	30	60	90	120	150	180	210	240	270	300	330	360	390	420	450	480	510	540	570	600
3	20	40	60	80	100	120	140	160	180	200	220	240	260	280	300	320	340	360	380	400
4	15	30	45	60	75	90	105	120	135	150	165	180	195	210	225	240	255	270	285	300
5	12	24	36	48	60	72	84	96	108	120	132	144	156	168	180	192	204	216	228	240
6	10	20	30	40	50	60	70	80	90	100	110	120	130	140	150	160	170	180	190	200
7	9	17	26	34	43	51	60	69	77	86	94	103	112	120	129	137	146	155	163	172
8	8	15	23	30	38	45	53	60	68	75	83	90	98	105	112	120	128	135	143	150
9	7	13	20	27	33	40	47	53	60	67	73	80	87	93	100	107	113	120	127	134
10	6	12	18	24	30	36	42	48	54	60	66	72	78	84	90	96	102	108	114	120
11	6	11	16	22	27	33	38	44	49	55	60	65	71	76	82	87	93	98	104	109
12	5	10	15	20	25	30	35	40	45	50	55	60	65	70	75	80	85	90	95	100
13	5	9	14	19	23	28	32	37	42	46	51	55	60	65	69	74	79	83	88	92
14	4	9	13	17	21	25	30	34	39	43	47	51	56	60	64	69	73	77	81	86
15	4	8	12	16	20	24	28	32	36	40	44	48	52	56	60	64	68	72	76	80
16	4	8	11	15	19	23	23	30	34	37	41	45	49	53	56	60	64	68	71	75
17	4	7	11	14	18	21	25	28	32	35	39	42	46	49	53	56	60	64	67	71
18	3	7	10	13	17	20	23	27	30	33	37	40	43	47	50	53	57	60	63	67
19	3	6	10	12	16	19	22	25	28	32	35	38	41	44	47	51	54	57	60	63
20	3	6	9	12	15	18	21	24	27	30	33	36	39	42	45	48	51	54	57	60
21	3	6	9	11	14	17	20	23	26	29	31	34	37	40	43	46	49	51	54	57
22	3	5	8	11	14	16	19	22	25	27	30	33	35	38	41	44	46	49	52	55
23	3	5	8	10	13	16	18	21	24	26	29	31	34	37	39	42	44	47	50	52
24	3	5	8	10	13	15	18	20	23	25	28	30	33	35	38	40	43	45	48	50
25	2	5	7	10	12	14	17	19	22	24	26	29	31	34	36	38	41	43	46	48
26	2	5	7	9	12	14	16	18	21	23	25	28	30	32	35	37	39	42	44	46
27	2	4	7	9	11	13	16	18	20	22	24	27	29	31	33	36	38	40	42	44
28	2	4	6	9	11	13	15	17	19	21	24	26	28	30	32	34	36	39	41	43
29	2	4	6	8	10	12	15	17	19	21	23	25	27	29	31	33	35	37	39	41
30	2	4	6	8	10	12	14	16	18	20	22	24	26	28	30	32	34	36	38	40
35	2	3	5	7	9	10	12	14	15	17	19	21	22	24	26	27	29	31	33	34
40	2	3	5	6	8	9	11	12	14	15	17	18	20	21	23	24	26	27	29	30
45	1	3	4	5	7	8	9	11	12	13	15	16	17	19	20	21	23	24	25	27
50	1	2	4	5	6	7	8	10	11	12	13	14	16	17	18	19	20	22	23	24
55	1	2	3	4	5	7	8	9	10	11	12	13	14	15	16	17	19	20	21	22
60	1	2	3	4	5	6	7	8	9	10	11	12	13	14	15	16	17	18	19	20

EXPLANATORY NOTE.—The following examples illustrate the construction of this table. If the reading at upper surface of emulsion is 50 c.c. and the elapsed time is 15 minutes, the rate of settling or demulsibility $= (60-50) \times \frac{60}{15} = 40$ c.c. per hour; for a reading of 45 c.c. and a time of 10 minutes, demulsibility $= (60-45) \times \frac{60}{10} = 90$ c.c. per hour.

watch. Withdraw the paddle from the cylinder and wipe it with the finger, returning to the cylinder the emulsion thus collected.

Allow the cylinder to stand in the water-bath at 130° F. and after one minute from the time the paddle was stopped, take a reading to the nearest c.c. mark, of the line of separation between clean oil at the top and the emulsion layer immediately below. To make this reading, remove the cylinder from the bath momentarily. Obtain from Table 10 the average rate of settling corresponding to the elapsed time (one minute) and the reading at upper surface of emulsion. Record this figure. Continue to take readings at intervals of one minute from the time of stopping the paddle and record the corresponding rate of settling taken from Table 10. When the oil has separated completely from emulsion or when the recorded rates of settling show a maximum, further readings are unnecessary, and the number representing the maximum rate of settling should be reported as the " Herschel Demulsibility " of the oil.

If the maximum rate of settling is not reached in one hour, readings should be discontinued and the average rate for sixty minutes should be reported. If no oil has separated in one hour, the demulsibility is 0. Oils which separate completely n one minute have a demulsibility of 1200, provided emulsion was actually formed. It is to be noted that an oil having a demulsibility of 300 separates completely in four minutes and should therefore be considered as a good oil so far as emulsification is concerned.

Determination of Carbon Residue

(Test Method No. 39)

This method was devised by Dr. P. H. Conradson and adopted by the American Society for Testing Materials [1] as a Standard Test under Serial Designation D 47–18. It has been widely used and is an official test of the National Petroleum Association and of the U. S. Committee on Standardization of Petroleum Specifications. The interpretation of the test is still a matter of discussion among chemists and lubrication engineers. In general it may be said that specifications for the permissible amount of carbon residue in a given product should be drawn so that they do not exclude products which have proved entirely satisfactory in practice.

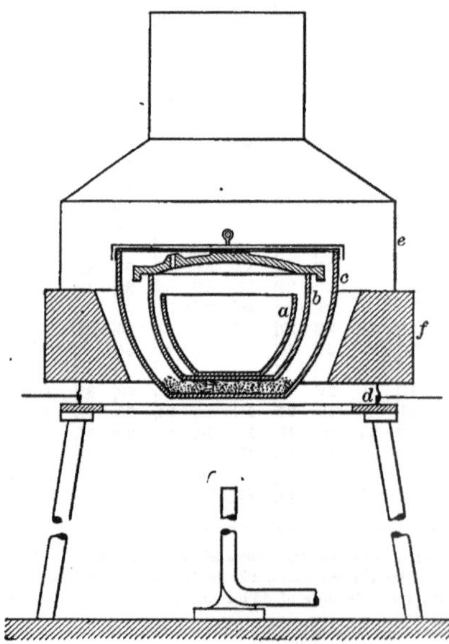

Fig. 46.—Carbon residue apparatus.

Apparatus.—(*a*) Porcelain crucible, wide form, glazed throughout, 25 to 26-c.c. capacity, 46 mm. in diameter.

(*b*) Skidmore iron crucible, 45-c.c. ($1\frac{1}{2}$ oz.) capacity, 65 mm. in diameter, 37 to 39 mm. high, with cover, without delivery tubes and one opening closed.

(*c*) Wrought-iron crucible with cover, about 180 c.c. capacity, 80 mm. diameter, 58 to 60 mm. high. At the bottom

[1] A. S. T. M. Standards, 1918, p. 620.

of this crucible a layer of sand is placed about 10 mm. deep or enough to bring the Skidmore crucible with cover on nearly to the top of the wrought-iron crucible.

(*d*) Triangle, pipe-stem covered, projection on side so as to allow flame to reach the crucible on all sides.

(*e*) Sheet-iron or asbestos hood provided with a chimney about 2 to 2½ inches high, 2⅛ to 2¼ inches in diameter to distribute the heat uniformly during the process.

(*f*) Asbestos or hollow sheet-iron block, 6 to 7 inches square, 1¼ to 1½ inches high, provided with opening in center 3¼ inches in diameter at the bottom, and 3½ inches in diameter at the top.

Method.—The test shall be conducted as follows:

Ten grams of the oil to be tested are weighed in the porcelain crucible, which is placed in the Skidmore crucible and these two crucibles set in the larger iron crucible, being careful to have the Skidmore crucible set in the center of the iron crucible, covers being applied to the Skidmore and iron crucibles. Place on triangle and suitable stand with asbestos block and cover with sheet-iron or asbestos hood in order to distribute the heat uniformly during the process.

Heat from a Bunsen burner or other burner is applied with a high flame surrounding the large crucible, until vapors from the oil start to ignite over the crucible, when the heat is slowed down so that the vapor (flame) will come off at a uniform rate. The flame from the ignited vapors should not extend over 2 inches above the sheet-iron hood. After the vapor ceases to come off the heat is increased as at the start and kept so for five minutes, making the lower part of the large crucible red hot, after which the apparatus is allowed to cool somewhat before uncovering the crucible. The porcelain crucible is removed, cooled in a desiccator and weighed.

The entire process should require about one-half hour to complete when the heat is properly regulated. The time will depend somewhat upon the kind of oil tested, as a very thin, rather low-flash-point oil will not take as long as a heavy, thick, high-flash-point oil.

Note.—It will assist in regulating the rate of heating if the apparatus is heated before the porcelain crucible containing the oil is introduced. It is essential that the heat be applied during the test in such manner that the oil does not boil over from the porcelain into the Skidmore crucible. If this happens, the determination must be disregarded and repeated.

Heat Test for Lubricating and Insulating Oils

(Test Method No. 40)

Heat tests are applied to lubricating oils and to transformer oils to determine their stability toward heat and also as a means of detecting incomplete or improper refining for specific purposes. As commonly carried out in the presence of air, they show the combined effect of heat and oxidation. The problem of standardizing an oxidation test so that it may be used quantitatively has never been satisfactorily solved although work now in progress at the Bureau of Standards on the Waters Oxidation Test [1] seems likely to culminate in a suitable method. The so-called "sludge tests" for transformer oils consist in bubbling air or oxygen through the heated oil, often in the presence of metallic copper. None of these tests can be considered as entirely satisfactory. At the present time, therefore, a heat test must be largely a qualitative and comparative test. The following procedure will be found to give valuable indications of the quality.

Pour 100 c.c. of the oil to be tested into a perfectly clean and dry 4-oz. oil sample bottle. Close the bottle with a cork having along one side a small groove to serve as a vent. Place the bottle in an oil-bath of suitable size and shape and raise the tem-

[1] Bureau of Standards Circular No. 99, Carbonization of Lubricating Oils, 1920, gives the latest published information.

perature to 450° F. Maintain at this temperature for any desired length of time. For most purposes one hour is sufficiently long and is recommended. Remove the bottle and allow to cool. Examine the resulting oil for color, transparency and odor. Allow it to stand for twelve to twenty-four hours and note the formation of any sediment. In laboratories where color measurements can be made, such measurements on the original and on the heated oil will prove convenient for purposes of record. By adopting a suitable standard rate, method and time of heating, this test may be rendered fairly duplicable.

The interpretation of this test is largely a matter of individual opinion. In general the oils which darken least are considered the best. All oils will show slight darkening and such darkening is of course most noticeable with the lightest colored oils. The formation of any appreciable amount of sediment is an even more reliable evidence of poor quality for special purposes according to present-day standards.

Corrosion of Metals

(Test Method No. 41)

This subject is of great importance in connection with petroleum products, most of which are used in contact with metal of some sort. The test described here was devised [1] in connection with a study of lubricating oils. It is quicker and more severe than the usual twenty-four hour test at room temperature.[2] The object of the test is to determine qualitatively the corrosive or staining properties of an oil toward polished copper as an indication of corrosive action on other metals.

Apparatus.—Strips of polished No. 10 gage copper, about 10 mm. wide and 125 mm. long.

4-oz. oil sample bottles and corks to fit. Each cork should have a small groove along one side for an air vent and should be split halfway up so that the end of a copper strip may be inserted. When the cork is pushed into the mouth of the bottle it will act as a clamp to hold the copper strip.

Water-bath equipped with a suitable heater and with a cover, perforated to receive the desired number of 4-oz. bottles.

Procedure.—Prepare the copper strips by heating to dull redness, chilling by immersion in cold water and then polishing thoroughly with very fine sandpaper. Pour 100 c.c. of the oil to be tested into a 4-oz. sample bottle, insert the end of the copper strip into the split cork and push the cork firmly into the mouth of the bottle. Place the bottle in the water-bath so that the shoulder is just above the perforated cover. Maintain the bath at the boiling point of water for five hours without disturbing the bottle. It is preferable to carry out the test on each oil in duplicate. The number of tests which can be run at one time is limited only by the capacity of the bath.

Interpretation.—Darkening of the copper on the submerged surface indicates corrosion due to sulphur or a corrosive sulphur compound.

Diminution of the polish or the appearance of a dull reddish color on the submerged surface of the copper indicates corrosion due to acids or acid compounds. In extreme cases, some pitting of the surface may be observed.

Darkening or corrosion of the emergent surface indicates formation of vapors containing acid or sulphur compounds.

High-grade oils should leave on the copper no more than the faintest indication of the depth to which it was immersed.

[1] By H. G. Smith, Chemical Engineer, The Atlantic Refining Co.

[2] Bulletin No. 5, p. 30, U. S. Committee on Standardization of Petroleum specifications.

Determination of Sulphur by Lamp Method

(Test Method No. 42)

This method is the one in common use for the determination of sulphur in illuminating oils and may be used for naphthas and gas oils. It has also been employed with modifications [1] for sulphur in lubricating and other viscous oils but cannot be recommended for this purpose. The results obtained on all oils are usually slightly low on account of retention of small amounts of sulphur compounds in the wick.

Apparatus.—See Fig. 47.—Chemically resistant glass is necessary.

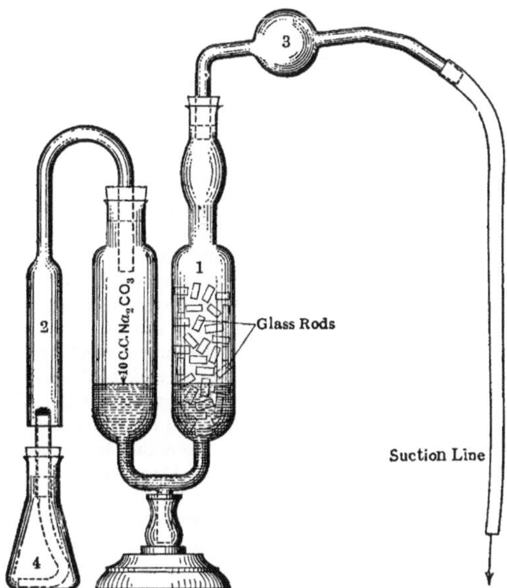

Fig. 47.—Sulphur by lamp method.

1. Absorber, about 150 c.c. capacity, containing glass beads or short pieces of glass rod in the suction side as shown.
2. Glass chimney connected with the absorber by a rubber stopper.
3. Glass spray trap connected with the absorber by a rubber stopper.
4. Small lamp of about 25 c.c. capacity. This lamp may conveniently consist of a 25–33 c.c. Erlenmeyer flask and a cork carrying a short section of glass tubing, about $\frac{1}{8}$ inch inside diameter for naphthas and lamp oils, $\frac{3}{16}$ inch for gas oils. The cork must be grooved along one side so that air may enter the flask while the oil is being consumed.
5. Ordinary cotton wicking.
6. Filter pump or other means for continuous suction and rubber tubing to connect with spray trap.

Solutions.—Standard hydrochloric acid solution containing 2.275 g. HCl per liter and carefully checked for accuracy.

[1] Compare Conradson, J. Ind. Eng. Chem., **4** (1912), 842.

Standard sodium carbonate solution containing 3.306 g. Na_2CO_3 per liter. Exactly 10.0 c.c. should be required to neutralize 10.0 c.c. of the hydrochloric acid solution.

Methyl orange solution in distilled water, containing 0.004 g. methyl orange per liter.

Procedure.—It is advisable to make a blank determination at the same time and under the same conditions burning sulphur-free alcohol in the lamp. This will furnish a fairly accurate correction for acid impurities in the atmosphere.

Prepare the lamp according to the kind of oil to be tested. For naphthas and lamp oils use the ⅛ inch wick tube and two strands of wicking; for gas oils use the $\frac{3}{16}$-inch wick tube and three strands of wicking. New wicking, about 4½ inches long, is required for each test and must be free from twists after it has been passed through the wick tube. It should be trimmed flat with very sharp scissors. Pour into the clean dry lamp about 20 c.c. of the oil to be tested, insert wick and cork and weigh the assembly with an accuracy of 0.001 g.

Rinse out the absorber containing the glass beads thoroughly with distilled water and add exactly 10.0 c.c. of the standard sodium carbonate solution from an accurately calibrated burette, allowing the burette to drain for three minutes before taking the reading. Rinse the chimney and the spray trap with distilled water, dry the chimney and connect both to the absorber as shown in Fig. 47. Set up the apparatus for the blank determination in exactly the same manner and using exactly 10.0 c.c. of the sodium carbonate solution. Apply gentle suction to both absorbers, light both the weighed oil lamp and the alcohol lamp and then place in position under the chimneys so that the tops of the wick tubes extend into the chimneys not more than $\frac{1}{16}$ inch. Adjust the wick height and the suction so that the flame is steady, free from smoke and approximately ¼ inch high. This requires that the wick be flush with the top of the wick tube for naphthas and a little higher for lamp and gas oils. The room must be free from drafts. The suction on the blank should be so adjusted that air is drawn through both determinations at the same rate. Continue burning for about two hours, or less if the sulphur content of the oil is high. During this time the oil should be consumed at the rate of about 1 gram per hour.

Stop the suction on both absorbers and extinguish the flames. Weigh the oil lamp immediately and calculate by difference the weight of oil consumed. Working with the blank first, disconnect the spray trap and chimney and wash them thoroughly with the methyl orange solution, using a wash bottle with a very fine jet and collecting the washings in the absorber. The amount of solution required for washing should not exceed 35 c.c. Carefully titrate the very faintly yellowish solution in the absorber with standard hydrochloric acid, added to the suction side of the absorber from an accurately calibrated burette. During this titration, the contents of the absorber should be agitated carefully, either by blowing through a rubber tube held between the operator's lips and connected at the other end with the chimney side of the absorber or else by the use of a suitable rubber syringe bulb. As the end-point is approached, draw the liquid back into the chimney side between each addition of acid and then blow it into the suction side, agitating as before. As soon as the first permanent pink color appears, the end-point has been reached. Read and record the volume of hydrochloric acid solution used.

Rinse the chimney and spray trap used in the actual determination into the absorber to which they were connected, exactly as prescribed for the blank. If the methyl orange solution in the absorber has a pink color, too much oil has been burned and the determination must be repeated, burning for a shorter time. Titrate just as in the blank, making sure that the absorber is cold. Read and record the volume of hydrochloric acid solution required.

Calculate the sulphur content of the oil by substituting the proper values in the following formula:

$$\text{Per cent sulphur} = \frac{(\text{c.c. HCl for blank} - \text{c.c. HCl for sample}) \times 0.1}{\text{grams of oil burned}}.$$

If a blank is not run, the formula is:

$$\text{Per cent sulphur} = \frac{(\text{c.c. } Na_2CO_3 - \text{c.c. HCl}) \times 0.1}{\text{grams of oil burned}}.$$

These formulæ are correct only for the standard solutions specified, 1 c.c. of each being equivalent to 0.001 gram of sulphur. The use of solutions of any other strength, such as tenth normal, involves more complicated calculation and is not advisable.

Determination of Sulphur by Bomb Method

(Test Method No. 43)

The method here described is the result of an extended investigation[1] of the sources of error in the determination. It has been in use for more than a year and will give accurate and reliable results if the directions are followed exactly as given. No other method for sulphur in petroleum products is as satisfactory, and the necessary apparatus should be part of the equipment of every laboratory making the determination. The method recently described by Dr. C. E. Waters of the Bureau of Standards [2] is recommended where a bomb is not available.

Bomb.—Accurate results cannot be expected unless a suitable bomb is used. There are several of these on the market at the present time. Among the characteristics which are considered essential are the following: The capacity should not be less than 300 c.c. and preferably about 375 c.c. The lining should be chemically and physically resistant. When closed and filled under pressure, the bomb should show no leak. The oxygen-charging valve is often faulty in this respect. The design and construction of the bomb should be such as to prevent exposure of the gasket to gases within the bomb.

Procedure.—Place 20 c.c. of distilled water in the bottom of the bomb. Use 0.5–1.0 gram of oil, weighed into the sample cup of the bomb, when the material is not volatile. For volatile materials use a very thin-walled glass bulb of the type used in the ultimate organic analysis of such liquids, and place a few drops of sulphur-free alcohol in the sample cup to start combustion. Arrange the ignition mechanism according to the bomb used and close the bomb tightly. Admit oxygen until a pressure of 40 atmospheres is reached. Ignite. Place the bomb in cold water for twenty minutes. Shake vigorously for twenty-five seconds and allow to drain for five minutes. Release the pressure rather slowly and open the bomb. Using distilled water in a wash bottle with a very fine jet, wash the wires and cover thoroughly, allowing the washings to collect in the bomb. In the same way wash the sample cup held by small tongs. Transfer the solution from the bomb to a 500 c.c. beaker and wash the inside of the bomb thoroughly. The total volume of solution thus obtained need not exceed 350 c.c. Avoid any loss of material by spattering or otherwise in the various washings.

Filter the solution through a washed filter paper into another beaker, of smaller size if possible. Wash the filter thoroughly. Add 2 c.c. of concentrated hydrochloric acid and 10 c.c. of saturated bromine water. Evaporate to about 75 c.c. over a hot plate or steam-bath. To the hot solution add 10 c.c. of a 10 per cent barium chloride solution, as hot as possible, in a very fine stream, or dropwise so that thirty to forty-five seconds are required. Stir vigorously with a glass rod during this addition and for

[1] In the Chemical Laboratory of The Atlantic Refining Company; report now in preparation.

[2] Technologic Paper No. 177, Sulphur in Petroleum Oils (1920).

four minutes afterward. Allow the precipitate to settle for one hour on a steam-bath. Cool and let stand for one hour (or longer) at room temperature. Filter carefully through a suitable ashless filter paper and wash the precipitate with hot water, first by decantation and then on the filter until free from chloride. Transfer the wet filter paper and precipitate to a weighed platinum crucible. Dry carefully over a low flame. Allow the filter paper to burn away and then ignite until the precipitate is just burned white. Cool in a desiccator and weigh. From the increase in weight, which is barium sulphate, calculate the percentage of sulphur as follows:

$$\text{Per cent of sulphur} = \frac{\text{grams } BaSO_4 \times 13.734}{\text{grams oil used}}.$$

Accuracy.—Analyses by this method on pure compounds of known sulphur content are in error by less than 1 per cent of the sulphur present. Duplicate determinations on petroleum products should agree within 1 per cent of the sulphur present except in the case of oils of very low sulphur content, where slightly larger variations are to be expected. The following results were obtained in routine work without special precautions:

	Per cent	Per cent
Sample 1	3.87	3.82
Sample 2	0.441	0.447
Sample 3	3.12	3.10
Sample 4	0.213	0.213
Sample 5	1.071	1.064
Sample 6	0.231	0.222

While such accuracy can be easily obtained by any careful operator, deviations from the directions have been found to give incorrect and discordant results.

Determination of Acidity of Petroleum Products

(Test Method No. 44)

The accurate determination of free acid in petroleum oils is considerably more difficult than is commonly believed. Oils which are absolutely neutral will show a slight apparent acidity when titrated in the usual way. This is due to the fact that petroleum hydrocarbons possess the power of adsorbing a small amount of alkali and hence appear to be acid unless special precautions are observed. The following directions are suitable for lubricating oils as received. Gasoline and naphtha should be distilled according to Test Method No. 23 to a 10 per cent residual, and kerosene should be distilled to a residual corresponding to a vapor temperature of 450° F. Reported acidity determined on these residuals should be on the basis of the weight of material used for the distillation.

Procedure.—Weigh accurately 10 to 12 grams of the material to be tested in a clean dry 250 c.c. Erlenmeyer flask. In the case of residuals from distillation, the weight used may be smaller if necessary. In another flask mix 50 c.c. of 60 per cent ethyl or denatured alcohol and one c.c. of a 1 per cent alcoholic solution of phenolphthalein. From a burette add just enough tenth-normal aqueous caustic soda to give a faint pink color. Transfer this neutralized solution to the Erlenmeyer flask containing the weighed petroleum oil. Heat just to boiling on a steam-bath or hot plate, swirling the contents frequently to insure thorough mixing. Titrate the mixture while hot with tenth-normal aqueous caustic soda, free from carbonate, to the first appearance of a pink color, agitating thoroughly between each addition of alkali. From the volume of tenth-normal alkali required the acidity is calculated and reported as percentage by weight of oleic acid, each c.c. of tenth-normal alkali being equivalent to 0.0282 gram of oleic acid.

The following example shows the data and calculation involved: 10.163 grams of lubricating oil required 0.15 c.c. of caustic soda solution whose strength is 0.107 normal. Then each c.c. of the alkali is equivalent to $0.0282 \times 1.07 = 0.0302$ gram of oleic acid. Total oleic acid equivalent of the sample is $0.0302 \times 0.15 = 0.00453$ gram. Acidity is therefore $(0.00453 \div 10.163) \times 100 = 0.045$ per cent calculated as oleic acid.

Occasionally, caustic potash is used in place of caustic soda and acidities are expressed as milligrams of caustic potash required for 1 gram of oil. For all practical purposes, the oleic acid percentage multiplied by 2 gives milligrams of potassium hydroxide per gram of oil.

Fatty-Oil Content

It should be clearly understood that the fatty-oil content of compounded oils is not necessarily proportional to the saponification numbers of these oils. Unfortunately this erroneous assumption is frequently made [1] and it seems necessary to recall the undisputed facts in the matter.

First. The per cent of fatty oil in a compounded oil is exactly equal to 100 times the quotient of the saponification number of the compound divided by the saponification number of the fatty oil used.

Second. The saponification numbers of fatty oils used for compounding may vary from 130 (sperm oil) to 225 (blown cottonseed oil).

Third. Hence it follows that the percentages of fatty oil in a series of compounded oils, all having the same saponification number, will be different if the fatty oils used have different saponification numbers.

Fourth. The fallacy of attempting to calculate the fatty-oil content of a compounded oil by assuming a saponification number of 195 for all fatty oils is strikingly shown in the following example: A compound of 300° Mineral Seal with 25 per cent by weight of either prime winter-strained lard oil (saponification number 195) or sperm oil (saponification number 130) will meet the current specification for Signal Oil Grade B, as to fatty oil. The saponification number of the lard oil compound will be 48.7; of the sperm oil compound 32.5. If the saponification number of the sperm oil is erroneously assumed to be 195, the calculated content by weight of fatty oil in the sperm oil compound will be only 16.7 per cent whereas the true content is 25 per cent.

Obviously fatty-oil content can be calculated from the saponification number of a compounded oil only when the saponification number or nature of the fatty oil is either known or determined. Determinations of this kind are difficult and require directions [2] beyond the scope of this chapter. The saponification numbers of the fatty oils used for compounding are as follows:

Lard oil	192–198
Tallow	193–198
Neatsfoot	193–204
Fish	140–193
Sperm	120–140
Castor	176–187
Rapeseed	170–179
Soya bean	189–197
Peanut	186–197
Cottonseed	191–197
Blown rapeseed	195–216
Blown cotton seed	210–225
Degras	110–210

[1] Compare Bulletin No. 5, U. S. Committee on Standardization of Petroleum Specifications, "Fatty Oil," p. 28.

[2] Consult Lewkowitsch, Chemical Technology of Oils, Fats and Waxes.

If the saponification numbers of both fatty oil and compounded oil are known, the following formula may be used:

$$\text{Per cent of fatty oil} = \frac{100 \times \text{Sapon. No. of compound}}{\text{Sapon. No. of fatty oil}}.$$

Determination of Saponification Number

(Test Method No. 45)

The saponification number is the number of milligrams of potassium hydroxide, KOH, required to saponify 1 gram of the material to be tested. Pure hydrocarbon mixtures contain no fat and therefore have a saponification number of 0. Compounded oils have saponification numbers whose magnitude depends partly on the nature and partly on the quantity of fatty oil mixed with the petroleum oil. The method here described is simpler than some others but has been found to be quite as accurate.

Solutions.—1. Exactly half normal hydrochloric acid solution, accurately standardized.

2. Approximately half normal alcoholic potassium hydroxide solution, which must be clear and free from any pink color. To prepare this solution, dissolve about 33 grams of potassium hydroxide sticks of high purity in 1000 c.c. of purified 95 per cent ethyl alcohol, allow the solution to settle and filter directly into a clean dry reagent bottle. Use a clean rubber stopper for the reagent bottle and keep the solution in a cool dark place. Alcoholic potassium hydroxide made from unpurified alcohol soon acquires a reddish color which renders it unfit for use. Purified alcohol may be prepared as follows: Dissolve about 6 grams of potassium hydroxide in 1000 c.c. of 95 per cent alcohol and either allow to stand for one week or else boil for two hours under a reflux condenser. Then distill off the alcohol on a steam bath, using a short Hempel column or equivalent.

3. Solution of phenolphthalein made by dissolving 0.50 gram in 500 c.c. of purified 95 per cent ethyl alcohol.

Procedure.—The amount of compounded oil used for the determination should be about 5 grams, unless the fatty-oil content is high, in which case a smaller quantity should be taken so that the half normal hydrochloric acid required for the titration is at least 13 c.c.

Weigh the proper amount of the oil to be tested in a perfectly clean and dry 250–300 c.c. Erlenmeyer flask. Use two other clean dry flasks, of the same size and shape, for blanks and follow the subsequent directions with exact duplication in each case. Perform each operation first on one blank, then on the determination proper and finally on the second blank.

From the same accurate burette, add exactly 25.0 c.c. of the alcoholic potassium hydroxide to each flask, allowing one minute for wall drainage. Connect the three flasks with reflux condensers and boil gently for two hours on an electric hot plate or equivalent, shaking the oil flask frequently. Remove all three flasks from the hot plate, add 10.0 c.c. of the 0.1 per cent phenolphthalein solution to each and cover the oil flask and second blank with watch glasses. Titrate the first blank while hot with half normal hydrochloric acid until the pink color is destroyed by a single drop and record the required volume. Titrate the saponification mix, swirling the flask after each addition of acid and tilting it slightly so that the color of the bottom layer may be observed. As soon as the red color begins to fade but before it has disappeared, heat the mixture just to boiling and mix by swirling several times. Then finish the titration as before, being careful not to pass the end-point. Record the required volume of acid. Then heat the second blank to boiling and titrate in exactly the same way as the first and record the required volume of acid.

Subtract the volume of exactly half-normal hydrochloric acid required for the determination proper from the average volume required for the two blanks. Multiply the difference by 28.05 and divide this product by the number of grams of oil used. The quotient is the saponification number of the oil.

NOTES.—For referee purposes two saponifications should be made but only two blanks are necessary.

Use of larger quantities of oil or of potassium hydroxide does not increase the accuracy of the determination. Sodium hydroxide should not be used in place of potassium hydroxide and no acid other than hydrochloric is satisfactory.

The addition of naphtha or benzene to the saponification mixture is unnecessary and makes the end-point harder to distinguish.

The prescribed heating before final titration of the determination proper is necessary for two reasons, first to eliminate adsorption of alkali by the oil, second,to decrease the viscosity of the oil.

The practice of adding a slight excess of hydrochloric acid, boiling and then titrating back with standard alkali is a waste of time and an additional source of error.

GASOLINE PRECIPITATION TEST FOR LUBRICATING OILS

(Test Method No. 46)

At the present time there is no standard method for the gasoline precipitation test and reported results show wide variations. The test is usually applied to black oils and unfiltered cylinder oils, occasionally to other products. Provided the same quality of gasoline is used in each test, fairly duplicable results can be obtained by a gravimetric method similar to Test Method No. 48.[1] Even by this method, determinations on the same sample in different laboratories vary considerably and the gravimetric method requires much more time and skill than its importance justifies, especially as the meaning of the test is obscure. The volumetric method by centrifuge is simpler and is entirely satisfactory for a test which at best is comparative only.

The gasolines prescribed in various precipitation methods include " 86/88 gasoline," " petroleum ether," " normal benzine " and others equally indefinite as to quality and frequently not obtainable. The method here described gives a practical specification for a product not dangerously volatile, of reasonably uniform quality and commercially available.

Apparatus.—Same as in Test Method No. 22 [2] except that a water-bath is not required.

Precipitation gasoline.—Petroleum naphtha of the following specification:

Specific gravity at 60° F.	0.695–0.705	
Initial boiling point, A. S. T. M., 113–131° F.	(45–55° C.)	
End-point, A. S. T. M. not higher than 248° F.	(120° C.)	

Procedure.—1. Exactly 10.0 c.c. of the oil to be tested shall be measured in each of the two clean and dry centrifuge tubes at room temperature. Each tube shall be filled to the 100 c.c. mark with the prescribed diluent and closed tightly with a softened cork (not a rubber stopper). Each tube shall then be inverted at least twenty times, allowing the liquid to drain thoroughly from the tapered tip of the tube each time. The tubes shall then be placed in a water-bath at 90–95° F. for five minutes. The corks shall be momentarily removed to relieve any pressure and each tube shall again be inverted at least twenty times exactly as before. The success of this method depends to a large degree upon having a thoroughly homogeneous mixture which will drain quickly and completely from the tapered tip when the tube is inverted.

[1] Page 715.
[2] Page 670.

2. The two centrifuge tubes shall then be placed in the centrifuge (1) on opposite sides and shall be whirled at a rate of 1400–1500 R.P.M. or equivalent for ten minutes. The volume of sediment at the bottom of each tube shall be read and recorded estimating to 0.05 c.c. if possible. The tubes shall then be replaced in the centrifuge, again whirled for ten minutes as before, and removed for reading the volume of the sediment as before. This operation shall be repeated until the volume of sediment in each tube remains constant for three consecutive readings. In general, not more than four whirlings are required.

3. The volume of the solid sediment at the bottom of each centrifuge tube shall be read estimating to 0.1 c.c. or closer if possible. If the two readings differ by not more than 0.1 c.c. the mean of the two shall be reported as the "Precipitation Number." If the two readings differ by more than 0.1 c.c. two more determinations shall be made and the average of the four determinations shall be reported.

Accuracy.—With care and proper attention to details, duplicate determinations of precipitation number by this method should not differ by more than 0.1 provided the centrifuge tubes are accurate and readable to this degree.

Tar Tests

(Test Method No. 47)

The various tar tests applied to lubricating, insulating and other petroleum oils fall in the same category as gasoline precipitation tests as to obscurity of meaning. The three described here are presented because they are frequently used, especially in connection with exported products. Of the three given, the Tar Value and Tar-forming Value are the more rational and reproducible though obviously not applicable to compounded oils.

A. **Kissling tar** value [1] is determined as follows: 50 grams of the lubricating oil are warmed in a flask with air condenser to about 80° C. (176° F.) with 50 c.c. of alcoholic sodium hydroxide (50 grams alcohol and 50 grams of a 7.5 per cent solution of NaOH in water) and then shaken vigorously and continuously for five minutes. The mixture is put into a separatory funnel, allowed to separate hot, and then after cooling, as large a part as possible of the alkaline solution containing the tarry matter filtered; the clear solution thus obtained is acidified in a separatory funnel and extracted with two 50 c.c. portions of benzol. The benzol is evaporated, the residue weighed and the weight thus obtained, calculated to percentage by weight, is the Kissling Tar value.

B. **Kissling tar** forming value is determined by heating the oil for fifty hours at 302° F. (150° C.) and then proceeding in exactly the same manner as for tar value.

C. Acid tar test.—The basis of this test is the fact that sulphuric acid by combination, polymerization and emulsification will remove various constituents of most petroleum oils. In different countries and for different products there are slight modifications as to strength and proportion of acid, temperature and time. The following method is typical:

Pour 50 c.c. of the oil to be tested and 50 c.c. of gasoline into a cylinder of about 125 c.c. capacity, provided with a ground-glass stopper and accurately graduated from the bottom up in 1 c.c. divisions to 100 c.c. Mix thoroughly and add 10 c.c. of 66° Bé. sulphuric acid. Shake the mixture vigorously for five minutes and allow to settle at room temperature for twenty-four hours. Read the volume of the bottom or acid layer. Subtract 10 c.c. from this volume. Multiply the remainder by 2 and record the product as "Acid Tar Test—per cent."

[1] Kissling, Chem. Zeit., **30** (1906), 932; **31** (1907), 328; **32** (1908), 938; **33** (1909), 521.

Solubility Tests

Solubility tests on petroleum products are restricted in most cases to asphaltic materials. Exceptions are the Gasoline Precipitation (Test Method No. 46) for cylinder and black oils and a rarely used test for paraffin wax depending on its relatively slight solubility in acetone.[1] The following methods are those devised for asphaltic products.[2]

Determination of Solubility

(Test Method No. 48)

This method consists in separating material insoluble in the solvent selected by means of a Gooch crucible and asbestos mat. The success of the method depends largely on the proper formation of the filter. The necessary apparatus is shown in Fig. 48, in which (*a*) is a 500 c.c. suction flask, (*b*) a rubber stopper, (*c*) a filter tube, (*d*) a section of suitable thin rubber tubing and (*e*) the Gooch crucible. Satisfactory dimensions for Gooch crucibles are: top diameter 44 mm.; bottom diameter 36 mm.; height 25 mm.

The asbestos used should consist of clean fibers $\frac{1}{8}$ to $\frac{3}{8}$ inch long, suspended in water. Freshly prepared asbestos should be washed by decantation until all fine particles are removed. With the apparatus assembled as shown, but without applying suction, fill the crucible with the water containing suspended asbestos. Allow to stand for about one minute, then apply very gentle suction until the water has been drawn through. Discontinue suction and again fill with asbestos suspension. Renew suction as before and repeat until a firm asbestos mat has been formed of thickness such that the position of the perforations in the crucible is barely discernible when the crucible is held between the eye and a very bright light. The proper preparation of such a filter should require at least three fillings of the crucible, which fact gives a definite idea of the amount of asbestos to be suspended in the water. The filter so prepared should be washed thoroughly with distilled water poured on in a manner to avoid disturbing the asbestos and using gentle suction. The crucible should then be removed, dried at 212° F. and then ignited at a low red heat. It should be cooled in a desiccator, weighed accurately and kept in the desiccator until needed.

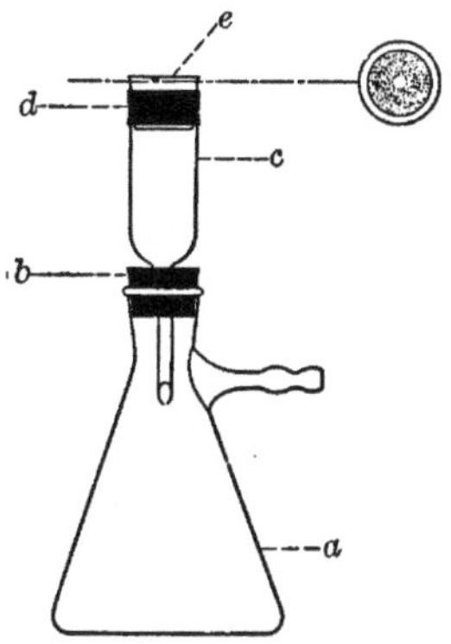

Fig. 48.—Suction Flask and Gooch filter assembly.

The solvents ordinarily used are carbon bisulphide, petroleum naphthas and carbon tetrachloride. The materials soluble in carbon bisulphide are reported as "Total Bitumen." The material soluble in carbon bisulphide but insoluble in petroleum naphtha is reported as "Asphaltenes." The material soluble in carbon bisulphide but insoluble in carbon tetrachloride is reported as "Carbenes." Carbon bisulphide and carbon tetrachloride are chemical compounds and should be used in pure form. Petroleum naphtha is an indefinite term applied to various mixtures of hydrocarbons. For this reason all determinations of solubility in naphtha must include a description of the kind of naphtha used, as to gravity at 60° F. and the Over- and End-Points according to Test Method No. 23. Two grades of petroleum naphtha often used are: $\frac{86}{88}$ Gas-

[1] U. S. Navy Dept. Specifications for Illuminating Wax.

[2] Compare Laboratory Manual of Bituminous Materials, Prévost Hubbard (Wiley, 1916).

oline, Sp. Gr. about 0.645; Over-Point about 77° F.; End-Point about 212° F. and 74/76 Gasoline, Sp. Gr. about 0.700; Over-Point about 122° F. and End-Point about 248° F. Both these products should show less than 4 per cent soluble in 98 per cent sulphuric acid.

Precautions.—Both naphtha and carbon bisulphude are very volatile and their vapors ignite very readily. Neither should be used in the vicinity of a flame or even of an electric hot plate. A fire extinguisher should be immediately available even where these precautions are observed.

Procedure is approximately the same regardless of the solvent used. Weigh in a 150 c.c. Erlenmeyer flask 1 to 2 grams of the material to be tested. This quantity may be varied in special cases but the weight of material dissolved in the solvent should be approximately 1 gram. Add exactly 100 c.c. of the solvent selected, cork the flask tightly with a previously softened cork (not a rubber stopper). Mix by swirling frequently at room temperature until the asphaltic material is thoroughly disintegrated. Set aside in a dark place for at least five hours and preferably over night. Place the Gooch crucible, prepared as described, in the filter tube and moisten the asbestos with a few drops of the solvent. Apply very gentle suction and carefully decant the solution in the 150 c.c. Erlenmeyer flask through the filter, taking care that the precipitate is not disturbed. As soon as the decanted liquid shows a tendency to carry precipitate to the filter, stop decantation and suction. Add 15 c.c. of fresh solvent to the 150 c.c. Erlenmeyer flask and mix by swirling. Allow to settle for a few minutes and decant as before. Repeat this procedure twice more and then transfer all the precipitate to the filter by means of more fresh solvent, using a glass rod if necessary. Wash the precipitate and the inside of the crucible with fresh solvent until the filtrate is colorless, using as little solvent as possible. Draw air through the crucible until the solvent odor has disappeared. If necessary, clean the outside of crucible. Dry the crucible and contents in an air-bath for thirty minutes at 212–220° F., cool in a desiccator and weigh. The increase in weight is the material insoluble in the selected solvent and should be calculated to a percentage basis in the usual way. The solubility is obtained by subtracting from 100 per cent the percentage insoluble.

EVAPORATION TESTS

Volatile petroleum products are often subjected to a test to determine the completeness or rapidity of evaporation. A typical method is as follows: Measure 50 c.c. or other suitable volume of the material to be tested into an accurately weighed dish. Allow to evaporate at any suitable desired temperature not higher than 104° F. (40° C.). Determine the weight of the residue when this becomes constant and report as grams per 100 c.c.

A rapid method for determining completeness of evaporation consists in saturating part of a piece of clean white filter paper with the material to be tested and allowing it to stand at room temperature for a specified time, as for example thirty minutes. Incomplete evaporation is indicated by an oily stain on the paper or even more delicately by the presence of an odor in cases where no stain is visible.

Comparative tests on two similar products are usually made by allowing equal volumes (2 c.c.) of each to evaporate from two watch glasses, side by side, and noting the time required for complete evaporation in each case.

Difficultly volatile petroleum products are subjected to evaporation tests for the purpose of determining the loss under specified conditions.[1] For example, the following method for lubricating oil was formerly recommended by the U. S. Committee on Standardization of Petroleum Specifications:[2] "Twenty grams of the oil are placed

[1] See P. H. Conradson, Proc. Am. Soc. Test. Materials, 18, I, 323 (1918).

[2] Bulletin No. 4, p. 3 (April, 1920).

in a weighed flat-bottom glass crystallizing dish having a diameter of approximately 3¾ inches. The dish is then placed in an oven at a temperature of 212° F. for two hours, cooled in a desiccator and weighed." The results obtained by this method are not concordant in many cases for the reason that many of the factors influencing results are not definitely described. Among the important factors affecting evaporation tests are the following: temperature and time of exposure, size of sample, size and shape of container and of oven, position of sample in oven, barometric pressure and ventilation of oven. The following method is therefore recommended as much more satisfactory than the older procedure.

Determination of Loss on Heating of Oils and Asphaltic Compounds

(Test Method No. 49)

This is the method adopted as Standard by the American Society for Testing Materials in 1920.[1] It has been revised several times since the adoption of the original form

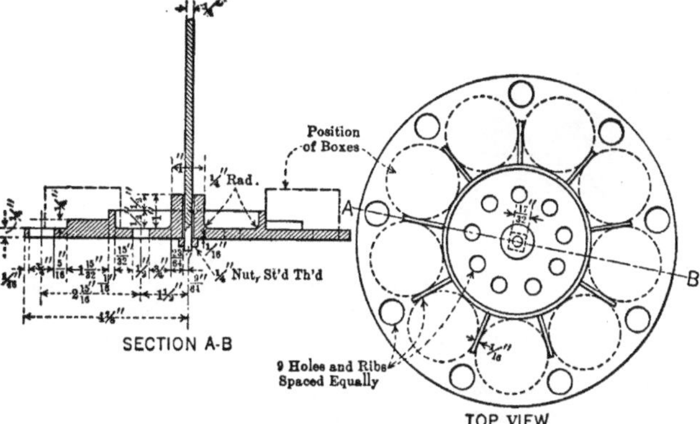

Fig. 49.—Details of revolving aluminum shelf.

in 1911 and may therefore be considered as reliable. For specific purposes in connection with petroleum products this method may be modified slightly in special cases but such modifications should be confined to changes in (*a*) weight of material used, (*b*) temperature of exposure and (*c*) time of exposure. No other variations are permissible.

1. This test is used to determine the loss in weight (exclusive of water) of oil and asphaltic compounds when heated as hereinafter prescribed. The material under examination must therefore first be tested for water and if water is found to be present, it must be removed by suitable methods of dehydration before the material is subjected to the loss on heating test; or another sample obtained which is free from water.

Apparatus.—2. *Oven.*—The oven may be either circular or rectangular in form and may be heated by either gas or electricity. Its interior dimensions shall be as follows: height, not less than 40.64 cm. (16 inches); width and depth or diameter, at least 4.08 cm. (2 inches) greater than the diameter of the revolving shelf.

It shall be well ventilated and shall be fitted with a window in the upper half of the door, so placed and of sufficient size to permit the accurate reading of the thermometer

[1] A. S. T. M. Standard Test Serial Designation D, 6–20.

without opening the door. It shall also be provided with a perforated circular shelf preferably of approximately 24.8 cm. (9.75 inches) diameter. This shelf shall be placed in the center of the oven and shall be suspended by a vertical shaft and provided with mechanical means for rotating it at the rate of 5 to 6 revolutions per minute. It shall be provided with recesses equidistant from the central shaft in which the tins containing the samples are to be placed. (A recommended form of aluminum shelf is shown in Fig. 49). A convenient oven of the electric type is shown in Fig. 50.

Fig. 50.—Electric oven with revolving shelf.

3. *Thermometer*.—The thermometer shall be between 12.7 cm. (5 inches) and 15.24 cm. (6 inches) in length and the mercury bulb shall be from 10 to 15 mm. (0.39 to 0.59 inch) in length. The scale shall be engraved on the stem, shall be clear-cut and distinct, and shall run from 150° to 175° C. (302 to 347° F.) in 1° C. divisions and shall commence substantially 3.81 cm. (1½ inches) above the top of the bulb. Every fifth graduation shall be larger than the intermediate ones and shall be numbered. The degrees shall be substantially 3.17 mm. (⅛ inch) apart. The thermometer shall be furnished with an expansion chamber at the top and have a ring for attaching tags. It shall be made of a suitable quality of glass and be so annealed as not to change its readings under conditions of use. It shall be correct to 0.25° C. (0.45° F.) as determined by comparison at full immersion with a similar thermometer calibrated at full immersion by the United States Bureau of Standards.

4. *Container*.—The container in which the sample is to be tested shall be of tin, cylindrical in shape, and shall have a flat bottom. Its inside dimensions shall be substantially as follows: diameter, 55 mm. (2.17 inches); depth, 35 mm. (1.38 inches). (A 3-oz. Gill style ointment box, deep pattern, fulfills these requirements.)

Preparation of sample.—5. The sample as received shall be thoroughly stirred and agitated, warming, if necessary, to insure a complete mixture before the portion for analysis is removed.

Testing.—6. Weigh 50 grams of the water-free material to be tested into a tared container conforming to the requirements of Section 4. Bring the oven to a temperature of 163° C. (325° F.), and place the tin box containing the sample in one of the recesses of the revolving shelf. The thermometer shall be immersed for the depth of its bulb in a separate 50-gram sample of the material under test, placed in a similar container, and shall be conveniently suspended from the vertical shaft. This sample shall rest in one of the recesses upon the same shelf and revolve with the sample or samples under test. Then close the oven and rotate the shelf 5 to 6 revolutions per minute during the entire test. Maintain the temperature at 163° C. (325° F.) for five hours, then remove the sample from the oven, cool and weigh, and calculate the loss due to volatilization.

7. *Permissible variation in temperature*.—During the five-hour period the temperature shall not vary more than 1° C. All tests showing a greater variation in temperature shall be rejected.

Accuracy.—8. Up to 5 per cent loss in weight the results obtained may be considered as correct within 0.5. Above 5 per cent loss in weight the numerical limit of error increases 0.01 for every 0.5 per cent increase in loss by volatilization as follows:

Volatilization loss, per cent	Numerical correction	True volatilization loss, per cent
5.0	±0.50	4.50 to 5.50
5.5	±0.51	4.91 to 6.01
6.0	±0.52	4.58 to 6.52
10.0	±0.60	9.40 to 10.60
15.0	±0.70	14.30 to 15.70
25.0	±0.90	24.10 to 25.90
40.0	±1.20	38.80 to 41.20

Precautions.—9. Under ordinary circumstances a number of samples having about the same degree of volatility may be tested at the same time. Samples varying greatly in volatility should be tested separately. Where extreme accuracy is required not more than one material should be tested at one time and duplicate samples of it should be placed simultaneously in the oven. Such duplicates shall check within the limits of accuracy given above. Results obtained on samples showing evidences of foaming during the test shall be rejected.

NOTE.—If additional periods of heating are desired, it is recommended that they be made in successive increments of five hours each.

DETERMINATION OF PENETRATION OF ASPHALTS

(Test Method No. 50)

The pentration test on solid and semi-solid petroleum asphalts serves not only as the basis for specifications but also as a valuable control in refining processes. Measurements of penetration at different temperatures permit accurate study and comparison of asphalts as to their susceptibility to temperature changes. While the following method is an A. S. T. M. Standard,[1] for routine work it is convenient to omit stirring the melted sample and to cool by partly immersing the container in water at 77° F. for thirty minutes, followed by complete immersion for one hour. The penetrometer shown in Fig. 51 is entirely satisfactory.

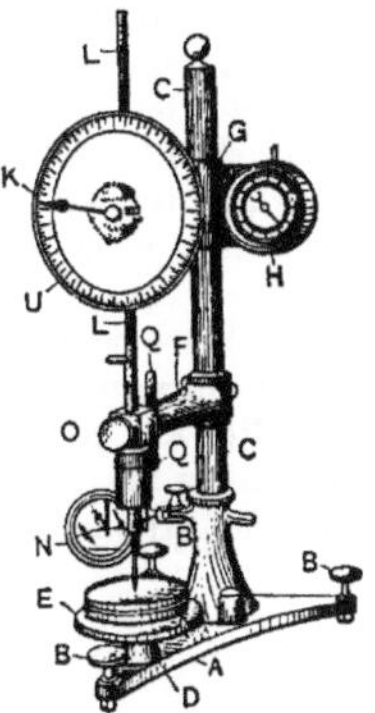

FIG. 51.—Penetrometer for asphalt.

Definition.—1. Penetration is defined as the consistency of a bituminous material, expressed as the distance that a standard needle vertically penetrates a sample of the material under known conditions of loading, time and temperature. Where the conditions of test are not specifically mentioned, the load, time and temperature are understood to be 100 grams, 5 seconds, 25° C. (77° F.), respectively, and the units of penetration to indicate hundredths of a centimeter.

Apparatus.—2. The container for holding the material to be tested shall be a flat-bottom, cylindrical dish, 55 mm. ($2\frac{3}{16}$ inches) in diameter and 35 mm. ($1\frac{3}{8}$ inches) deep.[2]

3. The needle [3] for this test shall be of cylindrical steel rod 50.8 mm. (2 inches) long

[1] Adopted under the Serial Designation D 5-16, A. S. T. M. Standards 1916, 530.

[2] This requirement is fulfilled by the American Can Company's Gill style ointment box, deep pattern, 3-oz. capacity.

[3] A modified blunt needle is now being considered; A. S. T. M. Proceedings, 20, Part I (1920), 435.

and having a diameter of 1.016 mm. (0.04 inch) and turned on one end to a sharp point having a taper of 6.35 mm. (¼ inch).

4. The water-bath shall be maintained at a temperature not varying more than 0.1° C. from 25° C. (77° F.). The volume of water shall not be less than 10 liters and the sample shall be immersed to a depth of not less than 10 cm. (4 inches) and shall be supported on a perforated shelf not less than 5 cm. (2 inches) from the bottom of the bath.

5. Any apparatus which will allow the needle to penetrate without appreciable friction, and which is accurately calibrated to yield results in accordance with the definition of penetration, will be acceptable.

6. The transfer dish for container shall be a small dish or tray of such capacity as will insure complete immersion of the container during the test. It shall be provided with some means which will insure a firm bearing and prevent rocking of the container.

Preparation of sample.—7. The sample shall be completely melted at the lowest possible temperature and stirred thoroughly until it is homogeneous and free from air bubbles. It shall then be poured into the sample container to a depth of not less than 15 mm. (⅝ inch). The sample shall be protected from dust and allowed to cool in an atmosphere not lower than 18° C. (65° F.) for one hour. It shall then be placed in the water-bath along with the transfer dish and allowed to remain one hour.

Testing.—8. (*a*) In making the test the sample shall be placed in the transfer dish filled with water from the water-bath of sufficient depth to completely cover the container. The transfer dish containing the sample shall then be placed upon the stand of the penetration machine. The needle, loaded with specified weight, shall be adjusted to make contact with the surface of the sample. This may be accomplished by making contact of the actual needle point with its image reflected by the surface of the sample from a properly placed source of light. Either the reading of the dial shall then be noted or the needle brought to zero. The needle is then released for the specified period of time, after which the penetration machine is adjusted to measure the distance penetrated.

At least three tests shall be made at points on the surface of the sample not less than 1 cm. (⅜ inch) from the side of the container and not less than 1 cm. (⅜ inch) apart. After each test the sample and transfer dish shall be returned to the water-bath and the needle shall be carefully wiped toward its point with a clean, dry cloth to remove all adhering bitumen. The reported penetration shall be the average of at least three tests whose values shall not differ more than four points between maximum and minimum.

(*b*) When desirable to vary the temperature, time and weight, and in order to provide for a uniform method of reporting results when variations are made, the samples shall be melted and cooled in air as above directed. They shall then be immersed in water or brine, as the case may require, for one hour at the temperature desired. The following combinations are suggested:

At 0° C. (32° F.) 200 gram weight, 60 seconds.
At 46.1° C. (115° F.) 50 gram weight, 5 seconds.

Determination of Asphalt Content

(Test Method No. 51)

This method is applicable to all asphaltic materials having a penetration greater than 100 at 77° F. It gives more information than an evaporation test for many purposes and is recommended for study of refinery yields of asphalt as well as for specification purposes. "Asphalt Content" is arbitrarily taken as percentage of 100 penetration material.

Apparatus.—1. A. S. T. M. penetration cups, flat-bottom, cylindrical form, $2\frac{3}{16}$ inches diameter and $1\frac{3}{8}$ inches deep.

2. Thermometer, of range including 400–500° F.

3. Electric hot plate, three heat type. (The oven described under Test Method No. 49 may be used but is not necessary.)

4. Balance, capacity 100 grams, sensitive to 0.01 gram.

5. Penetrometer.

Procedure.—Turn heat on the hot plate. Fill three weighed cups to about $\frac{3}{4}$ inch below the top with the material to be tested and find the weight of material in each cup. Fill a fourth cup with high-flash petroleum cylinder oil and place the four cups on the hot plate with the cylinder-oil cup in the center. Suspend the thermometer so that the bulb is completely immersed in this cylinder oil. If the material to be tested contains moisture or volatile solvents an asbestos board $\frac{1}{4}$ inch thick should be interposed between the hot plate and the cups until there is no danger of foaming. Hold the temperature as read on the thermometer at 450–500° F. and evaporate the samples to a penetration of 100 at 77°.

The point at which samples should be removed for trial of penetration is a matter of judgment and experience. The penetration should be taken as described in Test Method No. 50, cooling each time. Where the approximate percentage is known, a rough weighing will indicate whether evaporation is sufficient to justify a penetration test. Where a specification prescribes a minimum or maximum percentage of 100 penetration material, one cup should be evaporated to each percentage and the penetration taken. With experience it is possible to obtain the desired penetration without more than three trials. Any penetration between 90 and 110 may be considered as 100 because the evaporation loss is negligible between these figures. The residual material calculated to per cent of the original is the asphaltic content or "per cent of 100 penetration."

Determination of Ductility

(Test Method No. 52)

Ductility as determined by this method is the distance in centimeters to which an asphalt can be stretched without rupture under certain specified but arbitrary conditions. The test is usually applied to asphaltic cements as specified by the American Society of Municipal Improvements.

The apparatus consists of a ductility machine and mold (Fig. 52). The machine itself is a device for stretching the asphalt in the mold at a definite rate of speed, usually 5 cm. per minute, the asphalt being held in water during the operation at a specified temperature, usually 77° F.

Procedure.—The mold should be placed upon a brass plate. To prevent the asphalt from adhering to the plate and the inner side of the two removable pieces of the mold, they should be well amalgamated. The different pieces of the mold should be held together in a clamp or by means of an India rubber band. The material to be tested is poured into the mold while in a molten state, a slight excess being added to allow for shrinkage on cooling. After the briquette is nearly cool, it is smoothed off level by means of a heated palette knife. When cooled, the clamp is taken off and the two side pieces removed, leaving the briquette of asphalt firmly attached to the two ends of the mold, which thus serve as clips. The briquette should be immersed in water maintained at the required temperature for at least thirty minutes or until the whole mass of bitumen is at 77° F. It is then pulled apart at the required rate of speed in a suitable machine, the briquette being entirely immersed in water maintained at 77° F. during the entire operation of pulling. Any pieces of dirt, wood, or extraneous matter in the briquette may cause the fracture of the fine thread before the true maximum ductility of the material under examination has been reached. Great care should be observed, therefore, to avoid the presence of such foreign matter in the bitumen when it is poured into the mold. The average of at least two tests shall be recorded as the ductility of the

Fig. 52.—Ductility machine for asphalt.

sample under examination. These tests must not differ more than 20 per cent from their average.

Float Test

(Test Method No. 53)

The float test is a convenient method for control of refinery processes and for use in connection with specifications. It does not measure any single property but is an indication of consistency. It is commonly used on materials too soft for penetration but not sufficiently liquid for viscosity determination at the temperature involved. The usual temperatures employed are 32°, 122° and 150° F. according to the consistency of the material.

The essential parts of the apparatus are an aluminum float and a conical brass collar which can be screwed into an opening in the float. The combined weight of float and collar is exactly 50 grams and the dimensions are shown in Fig. 53.

In making the test the brass collar is placed with the small end down on a brass plate, which has been previously amalgamated with mercury by first rubbing it with a dilute solution of mercuric chloride or nitrate and then with mercury. A small quantity of the material to be tested is heated in a metal spoon until quite fluid, with care that it suffers no appreciable loss by volatilization and that it is kept free from air bubbles. It is then poured into the collar in a thin stream until slightly more than level with the top. The surplus may be removed, after the material has cooled to room temperature, by means of a spatula blade which has been slightly heated. The collar and plate are then placed in ice water maintained at 41° F., and left in this bath for at least fifteen minutes. Meanwhile a suitable container is filled about three-fourths full of water and placed on a tripod, and the water is heated to any desired temperature at which the test is to be made. This temperature should be accurately maintained, and should at no time throughout the entire test be allowed to vary more than 1° F. from the temperature selected. After the material to be tested has been kept in the ice water for at least fifteen minutes, the collar and contents are removed from the plate and screwed into the aluminum float, which is then immediately floated in the warmed bath. As the plug of bituminous material becomes warm and fluid, it is gradually forced upward and out of the collar, until water gains entrance to the saucer and causes it to sink

The time in seconds between placing the apparatus on the water and when the

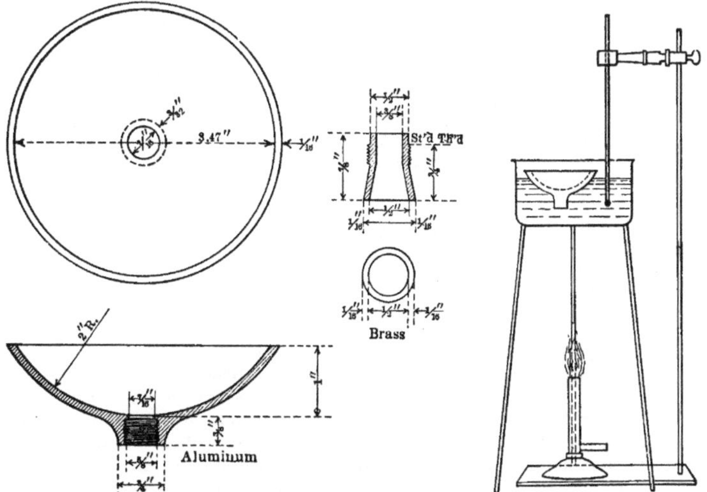

Fig. 53.—Float test apparatus.

water breaks through the bitumen is determined by means of a stop-watch and is taken as a measure of the consistency of the material under examination.

Fixed Carbon and Ash

(Test Method No. 54)

This method is usually applied to asphaltic petroleum products. While it is arbitrary as far as fixed carbon is concerned and not particularly accurate, it is of considerable value in distinguishing asphaltic materials from different sources. For fixed carbon in lubricating oils, Test Method No. 39 [1] is preferable. Ash in such oils should be determined by the modified method for ash as given.

Procedure.—One gram of the material is placed in a platinum crucible weighing from 20 to 30 grams and having a tightly fitting cover. It then is heated for seven minutes over the full flame of a Bunsen burner, as shown in Fig. 54. The crucible should be supported on a platinum triangle with the bottom from 6 to 8 cm. above the top of the burner. The flame should be fully 20 cm. high when burning freely, and the deter-

[1] Page 704.

mination should be made in a place free from drafts. The upper surface of the cover should burn clear, but the under surface should remain covered with carbon, excepting in the case of some of the more fluid bitumens, when the under surface of the cover may be quite clean.

The crucible is removed to a desiccator and when cool, is weighed, after which the cover is removed, and the crucible is placed in an inclined position over the Bunsen burner and ignited until nothing but ash remains. Any carbon deposited on the cover is also burned off. The weight of ash remaining is deducted from the weight of the residue after the first ignition of the sample. This gives the weight of the so-called fixed or residual carbon, which is calculated on a basis of the total weight of the sample exclusive of mineral matter. If the presence of a carbonate mineral is suspected, the percentage of mineral matter may be obtained most accurately by treating the ash with a few drops of ammonium carbonate solution, drying at 100° C., then heating for a few minutes at a dull red heat, cooling and weighing.

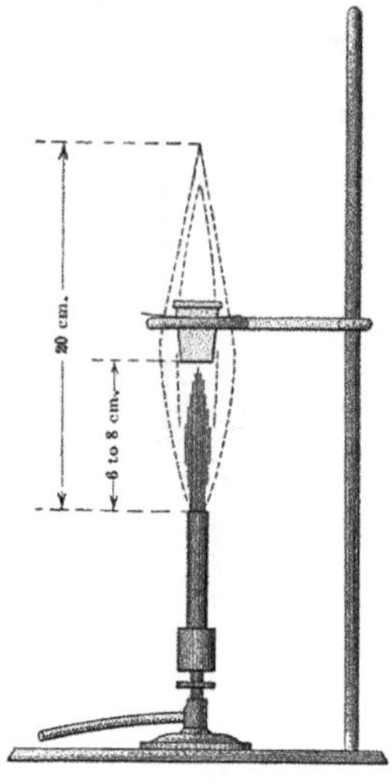

An excellent form of crucible for this test is shown in Fig. 54. It has a cover with a flange 4 mm. wide fitting tightly over the outside of the crucible, and weighs complete about 25 grams. Owing to sudden expansion in burning some of the more fluid bitumens, it is well to hold the cover down with the end of the tongs until the most volatile products have burned off.

Fig. 54.—Fixed carbon apparatus.

Some products, particularly those derived from Mexican petroleum, show a tendency to expand suddenly and foam over the sides of the crucible in making this determination, and no method of obviating this trouble without vitiating the result has thus far been forthcoming. Experiments in the laboratory of the Office of Public Roads and Rural Engineering indicate that the difficulty may be overcome by placing a small piece of platinum gauze over the sample and about midway of the crucible. The gauze should be so cut or bent as to touch the sides of the crucible at all points, and, of course, is weighed in place in the crucible before and after ignition.

Modified **ash method**.—Since the percentage of ash in lubricating oils is usually very small, it is necessary to use a fairly large sample where accuracy is desired. The following method will be found satisfactory: Weigh about 25 grams of the oil to be tested in a weighed platinum evaporating dish [1] of about 40 c.c. capacity. Set on a platinum triangle on a ring stand under a laboratory hood. Heat carefully until the oil takes fire and allow it to burn away without applying additional heat. Then ignite carefully at first and later with a full Bunsen flame, until all carbonaceous matter disappears. The ash is likely to be reddish on account of the presence of iron. Cool in a desiccator and weigh. Calculate to percentage by weight on the original oil.

Oil and Moisture in Paraffin Wax

(Test Method No. 55)

Accurate determination of "Oil and Moisture" in paraffin wax cannot be made by any known method. Reports made by some commercial laboratories showing percentage to the second decimal place are humorously inaccurate. The results obtained

[1] A platinum crucible is not as satisfactory as the dish.

by the method here described may be only comparative on some products but they are reproducible within 10 per cent of the "oil and moisture" found, except for waxes showing less than 1 per cent by the test. In this case, the allowable difference between duplicate determinations is about 0.2 per cent of the wax used.

If correct information as to moisture is required, water should be determined according to Test Method No. 20 [1] or by allowing it to settle out of a melted 100 c.c. wax sample in a B. S. tube. The "oil" in paraffin wax is arbitrarily assumed to be that part of it which is liquid at 60° F. The determination of oil and moisture in fully refined paraffin wax is of little value. Such wax should be free from oil and examinations for taste and odor of the melted wax are far more reliable than any pseudo-analysis. Absence of taste and odor in commercial paraffin may be confidently accepted as proof that no oil is present. The pressing method here described is suitable for semi-refined paraffin, scale wax and intermediates.

Fig. 55.—Wax-testing press.

Apparatus.—1. Wax-testing press, Fig. 55, capable of giving a pressure of about 900 pounds per square inch. The wax cup should consist of a flat circular base and a ring, 5⅝ inches inside diameter fitting into a depression on the base. A plunger having a flat circular face should fit into the wax cup with about 0.002 inch clearance.

2. Supply of high-grade, soft and very absorbent blotters, accurately cut into disks exactly 5⅝ inches diameter. Colored blotters are preferable.

3. Supply of thin but strong, tightly woven silesia or similar cloth free from nap. This cloth should be accurately cut into disks, 5⅝ inches diameter.

4. Set of analytical weights and balance sensitive to 0.005 gram.

5. Desiccator with calcium chloride or equivalent.

6. Dry refrigerator, temperature 40–50° F.

Sample. — Unless wax is entirely liquid, proper sampling requires considerable care and judgment, because wax in solid form, especially scale wax, is not homogeneous. A satisfactory method for sampling a solid cake is to remove a diagonal section of the full thickness of the cake. Very hard wax may be cut with a rip saw. Samples should always be placed and kept in glass or metal containers and never in any absorbent material such as paper or cardboard. Samples sent to the laboratory should be thoroughly representative of the material in question and should be kept in a cool place.

If the laboratory sample is solid, it may be used as received, cutting out a diagonal section of size suitable for the test. Samples of high oil-content should be melted in entirety. Liquid samples should be stirred and a quantity slightly greater than that required for the test should be poured into an evaporating dish or a beaker cover and allowed to solidify.

[1] Page 668.

Procedure.—It is desirable to have the room temperature at 60° F. or as near this as possible. Keep the cloth and blotter disks in a desiccator at room temperature. Keep the clean and dry wax-cup base, ring and plunger in the dry refrigerator so that they are cooled to 60° F. or lower if the room temperature is above 60° F. It is advisable to have the temperature of the refrigerator lower than 60° F. by about one-half the difference between room temperature and 60° F.

The original laboratory sample having been reduced to a solid test portion as described chill this test sample to 60° F. or even lower. According to the hardness of the chilled test sample, scrape or shave it into very small flakes or thin shavings. Do not attempt to grind the sample. Place two of the cloth disks on the same pan of the balance, counterpoise and leave them on the balance pan, weigh 30–35 grams (or 500 grains) on the top cloth, with an accuracy of 0.005 gram. Distribute the wax evenly over the disk, leaving bare a ½ inch ring around the outside of the disk. Lay the second cloth disk over the wax and carefully transfer the cloth-wax "sandwich" to the refrigerator together with a sufficient number of blotter disks. These disks should be separated to facilitate cooling. After ten minutes place the cloth-wax "sandwich" between blotters, using two or three on each side according to the oil content of the wax. Make sure that the edges of all the disks are even and push the assembly into the cooled wax cup. It is essential that the disks fit perfectly inside the ring without any humps or wrinkles to prevent the application of pressure to the wax.

Transfer the assembled apparatus to the wax-testing press at once and gradually apply a pressure of about 900 pounds per square inch, taking about thirty seconds for the operation. Press for five minutes and release. Quickly replace the used blotters with fresh cold ones and again press for five minutes as before. If the blotters next to the cloth show more than a trace of oil, replace them again and set the assembly including wax cup and plunger in the refrigerator. After fifteen minutes press again as before, repeating until the blotters next to the cloth show only the faintest trace of oil. Separate the cloth-wax "sandwich" from the blotters, picking off any adhering shreds, which are easily seen if colored blotters are used. Leave the "sandwich" in the desiccator at room temperature for fifteen minutes and then weigh.

Calculate the "oil and moisture" by substituting the proper values in the following formula,

$$\text{Per cent oil and moisture} = \frac{(B-A)\times 100}{W},$$

where A is the combined cloth-wax weight after pressing;
B is the combined cloth-wax weight before pressing;
W is weight of wax used.

Duplicate tests should be made for referee purposes and both results should be reported.

NOTE.—It is not advisable to attempt separation of wax from cloth disk before weighing as practiced by some laboratories. Results obtained thus are too high as may be demonstrated by trial on oil-free fully refined paraffin.

APPROXIMATE VAPOR PRESSURE OF GASOLINE

(Test Method No. 56)

According to the regulations of the Interstate Commerce Commission[1] liquid condensates from natural gas or casing-head gas require extra care in handling and are subject to certain definite special requirements depending on the vapor pressure of the liquid. The measurement of this vapor pressure may be made by the method here described.

[1] See I. C. C. Regulations, paragraph 1824 (k) as revised Dec. 15, 1919.

Apparatus.—1. Vapor pressure bomb, of cylindrical form as shown in Fig. 56, capacity about 400 c.c. It should be made of metal, preferably brass about $\frac{1}{4}$ inch thick and all joints should be well soldered. The pressure gage should be accurate to $\frac{1}{2}$ pound per square inch and should be soldered to the union connection.

2. Measuring cylinder or cup, suitable for measuring exactly one-tenth of the total capacity of the vapor pressure bomb as determined by trial.

3. Water-bath of size sufficient to permit immersion of the vapor pressure bomb assembly to the lower edge of the pressure gage. An ordinary 10-quart bucket is suitable.

4. Thermometer accurate to 1° F. at 70°, 90° and 100° F.

5. Small funnel with stem about 3 inches long and small enough to enter the mouth of the bomb.

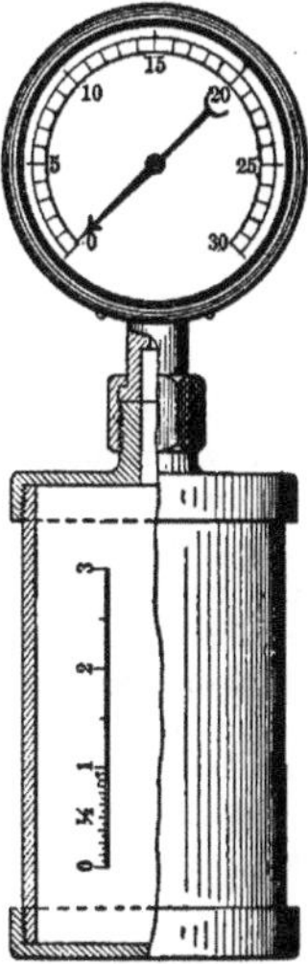

Fig. 56.—Vapor pressure bomb.

Procedure.—Have the bomb clean and dry. If possible, take the sample directly from the storage tank or tank-car by lowering the bomb, in a vertical position with pressure gage removed, into the gasoline and leaving it submerged for about four minutes. Withdraw the completely filled bomb and pour out exactly one-tenth its total capacity, measuring in the cylinder or cup. Take the temperature of the gasoline in the bomb and record as "filling temperature." Connect the pressure gage to the bomb, using a rubber gasket, and tighten the union connection with a short-handle wrench.

If samples for test are sent to the laboratory in any container other than the closed bomb, such containers must be perfectly gas-tight and the sample must be thoroughly representative of the material in question. In this case both bomb and sample should be placed in a refrigerator and cooled to about 65° F. before filling the bomb. Then pour about 200 c.c. of the sample into the bomb, using the small funnel, shake gently and pour out all the gasoline. Fill the bomb completely full with the test sample, pour out one-tenth of its total capacity and immediately close the bomb just as described above.

Immerse the vapor pressure bomb assembly to the lower edge of the pressure gage in water at 70° F. for five minutes, stirring the water constantly and maintaining its temperature within 1° of 70° F. Tap the pressure gage lightly with the finger tips, read and record the indicated pressure to the nearest $\frac{1}{2}$ pound mark. Remove the bomb from the water-bath and loosen the union connection until the gage shows zero pressure. After twenty seconds, tighten the union, using a short-handle wrench and immerse the bomb assembly to the lower edge of the pressure gage in water at the temperature required by the Interstate Commerce Commission, namely, 100° F., Mar. 1st to Nov. 1st and 90° F., Nov. 1st to Mar. 1st. Maintain the required temperature for ten minutes, stirring constantly during this period. Then tap the pressure gage lightly and read the indicated pressure as before. Correct this pressure according to the initial filling temperature of the gasoline in the bomb, as follows: Deduct 1 pound for filling temperature 50–59° F. inclusive; 2 pounds for 40–49° F. inclusive and 3 pounds for filling temperature below 40° F.

The final report should include:

Date of test.
Method of filling.
Filling temperature.
Pressure at 70° F. before venting.

Temperature of final pressure measurement.
Actual pressure at this temperature.
Corrected pressure.
Gravity of the gasoline at 60° F.

Precautions.—A high-class pressure gage is essential and even this must be frequently checked for accuracy. Proper corrections for scale error should be made when necessary.

The assembled apparatus must be perfectly tight and free from leaks. Such leaks will be indicated by the appearance of gas bubbles or gasoline when the bomb assembly is immersed in the water-bath.

The test must be applied to fresh samples only and the sample used for this test should not be used more than once.

After each test the bomb should be emptied and allowed to drain thoroughly.

Conclusion

While it is impossible in the space available to cover all the methods used in testing petroleum products, it is believed that all of the important ones have been presented in such form as to be workable. Some omissions may be noted, such as tests for lubricating greases, friction tests, sludge tests, paraffin scale determinations, unsaturated hydrocarbon measurements and others. Such omission has been intentional for the reason that satisfactory results cannot be obtained from any of these tests in the present state of knowledge.

In conclusion, the author desires to acknowledge his indebtedness for the many valuable criticisms and suggestions received from Messrs. I. K. Giles, E. M. Johansen, C. H. Osmond, D. Schultz, and E. A. Ritter of the Laboratory Department of The Atlantic Refining Company.

NATURAL-GAS GASOLINE

BY

H. C. COOPER [1]

I. PHYSICAL CHEMISTRY OF NATURAL GAS AND NATURAL-GAS GASOLINE

General description of natural gases.—In this chapter the term natural gas will be used to refer only to the inflammable hydrocarbon gases, which are produced from the earth in great volume, generally through drilled wells, and which form the basis of the natural-gas industry. Many gases of differing constitution have been found in nature, but the volumes produced have been so small when compared with those of the hydrocarbon gases that by common usage the name natural gas has come to be applied only to these hydrocarbon gases unless the contrary is stated.

Constitution.—Natural gas is regarded by chemists as a mechanical mixture of the several hydrocarbons of which the whole volume is composed, and hence natural gas follows all the gas laws which govern such mixtures of distinct gases. See gas laws in Section II.

TABLE 1

Analysis of Typical "Dry" and "Wet" Natural Gases

PERCENTAGE OF EACH GAS IN THE TOTAL VOLUME

Letter	Methane, per cent	Ethane, per cent	Propane, per cent	Butane, per cent	Pentane, hexane, etc., per cent	Nitrogen, per cent	Carbon dioxide, per cent	Quality of gas
A	84.7	9 4	3.0	1.3	......	1.6	......	Dry
B	84.1	6.7	......	......	......	8.4	0.8	Dry
C	36.8	32.6	21.1	5.8	3.77	......	......	Wet

A—Bureau of Mines Technical Paper No. 104, p. 16, Pittsburgh natural gas supply.
B—Bureau of Mines Technical Paper No. 158, p. 5, Kansas City natural gas supply.
C—Bureau of Mines Bulletin No. 88, p. 16.

Table 1 shows that natural gas may vary widely in composition. The individual hydrocarbons described in Table 2 are necessarily the same when they occur in any natural gas, but the proportions of the volume of the gas mixture formed by each hydrocarbon varies widely in the natural gas produced in different sections of the country, and frequently even in the same section.

[1] Credit is due the following for valuable assistance given in the preparation of this chapter: Roger Chew, H. M. Weir, Dr. R. P. Anderson, Dr. J. B. Garner, C. C. Reed and John Mosby.

TABLE 2

Physical Data of the Hydrocarbons of Natural Gas. Paraffin Series

BY DR. J. B. GARNER

Name and chemical symbol	State in which isolated at 60° F.[1]	Specific gravity, air = 1, at 60° F.	Specific gravity, water = 1[1]	Calculated volume gas in cubic feet from 1 gallon fluid 60° F. and 30 inches mercury [5]	Pounds per U. S. gallon at 60° F.[3]	Weight per 1000 cubic feet gas at 60° F. and 30 inches mercury [5]	**B.t.u. per cubic foot low value at 32° F. and 30 inches of mercury [4]**
Methane, CH_4.....	Gas	0.5547[1]	0.4150 at −263.2° F.	82.13	3.457	42.12	**1066**
Ethane, C_2H_6......	Gas	1.036[2]	0.446 at 32° F.	47.04	3.715	78.97	**1844**
Propane, C_3H_8.....	Gas	1.5204[1]	0.536 at 32° F.	38.46	4.455	145.8	**2654**
Butane, C_4H_{10}......	Gas	2.0100[2]	0.600 at 32° F.	32.73	4.998	152.7	**3447**
Pentane, C_5H_{12}.....	Liquid	2.493[5]	0.6337 at 60° F.	27.83	5.275	189.4	**4250**
Hexane, C_6H_{14}.....	Liquid	2.980[5]	0.6678 at 60° F.	24.55	5.558	226.4	**5012**
Heptane, C_7H_{16}....	Liquid	3.460[5]	0.6885 at 60° F.	21.78	5.732	263.2	
Octane, C_8H_{18}......	Liquid	3.950[5]	0.7074 at 60° F.	19.62	5.889	300.1	
Nonane, C_9H_{20}.....	Liquid	4.430[5]	0.7223 at 60° F.	17.84	6.014	337.0	

Name and chemical symbol	B.t.u. per pound low value at 32° F. and 30 inches of mercury [5]	Critical pressure, pounds per square inch	Critical temperature degrees F.	Specific heat at constant volume [2]	Specific heat at constant pressure [2]	Ratio of specific heats [2]
Methane, CH_4.....	25,280	736.5[2]	−139.9[2]	0.595	0.780	1.316
Ethane, C_2H_6......	23,560	665.8[2]	95.0[2]	......	....	1.220
Propane, C_3H_8.....	22,910	662.8[2]	206.6[2]	......	....	1.153
Butane, C_4H_{10}.....	22,600	522.0[4]	307.4[4]	0.530	0.587	1.108
Pentane, C_5H_{12}.....	23,430	486.6[2]	387.0[2]	0.512	....	....
Hexane, C_6H_{14}.....	21,630	441.9[2]	454.1[2]	0.527	....	....
Heptane, C_7H_{16}....	......	394.8[2]	512.3[2]	0.504	....	....
Octane, C_8H_{18}......	......	371.2[2]	565.5[2]	0.505	....	....
Nonane, C_9H_{20}.....	......	......	......	......	....	....

[1] Beilstein's Organische Chemie. [2] Landolt's Tables.
[3] U. S. Bureau of Standards Circular No. 57. [4] Burrell, G. A.: Gasoline Industry.
[5] Calculated by Dr. J. B. Garner. NOTE.—In the calculations, made by Dr. Garner the following equivalents were used:

1 pound..................................	453.59	grams
1 U. S. gallon............................	3.7853	liters
1 cubic foot..............................	28.317	liters
1 atmosphere..............................	14.73	pounds per square inch
Baumé modulus............................	140	

Table 2 shows the physical characteristics of the nine hydrocarbons, which in various proportions form the natural gas of commerce. Of these the first four hydrocarbons are permanent gases at atmospheric pressure and temperature, and each, if isolated, would follow the physical gas laws for a perfect gas. The remaining five hydrocarbons are liquids when isolated and reduced to atmospheric temperature and pressure, and each thus isolated follows the physical laws governing liquids. These five liquid hydrocarbons—pentane and higher, are the group which, when separated from the associated substances in the body of natural gas, form what is termed Natural-gas gasoline.

When any of these liquid hydrocarbons are present in the natural-gas mixture they are necessarily in the gaseous form. This condition is explained in Section II, Gas Laws, partial pressures and vapor tension of liquids.

The natural gases shown in Table 1 are placed according to their analysis in one of the two general groups " dry " and " wet." It is necessary to explain that these terms are entirely arbitrary and have come into use in the natural-gas gasoline industry to describe in a rough way the gasoline content of a natural gas.

The gases which contain gasoline constituents sufficient to yield 1 gallon or more per 1000 cubic feet of natural gas, and from which this gasoline content may be extracted by simple compression and cooling, are said to be " wet "; and in contrast those natural gases which contain little of the gasoline-making constituents and which require some absorption method for the extraction, are said to be " dry."

Casing-head gas.—This is the trade name given to the gas which issues from the casing-head of oil wells. It is generally a typical " wet " gas. (See Table 1.)

Gas is not usually allowed to accumulate pressure in the casing of oil wells. Such pressure tends to hold back oil production. The gas is either allowed to escape at atmospheric pressure or is gathered by a vacuum pump and delivered to a gas line. The gas, being in contact with oil at the bottom of the well, becomes saturated with vapor of the light constituents of the oil. These are pentane, hexane, and all the gasoline constituents. As the pressure of the gas in the casing head is lowered, the proportions, by volume, of these gasoline vapors in the mixture of gas increase; thus, under a vacuum a yield of several gallons of gasoline may be obtained from 1000 cubic feet of the mingled gas and vapor.[1]

Natural gas of commerce.—This name properly describes the gas obtained from wells which do not produce oil, and forms the great bulk of the gas supplied by natural gas companies to cities and towns for fuel. In commercial gas-producing wells the pressure of the gas is generally above atmospheric, and very little oil is present in the sand. Under such conditions the gas does not contain more than a trace of gasoline vapors and for that reason is termed " dry " gas.

The following description of the method of analyzing natural gas has been drawn up by Dr. J. B. Garner, director of chemical and engineering research, gas, oil and coal, Hope Natural Gas Company, Pittsburgh, Pa.

Gas Analysis

1. **Apparatus.**—The most convenient, as well as the most accurate type of apparatus for analysis is that invented by Colonel George A. Burrell, formerly of the United States Bureau of Mines. This apparatus consists of:

(1) 100-c.c. gas burette graduated in tenths;
(2) Four absorption pipettes;
(3) One combustion pipette (all pipettes are connected by a manifold) and,
(4) One copper-oxide tube, electrically heated.

[1] It was from these " wet " casing-head gases that the light gasoline condensates were first collected in 1904. See U. S. Bureau of Mines, Bulletin 88, p. 9.

All parts of the apparatus are conveniently arranged with reference to one another and are practically faultless in operation and manipulation. The apparatus can be purchased from practically any of the reliable chemical supply and apparatus companies.

2. **Sampling.**—Proper sampling is as important a factor in the analysis of any gas as the method of analysis, or apparatus used. Proper containers, i.e., ones which do not leak, as well as proper method, should be employed in collecting the sample. If possible the sample of gas should be taken by the displacement of mercury. If mercury is not available the gas sample should be collected over water. However, before the collection of the sample by displacement of either mercury or water, the line delivering the gas should have a rapid stream of gas delivered through it so as to insure complete removal of air. The importance of proper and correct methods of sampling is often overlooked and, as a consequence of not using care, inaccurate conclusions are drawn regarding not only the composition of the gas, but also its specific gravity and heating value.

3. **Procedure.**—Before proceeding with the analysis of any sample of gas, the analyst should examine the apparatus carefully to assure himself that it is in such condition that a correct analysis can be made. The mercury used in the analysis should always be perfectly clean, for if it is in any other condition it will adhere to the sides of the burette, and thus incorrect readings will be taken. It often happens that the mercury becomes contaminated with stopcock grease, or with sulphur compounds contained in the gas sample. If the mercury is not clean, it should be filtered through a 10 per cent solution of sodium hydroxide, then washed several times with distilled water, and finally filtered through a 25 per cent solution of nitric acid.

The gas burette and combustion pipette, if not clean, should be put in proper condition by the use of a mixture of concentrated sulphuric acid and potassium or sodium dichromate, then washed with distilled water and dried.

All stopcocks should be greased, but an excess of grease should be avoided as this may cause an obstruction to the flow of gas in the capillary tube of the manifold.

After it has been found that the mercury, burette, combustion pipette, and stopcocks are in proper condition, confirmation of this condition should be made by a positive test. This is done by expelling the air from the gas burette, then producing a partial vacuum on the manifold by lowering the level of the mercury to the 50 c.c. mark on the gas burette. The reading is taken and if the volume is the same after five minutes the level of the mercury is lowered to the 80 c.c. mark on the burette. If there is no change in the volume after five minutes the apparatus may be assumed to be leakproof.

Before transferring the gas sample to the burette for analysis, a sample of air should be drawn into the burette. The oxygen should be extracted from this by passing the air into the alkaline pyrogallol solution until the volume is the same after two successive treatments of the air with this solution. The residual gas is nitrogen. All the pipettes, the manifold, and the manometer should be washed at least twice with this sample of nitrogen. In transferring the gas which is to be analyzed to the burette, a small capillary glass tube, the bore of which is not greater than 1 millimeter, is connected to the sampling tube and the burette by short pieces of thick-walled rubber tubing. (The rubber tubing must be cleaned prior to use, with caustic soda solution and water.) The length of the rubber tubing should be such that as little of it as possible will be exposed to the gas. Rubber absorbs heavy hydrocarbons from gas rich in these constituents. The gas should be displaced from the container preferably by mercury, as water always contains a small amount of dissolved air. The burette and manifold should be washed several times with the gas before the sample to be used for the analysis is taken. The temperature of the water jacket and the barometric reading should be taken and recorded after each reading in the gas burette.

A sample of 95 to 100 c.c. should be taken for the analysis. This sample should be allowed to remain in the burette until the volume is constant, before it is measured against the mercury in the manometer at atmospheric pressure. The volume in the burette is measured at atmospheric pressure and is recorded as *the original* volume (*A*). The gas is then passed several times into the pipette containing potassium hydroxide solution, each time leaving the gas in contact with the solution for a period of two minutes. The volume of the residual gas must be constant after two consecutive readings before it may be assumed that the total content of carbon dioxide has been removed. Record this volume as (*B*). (If the gas contains hydrogen sulphide, this gas will be absorbed by the potassium hydroxide. All gas samples should be tested beforehand with silver nitrate and lead acetate solution to detect the presence of this substance.) When it has been shown that the carbon dioxide content has been removed from the gas, the difference between the original volume (*A*) and the resultant volume (*B*) is equal to the volume of carbon dioxide contained in the gas (hydrogen sulphide not being present). The gas isthen passed either into the pipette containing fuming sulphuric acid (20 per cent sulphur trioxide) or into the pipette containing bromine water. (Fuming sulphuric acid absorbs not only the unsaturated hydrocarbons—ethylene, propylene, butylene, etc., of the olefine series, but also pentane, hexane, heptane, etc., of the paraffin series. The decrease in volume which a sample of gas undergoes when treated with fuming sulphuric acid does not, therefore, accurately represent the unsaturated hydrocarbon content of the gas. Bromine water is better suited for an accurate determination of the amount of these in the gas. This reagent is prepared by diluting a saturated solution of bromine water with twice its volume of distilled water.) The analytical procedure is as follows:

The resultant gas (*B*), after carbon dioxide has been removed, is delivered to the bromine water pipette and permitted to remain in it for three minutes. The residual gas with some bromine in it is then passed into the potassium hydroxide solution for the removal of the bromine content and the volume of the residual gas is then measured in the gas burette. This procedure is repeated until the volume of the residual gas (*C*) is constant after two consecutive readings. The difference in volumes of the gas before (*B*) and after treatment with bromine (*C*) is recorded as the unsaturated hydrocarbon content of the gas.

The resultant gas (*C*) is passed from the burette into the pipette containing alkaline pyrogallol solution. (Care must be taken to use only a solution of alkaline pyrogallol which has been freshly prepared and is of the proper concentration.) The gas and the solution are permitted to remain in contact with one another for a period of five minutes. The gas and solution must be brought into contact a sufficient number of times until two consecutive readings are the same. Record this as volume (*D*). The difference between volumes (*C*) and (*D*) is due to oxygen. The residual gas (*D*) is transferred from the burette to the pyrogallol pipette for storage and subsequent use.

A sample of pure oxygen (about 75 c.c.) is drawn into the gas burette and measured accurately over mercury at atmospheric pressure (Volume *E*); then the leveling bulb should be lowered so as to cause a partial vacuum in the burette and on the manifold. The stopcock to the pyrogallol tube is carefully opened and about 20 c.c. of the gas sample drawn into the burette along with the oxygen. The volume of this mixture (Volume *F*) should be accurately determined at atmospheric pressure. The volume of gas used is the difference between the volume of the mixture (*F*) and the volume of the oxygen used (*E*). The electric furnace is lowered over the copper oxide tube and the temperature of this is raised to about 275° C. The "oxygen-gas" mixture (*F*) is then passed slowly back and forth through the copper-oxide tube for a period of fifteen minutes. The electric furnace is then raised and the copper-oxide tube is allowed to again assume the room temperature. When the volume of the resultant gaseous mixture has been shown to be constant, this volume is recorded as (*G*). The contraction (*H*) i.e. ($F-G=H$), has been due to hydrogen. The resultant gas is then passed into the potassium hydroxide pipette for storage and subsequent use.

Another sample of pure oxygen (about 20 c.c.) is collected in the burette and its volume carefully determined at atmospheric pressure (Volume I). The sample of oxygen (I) is then delivered into the combustion pipette by the displacement of mercury. The entire sample of gas (*G*) is drawn from the potassium hydroxide pipette and the volume of the mixture accurately determined (*J*). The contraction in volume (*K*) i.e., ($G-J=K$), of this sample of gas is due to carbon monoxide. The sample of gas $-(J)$ is then delivered into the combustion pipette and the mixture burned. The gas and apparatus are permitted to cool to room temperature and the resultant volume is measured (*L*). The first contraction (*M*) in the volume of the gas, i.e., $[(J+I)-L=M)]$ is recorded and the gaseous mixture resulting is then passed into the potassium hydroxide pipette. After three minutes contact with the solution contained therein, the volume of gas is again measured (*N*). The loss in volume ($L-N$) is due to "first" carbon dioxide. The residual gas is delivered the second time into the combustion pipette and burned and the same procedure followed as before until the volumes are constant after two successive treatments in the combustion and potassium hydroxide pipettes. The "several contractions" in volume and the "several carbon dioxides" are recorded.

The above data are utilized in the following manner for the calculation of analytical results:

I. Volumetric **composition**.—*A*. The percentages of carbon dioxide, unsaturated hydrocarbons, oxygen, hydrogen, and carbon monoxide are calculated by dividing the contraction in volume, due to each of these substances, by the volume of the original sample used (*A*) and multiplying the quotient by 100.

B. The percentages of methane and ethane are calculated by substituting the combustion results in the following equations:

(1) Total contraction $=2CH_4+2.5C_2H_6+1.5H_2$.
(2) Total carbon dioxide $=CH_4+2C_2H_6$.

The volume of hydrogen (*H*) in the sample of gas used for the combustion has already been determined and must be substituted in Equation (1). Equation (2) is then multiplied by 2 and Equation (1) subtracted. The number of c.c. then in the combustion sample are thus calculated:

(3) $2CH_4+4C_2H_6=$ total contraction.
$2CH_4+2.5C_2H_6+1.5H_2=$ total carbon dioxide.
Total contraction $-$ total carbon dioxide $=1.5C_2H_6-1.5H_2$.

This volume of ethane is then substituted in either one of the equations above and the volume of methane calculated. When the volumes of methane and ethane thus calculated are divided by the volume of the sample used and the quotient multiplied by 100, the products obtained will be the percentages of methane and ethane in the gas.

C. The percentage of nitrogen is calculated by subtracting from 100 the sum of the percentages of all the other substances.

II. **Heating value of the gas.**—The heating value of the gas is calculated by multiplying the percentage of each combustible substance in the gas by its heating value and summing up the products thus obtained.

Table of Heating Values per Cubic Foot

Hydrogen	341
Carbon monoxide	338
Methane	1066
Ethane	1844
Ethylene or unsaturated hydrocarbons	1700

III. **Specific gravity of gas.**—The specific gravity of the gas is calculated by multiplying the percentages of each of the substances by the specific gravity of the gas and then summing up the products thus obtained.

Table of Specific Gravities

Air = 1

Hydrogen	0.0696
Oxygen	1.0530
Nitrogen	0.9673
Carbon monoxide	0.9672
Carbon dioxide	1.5291
Methane	0.5576
Ethane	1.0750
Ethylene	0.9852

II. COMPRESSION AND COOLING METHOD FOR SEPARATING GASOLINE FROM NATURAL GAS

Field of application and physical laws which govern the process.—Gases are completely miscible—that is to say, they will mix in all proportions. In a confined space, such as a tank, a mixture of several gases will diffuse until each completely fills the tank as if it alone were present. This is true of all mixtures in which the various gases have no chemical action on each other. Natural gas is such a mixture of gases and vapors of liquids. Suppose a certain mass, or quantity, of natural gas is placed in a tight tank—then each of the gases of which the mixture is composed will expand until each completely fills the tank.

Consider the following example—The pressure, as indicated by a gage placed on a tank containing natural gas is 100 pounds absolute. Then if the gas is composed of methane 50 per cent and ethane 50 per cent each by volume, the pressure on the methane will be 50 pounds absolute, and likewise the pressure on the ethane will be 50 pounds absolute. That is to say, the total pressure, 100 pounds absolute, is the sum of the partial pressures of each component gas, which in this case would be 50 pounds on each. Thus it is seen that the partial pressure on each gas bears the same relation to the total pressure on the mixture that the volume of each gas bears to the total volume. This law is known as Dalton's Law of partial pressures and is stated as follows:[1] "The total pressure of a mixture of gases is the sum of the pressures which would be exerted by each of the components if it alone occupied the total volume."

The application of this law to either the compression or absorption method of separating gasoline vapors from natural gas is illustrated by the following example:[2]

"Assume that a natural gas contains 1.5 per cent by volume of hexane (which is approximately 0.5 gallon per 1000 cubic feet) at 80° F. The vapor pressure of hexane at 80° F. is 2.9 pounds per square inch, as is stated in Table 3. The pressure necessary

[1] G. Senter: Outline of Physical Chemistry, p. 81.

[2] Bureau of Mines Technical Paper 232, W. P. Dykema and Rqv O. Neal, p. 9.

before the hexane will start to precipitate is $(100 \div 1.5) \times 2.9$ or 193 pounds per square inch. The pressure necessary to cause the hexane to start to condense at other temperatures is shown by the curve (Fig. 1). If this natural gas is contaminated with an equal volume of dry gas or air, the hexane content will be decreased to 0.75 per cent, and a pressure of 386 pounds per square inch will have to be applied before the hexane will start to condense. The same reasoning applies to pentane, heptane, and other gasoline vapors in casing-head gas. Upon the partial vapor pressure of each component depends the pressure at which the compressor must be worked in order to precipitate the gasoline.

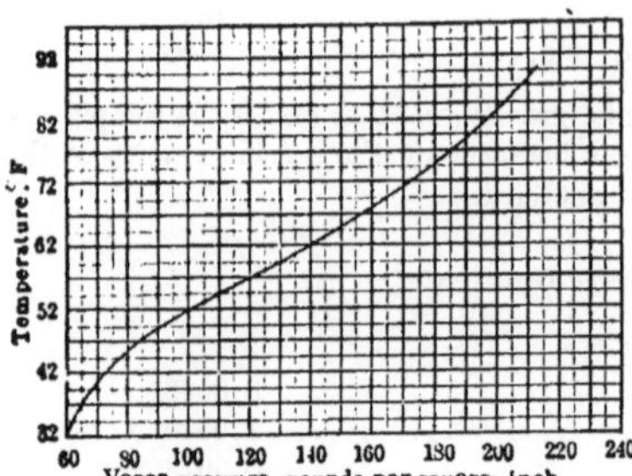

Fig. 1.—Curve showing pressures required at different temperatures to condense hexane from a mixture of gases, of which it forms 1.5 per cent by volume.

In the absorption process, partial pressures of the various recoverable constituents are equally important. Each particular hydrocarbon continues to dissolve in the absorption medium until its partial vapor pressure in the gas is equal to its partial vapor pressure from the absorption menstruum. For example, hexane in a mixture of gases will dissolve in 'mineral seal oil' until its partial vapor pressure from the oil is at equilibrium with its partial vapor pressure in the gas."

Table 3

Vapor Pressures and Corresponding Temperatures at which Certain Hydrocarbons will Condense *

Temperature (° F.)	Vapor Pressure (Pounds per Square Inch)				
	Propane (C_3H_8)	Butane (C_4H_{10})	Pentane (C_5H_{12})	Hexane (C_6H_{14})	**Heptane (C_7H_{16})**
32	72.5	15.9	3.5	0.9	**0.2**
40	84.6	19.9	4.4	1.0	**0.3**
50	100.7	24.1	5.4	1.4	**0.4**
60	115.4	30.0	6.8	2.0	**0.5**
70	130.1	36.7	8.2	2.5	**0.7**
80	147.2	44.1	9.9	2.9	**0.9**
90	165.0	53.2	11.9	3.2	**1.1**

* Diserus, P.: Recovery of gasolene from casing-head and natural gas, Jour. Engineers' Club of St. Louis, Jan.-Feb., 1918.

On this principle rests what is termed the compression method of extracting gasoline from natural gas.

Referring to Fig. 2.—The fresh gas is compressed to such a pressure that after it passes through the cooler a deposit of pentane and heavier hydrocarbons called gasoline will take place in the accumulator as shown.

The quantity of gasoline recovered from each 1000 cubic feet of fresh gas depends on the proportion by volume of the pentane and heavier gases in the fresh gas and the pressure produced on the fresh gas by the compressor and the temperature maintained in the cooler.

Referring to Table 1.—Wet gases, when treated with equal efficiency, will yield more gasoline than dry gases. This process may be successfully used in treating rich casing-head gas on account of the large proportion of vapors of pentane and the higher hydrocarbons in such " wet " gases. For rich gas a moderate pressure in the compressor (say 275 pounds) will exert sufficient partial pressure on these gasoline constituents to largely condense them, and this can occur at moderate temperatures (70° to 80° F.) in the cooler.

The temperature in the cooler should be the lowest that can be produced by the use of water from a cool stream or well, or of water cooled by atomizing in air if necessary. Table 3 shows that the vapor pressure of any of the hydrocarbons increases with temperature rise; hence it requires greater pressure to condense them, should the cooler temperature be 90° F., than if it were maintained at 70° F.

Consider the operation of this system. Suppose the pressure to be 250 pounds in the compressor, cooler and accumulator, and the temperature 70° F. in the cooler;

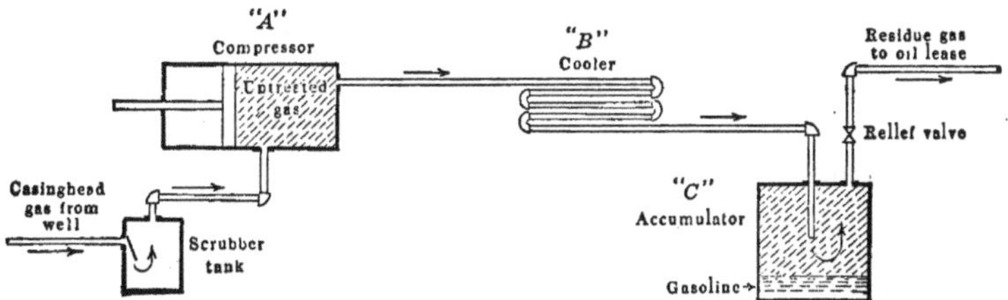

FIG. 2.—Diagram of compression and cooling process plant.

then under these conditions, if hexane, for example, is present in the fresh gas " *A*," it will begin to condense in " *B* " providing it is present in such a large percentage by volume in the mixture that its percentage share of the 250 pounds total pressure is above its condensing pressure at a temperature of 70° F. If there is sufficient hexane in the mixture to bring about condensation, it will condense and collect in " C " only until the vapor of hexane has decreased in volume to that percentage in the mixture at which its partial pressure is just equal to its condensing pressure at a temperature of 70° F.

Thus it is seen that all of the pentane, hexane and other gasoline constituents cannot be separated from the gas mixture by compression and cooling; but in the case of " wet " casing-head gases such a large amount of gasoline can be separated by this simple process and it can be done so cheaply that the method has become firmly established as a commercial process, with pressures of about 250 pounds and ordinary atmospheric temperatures in the cooler.

It is possible to increase the yield of gasoline by the use of excessively high pressures, and by artificial cooling by means of refrigerating apparatus.

The cost of this apparatus and the expense of operation has to be considered in connection with the probable quality of the increased output. Under high pressure and intense cold the gases propane and butane will be condensed temporarily in the gasoline and since they will evaporate rapidly when the gasoline is reduced to atmospheric pressure they are apt to carry off a part of the stable product, thus greatly increasing the real losses by evaporation. In this connection the all important matter to be established before installing a gasoline plant of any type, on either wet or dry gas, is the quantity and quality of the gas, and the assurance of a continuing supply. The engineer must determine these controlling facts beyond reasonable doubt. When this has been done, if the gas is sufficiently rich to justify a simple compression method of extraction, and if the quantity of the gas is large, it would be well to consider carefully the use of some absorption apparatus on the residual gas, when the cooler cannot be maintained at a reasonably low temperature (say 80° F. in summer) by natural means.

As a general rule, only casing-head gas is rich enough for the use of the compression method alone, and such gas is very apt to be uncertain in volume; hence only the simplest and most inexpensive apparatus should be used except in cases where the plant is operating on a large territory in which additional wells are constantly being drilled and an additional supply of gas is being made available.

III. ABSORPTION METHOD FOR SEPARATING GASOLINE FROM NATURAL GAS

Field of application and physical laws which govern the process.—Referring to Fig. 3.—The natural gas, having been compressed in *A*, for the purpose of transmission

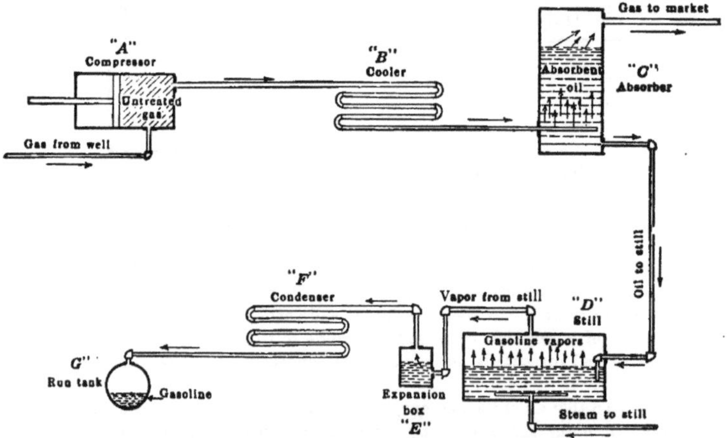

FIG. 3.—Diagram of absorption process plant.

to market, passes through cooler *B*, which is cooled by water, into an absorber *C* wherein the gas is thoroughly washed in a suitable oil. Thence the gas passes into the pipeline on its way to market.

If the oil is then removed from the absorber and examined it will be found that the Baumé gravity has been raised, and if the oil is placed in a still *D* and heated as with steam a vapor will pass over through the condenser *E* and deposit as a fluid in the container *G*. This fluid is a mixture of the hydrocarbons, pentane and higher, which is the natural-gas gasoline.

(1) The quantity and quality of gasoline fluid thus obtained depends upon (*a*)

the percentage of gasoline forming constituents in the gas, (b) the temperature of the oil and gas during the process of absorption in C, being greatest when cold, (d) the pressure of the gas during the process of absorption in C, being greatest when the pressure is high, (e) the quality of the oil used for this purpose, as the ability of the oil to absorb the gasoline vapors is a specific quality.

(2) In case of a dry gas (A, Table 1) the compression and cooling process can produce no gasoline commercially since the percentage by volume of the gasoline constituents is too small. Therein consists the usefulness of the absorption process, as the dissolving effect of oil brings about a separation, and a dense vapor consisting almost entirely of gasoline constituents is formed in the still. These vapors condense readily.

The physical laws which govern the absorption process are (a) the law of partial pressures which was discussed under the compression process and (b) the laws of the solution of gases in liquids. The latter is stated as follows.[1] "For gases which are not very soluble and do not enter into chemical combination with the solvent the relation between pressure and solubility is expressed by Henry's Law as follows. The quantity of gas taken up by a given volume of solvent is proportional to the pressure of the gas."

In the absorber C (Fig. 3) the action of absorption takes place between the several hydrocarbons forming the natural gas (Table 2) and the absorption oil. At any given temperature and pressure in the absorber the oil has a certain power of absorption for each hydrocarbon. Pentane, and those hydrocarbons which are liquids when separated from the natural gas, dissolve and are held in solution by this oil as liquids. At the same time true gases, butane, propane, etc., also enter the oil to some extent and are held by it, but it is not known whether they are then in the liquid state, or still in a gaseous condition. When the oil is withdrawn from the absorber C and subjected to heat in the still, the dissolved hydrocarbons are driven off from the oil and passed through a cooler in which those hydrocarbons which are liquid at atmospheric temperature and pressure condense, forming the mixture termed gasoline. At the same time those hydrocarbons, butane, propane, etc., which are normally gases, do not condense but pass off as gas, except for a small part which are held in solution in the gasoline. Thus it is seen that gases are absorbed in the oil at certain points in the process of absorption.

It is not possible to separate all of any one hydrocarbon from the natural-gas mixture by means of the process of absorption. This fact can be shown by considering the action of absorption in the case of one hydrocarbon, for example, hexane. As the molecules of this substance enter into solution with the oil the vapor tension of the mixture is raised; this action continues until the vapor tension of the mixture is equal to the vapor pressure of the hexane in the gas. That is to say, the molecules of hexane leave the mixture again as rapidly as they enter, hence a point of equilibrium is reached, leaving some hexane still in the gas in vapor form. This action is true for all the hydrocarbon gases forming the natural gas mixture. Those hydrocarbons, methane to butane (Table 2), having the highest vapor pressure in the gas are least readily absorbed by the oil, and the larger part of these substances which have been absorbed escape quickly at low temperature when the charged oil is withdrawn from the absorber. The heavy hydrocarbons, pentane and higher, form a more stable solution in the oil, and a much larger percentage of each of them is caught and held in the oil.

In the practice of the absorption process as applied to natural gas the following factors, which are of paramount importance, will be considered separately.

Quality of the absorbent oil.—A suitable oil should have the following general characteristics:

[1] G. Senter: Outline of Physical Chemistry, p. 82.

Petroleum oil, paraffin base.
Cold test, not above 32° F.
Flash point, not below 225° F. open.
Initial boiling point, not below 450° F.
Final boiling point, not above 675° F.

The oil must pass the following emulsification test:

Emulsification Test

Apparatus.—(1) Cylindrical 4-ounce oil sample bottle with cork; (2) Thermometer (3) Water-bath containing 3 to 4 inches of water.

Procedure.—Make sure that the 4-ounce bottle is perfectly clean, then put into it 1 ounce of distilled water and add an equal amount of the oil to be tested. Insert the cork and place the bottle and contents in the water-bath. Heat the bath until the temperature of the contents of the bottle reaches a temperature of 180° F. This point is determined by removing the cork and inserting the thermometer from time to time.

Remove the bottle and contents from the bath. Fit the cork tightly, and shake vigorously for five minutes. Allow to stand one hour and interpret results as follows:

(1) If the oil and water have entirely separated leaving a clear line of demarcation, and the oil and water layers are free from cloud, the emulsification is " good."

(2) If the oil and water have separated leaving a very thin layer between, which gives the appearance of an emulsion, and if the oil and water layers are only slightly clouded, the emulsification is " fair."

(3) If the oil and water have not separated or have only partially separated leaving an indistinct line of demarcation, or an emulsified layer has formed between the water and oil layers and the water and oil layers are considerably clouded, the emulsification is " poor."

Good = Passes.
Fair = Passes, but very close.
Poor = Does not pass.

The oil generally used is known as mineral seal oil and conforms to the following general specifications:

Viscosity	45 to 50 seconds at 100° F.
Flash	300° F.
Gravity	38 5 to 39.5 Bé. at 60° F.
Initial boiling point	518° F.
Final boiling point	666° F.
Cold test (important)	30° F.

The following specification is for an oil which is slightly heavier than mineral seal and has been successfully used:

Gravity	33.6° Bé.
Color	Dark straw
Flash	230° F. open

Distillation

Begins to boil at 470° F.

Per cent		Degrees F.	Per cent		Degrees F.
10	off at	490	10	off at	549
10	off at	496	10	off at	568
10	off at	503	10	off at	606
10	off at	512	10	off at	658
10	off at	521	—		
10	off at	534	100		

Final boiling point 658° F.

The following are typical mineral seal oils in successful use:

Gravity	40.0 Bé.	39.1 Bé.
Color	15	15
Open flash	260° F.	255° F.
Open fire	300° F.	300° F.
Pensky closed	260° F.	245° F.
Cold test	27 C.	28 C.
Viscosity	1300	1360
S.D.	0.035	0.035
Doctor test	Negative	Negative

Distillation

Cuts	Per cent	Begins to boil at 510° F.		Begins to boil at 490° F.	
		Boiling point, Degrees F.	Gravity Bé.	Boiling point, Degrees F.	Gravity Bé.
1	10	537	41.6	535	41.9
2	10	543	41.2	545	41.0
3	10	551	40.9	556	40.7
4	10	557	40.7	564	40.3
5	10	565	40.5	573	39.9
6	10	575	40.3	583	39.5
7	10	585	39.8	595	39.1
8	10	595	39.3	610	38.3
9	10	618	38.5	637	37.4
10	8.5	660	37.2	685	35.7
	1.5	Above 660° F.		Above 685° F.	
	100.0				

	Final 660° F.	Final 685° F.
Off at 520° F.	0.75 per cent	4.0 per cent
Off at 560° F.	43.00 per cent	35.5 per cent

Specific Heats of Absorption of Oils

On a basis of " water equals 1," the value for the oils usually used in the absorption process may be taken as .47. That is to say, .47 B.t.u. is required to raise the temperature of 1 pound of oil 1° F.

Pressure.—If 1000 cubic feet of natural gas containing gasoline constituents is passed through a certain quantity of oil, for example 10 gallons, as in the absorber *C* (Fig. 3), and if the temperature is kept constant, for example at 70° F., the quantity of gasoline constituents which will dissolve in the oil will vary with the pressure. That is to say, it will increase with increase of pressure and decrease with decrease of pressure. This fact accords with Henry's Law. However, with increasing pressure, and consequent increase of yield, the quality of the gasoline mixture also changes; it becomes more volatile or wild, containing more dissolved butane and propane; and conversely the quality is more stable as the pressure and resultant quantity of gasoline are diminished.

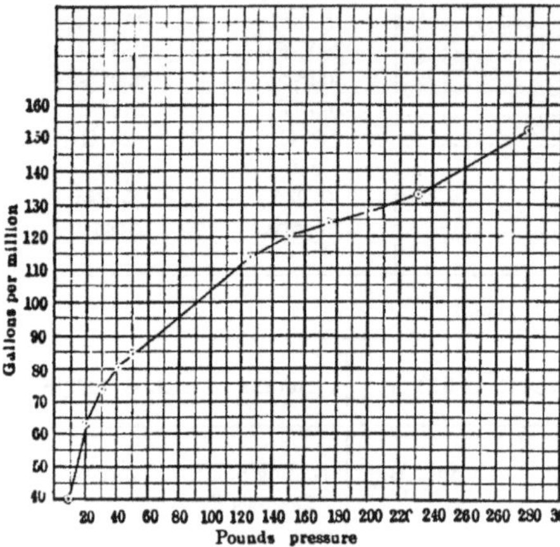

Fig. 4.—Curve showing the relationship of yield of gasoline to pressure.

The curve (Fig. 4) was plotted by Dr. J. B. Garner, from experiments made by him at Hastings, W. Va., on gas of that state. Experiments of a like nature made by the same chemist on gases of other fields, and similar experiments by other chemists have shown like results varying only in degree, dependent upon quality of the gas.

In practice, it is not commercially profitable to compress " dry " natural gas to a high pressure for the sole purpose of increasing the yield of gasoline. Gasoline plants should be located, if other factors permit, to take advantage of as high pressure as the transportation lines afford; but power costs are too great to compress large volumes of gas for any purpose other than transportation.

If in the treatment of casing-head gas, rich in gasoline, the absorption method is applied, then on account of the small volumes of gas and the comparatively small investment necessary for power, the gas may be compressed for the purpose of gathering gasoline and distributing the residue gas to the leases. For gases containing 1½ or more gallons of gasoline per thousand cubic feet, a pressure of 80 pounds for absorption has been found to work well.

Temperature.—At a fixed pressure, the quantity of gasoline which a given volume of oil will absorb from a given volume of natural gas will increase as the temperature is lowered and will decrease if the temperature is raised. This result is explained by the fact that as the temperature is lowered the vapor tension of each hydrocarbon in the natural gas is lowered and hence they do not have as great a tendency to leave the oil when once dissolved in it as they do when at a high temperature. But again the quality of the gasoline changes with change in the temperature of the absorption; the product becomes more volatile as absorption takes place at decreasing temperature, and becomes more stable as temperatures are raised. More butane and propane dissolve in the oil at low temperatures than can be absorbed at high temperatures.

The curve (Fig. 5) has been plotted by Dr. J. B. Garner from experiments made by him at Hastings, West Va. Similar observations have been made by other experimenters which support the foregoing statements.

Absorption gasoline plants have been equipped in some instances with refrigeration apparatus applied in various ways to bring about lower temperatures of absorption than can be obtained by cooling with water at atmospheric temperatures.

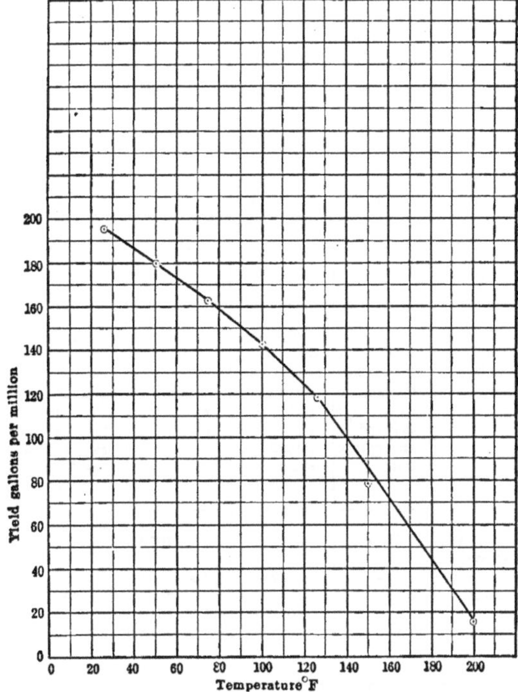

Fig. 5.—Relationship of yield of gasoline to temperature.

The factors which govern the advisability of such investment are (*a*) the probable duration of time each year when natural cooling is not sufficiently effective, and (*b*) the market which can be found for a volatile product, as it will be necessary to blend quite heavily in summer if the absorption is carried on at temperatures much below 70° F.

Practice in this respect must differ as between different climates; it must also vary with the particular product which it is proposed to make.

Ratio of quantity of oil to quality of gas.—Table 4 is given for the purpose of plant design. This table is based on the use of mineral seal oil or an oil of equal quality.

The table is based on the following assumptions, viz., (*a*) the plant design provides good contact between gas and oil, (*b*) the temperatures of the oil and gas range from 50° F. in winter to about 80° F. in summer as an average.

In designing a gasoline plant to be built on a natural gas transportation line carrying dry gas to market, the still and oil-circulating apparatus should be ample to circulate as much oil as is required for the maximum quantity of gas per hour in winter. If this is done a greater ratio of oil to gas can be circulated with the same equipment in summer when the rate of flow of gas per hour is less than the maximum, but temperatures are higher. Likewise as the field pressure declines the quality of the gas may increase in gasoline content, but under these conditions the quantity of gas per hour will decline; therefore the same circulating system will provide a sufficient ratio of oil to gas to meet these conditions.

In operation the degree of saturation of gasoline in oil should not be carried so far that the Baumé rise in the mixture with oil will be above 3° in winter when the oil is 50° F. to 60° F., or 1½° in summer at a temperature of 80° F. These are limits found by experience to give efficient absorption in plant operation.

TABLE 4

Oil Circulation Required in Practice for Absorption of Gasoline at Differing Pressures and Differing Qualities

1	2	3	4	5	6	7
General quality of the gas undergoing treatment	Quality of oil used as an absolvent	Contents of gasoline gallons per thousand cubic feet as found by absorption test	Average pressure under which absorption will take place. Pounds per square inch	Average temperature at which absorption will take place	Average quantity of oil in gallons required per thousand cubic feet of gas treated. Good contact assured	Capacity of stills, circulating pumps, etc., in gallons per thousand cubic feet of gas treated
Dry	Mineral seal or equivalent	0.15–0.3	100–250	60° F.–70° F.	7	12.0
Dry	ditto	0.4 –0.6	100–250	60° F.–70° F.	9–14	14–18
Dry	ditto	0.15–0.3	10– 20	60° F.–70° F.	30	40
Wet	ditto	1.0	80–100	60° F.–70° F.	25	35
Wet	ditto	2.0	80–100	60° F.–70° F.	50	65

NOTE.—Column 7 shows the capacity of oil circulation (stills, pumps, etc.) which should be provided to allow for changing conditions of temperature, pressure, quality of gas, rate of flow, etc., which take place in any gasoline plant throughout the year. If provision for ample oil circulation is made when designing the plant then the most economical and efficient circulation can be maintained for each operating condition. Such provision will add but little to the initial expense.

IV. PROCESS OF SEPARATION OF GASOLINE CONSTITUENTS FROM NATURAL GAS BY MEANS OF SOLID SUBSTANCES

Solid substances tend to condense on their surfaces any gas or vapor with which they come in contact. This is a specific quality for each substance, that is to say, the amount of such condensation which will take place depends upon the nature of the solid and the nature of the gas. Condensation increases as the temperatures of the gas and solid substance are lowered; it also increases as pressure on the gas is increased.

Various solid substances have been suggested for separating gasoline from natural gas, but up to the present time charcoal, properly prepared, has been experimented with the most widely. Some of the gels are also being brought forward. These processes are all protected by patents.

Fig. 6 is a diagram showing the path of natural gas when passing through a solid substance, as charcoal in this process.

The gas passes first through one absorber, until the proper concentration of gasoline in the charcoal is reached, whereupon the gas stream is switched to the other absorber while the absorbed gasoline is distilled from the charcoal in the first absorber. The process is then reversed, each absorber being alternately a still and an absorber. Plans for permanent plants of this type include three absorbers, so that between the periods of absorption and distillation, the absorber and its contained charcoal may undergo a period of cooling during which the gas is passed through it after the gasoline has been extracted.

The following description of the adsorption process, which employs charcoal in treating natural gas, is written by Dr. R. P. Anderson:

It has long been known that charcoal adsorbs gases. The amount of any particular

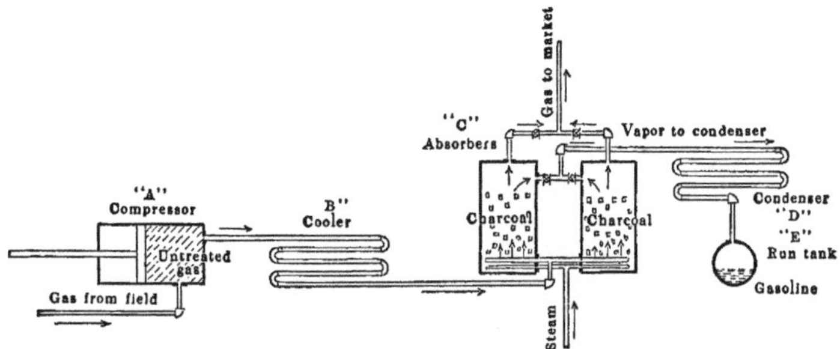

Fig. 6.—Diagram showing the path of natural gas when passing through charcoal.

gas adsorbed under given conditions depends upon the nature of the material from which the charcoal is produced, upon the method of charring, and upon the method of activation. The activation methods were developed during the World War in the attempt to increase the value of charcoal for use in gas warfare, and as a result of their development, the adsorptive capacity of this material has been greatly increased.[1]

When natural gas is passed through activated charcoal each of its constituents is absorbed to some extent, and under proper conditions a fairly sharp separation of the gasoline-forming constituents from the other constituents may be effected. This fact has been utilized in the charcoal method for the approximate determination of the gasoline content of natural gas [2] and later in the charcoal process for the ex raction of gasoline from natural gas.[3] The rights of the charcoal process are owned by the Gasoline Recovery Corporation, 62 Cedar Street, New York City. The principal features of the method will be considered in the following discussion.

Removal of gasoline vapor from gas.—To remove gasoline vapor from the natural gas to be treated, the gas is passed through an absorber containing the charcoal and thence to the point of distribution. After the proper degree of saturation of the charcoal

[1] See N. K. Chaney, The Activation of Charcoal, Trans. Am. Elect. Soc., **36**, 1919.

[2] Oberfell, Shinkle and Meserve: Jour. Ind. Eng. Chem., **11**, 197, 1919; Oberfell and Burrell: Gas Record, 2, **16**, 45, 1919; Anderson and Hinckley: Jour. Ind. Eng. Chem., **12**, 735, 1920.

[3] Burrell, Oberfell and Voress: Chem. and Met. Eng., **24**, No. 4, 1921.

with gasoline has been obtained, the gas is diverted to a second absorber, and the charcoal in the first is ready for distillation.

One of the most important features in connection with plant operation is the control of the degree of saturation of the charcoal. That this is true can be realized only after the mechanism of the process of saturating the charcoal with gasoline is understood. When natural gas is first passed through unsaturated, activated charcoal, rapid adsorption takes place, as evinced by the heat that is liberated. The heat wave advances gradually through the entire depth of the charcoal and finally the temperature of the charcoal drops to approximately that of the entering gas. The progress of the heat wave marks the saturation of the charcoal with the constituents of the natural gas, but this takes place in such a fashion that the separation of the gasoline-forming constituents of the natural gas from the other constituents is slight. The charcoal absorbs a little methane, more ethane, still more propane and butane, and practically all of the vapors of pentane, hexane, heptane, etc. When these substances are distilled from the charcoal, the percentage of gasoline vapor (pentane, hexane, heptane, etc.) in the gas obtained is small and it is impossible to condense it efficiently. Most of the gasoline vapor will pass off with the methane, ethane, propane and butane.[1]

To obtain the proper separation of the gasoline-forming constituents of natural gas from the other constituents, the passage of gas through the charcoal must be continued for some time after the first saturation point just mentioned. During this time, the gasoline-forming constituents are largely retained by the charcoal, displacing the other constituents, until finally the gas liberated from the charcoal on distillation contains such a high percentage of gasoline vapor that it can be condensed with but little loss. It is probable that the saturation of the charcoal which gives the largest yield of gasoline from the gas is one in which the passage of gas has been continued beyond the point where the charcoal is capable of removing completely the gasoline-forming constituents from the gas. If this is true, the loss due to incomplete adsorption is more than offset by the gain in the efficiency of condensation of gasoline from the vapors driven from the charcoal.

The saturation that should be obtained for the largest yield can be determined only by experiment. It depends upon the nature of the charcoal, the nature of the gas, and upon operating conditions. At one plant a saturation of 20 per cent by weight has been found satisfactory.

Removal of gasoline vapor from charcoal.—The gasoline-forming constituents of the natural gas are presumably present in the pores of the charcoal in liquid form,[2] and this liquid is removed from the charcoal by distillation with superheated steam. After the natural gas is diverted to another absorber, the pressure on the first absorber is reduced to atmospheric and superheated steam is admitted.

The first effect of the steam heating is to liberate the lighter constituents of the gas that have been retained by the charcoal. As the heating progresses, the amount of gas decreases and gasoline vapor and steam begin to come over. Complete removal of the gasoline vapor from charcoal cannot be effected by the use of saturated steam, and in the earlier stages of the development of the process superheated steam was always used. It was later ascertained, however, that almost all of the gasoline could be recovered from the charcoal by the use of saturated steam, and that the residual gasoline content of the charcoal had no deleterious effect upon the absorptive characteristics of the charcoal that could be detected in the yield of gasoline obtained from the gas.

Cycle of operations.—To treat a given quantity of natural gas continuously, at least two absorbers must be provided. During the saturation of the charcoal in one

[1] For explanation of this, see under compression method.

[2] Lamb and Coolidge: Jour. Am. Soc., 42, 1146, 1920.

absorber, the gasoline is distilled from the charcoal in the other. Where but two absorbers are used, the absorber in which a distillation has just been carried out will not have cooled to room temperature before being used again. A slight reduction in yield results from passing gas through a hot absorber, and for this reason it seems desirable to provide three absorbers.

With three absorbers, the gas is passed through two in series while distillation is taking place in the third, and by connecting the absorber from which the gasoline has just been distilled as the second of the series for the next absorption, the gas which is being treated may be used to cool the hot absorber without any reduction in efficiency. The use of two absorbers in series is also valuable in preventing loss of gasoline during the absorption process, since after the second absorber is cooled it will retain the greater part of the gasoline which may pass through the first absorber while the proper concentration is being reached. Where the saturation period is considerably longer than the distillation period, two absorbers can be used efficiently by connecting the absorber from which the gasoline has just been distilled in series with the first absorber until the proper saturation has been built up in the first absorber.

Nature of the product.—In general, gasoline produced by the charcoal process has a lower vapor pressure for a given gravity than gasoline produced from natural gas by any other method. This is due partly to the relatively sharp separation of the gasoline-forming constituents of the natural gas from the other constituents during the adsorption and partly to the additional separation that is effected during the distillation, as a result of its being a "batch" process. Also charcoal-process gasoline is not contaminated with a small percentage of oil, as is the case with gasoline produced by the oil-absorption method.

Applicability of method.—The use of charcoal for the approximate determination of the gasoline content of various natural gases has brought out the fact that it is especially suited to testing lean gas, that it is fairly satisfactory on moderately rich gas, and that on rich gas the results are noticeably low. On lean gas, the results obtained with the charcoal method of testing are usually considerably higher than those obtained by the oil-absorption method; while on rich gas, the results are usually lower than the actual plant yield by the compression method. It would appear, therefore, from a consideration of yield, that the logical field of development of the charcoal process should be in treating dry gas directly or in treating residue gas from absorption or compression plants.

The following study of the general subject of gaseous adsorption by solids has been made by Mr. H. M. Weir, B.Ch.E.

Mr. Weir develops the general state of knowledge of this subject, drawing particular attention to the specific quality of the action of adsorption.

In the recent literature dealing with the possibilities offered by such solids as activated charcoal and silica gel in the recovery of gasoline from natural gas, the specific or "selective" nature of this adsorption of vapors has been stressed considerably. The reason for such emphasis is apparent from a study of results obtained in even the crudest experiments. The number and complexity of the compounds in a mixture of natural gas and gasoline vapors makes a study of the fundamental factors influencing adsorption more difficult than need be, and it is interesting in this connection to consider results which have been recorded in the past where simple gases were involved. All of the earliest work on gaseous adsorption by solids was done by means of charcoal, since silica gel was unknown.

Adsorbing power of charcoal.—It is evident from James Boyle's writings that he knew that some, if not all, solids tend to condense on their surface any gases or vapors with which they are in contact. It was predicted very early that charcoal, which offers a large surface relative to its mass, would exhibit a large adsorptive effect.

Probably the first intensive study of the power possessed by charcoal made from different woods to adsorb various gases was that of Hunter published from 1862 to 1872.[1] Some of his results are tabulated below.

TABLE 5.

Charcoal from	Volume of Gas Adsorbed per Volume of Charcoal (0° C., 760 mm. Hg)		
	NH_3	CO_2	$(CN)_2$
Logwood	111	55	87
Ebony	107	47	90
Lignum vitæ	89	47	....
Cocoanut shell *	176	71	114
Boxwood	86	31	29

* Summarized from results published at a later date than others tabulated.

The relative adsorptive value of different charcoals as shown above was fairly well borne out by government investigation for gas mask fillers. It was found that charcoal from cocoanut shell, properly treated, was the best adsorbent.

Effect of pressure, temperature, and boiling point of a gas or vapor undergoing adsorption.—It is often stated that with the same solid and the same gas, the amount of adsorption is greater the higher the pressure and the lower the temperature, and also that of two gases or vapors the one having the higher boiling point is the more easily adsorbed.

The first generalization is confirmed by a great many published determinations, but Dewar[2] obtained some results indicating that it may not always apply.

TABLE 6.

Pressure atmospheres	Hydrogen adsorbed by charcoal at −185° C. Volume in c.c. adsorbed in 6.7 grams
1	620
5	925
10	1050
15	1000
20	975
25	925

Dewar expressly stated that he did not consider the lower values obtained at 20 and 25 atmospheres to be due to experimental error.

There are a number of determinations on record which do not bear out the statement that a gas or vapor is adsorbed more strongly the higher its boiling point. Among

[1] Phil. Mag., 25, 364; also Jour. Chem. Soc., Vols. 18, 20, 21, 23, 24 and 25.
[2] Proc. Royal Institution, 18, 437.

the fixed gases Hempel [1] found that nitrogen is absorbed more than argon, hydrogen sulphide more than carbon dioxide, and ethane more than carbon dioxide, all being adsorbed at 20° C. In every case the compound adsorbed to a lesser extent has the higher boiling point. In the case of true vapors Dewar [2] found that ethyl alcohol, ether, benzene and aldehyde were all adsorbed more than water vapor though the boiling point " generalization " would require the reverse.

The results cited make it evident that the adsorption of gases by charcoal is really specific since it varies with the nature of both gas and solid, and in a manner which can not be predicted with certainty from knowledge of pressures and boiling points of the gases involved. It is not even true that the most porous charcoal is the best adsorber; cocoanut charcoal is quite firm and has a metallic luster in fracture.

Adsorption applied to a mixture of gases, as in natural gas.—Very little work has been done on the problem of adsorption of known mixtures of gases, but it represents a field that should be investigated and which would yield the most interesting results in connection with gasoline recovery from natural gas. A general rule seems to be that the more readily adsorbed gas displaces the other to some extent. Again, however, this does not always occur, as evidenced by Bergter's [3] results. He showed that at pressures of 0.5 mm. to 10 mm. oxygen is adsorbed more than 30 times as strongly as nitrogen, but that under these conditions the presence of oxygen increases the amount of nitrogen over that adsorbable alone. In the case of readily condensable vapors it is possible that the condensation of one vapor helps to carry down the other.

State of the gas or vapor after adsorption.—The suggestion of condensation of vapors makes it interesting to inquire just what evidence there is to show that adsorbed gases may actually be condensed in the adsorbing solid. In 1906 Dewar[4] calculated the apparent density of some gases absorbed by charcoal at low temperatures and found that in some cases the density was even greater than that of the liquefied gas. A tabulation of some of Dewar's results is given here.

TABLE 7.

Gas	Density in Grams per Cubic Centimeter		
	Temperature (C.) of adsorption	Density of adsorbed liquid	Density of liquefied gas
CO_2	+ 15	0.70	0.80
O_2	−183	1.33	1.12
N_2	−193	1.00	0.84

It may be well to note here that the vapor pressure of a film of apparently liquefied gas after adsorption bears no relation to the vapor pressure of a mass of the liquefied gas. The phase rule holds only when the effects of capillary tension are eliminated. It is well known that the vapor tension of a liquid in a capillary tube is less than that of the liquid in mass at the same temperature.

Heat of liquefaction and adsorption.—Adsorption is accompanied by a marked evolution of heat. The accompanying table relates the molecular heat of adsorption

[1] Zeit. Elektrochemie, 13, 724.
[2] Proc. Royal Institution, 18, 179.
[3] Drude's Ann., 37, 480.
[4] Proc. Royal Institution, 18, 438.

with the heat of liquefaction of several gases. The work was done by Dewar[1] with charcoal as the adsorbent.

TABLE 8.

Gas	Molecular heat of adsorption	Molecular heat of liquefaction
H_2	1600	238
N_2	3684	1372
O_2	3744	1664

The figures are not surprising since it is well known that the wetting of a porous body by a liquid will evolve heat. Consequently if the gas is actually condensed as seems likely from these figures, it is probable that the heat generated on liquefaction would be increased by that given off by the liquid wetting the charcoal.

In 1913 Gaudechon [2] obtained some figures on the heat evolved on wetting wood charcoal with various liquids. Some of his results are given here.

TABLE 9.

Liquid	Gram calories per gram of charcoal
"Pentane" ($C_5H_{12}+C_6H_{14}$)...	0.4
Chloroform................	2.3
Amyl alcohol..............	3.7
Water......................	3.9
Ethyl alcohol..............	6.9
Methyl alcohol.............	11.5

An interesting calculation was made by Melsens [3] in 1874. Assuming that the heat on wetting charcoal was due to compression of the liquid he calculated that since the temperature rise of water was about 1° centigrade when suddenly compressed to 770 atmospheres, the temperature rise of 1.16° when charcoal was wetted corresponded to a pressure of 893 atmospheres.

Lamb and Coolidge [4] showed that the difference between the heat of adsorption and heat of liquefaction was practically constant for several liquids studied when referred to equal volumes. Assuming this heat difference to be the heat of compression the force was estimated as 37,000 atmospheres.

From the references given it is evident that though a considerable amount of work has been done, the phenomena of adsorption by porous media such as charcoal (or silica gel) have not been correlated so that the behavior of a given simple gas or simple mixture of gases can be predicted. Assuredly it is impossible to speak in other than a roughly qualitative way when it comes to adsorption of gas and vapor from such complex mixtures as occur in " wet " natural gas.

[1] Proc. Royal Institution, 18, 183.
[2] Compt. Rend., 157, 209.
[3] Ann. Chim. Phys. (5), 3, 522.
[4] Jour. Am. Chem. Soc., 42, 1146.

In recent numbers of Chemical and Metallurgical Engineering [1] articles have appeared dealing with adsorption of gasoline vapors by both charcoal and silica gel. Results given for the "Charcoal process" were obtained by the use of natural gas which carried gasoline vapors. The results cited in favor of silica gel were obtained by passing air saturated by pentane vapors through the gel. In view of the variations in adsorption as shown by the references cited it seems unlikely that the results given are of much value in comparing the operation of the two processes.

V. PHYSICAL CHARACTERISTICS OF NATURAL GAS AND NATURAL-GAS GASOLINE

Specific gravity of natural gas.—The density or weight of a unit volume of natural gas is usually expressed in terms of its relation to the density or weight of a like volume of air. The term "specific gravity" is given to this relation, and the expression is written thus: specific gravity, air equals 1.

The effusion method and apparatus for determining specific gravity of gas is here described. The results give the comparison between the densities of gas and air, and are based on the fact that when a definite volume of gas and an equal volume of air are taken, both at the same temperature and pressure, if each is allowed to escape into the air through the same small orifice the relation between the time taken by the various gases to escape will be the same as the relation of the square roots of their respective densities.

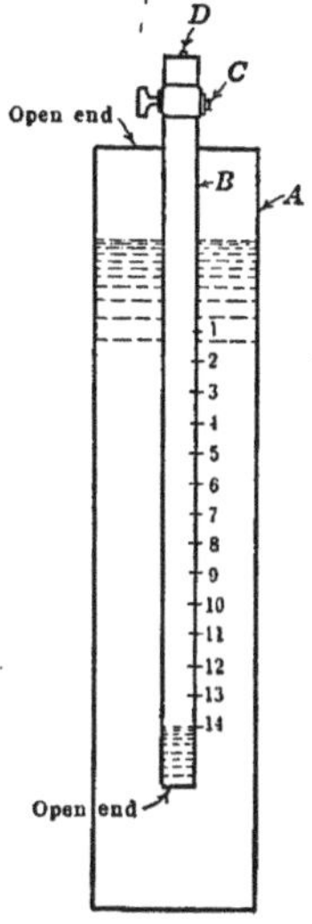

Fig. 7.—Diagram of apparatus for determining the specific gravity of natural gas.

An apparatus for making these determinations is described as follows:

A is a glass jar open at the top. *B* is a glass tube open at the bottom, and supplied with a valve *C* near the top. *D* indicates an orifice opening in the top of *B* which is closed except for this orifice.

The jar *A* is filled with water to a point near the top. Tube *B* is withdrawn from its position shown in *A*, and allowed to fill with air. Valve *C* is then closed and tube *B* again placed in *A*. The water compresses the air in *B* to some point such as 14. Valve *C* is suddenly opened and the air is forced through the orifice by the head of water. The time required for the water to pass from one mark, such as 12 to another mark, such as 3, is taken in seconds by use of a stop-watch.

In like manner the gas which is to be compared with air is passed by means of a rubber hose into the inner tube *B*, and allowed to fill it, and thereupon the time required for the water to rise from the same mark 12 to the mark 3 is noted.

The relation between the densities of the gas and the air are then calculated as follows:

Let T = time required by air to pass through orifice;
T_1 = time required by gas to pass through orifice.

Then $$\frac{T_1^2}{T^2} = \text{specific gravity of gas, air} = 1.$$

[1] Chem. and Met. Engineering, Vol. 23, pp. 1155, 1219 and 1251; also Vol. 24, p. 156.

Example.—Let T for air = 52 seconds;
and T_1 = 44 seconds.

Then
$$\frac{44^2}{52^2} = .64,$$
which is the specific gravity of the gas.

This is a delicate instrument. A large number of runs with air and gas should be made and the average time for a like volume of each to pass through the orifice should be used for calculation. Care should be taken that no water is lost during the experiments as the same head of water must act on both the gas and air in forcing each through the orifice. The air, gas and water should be at the same temperature.

The air surrounding the instrument should be perfectly pure and quiet. A wind blowing over the orifice will affect the readings; or if the experiments are being conducted indoors the accumulation of gas in the atmosphere may prevent true readings. The experiments are most accurate when conducted in a sheltered location out of doors.

NOTE.—Several methods of actually weighing gases have been used. A description of one such piece of apparatus is contained in Technologic Paper No. 89, Bureau of Standards, A Specific Gravity Balance for Gases, by Junius David Edwards.

Specific gravity of gasoline.—The methods of taking the specific gravity of liquids are presented in the chapter on methods of testing petroleum products, and tables of the relation of Baumé gravity to specific gravity, etc., were given in the general tables.

TABLE 10

Latent Heats of Evaporation—B.t.u. per Pound (Avoirdupois)

(Compiled by Dr. J. B. Garner)

Letter	Name of substance	B.t.u. per pound avoirdupois
	Methane	Not known
	Ethane	Not known
	Propane	Not known
A	Butane	197.0
B	Pentane	134.8
C	Hexane	142.9
D	Heptane	133.2
E	Otcane	127.7
F	89° Bé. gasoline	100.2
G	65° Bé. gasoline	100.6
H	56° Bé. naphtha	103.5
I	43° Bé. kerosene	105.4

A—Refrigeration experiments by Dr. J. B. Garner.
B—Smithsonian Tables—John.
C—Chemiker Kalender—Mabery.
D—Chemiker Kalender—Mabery.
E—Chemiker Kalender—Longinnin. Also Landolt's tables, pp. 836 to 843.
F. G. H. I—Bureau of Mines, Bulletin 88, p. 49.

Heat value of natural-gas gasoline.—The heating value and volume of vapor per gallon for natural-gas gasoline of various specific gravities, are matters of importance

TABLE 11

Properties of Natural Gas Hydrocarbons

Formula	Calories per gram Mol.[5]	B.t.u. per pound[1]	Pounds per gallon at 60° F.	B.t.u. per gallon	B.t.u. per cubic foot 60° F.—30″ Mercury[2]	Melting point		Boiling point		Specific gravity		Gravity, Baumé Scale	Cubic feet of vapor per gallon
						° C.	° F.	° C.[3]	° F.	Gas, air = 1	Liquid, water · 1[4]		
CH_4	211,900	23,790	....	....	1008.4	−184.0	−299.0	−160.0	−256.0	0.554			
C_2H_6	370,400	22,200	....	....	1763.6	−172.1	−277.8	− 84.1	−119.4	1.038			
C_3H_8	429,200	21,620	....	....	2519	−187.8	−306.2	− 44.1	− 47.4	1.523			
C_4H_{10}	....	21,320	....	....	3274	−135.0	−211.0	+ 0.3	+ 32.5	2.007	0.600	103.3	
C_5H_{12}	....	21,130	5.214	110.200	4029	−130.8	−203.4	36.4	97.7	2.491	0.626	93.6	27.3
C_6H_{14}	....	21,010	5.519	116.000	4784	− 94.0	−137.2	69.0	156.2	2.975	0.663	81.2	24.2
C_7H_{16}	....	20,920	5.727	119.800	5540	− 97.1	−142.8	98.4	209.2	3.459	0.688	73.5	21.6
C_8H_{18}	....	20,850	5.885	112.700	6295	− 56.6	− 69.9	125.5	257.9	3.944	0.707	68.0	19.5
C_9H_{20}	....	20,800	6.014	125.100	7050	− 51.0	− 59.8	150.5	302.9	4.428	0.722	639	17.7
$C_{10}H_{22}$	....	20,760	6.114	126.900	7805	− 32.0	− 25.6	173.0	343.4	4.912	0.734	60.7	16.2
$C_{11}H_{24}$	....	20,730	6.185	128.200	8560	− 25.6	− 14.1	194.5	382.1	5.396	0.743	58.4	15.0

[1] Computed with formula derived by Dr. Anderson as explained in his original text, B.t.u. per lb. $= \frac{20{,}375+6832}{n+0.1438}$.

[2] Computed with formula derived by Dr. Anderson as explained in his text, H.V. $= 755.2n + 253.2$.

[3] Boiling point of ethane was determined at a pressure of 749 mm., and that of nonane at 759 mm.; otherwise the figures are for 760 mm.

[4] The values are the ratio of the weights of the hydrocarbon and water at 60°. No correction has been made to the value of butane which was determined at its boiling point.

[5] Thomsen's figures for the hydrocarbons, methane, ethane and propane. (Landolt-Börnstein, "Physikalisch-chemische Tabellen," 1912, p. 909.)

in calculations involving the reduction, in heat value and volume, of natural gas through the removal of all or a part of its gasoline content, and certain combustion problems involve specific heat and heat of vaporization for gasoline.

For the purpose of such engineering calculations the following quotations, tables and figures as marked are taken from articles by R. P. Anderson in the Journal of Industrial and Engineering Chemistry.

Table 11 was compiled from two tables by Dr. R. P. Anderson, one marked 1 in Journal of Industrial Chemistry, Vol. 12, No. 9, page 852, September, 1920 and table marked 1 in same journal Vol. 12, No. 6, page 547, June, 1920. This table was calculated by Dr. Anderson with formulae deduced by him on the assumption, as stated in the text of his article (Vol. 12, No. 9, page 852) "that the relationship between the heating value of normal paraffin hydrocarbons, expressed in calories per gram-molecule, and the number of carbon atoms per molecule is a linear one." The formulae are not copied here but are clearly worked out in the text.

Table 11 shows physical values for the hydrocarbons which differ only slightly from those shown in Table 2, compiled by Dr. J. B. Garner.

The heating values given in Table 11 are for hydrocarbons in vapor form; therefore the latent heat of vaporization must be subtracted from these values to obtain heating values of the liquids. Further, Dr. Anderson shows the relation which exists between the specific gravity (G) and B.t.u. per pound of any of the hydrocarbons, and deduces the following equations.

$$\text{B.t.u. per pound} = 23{,}330 - 3500G. \qquad (1)$$

If the values for specific gravity given in Table 10 be substituted for G in this equation the results will agree within .03 per cent with those given in Table 10.

$$\text{B.t.u. per gallon at } 60° \text{ F.} = (23{,}330 - 3500G)\, 8.328G, \qquad (2)$$

or

$$\text{B.t.u. per gallon at } 60° \text{ F.} = 14{,}200 + 153{,}500G. \qquad (3)$$

Dr. Anderson states in the same article that Equations (1) and (2) are directly applicable to gasoline which is a mixture of several hydrocarbons "providing the assumption be made that it consists only of the normal paraffin hydrocarbons, pentane to undecane, inclusive. In other words, the formulae do not hold merely for the various individual hydrocarbons, but also for the various mixtures of these hydrocarbons which might exist as gasoline. The composition of gasoline of a certain specific gravity might vary considerably as to the number and relative proportions of the various constituents; but, under the assumption that has been made, this would have no material effect upon the heating value of the gasoline expressed either in B.t.u. per pound, or in B.t.u. per gallon.

"On the other hand, the specific gravity of a gasoline does not define exactly the average number of carbon atoms per molecule, the specific gravity of its vapor, the heating value of its vapor in B.t.u. per cubic foot, or the volume of vapor required to produce one gallon of gasoline, since the relationships between these characteristics and specific gravity are curvilinear."

When the hydrocarbons forming gasoline are separated from the body of natural gas, Dr. Anderson has this to say concerning the effect on the quality of the remaining gas: "The reduction in volume due to the removal of gasoline vapor from natural gas depends primarily upon the yield and specific gravity of the gasoline that is produced; the reduction in heating value primarily upon the heating value of the gas, and upon the yield and specific gravity of the gasoline removed; and the reduction in specific gravity primarily upon the specific gravity of the gas, and upon the yield and specific gravity of the gasoline removed. It is not a difficult matter to determine experimentally what the change in volume, and specific gravity amounts to in any given case, unless the change is extremely small, but it is frequently not convenient to do so. It is also unnecessary, since this change can be computed with sufficient accuracy for all ordinary purposes if the necessary information is available.

"In making the computation, the relationship between the specific gravity of the gasoline and (a) the specific gravity of its vapor, (b) the quantity of vapor required to produce a gallon of gasoline, and (c) the heating value of this vapor in B.t.u. per cubic foot must first be established."

By means of curves made by plotting the values of specific gravity of vapor against specific gravity of liquid for pentane, hexane and heptane, Dr. Anderson has worked out Tables (12 and 13), copied here from the text of his article Vol. 12, No. 9, page 852, which shows probable composition of natural-gas gasoline of different Baumé gravities.

TABLE 12

Gravity, ° Bé.	Average of maximum and minimum, specific gravity of vapor	Maximum error of specific gravity, per cent	Percentage composition corresponding to specific gravity			Vapor per gallon of mixture, cubic feet	Maximum error of volume, vapor, per cent	
			C_5H_{12}	C_6H_{14}	C_7H_{16}			
90	2.64	±0.5	77.6	14.1	8.3	26.4	±0.2	
85	2.85	±1.1	45.4	34.3	20.3	25.1	±0.4	C_5H_{12}, C_6H_{14}
80	3.09	±1.2	17.0	42.0	41.0	23.7	±0 5	C_7H_{16} mixtures
75	3.37	±0.2	3.9	9.6	86.5	22.1	±0.1	

TABLE 13

Gravity, ° Bé.	Average of maximum and minimum specific gravity of vapor	Maximum error of specific gravity	Percentage composition corresponding to specific gravity			Vapor per gallon of mixture, cubic feet	Maximum error of volume, vapor	
			C_4H_{10}	C_5H_{12}	C_6H_{14}			
90	2.66	±2.1	15.0	35.9	49.1	26.4	±0.2	C_4H_{10}, C_5H_{12}
85	2.83	±0.4	6.6	15.7	77.7	25.1	±0.1	C_6H_{14}, mixture
			C_6H_{14}	C_7H_{16}	C_8H_{18}			
80	3.06	±0.2	87.5	8.0	4.5	23.8	0.0	C_6H_{14} C_7H_{16}
75	3.39	±0.8	36.6	40.4	23.0	22.1	±0.1	C_8H_{18} mixtures

Dr. Anderson states in the same article: "The figures given in Table 14 for specific gravity of vapor and cubic feet of vapor per gallon represent the average of the values given in Tables 12 and 13. The heating values in the table have been computed from the specific gravities of the vapor by means of the formula

$$\text{B.t.u. per cubic foot at } 60°\text{, 30 inches mercury, dry} = 1559.7\,Gv + 144.6,$$

this procedure being permissible on account of the linear nature of the heating value, specific gravity relationship (Gv equal specific gravity of vapor).

TABLE 14

Gravity, ° Bé.	Probable specific gravity of vapor	Probable heating value B.t.u. per cubic foot	Probable volume vapor per gallon, cubic feet
90	2.65	4278	26.4
85	2.84	4574	25.1
80	3.07	4933	23.7
75	3.38	5416	22.1

It seems reasonable to assume that the values in Table 14 apply quite closely (probably within 1 per cent) to the ordinary natural-gas gasoline. With these values established the computation of the change in heating value or specific gravity of any natural gas may easily be carried out.

The relationship between heating value of gas before and after gasoline extraction is given by the formula

$$\text{H.V. untreated gas}\times 100=\text{H.V. treated gas}\times(100-V)+\text{H.V. gasoline vapor}\times V,$$

or $$\text{H.V. treated gas}=\frac{\text{H.V. untreated gas}\times 100-\text{H.V. gasoline vapor}\times V}{100-V},$$

where V represents the volume of gasoline vapor removed.

Similarly the specific gravity of treated gas

$$=\frac{\text{Sp. gr. of untreated gas}\times 100\text{-sp. gr. gasoline vapor}\times V}{100-V}.$$

In Technical Paper 253, Department of the Interior, Bureau of Mines, by Donald B. Dow, the following statement concerning the reduction in heat value of natural gas through the removal of gasoline vapors, is made:

"The percentage loss can be calculated in a more or less arbitrary manner on the basis of a gallon of gasoline being equal to 30 cubic feet of vapor, and the calorific value of the gasoline recovered being 119,400 B.t.u. per gallon. This figure is based on the average absorption gasoline which has a gravity of about 80° Bé. and a heating value of 21,500 B.t.u. per pound. The heating value of gasoline is rarely determined, as this property is not considered important in judging the quality of gasoline, consequently few figures on B.t.u. can be found. Those at hand show values ranging from 20,097 B.t.u. per pound in gasoline having a gravity of 58° Bé. to 21,269 B.t.u. per pound in gasoline having a gravity of 64° Bé. These figures differ with different methods of production and, as regards refined gasoline, differ with the crude used. However, in general, as the gravity of the gasoline rises the heating value also rises. This is due to the fact that it will take a larger volume of very light gasoline to weigh a pound than of a heavier or lower-gravity product. Hence it will be seen that the figure 21,500 B.t.u. per pound is logical for gasoline of the gravity obtained in the absorption process, and no appreciable difference in percentage loss would occur should this figure vary slightly.

"These factors are practically correct when applied to the average commercial gas. The percentage loss in heating value is then equal to

$$\frac{B-\left(\dfrac{B-(A\times 119{,}400)}{1-(A\times 30)}\right)}{B},$$

where A = gasoline extracted per cubic foot of gas;
B = B.t.u. value of gas before treatment in gasoline plant;
119,400 = B.t.u. per gallon of gasoline.

"This formula is developed as follows: A times 30 represents the quantity of gas equivalent to that of the gasoline removed; subtracting this quantity from 1 gives the quantity of gas left after the gasoline has been removed from 1 cubic foot. A times 119,400 represents the heating value of this quantity of gasoline, and the difference between this product and B represents the heating value of the residual gas; hence this latter figure divided by the actual quantity of gas present gives the heating value of a cubic foot of the gas. Then by subtracting this figure from B, the original heating value of the gas, the loss is obtained, and this in turn divided by B gives the percentage loss."

VI. METHODS OF BLENDING, SHIPPING, STORING, AND TRANSFERRING GASOLINE

If a quantity of natural-gas gasoline is placed in a closed vessel, its vapor will exert a pressure in the vessel which is proportional to the temperature. This pressure is termed the vapor tension of the gasoline.

Table 3 gives the vapor pressure or tension at different temperatures for several of the hydrocarbons which are generally components of the gasoline mixture, when these hydrocarbons are isolated and in their pure state.

But a number of samples of natural-gas gasoline, all having the same Baumé gravity may have widely differing vapor tensions, since, as is shown by Tables 12, 13, and 14 the components in the several mixtures having the same gravities may differ in their proportion in the mixture.

Hence, in shipping natural-gas gasoline, it is a matter of importance to test the gasoline for vapor tension whether in a raw state or blended with naphtha, to establish its safety and prevent losses by evaporation.

Shipment in tank cars.—For purposes of safety, the Interstate Commerce Commission has issued rules governing shipments of explosives, and in its publication of the regulations has set rules for making vapor-tension tests of gasoline after it has been placed in the cars. These rules are subject to change. At present they are as shown in the chapter on testing methods.

Blending gasoline.—Natural-gas gasoline in its raw state is too volatile for direct use as a motor fuel. By blending the gasoline with the light distillates of crude oil, which distill off just after the straight-run gasoline has been taken out of the crude, a satisfactory motor fuel is obtained.

The following tables Nos. 15 and 16, showing the physical characteristics of blends of gasoline and naphtha have been compiled by Mr. C. C. Reed, C.E.

The per cent of each constituent in the blend is governed by the Baumé gravity of each and the desired Baumé gravity of the product.

The important matter in blending is to obtain a thorough mixture. When all gas, propane, butane, etc., has been allowed to escape from the gasoline, and the latter in its truly liquid state is carefully mixed with naphtha, it forms a stable blend. If the blend is exposed to the atmosphere a gradual evaporation of the blend takes place lowering the Baumé gravity, but no separation takes place in the remaining fluid.

Blending may be performed by injecting the naphtha into the coils of the coolers, or into the make tanks or into the stock tanks in compression plants. In absorption plants naphtha may be injected into the condenser, or in the run tanks. Blending may be accomplished by pumping the naphtha and gasoline together into the storage tanks. The result from any of these methods is the same when properly carried out.

TABLE 15

Gasoline Naphtha Blends

Gasoline, Baumé gravity 60° F.	Naphtha, Baumé gravity	Per cent of gasoline	Per cent of naphtha	Blend, Baumé gravity 60° F.
80.5	57.3	90.0	10.0	77.9
80.5	57.3	80.0	20.0	74.9
80.5	57.3	70.0	30.0	72.4
80.5	57.3	60.0	40.0	71.0
80.5	57.3	50.0	50.0	67.2
80.5	57.3	40.0	60.0	65.1

TABLE 16

Gasoline Naphtha Blends

Gasoline, Baumé gravity 60° F.	Naphtha, Baumé gravity	Per cent of gasoline	Per cent of naphtha	Blend, Baumé gravity 60° F.
80.0	59.0	50.0	50.0	68.4
80.0	59.0	52.0	48.0	69.0
80.0	59.0	54.0	46.0	69.2
80.0	59.0	56.0	44.0	69.6
80.0	59.0	58.0	42.0	70.0
80.0	59.0	60.0	40.0	70.4
80.0	59.0	48.0	52.0	68.0
80.0	59.0	46.0	54.0	67.6
80.0	59.0	44.0	56.0	67.0
80.0	59.0	42.0	58.0	66.5
80.0	59.0	40.0	60.0	66.0

Claims have been made that heavy blending of the freshly made gasoline in the accumulator tanks of compression plants and the run tanks of absorption plants results in increased yields of gasoline. The quantity of liquid first gathered contains much butane and propane, and these gases, being only temporally in the liquid state, rapidly escape, thus reducing the quantity of final product. When mixed with naphtha these temporary liquids are held for a longer period of time, leave the mixture slowly and thus do not carry off mechanically so much pentane and other true liquids. Thus an excessive amount of blending does result in a greater apparent temporary yield, but the butane and pentane eventually do escape from the mixture. The net result is a slightly larger saving of the pentane and heavier liquids, but practically none of the butane and other gases are permanently held.

For records of evaporation losses in blending see Bulletin 88, Bureau of Mines, by George A. Burrell, F. M. Seibert and G. G. Oberfell, page 76.

A typical specification for naphtha for use with natural-gas gasoline is as follows:

TABLE 17

Blending Naphtha Specifications (for Cutting Vapor Pressure and Gravity)

Initial boiling point not less than	190° F.
Not more than 10 per cent over at	248° F.
Not less than 90 per cent over at	374° F.
End	437° F. or lower
Color	Water-white
Odor	Sweet

TABLE 18

Blending Naphtha Specifications (for Finished Motor Gasoline)

Not more than 10 per cent over at	194° F.
Not less than 40 per cent over at	284° F.
End	446° F. or lower
Color	Water-white
Odor	Sweet
Straight-run	

All grades of petroleum distillates can be used to make blends, and are so used, depending upon the specifications of the ultimate product.

Pipe-lines.—Pipe-lines for the transportation of gasoline should be very carefully laid, preferably of screw pipe. The collars should be taken off the pipe when received, the threads cleaned and coated with shellac and pipe screwed together very tightly. Ditch the line about 2 feet deep and cover thoroughly; this is important. Give the line plenty of slack.

Losses due to agitation in pumping through pipe-lines cannot be predicted; however, pipe-lines are usually not used except for transportation from distant localities to the railroad, in which case this method is generally more economical than hauling in drums.

The following notes are made covering a 2-inch line transporting gasoline from a large absorption plant in Pennsylvania. The length of the line is 26.5 miles. The quality of gasoline transported over a period of one year was 83.2 Bé. gravity. The total quantity transported in one year was 3,273,130 gallons.

The gasoline was measured in the storage tank at the plant. At the receiving end it was measured as shipped in tank cars; thus the transportation losses through the line included all handling at the receiving end of the line. The measured loss was 3.45 per cent, but in general, losses in loading lines very much exceed this percentage for the reason that they are only used intermittently. This particular line was constantly in use; hence the escape from it, due to the regasifying of some of the gasoline, was very small in proportion to the amount transported. The gasoline made at this particular plant is very stable although of high gravity.

VII. TESTING NATURAL GAS FOR GASOLINE CONTENT

General observations.—When it is proposed to build a plant for extracting gasoline from natural gas, a test of the gas should be made by a method identical in operation with that of the plant. Thus, for wet gases from oil wells, if a plant using the compression and cooling process is to be installed, each well which is to contribute to the gas supply should have its gas thoroughly tested by compressing and cooling at pressures and temperatures such as will obtain in plant operation. Similarly in the case of dry natural gases, for plants planned for the absorption process, tests should be made of the gas to be treated by the use of oil, and under pressure and cooling conditions such as will obtain in plant operation. And further, should it be planned to construct a plant using charcoal or other solid adsorbent, tests using those substances should be made.

The object of "testing" gas is not alone to determine the presence of gasoline, but also to show the recovery which may be expected from a plant operating under the method employed in the test. All such tests must be carried on for a sufficient length of time to obtain average conditions, as the quality of gas changes frequently, and further, tests should be made on a sufficiently large scale, so that the product may be thoroughly examined to determine the actual yield of gasoline of salable quality, after all temporary fluids have been separated from the condensate.

Compression tests.—The apparatus necessary for such tests may be put in many forms, but for convenience it should be portable and is generally constructed on a truck or wagon which may be transported readily to the wells.

The gas to be tested may be conducted from the well through a small pipe to a suction tank of 3 to 5 cubic feet capacity, conveniently set on the truck body. This tank is provided for the purpose of collecting any water or oil coming from the well,

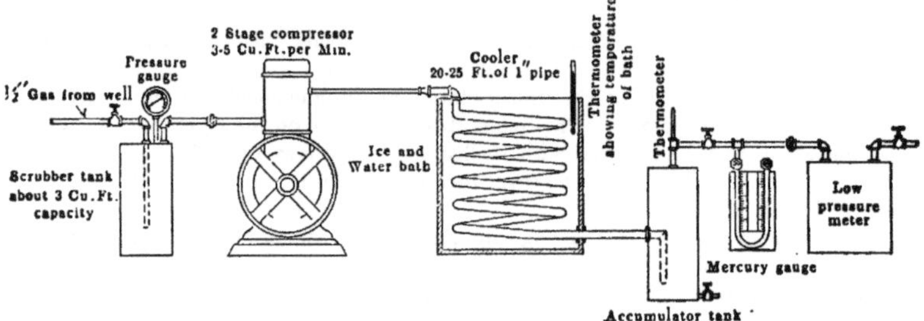

FIG. 8.—Diagram of apparatus for testing gas by compression and cooling.

thus preventing such fluids from entering the compressor. From the suction tank a pipe conveys the gas to the compressor, which should preferably be of two-stage design and capable of compressing to 250 pounds gage pressure. A capacity of 3 to 5 cubic feet per minute is a suitable size for this purpose. A gage or mercury column arranged to indicate suction pressure should be attached to the suction tank. A small gas or gasoline engine may be belted to the compressor for driving it. The compressed gas passes through cooling coils conveniently made of 1-inch pipe immersed in a tank of water in which ice can be placed. The tank should be large enough to accommodate about 20 to 25 lineal feet of the pipe arranged in a coil, and at the end of the coil a drip pipe should be connected to collect the condensate. A thermometer well should be placed in the pipe leaving the drip pot to indicate condensing temperatures. The meter is placed on the outlet gas after the pressure has been reduced. Additional cooling may be obtained by passing the gas from the cooling tank into an expanding coil made of 2-inch pipe having a 1-inch pipe passing through it. The drip is then placed on the end of the 2-inch pipe and the gas passes from the tank cooler through the space between the 2-inch and 1-inch pipes, after which it is expanded through the 1-inch pipe, thus cooling the gas in the annular space. The meter is then placed on the outlet of this expander cooler. If extreme cooling is carried out in a testing apparatus, the results may indicate yields which may not be attainable in a plant operating under less favorable conditions. In operation the pressure on the gas is reduced to practically atmospheric before entering the meter and a mercury gage is placed on the discharge pipe at the meter to indicate the exact pressure of the gas for the purpose of calculating the volume of gas remaining after passing through the testing apparatus.

In making a test the pressure and condensing temperature should be observed, and the test should be continued until a sufficiently large quantity of condensate is obtained so that a distillation may be made and the Baumé gravity of the product after weathering obtained.

Absorption Tests

The apparatus shown in Fig. 9 has been extensively used for the purpose of determining the quantity of gasoline which may be recovered from dry or wet natural gas by plants constructed to use the absorption process with oil. In making a test with this apparatus the oil which will be used in the finished plant should be used.

The natural gas undergoing test passes from the inlet header (Fig. 9) through the oil in the two autoclaves in series. The number of autoclaves to be used in series depends

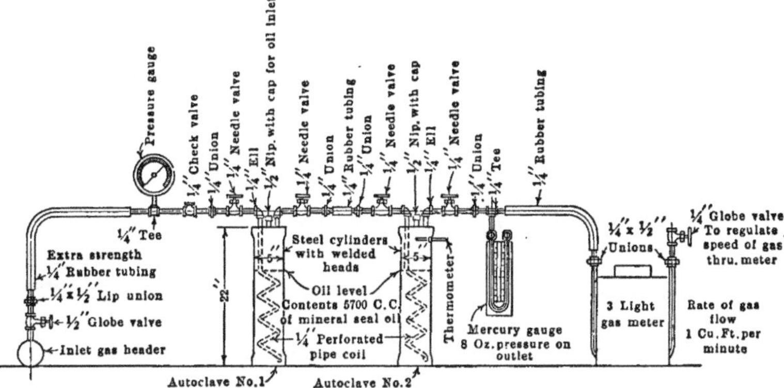

Fig. 9.—Diagram of apparatus used for testing gas by the absorption process.

upon the richness of the gas in gasoline content. A sufficient number should be used to leave only a small quantity of gasoline in the gas when it passes through the final autoclave in the series, and the test should be stopped when the Baumé gravity in the final autoclave has risen not more than one degree above the original oil.

In conducting a test the gas should be passed through the autoclaves at a rate not exceeding 1½ cubic feet per minute. A slower rate is preferable. A number of check runs should be made, and for important tests, four autoclaves should be used in series, so that the total quantity of gasoline gathered in the autoclaves will be sufficient, when distilled, to permit thorough examination for quality. The errors in absorption tests lie in conducting the test too hurriedly and on too limited a scale, so that the quality of the product is not fully determined.

It is important that the conditions of pressure and temperature during the test shall be those which can be approximated in the finished plant. It is the purpose of these tests to indicate what may be accomplished in the oil absorption plant when operated in accordance with the test conditions. Therefore, all observation should be made with this in view.

When the absorption in the autoclaves has reached the predetermined point in the final autoclave, absorption is stopped and the oil is withdrawn and measured in each autoclave separately. The increased volume thus observed is due to absorbed gasoline and dissolved gases. The Baumé gravity for each autoclave is noted.

The entire body of oil may be mixed together and distilled, and for this purpose a small steam still using steam at atmospheric pressure is suitable. The distillation apparatus shown in Fig. 10 indicates diagrammatically a steam still and a suitable condenser and collecting graduate for the distillate. The distillation should be carried on very slowly so that no oil will be carried over with the gasoline. The condenser coil is surrounded by a bath of ice-water. By these means the conditions which obtain in a plant are approximated, for while in the plant the condensing temperature is not as low as in the test apparatus, it is possible in the plant to arrange for compressing uncondensed vapors from the condenser tail pipes, or the pressure in the condenser itself may be raised to accomplish this condensation. Therefore, in general, the test condenser using the ice-water bath will fairly approximate plant conditions.

The Baumé gravity of the product as condensed should be taken and a careful

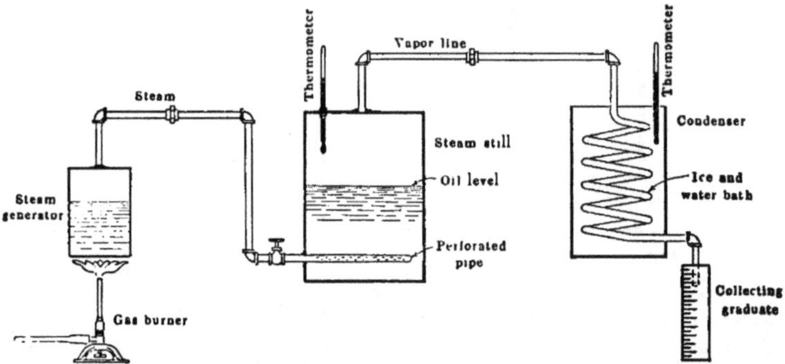

FIG. 10.—Diagram of apparatus used for recovering gasoline from oil. Used in absorption tests.

distillation should be run on the product to determine its quality and to arrive at the quantity of stable product, because the raw condensate will usually contain dissolved gases and may also contain a quantity of oil carried over from the still. It will generally be of too light and volatile a nature to meet commercial requirements. The distillation of the condensate, therefore, will determine the amount of merchantable product which has been obtained, and from this and the quantity of gas which has passed through the absorption apparatus, the yield of merchantable gasoline in gallons per thousand cubic feet of gas treated may be determined.

The observations to be made when making an absorption and distillation test are as follows:

A. Record should be kept of the quality and kind of oil used in the test.

B. Record should be made of the source of gas supply, and in testing casing-head gas from oil wells a complete record of all wells should be made for the purpose of determining the probable life of the gas supply.

C. Record the volume of oil used in each autoclave. This should be about 5700 cubic centimeters.

D. Take the Baumé gravity of oil corrected to 60° F. before the test is begun.

E. Take reading of the meter at beginning and ending of test, also pressure, if any, on the meter. From this the calculation of the total gas treated is made.

F. Take the time at beginning and ending of the test for determining the rate at which the gas was passed through the oil.

G. Record the pressure of the gas and the temperature of the oil and gas in the autoclaves while the test is in progress. This information is to guide the plant designer in determining the size of absorbers and coolers necessary in the finished plant.

H. Observe the increase in volume of oil in each absorber.

I. Observe the gravity of oil as taken from each absorber in degrees Baumé, corrected to 60° F.

J. Record the total volume of raw condensate obtained in the test.

K. Calculate from above data the gallons of raw condensate obtainable per thousand cubic feet of gas as treated.

L. From result of distillation of the condensate, calculate the gallons of gasoline of stable quality obtainable per thousand cubic feet of gas as treated.

Engineers have constructed autoclaves and washing bottles of various designs in which to make absorption tests. In like manner distillation apparatus of various kinds are successfully used. Dry distillation is used where steam equipment is not available. Much of this apparatus has been described in Bureau of Mines Bulletin 151, by W. P. Dykema. The same principles govern all tests in oil similar to the autoclave system described.

The following methods of testing natural gas for gasoline content are described in Bureau of Mines Technical Paper 87, by G. A. Burrell and G. W. Jones.

1. Solubility tests, in which the methods of testing a natural gas to determine its degree of solubility in different oils are described.

2. Testing by the absorption process.

3. Laboratory method, in which, by low temperatures, the natural gas is separated into two parts, gasoline and permanent gases.

Use of charcoal in determining the gasoline content of natural gas.—For this purpose a tube of metal or glass may be partially filled with activated charcoal which passes 8 to 14 mesh, and arranged to pass the natural gas continuously through the tube. This is a very delicate experiment and requires careful operation with regard to the following points:

Rate of flow.—This depends upon the quality of the gas undergoing test and numerous tests must be made to determine this rate. Since the gasoline constituents are rapidly adsorbed by the charcoal, heat is generated so fast that the temperature rise may liberate part of the gasoline after it has been adsorbed. A rate which proves, after trial, to give small temperature rise must be adopted.

Quantity of gas treated.—If too small an amount of gas is treated, it is difficult to entirely condense the gasoline when extracted from the charcoal. If too large a volume of gas is treated a part of the gasoline content will pass through the charcoal and not be absorbed. Numerous trials must be made to determine the quantity to be passed through the charcoal to obtain the greatest yield of gasoline per unit of gas treated.

Temperature of gas and charcoal.—Experience has demonstrated that the results are best when experiments are conducted at ordinary atmospheric temperatures. At high temperatures the recovery of gasoline falls off.

Method of recovering gasoline from charcoal.—The most satisfactory method of recovering the gasoline from charcoal is to place the charcoal in a distillation flask, or other suitable vessel and cover it with mineral seal oil. The gasoline will be liberated from the charcoal and become dissolved in the oil. Upon the application of heat to the flask the gasoline will distill from the oil and can be condensed in the usual manner. The charcoal is of no further value after this operation.

On account of the small scale of these experiments, the quantities of gasoline recovered do not permit of examination for quality.

In the Journal of Industrial and Engineering Chemistry, Vol. 12, No. 8, page 735, August 1920, paper by R. P. Anderson and C. E. Hinckley, "Use of Charcoal in Determining the Gasoline Content of Natural Gas," a description is given of a portable apparatus which the authors have used for making tests.

VIII. CONSTRUCTION OF GASOLINE EXTRACTION PLANTS

Compression plants (Fig. 11) may be arranged in any convenient manner to suit local conditions. The field of application is limited to the treatment of wet gases, usually gathered from oil-well casing heads. No plant should be built unless the supply of gas is proved to be enduring and of sufficient volume to warrant expenditure and give necessary earnings; too few wells should not be depended upon for a gas supply. The producing property should afford room for additional drilling and should be in the

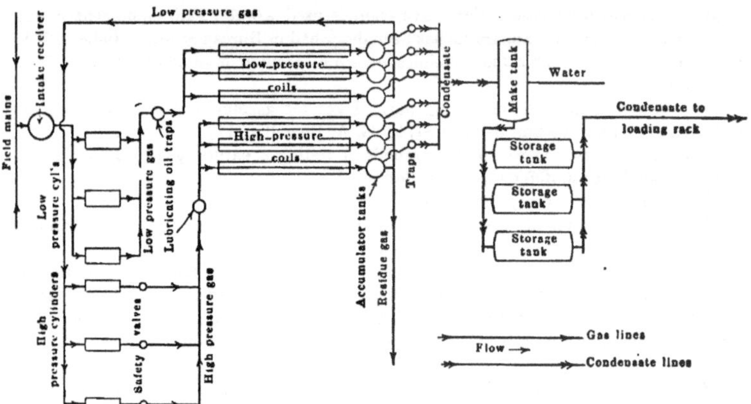

FIG. 11.—Diagram of a compression plant.

hands of operators who will take care of the wells. When the gas supply is adequate and enduring the compression plant is profitable and economical in operation.

Absorption plants (Fig. 12) are susceptible of wide variation in design. The methods and apparatus for absorption, distillation, blending and storage are very numerous.

The field of application may be either wet or dry gases. For wet gases the absorption method does not present any commercial advantage over the compression method, except in a few unusual cases. The cost of absorption plants is greater than that of compression plants and the efficiency of the former in gasoline production from wet gas is not sufficiently in excess of the latter to warrant the installation of an absorption plant unless the volume of wet gas to be treated is very large, under which circumstances the cost of the two types of plant is nearly equal and operating costs are comparable. Earnest thought should be given to any proposition contemplating the installation of an absorption plant on a small volume of casing-head gas. The salvage value of small absorption plants is not so great a proportion of the initial investment as is the case in compression plants.

For dry-gas treatment a study should be made to establish the certainty of a continuous volume of gas supply. The variation in volume as between winter and summer, and the conditions of cooling-water supply, markedly affect the expense of onstruction. When either a compression plant or an absorption plant is to be designed the character of the product which it is intended to produce should be determined upon. The investment in tankage, loading facilities and transfer apparatus is affected by the matter of blending, or redistilling. Original estimates of investment and operation costs may be far from what is developed if these matters are not given full consideration.

Gas compressors.—The two tables, Nos. 19 and 20, are given to indicate the approximate power required to compress 100,000 cubic feet of natural gas to pressures which may be found applicable to the wet gases used in the compression process.

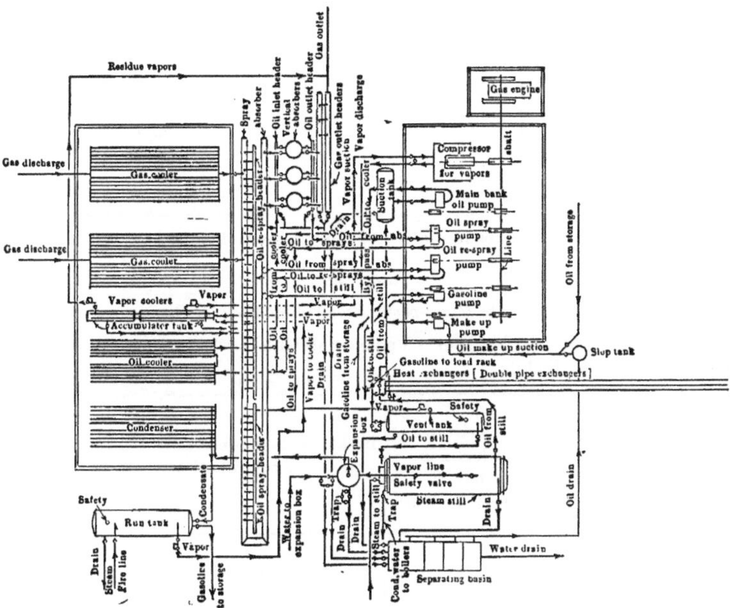

Fig. 12.—Diagram of an absorption plant.

Gas expanders.—Properly constructed steam engines may be operated with gas instead of steam. The power developed may be employed to drive pumps, or a gas compressor in compression plants. The expander engines may be designed either to be direct-connected to the compressors or operated by belt.

The gas which passes through the steam engines doing work is cooled in proportion to the quantity of the work performed, so that, upon exhausting, the temperature of the gas is 50° or 60° F. in many cases. This cold gas may be applied to cool water or gas, and then returned to the field lines. When a very large volume of casing-head gas is available for treatment, and the supply is assured, the investment may be justified, providing the cooling effect will produce a sufficient result in increasing the yield of salable gasoline. Climatic conditions govern this matter; in the middle and northern states artificial cooling can be effective only during the few hot months, and increased earnings are therefore limited in duration.

TABLE 19

Compression of Natural Gas

Compressors of two-stage type, gas-cooled between stages, suction pressure atmospheric, 14.7 pounds absolute.

Volume delivered per 24 hours in cubic feet	Discharge pressure high stage in pounds gage	Brake horse-power required at belt wheel of compressor or at rod in compressor of direct-connected type
100,000	300	23.0
100,000	250	21.6
100,000	225	20.7
100,000	200	20.25
100,000	125	16.7

Horse-power in direct ratio to volume compressed

TABLE 20

Brake Horse-power Required by Vacuum Pumps

Suction pressure at peak, about 4.5 pounds absolute discharge pressure at atmosphere, 14.7 pounds absolute.

Volume delivered per 24 hours in cubic feet at atmospheric pressure	Brake horse-power required at belt wheel of compressor
100,000	10.0

Horse-power in direct ratio to volume compressed

The following examples showing the work delivered and temperatures of gas attained by expanders are given by the Ingersoll-Rand Company as typical of apparatus built by them.

Example No. 1.—Natural gas expander with direct-driven compressor for recompressing this gas.

Working conditions:

Initial gas pressure to high-pressure expander cylinder, 225 to 275 pounds gage.
Exhaust gas pressure from low-pressure expander cylinder, 5 pounds gage.
Intake gas pressure to compressor cylinder, 2 to 3 pounds gage.
Discharge gas pressure from compressor cylinders, 35 pounds gage.
Gas available, 1,530,000 cubic feet of free gas per day.

Machine recommended for these conditions.

Ingersoll-Rand Company, 8 and 15 and 15×14 XPV–4.
Compound expander, high-pressure cylinder 8-inch diameter×14-inch stroke.
low-pressure cylinder 15-inch diameter×14-inch stroke.
Duplex compressor, 2 cylinders, 15-inch diameter×14-inch stroke.
P. D. 970 cubic feet, minimum.
R.P.M., 170.
Cut-off, $\frac{3}{8}$-stroke.
I.H.P. developed, 97.
Exhaust temperature of gas from low-pressure expander cylinder 60° to 70° F.
Initial temperature of gas from high-pressure expander cylinder, 60° F.

Example No. 2.—Expander for natural gas with two-stage compressor for recompressing this gas to 300 pounds discharge.

Working conditions:

Initial gas pressure to expander cylinder, 250 pounds gage.
Exhaust gas pressure from expander cylinder, 5 pounds gage.
Gas available, 2,200,000 cubic feet per day of free gas.

Machine recommended for these conditions.

Ingersoll-Rand Company, 9 and 18 and 16 and 8×16, XPV–3.
Compound expander cylinders:

High-pressure cylinder, 9-inch diameter×16-inch stroke.
Low-pressure cylinder, 18-inch diameter×16-inch stroke.

Compound compressor cylinders:

Low-pressure cylinder, 16-inch diameter×16-inch stroke.
High-pressure cylinder, 8-inch diameter×16-inch stroke.
P.D., 533 cubic feet per minute.
Discharge pressure, 300 pounds gage.

R.P.M., 150.
I.H.P., 140.
Initial gas temperature to expander cylinders, 80° F.
Exhaust temperature from low-pressure expander, 70° F.

Example No. 3.—Expander for natural gas with two-stage compressor for recompressing this gas to 300 pounds gage discharge.

Working conditions:

Initial gas pressure to expander cylinders, 125 pounds gage.
Exhaust gas pressure from expander cylinders, 5 pounds gage.
Gas available, 2,850,000 cubic feet of free gas per day.
Intake to compressor cylinders, 2½ pounds gage.
Discharge from compressor cylinders, 300 pounds gage.

Machine recommended for these conditions:

Ingersoll-Rand Company, 13 and 21 and 16 and 8×16, XPV–3.
Compound expander cylinders:

High-pressure cylinder, 13-inch diameter×16-inch stroke.
Low-pressure cylinder, 21-inch diameter×16-inch stroke.

Compound compressor cylinders:

Low-pressure cylinder, 16-inch diameter×16-inch stroke
High-pressure cylinder, 8-inch diameter×16-inch stroke.
P.D. 533 cubic feet per minute.

R.P.M., 150.
I.H.P., 140.
Initial, gas temperature to expander cylinders, 80° F.
Exhaust, gas temperature from expander cylinders, 35° F.

Example No. 4.—Expander for natural gas—power only.

Working conditions:

Initial gas pressure to expander cylinder, 125 pounds gage.
Exhaust gas pressure from expander cylinder, 5 pounds gage.

Gas available, 1,440,000 cubic feet per day of free gas.

Machine recommended for this service:
Ingersoll-Rand Company, 12×12 F.P.-1.
R.P.M., 175.
B.H.P., 55.
Initial temperature of gas to expander cylinder, 80° F.
Exhaust temperature of gas from expander cylinder, 25° F.

Absorbers.—Any apparatus constructed to provide means for bringing the natural gas and absorbing oil together and making contact between them is termed an absorber. The many forms of apparatus which have been designed for this purpose may be classified as follows:

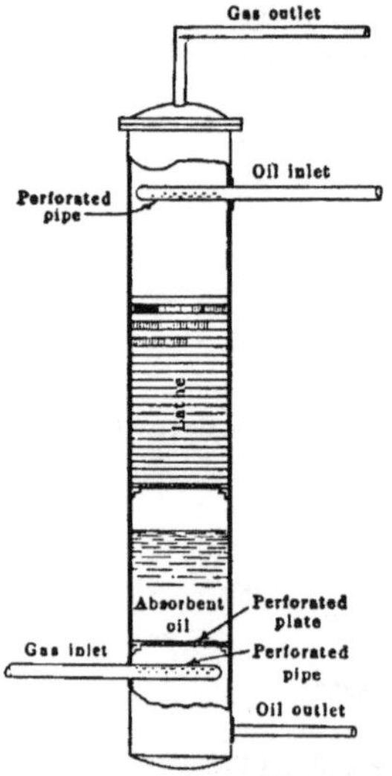

(*a*) Tanks in which a body of oil is maintained continuously, the gas passing up through it in fine streams of bubbles.

(*b*) Vertical towers in which the gas rises from the bottom to the top while the oil descends through it, being spread out over baffles to provide contact surface.

(*c*) Large pipe-lines through which the gas passes, the oil being sprayed into the gas and then collected.

The general principle to be observed in design is that the gas and oil must flow in counter current. Thus the fresh oil meets the gas just leaving the absorber and advances through the gas toward the point where the gas is entering.

Experience at present indicates that for purposes of efficiency and economy of cost a vertical chamber, as illustrated in Fig. 13, is the most satisfactory. Illustrations of such absorbers are shown in Bulletin 176, Bureau of Mines, by W. P. Dykema.

Fig. 13.—Diagram of vertical absorber. For capacities as per Table 21, oil level should be kept below gas inlet.

In a vertical absorber of given diameter a certain volume of gas can pass at a given pressure in twenty-four hours without carrying over any oil in suspension.

Table 21 gives the quantity in millions of cubic feet of free gas per day for different diameters of absorbers. Note that the space of 8 feet 5 inches is left between the upper row of lath and the top of the absorber for a separating space.

TABLE 21

Capacity of Free Gas in Millions of Cubic Feet per 24 Hours

Clear space on top of lath sections about 8 feet 5 inches. Ample space between gas inlet and bottom of lath section. Oil level kept below gas inlet.

Gage pressure in pounds	Inside Diameter of Absorbers												
	8 inches	10 inches	12 inches	16 inches	20 inches	24 inches	30 inches	36 inches	42 inches	48 inches	60 inches	72 inches	84 inches
0	0.019	0.030	0.045	0.080	0.128	0.182	0.285	0.410	0.560	0.730	1.142	1.672	2.237
25	0.051	0.081	0.121	0.215	0.342	0.487	0.761	1.095	1.493	1.948	3.047	4.460	5.967
50	0.825	0.133	0.197	0.350	0.556	0.792	1.237	1.780	2.426	.166	4.952	7.248	9.697
75	0.114	0.184	0.273	0.485	0.770	1.096	1.713	2.465	3.360	.384	6.856	10.036	13.428
100	0.146	0.236	0.350	0.620	0.985	1.401	2.190	3.150	4.293	3.603	8.761	12.824	17.158
150	0.209	0.340	0.502	0.890	1.414	2.010	3.142	4.520	6.160	8.040	12.571	18.400	24.618
200	0.273	0.443	0.655	1.160	1.842	2.620	4.095	5.890	8.026	10.476	16.380	23.976	32.079
250	0.336	0.546	0.807	1.430	2.271	3.230	5.047	7.260	9.893	12.913	20.190	29.552	39.539
300	0.400	0.650	0.960	1.700	2.700	3.840	6.000	8.640	11.760	15.350	24.000	35.130	47.000
350	0.463	0.753	1.112	1.970	3.128	4.449	6.952	10.010	12.626	17.786	27.809	40.704	54.460
400	0.527	0.856	1.265	2.240	3.557	5.058	7.905	11.380	14.493	20.222	31.619	46.280	61.921
450	0.590	0.960	1.417	2.510	3.985	5.668	8.857	12.750	16.359	22.659	35.428	51.856	69.381
500	0.654	1.063	1.570	2.780	4.414	6.278	9.810	14.120	18.226	25.096	39.238	57.432	76.842

Capacity of vertical absorbers.

Laths make the best baffles. They should be of soft wood, about ½ inch thick and 3 inches wide. A clear space of 50 per cent of the absorber area should be left through the baffles.

Vertical absorbers for high pressure, 100 pounds or more, may be about 30 feet high, and have about 9 feet of baffles.

The oil should not be sprayed in violently, but should be allowed to flow quietly over the top row of baffles and spread over them.

The absorption efficiency of the absorber depends on the time of contact between the oil and gas, and the thoroughness with which they are mixed together. At widely differing rates of flow of gas through a given absorber, channeling effects take place, and the gas and oil do not always make good contact. To overcome this, very high towers have been proposed. Another means of overcoming this tendency is to place several vertical absorbers in series, the oil being pumped from one to the other, counter current to the gas.

An arrangement shown in Fig. 12 has been found efficient. A sufficient number of vertical absorbers are placed in a parallel bank to provide separating capacity for the gas and oil; then a single-spray absorber is used, through which all the gas passes on its way to the vertical absorbers; the oil is collected from the bottoms of all the vertical absorbers and is sprayed and resprayed four to six times over and over in the spray absorbers. A very thorough contact is thus obtained. (See oil circulation.)

Vertical absorbers are efficient if they are made without baffles and contain instead a deep body of oil. They should be combined with an efficient spray absorber. A plant thus equipped, which may be taken as an example of this type of vertical absorber, has absorbers 4 feet in diameter, 25 feet high, and the gas enters 20 inches above bottom in perforated pipes. Oil enters 12 feet from the top of the absorber; gas leaves at top of the absorber. The oil level is about 5 feet above the gas inlet. The gas capacity of this absorber is 4.5 million cubic feet per day at 240 pounds gage pressure, showing that a body of oil above the gas inlet decreases the capacity of a vertical absorber.

Spray absorbers for high-pressure gas (Fig. 14) are laid horizontally and are actually nothing more than an enlargement of the gas line as it leads up to the vertical absorbers. The spray nozzles should be placed about 24 to 26 inches apart and their number is governed by the volume of oil to be handled. This spray absorber should be long enough to allow the oil to separate from the gas and settle in the collecting leg. Any oil which does not so settle is separated in the vertical absorber.

Spray absorbers 16 inches to 20 inches in diameter work well. The length should be 150 to 200 feet. For volumes of gas up to 20 million feet per twenty-four hours at 150 pounds pressure, one 20-inch diameter line can be used to form such an absorber. For larger volumes use absorbers in parallel. A large plant passing 140 million cubic feet per day at 250 pounds pressure uses two spray lines in parallel, 30 inches in diameter and 200 feet long. The oil drip should be ample in size with a drip seal for the line leading to the still.

Spray nozzles may be ½-inch, ¾-inch or 1-inch and so connected that they may be back-blown for cleaning and be withdrawn from the absorber if necessary. The capacity of each nozzle in gallons per minute must be obtained from the maker of the particular nozzle chosen, also the differential pressure required—15 to 30 pounds. Any first-class nozzle such as used for atomizing water will serve for this oil-spray purpose.

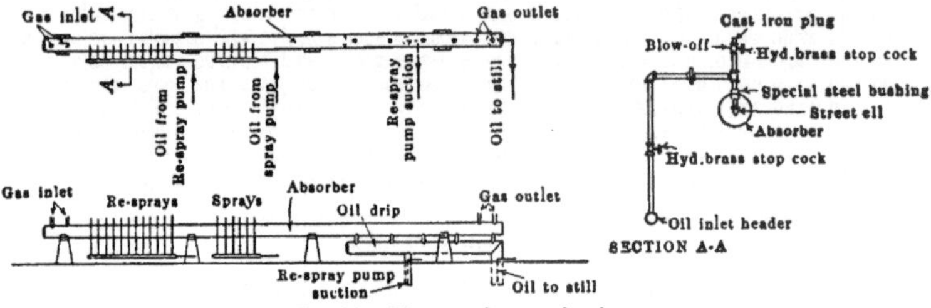

Fig. 14.—Diagram of spray absorber.

TABLE 22

Capacity of Free Gas in Millions of Cubic Feet per 24 Hours

Absorbers are assumed to be half full of oil and with proper oil and gas connections to and from absorbers. No oil will be carried over with the gas if capacity is kept within limits given in this table.

Gage pressure in pounds	Inside Diameter of Absorbers												
	8 inches	10 inches	12 inches	16 inches	20 inches	24 inches	30 inches	36 inches	42 inches	48 inches	60 inches	72 inches	84 inches
0	0.021	0.027	0.033	0.046	0.059	0.076	0.105	0.133	0.174	0.214	0.333	0.482	**0.656**
25	0.056	0.072	0.088	0.123	0.158	0.203	0.280	0.355	0.460	0.571	0.889	1.285	**1.750**
50	0.091	0.117	0.144	0.200	0.257	0.330	0.454	0.577	0.745	0.928	1.444	2.089	**2.843**
75	0.127	0.162	0.199	0.277	0.356	0.457	0.629	0.799	1.031	1.285	2.000	2.892	**3.937**
100	0.163	0.208	0.255	0.354	0.456	0.584	0.803	1.022	1.316	1.643	2.555	3.696	**5.031**
150	0.234	0.298	0.366	0.508	0.654	0.838	1.152	1.466	1.887	2.357	3.667	5.303	**7.218**
200	0.306	0.389	0.477	0.662	0.853	1.092	1.501	1.911	2.458	3.071	4.778	6.910	**9.405**
250	0.378	0.479	0.588	0.816	1.051	1.346	1.850	2.356	3.029	3.786	5.888	8.517	**11.592**
300	0.450	0.570	0.700	0.970	1.250	1.600	2.200	2.800	3.600	4.500	7.000	10.152	**13.780**
350	0.522	0.660	0.811	1.124	1.448	1.854	2.548	3.245	4.171	5.214	8.111	11.732	**15.967**
400	0.593	0.751	0.922	1.278	1.646	2.108	2.897	3.689	4.742	5.928	9.222	13.339	**18.154**
450	0.665	0.841	1.033	1.432	1.844	2.362	3.246	4.134	5.313	6.643	10.333	14.946	**20.342**

Capacity of horizontal absorbers 20 feet long.

Should horizontal absorbers be used instead of the vertical type, refer to Table 22 for capacity recommended. The spray absorber should be used in conjunction with the horizontal absorber in a complete plant. In operation it is difficult to distribute the gas equally to a bank of absorbers, either of the vertical or the horizontal type, and for this reason the gas-distributing heads up to each absorber should be of large size and all perforated distributing pipes should be arranged for back-blowing. By making the spray absorber of ample size and by making the respray pump large, as recommended, any inefficiency in the vertical or horizontal absorber can be made up.

Oil circulation, pumps, etc.—The quantity of oil to be circulated per hour in any given plant is determined by the volume of gas to be treated per hour, its gasoline content and pressure. (See description of process, Table 4.)

For example, assume that a circulation of 6000 gallons of oil per hour will be required. Then the pumping equipment for an arrangement of vertical and spray absorbers as per Fig. 13 will be as follows:

(*a*) One pump of 6000 gallons capacity per hour, designed to take oil at no head and deliver against full gas pressure into the tops of the whole bank of vertical absorbers.

(*b*) One pump of 6000 gallons capacity per hour, designed to take the oil from the bottom of the vertical absorber at a suction pressure about equal to the full gas pressure in these absorbers and raise this pressure about 30 pounds for the purpose of atomizing and delivering the oil into the spray absorber. (Figs. 12, 13 and 14.)

(*c*) One pump of four times the capacity of the given oil circulation or in this case 24,000 gallons per hour, arranged to take suction at the full gas pressure in the spray absorber and respray the oil gathered from the leg, as shown (Fig. 14), back through the spray absorber.

For convenience all pumps should be grouped in one building, both those used for oil circulation and for all other service. Pumps may be steam-operated, if use can be made of exhaust steam, or they may be power pumps driven by belts from a line shaft, or be direct-driven by electric motors.

Fig. 12 illustrates an absorption plant using power pumps belt-driven from the line shaft and operated by a gas engine. The rate of oil circulation may be altered by having the main high-pressure pump and first spray pump arranged with by-passes or by altering the pulley sizes. Changes in circulation rate seldom have to be made hurriedly in plants. Electric motor drive may also be used with a line shaft. The main high-pressure pump delivering to vertical absorbers may be of plunger type or multistage centrifugal type, but pumps for spray absorbers should be centrifugal.

If steam pumps are used they may be of direct-acting type or steam turbine driven. The choice of pumps rests with the designer, but unless a method of distillation is chosen in which use can be made of all exhaust steam from pumps, it appears best to use power pumps and drive with belts from a line shaft using a gas engine for power.

The steam consumption of ordinary duplex steam pumps using 100 pounds steam pressure is 130 pounds of steam per horse-power per hour. On account of the many pumps used at a gasoline plant, also tail gas compressors, the stilling operation generally does not use all the exhaust steam, and if no other use exists for the steam, it must be wasted, thus increasing the fuel charge. The still should be treated with live steam (see Stills, Fig. 16) and power pumps should generally be used.

For information as to efficiency and power consumption of pumps the designer is referred to " Mechanical Engineers' Handbook " by Lionel S. Marks, and to the builders of such equipment.

The following Table 24 is given to show the percentage of the power delivered to the belt wheel of a power pump which is delivered by the pump in useful work, the difference being mechanical friction.

TABLE 23

Steam Consumption of Simple Duplex Pumps

(National Transit Co.)

Pounds of steam per pump horse-power-hour

Stroke in inches	Gage Pressure in Pounds at Steam Inlet						
	60	70	80	90	100	120	140
4	200	189	185	180	177	171	165
6	165	158	152	147	141	137	137
8	142	137	132	128	127	120	116
10	124	117	116	113	110	107	101
12	113	110	104	104	101	96	94
18	99	94	88	88	86	84	81
24	90	86	81	81	79	76	73

TABLE 24

Mechanical Efficiencies of Direct-acting Pumps

(From L. S. Marks)

Stroke in inches	Plunger pumps	Piston pumps	Outside packed plunger pumps	Pressure pumps
3	0.50	0.50	0.47	0.45
4	.55	.55	.52	.50
5	.60	.60	.57	.54
6	.65	.65	.61	.58
8	.70	.70	.66	.63
10	.75	.75	.71	.67
12	.775	.775	.74	.70
15	.80	.80	.76	.72
18	.825	.825	.78	.74
24	.85	.85	.81	.77
36	.875	.875	.83	.79

Heat exchangers.—The double-pipe form of heat exchanger has given satisfactory results, and while various forms may be used the double-pipe type alone will be described here, since it meets theoretical and practical requirements.

Properly proportioned heat exchangers are vital to the successful operation and economy of absorption plants. Complete transfer of heat between the hot oil leaving the still and the cold oil leaving the absorber of course cannot be economically attained, but an exchange of 80 per cent of this heat can be obtained by proper proportions. As the heat exchanger is increased in size the exchange of heat is increased also, but the friction increases rapidly. Friction assists in transfer of heat, but when a point is reached requiring more than a certain amount of pumping work to be done in circulating the oil through the pipes, the limit of economy is passed. This point has to be carefully determined by the designer.

The exchangers are best arranged with the hot oil leaving the still in the outside pipe (Fig. 15). The advisability of covering the exchanger with insulation depends upon local conditions. If the exchanger is insulated more heat will be saved in the still, but more water will be required for cooling the warm oil leaving the exchanger, and this will require power. In temperate climates the oil coolers should be made ample for summer conditions (see Coolers, Fig. 22) and the exchanger should be covered in winter. The cooler can then take any residual heat from the oil, and in summer a few of the final exchanger coils may be uncovered to allow heat to radiate into the air, thus aiding the water coolers. The exchanger coils should be spaced well apart and the most economical arrangement of exchanger insulation should be determined by experiment for each particular plant.

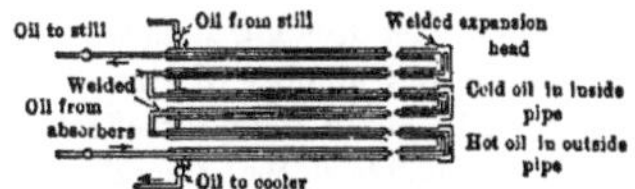

Fig. 15.—Double-pipe form of heat exchanger.

Table 22 gives dimensions of double-pipe heat exchangers in use. These exchangers are entirely covered in winter.

TABLE 25

Tabulation of Double-pipe Counter-flow Heat Exchangers (Not Insulated)

Cold oil in inside pipe, hot oil in outside pipe

Number of plant	Sizes of pipes in exchanger	Length of exchanger in feet	Number of pipes high	Total number of stands in exchanger	No. of stands in parallel	Number of series	Total heat exchanger area in square feet	Gallons of oil circulated per 24 hours	Exchanger area in square feet per 1000 gallons oil per 24 hours
1	3" ×5"	40	6	6	1	6	1230	90,000	13.6
2	3" ×5"	63	6	4	1	4	1207	75,000	16.1
3	2½"×4"	54	6	6	3	2	1350	90,000	15.0
4	2" ×3"	20	6	8	4	2	555	45,000	12.3
5	2" ×3"	40	6	4	2	2	555	45,000	12.3
6	3" ×5"	63	8	6	1	6	2685	120,000	22.4
7	2½"×4"	60	6	6	3	2	1500	120,000	12.5
8	2½" ×4"	120	6	3	3	1	1500	102,800	14.6

Number of plant	Average velocity of oil in feet per minute	Average Oil Temperatures in Degrees F.—Month of December, 1920						
		Cold oil from absorbers			Hot oil from still			Per cent of heat recovered
		Enter exchanger	Leave exchanger	Heating effect	Enter exchanger	Leave exchanger	Cooling effect	
1	138	57°	155°	98°	204°	82°	122°	0.760
2	116	57	232	175	284	93	191	.817
3	74	56	174	118	253	108	144	.688
4	50	56	180	124	255	119	136	.706
5	96	62	210	148	266	113	153	.789
6	194	65	224	159	266	68	198	.842
7	98	72	203	131	263	114	149	.772
8	85	76	251	175	315	127	188	.797

For general design the following proportions generally give satisfactory results in operation and investment charges are reasonable.

Operating method counter flow.—The area of the external surface of the inside pipe is about 14 square feet per 1000 gallons of oil circulated per twenty-four hours.

The oil velocity is 135 lineal feet per minute, about equal for both hot and cold oil.

The drop in pressure through an exchanger of this type should be about 4 pounds. Therefore, after deciding upon the sizes of pipes to be used and the area required, arrange the length and number of stands to give about this drop in pressure. The velocity will then be about right for effective exchange. Keep the number of stands of pipe as few as possible so that the oil will distribute.

Special examples of heat exchangers.

No. 1

Capacity, gallons per twenty-four hours	30,000
Number of stands of pipe	1
Size of pipes	2 inches×3 inches
Number of passes in stand	6
Length of exchanger	80 feet
Oil velocity	130 feet per minute
Friction loss	3 pounds
Area of exchanger	270 square feet
Internal area 2-inch pipe	3.36 square inches
Area between 2-inch and 3-inch pipe	2.96 square inches

No. 2

Capacity, gallons per twenty-four hours	50,000
Size of pipes	2½-inches×4 inches
Number of stands of pipe	1
Number of passes in stand	6
Length of exchanger	120 feet
Oil velocity	120 feet per minute
Friction loss	4 pounds
Area of exchanger	500 square feet

Instead of one stand of 2½-inch by 4-inch pipe for this capacity, an exchanger of two stands of 2-inch by 3-inch pipe would give equally good results.

No. 3

Capacity, gallons per twenty-four hours	85,000
Number of stands of pipe	1
Size of pipes	3 inches×5 inches
Number of passes in stand	6
Length of exchanger	120 feet
Oil velocity	140 feet per minute
Friction loss	4 pounds
Area of exchanger	616 square feet

These exchangers are used for very large plants.

The exchangers, coolers and stills are all interdependent and should be studied and considered together in designing a plant. If the plant is designed so that the stills operate at 15 pounds to 35 pounds pressure the oil will flow through the exchangers witnout pumping, which is a marked advantage in plant design, as hot oil is difficult to pump.

Distilling apparatus.—The quantity of oil circulated each twenty-four hours in any given plant should be adjusted so that samples of oil drawn from the vertical absorbers show a rise in gravity of not more than 1° Baumé.

The gas volumes passing through the plant vary widely from hour to hour, but oil rates cannot be varied rapidly; therefore plant operation can only make gradual changes depending upon the indications of the absorption in the oil samples. In summer the rate of circulation will tend to decrease if the gas flow decreases, but not in direct proportion, for when the absorbing oil is warm, 80° F., more is required per thousand feet of gas treated than is needed in winter with temperatures of 50° F.

Oil from the spray absorber, Fig. 13, passes first to a primary still or vent tank. Distillation is carried on continuously and in two stages, for the following reasons:

(*a*) The charged oil should not be distilled in batches but rather in a continuous process. This is because the very volatile fractions come off rapidly in the early stages of distillation; hence in a batch process the percentage of non-condensable fractions in the mixture of vapors coming off the still is very large in the early stages, and these gases carry off in suspension a large part of the merchantable gasoline. On the other hand, in a continuous process a smaller proportion of non-condensable gases is coming off at all times, and better condensation results.

(*b*) The charged oil leaving the absorber contains a large amount of butane and propane, which if allowed to enter the still will interfere greatly with the condensation. To avoid this effect these gases are liberated in a primary still or vent tank continuously and in such a manner as not to carry off any merchantable gasoline from the oil. After passing this point the oil passes into the main still where the gasoline is liberated from the oil.

This first distillation may be accomplished in two ways: (*a*) Reduce the pressure on the oil as it leaves the absorbers, allowing it to pass continuously into a suitable tank maintained at a level half full. Then at ordinary temperatures, 50° to 80° F. and at pressures of 0 pound to 10 pounds gage, the butane and propane will be liberated, leaving the gasoline constituents still absorbed in the oil. This method requires that the oil from the tank shall be pumped through the exchangers into the still. Further, all of the gases liberated from the oil must be compressed from the low-venting pressure up to the discharge pressure again. These compressed gases should be cooled and thoroughly dripped to recover gasoline, should any be carried off.

(*b*) If the pressure on the absorbers is 50 pounds or over, then the oil may be put through the heat exchangers by this direct pressure. At the end of the exchangers the temperature of the oil will be from 150° to 210° F., depending on the still temperatures. The oil then passes continuously into a tank called a primary still at 40 pounds pressure or higher. By proper adjustment of temperature and pressure a point can be reached where very little gasoline will be carried out of the oil. The vapors leaving the primary still should pass through a cooler and settling tank at the pressure on the primary still, so as to collect any fluid. Should any be found it should be passed into the main still and run through the condenser. These vapors being under high pressure will usually return readily to the suction of the compression station if the plant is located where this can be done; but if it is not so located the gases can be compressed into the discharge again without a great expenditure of power, as the pressure in the primary still is high.

This matter of primary distillation is of great importance but is susceptible of many variations of method and apparatus.

The primary still at high temperature and pressure is simpler in operation than the vent tank method, as it requires less compressing of liberated vapors and no pumping of oil into still. But when the absorption is being conducted at low pressure a compressor should recompress the vapors from the primary still, as the oil may be hot enough after leaving the exchanger to liberate some gasoline in the primary still should the pressure be low. The compression will condense such gasoline.

Table 23 gives combinations of temperature and pressure which have operated satisfactorily on some plants, but it is necessary to determine this combination for each plant by testing the gas escaping from the primary still.

TABLE 26

Primary still conditions: Temperature of entering oil, degrees F.	Approximate corresponding pressure pounds
185	40
215	60
225	70

The following Table 24 gives sizes which have proved satisfactory for vent tanks. The tanks should be safe at 75 pounds pressure, and be equipped with a back-pressure valve for vapor escape.

TABLE 27

Dimensions of Vent Tanks

(By John Mosby)

Volume of oil circulated in gallons per 24 hours	VENT TANK DIMENSIONS	
	Diameter	Length, feet
350,000	7 feet	20
240,000	6 feet	20
140,000	5 feet	20
110,000	4 feet	20
55,000	30 inches	20
35,000	20 inches	20

Main still.—The function of a main still is to provide an apparatus for adding heat to the oil in addition to that obtained from the exchanger, sufficient to distill the gasoline vapors from the oil, and also to provide a suitable space where the vapors can separate from the oil and collect continuously.

Heating is usually accomplished by the use of steam, as the temperatures required are not generally above 275° F. Temperature regulation may be accomplished readily by the use of steam, and, as will be shown, the presence of steam in the oil assists the distillation of the gasoline vapors.

Stills built to use direct heat instead of steam have been proposed and built. Their purpose is to eliminate the boiler installations, and it has also been stated that heat can be saved through this arrangement. The data available on this type of still are not as yet sufficient to demonstrate its true efficiency, fuel consumption, radiation losses, etc., or to show how well it can be regulated in operation.

Two types of steam stills are shown, Figs. 16 and 17.

Referring to Fig. 16, the horizontal tank is maintained about half full of the oil, and thus affords a liberating surface for the vapors. The oil is fed in continuously near the top of the tower, which is baffled, generally with hard cobblestones. The oil descends among the stones and is delivered from the tower into one end of the horizontal tank, whence it passes gradually through the length of the tank, all the time liberating the gasoline vapors, and is finally withdrawn at the outlet end of the tank, whence it is pumped through the exchangers and coolers and into the absorbers. The vapors are continuously liberated in the tank and pass upward through the tower, and thence through the vapor line to the separating box, or auxiliary condenser on to the main condenser. The heating is done by steam admitted through perforated pipe in the bottom of the tank. This may be exhaust steam from pumps if available, together with live steam direct from boilers.

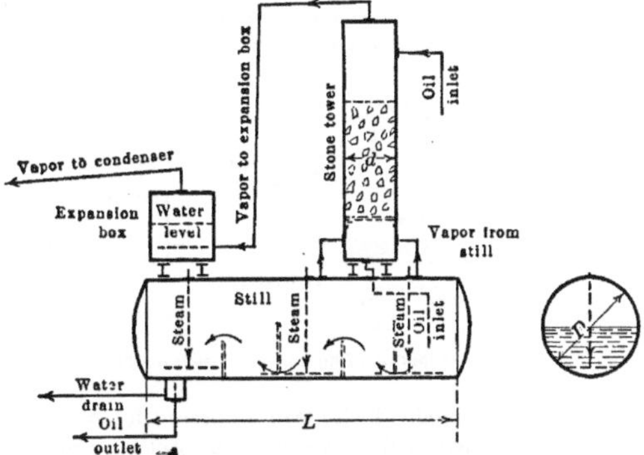

FIG. 16.—Diagram of steam still with tower, operated at low pressure.

The space above the oil level in the tank is filled with a mixture of gasoline vapors and water vapor and small quantities of oil in suspension. These mingled vapors on rising through the tower exchange heat with the entering oil which is cooler, and thus some of the light vapors of the oil are carried over to the condenser without descending to the tanks, while at the same time oil suspended in the rising vapors from the tank is caught and returned. Stills of this type have been widely used. They are made of light material and are worked at pressures close to atmospheric. The tower necessitates the use of a considerable quantity of steam for the single purpose of lifting the heavy gasoline vapors up through its length. The entire still is delicate in operation, but works well when operated at its best rate and proper temperature. It is apt to throw oil violently over into the condenser if attemps are made to force it, as the increase in volume of water vapor will choke the tower.

The capacity of the still may be taken as follows:

$L \times D \times 30 =$ gallons of gasoline per twenty-four hours;
$L =$ length of still in feet;
$D =$ diameter of still in feet.

Experience has shown that with a properly proportioned tower enough vapors may be liberated in twenty-four hours to form gasoline in the amount given by the formula. Thus a still 10 feet in diameter and 30 feet long may be worked at a rate of 9000 gallons of gasoline production per day. It is here assumed that the oil will be reasonably charged with gasoline, say 4 or 5 per cent, for if only a trace of gasoline were contained in the oil the still would be unable to heat sufficient oil to give this output in twenty-four hours.

The tower for a 10 by 30 foot still should be about 5 feet 6 inches in diameter, and 35 feet high. Two towers may be used to give free escape for vapors, but the oil and vapors are not apt to be distributed as well as in the single tower. The following formula has been developed by Mr. John Mosby:

$$\text{Diameter of tower in feet} = \sqrt{\frac{L \times D}{10}},$$

where L and D are respectively the length and diameter of the tank in feet.

The oil should enter the tower 8 feet below the vapor outlet. Vapors enter the tower at the bottom and a perforated plate above the vapor inlet supports the stones.

The still, Fig. 17, is built to be worked at vapor pressures as high as 35 pounds if

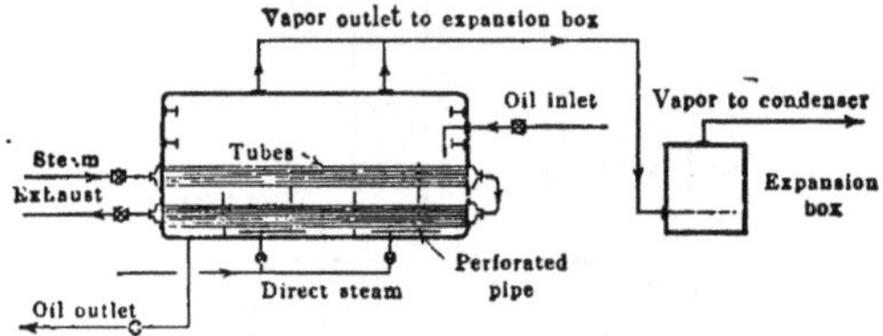

FIG. 17.—Diagram of high-pressure steam still.

desired. Tubes are shown through which steam is passed for heating the oil, and in addition some steam is admitted to the interior of the still for agitation and to assist distillation.

No tower is shown, since if the oil heat exchangers are ample very little added effect can be obtained from a tower.

The purpose of the tubes is to permit of limiting the amount of water which would otherwise enter the still and later have to be separated from the oil. Further, the tubes provide good distribution of heat. The latest stills built for pressure are designed without the tubes, all the steam entering the still direct.

A back-pressure trap is placed on the outlet of the tubes in these stills, and all steam which enters the tubes is reduced to water before being withdrawn, thus liberating all its latent heat in useful work.

With stills worked at 20 pounds pressure the gasoline vapors are more completely condensed in the condenser than can be accomplished with atmospheric pressure stills. In the case of the latter it is necessary to compress all the vapors from the condensers to gather the gasoline which would otherwise be lost as only partial condensation takes place at atmospheric pressure in this process. Experience has shown that by putting pressure on the condensers practically all of the merchantable gasoline is recovered without the necessity of compressing the vapors for that purpose. The still pressure also permits the oil to flow through the heat exchangers without the necessity of pumping. Plant operation as a whole is simplified by arranging that a pressure may be carried on the still and condenser.

The capacity of pressure stills, Fig. 17, may be taken as follows:

$L \times D \times 45$ = gallons gasoline per twenty-four hours;
L = length of still in feet;
D = diameter of still in feet.

It is here assumed that the entering oil contains 4 to 5 per cent of gasoline, and the still is worked at about 30 pounds pressure.

The diagram, Fig. 18, shows an expansion box of simple design. The vapors from the still enter it below the water line as shown, pass through the body of water and out from the top of the box to the condenser. The water level is maintained about 2 feet below the top of the box, the water entering at the bottom and being drained off at the highest level. A thermostat in the top of the box controls the admission of cool water so as to maintain a temperature found experimentally generally about 180° F., at which temperature any oil suspended in the gasoline vapors will settle out and collect on the

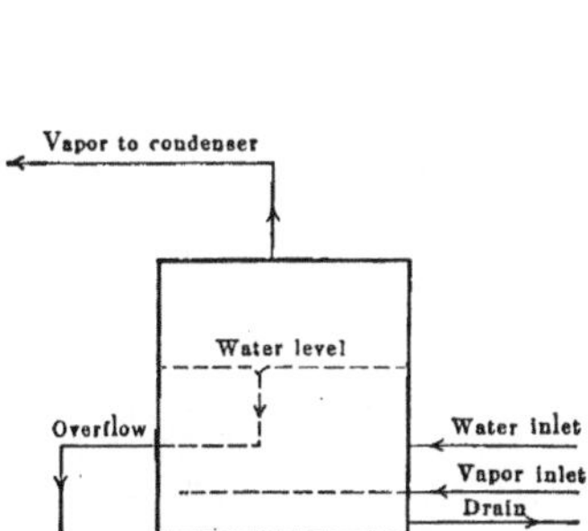

Fig. 18.—Expansion box of simple design.

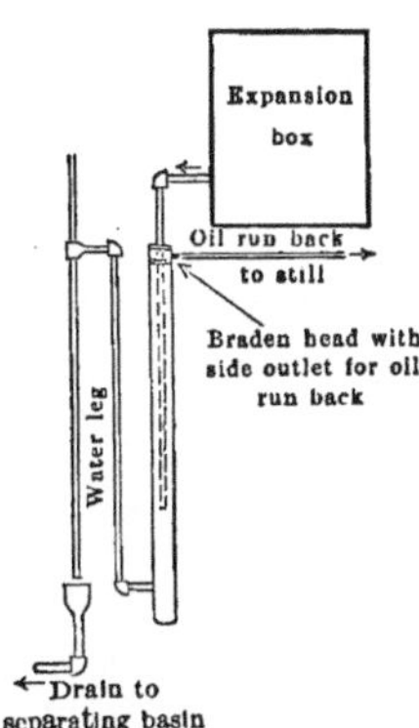

Fig. 19.—Automatic trap for taking the oil and water from an expansion box operating at low pressure.

surface of the water. A trap automatically drains off the oil and water, preventing the escape of vapor and maintaining the water level in the box.

Fig. 19 illustrates diagrammatically an automatic trap for taking the mingled oil and water from a still operating at low pressure.

An expansion box or some form of condensing apparatus with temperature control is necessary to rid the gasoline vapors of oil and produce products of narrow distillation range.

The principles governing distillation of gasoline from absorbent oil in any steam still are as follows:

The space above the oil is filled with a mixture of gasoline vapor and water vapor. If the still is being operated at 20 pounds pressure, and the water vapor forms 75 per cent of the total volume of the mixture, the water vapor will bear 15 pounds of the total pressure and the gasoline vapor 5 pounds, according to the law of part al pressures. Hence, with increase in the proportion of water vapor in the mixture, the ease and completeness of gasoline separation will increase. But the cost in heat will also increase enormously, for all the water vapor carries off its latent heat of evaporation to the condenser and very little of it can be recovered in vapor exchangers as the oil from the heat exchangers enters the still at nearly still temperature.

Referring to Fig. 20 suppose the still is operating at 20 pounds gage pressure. The temperature of steam of that pressure is about 258° F. Hence if the temperature of the still is held at 258° F., all the water in the still will be evaporated, at great expense of heat; but if the temperature is lowered to 250° F. then, as shown by the curve, for every pound of gasoline distilled from the oil, only 2 pounds of water will be carried through the still as steam.

In operating a steam still in a gasoline plant it is important to experiment and reduce the temperature below the evaporating temperature of water at the still pressure, and to continue to reduce it until sample distillations of the oil leaving the still show that gasoline in appreciable quantity is being left in the oil. These studies should be made on stills operating at either high or low pressures. In a plant equipped with low-pressure still and steam pumps passing exhaust steam into the still, if the temperature has to be raised in order to increase the rate of distillation, care should be taken to raise the

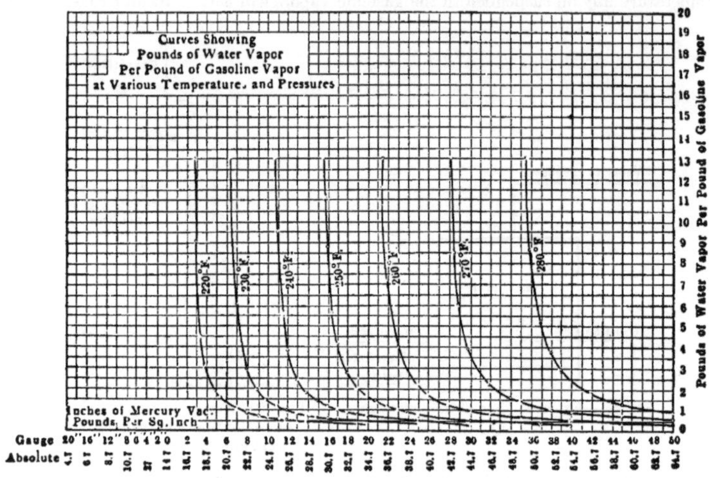

FIG. 20.—Curves showing pounds of water vapor per pound of gasoline vapor at various temperatures and pressures.

pressure above the evaporating point of the water. If this is not done an additional amount of hot live steam will be required to supply heat to evaporate the water. This operation overtaxes the tower and condenser and puts extra demands on the boilers. As low-pressure stills are usually not strong enough for more than 5 pounds gage pressure, the temperature should not be above 218° F. and the pressure should be about 3 pounds, as shown by the curves, Fig. 20.

Condensers and coolers: Absorption plants.—Gasoline plants are usually located where the supply of cooling water is limited in summer, and frequently the water is of a quality which produces a coating of slime and scale on hot pipes.

For this reason condensers and coolers of the submerged type are not being built, but are giving way to single exposed pipe coils so arranged that they can be readily kept clean. Fig. 12 shows a convenient arrangement for an absorption plant. A large cooling basin is built and all of the cooling pipes are placed over this basin. A wooden shutter work surrounds the basin and the cooling water is atomized in this enclosure over the coils.

Fig. 21 illustrates a typical vapor condenser. The size of vapor line from still to condenser coils should be about as shown in the following Table 28.

TABLE 28

Size of Vapor Line from Still to Condenser

Gallons of gasoline produced per 24 hours	Size of vapor line, inches
10,000 to 12,000	10
5,000 to 7,000	8
4,000 to 5,000	6
1,000 to 2,000	4
250 to 500	2

The number of pipes in one stand in the vapor condenser is varied, but good practice is to make it ten pipes high. Use very wide sweep bends so pipes will be kept wide apart for air circulation and to permit cleaning. Vapor enters the tops of these stands. All the stands are grouped together in parallel and should cover a combined sectional area equal to that of the vapor line or slightly larger. The total surface of the combined stands of pipe should be sufficient to give 0.2 to 0.3 square foot area for each gallon of gasoline to be condensed per twenty-four hours.

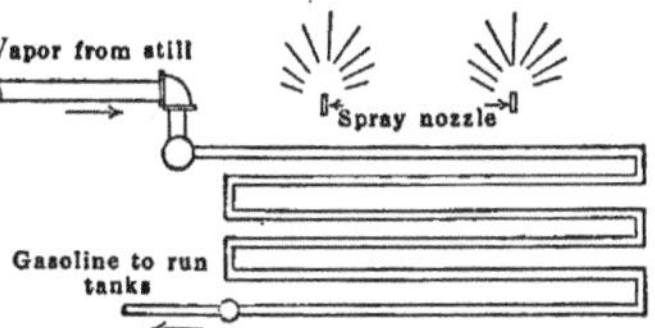

FIG. 21.—Typical vapor condenser.

The length of the condenser is varied to suit conditions. Thirty to 40 feet is a convenient length for the coils. The cooling basin should be built to accommodate this length. When the main vapor pipe is 6 inches in diameter or larger, use 3-inch pipe in the upper passes of the condenser, decreasing to 2-inch at the bottom. When vapor pipes are less than 6 inches in diameter, use 2-inch pipe throughout.

EXAMPLE NO. 1

Still 8 by 30 feet.

Maximum yield 12,000 gallons daily in winter when cooling water is 50° to 60° F.

Total condenser surface, square feet	1986
Square foot per gallon	0.16
Number of stands	9
Pipes in each stand	8
Length, feet	40

Size of pipe, 3-inch for the three top rows and 2-inch for the bottom rows.

This condenser gives satisfactory results, when the pressure on the still and condenser about is 20 to 40 pounds and water is limited in summer. This condenser is cooled by sprays of water over a cooling basin.

EXAMPLE NO. 2

This is a submerged coil of the refinery type. The still and condenser operate at atmospheric pressure. The cooling water in summer varies from 75° to 85° F., entering the condenser box. In winter this temperature averages about 55° F. The still is worked slowly in summer but the condensation at atmospheric pressure is very poor, so that the escaping vapors have to be compressed and cooled as in a compression plant. Fully half the total yield is obtained in this way in summer. In winter this condenser produces about 6000 to 7000 gallons of liquid and the vapors which escape are compressed and cooled, yielding about 3000 gallons of high-gravity product. This type of condenser is not so efficient as the exposed-coil type, for with the same water supply a condenser cannot be kept as cool in a body of water as by spraying the water in air and then letting it evaporate from the surface of the condenser coils.

Size of still, 10×30 feet.

Total production in winter, 10,000 gallons maximum, twenty-four hours.

Area of condenser, square feet	1802
Square foot per gallon	0.18

Oil coolers.—The following matters covering design and operation of the oil coolers are of importance. See Fig. 22.

(*a*) Coolers should be of the exposed type, as they are more effective in this arrangement than when submerged in a pit or tank.

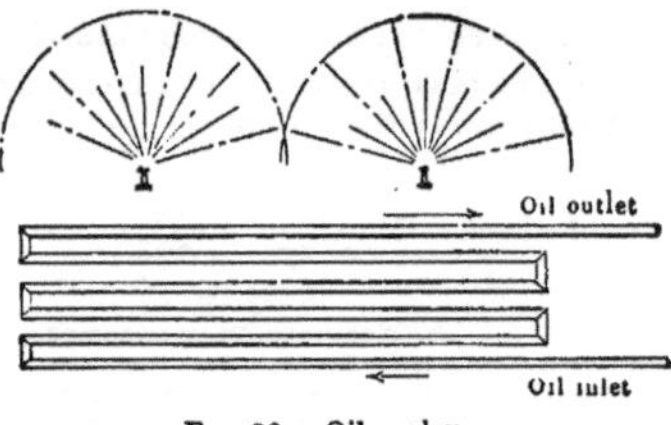

Fig. 22.—Oil coolers.

(*b*) The surface exposed in the cooler should be 15 to 20 square feet per 1000 gallons of oil circulated per twenty-four hours.

(*c*) The oil should circulate from bottom to top through the coils.

(*d*) The number of stands of coils should be such as to give the oil a maximum velocity of about 120 feet per minute through the cooler in order to keep down excessive friction loss.

(*e*) In operation, 2-inch pipe is the most satisfactory size. Care must be used to arrange the headers, to which the stands of pipe are connected top and bottom, in such a manner that the oil will divide and be distributed through all the stands. The headers should be very large.

(*f*) The number of passes in each stand is determined by the designer. Ten passes work well.

(*g*) Wide-sweep bends should be used for the coils, and the stands should be placed at least 10 inches apart; a greater distance is better. Air must have free circulation through the coils and water must touch every coil. Every coil must be constantly cleaned as all inland waters coat the surface of the coolers.

(*h*) Length of cooler is determined by dimension of cooling basin. Forty feet is a satisfactory length.

Oil circulation should be taken as maximum rate. In absorption plants on dry gas of natural-gas companies, this will occur in winter. A cooler based on winter rate for oil will generally be large enough for summer cooling in temperate climates as the oil rate is usually less in that season.

For refrigeration applied to oil in this process, each case becomes a separate study. The builders of refrigerating machines can supply data for power required and cost of equipment. Oil should not be cooled below ordinary water temperatures unless:

(*a*) The increased product, which will be volatile, can be marketed profitably.

(*b*) The season of the year during which a gain can be made over ordinary water cooling will be sufficiently long to make the increased yield pay the investment.

Gas coolers for compression gasoline plants.—(*a*) Experience indicates that for cooling the compressed gas to water temperatures, the most efficient coolers are those of similar general construction to the oil coolers described for absorption plants.

(*b*) The pipes are usually 2 inches in diameter, connected by return bends, thus forming coils. Several coils placed vertically form a stand. Several stands placed side by side, connected top and bottom by headers, complete the cooler.

(*c*) The gas should enter the header at the top of the coils and pass downward, carrying with it the fluid as it condenses. Particular attention should be given to drainage of the coils. The bottom pass of each stand should drain by gravity into the accumulator tank above the fluid level in the tank.

(*d*) The headers at top and bottom of the coils should be large in diameter, thus forming a reservoir of gas to help in distributing the flow evenly through each coil. If a large number of stands of coils are connected together, then the header should be fed with gas at several points. From three to twelve stands are frequently grouped together. Length of coils is 20 to 40 feet. The number of pipe passes in the stands varies from four up to sixteen.

(*e*) In cooling gas compressed by a two-stage compressor, coolers of equal size should be provided for both the low-stage and high-stage discharge.

(*f*) Experience indicates that about 0.8 square foot of surface should be provided in both low-stage and high-stage coolers for each 1000 cubic feet of gas compressed per twenty-four hours. No exact rule can be given for this surface, but it is here assumed that the coils are placed well apart, both vertically and horizontally, to permit of good air circulation, and further, the coils must be kept clean. The water should be thoroughly atomized and thus reduced to nearly atmospheric temperature. This assumes that the water is being used over and over again, as in a cooling basin or pond. If a continuous supply of fresh water, as from a well, is being used, atomizing is not necessary. In Bulletin 151 of the Bureau of Mines, data concerning cooling areas are presented.

An accumulator tank should be set to receive the mingled gas and fluid discharged from each cooler. If several two-stage compressors are operated in a plant the low-stage coolers are sometimes grouped together, as are the high-stage coolers. This is a matter for the designer to decide, but the main point to keep in mind is gas distribution through the coolers, and the provision of ample space in the accumulators for the gas to separate from the fluids. A capacity of 40 to 45 cubic feet for each 300,000 cubic feet of gas passing through the accumulators per twenty-four hours at 250 pounds pressure has been found to give satisfactory effect. Accumulators for high- and low-stage coolers are usually the same size.

Accumulator tanks are usually set vertically and vary from 3 to 4 feet in diameter and from 6 to 10 feet in length. They should be set low enough to permit about 2 feet of fluid to accumulate in the bottom, and the discharge of mingled fluid and gas from the cooler should be above this level. The gas should leave at the top of the accumulator.

The pressure in the low-stage cooler and accumulator is usually from 30 to 40 pounds gage and that in the high-stage cooler and accumulator 250 to 275 pounds gage. The temperatures of discharge from both stages of compression will be 200° to 250° F. at these pressures.

Accumulators are sometimes placed horizontally. The level of liquid is held near the center line of the tank. The cooler discharge enters at one end at the top, and a dry pipe, perforated and placed close to the top, collects the gas for egress from the tank. The same cubical capacity can be used for horizontal or vertical accumulators.

Cooling by expansion of gas.—For cooling the gas below temperatures which can be produced by the water-cooled coils described in the foregoing, the gas leaving the high stage accumulator, may be subjected to refrigeration as follows:

A heat exchanger of double-pipe coils is constructed. The gas for treatment is passed through the annular space between the pipes into an accumulator tank. Thence the gas passes into an expander engine (see description of expander engine) and the cold exhaust from the engine passes counter current through the inner pipe of the coil. By this means condensates which were not precipitated in the water cooler may be obtained from the gas. These fluids are then trapped to the " make tank." If the water coolers are ample in size, and the accumulator tank large, and if the water has been cooled to atmospheric temperature, it is likely that any further fluids obtained through refrigeration will be highly volatile.

Careful study should be given this subject by the designer, to determine the commercial value of such products. The experience of other plants operating in the same field is the best guide. The gas supply for the plant, if at all uncertain in volume, should be taken into account, as refrigerating machinery adds greatly to the cost of the plant.

In Bulletin 151 of the Bureau of Mines, page 43, a description is given of heat exchangers of various sizes of pipes. The general average gives a cooling surface of 0.563 square foot per 1000 cubic feet of gas treated per day. On page 46 of the same bulletin, reference is made to two plants in Oklahoma, in which the area of surface is 0.22 square foot per 1000 cubic feet of gas treated. Refrigeration is more frequently practiced in the Mid-West and California fields than in the Eastern fields.

The quantity of production made in the accumulators of the refrigerating coils, stated in per cent of the total production of the plants, varies widely, ranging from 10 to 50 per cent. It appears likely that with efficient water-cooling the percentage of merchantable product gained would be near the low figure.

General piping-compression plants.—A reservoir of pipe (or separator) should be set in each discharge pipe before reaching the cooler, to collect the oil and prevent it entering the cooler. The slight cooling of the gas by the air is generally sufficient to allow the oil to precipitate before any condensation of gasoline takes place. Ample relief valves should be placed in the discharge from each compressor cylinder attached between the discharge valve and the compressor, so that it cannot be cut off from the cylinder.

The gasoline which collects in each accumulator tank should be continuously withdrawn by an automatic tap, arranged to maintain a fixed level of fluid in the accumulator tank. This arrangement prevents the escape of gas and also releases the pressure on the gasoline as it flows to the storage tank. If it is arranged that the discharge from the high-pressure accumulator is delivered into the low-pressure accumulator, any gas liberated by this reduction of pressure is carried back through the high-stage compressor, the trap on the low-pressure accumulator thus delivering all of the gasoline to the storage tank, or make tank.

In Bulletin 151, previously referred to, attention is called (page 31) to the advantages of cooling the gas entering the plant for low-stage suction. This is important in warm climates if the suction lines are so exposed that the gas is found to be very warm. The plant capacity for gas is thus cut down, as the gas expands by heat. The suction lines should be buried and scrubber tanks should be sheltered from the sun. The bulletin above referred to contains typical plans of compression plants, showing the arrangements of coolers and accumulators.

Distillation of product.—The gasoline produced from casing-head gas by the compression process at high pressures—250 pounds—and water cooling, if too high in Baumé gravity for marketing in its raw state, may be reduced by natural weathering, or by steam distillation. The latter method should be carried only in a continuous process, and not in a batch process. In the continuous process the percentage of gas leaving the still with the gasoline vapors and entering the condenser, is so small that it does not carry away as large a proportion of true gasoline as when distillation is conducted in batches. In the latter case the rush of gas at the beginning of distillation carries away much gasoline, and thus the losses are necessarily large.

Scrubbing tanks should be placed on the incoming gas to collect oil and water, preventing them from entering the compressors. Many types of these tanks have been designed with different plans for baffles. The principal point is to have the tank large enough to bring the gas to rest. Let the gas enter in a pipe pointing downward near the middle of the tank. Set the tank vertically, letting the gas issue from the top. Scrubbers are usually about 4 feet in diameter and 10 to 12 feet high. These tanks must be heavy and tight so that vacuum may be produced in them.

Fuel consumption in compression and absorption plants.—Fuel consumption of gas engines used in gasoline plants varies with the type and condition of the engine. For two-cycle engines of about 80 B.H.P. the consumption of natural gas of 1000 B.t.u. can be fairly taken at 15 cubic feet per B.H.P. per hour. When larger engines are used they are generally of the four-cycle type and the fuel consumption per B.H.P. per hour may be taken at 10 cubic feet. It is to be remembered that the casing head gas in compression plants is high in heat value, so that the figure given above is not exact. The consumption of fuel for auxiliary machinery, pumps, electricity, etc., must be added to the consumption of the compressing engines or vacuum pumps.

Boiler power.—The boiler capacity required for a gasoline absorption plant includes the following:

Distillation of oil.
Heating of buildings.
Operation of pumps, if that type of equipment is used.
Operation of any other kinds of apparatus.

The design of the plant affects all of these items.

In the item of distillation, the theoretical consumption of heat by the oil will be only that given by the formula.

$$\text{Boiler horse-power} = \frac{Oh \times 7.1 \times 0.48 \times T}{33479},$$

where Oh = amount of oil passing through still in gallons per hour;
7.1 = weight of 1 gallon of oil in pounds;
0.48 = specific heat of oil;
T = number of degrees F. difference between still temperature and temperature at which oil enters still;
33479 = B.t.u. in one boiler horse-power.

But additional steam must be supplied for distillation purposes and passes to the condenser (see stills). Also the gasoline contained in the oil must be evaporated. All of the three distillation items are further affected by the rate at which the distillation process is carried out, and by radiation losses. If, therefore, an absorption plant is designed so that only the distilling operation requires steam, then about 50 per cent more boiler horse-power than theoretically necessary for distillation purposes alone should be provided. This arrangement will take care of all the losses and afford a stand-by unit.

The boiler plant should be subdivided so that boilers may be shut down when the plant operations permit. Boilers should always be worked at or somewhat below rating. It is uneconomical of fuel to force boilers when gas-fired. Information on boilers, steam pumps, and steam engines should be taken from Kent's "Mechanical Engineers' Handbook."

CLASSIFICATION OF NATURAL-GAS GASOLINE BY PRINCIPAL METHODS OF MANUFACTURE[1]

Gasoline Produced by Compression and by Vacuum in 1920

State	Number of plants.	Gasoline Produced			Gas Used	
		Quantity, Gallons.	Value.	Average price, Cents.	Estimated volume, M. cu. ft.	Average yield, Gallons
Oklahoma	266	162,761,829	$28,233,143	17.34	48,363,205	3.37
California (a)	44	35,347,691	6,619,893	18.72	27,856,279	1.27
Texas	35	30,144,880	5,272,276	17.48	10,098,420	2.99
West Virginia (b)	163	15,972,833	3,169,859	19.84	11,605,174	1.38
Pennsylvania	279	10,981,461	2,128,774	19.38	5,391,467	2.04
Wyoming	4	8,175,825	1,609,762	19.68	2,345,048	3.49
Louisiana	18	6,077,093	831,086	13.67	1,917,159	3.17
Illinois	92	6,054,916	1,307,980	21.6	2,889,334	2.10
Ohio	47	2,294,996	466,747	20.3	916,075	2.51
Kansas	7	1,574,482	315,906	20.1	780,820	2.02
New York	4	411,078	75,576	18.38	162,463	2.53
Kentucky	6	182,927	41,997	22.95	254,091	0.72
Total, 1920	965	279,980,011	$50,072,999	17.88	112,579,535	2.486
Total, 1919	1025	261,157,587	45,563,458	17.4	117,669,332	2.22

(a) Includes four combination compression and absorption plants.
(b) Includes seven combination compression and absorption plants.

Gasoline Produced by Absorption in 1920 (a)

State.	Number of plants.	Gasoline Produced			Gas Used	
		Quantity, Gallons.	Value.	Average price, Cents.	Estimated volume, M. cu. ft.	Average yield, Gallons
West Virginia	48	42,968,655	$9,879,692	22.99	162,714,884	0.26
Oklahoma	46	14,662,995	2,847,347	19.41	36,257,061	.40
California	26	12,860,285	1,703,926	13.24	15,916,116	.81
Pennsylvania (b)	27	(a)10,169,674	2,253,606	22.16	55,560,230	.18
Ohio	12	7,720,642	1,727,811	22.37	39,299,254	.20
Louisiana	13	4,532,536	881,527	19.4	35,836,884	.13
Kentucky	3	4,314,393	1,029,631	23.86	18,685,194	.23
Texas	7	2,811,148	498,533	17.73	5,753,793	.49
Kansas	3	} 3,291,478	639,047	18.6	13,280,749	.25
Wyoming	1			23.55		.22
Total, 1920	186	103,331,806	$21,461,120	20.76	383,304,165	.269
Total, 1919	166	90,377,439	18,633,305	20.6	374,928,966	.24

(a) Includes drip gasoline.
(b) Includes 1650 gallons of drip gasoline, valued at $240.

[1] U. S. Geological Survey.

ASPHALT

BY

R. G. SMITH

ASPHALT, or asphaltum, is a black or brownish-black product having cementitious, waterproof, and usually resilient properties. Its technical definition is: [1] Solid or semi-solid native bitumens, solid or semi-solid bitumens obtained by refining petroleum, or solid or semi-solid bitumens which are combinations of the bitumens mentioned with petroleums or derivatives thereof, which melt upon the application of heat, and which consist of a mixture of hydrocarbons and their derivatives of a complex structure, largely cyclic and bridge compounds.

The definition of bitumen is: A mixture of native or pyrogenous hydrocarbons and their non-metallic derivatives which are soluble in carbon disulphide.

CLASSIFICATION

Asphalts may be roughly classified as petroleum asphalts and native asphalts.

Petroleum asphalt.—Quantities of asphalt are manufactured at the present time from crude petroleum oils of a so-called asphalt base. These crude petroleums are now obtained from California, Texas, and the Mid-continent fields of the United States, and from Mexico.

THE REFINING OF ASPHALT

Asphalts are refined from crude petroleum oils by three principal processes which result in different characteristics. The asphalts refined by these processes are known as: (1) Steam-refined, (2) Air-refined, (3) Combination steam- and air-refined.

Steam-refined process.—The largest amount of asphalt is made by this process. The lighter portions of the crude oil are distilled by heat until the residue reaches the required consistency, being agitated all the time by steam through the agitator pipes at the bottom of the still.

The action of the steam is based upon the physical law that the total vapor pressure is equal to the sum of the vapor pressures of the components of two immiscible liquids, and that when distillation takes place the vapors will pass over in the ratio of their respective vapor pressures, the boiling points of the mixtures being the temperature at which the sum of the vapor pressures of the two liquids is equal to the pressure of the atmosphere. Owing to this fact, the distillation takes place at a materially lower boiling point than if steam were not used, the asphalt is freed from undesirable compounds and the formation of free carbon or coke is prevented.

This process gives a ductile asphalt with highly cementitious qualities.

Air-refined process.—After the crude oil is heated to a certain temperature air is introduced through pipes at the bottom of the still, the fire is drawn, and heat is supplied

[1] A. S. T. M. Standard, Serial Designation D8–18.

through the exothermic reaction of the air and crude petroleum oil. Care must be taken that the reaction takes place at a proper rate. The chemical reactions of air on crude petroleum oil in this process are not exactly known but it is believed that they lead to the formation of condensation products having an exceedingly complex nucleus. The asphalt obtained from this process has a higher melting point and less susceptibility to temperature changes than the steam-refined asphalt but has less ductility.

Combination steam- and air-refined process.—The crude oil is agitated with steam for a period and its refining is then finished with air agitation. Asphalt obtained by this process shows characteristics varying between those of steam-refined and air-refined asphalt, depending on the length of treatment with steam before the air is applied.

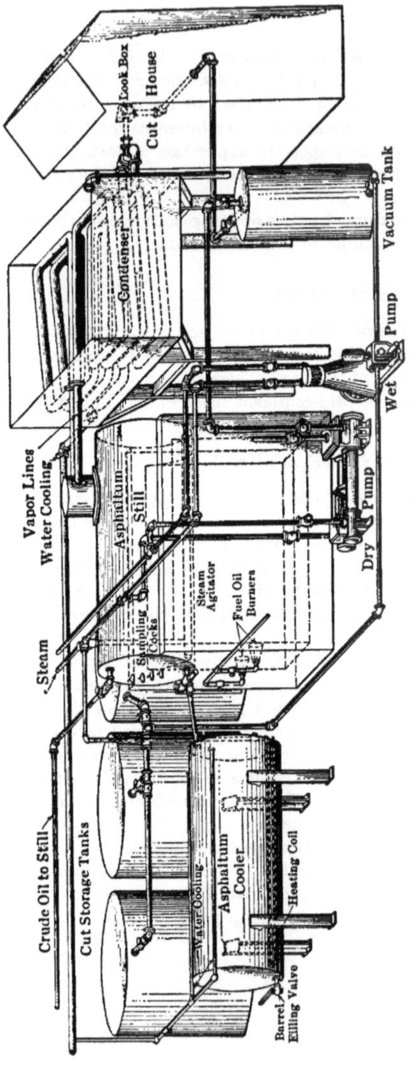

Fig. 1.—Batch method of refining asphalt.

Methods of Refining

There are two principal methods of refining crude petroleum oils into asphalt, known as:

(1) The batch method,
(2) The continuous method.

For Steam-refined Asphalt

The Batch Method.—Fig. 1 shows a diagrammatic sketch of the operation. The crude oil is placed in the asphalt still. The steam agitator pipes, shown in the bottom of the still, are placed so that the agitating vapor is directed against the shell of the still. Heat is supplied to the still by means of fuel oil or other available fuel. In modern practice, the oil is distilled under a vacuum so that the vapors will distill at a lower temperature. The vapor lines to a condenser are shown. The distillates are received at the cut house from which the different cuts are sent to cut storage tanks. After an asphalt has attained the desired consistency, it is withdrawn from the still and placed in an asphalt cooler where it is cooled to a suitable temperature by applying water on the outside of the metal cooler. It is then filled into tank cars or barrels.

In order to test for the consistency required a small sample of asphalt is taken from the sampling cocks, shown on the still, usually placed on a small piece of wood,

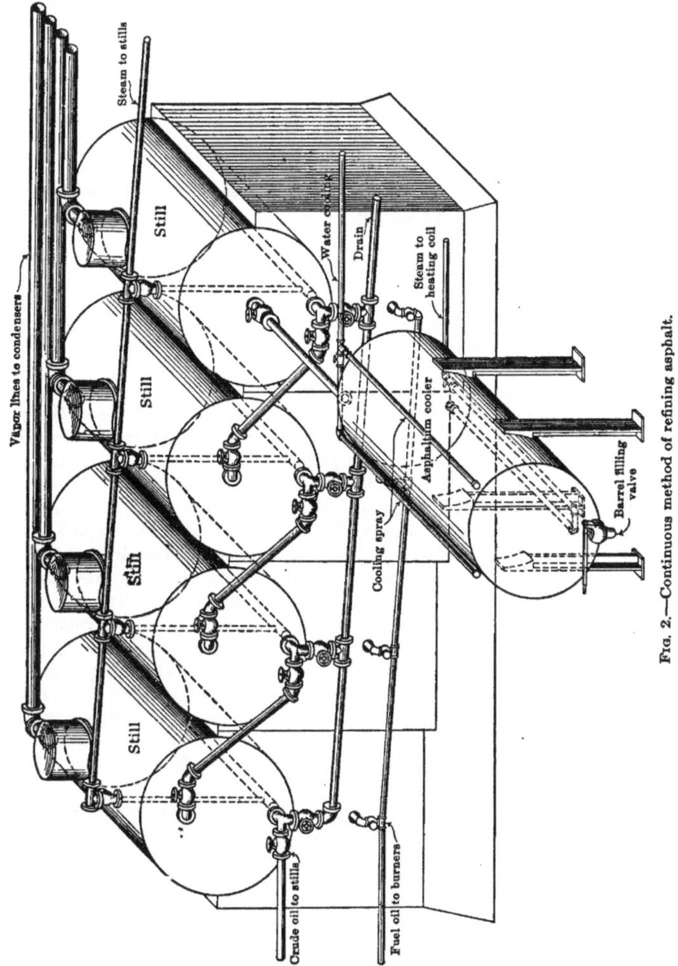

FIG. 2.—Continuous method of refining asphalt.

cooled rapidly in the air, and chewed by the "still" man to determine its consistency. When the proper consistency is attained, the fires are drawn and the asphalt allowed to flow to the cooler. There is no time to take the sample from the still to a laboratory and determine its penetration, as by the time such a determination could be made the asphalt in the still would have reached a harder consistency.

For Air-refined Asphalt

The Batch Method.—After the crude oil is heated to a desired temperature, air is applied through the agitator pipes. The fires are drawn and heat is supplied by the reaction until the desired consistency of the product is obtained. It is then drawn to the cooler.

For Steam-refined Asphalt

The continuous method.—In this method the crude oil is pumped through a series of stills, instead of remaining in one still until asphalt of the desired consistency is obtained. The first still takes off the lightest portions, the second still, those which are somewhat heavier, and so on until at the last still asphalt of the desired consistency is obtained. A diagrammatic sketch of this system is shown in Fig. 2. This system is applicable only to steam-refined asphalt.

NATIVE ASPHALTS

Native asphalts are found in varying quantities in all parts of the known world. Table A shows the native bitumens found in the United States, and Table B the native bitumens found outside of the United States.

The character of native asphalts is as follows:

Gilsonite.—This is an asphaltite derived from petroleum and one of the purest known hydrocarbons, is a hard, black substance of homogeneous texture which softens and flows at high temperature. Exceedingly brittle, it breaks with a lustrous conchoidal fracture which soon becomes dead black on exposure to the weather. Gilsonite is sold as " firsts " and " seconds," the higher grade being the larger lumps obtained from the center of the vein, and the second grade that obtained near the vein walls. The latter is generally accompanied by considerable powdered material of the same nature. This asphaltite has a wide variety of uses in the industries. Properly fluxed with a heavy asphaltic oil, it forms an asphaltic cement of peculiar rubbery texture little susceptible to temperature changes. It is also valuable as a base for marine paints and protective coatings for steel work, for use in the manufacture of automobile tires and other rubber products, for insulation, and for roofing materials.

Grahamite.—This is a comparatively pure, brittle, black bitumen which does not melt at a high temperature but merely intumesces. It is rarely of compact structure, often containing a large proportion of inorganic impurities, and when fractured the break is irregular or hackly and the broken surface dull. It is slightly heavier than gilsonite, but, like gilsonite, is an asphaltite.

Commercially, grahamite has been used successfully in the roofing industry, for when mixed with heavy asphaltic fluxes it is very rubbery and elastic and non-susceptible to heat changes. It is used as a substitute for rubber, as a filler between stone and brick blocks, and in the manufacture of flooring, varnishes and paints.

Trinidad asphalt.—Trinidad asphalt, the heaviest of the native asphalts, is a dull black material which melts at a relatively low temperature. It breaks with a conchoidal fracture.

This asphalt is known commercially as lake and land asphalt, depending on whether it is obtained from the large lake of the material located on the Island of Trinidad, or from the overflow covering the surrounding shores.

The native asphalt, as found, is thoroughly mixed with organic salts and mineral clay and contains variable quantities of water. When refined it is used as a paving material.

TABLE A

Native Bitumens Found in the United States

Region and variety	Description	State	County
EASTERN Bituminous sands and limestones	Native asphalt associated with mineral matter	Kentucky	Edmondson, Carter, Boyd, Breckinridge, Grayson, Warren, Logan
Native asphalt	Occurring free	Kentucky	Breckinridge, Grayson
Grahamite	Asphaltite	West Virginia	Ritchie
MID-CONTINENT Bituminous limestone and sands	Native asphalt associated with mineral matter	Oklahoma Texas Arkansas Louisiana Missouri	Murray, Jackson, Pontotoc, Comanche, Jefferson, Stephens, Garvin, Carter, Love, Marshall, La Flore, Atoka, McCurtin, Johnson Uvalde, Montague, Burnet, Anderson, Jasper, Cook Pike, Sevier Lafayette Lafayette
Grahamite	Asphaltite	Oklahoma Texas	Pushmataha, Atoka, Stephens Fayette, Webb
Impsonite	Asphaltic pyrobitumen	Arkansas Oklahoma	Scott Dougherty, Murray, La Flore
Native asphalt	Occurring free	Arkansas Oklahoma	Pike Carter, Murray
ROCKY MOUNTAIN Bituminous sands and limestone	Native asphalt associated with mineral matter	Utah	Grand, Carbon, Uinta
Native asphalt	Occurring free	Utah	Grand
Gilsonite	Asphaltite	Utah Colorado	Uinta Rio Blanco
Grahamite	Asphaltite	Colorado	Grand
Wurtzilite	Asphaltic pyrobitumen	Utah	Uinta
Ozokerite	Native mineral wax	Utah	Emery, Uinta, Wasatch
PACIFIC COAST Bituminous sands and limestones	Native asphalt associated with mineral matter	California	Santa Cruz, Santa Barbara, Orange, San Luis Obispo, Mendocino, Monterey
Native asphalt	Resembling gilsonite or an asphaltite	California	Kern
Glance pitch	Asphaltite	Oregon	Coos
Impsonite	Asphaltic pyrobitumen	Nevada	Eureka

Bermudez asphalt.—Bermudez asphalt is a relatively pure bitumen, black in color, uniform in structure, and possessing a bright luster. It is soft, breaks with a conchoidal fracture, and melts at a low temperature. It takes its name from the state of Bermudez in Venezuela where two large deposits have been found.

The chief use of the Bermudez asphalt is in the paving industry.

Cuban asphalt.—Cuban asphalt, as it is commercially known, is a native bitumen found near the village of Bejucal on the island of Cuba. There are many forms of native bituminous material throughout the island but only the Bejucal asphalt has been found in sufficient quantities and with the proper characteristics to warrant commercial development. It is dull black in color, melts at a high temperature, and breaks with a conchoidal fracture, in these respects being similar to the refined asphalt of Trinidad.

Table B

Native Bitumens Found Outside of the United States

Region and variety	Description	Country	State or Province
North America			
Bituminous sands and limestones	Native asphalts associated with mineral matter	Canada	Alberta, Mackenzie
		Mexico	Tamaulipas, Vera Cruz
		Cuba	Matanzas, Pinar del Rio, Havana, Camaguey
Glance pitch or manjak	Asphaltite	Mexico	
		West Indies	Barbados, Santo Domingo,
		Trinidad	St. Patrick
Grahamite	Asphaltite	Mexico	Tamaulipas, Vera Cruz
		Cuba	Pinar del Rio, Havana, Santa Clara
		Trinidad	San Fernando
Albertite	Asphaltic pyrobitumen	Canada	New Brunswick, Nova Scotia
		Cuba	
		Mexico	
South America			
Bituminous sands and limestones	Native asphalts associated with mineral matter		St. Patrick
		Argentina	Jujuy, Chubut
		Venezuela	Bermudez
		Peru	Luya
Glance pitch or manjak	Asphaltite	Colombia	Tolima
Europe			
Bituminous sands and limestones	Native asphalt associated with mineral matter	France	Landes, Gard, Haute-Savoie, Ain, Alsace-Lorraine, Basses-Alpes
		Switzerland	Val de Travers, Vaud
		Germany	Hanover, Westphalia, Hesse
		Italy	Marches, Abruzzi e Molise, Calabria, Campania, Sicily, Tyrol, Istria
		Hungary	Hungary
		Jugo-Slavia	Dalmatia, Herzegovina, Bosnia
		Czecho-Slovakia	Silesia and Moravia, Bihar Transylvania, Walachia
		Greece	Triphily, Argolis, Salonika
		Portugal	Estremadura
		Servia	
		Spain	Santander, Alava, Navarra, Tarragona, Soria
		Russia	Terek, Simbirsk
		Albania	Albania
Native asphalt	Occurring free	Switzerland	Neuchâtel
		France	Landes
		Greece	Zante
		Austria	Styria
		Jugo-Slavia	Croatia. Slavonia
		Montenegro	
		Russia	Vitebsk
		Georgia	Kutais
Elaterite	Asphaltic pyrobitumen	England	Derbyshire
		Spain	Albacete
Albertite	Asphaltic pyrobitumen	Scotland	Hoy

TABLE B—*Continued*

Region and variety	Description	Country	State or Province
ASIA Bituminous sands and limestones	Native asphalts associated with mineral matter	Russia Japan Arabia Mesopotamia Palestine Armenia India Persia Syria	Uralsk Ugo Wadi-Gharandel Dead Sea, Beirut
Native asphalt	Occurring free	Russia	Western Siberia
Glance pitch or manjak	Asphaltite	Syria	Hasbaya, Dead Sea
Elaterite	Asphaltic pyrobitumen	Russia	Semiryechensk
AFRICA Bituminous sands and limestones	Native asphalt associated with mineral matter	Algeria Nigeria Rhodesia Egypt	Oran Arabian Desert
Glance pitch or manjak	Asphaltite	Egypt	Arabian Desert
OCEANIA Native asphalts Rock asphalt	Occurring free	Timor	Portuguese
Elaterite	Asphaltic pyrobitumen	Australia	South Australia, New South Wales
Albertite	Asphaltic pyrobitumen	Australia	Tasmania, Victoria

NOTE.—From the foregoing it will be noted that native bituminous materials are well distributed in some form or other throughout the entire world.

Cuban asphalt is relatively much harder than Trinidad asphalt and contains a high percentage of fixed carbon. It very closely resembles grahamite and may even be classed as an asphaltite rather than an asphalt.

The hard Cuban asphalt is shipped to the United States where it must be mixed with a proper asphaltic flux before use in the paving industry.

Elaterite, wurtzilite, albertite and impsonite.—Elaterite, wurtzilite (also commercially known as elaterite), albertite, and impsonite are natural asphaltic substances called asphaltic pyrobitumens. They generally contain less than 10 per cent of mineral matter, and are characterized by their hardness and infusibility and by the bitumen yielded on destructive distillation. Black in color, these four classes of substances merge or grade into one another, so that they may be roughly distinguished from each other by their bright to dull surfaces, and their conchoidal to hackly fractures.

The principal uses of refined wurtzilite are in the making of marine and iron paints and other protective coatings, and in the manufacture of insulating materials.

Asphaltic sands.—Asphaltic sands are composed of asphalts found mixed in nature with varying proportions of loose sand grains. The quantity of bituminous cementing material extracted from the sand may run as high as 12 per cent and this bitumen is composed of a soft asphalt which rarely has a penetration as low as 60°.

Asphaltic sands with the addition of hard asphalt or a certain proportion of filler, and sometimes stone screenings, have frequently been used as a paving material.

Asphaltic limestones.—Limestones which have become impregnated with asphalt in varying proportions are termed asphaltic limestones. The percentage of asphalt is in some deposits as high as 12 to 14 per cent.

Asphaltic limestones, properly pulverized and mixed with sand, have been used in limited quantity for paving purposes.

TABLE C

Characteristics or Typical Tests

Characteristics	Asphalts				Asphaltites		Asphaltic Pyrobitumens			
	Trinidad Refined		Bermudez refined	Cuban (Bejucal)	Gilsonite or Uintaite	Grahamite	Elaterite	Wurtzilite	Albertite	Impsonite
	Lake	Land								
Penetration at 77° F. in degrees	6–10	0	20–30	0	0	0	...	0	0	...
Specific gravity at 77° F.....	1.40–1.42	1.40–1.45	1.05–1.08	1.30–1.35	1.05–1.10	1.15–1.20	0.90–1.05	1.05–1.07	1.07–1.10	1.10–1.25
Ductility at 77° F..........	...	...	11	...	0	...	...	...	...	...
Melting-point (R. and B.) °F.	200–210	220–230	180–200	240–260	250–350	Cokes	Inf.	Inf.	Inf.	Inf.
Soluble in CS_2 (per cent).....	56–57	54–55	93–97	70–75	98–100	45–100	10–20	5–10	2–10	1–6
Soluble in CCl_4 (per cent)....	56–57	...	92–96	...	99	20–99	...	...	...	...
Proportion soluble in 86°–88° Bé. naphtha (per cent)....	62–64	33–34	65–75	32–50	20–60	Tr.–50	5–10	Tr.–2	Tr.–2	Tr.–2
Volatile at 325° F. for 5 hrs. (per cent)..............	1.0–1.6	...	3–6	Less than 1	Less than 2	Less than 1	...	1–3	...	...
Fixed carbon (per cent).....	10–12	12–14	13–14	17–25	10–20	30–35	2–5	5–25	25–50	50–55

PRODUCTION OF ASPHALT

The production of asphalt from various sources is shown in the following tables, taken from Mineral Resources, 1918, United States Geological Survey, published March 31, 1920.

TABLE D

Asphalt Manufactured from Domestic Petroleum and Sold at the Refineries

Year	Quantity (short tons)	Value
1902	20,826	$303,249
1903	46,187	522,164
1904	44,405	459,135
1905	52,369	452,911
1906	64,997	615,406
1907	137,948	1,898,108
1908	119,817	1,540,396
1909	129,594	1,565,427
1910	161,187	2,225,833
1911	277,192	3,173,859
1912	354,344	3,755,506
1913	436,586	4,531,657
1914	360,683	3,016,969
1915	664,503	4,715,583
1916	688,334	6,178,851
1917	701,809	7,734,691
1918*	604,723	8,796,541
1919*	614,692	8,727,372
1920†	700,496	11,985,457

Asphalt Manufactured from Mexican Petroleum and Sold at the Refineries

Year	Quantity (short tons)	Value
1913	114,437	$1,743,749
1914	313,787	4,131,153
1915	388,318	3,730,436
1916	572,387	6,018,851
1917	645,613	7,441,813
1918*	597,697	9,417,818
1919*	674,876	7,711,510
1920†	1,045,779	14,272,862

* From Mineral Resources, 1919, U. S. G. S., published July 18, 1921.
† From U. S. Geo. Sur. Press. Bull. No. 476, August, 1921.

TABLE E

Native Asphalt and Bituminous Rock Produced in Principal Producing Countries

Year	United States		Trinidad *		Germany	
	Quantity (short tons)	Value	Quantity (short tons)	Value	Quantity (short tons)	Value
1906	73,062	$674,934	150,373	$832,964	129,388	$268,631
1907	85,913	928,381	171,271	832,274	139,567	264,494
1908	78,565	517,485	143,552	403,023	98,088	188,334
1909	99,061	572,846	159,416	459,446	85,446	176,897
1910	98,893	854,234	157,120	421,419	89,491	152,565
1911	87,074	817,250	†201,284	†603,800	90,256	154,938
1912	95,166	865,225	†212,236	†742,800	105,950	200,743
1913	92,604	750,713	‡253,830	‡733,187	116,294	188,654
1914	79,888	642,123	123,524	‡362,754	90,169	145,302
1915	75,751	526,490	6,172	‡26,819	35,715	62,647
1916	98,477	923,281	144,234	408,246	......	......
1917	81,604	773,424	147,261	421,867	13,382	......
1918	60,034	780,808	79,781	216,657	......	......
1919	88,281	682,989	105,063	275,858	......	......

* Includes small quantity of manjak, produced in Barbados.
† Exports. Figures for production not available.
‡ Fiscal year, April 1 to March 31.

Year	Cuba		France		Italy *	
	Quantity (short tons)	Value	Quantity (short tons)	Value	Quantity (short tons)	Value
1906	5,717	$26,605	216,405	$345,599	144,802	$349,926
1907	5,571	37,594	195,136	330,065	178,127	442,014
1908	6,875	31,574	188,616	264,188	148,433	368,306
1909	11,900	48,246	186,298	269,161	123,361	305,159
1910	2,320	13,685	187,085	277,210	179,261	452,911
1911	3,638	21,928	187,006	261,743	207,926	591,550
1912	17,260	87,500	343,656	393,994	200,560	581,383
1913	†1,749	30,935	45,714	129,809	188,602	521,398
1914	† 969	19,491	39,193	......	132,115	400,164
1915	† 486	11,247	12,905	......	52,525	184,621
1916	† 539	12,486	15,852	......	18,556	68,301
1917	521	13,191	13,303	......	9,529	38,965
1918	......	......	11,183	......	24,591	128,134
1919	......	......	19,634	225,772	85,980	587,563

* Only about 7 per cent of the quantity given represents asphalt, the remainder being bituminous sandstone and limestone.

† Figures shown indicate exports.

TABLE E—(*Continued*)

Native Asphalts and Bituminous Rock Produced in Principal Producing Countries

Year	Spain		Japan		Austria-Hungary	
	Quantity (short tons)	Value	Quantity (short tons)	Value	Quantity (short tons)	Value
1906	8,587	$17,130	43	$3,572	10,633	$778,781
1907	9,057	16,001	644	5,436	11,335	727,892
1908	13,635	24,084	2,650	25,564	12,239	768,162
1909	5,822	10,282	4,614	45,205	11,179	663,246
1910	7,072	18,308	526	29,004	9,070	702,022
1911	*4,124	8,754	1,389	13,728	†8,312	652,603
1912	5,938	13,003	3,199	32,518	†11,439	664,778
1913	*6,153	13,402	2,491	27,242	¶33,354	......
1914	6,355	13,847	2,212	25,836	¶21,289	......
1915	4,984	10,706	2,177	......	¶441	......
1916	8,064	......	2,538	......	......	......
1917	2,003	4,124	4,269	......	......	......
1918	4,070	8,586	3,304	52,210	......	......
1919	5,031	9,735	7,337	173,928	......	......

Year	Russia		Venezuela §		Mexico	
	Quantity (short tons)	Value	Quantity (short tons)	Value	Quantity (short tons)	Value
1906	‡12,517	$110,294	24,783	$98,250	1,531	$17,174
1907	‡14,116	101,705	42,153	167,938	4,945	182,265
1908	‡24,961	491,302	35,324	141,912	5,811	330,903
1909	† 2,665	4,599	41,767	180,061	6,031	106,484
1910	†27,544	176,518	35,717	‖151,000	3,140	39,681
1911	......	......	56,183	‖238,000	8,912	125,322
1912	......	......	73,780	‖312,000	33,611	462,230
1913	......	......	93,884	‖400,000		
1914	......	......	49,941	......		
1915	......	......	31,949	......	428,047	3,733,000
1916	......	......	49,176			
1917	......	......	54,410			
1918	......	......	47,314	.		
1919	......	......	41,582	......	......	......

* Figures for 1911 do not include 7165 tons of bituminous rock for which no value was reported. Figures for 1913 do not include 5112 tons of bituminous rock, valued at $5833.

† Includes ozokerite.

‡ Includes mineral pitch.

§ Exports. Presented through courtesy of the Barber Asphalt Co.

‖ Estimated.

¶ Austria. Figures for Hungary not available.

CHARACTERISTICS OF ASPHALT

The characteristics of an asphalt suitable for a particular use depend upon the nature of that use and the conditions to which it gives rise. In order to obtain an indication of the characteristics which will be suitable for certain conditions, tests on the different properties of asphalt are made, and a series of standard tests have been developed to give an indication of the qualities that are necessary for a particular use. The standard tests of asphalt may be divided into two classes:

Physical.—Consistency or penetration test,
Specific gravity,
Ductility,
Melting point.

Chemical.—Solubility in carbon disulphide (CS_2),
Solubility in carbon tetrachloride (CCl_4),
Solubility in 86° Baumé naphtha,
Volatilization test,
Flash point,
Fixed carbon or residual coke.

Details of these tests are as follows:

Consistency or Penetration Test

The consistency of asphalt is determined by means of a penetrometer, which measures the distance a standard needle will vertically penetrate a sample of asphalt, when weighted with 100 grams (3.53 ounces) for five seconds at 25° C. (77° F.). The penetration test in its present form was adopted as standard by the American Society for Testing Materials in 1916 and given the serial designation D 5–16.

A penetrometer is shown in Fig. 3. The needle for the test is a cylindrical steel rod 50.8 mm. (2 inches) long, has a diameter of 1.016 mm. (0.04 inch) and is turned on one end to a sharp point having a taper of 6.35 mm. ($\frac{1}{4}$ inch).

A sample of the asphalt to be tested is completely melted at the lowest possible temperature and stirred thoroughly until it is homogeneous and free from air bubbles. It is then poured into a flat-bottom cylindrical container—55 mm. ($2\frac{3}{16}$ inches) in diameter and 35 mm. ($1\frac{3}{8}$ inches) deep—to a depth of not less than 15 mm. ($\frac{5}{8}$ inch). The sample should be protected from dust and allowed to cool in an atmosphere not lower than 18° C. (64.4° F.) for one hour, when it should be placed in the water-bath at 25° C. (77° F.), and allowed to remain for one hour.

When the test is being made the sample is submerged in a water-bath, the temperature of bath and sample being kept constant at 25° C. (77° F.). The needle is adjusted to a delicate contact with the surface of the asphalt and the position of the dial needle on the penetrometer noted. The needle and its superimposed weight are then released for five seconds, after which the dial needle is again read; the difference between the " before " and " after " readings is the " penetration."

The needle and weight are released and arrested by means of a spring and the distance penetrated read off on the dial. Each ten divisions or so-called " degrees " indicates a distance of 1 mm. penetrated by the needle.

A metronome should be used for timing the test. The accuracy of the determination is dependent upon the method of preparing the samples for the test and the care taken in controlling the temperature during the test.

Generally three tests are made on each sample, each test being made at points on the surface of the sample not less than 1 cm. (0.3937 inch) from the side of the container, not less than 1 cm. apart. The determined "penetration" is an average of the three tests.

Specific Gravity Test

Gravity is the weight of the substance per unit volume, and asphalt is usually measured by specific gravity. Specific gravity is the ratio of the weight of asphalt to the weight of an equal volume of water.

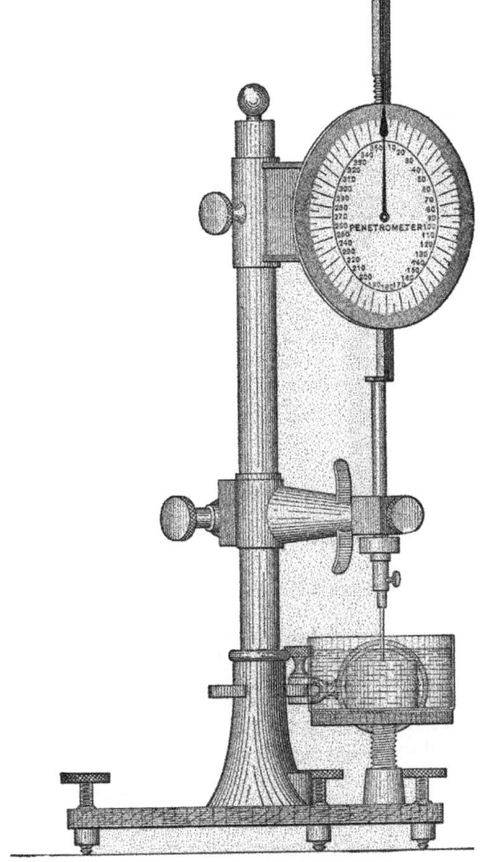

Fig. 3.—Penetrometer.

The specific gravity test is an indication of the identity of the asphalt, and is used in the calculation of weights for shipment or storage and for the determination of volume.

Methods of test.—On solid asphalts a sample of the material to be tested is suspended by means of a silk thread from the hook of one of the pan supports of a balance, and a short distance above the pan, and weighed. (See Fig. 4.) This weight is "a" The fragment is then immersed in distilled water at 25° C. (77° F.) and suspended, care being taken that the water contained does not touch the balance, and weighed again. This weight is "b." The specific gravity is then determined from the formula

$$\frac{a}{a-b}.$$

Another method is known as the "Pycnometer method." Pycnometers are various-shaped glass bottles, each of which has special advantages. They are fitted with ground-glass stoppers.

The pycnometer used for asphalt consists of a straight-walled glass tube approximately 70 mm. (2.76 inches) long and 22 mm. (.87 inch) in diameter, carefully ground to receive an accurately fitting solid glass stopper with a hole of 1.6 mm. (.063 inch)

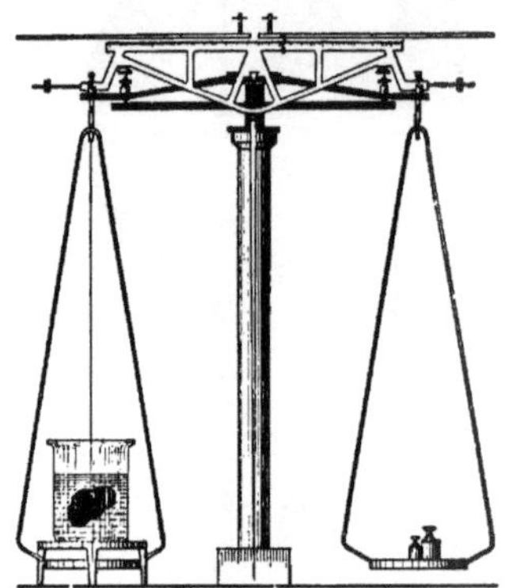

Fig. 4.—Specific gravity test.

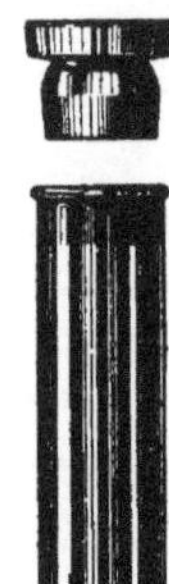

Fig. 5.—Pycnometer.

bore. The lower part of the stopper is made concave, in order to allow all air bubbles to escape through the bore.

The depth of the cup-shaped depression is 4.8 mm. (.19 inch) at the center. The stoppered tube has a capacity of about 24 c. c. (1.464 cubic inches) and when empty weighs about 28 grams (.0617 pound). (See Fig. 5.)

The procedure of obtaining the specific gravity is as follows: The pycnometer is weighed empty and this weight is called "a." It is filled with freshly distilled water at 25° C. (77° F.) and the weight is again taken and called "b." The sample of the asphalt is heated and poured into the dry pycnometer to about half the capacity of the latter. The tube and contents are allowed to cool to 25° C. (77° F.) after which the tube is carefully weighed. This weight is called "c." Distilled water 25° C. (77° F.) is then poured in until the pycnometer is full. After this, the stopper is inserted, and the whole cooled to 25° C. (77° F.) by thirty-minute immersion in a beaker of water maintained at this temperature. The pycnometer and contents are then weighed. This weight is called "d." The specific gravity of the asphalt is determined from the following formula:

$$\text{Specific gravity at 25° C. (77° F.)} = \frac{c-a}{(b-a)-(d-c)}.$$

Ductility Test

The ductility of asphalt is an indication of the value of its binding qualities and of its resistance to shock.

The test is made by pulling apart a briquet of standard size, 1 sq. cm. (0.155 square inch) in cross-section at the standard temperature of 25° C. (77° F.), and at the standard rate of 5 cm. (1.96 inches) per minute. The distance it is extended

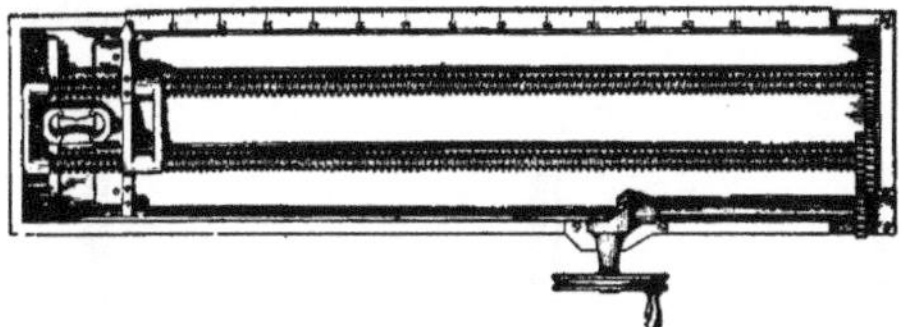

Fig. 6.—Ductility machine.

before breaking is measured in centimeters and is the measure of its ductility. A ductility machine is shown in Fig. 6.

Melting Point

There is no true melting point of asphalt, but various methods are in vogue for determining the softening point, or so-called melting point. This so-called melting point will vary with each particular method and an attempt has been made to gather the different methods that have been in use and to determine the differences in readings of each method.

The following methods are described:

Ring and ball,
Hook and cube,
Kramer and Sarnow,
General Electric,
Richardson,
Pohl.

The description of each test follows:

Test for melting point. Ring-and-ball method.—This is the standard method for the determination of the softening point of bituminous materials other than tar products as adopted by the American Society for Testing Materials in 1919, and given the serial designation D 36–19.

The ball is of steel 9.53 mm. ($\frac{3}{8}$ inch) in diameter and weighing 3.45 to 3.55 grams. The ring is of brass 15.875 mm. ($\frac{5}{8}$ inch) inside diameter, 6.35 mm. ($\frac{1}{4}$ inch) high and with walls 2.38 mm. ($\frac{3}{32}$ inch) thick supported at right angles to a thin steel rod.

A small quantity of the sample is slowly melted in a tablespoon, stirred to remove any air bubbles and then poured into a brass ring which has previously been set on a brass plate, the latter amalgamated to prevent sticking of the material thereto. When the ring cools, the excess material that should always be poured is removed with a hot knife.

The ball is then set on the asphalt and the whole, together with a thermometer, suspended in a 600-c.c. beaker filled with distilled water at 5° C. (41° F.) so that the bottom of the ring and bulb of the thermometer are exactly 2.54 cm. (1 inch) above the bottom of the beaker. The thermometer bulb must be within 0.635 cm. (¼ inch) of the ring but not touching.

After allowing to stand fifteen minutes, heat is applied in sufficient quantity to raise the temperature exactly 5° C. (9° F.) each minute.

As the bath becomes heated, the ball sinks into the asphalt forcing it out of the ring, and that temperature at which the sample touches the bottom of the beaker is taken as the melting-point temperature of the asphalt.

Test for melting point. Hook-and-cube method.—This method was first used in asphalt work by the Office of Public Roads, U. S. Department of Agriculture, but

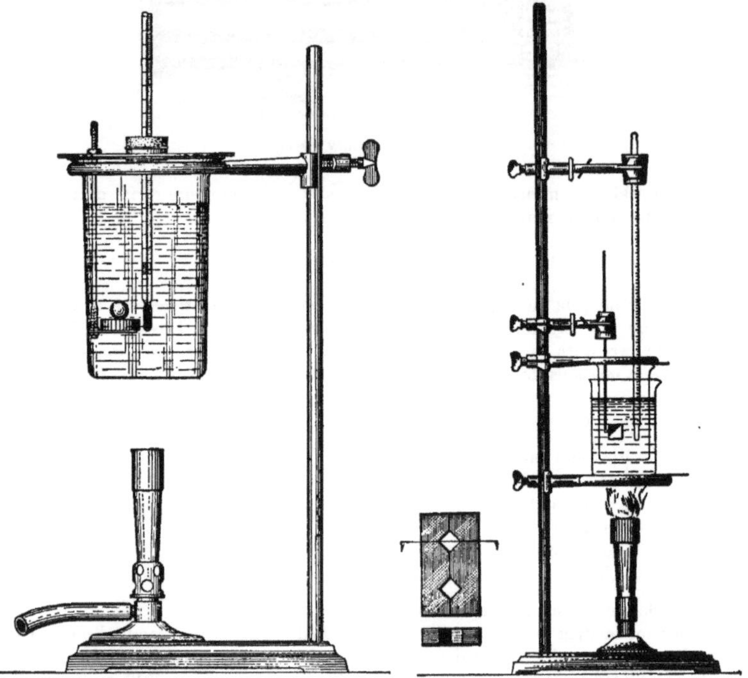

Fig. 7.—Ring-and-ball test for melting point. Fig. 8.—Melting point, hook-and-cube method.

has since been generally superseded except in its application to tar products. With this limitation it was adopted by the American Society for Testing Materials in 1920, and given the serial designation D 61-20.

For making the cube required in this test, a 12.7 mm. (½ inch) amalgamated brass cubical mold is used. This is placed on an amalgamated brass plate, the amalgamation being necessary in both cases to prevent sticking of the material to the metal.

The hook is made from a No. 12 B. & S. gage copper wire (diameter 2.05 mm. = 0.0808 inch) and bent at right angles.

A quantity of the sample to be tested is slowly melted on a tablespoon and from this an excess is poured into the prepared mold. Upon cooling a hot knife is used to cut away the surplus material, the cube is removed from the mold and placed on the hook with the wire passing through the center of two opposite faces.

The wire and a thermometer are then suspended through a cover over the top into a 400-c.c. tall-form Jena-glass beaker so that the bottoms of the cube and bulb of the thermometer are exactly 2.54 cm. (1 inch) above the bottom of the beaker, and equally spaced from each other and the sides of the beaker. The beaker is immersed in a bath of water or glycerin in a low-form 800-c.c. Jena-glass beaker to within about 1.905 cm. (¾ inch) of the bottom. The entire apparatus is then set up as shown in the diagram to permit heating with a Bunsen burner.

The temperature is raised at a uniform rate of 5° C. (9° F.) per minute.

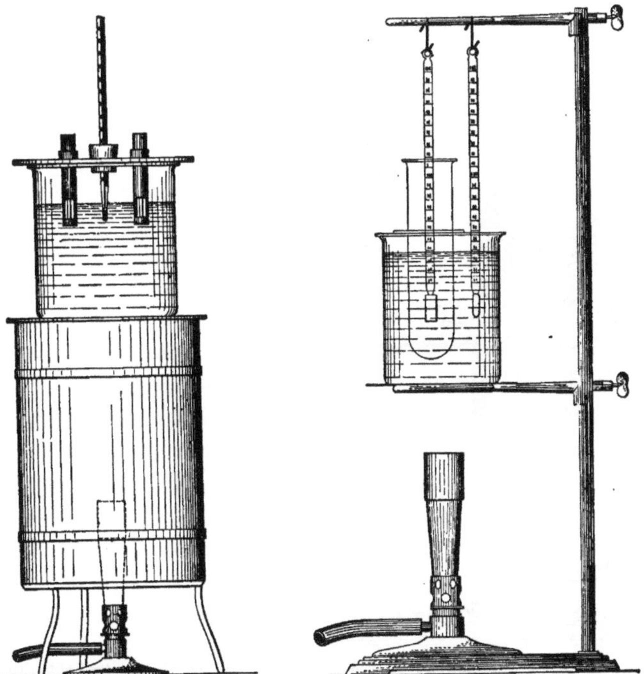

Fig. 9.—Melting point, Kramer and Sarnow method.

Fig. 10.—Melting point, General Electric method.

As the sample softens it gradually settles from the hook toward a paper set in the bottom of the beaker and that temperature at which the material touches this paper is considered the melting-point temperature of the sample.

Test for melting point. The Kramer and Sarnow method.—The apparatus consists of a 600-c.c. beaker filled to within 3.18 cm. (1¼ inch) of the top with some heating medium such as water, glycerin or paraffin and supported on a wire gauze over a Bunsen burner. Care should be taken to prevent fluctuations in temperature, due to sudden drafts by some such method as is shown in the accompanying diagram.

Just enough of the sample to be tested to reach a depth of 5 mm. (0.197 inch) when melted should be placed in a flat dish. Into this are set vertically two open glass tubes 6 to 7 mm. (.235 to .275 inch) internal diameter and about 8 cm. (3.15 inches) long, the bituminous sample in the dish rising inside the tubes to the predetermined depth of 5 mm. (0.197 inch) and forming thereby solid plugs within the tubes when set. On cooling the tubes are carefully removed, all excess material scraped away and on top of each of these plugs is placed 5 grams of mercury.

The plugged tubes are then immersed in the heating solution, the level of this medium reaching about the center of the mercury column in each tube.

These tubes are supported by corks set in a cardboard cover, the latter used to pre-

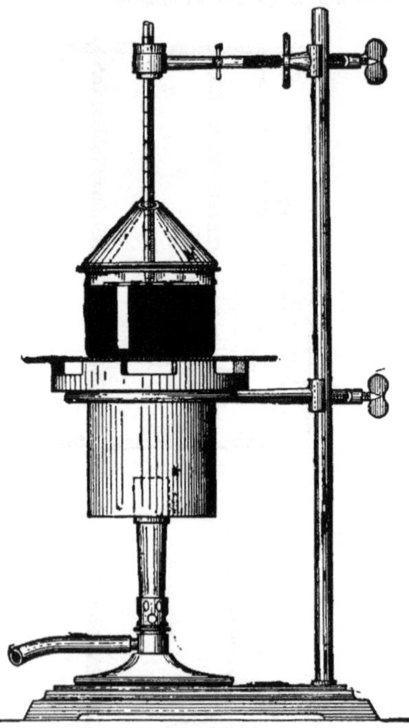

Fig. 11.—Richardson over-mercury method of testing melting point.

Fig. 12.—Pohl method of testing melting point.

vent cooling drafts striking the tubes. Through a third hole in the cover, halfway between the two tubes, is supported a thermometer, its bulb submerged in the liquid.

Heat is applied, at a uniform rate of 2¼° C. (4° F.) per minute. As the temperature rises the mercury gradually sinks and disappears, forcing the asphaltum out of the tube and finally breaking through and falling to the bottom of the beaker. The temperature at which it breaks through and falls to the bottom is taken as the melting point.

Test for melting point. General Electric method.—In this method a test-tube about 15.87 cm. (6¼ inches) long and 3.18 cm. (1¼ inch) in diameter, is immersed in a 600-c.c. beaker to within 1.90 cm. (¾ inch) of the bottom of the beaker and the beaker then filled with glycerin or water. On the test-tube three horizontal lines are scratched by means of a file, the lower line 1.90 cm. (¾ inch) from the bottom of the test-tube, the second 0.508 cm. (0.200 inch) above this first line, and the third 0.127 cm. (0.050 inch) above the second.

Suspended over this beaker are two 200° C. (392° F.) thermometers, one extending into the test-tube and the other into the glycerin bath. The mercury bulb in the thermometer of the test-tube should be of very uniform cross-section and so suspended that the bottom of the bulb is just even with the highest scratched line on the test-tube. The thermometer suspended in the glycerin, should be at the same height as the other.

After the average diameter of the bulb in the thermometer has been determined with the aid of a pair of micrometers, the sample to be tested is heated and gradually built up on this bulb until the new diameter is 0.254 cm. (0.100 inch) greater than the diameter of the bulb. A micrometer should also be used in this measuring. The sample should further extend below the bulb to a point just even with the second scratched line, thereby giving 0.127 cm. (0.050 inch) of sample below the end of the thermometer and around the bulb. The thermometer is then suspended in the test-tube.

The illustration, Fig. 10 (page 803), shows how the sample should cover the mercury bulb of the thermometer, and also the complete set-up of the apparatus.

With the aid of a Bunsen burner placed directly under the center of the beaker, heat is so applied that the temperatures, as recorded by both thermometers, rise gradually and evenly. This heat should always be regulated by increasing or decreasing the gas flame and never by moving the burner, and the rate of the temperature rise should be such that the difference in readings between the two thermometers is not greater than 15° C. (27° F.) and not less than 12° C. (21.6° F.) All drafts should be avoided.

With the increase in temperature the sample softens and drops toward the lowest scratched line on the test-tube. That temperature at which the sample is just even with this line is considered the melting-point temperature of the sample.

Test for melting point. The Richardson or over-mercury method.—In this method it is necessary that the sample of asphalt to be tested be rolled into a sphere about the size of a French pea, and this sphere then placed on a No. 2–0 microscope slide. It is frequently necessary to heat the sample slightly to free it from all nicks and irregularities.

As shown in the illustration Fig. 11, the apparatus consists of a 150-c.c. beaker nearly filled with mercury and supported on an iron stand over a 20-mesh wire gauze. The microscope slide containing the asphalt is placed on the surface of the mercury. A glass funnel, from which the stem has been cut, rests on top of the beaker, and through the hole in the funnel a thermometer is suspended with its bulb immersed in the mercury.

The flame from a Bunsen burner is then applied gradually, care being taken to raise the temperature at the rate of 5° C. (9° F.) per minute.

By placing the microscope slide on the surface of the mercury a mirror is formed, and it is upon this mirror-like action that the test depends. At the beginning of the test there appear to be two balls of the sample, one under the other and touching at their point of contact. Gradually as the temperature increases the sample melts until the two balls settle and converge into one. This temperature at which there appears to be but one perfect sphere is considered the melting point of the sample.

Test for melting point. Pohl method.—A test-tube 15.87 cm. (6¼ inches) long and 3.18 cm. (1¼ inches) diameter is immersed in a bath of glycerin or water to within 1.90 cm. (¾ inch) of the bottom of a 600-c.c. beaker.

The sample to be tested is first melted and poured into a metallic tube 2.54 cm. (1 inch) long and 0.635 cm. (¼ inch) inside diameter. The tube is then attached with a rubber band to the bulb end of a thermometer so that the center of the mercury bulb is even with the center of the metallic tube, after which the thermometer is suspended in the test-tube with the bottom of the bulb about 2.54 cm. (1 inch) above the bottom of the test-tube.

A Bunsen burner is placed under the beaker and the temperature raised at the uniform rate of 1½° C. (2½° to 3° F.) per minute.

The asphalt becomes softer as the temperature increases and the melting point is taken as that temperature at which the first drop of the sample falls from the metallic tube.

Fig. 13 shows in graphic form the determination of melting point by the different methods on a California steam-refined asphalt. Fig. 14 shows in graphic form the

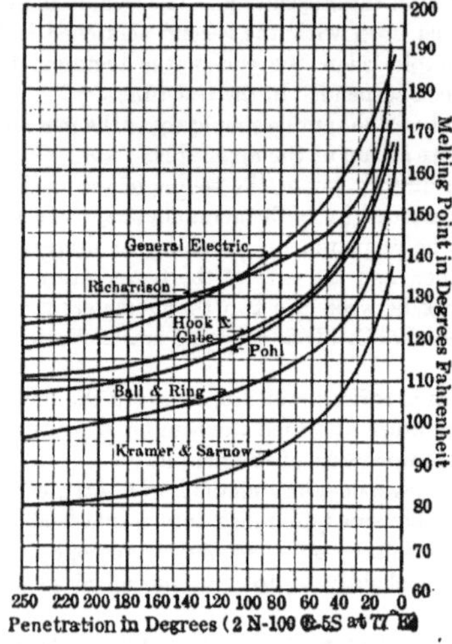

Fig. 13.—Determination of melting point by the different methods on a California steam-refined asphalt.

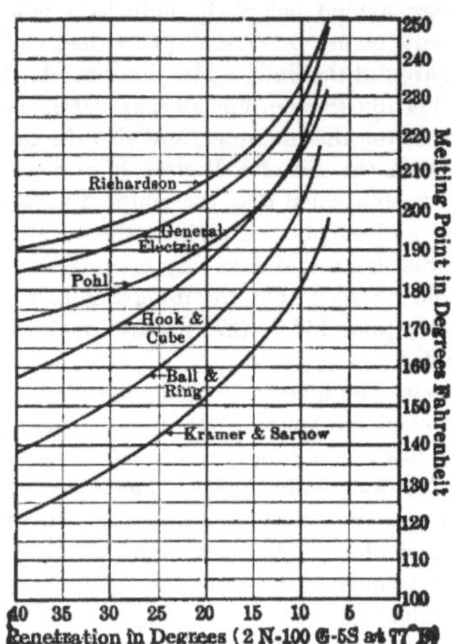

Fig. 14.—Determination of melting point by the different methods on a California air-refined asphalt.

determination of melting point by the different methods on a California air-refined asphalt.

Solubility in Carbon Disulphide (CS_2)

This test indicates the purity of the asphalt. The test as described below was developed and is used by the U. S. Office of Public Roads. It is not the standard solubility test of the American Society for Testing Materials. The standard solubility test adopted by the Society and which bears the serial designation D-4 was adopted in 1911 but is not advocated by the Society as the best test for general use, as it is longer and in many cases gives no better results than other and more expeditious tests. The test recommended by the Society is intended to be resorted to in cases of dispute only.

The usual solubility determination is made by dissolving the asphalt in carbon disulphide (CS_2) and recovering any insoluble matter by filtering the solution through an asbestos felt mat in a Gooch crucible.

The test is made by placing 1 to 2 grams (15.4 to 30.8 grains) of the asphalt to be tested in a previously weighed Erlenmeyer flask, then pouring over it in small portions 100 c.c. (.211 pint) of pure carbon disulphide, meanwhile continually agitating the flask until all lumps disappear and nothing adheres to the bottom. The flask is then corked and set aside for fifteen minutes.

A Gooch crucible is fitted with an asbestos filter mat about ⅛ inch thick. The crucible and filter are thoroughly dried, then brought to a red heat over a Bunsen burner, then cooled in a desiccator and weighed.

After weighing, the crucible is set up over the dry pressure flask as shown in Fig. 48, chapter on "Petroleum Testing Methods," and a suction hose fitted to the flask. The solution of asphalt in carbon disulphide is decanted through the asbestos filter without suction, by gradually tilting the flask. The contents of the crucible are washed with carbon disulphide until the washings run colorless. Suction is then applied until there is practically no odor of carbon disulphide in the crucible.

The crucible and contents are dried in the hot-air oven at 100° C. (212° F.) for about twenty minutes, cooled in a desiccator, and weighed.

This will give the amount of insoluble material, which may be organic or mineral matter. To separate the organic from the inorganic matter, the former is burned off by ignition at red heat, thus leaving the mineral matter in the form of ash which can be weighed on cooling.

Solubility in Carbon Tetrachloride (CCl_4)

This test is made when the identity of the asphalt is to be determined, or when there is any reason to suspect that the asphalt has been injured by overheating during the process of manufacture.

The determination is made by dissolving 1 or 2 grams of asphalt with carbon tetrachloride (CCl_4) and is conducted similarly to the carbon disulphide test, with the exception that the sample is allowed to stand in a dark place and not in the light, as insoluble compounds are formed by the action of light.

The difference between the amount of asphalt soluble in carbon disulphide and carbon tetrachloride is known as "carbenes."

Solubility in 86° Baumé Naphtha

The test is conducted similarly to the test with carbon disulphide, except that great care is taken to see that the asphalt is completely broken up and exposed to the solvent.

That portion of the asphalt which is dissolved by the naphtha is called "petrolenes"; the insoluble portion is "asphaltenes." The proportion of the former to the latter will indicate the physical properties which may be expected to be developed by the asphalt when in service.

If the petrolene content is too low, the asphalt will be brittle and will have low binding power; if too high, the asphalt will be very susceptible to changes of temperature, although of good binding power.

Volatilization Test

This test is used to determine the loss in weight of asphalt when subjected to a standard temperature under standard conditions and also to ascertain any changes in the character of the material due to such heating.

This test is used by the U. S. Office of Public Roads, and was adopted as standard in its revised form in 1920 by the American Society for Testing Materials. It is identified by the Society's serial designation D 6–20.

The determination when applied to asphalts alone, is made by placing 50 grams (1.765 oz.) of water-free asphalt in a flat-bottom dish, the inside dimensions of which are approximately 55 mm. ($2\frac{3}{16}$ inches) in diameter and 35 mm. about $1\frac{3}{8}$ inches) deep, and placing same in an oven (see Fig. 15), which has previously been brought to the required temperature of 163° C. (325.4° F.). The dish is maintained at this temperature for five hours. The temperature of the sample shall be considered as that of a similar quantity of the asphalt immediately adjoining it in the oven, in which the bulb of a standardized thermometer is immersed.

It is important that the standard dish be used in this test, for the reason that the evaporation or volatilization varies as the square of the diameter of the dish. It is also important that a standard oven be used in which an even temperature may be maintained.

Such an oven may be circular or rectangular in shape and the source of heat either

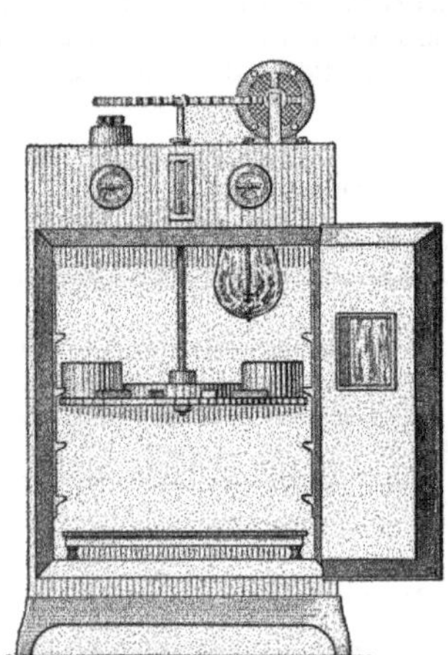

Fig. 15.—Volatilization test.

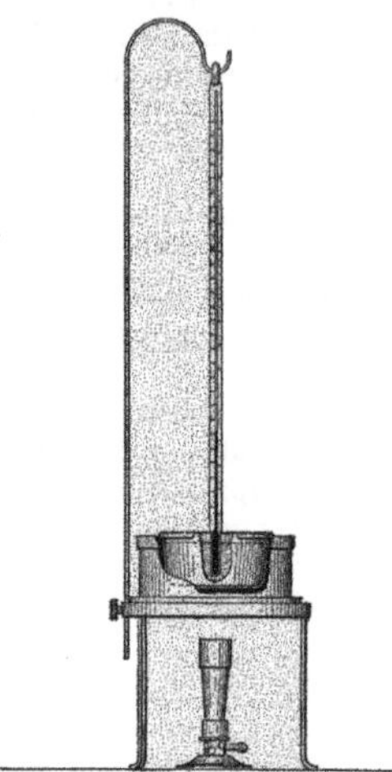

Fig. 16.—Flash-point tester.

gas or electricity. The height of the interior of the oven must be at least 40.64 cm. (16 inches) and the width or diameter not less than 5.08 cm. (2 inches) greater than the perforated circular shelf. Provision for proper ventilation must be made, and in the upper half of the door a window must be so placed that the thermometer may be accurately read without the opening of the door. The circular shelf should be approximately 24.8 cm. (9.75 inches) in diameter, and should be suspended by a vertical shaft in the center of the oven and revolved by a motor or other mechanical means, at the rate of five to six revolutions per minute.

Dishes of the samples to be tested are placed on the shelf, equally distant from the center of the shaft in recesses provided and the shelf revolved, counteracting thereby variations in temperature at various positions in the oven.

The penetration test on the residue is then made, as described under "Penetration test," page 799.

Flash Point

The flash-point test indicates the maximum safe temperature to which asphalt can be heated. This test is made in an open cup, shown in Fig. 16.

The temperature of the material in the cup is raised at the uniform rate of 5° C. (9° F.) per minute. From time to time the testing flame is brought almost in contact with the surface of the asphalt in the cup. A distinct flicker or flash over the entire surface of the cup shows that the flash point is reached and the temperature at this point is noted.

Fixed Carbon

This test is not now usually required in asphalt specifications. It is a means of identification of the asphalt and indicates properties which are more directly measured by the other standard tests. The method usually used is that recommended by the committee on coal analysis of the American Chemical Society, published in volume 21 (1899), page 1116, and may be briefly described as follows:

A ½ gram sample of the bitumen to be examined is weighed in a 20-gram platinum crucible provided with a close-fitting cover. This is heated for exactly seven minutes on a platinum triangle in the full flame of a Bunsen burner which has been carefully regulated so as to give a flame 6–8 cm. high to the top of the inner cone, and fully 20 cm. high to the top. The platinum crucible should be placed ½ cm. above the top of the inner cone. Care should be taken to prevent drafts as a steady even flame is required.

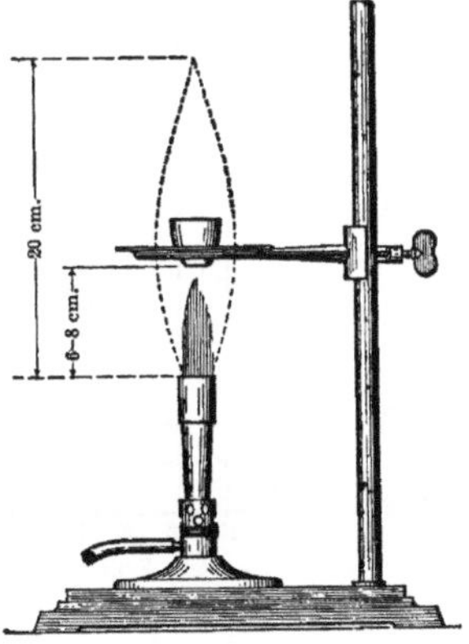

Fig. 17.—Fixed-carbon test.

The crucible while still covered is removed from the flame, placed in a desiccator, cooled and weighed. The cover is then removed, the crucible returned to the flame, placed on its side to give access to air, and ignited until only the ash remains. The cover is also heated to remove any carbon that may have been deposited on its under side. After being cooled again, the crucible and cover are weighed, this latter weight being subtracted from that previously obtained. This loss is the fixed carbon, the remainder is the ash.

When mineral matter containing carbonates is carried in the bitumen this ash is treated with a few cubic centimeters of a saturated solution of ammonium carbonate ($N_2H_8CO_3Aq$) and dried. It is then ignited at a low heat to remove excess of carbonate and the corrected weight of the ash is obtained.

CHARACTERISTICS OF PETROLEUM ASPHALT

Table F shows the characteristics of California steam-refined asphalt, and California air-refined asphalt. Figs. 18 and 19 show these characteristics in graphic form.

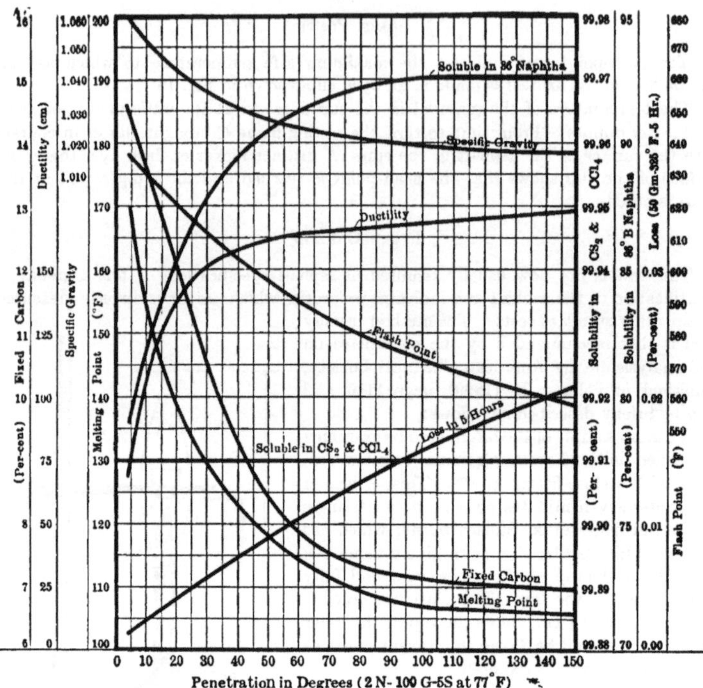

Fig. 18.—Characteristics of California steam-refined asphalt.

Table F

Characteristics of Asphalts from California Petroleum Oil

Steam-refined

Penetration (77° F., 100 g. 5 sec.).	10	20	30	40	60	80	100	150
Specific gravity	1.052	1.043	1.036	1.031	1.025	1.021	1.020	1.017
Ductility (cm.)	100	135	150	150+	150+	150+	150+	150+
Melting point (R. and B.) ° F.	155	139	130	123	114	109	107	106
Solubility in CS_2 (per cent)	99.91	99.91	99.91	99.91	99.91	99.91	99.91	99.91
Solubility in CCl_4 (per cent)	99.91	99.91	99.91	99.91	99.91	99.91	99.91	99.91
Solubility in 86° Bé. naphtha per cent)	82.0	85.3	87.5	89.5	91.2	92.0	92.6	92.6
Volatility 50 g., 325° F., 5 hr. (per cent)	0.00	0.00	0.00	0.01	0.01	0.01	0.02	0.02
Flash point °(F.)	630	620	611	603	590	579	571	558
Fixed carbon (per cent)	13.7	12.1	10.5	9.3	7.9	7.3	7.1	7.0

AIR-REFINED

Penetration (77° F., 100 g., 5 sec.)	5	10	15	20	30	40
Specific gravity	1.055	1.049	1.045	1.042	1.036	1.033
Ductility (cm.)	0.5	1.5	2.5	3.5	8.0	17.0
Melting point (R. and B.) ° F.	237	210	188	174	153	143
Solubility in CS_2 (per cent)	99.89	99.89	99.89	99.89	99.89	99.89
Solubility in CCl_4 (per cent)	99.88	99.88	99.88	99.88	99.88	99.88
Solubility in 86° Bé. naphtha (per cent)	57.6	58.5	59.5	60.5	62.6	65.0
Volatility 20 g., 325° F., 5 hr. (per cent)	0.6	1.0	1.4	1.8	2.4	2.7
Flash point (° F.)	394	392	389	387	384	382
Fixed carbon (per cent)	15.5	14.9	14.3	13.8	12.8	12.0

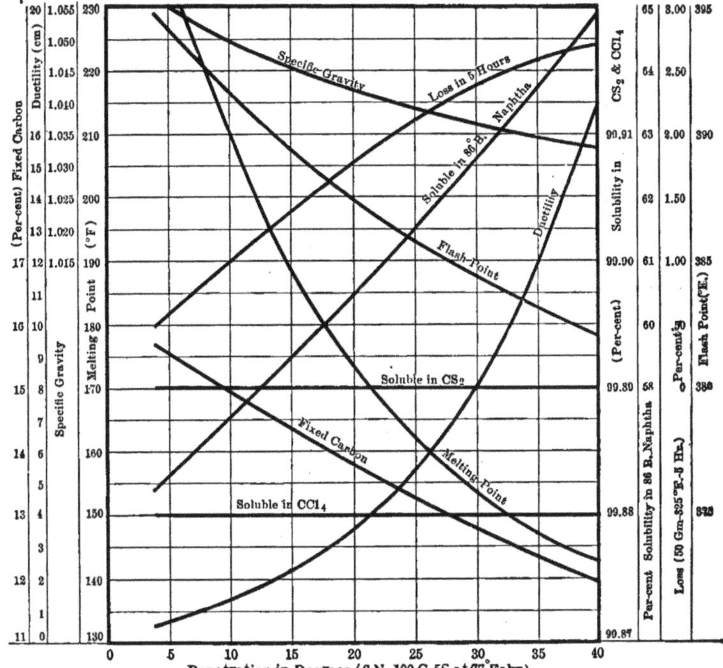

FIG. 19.—Characteristics of California air-refined asphalt.

Table G shows the characteristics of Mexican steam-refined and air-refined asphalt. Figs. 20 and 21 show these characteristics in graphic form.

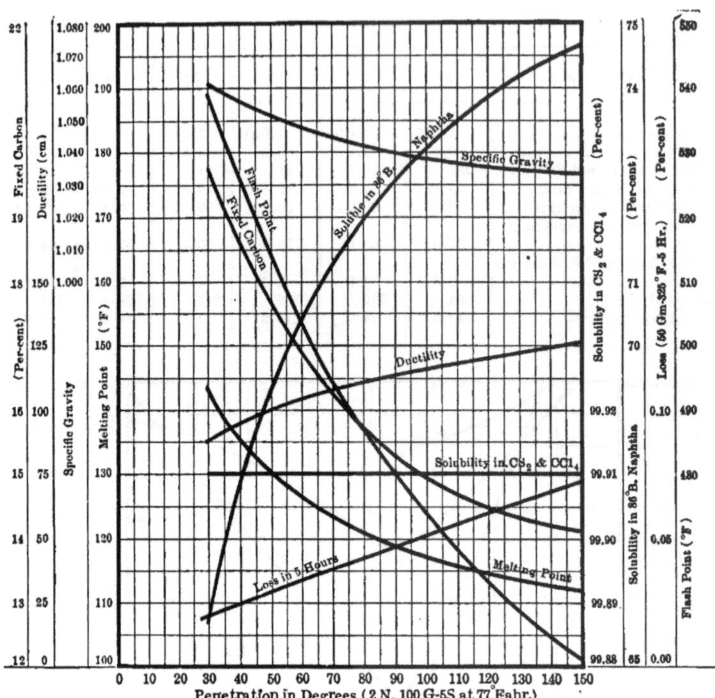

Fig. 20.—Characteristics of Mexican steam-refined asphalt.

Table G

Characteristics of Asphalts from Mexican Petroleum Oil

Steam-refined

Penetration (77° F., 100 g., 5 sec.)	30	40	60	80	100	150
Specific gravity	1.061	1.056	1.048	1.042	1.038	1.033
Ductility (cm.)	87	95	105	110	115	125
Melting point (R. and B.) ° F.	143	135	126	121	117	112
Solubility in CS_2 (per cent)	99.91	99.91	99.91	99.91	99.91	99.91
Solubility in CCl_4 (per cent)	99.91	99.91	99.91	99.91	99.91	99.91
Solubility in 86° Bé. naphtha (per cent)	65.8	67.8	70.4	71.9	73.0	74.7
Volatility 50 g., 325° F., 5 hr. (per cent)	0.02	0.02	0.03	0.04	0.05	0.07
Flash point (° F.)	540	526	504	487	474	451
Fixed carbon (per cent)	19.7	18.6	16.9	15.7	15.0	14.1

AIR-REFINED

Penetration (77° F., 100 g., 5 sec.)	10	20	30	40
Specific gravity	1.053	1.044	1.037	1.033
Melting point (R. and B.) ° F.	241	196	174	162
Solubility in CS_2 (per cent)	99.89	99.89	99.89	99.89
Solubility in CCl_4 (per cent)	99.88	99.88	99.88	99.88
Solubility in 86° Bé. naphtha (per cent)	54.1	56.9	59.9	63.0
Volatility 50 g., 325° F., 5 hr. (per cent)	0.01	0.02	0.03	0.04
Flash point (° F.)	485	460	438	420
Fixed carbon (per cent)	22.7	20.4	19.0	18.2

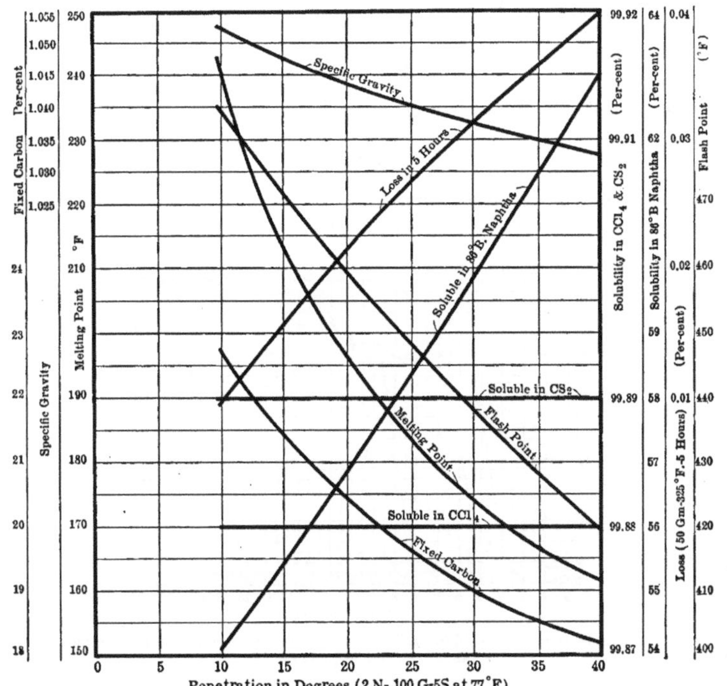

FIG. 21.—Characteristics of Mexican air-refined asphalt.

The differences in the characteristics of asphalt manufactured from Mexican and California crude oil are a means of identification only and do not prove any superiority of one product over the other. Both of these products have given excellent results in their service.

The greatest difference between the two is in the fixed-carbon content, which is approximately 7 to 9 per cent more in Mexican steam-refined than in California steam-refined asphalt.

The melting points of Mexican steam-refined asphalt are approximately 13° F. higher at 30° penetration and approximately 6° F. higher at 150° penetration, than those of California steam-refined asphalt.

The specific gravity of Mexican steam-refined asphalt is approximately .012 more at 30° and .016 more at 150° penetration than that of California steam-refined asphalt.

Another noticeable characteristic is that the solubility in 86° Bé. naphtha is approximately 20 per cent greater in California steam-refined asphalt than in Mexican steam-refined asphalt.

The ductility of California steam-refined asphalt is usually somewhat higher than that of Mexican steam-refined asphalt.

The differences between air-refined asphalt manufactured from Mexican and California petroleum oils are noticeable in melting point and fixed carbon but not in specific gravity. Also, the difference between the solubility of these two products, in 86° Bé. naphtha is very much less than with the steam-refined asphalts.

PACKAGES

Asphalt is marketed in tank cars and in wood and steel barrels of various types.

Tank Cars

Asphalt of 30° penetration and softer may be shipped in tank cars. This is the best and most economical method of shipping asphalt. Tank cars usually have a capacity

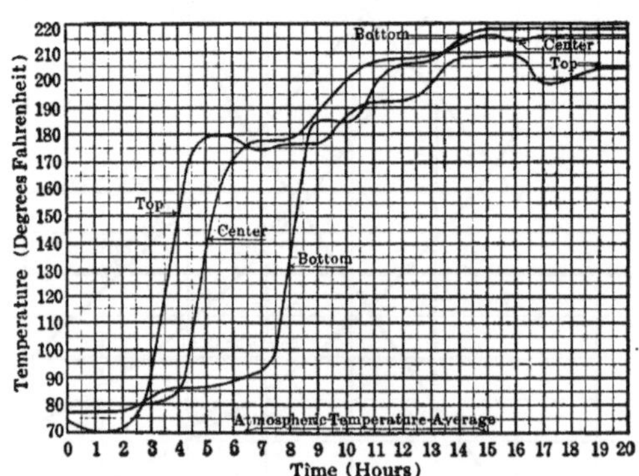

Fig. 22.—Temperature time curve of 10,000 gallon tank car, having 300 square feet of heating surface, steam at 100 pounds pressure.

of approximately 10,000 gallons, and are supplied with steam-heating coils having a heating surface of between 250 and 400 square feet. The usual unloading temperature is reached from an atmospheric temperature of 70° in from 20 to 30 hours, depending on the amount of surface of the steam coils, and the pressure of the steam. A 10 H.P. boiler is sufficient to develop the amount of steam required.

Fig. 22 shows a temperature time curve on the heating of a 10,000-gallon tank car, having approximately 300 square feet of heating surface, with steam at 100 pounds pressure. The temperature was taken at the top, center, and bottom of the car. It will be noted that the bottom of the car takes the longest time to heat, although the steam coils are placed in the bottom. This is due to the convection of the hot asphalt to the top of the car. It is therefore necessary to have the whole car heated before attempting to unload the asphalt.

Fig. 23 shows a temperature time curve on the cooling of asphalt in a 10,000-gallon tank car. It will be noted that considerable time is taken to cool asphalt in a tank car and that it may arrive at its destination without being cooled to atmospheric temperature.

The radiation in B.t.u.'s per square foot per hour per average degree Fahrenheit

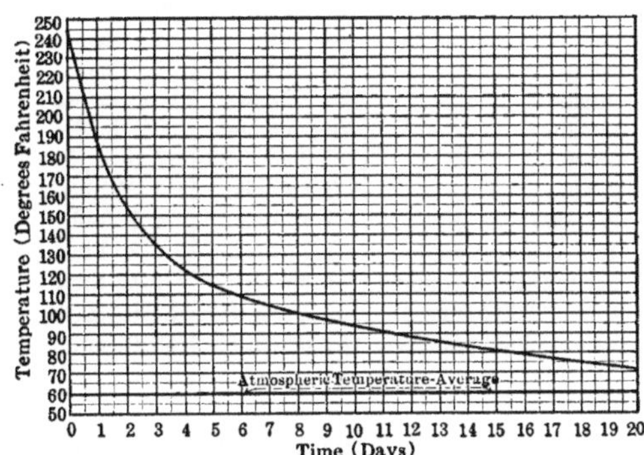

Fig. 23.—Temperature time curve of cooling of asphalt in a 10,000 gallon tank car.

difference of temperature is approximately 0.3. The specific heat of asphalt is approximately 0.46.

The economy of shipping in tank cars may be stated as follows:

(1) The cost of the barrel is saved.

(2) There is a saving in freight, due to the fact that freight is paid on net weight of asphalt shipped in tank cars, and on gross weight shipped in packages.

(3) The cost of unloading and preparing for use is much less in tank cars than in packages.

(4) There is less waste of material, due to asphalt sticking to packages.

Barrels

Wood is used practically exclusively for shipping asphalt in barrels from California. There are three types of packages.

(1) Package used for shipments by rail in the United States, or by coastal steamers, and known to the trade as "domestic asphalt barrel," in which asphalt harder than 130° penetration is shipped.

(2) Export wood barrels. A package used for shipments via steamer to foreign countries, of asphalts harder than 130° penetration.

(3) Heavy wood barrel. A package used for asphalt softer than 130° penetration, in both domestic and export shipments.

The dimensions of the three types of barrels are given in the following Table H.

TABLE H

	Domestic wood barrel, one end open	Export wood barrel, both ends closed	Heavy wood barrel, both ends closed
Thickness of stave	7/16 inch	9/16 inch	3/4 inch
Number of hoops	4	6	6
Height	34 inches	34 inches	34 inches
Packing capacity	52½ gal.	50 gal.	48 gal.
Weight with 1 head	23 lbs.	31 lbs.	
Weight with 2 heads	26 lbs.	35 lbs.	56 lbs.
Diam. at bilge—outside	25 inches	25 inches	25 inches
Diam. at bead—outside	20½ inches	20¼ inches	20½ inches
Over-all cubical measure	12.3 cu. ft.	12.3 cu. ft.	12.3 cu. ft.
Barrels per ton (40 cu. ft.)	3.25 bbls.	3.25 bbls.	3.25 bbls.
Barrels per net ton (2000 pounds)	4.4 bbls.	4.7 bbls.	5 bbls.
Barrels per gross ton (2000 pounds)	4.2 bbls.	4.3 bbls.	4.4 bbls.
Tons (40 cu. ft.) per ton (2000 pounds)	1.35 tons	1.45 tons	1.54 tons
Cubic feet per net ton (2000 pounds)	54.1 cu. ft.	57.8 cu. ft.	61.5 cu. ft.
Cubic feet per gross ton (2000 pounds)	52.0 cu. ft.	53.0 cu. ft.	54.0 cu. ft.

In the Eastern part of the United States the metal drum and wood barrel are both used, the former more extensively. The metal drum has the following dimensions:

Height	34⅞ inches
Diameter	21⅞ inches
Tare	18 pounds
Gage of iron	26
Average filling capacity	51.2 gallons

Method of lining barrels.—In order to prevent the asphalt from sticking to the staves of wood-barrels, the inside of the barrel is treated with a clay wash, and allowed to dry before the asphalt is placed in the barrel. By this means, the staves can be stripped from the barrel with very small loss of asphalt.

Loading box cars. The dunnage required for a railroad car approximately 36 feet in length, 9 feet in width, and 7 feet in height, is about 260 to 300 feet B. M.

USES OF ASPHALT

STREET AND HIGHWAY CONSTRUCTION

The primary use of asphalt is in street and highway construction. There are various classes of asphaltic pavements, the most important being that in which the asphalt is mechanically mixed with a mineral aggregate, this mixture being laid upon a foundation and rolled to the desired thickness.

There are eight well-recognized classes of paving mixtures of this type, the classification depending upon the size and proportions of the mineral aggregates.

Asphaltic concrete, coarse aggregate type.—In this type about 60 per cent of mineral aggregate is retained on a 10-mesh screen, the remainder passing through this screen.

The largest size of rock used in the mineral aggregate is usually approximately one-half of the thickness of the pavement.

Where asphaltic concrete is used as the foundation or base, the aggregate proportioned as above passes through a 2½-inch, a 2-inch, or a 1½-inch screen, depending on the thickness of the base. Where it is used for an asphaltic concrete surface, the aggregate usually passes a 1½-inch or a 1-inch screen, depending upon the thickness of the surface.

This pavement when laid does not present a close texture and a seal coat is usually applied to make it impervious and of a close, smooth-wearing texture. This seal coat usually consists of an asphalt applied at the rate of approximately one-third of a gallon per square yard, upon which are thrown fine screenings or coarse sand. The pavement is then rolled. Another type of seal coat is the application of a mixture of 90 per cent sand and 10 per cent asphalt which is placed on the pavement to a depth of approximately ¼ inch before the final rolling.

The principle of this type of pavement is that the coarse aggregate will interlock and give it stability. The fine aggregate fills the voids in the pavement, increasing the density and adding resistance to lateral displacement by the friction of its particles.

Asphaltic concrete, fine aggregate type.—In this type about 30 per cent of the mineral aggregate is retained on a 10-mesh screen and the largest size of the mineral aggregate passes through a ½-inch screen. This type presents a close textured surface and no seal coat is used in its construction. This pavement depends for its stability on the frictional resistance of the sand particles coated with asphalt, the rock adding to the density of the pavement.

Asphaltic concrete, open-binder type.—In this type 90 to 95 per cent of the mineral aggregate is retained on a 10-mesh screen, the largest size usually passing a 1-inch screen; the pavement depends solely for its stability on the interlocking of the mineral aggregate. This type was formerly used in what was known as "open binder," upon which was placed a sheet asphalt surface but it has been generally discarded in favor of the asphaltic concrete, coarse aggregate type, in which the voids are filled with fine aggregate.

Sheet asphalt.—This type is a mixture of a definitely graded mineral aggregate, usually quartz sand, all of which passes a 10-mesh screen. It depends for its stability upon the frictional resistance of the sand particles which are coated with asphalt. Practice has determined certain set gradings, and should be followed to insure the best results.

Asphalt sand mixture.—In many localities it is not possible to obtain the standard sheet asphalt mixture and an asphalt pavement is laid with the sand available. In many cases chemicals, such as copper sulphate and aluminum sulphate, are added in small amounts to the mixture with the intention of hardening it and increasing its stability under traffic.

Asphalt mastic pavements.—Mastic pavements differ from asphaltic concrete pavements or sheet asphalt pavements in that they contain much more filler, or fine material which passes a 200-mesh, and much more asphalt.

The mastics are generally intimately mixed for a very much longer period of time, such as twenty to forty minutes, so that they will pour and set in place without rolling. In order to obtain a smooth surface they are usually troweled with a wooden trowel while still warm, or "struck off" to the right thickness and grade and allowed to set.

Asphalt earth pavements.—Asphalt earth pavements have been laid in various parts of the eastern United States, usually under the name of "National" pavements. In this particular pavement the soil, of which approximately 75 per cent passes a 200-mesh, is mixed with approximately 14 to 18 per cent of asphalt and is generally laid on existing macadam bases. It depends for its stability on the high frictional resistance due to the large surface area of the earth particles. A softer asphalt is used in this type of pavement than in the sheet asphalt type in order to avoid the tendency to crack in cold weather.

Natural rock asphalt pavement.—Another type of pavement depending on the high frictional resistance and absorbent quality of mineral aggregate, is the natural rock asphalt pavement in which the mineral matter is a fine limestone, the greater portion of which passes a 200-mesh.

Table I gives the grading and characteristics of typical asphalt paving mixtures. It will be noted from this table that the percentage of asphalt increases with the decrease in the size of the mineral aggregate, or increase in the surface area of the aggregate. The percentage varies in the various types from 6 to 18 per cent.

Other types of paving in which the asphalt is not mechanically mixed with the mineral aggregate are used quite extensively and may be classified as follows:

Asphalt macadam.—This designation is given to a type of road where the asphalt is applied to the wearing course of crushed rock by means of a power pressure sprinkler, or by hand-pouring pots or other available means, after which it is covered with fine screenings and rolled. Generally this process is repeated two or three times in order to obtain thorough incorporation of the asphalt with the rock and screenings.

This type is used because of its lower cost of construction as compared with mechanically mixed pavements, but is not as satisfactory.

Asphalt surface treatments.—A treatment of the surface of an old macadam or concrete road by a thin carpet approximately $\frac{3}{8}$ inch thick of asphalt and screenings sometimes adopted. This treatment should not be confounded with asphalt macadam.

Foundations of pavements.—The above types of pavements are placed upon suitable foundations. These foundations are usually asphaltic concrete, hydraulic concrete, asphalt sand, water macadam, crushed rock, brick, or stone blocks.

The two principal foundations for new construction, especially where the mechanically mixed types of asphaltic wearing surfaces are to be used, are asphaltic concrete and hydraulic concrete.

Up to a recent period, because of its lower cost, hydraulic concrete has been the generally favored foundation, but commercial conditions, especially in the western United States in 1920, have favored asphaltic concrete for this purpose.

The asphaltic concrete foundation has been particularly featured in the western United States and there are upwards of 100,000,000 square feet of pavements of this character which have been laid by cities, counties, and states, and have given splendid satisfaction. Records of use up to twenty-five years with practically no maintenance have been obtained.

Asphaltic concrete as a foundation has many advantages, among which may be mentioned the fact that the asphaltic wearing surface will bond to the base, making a homogeneous pavement, and will thus avoid a frequent cause of waving, which may occur when asphalt surfaces of insufficient thickness are placed on hydraulic concrete bases without sufficient bond. It has good durability, and records show a low maintenance cost. It can be laid quickly, and can be used a few days after being laid as it requires no curing.

Asphaltic concrete is especially resistant to impact and tests of impact have shown much greater resistance than hydraulic concrete.

Specifications for asphalt for asphaltic concrete pavement.—The specifications on page 820 are suitable for an asphalt in coarse or fine asphaltic concrete and sheet asphalt mixtures.

TABLE I

Typical Paving Mixtures

Size of Aggregate		Asphaltic Concrete					Sheet asphalt mixture	Asphalt sand mixture	Asphalt mastic mixture	Asphalt earth pavement mixture
		Coarse Aggregate Base Course		Coarse Aggregate Surface Course		Fine aggregate surface course				
		No. 1	No. 2.	No. 1	No. 2					
Passing 2½-inch	Retained on 2-inch	10								
Passing 2-inch	Retained on 1½-inch	12								
Passing 1½-inch	Retained on 1-inch	15	18	18						
Passing 1-inch	Retained on ¾-inch	8	11	11	13					
Passing ¾-inch	Retained on ½-inch	10	13	13	16					
Passing ½-inch	Retained on ¼-inch	13	17	17	21	8				
Passing ¼-inch	Retained on 10 mesh	12	15	15	18	20	0	..	11	
Passing 10 mesh	Retained on 20 mesh		8	8	9	11	3	..	12	
Passing 20 mesh	Retained on 30 mesh	10	2	2	5	4	6	..	13	
Passing 30 mesh	Retained on 40 mesh		3	3	2	2	9	..	6	
Passing 40 mesh	Retained on 50 mesh	3	2	2	2	10	12	2	6	3
Passing 50 mesh	Retained on 80 mesh		2	2	2	17	27	9	6	3
Passing 80 mesh	Retained on 100 mesh	3	2	1	2	6	14	42	5	3
Passing 100 mesh	Retained on 200 mesh		2	2	3	13	15	39	23	16
Passing 200 mesh	Mineral filler	4	5	6	7	9	14	8	18	75
Total aggregate (per cent)		100.0	100.0	100.0	100.0	100.0	100.0	100.0	100.0	100.0
Approximate per cent asphalt		6.0	6.0	6.5	7.0	8.0	10.0	12.0	16.0	18.0

Specifications for Asphalt for Concrete Pavement

Penetration, 100 grams, five seconds, at 77° F. 40–90*
Specific gravity .... Not less than 1.000
Ductility of a sample briquet 1 sq. cm. in cross-section, elongated at 5 cm. per min., at 77° F. .... Not less than 30
Solubility in carbon disulphide (CS_2) .... Not less than 99 per cent
Solubility in carbon tetrachloride (CCl_4) .... Not less than 98.5 per cent
Evaporation from 50 grams in five hours at 325° F. in a dish 5.5 cm. (2⅛ inches) diameter and 3.5 cm. (1⅓ inches) deep ..... Not more than 3 per cent
Penetration of sample after evaporation test ..... Not less than 50 per cent of original penetration
Flash test .... Not less than 350° F.

* The exact limits of penetration specified depend upon the types of pavement, traffic, and the climatic conditions and should have a limiting range of 10° in penetration.

The following table gives a suggestion for penetrations under various conditions:

Table J

Type of pavement	Traffic	Climate		
		Cold	Moderate	Warm
Asphaltic concrete (coarse aggregate)	Light	70–80	70–80	60–70
	Moderate	70–80	70–80	60–70
	Heavy	60–70	60–70	50–60
Asphaltic concrete (fine aggregate)	Light	60–70	60–70	50–60
	Moderate	60–70	50–60	50–60
	Heavy	50–60	50–60	50–60
Sheet asphalt	Light	50–60	50–60	40–50
	Moderate	50–60	50–60	40–50
	Heavy	40–50	40–50	40–50

For the asphalt macadam type, the specifications for asphalt would be similar to the previous specification, except for penetration, which would be as follows:

Traffic	Climate		
	Cold	Moderate	Warm
Light	120–150	120–150	90–120
Moderate	120–150	90–120	90–120
Heavy	120–150	90–120	90–120

For the asphalt surface treatments, somewhat softer asphalts are used than for the asphalt macadam type, and in practice there is a great variation in degree of softness. A satisfactory treatment can be given with an asphalt of about 150° to 250° penetration.

For mastic pavements an air-refined asphalt having the following characteristics is generally used:

Penetration, 100 grams, five seconds, at 77° F. 15–40
Specific gravity . . . . Not less than 1.000
Ductility of a sample briquet 1 sq. cm. in cross-section, elongated at 5 cm. per min., at 77° F. 2–15
Melting point (ring and ball) . . . . Not less than 150° F.
Solubility in carbon disulphide (CS_2) . . . . Not less than 99 per cent
Solubility in carbon tetrachloride (CCl_4) . . . . Not less than 98.5 per cent
Solubility in 86° Bé. naphtha . . . . Not more than 75 per cent
Evaporation from 50 grams in five hours at 325° F. in a dish 5.5 cm ($2\frac{1}{8}$ inches) diameter and 3.5 cm. ($1\frac{3}{8}$ inches) deep . . . . Not more than 3 per cent
Penetration of sample after evaporation test . . . . Not less than 50 per cent of original penetration
Flash test . . . . Not less than 350° F.

The penetration and melting point will depend upon the temperature conditions under which the mastic is to be used. If laid on bridges or platforms under moderate to high temperatures, the asphalt should be 15° to 25° penetration with a melting point of not under 175° F. When laid inside of buildings, under even temperature conditions, the penetration should be 30° to 40°, and the melting point not less than 150° F.

Asphalt fillers.—Asphalt is used for filling the joints and cracks and for taking care of expansion in brick, stone-block, and hydraulic cement concrete pavements. There are three kinds of fillers:

(1) Asphalt heated and poured into the joints or expansion cracks.

(2) An asphalt grout in which sand is mixed with the asphalt and broomed or "squeegeed" into the joints.

(3) The prepared asphalt joint, in which the asphalt mixed with limestone, sand, or sawdust which may be placed between asphalt-saturated fabrics is prepared in pre-molded strips.

For this use, an asphalt of the following characteristics would be satisfactory:

Penetration, 100 grams, five seconds, at 77° F. . . . . 15 to 30
Specific gravity . . . . Not less than 1.000
Ductility of a sample briquet 1 sq. cm. in cross-section, elongated at 5 cm. per min. at 77° F. 2–15
Melting point (ring and ball) . . . . Not less than 150° F.
Solubility in carbon disulphide (CS_2) . . . . Not less than 99 per cent
Solubility in carbon tetrachloride (CCl_4) . . . . Not less than 98.5 per cent
Solubility in 86° Bé. naphtha . . . . Not more than 75 per cent
Evaporation from 50 grams in five hours at 325° F. in a dish 5.5 cm. ($2\frac{1}{8}$ inches) diameter and 3.5 cm. ($1\frac{3}{8}$ inches) deep . . . . Not more than 3 per cent
Penetration of sample after evaporation test . . . . Not less than 50 per cent of original penetration
Flash test . . . . Not less than 350° F.

Roofing

The second largest use for asphalt is in the roofing industry. The principal products of this industry are:

(1) Asphalt saturated felt.
(2) Prepared roofing.
(3) Asphalt shingles.

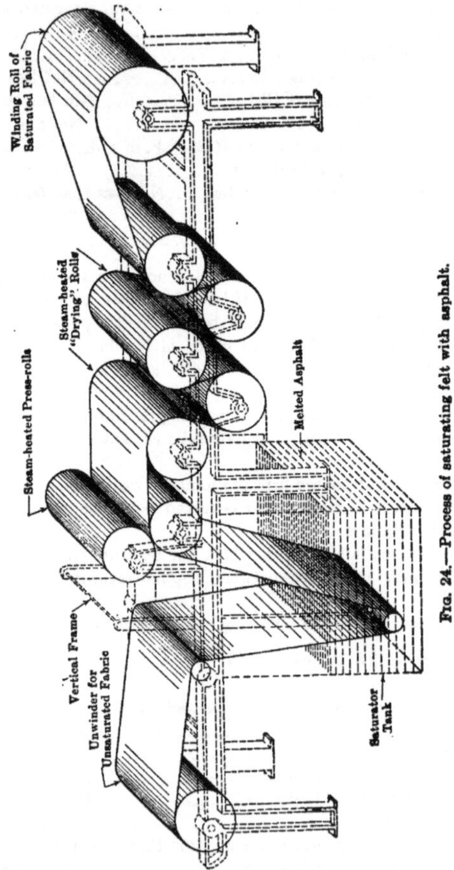

Fig. 24.—Process of saturating felt with asphalt.

Asphalt saturated felt.—In the manufacture of this product, the felt is first manufactured from rags, wool, etc., and then passed through a saturating tank containing asphalt. The process is illustrated in Fig. 24.

The felt is passed through the tank containing heated asphalt, then pressed between two rolls, and followed by a number of steam-heated rolls which leave the felt in what is termed a " dry " condition. It is then wound in large rolls by the winder. The

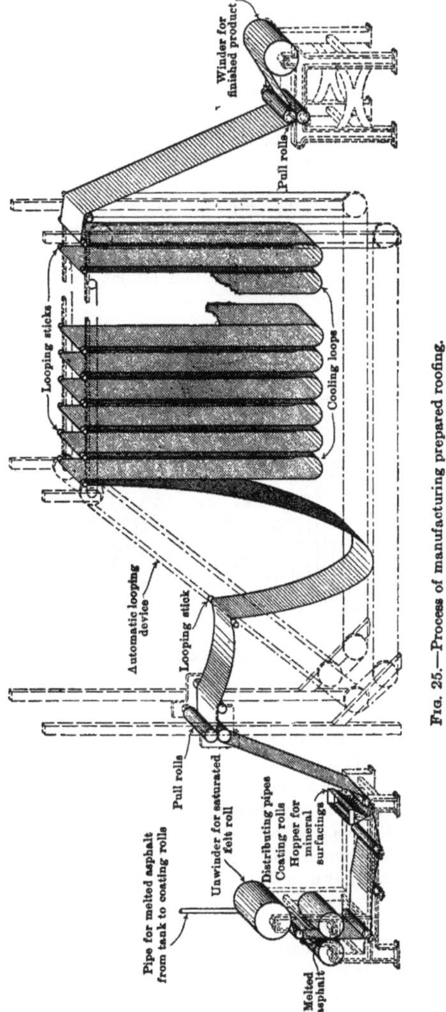

Fig. 25.—Process of manufacturing prepared roofing.

felt is saturated with not less than 60 per cent of its weight of asphalt and must be pliable. The asphalt used as a saturant varies between 90° and 250° penetration.

Prepared roofing.—If prepared roofing is to be manufactured, the asphalt-saturated roll is given a coating of harder asphalt, either by passing through a tank or passing through steam-heated rolls which distribute a coating fed from an overhead storage tank. This process is shown in Fig. 25.

The penetration of the asphalt varies for different conditions but is usually between 10° and 20° penetration, with a melting point of approximately 200° F.

The coating while still hot receives a dusting of mineral matter, usually talc or sand or slate, after which it is cooled by an automatic looping device. This looping device gives capacity to the plant and allows the coating to take place continuously though there be a temporary stoppage of the winder.

From the cooling loops the prepared roofing now passes to a winder where it is made into definite sized rolls for market.

The two processes as shown in the figures are called the intermittent method. Some plants operate without any winding of the asphalt-saturated felt, in which case the process is known as the continuous method.

Asphalt shingles.—If shingles are to be manufactured, a machine cuts the prepared

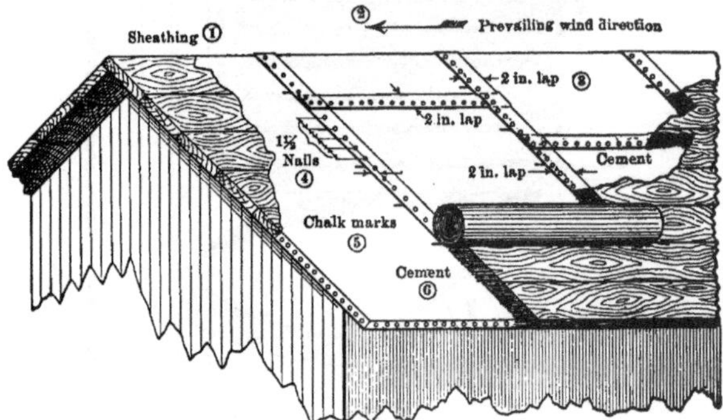

FIG. 26.—Method of applying prepared roofing.

roofing to the proper size. Usually the coating is a mineral matter of a red, green, or white color. The prepared roofings and shingles have proved very satisfactory, as they are low in first cost and can be applied to a roof at a low cost of labor.

The method of applying prepared roofing is well shown in Fig. 26.

Asphalt and gravel roofs.—Where built-up asphalt and gravel roofs or waterproofing membranes are required, the asphalt-saturated felt without the hard coating is used and is applied as shown in Fig. 27.

Usually three, four, or five plies of asphalt-saturated felt are used, the plies being firmly cemented together by asphalt. When a five-ply roof is built, it may be put on five separate plies, or in combination of three and two, as shown in the drawings. When the roofing is applied over wood a dry sheet or sheeting paper should be first applied before the application of the roofing.

Where a waterproofing membrane is required the gravel is not placed on the membrane, as tiling or concrete is sometimes placed over the waterproofing membrane.

The characteristics of an asphalt that has proved satisfactory for built-up roofs are given on page 826.

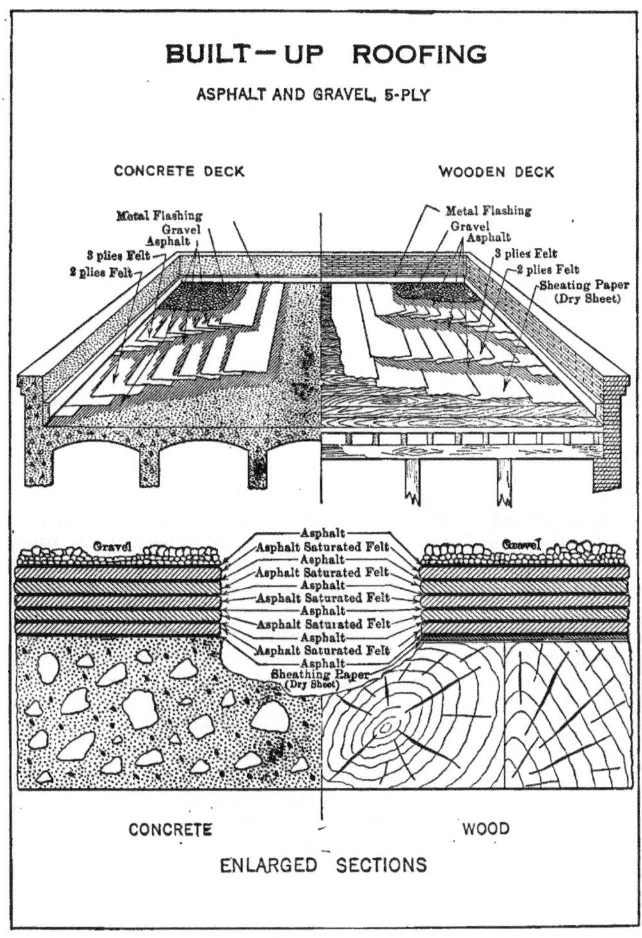

Fig. 27.—Method of applying built-up roofs.

Characteristics of Asphalt for Built-up Roofs

Penetration, 100 grams, five seconds, at 77° F. 15–40
Specific gravity. . . . . .Not less than 1.000
Ductility of a sample briquet 1 sq. cm. in cross-section, elongated at 5 cm. per min., at 77° F. 2–15
Melting point (ring and ball). . . . . .Not less than 150° F.
Solubility in carbon disulphide (CS_2). . . . . .Not less than 99 per cent
Solubility in carbon tetrachloride (CCl_4). . . . . .Not less than 98.5 per cent
Solubility in 86° Bé. naphtha. . . . . .Not more than 75 per cent
Evaporation from 50 grams in five hours at 325° F. in a dish 5.5 cm. ($2\frac{1}{8}$ inches) diameter and 3.5 cm. ($1\frac{1}{3}$ inches) deep. . . . . .Not more than 3 per cent
Penetration of sample after evaporation test. . . .Not less than 50 per cent of original penetration
Flash test. . . . . .Not less than 350° F.

The penetration and melting point will depend upon the temperature conditions under which the roof is to be used. Where the temperatures are moderate 25° to 40° penetration is used, and in warmer regions such as the hot valleys of the interior of California 15° to 25° penetration meets the conditions more satisfactorily.

Floor Coverings

An asphalt-saturated felt is the foundation upon which is printed a suitable design for floor coverings. They are economical and are proving satisfactory.

Pipe Dipping

Asphalt is used as a coating for both steel and wooden pipe. It prevents the corrosion of steel pipe and of the steel bands on wooden pipe, besides giving a wearing coating. The thickness of asphalt left on the pipe depends upon the temperature of the asphalt in the dipping tank, and the pipes should be left in the tank for a sufficient length of time in order that a good bond may be obtained between the asphalt and the pipe material.

Where steel pipes are dipped they are usually left in the dipping vat from ten to twenty minutes. They are then hung above the vat in a sloping position to drip and cool. Wooden pipe is not dipped but is placed on revolving rollers in a bath of hot asphalt at such a height that asphalt will not enter the inside of the pipe. It usually takes from three to eight minutes to coat wooden pipe by this process. The pipe is rolled in sawdust immediately after coating. At times when it is necessary to ship pipes directly after the coating with asphalt and sawdust they are subjected to a spray of cold water to chill the surface so as to facilitate handling and loading.

A suggestion for a pipe-dipping plant is shown in Fig. 28. A melting tank is provided in which the asphalt is heated by means of a steam-heating coil, so that it will flow readily. From the melting tank the asphalt flows to the dipping tank which is likewise equipped with steam coils. From the dipping tank the asphalt is circulated through an asphalt heating retort which heats it to the required temperature. It flows back to the dipping tank and is kept circulating so that the temperature is constant in the dipping tank. The asphalt in the dipping tank loses temperature through the cooling effect of the mass of cold pipe being dipped and through loss of heat by radiation.

The retort usually consists of six to eight lengths of 3-inch pipe so arranged that 6 lineal feet of each pipe are exposed to the hot gases of the heating chamber, care being taken by means of baffles to prevent direct contact with the flame. This diameter of pipe is recommended as smaller sizes have a tendency to burn out or overheat the asphalt and uniform heating with a larger pipe is more difficult.

All fittings and return bends are to be outside the retort and fitted with flanged connections. Pipe coils are to be inclined to the front to allow the drawing off of the asphalt when the plant is shut down, through small drain cocks tapped in the return bends. Usually a steam-jacketed rotary pump is satisfactory for the circulating pump.

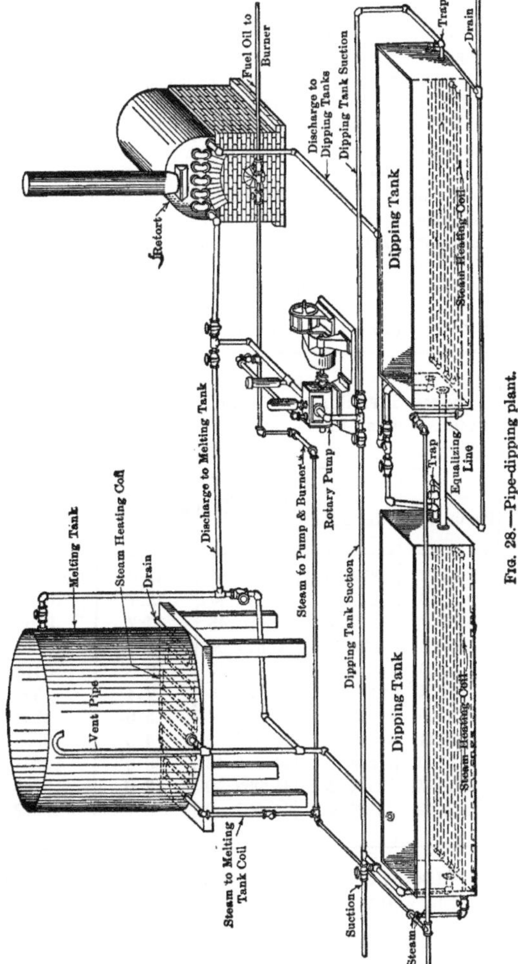

Fig. 28.—Pipe-dipping plant.

Pipe-Line Coating

It is not always advisable to have the coating placed on pipes by means of dipping, as part of it is likely to be broken off by shipping and handling, and a coating may be placed on pipe when in place.

For old pipe-lines now in use, the following procedure for coating with asphalt has been found to be satisfactory:

(1) The pipe is thoroughly cleaned of all scale and dirt, generally with a wire brush.

(2) The pipe is given a priming coat of a bituminous priming solution, usually with a flat paint brush.

(3) Two coatings of asphalt are applied at a temperature between 300° and 350° F. One coat is applied by the gang working in one direction and the other coat by the same gang reversing the direction and retracing its steps. In this way a coating of asphalt averaging ⅛ inch or more in thickness is readily applied to the pipe.

The coat is applied by means of a canvas swab 30 inches wide and made of 6 feet of 16-ounce canvas with a wooden stick at each end for a handle. Two men use this swab while one man pours the asphalt on the pipe. The purpose of the canvas swab is to obtain a uniform thickness.

Sometimes on hot pipes a wrapping of burlap is placed around the asphalt in order to hold the latter in place on the pipe.

The following table shows the covering capacity of asphalt ⅛ inch thick on pipes:

TABLE K

Covering Capacity of Asphalt, 1 Coat, ⅛ Inch Thick

Nominal diameter of pipe in inches	External Pipe Surface in Square Feet		Weight of pipe, including couplings in pounds per mile	One U. S. gallon—231 cu. in. asphalt ⅛ inch coat will cover. Feet	Asphalt Required (in gallons) when Applied ⅛ inch Thick	
	Per lin. foot	Per mile			Per lin. foot	Per mile
1	0.3442	1817.376	8912.64	37.3	0.0268	142.61
1½	0.4976	2627.328	14509.44	25.7	.0388	202.72
2	0.6217	3282.576	19620.48	17.7	.0562	255.87
3	0.9163	4838.064	40524.00	14.0	.0714	377.98
4	1.1780	6219.840	57974.40	10.8	.0918	464.64
6	1.7344	9157.632	102257.76	7.4	.1351	713.54
8	2.2580	11922.240	154244.64	5.6	.1759	928.96
10	2.8144	14860.032	219880.32	4.5	.2203	1158.67
12	3.3790	17841.120	268836.48	3.8	.2633	1391.15
15	4.1887	22116.336	342962.40	3.0	.3265	1723.25
18	4.7124	24881.472	342962.40	2.7	.3672	1939.83

Characteristics.—Usually an air-refined asphalt having the following characteristics is used for pipe-dipping work:

Penetration, 100 grams, five seconds, at 77° F. 10–40
Specific gravity . . . . Not less than 1.000
Ductility of a sample briquet 1 sq. cm. in cross-section, elongated at 5 cm. per min., at 77° F. 2–15
Melting point (ring and ball) . . . . 150°–250° F.
Solubility in carbon disulphide (CS_2) . . . . Not less than 99 per cent
Solubility in carbon tetrachloride (CCl_4) . . . . Not less than 98.5 per cent

Solubility in 86° Bé. naphtha . . . . Not more than 75 per cent
Evaporation from 50 grams in five hours at 325° F. in a dish 5.5 cm. (2⅛ inches) diameter and 3.5 cm. (1⅜ inches) deep . . . . Not more than 3 per cent
Penetration of sample after evaporation test . . . . Not less than 50 per cent
Flash test . . . . Not less than 350° F.

The penetration and melting point will depend upon the use to which the pipe is to be put. If it is to carry cold liquids, such as water, at atmospheric temperatures, 30 to 40 penetration can be used with a melting point not more than 150° F. Where asphalt is to be used on pipe-lines carrying hot liquids, 10 to 15 penetration is used, with a melting point of not less than 200° F. In many cases where hot liquids are to be carried it will be necessary to wrap burlap around the covering to prevent the gradual falling off of the coating.

Briquetting

Coal-dust is made readily available for commercial use by binding the dust with asphalt to make briquets. Asphalt is the best-known binder and is used at the rate of 4 to 6 per cent by weight of the coal-dust. It is usually heated and intimately mixed in a mixer after which the dough is pressed into molds, compressed into briquets and cooled.

The characteristics of an asphalt for this purpose are as follows:

Penetration, 100 grams, five seconds, at 77° F. 20–30
Specific gravity . . . . Not less than 1.000
Ductility of a sample briquet 1 sq. cm. in cross-section, elongated at 5 cm. per min., at 77° F. 2–15
Melting point (ring and ball) . . . . Not less than 160° F.
Solubility in carbon disulphide (CS_2) . . . . Not less than 99 per cent
Solubility in carbon tetrachloride (CCl_4) . . . . Not less than 98.5 per cent
Solubility in 86° Bé. naphtha . . . . Not more than 75 per cent
Evaporation from 50 grams in five hours at 325° F. in a dish 5.5 cm. (2⅛ inches) diameter and 3.5 cm. (1⅜ inches) deep . . . . Not more than 3 per cent
Penetration of sample after evaporation test . . . . Not less than 50 per cent
Flash test . . . . Not less than 350° F.

Manufacture of Paints

Asphalt, both of the natural and petroleum classes, is used as the basis of paints. When used for damp-proofing concrete or masonry walls, the paint usually consists of an asphalt and a solvent or thinner in such proportions that it may be easily brushed out. These proportions are 30 to 50 per cent of asphalt by volume.

When used as a roof paint the consistency is usually somewhat greater and there is 50 to 70 per cent of the base. In roofing paint, filler is sometimes added and the consistency is such that the paint has to be troweled on.

An asphalt for the above process usually has a penetration between 15 and 40 and a melting point between 150° and 220° F.

Asphalt paint used as a protection for metal consists of an asphalt base, a thinner, and a vegetable oil to produce a toughening effect upon the coating. The asphalts used for this purpose include a wide variety or combination of varieties of natural and petroleum asphalts many of which have given excellent service as a metal protection.

It has been suggested that an asphalt paint should be used to coat the steel in reinforced concrete. If a paint of the proper consistency is chosen, the natural bond between the concrete and the steel will be increased and the paint will prevent the corrosion due to dampness or electrolytic effects, and thus the subsequent cracking of reinforced concrete, due to the oxidation and increase in volume of the steel will be prevented.

In the manufacture of paints, the base is usually brought to a fluid by heating and the solvent added, a small amount at a time, while the paint is kept constantly stirred.

Electrical Insulation

Asphalt is used in combination with animal and vegetable oils, rubbers, sulphur, etc., in the manufacture of compounds for electrical insulation purposes, including the insulation of cotton-covered transmission wires, the field and armature windings of motors, dynamos, transformer coils, and magnetos, and for a sealing and insulating compound in conduits for lead-cable covered wires.

The electrical resistance of asphalt is high. The following table shows the resistance of California air-refined asphalt:

	Resistance in ohms per cubic centimeter
50° F.	$27,540\times10^9$
100° F.	139×10^9
150° F.	1.26×10^9

OIL SHALE

BY

DAVID ELIOT DAY[1]

GENERAL

Introduction.—This chapter is offered by the writer with a full realization of its limitations and incompleteness. It is difficult to discuss the technology of a subject when this technology is in a formative stage, when many of the fundamental data are lacking and when even the basic principles are the subject of controversy. Until careful investigators have given the subject the full treatment it deserves, a full and reliable presentation of the data on oil-shale engineering is impossible. In this chapter certain very general principles are given, together with a compilation of existing data. It will be noticed that no attempt has been made to discuss the details of the industry as it exists in other countries. This omission was made purposely as the value of the technology developed in other countries, as far as practical application to the problems to be encountered in the United States is concerned, is at least doubtful. Furthermore, the Scottish and French oil-shale industries have been thoroughly and excellently treated by other writers, so that a repetition here is neither necessary nor advisable. No history has been included, as history has no place in a reference handbook.

All analyses, physical and chemical data, and cost figures, where not original, were taken from sources of established accuracy, Government figures being used wherever possible. This course has certain disadvantages, as in the case of Colorado and Utah analyses. Many of the figures given under this head were taken from U. S. G. S. Bull. 641, "Oil Shales of Northwestern Colorado," by Dean E. Winchester, and it is rather certain that these figures, as a whole, are misleading both because oil-shale analysis was not fully understood when Mr. Winchester made his examination and because the figures given include many beds which were known to be lean in oil, but which were examined for scientific reasons. However, so much that was misleading in the other extreme has been published that an unavoidable understatement should only have the effect of a mild corrective.

[1] **Acknowledgments.**—In the preparation of this chapter the writer has drawn freely from various published sources of information. Under "Occurrence" particularly, the data were compiled almost entirely from the literature on the subject. Bulletins of the United States Geological Survey, the United States Bureau of Mines, and the various State Surveys were particularly valuable in preparing the tables of analyses and technical data. In addition to the sources to which reference is made in the text, the author is particularly indebted to Dr. David T. Day, who edited the entire article and provided unpublished data of great value, and to Florence P. Day and A. H. Heller, who were responsible for portions of the article, and whose perseverance in consulting references on the occurrence of oil shale made possible the wide range of this section.

Definition of oil shale.—Any hard-and-fast definition of oil shale will be subject to geographical limitations. Thus the definition of an observer whose experience has been limited to Scotch shale will not apply to Colorado shale. Nor will a close definition of Colorado shale apply to Tennessee shale, Nevada shale, or California shale. Mr. Cunningham Craig has defined oil shale as "An argillaceous or shaly deposit from which oil may be obtained by distillation, but not by trituration or treatment with solvents." This definition is not technically correct even for Scotch shales, as recent experiments have shown that some oil may be extracted from practically every shale by the use of suitable solvents under certain physical conditions. As discussed in this chapter the term oil shale will include all rocks which, because of their chemical or physical properties or owing to the conditions of sedimentation under which they were formed, may be petrographically classed as shale and which will yield volatile hydrocarbons by some method of treatment. This definition is purposely broad and inclusive, covering the entire range of commercially important shales from those closely related to cannel coals and lignites which must be destructively distilled, to those of the California type from which the entire oil content may be obtained by treatment with solvents.

Classification

By this definition, the materials from which oil may be produced by destructive distillation or by the action of solvents may be divided under three heads:

(1) True coals, cannels, and lignites.
(2) Tar sands, oil sands, and similar substances.
(3) Oil shales.

Oil shales may be further classified by the following criteria:

(1) Chemical composition.
  a. Ratio of volatiles to fixed carbon to ash.
  b. Ratio of total bituminous content to amount recoverable by solvents.
  c. Analysis of recovered hydrocarbons.
  d. Analysis of mineral ash.
(2) Physical properties—Color, hardness, specific gravity, cleavage, etc.
(3) Geology.
  a. Conditions governing formation.
  b. Geologic age.

These classifications are merely suggested, for by no means enough research work has been done to provide the necessary data for classifying in detail. The classification based on the ratio of volatiles to fixed carbon to ash has been enlarged on as shown in the accompanying triangular diagram (Fig. 1). From this it is seen that the fixed-carbon content of oil shale is fairly constant and relatively low, never more than 20 per cent, and that the variation from low-grade to high-grade shale is due to increase of volatiles at the expense of ash. Coal, on the other hand, has a uniformly low and constant ash content, the variation from anthracite to cannel coal being due to an increase in volatiles at the expense of fixed carbon. The only point of contact between the coals and shales is at the high-volatile, low-ash, and low-fixed-carbon corner of the triangle, where the very rich torbanite, albertite, and similar extreme occurrences of shale (really not true shales in a geological sense) have similar characteristics to cannel coals. The low-fixed-carbon content, therefore, seems to be the outstanding point of difference between the average shale and coal.

A very interesting classification of oil shale is possible by determining the ratio of the total bituminous content to the amount recoverable by solvents. Sufficient work has been done on this subject to show that some oil can be recovered from any shale by the use of various solvents under various conditions of temperature and pressure. The total bituminous content may be recovered by treating California shales with solvents, 25 to 30 per cent from Colorado and Utah shale, as high as 60 per cent from

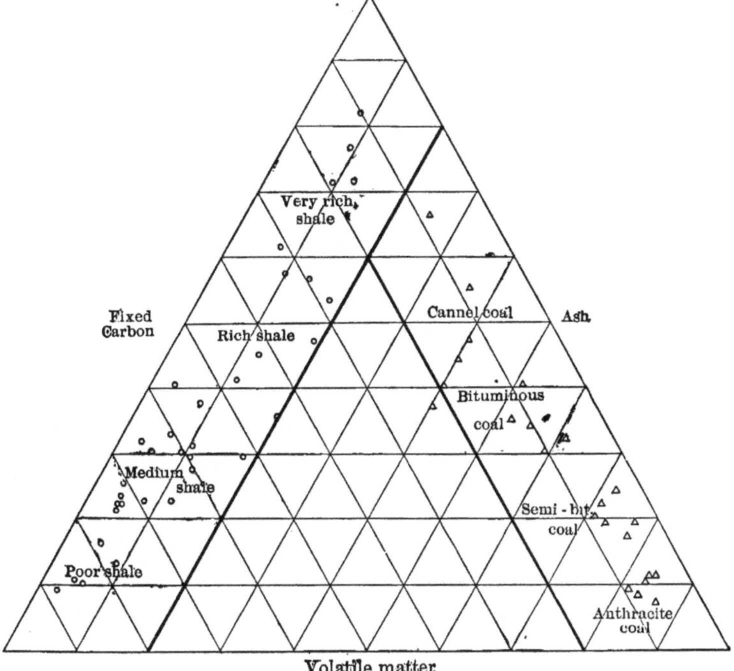

FIG. 1.—Triangular diagram; volatile to free carbon to ash.

Nevada shale, and a fairly low per cent from shale from the eastern states. The rôle that capillarity plays in retaining the hydrocarbons in the fine pores of the shale is very important and has led many investigators to underestimate the amount of oil which may be recovered by solvents.

Analysis of the recovered hydrocarbons and of the mineral ash does not offer a very definite basis for classification, nor do the physical properties of the shale. Data on the origin of the shale and the conditions affecting its deposition, would offer a very excellent basis for a comprehensive classification, but so far such data are lacking. A classification of oil shale on the basis of geologic age is given in the following table:

GEOGRAPHICAL AND GEOLOGICAL CLASSIFICATION OF OIL SHALE

Miocene—California.

Oligocene, Eocene —Colorado, Utah, Wyoming, Brazil, Spain, Serbia, Italy.

Cretaceous—Saskatchewan, Brazil, England, South Africa.

Jurassic—England, Scotland.

Triassic—South Africa, Germany.

Permian—Montana, France, New South Wales.

Pennsylvanian—Illinois, Pennsylvania.

Mississippian—New Brunswick, England, Scotland, Tasmania.

Devonian—Indiana, Ohio, Kentucky, Tennessee, New Brunswick.

Origin of oil shale.—Little is definitely known concerning the origin of oil shale and the nature of the oil-producing substances in the shale. There are those who contend that the oil is indigenous to the shale itself and is present not as liquid hydrocarbons nor as the solid residue from some previously existing liquid hydrocarbons but as bituminous or carbonaceous matter forming an integral part of the shale as originally deposited; others advance the theory that the oil is not indigenous to the shale itself but has migrated from some other source, has saturated the pores of the shale, and has evaporated to leave a comparatively insoluble residue.

The facts in the case, as so far determined, bear out both hypotheses and suggest that either may be correct in individual instances, or that any combination of the two may be present. Thus on the one hand there are the coals which contain large volatile and ash percentages with little fixed carbon. These coals, which may grade down to a true oil shales by the previous definition, contain little if any oil in a soluble state and are undoubtedly the original source of the oil produced from them. On the other hand, the diatomaceous shales of California carry all their oil in a completely soluble form, and their degree of productiveness is governed by the ordinary factors of oil accumulation, proving that the oil has migrated, if not from some other formation, certainly within the shale itself. A mean between these two types of shale is that found in Utah and Colorado. Experiments have definitely proved that a large portion of the oil recoverable from this shale exists as a mixture of difficultly soluble hydrocarbons, in all probability the evaporated remains of a once liquid oil. On the other hand, there are included in the rich shales of this region, porous sandstone lentils which are often entirely free from any traces of oil. Microscopic examination of thin sections of rich shale from this area has determined the presence of sufficient organic residue, such as the remains of algæ, spores, pollen, etc., to support the theory that the oil was formed by the destructive distillation of this organic matter. It is probable then that the source of this oil was the organic matter in the shale itself, that destructive distillation has gone far enough to convert at least part of this organic matter into soluble hydrocarbons, and that these soluble hydrocarbons have been retained by capillarity in the pores of the shale without any general migration.

Artificial oil shale has been made by both methods. In one set of experiments, a mixture of fuller's earth and spores was distilled and yielded oil very similar to that produced from Scotch oil shales. In another experiment, diatomaceous shale from California was soaked in the heavy oil produced in that state, and the resulting shale was identical with that found in the oil-bearing localities.

A comparison of the ratio of volatiles to ash to fixed carbon rather supports the soluble hydrocarbon theory. If oil shale is considered to contain no soluble hydrocarbons, but yields oil only by the destructive distillation of organic or bituminous matter, its origin must be very similar to that of coal, and the ratio of volatiles to fixed carbon to ash ought to be quite similar to the ratio in various types of coal. To amplify this idea, a chart (Fig. 2) was drawn, on which were plotted points representing the ratio: per cent volatiles to per cent ash. The points on this chart representing shales of various degrees of richness, from various parts of the world, indicate a fairly definite line (*AB* on the chart). Points representing coals also indicate a line (*CD* on the chart) which has absolutely no relation to line *AB*. A third line (*EF* on the chart) was drawn to represent the values obtained if a porous shale or fuller's earth was saturated with

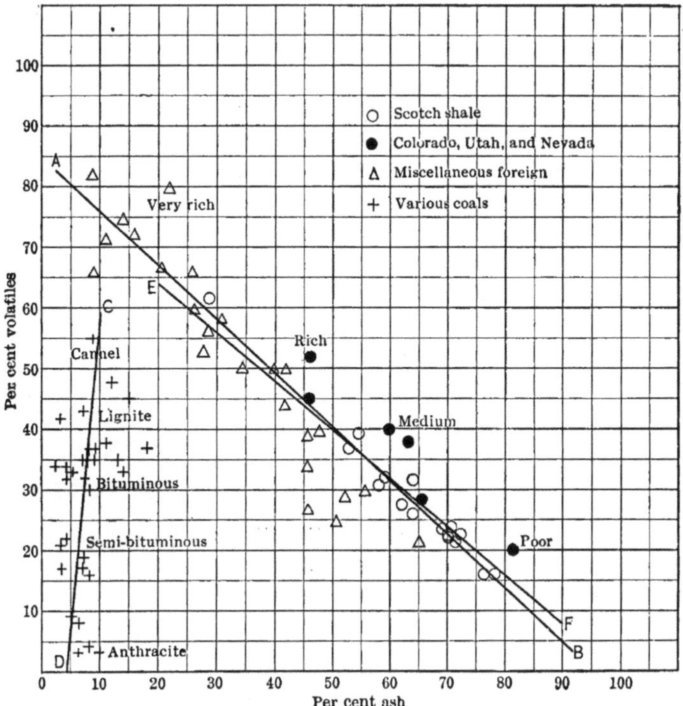

Fig. 2.—Relation of per cent volatile to per cent ash.

different amounts of oil and then subjected to the same process of distillation employed in obtaining the results for natural shales and coals. This line, as seen, nearly coincides with the line *AB* drawn for natural shales. This evidence favors the theory of origin of oil shale by saturation rather than the coal-like origin necessary to substantiate the destructive distillation theory.

OCCURRENCE

Some idea of the occurrence and importance of oil shale in various parts of the United States and in foreign countries may be had from the following tables, which are mainly abstracts from technical papers which have appeared in the various mining and petroleum periodicals.

OCCURRENCE OF OIL SHALE IN THE UNITED STATES

California

Location	Thickness sampled, feet	Oil, gallons per ton	Gravity, Baumé	Gas, cubic feet per ton	Ammonia sulphate, pounds per ton
Santa Barbara Co. Casmalia	27	25.5	19.7	1500	....
Casmalia	11	23.8	21.0	1300	....
Casmalia	31	25.8	21.5	1300	....
Casmalia	55	27.4	22.0	1400	....
Graciosa Ridge	....	52	21.5	....	....
Casmalia	....	55	....	1425	....
*Sisquoc Ranch	45	36	25.0	1700	....
*Carpenteria	14	42	23.6	1650	....
Kern County McKittrick	....	62	....	....	....
Riverside Co. Elsinore	5	65	36.0	....	....

REMARKS

Counties—Santa Barbara, Ventura, Los Angeles, Kern and San Bernardino.
Age—Miocene. Formation—Monterey.
Area in State—100 to 150 square miles.
Thickness of formation—100 to 5000 feet.
Analyses and samples by Turner, Geo. W., Day, David E., and Day, D. T.
Typical California shale is a porous diatomaceous shale more or less saturated with petroleum. Deposits are generally related to producing fields and are found near the crests of large anticlines. The Elsinore deposit is an exception to this rule, the shale being quite similar to that from Elko, Nevada.

* Saturated sands.

Colorado

Location	Thickness sampled, feet, inches	Oil, gallons per ton	Gravity, Baumé	Gas, cubic feet per ton	Ammonia sulphate, pounds per ton
Conn Creek, Sec. 1, T.7S., R.98W.	1 4	61–68	17–24	....	22.5 (1)
Kimball Creek, Sec. 5, T.7S., R.100W.	6 0	26–31	20.5	....	.... (1)
Parachute Creek, Sec. 29, T.5S., R.95W.	5 10	20	21	....	.... (1)
Parachute Creek, Sec. 36, T.5S, R.96W.	6 4	18	27.1	....	8.7 (2)
Roan Creek, Sec. 1, T.7S., R.98W.	....	65.3	23 6	4549	7.0 (2)
Book Cliffs, Sec. 6, T.6S., R.94W.	10 0	40.6	29.2	1916	4.3 (2)
Book Cliffs, Sec. 22, T.6S., R.95W.	8 0	20.7	33.1	....	.... (2)
Dry Fork Creek, Sec. 7, T.7S., R.100W.	5 0	38	27.5	....	7.4 (2)

Counties—Garfield, Rio Blanco, Moffat, and Mesa.
Age—Eocene. Formation—Green River.
Area in State—2500 square miles.

General averages, U. S. G. S. samples of Colorado shale:
Samples examined: 65.
Average oil content: 19.05 U. S. gallons per short ton.
Average gas yield: 2705 cubic feet per short ton (20 samples).
Average yield ammonia sulphate: 7.3 pounds per short ton (36 samples).
Percentage of samples examined which were average or above: 37 per cent.
Average thickness of beds represented by these samples: 4 feet 6 inches.
Average oil content of these commercial beds: 29.8 U. S. gallons per ton.

Indiana

Location	Thickness sampled, feet	Oil, gallons per ton	Gravity, Baumé	Gas, cubic feet per ton	Ammonia sulphate, pounds per ton
New Albany	....	9.1	....	958	....
New Albany	....	11.2	....	2043	0.15 (3)
New Albany	....	11.9	....	1197	0.0 (3)
Boonville	....	14.0	....	2522	0.97 (3)
Boonville	....	15.4	....	2922	0.61 (3)
Miscellaneous	....	10.0	....	997	28.5
Miscellaneous	....	12.75	....	912	22.2
Miscellaneous	....	18.00	....	1836	20.4
Miscellaneous	....	20.00	....	....	....
Miscellaneous	....	30.00	....	....	....

Counties—Floyd, Clark, Scott, Jefferson, Jennings, and Jackson.
Age—Devonian. Formation—New Albany.
Area in State—500 square miles.
General averages, U. S. G. S. and Indiana G. S. samples:
Samples examined: U. S. G. S., 8; Indiana G. S., 14.
Average oil content: U. S. G. S., 10.0 gallons per ton; I. G. S., 14 6 gallons per ton.
Average yield gas: U. S. G. S. 1400 cubic feet per ton; I. G. S., 1076 cubic feet per ton.
Average yield ammonia sulphate: Indiana G. S. 27.3 pounds per ton.
Average thickness of formation: 70 to 140 feet.

[1] Woodruff, E. A., & Day, D. T., U. S. G. S., Bull. 581.
[2] Winchester, Dean E., U. S. G. S., Bull. 641.
[3] Determinations by Ashley, G. H., U. S. G. S. Bull. 641.
Miscellaneous samples, representing average oil content of Albany shale throughout Indiana

Illinois

Location	Thickness sampled, inches	Oil, gallons per ton	Gravity, Baumé	Gas, cubic feet per ton	Ammonia sulphate, pounds per ton
Sangamon Co.	....	11.9	....	2186	.65
Gallatin Co.	....	16	....	....	3.44
Johnson Co.	....	36.6	....	6389 (*a*)	
Johnson Co.	....	45.1	....	5860 (*a*)	
(*b*) McLean Co.	6	....	....	....	....
(*b*) LaSalle Co.	10	....	....	....	....
(*b*) Illinois Bituminous	....	2.5	....	....	9
(*c*) Average	....	14	....	....	....
(*d*) Average	....	25	....	8665	30

REMARKS

Counties—Sangamon and Gallatin.
Age—Pennsylvania series.

Ashley, G. H.: U. S. G. S., Bull. 641.

(*a*) Ozark Shale Analysis by Illinois State Survey.
(*b*) Cannel coal by G H. Ashley, U. S. G. S., Bull. 659.
(*c*) 2 samples.
(*d*) 7 samples of coal.

Kentucky

Location	Thickness	Oil	Gravity	Gas	Ammonia sulphate
Louisville	feet 8	11.2	....	1016	....
Crittenden Co.	....	14.0	....	....	2.58
Estell Co.	....	23-30	....	....	40.2
Powell Co.	....	22.7	....	4167	97.8
Rockcastle Co.	....	8.0	....	3000	....
Taylor Co.	....	27.75	....	....	....
Fleming Co.	....	21.5	....	....	....
* Breckenridge	....	13.0	....	....	....
Garrard	....	21.0	....	....	....

Counties—Crittenden, Louisville, Caseyville, Clay City.
Age—Devonian. Formation—Chattanooga shale.

Ashley, G. H.: U. S. G. S., Bull. 641.

* Cannel coal by G. H. Ashley, U. S. G. S., Bull. 659.

Montana

Location	Thickness, feet	Thickness, inches	Oil	Gravity	Gas	Phosphate per cent
Daly spur O.S.L. R.R., Sec. 2, T.9S., R.10W.	4	8	14	....	....	3.26
Daly spur O.S.L. R.R., Sec. 2, T.9S., R.10W.	4	7	0	....	....	19.41
Daly spur O.S.L. R.R., Sec. 2, T.9S., R.10W.	14	0	17	....	....	1.72
Daly spur O.S.L. R.R., Sec. 2, T.9S., R.10W.	10	0	13	....	....	....
Small Horn Canyon, Sec. 14, T.9S., R.9W.	5	6	21	....	....	0 63
Small Horn Canyon, Sec. 14, T.9S., R.9W.	17	8	17	....	....	....
Dry Canyon near Dell, Sec. 12, T.13S., R.10W.	8	0	17	....	....	5.94
Little Sheep Creek, Sec. 4, T.15S., R.9W.	14	0	16	....	....	5.57
Little Sheep Creek, Sec. 4, T.15S., R.9W.	4	0	12	....	....	10.28
Little Sheep Creek, Sec. 4, T.15S., R.9W.	2	6	12	....	....	10.28

Counties—Beaverhead.
Age—Permian. Formation—Phosphoria. Steep dip, faults.
Area in State—500 square miles.

General Averages, U. S. G. S. samples of Montana shale:
Samples examined: 15.
Average oil content: 12.25 S. U. gallons per short ton.
Average yield of phosphoric acid in oil bearing beds: 4 per cent.
Average yield of phospho ic acid in related beds: 15.77 per cent.
Percentage of samples examined which were average or over: 60 per cent.
Average thickness of beds represented by these samples: 10 feet 3 inches.
Average oil content of these commercial beds: 16 gallons per ton.

OCCURRENCE OF OIL SHALE IN THE UNITED STATES

Nevada

Location	Thickness sampled, feet, inches		Oil, gallons per ton	Gravity, Baumé	Gas, cubic feet per ton	Ammonia sulphate, pounds per ton	REMARKS
Elko at Catlin plant	5	0	50	....	....	....	County—Elko.
Elko, Nev.	..	..	86 8	28	3891	6.0	* Determination by Day, David T.
Elko, Nev.	2	4	50.0	35.8	....	4.5–8.4	† Determination by Day and Winchester, U. S. G. S., Bull. 641.
*Elko, Southern Pacific	1	0	80	30	....	....	‡ Determination by Winchester, Dean E., U. S. G. S., Bull. 641.
†Elko, Southern Pacific	6	0	30	28	....	....	(a) Surface shale.
‡Elko, Southern Pacific	4	0	30	25	....	....	(a) Average.
Elko, Southern Pacific	12	0	15	28	....	....	(a) Paraffin wax, 26 per cent.
Elko near Carlin	7	0	35	....	....	....	(a) Large bed.
Elko, Berner basin	8	0	20	35	....	....	
North of Elko	10	0	30	30	....	....	

Ohio

Location	Thickness sampled		Oil, gallons per ton	Gravity, Baumé	Gas, cubic feet per ton	Ammonia sulphate, pounds per ton	REMARKS
(a) Glen Mary	No. 1 5 ft. No. 2 3 ft.		7.7 5.6	....	1200 958	....	Location sampled—Glen Mary, Columbus, Gahanna.
(a) Columbus	....		4.0	....	....	....	Age—Devonian. Formation—Ohio (Chattanooga or New Albany) shale.
(a) Gahanna	feet, 6	inches 8	11.0	....	....	....	(a) Ashley, G. H.: Black shales of the Eastern States, U. S. G. S., Bull. 641.
(b) Average	....		7.1	....	....	....	(b) 4 samples.
(c) Mahoning Co.	5	0	....	....	....	....	(c) Cannel coal by G. H. Ashley, U. S. G. S. Bull., 659.
(c) Licking Co.	....		....	....	10,080	....	(d) Coal, Geol. Sur. of Ohio, Bull. 20, 4th Series, 112.
(c) Wayne Co., Sterling	....		....	....	11,782	....	
(c) Coshocton	....		74	....	....	....	
(c) Hocking Valley	....		....	....	8,960	....	
(d) Jackson Co.	2	0	....	....	6,200	....	
(d) Jackson Co.	2	0	....	....	5,960	....	

Pennsylvania

Location	Thickness sampled	Oil, gallons per ton	Gravity, Baumé	Gas, cubic feet per ton	Ammonia sulphate, pounds per ton	REMARKS
Cannelton	Feet 3–5	27.3	....	2905	....	Location sampled—Cannelton, Clay Township; Queen Junction, Butler, Zelienople Quadrangle
Clay Township	1	45.0	....	....	3.72	Age—Pennsylvania series. (Associated with cannel coal.)
Queen Junction	2	43.0	....	....	4.99	Area in State—Several thousand acres.
Clay Township	1	34.0	....	....	5.21	Ashley, G. H.: U. S. G. S., Bull. 641.
Butler	1	24.0	....	....	9.43	* Cannel coal, G. H. Ashley, U. S. G. S. Bull. 659.
Zelienople Quadrangle	6	18.0	....	....	2.53	
* Armstrong Co.	9	33.6	....	5029	5.37	
* Beaver Co.	6	37.3	....	5268	2 24	
* Indiana Co.	8	20.3	....	4790	5.57	
* Center Co.	....	10.7	....	4071	1.80	

OCCURRENCE OF OIL SHALE IN THE UNITED STATES

Tennessee

Location	Thickness sampled, feet	Oil, gallons per ton	Gravity, Baumé	Gas, cubic feet per ton	Ammonia sulphate, pounds per ton
(a) Rockwood No. 1	2	7.7	....	1488	....
No. 2	5	5.5	....	1128	....
(a) Baker's Station No. 1	5	9.1	....	1916	....
No. 2	5	9.1	....	1485	....
No. 3	5	6.3	....	1557	....
(a) Newcomb	feet, inches 0 6	21	....	....	....
(b) Dickson Co.	45 0	12–18	....	1000–2000	....
(c) Campbell Co.	3 0	....	....	....	....
(c) Newsom	....	4.2	....	835	.21
(d) Average	....	5.2	....	....	....

REMARKS

Location sampled—Cumberland Gap, Rockwood, Baker's Station, Newson, Newcomb, Dickson.
Age—Devonian. Formation—Chattanooga shale.

(a) Ashley, G.H.: U. S. G. S., Bull. 641.
(b) Day, David E.
(c) Cannel coal—G. H. Ashley, U. S. G. S., Bull. 659.
(d) 13 samples.

Utah

Location	Thickness sampled (feet, inches)	Oil, gallons per ton	Gravity, Baumé	Gas, cubic feet per ton	Ammonia sulphate, pounds per ton
Soldier's Summit	6 3	16.8	26 6	2864	3.5
Soldier's Summit (2 miles north)	6 11	24	24.7	....	12.21
Juab	2 9	11.9	25.6	1198	2.5
Watson, Sec. 27, T.11S., R.25E.	6 10	37	30.42	....	7.8
Hell's Hole Canyon, Sec. 15, T.10S., R.25E.	4 8	45	26	....	9.2
Dragon, Sec. 4, T.12S., R.25E.	4 4	30	23.2	....	7.0
Bitter Creek, 12 miles S.W. Dragon	5 9	43	27.1	....	5.2
Willow Creek, 28 miles W. Dragon	5 2	15	28.4	....	2.66
Soldier's Summit (6 miles north)	12 6	19	28.5	....	4.4
Colton (4 miles east)	5 0	11	27.5	....	3.6

REMARKS

Counties—Carbon, Duchesne, Uinta and Utah.
Age—Eocene. Formation—Green River. Structure—Synclinal basin.
Area in State—3500 square miles.
Analyses from U. S. G. S., Bull. 691B: Winchester, Dean E.
General averages, U. S. G. S. samples of Utah shale:
Samples examined: 94.
Average oil content: 23.8 gallons per short ton.
Average gas yield:
Average yield ammonia sulphate: 4.9 pounds per ton.
Percentage of samples examined which were average or above: 39 per cent.
Avera-e thickness of beds represented by these samples: 4 feet 3 inches.
Average oil content of these commercial beds: 36 gallons per ton.

OCCURRENCE OF OIL SHALE IN THE UNITED STATES

Wyoming

Location	Thickness sampled, feet, inches		Oil, gallons per ton	Gravity, Baumé	Gas, cubic feet per ton	Ammonia sulphate, pounds per ton
Red Desert Basin, Sec. 9, T.13N., R.99W.	5	0	30	30.79	....	3.94
On Green River, Sec. 27, T.13N., R.108W.	3	3	19	....	....	9.32
On Green River, Sec. 36, T.16N., R.108W.	6	0	13	26.86	....	5.06
S. of town of Green River, Sec. 27, T.16N., R.106W.	8	1	19	25.50	....	8.68
At town of Green River, Sec. 8, T18N., R107W.	1	8	29	22.47	....	11.71
Fossil Buttes, Sec. 5, T.21N., R.107W.	2	0	50	27.49	....	2.0
*Green River, Sec. 25, T.17N., R.107W.	35	0	28	....	1200	13.8
*Green River, Sec. 25, same Twp.	26	0	34	....	1200	16.2
*Sec. 27, T. and R. as above	30	0	33	....	1420	14.8
*Green River, Sec. 22, T.16N., R.107W.	3	0	75	....	2600	29.9

REMARKS

Counties—Lincoln, Uinta and Sweetwater.

Age—Eocene. Formation—Green River.

Area in State—3000 to 7000 square miles.

General averages, U. S. G. S. samples of Colorado shale:

Samples examined: 41. Analysis from U. S. G. S., Bull 641.

Average oil content: 14.5 U. S. gallons per short ton.

Average gas yield:

Average yield ammonia sulphate: 6.2 pounds per ton.

Percentage of samples examined which were average or above: 37 per cent.

Average thickness of beds represented by these samples: 4 feet 2 inches.

Average oil content of these commercial beds: 24.0 gallons per ton.

General averages, private samples of Wyoming shale: (Samples marked by * taken and examined by Erb Wuensch for American Shale and Petroleum Co.)

Samples examined: 15.

Average oil content: 36 U. S. gallons per short ton.

Average thickness of beds: 17 feet.

Average gas yield: 1350 cubic feet per short ton.

Average ammonia sulphate yield, dry: 6.3 pounds per short ton.

Average ammonia sulphate yield, with steam: 16 pounds per ton.

OCCURRENCE OF OIL SHALE IN CANADA

British Columbia

Location	Thickness sampled, feet, inches	Oil, gallons per ton	Gravity, Baumé	Gas, cubic feet per ton	Ammonia sulphate, pounds per ton	REMARKS
						County—Ashcroft.
Ashcroft shales	....	50 Some 65	....	....	....	

Newfoundland

Location	Thickness sampled, feet, inches	Oil, gallons per ton	Gravity, Baumé	Gas, cubic feet per ton	Ammonia sulphate, pounds per ton	REMARKS
Average †	....	40	....	....	....	Location—*North side Notre Dame Bay, Cape Rouge Peninsula, vicinity of Deer and Grand Lakes.
						Area in province—150 square miles.
						Fixed carbon in shale, 35 per cent.
						Ash in shale, 29 per cent.

* Abraham, Asphalt and allied substances, p. 162. † Petroleum Review, Vol. 33, p. 209.

New Brunswick

Location	Thickness sampled, feet, inches	Oil, gallons per ton	Gravity, Baumé	Gas, cubic feet per ton	Ammonia sulphate, pounds per ton	REMARKS
Albert Mines 1	6 6	58	26.5	....	83	Location—Albert County and Westmoreland County at Albert Mines, Taylorville and Baltimore—Hillsboro and Coverdale, along Petitcodiac River, between Memracook and College Bridge Station.
2	3 6	47	26.9	....	60	
3	5 0	54.6	27.0	....	48	
Baltimore 1	6 0	65	26.6	....	110	
2	5 0	59	26.9	....	67	
3	.. ..	61	23.7	....	111	Age—Devonian. Formation—(Albert shale) and Lower Carboniferous (Horton).
Taylorville 1	2 2	51	25.4	....	93	
2	3 0	58	23.7	....	98	
3	5 0	61	26.5	....	110	
Hayward Brook	.. ..	35	21	....	75	Ells, R. W.: Report on the bituminous shales of New Brunswick and Nova Scotia.
Albert County	.. ..	90–120	....	....	80–90	
†Average	.. ..	40.09	....	....	79.94	† Gray—The coal fields and coal industry of Eastern Canada. Canada Dept. of Mines, Mines Branch, Bull. 14, 1917.
Posser Brook	.. ..	35	26.4	....	75	
Miscellaneous	.. ..	14.6	34.6	....	80	
Irvings	.. ..	48	22.1	....	71	

Nova Scotia

Location	Thickness sampled, feet, inches	Oil, gallons per ton	Gravity, Baumé	Gas, cubic feet per ton	Ammonia sulphate, pounds per ton	REMARKS
Antigonish Co., Sawmill Brook, surface of banks	100	13.2	25	....	22	Counties—Antigonish, Pictou. Age—Horton series—Carboniferous (Possibly Devonian).
Antigonish Co., bed of Sawmill Brook	100	12	....	....	38	* Abraham: Asphalt and Allied Substances, p. 220.
Antigonish Co., Sawmill Brook, plain shales	100	12	....	....	34	Analyses by Ells, R. W.
McLellans Brook near New Glasgow, Pictou Co.	....	50.5	....	....	41	
McLellans Brook—Black's Old Mill Site, Pictou Co.	....	17.5	25.5	....	35	
Marsh Brook, Pictou Co.	....	16.8	....	....	....	
Woodbrow Sta.—1 mile west. Black shale	10	17.1	....	....	....	
6 miles S. Sydney	10 inches (exposed)	30	24	3500	0.72% nitrogen	
Macadam Lake	feet 2–10	15–20	....	....	....	
Arcadian shale *	.. ..	4–23	....	....	9–40	

OCCURRENCE OF OIL SHALE IN CANADA

ONTARIO

Location	Thickness sampled, feet	Oil, gallons per ton	Gravity, Baumé	Gas, cubic feet per ton	Ammonia sulphate, pounds per ton	REMARKS
	10-35	12	....	....	....	Age—Ohio series.

QUEBEC

Location	Thickness sampled, feet	Oil, gallons per ton	Gravity, Baumé	Gas, cubic feet per ton	Ammonia sulphate, pounds per ton	REMARKS
Gaspé (3 samples)	....	38.6	18	....	47.3	Location—*Along York and Dartmouth Rivers, along St. Lawrence from Quebec to Montreal, along Ontario Peninsula near Port Hope, Lake Huron shore near Georgian Bay, near Collingwood. Age†—At Gasp'—Devonian.

* Abraham: Asphalt and allied substances, p. 162. † Petroleum Review, Vol. 33, p. 209.

SASKATCHEWAN

Location	Thickness sampled, feet	Oil, gallons per ton	Gravity, Baumé	Gas, cubic feet per ton	Ammonia sulphate, pounds per ton	REMARKS
Chance sample at Carrot River	150	7	....	....	22.5	Location—Pasquia Hills near Carrot River—Uabi River. Age—Niobrara Cretaceous.
Random sample	35	48	....	....	33.5	Mansfield: Petroleum Review, Vol. 34, p. 159, 1916.

OCCURRENCE OF OIL SHALE IN SOUTH AMERICA

ARGENTINA

Location	Thickness sampled, feet	Oil, gallons per ton	Gravity, Baumé	Gas, cubic feet per ton	Ammonia sulphate, pounds per ton	REMARKS
Argentina	100	80	....	....	....	Location—Area at Rio Grande, 240 miles southwest of Alvear. Area—1½ miles by 20 miles.

BRAZIL

Location	Thickness sampled, feet	Oil, gallons per ton	Gravity, Baumé	Gas, cubic feet per ton	Ammonia sulphate, pounds per ton	REMARKS
Barreira do Boqueirao *	2-4 meters	....	....	....	....	Location—*States of Pernambuco, Parahyba, Rio Grande do Norte, Espirito Santo, Sergipe at Cape Santo Agostinho. Rio Formosa, Tamandaré, Abren de Una, Maragogy, Sao Paulo, Rio Grande do Sul, Maranhao.
Raicho Doce shales *	....	44.73	....	90 60 candle gas	....	Age—* Tertiary—Eocene. Synclinal.
Maranhao ‡	....	100	....	.	....	† N.E. coast of Brazil—Cretaceous. (Permian?)
Bahia, Marahu ‡	....	60	....	....	....	
Marahu ‡	feet 9-12	100	30-29	....	....	
Sa Pa lo ‡	....	35	....	....	....	* Branner: Trans. Am. Inst. Min Eng., Vol. 30, p. 537, 1901.
Alagvas	....	44.73	....			† Mansfield: Petroleum Review, Vol. 39, p. 159, 1916.
Bahia	....	113.00	29-30			‡ Williams, H. E.: Oil shales and petroleum prospects in Brazil.
Sao Paulo	....	27.00	....			

OCCURRENCE OF OIL SHALE IN CONTINENTAL EUROPE

Bulgaria

Location	Thickness sampled, feet	Oil, gallons per ton	Gravity, Baumé	Gas, cubic feet per ton	Ammonia sulphate, pounds per ton	REMARKS
Deposits, average	8	32	....	....	....	Location—Brjeznik, Radoriew, Kazanlik, Sirbinova. Poportzi. Tonnage—30,000 tons available, yield 32 gallons.
Sirbinova	80	....	....	....	....	
Brjeznik	1 0	....	....	....	....	

Esthonia

Location	Thickness sampled, feet	Oil, gallons per ton	Gravity, Baumé	Gas, cubic feet per ton	Ammonia sulphate, pounds per ton	REMARKS
* Northern	....	41	....	5,300	....	Location—Along Gulf of Finland. On railroad between Johwi and Rakwere. Tonnage—1,500,000,000. Volatile matter, 70.06 per cent; coke, 11.08 per cent.; ash, 18.86 per cent.

* Pet Times, April 17, 1920, 412.

France

Location	Thickness sampled, feet	Oil, gallons per ton	Gravity, Baumé	Gas, cubic feet per ton	Ammonia sulphate, pounds per ton	REMARKS
Autun	5	1`	....	....	25–30	Location—*Autun, Buxiere les Mines, the Midi, the Central Plateau and Vendée. In the Riviera. Age—† Permo-carboniferous. Area—* 45,000 acres.
Buxière les Mines	8	16–22	23	....	10–15	
Province of Var	3–11	60–70	....	....	....	
Buxière les Mines	....	20	21.5	....	17	
Buxière les Mines	....	11	20.3	....	13	* Scheithauer: Shale Oil and Tars, p. 9.
Buxière les Mines	....	10	18.9	....	12	† Mansfield: Petroleum Review, Vol. 34, p. 159, 1916.
Buxière les Mines	....	16	19.5	....	15	
Autun	....	21	24.7	....	....	
Bourbon Saint-Hilaire	....	20	24.7	....	....	
Faymoreau	....	19	31.5	....	....	

Germany

Location	Thickness sampled, feet	Oil, gallons per ton	Gravity, Baumé	Gas, cubic feet per ton	Ammonia sulphate, pounds per ton	REMARKS
Wurtemberg	....	18–25	....	3200–5000	....	Location — * Rhenish Prussia, Saxony at Halle, Zeitz, Weissenfels, Eisleben, Aschersleben—Hesse at Messel—Bavaria, Wurtemberg, Westphalia. Age—† Liassic a e at Wurtemberg. Messel "shalc" probably lignite.
Messel	....	32	....	1900	71 gals. ammo. sulph. water	
Wu:temberg	....	19	....	2984	....	* Abraham: Asphalt and allied substances, p. 164 † Mansfield: Petroleum Review, Vol. 34, p. 159, 1916.

OCCURRENCE OF OIL SHALE IN GREAT BRITAIN

England

Location	Thickness sampled, feet	inches	Oil, gallons per ton	Gravity, Baumé	Gas, cubic feet per ton	Ammonia sulphate, pounds per ton	REMARKS
b Norfolk Co., Smith series	6	0	50	....	....	....	Location—*a* County of Dorset at Kimmeridge, Ringstead Bay, Osmington Mills, extending to Portesham and Abbotsbury—North end of Island of Portland, Portland Harbor, near Shaftesbury; County of Norfolk near King's Lynn; County of Lincoln at Humber, near Caister; County of York near Filey; Counties of Sussex near Battle; Counties of Somerset, Wilts, Gloucester, Berks, Oxford, Buckingham, Bedford, Cambridge, Essex, Kent, Hereford, Suffolk, and Flint. Age—*a* Carboniferous (County of Flint). Jurassic and Carboniferous systems—Kimmeridge clay of Upper Oolites—Faults Anticline. (*a*) Mansfield: Petroleum Review, Vol. 34, pp. 159, 179, 1916. (*b*) Forbes-Leslie: Petroleum Review, Vol. 35, pp. 327, 348, 367, 1916. (*c*) Sir A. Strahan: Imperial Geol. Survey. (*d*) Cannel coal, G. H. Ashley, U. S. G. S., Bull. 659.
Puny drains series	7	0	50–51	17–18	25,000	66	
b On English Oil Fields, Ltd., property	8	0	39.6	....	....	22.6	
Dorset Co., (*c*) Kimmeridge shale, Blackstone bed	No. 1 2	6	40.6	13	....	22.7	
	No. 2 2	6.5	38.4	....	....	32.4	
(*c*) Kimmeridge shale	No. 1 2	0	25.5	10	....	28.5	
	No. 2 2	0	28.1	....	....	14.6	
	No. 3 1	9	29.9	....	....	25.3	
(*d*) Derbyshire	..	..	82	....	....	....	
(*d*) Wigan Cannel	..	..	74	....	....	....	
(*d*) Newcastle	..	..	68	....	1,400	....	
Wales, Flintshire	..	..	37	....	....	....	

OCCURRENCE OF OIL SHALE IN GREAT BRITAIN

Scotland

Location	Thickness sampled, feet	Oil, gallons per ton	Gravity, Baumé	Gas, cubic feet per ton	Ammonia sulphate, pounds per ton
b Isl nd of Raasay	feet 7–10	12	....	....	6.2
b Skye Coast, Holm and Prince Charles' Cave	....	12.8 12.8	....	....	7.4 7.4
c Lothian shales	....	25 2 gals. light naphtha	....	3000	45
c Lothian shales	....	10–55	....	....	6.70
d West Calder district, Dunnet shale	10–15	30	....	....	25
d West Calder district, Broxburn shale	3	20	....	....	....
d Addiewell, Fells shale	2.5–3.5	28	....	....	10–14
d Pumpherston shale	92	20	....	....	50–60
d Dalmeny shale field	7	30	....	....	35
Oakbank, Dunnet shale	4–12	24–33	....	....	14–34
(*f*) Broxburn curly seam	2 3	26.7	....	....	39.4
(*f*) Broxburn grey seam	2–3	23.7	....	....	39
(*f*) Sa dhole Pit	2 4	35.6	26–27	....	20–35
(*f*) Pentland field	2 6	32–34	25.9 27.4	....	19–20
(*f*) Newliston	5 10	33.6	29.8	....	24.2
(*f*) Crossgreen	5 6	25.1	30.5	....	35.4
(*f*) Duddingston	15 3	21	....	....	....
(*f*) Oakbank, McLean shale	5	38.4	....	....	....
(*f*) Oakbank, Lower Wild shale	5 6	19.3	....	....	....
(*f*) Kiltongue Coal seam, Aird ie shale	....	33	16.2	....	12
(*f*) Gunsgreen, Mungle shale	3 1	36	....	....	30
(*f*) Westwood, Addiewell grey shale	....	28.3	29.2	....	13

REMARKS

Location—*a* West Lothian, Midlothian, Fife Counties; Islands of Raasay and Skye.
e Counties of Linlithgow, Peebles at West Calder, Uphall, Queens Ferry, Loanhead.
c Burntisland, Broxburn, Dalmeny, Bathgate, Philipstoun, Forkne k, Addiewell, Hopetown, Westwood, Deans, Seafield, Dewliston, Lanackshire at Tarbrax, Cobbinshaw, Stirlingshire at Blackriggs Renfrewshire, Ayrshire.

Age—*a* Shale seams in the Calciferous sandstone at base of Carboniferous rocks. *b* Jurassic shales on islands of Raasay and Skye at base of great Estuarine series.

(*f*) Memoirs of the Geological Survey, Scotland, 1912.

a Mansfield: Petroleum Review, Vol. 34, p. 159, 1916.
b Forbes-Leslie, Vol. 35, pp. 327, 348, 367, 1916.
c Lee: Nature, Vol. 92, p. 169, 1913.
d Abraham: Asphalt and allied substances, pp. 163, 220, 221.
e Trans: Edinburgh Geol. Survey, Vol. 8, p. 120, 1898–1905.

12 gallons of light naphtha.

OCCURRENCE OF OIL SHALE IN CONTINENTAL EUROPE

Italy

Location	Thickness sampled, feet	Oil, gallons per ton	Gravity, Baumé	Gas, cubic feet per ton	Ammonia sulphate, pounds per ton	REMARKS
						Location—Province of Chieti, town of Valentino. Age—Eocene.
Province of Chieti	10–100 aver. 60	12–15	21	....	....	Murrie: Journal of Society of Chemical Industry, Vol. 4, p. 182, 1885.

Serbia

Location	Thickness sampled, feet	Oil, gallons per ton	Gravity, Baumé	Gas, cubic feet per ton	Ammonia sulphate, pounds per ton	REMARKS
Alexinatz		43-46			23.6	Location—*West Serbia near river Golabara. †At Alexinatz. ‡At Subotini. Age—* Eocene. Area—* 30 square miles.
Subotini	....	85	....	4600	....	

* 2 per cent semi-solid hydrocarbon; 3.02 per cent of water combination.
‡ From Subotini shale: 8 per cent water; 29¼ per cent ash; carbon (in ash), 17.3 per cent.
* Griffiths: Chemical News, Vol. 49, p. 107, 1884.
† Mansfield: Petroleum Review, Vol. 34, p. 159, 1916.
‡ Mills: Manual of Destructive Distillation.

Spain

Location	Thickness sampled, feet	Oil, gallons per ton	Gravity, Baumé	Gas, cubic feet per ton	Ammonia sulphate, pounds per ton	REMARKS
Castellon ‡	4	15.7	13	....	9.2	Location—*Ronda district in Southern Spain. †North of Barcelona. At Castellon Conil. Age—Conil shales—Eocene.
Burgos Spain, § S. of Europe Exploration Co.	....	....	21	....	....	

* Abraham: Asphalt and Allied Substances, p. 164.
† Mansfield: Petroleum Review, Vol. 34, p. 159, 1916.
‡ A. B. Thompson: Petroleum Mining.
§ Neuburger and Noalhat: Petroleum Technology.

OCCURRENCE OF OIL SHALE IN AFRICA

Africa

Location	Thickness sampled, feet, inches	Oil, gallons per ton	Gravity, Baumé	Gas, cubic feet per ton	Ammonia sulphate, pounds per ton	REMARKS
† Ermelo	2 0	30–34	....	....	....	Location—†Middelburg, Ermelo and Wakkerstroom districts, Transvaal Gaza Land, Mozambique, Portuguese E. Africa, Orange River Colony. Age—*Triassic. Karroo formation. Gaza Land—Cretaceous. Area—†Ermelo district—few hundred acres; Wakkerstroom district—few hundred acres. African Oil Corp.; 750 acres.
† Wakkerstroom	0 9	up to 90	....	....	....	
African Oil Corp., top shale	1 9.1	32.7	30	....	....	
African Oil Corp., middle shale	1 9.6	29.2	30	....	....	
African Oil Corp., bottom shale	0 9	19.4	30	....	....	* Mansfield: Petroleum Review, Vol. 34, p. 159, 1916. † Mining World, Vol. 34, p. 74, 1911. ‡ Sir John Cadman: Oil Resources British Empire.
‡ Wakkerstroom	1 0	22	....	....	....	
Transvaal	....	40	....	....	....	Pet. Rev. **38**, 373, 374, 39, South African Min. and Eng. Jour. **29**: 2, March 6, 1920, 9–10.
Middleburg coal district	6 0	30–40	....	....	....	
South Africa	5.6	27.2	....	....	....	
Congo (*a*)	27	58	....	....	....	

a Mining World, Vol. 34, p. 1182, 1911

OCCURRENCE OF OIL SHALE IN AUSTRALIA

New South Wales

Location	Thickness sampled, feet	Oil, gallons per ton	Gravity, Baumé	Gas, cubic feet per ton	Ammonia sulphate, pounds per ton
Aver ge of New South Wales	1-2.5	100-150	....	18,000 39 candle gas	....
General average	....	80-100	....	....	....
Deposit of Newness	....	14	....	....	....
Harkey Vale	....	60	....	....	....
Torbane	(thin)	150	....	....	....
† Coorangiti shale	....	14-150	....	....	20-30
‡ Blue Mountain	2-4.5	100-120	....	....	70
§ Capertee	....	125	....	....	....
Miscellaneous	5	....	....	....	....
Blue Mountains	....	100	....	17,600 49 c.p.	70

REMARKS

Location—Cumberland County at Lake Macquarrie, Greta; Cook County at Mount York, Mount Megalong; Camden County at Joadga Creek, Cambewarren Ranges, Broughton Creek; Tonali River and Burragorang; in the Blue Mountains at Blackheath; vale of Harkey. Murruruidi, 75 miles N.W. of Newcastle, Wolgan Valley. Caperte , 100 miles N.W. of Sydney.
Age—Permo-Carboniferous at Blue Mountain.

*Mills: Manual on Destructive Distillation, p. 16.
† Abraham: Asphalt and Allied Substances.
‡ Jene, H. L.
§ Heller, A. H.

Queensland

Location	Thickness sampled, feet	Oil, gallons per ton	Gravity, Baumé	Gas, cubic feet per ton	Ammonia sulphate, pounds per ton
Baffle Creek	9-10	38	....	....	29.6
Narrows, Port Curtis district	....	28	....	....	47
Duaringa	....	31	....	....	....
* Shale Gully	8	11	....	....	29.6
* Munduran Creek	....	12.5	....	....	....
* Munduran Creek	....	28	....	....	47
* Munduran Creek	....	21.4	....	....	....
* Munduran Creek	....	8.9	....	....	....
* Harpur Creek	....	12	24.7	....	18.29
* Harpur Creek	....	12	....	....	38

Location—Baffle Creek; Narrows (Port Curtis district).

* Geol. Sur. of Queensland Pub. 249.

Tasmania

Location	Thickness sampled, feet	Oil, gallons per ton	Gravity, Baumé	Gas, cubic feet per ton	Ammonia sulphate, pounds per ton
Latrobe and Railton deposits—called Tasmanite	4	40	....	....	....
Mersey	1-12	40	....	....	....
Mersey River Valley	4	....	....	....	....
Miscellaneous	....	40-65	....	....	....

Location—Latrobe and Railton Devon County, East of Table Cape. Mersey.
Age—Carboniferous.
Area—300 acres. 12,000,000 tons in Mersey district.

OCCURRENCE OF OIL SHALE IN AUSTRALIA

New Zealand

Location	Thickness sampled, feet	Oil, gallons per ton	Gravity, Baumé	Gas, cubic feet per ton	Ammonia sulphate, pounds per ton	REMARKS
Orepuki	....	20–40	....	....	....	Location—Orepuki, Anstren, Waiwali, Aukland, D'Orville Island, Chatham Island, Mongonui, Waikaia. Area—1,000,000 tons in Waikaia district. Abraham: Asphalt and Allied Substances. Trans. and Proc. of New Zealand Inst. 52 (1920), 27–29.
Orepuki	4	....	....	....	....	
Orepuki	4.5	42	26	....	....	
Waikaia	....	38.41	....	....	19.12	
Waikaia	....	38	15.7	4,000	8	
Waiarewau	8	28.3	....	....	....	

MISCELLANEOUS

Oil shale, concerning which no accurate information is available, is reported at the following places:

United States—Alabama.
Maryland.
Texas.

Mexico—Tolnea, Soledad, Mendy.
Peru—Tertiary shales near Cerro de Pasco, fairly rich.

EUROPE

Great Britain—Wales, in Flintshire coal measures.
Austria—Moravia, Bohemia, Tyrol.
Portugal—Near Monte Real; Leiria, north of Alcobara, near Torres Vedras.
Spitzbergen—Capes Thorsden, Staratsehen, and King's Bay.
Sweden—Kinnekulle, Nerike, and Ostergötland.
Norway—Shale associated with coal on Island of Anden.
Montenegro—Triassic shale at Bukowich.

ASIA

Turkey—Muchado Plain.

AFRICA

Gaza Land—Mozambique, Orange River Colony.

COMMERCIAL CONSIDERATIONS

General.—There is little doubt that there will eventually be an oil-shale industry of great importance in this country. Figures showing the production and consumption of petroleum and its products, the decline in production of old oil-fields, and the potentialities of new fields demonstrate this fact clearly. The question is, "When will conditions be such as to warrant the investment necessary to put oil-shale development on a commercial basis?" Opinion is divided on this question. Many careful investigators have concluded that the production of oil from shale at a profit will not be possible for ten or even twenty years. Others are of the opinion that profitable operations may be carried on at present. Any broad prediction as to the industry in general is difficult. The costs of producing oil from shale are not known, nor can they be closely approximated, nor is the value of the oil recovered definitely known. It seems logical to assume that this industry will not be developed into its full importance over night. Certain deposits of very rich shales with particularly favorable mining and transportation facilities will be worked first, and even now one such deposit in Nevada is being operated on a commercial scale.

Sampling and analysis.—Inasmuch as oil-shale development will be, in the strictest sense, a mining and manufacturing enterprise, its success will depend primarily on operating costs, and the data upon which these costs will be based should, of course, be carefully collected before operations are commenced. In discussing these data, it is essential to separate *facts* from *approximations* and to list them under these two categories. Thus we have as *facts* which can be absolutely determined in advance:

(*a*) Amount of commercial shale in a given property.

(*b*) Amount of recoverable oil in above.

(*c*) Units of shale which must be mined and treated to produce one unit of oil.

(*d*) Character and value of this oil.

(*e*) Distance and cost of transportation of above to a suitable point of market.

The following data cannot be stated as *facts* but can be more or less closely *approximated*:

(*a*) Cost of mining sufficient units of shale to produce a unit of oil.

(*b*) Cost of treating the above amount of shale to produce oil therefrom.

(*c*) Cost of further treating this oil to produce a marketable product.

It is the duty of the oil-shale engineer accurately to collect the *facts* listed, and from his knowledge of mining in analogous situations, of retorting in this and similar industries, and of refining treatment and costs in the petroleum industry, closely to *approximate* what the remaining data will be.

Sampling the property.—Aside from the preliminary examination of a property before acquisition, in which hand sampling may be justifiable, the most satisfactory method of examining a property is by means of a core drill. The advantages of this method may be enumerated as follows:

(*a*) It is the only method which will recover acceptable samples from beds which do not outcrop on the property.

(*b*) The core forms an absolutely reliable continuous sample of all the beds encountered on the property.

(*c*) This sample may be readily and accurately subdivided into samples of retorting size.

(*d*) The samples recovered are at depth, where the actual mining will take place, not at the surface, where a sample may or may not represent what exists below.

(*e*) Many facts concerning water conditions, mining methods employable, hardness of formations encountered, etc., are obtained at no extra cost.

The cost of core drilling, although considerably greater than that of haphazard hand sampling, is a very small item when considered in its relation to the tonnage thus proved, and the results derived. A series of three holes, for instance, should prove up at least 160 acres each or 500 acres in all. If this acreage contains a thickness of 100 feet of commercial shale in all, roughly 121,000,000 tons will thus be proved. At a cost of $5.00 a foot for drilling, and assuming the necessary depth of the holes to average 500 feet each, the total cost for drilling will amount to $7500 or about $\frac{5}{1000}$ of a cent per ton. The recovered core should be carefully labeled and sent to the laboratory for analysis.

Examining the shale.—The chemical engineer in charge of the laboratory examination of oil shale and shale oil has two distinct and important objects as a basis for his examination. First is the determination of the economic aspects of the deposits including:

(*a*) Amount of commercial shale.

(*b*) Quality of commercial shale.

(*c*) Total amount of oil available.

(*d*) Character and value of oil.

(*e*) Character and value of by-products.

Second, he should determine the conditions and economies which will affect the commercial treatment of this particular type of shale, and should therefore seek to establish:

(*a*) The temperature at which the retort should be operated as regards:

1. Quantity of oil.
2. Quality of oil.
3. Economy of operation.

(*b*) Percentage of the total oil which may be economically recovered from the shale.

(*c*) Time period necessary for this recovery at the recommended temperature.

(*d*) Special conditions, such as the use of steam or vacuum, which will favorably affect the operation.

(*e*) Amount of fuel necessary to the operation over the gas produced during the operation.

To achieve highest efficiency, the work of examining the cores from diamond drilling can be best divided into three operations, as shown in Fig. 3, all of which may progress simultaneously in the well-arranged laboratory.

Operation "A."—Finely powdered, gram samples from each foot of the core should first be examined in test-tubes over a Bunsen flame. This is an exceedingly rapid and delicate method of eliminating beds which do not carry at least 10 gallons of oil to the ton. The test-tube containing the powdered shale is held in a horizontal position with its closed end in the hottest part of the flame, and is heated until the glass is fused, the amount of vapors coming off being carefully noted. If no oil vapor is given off during the operation, the shale may be classed as barren and eliminated from further consideration. If yellowish vapors, smelling of petroleum, are given off, but no drops of oil condense on the cool surfaces of the tube, the shale may be said to contain a trace of oil, but no commercial quantity, and may also be eliminated. When small drops of oil form on the sides of the tube, the sample is checked for examination under operation "B." In the operation chart, "I" represents the total core, tested by this method, "II" the material containing interesting quantities of oil reserved for further test, "III" material containing less than 10 gallons of oil, which is eliminated by this method.

Operation "B."—The core "II" determined by operation "A" to contain interesting amounts of oil is now divided by a core cutter or other means into two or more equal vertical parts. Samples representing each 5 feet of this core are quartered down to 8½ ounces, and these samples are tested for oil, gas, and (if desired) ammonia content. This test is made in a battery of small mercury retorts with suitable condensers, collectors, and means of heating, and the results are noted and kept for checking and correction by the data subsequently obtained. This operation is not intended to determine absolutely the quantity or quality of the oil, but it is a fast and efficient method of calculating the number, thickness, and attitude of the commercially important beds. The results from operations " A " and " B " should be expressed in one or more charts similar to chart AB (Fig. 4). The remaining core falls into two classes, commercial shale " IV " and non-commercial shale " V."

Operation " C."—Representative 50-pound samples of rich shale produced by the quartering in the above operation are now tested in a research retort which should be provided with means for mixing the shale during heating, pyrometers in the retort and in the fines, ports for introducing steam and for vacuum, and apparatus for measur-

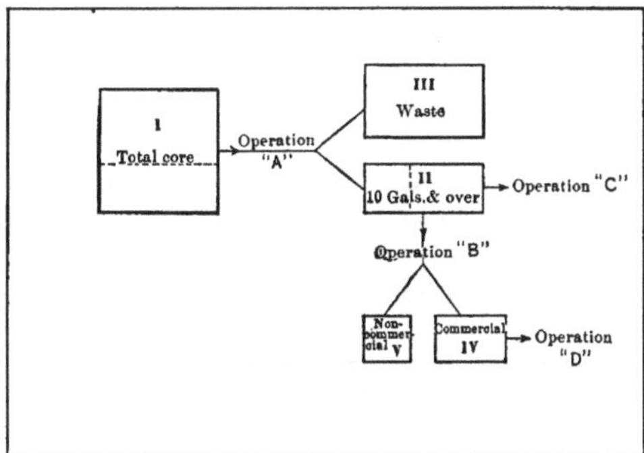

Fig. 3.—Operation chart for examination of diamond-drill core.

ing the gas formed. The retort is slowly heated, the temperature being noted every ten minutes and the following data recorded:

(*a*) Time required for first drop of water to come over.
(*b*) Temperatures at that point in retort and fines.
(*c*) Time required to eliminate all free water.
(*d*) Temperatures at that point.
(*e*) Time required for first drop of oil to come over.
(*f*) Temperatures at that point.

Holding the temperature at that point, all possible oil is taken off, and the time and the amount and character of the oil obtained are noted. The temperature is then gradually raised 20° and the above process is repeated. This is done at 20° intervals until all the oil is taken off. The retort is then cooled, and the residue is tested for oil, carbon, and (if desired) ammonia content.

The above operation is then repeated in half the time and the results noted as above and compared. If advisable, it should be repeated in still less time. It should then be repeated using steam at the temperatures at which high percentages of unsaturates are formed, and under different amounts of vacuum, the results being noted and compared as above. The results from these tests should be recorded on charts similar to CD, CS, CV (Fig. 5). The above operation will determine the temperature, time, and conditions most suitable to efficient retorting. It should be carried on in a retort which closely simulates the conditions obtaining in a commercial plant. Although there are other installations which serve effectively, Fig. 6 shows a type of apparatus which has given very satisfactory service in the laboratories of The Day Company. It is a miniature reproduction of the retort used in Germany in obtaining oil from lignite, the only large-capacity, low-temperature method in successful operation at the present time. The laboratory plant as shown in the illustration is composed of a cast-iron cylindrical retort, closed at the bottom, and fitted at the top with a flanged

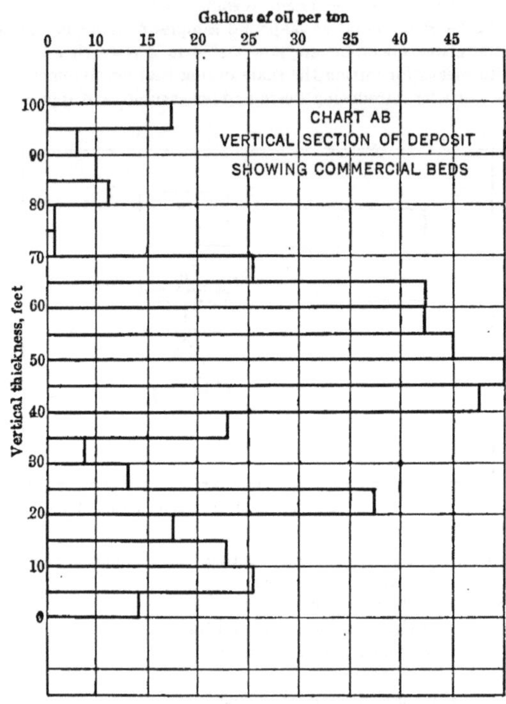

FIG. 4.—Method of tabulating results of oil shale cores.

cover. Inside the retort is a perforated pipe on which are several conical baffles. This pipe is open at the top but is closed at the bottom by a circular sheet-iron plate, to which is affixed a shaft which protrudes through the bottom of the retort by means of a stuffing box and terminates in a crank for rotating the entire assembly. The whole is enclosed in a suitable furnace. In practice, the cover plate is removed and the 50-pound sample of shale is evenly distributed between the baffle plates and the retort wall. The cover is then replaced and the retort heated. As the vapors are formed, they enter the pipe through the perforations below the baffles and pass off through the vapor line to the condenser. Pyrometer ports in the retort and the flues provide for a thorough measurement of temperature. Connections are also provided for the introduction of steam, or the use of vacuum. During the heating the interior assembly is slowly rotated by the crank, keeping the shale thoroughly mixed. The "sticking-point" may be noted by this means. When the distillation is completed, the flanged cover containing the vapor line is removed, and the interior assembly is taken out through the top, together with the spent shale.

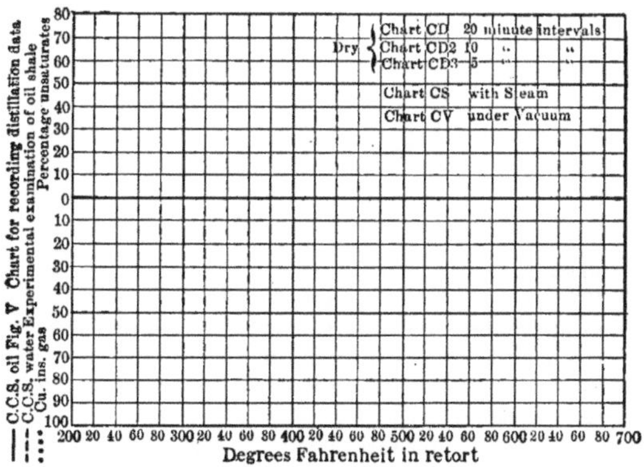

Fig. 5.—Chart for recording distillation data. Experimental examination of oil shale.

Operation "D."—Operation "B" has determined the beddings which will be commercially worked, and operation "C" the most favorable method and conditions of working. In operation "D," 50-pound representative samples of each of these commercial beddings are tested according to this method. The results will give the

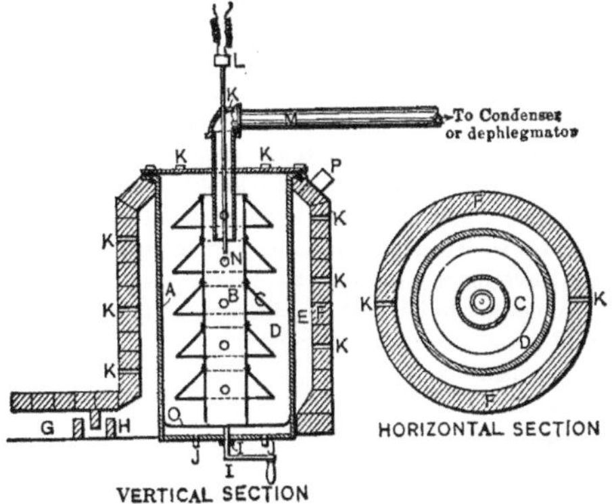

Fig. 6.—Sketch of experimental retort. Experimental examination of oil shale.

amount of oil to be expected from the commercial operations and will at least throw some light on what the cost of these operations will be.

Analysis of the oil.—The analysis of the oil will follow the practice now employed in progressive petroleum laboratories. The crude oil from each of the tests previously described is first examined and the water and sediment content, gravity, flash and boiling point determined. The crude is then fractionally distilled into 10-per cent fractions and the data just listed are obtained for each of these fractions. The fractions are then grouped into the following commercial products: gasoline and motor spirit, illuminating oil, gas oil, fuel oil, lubricating oil, paraffin wax, and residue. These products are further tested to determine:

(1) The amount, method, and cost of treatment necessary to put them into marketable shape.

(2) Their market value.

Mining Methods

By A. H. Heller

General considerations.—If this were merely a discussion of mining oil shale in the Colorado–Utah district, no doubt the largest oil-shale region in the world, it would be possible to concentrate on three or four methods of mining at the most, and one or more of them would be applicable to all deposits in this area. Taking oil-shale deposits as a whole throughout the world, their modes of occurrence are so varied that many more methods of mining must be reviewed.

Most articles so far written on this subject have taken coal-mining methods as a basis for determining estimates of cost, and for working out the systems that may be used for the most economical mining of oil shale. Although certain ideas will be borrowed from coal mining, it is no doubt far better to use our present metal-mining methods as a basis, for the following reasons:

(1) Very little shale is friable and as brittle as coal, the majority being very tough, and the rate of progress of drilling comparing favorably with the average metallic ore.

(2) Experience in Scotland and thorough tests conducted by the U. S. Bureau of Mines have determined that the dust from shale mines is non-explosive, a circumstance which eliminates the necessity for the elaborate and costly systems of ventilation now practiced in coal mines. For this reason the room-and-pillar method of mining coal is much more expensive than the same method (called breast-and-bench mining) used in metal mines. On account of the explosive nature of the gas and dust in coal mines, they are also limited to the use of " permissible " explosives.

(3) Very few detailed coal-mining costs are available, and these few usually include crushing, washing, screening and handling, besides sundry miscellaneous costs, which would be distributed in a shale mine and plant. For metal mining we have a wealth of costs to draw from, covering a great variety of methods.

In determining the method to be used for mining a certain bed of oil shale, the following factors must be taken into consideration:

(1) Thickness of the bed.
(2) Dip of the bed.
(3) Strength of walls or floor and roof.
(4) Strength of shale.
(5) Area of shale bed.
(6) Amount of overburden.
(7) Location of deposit in relation to normal ground level.
(8) Value of the material to be mined.
(9) Cost of labor and supplies.
(10) Uniformity of grade of shale throughout the deposit.
(11) Daily tonnage.

Classes of shale deposits.—*Thin, flat or slightly dipping beds with little or no overburden.* This type of deposit is relatively uncommon and can be mined by open-cut methods with steam shovels, excavators, or scrapers.

Thick, flat or slightly dipping deposits, with little or no overburden. Open-cut methods are the most economical for this type of deposit. Where the area of the deposit is limited, or where the shale pitches slightly, and will soon be covered by overburden, underground methods will have to follow open-cut operations.

Thin, flat or slightly dipping beds, with excessive overburden. This type of deposit is rather common in the Rocky Mountain region. Taken in order of cheapness the methods most applicable are breast and breast-and-bench systems of metal-mining methods, and longwall and room-and-pillar systems of coal-mining. The breast, and breast-and-bench systems of metal mining are essentially room-and-pillar methods, but differ from the latter in that the pillars are not removed, with resulting lowered cost.

Thick, flat or slightly dipping beds or masses, with excessive overburden. This is a very common type of occurrence of shale having a low oil content. The most feasible mining methods are underground glory-hole methods, or caving methods which would include top-slicing, sub-level caving, and block-caving.

Steeply inclined deposits. This type is unusual in the Rocky Mountain region, but typical of the Scottish and many other deposits. With a very thick bed, caving systems may be used, just as for flat deposits. Room-and-pillar methods are applicable on the thinner beds. Open-stope methods may be used, providing the ore and walls are strong, and the dip of the bed less than 45°. With a dip greater than 60° shrinkage stoping may be used.

Description of Methods

Breast and Breast-and-bench methods.—These methods are extensively used in metal mining, and include breast stoping for beds of less than 12 feet in thickness, and breast and bench stoping for beds of greater thickness. With the latter method, the upper part is usually mined by breast holes, and the lower part by vertical holes. Where the upper part is very tough and the lower part breaks very easily, it is cheaper to run a heading in the lower part, e.g., 100 feet, followed by breaking down the upper part. Where benches are used their slope should be steep enough to minimize bedding. Ore is mined by these methods in the southeast and southwest Missouri and the Wisconsin zinc districts, and about 5 tons is broken per man per day.

In the mining of shale it will be desirable to have a systematic arrangement and uniform size of pillars. Pillars are usually 10 to 20 feet in diameter and 20 to 50 feet apart. Ten to thirty-five per cent of the shale is ordinarily left in the pillars, depending upon the thickness of the shale body, regularity of occurrence, character of walls, strength of shale and depth at which it occurs.

The breast methods are to be advocated for shale mining, for they result in cheap mining although there is an incomplete extraction of the shale.

Room-and-pillar method where pillars are removed.—This method is used extensively in coal mining, the value of the coal in the pillars being greater than the expense of mining them. This type of mining is also used for oil shale in Scotland, and has been one of the few methods so far advocated for the mining of shale deposits. It is also being used in a simplified form in the mining of salt and iron ores. Although this is a more systematic method than breast stoping, the principle is the same. It is applicable to both flat and steep beds, providing the shale is abundant and the walls and shale are strong. Notwithstanding that pillars are removed in Scotland, they are able to mine an average of 6 to 8 tons per man.

Longwall method.—This is also a coal-mining method, but has been used very little in the United States. It is best for a strong roof pressure, and on friable shale if the roof permits. The longwall system is cheaper than the room-and-pillar but the requirements are more rigid.

Underground glory-hole method.—Application of this method is limited to large bodies of shale, such as masses or thick beds, with both strong walls and ore. At Phoenix, B. C., where this method is used in metal mining over 8 tons of ore is produced per man underground.

Top-slicing caving method.—This method is not to be recommended for shale mining generally, on account of the poor ventilation, and high timbering cost, but it is the only economical caving system where the overlying ground, shale and walls are weak.

Sub-level caving.—This is a cheaper method than top-slicing, less timber being required. It also allows better ventilation.

Block-caving.—This method is the cheapest underground mining method that can be employed, but large-scale systematic work and good judgment are necessary if it is to be used successfully. This is a type of mining which has been given no publicity in regard to shale work, perhaps because no large-scale mining of oil shale, to which the method is adaptable, has been carried on, and also because oil-shale mining methods so far suggested have followed along the lines of Scottish practice. In the United States there are certain oil-shale deposits to which this method of mining may be applied. The requirements are that the beds be more than 70 feet thick, and that the shale should be neither very soft nor in tough solid masses, but should be sufficiently fractured to cave easily when blasted in blocks. With this system the ventilation is good, and the extraction of shale will range from 70–90 per cent. The Ohio Copper Company of Utah had such success and such low costs with this method on a friable quartzite ore, that the Inspiration Copper Company adopted it. It is also used on iron ores in Michigan.

Shrinkage stopes.—In this method the ore accumulates until the stope is filled. It is applicable to steeply dipping deposits, exceeding 60°, having strong ore and strong walls.

Open-cut methods.—There are two general types of open-cut methods, namely, glory-hole and steam-shovel operations. In the first the shale is drawn out below the deposit. This method would be applicable to certain thick beds, especially those elevated at any distance above the normal ground level, raises being run up to the deposit for drawing out the ore. Steam-shovel methods are the only economical ones applicable to shallow beds having little or no cover. The general features appear simple, but good management, close attention to details, and systematic work are essential to success.

Where overburden occurs, it is a question to be solved with each individual mine, whether it is more profitable to operate by stripping with steam shovels, or to use underground methods directly. In a certain anthracite mine in Pennsylvania, it was found more profitable to strip 2.75 cubic yards of overburden for every ton of coal recovered, than to employ underground systems of mining.

All open-cut methods have a disadvantage as compared with underground mining, in that operations must cease entirely in wet weather, or a wet sticky material must be handled which causes trouble in subsequent operations. This is a very important factor, especially with Colorado–Utah deposits.

Costs

Coal-mining costs will not be taken as a basis for estimating those of shale mining, because very few detailed costs are available, and the total given usually includes items which should not be included under mining. Coal-mining method requirements are also more rigid than those which would ordinarily be practiced in shale mining.

In comparing metal and shale mining, the following factors, which tend to make the mining of shale a cheaper operation, must be considered:

(1) Greater regularity of shale deposits.

(2) Smaller development expense.

(3) No hoisting in most cases.

(4) Very little pumping in the Rocky Mountain region.

(5) Greater progress in shale than in metallic ores.

(6) Less timbering.

For steam-shovel open-cut methods, D. C. Jackling has estimated a cost of $0.50 per ton for a deposit of oil shale 50 feet thick with no overburden, and the Stimpson Equipment Company made an estimate of less than $0.40 per ton for a similar deposit. On the Mesabi Range with iron ores, mining costs from $0.25 to $1.00 per ton, including stripping and capital charges. The direct mining cost at the Utah Copper Mines is about $0.25 per ton, not including a stripping cost of about $0.10 per ton. At a certain Pennsylvania anthracite mine stripping operations cost from $0.20 to $0.50 per ton, and the underlying coal is mined for less than $0.30 per ton.

For the block-caving system, the Ohio Copper Company undoubtedly operates at the lowest cost, having an ideal ore for this type of mining. In 1916 they were mining for $0.25 per ton, which included superintendence, office, engineering, blacksmith, development, mining, timbering, loading and supplies, but does not include depreciation, amortization and taxes. This was on a 2500-ton per-day basis. In 1918 the expense for the same items, including taxes, was $0.35 per ton. At the Inspiration Copper Mine the costs, excluding development, are about $0.40 per ton.

Top-slicing methods on the Mesabi Range cost about $2.00 per ton. This is the most expensive caving system that can be employed.

Shale is being mined underground at Elko, Nevada, on an 8-foot, 30-degree vein for $1.50 per ton.

The breast-and-bench methods of mining in southeast and southwest Missouri and in Wisconsin, are at present costing about $1.25 per ton.

In general, it may be said that very little shale mining will be done under $0.50 per ton, and then only if open-cut methods or block-caving systems are employed. Breast-and-bench mining will cost from $0.75 to $1.25 per ton, at a sacrifice of a considerable amount of the shale. By the use of the longwall and room-and-pillar systems, and more expensive forms of mining, the cost will no doubt run from $1.00 to $2.00 per ton, with an average for these latter systems of about $1.50.

Crushing

General considerations.—From the standpoint of its crushing properties, all oil shale will fall into one of the three following classes:

(1) Brittle or splintery shale, easily crushed.

(2) Massive, dry shale, with enough fractures and cleavages not to offer too strong a resistance to crushing.

(3) Massive tough shale, much of which has a tendency to stick.

As a general rule the higher the oil content of the shale, the greater resistance it offers to crushing. In a good many instances metallic ores have crushing properties similar to those of shale of classes 1 and 2, and there are a few cases of tough schistose ores that are similar to shales of class 3. It would be extremely unusual to find a shale that would crush as easily as a coal; yet, notwithstanding this fact, most of the machinery so far used in shale crushing has been borrowed from coal-preparation practice.

Layout and design of crushing plant.—In order to design and lay out a crushing plant, and to choose the proper equipment the following factors must first be known:

(1) Size of material to be sent to crushing plant.

(2) Daily tonnage required.

(3) Uniformity and size of product desired.

(4) Nature of material to be crushed, which would include degree of hardness, tenacity, compressibility and moisture and oil content.

(5) Type of power available.

(6) Class and type of construction desired.

(7) Sites available and their location with relation to mine and retorting plant.

Knowing these factors the following items, as regard the layout and design of the plant, can then be determined:

(1) Stages of crushing to be employed.

(2) Type and make of crushers to be used for each stage.

(3) Methods of conveying shale to and from each stage.

(4) Type and make of screening apparatus.

(5) Type and make of feeders.

(6) Location, types of bins and their capacities.

(7) Kind of plant, whether single or multi-floor.

(8) Location of power and transmission equipment.

(9) Lighting, heating and ventilating of building.

(10) General considerations, which would have to do with the mechanical efficiency of the plant, accessibility to all moving parts of machines and economy of operation.

Stages of crushing to be used.—Single-stage crushing is that in which only one reduction is made: primary and secondary crushing is that which consists of more than one crushing operation. A combination of grinding with single-stage or primary and secondary crushing is also employed. The stages of crushing employed are directly dependent upon the size and uniformity of both the feed and the product. The feed to the crushing plant will range from a maximum of 6 inches to a maximum of 24 inches in diameter, depending upon the method of mining employed, and the amount of sledging done in the mine, the average maximum to most crushing plants being about 12 inches.

Single-stage crushing will rarely be used in the crushing of shale, because the product of any standard single-stage or primary crusher will be too large for economic retorting. Crushers have been built that will reduce coal from run-of-mine size to one suitable for retorting, and while they may be economical on coal, they are inefficient when applied to the reduction of shale.

Primary and secondary crushing is the most economical practice, and it has been developed to a high point of efficiency in the reduction of metallic ores, and will no doubt be the most generally used system for shale reduction. The primary types of crushers to be recommended are either gyratory or jaw, their size depending upon the capacity of the plant, units desired, and size of feed and product. In secondary crushing one may employ smaller types of gyratories or plain, corrugated or spike-toothed rolls. Jaw types, although at one time used extensively for secondary crushing, have been replaced by other types. A combination of two or more of the above types of crushers may be worked out in order to obtain the most efficient practice. The product obtained may range from a maximum of $\frac{1}{8}$ inch to a maximum of 2 inches according to the amount and kind of crushing used.

Grinding in combination with primary or primary and secondary crushing.—The use of a finely ground product is being advocated by some engineers interested in shale-oil development. Material of this size in the proper form of retort, will undoubtedly lessen the time and the cost of retorting, but this saving hardly offsets the increased cost of reduction. If, with certain shales, it is found that the residue from retorting may in some way or for some purpose be used to a profit in pulverized form, very fine reduction before retorting may be more economical.

Type and make of crushers to be used for each stage.—The kinds of crushers applicable to each type of crushing operation, may be classified as follows:

(1) Jaw crushers for primary crushing.
(2) Gyratory crushers for either primary or secondary crushing.
(3) Plain, corrugated or spiked rolls, for secondary crushing.
(4) Miscellaneous crushers for secondary crushing.
(5) Coarse-grinding ball or rod mills.
(6) Fine-grinding ball or rod mills.
(7) Miscellaneous fine-grinding mills.

Jaw crushers, of which the Blake is the most common type, work very well as primary crushers on most shales, provided that the jaws are corrugated and that they have a small angle of nip. If the mine contains both dry and sticky types of shale, they should be well mixed before crushing. The average proportion of reduction on shale will be a little less than 3 to 1. For primary crushing both jaw and gyratory types are now made in such sizes that they will take the largest run-of-mine pieces that can be conveniently handled, and this has tended to eliminate hand sledging in the mine.

Gyratory crushers have a larger capacity than the jaw types for the same size jaw opening, the proportion of reduction being about the same. If the gyratory can be fitted with corrugated liners, any type of shale may be handled. Common types of gyratory crushers are the Telsmith, Gates, and McCully, the first being particularly adapted to secondary crushing.

Various types of rolls.—These are essentially secondary crushers, although the spike-toothed type of roll has been used for primary work in some instances in Scotland. It is of rigid construction, and when fitted with solid alloy steel teeth it becomes an economical machine for the reduction of shale. It has an advantage over the other types of rolls in that its nip is positive, and the character of the shale in no way affects its operation. Of the flat and corrugated types the latter is best suited to shale in that it makes less slabs, but the upkeep is higher as a worn corrugated roll must be discarded, whereas a worn smooth-faced roll may be turned down. Most smooth-type rolls are provided with a regulation for side movement of the roll, making for an even wear of the shell, but this regulation is impossible with the corrugated roll.

Miscellaneous crushers for secondary crushing.—The disk crusher of the Symons type requires a brittle material and is not adaptable to tough shale, or to that which has a tendency to slab, and for these reasons will not be generally applicable to the crushing of shale. There are a good many ring-and-hammer, and ring-and-disk types of crushers on the market, most of which have been successfully applied to coal breaking. As a rule most of these crushers attempt too great a reduction in the one machine, and have a multiplicity of parts, which should be avoided in any crushing apparatus. Also very few of these crushers have been built to resist abrasion to any extent.

Grinding.—Economical grinding of shale requires a preliminary crushing. Any method of dry grinding will cost from two to three times as much as the preliminary crushing, depending upon the reduction and the type of apparatus used, and for this reason the grinding of shale will be practiced but little. For grinding to not finer than

35 mesh, comminuters as used at Golden Cycle Mill in Colorado, Krupp Mills as used in Australia, or a rod mill of the Forrester type may be used. The latter with proper modifications could be made into a very efficient mill, on account of its low power consumption. With metallic ores any method of wet grinding is much cheaper than dry grinding, but wet grinding of shale is to be avoided, as filtering and drying would have to be resorted to before retorting.

For grinding to very fine sizes, standard ball or rod mills could be used, with a suction provided for removing the material when fine enough. Such methods, however, would prove impractical on wet or sticky shale.

Methods of conveying shale in crushing plants.—For conveying very coarse or run-of-mine ore, pan conveyors are being used, but their cost of upkeep is very high, and any conveying of such size of material is not to be recommended. For conveying the crushed materials the shaking and apron types of apparatus do not compare favorably with the belt conveyors, and if any elevation of the shale is desired it should be carried on as far as possible through the use of the latter. The greatest angle at which they will convey shale is 30°, and then only when the belt has a high rate of speed. Elevation and conveying by means of elevators should not be used unless absolutely necessary. The only practical form of elevator is the belt and bucket, as the sprocket and chain type, although more positive, requires more maintenance. Belt and bucket elevators are made in sizes up to 48 inches in width, and elevate materials as large as 8 inches in diameter. Even the best designed elevators are a source of trouble, and the only way to be assured of continuous running is to have all elevators in duplicate. Automatic skips have been tried in ore-crushing plants, and may be used where upwards of 1000 tons are handled per day.

Screening apparatus.—The feed to all crushers should be screened, so that material under the size of the crusher product may by-pass the crusher. This reduces the tonnage to the crusher and consequently the cost of crushing. Grizzlies are adaptable for use ahead of primary crushers. For secondary products high-speed vibrating screens or trommels may be used. Shaking screens and shaking grizzlies are not to be recommended as the shale tends to clog them up, unless an angle is used which is much greater than that usually employed.

Feeders.—Mechanical feeders for the crushers should always be used, the one giving the least trouble being the pulley or drum type.

Bins.—There are three standard forms of bins, round with conical bottom, rectangular with sloping bottom, and rectangular with flat bottom. The round bin with conical bottom, usually built of steel or reinforced concrete, is no doubt the best as there are no corners in which the shale can hang up, and the bin may be emptied completely without the use of hand labor. Rectangular bins may be built either of wood or steel and concrete. A slanting bottom is preferable to the flat bottom, as the extra storage obtainable from the flat bottom, and usually held in reserve, may be secured in the slanting-bottom bin by building it a little larger. In all rectangular bins, there is a tendency for the shale to build up in steep walls, so that the tonnage of free-flowing material to the crusher is at all times very small. In some instances with a 500-ton bin, less than one-fifth of this would flow freely to the bin gate. Primary bins should be large enough or numerous enough to contain enough shale to run the crushing plant from twenty-four to forty-eight hours, depending upon the type of mining and the transportation used. Secondary bins for crushed shale should hold at least a twenty-four hour run of the crushing plant, as unforeseen difficulties often occur which may necessitate a shut-down in the crushing plant.

Kind of plant.—The tendency in the design of crushing plants has been to make them of the multi-floor type, with the various stages of crushing and screening on separate floors, fall being obtained through gravity and the use of elevators. Not only is the use of elevators attended with operation difficulties, but the placing of the machinery on various floors adds to the cost of attendance and supervision. A much better plan would be to convey the shale by belt conveyors from one crushing unit to the next, and, if fall is available, to utilize it in the placing of bins so that elevators or long conveyors will not be necessary. In this way all of the machinery will be on one floor, attendance and the initial installation and building cost will be decreased, and one crane may handle all machinery for repairing and adjustments.

Power and transmission equipment.—Each crushing unit should have separate motors. A feeder should, wherever possible, be driven from the same motor which drives the crusher it feeds, with suitable clutches provided. All transmission machinery should be easily accessible.

Costs.—The cost of crushing will depend upon the size of feed and size of product, daily tonnage, character of shale, kind of machinery used and the cost of labor and supplies. The following are estimates of cost per ton on average shale for various degrees of crushing and various tonnages:

Crushed to pass		Daily Tonnage		
	250 tons	500 tons	1000 tons	2000 tons
2 inch	$0.10	$0.08	$0.06	$0.04
1 inch	.14	.12	.10	.08
½ inch (No. 2)	.20	.17	.15	.13
¼ inch	.30	.27	.24	.21
45-mesh	.50	.45	.40	.35
80-mesh	.70	.62	.56	.50

Retorting Methods

General considerations.—In general, the only commercial method of recovering oil from shale is a regulated heat treatment in some form of closed retort. Many retorts have so far been designed for this purpose and of these several have demonstrated their fitness for commercial operations. In the following discussion the principles of recovering oil from shale by treatment in a retort are stated, and some suggestions are made as to the apparatus necessary for this purpose.

Principles of shale retorting.—By the "principles of retorting shale for oil" are meant those which are involved in producing oil as the main and most important product. By-products introduce separate factors which necessitate separate discussion. In this article the oil-shale retort will be treated as a machine for producing oil, and oil alone. As determined by research so far completed, the factors governing the efficiency, economy, and oil recovery of a retort, considered from the standpoint of the retort alone and eliminating the factors introduced by the type of shale or its preparation before retorting, are as follows:

(1) Factors affecting efficiency.
- (*a*) Thin shale layer.
- (*b*) Mechanical agitation.
- (*c*) Complete heat control.
- (*d*) Efficient application of heat.

(e) Efficient furnace design.
(f) Simple mechanism.
(g) Large unit capacity.

(2) Factors affecting economy.
(a) Continuous operation.
(b) Minimum number of moving parts.
(c) Maximum use of automatic machinery.
(d) Minimum fuel requirements.
(e) Accessibility of parts.
(f) Mechanical strength and endurance.

(3) Factors affecting the quantity and grade of oil recovered.
(a) Temperature control.
(b) Even heat application.
(c) Progressive heating.
(d) Protection of vapors from abnormal temperatures.
(e) Free passage of vapors.
(f) Freedom from dust.

The importance of many of these factors is self-evident, but in order that their bearing on the subject and their relation to each other may be fully understood, they will be discussed individually.

Factors affecting efficiency.—*Thin shale layer.*—The importance of having a thin shale layer in the retort is brought forward most strongly by a consideration of the insulating qualities of the spent shale. Due to its high porosity and its carbon content, spent shale is one of the best insulating materials that can be imagined. A thickness of 10 inches of spent shale is practically equal in insulating qualities to a two-course wall of fire-brick. If a thick layer of shale is to be heated, that portion of the layer lying nearest to the heating surface will be distilled quickly and easily, but this spent layer will effectively insulate the portion of the shale lying farther from the heat, and will make its distillation exceedingly slow and difficult. This explains one of the faults of the Pumpherston retort in its application to American shale. This retort is circular in cross-section with a diameter of 22 inches at the top and 26 inches at the bottom, and is heated from the outside. The minimum shale layer is therefore 11 inches in thickness. Experiments showed that it was relatively easy to distill the oil from the 4-inch zone nearest to the retort, but that the complete distillation of the material lying within this zone took nearly four times the period necessary for the outer zone and necessitated a correspondingly large amount of heat. A thick shale layer also affects the quantity and grade of oil recovered and will be discussed under that head .

Mechanical agitation.[1]—The agitation of the material as it passes through the retort is important both from the standpoint of heat efficiency and operating efficiency. It is obvious that a greater shale layer may be retorted with the same amount of heat and in the same time if this layer is constantly stirred up and agitated, than would be possible if this layer remained stationary. The effect of this agitation is constantly to remove the spent material and to substitute fresh material in its place, by this means avoiding the insulating effect of a stationary layer of spent material in contact with the zone of heating.

Furthermore, most rich shales tend to coke in the retort unless they are subjected to constant agitation. In the retorts of the Scotch type, where no mechanical agitation is used and gravity alone is relied upon to carry the shale through the retort, rich

[1] "Agitation," in the sense used in this discussion, means slow, irregular movement, both lateral and vertical, rather than violent motion.

shales coke and adhere to the sides of the retort, clog the apparatus and cause expensive shut-downs and repairs.

Mechanical agitation has a decided effect on the quantity and grade of oil recovered and will be discussed under that head.

Complete heat control.—The importance of complete heat control is so obvious as to require little discussion. If too little heat is applied the time of retorting will be uselessly increased, with a loss in retort efficiency. If too much heat is applied some of the oil will be lost as permanent gas, and heat will be wasted, thus reducing both operating and heat efficiency.

Correct application of heat.—Not only must the heat be under complete control but it must be correctly applied. To achieve high efficiency, shale should be heated progressively to higher and higher temperatures. Consequently the greatest heat should be applied at the discharging end of the retort and should decrease gradually toward the charging end. Furthermore, heat should be applied so that every particle of shale receives the same heat treatment given every other particle. Careful retort design is necessary to produce this result. Another question worthy of consideration under this heading is the relation of the number of square feet of effective heating surface to the total surface of the retort and to the tonnage treated in one charge. Many of the retorts designed recently are inefficient because the *effective* heating surface is only a small percentage of the surface actually heated. This is particularly true of horizontal cylindrical retorts where heat is applied to the entire circumference of the retort, whereas the effective heating area is limited to a relatively small arc on the bottom.

Correct furnace design.—Too little attention has been given to the furnace in which the retort is placed. This furnace should not only be designed to regulate and apply the heat as discussed under the last three headings but should also be designed with attention to the general principles of furnace efficiency.

Simple mechanism.—The simpler the mechanism, the lower will be the cost of installation and of upkeep and operation. Many retorts which produce splendid results in the laboratory are totally unfitted for commercial operations because of their complicated mechanism. It must be remembered that a commercial shale plant must be capable of handling large tonnages continuously. Any retort whose mechanism is complicated cannot escape repairs, shutdowns and resulting loss. Moreover, the retort will be subjected to a temperature at which iron will not withstand great stresses, so that metal moving parts subjected to heat cannot be expected to stand up under the normally severe operating conditions.

Large capacity.—In designing an oil-shale retort, or in considering the adoption of a process, the primary consideration should be the adaptability of the retort to large-scale operations. The oil-shale industry is fundamentally a low-grade mining and manufacturing enterprise. To be commercially successful it must handle large tonnages at low costs. For this reason any successful retort must not only be adaptable to large-scale development but must be capable of attaining this development without too great duplication of units. It has been emphasized before that the construction of a successful experimental shale retort is not a difficult matter, but that duplicating the results on a large scale is another problem entirely. It is not beyond the powers of a good engineer to quickly grasp the potentialities of a retort from this point of view, and it is surprising how often this fundamentally important factor is overlooked.

Factors affecting economy.—All of the factors previously discussed have a direct bearing on economy of operation. Other factors which affect economy but which are not directly concerned with the efficiency of the retort will now be discussed.

Continuous operation.—Continuous operation is essential not only from a standpoint of efficiency but to secure economical operation. No matter how ideal a "batch" type of retort may be from an efficiency standpoint, it can never be as economical as a correctly designed continuous retort. The time required to empty and recharge a retort of the non-continuous type is a total loss and even if this period consumes only ten minutes of every hour, such a retort would be idle $16\frac{2}{3}$ per cent of the time or fifty days out of a three-hundred day working year. Heat is wasted during the idle period if the retort is kept hot, or an excess of heat is required in starting up if it is allowed to cool down. A non-continuous type of apparatus cannot be made as nearly automatic as a continuous retort and consequently requires a greater labor and superintendence charge per ton capacity.

Minimum number of moving parts.—The number of moving parts required should be kept at a minimum, not only for the reasons discussed under efficiency but also for reasons of economy. In the first place, these parts require power to operate them, which means expense. In the second place, moving parts are always subject to wear and in course of time will need replacing with the attendant costs both for new material and for non-productive time. The relation of moving parts to high temperatures, as already discussed, is a very important consideration.

Maximum use of automatic machinery.—The effect of automatic machinery on the economy of operations is too well known to require any discussion. This factor has perhaps been over emphasized by many inventors, who in their attempt to make an entirely automatic retort have defeated their own purpose by introducing an impossible amount of complicated machinery.

Minimum fuel requirements.—It is obvious that to obtain operating economy, fuel requirements should be kept at a minimum. For this reason full use should be made of all possible heat applied (1) by designing an efficient furnace, (2) by controlling the heat applied, and (3) by at least recovering some of the heat from the spent shale and the vapors. It is, of course, impossible to recover all the heat applied during the operation but it is equally unnecessary to waste it all. The heat in the spent shale may be used to pre-heat air for combustion or in the production of producter gas, by the combustion under the retort of the coke in the spent shale. The heat in the vapors may be recovered to some extent, by the use of heat exchangers, in pre-heating the incoming shale, etc. Very little thought has been given to this subject, which will have a really important bearing on economical operations.

Accessibility of parts.—The entire retort should be completely accessible. Shale may tend to coke or clog the retort, or carbon may form on the walls. Consequently access to the interior should be provided so that the walls may be scraped or cleared if necessary. The furnace should be so designed that damaged or burnt-out retorts may be replaced without tearing down the brickwork. Finally, if any dust is formed in the retort, it may be carried over by the vapors and deposited in the vapor lines. For this reason these lines should be arranged for easy cleaning.

Mechanical strength and endurance.—The last and possibly the most important factor is the mechanical strength and endurance of the retort. No matter how sound and perfect a retort may be from a theoretical point of view, it is worthless unless it can stand up under the rigors of continuous service without excessive repairs and replacements. To accomplish this result parts subject to stress must be protected from heat, moving parts and complications must be reduced to a minimum and the whole structure must be designed with a clear understanding of operating conditions and a knowledge of sound mechanical construction.

Factors affecting the quantity and grade of oil recovered.—These, the technical factors affecting the retort, may be considered as variables inasmuch as they may change with the raw material and with the desired product. However, certain principles govern the treatment of shale of any type with the production of any desired product. The essential factors for the application of these principles follow.

Temperature control.—If we adopt the theory that oil is produced from shale by "cracking" complex hydrocarbons already existing in the shale, the importance of complete temperature control is at once evident. Experiments have proved that radically different results are obtained when shales are treated at different temperatures. When very high temperatures are used a large amount of light oil is produced at the expense of a lessened total recovery. When low temperatures are used, a maximum amount of oil is recovered, but this oil has not the large percentage of low-boiling fractions. Market conditions will determine the product from which the greatest net return can be expected. To recover this product, the range of temperature necessary must be determined and the retort must be operated constantly at these temperatures. Absolute temperature control is essential in accomplishing this result.

Even heat control.—Inasmuch as the use of different temperatures will result in the production of different grades of oil from the same shale, similar heat treatment must be given each particle of shale to produce a constant product and to avoid losses. This will necessitate:

(1) A thin shale layer.
(2) Mechanical agitation.
(3) Continuous treatment.

The thin shale layer avoids overheating of the shale nearest to the retort wall in the attempt to heat sufficiently the shale farther removed. Mechanical agitation further reduces the effective insulating property of the shale layer and makes even treatment of the individual particles possible. Uniform heating conditions can only be maintained by continuous treatment.

Progressive heating.—Experimental work and research have shown that to obtain a maximum yield of the most desirable product, the shale must be heated progressively. In practice the temperature of the shale should be brought up gradually to the point at which oil vapors are formed and then increased uniformly until all the oil is driven off. To accomplish this not only necessitates perfect control of the heat applied, but also involves the application of the most intense heat at the discharging end of the retort with an even decrease to the charging end.

Protection of vapors from abnormal heat.—Shale vapors as formed are composed of rather complex molecules which are easily broken down if exposed to temperatures higher than those at which these vapors were formed. Some of this breaking down or "cracking" may be desirable for the production of light oils, but in any case it must be carefully controlled. In general, it seems essential to withdraw the vapors as rapidly as formed and to protect them from high temperatures. If, on the other hand, the vapors come in contact with a layer of relatively cool shale and are condensed and revaporized, a very large loss will result as well as the production of undesirable hydrocarbons. Except in special cases, therefore, the vapors should be withdrawn without overheating or redistillation.

Free passage of vapors.—One of the most essential factors in obtaining the result discussed above is a free passage for the vapors, not only in the retort itself but also from the retort to the condenser. Pressure in the retort is probably undesirable from the standpoint of products formed and it certainly increases the difficulty of preventing leaks, particularly in a continuous apparatus. The use of vacuum, on the other hand, although theoretically desirable, introduces mechanical complications which should be avoided if possible. The most practical plan is to provide sufficiently large passages for the vapors to allow them to pass out freely without creating pressure in the retort and without necessitating the use of a vacuum.

Freedom from dust.—Although, as already discussed, mechanical agitation is essential to successful retorting, this agitation should not produce dust by grinding the shale, nor should it tend to stir up the dust that may be present with the crushed shale. Shale dust, if given the opportunity, will go over into the vapor lines and condenser with the vapors and will absorb oil to form a thick paste, which not only clogs up the lines and causes countless mechanical difficulties, but also wastes a large amount of oil. The writer has inspected processes which, on account of dusting, yielded a product containing nearly as much mineral matter as the original shale.

Apparatus.—The types of retort which have been applied to oil shale are as follows:

I. Vertical retorts.
   a. Gravity feed.
   b. Mechanical feed.
   c. Combination method.

II. Horizontal retorts.
   a. Stationary retort with shale conveyor.
   b. Rotary retort.

III. Inclined retort.
   a. Stationary retort with shale conveyor.
   b. Rotary retort.

The Pumpherston process, extensively used in Scotland, is an example of class I–*a*. The retort used is vertical, circular in cross-section and tapered slightly from the bottom to the top. Shale is fed into a hopper at the top of the retort, and is removed constantly at the bottom. Its passage through the retort is effected by gravity alone. This retort is of the dual type, the upper cast-iron section being used for the production of oil, while the lower fire-brick portion serves as a gas producer and to recover ammonium sulphate. The advantages of this retort are its extreme simplicity, the absence of mechanical complications, and its strong and durable construction. Its disadvantages are its low heat efficiency, the lack of means to prevent mass carbonization, and the large initial and operating cost.

Type I–*b* is illustrated by the Galloupe process, an American invention. This process is vertical and cylindrical, and has mechanical means to provide positive movement of the shale through the retort. This type of retort embraces many of the factors essential to successful shale distillation and has given good results in experimental tests.

Type I–*c*, of which the Day–Heller retort is an example, uses both gravity and mechanical means to keep the shale moving. The retort shown in the illustration employs slanting hearths down which the shale moves by gravity. A revolving inner member serves to keep the shale agitated and prevents sticking and clogging. This type of retort includes most of the factors essential to perfect retorting and shows promise of commercial success.

Type II–*a* has been the basis for several retorts such as those used in the Ginet, the Day, the Jensen, the Johns and other processes. This type involves the use of a stationary horizontal retort through which the shale is moved by mechanical means. Screw conveyors in cylindrical retorts and rabble arms in flat-bottom retorts are the means most commonly employed in conveying the shale. The advantages of this type of retort are that it is positive and easily regulated and has good heat control. The disadvantages lie in the large number of moving parts which produce mechanical difficulties and operating expense, low heating efficiency due to the small percentage of effective heating surface, and the structural weaknesses inherent in this type of construction.

Class II–*b*, as illustrated by the Randall process, is a rotary retort not unlike a cement kiln. If, as in cement-kiln practice, internal heating could be used, this retort would

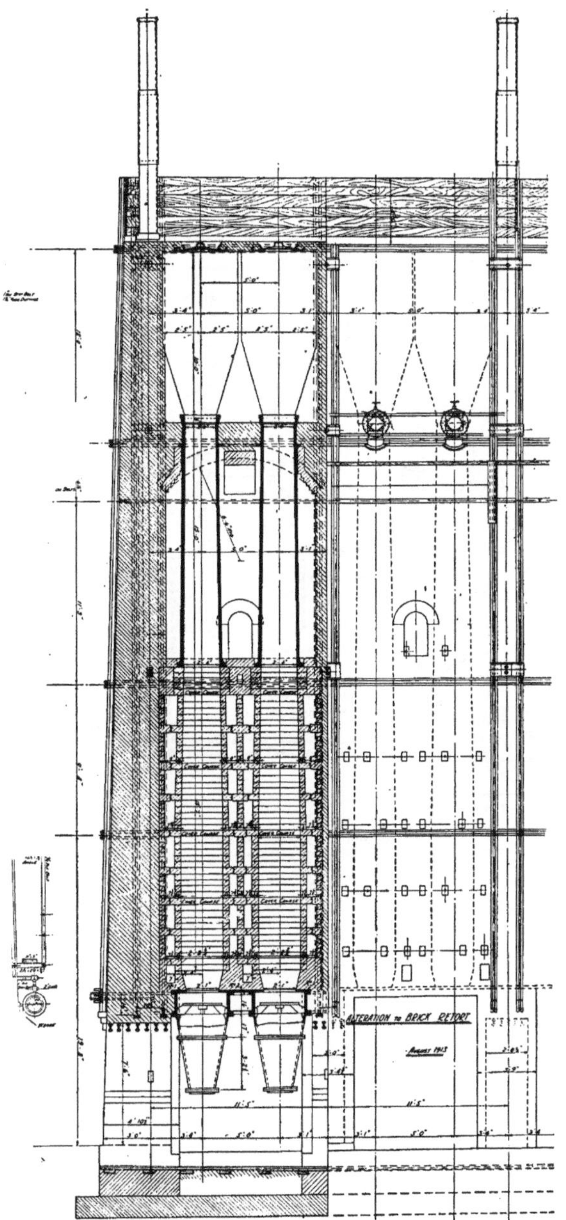

FIG. 7.—Sketch of Pumpherston Retort.

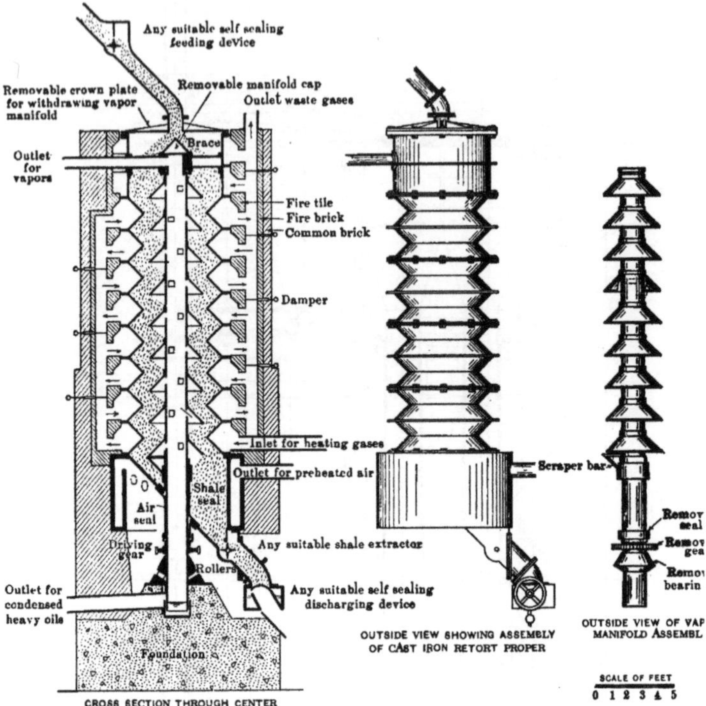

Fig. 8.—Sketch of Day-Heller Retort.

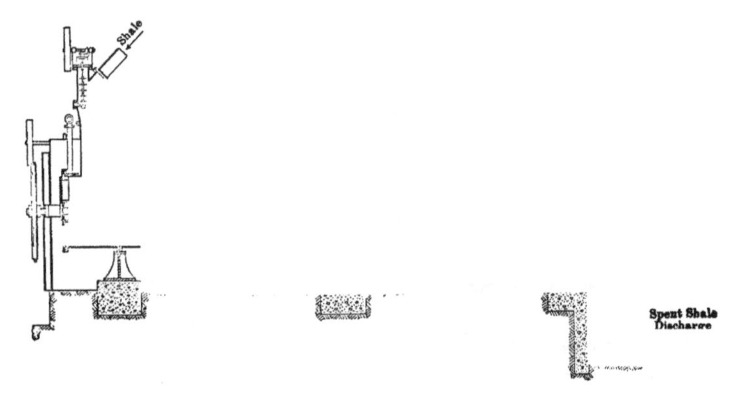

Fig. 9.—Sketch of Ginet Retort.

no doubt provide a very desirable method of oil-shale distillation. The disadvantages of internal heating with fully reduced flue gases or with an inert gas are, first, the expense of heating such an inert gas, and second, the difficulty of separating a small amount of condensable vapor from a very large amount of permanent gas. If internal heating is not used the rotary kiln loses its efficiency and presents serious mechanical and structural difficulties.

Classes III-*a* and III-*b* do not differ essentially from the corresponding classes II-*a* and II-*b* and have the same good and bad points.

Construction and arrangement of retorts.—The materials from which oil-shale retorts may be constructed are cast iron, cast steel, malleable iron, steel, and fire-brick. Cast iron and cast steel offer greater resistance to high temperature than do ordinary iron or steel, but they are also more apt to crack and break if subjected to sudden changes in temperature or uncompensated expansion. Fire-brick, of course, offers splendid resistance to high temperatures, but it is difficult to construct or to maintain a fire-brick retort that will be leak-proof. The heat conductivity of fire-brick is very poor. The use of Monel metal, calorized iron and other metals capable of resisting high temperatures is worthy of careful consideration.

Fig. 10.—Sketch of Randall Retort.

In general, the vertical type of retort will be the most satisfactory. This type of retort is exceedingly strong structurally; it occupies a minimum of ground space, and requires a minimum number of moving parts. To gain full efficiency from this retort, however, the furnace surrounding it must be carefully designed. Lateral flues are more efficient than vertical flues and should be employed where possible.

Vertical or horizontal retorts should always be built in "benches," so that heat radiation may be reduced and operations centralized. Each retort, however, should be entirely independent of the other retorts in the same bench, in order that any retort may be "cut out" for repairs without shutting down the entire bench. Positive and automatic means of charging and discharging the shale should be provided.

The well-designed oil-shale plant is laid out according to a carefully considered plan which should provide for a minimum of supervision, and should recognize the possibility of future enlargement of the plant.

Practices which make for economy and efficiency.—Economy and efficiency in retorting oil shale may be aided by the following practices:

(1) Preheating the incoming shale;

(2) Recovering heat from the spent shale;

(3) Using the carbon content of the spent shale in the formation of producer gas;

(4) Fractional condensation of the vapors.

Preheating the incoming shale will result in very important economies. The retort itself will be designed to produce oil by destructive distillation. The shale must be thoroughly dried and heated to a sufficient temperature before destructive distillation will take place, and the further this preheating can be accomplished before the shale enters the retort, the greater the capacity of the retort will be. Preheating can be accomplished by the direct heat of the waste flue gases on the incoming shale in any desired form of dryer, by the indirect heat of the outgoing oil vapors in a suitable heat exchanger, or by a combination of the two methods.

Heat may be recovered from the spent shale by placing an air jacket around the discharge opening of the retort. By this means the spent shale may be used to heat air for combustion.

It is possible to make use of the carbon content in the spent shale by treating while hot in a gas producer. The Scotch make use of this principle in their retort, the bottom half of which is designed primarily as a gas producer. Present practice suggests that the gas producer should not be made a part of the retort itself but should be provided as a separate piece of apparatus. The gas produced by this means, added to the fixed gas produced in the retort, should furnish more than enough heat for the retorting operation. If additional gas is required for any purpose, it may easily be produced by adding a little fresh shale to the spent material in the producers.

Fractional distillation of oil shale is physically impractical. This is perfectly clear if the cracking theory of producing oil from shale is accepted. It is well known that under any constant conditions of temperature and pressure, the complete condensates of the vapors formed at different stages of the " cracking " reaction do not differ essentially. This is true also of oil-shale distillation. The vapors produced contain a mixture of hydrocarbons of different boiling points, which does not vary markedly from the beginning of the reaction to the end. It is, however, quite possible to fractionally condense these vapors and to separate out, by differential cooling, oil mixtures of different boiling points. This is standard refinery practice in continuous refining as practiced by the Shell Oil Co. and others. Several excellent types of dephlegmators have been designed which accomplish the desired result.

In its simplest form, fractional condensation consists in passing the vapors through a series of condensers, maintained at fixed temperatures. In the first and warmest condenser, heavy hydrocarbons which have high boiling and condensing points are liquefied and collected; in the second condenser, which is maintained at a temperature several degrees cooler than that of the first, the next fraction condenses and so forth. By this means a direct rough separation is made from the vapors and further refining is much reduced and simplified.

Refining

Little progress has been made in determining the exact practice which will be required in refining shale oil. It is clearly recognized that shale oil will yield products which are different from, but not necessarily superior or inferior to, the corresponding products from ordinary petroleum. The difference in chemical composition of these shale-oil products from the ordinary petroleum products will involve some changes in refining methods, particularly in acid and soda treatment or "agitation." The present types of agitators are designed to remove all the unsaturated compounds in the oil treated. Inasmuch as shale oils contain from 50 to 90 per cent of unsaturated compounds, and as the presence of at least a portion of the unsaturates in the refined oil will not decrease its value, present methods of treating cannot be applied without serious losses.

Moreover, it is difficult to redistill shale oil without cracking and breaking down the heavier hydrocarbon molecules. For this reason, the redistillation of shale oil must be conducted with great care, and may even necessitate methods which, although known, have not been generally applied to ordinary petroleum distillation.

Recent great advances have been made in the filtration of oils for the selective absorption and adsorption of undesirable ingredients. This has been achieved primarily by the discovery of clays with higher absorbing value than the fuller's earth previously used. Next, the efficiency of clays has been greatly increased by treatment with sulphuric acid. Silica gels give promise of greater efficiency. Finally, researches into the use of various forms of charcoal for absorption in gas-mask work during the war has been extended to the use of improved, especially of "activated," charcoals for oil filtration. It is claimed by good authority that for some purposes activated charcoals have given at least twenty-five times the efficiency of standard fuller's earth. In all probability the separation of undesirable ingredients from shale-oil fractions will be effected by a preliminary treatment with dilute sulphuric acid for the removal of oily bases, then by a slight treatment with strong acid followed by a "clay wash" or filtration.

By-products.—Products generally recoverable in oil-shale distillation are oil, fixed gas, spent shale, and ammonium sulphate. Considering the oil as the main product, and assuming that the fixed gas will be consumed during the operation, there remain as by-products the spent shale and the ammonium sulphate. In certain instances, potash and phosphates may be commercially recovered. It is doubtful that the spent shale, with the exception of that from California, will have any great market value. By this it is not meant that this spent shale will have no commercial uses; undoubtedly it will; but with the tremendous production of spent shale which must accompany any important production of shale oil it will be next to impossible to sustain a high market value for the material, no matter how desirable it may be. Ammonium sulphate should command a good market and will certainly add perceptibily to the profits made from working a shale which will yield sufficient ammonium sulphate to justify its recovery. By no means all shales contain such amounts. Potash and phosphates are not generally recovered from the oil shale itself, but are sometimes found in associated beds which may be included in the mining operations. Deposits of this type are found in Tennessee, Montana and a few other localities.

THE KNOWN SHALE AREAS OF THE UNITED STATES—OCCURRENCE, GEOLOGY AND CHARACTERISTICS

Eastern States

"The black shales of the Eastern United States are mainly at one general horizon, in the upper Devonian or possibly in the lower Carboniferous, which extends from New York to Alabama and westward to the Mississippi. Other extensive deposits of black shale occur at one or more horizons in the lower Devonian and at one horizon in the Ordovician. In addition, black shales overlie some of the coal beds, especially certain beds in the eastern interior coal field.

"The principal body of black shale is known as the Chattanooga, New Albany, or Ohio shale. This bed underlies the eastern coal fields, and crops out in a long line from central Alabama northeastward through Tennessee and Virginia and all around the Nashville Basin in central Tennessee. West of the Appalachian coal field its outcrop extends from north to south across central Ohio, passing close to Columbus, and reaching the Ohio River near Vanceburg. Thence the outcrop makes a loop through central Kentucky, past Lebanon and northward to Louisville, from which it stretches in a broad belt northwestward across Indiana, past Indianapolis nearly to Chicago. From this western belt of outcrop the shale extends eastward under eastern Ohio and underlies nearly all of Kentucky except the area within the loop described and all of Indiana west of the outcrop."[1]

[1] Ashley, G. H.: "Black Shales of the Eastern States," U. S. Geol. Survey, Bull. 641.

Characteristics of Eastern Shales

The shales of the eastern states belong to one general type which may be des as follows:

Color:	
Fresh surface	Black
Weathered surface	Blue to gray
Character:	
Fresh	Generally massive
Weathered	Fissile
Streak	Brown
Hardness	3.0 to 4.0
Texture	Fine-grained
Elasticity	Tough to brittle
Specific gravity	2.0 to 2.7

TABLE I

Raw Shale—Chemical Composition—(Proximate) *

	Sample I	Sample II	
Volatile hydrocarbons	12.84	13.14	22
Fixed carbons	6.25	8.50	6.
Ash	80.91	78.36	70.
Total	100 00	100.00	

* Indiana shale. Analysis by John R. Reeves.

TABLE II

Spent Shale—Analysis of Ash *

	Per cent
Carbon	11.0
SiO_2	59.2
FeS	12.2
Al_2O_3	17.5
CaO	0.1
MgO	0.14
Total	100.14

* Indiana shale. Analysis by Hans Duden.

TABLE III

Distillation Tests of Crude Shale Oil *

Color of oil: Green, changing on exposure to deep red-black.

	Cubic centimeters per hundred
To 150° C	17.75
150°–200° C	11.75
200°–250° C	17.75
250°–300° C	16.00
Above 300° C	8.50
Coke and tar	28.25
Percentage of unsaturated hydrocarbons	55
Percentage of pyridines or pyridine compounds	2–5
Percentage of creosote	1–2
Percentage of phenol	0–0.5
Sulphur compounds	2–3
Specific gravity of crude oil	0.8900
Heating value, crude oil, B.t.u	19,200

* Figures are from tests on Indiana shales made by John R. Reeves, Indiana Geol. Survey.

Representative samples, showing the oil, gas, and ammonia content of these shales will be found in tables for the respective states under the general head of occurrence. However, most of the samples were examined by the United States Geological Survey in connection with a report by G. H. Ashley, published in 1916. At that time the technology of examining shale samples was not very well worked out and the results obtained by the Government investigations are by no means consistent. It is probable, then, that the highest results quoted are nearer correct than the low ones. This is particularly true of the determinations of ammonium sulphate. The black shales of the eastern states should yield, on the average, when retorted with steam, 15 to 25 gallons of oil, 2000 cubic feet of permanent gas, and 25 to 49 pounds of ammonium sulphate per ton.

The shale in the eastern states is generally uniform, is quite thick (40 to 100 feet), is readily accessible to transportation and to markets, and is often well located for inexpensive mining. It is sometimes found associated with cannel coals which may be mined with it and which will greatly increase the total yield of oil. In some instances it is found in contact with phosphate beds which may yield commercial amounts of phosphate as a by-product to the shale enterprise. For the various reasons noted, this shale is commercially interesting in spite of its low yield in oil.

COLORADO, UTAH, AND WYOMING

" In Colorado, Utah and Wyoming, the oil-yielding shale is confined almost entirely to the middle part of the Green River formation of early Tertiary (Eocene) age. The occurrence of this formation, and the points at which rich shale has been sampled are shown in the accompanying map. In northwestern Colorado the Green River formation is the youngest present, but north of White River and only a few miles west of the Colorado–Utah line the Bridger formation rests unconformably on the Green River as well as older formations, and along the northern edge of the Uinta Basin in Utah the Bridger obscures the entire outcrop of oil shale. The Bridger formation also occupies the central part of the Green River Basin in southwestern Wyoming, and west of Burnt

Fork and south of Carter overlaps the Wasatch formation, covering the outcrop of the Green River.

"The Green River is underlain by the Wasatch and this in turn by the Mesaverde, which in Colorado and Utah is coal bearing. The Wasatch in Wyoming contains many valuable beds of coal which may be needed for fuel when the oil-shale industry is developed.

"The Green River formation has a maximum thickness near the mouth of Piceance Creek of about 2600 feet and may be separated there on the basis of the presence or absence of oil-yielding shale into three fairly distinct parts. The upper and lower parts of the formation are practically barren, but the middle member of the formation contains, at every locality examined, beds of shale that will yield considerable oil. The section measured at Piceance Creek shows the oil-yielding formation to be 1550 feet thick, whereas the lower barren part is only 342 feet thick, and the upper barren part is 716 feet thick. According to measurements made at Morris Station on the Book Cliffs, the upper 595 feet of the section there exposed is oil-yielding, but the underlying part, 1487 feet thick, includes no beds which will yield a considerable amount of oil. The great thickness of the barren beds at Morris corresponds very closely with the thickness of the lower part of the formation as described by Woodruff in a section measured along the Mount Logan trail in Sec. 26, T. 7 S., R. 97 W., only a few miles to the west and on the same general cliff. In general, the lower member of the Green River is very variable both in thickness and in character. Along Evacuation Creek near Dragon, Utah, this member includes about 600 feet of coarse sandstone, oolite, and shale with no persistent bed, and some very remarkable lenticular beds. Only a few miles away in Hell's Hole Canyon, northwest of Watson, the lower part of the Green River formation consists largely of shale with comparatively thin sandstone beds and only a little oolite.

"In contrast to this extreme irregularity in the lower part of the formation the thin beds of the oil-yielding portion are remarkably persistent. At three places sampled in eastern Utah, separated from one another by 5 and 7 miles, there are three thin beds of sandstone which are remarkably regular in thickness, while the interval between them varies only slightly from place to place.

"The line between the Wasatch and the Green River formations is very difficult to follow because of lack of exposures and is very hard to identify accurately from place to place. There seems to be a general gradation from the upper part of the Wasatch formation into the lower part of the Green River.

"The Green River formation consists principally of shale, but contains, especially in its lower part, beds of sandstone, many of which are ripple-marked. Most sections show one or more beds of oolite and some conglomerate or conglomeratic sandstone. Near the old Black Dragon Mine, Utah, however, the lower part of the formation, according to measurements made by Woodruff, contains oolite and sandstone equal to more than half of the exposed thickness of the beds (529 feet). There is at the base of the upper part of the formation on Yellow and Piceance Creeks a bed of massive sandstone which may be equivalent to the Tower sandstone of Powell in southwestern Wyoming. As is shown by the section, pages 884–887, there are beds in nearly every section which yield at least 15 gallons of crude oil to the ton of shale, but the correlation of beds from one measured section to another is very uncertain, although the sections may be only a few miles apart. A careful study of the strata exposed in a continuous cliff face a m le or more in extent shows that although the formation appears to be remarkably regular in thickness, individual beds vary greatly from place to place. A single massive bed, 5 feet thick, may change within less than half a mile to a comparatively thin-bedded shale. Study of any single bed at several places along its outcrop, to determine its variability in thickness, bedding, mode of weathering, and value as a source of oil, was made impossible by lack of exposures, except near the mouth of Piceance Creek in Colorado and along the west side of the Green River in southwestern Wyoming. Beds carefully measured and sampled in these two localities yielded rather inconsistent results."[1]

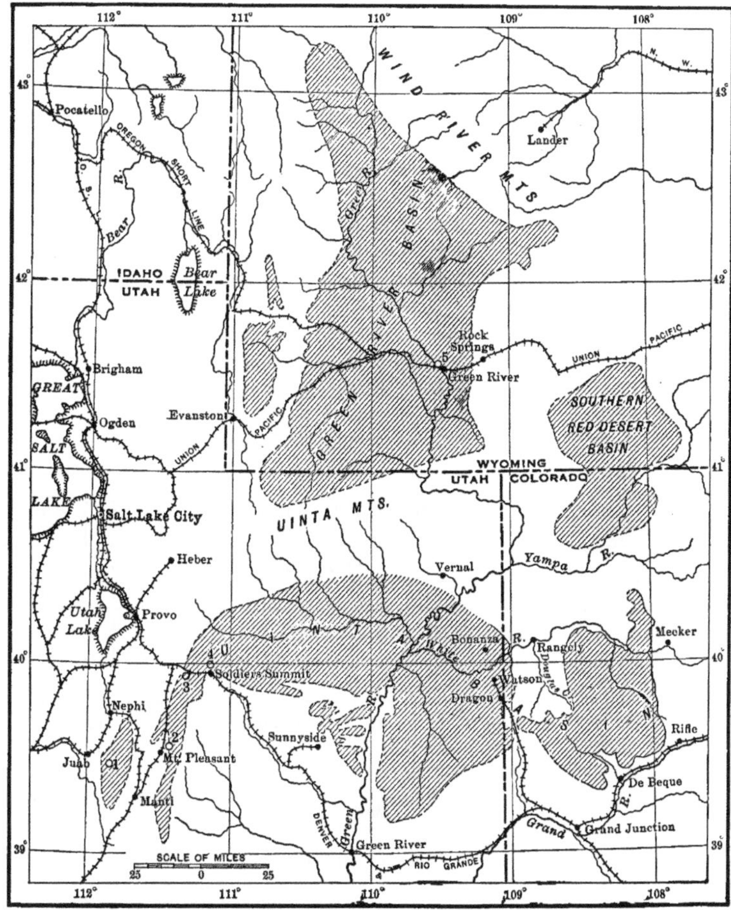

Fig. 11.—Map of oil shale deposits of Colorado, Utah, and Wyoming.

As has been stated previously, the methods used when these samples were tested had not reached a high state of perfection with the result that the oil yields are rather uniformly low. Several sections made by the author along Evacuation Creek, south of Watson, Utah, indicated that there was a thickness of 100 feet of shale in that area which would

[1] Winchester, Dean E.: Oil Shale in Northwestern Colorado, U. S. Geol. Survey, Bull. 641.

average 28 gallons of crude oil to the ton, from the top to the bottom, that 40 feet of this layer averaged 45 gallons to the ton, and that two ten-foot shale members in the same layer averaged 60 gallons to the ton. This is perhaps the best exposure personally examined in the Colorado–Utah–Wyoming field.

The shales from the Green River formation have the following general characteristics:

Color
- Fresh surface........................... Brown to black
- Weathered surface...................... Brown, gray to blue white

Character
- Fresh.................................. Generally massive
- Weathered............................. Fissile, curly, papery

Streak................................... Brown
Hardness................................. 3 to 4
Texture.................................. Fine-grained
Elasticity............................... Very tough and elastic
Specific gravity......................... 1.52 to 2.06

TABLE IV

Ash Analyses, and Relation of Ash to Volatiles of Green River Shales *

Sample	SiO_2	$Fe_2O_3+Al_2O_3$	CaO	MgO	Volatile plus fixed carbon	Oil content gallons per ton
De Beque, Colo.	44.7	25.6	17.6	5.28	40.0	42.7
Dragon, Utah	45.8	16.4	23.9	7.9	42.2	41.6
Dragon, Utah	46.8	17.5	23.9	8.9	34.3	21.7
Green River, Wyo.	38.9	12.4	38.3	4.9	34.9	23.4
Green River, Wyo.	41.9	18.8	17.6	10.9	48.1	58.7

* Gavin, M. J., and Sharp, L. H.: Reports of Investigations, U. S. Bureau of Mines.

TABLE V

*Physical Properties**

Sample	Specific gravity	Specific heat	Heat conductivity	Heat of combustion
De Beque, Colo.	1.92–2.06	0.265	0.00314–0.00518	2460 cal. per gram 4428 B.t.u. per pound
De Beque, Colo., spent shale	....	0.223	....	600 cal. per gram 1080 B.t.u. per pound

*Gavin, M. J., and Sharp, L. H.: Reports of Investigations, U. S. Bureau of Mines.

TABLE VI

Relation of Oil Content to Heating Value * (See Chart)

Sample	Oil, gallons per ton	Volatiles plus fixed carbon, per cent	Heating value, B.t.u. per pound
Soldier's Summit, Utah	16.8	33.55	2266
On White River, Colo.	8.4	20.19	1195
On White River, Colo.	40.6	38 07	4030
Book Cliffs, Colo.	28 0	40 02	3068
Conn Creek, Colo.	65.3	52.04	7036
† De Beque, Colo.	42.7	40.00	4428

* Winchester, Dean E.: Oil Shales in Northwestern Colorado, U. S. Geol. Survey, Bull. 641.
† Gavin, M. J. & Sharp L. H.—U. S. Bureau of Mines. Reports of Investigations.

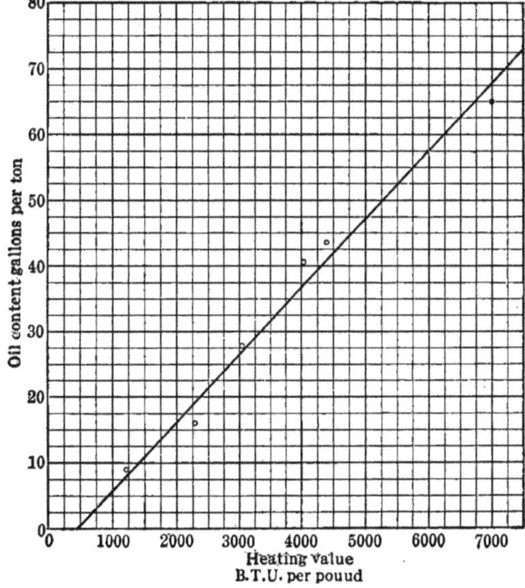

FIG. 12.—Chart showing relation of oil content to heating value for various shales.

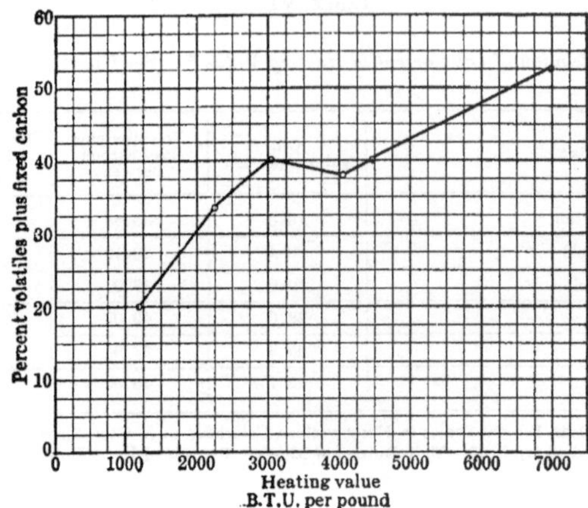

FIG. 13.—Chart showing relation of volatiles plus fixed carbon to heating value for various shales.

PHYSICAL AND CHEMICAL PROPERTIES OF SHALE OIL FROM COLORADO, UTAH AND WYOMING

TABLE VII

Physcial Properties of Crude Shale Oil *

(Before and after treatment with sulphuric acid)

Sample	Gravity at 60° F.		Color
	Specific	Degrees Bé.	
Kimball Creek, Colo.	0.9302 †0.8950	20.5 26.5	Dark brown Dark red
Parachute Creek, Colo.	0.9271 †0.8805	21 29	Dark brown Dark red
Conn Creek, Colo.	0.9103 †0.8695	24 31	Dark brown Red

* Woodruff, E. G., and Day, D. T.: Oil Shale of Colorado and Utah, U. S. Geol Survey, Bull. 581.
† After treatment with sulphuric acid.

Table VIII

Distillation Tests, Crude Shale Oil *

Sample No.	1	2	3	4	5	6
Initial B.P., ° C.	70	80	54	32	25	72
Distillation	Cubic centimeters per hundred					
To 100° C.	2.5	2.5	4	5.8	4.1	2.
100°–125° C.	1	3.5	2	4.8	3.6	1.
125°–150° C.	2.5	4.5	3	5.6	5.5	8.
150°–175° C.	6	6	5	6.4	5.8	6.5
175°– 00° C.	5	5	4.5	6.6	6.7	5.
200°–225° C.	5	6	5	6.8	6.1	5.
Total to 225° C.	22.0	27.5	23.5	36.0	31.8	27.5
225°–250° C.	5	8	5	7.2	6.4	5.
250°–275° C.	7	8.5	7	9.4	7.2	7.
275°–300° C.	7	9	12	8.7	9.0	7.
Total 225°–300° C.	19.0	25.5	24.0	25.3	22.6	19.
Total distillate.	41	53.0	47.5	61.3	54.4	46.5
Total residuum.	59	47	52	38.5	43	53.5
	Specific gravity at 60° F.					
Crude.	0.9290	0.8838	0.9126	....	....	0.9327
Fraction to 150°	0.7974	0.7568	0.7605	....	....	.8202
Fraction, 150°–300° C.	0.8742	0.8524	0.8538	....	....	.8876
Residuum	0.9894	0.9368	0.9628	....	....	1.0160
Percentage of asphalt.	4.10	0.47	1.03	....	....	3.62
Percentage of paraffin.	3.72	4.70	4.00	....	....	1.63
Percentage of sulphur.	....	1.42	0.69	....	....	....
Percentage of nitrogen.	1.549	1.849	2.135	....	....	1.643
	Unsaturated hydrocarbons					
Percentage in crude.	86	72	Undetermined	....	....	81.6
Percentage in 300° fraction.	71	57	58	....	....	71.

Sample 1. White River, Rio Blanco County, Colo. Steam used. Oil content 22.88 gallons per ton.
Sample 2. White River, Rio Blanco County, Colo. Oil content 40.6 gallons per ton.
Sample 3. Conn Creek, Garfield County, Colo. Oil content 65.3 gallons per ton.
Sample 4. White River, Uinta County, Utah. Oil content 33.3 gallons per ton.
Sample 5. Duchesne County, Utah. Oil content, 39 gallons per ton.
Sample 6. Same as sample 1, distilled dry.

* Winchester, Dean E.: Oil Shales in Northwestern Colorado, U. S. Geol. Survey, Bull. 641.

Experiments on cracking shale-oil residuum.—That the residue above 175° C. from the ordinary distillation of shale oil may be cracked to yield additional gasoline was determined by experiments made in a Rittman furnace. In these tests shale oil from White River and Ninemile Creek, Utah, was distilled by ordinary methods and the residuum was then run through the Rittman process. The recovery by ordinary distillation is given in columns 4 and 5, Table VIII. The cracking tests follow:

TABLE IX

Rittman Furnace Tests on Residuum Over 175° C. Obtained from Ordinary Distillation of Shale Oil *

Pressure used: 150 pounds

	Sample A	Sample A'	Sample B	Sample B'
Temperature used, °C.	525	550	525	600
Specific gravity of residuum	0.920	0.920	0.957	0.957
Specific gravity of recovered oil	.901	.902	.929	.959
Percentage recovery	79	79	82	70

* Winchester, loc. cit.

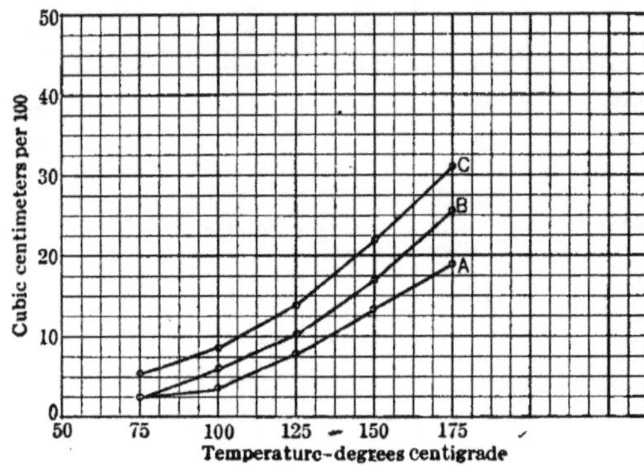

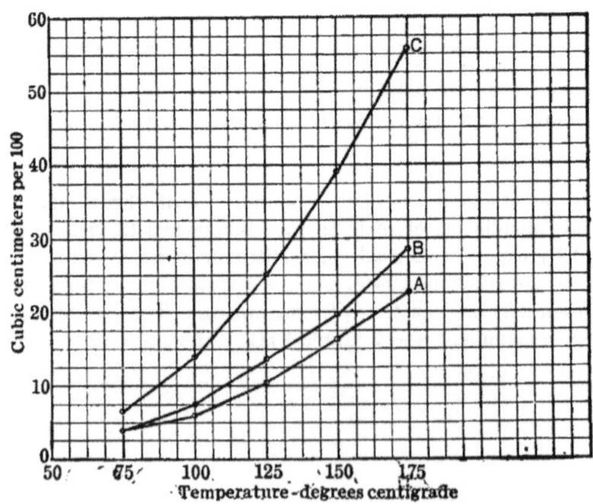

FIGS. 14 and 15.—Charts showing increase in low boiling fractions obtained by cracking shale oil residuum.

TABLE X

Distillation of Oil Recovered in Above Tests

(See Charts 14 and 15)*

	Sample A	Sample A′	Sample B	Sample B′
Specific gravity of recovered oil......	0.901	0.902	0.929	0.959
Distillation (per cent by weight)				
To 75° C.	....	4.2	....	3.8
75°–100° C.	2.2	6.2	2.8	1.8
100°–125° C.	1.1	8.5	1.1	2.5
125°–150° C.	1.4	10.7	1.6	2.7
150°–175° C.	2.7	13.8	3.0	4.2
175°–200° C.	7.0	20.0	6.3	6.9

* Winchester, loc. cit.

Pyridine bases in shale oil.—Woodruff and Day [1] noted the odor of pyridine compounds in the shale oil collected by them from Colorado, Utah, and Wyoming shales. Determinations of the amount of pyridine compounds existing in shale oil from various localities resulted as follows:

TABLE XI

Pyridine Compounds in Dry Distilled Shale Oil *

Location of bed sampled	Oil yield, gallons per ton	Nitrogen in shale (original) per cent	Specific gravity of oil	Pyridine compounds percent by weight
Near Watson, Utah	32	0.53	0.8998	3.88
Near Watson, Utah	15	.35	.8870	4.28
Temple Station, Utah	90	1.30	.8745	5.99
Sweetwater County, Wyo.	34	.68	.8994	2.26
Fossil Butte, Wyo.	8	.10	.8705	8.91

* Winchester, Dean E.: Oil Shales in the Uinta Basin, U. S. Geol. Survey, Bull. 691B.

[1] U. S. G. S. Bulletin 581.

Amount and Character of Fixed Gases Recovered in Oil Shale Distillation

Table XII

Cubic Feet of Fixed Gases Produced in the Distillation of Shales of Different Oil Content *

Location sampled	Oil content, gallons per ton	Specific gravity, oil	Fixed gas, cubic feet per ton
On White River, Colo., near Utah line....	11.9	0.9010	**2395**
Cathedral Bluffs, Colo..................	33.6	.8919	**3034**
On White River, Colo., Wilson Ranch.....	40.6	.8838	**3832**
Book Cliffs, Colo.......................	40.6	.8790	**1916**
Conn Creek, Colo........................	65.3	.9115	**4549**
Green River, Wyo........................	29.4	.9130	**2978**

* Winchester, loc. cit.

Table XIII

Analyses of Fixed Gases Produced from Shale *

Stage of distillation	Carbon dioxide (CO_2)	Carbon monoxide (CO)	Oxygen	Unsaturated hydro-carbons	Other hydro-carbons and nitrogen
First..........	12.70	1.56	1.56	11.10	73.08
Second........	5.20	7.50	1.50	9.50	76.30
Third.........	3.20	6.20	2.00	10.00	78.80
Fourth........	4.00	1.60	1.00	12.40	81.00
Last..........	12.40	9.60	0.60	6.40	71.00

* Woodruff and Day, loc. cit.

Note.—The tests were made by Woodruff and Day on shale from Conn Creek, Colo., oil content 68 gallons to the ton. The shale was heated for six hours in a closed retort, without steam, and the gas was collected at regular intervals. The high yield of CO_2 in the last stages of distillation is considered due to the dissociation of the calcium carbonate in the shale. At the highest temperatures, some of this CO_2 was reduced to CO.

Ammonium Sulphate and Other By-products

Table XIV

Comparison of Amounts of Ammonium Sulphate Recovered by Steam and by Dry Methods of Distillation *

Sample	Oil				Ammonium Sulphate		
	With steam		Without steam		Theoretical yield, pounds per ton	Yield as determined	
	Yield, gallons per ton	Specific gravity	Yield, gallons per ton	Specific gravity		With steam, pounds per ton	Without steam, pounds per ton
Wilson Ranch, Colo.	10	0.9135	8.4	0.8946	43.2	29.9	18.3
Wilson Ranch, Colo.	44	.9630	40.6	.8838	50.8	34.0	8.5
W. of Rifle, Colo.	39	.9234	28.0	.9126	43.2	15.8	7.3
Near Watson, Utah	55	.9286	55.0	.9052	75.4	23.1	9.6
Soldier's Summit, Utah	23	.9346	16.8	.8937	36.6	13.4	3.5

Winchester, loc. cit.

Note.—The steam distillation was carried on in the same type of retort, and, as nearly as possible, under the same conditions as the dry distillation, the only difference being that superheated steam was injected during the entire period of distillation. This has had the effect of: (1) increasing the yield of oil in almost every case; (2) lowering the gravity and the quality of the oil; and (3) doubling and sometimes tripling the yield of ammonium sulphate.

SECTION OF GREEN RIVER FORMATION IN NORTHWESTERN COLORADO

Forks of Parachute Creek *

About 10 miles north of Grand Valley, about Sec. 36, T.5S., R.96W.

	Feet.
Interval to top (estimated)	150
Shale (estimated average 15 gallons)	125
Shale (estimated average 30 gallons)	75
Shale, rich, hard (5 feet near middle of bed; 42 gallons)	25
Shale (average probably 30 gallons)	30
Covered and lean shale	86
Shale (estimated 30 gallons)	45
Shale (as good or better than 36 gallons)	4.5
Shale, sandy, lean	6
Shale (estimated 25 gallons)	4
Shale (nearly 36 gallons)	5
Shale, medium	15
Shale, hard, blue (36 gallons)	3
Shale, medium	9
Shale, hard, blue (36 gallons)	2
Shale, medium	9
Shale, hard, blue (36 gallons)	2.5
Interval, probably 40 per cent shale in beds 3 feet or more thick which will test 25 gallons or more to the ton	175
Shale (19 gallons)	2
Shale, medium rich	5.5
Shale (19 gallons)	3
Shale, medium rich	11
Shale (19 gallons)	2
Shale, lean	6.5
Shale, hard, blue (19 gallons)	5.3
Shale, sandy, yellow	5
Shale (estimated 25 gallons)	18
Shale, slightly less rich than 18 gallons	12
Shale, hard (18 gallons)	6.3
Sandstone	0.2
Shale, lean	.....
	847.8

* Winchester, Dean E.: Results of dry distillation of miscellaneous shale samples, U. S. Geol. Survey, Bull. 691 B.

Sections in Northwestern Utah

Location Q, along Evacuation Creek between Temple Switch and Dragon, Utah

		Feet.
Shale, thin bedded, lean to barren		40
Shale, hard, dark (estimated yield, 20 gallons)		1
Shale, lean to barren, thin bedded; a few rich layers less than 1 inch thick (Diptera larvae)		155
Sandstone		1
Shale, platy, lean to barren; two or three rich beds about 1 inch thick		36
Shale, thin bedded, rich		1
Shale, lean		6
Sandstone, persistent		0.4
Shale, lean, thin bedded		4
Shale, hard, dark brown, rich	(Sample yielded 31 gallons.)	1.8
Shale, hard, light brown, rich		2.1
Shale, hard, dark brown, rich		0.5
Shale, sandy (not in sample)		......
Shale, hard, dark brown, rich		.8
Shale, hard, dark brown, rich		.66
Shale, dark, tough	(Sample from richest part of upper bench; 90 gallons. Sample from whole bed at surface; 32 gallons.)	1.33
Shale, dark, platy		1.4
Shale, dark, hard, rich		0.24
Shale, soft, dark brown		0.25
Shale, hard, dark, rich		0.25
Shale, soft, dark brown		0.4
Shale, hard, dark, rich	(Sample from whole bed 1½ feet back from outcrop; 55 gallons.)	0.3
Shale, soft, dark brown		0.8
Shale, hard, dark, rich		1.5
Shale, thin-bedded, platy	(Sample yielded 15 gallons.)	4.2
Shale, rather lean and papery		2
Shale, hard, dark brown to black (32 gallons)		4.25
Shale, hard, lean, some thin sandstone layers		3.8
Shale, hard, rich (two samples, one yielding 23 gallons and the other yielding 18 gallons)		3.9
Shale, minutely banded, some rich layers (10 gallons)		6.5
Shale, lean to barren, with two bands of small dark sandstone lenses		4
Sandstone, hard, quartzitic, persistent		0.4
Shale, sandy, barren; thin beds of sandstone		3.1
Shale, brown and black, rich	(12 gallons.)	.6
Shale, hard, weathers green		
Shale, sandy; weathers greenish-gray; lean to barren		2
Shale, hard (9 gallons)		1
Sandstone, rough, coarse, containing asphalt; top and bottom surfaces irregular, with shale conforming to the irregularities		1.8
Shale		7
Sandstone, persistent		0.25
Shale, lean, sandy, gray to reddish, with several thin layers of sandstone		13.7

SECTIONS IN NORTHWESTERN UTAH—*Continued*

		Feet.
Shale, hard, rich	(32 gallons.)	2
Shale, soft		2
Shale, hard, rich		1.75
Sandstone		.25
Shale, hard	(Sample yielded 6 gallons.)	.27
Sandstone (not included in sample)		.1
Shale, hard		1.6
Sandstone, persistent (not included in sample)		.2
Shale, hard		.9
Shale, clayey		.4
Shale, hard, mostly lean, with thin beds of richer shale		1.5
Sandstone, persistent		.25
Shale, hard, rich	(7 gallons.)	2.75
Sandstone		.1
Shale, hard, rich		.9
Sandstone		.05
Shale, hard, rich		2.5
Sandstone		.1
Shale, hard, dark (7 gallons)		.7
Sandstone, persistent		.5
Shale, hard, rich to lean	(Sample yielded 14 gallons.)	3.25
Horizon of sandstone lenses, none of which occurred where sample was taken		
Shale, hard, rich		2.5
Sandstone		.1
Shale, probably lean	(Sample yielded 19 gallons.)	.6
Sandstone, bearing gypsum (not included in sample)		.4
Shale, hard, dark, rich; some gypsum near top		1.8
Shale, with considerable gypsum		.33
Shale, very dark brown, rich		.4
Sandstone, brownish, shaly		23.
Shale, papery, lean		1.5
Shale, rich; weathers blue		.5
Shale, sandy, and barren shaly sandstone		10.
Shale, rich, papery		.5
Shale, sandy, barren		17.5
Sandstone, brown, massive		.9

* From Oil Shale of the Uinta Basin, Northern Utah, by Dean E. Winchester, U. S. Geol. Surv Bull. 691-B.

Sections of Parts of Green River Formation in Southwestern Wyoming

T. 14*N.*, R. 99W.*

		Feet.
Sandstone, coarse-grained, not massive		50
Sandstone, containing fossil shells		.33
Sandstone, coarse-grained, thin-bedded		10
Covered, probably sandy shale		35
Sandstone, coarse		8
Covered, mostly shale		30
Shale, papery, drab, lean		5
Shale, thin, barren, and sandstone		72
Shale, drab, thin, lean		3
Shale, thin, drab, barren		20
Shale, thin, lean		30
Sandstone, concretionary		1
Shale, thin, lean		14
Oolite and chert		
Shale, thin-bedded, lean		14.5
Shale, thin-bedded; weathers blue, rich	(Sample yielded 30 gallons.)	2
Shale, gray, sandy (not included in sample)		1.6
Sandstone, yellow (not included in sample)		.1
Shale, thin-bedded; weathers blue; rich		3
Shale, yellow, sandy		28
Shale, papery, lean		40
Shale, drab, fissile		10
Sandstone, concretionary		1
Shale, drab, papery		13
Oolite		0.5
Shale, drab, papery		10
Sandstone, oolitic		0.33
Shale, drab, fissile		12.5
Sandstone, micaceous		1
Sandstone, yellowish		3
Shale, drab, thin sandstone lenses		26
Sandstone, shaly, yellowish		1
Shale, drab, papery, barren		5
Sandstone, shaly, yellowish		1.5
Shale, greenish drab		37
Maroon clay shale (probably Wasatch)		......
		489.03

* From Oil Shale in Northwestern Colorado and Adjacent Areas, by Dean E. Winchester: U. S. Geol. Survey, Bull. 641.

Commercial possibilities and development.—Colorado, Utah, and Wyoming may be classified as to commercial importance as follows:

	Colorado	Utah	Wyoming
Transportation	1	3	2
Mining conditions	2	1	2
Richness of shale	2	1	3
Quality of oil	1	1	2

The Denver and Rio Grande Railway between Rifle and De Beque, Colorado, affords transportation facilities for the rich shale deposits on Conn, Roan, Parachute, and Kimball Creeks. Wyoming has the Union Pacific Railway, which at the town of Green River affords access to a portion of the shale area. Utah unfortunately has only the narrow-gage road of the Uinta Railway, with terrific grades to be encountered before reaching the main line of the Denver & Rio Grande Railway. Even the Uinta reaches only a very small portion of the rich Utah shales.

The richest beds of shale, according to experiments made thus far, are in the Uinta Basin of Utah. Thirty-one out of eighty-eight samples from this area (35 per cent), examined by the United States Geological Survey, yielded 30 or more gallons of oil to the ton against eight out of fifty-five (14.5 per cent) in Colorado and four out of forty (10 per cent) in Wyoming. Furthermore, the rich beds in Utah are often thick and well located for cheap mining. On White River and along the Uinta Railway in Utah, the writer has examined beds of shale, averaging 25 to 35 gallons for a thickness of 100 feet, which have little overburden and could be mined with open-pit methods. Many locations in this area have abundant water, good plant sites, and splendid facilities for disposing of the spent shale.

Colorado is not quite as fortunate in this respect. Although there are rich beds of shale along Conn, Roan, Kimball, and Parachute Creeks, the beds are not as thick as those in Utah, and a thickness of 100 feet would contain a large proportion of lean shales. Furthermore, the shale deposits nearest to transportation are so high upon the faces of almost perpendicular cliffs as to be practically inaccessible and generally have a heavy overburden of lean shale. In Wyoming the rich shale is generally in thin beds with a heavy overburden.

Shale oil from Utah contains a slightly higher percentage of light products than Colorado shale oil, but in other respects the two oils are very similar. Wyoming oils generally contain a fairly high sulphur percentage and are less valuable for this reason.

Good transportation facilities have centered the early commercial development of oil shale at Debeque and Rifle, Colorado. So far the development work has not progressed beyond the experimental stage. The industry has suffered from the lack of good financing and good engineering ability. It is unfortunately true that the actual results obtained represent but poorly the amount of money spent in producing them. A hopeful sign for the future is that large companies, with abundant capital for developing the industry and engineering skill to direct this development along proper lines, are entering the field and are carefully and systematically carrying out research work preparatory to large operations.

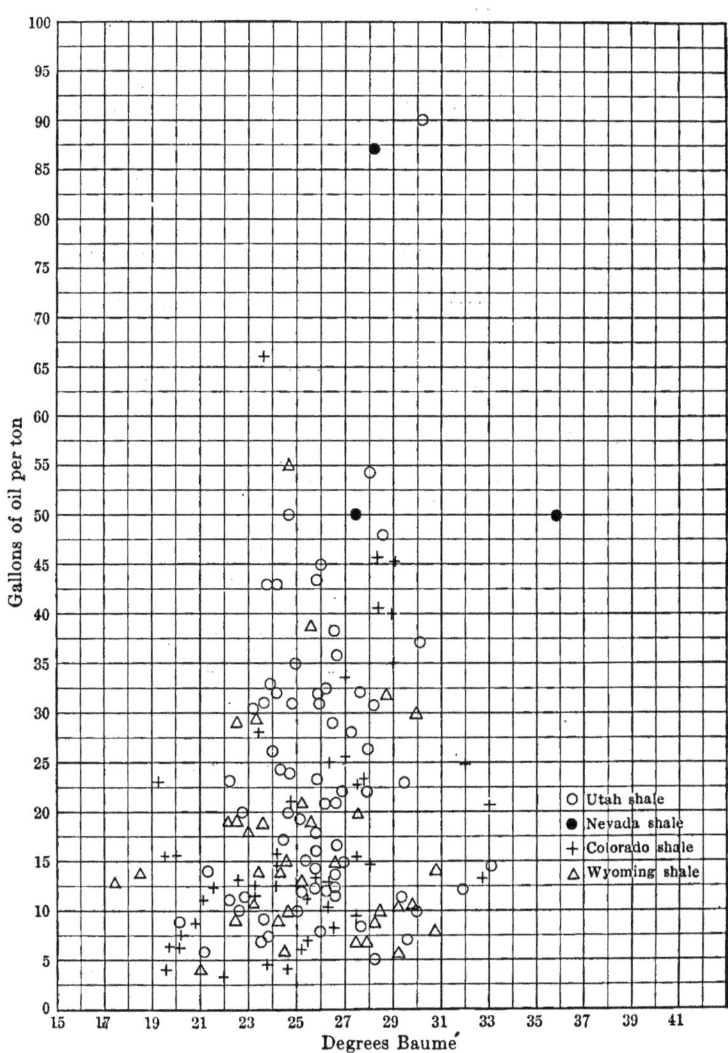

FIG. 16.—Chart showing relation of oil content in gallons per ton to gravity of recovered oil for various Colorado, Utah and Wyoming shale samples.

Nevada

The oil-snale deposits of Nevada, thus far determined, are confined to a relatively small area near Elko and Carlin, near the main line of the Southern Pacific Railway. The shale differs physically, chemically, and geologically from that of the Colorado–Utah–Wyoming area. Geologically the shale is of Tertiary age. The rich beds are few in number and are seldom over 5 feet in thickness. This area has been subjected to considerable geologic disturbance, and the shale beds dip at angles of 30° or more and are often faulted.

Characteristics of Nevada Shale

Color:
- Fresh surface........................ Brown to black
- Weathered surface.................... Gray to blue white

Character:
- Fresh................................ Generally fissile
- Weathered............................ Generally fissile

Streak................................. Brown
Hardness............................... 3.0 to 3.5
Texture................................ Fine-grained
Elasticity............................. Tough
Specific gravity....................... 1.9 to 2.5

Table XV

Ash Analyses and Relation of Ash to Volatiles

Sample	SiO_2	Fe_2O_3 Al_2O_3	CaO	MgO	Volatile and fixed carbon	Oil content gallons per ton
Elko, Nev.*	65.5	25.5	0.6	0.8	33.5	32.5
Elko, Nev.†	....	....	....	....	45.52	86.8

* Gavin, M. J., and Sharp, L. H.: Reports of investigations, U. S. Bureau of Mines.
† Winchester, Dean E.: Oil Shale in Northwestern Colorado, U. S. Geol. Survey, Bull. 641.

Table XVI

Relation of Oil Content to Heating Value *

Sample	Oil, gallons per ton	Volatiles plus fixed carbon, per cent	Ash, per cent	Heating value, B.t.u.
Elko, Nev	86.8	45.52	46.21	7796

* Winchester, Dean E.: Oil Shale in Northwestern Colorado, U. S. Geol. Survey, Bull. 641.

Table XVII

Physical Properties of Crude Shale Oil from Elko, Nevada †

Sample	Gravity at 60° F.		Color
	Specific	Baumé	
Elko, ev	0.8850	28.2	Greenish-brown
Elko, Nev	.8449	35.70	Greenish-brown

† Winchester, loc. cit.

TABLE XVIII

Distillation Tests, Crude Oil Shale, from Elko, Nevada

Initial B.P., ° C.	52*	90†	80°	70°		
Distillation	Cubic centimeters					
To 100° C.	10					
100°–125° C.	0.5					
125°–150° C.	1.5					
150°–175° C.	2					
175°–200° C.	4					
200°–225° C.	4					
Total to 225° C.	22.0	12	12	28		
225°–250° C.	6					
250°–275° C.	6					
275°–300° C.	10					
Total 225°–300° C.	22.0	10	12	22		
Total distillate	44.0	22	24	50		
Total residuum	56.0	78	76	50		
	Specific gravity at 60° F.					
Crude	0.8850	0 933	0.870	.800 to .950		
Fraction to 150° C.	0.7769					
Fraction, 150°–3.0° C.	0.8466					
Residuum	0.9643					
Percentage of asphalt	0.82	0	0	....		
Percentage of paraffin	6.93	....	26	....		
Percentage of sulphur	1.06	....	....	....		
Percentage of nitrogen	0.887	....		2 to 4‡		
	Percentage of unsaturated hydrocarbons					
150°–300° C.	55.0	11				

* Compiled from U. S. Geol. Survey, Bull. 641, Oil Shale in Northwestern Colorado, by Dean E Winchester.

† Unpublished anal. by D. T. Day. ‡ Pyridine bases.

TABLE XIX

Cubic Feet of Gases Produced in Shale Distillation

Location sampled	Oil content, gallons per ton	Specific gravity	Fixed gas, cubic feet per ton
Elko, Nev.	86.8	0.8850	3891

TABLE XX

Yield of Ammonium Sulphate by Dry and Steam Distillation

Sample	OIL				AMMONIUM SULPHATE		
	With steam		Without steam		Theoretical yield	Yield as determined	
	Yield, gallons per ton	Specific gravity	Yield, gallons per ton	Specific gravity		With steam	Without steam
Elko, Nev.	50.0	0.9109	50.0	0.8449	80.1	8.4	4.5
Elko, Nev.	....	....	86.8	.8850	36.6	....	6.0

Commercial considerations.—Splendid transportation facilities have aroused very general interest in the oil-shale deposits near Elko and Carlin, Nevada. Both of the towns named are on the main line of the Southern Pacific, and the shale beds are only a few miles from the railway. The shale itself is sometimes exceptionally rich, ranging from 18 to more than 85 gallons to the ton, and there are at least two 5-foot beds which will yield 50 gallons to the ton. As these beds dip into the hill at angles of about 30° underground mining will be necessary. Very good plant sites can be found, and water is plentiful. The oil from Nevada shale is exceptionally high in paraffin and is very valuable for that reason. The chief objections to the locality are: (1) the rather limited tonnage of rich shale, (2) the necessity for underground mining, to some extent at least, at considerable depth, and (3) the irregular and faulted nature of the shale.

California

The oil shales of California are found at one general horizon, the upper member of the Monterey formation (Miocene). In this state the oil shales are the result of the saturation of a porous shale body with liquid oil and the oil-bearing quality of the Monterey formation is determined by general conditions governing the accumulation of oil. Thus, although the Monterey formation is present over large areas in California, it is oil-bearing only in certain restricted areas where the porous member of the Monterey is found capping a productive anticline. The characteristic occurrence of the shale, therefore, is not a series of alternately rich and lean beds of great lateral area as in Colorado and Utah, but usually takes the form of a block of shale, fairly homogeneous as to oil content, which may be several hundred feet thick, from a few hundred feet to a mile wide, and several hundred feet to several miles long. Deposits of this type are known to exist in Santa Barbara County near Santa Maria and Casmalia, in Ventura County and Kern County, and probably exist in Fresno, Los Angeles, and San Luis Obispo Counties.

The shale is peculiar both chemically and physically, inasmuch as it is made up almost entirely of the remains of diatoms. These diatomaceous remains are siliceous and the shale, where very pure, contains 90 to 99 per cent of amorphous silica. The characteristics of the shale are as follows:

Color:	
Fresh surface	Chocolate brown
Weathered surface	Light brown to tan
Character:	
Fresh	Massive
Weathered	Massive
Streak	Brown
Hardness	2 to 3
Texture	Fine-grained
Elasticity	Tough
Specific gravity	1.0 to 2.0

TABLE XXI

Ash Analyses and Relation of Ash and Oil Content

Sample	SiO_2	Fe_2O_3 Al_2O_3	CaO	MgO	Volatile and fixed carbon	Oil content gallons per ton
* Casmalia, California.......	75.8	19.1	1.4	0.9	....	18.0
† Casmalia, California.......	83.1	12.6	3.4	0.7	....	25 0

* Bureau of Mines, T. B. Brighton, analyst.
† Private work, George W. Turner, analyst.

PHYSICAL AND CHEMICAL PROPERTIES OF CALIFORNIA SHALE OIL

TABLE XXII

Physical Properties of Crude Shale Oil

Sample	Gravity at 60° F.		Oil content, gallons per ton	Color
	Specific	Baumé gravity		
Casmalia, Sta. Barbara County, Calif	0.9352 .926 .923	19.7 21.0 21.5	25.5 23.8 25.8	Brown Red-brown Red-brown
Graciosa Ridge, Sta. Barbara County.	.923	21.5	52.0	Brown
Elsinore, Riverside County	.846	36 0	65.0	Greenish-brown

TABLE XXIII

Distillation Tests, Crude Shale Oil

Sample No.	1	2	3	4
Initial B. P. Centigrade..........	40	60	85	72
To 85° C	5.2	1.5	......	2.1
85°-115° C....................	6.5	3.5	3.5	3.5
115°-130° C...................	.8	1.5	4.6	2.1
130°-150° C...................	4.1	1.8	2.3	4.2
150°-190° C...................	4.1	8.2	9.3	9.7
Distillate....................	20.7	16.5	19.7	21.6
Residue.......................	79	83	80	78

AMOUNT AND CHARACTER OF FIXED GASES FROM CALIFORNIA OIL SHALE

TABLE XXIV

Cubic Feet of Fixed Gas Produced from Various Grades of Shale

Location sampled	Oil content, gallons per ton	Gravity oil, Baumé	Fixed gas, cubic feet per ton
Casmalia....................	25.5	21.2	958
Casmalia....................	29.6	20.0	900

By-products.—Tests made without the use of steam failed to recover more than 4 or 5 pounds of ammonium sulphate per ton of shale, and for that reason it is doubtful if the shale in this region will be worth treating for its ammonia content. The spent shale, however, has interesting possibilities as in insulating material, for decolorizing and filtering purposes, and for the manufacture of light-weight bricks.

Commercial possibilities.—Considered from the point of view of commercial operations, California shales are exceptionally interesting. Transportation facilities are ideal, living and working conditions are splendid, and water and power are generally abundant. The type of shale deposit, as previously discussed, presents unusual opportunities for cheap, large-scale mining. Furthermore, as these are saturated shales, the oil recovered is not at all different from ordinary petroleum, and offers none of the difficulties in refining common to shale oil from other localities. The products from the oil have a ready present market and their sale will not require the "campaign of education" often considered essential in the marketing of shale-oil products from Utah–Colorado–Wyoming shales. Few samples show the exceptional oil content recovered from certain rich beds in the Uinta Basin, but the average from large tonnages should be quite as high as that from any shale area in the country.

INDEX

A

Abadan, Persia, oil refinery, i, 142
Abandoning oil wells, i, 290, 299, 300
Abel-Pensky tester, i, 624, 638, 639
Abilene oil field, Texas, i, 113
Abney level, i, 179
Abnormal dips, geology, i, 171
Abrasion, frictional, in lubrication, ii, 637
Absolute pressures, natural gas, table, i, 456
Absolute viscosity, i, 640, ii, 626
Absolute zero, ii, 807
Absorbers, natural-gas gasoline, i, 768
Absorption method, natural-gas gasoline, i, 738, 742, 764
  production by states, i, 786
Absorption of oils, i, 742
Absorption oil, characteristics, i, 739
Absorption plants, i, 780
Absorption system, wax plant, ii, 192
Absorption temperature, natural-gas gasoline i, 742
Absorption test, gasoline in natural gas, i, 761
Acadia Parish, La., oil, i, 95
Acala, Mex., oil seepage, i, 64
Accounting, refinery, ii, 424
Acenaphthalene, i, 508
Acenaphthene, i, 536
Acetic acid, melting point, ii, 806
  specific heat, ii, 811
Acetylene, i, 478, 528
  specific gravity, ii, 832
Acetylene, series of hydrocarbons, i, 478
Acid blow cases, ii, 40, 76
Acid coke, ii, 371
Acid heat test, petroleum products, i, 588, 690
Acid recovery, ii, 402
Acid sludge, ii, 369
Acid tank, ii, 39, 70
Acid tar test, i, 714
Acidity test, petroleum products, i, 588, 710
Aclinal structure, i, 45, 46
Adair County, Mo., oil, i, 97
Adair oil field, Okla., i, 104
Adams County, Ind., oil, i, 88
Adams County, Mo., oil, i, 97
Adiabatic line, ii, 525
Adsorption, gaseous by solids, i, 747
Adsorption method for natural-gas gasoline i, 745
Adsorption oil, i, 534
Aeroplane, engine, ii, 618, 619
Aeroplane gasoline, ii, 361
Ætna oil burner, ii, 477
Africa, asphalt, i, 157
  bitumen, i, 793
  oil, i, 75, 156
  oil shale, i, 846, 848
Aggregates, for concrete construction, ii, 129 249
Agitators, lubricating oil, ii, 68
  refined oil, flux and lubricants, ii, 62
  refinery, ii, 62, 92, 186, 221, 367
  skimming plant, ii, 19, 30
Ain, France, bitumen, i, 792
Air, ii, 786–789
  specific gravity, ii, 832
  specific heat, ii, 811
  standard cycle in internal combustion, ii, 526
Air and vapor mixture, table, ii, 789
Air-blast injection, engines, ii, 571
Air compressors, Diesel engine, ii, 567, 581
  rule for capacity, ii, 301
Air cooling, ii, 829
Air drive, i, 315
Air flooding, in oil production, i, 315
Air furnace, oil fired, ii, 507
Air inlet pipes, Diesel engine, ii, 577
Air inlet silencer, Diesel engine, ii, 579
Air lift, oil pumping, i, 315
Air pressures, conversion table, ii, 787
Air-refined asphalt, process, i, 790
Air supply, internal combustion engine, ii, 553
Air tanks, Diesel engine, ii, 576
  refinery, ii, 78
Alabama; Mississippian series, i, 29
  natural gas, i, 40, 79
  natural gas production and consumption, statistics, i, 352, 354
  oil, i, 40, 80
  oil shale, i, 848
  Selma chalk, i, 19
Alameda County, Calif., oil, i, 539
Alamo-Chapopote oil field, Mex., i, 124, 125

Alaska, oil, i, 78, 80, 539
oil production statistics, i, 340
Alazan-Potrero del Llano oil field, Mex., i, 124, 125
Albania, Turkey, asphalt and bitumen, i, 137, 792
Albert County, New Brunswick, albertite dikes, i, 6, 68
oil, i, 121
Alberta, Canada, Cambrian system, i, 41
bitumen, i, 792
Dakota formation, i, 5
oil analyses, i, 575
oil and gas, i, 123
overthrust faults, i. 66
tar sands, i, 5
Albertite, i, 793
Australia, i, 793
Europe, i, 792
New Brunswick, i, 6, 68
North America, i, 792
Oceania, i, 792
tests and characteristics, i, 794
Alcohol, boiling point, ii, 806, 807
expansion by heat, ii, 812
freezing point, ii, 807
hydrocarbon solvent, i, 536
latent heat of evaporation, ii, 823
specific heat, ii, 811, 812
Aleppo oil field, Pa., i, 111
Alexandria oil field, Ind., i, 89
Algeria, oil, i, 157
average specific gravity, i, 322
production statistics, i, 325
Algonkian age, i, 2, 9
Alidade, open sight, i, 179
telescopic, i, 180
Aliphatic hydrocarbons, specific heat, i, 531
Alkali, recovery, ii, 404
Alkyl sulphides, in oil, i, 529
Allegany County, N. Y., oil, i, 99, 107, 551
Allegheny County, Pa., gray sand, i, 35
oil, i, 558
Allegheny formation, gas sands, i, 24, 108, 116
Allen County, Kans., oil, i, 90, 544
Allen County, Ky., gas, i, 34
oil fields, i, 92, 546
Allen County, Ohio, oil, i, 102, 551
nitrogen in, i, 530
sulphur in, i, 525
Allen County oil field, Ind., i, 89
Allen oil field, Okla., i, 105
Allen oil field, Tex., i, 113
Allison vanishing thread tubing, dimensions, ii, 561
Alloy (tin-lead), melting point, ii, 807
Alloys, refining construction, ii, 281
Alluvium, i, 12, 83, 160
Alluwe oil field, Okla., i, 104, 105
Allyl benzol, i, 484
Allylene, i, 478
Alsace, France, bitumen, i, 792
oil, i, 133, 325
analyses, i, 578
average specific gravity, i, 322
oxygen in, i, 524
Alta Vista oil fields, Tex., i, 113
Alternating current motors, ii, 295
Alternators, Diesel engine, parallel operation, ii, 582
Altoona oil field, Kansas, i, 91
Alumina, specific heat, ii, 811
Aluminum, melting point, ii, 806, 807
refinery use, ii, 279
Amarillo gas and oil field, Tex., i, 21, 113, 114
American Chemical Society, fixed carbon test for asphalts, i, 809
"American Medical oil," i, 359
American melting point, wax, i, 679, 683
American Society for Testing Materials, carbon residue test, i, 704
determination of loss on heating oils and asphaltic compounds, i, 717
emulsification tests, i, 695, 699
melting point test, i, 802
penetration tests, asphalts, i, 719
viscosity of lubricants, standard test, i, 644
volatilization test, asphalt, i, 802
American Society of Municipal Improvements, ductility test, asphalt, i, 721
Ames limestone, i, 23, 32
Ammonia, aqua, gravity and strength, table, ii, 840
boiling point, ii, 806, 807
specific gravity, ii, 832
specific heat, ii, 811
temperature correction, table, ii, 840
vapor, saturated, ii, 839
Ammonia pumps, wax plant, ii, 194, 196
Ammonium hydrate, refinery use, ii, 368, 371
Ammonium sulphate in shale, i, 873, 883, 891
Amorphous silica, California shale, i, 892
Amorphous wax, ii, 359
Amsden sand, i, 98
Amsterdam oil field, Ohio, i, 101
Amyl benzol, i, 488
Amyl butylcaproyle, i, 468
Amylene, i, 472
Analyses, ash, California shales, i, 893
Green River shales, i, 876
Nevada shales, i, 890
petroleum products, i, 588
Analyses, boiler steel, ii, 144
crude oil, ii, 325, 331, 538

Analyses, fixed gases from oil shale, i, 882
natural gas, i, 729
oil shale, i, 836, 872
pipe steel, ii, 551
still gases, ii, 338
structural steel, ii, 116, 122
Analysis, gas, method of, i, 731
oil shale, method of, i, 849
shale oil, method of, i, 853
Andean oil district, Argentina, i, 140, 149
Peru, i, 148
Ada gas field, Oklahoma, i, 105
Anderson County, Tex., bitumen, i, 791
Aneroid barometer, i. 181
Angles, refinery construction, ii, 30, 103, 107
Angle bracing, refinery construction, ii, 28, 52–64
Aniline, boiling point, ii, 806
Annealing furnace, ii, 505, 508, 514
Annealing, pipe, ii, 502
Annona chalk, i, 19, 95
Anse la Butte oil field, Louisiana, i, 550
Anthracene, i, 514, 536
Anticlinal bulges, i, 59
Anticlinal structures, i, 45, 47
India, i, 141
Anticlinal theory, oil occurrence, i, 44, 73
Anticlines, Cincinnati, i, 102
United States oilfields, i, 79
Antimony, melting point, ii, 806, 807
specific heat, ii, 811
Appalachian Basin, i, 21, 140
Appalachian field, deliveries, petroleum, i, 345, 358
oil production, quantity and value of since 1859, i, 330, 356
stocks, i, 343, 357
Appalachian geosyncline, i, 102
Appalachian Mountain system, oil, i, 9
Appalachian oil district, Ohio, i, 102
Appalachian oil fields, i, 78
anticlinal and synclinal structures, i, 47
Keener sand, i, 31
Mississippian series, i, 31
sand stones, i, 73
structural types, i, 79
trend of direction, i, 9
Apollo oil field, Pennsylvania, i, 111
Apsheron Peninsula, Russia, mud volcanoes, i, 6
oil fields, i, 127
Arabia, bitumen, i. 793
oil, i, 145
Arc of contact, lubrication, ii, 631
Archean age, i, 2, 9
Archer County, Tex., oil, i, 114
Archers Fork oil field, Ohio, i, 101
Area, measures, ii, 762, 766
Argentina, bitumen, i, 792
oil analyses, i, 577
oil, average specific gravity, i, 322
oil, gas, and asphalt, i, 148
oil production statistics, i, 150, 325
oil shale, i, 842
oil, unit of measure, ii, 761
stratigraphic occurrence of petroleum, i, 11
Argentine gas pool, Kansas, i, 91
Arizona, oil, i, 80
Arkadelphia formation, i, 18, 95
Arkansas, bitumen, i, 791
Mississippian series, i, 29
natural gas production and consumption, statistics, i, 352, 354
oil, i, 80
oil analyses, i, 539
oil, production statistics, i, 356
Arkansas City oil field, Kansas, i, 91
Armenia, asphalt, oil, maltha, and bituminous springs, i, 145
Armstrong County, Pa., "Fourth" sand, i, 35
Armstrong Mills oil field, Ohio, i, 100
Armstrong Run oil field, Pa., i, 111
Aromatic distillate stock, ii, 369
Aromatic series of hydrocarbons, i, 480
Arrested anticline, oil geology, i, 55, 73
Artificial oil shales, i, 834
Asbestos, grease filler, ii, 674
packing, refinery, ii, 285, 286
refinery insulating, ii, 282, 286
Ash analyses, California shale, i, 893
Green River shale, i, 876
Nevada shale, i, 876
petroleum products, i, 588, 723
Ashland County, Ohio, Clinton sand, i, 39
Ashland oil field, Ohio, i, 101
Ashtabula County, Ohio, gas, i, 33
Asia, oil, i, 75, 137
oil shale, i, 848
Asia Minor, oil and gas, i, 145
Asphalt, i, 787–829
Albania, i, 137
Angola, i, 157
Arkansas, i, 80
Argentina, i, 150
Armenia (Turkish), i, 145
Asia, i, 792
briquette binder, i, 829
British Guiana, i, 155
Canada, i, 121
characteristics, i, 798, 809
Chile, i, 155
classification, i, 787
composition, i, 524
Cuba, i, 159

Asphalt, determination of ductility, i, 721
determination of penetration, i, 717
electrical insulation, i, 830
Europe, i, 792
fixed carbon test, i, 809
flash point test, i, 809
floor covering, i, 826
Germany, i, 134
in oil, i, 459
marketing, i, 814
melting point, methods of testing, i, 685, 687, 802
Mesopotamia, i, 141
Mexico, i, 124
Nevada, i, 98
Oceania, i, 793
Oklahoma, i, 22
paints, i, 829
Palestine, i, 144
Pitch Lake, Trinidad, i, 6, 159
processes of refining, i, 787
producing countries, i, 796
production from petroleum, i, 795
Rancho La Brea. i, 16
roofing, i, 822
saturated, felt roofing, i, 822
sheet, i, 817
solubility tests, i, 806
Spain, i, 136
surface indication of oil, i, 2, 3, 5, 9, 167
tests and characteristics, i, 794
Trinidad, i, 159
United States, i, 790
uses, i, 816
Venezuela, i, 151
volatilization test, i, 807
Wyoming, i, 67
Asphalt base oils, ii, 453
Asphalt content, determination of, i, 588, 721
Asphalt earth pavements, i, 817
Asphalt fillers, kinds, i, 821
Asphalt flux, ii, 365
Asphalt grout, i, 821
Asphalt macadam, i, 818
Asphalt mastic pavements, i, 817
Asphalt paving mixture, i, 818
Asphalt pipe coating, i, 826
Asphalt sand mixture, i, 817
Asphalt shingles, i, 824
Asphalt surface treatment, i, 818
Asphaltenes, i, 715, 807
Asphaltic-base crudes, refining, ii, 355
Asphaltic compounds, determination of loss on heating, i, 717
Asphaltic concrete, i, 816, 817, 818
Asphaltic flux oil, methods of testing, i, 588
Asphaltic limestones, i, 793
Asphaltic oils, effect on oil migration, i, 74
Asphaltic pavement, foundations, i, 818
Asphaltic petroleum products, ash test, i, 723
fixed carbon test, i, 723
pyrobitumens, tests and characteristics, i, 794
Asphaltic sands, description and use, i, 793
Asphaltites, i, 794
Assam, India, oil fields, i, 140
Assay distillation, petroleum products, i, 588, 678
Asymmetrical anticlines, i, 47
India, i, 141
Atascosa County, Tex., oil, 114
Aten oil field, Pennsylvania, i, 108
Athabaska, Can., tar, i, 5, 121, 123
Athens County, Ohio, Berea sand, i, 33
Athens oil field, Ohio, i, 101
Atmospheric pressure, ii, 786
Atoka County, Okla., bitumen, i, 791
Atlantic gravity temperature corrector, description and use, i, 601
Atlantic lubricator, i, 696
Atlantic **Refining Co.**, bomb method sulphur determination, i, 709
burner description, i, 677
Atmospheric air, expansion by heat, ii, 812
Atomization, Diesel engine, ii, 575
liquid fuel, ii, 448
lubricants, ii, 645, 663
Atomizer, mechanical, oil heaters, ii, 481
Atomizing burner, ii, 449
Attachment angles, oil tankers, i, 377
Atterite, refinery use, ii, 279
Atwood oil field, Pennsylvania, i, 111
Auglaize County, O., oil, i, 102
nitrogen in, i, 529
oil, sulphur in, i, 525
Augusta oil field, Kan., i, 91, 545
Austin Group, i, 19, 95
Australia, albertite, i, 793
bitumen, i, 793
elaterite, i, 793
oil, i, 166
oil and gas, i, 164
oil shale, i, 164, 847
Austria, bitumen, i, 792
oil shale, i, 848
Austria-Hungary, asphalt, i, 797
Automobile engines, ii, 620
lubrication, ii, 666
Automobile gasoline, ii, 361
Automobile oils, characteristics, ii, 677
Automobiles, number in United States, ii, 623
Avalon oil field, Pa., i, 110
Avant oil field, Oklahoma, i, 105
Aviation gasoline, ii, 361

Avis oil field, Texas, i, 113
Axle grease, ii, 675

B

Babcock and Wilcox boilers, ii, 467, 481
Bacon, Brooks and Clark cracking process, ii, 435
Bagasse, use with oil, ii, 477
Bailers, standard drilling, i, 248
Bailey oil field, Texas, i, 113
Bailing, in oil wells, i, 313
  tools, standard, i, 228
Bakersfield, Calif., pipe line heating, i, 405
Bakerstown oil field, Pennsylvania, i, 111
Baku, Russia, mud volcanoes, i, 6
  natural gas springs, i, 42
  oil analysis, i, 581
  oil fields, i, 50, 127, 142
  oil, mercaptans, i, 527
Balakhani, Russia, oil analysis, i, 581
  oil fields, i, 127
Balancing, Diesel engines, ii, 565
Balcones fault zone, Texas, i, 114
Bald Hill oil field, Oklahoma, i, 105, 556
Baldwin oil field, Kansas, i, 91
Balltown oil field, Pennsylvania, i, 111
Bangs oil field, Texas, i, 113
Barbacoas formation, Venezuela, i, 152
Barbados, W. I., glance pitch (manjak), i, 792
  oil, i, 161
Barbey ixometer, i, 643
Barbour County, W. Va., oil, i, 116
Barlow's formula, internal fluid pressure, ii, 519, 521, 527, 528
Barnes oil field, Oklahoma, i, 105
Barnesville oil field, Ohio, i, 101
Barograph, i, 181
Barometer, aneroid, i, 181
Barometric method, mapping, i, 177, 183
Barrels, asphalt, i, 815
  refinery, manufacture, ii, 226
  steel, ii, 566
Barren County, Ky., "Deep" sand, i, 94
  gas, i, 34
  High Bridge formation, i, 40
  natural gas, i, 40
  oil, i, 92, 94, 546
Barrier oil field, Kentucky, i, 93
Bars, weight and areas, table, ii, 781
Bartlesville oil field, Okla., i, 104–105, 556
Bartlesville sand, i, 28, 29, 30, 106
Barytes, grease filler, ii, 674
Basalt crevices, oil in, i, 67
Basalt plugs, i, 59, 63
Basaltic dikes, Mexico, i, 67
Base maps, i, 177
Base oils, cracking, ii, 430
Basin oil field, Wyoming, i, 118, 120
Basses-Alpes, France, bitumen, i, 792
Batch process, oil treating, ii, 369
  refining asphalt, i, 788, 790
Batch stills, ii, 324, 333
Bates County, Mo., oil, i, 97
Bath County, Ky., oil, i, 92, 546
Batson oil field, Texas, i, 113, 564
Baumé gravity, Bureau of Standards, i, 594
  correction for temperature, table, ii, 836
  oil equivalents, ii, 833, 834
  Tagliabue, i, 595
  use of correction tables for temperature, i, 601
Bauxite, refinery use, ii, 368
Bavaria oil field, Germany, i, 134
Bayard sand, i, 36, 108, 116
Bays Fork oil field, Kentucky, i, 93
Beagle oil field, Kansas, i, 91
Beall oil field, Pennsylvania, i, 110
Beallsville oil field, Ohio, i, 100
Beaman stadia arc, i, 180, 195
Bear Creek gas field, Kentucky, i, 93
Bear Creek oil field, Pennsylvania, i, 110
Beardmore engine, ii, 537
Bearing area, in lubrication, ii, 641
Bearing design, in lubrication, ii, 668
Bearing greases, ii, 674
Bearing oils, ii, 648
Bearing plates, ii, 246
Bearing pressures, in lubrication, ii, 544, 633, 635, 642, 666
Bearing temperatures, in lubrication, ii, 640
Bearings, metal, lubrication, ii, 631
  oil grooves in, ii, 668
Beaumont oil field, Kansas, i, 91
Beaumont oil field, Texas, rotary drilling system, i, 251
  standard casing requirements, i, 221
Beaver Creek oil field, Ky., i, 34, 92, 93, 94
Beaver Dam oil field, Ohio, i, 101
Beaver Falls oil field, Pennsylvania, i, 110
Beaver sand, i, 25, 92, 94
Bedford "red rock," i, 33
Bedford shale, i, 32, 34
Beds, geologic mapping, i, 169
Beehunter oil field, Indiana, i, 89
Beeswax, latent heat of fusion, ii, 823
  melting point, ii, 807
Beggs oil field, Oklahoma, i, 105, 556
Bejucal asphalt, i, 791, 794
Belgium, oil, i, 136
Bell topping retort, ii, 238
Bellair oil field, Illinois, i, 87
Belle Isle oil field, Louisiana, i, 63
Belle River oil pool, Ontario, Can., i, 122, 123
Bellevernon oil field, Pennsylvania, i, 110

Bellevue oil field, Pennsylvania, i, 110
Belmont County, Ohio, oil, i, 102
nitrogen in, i, 530
Belridge oil field, California, i, 16, 81, 82, 83, 541
Belt conveyors, for filter plant, ii, 201
horse power for, ii, 305
Belting, link, ii, 306
Belts, horse power, ii, 303
Bending movements, reinforced concrete, ii, 257
Bend limestone, i, 13, 31, 114
Bend tests, steel, ii, 117, 123, 144
Benedict oil field, Kansas, i, 91
Bengal, India, oil, i, 141
Benson's formula, heat transmission, air to water, ii, 829
Benton cracking process, ii, 426
Benton shale, i, 18, 119
Benzene (benzol), i, 480
hydrocarbon solvent, i, 536
Benzerythrene, i, 518, 536
Benzine, ii, 338, 363, 423
boiling point, ii, 806
condenser coil data, ii, 172
specific heat, ii, 812
treating, ii, 368
Benzol (benzene), i, 480, 536
heat of combustion, ii, 809
hydrocarbons, i, 480
Benzophenone, boiling point, ii, 518
Berea grit, i, 32, 33
Berea oil field, Ohio, i, 100
Berea sand, i, 15, 25, 32, 59, 92, 94, 103, 108, 116
Bermudez, Venezuela, asphalt, i, 791, 794
bitumen, i, 792
Bernadillo County, N. Mex., oil, i, 99
Bernoulli's theorem, i, 380
Bessemer engine, ii, 539
Bessemer oil field, Pa., i, 110
Bessemer steel pipe, ii, 501
Best, W. N., ii, 445
Bexar County, Tex., oil, i, 114, 565
Bibi-Eibat oil field, Russia, i, 47, 68, 127
oil analysis, i, 581
Big Dunkard sand, i. 24
Bigheart oil field, Oklahoma, i, 104, 556
Big Horn County, Mont., oil, i, 97
Big Horn County, Wyo., oil, i, 570
Big Injun sand, i, 15, 31, 32, 33, 94, 103, 108, 116
Big Lime, i, 21, 26, 27, 29, 31, 37, 38, 39, 108
(Greenbrier limestone), W. Va., i, 15, 31, 116
(Maxville limestone), Ohio, i, 15, 25, 31, 103
Big Muddy oil field, Wyo., i, 118, 120, 571
Big Sinking oil field, Kentucky, i, 92, 93, 547
Billet heating furnaces, oil fired, ii, 493
Billings oil field, Oklahoma, i, 21, 25, 104
Binagadi oil field, Russia, i, 127
Bins, oil shale plants, i, 860
Bird Creek-Flat Rock-Skiatook oil field, Oklahoma, i, 105
Bird Creek oil field, Oklahoma, i, 554
Birmingham gage, sheets, ii, 773, 774
Birnie's formula, internal fluid pressure, ii, 519, 526, 527, 528
Bismuth, latent heat of fusion, ii, 823
melting point, ii, 806
(melted) specific heat, ii, 812
Bisulphide of carbon, latent heat of evaporation, ii, 823
Bit hook, i, 282
Bits, dressing of, i, 243
rotary, i, 261
standard, i, 225, 227
Bittenbender oil field, Pennsylvania, i, 110
Bitumen, i, 787
Albania, i, 137
Arabia, i, 145
Brazil, i, 154
Cuba, i, 159
East Africa, i, 158
England, i, 135
European Turkey, i, 134
Falkland Islands, i, 156
foreign, i, 792
Gold Coast, i, 158
in igneous rocks, i, 66
Mesopotamia, i, 141
Nevada, i, 98
Nigeria, i, 158
United States, i, 790
Victoria, i, 166
Bituminous dikes, i, 5, 6, 71
Bituminous lakes, i, 3, 5
Borneo, i, 5
Bituminous lignite, Porto Rico, i, 161
Bituminous limestone, i, 3, 4, 792
melting point, i, 685
Bituminous mixtures, protective coatings, ii, 572
Bituminous products, method of determination of i, 715, 806
Bituminous rock, Philippine Islands, i, 163
production, i, 796
Bituminous sands, i, 3, 4, 792
Bituminous seals, oil sands, i, 46, 68, 74
Bituminous seepages, i, 3, 5
Bituminous shale, i, 9
Bulgaria, i, 137
Scotland, i, 135
Wales, i, 135

Bituminous springs, Armenia (Turkish), i, 145
Bixby oil field, Oklahoma, i, 104
Bixler sand, i, 28
Black Creek, Ontario, Canada, oil, i, 121
Black oil, ii, 356
methods of testing, i, 588
Black oil field, Ohio, i, 100
Black oil sand, i, 35
Black Ranch oil field, Texas, i, 113
Black Run oil field, Pennsylvania, i, 110
Black shale, i, 871
Black steel sheets, weight, table, ii, 778
Blackford County, Ind., oil, i, 88
Blackshire oil field, Pennsylvania, i, 110
Blacksmith tools, drilling, i, 217
Blackstone engine, ii, 533
Blackwell oil field, Oklahoma, i, 21, 104
Blackwell, Okla., crude, characteristics, ii, 13
Blackwell sand, i, 21, 25, 106
Bladensburg oil field, Ohio, i, 100, 552
Blaine-Laurel Creek oil field, Kentucky, i, 93
Blair oil field, Pennsylvania, i, 110
Blast furnace gas, fuel oil equivalent, ii, 476
Bleacher, lubricating oil, ii, 73
refined oil, ii, 72
skimming plant, ii, 19, 37
Blending, natural-gas gasoline, i, 757, 758
Block-caving method, oil shale mining, i. 856
Blood char, refinery use, ii, 368, 379
Blossom sand, i, 12, 19, 95
Blowouts and their prevention, i, 268
Blow cases, acid, refinery, ii, 76
Bluck oil field, Ohio, i, 101
Blue Creek oil field, West Virginia, i, 117, 569
Blue Monday sand, i, 35
Board Tree oil field, Pennsylvania, i, 110
Boardtree oil field, W. Va., i, 117
Boat transportation, oil, i, 370
Bodmer's formula, condensation in cylinders, ii, 659
Boiler-house, refinery, ii, 229
Boiler, oil country specifications, i, 224
Boiler manufacturing, oil fuel, ii, 497
Boiler-plate, refinery use, ii, 288
strength, at high temperatures, ii, 815
Boiler power, gasoline absorption plant, i, 785
Boiler shells, refinery, ii, 288
Boiler steel, standard specifications, ii, 143
Boiler tubes and flues, internal fluid pressure, ii, 534
Boilers, oil fired, ii, 463
refinery, ii, 92, 191, 287
tubular, California type, i, 225
tubular, oil fired, ii, 464
Bolinder engine, ii, 537
Boiling point, adsorption, relation to temperature and pressure, i, 748
hydrocarbons, i, 461
various liquids, ii, 806
water, table, ii, 790
Bolivia, oil, i, 153
oil, analysis, i, 577
oil, unit of measure, ii, 761
Bolted tanks, refinery, ii, 212
Bolton Creek oil field, Wyoming, i, 120
Bolts, connecting rod, internal combustion engine, ii, 543
cylinder heads, Diesel engine, ii, 563
refinery use, ii, 268
Bomb method, determination of sulphur in oils, i, 709
Bond County, Ill., oil, i, 86
Bone char, filter plant use, ii, 201, 202
refinery use, ii, 368, 377
Bonner Springs oil field, Kansas, i, 91
Booch sand, i, 30, 31, 106
Boone chert, i, 30, 34
Boone County, W. Va., oil, 116
Boone Dome oil pool, Wyoming, i, 120
Borneene, i, 512
Borneo, oil, i, 161, 162
oil analyses, i, 583
Bornylene, i, 512
Bossier Parish, La., oil, i, 95
Boston casing, inserted joint, weights and dimensions, ii, 559
weights and dimensions, ii, 556
Boston oil field, Oklahoma, i, 104
Bothwell oil field, Ontario, Can., i, 121, 122, 576
Bosworth oil field, Ohio, i, 100
Bottineau County, N. D., gas, i, 102
Bottom hole packer, i, 291, 293
Bottom outlet valve, tank cars, i, 363
Bottom settlings, i, 690. See also B. S.
Boulder County, Colo., oil, i, 542
Boulder oil field, Colorado, i, 84, 85, 542
Boulder sand, i, 35
Bourbon County, Ky., natural gas, i, 40
Bowerston oil field, Ohio, i, 100
Bowl, fishing, i, 289
Bowling Green oil field, Kentucky, i, 93
Box cars, asphalt, i, 816
Boyd County, Ky., bitumen, i, 791
Boyle's law, gases, ii, 525
Boynton oil pool, Oklahoma, i, 30, 104, 105
Boynton sand, i, 30
Bradford oil field, Pa., i, 37, 110
Bradford sand, i, 36, 108
Bradford third sand, i, 38
Brady Run oil field, Pennsylvania, i, 110
Brady's Bend oil field, Pennsylvania, i, 110

Brake horse-power, Diesel engines, ii, 568
internal combustion engines, ii, 530
Brant County, Ontario, Can., oil, i, 121, 123
Brass, Crane special, tensile strength, ii, 818
elongation, ii, 821
radiation and reflection, ii, 824
refinery use, ii, 279
rolled rod, elastic limit, ii, 820
rolled rod, tensile strength, ii, 820
specific heat, ii, 811
Bratcher Hollow oil field, Kentucky, i, 93
Braxton County, W. Va., gas, i, 116
Brazil, oil, i, 154
oil analysis, i, 577
oil shale, i, 834, 842
oil, unit of measure, ii, 761
Brazoria County, Tex., oil, i, 114, 566
Brea, California, i, 68
Wyoming, i, 67
Breakdown pressure, lubricants, ii, 633
Breakdown viscosity, lubrication, ii, 632
Breast and bench method, oil-shale mining, i, 855
Breast method, oil shale mining, i, 855
Breathitt County, Ky., oil, i, 92
Brecciation, i, 64
Breckenridge County, Ky., bitumen, i, 791
Breckenridge (Parks), oil field, Texas, i, 31, 113
Bremen oil field, Ohio, i, 53, 59, 100, 552
Brenner oil field, Pennsylvania, i, 110
Bricker-Snyder oil field, Ohio, i, 100
Brickwork, refinery construction, ii, 248
Brickwork and masonry, specific heat, ii, 811
Bridge, oil tanker, i, 376
Bridgeport gas field, West Virginia, i, 117
Bridger formation, oil shale, i, 873
Bridgeville oil field, Pennsylvania, i, 110
Brigg's standard, pipe, ii, 546
Bright stocks, ii, 359
Brine, refinery, ii, 372
sodium chloride solutions, ii, 848
Brine pump, wax plant, ii, 194
"Bringing in" a well, i, 252
Brinkhaven oil field, Ohio, i, 100
Briquets, asphalt, binder, i, 829
Bristoria oil field, Pennsylvania, i, 110
Bristow oil field, Oklahoma, i, 104
British Borneo, oil, i, 163
British Colombia, oil, i, 41, 68
oil shale, i, 841
British Guiana, oil, i, 155
British Honduras, oil, i, 126
British Isles, oil, i, 23
British thermal unit, ii, 446, 808
Broad Ripple oil field, Indiana, i, 89
Broken Arrow oil field, Oklahoma, i, 104
Bromine, boiling point, ii, 806
melting point, ii, 806
specific heat, ii, 812
Bronson oil field, Kansas, i, 91
Bronze, melting point, ii, 806
refinery use, ii, 279
Brooke County, W. Va., oil, i, 116
Brooklyn oil field, Ohio, i, 100
Brookville oil field, Pennsylvania, i, 110
Brown-Pickering topping plant, ii, 83
Brown County, Tex., oil, i, 28, 114, 559
Brown sediment. See B. S.
Brownstone marl, i, 19, 95
Brownwood oil field, Texas, i, 113, 559
Brunton compass, i, 178
Brush Creek oil field, Pennsylvania, i, 110
Brushy Creek oil field, Ohio, i, 100
Brushy Mountain gas field, Oklahoma, 104
B. S., i, 432, 588, 670, 690; ii, 359
B. S. and water, fuel oil, ii, 340
separation by centrifuge method, i, 588
B.t.u., ii, 446, 791, 808
Buck Creek, Ky., oil field, i, 92, 93
Buck Creek oil field, Okla., i, 104
Buckeye engine, ii, 540
Buckeye pipe laying machine, i, 412, 413
Buck Run oil field, Ohio, i, 100
Buffalo Basin gas pool, Wyoming, i, 120
Buffalo line test, Pitot tube, i, 436
Buffalo oil field, Kan., i, 91.
Buffalo oil field, Pa., i, 110
Building materials, spec. grav., table, ii, 831
Buildings, construction data, ii, 244
refinery, ii, 228
Built-up roofing, asphalt, i, 825
Bulb immersion thermometer, i, 592
Bulgaria, oil shale, i, 137, 843
Bulldog, fishing tool, i, 211, 278
Bullion oil field, Pennsylvania, i, 110
Bumper, fishing tool, i, 278
Bunkering wharf, ii, 241
Bureau of Standards, Baumé gravity, i, 594
determination of sulphur in oils, i, 709
emulsification of oils, i, 701
Waters oxidation test, i, 705
Burgess sand, i, 30, 106
Burgettstown oil field, Pennsylvania, i, 110
Burgoon sandstone, i, 32
Burkburnett oil field, Texas, i, 22, 28, 113, 114, 562
drilling methods, i, 202
gas pumps in, i, 320
standard casing requirements, i, 221
Burkett oil field, Texas, i, 113
Burkhart oil field, Ohio, i, 100
Burlap packer, i, 293

Burma, India, oil, i, 6, 50, 148, 149
oil analyses, i, 581
oil, unit of measure, ii, 761
Burmeister and Wain engine, ii, 598
Burmeister and Wain sprayer, ii, 574
Burners, oil, ii, 447
Burnett County, Tex., bitumen, i, 791
Burning point, i, 623
Burning Springs oil field, Kentucky, i, 93
Burning Springs oil field, W. Va., i, 117
Burnt clays, surface indications of petroleum, i, 3, 7
Bursting pressure, line pipe and tubing, formula and tables, i, 301
tank cars, i, 361
Bursting tests, tubes and pipes, ii, 527, 529
Burton cracking process, ii, 428, 431
Burton still, ii, 326
Busch-Sulzer Diesel engine, ii, 584, 613
Busseyville oil field, Kentucky, i, 93
Bustenari oil field, Rumania, i, 130
Butadiene 1-3, i, 478
Butane, i, 460
Butler County, Kan., oil, i, 90, 545
Butler fourth sand, i, 35
Butler gas sand, i, 33, 108
Butler oil field, Ohio, i, 100
Butler oil field, Pennsylvania, i, 110
Butler "second" sand, i, 34, 35
Butler third sand, i, 35
Butler third stray sand, i, 35
Butler 30-foot sand, i, 33
Butlersville oil field, Kentucky, i, 93
Butt and double strap joint, ii, 265
Butt straps, boiler, ii, 289
Butt-weld process, pipe, ii, 503
Butyl acetylene, i, 478
Butyl benzol, i, 488
Butyl pentyl sulphide, i, 529
Butyl sulphide, normal, i, 529
Butylene, i, 472
By-products, oil shale distillation, i, 871
Byron oil field, Wyoming, i, 118, 120

C

Cabell County, W. Va., gas, i, 116
Cabin Creek oil field, West Virginia, i, 117, 569
Cable chain, refinery use, ii, 271
Cable guide, tanks, ii, 111
Cable-tool drilling system, i, 202
Cable tools, fishing, i, 275
Cacheuta oil field, Argentina, i, 149
Caddo County, Okla., oil, 22, 103
Caddo Parish, La., oil, i, 95, 548
Caddo oil field, Louisiana, i, 7, 47, 96
Caddo oil field, Texas, i, 31, 60, 114, 560
"Caddo" sand, i, 18
Cadiz-Maxwell oil field, Ohio, i, 101
Cadmium, melting point, ii, 806
Cairo oil field, West Virginia, i, 117
Cairo oil sand, i, 68
Calcasieu Parish, La., oil, i, 95, 549
Calcium, melting point, ii, 806
Calcium chloride solutions, ii, 847
Calcium hypochlorite, refinery use, ii, 368
Calcium sulphate, occurrence in oil field, i, 8
Calf wheel parts, i, 210
Calhoun County, W. Va., oil, i, 116
California, alluvium, i, 12
anticlinal and synclinal structures, i, 47
bitumen, i, 791
bituminous lake, i, 5
brea, i, 68
burnt clay, i, 7
Chico formation, i, 17
Cretaceous system, i, 12, 17, 83
deliveries of petroleum, i, 345, 358
diatomaceous shale, i, 16, 834, 892
Eocene series, i, 12, 17, 83
Fernando formation, i, 16, 83
Lake View gusher, i, 248
Miocene series, i, 12, 16, 83
Monterey shale, i, 16, 68, 83, 892
natural gas, i, 53
natural gas, production and consumption, statistics, i, 352, 354
natural-gas gasoline, statistics, i, 786
oil, i, 78, 80
oil analyses, i, 539
oil, nitrogen in, i, 530
oil production, per well per day, i, 349
oil production statistics, i, 82, 340, 356
oil, refining, ii, 79
oil, sulphur in, i, 526
oil, vaporizing point, ii, 497
oil seepages, i, 5
oil shale, i, 834, 836
oil shale, ash analyses, i, 893
oil shale, fixed gases from, i, 894
oil strata, i, 12, 83
oil wells, number of, i, 349
Oligocene series, i, 12, 17, 83
outcrops closed by brea, i, 68
overthrust faults, i, 68
overturned folds, i, 53
Pliocene series, i, 12, 16, 83
Quaternary system, i, 12, 16, 83
rotary drilling system, i, 251
San Pedro formation, i, 16, 83
Santa Margarita formation, i, 16, 68
Sespe formation, i, 12, 17, 83
standard casing requirements, i, 221
stock of petroleum, i, 343, 357
structural types, i, 79

California, Tejon formation, i, 12, 17, 83
Tertiary system, i, 12, 16, 83
Vaqueros formation, i, 17, 83
California air-refined asphalt, characteristics, i, 811
California casing, collapsing pressure of, table, i, 304
table of weights and dimensions, i, 302
California Creek oil field, Oklahoma, i, 105
California fuel oil, specifications, ii, 341
California oil field, stocks, i, 343
California pipe-laying machine, i, 412, 413
California regular rig irons, i, 208, 210
California standard steel drilling rigs, specifications, i, 211, 213
California steam refined asphalt, characteristics, i, 809
Callahan County, Tex., oil, i, 31, 114
Callery oil field, Pennsylvania, i, 110
Calorie, ii, 808
Calorific tests, internal-combustion engines, ii, 549
Calorific value, coal and oil, ii, 341, 477, 480, 487, 497, 505
Cam shaft, Diesel engine, ii, 569
Cambrian system, Alabama, i, 80
Canada, i, 121
geographic distribution, i, 11, 40
oil strata, i, 13, 15
United States oil fields, i, 79
Cambridge gas pool, Ohio, i, 101
"Cambro-Ordovician" system, i, 40
Cameron oil field, Ohio, i, 101
Cameron oil field, West Virginia, i, 117
Cammellaird-Fullegar marine Diesel engine, ii, 596
Camphane, i, 522
Camphene, i, 512
Campton oil field, Kentucky, i, 92, 93
Campton sand, i, 94
Canada, albertite, i, 6, 68
asphalt, i, 121
bitumen, i, 792
Dakota sand, i, 20
faults, i, 68
geanticlinal structures, i, 53
greenstone dike, Tar Point, Ont., i, 67
natural gas, i, 53, 65, 121
New Brunswick oil field, i, 78
oil, i, 121
oil, analyses, i, 576
oil, average specific gravity, i, 322
oil, nitrogen in, i, 530
oil, oxygen in, i, 524
oil, production statistics, i, 123, 324
oil, sulphur in, i, 526, 528
oil, unit of measure, ii, 761
Canada, oil shale, i, 841
oil strata, i, 12, 121
Ontario oil field, i, 78
sedimentary rocks, i, 65
stratigraphic occurrence of oil, i, 11
"Tar sands," i, 5
Canadian distillates, sulphur in, i, 527, 528
Canary-Copan oil field, Oklahoma, i, 104
Candor oil field, Pennsylvania, i, 111
Caney oil field, Kansas, i, 91
Caney oil field, Oklahoma, i, 104
Caney sand, i, 39, 94
Caneyville oil field, Kentucky, i, 93
Cannel City oil field, Kentucky, i, 92, 93
Canonsburg oil field, Pennsylvania, i, 110
Cans, refinery products, manufacture, ii, 225
Canvas packer, i, 293
Canyon formation, i, 28
Capacho formation, Venezuela, i, 152
Capacities, discharging, pipe, ii, 568
Capacity, measures, ii, 762, 764, 767
oil tankers, i, 374
pipe line, computations, i, 396
tanks, formula, ii, 99
tanks and pipes, table, ii, 770
thermal, ii, 810
Capillarity, in lubrication, ii, 637, 641
oil, i, 534
relation to oil occurrence, i, 45
Capping, standard drilling, i, 248
"Cap rock," nature of, i, 42
Caproylhydride, i, 462
Car repair shop, refinery, ii, 232
Carbenes, i, 715, 807
Carbon, heat of combustion, ii, 808, 809
in cracking, ii, 431, 434
melting point, ii, 807
oil heated furnace, ii, 484
Carbon bisulphide, boiling point, ii, 806
Carbon content, spent shale, i, 870
Carbon County, Mont., oil, i, 97
Carbon County, Utah, bitumen, i, 791
Carbon County, Wyo., oil, i, 572
Carbon dioxide, expansion by heat, ii, 812
specific gravity, ii, 832
sublimation in inert liquid, ii, 807
Carbon dioxide recorder, oil heater, ii, 475
Carbon disulphide, asphalt solubility test, i, 806
Carbon monoxide, expansion by heat, ii, 812
heat of combustion, ii, 809
specific gravity, ii, 832
Carbon ratios in coal, relation to petroleum, i, 10
Carbon residue, oil, test, i, 704
Carbon tetrachloride, fire extinguisher, ii, 317
asphalt solubility test, i, 807

Carbonate of lead, radiation and reflection, ii, 824
Carbondale chilling machine wax plant, ii, 193
Carbondale soft wax press, ii, 198
Carbonic acid, fire extinguisher, ii, 317
melting point, ii, 806
specific gravity, ii, 811
Carbonic oxide. *See* carbon monoxide.
Carboniferous sediments, Oklahoma, i, 20
Carboniferous shales, "cover" rocks, i, 42
Carboniferous system, Alabama, i, 80
England, i, 135
geographic distribution, i, 11
Kentucky, i, 92, 94
Mississippian series, i, 29
Montana, i, 98
oil strata, i, 13, 15, 21
oil shale, i, 841, 844, 845, 847, 871
Oklahoma, i, 106
Pennsylvanian series, i, 22
Permian series, i, 21
Scotland, i, 135
Tasmania, i, 166
United States, i, 79
Wyoming, i, 119
Carbopetrocene, i, 518
Carburetter, Claudel-Hobson, ii, 620
gasoline engines, ii, 615
Caribbean oil district, Colombia, i, 151
South America, i, 146
Venezuela, i, 150
Carlinville oil field, Illinois, i, 86, 87
Carlyle oil field, Illinois, i, 86, 87
Carnot cycle, ii, 526
Carophyllene, i, 514
Carriers, common, pipe lines as, i, 379, 428
handling of tank cars by, i, 366
Carroll County, Ohio, Berea sand, i, 33
oil, i, 102
"salt sand," i, 25
Carrolltown oil field, Pennsylvania, i, 110
Cars, tank, oil transportation, i, 360
Carson County, Tex., oil, i, 114
Carson oil field, Pennsylvania, i, 110
Carson oil field, West Virginia, i, 117
Carter County, Ky., bitumen, i, 791
Carter County, Okla., bitumen, i, 791
grahamite, i, 71
oil, i, 54, 103,
Permian series, i, 22
Casey oil field, Illinois, i, 87
Casey oil field, Oklahoma, i, 104
Casiano oil field, Mexico, i, 124
Casing, collapsing pressure formula and table, i, 301
cylinder, Diesel engine, ii, 562
Casing, fishing for, i, 286
fishing with, i, 284
oil well, i, 300
perforated, i, 297
standard, drilling, i, 219, 235
weights and dimensions, i, 302, ii, 556
Casing anchor packer, i, 291, 293
Casing apparatus and shoes, i, 244, 245
Casing cutter, i, 287, 288
Cashing head, control, i, 312
Casing head gas, i, 731
Casing head gasoline, i, 320
Casing lines, specifications, i, 234
Casing spear, fishing tool, i, 287
Casing splitter, i, 288
Casing swedge, i, 288
Casing tools, standard drilling outfit, i, 217
Casmalia oil field, California, i, 540
Cass County, Mo., oil, i, 97, 551
Cast iron, Crane soft, tensile strength, ii, 818
latent heat of fusion, ii, 823
fittings, ii, 277
melting point, ii, 807
pipes, standard dimensions, ii, 274
radiation and reflection, ii, 824
Castings, iron, refinery, ii, 273
steel, refinery, ii, 279
Castle Shannon oil field, Pennsylvania, i, 110
Castor oil, saponification number, i, 711
Cat Canyon oil field, California, i, 81
Cat Creek oil field, Montana, i, 97
Cat Creek sand, i, 20
Catalysts, cracking, ii, 435
Catfish Run oil field, Pennsylvania, i, 110
Catoosa oil field, Oklahoma, i, 104
Catskill formation, i, 34, 108, 116
Cattaraugus County, N. Y., oil, i, 99
Caucasus Mountains, oil, i, 127, 131
Caulking, acid blow case, ii, 40, 77
agitator, ii, 30, 63, 64
bleachers, ii, 38, 72, 73
condenser boxes, ii, 53, 54, 56, 58, 59
filters, ii, 65
pumping-out pan, ii, 30, 60
stills, ii, 25–27, 43–51
tank car, i, 362
tanks, ii, 33–40, 60–78, 103
Caustic soda, refinery use, ii, 368
Cecil oil field, Pennsylvania, i, 110
Cedar Creek anticline, Montana, i, 97
Celebes, oil, i, 162
Celsius thermometer scale, ii, 832
Centigrade thermometer scale, ii, 832
Cement, for concrete tanks, ii, 129
refinery construction, ii, 249
Cement kilns and dryers, oil fuel, ii, 520
Cement oil field, Oklahoma, i, 22, 104

Cementing oil wells, i, 290, 291, 295
Cenozoic period, California oil fields, i, 83
geographic distribution, i, 11
oil strata, i, 12, 14
Centering, plane table, i, 193
Centerville oil field, Kansas, i, 91
Centerville oil field, Butler County, Pa., i, 110
Centerville oil field, Crawford County, Pa., i, 110
Centerville oil field, Washington County, Pa., i, 110
Central America, oil, i, 126
Central Andean oil district, Peru, i, 148
South America, i, 146
Central Ohio oil and gas field, i, 102
Central Texas oil field, i, 114
Centrifugal pumps, refinery, ii, 207, 208
Centrifuge, determination, B. S. and water, i, 670
Centrifuge stock, ii, 359
Centrifuging, ii, 389
Ceram, oil, i, 161
Cerotene, i, 476
Cerro Azul oil field, Mex., i, 20, 124, 125, 248
Cerro de Oro formation, Venezuela, i, 152
Cervantes, Mexico, oil seepages, i, 64
Ceryl paraffin, i, 472
Cetene, i, 476
Chain hydrocarbons, i, 460
Chains, refinery use, ii, 271
Chalcacene, i, 516
Chalk, specific heat, ii, 811
Chanute oil field, Kansas, i, 91, 544
Chapapote, Mexican oil seepage, i, 43
occurrence, i, 5
Chapopote oil field, Mexico, i, 124, 125
Characteristics, asphalts, i, 794
oil, i, 457
Charcoal, gasoline extraction, i, 745
oil adsorption, i, 534
specific heat, ii, 811
Charcoal test, gasoline in natural gas, i, 763
Charles' law, gas expansion, ii, 524
Chars, refinery use, ii, 368
Chartiers oil field, Pennsylvania, i, 110
Chaser, pipe threading, ii, 535
Chaseville oil field, Ohio, i, 100
Chassis, automobiles, ii, 621
Chattanooga oil field, Ohio, i, 100
Chattanooga shale, i, 30, 34, 37, 42, 94, 871
Chautauqua County, Kans., oil, i, 90, 545
Chautauqua-Peru-Sedan oil field, Kans., i, 91
Chaves County, N. Mex., oil, i, 99
Cheese box stills, ii, 233
Chelsea oil field, Oklahoma, i, 104, 557
Chemical characteristics, cylinder oils, ii, 661
Chemical composition, pipe line, ii, 550
Chemical Construction Co., acid recovery process, ii, 403
Chemical fire extinguishers, ii, 316
Chemical properties, boiler steel, ii, 143
steel pipe, ii, 501
structural steel, ii, 116, 122
Chemical treatment, refinery, ii, 367
Chemistry of natural-gas gasoline, i, 729
Chemung formation, i, 13, 15, 36, 107, 108, 116
Cherokee shales, i, 28, 29
Cherry Grove oil field, Pennsylvania, i, 110
Cherry picker, fishing, i, 277
Cherryvale oil field, Kans., i, 91, 546
Chester group, i, 31
Chester Hill oil field, Ohio, i, 100
Chicken Farm oil field, Okla., 104
Chico formation, i, 12, 17, 83
Chijol oil field, Mexico, i, 124, 125
Childers oil field, Oklahoma, i, 104, 555
Chile, oil, gas, and asphalt, i, 155
Chilling, refinery, ii, 372
Chilling machines, wax plant, ii, 194
Chimney, oil fuel, table, ii, 309
refinery, ii, 307
steam boilers, size, ii, 308
China, oil, oxygen in, i, 524
oil and gas, i, 137, 143
Pennsylvanian series, i, 23
Permian rocks, i, 21
stratigraphic occurrence of oil, i, 11
China clay, grease filler, ii, 674
Chip space, threading die, ii, 536
Chipping machine, wax packing, ii, 375
Chlopin, nitrogen compounds, ii, 367
Chloride of lime, refinery use, ii, 368
Chloroform, boiling point, ii, 806
specific heat, ii, 811
Choctaw County, Ala., oil and gas, i, 80
Chromic acid, refinery use, ii, 372
Chromium, fusing, ii, 807
Chrysene, i, 516, 536
Chugwater formation, i, 12, 22, 119
Cincinnati formation, i, 94
Cincinnati geanticline, in United States oil fields, i, 53, 79
Cincinnati shales, Kentucky, natural gas, i, 39
Cincinnati uplift, i, 72
Cincinnatian series, i, 39, 94
Cinnamene, i, 482
Circular mill gage, wire, ii, 272
Circumferential seam, boiler shell, ii, 288
Cisco formation, i, 28
Citrene, i, 510
Claggett oil field, Oklahoma, i, 104
Claiborne Parish, La., Nacatoch sand, i, 18
oil, i, 95, 549
Clapp, F. G., i, 1, 191

Claremore oil field, Oklahoma, i, 104
Claremore sand, i, 34
Clarendon oil field, Pennsylvania, i, 110
Clarion County, Pa., oil, i, 35
Clarion oil field, Pennsylvania, i, 110
Clarion sandstone, i, 25
Clark County, Ill., oil, i, 86, 543
Clarksville gas field, West Virginia, i, 117
Claudel-Hobson carburetter, ii, 620
Clavarino's formula, internal fluid pressure, ii, 519, 522, 526, 527, 528,
Clay, as "cover" rock, i, 42
Clay County, Tex., Cisco formation, i, 28
oil, i, 114, 559
Permian series, i, 21, 22
Clay County, W. Va., oil, i, 116
Clay wash in refining oils, i, 534
Clays, United States oil fields, i, 79
Clearance, lubrication, ii, 627, 631, 640, 647
pipe threading, ii, 537
Clerk, specific heat of gases, table, ii, 528
Cleveland gas field, Ohio, i, 100
Cleveland, Ohio, oil sand, i, 39
Cleveland oil field, Oklahoma, i, 104, 556
Cleveland open cup method, flash and fire test, i, 624, 626, 627
Cleveland sand, i, 27, 106
Clifton oil field, Pennsylvania, i, 110
Clinometer, i, 178
Clinton County, Ky., oil, i, 40, 94
Clinton formation, i, 15, 39, 94
Clinton limestone formation, i, 15, 30, 121
Clinton, Ohio, limestone, i, 39
Clinton sand, i, 15, 37, 39, 42, 55, 59, 73, 94, 102, 103
Clinton shale, as "cover" rock, i, 12
Clintonville oil field, Pennsylvania, i, 110
Clovene, i, 514
"Close" sand, i, 41
Cloud test, petroleum products, i, 583
Cloud point, petroleum, i, 658, 659, 660
Cloverly sand, i, 119
Cloverport gas field, Kentucky, i, 93
Coal, specific heat, ii, 811
Coal beds, relation to commercial oil fields, i, 10
Coal County gas pool, Oklahoma, i, 104
Coal equivalent, fuel oil, ii, 476, 480, 487, 497, 505
Coal tar, as fuel, ii, 476
Coalinga oil field, California, i, 17, 50, 68, 81, 82, 83, 541
Coastal Plain, Mexico, i, 63
Coatings, protective, ii, 570, 572
Cobalt, fusing, ii, 807
Cock, oil regulating, ii, 462, 492
Cock taps, charging, ii, 26, 27
Cocinylene, i, 476
Coefficient of heat resistance, insulating materials, ii, 827
expansion, ii, 812
friction, lubricants, ii, 633
lateral contraction, tubes, ii, 522
thermal capacity, ii, 810
Cofferdams, oil tankers, i, 376
Coffey alcohol still towers, ii, 343
Coffeyville formation, i, 26
Coffeyville oil field, Kansas, i, 91, 546
Coil condenser data, ii, 164–179
Coke, acid, ii, 371
method of testing, i, 588
specific heat, ii, 811
Coke oven gas, fuel oil equivalent, ii, 477
Coking, in refining oils, ii, 323, 329
Coking still, refinery, ii, 150
Colbert sand, i, 30
Cold-pressing, refinery, ii, 372
Cold-settled wax, ii, 359
Cold-settling, refinery, ii, 376
Cold tests, oil, i, 588, 657, 658
Coleman County, Tex., oil, i, 31, 114, 559
Colerain oil field, Ohio, i, 100
Coles County, Ill., oil, i, 86
Colfax County, N. Mex., oil, i, 99
Collapsing pressure, line pipe and tubing, formula and tables, i, 301
pipes, tables, ii, 507
steel pipes, ii, 505
"Collar buster," i, 288
Collinsville oil field, Oklahoma, i, 30, 104, 555
Colloidal suspension, oils, i, 534
Colombia, bitumen, i, 792
mud volcanoes, i, 6
oil, i, 151
oil analyses, i, 577
stratigraphic occurrence of oil, i, 11
Color determination, oils, i, 188, 662
Color readings, oil, comparison, i, 663
Color standards, petroleum and products, ii, 356
Colorado, bitumen, i, 791
Cambrian sediments, i, 40
Dakota sand, i, 20
Florence oil field, i, 67
Green River formation, i, 884
natural gas, production and consumption, statistics, i, 352, 354
oil, i, 84, 878
oil analysis, i, 542
oil in dikes, i, 67
oil production per well per day, i, 349
oil production statistics, i, 84, 338, 356
oil shale, i, 84, 834, 836, 873, 876, 878, 888
oil strata, i, 14

Colorado, oil wells, number of, i, 349
Colorado group, i, 18, 98, 118, 119
Colorimeters, i, 662
Columbus formation, i, 37
Column reinforcement, ii, 257
Comanche County, Okla., bitumens, i, 791
  oil, i, 103
Comanche County, Tex., oil, i, 31, 114
Comanche oil field, Oklahoma, i, 104, 557
Comanche series, i, 12, 14, 20, 95
Combination socket, i, 288
Combination system, drilling, i, 269
Combustion, fuel oil, ii, 445
  heat of, ii, 809
Combustion chamber, fuel oil, ii, 446
Combustion engineering, ii, 445
Comodoro Rivadavia oil field, Argentina, i, 149, 150, 576
Compartment cars, oil transportation, i, 360
Compass, use in oil geology, i, 178
Compass and clinometer method, mapping, i, 177
Compounded oils, fatty content, i, 712
  in lubrication, ii, 658
Compounding oils, ii, 658
Compressed air, ii, 297, 298
  oil-well pumping, i, 314
Compression, natural gas, i, 447, 735
  power required, i, 450
Compression cycles, ii, 526
Compression method, natural-gas gasoline, production by states, i, 786
Compression of fluids, ii, 832
Compression plant, natural-gas gasoline, i, 764
Compression system, wax plant, ii, 192
Compression test, gasoline in natural gas, i, 760
Compressor gas, volumetric efficiency, i, 450, 452
Compressors, natural gas, i, 765
  refinery, ii, 297
Concordia Parish, Louisiana, oil, i, 550
Concrete, asphaltic, i, 817
  crushing strength, ii, 245
  protective covering, ii, 573
  refinery construction, ii, 248, 252
  reinforced, working stresses, ii, 245
  tankage, refinery, ii, 86
  tanks, storage, ii, 127
Condensation, fractional, cracking, ii, 432
Condensation temperature, hydrocarbons, i, 736
Condenser boxes, refinery, ii, 28, 52
Condensers, absorption plants, i, 780
  refinery, ii, 91, 152, 216, 238
Condensing capacity, steam still, ii, 15
Condensing surface, condenser, computations, ii, 159
  steam still, ii, 15
Conduction of heat, ii, 825, 826
Conglomerates, United States oil fields, i, 79
Connecting rod and bolts, internal combustion engine, ii, 543
Connemaugh formation, i, 23, 24, 107, 116
Conradson carbon test, petroleum products, i, 588
  emulsification tests, petroleum products, i, 588, 695, 699
Consistency test, asphalt, i, 798
Constant pressure, engine, ii, 559
  internal combustion engines, ii, 526, 529
Constant volumes, internal combustion engines, ii, 526, 529
Construction, pipe lines, i, 406
Consumption, domestic, of petroleum, i, 347, 358
Contact, arc of, lubrication, ii, 631
Contact of sedimentary and igneous rocks, i, 45, 64
Continental Europe, oil shale, i, 843, 846
Continuous distillation, ii, 326
Continuous furnaces, oil fired, ii, 493
Continuous method, prepared roofing, i, 825
  steam-refined asphalt, i, 790
Continuous treatment, refinery, ii, 369
"Contour interval, i, 172
"Contours," 171
Contra Costa County, Calif., oil, i, 539
Control casing head, i, 312
Controls, mapping, i, 176
  refinery, ii, 307
Convection, heat loss by, ii, 830
Convection of heat, ii, 825, 826
Converse County, Wyo., oil, i, 571
Conversion tables, air pressures, ii, 787
  measures, ii, 762–768
  temperature, ii, 802
  volume, weight, energy, ii, 803
  weights, ii, 764–768
Conveyors, filter plant, ii, 201
  oil shale plants, i, 860
Conylene 1, 4-octadiëne, i, 478
Coodys Bluff-Alluwe oil field, Okla., i, 104
Cook County, Tex., bitumen, i, 791
Cookers, acid recovery, ii, 403
Cookson oil field, Pennsylvania, i, 110
Coolers, absorption plants, i, 780
Cooling, Diesel engine, ii, 577
Cooling method, natural-gas gasoline, i, 735
Cooling of air, ii, 829
Cooling surface, condenser, computations, ii, 159
Cooling towers, refinery, ii, 311

Coons, Anne B., i, 321
Cooper, H. C., i, 729,
Cooper oil field, Kentucky, i, 93, 547
Cooper oil field, Ohio, i, 100
Cooper sand, i, 33, 94
Cooperage buildings, refinery, ii, 233
Coos County, Ore., bitumens, i, 791
Copan oil field, Oklahoma, i, 104
Copper, melting point, ii, 806, 807
radiation and reflection of heat, ii, 824
refinery use, ii, 279
specific heat, ii, 811
Copper-matting and refinery furnace, oil fuel, ii, 516
Copper oxide, oil sweetening, ii, 334
Coraopolis oil field, Pennsylvania, i, 110
Cordage, drilling, i, 229
standard drilling outfit, i, 217
Corder oil field, Kentucky, i, 93
Core drill, oil shale sampling, i, 849
Cork, refinery insulating, ii, 282, 286
Corniferous limestone, i, 13, 15, 37, 88, 92, 94, 111
Corning oil field, Ohio, i, 100
Cornplanter oil field, Pennsylvania, i, 110
Correction curves, barometric readings, i, 185, 187
Correction factors, gas flow, i, 441
Corrosion, iron and steel, ii, 570
of metals, tests with petroleum products, i, 588, 706
Corrosion test, gasoline and naphtha, i, 669
petroleum products, i, 588
Corrosive substances, in grease lubricants, ii, 674
Corrugated sheets, weight, table, ii, 780
Corsicana oil field, Texas, i, 60, 113, 114, 560
Corsicana, Texas, rotary drilling, i, 251
Corundum, specific heat, ii, 811
Cosmoline, ii, 380
Cost, distillation by Gray's tower, ii, 357
oil engine operation, California, ii, 611
oil shale crushing, i, 861
operation Diesel engine, ii, 582, 583
refinery construction, ii, 6
refinery operation, ii, 421
shale mining, i, 857
steel tanks, ii, 126
Costa Rica, oil, i, 126
Cotton County, Okla., oil, i, 103
Cottonseed oil, saponification number, i, 711
Cottonwood oil field, Wyoming, i, 119, 120
Counter-current pipe, heat exchanger, ii, 17
Couplings, pipe, ii, 550–565
casing and pipe, weights and dimensions, tables, i, 302
natural-gas pipe lines, i, 453
Couplings, rotary drill pipe, i, 261
"Cover" rocks, nature, i, 42
Cowanshannock oil field, Pennsylvania, i, 110
Coweta oil field, Oklahoma, i, 104
Cowley County, Kans., oil, 90
Cow Run-Newells Run-Bosworth oil field, Ohio, i, 100
Cow Run sand, i, 23, 103, 108, 116
Coyote Hills oil field, California, i, 81
Cracked gasoline, ii, 430
Crackene, i, 518
Cracking, ii, 336, 355, 385
distillates, ii, 430
oil shale, i, 865
refinery, ii, 323, 324
shale oil residuum, i, 879
Cracking processes, ii, 427–444
Crafton oil field, Pennsylvania, i, 110
Craig oil field, Pennsylvania, i, 110
Craig sprayer, ii, 575
Crankshaft, Diesel engine, ii, 563
internal combustion engine, ii, 543
Crawford County, Ill., oil, i, 86, 543
Crawford County, Kans., oil, i, 545
Creek County, Okla., oil, i, 103, 554
Creekside oil field, Pennsylvania, i, 110
Cretaceous, system, Alabama, i, 80
Arabia, i, 145
Argentina, i, 150
Armenia, i, 145
Bolivia, i, 154
Brazil, i, 154
California, i, 83
Canada, i, 123
Cyprus, i, 133
geographical distribution, i, **11**
Germany, i, 134
Greece, i, 133
Jugo-Slavia, i, 134
Louisiana, i, 92
Mesopotamia, i, 141
Mexico, i, 124
Montana, i, 98
North Dakota, i, 102
oil shale, i, 834, 842, 846
oil strata, i, 12, 14, 17
Oklahoma, i, 106
Palestine, i, 144
Rumania, i, 131
saline domes, i, 61
Tennessee, i, 111
Texas, i, 114
Tunis, i, 157
United States oil fields, i, **79**
Venezuela, i, 152
volcanic plugs, i, 63
Washington, i, 115

Cretaceous, system, West Africa, i, 157
Wyoming, i, 119
Crevice oil, i, 67
Crevices, igneous rocks, i, 45, 66
sedimentary rocks, i, 46, 67
Crider oil field, Pennsylvania, i, 110
Crinoidal fossils, Trenton limestone, i, 40
"Crinoidal" limestone, i, 23
Critical viscosity, lubrication, ii, 632
Crook County, Ore., oil, i, 107
Crooksville oil field, Ohio, i, 100, 553
Cross anticline, i, 59
Cross bedding, i, 171
Cross Belt oil field, Pennsylvania, i, 110
Cross head, Diesel engine, ii, 567
lubrication, ii, 639
Cross rods, condenser box, ii, 52, 54, 56, 57
Cross ties, condenser box, ii, 28, 29, 59
flow tanks, ii, 60, 61, 62
pumping-out pans, ii, 30, 60
Crossings, pipe line, i, 415
Crossley internal combustion engine, ii, 532, 542
Crotonylene, i, 478
Crows Run oil field, Pennsylvania, i, 110
Cruce gas field, Oklahoma, i, 104
Crude delivery, at refinery, ii, 82
Crude oil. See also under Oil and Petroleum.
for cracking, ii, 431
for refining, ii, 12
typical distillations, ii, 325, 331, 354, 355
Crude oil still, condenser coil data, ii, 164–179
refinery, ii, 42, 140
skimming plant, ii, 25
Crude still condenser box, ii, 52, 54
Crude still products, ii, 338
Cruse sand formation, Trinidad, i, 160
Crushers, oil shale mining, i, 859
Crushing plant, oil shale, i, 858
Crushing properties, oil shale mining, i, 857
Crushing strength, concrete, ii, 245
Cuba, asphalt, i, 159
asphalt production, i, 796
bitumen, i, 159, 792
oil, i, 159
oil analyses, i, 575
oil production statistics, i, 161, 325
Cuban asphalt, i, 793, 794
Cube method, melting point determination, asphalts, i, 687
Cuddy oil field, Pennsylvania, i, 110
Culbertson oil field, Texas, i, 113
Cumberland County, Ill., oil, i, 86, 543
Cumberland County, Ky., oil, i, 40, 94, 546
Cumberland shallow sand, i, 94
Cumol (cumene), i, 486
Cunningham tank protective system, ii, 316
Cup greases, ii, 674
Cushing oil field, Oklahoma, i, 104, 556
oil sands, i, 26
standard casing requirements, i, 221
Cushing townsite oil field, Oklahoma, i, 104
Cutter, casing, i, 287, 288
Cuyahoga County, Ohio, oil, i, 39
Cyclene, i, 512
Cycles, compression, ii, 526
internal-combustion engine, ii, 529
Cyclic hydrocarbons, i, 460
Cyclobutane, i, 490
Cyclobutene, i, 490
Cyclobutyl diethyl methane, i, 492
Cycloheptane, i, 502
Cyclohexane, i, 494
Cyclo nonomethylene, i, 502
Cyclononane, i, 502
Cyclo-octadiëne, i, 520
a-Cyclo-octadiëne, i, 520
b-Cyclo-octadiëne, i, 520
Cyclo-octane, i, 502
Cyclo-octene, i, 520
Cyclopentadiëne, i, 520
Cyclopentane, i, 492
Cyclopropane, i, 490
Cylinder, automobile, ii, 621
internal combustion engine, ii, 542
internal fluid pressure, ii, 519
Cylinder casing, Diesel engine, ii, 641
Cylnder cover, Diesel engine, ii, 563
Cylinder head bolts, Diesel engine, ii, **565**
Cylinder liner, Diesel engine, ii, 562
Cylinder lubrication, ii, 650
Cylinder oil, ii, 651
chemical characteristics, ii, 661
method of testing, i, 588
viscosity temperature curves, ii, 654
Cylindrical tanks, capacity and dimensions, ii, 772
Cymogene, ii, 361
Cymol (cymene), i, 486
(meta), i, 488
(ortho), i, 486
(para), i, 486
Cyprus oil, i, 133
Czecho-Slovakia, oil, i, 134
oil production statistics, i, 134

D

Dague oil field, Pennsylvania, i, 110
Dakota formation, i, 5
Dakota group, i, 98
Dakota sand, i, 20, 119, 123
Dakota sandstone, i, 12, 119
Dallas County, Iowa, natural gas, i, 90

Dallas oil field, Kansas, i, 91
Dallas oil field, Wyoming, i, 120, 571
Dalton's law, gas pressure, i, 735
Damon Mound oil field, Texas, ii, 113, 566
d'Arcy equation, i, 382
Davis oil field, Pennsylvania, i, 110
Dawson County, Mont., gas, i, 97
Day, David Eliot, i, 831
Day, David T., i, 1, 67, 74, 84, 457, 831, 836, 838, 878, 882, 891
Day-Heller retort, oil shale, i, 866
Day, Roland B., ii, 427
Dayton oil field, Texas, i, 113
Deadweight, oil tanker, i, 374
Deadwood, tanks, ii, 135
Dean and Stark, water determination in petroleum, i, 668
Deaner oil field, Oklahoma, i, 104
De Beque oil field, Colo., i, 84, 85, 542
Decahydronaphthalene, i, 520
Decanaphthene, i, 500
Decane, i, 468, 533
Decker oil field, Ohio, i, 100, 552
Declination line, i, 192
Decolorization, petroleum products, ii, 377
Decomposition, mineral oils, ii, 663
Decylene, i, 474, 533
Decylthiophane, i, 528
De Golyer's classification of seepages, i, 4
Degras, in oil compounding, ii, 661
Degras oil, saponification number, i, 711
De la Vergne cooling system, ii, 578
De la Vergne engine, ii, 531, 532, 535, 542, 591, 610
De la Vergne sprayer, ii, 573
Delaware-Childers oil field, Okla., i, 104, 555
Delaware County, Ind., oil, i, 88
Delaware formation, i, 37
Delbridge, T. G., i, 587
Deliveries of domestic petroleum, i, 345, 358
Delivery of crude oil, at refinery, ii, 82
Demulsification, lubricants, i, 701, ii, 645
Demurrage, tank car, i, 366
Denbow-Wahl-Black-Whitterbrook oil field, Ohio, i, 100
Dennis Run oil field, Pennsylvania, i, 110
Density, gases, adsorption, i, 749
  oil in pipes, i, 389
  oils, relation to viscosity, i, 391
Dephlegmating towers, ii, 219, 343
Dephlegmating tube still, ii, 148
Derbyshire, England, elaterite, i, 792
  oil, i, 135
Derrick, rotary, in oil drilling, i, 253
Derrick plan, steel, i, 215
Desdemona oil field, Texas, i, 31, 113
De Soto Parish, La., oil, i, 95, 550
De Soto-Red River oil field, Louisiana, i, 96
Detail mapping, oil geology, i, 174
Deutz sprayer, engines, ii, 572
Devil's Basin oil field, Montana, i, 97
Devil's Den oil field, California, i, 81
Devonian system, Canada, i, 121
  Kentucky, i, 92, 94
  New Brunswick, i, 71
  oil shale, i, 836, 837, 838, 839, 841, 842, 871
  oil strata in, i, 13, 15, 34, 42
  Oklahoma, i, 103, 106
  Pennsylvania, i, 108
  South Africa, i, 158
  stratigraphic occurrence of petroleum, i, 11
  West Virginia, i, 116
Dewar cracking process, ii, 428
Dewdrop oil field, Pennsylvania, i, 110
Dewey-Bartlesville oil field, Okla., i, 104
Dexter oil field, Kansas, i, 91
Diacetylene series of hydrocarbons, i, 481
Diallyl-1-5 hexadiëne, i, 478
Diallylene, i, 478
Diamond B. X. casing, weights and dimensions, ii, 557
  drive pipe, weights and dimensions, ii, 563
Diamond Springs gas field, Kentucky, i, 93
Diamylene, i, 474
Dianthryl, i, 518
Diapir structures, i, 63
Diastrophism, relation to oil occurrence, i, 45
Diatomaceous earth, refinery insulating, ii, 282
Diatomaceous shale, California, i, 16, 17, 834, 892
Dibenzyl, i, 506
Dicaryophyllene, i, 514
Dickson County, Tenn., oil, i, 111,
Dicyclic diolefines, i, 520
  hydrocarbons, i, 508, 520
Dicyclo-o, 4-, 4-decane, i, 520
Dicyclododecatriëne, i, 520
Dicyclononane, i, 520
Dicyclo-octadiëne, i, 520
Dicyclo-octene, i, 520
Dicyclopentadiëne, i, 520
Dies, pipe threading, ii, 535
Diesel engine, ii, 294, 523, 556, 563, 567
Diethyl, i, 460
Diethyl benzol, i, 486
  cyclobutane, i, 492
  -1-3-cyclohexane, i, 498
  toluol, i, 488
Dietz cracking process, ii, 425
Diffusion, oils through fuller's earth, i, 74, 535, ii, 378
Dihexanaphthene, i, 502

Dihydro-anthracene, i, 516
Dihydrobenzol, i, 488
Dihydrobetulene, i, 504
Dihydrodimethyl naphthalene, i, 508
Dihydromyrcene, i, 510
Dihydronaphthalene-1-4, i, 508
Dihydropinene, i, 520
Dihydrotoluol-Δ-, 1-, 3-, i, 490
Dihydroxylol, i, 490
Di-isoamyl, i, 468
Di-iso-butyl, i, 466
Di-isobutylene, i, 474
Di-isohexyl, i, 468
Di-iso-propenyl, i, 478
Di-isopropyl, i, 462
Dikes, i, 6, 67, 68
  in contact with sedimentary rocks, i, 64
Dimethyl acetylene, i, 478
  2-3-, anthracene, i, 514
  2-4-, anthracene, i, 514
  benzene, i, 480
  β, y-, butadiëne, i, 478
  2, 3, butane, i, 464
  1-2-cycloheptane, i, 502
  cyclohexane (ortho), i, 496
  1, 1-, cyclohexane, i, 496
  1-, 2-, cyclohexane, i, 494
  cyclo-octadiëne, i, 502
  1, 1-cyclopentane, i, 492
  1, 2, cyclopentane, i, 492
  1, 3, cyclopentane, i, 492
  1-, 2-, cyclopropane, i, 490
  diacetylene, i, 480
  diethyl methane, i, 464
  1-, 3-, 4-, ethyl benzol, i, 484
  2-, 3-, 5-, ethyl benzol, i, 484
  1-, 3-, 5-ethyl cyclohexane, i, 500
  ethyl ethylene, i, 474
  ethylene, i, 472
  fulvene, i, 518
  heptane, i, 496
  2-, 4-, heptane, i, 468
  2-, 5-, heptane, i, 468
  2-, 6-, heptane, i, 468
  2,5-, 1,5-hexadiëne, i, 478
  1-, 1-, hexamethylene, i, 494
  1-, 3-, hexamethylene, i, 496
  1, 4, hexamethylene, i, 496
  2, 3, hexane, i, 466
  2, 4, hexane, i, 466
  2, 5, hexane, i, 466
  3, 4, hexane, i, 466
  1-3-5-isobutyl cyclohexane, i, 500
  iso-propyl methane, i, 462
  metacyclohexane, i, 496
  naphthalene β, i, 508
  2-, 6-, octane, i, 468
Dimethyl pentane, i, 464
  2, 4-, pentane, i, 464
  pentene, 2 (2, 3), i, 474
  pentene, 2 (2, 4), i, 474
  propyl methane, i, 462
  trimethylene, i, 490
  1, 1, trimethylene, i, 490
  1-, 2-, trimethylene, i, 490
Dimyricyl, i, 472
Dinaphthyl, I, 516, 536
  methane, α-, i, 516
  methane β, i, 516
Diolefines, i, 478
Diolefines, cyclic, i, 518
Dip, determination, i, 168, **171**, **182**
Dip arrow maps, i, 171, 172, 176
Dipentene, i, 510
Diphenyl, i, 506, 536
  benzol (para), i, 516
  diacetylene, i, 516
  ethane, i, 506
  fulvene, I, 516
  -m-tolyl methane, i, 516
  methane, i, 506, 536
Dipropargyl, i, 480
Direct current motors, ii, 295
Discharge head, pumps, ii, 210
Disk wall packer, i, 293, 294
Displacement tonnage, oil tankers, i, 374
Distillate fuel, testing method, i, 588
Distillates, for cracking, ii, 430
Distillation, costs, ii, 357
  determination of water in petroleum, i, 668
  dry heat, ii, 323
  end point method, i, 667, 673, 715
  fractional, ii, 323
  gasoline, naphtha, and kerosene, i, 673
  natural-gas gasoline, i, 784
  oil shale sample, i, 851
  steam, ii, 323, 343
Distillation record, stripping ii, 154
Distillation tests, oil, i, 873
  shale oil, i, 879, 891, 893
Distilling apparatus, gasoline extraction plant, i, 775, See also Stills
Disuberyl, i, 504, 506
Ditching, pipe line, i, 415
Ditolyl, i, 506
Divinyl, i, 478
Dixmont oil field, Pennsylvania, **i, 110**
Docosane, i, 470
" Doctor solution," i, 689, ii, 334, 368
" Doctor " test, petroleum products, i, 588, 689
" Doctor " treatment, petroleum products, ii, 14
Doddridge County, W. Va., oil, i, 116, 568

Dodecanaphthene, i, 500
Dodecane, i, 468
Dodecylene, i, 474
Dodecylthiophane-sulphone, i, 528
Dog-house, steel foundry, ii, 505
Dolomite, oil occurrence in, i, 41
Dome cover, tank cars, i, 367
Dome, stills, ii, 25, 26, 42, 44, 45, 47
tank cars, i, 362
Domes, geologic, i, 45, 59, 79
Dorseyville oil field, Pennsylvania, i, 110
Dos Bocas oil well, Mexico, i, 20, 124, 125, 248
Dotriacontane, i, 472
Double extra strong pipe, internal fluid pressure, ii, 533
weights and dimensions, ii, 555
wrought, dimensions, table, ii, 784
Double spray system, water cooling, ii, 315
Dougherty County, Okla., bitumen, i, 791
Douglas oil field, Wyoming, i, 118, 120
Douglass oil field, Kansas, i, 91
Dover West oil field, Ontario, Can, i, 122
Dowling oil field, Ohio, i, 100
Draft gage, oil heaters, ii, 472
Drains, refinery, ii, 242
Drake well, Pennsylvania, i, 107, 121, 201
Draught, chimneys, ii, 310
Draw-bar rating, tractors, ii, 623
Draw-off plugs, condenser boxes, ii, 53-59
pumping out pans, ii, 60
Dresser coupling, natural-gas pipe lines, i, 453
Drilling, cost, standard system, i, 250
dry hole method, i, 243
systems of, i, 201, 252, 267
Drilling lines, i, 231, 234
Drilling machines, i, 250
Drilling outfit, standard, specifications, i, 216
Drilling reports, typical, i, 238
Drilling tools, standard, i, 216, 233
Drill pipe, rotary, i, 256
weights and dimensions, ii, 565
Drive pipe, oil wells, i, 301
weights and dimensions, i, 305, ii, 563, 564
Drive pipe casing, standard, i, 236
Drooling burner, ii, 450
Drop forging furnace, oil fired, ii, 489, 494
Drop-out box condenser, ii, 218
Drumright oil field, Oklahoma, i, 104, 105
Drums, oil, manufacture, ii, 226
Dry coil still, ii, 345
Dry Fork oil field, Kentucky, ii, 93
Dry heat distillation, ii, 323
"Dry-hole" drilling, i, 243
Dry measure, ii, 764, 767
"Dry" natural gas, i, 731
Dryer, filter plant, ii, 206
Dual drilling system, i, 202
Dubois County, Ind., oil, i, 88, 544
Ductility, gun metals, ii, 815
Ductility test, asphalt, i, 721, 801
petroleum products, i, 588
Duff City oil field, Pennsylvania, i, 110
Duke sand, Texas, i, 31
Dulong's formula, heat value, ii, 810
Dumas Station oil field, Pennsylvania, i, 110
Dumbaugh oil field, Pennsylvania, i, 110
Dunbridge oil field, Ohio, i, 100
"Duncan gas field," Oklahoma, i, 104
Duncan oil field, Oklahoma, i, 557
Duncan oil field Pennsylvania, i, 110
Dunkard sand, i, 15, 24, 108
Dunn oil field, Pennsylvania, i, 110
Duplex lamp, Dual burner, i, 691, 693
Duplex pumps, refinery, ii, 207
Duplex screw-down liner packer, i, 293
Durand's theory, flow of viscous fluids, i, 381
Duriron, refinery use, ii, 279
Durol, i, 484
Dutch East Indies, oil, i, 78, 161
average specific gravity, i, 322
production statistics, i, 161, 325
unit of measure, ii, 761
Dutch Guiana, oil, i, 155
Dutch New Guinea, oil, i, 162
Dutcher sand, i, 30, 106
Dutton cracking process, ii, 425
Dutton oil pool, Ontario, Can., i, 122, 123
Duval County, Tex., oil, i, 114
Dynamic meter, natural gas, i, 433
Dynamometer, oil, ii, 630
water, ii, 547

E

Eagle Aero gasoline engine, ii, 619
Eagle Ford clay formation, i, 19, 95
Earth-balsam, i, 131
East Africa, oil, i, 158
East Indies, oil, i, 11, 75
East Liverpool gas pool, Ohio, i, 100
East Toledo oil field, Ohio, i, 100
Eastland County, Tex., oil, i, 31, 114, 559
Ebano oil field, Mexico, i, 64, 124, 125, 574
Echigo oil district, Japan, i, 143
Economic boiler, oil fired, ii, 471
Economizers, refinery, ii, 292
Economy-Legionville-Harmony oil field, Pennsylvania, i, 110
Ecuador, oil, i, 11, 154
oil analyses, i, 577
Eddy County, N. Mex., oil, i, 99, 551
Edeleanu process, distillation, ii, 391
Edgar County, Ill., oil, i, 86
Edgerly oil field, Louisiana, i, 96, 549
Edgerton oil field, Kansas, i, 91

Edinburg oil pool, Pennsylvania, i, 110
Edison wire gage, ii, 272
Edmondson County, Ky., bitumen, i, 791
Edna oil field, California, i, 81
Edwards limestone, i, 12
Efficiency, Diesel engine, ii, 560, 561, 582
engines, ii, 609
gas compressor, volumetric, i, 450
internal combustion engines, ii, 530, 549
refinery, i, 421
riveted joints, ii, 265
Egbell, Czecho-Slovakia, oil field, i, 134
Egger oil field, Ohio, i, 100
Egypt, bitumen, i, 793
oil, i, 78, 156
oil analyses, i, 583
oil, average specific gravity, i, 322
oil production statistics, i, 157, 325
oil, unit of measure, ii, 761
Eicosane, i, 470
Eicosylene, i, 476
"1800-foot sand," i, 19
Elastic limit, metals, ii, 820–822
Elaterite, occurrence, i, 792, 793, 794
Elbing oil field, Kansas, i, 91
Eldersville oil field, Pennsylvania, i, 110
Eldorado oil field, Arkansas, i, 539
Eldorado oil field, Kansas, i, 91, 545
casing requirements, i, 221
Electra oil field, Texas, i, 21, 28, 47, 113, 563
Electric heating, cracking, ii, 433
Electric meter, natural gas, i, 433
Electric power, refinery, ii, 287
Electrical equipment, oil tankers, i, 377
Electrical method, specific heat determination, ii, 810
Electrical units, equivalent values, ii, 769
Electrolysis, pipe lines, i, 453, ii, 574
Electromagnets, in fishing, i, 290
Elevations, mapping, i, 193
Elevator casing, rotary drilling, i, 254
Elgin County, Ont., Can., natural gas, i, 123
Elgin sandstone, i, 25
Elida oil field, Ohio, i, 100
Elizabeth sand, i, 36, 108, 118
Elk Basin oil field, Montana-Wyoming, i, 97, 118, 120, 570
Elk City gas pool, Kansas, i, 91
Elk City oil field, Pennsylvania, i, 110
Elk County, Kans., oil, i, 90, 545
Elk Falls oil field, Kansas, i, 91
Elk Hills oil field, California, i, 81
Elk Run oil field, Ohio, i, 100
Elko oil shale, Nevada, i, 98, 890
Elliott tester, oils, i, 624, 632
Ellis formation, i, 98
Elmore oil field, Kansas, i, 91
Elongation, brass and Monel metal, ii, 821
Elyria gas field, Ohio, i, 100
Emba oil field, Russia, i, 127
Embar sand, i, 13, 22, 67, 98, 119
Emery County, Utah, bitumen, i, 791
oil, i, 115
Emlenton-Ritchie Run oil field, Pa., i, 110
Empire oil field, Oklahoma, i, 104
Emulsification tests, oil, i, 695, 740
Emulsions, in pipe lines and tanks, i, 431
Endeavor oil field, Pennsylvania, i, 110
End point method, distillation, i, 667, 673, 715
Energy, weight and volume equivalents, ii, 803
Engine distillate, ii, 339
Engine efficiency, ii, 527, 609
Engine oil, characteristics, ii, 677
methods of testing, i, 588
Engines, aeroplane, ii, 618, 619
automobile, ii, 620
internal combustion, ii, 294, 523, 530, 542,
oil, ii, 523
stationary, ii, 531
steam, refinery, ii, 293
England, bitumen, i, 792
gas and oil, i, 7, 11, 135
oil analyses, i, 579
oil, average specific gravity, i, 322
oil production statistics, i, 325
oil shale, i, 834, 844
Engler distilling flask, i, 673
Engler viscosimeter, i, 643, 655, 656
English method, determination, melting point of paraffin wax, i, 679
English, oil field, Pennsylvania, i, 110
Enniskillen oil field, Ontario, Can., i, 121
Enterprise oil field, Pennsylvania, i, 110
Eocene series, Arabia, i, 145
Armenia, i, 145
Belgium, i, 136
California, i, 83
Colorado, i, 84
Florida, i, 84
Galicia, i, 132
Germany, i, 134
India, i, 141
Italy, i, 132
Jugo-Slavia, i, 134
Madagascar, i, 158
Mesopotamia, i, 141
Mexico, i, 124
oil shale, i, 834, 836, 839, 840, 842, 846, 873
oil strata, i, 12, 14, 17
Rumania, i, 131
Spain, i, 136
Trinidad, i, 160

Eocene series, Venezuela, i, 152
Epperson sand, i, 92
Equivalent of heat, mechanical, ii, 524, 808
Erie oil field, Kansas, i, 91, 546
Erie oil field, Pennsylvania, i, 110
Erosion, oil geology, i, 170
Erythrene, i, 478
Escambia County, Fla., oil, i, 84
Essex County, Ontario, Can., oil, i, 121, 123
Esthonia, oil shale, i, 843
Estill County, Ky., oil, i, 92, 548
Ethane, i, 460, 730, 752
Ether, boiling point, ii, 807
  expansion by heat, ii, 812
  hydrocarbon solvent, i, 536
  latent heat of evaporation, ii, 823
  specific heat, ii, 811, 812
  sulphuric, boiling point, ii, 806
Ethyl acetylene, i, 478
  alcohol, ii, 401, 806, 807, 823, 830
  anthracene, i, 514
  benzol, i, 482
  cyclobutane, i, 492
  cycloheptane, i, 502
  cyclohexane, i, 498
  ethylene, i, 472
  hexamethylene, i, 498
  hexane, i, 466
  isoamyl, i, 464
  isobutane, i, 464
  isopropane, i, 460
  methyl benzol, i, 484
  naphthalene i, 508
  naphthene, i, 498
  pentyl sulphide, i, 529
  phenyl acetylene, i, 480
  propyl sulphide, i, 529
  sulphide, i, 529
  tetramethylene, i, 492
  toluol, i, 484
Ethylene, i, 458, 472
  heat of combustion, ii, 809
  in refinery gases, ii, 400
  specific gravity, ii, 832
Ethylene benzol, i, 482
Ethylene hydrocarbons, specific heats, i, 532
Euphemia oil field, Ontario, Can., i, 121
Eureka County, Nev., bitumen, i, 791
Eureka oil field, Kansas, i, 91
Eureka oil field, West Virginia, i, 117
Eureka-Volcano-Burning Springs anticline, West Virginia, i, 47, 117
Europe, bitumen, i, 792
  oil, i, 75, 127
  oil shale, i, 848
Evans City oil field, Pennsylvania, i, 110
Evans cracking process, ii, 438
Evaporating towers, refinery, ii, 184, 219, 243
Evaporation, factors, steam, ii, 799
  latent heat of, table, i, 752, ii, 823
  lubricants, ii, 645
  tests, petroleum products, i, 588, 716
  total heat of, ii, 824
  water, ii, 313, 314
Ewings Mill oil field, Pennsylvania, i, 110
Exchangers, condenser, ii, 181
Exhaust parts, Diesel engine, ii, 563, 577
Expanders, natural gas, i, 765
Expansion, gases, ii, 812
  heat, ii, 812
  liquids, ii, 812
  solids, ii, 813
  steam pipes, ii, 798, 802
  trunks, oil tankers, i, 376
Explosion hatch, tank, ii, 111
Exports, petroleum, i, 349
External upset tubing, weights and dimensions, ii, 562
Extra heavy fittings, dimensions, ii, 278
Extra strong pipe, internal fluid pressure, ii, 532
  weights and dimensions, ii, 555
  wrought, dimensions, table, ii, 783
Extraction plant, natural-gas gasoline, i, 764

F

Fabric, protective covering, ii, 573
Fagundus oil field, Pennsylvania, i, 110
Fahrenheit thermometer scale, ii, 832
Fairbanks-Morse internal combustion engine, ii, 537
Fairfield County, Ohio, oil and gas, i, 39, 102, 552
Falkland Islands, oil, i, 156
Fallsburg oil field, Kentucky, i, 93
Fan-tailed burner ii, 451
Fatty oils, saponification numbers, i, 712
Faults, closed, i, 68
  overthrust, i, 68
  United States oil fields, i, 46, 79
Fayette County, Ala., oil and gas, i, 79
Fayette oil field, Pennsylvania, i, 110
Fayetteville formation, i, 30
Feed, lubrication, ii, 634, 646, 662
  oil and water, cracking, ii, 433
  oil lubrication, ii, 638, 640
Feeders, oil shale plants, i, 860
Feeding, grease lubrication, ii, 674
  methods, lubricants, ii, 660
Feed-pump control, cracking, ii, 433
Feed water heaters, refinery, ii, 292
Felt, asphalt saturated, roofing, i, 822
Felts, refinery insulating, ii, 282
Fenchene, i, 514

Ferghana, Russia, oil analyses, i, 582
oil field, i, 127, 145
Fergus County, Mont., oil, i, 97, 551
Fernando formation, i, 12, 16, 83
Ferris oil field, Wyoming, i, 120, 572
Ferrosteel, Crane, tensile strength, ii, 819
Fichtelite, i, 522
Fiddlers Run oil field, Pennsylvania, i, 110
Field mapping, i, 168, 172, 173
Field methods, oil geology, i, 167
Field work, oil geology, i, 175
Fifteen oil field, Ohio, i, 100, 553
Fifth sand, i, 35, 36, 108, 116
50-foot sand, i, 34, 35, 108, 116
Fillers, asphalt, i, 821
automatic, refinery, ii, 225
lubricating greases, ii, 674
Filling, pipe line, i, 415
Filling tanks, refinery, ii, 75
Fillmore oil field, California, i, 81
Film, as oil indication, i, 168
Filter, refinery, ii, 65, 201
Filter house, refinery, ii, 92, 200, 231
Filter wash stills, refinery, ii, 51
Filtering, refinery, ii, 377
fractional, i, 74, 535
Filtration, fire hazards, ii, 412
Findlay oil field, Ohio, i, 55, 100
Finland, oil, i, 131
Finleyville oil field, Pennsylvania, i, 110
Fire brick, furnaces, ii, 463
refinery construction, ii, 259
Fire extinguishers, refinery, ii, 415, 418
Fire-extinguishing, oil tankers, i, 377
refinery, ii, 405, 418
Fire fighting equipment, refinery, ii, 315
Fire hazards, refinery, ii, 319, 411
Fire point, i, 623
Cleveland open cup tester, i, 629
Elliott tester, i, 632
Pensky-Martens closed tester, i, 638
Fire prevention, refinery, ii, 315, 405
Fireproof construction, refinery, ii, 229
Fire protection, pipe lines, i, 423
Fire test, i, 623, 624
Fire Worshipers, i, 4, 127
First Bradford sand, i, 36
First Cow Run sand, i, 23, 25
First Mountain sand, i, 25
First Venango sand, i, 35
Fisher oil field, Ohio, i, 100
Fishing, oil well, i, 274
Fishing tools, i, 218, 274
Fish oil, saponification number, i, 711
Fissures in shale, oil, i, 67
Fitting, steel plate work, ii, 42
Fittings, agitator, ii, 63, 64
Fittings, cast-iron, ii, 277
pipe line, i, 417
standard drilling outfit, i, 217
Five Islands, Louisiana, salt domes, i, 61
Five Points-Florence oil field, Pa., i, 110
Fixed carbon test, asphalt, i, 809
asphaltic petroleum products, i, 723
petroleum products, i, 588
Fixed gases, oil shale distillation, i, 882
Flat Rock oil field, Oklahoma, i, 105
Flat rolled iron, weight, table, ii, 776
steel, weight, table, ii, 776
Flattening test, pipes, ii, 551
Flanges, acid blow case, ii, 40, 77
bleachers, ii, 38, 72, 73
condenser boxes, ii, 29, 53, 55, 56, 58, 59
filters, ii, 65
pumping-out pan, ii, 30, 60
refinery use, ii, 273
standard dimensions, ii, 277, 280
stills, ii, 25, 43, 45, 46, 47, 50, 51
still towers, ii, 26, 27, 48, 49
tanks, ii, 32-39, 60-78, 107
Flanging, steel plate work, ii, 42
Flash point, asphalt, i, 809
fuel oil, ii, 340
Flash point of oils i, 623
Abel tester, i, 638
Elliott tester, i, 632
Pensky-Martens tester, i, 633, 634, 635
Tag closed tester, i, 629
testers used, i, 588
Flash test, oil, Cleveland open cup method, i 626
comparison by different instruments, i, 62
Flat Rock oil field, Oklahoma, i, 105
Flax, packing in refinery construction, ii, 28
Flints Mills oil field, Ohio, i, 100
Float test, petroleum products, i, 586, 722
Floc, refinery, ii, 370
Floc test, petroleum products, i, 588, 691
Floor beams, concrete construction, ii, 252
Floor coverings, asphalt, i, 826
Floor slabs, reinforced concrete, ii, 252
Florence oil field, Colorado, i, 67, 84, 85, 54
Florence oil field, Kansas, i, 91
Florence oil field, Ontario, Can., i, 122
Florence oil field, Pennsylvania, i, 110
Florida, oil, i, 19, 84
Flow, steam, in pipes, table, ii, 797
oil through pipes, i, 380
Flow capacity, chart, i, 401
in pipe lines, i, 396
Flow tanks, refinery, ii, 60
Flowing wells, location and output, i, 310
Floyd County, Ky., oil and gas, i, 31, 34, 92 94, 547

Fluid friction formula, i, 439
Fluid injecting, apparatus, oil heater, ii, 475
Fluid pressure, pipes, internal, ii, 519
Fluoranthene, i, 516, 518, 536
Fluorene, i, 506, 536
Flush joint tubing, weights and dimensions, ii, 560
Flushing oil field, Ohio, i, 100
Flux, refinery, ii, 41
Flywheel, internal combustion engines, ii, 546
  Diesel engine, ii, 569
Foam lines, refinery, ii, 243
Foaming solution, fire extinguishing, ii, 420
Foamite system, fire extinguishing, ii, 317
Fonner oil field, Pennsylvania, i, 110
Foot-pound, ii, 808
Foot valve and strainer, oil heaters, ii, 461
Foots oil, ii, 374
Ford City oil field, Pennsylvania, i, 110
Foreign oil units, weights and measures, ii, 761
Forest clay formation, Trinidad, i, 160
Forging, oil fuel, ii, 487
Forging furnaces, ii, 489
Formation maps, oil geology, i, 171, 172, 176
Formosa, Japan, oil, i, 78, 143
  oil production, statistics, i, 143
Forms, concrete, ii, 251
  concrete tank construction, ii, 132
Ford Bend County, Tex., oil, i, 114
Fort Scott limestone, i, 26, 27
Fortuna sand, i, 22
40-foot sand, i, 108
Forward oil field, Pennsylvania, i, 110
Fossiliferous shales, i, 40
Fossils, vertebrate, in asphalt, i, 16
Foster tester, for oil, i, 623
Foundation materials, ii, 247
Foundations, refinery, ii, 237, 244
Four-cycle engine, Busch-Sulzer, ii, 584
  Diesel, ii, 560
  horizontal stationary, ii, 531
  internal combustion, ii, 529
  Snow Diesel, ii, 591
Fourth sand, i, 35, 36, 108, 116
Fourth sand oil field, Pennsylvania, i, 110
Fox-Bush oil field, Kansas, i, 91
Fox oil field, Oklahoma, i, 104
Fractional condensation, cracking, ii, 423
  oil shale vapors, i, 870
Fractional distillation, ii, 323
Fractional fusion, refinery, ii, 373
Fractionation, oils, i, 534
France, asphalt, i, 796
  bitumen, i, 792
France, oil, i, 133
  analyses, i, 578
France, production statistics, i, 325
  shale, i, 834, 843
  unit of measure, ii, 761
Francis gas field, Oklahoma, i, 104
Franklin County, Kans., oil, i, 90, 545
Franklin oil field, Pennsylvania, i, 110
Frasch sweetening process, ii, 324, 334, 392
Frazeysburg oil field, Ohio, i, 100
Fredonia oil field, Kansas, i, 91
Freezing, in lubrication, ii, 664
  oil well, i, 220, 246, 262, 286, 298
Fremont County Colo., i, 542
Fremont County Wyo., oil, i, 119, 571
Fremont oil field, Ohio, i, 100
French Guiana, oil, i, 156
French thermal unit, ii, 808
Fresno County, Calif., oil, i, 80, 541
  nitrogen in, i, 530
  sulphur in, i, 526
Friction in lubrication, coefficient of, ii, 633
Friction relations of viscosity in lubricants, ii, 630
Friction socket, i, 278, 289
Friction tests, lubricants, ii, 631
Frictional force, lubrication, ii, 627, 629
Frictional loss, in lubrication, ii, 638
Frictional resistance, flow of liquid, i, 381
  piston lubrication, ii, 656
Frogtown-Greenville oil field, Pa. i, 110
Frontier formation, i, 17, 97, 98, 118
Froth-producing mixtures, fire extinguishing, ii, 418
"Frozen" casing, i, 249
Frozen Creek oil field, Kentucky, i, 93
Frozen grease, lubricant, ii, 673
Fryburg oil field, Pennsylvania, i, 110
Fuel consumption, Busch-Sulzer Diesel engine, ii, 590
  curves, internal combustion engine, ii, 533, 535
  extraction plants, natural-gas gasoline, i, 785
  Snow Diesel engine, ii, 591
Fuel economy, stills, ii, 333
Fuel injection, engines, ii, 571
Fuel oil, ii, 424
  British Admiralty specifications, ii, 340
  California, specifications, ii, 341
  condenser coil data, ii, 170, 178
  fuel equivalents, ii, 476, 480, 487, 497, 505
  locomotives, ii, 484
  Navy specifications, ii, 340
  tankers, i, 378
  use of, ii, 445
Fuel oil stock, ii, 370
Fuel pump, Diesel engine, ii, 575
Fuels, mixed, heating value, ii, 808, 810

Fuller's earth, diffusion of oil, through, i, 74, 534
refinery use, ii, 201, 202, 368, 377, 418
Fullerton oil field, California, i, 16, 81, 82, 83
Fulton County, Ind., oil, i, 88
Fuming sulphuric, acid, ii, 411
content of SO_3, table, ii, 845
Fuqua topping plant, ii, 82
Furbero oil field, Mexico, i, 64, 124, 125, 126, 574
Furnace, cracking, ii, 432
forging, ii, 489
revivifying, filter plant, ii, 203
Fusion, fractional, refinery, ii, 373
latent heat of, ii, 823
Fusion points, seger cones, ii, 804
Fyzabad pool, Trinidad, i, 159

G

Gage, sheet and wire, ii, 773, 774,
tin plate, table, ii, 779
wire, ii, 272, 773, 774
Gage pressures, compressed air, ii, 299
Gage table, tanks, ii, 135
Gaging, concrete tanks, ii, 133, 134
pipe line, i, 426
refinery, ii, 425
Gaines oil field, Pennsylvania, i, 110
Gainesville oil field, Kentucky, i, 93
Galicia, oil, i, 47, 67, 78, 131
analyses, i, 580
oxygen in, i, 524
production statistics, i, 132, 324
Gallipolis oil field, Ohio, i, 100
Galloupe process, oil shale, i, 866
Galvanized sheets, weights, table, ii, 779, 780
Gantz sand, i, 34, 108, 116
Garber oil field, Oklahoma, i, 104
Garbutt rod, i, 318
Garfield County, Mont., oil, i, 551
Garfield County, Okla., oil, i, 103, 555
Garnett oil field, Kansas, i, 91
Garrett Run oil field, Pennsylvania, i, 110
Garrison formation, i, 21
Garrison oil field, Pennsylvania, i, 110
Garvin County, Okla., bitumen, i, 791
Garvin oil field, Pennsylvania, i, 110
Gas, natural, compression, i, 447
gasoline from, i, 729
production statistics, i, 352
Gas, tests of pipe line flow, i, 446
Gas analysis, method of, i, 731
Gas anchor packer, special, i, 293
Gas engine fuel, ii, 361
Gas engines, refinery, ii, 287
Gas expansion, internal combustion engines, ii, 526
Gas-oil, ii, 339, 363, 424
condenser coil data, ii, 178
sulphur test, i, 707
test methods, i, 588
Gas pipe, standard dimensions, table, ii, 78 784
Gas pumps, for oil wells, i, 320
Gas sand, i, 15, 25, 108
Gas seepages, as indications of oil, i, 167
Gas Springs, Argentina, i, 150
Asia Minor, i, 145
France, i, 133
Rumania, i, 131
Gas tanks, oil well, i, 313
Gaseous adsorption, by solids, i, 747
Gases, adsorption density, i, 749
expansion, ii, 812
heat of liquefaction, i, 749
specific gravity, ii, 832
specific heat, ii, 524, 528, 811
Gasoline, ii, 338, 360
acid heat test, i, 690
average distillations, ii, 350
corrosion and gumming test, i, 689
cracked, ii, 430
determination of vapor pressure, i, 726
"doctor test," i, 689
end point method of distillation, i, 673
gravities, ii, 346, 349, 352
natural gas, i, 729
Panuco crude, ii, 431
refinery production, United States, ii, 34
testing methods, i, 588
treating, ii, 368
volatilities, ii, 346, 352
Gasoline content, natural gas, i, 759
Gasoline distillate, condenser coil data, ii, 1
Gasoline engines, ii, 523, 615
Gasoline precipitation test for lubricati oils, i, 713
Gasoline vapor, removed from gas, i, 745, 7
Gasometers, refinery, ii, 125
Gaspé Peninsula, Quebec, Can., oil, i, 41, 121, 576
Gate houses, pipe line, i, 417
Gate valve, use in gushers, i, 312
Gathering lines, oil, i, 428
Gaujene, i, 514
Gaza Land, Africa, oil shale, i, 848
Gbely oil field, Czecho-Slovakia, i, 134
Geanticlinal folds, i, 53
Geanticline, United States oil fields, i, 79
Gear greases, ii, 674
Gels, adsorption method, gasoline extracti i, 745
silica, absorbent, ii, 368
Gemsah, Egypt, oil, i, 156

Gemsah, Egypt, oil analysis, i, 583
General Electric method, melting point determination, asphalt, i, 803
Generators, refinery, ii, 287, 295
Genesee shale, i, 34
Geographic classification, oil shale, i, 834
Geologic ages, oil formations, Canada, i, 121
  Kentucky, i, 14, 94
  Louisiana, i, 12, 95
  North America, i, 12, 15
  Ohio, i, 14, 103
  Oklahoma, i, 12, 106
  Pennsylvania, i, 118
  United States, i, 79
  Wyoming, i, 12, 119
  West Virginia, i, 14, 116
Geologic field work, oil, i, 167
Geologic mapping, oil, i, 168
Geologic maps, oil, i, 171
Geological classification, oil shales, i, 834
George oil field, Pennsylvania, i, 110
Georges Creek oil field, Kentucky, i, 93
Georgia, oil, i, 84
Geosyncline, Appalachian, i, 102
  Great Plains, i, 72
Germany, asphalt, i, 796
  average specific gravity, i, 322
  bitumens, i, 792
  gas and oil shale, i, 11, 78, 133
  oil analyses, i, 579
  oil production statistics, i, 134, 324
  oil shale, i, 834, 843
  oil, unit of measure, ii, 761
Gibson County, Ind., oil, i, 88
Gibson oil field, Kentucky, i, 93
Gibsonburg-Helena oil field, Ohio, i, 100
Gibsonia oil field, Pennsylvania, i, 110
Gibsonville gas field, Ohio, i, 100
Gill, John D., ii, 625
Gilmer County, W. Va., oil, i, 116
Gilsonite, i, 71, 790, 794
Girders, reinforced concrete, ii, 252
Girty oil field, Pennsylvania, i, 110
Glade oil field, Pennsylvania, i, 110
Glade Run oil field, Pennsylvania, i, 110
Glance pitch (manjak), i, 792
Glass. radiation and reflection of heat, ii, 824
  specific heat, ii, 811
Glass industry, oil fuel, ii, 516
Glauconite, i, 19
Glenhazel oil field, Pennsylvania, i, 110
Glenn Pool oil field, Oklahoma, i, 104, 554
Glenn sand, i, 28
Glenns Run oil field, West Virginia, i, 117
Glenoak oil field, Oklahoma, i, 104
Glensfield oil field, Pennsylvania, i, 110
Glue solutions, fire extinguishing, ii, 318
Glycerin, as lubricant, ii, 643
  boiling point, ii, 807
"Go devil," i, 432
Gold, melting point, ii, 806, 807
  radiation and reflection of heat, ii, 824
  specific heat, ii, 811
Gold Coast, oil, i, 158
Goldingham, Arthur H., ii, 523
Goose Creek oil field, Texas, i, 113, 564
Goose Run and Mitchell oil field, Ohio, i, 100
Gordon sand, i, 31, 35, 36, 108, 116
Gordon stray sand, i, 35
Gotebo oil field, Oklahoma, i, 104, 555
Gould oil field, Ohio, i, 100
Governor, Busch-Sulzer Diesel engine, ii, 586, 590
  pendulum, internal combustion engine, ii, 547
Grab, fishing tool, i, 282, 286
Grady County, Okla., i, 103
Graham oil field, Texas, i, 113, 563
Graham-Fox oil field, i, 104
Grahamite, i, 790, 792, 794
Grahamite dike, i, 6, 7, 68
Grand County, Colo., bitumen, i, 791
Grand County, Utah, bitumen, i, 791
  oil, i, 115
Grand Valley oil field, Pennsylvania, i, 110
Granite oil field, Okla., i, 104
Grant County, Ind., oil, i, 88, 544
Grantsville oil field, West Virginia, i, 117
Grapeville oil field, Pennsylvania, i, 110
Graphite, filler for lubricating grease, ii, 674
  specific heat, ii, 811
Graphitic coal, ii, 520
Grass Creek oil field, Wyoming, i, 118, 120, 572
Gratings, filters, ii, 65
  steam still towers, ii, 20, 50
Gravel and asphalt roofs, i, 825
Gravimetric method of determination of B. S. i, 670
Gravitation, relation to oil occurrence, i, 45
Gravities, gasoline, ii, 346, 349, 352
  stripping, ii, 327, 328
Gravity, comparison of scales, i, 596
  correction for temperature, i, 600
  natural-gas gasoline, relation to composition, i, 755
  oil, relation to viscosity, i, 391
  "observed," i, 599
  temperature corrector, Atlantic, i, 601
  test for, i, 594
Gravity of Liquids, hydrometer method, i, 597
  weighing method, i, 618
Gravity of solids, hydrometer method, i, 588, 621
  weighing method, i, 588, 621

Gravity drop furnace, filter plant, ii, 203
Gray chilling machine, wax plant, ii, 195
Gray dephlegmating towers, ii, 221, 357
Gray oil field, Tex., i, 113
Gray sand, i, 35
Gray wax-moulding press, ii, 375
Grays Park oil field, Pennsylvania, i, 110
Grayson County, Ky., bitumen, i, 791
oil, i, 92
Graysville oil field, Ohio, i, 100, 552
Grease lubrication, ii, 671
Greases, lubricating, ii, 672
Great Britain, oil analyses, i, 579
oil and gas, i, 135
oil shale, i, 135, 844, 848
oil, unit of measure, ii, 761
Great Plains, geosyncline, i, 72
Cambrian system, i, 40
Greece, bitumen, i, 792
oil, i, 137
Greeley oil field, Kansas, i, 91
Greenbrier limestone, i, 15, 29, 31, 116
Greene County, Pa., oil, i, 35, 558
Green Hill oil field, Kentucky, i, 93
Green River formation, oil shale, i, 873, 884
Green River shales, ash analyses, i, 876
Greenville gas pool, Illinois, i, 87
Greenville oil field, Pennsylvania, i, 110
Greenwood County, Kan., oil, i, 90, 545
Greybull oil field, Wyoming, i, 118, 120, 570
Griffin oil field, Kentucky, i, 93
Griffith oil field, Ohio, i, 552
Griffithsville oil field, West Virginia, i, 117
Grinding, oil shale, i, 859
Grizzlies, oil shale plants, i, 860
Ground cork, grease filler, ii, 674
Ground joint pipe, ii, 275
Grozni, Russia, oil analysis, i, 581
oil field, i, 47, 127
Gross register tonnage, oil tankers, i, 374
Grubbs oil field, Pennsylvania, i, 110
Guelph limestone, i, 15, 121
Guernsey County, Ohio, oil, i, 33, 102
Guilford oil field, Kansas, i, 91
Gulf Coast oil field, i, 19, 78, 79, 356
deliveries of petroleum, i, 349, 358
quantity and value oil production since 1889, i, 337, 348, 356
saline domes, i, 60
stocks of petroleum, i, 343, 357
Texas, i, 114
Gulf Coastal Plain, Mexico, i, 124
Gulf series, i, 18, 95
Gum, tar stills, ii, 329
"Gumbo," drilling bit used, i, 262
Gum lac, radiation and reflection of heat, ii, 824
Gumming tests, petroleum products, i, 588, 689
Gun metals, ductility, ii, 815
Gurley compass, i, 178
Gusher, description of, i, 310
Gypsum, relation to oil occurrence, i, 8, 61
grease filler, ii, 674
Gyratory crushers, oil shale mining, i, 859

H

Hadenville oil field, Pennsylvania, i, 110
Haiti, oil, i, 161
Haldinand County gas fields, Ontario, Can., i, 122, 123
Hale oil field, Kansas, i, 91
Hallett oil field, Oklahoma, i, 104
Halltown oil field, Pennsylvania, i, 110
Hamilton County, Iowa, gas, i, 90
Hamilton formation, i, 38, 94, 121
Hamilton oil field, Wyoming, i, 119, 120
Hamilton Switch oil field, Oklahoma, i, 104
Hammerschmitt oil field, Pennsylvania, i, 110
Hammer-weld process, pipes, ii, 504
Hammond coupling, natural gas pipe lines, i, 453
Hancock County, Ohio, oil, i, 102, 552
nitrogen in, i, 529
sulphur in, i, 525
Hancock County, W. Va., oil, i, 116, 568
Hand-fired boiler, oil and coal, ii, 472
Hand level method, mapping, i, 177, 182
Hand lens, i, 181
Handling tank cars, i, 366
Hardin County, Tex., oil, i, 114, 564
Hard press, wax plant, ii, 199
Harmony oil field, Pennsylvania, i, 110
Harper oil field, Ohio, i, 100
Harris County, Tex., oil, i, 114, 564
Harrison County, Ohio, oil, i, 102
Harrison County, W. Va., oil, i, 116, 568
Harrisville oil field, Pennsylvania, i, 110
Harrisville oil field, West Virginia, i, 117
Hartford oil field, Kentucky, i, 93
Haskell oil field, Oklahoma, i, 104
Hatch, explosion, tanks, ii, 111
Hatches, bleacher, ii, 38, 73, 74
Haysville oil field, Pennsylvania, i, 110
Hazelhurst oil field, Pennsylvania, i, 110
Headache post, i, 272
Heads, boiler, refinery, ii, 289
filter, ii, 65
pump, ii, 210
stills, ii, 25–27, 45–51
water, comparison with various units, table, ii, 792
Healdton oil field, Okla., i, 59, 104, 554
Heat, conduction, ii, 825, 826

Heat, convection, ii, 825, 826
effect on hydrocarbons, i, 458
expansion, ii, 812
latent, ii, 822
loss by convection, ii, 830
quantitative, measurement, ii, 808
radiation of, ii, 824
reflection of, ii, 824
sensible, ii, 824
specific, ii, 810
transmission, ii, 828, 829
Heat application, oil shale retorting, i, 863
Heat balance, engines, ii, 549, 556, 584
Heat consumption, engines, ii, 549, 583
Heat control, "cracking" oil shale, i, 865
oil shale retorting, i, 863
Heat exchange, condensers, ii, 182
data, Trumbull process, ii, 388
steam still, ii, 17
Heat exchangers, gasoline extraction plants, i, 772
Heat-exchanging stills, refinery, ii, 50
Heat losses, engines, ii, 527
furnace walls, ii, 237
uninsulated surfaces, ii, 283
Heat of adsorption, i, 749
Heat of combustion, ii, 809
Heat of evaporation, total, ii, 824
Heat of liquefaction, gases, i, 749
Heat of vaporization, steam, ii, 794
Heat recovery, spent shale, i, 869
Heat test, lubricating and insulating oils, i, 705
petroleum products, i, 588
Heat-treating furnaces, ii, 505
Heat unit, ii, 808
Heat units, equivalent values, ii, 769
water, ii, 791
Heat value, natural gas, i, 734, 756
natural-gas gasoline, i, 752
shales, relation to oil content, i, 877
Heater, for pipe lines, i, 403
oil burning, ii, 447
Heater box, oil heaters, ii, 479
Heater pipes in cars for transportation of petroleum, i, 360
Heavy foots oil, sweating cuts, ii, 374
Heavy standard rig specification, i, 206
Heidelberg oil field, Ky., i, 92
Heidelberg oil field, Pa., i, 110
Heine boiler, oil fired, ii, 469
Helena oil field, Ohio, i, 100
Helicoid conveyor, capacity and horsepower, ii, 307
filter plant, ii, 201
Hemellithol, i, 482
Hendershot oil field, Ohio, i, 100
Hendersonville oil field, Pa., i, 110
Heneicosane, i, 470
Henry's law, gases, i, 739
Hentriacontane, i, 472
Hepler oil field, Kans., i, 91
Heptacosane, i, 472
Heptadecane, i, 470
Heptadecylene, i, 476
Heptamethylene, i, 502
Heptanaphthene, i, 494
Heptane, i, 464, 533
Heptine, i, 478
Heptylene, i, 474, 533
Heptylthiophane, i, 528
Heptylthiophane-sulphone, i, 528
Herschel demulsibility test, petroleum products, i, 588, 695, 701
Hershey on friction, ii, 633
Hesperidene, i, 510
Hess-Ives tint photometer, use in oil tests, i, 662
Hesselman sprayer, engines, ii, 571
Hewins oil field, Kans., i, 91
Hewitt oil field, Okla., i, 104, 554
Hexacontane, i, 472
Hexacosane, i, 470
Hexadecane, i, 470, 533
Hexadecyl acetylene, i, 478
Hexadecylene, i, 476
Hexadiëne, 1, 5, i, 478
Hexaethyl benzol, i, 488
Hexahydro-anthracene, i, 516
Hexahydrobenzol, i, 494
Hexahydro-4-cumol, i, 500
Hexahydrocumol, i, 500
Hexahydrocymeme (para), i, 500
Hexahydro-isoxylol, i, 496
Hexahydromesitylene, i, 498
Hexahydrometaxylol, i, 494
Hexahydro 1-, 3-, metaxylol, i, 494
Hexahydroorthoxylol, i, 496
Hexahydroparaxylol, i, 498
Hexahydropropyl benzol, i, 500
Hexahydro-pseudocumol 1,- 2-, 4, i, 496
Hexahydro-pseudocumol 1, 2, 4 (para), i, 496
Hexahydrotoluene, i, 494
Hexamethyl benzol, i, 488
Hexamethylene, i, 494
Hexanaphthene, i, 494
Hexane, i, 460, 462, 533
Hexatriacontane, i, 472
Hexyl acetylene, i, 478
Hexylene, i, 474, 533
Hexyl sulphide, i, 529
Hexylthiophane, i, 528
Hexylthiophane-sulphone, i, 528

Hickory Creek-Pond Creek oil field, Okla., i, 104
Hickory oil field, Pa., i, 110
Hidden Dome gas pool, Wyo., i, 120
High Bridge formation, i, 40, 94
High dip, relation to occurrence of petroleum, i, 168
High pressure burner, ii, 449
High pressure gas-pipe lines, ii, 452
High pressure internal combustion engines, i, 542
Highway construction, asphalt, i, 816
Hiseville oil field, Ky., i, 93
Hocking County, Ohio, oil, i, 39, 102
Hoffman-Moore-Burkhart oil field, Ohio, i, 100
Hog still, ii, 345
Hogshooter oil field, Okla., i, 104
Hogshooter sand, i, 29, 30, 106
Hohman oil field, Ohio, i, 100
Hoid-Brons sprayer, ii, 575
Hoist or draw works, rotary drilling, i, 253
Holland, oil, i, 136
  unit of measure, ii, 761
Holliday oil field, Tex., i, 113
Hollidays Cove oil field, W. Va., i, 117
Homer gas field, Ohio, i, 100
Homer oil field, La., i, 18, 96, 550
Homer oil field, Okla., i, 104
Homewood oil field, Pa., i, 110
Homewood sandstone, i, 25
Homeworth oil field, Ohio, i, 100
Hominy oil field, Okla., i, 104
Hominy townsite oil field, Okla., i, 104
Homocline, oil geology, i, 53
Honduras, oil, i, 126
Hook, threading die, ii, 536
Hook wall packer, i, 294
Hook and cube method, melting point determination, asphalt, i, 802
Hookstown oil field, Pa., i, 110
Hopewell oil field, Ohio, i, 100
Hopper oil field, Pa., i, 110
Horizontal distances, mapping, i, 193
Horizontal engines, Diesel, ii, 561
  stationary, ii, 531
  two-cycle, ii, 537
Horizontal retorts, oil shale, i, 866
Horizontal stills, refinery, ii, 140
Hornsby-Akroyd engine, ii, 523, 531
Hornsby 1915 "R" type internal combustion engine, ii, 533
Horse-power, air compression, ii, 301
  belts, ii, 303
  belt conveyors, ii, 305
  formula, N. A. C. C., ii, 622
  formula, tractor engines, ii, 622
Horse-power, formulas, Society of Automotive Engineers, ii, 621
  helicoid conveyors, ii, 307
  internal combustion engines, ii, 530, 549, 568
  steam engines, ii, 294
  steel shafting, tables, ii, 302
Horton sand, i, 25, 92, 94
Hoskins Mound oil field, Tex., i, 113
Hot oil system, pipe lines, i, 406
Hot Springs County, Wyo., oil, i, 119, 572
Hot surface type internal combustion engine, ii, 530, 542
Howard oil field, Kans., i, 91
Hudson oil field, Wyo., i, 120
Hudson River shales, i, 39
Hull Creek oil field, Tex., i, 565
Hull oil field, Kans., i, 91
Hull oil field, Tex., i, 113, 565
Humble oil field, Tex., i, 113, 564
Humboldt oil field, Kans., i, 91, 544
Humidity of air, relative, table, ii, 787
100-foot sand, i, 34, 108, 116
Hungary, bitumen, i, 792
Huntington County, Ind., oil, i, 88
Hurghada, Egypt, oil analysis, i, 583
  oil field, i, 157
Huron sandstone, i, 88
Hurry-up sand, i, 24
Hydraulic circulating system, i, 247, 248
Hydraulic concrete, pavement foundations, i, 818
Hydraulic fittings, general dimensions, ii, 278
Hydraulic flow, i, 381
Hydraulic press, wax plant, ii, 199
Hydraulic pressure, relation to petroleum occurrence, i, 45
Hydrindene, 1, 2-, i, 484
Hydrocarbons, condensation temperature, i 736
  natural gas, i, 730, 753
  oxidation, i, 458
  paraffin series, i, 730
  petroleum, i, 457
  soluble, theory of oil shales, i, 834
  specific heats, i, 531
  vapor pressures, i, 736
  Viscosities, ii, 647
Hydrochloric acid, expansion by heat, ii, 812
  specific heat, ii, 812
Hydrogen, boiling point, ii, 807
  expansion by heat, ii, 812
  heat of combustion, ii, 809
  heating value, ii, 808
  specific gravity, ii, 832
  specific heat, ii, 811

Hydrogen sulphide, occurrence with petroleum, i, 8, 527
removal, ii, 400
Hydrometer, gravity determination, liquids, i, 597
gravity determination, solids, i, 621
method of reading, i, 598, 599
use, i, 596
Hydrometer readings, correction for meniscus, i, 599
Hydropinene, i, 512
Hydrostatic pressure, line pipe and tubing, formula and tables, i, 301
Hydrostatic test, pipes, ii, 551
Hyponitric acid, melting point, ii, 806

I

Ice, latent heat of fusion, ii, 823
melting point, ii, 806
radiation and reflection, ii, 824
specific heat, ii, 811
Idaho, oil and gas, i, 86
Igneous beds, Mexico, i, 126
Igneous intrusions, i, 64
Igneous rocks, i, 65, 66
Ignition, internal combustion engines, ii, 553
Ignition chamber, boilers, ii, 463, 467
Illinois, anticlinal and synclinal structures, i, 47
deliveries of petroleum, i, 345, 358
La Salle anticline, i, 47
Mansfield sandstone, i, 29
Mississippian series, i, 29
natural gas, production and consumption, statistics, i, 352, 354
natural-gas gasoline statistics, i, 786
oil, i, 86
oil analyses, i, 543
oil production statistics, i, 86, 335, 356
oil, still gas from, ii, 338
oil fields, i, 78, 79
oil seepages, i, 4
oil shale, i, 834, 837
oil strata, i, 14, 15
Pennsylvanian series, i, 22
stocks of petroleum, i, 343, 357
Illuminating oil, doctor test, i, 689
floc test, i, 691
lamp test, i, 692
sulphur test by lamp method, i, 707
testing methods, i, 588
Imperial Ideal rig iron, specifications, i, 207, 209
Imperial oil field, Pa., i, 110
Impervious covers, effect on oil migration, i, 74
Imports, petroleum, i, 347, 358
Impression block, i, 278, 284
Impsonite, i, 791, 793, 794
Inclined retort, oil shale, i, 866
Independence oil field, Kans., i, 91
Index of refraction, oils, i, 532
India, bitumen, i, 793
mud volcanoes, i, 6
oil, i, 137, 140
oil analyses, i, 581
oil, oxygen in, i, 524
oil production statistics, i, 140, 325
Indiana, anticlinal and synclinal structures, i, 47
Corniferous limestone, i, 38
Mississippian series, i, 29
natural gas, production and consumption statistics, i, 352, 354
New Albany shale, i, 34
oil, i, 88
oil production statistics, i, 88, 334, 356
oil, still gas from, ii, 337
oil, sulphur in, i, 526
oil shale, i, 834, 836
oil shale analyses, i, 872
oil strata, i, 14, 15
Pennsylvanian series, i, 22
tar distillation, ii, 331
Trenton limestone, i, 40
Indicated horse-power, internal-combustion engines, ii, 530
Inez gas field, Ky., i, 93
Ingalls oil field, Okla., i, 104
Ingomar oil field, Pa., i, 110
Inola oil field, Okla., i, 104
Inorganic theory, oil occurrence, i, 44
Inserted joint, Boston casing, ii, 559
Inspection reports, refinery, ii, 417
Installation, hot surface type engine, ii, 555
Instruments, geologic field work, i, 178
Insulated cars, oil transportation, i, 360
Insulating materials, heat conduction and resistance, ii, 827
refinery, ii, 282
Insulating oils, heat test for, i, 705
Insulation, pipe lines, i, 453, ii, 573
refinery, ii, 237, 282
tanks, ii, 124
use of asphalt for, i, 830
Insurance rates, refinery, ii, 319
tank spacing, ii, 85
Intake pressure, oil in pipe, i, 397
Interior upset rotary pipe, weights and dimensions, ii, 563
Intermittent method, prepared roofing, i, 825
Internal combustion engines, ii, 523
fuel consumption curves, ii, 533, 535
lubrication, ii, 663

Internal combustion engines, refinery, ii, 294
Internal fluid pressure, formulas, comparative tables, ii, 524
tubes, pipes and cylinders, ii, 519, 531
Internal heat, relation to occurrence of petroleum, i, 45
Internal mixing, centrifugal burner, ii, 451
Internal pressure test, pipe, ii, 550
Internal thermal resistance, ii, 825
Intersection, in mapping, i, 195, 196
Interstate Commerce Commission, on shipment of explosives, i, 757
regulations regarding vapor pressure of gasoline, i, 726
tank car regulation, i, 359, 363, 364, 366, 428
Intrusive beds, i, 64
Invar, linear expansion, ii, 813
Iodine, melting point, ii, 807
Iola gas field, Kans., i, 91
Iowa, Mississippian series, i, 29
natural gas, production and consumption statistics, i, 352, 354
oil and gas, i, 90
Iowa Park oil field, Tex., i, 113
Ireland and Hughes joint, i, 229
Iridium, fusing point, ii, 807
Iron, Crane malleable, tensile strength, ii, 819
Crane soft cast, tensile strength, ii, 818
effect of cold, ii, 816, 817
flat rolled, weight, table, ii, 776
latent heat of fusion, ii, 823
melting point, ii, 807
protective coatings, ii, 570
tensile strength at high temperatures, ii, 816
wrought, melting point, ii, 807
wrought, specific heat, ii, 811
Iron Creek oil field, Wyo., i, 118, 120
Iron foundry, oil as fuel, ii, 507
Iron plate, weight, table, ii, 775
Iron scum, i, 3
Iron sheets, weights of, ii, 773
Iron tubes, thermal expansion, ii, 814
Irvine oil field, Ky., i, 92, 93, 546
Irvine sand, i, 37, 94
Isatine reaction, sulphur compounds, i, 527
Isherwood system, tank vessel construction, i, 373
Island Creek oil field, Ky., i, 93
Island Creek oil field, Ohio, i, 100
Isoamyl-a-dihydrophellandrene, i, 512
Iso-amyl benzol, i, 483
Iso-amyl menthane, i, 512
Isogonic lines, i, 192
Iso-anthracene, i, 514
Isobutane, i, 460
Isobutyl benzol, i, 488
Isobutyl cyclohexane, i, 500
Isobutyl sulphide, i, 529
Isobutylene, i, 472
Isodurol, i, 484
Isohexane, i, 462
Isomeric changes, hydrocarbons, i, **458**
Iso-octane, i, 466
Iso-octylthiophane, i, 528
Isopentane, i, 460
Isoprene, i, 478
Isopropyl acetylene, i, 478
Isopropyl alcohol, ii, 399, 401
Isopropyl benzol, i, 484
Isopropyl ethylene, i, 472
Isopropyl toluene (para), i, 488
Isothermal line, ii, 525
Isovols, i, 10
Italy, asphalt and bituminous rock, i, **796**
bitumen, i, 792, 796
oil, i, 132
oil analyses, i, 580
oil production statistics, i, 132, 324
oil, unit of measure, ii, 761
oil shale, i, 834, 846
stratigraphic occurrence of oil, i, 11
Ivory, radiation and reflection, ii, 824
Ivyton oil field, Ky., i, 93

J

Jack County, Tex., oil, i, 114, 559
Jacket wall, Diesel engine, ii, 562
Jackson County, Ky., oil, i, 548
Jackson County, Mo., oil, i, 97
Jackson County, Okla., bitumen, i, 791
Jackson County, W. Va., oil, i, 116
Jackson Ridge oil field, Ohio, i, 100
Jacksonville gas pool, Ill., i, 87
Japan, asphalt, i, 797
bitumen, i, 793
natural gas, i, 137, 143
oil, i, 135, 141
oil analyses, i, 582
oil production statistics, i, 143, 324
oil, stratigraphic occurrence, i, 11
oil, unit of measure, ii, 761
Jars, fishing, i, 277
Jars and mandrell, i, 285, 288
Jasper County, Ind., oil, i, 88, 89, 544
Jasper County, Tex., bitumen, i, 791
Jasper oil field, Ind., i, 89
Java, oil, i, 161, 162
oil analyses, i, 584
oil, oxygen in, i, 524
Jaw crushers, oil shale mining, i, 859
Jay County, Ind., oil, i, 88
Jefferson County, Ohio, oil, i, 57, 102

Jefferson County, Okla., bitumen, i, 791
Jefferson County, Texas, oil, i, 114, 565
  nitrogen in, i, 530
Jefferson Davis Parish, La., oil, i, 95, 550
Jefferson oil field. Pa., i, 110
Jenks oil field, Okla., i, 104
Jennings gas field, Tex., i, 113
Jennings oil field, La., i, 96, 550
Jennings oil field, Okla., i, 104
Jet, radiation and reflection, ii, 824
Jewell oil field, Ky., i, 93
Jewett oil field, Ohio, i, 100
Johnson County, Ky., oil, i, 92
Johnson County, Okla., bitumen, i, 791
Johnson Fork oil field, Ky., i, 93, 547
Joint cracks, United States oil fields, i, 79
Joints, bell and spigot, ii, 275
  casing, ii, 559, 560
  concrete tanks, ii, 132
  double swivel, swing pipe, ii, 107
  pipe, ii, 566
  riveted, ii, 265
  rotary tool, i, 259
  threaded, lubrication of, ii, 548
  tool, standard, i, 225, 230
Jones sand, i, 26, 31, 92
Joule's equivalent, ii, 524, 808
Journal bearings, lubrication, ii, 636, 639
Juan Casiano oil well, Mexico, i, 20
Jugo-Slavia, oil and gas, i, 134
Junction City oil field, Ohio, i, 59, 100
Junker's Diesel engine, ii, 594
Jurassic system, Alaska, i, 80
  Germany, i, 134
  Madagascar, i, 158
  Montana, i, 98
  oil shale, i, 834, 844, 845
  oil strata, i, 11, 12, 14, 20
  Wyoming, i, 118, 119

K

Kanawha County, W. Va., oil, i, 116, 569
Kane oil field, Pa., i, 110
Kane sand, i, 108
Kansas, anticlines in, i, 47
  Big Lime sand, i, 26
  Cambrian sediments, i, 40
  lenticular sands, i, 57
  limestone, i, 29
  Mississippi lime sand, i, 29
  monoclinal bulges, i, 59, 72
  natural gas, consumption and production statistics, i, 352, 354
  natural-gas gasoline, statistics, i, 786
  oil, i, 90
  oil analyses, i, 544
  oil production statistics, i, 90, 336, 356
Kansas, oil, still gas from, ii, 338
  oil strata, i, 12
  Oswego lime, i, 27, 30
  Pennsylvanian series, i, 22
  Permian rocks, i, 21
  Peru oil field, i, 91
  Peru sand, i, 27
  salt unassociated with oil, i, 7
  standard casing requirements, i, 221
  structural terraces, i, 57
Karns City oil field, Pa., i, 110
Kay County, Okla., oil, i, 103, 555
Keener sand, i, 14, 31, 94, 103, 116
Kelleyville oil field, Okla., i, 104
Kendall oil field, Pa., i, 110
Kent County, Ontario, Canada, oil, i, 121, 123
Kentucky, Berea sand, i, 32
  Beaver sand, i, 25, 33
  Big Injun sand, i, 32
  Big Lime, i, 31
  bitumen, i, 791
  Caney sand, i, 39
  Chattanooga shale, i, 34, 37
  Cincinnati series, i, 39
  Clinton formation, i, 39
  Cooper sand, i, 33
  Corniferous limestone, i, 37, 38
  High Bridge formation, i, 40
  Horton sand, i, 35
  Irvine sand, i, 37
  Knox dolomite, i, 40
  Lexington formation, i, 40
  Mauch Chunk shale, i, 31
  Maxon sand, i, 31
  Mississippian series, i, 15, 29, 31
  Mt. Pisgah sand, i, 33
  natural gas, i, 31, 34, 40
  natural gas consumption and production statistics, i, 352, 354
  natural-gas gasoline statistics, i, 786
  Niagara formation, i, 38
  oil, i, 92, 94
  oil analyses, i, 546
  oil production statistics, i, 92, 331, 356
  oil shale, i, 834, 837
  oil strata, i, 14
  Onondaga limestone, i, 37
  Otter sand, i, 33
  Pennsylvanian series, i, 22, 25
  Pike sand, i, 25
  Pottsville conglomerate, i, 25
  Ragland sand, i, 37
  Silurian system, i, 38
  Silurian-Devonian system, i, 37
  Slickford sand, i, 33
  Ste. Genevieve limestone, i, 31
  St. Louis limestone, i, 31

Kentucky, Stray sand, i, 33
Trenton limestone, i, 40
Upper Sunnybrook sand, i, 39
Keown oil field, Pa., i, 110
Kern County, Calif., bitumen, i, 791
oil, i, 80, 541
Kern River oil field, Calif., i, 16, 81, 82, 83, 541
Kerosene, base for cracking, ii, 430
end point method, distillation, i, 673
Kerosene distillate, ii, 339, 363, 368, 424
condenser coil data, ii, 168, 176
Kerosene stock, ii, 339
Key bed, geology, i, 169, 171
Key horizon, geology, i, 171
Kier's Rock oil, i, 359
Kifer oil field, Pa., i, 110
Kilbuck oil field, Ohio, i, 100
Kilgore oil field, Ohio, i, 100
Kilgore oil field, Okla., i, 104
Kinematic viscosity, i, 640
Kinks, wire ropes, i, 232
Kinney rotary pump, ii, 210
Kiowa County, Okla., oil, i, 555
Kissling tar value, petroleum products, determination, i, 714
Kittanning coal beds, i, 25
Kittanning sandstone, i, 25
Klondike oil field, Ohio, i, 100
Knocking, engines, ii, 553, 581
Knott County, Ky., oil, i, 31, 92, 94
Knox County, Ky., oil, i, 92
Knox County, Ohio, oil, i, 39, 102, 552
Knox dolomite, i, 40, 94
Knoxville oil field, Ohio, i, 100
Kootenai formation, i, 20, 98
Kormann cracking process, ii, 433
Kossuth oil field, Pa., i, 110
Kramer and Sarnon melting point method, asphalt, i, 803

L

Labette County, Kans., oil, i, 90
Labette formation, i, 27
La Brea asphalt deposit, Calif., i, 16, 83
La Brea sand, Trinidad, i, 159, 160
Laccoliths, i, 64, 126
La Cygne oil field, Kansas, i, 91
Lafayette County, La., bitumen, i, 791
Lafayette County, Mo., bitumen, i, 791
La Flore County, Okla., bitumen, i, 791
La Fourche Parish, La., gas, i, 95
Lagonia oil field, Pa., i, 110
Lagunillas series, Venezuela, i, 152
Lahee, Frederic H., i, 167
Lake County, Ohio, gas, i, 33
Lake County oil field, Ind., i, 89
Lake Creek oil field, Pa., i, 110
Lake Maracaibo, Venezuela, oil near, i, 150
Lake View gusher, Calif., i, 248
Lambton County, Ontario, Can., oil, i, 121
Lamë's formula, internal fluid pressure, ii, 519, 522
Lamoure County, N. Dak., natural gas, i, 102
Lamp black, radiation and reflection, ii, 824
Lamp burning test, oil, i, 691, 693
petroleum products, i, 588
Lamps used in oil tests, i, 691, 693
Lamp method test, sulphur in oils, i, 707
Lance Creek oil field, Wyo., i, 118, 120, 572
Lander oil field, Wyo., i, 22, 118, 119
Landolt's tables of organic compounds, i, 730
Lantz oil field, Pa., i, 110
Lap joints, riveted, ii, 265
Lap-weld process, pipes, ii, 502
Lap-welded, pipe, ii, 334, 565
Lardintown oil field, Pa., i, 110
Lard oil, saponification number, i, 712
Laredo gas field, Tex., i, 114
La Salle anticline, Ill., i, 86
Latch jack, fishing tool, i, 277
Latent heat, ii, 822
of evaporation, table, i, 752, ii, 823
of fusion, ii, 823
of steam, ii, 794
Lateral contraction, tubes, ii, 522
Latrobe oil field, Pa., i, 110
Laurel Creek oil field, Ky., i, 93
Laurene, i, 512
Laurylene, i, 476
Lawrence County, Ill., oil, i, 86, 543
Lawrence County, Ky., oil, i, 92, 547
Laws of lubrication, ii, 629
Lawton oil field, Okla., i, 104
Laying pipe line, i, 408
Layne centrifugal pump, ii, 210
Layton sand, i, 26, 27, 106
Lead, latent heat of fusion, ii, 823
melting point, ii, 806, 807
refinery use, ii, 279
specific heat, ii, 811, 812
Lead for lining agitators, ii, 187
Lead in cast-iron pipe, ii, 275
Lead threading die, ii, 539
Leaks, tank cars, i, 369
Leamington gas field, Ontario, Can., i, 122
Leatherwood oil field, Pa., i, 110
Lee County, Ky., oil, i, 92, 547, 548
Leechburg oil field, Pa., i, 110
Leetsdale oil field, Pa., i, 110
Legionville oil field, Pa., i, 110
Leidecker sand, i, 30
Leman condenser, ii, 216
Length, measures, ii, 762, 766

Lenhart oil field, Ky., i, 92
Lenticular sands, i, 53, 57
Leonard oil field, Okla., i, 104
Leucacene, i, 516
Leveling, plane table, i, 193
Levels, geologic field work, i, 179
Lewis County, W. Va., oil, i, 116, 568
Lewisville-Decker-Fisher oil field, Ohio, i, 100
Lexington formation, i, 40, 94
Liassic system, Belgium, i, 136
  England, i, 135
  oil shale, i, 843
  Sweden, i, 136
  Switzerland, i, 136
Liberty County, Tex., oil, i, 114, 565
Liberty gasoline engine, ii, 618
Licking County, Ohio, oil, i, 39, 102
Lickskillit oil field, Pa., i, 110
Lietzenmayer sprayer, engines, ii, 571
Light end distillate, ii, 339
Lighting plant, standard drilling system, i, 251
Light oil agitators, ii, 186
Lighting equipment, oil tankers, i, 378
Lily white oil, registered by different instruments, i, 663
Lima crude oil, refining of, ii, 79
  typical distillation, ii, 325, 331, 332
Lima district, Ind., oil, i, 544
Lima-Indiana oil field, i, 78
  deliveries of petroleum, i, 345, 358
  geologic structures, i, 79
  Indiana, i, 88
  Ohio, i, 102
  oil production, quantity and value since 1886, i, 334, 356
  stocks of petroleum, i, 343, 357
  Trenton limestone, i, 40
Lima oil field, Ohio, i, 100, 551
  oil, nitrogen in, i, 530
  oil production statistics, i, 334, 356
Lime, refinery use, ii, 372
Lime kiln, oil fuel, ii, 520
Lime-soap, in lubricating greases, ii, 673
Limestone, Ames, i, 23
  Bend, i, 13, 31
  Clinton, i, 15
  Corniferous, i, 37, 92
  crinoidal, i, 23
  Edwards, i, 13
  Fort Scott, i, 27
  Guelph, i, 15, 121
  Hamilton, i, 121
  Marble Falls, i, 31
  Maxville, i, 25, 29, 31, 103
  Mississippi, i, 13, 29
  Niagara, i, 15, 39
Limestone, oil productive, i, 12, 13, 14, 41
  Onondaga, i, 37, 92, 121
  Oologah, i, 26
  Ste. Genevieve, i, 31
  St. Louis, i, 31
  Tamasopo, i, 12, 20, 63, 124, 126
  Trenton, i, 15, 39, 40, 79, 86, 88, 94, 102, 103, 121
  United States oil fields, i, 79
Limestone County, Tex., oil, i, 560
Limestone oil field, Pa., i, 110
Limestone Run oil field, Pa., i, 110
Limmer asphalt, i, 134
Limonene, d or l, i, 510
Limonene-e, i, 510
Lincoln County, Ky., oil, i, 92
Lincoln County, Ontario, Can., gas, i, 123
Lincoln County, W. Va., oil, i, 116
Linden oil field, Pa., i, 110
Liner, cylinder, Diesel engine, ii, 562
Line pipe, oil wells, i, 301
  specifications, ii, 550
  weights and dimensions, i, 305, ii, 554
Line pipe casing, standard, i, 236
Lines, drilling, i, 231
  refinery, ii, 242
Link belting, ii, 306
Linseed oil, boiling point, ii, 806
  expansion by heat, ii, 812
Lintels, ii, 245
Lip, threading die, ii, 535
Liquefaction, gases, heat of, i, 749
Liquid air, boiling point in oxygen, ii, 807
Liquid exchangers, condensers, ii, 181
Liquid measures, ii, 764, 767
Liquid-phase process, cracking, ii, 428, 436
Liquids, compression, ii, 832
  expansion, ii, 812
  specific gravity, table, ii, 830
  specific heats, ii, 812
Lisbon oil field, Ohio, i, 100
Litchfield gas field, Ky., i, 93
Litchfield oil field, Ill., i, 87
Litchfield oil field, Ohio, i, 100
Litharge, refinery use, ii, 368
Lithological character of oil strata, i, 41
Little Buffalo Basin oil field, Wyo., i, 118
Little Dunkard sand, i, 24
Little Hominy oil field, Okla., i, 104
Little Poland, oil, i, 131
Little Richmond Creek oil field, Ky., i, 93
Live load, lubrication, ii, 631
Livingston distillation process, ii, 324
Llano Burnett uplift, Tex., i, 114
Llanos formation, Trinidad, i, 160
Lloyds Register of Shipping, Ship Classification rules, i, 376

Load displacement, oil tankers, i, 374
Load, water line, oil tankers, i, 374
Loading racks, refinery, ii, 225
Lobitos oil field, Peru, i, 53, 148, 578
Locke level, geologic field work, i, 179
Lockwood oil field, Pa., i, 110
Loco oil field, Okla., i, 104
Locomotive boiler, oil fired, ii, 473
Locomotive fuel, oil, ii, 485
Locust Grove oil field, Ohio, i, 100
Lodi gas pool, Ohio, i, 100
Logan County, Ky., bitumen, i, 791
Logan County, Ky., oil, i, 547
Logan County, W. Va., gas, i, 116
Logs, well, i, 238
Lohn oil field, Tex., i, 113
Lompoc oil field, Calif., i, 17, 68, 81, 82, 540
Long measure, ii, 760
Long wall method, oil shale mining, i, 856
Longitudinal framing system, tank vessel construction, i, 373
Longton oil field, Kans., i, 91, 546
Loops, pipe lines, i, 400
Los Angeles City oil field, Calif., i, 82, 540
Los Angeles County, Calif., oil analyses, i, 539
oil, nitrogen in, i, 530
oil, sulphur in, i, 526
San Pedro formation, i, 16
Los Angeles oil field, Calif., i, 50, 68
Los Angeles-Sour Lake oil field, Calif, i, 81
Los Naranjos oil field, Mexico, i, 124
Losses, heat, refinery, ii, 283
Lost Hills oil field, Calif., i, 16, 81, 82, 83, 541
Lost Soldier oil field, Wyo., i, 120, 572
Louisa County, Iowa, gas, i, 90
Louisiana, Annona chalk, i, 19
anticlinal and synclinal structures, i, 47
Austin group, i, 19
bitumen, i, 791
Brownstone marl, i, 19
Blossom sand, i, 19
Caddo sand, i, 18
Eagle Ford clay, i, 19
Homer pool, i, 18
Marlbrook marl, i, 19
mud volcanoes, i, 7
Nacatoch sand, i, 18
natural gas, i, 18
natural-gas gasoline statistics, i, 786
natural gas production and consumption statistics, i, 356, 358
northern, oil production statistics, i, 336, 356
oil analyses, i, 548
oil and gas, i, 92
oil associated with sulphur and salt, i, 8
oil production statistics, i, 341, 356
Louisiana, oil strata, i, 12
oil, sulphur in, i, 526
Quaternary system, i, 16
rotary drilling system, i, 251
saline domes, i, 59, 60, 61
Shreveport gas sand, i, 18
Vivian gas sand, i, 18
Woodbine sand, i, 19, 20
Love County, Okla., bitumen, i, 791
Lovibond glasses, Union Petroleum colorimeter, i, 665
Lovibond tintometer, i, 662, 663, 666, 667, 668
Lovi oil field, Pa., i, 110
Low dip, relation to occurrence of petroleum, i, 168
Lower Coroni formation, Venezuela, i, 152
Low pressure burner, ii, 452
internal combustion engines, ii, 542
Lower Cretaceous series, Galicia, i, 132
Louisiana, i, 20, 95
Mexico, i, 126
Montana, i, 98
oil shale, i, 12, 14, 20
Oklahoma, i, 106
Rumania, i, 131
Venezuela, i, 152
Wyoming, i, 118, 119
Lower Miocene series, Calif., i, 83
Lower Mississippian age, i, 31
Lower Neocene series, Russia, i, 65
Lower Trenton sand, i, 94
Lowerators, refinery, ii, 227
Lowering, pipe line, i, 415
Loyalty Islands, oil, i, 164
Lubricants, corrosion effect on metals, i, 706
Diesel engine, ii, 581
pipe threading, ii, 541
refinery, ii, 41
selection of, ii, 625, 643
threaded joints, ii, 548
Lubricating greases, ii, 672
Lubricating oils, agitators, ii, 63, 186, 190
bleachers, refinery, ii, 73
characteristics, ii, 677
condenser coil data, ii, 170, 178
gasoline precipitation test, i, 713
heat test for, i, 705
natural, ii, 323
physical tests, ii, 676
reduction, ii, 357
specifications, ii, 675
treating, ii, 368
Lubrication, ii, 625–677
Diesel engine, ii, 579
laws of, ii, 629
pipe joints, ii, 566
threaded joints, ii, 548

Lubrication systems, ii, 668
Luchaire tester, oil, i, 624
Lugs, filters, ii, 65
stills, ii, 25, 43, 44, 45, 47
still towers, ii, 26, 27, 48, 49
Lupton sand, i, 98
Lye tanks, refinery, ii, 40, 77, 78
Lyon County, Kans., oil, i, 90

M

Macadam, asphalt, i, 818
MacCamy condenser, i, 216
Machine oils, characteristics, ii, 677
Machinery, standard drilling outfit, i, 216
Machines ditching, for pipe lines, i, 415
pipe laying, i, 411, 415
Mackenzie, Canada, bitumen, i, 792
Mackenzie River oil field, Canada, i, 123, 576
Macksburg, 140 foot sand, i, 24
Macksburg 800 foot sand, i, 103
Macksburg 500 foot sand, i, 25, 103
Macksburg 300 foot sand, i, 103
Macksburg oil field, Ohio, i, 57, 100, 553
sulphur in, i, 526
Macoupin County, Ill., oil, i, 543
Madagascar, oil, i, 158
Madill oil field, Okla., i, 20, 104, 106, 555
Madison County, Ark., asphalt, i, 80
Madison County, Ind., oil, i, 88
Madison County, Iowa, oil, i, 90
Madison oil field, Kans., i, 91
Madison oil field, Pa., i, 110
Magnesia, specific heat, ii, 811
Magnesian limestone, specific heat, ii, 811
Magnesium, melting point, ii, 806
Magnesium carbonate, refinery insulating, ii, 282, 286
Magoffin County, Ky., oil, i, 92
Magowan Lamp No. 1, Sunhinge burner, i, 691, 693
Mahaffey oil field, Pa., i, 110
Mahoney gas pool, Wyo., i, 120
Mahoning County, Ohio, oil, i, 102
Mahoning sandstone, i, 24
Maikop oil field, Russia, i, 65, 127
Malaga-Brushy Creek-Harper oil field, Ohio, i, 100
Malheur County, Ore., oil, i, 107
Maline oil field, Kans., i, 91
Maltha, Turkish Armenia, i, 145
Manganese, fusing, ii, 807
Manheads, stills, ii, 27, 42, 51
Manholes, tanks, ii, 40, 76, 79
bleacher, ii, 38, 72, 74
tanks, ii, 32, 37, 68, 78, 107, 133
Manifold sand, i, 32
Manifolds, refinery, ii, 224
Manila cable or rope, weight and strength, i, 231
Manjak, Africa, i, 792
Asia, i, 792
Barbados, i, 161
North America, i, 792
South America, i, 792
Trinidad, i, 159
United States, i, 792
Mannecks, filters, ii, 65
stills, ii, 26, 27, 43, 51, 52
still towers, ii, 28, 50
Mannington oil field, W. Va., i, 117
Manor oil field, Pa., i, 110
Mansfield sandstone, i, 29
Manufacturers' standard wire gage, ii, 272
Mapleton oil field, Kans., i, 91
Mapping, petroleum, i, 168, 172
Maps, petroleum geology, i, 171
Maracaibo oil district, Venezuela, i, 151, 152
Maracaibo series, Venezuela, i, 152
Marble, radiation and reflection, ii, 824
specific heat, ii, 811
Marble Falls limestone, i, 31
Marcos shale, i, 14
Margaric acid, melting point, ii, 806
Margarylene, i, 474
Marine boilers, oil fired, ii, 480
Marine Diesel crankshafts, ii, 564
Marine engines, Diesel, ii, 561, 562, 594, 599
Marion County, Fla., oil, i, 84
Marion County, Ill., oil, i, 543
Marion County, Ind., oil, i, 88, 89
Marion County, Kans., oil, i, 90
Marion County, Tex., oil, i, 560
Marion County, W. Va., oil, i, 116
Markham oil field, Tex., i, 113
Markham sand, i, 27
Marlbrook marl, i, 19, 92
Mars oil field, Pa., i, 110
Marsh gas, i, 460
heat of combustion, ii, 809
specific gravity, ii, 832
Marshall County, Okla., bitumen, i, 791
oil, i, 20, 103, 106, 555
Marshall County, W. Va., oil, i, 116
Martin County, Ind., oil, i, 88
Martin County, Ky., oil, i, 31, 94
Martinez formation, i, 83
Maryland, natural gas, production and consumption statistics, i, 354, 358
oil shale, i, 848
Mason County, Ky., oil, i, 92
Masontown oil field, Pa., i, 110
Mass, measures of, ii, 764, 767
Mastic asphalt pavements, i, 817
Matagorda County, Tex., oil, i, 114

Mauch Chunk formation, i, 94, 108, 116
Mauch Chunk shales, i, 29, 31
Maud oil field, Okla., i, 104
Maverick Springs oil field, Wyo., i, 120, 571
Mawhinney oil field, Pa., i, 110
Maxon sand, i, 31, 94
Maxton sand, i, 25, 31, 103, 108, 116
  relation to Maxon sand, i, 31
Maxville limestone, i, 15, 25, 29, 31, 103, 116
Maxwell oil field, Ohio, i, 101
McArthur oil field, Ohio, i, 100
McCloskey oil field, Ill., i, 87
McComb cracking process, ii, 435
McConnellsville gas field, Ohio, i, 100
McCormick oil field, Pa., i, 110
McCreary County Ky., oil, i, 92, 94
  Trenton limestone, i, 40
McCulloch County, Tex., oil, i, 114
  Canyon formation, i, 28
McCurdy oil field, Pa., i, 110
McCurtain County, Okla., bitumen, i, 791
McDonald oil field, Pa., i, 110
McDonald sand, i, 36
McIntosh County, Okla., oil, i, 103
McIntosh and Seymour Diesel engine, ii, 592, 614
McIntyre oil field, Ohio, i, 100
McKean County, Pa., oil, i, 558
McKeesport oil field, Pa., i, 110
McKinley County, N. Mex., oil, i, 99
McKittrick formation, i, 83
McKittrick oil field, Calif., i, 16, 68, 81, 82, 83, 541
McLennan County, Tex., oil, i, 560
McMurry oil field, Pa., i, 108
Meade County, Ky., gas, i, 34
Meadows sand, i, 30
Mean specific heats, gases, ii, 529
Measures and weights, ii, 760
  foreign oil units, ii, 761
  petroleum, ii, 761, 768
Mecca oil field, Ohio, i, 100, 530
Mechanical efficiency, ii, 530
  Diesel engine, ii, 561
Mechanical equivalent, heat, ii, 808
Mechanical units, equivalent values, ii, 769
Medina County, Tex., oil, i, 567
Medina formation, i, 102
Medina oil field, Ohio, i, 39, 100, 102
Medina sands, i, 15, 39, 121
Mediterranean formation, Rumania, i, 131
Meigs County, Ohio, Berea sand, i, 33
Melene, i, 476
Melting point, American wax method for petroleum products, i, 588
  asphalt, i, 801, 802
Melting point, Bureau of Standards wax method for petroleum products, i, 588
  cube method for petroleum products, i, 588
  hydrocarbons, i, 460
  paraffin wax, i, 679, 681
  petrolatum, i, 684
  ring and ball method for petroleum products, i, 588
  Saybolt petrolatum method for petroleum products, i, 588
  Saybolt wax method for petroleum products, i, 588
  various substances, ii, 806
Menden oil field, Ohio, i, 100
Mendocino County, Calif., bitumen, i, **791**
Menifee County, Ky., oil, i, 92, 547
Menifee gas field, Ky., i, 93
Menthane, i, 500
Menthene, i, 500, 512
Mento naphthene, i, 498
Mercaptans in petroleum, i, 527
Mercer County, Ohio, oil, i, 102
Mercer County, Pa., oil, i, 558
Mercer oil field, Pa., i, 110
Mercury, boiling point, ii, 806, 807
  expansion by heat, ii, 812
  freezing point, ii, 807
  melting point, ii, 806
  radiation and reflection, ii, 824
  specific heat, ii, 812
Mesa County, Colo., oil, i, 542
Mesitylene, i, 482
Mesopotamia, oil, i, 141
Mesozoic period, California oil fields, i, 83
  Madagascar rocks, i, 158
  North Carolina, i, 102
  oil strata, i, 11, 12, 14
Metal packing, refinery, ii, 285, 286, 287
Metals, heat-conducting power, ii, 825
  non-ferrous, refinery use, ii, 281
  tenacity, ii, 815
Metamorphism, relation to oil occurrence, i, 2, 9, 45, 64
Meta-styrol, i, 482
Metaxylol, i, 482, 528
Meters, natural gas, i, 432
  refinery, ii, 307
Methane, i, 460
  heat of combustion, ii, 809
  specific gravity, ii, 832
Methylacetylene, i, 478
Methyl allylene, i, 478
Methyl amyl acetylene, i, 478
Methylanthracene, i, 536
Methyl-α anthracene, i, 514
Methyl-β anthracene, i, 514

Methyl benzol, i, 480
Methyl-l-bicyclo-1-3-3-nonane, i, 522
Methyl-α butadiëne, i, 478
Methyl-β butadiëne, i, 478
Methyl cyclobutane, i, 490
Methyl cyclohexane, i, 494
Methyl cyclopentane, i, 492
Methyl-1-2 cyclopentane, i, 492
Methyl cyclopropane, i, 490
Methyl decyl acetylene, i, 478
Methyl diethyl methane, i, 462
Methyl dihexanaphthene, i, 502, 504
Methyl-1, di-iso-propyl 3, 5 benzol, i, 486
Methyl ethyl acetylene, i, 478
Methyl ethyl benzol, i, 484
Methyl-l-ethyl-2-cyclohexane, i, 498
Methyl-l-ethyl 3-cyclohexane, i, 498
Methyl-l-ethyl-2-cyclopentane, i, 494
Methyl-ethyl-3-cyclopentane, i, 494
Methyl ethyl ethylene, i, 472
Methyl-1-ethyl-3-isopropyl-4-cyclohexane, i, 500
Methyl ethyl iso-propyl methane, i, 464
Methyl-2-ethyl-3-pentane, i, 466
Methyl ethyl propyl methane, i, 464
Methyl-2 heptane, i, 466
Methyl-3 heptane, i, 466
Methyl-4-heptane, i, 466
Methyl hexamethylene, i, 494
Methyl-2-hexane, i, 464
Methyl isopropyl benzol (meta), i, 486
Methyl-isopropyl-4-cyclohexane, i, 500
Methyl naphthalene, β, i, 508
Methyl naphthalene, i, 508
Methyl-4 octane, i, 466
Methyl pentamethylene, i, 492
Methyl phenathrene, i, 516
Methyl phenyl fulvene, i, 490, 516
Methyl propyl acetylene, i, 478
Methyl propyl benzol, i, 486
Methyl propyl tetramethylene, i, 492
Methyl sulphide, i, 529
Methyl tetramethylene, i, 490
Methyl trimethylene, i, 490
Metric system, weights and measures, conversion tables, ii, 766
Mexia oil and gas field, Tex., i, 20, CO, 113, 114, 560
Mexican air-refined asphalt, characteristics, i, 813
Mexican crude oil, refining, ii, 80
Mexican steam-refined asphalt, characteristics, i, 812
Mexico, anticlinal and synclinal structures, i, 47
  asphalt, i, 797
  basaltic dikes, i, 64
  basalt plugs, i, 59
  bitumen, i, 792
  bituminous seepages, i, 4
  Cerro Azul oil well, i, 20, 248
  Dos Bocas gusher, i, 20, 248
  drilling methods, i, 202
  Ebano oil field, i, 64
  Furbero oil field, i, 64
  Juan Casiano well, i, 20
  laccoliths, i, 64
  oil analyses, i, 574
  oil and asphalt, i, 78, 124
  oil production statistics, i, 124, 325
  oil seepages, i, 64, 67
  oil shale, i, 848
  oil springs along fault lines, i, 68
  oil strata, i, 11, 12
  oil, unit of measure, ii, 761
  Panuco crude, ii, 431
  Panuco oil field, i, 20
  pipe lines, i, 380
  Potrero del Llano well, i, 20, 248
  quaquaversal structure, i, 59
  rotary drilling system, i, 251
  San Felipe formation, i, 63
  sedimentary and igneous rocks, i, 64
  standard casing requirements, i, 221
  Tamasopo limestone, i, 20, 63
  volcanic plugs, i, 63
Meyers oil field, Okla., i, 104
Miami County, Kans., oil, i, 90, 546
Mica, filler for lubricating grease, ii, 674
Michigan, Mississippian series, i, 29
  natural gas, production and consumption statistics, i, 352, 354
  oil, i, 95
  oil analyses, i, 551
  oil production statistics, i, 340
Mid-Continent crude, in refining, ii, 13
Mid-Continent oil field, deliveries of petroleum, i, 345, 358
  quantity and value, oil production since 1889, i, 336, 356
  stocks of petroleum, i. 343, 357
Mid-Continent oil fields, i, 78
  anticlinal and synclinal structures, i, 47
  geologic structures, i, 79
  Kansas, i, 90
  Louisiana, i, 95
  Oklahoma, i, 103
  Texas, i, 114
Middle carboniferous system, China, i, 144
Middle Miocene series, i, 68
  Java, i, 162
Midway oil field, Calif., i, 16, 81, 82, 83, 542,
Mietz internal combustion engine, ii, 540, 542
Mifflin oil field, Pa., i, 110

Migration, oil, i, 64, 67, 73, 74
Mileage, allowance, tank cars, i, 366
Miller's Creek, oil field, Ky., i, 92, 93
Millsap oil field, Tex., i 113
Millstone oil field, Pa., i, 110
Minatitlan, Mexico, oil and oil refinery, i, 124
Mineral seal distillate, ii, 363
Mineral seal oil, as absorption oil, i, 741, 743
Mineral wax, Scotland, i, 67
Mineral Wells gas field, Tex., i, 113
Mingo oil field, Ohio, i, 100
Mingo County, W. Va., gas, i, 116
Minnesota, oil and gas, i, 96
Miocene series, Algeria, i, 157
  Barbados, i, 161
  Borneo, i, 162
  Czecho-Slovakia, i, 134
  Egypt, i, 156
  France, i, 133
  Galicia, i, 132
  Greece, i, 137
  India, i, 141
  Italy, i, 132
  Jugo-Slavia, i, 134
  Mesopotamia, i, 141
  New Guinea, i, 163
  New Zealand, i, 164
  oil shale, i, 834, 836, 892
  oil strata, i, 12, 14, 16
  Philippine Islands, i, 163
  Poland, i, 132
  Rumania, i, 131
  Russia, i, 127
  South Australia, i, 166
  Sweden, i, 136
  Switzerland, i, 136
  Trinidad, i, 160
  Turkey, i, 134
  Venezuela, i, 152
Miola oil field, Pa., i, 110
Mississippi, oil and gas, i, 96
Mississippi lime, i, 29, 34
Mississippi sand, i, 13, 29
Mississippi Valley, Cambrian sediments, i, 40
  Mauch Chunk red shales, i, 31
Mississippian limestone, i, 13, 29
Mississippian series, Appalachian oil fields, i, 31
  Arkansas, i, 80
  Canada, i, 121
  England, i, 135
  Illinois, i, 86
  Indiana, i, 88
  Kansas, i, 90
  Kentucky, i, 92, 94
  Mid-Continent oil fields, i, 29
  New Brunswick, i, 71, 123
Mississippian series, Ohio, i, 102, 103
  oil shale, i, 834
  oil strata, i, 13,15, 29
  Oklahoma, i, 22, 106
  Pennsylvania, i, 34, 108
  Tennessee, i, 111
  Texas, i, 114
  West Virginia, i, 116
Missouri, bitumen, i, 791
  Mississippian series, i, 29
  natural gas, production and consumption statistics, i, 356, 358
  oil, i, 97
  oil analyses, i, 551
  oil production statistics, i, 340
  Pennsylvanian series, i, 22
Mitchell and Dunn patents, for oil fields, i, 315
Mitchell oil field, Ohio, i, 100
Mixture method, determination of specific heat, ii, 810
Mobility, of lubricants, ii, 644
Moisture in paraffin wax, determination of, i, 725
Molecular attraction, relation to petroleum occurrence, i, 45
Moncton, New Brunswick, Canada, gas, i, 123
Monel metal, elastic limit, ii, 821
  elongation, ii, 821
  refinery use, ii, 279
  tensile strength, ii, 818, 821
Monoclinal accumulations, i, 57
  bulges, i, 59
  noses, i, 55
  ravines, i, 55
  structures, i, 45, 53
Monoclines, United States oil fields, i, 79
Monocyclic terpenes, i, 510, 512
Monoethyl trimethyl benzol, i, 488
Monongalia County, W. Va., oil, i, 116, 568
Monroe County, Ill., oil, i, 86
Monroe County, Mich., oil, i, 95
Monroe County, Ohio, Keener sand, i, 31
  Maxton sand, i, 25
  oil, i, 100, 102, 552
Monroe formation, i, 37
Monroe gas field, La., i, 96
Montague County, Tex., bitumen, i, 791
Montana, Cat Creek sand, i, 20
  natural gas, production and consumption statistics, i, 352, 354
  oil analyses, i, 551
  oil production statistics, i, 338, 356
  oil, refractive index, i, 532
  oil and gas, i, 97
  oil shale, i, 834, 837

Montana group, i, 98, 118, 119
Montebello oil field, Calif., i, 81, 82, 540
Montenegro, bitumen, i, 792
oil, i, 134
oil shale, i, 848
Monterey County, Calif., bitumen, i, 791
oil, i, 80
Monterey shale formation, i, 16, 68, 83, 892
resin, i, 525
Montgomery County, Ill., oil, i, 86, 544
Montgomery County, Kans., oil, i, 90, 546
Moon oil field, Pa., i, 110
Moon Run and Neville Island oil field, Pa., i, 110
Moorecroft oil field, Wyo., i, 118
Moore oil field, Ohio, i, 100
Moore soft wax press, ii, 199
Moores Junction oil field, Ohio, i, 100
Moose Ridge oil field, Ohio, i, 100
Moran oil field, Kans., i, 91, 544
Moran oil field, Tex., i, 113, 561
Morehouse Parish, La., oil, i, 95
Morgan County, Ky., Caney sand, i, 39
oil, i, 92, 94, 547
Morgan County, Ohio, Berea sand, i, 33
Clinton sand, i, 39
Cow run sand, i, 23
oil, i, 102, 553
Morison Run oil field, Pa., i, 110
Morocco, oil, i, 157
analysis, i, 583
Morris oil field, Okla., i, 104, 556
Morris oil field, Tex., i, 113
Morris pool, Okla., i, 28, 30
Morris sand, i, 30
Morrison formation, i, 20, 119
Morse oil field, Okla., i, 104
Morse wire gage, ii, 272
Mortar, refinery construction, ii, 251
Mosa oil pool, Ontario, Can., i, 122, 123
Moscow oil field, W. Va., i, 117
Motor gasoline, ii, 361, 362
gravities, ii, 346
Motor ship Narragansett, engine equipment and records, ii, 599
Motor vehicle registration, U. S., ii, 623
Motors, refinery, ii, 295
Moulder oil field, Ky., i, 93
Mounds sand, i, 30, 106
Moundsville oil field, W. Va., i, 117
Mt. Morris oil field, Pa., i, 110
Mt. Nebo oil field, Pa., i, 110
Mt. Perry oil field, Ohio, i, 100
Mt. Pisgah oil field, Ky., i, 93
Mt. Pisgah sand, i, 33, 94
Mt. Vernon oil field, Ohio, i, 100
Mountain sand, i, 32, 108
Mowry shale, i, 98, 119
Mud lubricator for oil wells, i, 250
Mud volcanoes, general characteristics and occurrence, i, 6
New Guinea (Papua), i, 162
relation to occurrence of petroleum, i, 3
Mule Creek oil field, Wyo., i, 120
Multiwhirl condenser, ii, 217
Muncie oil field, Ind., i, 89
Murdocksville oil field, Pa., i, 110
Murray County, Okla., bitumen, i, 791
Murraysville sand, i, 33, 108
Murryville oil field, Pa., i, 110
Muscatine County, Iowa, natural gas, i, 90
Muskingum County, Ohio, Clinton sand, i, 39
oil, i, 102
Muskogee County, Okla., Booch sand, i, 30
Boynton pool, i, 30
Leidecker sand, i, 30
oil, i, 103
Muskogee oil field, Okla., i, 104, 556
Muskogee sand, i, 30, 106
Musselshell County, Mont., oil, i, 97, 551
Myricyl paraffin, i, 472

N

Nacatoch sand, i, 12, 18, 95
Nacogdoches County, Tex., oil, i, 560
Nansen oil field, Pa., i, 110
Naperian clay formation, Trinidad, i, 160
Naphtha, ii, 338
acid heat test, i, 690
corrosion and gumming test, i, 689
doctor test, i, 689
end point method, distillation, i, 673
gasoline blends, i, 758
solubility test, asphalt, i, 807
sulphur test, lamp method, i, 707
testing methods, i, 588
treating, ii, 368
Naphtha distillate, condenser coil data, ii, 164, 166
Naphthalene, i, 508, 536
boiling point, ii, 806, 807
Naphthalene α phenyl, i, 508
Naphthalene tetrahydride, i, 508
Naphthene series of hydrocarbons, i, 490
Narragansett, Diesel engine equipment, ii, 599
Nashport oil field, Ohio, i, 100
National line pipe, specifications, ii, 550
weights and dimensions, ii, 554
"National" pavements, i, 817
National Petroleum Association, carbon residue test, i, 704
colorimeter, i, 665
color test, oils, i, 662

National Petroleum Association, floc test, i, 691
Native asphalts, classification, i, 790
Native bitumens, United States, i, 790
Natrona County, Wyo., oil, i, 17, 118, 572
Natural gas, Ala., i, 79
analyses, i, 729
Argentina, i, 149
Arkansas, i, 80
Asia Minor, i, 145
Bayard sand, i, 36
Big Lime, i, 31
Borneo, i, 162
California, i, 53
Cambrian system, i, 41
Canada, i, 53, 121
Chattanooga shale, i, 34
Chile, i, 155
China, i, 144
Cincinnati shale, i, 40
Clinton sand, i, 39
commercial, i, 731
compression, i, 450
compressors, i, 765
Dalton's Law of pressure, i, 735
dry, i, 731
England, i, 135
expanders, i, 765
gasoline separation by solids, i, 744
Germany, i, 134
heat value, i, 756
Hogshooter sand, i, 29
Idaho, i, 85
Iowa, i, 90
Japan, i, 143
Jugo-Slavia, i, 134
Kentucky, i, 34, 40
Knox dolomite, i, 40
lenticular sands, i, 59
Louisiana, i, 7, 95
Mesopotamia, i, 141
meters, i, 432
Minnesota, i, 96
Mississippi, i, 96
monoclinal structures, i, 53
monoclinal noses, i, 55
Montana, i, 97
Murraysville sand, i, 33
Muskogee sand, i, 30
New Brunswick, Can., i, 71, 123
New Jersey, i, 98
New South Wales, i, 164
New York, i, 99
North Dakota, i, 102
occurrence with petroleum, i, 1, 73
Ohio, i, 102
Ohio shale, i, 33
Natural gas, Oklahoma, i, 20, 30
Ontario, Can., i, 123
Oswego lime, i, 27
Pennsylvania, i, 107
Permian strata, i, 21
Philippine Islands, i, 163
physical characteristics, i, 751
pipe line formula development, i, 438
pipe line storage capacity, i, 446
pipe line transportation, i, 452
pressure, i, 448
production and consumption statistics, i, 352, 354
Queensland, i, 166
Rumania, i, 131
Russia, i, 4, 127
Santo Domingo, i, 161
sedimentary and igneous rocks, i, 65
South Dakota, i, 108
Speechley sand, i, 36
Squirrel sand, i, 38
structural terraces, i, 57
Sweden, i, 136
Switzerlnd, i, 136
test, gasoline content, i, 759
Texas, i, 114
Tiona sand, i, 36
transportation of, i, 432
Trinidad, i, 159
Washington, i, 115
West Virginia, i, 116
wet, i, 731
Natural gas engines, ii, 523
Natural gas hydrocarbons, properties, i, 753
Natural-gas gasoline, i, 729, 786
production by states and methods, statistics, i, 786
production statistics, i, 351
Natural gas springs, i, 3, 4
Natural lubricating oils, ii, 323
Natural rock asphalt pavement, i, 818
Nautical measure, ii, 760
Navarro County, Tex., oil, i, 114, 560
Navy gasoline, ii, 361, 362
Neatsfoot oil, saponification number of, i, 711
Nebraska, Cambrian sediments, i, 40
oil, i, 97
Permian rocks, i, 21
Neck grease ii, 675
Neocomian strata, i, 126, 131
Neodesha oil field, Kans., i, 91, 546
Neosho County, Kans., oil, i, 90, 546
Neosho Falls oil field, Kans., i, 91
Net register tonnage, oil tankers, i, 376
Neutral stock, ii, 370, 381
Neutralization, refinery stock, ii, 369, 370, 371, 372,

Nevada, asphalt, i, 98
bitumen, i, 791
oil, 98
oil shale, i, 98, 838, 890
Neville Island oil field, Pa., i, 100
New Albany oil field, Kans., i, 91
New Albany shale, i, 34, 871
New Bethlehem oil field, Pa., i, 110
New Brunswick, Can., albertite, i, 6, 68, 792,
natural gas, i, 123
oil, i, 78
oil analysis, i, 576
oil shale, i, 834, 841
oil strata, i, 15, 121
New Caledonia, oil, i, 164
New Castle oil field, Wyo., i, 570
New Century dryer, filter plant, ii, 206
Newells Run oil field, Ohio, i, 100
New Freeport oil field, Pa.,i, 110
New Gallilee oil field, Pa., i, 110
New Geneva oil field, Pa., i, 110
New Guinea, oil, i, 161, 162
oil analyses, i, 584
New Jersey, oil and gas, i, 98
Newkirk sand, i, 25
New Mexico, Cambrian sediments, i, 40
oil, i, 99
oil analyses, i, 551
oil production statistics, i, 340
Permian series, i, 21
New Salem oil field, Kans., i, 91
New Salem oil field, Pa., i, 110
New Sheffield oil field, Pa., i, 110
New South Wales, bitumen, i, 793
oil, i, 164
oil shale, i, 834, 847
New Straitsville oil field, Ohio, i, 59, 100, 553
New Waterford oil field, Ohio, i, 100
New York, Allegany County, early production, i, 107
Cambrian system, i, 41
Corniferous outcrops, i, 38
Genesee shale, i, 34
natural gas, i, 41, 65
natural gas consumption and production, statistics, i, 352, 354
natural-gas gasoline, statistics, i, 786
oil analyses, i, 551
oil and gas, i, 99
oil production statistics, i, 330, 356
oil strata, i, 15
New York State closed tester, i, 632
New Zealand, oil, i, 164
oil analyses, i, 584
oil shale, i, 848
Newark gas field, Ohio, i, 100
Newbury oil pool, Ohio, i, 39
Newburg sand, i, 15, 37, 39
Newcastle oil field, Ind., i, 89
Newcastle oil field, Wyo., i, 120
Newfoundland, Canada, oil, i, 124
oil analysis, i, 576
oil shale, i, 841
Newhall oil field, Calif., i, 51, 82, 540
Newkirk oil field, Okla., i, 104, 555
Newkirk sand, i, 25, 106
Niagara formation, i, 13, 15, 37, 38, 94, 103
Niagara limestone, i, 15, 39
Niagara sand, i, 94
Niagaran series, i, 94
Nichols gravity apparatus, i, 599
Nichols' wax melting point recorder, i, 680
Nickel, melting point, ii, 806, 807
Nineveh oil field, Pa., i, 110
Nineveh sand, i, 116
Nineveh 30-ft. sand, i, 35 108
Niobrara County, Wyo., oil, i, 572
Niobrara shale, i, 98, 118
Nipple, steel die, fishing tool, i, 286
Nitric acid, boiling point, ii, 806
expansion by heat, ii, 812
Nitrogen, boiling point, ii, 807
expansion by heat, ii, 812
in petroleum, i, 458, 528
specific gravity, ii, 832
specific heat, ii, 811
Nitroglycerin, melting point, ii, 806
shooting oil wells with, i, 297
Noble County, Ohio, oil, i, 102, 553
Noble County, Okla., oil, i, 103
Nonacosane, i, 472
Nonadecane, i, 470
Nononaphthene, i, 496, 502
Nonane, i, 466, 468, 533
Non-conducting materials, relative values, ii, 284
Non-emulsifying lubricants, ii, 644
Non-ferrous metals, refinery construction, ii, 281
Nonylene, i, 474, 533
Nonylthiophane, i, 528
Nonylthiophane-sulphone, i, 528
Nordhoff oil field, Calif., i, 81
Norfolk County, Ont., Can., gas, i, 122, 123
Normal dips, geology, i, 171
North America, bitumen, i, 792
Cambrian sediments, i, 40
oil production, i, 75, 78
oil strata, i, 12, 14
North Baltimore oil field, Ohio, i, 100
North Carolina, oil, i, 102
North Dakota, natural gas consumption and production, statistics, i, 352, 354
oil and gas, i, 102

North Oregon Basin gas pool, Wyo., i, 120
North Petroleum oil field, Ky., i, 93
North Warren oil field, Pa., i, 110
North Warren shale, i, 36
Norway, oil shale, i, 848
Noses, structural, i, 53, 55
Notch, structural, i, 53
Nova Scotia, Can., albertite, i, 792
oil shale, i, 841
Pennsylvanian series, i, 23
Nowata-Claggett oil field, Okla., i, 104
Nowata County, Okla., oil, i, 103, 555
Wheeler sand, i, 28
Nowata formation, i, 26
Nozzles, steel tanks, ii, 107
Nyassaland, Portuguese, oil, i, 158

O

Oakgrove oil field, Ohio, i, 100
Oakland City oil field, Ind., i, 89
Observed gravity, i, 599
Observed temperature, i, 598
Oceania, bitumen, i, 793
oil, i, 161
Ochelata-Ramona oil field, Okla., i, 104
Ochelhäuser piston, two-cycle marine Diesel engine, ii, 596
Octane, i, 464, 533
Octocosane, i, 472
Octodecane, i, 470, 533
Octodecylene, i, 476
Octodecylthiophane, i, 528
Octohydrophenanthrene, i, 516
Octomethylene, i, 502
Octotriacontane, i, 472
Octylene, i, 474, 533
Octylthiophane, i, 528
Octylthiophane-sulphone, i, 528
Odometer, i, 182
Oenanthylene, i, 474
Ohio, Ames limestone, i, 32
anticlinal bulges, i, 59
Berea sand, i, 32, 33
Big Injun sand, i, 32
Big Lime, central Ohio, i, 37
Big Lime (Maxville) sand, i, 25, 31
Bremen oil field, i, 53, 59
Clinton limestone, i, 39
Clinton sand, i, 39, 55, 59
Connemaugh formation, i, 23
Cow Run sand, i, 23
crevice oil, i, 67
First Cow Run sand, i, 25
Junction City oil field, i, 59
Keener sand, i, 31
Maxton sand, i, 25, 31
Medina series, i, 39
Ohio, Mississippian series, i, 15, 29
monoclinal structures, i, 53, 55, 59, 72
natural gas, i, 33, 39, 102
natural gas, consumption and production, statistics, i, 352, 354
natural-gas gasoline, statistics, i, 786
natural gas springs, i, 4
Newburg oil pool, i, 39
Newburg sand, i, 39
New Straitsville oil field, i, 59
Ohio shale, i, 33, 37
oil, i, 102
oil, alkyl sulphide in, i, 529
oil analyses, i, 551
oil, nitrogen in, i, 529
oil, oxygen in, i, 524
oil production statistics, i, 341, 356
oil, sulphur in, i, 525
oil seepages, i, 4
oil shale, i, 834, 838
oil strata, i, 14, 103
140-ft. sand, i, 24
Pennsylvanian series, i, 22
Pittsburgh coal, i, 23
salt sands, i, 25, 33
salt water, i, 25, 33
salt with oil, i, 7
Second Cow Run sand, i, 25
Silurian system, i, 38
Silurian-Devonian system, i, 37
Southeastern and central, oil production statistics, i, 332
squaw sand, i, 32
Stadler sand, i, 39
structural terraces, i, 55
Trenton limestone, i, 40
Upper Freeport coal, i, 24
Wooster oil fields, i, 59
Ohio County, Ky., oil, i, 92
Ohio County, W. Va., oil, i, 116
Ohio-Indiana oil fields, anticlinal and synclinal structures, i, 47
Ohio shale, i, 32, 34, 37, 871
natural gas in, i, 33
Ohio series, oil shale, i, 842
Oil, asphalt base, ii, 453
compounding, i, 712
cylinder. *See* Cylinder Oils.
fuel, use of, ii, 445
occurrence. *See* under separate States and Counties and also under "Occurrence of Petroleum" in Table of Contents.
piston lubrication, ii, 655
road, ii, 365
Oil. *See also* Petroleum.
Oil and moisture test, petroleum products, i, 588

Oil and moisture testing press, wax plant, ii, 224
Oil bearing formations, California, i, 83
Canada, i, 121
India, i, 141
Kentucky, i, 94
Louisiana, i, 95
Montana, i, 98
North America, i, 12, 14
Ohio, i, 103
Oklahoma, i, 106
Pennsylvania, i, 108
Trinidad, i, 160
United States, i, 70
Venezuela, i, 152
West Virginia, i, 116
world, i, 11
Wyoming, i, 119
Oil burners, ii, 447
Oil City oil field, Ky., i, 93
Oil City oil field, Pa., i, 110
standard casing requirements, i, 220
Oil Creek, early oil production, i, 34, 107
Oil distillation, ii, 323
Oil drilling systems, i, 201
Oil engines, ii, 523
operating costs, ii, 612
operating tests, ii, 610, 613, 614
Oil equivalents, weight and volume, Baumé and specific gravities, ii, 833, 834
Oil field development, i, 201
Oil fields, aclinal or sub-aclinal structure, i, 46
anticlinal and synclinal structure, i, 47
relation to coal beds, i, 10
structural habits, i, 72
Oil film, lubrication, effect of velocity, ii, 637
Oil film pressure, lubricants, ii, 633
Oil fuel, chimney, table, ii, 309
oil tankers, i, 378
Oil fuel. *See also* Fuel Oil.
Oil grooves, for lubrication, ii, 668
Oil heaters, ii, 447
Oil in paraffin wax, i, 726
"Oil lines," value, i, 9
Oil lubricator, rotary, i, 262
Oil Mountain oil field, Wyo., i, 120
Oil pipe lines, common carriers, i, 428
Oil pump rooms, oil tankers, i, 377
Oil pumping systems, burners, ii, 453
Oil pumps, internal combustion engines, ii, 545
oil tankers, i, 376
Oil refining methods, ii, 323
Oil reservoirs, formed by brecciation and metamorphism, i, 64
Oil sands, porosity, i, 42
sealed by bituminous deposits, i, 68
Oil seepages, i, 3, 167
Alabama, i, 40, 80
Alaska, i, 80
Argentina, i, 150
Australia, i, 166
basaltic dikes, i, 64
Bolivia, i, 154
British Columbia, i, 68
Canada, i, 68, 121
China, i, 143
Colombia, i, 151
De Golyer's classification, i, 4
Dutch Guiana, i, 156
Egypt, i, 156
fault lines, i, 68
France, i, 133
French Guiana, i, 156
Georgia, i, 84
Greece, i, 137
Jugo-Slavia, i, 134
Knox dolomite, i, 40
Mesopotamia, i, 141
Mexico, i, 64, 67, 68, 124
New Guinea, i, 162
New South Wales, i, 164
Palestine, i, 144
Papua, i, 162
Philippine Islands, i, 163
Sakhalin, i, 143, 145
Spain, i, 136
Sundance formation, i, 20
Switzerland, i, 136
Trinidad, i, 4, 159
Tunis, i, 157
Victoria, i, 166
volcanic plugs, i, 63, 64
Washington, i, 115
Wyoming, i, 20, 68
Oil shale, i, 831–894
analyses, i, 836, 872
areas, United States, i, 871
Brazil, i, 154
Bulgaria, i, 136, 137
classification, i, 832
Colorado, i, 84
Cuba, i, 161
deposits, classes, i, 855
distillation, fixed gases, i, 882
Germany, i, 134
Nevada, i, 98
New South Wales, i, 164
Queensland, i, 166
resins in, i, 524
Scotland, i, 67, 135
South Africa, i, 158
Tasmania, i, 166
Wales, i, 135

Oil shale vapors, i, 870
Oil spray, internal combustion engine, ii, 553
Oil springs. *See* Oil Seepages.
Oil Springs oil field, Ontario, Can., i, 121, 122, 123, 576
  nitrogen in, i, 530
  sulphur in, i, 526
Oil stock, condenser coil data, ii, 178
Oil supply, internal combustion engines, ii, 553
Oil, synonymous with petroleum, i, 1
Oil tankers, i, 371
Oil tar, Wyoming, i, 67
Oilton oil field, Okla., i, 104
Oil Valley oil field, Ky., i, 547
Oil well tubing, weights and dimensions, ii, 562
Oil wells, burning, control of, i, 312
  flowing, i, 310
  U. S., number of, i, 309, 349
Okemah oil field, Okla., i, 104
Okesa oil field, Okla., i, 104
Okfuskee County, Okla., oil, i, 103
Oklahoma, anticlinal and synclinal structures, i, 47, 159
  asphalt seepages, i, 22
  Bartlesville sand, i, 28
  Big Lime, i, 26
  Billings oil field, i, 21, 25
  bitumen, i, 791
  Bixler sand, i, 28
  Blackwell crude, ii, 13
  Blackwell pool, i, 21, 25
  Booch sand, i, 30
  Boone chert, i, 30, 34
  Boynton pool, i, 30
  Burgess sand, i, 30
  Cambrian sediment, i, 40
  Carboniferous system, i, 21
  Cement oil field, i, 22
  Chattanooga shale, i, 30, 34
  Cherokee shale, i, 28, 29
  Claremore sand, i, 34
  Cleveland sand, i, 27
  closed faults, i, 68
  Coffeyville formation, i, 26
  Colbert sand, i, 30
  Collinsville oil field, i, 30
  Cushing oil field, i, 26, 28
  domes, i, 59
  Dutcher sand, i, 30
  Elgin sandstone, i, 25
  Fayetteville formation, i, 30
  Ft. Scott limestone, i, 26
  Fortuna sand, i, 22
  Garrison formation, i, 21
  Glenn sand, i, 28
Oklahoma, grahamite dikes, i, 6, 71
  Healdton oil field, i, 59
  Hogshooter sand, i, 29, 30
  Jones sand, i, 26
  Layton sand, i, 26, 27
  Leidecker sand, i, 30
  Lower Cretaceous series, i, 20
  Madill oil pool, i, 20
  Markham sand, i, 27
  Meadows sand, i, 30
  Mississippi lime, i, 29, 34
  Mississippian series, i, 13, 29
  monoclinal bulges, i, 59
  monoclinal noses, i, 55
  Morris sand, i, 28, 30
  Mounds sand, i, 30
  mud volcanoes, i, 7
  Muskogee County, i, 30
  natural gas, i, 29
  natural gas, consumption and **production** statistics, i, 352, 354
  natural-gas gasoline statistics, i, 786
  Newkirk oil field, i, 25
  Newkirk sand, i, 25
  oil, i, 52, 103
  oil analyses, i, 554
  oil, production statistics, i, 336, 356
  oil strata, i, 12, 106
  Oswego limestone, i, 21, 25, 26, 27, 28, 29, 30
  Pennsylvanian series, i, 22
  Permian series, i, 21
  Peru sand, i, 27
  Pitkin limestone, i, 30
  Ponca City oil field, i, 21, 25
  Ponca City sand, i, 25
  Pre-carboniferous rocks, i, 22
  quaquaversal structure, i, 59
  Red Fork sand, i, 28
  relation of geosyncline to oil fields, i, 72
  rotary drilling system, i, 251
  Sapulpa sand, i, 30
  Second Booch sand, i, 30
  Skinner sand, i, 28
  Squirrel sand, i, 28
  standard casing requirements, i, 221
  Trinity sand, i, 20
  Tucker sand, i, 30
  Tulsa County, i, 27, 29, 30
  Wagoner County, i, 30
  Washington County, i, 30
  Wheeler oil pool, i, 59
  Wheeler sand, i, 21, 25, 26, 27, 28, 30
  Wilcox sand, i, 34
Oklahoma crude, fuel oil yield, ii, 342
  fuel oil, vaporizing point, ii, 497
  gravities, continuous stills, ii, 327

Oklahoma crude, refinery running time and fuel consumption, ii, 333
typical distillation, ii, 355
Oklahoma oil fields, disposition of oil and water in, i, 73
lenticular sands, i, 59
structural habits, i, 72
water with oil, i, 73
Okmulgee County, Okla., Booch sand, i, 30
Morris pool, i, 30
Morris sand, i, 30
Mounds sand, i, 30
oil, i, 103, 106, 556
Wilcox sand, i, 34
Oldham County, Ky., gas, i, 40
Old shore lines, relation to oil occurrence, i, 41, 65
Olefiant gas, heat of combustion, ii, 809
Olefinacetylene, i, 478
Olefine series, hydrocarbons, i, 472
Olefines in refinery gases, ii, 400
Oligocene series, Calif., i, 83
Czecho-Slovakia, i, 134
Galicia, i, 132
India, i, 141
Jugo-Slavia, i, 134
oil strata in, i, 12, 14, 17
oil shale, i, 834
Poland, i, 132
Rumania, i, 131
Switzerland, i, 136
Trinidad, i, 160
Venezuela, i, 152
Olive oil, expansion by heat, ii, 812
specific heat, ii, 812
Olympia oil field, Ky., i, 92, 93, 546
100-foot sand, i, 34, 108
140-foot sand, i, 24, 103
Onondaga limestone, i, 37, 92, 94, 121
Onondaga oil field, Ontario, Can., i, 122, 123
Ontario, Can., "Clinton" shales, i, 39
Corniferous limestone, i, 38
Guelph limestone, i, 15, 121
Hamilton limestone, i, 121
Medina sands, i, 15, 39, 121
natural gas, i, 41, 65, 123
oil, i, 14, 121, 123
oil analyses, i, 576
oil, nitrogen in, i, 530
oil shale, i, 842
salt unassociated with oil, i, 7
Oologah limestone, i, 26
Oolite, Green River formation, i, 874
Open cut methods, oil shale mining, i, 856
Open-hearth steel pipe, ii, 501
Open sand, i, 41
Optical activity, oil, i, 532
Orange County, Calif., bitumen, i, 791
oil, i, 80
Orange oil field, Tex., i, 113, 114
Ordovician system, Can., i, 121
geographic distribution, i, 11, 39
Indiana, i, 88
Kentucky, i, 92, 94
Ohio, i, 102, 103
oil shale, i, 871
oil strata, i, 13, 15
United States oil fields, i, 79
Oregon, bitumen, i, 791
natural gas consumption and production statistics, i, 353, 354
oil, i, 67, 107
Oregon Basin oil field, Wyo., i, 118
Organic theory, oil occurrence, i, 44
Orientation, plane table, i, 193
Orifice, natural gas meter, i, 433
Origin of oil shales, i, 834
Orinoco oil district, Venezuela, i, 146, 151
Oriskany sandstone, i, 121
Ormsby oil field, Pa., i, 110
Osage County, Okla., Cleveland sand, i, 27
oil, i, 103, 556
Peru sand, i, 27
squirrel sand, i, 28
Osage Junction oil field, Okla., i, 104
Osage oil field, Wyo., i, 120, 570
Oskamp oil field, Ky., i, 93, 546
Osawatomie oil field, Kans., i, 91
Oswego lime, i, 21, 25, 26, 27, 28, 29, 30
Otsego oil field, Ohio, i, 100
Ottawa County, Ohio, oil, i, 102
Otter sand, i, 33, 94
Otto oil field, Kans., i, 91
Ouachita Parish, La., oil, i, 95
Outage, minimum, tank cars, i, 365
Outcrops, bituminous, i, 4
oil geology, i, 169, 170
Outlet plug, acid tanks, ii, 39, 76
Oven, wax plant, ii, 199
Over and short accounts, i, 426
Overshot, fishing tool, i, 284
Overturned folds, oil and gas in, i, 53
Owasso oil field, Okla., i, 104
Ownership, tank cars, i, 366
Oxidation, hydrocarbons, i, 458
lubricants, ii, 645
mineral oils, ii, 663
Oxygen, in oils, i, 458, 524
specific gravity, ii, 832
specific heat, ii, 811
Ozokerite, Philippine Islands, i, 163
United States, i, 791

P

Pace counter, i, 182
Pacific oil district, Colombia, i, 151, 153
  Peru, i, 148
  South America, i, 146
Package transportation, petroleum, i, 359
Packer method, water exclusion, i, 290, 291
Packing, refinery, ii, 285
Packingless type pump, internal combustion engines, ii, 546
Paden oil field, Okla., i, 104
Paint Creek oil field, Ky., i, 93
Paints, asphalt bases, i, 829
  protective, ii, 124
Painting, tank, ii, 111, 124
Pale oils, filtering, ii, 381
Paleography, relation to occurrence of oil, i, 42
Paleozoic period, Cambrian rocks, i, 41
  oil, strata, i, 13, 15
  stratigraphic occurrence of petroleum, i, 11
  Venezuela, i, 152
Palestine, oil and asphalt, i, 144
Palladium, melting point, ii, 807
Palo Pinto County, Tex., oil, i, 114, 561
  Strawn formation, i, 29
Panama, oil, i, 126
  pipe lines, i, 379
Panama Canal, oil tanker, tonnage, i, 376
Panola County, Tex., oil, i, 561
Panuco crude, gasoline from, ii, 431
Panuco oil field, Mexico, i, 20, 124, 125
  drilling methods, i, 202
  standard casing requirements, i, 221
Paola-Rantoul oil field, Kans., i, 91, 546
Paper mill board, refinery packing, ii, 285, 287
Paper stock, ii, 359
Papoose sand, i, 32
Papua, oil, i, 161, 162
Paraffin, Scotland, i, 67
  latent heat of fusion, ii, 823
Paraffin base oils, ii, 453
Paraffin distillate, ii, 341, 355
Paraffin hydrocarbons, i, 460, 730
  specific heats, i, 531
Paraffin jelly, ii, 380
Paraffin series of hydrocarbons, i, 460, 730
Paraffin stock, ii, 370
Paraffin wax, melting point determination, i, 679
  moisture determination, i, 725
  oil determination, i, 724–725
  oxidation, i, 459
  refined, ii, 375
  solubility, i, 538
  tests, i, 588
  treating, ii, 368
Parallel operation, Diesel engine, ii, 582
Park County, Wyo., oil, i, 570
Parmleysville oil field, Ky., i, 93, 547
Partial immersion thermometers, i, 592
Passivity, lubricants, ii, 644
Patents, acid recovery, ii, 402
  cracking petroleum oils, ii, 440
Pawhuska oil field, Okla., i, 104
Pawnee County, Okla., Cleveland sand, i, 27
  oil, i, 103, 556
Payette, Ida., oil, i, 86
Payne County, Okla., oil, i, 103, 556
Peabody oil field, Kans., i, 91
Peace River region, Can., oil, i, 123
Peanut oil, saponification number, i, 711
Pearson oil field, Okla., i, 104
Pearson, S. and Son, oil activities in Mexico, i, 124
Peay sandstone, i, 18, 98
Pechelbroun, Alsace, oil analyses, i, 578
  wells and refinery, i, 133
Péclet's formula, external conduction, heat, ii, 826
Pecos County, Tex., oil, i, 567
Pecos Valley oil field, Tex., i, 114
Pelican oil field, La., i, 96
Penetration test, petroleum products, i, 588, 719, 798
Penetrometer, i, 588, 719, 799
Pennsylvania, anticlinal and synclinal structures, i, 47
  anticlinal bulges, i, 59
  Armstrong County, oil, i, 35
  Bayard sand, i, 36
  Bedford shale, i, 32
  Berea sand, i, 32
  Big Injun sand, i, 15, 33
  Big Lime (Maxville), i, 31
  Blue Monday sand, i, 35
  Boulder sand, i, 35
  Bradford sand, i, 36
  Bradford Third sand, i, 38
  Butler Fourth sand, i, 35, 36
  Butler gas sand, i, 33
  Butler Second sand, i, 34
  Butler Third sand, i, 35
  Butler Third Stray, i, 35
  Chemung formation, i, 36
  Clarion County, oil, i, 35
  Connemaugh formation, i, 23
  Cow Run sand, i, 23
  Drake well, i, 201
  Dunkard sand, i, 15, 24
  Elizabeth sand, i, 36
  Fifth sand i, 36
  50-foot sand, i, 35
  First Mountain sand, i, 25

Pennsylvania, Fourth sand, i, 35, 36
Gantz sand, i, 34
gas sand, i, 15, 25
Gordon sand, i, 35, 36
Gordon stray sand, i, 35
Gray sand, i, 35
Manifold sand, i, 32
Mauch Chunk red shales, i, 31
Maxton sand, i, 25, 31
McDonald sand, i, 36
Mississippian series, i, 29, 31
Mountain sand group, i, 32
Murraysville sand, i, 33
natural gas, i, 33, 36, 107
natural gas production and consumption statistics, i, 352, 354
natural-gas gasoline statistics, i, 786
Nineveh 30-foot sand, i, 35
North Warren shale, i, 36
oil, i, 107
oil analyses, i, 558
oil, oxygen in, i, 524
oil production statistics, i, 331, 356
oil, sulphur in, i, 525
oil shale, i, 834, 838
oil strata, i, 14, 108
100-foot sand, i, 34
Permian series, i, 21
Pithole grit, i, 33
Pittsburgh coal, i, 23, 35, 36
Pocono formation, i, 34
salt brine, occurrence, i, 8
salt sand, i, 15, 25
Sixth sand, i, 36
Snee sand, i, 35
Speechley sand, i, 36
standard casing requirements, i, 220
Stoneham sandstone, i, 36
Third mountain sand, i, 33
Tiona sand, i, 36
Upper Freeport coal, i, 24
Venango First sand, i, 34
Venango sand group, i, 34, 35
Venango Second sand, i, 35
Venango Third sand, i, 36
Warren First sand, i, 36
Warren sand group, i, 36
Warren Second sand, i, 36
Pennsylvania crude, typical distillation, ii, 325, 354
Pennsylvania Rock-oil Co., i, 107
Pennsylvanian series, Appalachian oil fields, i, 23
Illinois, i, 86
Kansas, i, 90, 125
Kentucky oil fields, i, 25, 92, 94
Montana, i, 98
Pennsylvanian series, Northern and Central Texas, i, 28
Ohio, i, 102, 103
oil shale, i, 834, 837, 838
oil strata, i, 13, 15, 22
Oklahoma, i, 25, 103, 106
Pennsylvania, i, 107, 108
relation to Mississippian series, i, 29
relation to Permian series, i, 21
Texas, i, 114
West Virginia, i, 116
Wyoming, i, 118
Pensky-Marten closed up tester, i, 624, 637, 638
Pentacosane, i, 470
Pentadecanaphthene, i, 502
Pentadecane, i, 470
Pentaethyl benzol, i, 488
Penta methyl benzol, i, 488
Pentamethylene, i, 492
Pentane, i, 460, ii, 361
Pentatriacontane, i, 472
Pentyl sulphide, i, 529
Percussion, drilling system, i, 203, 242
Perforating oil wells, i, 290, 297
Perhydroacenaphthene, i, 522
Perhydroanthracene, i, 522
Perhydrofluorene, i, 522
Perhydrophenanthrene, i, 522
Perhydropicene, i, 522
Perhydrorecene, i, 522
Perkins one-plug method, oil well cementing, i, 296
Permian series, Brazil, i, 154
Montana, i, 98
natural gas, i, 21
oil shale, i, 834, 837, 842
oil strata, i, 13, 15, 21
Oklahoma, i, 21, 103, 106
"Red beds," i, 21
Texas, i, 21, 114
Wales, i, 135
Wyoming, i, 21, 67, 119
Permo-Carboniferous series, i, 843, 847
Permo-Carboniferous shales, Australia, i, 166
Perry County, Ohio, oil, i, 102, 553
Perry County, Pa., oil, i, 558
Perrysville oil field, Pa., i, 110
Pershing oil field, Okla., i, 104
Persia, bitumen, i, 793
oil, i, 137, 142
oil analyses, i, 582
oil production statistics, i, 142, 325
Peru, anticlinal and synclinal structures, i, 53
bitumen, i, 792
oil, i, 146, 148
oil analyses, i, 578
oil production statistics, i, 148, 325

Peru, oil shale, i, 848
oil, unit of measure, ii, 761
stratigraphic occurrence of petroleum, i, 11
Peru oil field, Kans., i, 91
Peru sand, i, 27, 106
Petersburg oil field, Ind., i, 89
Petrocene, i, 518
Petrohol process, ii, 399
Petrol-Alcohol process, ii, 399
Petrolatum, ii, 359
as lubricant, ii, 644
lubricant grease, ii, 673
melting point, determination, i, 684
method of testing, i, 588
white (album), ii, 380
Petrolatum greases, ii, 674
Petrolatum stock, ii, 359, 364, 380
Petrolenes, i, 807
Petroleum, analyses, i, 538
asphalt, i, 524
average production per well in U. S., i, 309, 349
carbon residue determination, i, 704
characteristics, i, 457
cloud point, i, 658, 659, 660
cold test, i, 658, 659
color determination, i, 662
description, i, 457
elements in, i, 458
emulsification tests, i, 695
foreign oil units of measure, ii, 761
lamp burning tests, i, 691
nitrogen content, i, 529
occurrence. *See* under separate States and Countries and also under "Occurrence of Petroleum" in Table of Contents.
optical activity, i, 532
oxygen in, i, 524
physical properties, i, 531
pour point, i, 660, 661
production methods, i, 309
production statistics, i, 321
radioactivity, i, 535
related products, diagram, ii, 382, 383, 384
refining, ii, 1–426
refining to asphalt, i, 787
resins, i, 524
solubility, i, 535
specific heat, i, 531
sulphur in, i, 525
surface indications, i, 167
temperature corrections for Baumé and specific gravities, ii, 835, 836, 837, 838
testing methods, i, 587
United States production and value, i, 328
units of measure, ii, 768
Petroleum, water determination in, i, 668
world's production, i, 324
Petroleum. *See also* Oil.
Petroleum asphalt, characteristics, ii, 809
Petroleum distillation, ii, 323
modified assay method, i, 678
Petroleum engines, internal combustion, ii, 523
Petroleum geology, field methods, i, 167
Petroleum jelly, ii, 360, 380
Petroleum production, i, 201
Petroleum products, acidity of, i, 710
ash test, i, 723
calorific values, ii, 342
decolorization, ii, 377
diagram showing relations, ii, 382, 383, 384
doctor treatment, ii, 14
evaporation tests, i, 716
fatty oil content, i, 711
fixed carbon test, i, 723
float test for, i, 722
heat loss determination, i, 717
solubility tests, i, 715
sulphur determination, i, 709
tar tests, i, 588, 714
transportation, i, 359
Petroleum residuum, as grease lubricant, ii, 673
Petroleum testing laboratory, functions of, i, 587
Petroleum Testing Methods, i, 587–728
Petroleum Center oil field, Pa., i, 110
Petrolia oil field, Can., i, 122, 576
nitrogen, i, 121, 123, 530
sulphur, i, 526
Petrolia oil field, Pa., i, 110
Petrolia oil field, Tex., i, 22, 59, 113, 559
Petroliferous rocks, tests for oil in, i, 167
Petroliferous sands, as indications of petroleum, i, 167
Petter internal combustion engine, ii, 537, 542
Phellandrene α, i, 510
Phellandrene β, i, 512
Phenanthrene, i, 516, 536
Phenol acetylene, i, 484
Phenol ditolyl methane, i, 506
Phenol ethane, i, 482
Phenols, in crude oil, i, 459
Phenyl anthracene, i, 516
Phenyl ditolyl methane, i, 516
Phenyl naphthalene, i, 536
Phenyl pentane, i, 488
Phenyl toluol, i, 506
Philadelphia Road oil field, Ohio, i, 100
Philippine Islands, oil analyses, i, 585
oil and gas, i, 163
Phillips oil field, Pa., i, 110

Phosphorus, boiling point, ii, 806
latent heat of fusion, ii, 823
melting point, ii, 806
specific heat, ii, 811
Physical characteristics, natural-gas and natural-gas gasoline, i, 751
Physical properties, boiler steel, ii, 144
shale oil, i, 878, 890, 893
steel pipe, ii, 501
structural steel, ii, 116, 122
Physical tests, lubricating oils, ii, 676
Picene, i, 516, 518, 536
Piceneikosihydride, i, 506
Pickett County, Tenn., oil, i, 108
Pico oil field, Calif., i, 81, 530
Piedras Pintas oil field, Tex., i, 113, 114
Pierre formation, i, 14, 17, 98, 119
Pike County, Ark., asphalt, i, 80
bitumen, i, 791
Pike County gas pool, Ill., i, 87
Pike County, Ind., oil, i, 88
Pike sand, i, 25, 92, 94
Pilot Butte oil field, Wyo., i, 118, 120, 571
Pilot light, oil burner, ii, 492
Pinane, i, 520
Pincher Creek oil district, Can., i, 41, 68
Pine Dome gas pool, Wyo., i, 120
Pine Grove oil field, Pa., i, 110
Pine Prairie oil field, La., i, 96
Pine Run oil field, Pa., i, 110
Pinene, i, 512
Pinion grease, ii, 674
Piolett oil field, Pa., i, 110
Pipe, ii, 501
asphalt coating, i, 826
capacity, table, ii, 770
cast iron, standard dimensions, ii, 274
compressed air, diameter, relation to volume, ii, 300
covering capacity, i, 828
Diesel engine, ii, 577
dimensions and capacity, table, ii, 771
flow of oil through, i, 380
perforated, i, 297
refinery, ii, 242
rifled, i, 406
steam, expansion, ii, 798
steam, gas, and water, standard dimensions, ii, 782–784
volume of flow, table, i, 395
water, discharging capacities, ii, 568
yield point tests, ii, 526
Pipe and machine shop, refinery, ii, 228
Pipe cutting and threading outfit, standard drilling, i, 217
Pipe dipping, asphalt, i, 826
Pipe fittings, concrete tanks, ii, 133
Pipe line, asphalt coating, i, 826
computations for capacity, i, 396
construction, i, 406
flow, i, 380
flow, gas, i, 439, 446
gas, tests of, i, 445
formula, application of, i, 442, 444
heating, i, 403
loops, i, 400
natural-gas gasoline, i, 759
operation, i, 424
storage capacity, natural gas, i, 446
transportation, natural gas, i, 432
transportation, petroleum, i, 359, 379
Pipe standards and use of pipe, ii, 501–574
Pipe thread formula, ii, 546
Pipe threading, ii, 535
Piperylene, i, 478
Piping, compression plants, i, 784
swivel joint burner, ii, 479
Piru oil field, Calif., i, 81
Piston, Diesel engine, ii, 566
internal combustion engine, ii, 543
lubrication, ii, 652, 665
Piston blowing, internal combustion engine, ii, 554
Piston cooling, Diesel engines, ii, 567, 589
Piston speed, internal combustion engines, ii, 543
Pitch, general occurrence, i, 5
Pitch dike, Scotland, i, 67
Pitch Lake, Trinidad, i, 6, 68, 159
Pitch Lake, Venezuela, i, 68
Pit hole grit, i, 33
Pitkin limestone, i, 29, 30
Pitot tube, natural gas, i, 433
Pittsburgh coal bed, i, 123
Pittsburgh coal bed, relation to Bayard sand, i, 36
Berea sand, i, 32
Big Injun sand, i, 31
Bradford sand, i, 36
Cow Run sand, i, 23
Dunkard sand, i, 24
Elizabeth sand, i, 36
Fifth sand, i, 36
Fourth sand, i, 36
gas sand, i, 25
Gordon sand, i, 36
Gray sand, i, 35
Macksburg 500-ft. sand, i, 25
Nineveh 30-ft. sand, i, 35
Salt sands, i, 25
Second Cow Run sand, i, 25
Sixth sand, i, 36
Speechley sand, i, 36
Tiona sand, i, 36

Pittsburgh coal bed, relation to Upper Freeport coal, i, 24
Venango First sand, i, 34
Warren sand, i, 36
Pitt sweetening process, ii, 336
Placard, tank car, i, 365, 367
Plane table, geologic mapping, i, 177, 179, 191
Plant design, refinery, ii, 4
Plate-heating furnace, ii, 497
Plates, boiler, refinery, ii, 288
iron and steel, weight, table ii, 775, 778
refinery construction, ii, 261
tanks, ii, 93, 103
tin, weights and gages, table, ii, 779
Platinum, fusing, ii, 807
melting point, ii, 807
radiation and reflection of heat, ii, 824
Pleasants County, W. Va., oil, i, 116, 117, 569
Pleasantville oil field, Ohio, i, 100, 552
Pleistocene series, California, i, 83
oil strata, i, 12
Trinidad, i, 160
Pliocene series, Albania, i, 137
California, i, 83
Italy, i, 132
Japan, i, 143
Jugo-Slavia, i, 134
oil strata, i, 12, 16
Rumania, i, 131
Russia, i, 127
Trinidad, i, 160
Plotting scale, oil mapping, i, 180
Plugging, oil heater atomizer, ii, 484
oil wells, i, 290, 299
Plum Creek oil field, Pa., i, 110
Plumb Run oil field, Ohio, i, 100
Plum Run oil field, Pa., i, 110
Plumbus oxide, refinery use, ii, 368
Plunger pump, i, 317
Pluymert, N. J., i, 370
Pluymert conversion table, tanker tonnage, i, 375
Plymouth oil field, Ill., i, 87
Plympton-Sarnia oil field, Ont., Can., i, 122
Pocket transit, i, 178
Pohl method melting point determination, asphalt, i, 805
Point Lookout oil field, Pa., i, 110
Poison Spider oil field, Wyo., i, 118, 120
Poland, mud volcanoes, i, 6
oil, i, 131
oil analyses, i, 580
oil production statistics, i, 132, 324
overturned folds, i, 53
stratigraphic occurrence of oil, i, 11
See also Galicia.
Polk County, Iowa, i, 90
Ponca City oil field, Okla., i, 21, 25, 104
Ponca City sand, i, 25
Pond Creek oil field, Okla., i, 104
Pontotoc County, Okla., bitumen, i, 791
oil, i, 103
Pooleville oil field, Okla., i, 104
Poop, oil tanker, i, 376
Porosity, oil sand, i, 42
Portable drilling machines, i, 250
Portable furnaces, oil fired, ii, 489, 493
Porter County oil field, Ind., i, 89
Porter oil field, Okla., i, 104
Portland cement, refinery construction, ii, 249
Porto Rico, oil, i, 161
Portugal, bitumen, i, 792
oil, i, 136
oil shale, i, 848
Potassium, melting point, ii, 806
Potassium bichromate, color readings by, i, 663
Potassium permanganate, refinery use, ii, 372
Potassium sulphate, melting point, ii, 806
Poteau gas field, Okla., i, 104
Potrero del Llano oil field, Mexico, i, 20, 124, 248
Potsdam sandstone, i, 15, 21
Pottsville formation, i, 25, 94, 108, 116
Potwin oil field, Kans., i, 91
Pound calorie, ii, 808
Pour method, petroleum products, i, 588
Pour point, oil, i, 660, 661
Powder River oil field, Wyo., i, 21, 118
Powell County, Ky., oil, i, 92
Power equipment, oil shale plant, i, 861
Power generators, refinery, ii, 287
Power-house, refinery, ii, 229
Power losses, lubrication, ii, 640
Power plant, pipe line, i, 417
standard drilling system, i, 222
Power producers, comparison of average power, ii, 296
Power transmission, refinery, ii, 298
Powers, for multiple pumping, i, 319
Pre-Cambrian system, Venezuela, i, 152
Pre-Carboniferous rocks, Okla., i, 22
Pre-Cretaceous system, Venezuela, i, 152
Precipitation in gasoline specifications, i, 713
Preheaters, cracking, ii, 432
Preheating, oil shale, i, 869
Prehnitol, 1, 2, 3, 4, i, 484
Prepared roofing, asphalt, i, 824
Presses, wax plant, ii, 198, 223
Pressing, refinery, ii, 372
Pressure, adsorption, relation to temperature and boiling point, i, 748
atmospheric, ii, 786
bearing lubrication, ii, 642

Pressure, bearings, internal combustion engine, ii, 544
collapsing, steel pipes, ii, 505
compressed air, relation to volume, ii, 298
gas, Dalton's law, i, 735
heads of water, table, ii, 792
in cracking, ii, 435
masonry, ii, 245
mean effective, air compression, ii, 301
natural gas, i, 448
natural gas pipe lines, table, i, 455, 456
pipe line, i, 401
relation to yield ol gasoline, i, 742
saturated ammonia vapor, ii, 839
tubes, pipes, and cylinders, internal fluid, ii, 519
working, boilers, ii, 288, 290
Pressure curves, natural gas, i, 437
Pressure friction experiments, lubricants, ii, 631
Pressure relief valves, ii, 461
Pressure-still, cracking, ii, 434
Pressure variation, barometric readings, i, 185
Preston oil field, Okla., i, 104
Price cracking process, ii, 825
Primary and secondary crushing, oil shale, i, 848
Prime white distillate, ii, 339
Prime white oil, i, 663
Princeton oil field, Ind., i, 88, 89
Processes, refining, ii, 322
Production of petroleum, i, 321
Production statistics, natural gas, i, 351, 354
natural-gas gasoline, i, 351, 756
petroleum, i, 321
refinery gasoline, ii, 346
Profile method of recording, i, 184
Propane, i, 460
Propanol, ii, 401
Propine, i, 478
Proportional meter, natural gas, i, 433
Propylacetylene, i, 478
Propyl alcohol, ii, 401
Propyl benzol, i, 482, 484
Propyl cyclohexane, i, 498
Propyl dimethyl methane, i, 462
Propylene, i, 472, ii, 401
Propyl hexamethylene, i, 500
Propyl iso-propyl, i, 462
Propyl naphthalene, i, 508
Propyl sulphide, normal, i, 529
Protractor, i, 181
Pseudobutylene, i, 472
Pseudocumol, i, 482
Pseudopropyl naphthalene, i, 508
Puente Hills oil fields, Calif., i, 16, 17, 81, 83, 540
Puente Hills oil fields, Calif., nitrogen in oil, i, 530
sulphur in oil, i, 526
"Puking," refinery, ii, 407
Pulley diameters, determination, ii, 303
Pulsometer, oil heater, ii, 462
Pulverizer, internal combustion engines, ii, 544, 571
Pump, fuel, Diesel engine, ii, 575
gasoline extraction plants, i, 771
oil, internal combustion engine, ii, 545
oil heaters, ii, 453
pipe line, i, 421
refinery, ii, 92, 206
refinery water supply, ii, 241
wax plant, ii, 194, 196
Pump house, refinery, ii,229
Pump speed regulator, oil burning system, ii, 461
Pump stations, pipe line, i, 417, 418
Pumpherston process, oil shale, i, 866
Pumping, oil well, i, 317
Pumping-out pan, refinery, ii, 59
skimming plant, ii, 29
Punching, steel plate work, ii, 42
steel tanks, ii, 101
Punjab oil fields, India, i, 141
Pushmataha County, Okla., grahamite (asphaltite), i, 791
Putnam County, W. Va., gas, i, 116
Putnam oil field, Tex., i, 113
Puzzolan cement, refinery construction, ii, 249
Pycnometer method, i, 800
Pyrene, i, 516, 536
fire extinguisher, ii, 317
Pyridine bases, shale oil, i, 881
Pyridine, in oil, i, 528
Pyrobitumens, asphaltic, i, 794
Pyrrolylene, i, 478

Q

Quadrant formation, i, 22, 98
Quadrant, fireman's regulating, ii, 492
Quaquaversal structures, i, 45, 59
Quartdecylthiophane, i, 528
Quartz, specific heat, ii, 811
Quaternary system, California oil fields, i, 83
oil strata, i, 11, 12, 16
Texas, i, 114
Trinidad, i, 160
United States oil fields, i, 79
Venezuela, i, 152
Quay-Yale oil field, Okla , i, 104
Quebec, Can., Cambrian system, i, 41
closed faults, i, 68
greenstone dike, i, 67

Quebec, Can., natural gas, i, 65
oil shale, i, 842
oil strata, i, 15
sedimentary and igneous rocks, i, 65
Quebec group, Can., i, 15, 121
Queensland, natural gas, i, 166
oil shale, i, 166, 847
Quicklime, specific heat, ii, 811
Quicksilver. *See* Mercury.
Quinton gas pool, Okla., i, 104

R

Radiation, in mapping, i, 195
Radiation loss, bare steam lines, ii, 660
Radiation of heat, ii, 824
Radio activity, oil, i, 535
Ragland oil field, Ky., i, 92, 93, 546
Ragland sand, i, 37, 94
Railroad connections, refinery, ii, 9
Railroad rules, tank cars, i, 366
Railroad transportation, oil, i, 359
Rake, threading die, ii, 536
Ramona oil field, Okla., i, 104
Ramsey oil field, Pa., i, 110
Rancho la Brea, asphalt seepage, i, 16
Randall process, oil shale, i, 866
Randolph County, Ill., oil, i, 544
Randolph County, Ind., oil, i, 88
Rangely oil field, Colo., i, 84, 85, 542
Ranger oil field, Tex., i, 31, 113, 114, 562
Rankine cycle, ii, 526
Rankine's formula, external conduction of heat, ii, 826
internal conduction of heat, ii, 825
Rantoul oil field, Kans., i, 91, 545
Rasp, rotary, i, 282
"Rat-hole," rotary drilling, i, 267, 268
Rattlesnake oil field, Pa., i, 110
Ravenna oil field, Ky., i, 93
"Ravine," monoclinal, i, 53
Raymilton oil field, Pa., i, 110
"R. E." (resistance to emulsification) test, oils, i, 588, 696
Reaming, pipe, ii, 550, 552
Réaumur thermometer scale, ii, 832
Reconnaissance mapping, oil geology, i, 174, 190
Records, oil testing methods, i, 590
Recovery house, refinery, ii, 230
Rectangular tank, capacity and dimensions, table, ii, 772
Red beds, i, 21
Red Fork oil field, Okla., i, 104, 558
Red Fork sand, i, 28
Red Medina sand, i, 15, 39, 121
Red oils, filtering, ii, 381
Red River oil field, La., i, 96
Red River Parish, La., oil, i, 95, 550
Reduced crude stock, ii, 370
Reduced wax distillate, ii, 364
Reducing, refinery, ii, 323
Reducing still condenser boxes, refinery, ii, 55
Reducing stills, ii, 46, 213
products, ii, 360
refinery, ii, 45, 150
Redwood admiralty viscometer, i, 643, 654, 655
Redwood viscosimeter, i, 643, 654
Reeves County, Tex., oil, i, 114, 567
Refined oil agitators, ii, 62
Refined oil bleachers, ii, 72
Refinery, buildings, ii, 88
construction, ii, 6, 227
engineering, ii, 1
equipment, ii, 227
flux and lubricants, ii, 41
insulation, ii, 282
operation, ii, 1, 319, 421
organization, ii, 321
products, ii, 338
products, diagram, ii, 382, 383, 384
site, ii, 1, 6
Refining apparatus, standard, ii, 91
Refining general scheme, ii, 79
oil, ii, 1
processes, ii, 12, 322, 385
shale oil, i, 870
Reflected light, oil examination, i, 662
Reflection of heat, ii, 824
Refractive indices, petroleums and hydrocarbons, i, 532
Refrigerating units, wax plant, ii, 192
Refrigeration oils, characteristics, ii, 677
Regional dips, oil geology, i, 171
Register tonnage, oil tankers, i, 374
Regnault, specific heats of gases, ii, 525
Reinforced concrete, refinery construction, ii, 252
working stresses, ii, 245
Reinforcement, concrete tanks, ii, 130, 131
Reinforcing bars, spacing of, ii, 252
Reiser gas field, Tex., i, 113
Repair shops, oil fuel, ii, 497
Rerun benzine, condenser coil data, ii, 172
Rerunning, refinery, ii, 323, 334
Rerunning still, refinery, ii, 150
Resection, in mapping, i, 195, 196
Residual fuel, testing methods, i, 588
Residuum, petroleum, as grease lubricant, ii, 673
shale oil, i, 879
steam distillation, ii, 354
Resins, i, 524
Resistance, lubrication, ii, 627, 656

Retendodecahydride, i, 522
Retene, i, 516, 536
Retorting, oil shale, i, 861
Retorts, oil shale, i, 866
  refinery, ii, 238
Reversals, oil geology, i, 171
Reversing, marine Diesel engine, ii, 596, 604
Revivification, filters, ii, 379
Revivifying furnace, filter plant, ii, 203
Rexville oil pool, N. Y., i, 99
Rhigolene, ii, 361
Rhinyl allylene, i, 484
Rhodacene, i, 516
Rhodesia, Africa, bitumen, i, 793
Richardson ball method, melting point determination, asphalt, i, 805
Richay furnace, filter plant, ii, 206
Richhill oil field, Pa., i, 110
Richland County, Ohio, oil, i, 39
Richmond oil field, Ind., i, 89
Richmond oil field, Ohio, i, 100
Ridgetop shale, i, 34
Ridgway oil field, Pa., i, 110
Rifled pipe, i, 406
Rig, California standard steel, drilling specifications, i, 211, 213
  combination, specifications, i, 271
  construction, standard, i, 223
  rotary, i, 263
  standard, i, 202, 205
Rig iron, California, regular, i, 208, 210
  imperial ideal, i, 207
  standard, i, 207
Right of way, pipe line, i, 406–407
Riley oil field, Ind., i, 89
Rimersburg oil field, Pa., i, 110
Rinards Mills oil field, Ohio, i, 100
Ring and Ball method, melting point determination, bituminous products, i, 685, 802
Rio Blanco County, Colo., bitumen, i, 791
  oil, i, 542
Ripley oil field, Okla., i, 104
Ritchie County, W. Va., grahamite (asphaltite), i, 6, 68, 791
  oil, i, 116, 569
Ritchie Mines, W. Va., grahamite, i, 6, 68, 791
Ritchie Run oil field, Pa., i, 110
Riverton oil district, Tenn., i, 108
Riveted casing, standard, i, 237
Riveted joints, ii, 265
Riveting, acid blow case, ii, 40, 77
  agitator, ii, 30, 62, 64
  bleachers, ii, 38, 72, 73
  boiler sheets, ii, 288
  condenser boxes, ii, 28, 29, 53, 54, 56, 58, 59
  filters, ii, 65
Riveting, oil tankers, i, 376
  pumping-out pan, ii, 30, 60
  steel plate work, ii, 42
  still, ii, 25, 26, 43, 44, 46, 47, 48, 50, 51
  still towers, ii, 27, 49
  tanks, ii, 31–40, 60–78, 93, 101
  tank cars, i, 361
Rivet-making furnace, ii, 497
Rivets, refinery construction, ii, 261
Road oil, ii, 365
  methods of testing, i, 588
Roane County, W. Va., oil, i, 116
Roaring Run oil field, Pa., i, 110
Robertson County, Tenn., oil, i, 111
Rochester oil field, Ind., i, 89
Rock County, Neb., oil, i, 97
Rock Haven gas field, Ky., i, 93
Rock pressure, in oil occurrence, i, 45
Rockville oil field, Pa., i, 110
Rocky Branch oil field, Ky., i, 93, 547
Rocky Mountain oil field, oil strata, i, 79
  deliveries of petroleum, i, 345, 358
  quantity and value of oil production since 1887, i, 338, 356
  stocks of petroleum, i, 343, 357
Rod packing, refinery, ii, 286
Rod wax, ii, 359
Rodemer oil field, Ky., i, 93
Rods, geologic field work, i, 179
  sucker, fishing for, i, 288
Rogers County, Okla., oil, i, 103, 554
Rolling steel plate work, ii, 42
Romney oil field, Ontario, Can., i, 122, 123
Roof, agitator, ii, 31, 63, 64
  steel tanks, ii, 107
  water-top, tanks, ii, 95
  wood, for tanks, ii, 95
Roof supports, bleacher, ii, 38, 73, 74
  tanks, ii, 32–37, 66–78, 97, 107
Roofing asphalt, i, 822
Room-and-pillar method, oil shale mining, i, 855
Rope knives, i, 282
Ropes, drilling, i, 229
  standard steel hoisting, specifications, i, 234
  steel sand lines, specifications, i, 234
Ropeseed oil, saponification number, i, 711
Rosenberry oil field, Pa., i, 110
Rosin, in grease lubricants, ii, 674
Ross and Somerset Twps. oil field, Pa., i, 110
Rose burner, ii, 450
Ross Creek oil field, Ky., i, 92, 93, 548
Rotary burner, oil, ii, 477
Rotary drilling system, i, 201, 251
Rotary fishing, i, 284
Rotary kiln furnace, filter plant, ii, 206
Rotary machine, i, 254

Rotary pipe, special upset, weights and dimensions, table, i, 306
weights and dimensions, ii, 563, 565
Rotary pumps, ii, 207, 209
Rotary retort, oil shale, i, 866
Rotary shoe, i, 286
Rotary special pipe, weights and dimensions, table, i, 306
Round reamer, i, 282
Rowan County, Ky., oil, i, 92
R. S. A. semaphore lamp, i, 692
R. S. A. semaphore lamp No. 3, Dressel burner, i, 693
R. S. A. signal hand lantern, i, 692
R. S. A. signal hand lantern, No. 1 Adlake, i, 693
Rubber packing, refinery, ii, 285, 286
Rubber solvent, ii, 361
Rumania, bailing in oil wells, i, 313
bitumen, i, 792
Bustenari, i, 130
Campina oil field, i, 47
mud volcanoes, i, 6, 131
natural gas, i, 131
oil, i, 131
oil analysis, i, 581, 582
oil, oxygen in, i, 524
oil production statistics, i, 131, 324
oil, unit of measure, ii, 761
overthrust faults, i, 68
overturned fold, i, 53
quaquaversal structures, i, 59
saline domes, i, 59, 63
salt associated with oil, i, 8
stratigraphic occurrence of oil, i, 11
volcanic plugs, i, 63
Run-down lines, refinery, ii, 242
Run-down tanks, refinery, ii, 36, 74
Russia, anticlinal and synclinal structures, i, 47
asphalt, i, 797
bailing in oil wells, i, 313
Baku, i, 50
Bibi-Eibat oil field, i, 47,68
bitumen, i, 793
bituminous lakes, i, 5
closed faults, i, 68
Grosni oil field, i, 47
laccoliths, i, 65
Lower Neocene series, i, 65
Maikop oil field, i, 65
monoclinal unconformity, i, 65
mud volcanoes, i, 6
natural gas springs, i, 74
oil, i, 127, 145
oil analyses, i, 581
oil production statistics, i, 127, 324
Russia, oil, unit of measure, ii, 761
Permian series, i, 21
rotary drilling system, i, 251
stratigraphic occurrence of oil, i, 11
Surakhany oil field, i, 50
Tcheleken, oil, i, 65
Upper Oligocene series, i, 65
Ruston and Hornsby sprayer, ii, 573
Ruston and Proctor internal combustion engine, ii, 531
Ruston modern high compression engine, ii, 534, 542
Rutylene, i, 474

S

Sabathé sprayer, engines, ii, 572
Sabine uplift, La., i, 92
Sabinene, i, 514
Sabunchi oil field, Russia, i, 127
Sac County, Iowa, gas, i, 90
Safety factors, lubrication, ii, 638, 646, 649
Safety regulations, railroad transportation of oil, i, 363
Safety valve, Diesel engine, ii, 576
steam, ii, 798
Sage Creek oil field, Wyo., i, 120
Saginaw County, Mich., oil, i, 95, 551
St. Catherine, Ontario, Can., oil, i, 121
St. Clair County, Mich., oil, i, 95
Ste. Genevieve formation, i, 31, 94
St. John's natural gas meter, i, 433
St. Louis formation, i, 94
St. Louis limestone, i, 31, 84, 108
St. Martin Parish, La., oil, i, 95, 133
St. Mary's oil field, Ohio, i, 100
St. Mary's oil field, W. Va., i, 117
St. Quirinus oil, i, 133
Sakhalin, Island of, oil, i, 143, 145
Salem oil field, W. Va., i, 117
Salina formation, i, 37
Saline domes, closed, i, 59
Louisiana, i, 60, 95
perforated, i, 63
Rumania, i, 131
Texas, i, 114
United States oil fields, i, 79
Salt, in Big lime, i, 37
melting point, ii, 806
relation to oil occurrence, i, 37
solidification point, ii, 807
Switzerland, natural gas in, i, 136
Salt Creek oil field, Wyo., i, 17, 18, 118, 1 572
Salt domes, i, 59
Salt Lake oil field, Calif., i, 16, 82, 83
Salt Run oil field, Ohio, i, 100
Salt sand, i, 15, 25, 33, 92, 103, 108, 116

Salt water, associated with oil, i, 7, 134
in Berea sand, i, 33
in Big lime, i, 37
in Blossom sand, i, 29
in Mexican oil fields, i, 126
in Mississippi lime, i, 29
in Nacatoch sand, i, 19
in Philippine Islands, i, 163
in Texas oil fields, i, 114
in Wheeler sand, i, 28
Salt wells, associated with oil, China, i, 144
Salta series, Bolivia, i, 154
Saltsburg sandstone, i, 24
"Salvy" stocks, ii, 359
Samaratian formation, i, 131, 134
Samples, oil, marking of, i, 592
Sampling, gas, i, 731
oil, i, 590
oil shale, i, 849
petroleum products, i, 592
value of in petroleum testing methods, i, 590
San Antonio oil field, Tex., i, 114, 565
San Cristobal formation, Venezuela, i, 152
San Cristobal oil field, Mexico, i, 124
San Felipe shales, Mexico, i, 63, 124
San Juan County, Utah, oil, i, 115, 567
San Luis Obispo County, Calif., bitumens, i, 791
oil, i, 80
San Marco oil field, Mexico, i, 124
San Patricio County, Tex., oil, i, 114
San Pedro formation, i, 16, 83
San Pedro oil field, Mexico, i, 124
Sand, in fire extinguishing, ii, 318
refinery use, ii, 418
specific heat, ii, 811
Sand Draw gas pool, Wyo., i, 120
Sand Hill oil field, Ohio, i, 100
Sand lines, specifications, i, 234
Sand mixtures, asphaltic, i, 817
Sand reels, i, 209
Sand rock, i, 19
Sand screens, in drilling, i, 298
Sandle oil field, Pa., i, 110
Sandoval oil pool, Ill., i, 86, 87
Sands, Louis C., i, 201
Sandstone, bituminous outcrop, i, 3, 4
Canada, i, 12
Dakota, i, 12, 20
Eagle, i, 12
Embar, i, 13, 67, 98, 119
Green River formation, i, 874
Huron, i, 88
Mississippian series, i, 29
oil occurrence in, i, 41
Oriskany, i, 121
Sandstone, Peay, i, 98
Pennsylvanian series, i, 23
Potsdam, i, 15, 121
Saltsburg, i, 24
Shannon, i, 17, 119
Stoneham, i, 36
Tensleep, i, 13, 22, 98
Torchlight, i, 98
United States oil fields, i, 79
Virgelle, i, 12, 98
Wall creek, i, 17, 119
Sandusky County, Ohio, oil, i, 102
oil, nitrogen in, i, 530
oil, sulphur in, i, 525
Sandy Creek oil field, Pa., i, 110
Sanilac County, Mich., oil, i, 95
Sanpete County, Utah, oil, i, 115
Santa Anna oil field, Tex., i, 113
Santa Barbara County, Calif., bitumen, i, 791
oil, i, 80, 540
oil, nitrogen in, i, 530
oil, sulphur in, i, 526
Santa Clara, Mex., bitumen, i, 792
Santa Clara County, Calif., oil, i, 80, 540
Santa Clara Valley oil field, Calif., oil, i, 17, 83
Santa Cruz County, Calif., bitumen, i, 791
Santa Margarita formation, i, 16, 68, 83
Santa Maria oil field, Calif., i, 17, 81, 82, 83, 540
Santa Paula oil field, Calif., i, 81, 82, 541
Santene, i, 510
Santo Domingo, West Indies, glance pitch (manjak), i, 792
oil and gas, i, 158, 161
Saponification number of oils, method of determination, i, 588, 712
Sapulpa sand, i, 30
Saratoga chalk, i, 19
Saratoga oil field, Tex., i, 113, 564
Sarawak, British Borneo, oil, i, 162
Sarnia oil field, Ontario, Can., i, 122, 123
Saskatchewan, Can., oil shale, i, 834, 842
Saturated ammonia vapor, pressure and temperature, ii, 839
Saturated brine, boiling point, ii, 806
Saturated hydrocarbons, i, 520
Saturated polycyclic hydrocarbons, i, 522
Saturated steam, properties, table, ii, 793
Saturation, air, table, ii, 787
Say oil field, Pa., i, 110
Saybolt chromometer, color readings, i, 662, 663, 664
Saybolt furol viscosimeter, i, 643, 648, 654, 658
Saybolt improved wax tester, i, 679

Saybolt lamp, No. 2 Sunhinge burner, i, 691, 693
Saybolt petrolatum method, melting point determination, i, 684
Saybolt standard viscosimeter, i, 645
Saybolt thermo viscosimeter, i, 643, 644, 648, 649
Saybolt universal flash tester, i, 624
Saybolt universal viscosimeter, i, 643, 644, 645, 654, 661
Saybolt viscosity, relation to absolute, ii, 628
Saybolt wax method, melting point determination, i, 679
Saxonburg oil field, Pa., i, 110
Scale wax, ii, 372
sweating, ii, 374
Scarfing, steel plate work, ii, 42
Schmid's formula, cylinder oil feed, ii, 663
Schriver oil field, Ohio, i, 100
Schulter oil field, Okla., i, 104
Scio oil field, Ohio, i, 100
Scotland, bitumen, i, 67, 135, 792, 834, 835
oil and oil shale, i, 67, 135, 834, 835
pipe line in, i, 380
shale oil analyses, i, 579
Scotland beds, Barbados, i, 161
Scott County, Ark., asphalt, i, 80
impsonite, i, 791
Scott County, Tenn., oil, i, 108, 558
Scott-Griffith oil field, Ohio, i, 100
Scott method, cementing oil wells, i, 296
Scott stilling system, ii, 345
Scottsville oil field, Ky., i, 93
Screening oil wells, i, 297
Screens, oil shale crushers, i, 860
Screw threads, bolts, ii, 269
Screwed fittings, general dimensions, ii, 278
Scum, as surface indication of oil, i, 3
Sea lines, oil transportation, i, 416
Sea water, boiling point, ii, 806
Seamless interior upset drill pipe, weights and dimensions, ii, 565
Seamless pipe, ii, 502, 504
Seamless steel tubing, weight, table, ii, 785
Sebastian County, Ark., gas, i, 80
Second Booch sand, i, 30
Second counters, use of with viscosimeters, i, 647
Second Cow Run sand, i, 25
Second gas sand, i, 19
"Second" sand, i, 34, 35
Second Wall Creek sand, i, 119
Sedan oil field, Kans., i, 91
Sedimentary rocks, oil in, i, 64, 67
Seepages. *See* Oil seepages.
Seepages, relation to oil occurrence, i, 3, 47
Seger cones, fusion points, ii, 804
Selma Chalk, i, 12, 19
Seminole County, Okla., oil, i, 557
Seneca County, Ohio, oil, i, 102
Seneca oil, i, 359
Sensible heat, ii, 824
Separating towers, refinery, ii, 184
Separators, refinery, ii, 243
Serbia, bitumen, i, 792
oil shale, i, 834, 846
Sespe formation, i, 12, 17, 83
Sespe oil field, Calif., nitrogen in, i, 530
sulphur in, i, 526
Sesquiterpene, i, 514
Sesquiterpene cardinene, i, 514
Sesquiterpene (tricyclic), i, 514
Setting, concrete, ii, 251
"Setting up," plane table, i, 193
Settling, strata, oil geology, i, 170
Seven Lakes oil field, New Mexico, i, 99
Sevier County, Ark., bitumen, i, 791
gas, i, 80
Sevins oil field, Pa., i, 110
Sewers, refinery, ii, 242
Sexdecylthiophane, i, 528
Shackelford County, Tex., Cisco formation, i, 28
oil, i, 561
Strawn formation, i, 29
Shafting, cold rolled, tensile strength, elastic limit, elongation, ii, 822
horsepower, tables, ii, 302
Shale. *See* Oil shale.
oil in, i, 67
United States oil fields, i, 79
Shale oil, analyses, i, 579
distillation tests, i, 879, 891, 893
physical properties, i, 878, 890, 893
pyridine bases, i, 881
refining, i, 870
relation of yield to gravity, chart, i, 889
residuum, cracking, i, 879
Shallow oil field, Ill., i, 86
Shambur oil field, Pa., i, 110
Shamrock oil field, Okla., i, 104
Shannon oil field, Wyo., i, 12, 17, 120, 572
Shannopin oil field, Pa., i, 110
Sharples super centrifuge, ii, 377, 389
Shear reinforcement, ii, 252
Shearing, acid blow case, ii, 40
acid tank, ii, 39
bleachers, ii, 38
steel plate work, ii, 42
tanks, ii, 32, 33, 34, 36, 37
Sheet asphalt, i, 817
Sheet packing, refinery, ii, 285
Sheets, black steel, table of weights, ii, 778
corrugated, table of weights, ii, 780

Sheets, iron and steel, weights, ii, 773
standard galvanized, table of weights, ii, 779
Sheets Run oil field, Ohio, i, 100
Sheffield oil field, Pa., i, 110
Shelburn oil field, Ind., i, 89
Shelbyville oil field, Ind., i, 89
Shell, boiler, refinery, ii, 288
filter, ii, 65
still, ii, 25–27, 43–51, 107
Shellhammer oil field, Pa., i, 110
Shensi series, China, i, 144
Shepard oil field, Okla., i, 104
Sherad oil field, Wyo., i, 120
Shetland oil field, Ontario, Can., i, 122
Shingles, asphalt, i, 824
Shinglehouse oil field, Pa., i, 110
Shinnston oil pool, W. Va., i, 117, 568
Ship classification rules, oil tankers, i, 376
Shipment, natural-gas gasoline, i, 757
refined products, ii, 12
Shipper, tank car, obligations of, i, 367
Shipping, fire risks, ii, 413
Shipping buildings, refinery, ii, 232
Shoe, fishing tool, i, 289
Shoes, rotary, i, 262, 286
Shooting oil wells, i, 290, 297
Shore lines, relation to occurrence of petroleum, i, 45, 65
Shreveport gas field, La., i, 96
Shreveport sand, i, 18
Shrinkage stopes method, oil shale mining, i, 856
Siam, oil i, 145
Siberia, oil, i, 137
Sicily Island oil field, La., i, 96
Sicily, Italy, bitumen, i, 792
Siggins oil field, Ill., i, 87
Silencer, Diesel engine, ii, 579
Silica, specific heat, ii, 811
Silica gels, refinery use, ii, 368
Silliman's discovery, cracking, ii, 324
Silocel, insulating brick, ii, 282
Silurian system, Canada, i, 121
geographic distribution, i, 11
Kentucky, i, 94
Ohio, i, 102, 103
oil-bearing formations in, i, 13, 15, 38
Sweden, i, 136
United States oil fields, i, 79
Silurian-Devonian system, oil strata, i, 37
Silver, latent heat of fusion, ii, 823
melting point, ii, 806, 807
specific heat, ii, 811
Silver leaf, radiation and reflection of heat, ii, 824
Simplex refining process, ii, 385
Sinking oil field, Ky., i, 93, 548
Single stage crushing, oil shale, i, 858
Sistersville oil field, Ohio, i, 100
Sistersville oil field, W. Va., i, 117
Sixth sand, i, 36, 108, 116
Skelp, pipe, ii, 502, 503
Skiatook oil field, Okla., i, 105, 554
Skimming, refinery, ii, 323, 324, 326
Skimming plant, ii, 18
Skinner sand, i, 28
Slack wax, ii, 372
barrel, transportation, ii, 375
sweating cuts, ii, 374
Slatelick oil field, Pa., i, 110
Slickford oil field, Ky., i, 33, 93, 94
Slide rule, for flow capacity, i, 403
Sligo oil field, Pa., i, 110
Slip socket, i, 278
Slippery Rock oil field, Pa., i, 110
"Sludge" acid, ii, 368, 402
Sludge acid residues, utilization, ii, 404
Sludge tests for transformer oils, i, 705
Slush pump, rotary, i, 258
Sluss oil field, Kans., i, 91
Smethport oil field, Pa., i, 110
Smith, A. D.,ii, 1
Smith, R. G., i, 787
Smith-Dunn process of air lift for oil wells, i, 315
Smiths Falls oil field, Ontario, Can., i, 122
Smiths Ferry oil field, Pa., i, 110
Smock oil field, Kans., i, 91, 545
Smudge oil, ii, 342
Snee sand, i, 35
Snow-Diesel engine, ii, 591
Snyder oil field, Ohio, i, 100
Soap Creek oil field, Mont., i, 97
Soap stock, mineral, ii, 359
Socket, combination, i, 288
friction, i, 289
Soda ash, refinery, ii, 368
Soda soap, in lubricating greases, ii, 673
Sodium, melting point, ii, 807
Sodium carbonate, refinery use, ii, 368
Sodium chloride solutions, ii, 848
Sodium hydrate, refinery use, ii, 368
Sodium hydroxide solution, concentration, table, ii, 846
Sodium plumbate for sweetening, ii, 336
refinery use, ii, 334, 368
Sodium silicate, refinery use, ii, 372
Soda, specific heat, ii, 811
Sodom oil field, Pa., i, 110
Soft paraffin, ii, 380
Soft sand, i, 41
Soft wax press, ii, 199
Solar oil, condenser coil data, ii, 178

Solid asphalt, method of testing, i, 588
Solid injection, fuel engines, ii, 571, 572
"Solid" point, oil, i, 661
Solidification method, melting point determination. i, 681
Solids, expansion, ii, 813
specific heat, ii, 811
Solubility, paraffin wax, i, 538
petroleum, i, 535
solid hydrocarbons, i, 536
Solubility in CS_2, petroleum **products**, i, 588
Solubility tests, asphalt, i, 806
petroleum products, i, 715
Solvent, rubber, ii, 361
Somaliland, oil, i, 158
Somerfeld on friction, ii, 633
Somerset oil field, Tex., i, 113, 114
Somerset township oil field, Pa., i, 110
Sour distillates, ii, 334
Sourlake oil field, Tex., i, 113, 564
South Africa, oil, i, 158
oil shale, i, 834
South America, bitumen, i, 792
oil, i, 146
oil shale, i, 842
pipe lines, i, 380
rotary drilling system, i, 251
South American oil fields, anticlinal and synclinal structures, i, 47
South Atlantic oil district, Argentina, i, 149, 150
South America, i, 146
South Bosque oil field, Tex., i, 113, 560
South Dakota, natural gas production and consumption statistics, i, 354, 356
oil and gas, i, 111
South Elgin oil field, Okla., i, 104
South Oregon Basin gas pool, Wyo., i, 120
"South Islands," La., salt domes, i, 61
Southern Nigeria, oil, i, 158
South Penn casing, i, 236
weights and dimensions, ii, 558
South Petroleum oil field, Ky., i, 93
Southern oil fields, Mexico, standard casing requirements, i, 221
Soya bean oil, saponification number, i, 711
Spain, asphalt and bituminous rock, i, 797
bitumen, i, 792
oil analysis, i, 581
oil and asphalt, i, 136
oil shale, i, 834, 846
Sparta oil field, Ill., i, 87
Spear, casing, i, 287
Specific gravities of hydrocarbons, i, 460
Specific gravity, ii, 594, 830
asphalt, i, 799
Specific gravity, corrections for temperature, table, ii, 835, 837
gasoline, determination, i, 752
natural gas, determination, i, 751
oil equivalents, weight and volume, ii, 833, 834
use of correction tables for temperature, i, 601
Specific heat, ii, 524, 810
gases, ii, 528, 811
liquids, ii, 812
of absorption of oils, i, 742
oil, i, 531
solids, ii, 811
woods, ii, 811
Speechley sand, i, 36, 108
Speed-friction experiments, lubricants, ii, 631
Speed regulation, internal combustion engine, ii, 547
Speed regulator, pump, ii, 461
Speller, F. N., ii, 681
Spencerville oil field, Ohio, i, 100
Spent shale, analysis of ash, i, 872
heat recovery, i, 870
use of carbon content, i, 870
Sperm oil, saponification number, i, 711
Spermaceti, latent heat of fusion, ii, 823
melting point, ii, 806
Spindle oils, methods of testing, i, 588
characteristics, ii, 677
Spindletop oil field, Tex., i, 59, 113, 114, 252, 565
Spindletop oil field, Wyo., i, 120
Spitzbergen, oil shale, i, 848
Splicing, cordage, i, 235
Split gland, cast iron, ii, 108
Spray ponds, refinery, ii, 311
Sprayers, internal combustion engines, ii, 530, 544, 571
Spring Valley oil field, Wyo., i, 120, 573
Spud, i, 282
Spudders, standard drilling, i, 250
Spudding, i, 240
Spurrior oil district, Tenn., i, 108
St. Clair County, Mich., oil, i, 551
Squaw sand, i, 32, 94, 116
Squirrel sand, i, 28, 106
Stability, lubricants, ii, 643, 672
Stadia reduction tables, i, 180
Stadia rod, i, 179
Stadia slide rule, i, 180
Stadia traversing method, mapping, i, 197
Stadler sand, i, 39
Stairway, tank, ii, 107
Standard buildings, refinery, ii, 228
Standard drilling outfit, specifications, i, 216
Standard drilling system, i, 203

Standard fittings, dimensions, ii, 278
Standard marine gasoline engine, ii, 615
Standard pipe, internal fluid pressure, ii, 531
  weights and dimensions, ii, 553
Standard rig, construction plan, i, 204
Standard rig iron, specifications, i, 207
Standard steel tankage, ii, 87
Standard-tool drilling outfit, i, 202
Standard welded pipe, specifications, ii, 549
Standard white distillate, ii, 339
Standard white oil, i, 663
Standard wrought pipe, dimensions, table, ii, 782
Standardizing specifications, combination rig, i, 270
Starr oil field, Ohio, i, 100
Stationary boiler, oil equipment, ii, 463
Station Camp oil field, Ky., i, 92, 93
Station, in mapping, i, 176, 195
Stationary engine, four-cycle horizontal, ii, 531
  Diesel, ii, 576
  two-cycle, internal combustion, ii, 537
Stationary retort, oil shale, i, 866
Statistics of petroleum and natural gas production, i, 321–358
Staunton gas pool, Ill., i, 87
Steam, ii, 793–799
Steam and air refined asphalt, combined process, i, 788
Steam boilers, chimneys for, ii, 308
  refinery, ii, 287
Steam condensation, Bodmer's formula, ii, 659
Steam cylinder lubrication, ii, 651
Steam distillation, ii, 343
Steam engine, refinery, ii, 293
Steam flow meter, oil heater, ii, 475
Steam lines, refinery, ii, 242
Steam pipe, standard dimensions, table, ii, 782–784
Steam-refined asphalt, batch method, i, 788
  processes, i, 787, 790
Steam-refined stock, ii, 364, 380
Steam reducing, refinery, ii, 323, 354
Steam requirements, skimming plant, ii, 18
Steam still, refinery, ii, 26, 47, 151
Steam still condenser box, refinery, ii, 57
Steam still products, ii, 360
Steam still towers, refinery, ii, 27, 49, 184
Steam stilling, ii, 14, 323
Stearic acid, melting point, ii, 807
Stearine, melting point, ii, 806
Stebinger drum, i, 180, 195
Steel, corrosion, ii, 570
  Crane cast, elastic limit, ii, 820
  Crane cast, tensile strength, ii, 820
Steel, effect of cold, ii, 816, 817
  expansion, ii, 814
  flat rolled, weight, table, ii, 776
  melting point, ii, 807
  protective coatings, ii, 570
  radiation and reflection of heat, ii, 824
  specific heat, ii, 811
  tensile strength at high temperature, ii, 816
Steel and tower specifications, skimming plant, ii, 24
Steel castings, refinery, ii, 279
Steel foundry, oil fuel, ii, 505
Steel pipes, ii, 501
Steel plate, refinery construction, ii, 261
  weight, table, ii, 775, 778
Steel plate construction, skimming plant, ii, 23
Steel plate work, flux and lubricants, refinery ii, 41
Steel sheets, weights, ii, 773
Steel specifications, tanks, ii, 116
Steel tanks, costs, ii, 126
  refinery, ii, 92
Steel tubes, thermal expansion, ii, 814
Steel tubing, seamless, weights, table, ii, 785
Steffy oil field, Ky., i, 93
"Step," geologic field work, i, 182, 195
Stephens County, Okla., bitumen, i, 791
  grahamite, i, 71
  oil, i, 103, 557
Stephens County, Tex., Bend limestone, i, 31
  oil, i, 114, 561
Steuben County, N. Y., oil, i, 99
Steubenville oil field, Ky., i, 93
Stewart's formulæ, collapsing pressures, pipes, ii, 505
Stickers, pipe threading, ii, 536
Stilbene, i, 516, 536
Still, crude, ii, 44
  dry coil, ii, 345
  filter wash, ii, 51
  gasoline extraction plants, i, 775
  heat exchanging, ii, 50
  hog, ii, 345
  refinery, ii, 25, 42, 44, 91, 140, 213, 232
  sweetening, ii, 334
  tar, ii, 329
  tower, ii, 325
Still coke, ii, 330
Still condenser box, skimming plant, ii, 28
Still gas, ii, 337, 338
Still products, reducing, ii, 360
  steam, ii, 360
Still wax, ii, 329
Stilling, fire precautions, ii, 406
Stillwater oil field, Ky., i. 92

Stocks, held by pipe line and marketing companies, i, 349
used for refinery products, ii, 359
Stoker setting, refinery, ii, 235
Stollmeyer sand, Trinidad, i, 160
Stollmeyer Cruse shale, Trinidad, i, 160
Stone, specific heat, ii, 811
Stone Bluff oil field, Okla., i, 104
Stone work, refinery construction, ii, 248
Stoneham sandstone, i, 36
Storage, natural-gas gasoline, i, 757
refinery yard, ii, 84
skimming plant, ii, 19
Storage tank, concrete, ii, 127
refinery, ii, 31, 32, 34, 35, 65, 66, 67, 68, 69
Storehouse, refinery, ii, 228
Story County, Iowa, gas, i, 90
Stove distillate, ii, 340
"Stovepipe" casing, standard, i, 236, 237
Straight shot burner, ii, 450
Strainers, in drilling, i, 298
Strapping, tank, ii, 134
Stratigraphic intervals, i, 170
Stratigraphic occurrence of petroleum, i, 11
Stratigraphy, relation to occurrence of petroleum, i, 11
Strawn oil field, Tex., i, 29, 113
Stray sand, i, 34, 92, 94
"Stray" Third sand, i, 35
Street construction, asphalt, i, 816
Strength, tubes, pipes, cylinders, ii, 519
Stress, steel pipes, ii, 519
Strike, determination of, i, 182
Stringing, pipe line, i, 408
Stripping, refinery, ii, 323, 324, 326
Stroke, internal combustion engine, ii, 543
Stroud oil field, Okla., i, 104
Structural classification, oil and gas, i, 45
"Structural contour lines," oil geology, i, 9
Structural habits, oil and gas fields, i, 72
Structural steel, specification for building, ii, 116, 121
Structural theory, oil occurrence, i, 44
Structural types, United States oil fields, i, 79
Structure contour maps, i, 171
Stub's steel wire gage, ii, 272
Stuffing box, cable, ii, 111
Stumptown oil field, Ohio, i, 100
Styrol, i, 482
Sub-aclinal structures, i, 45, 46
"Sub-Clarksville" sand, i, 19
Suberane, i, 502
Sub-level caving method, oil shale mining, i, 856
Submarines, Diesel engine and tests, ii, 602
Subsurface mapping, oil geology, i, 172
Subsurface structure, oil geology, i, 174
Sucker Rod oil field, Pa., i, 110
Sucker rods, fishing for, i, 288
Suction lift, pumps, ii, 210, 212
Suction pipe, oil heater, ii, 460
Suez Canal, oil tanker tonnage, i, 376
Sugar, as lubricant, ii, 643
Sugar Creek oil field, Pa., i, 110
Sugar Grove-Gibsonville gas field, Ohio, i, 100
Sugar refinery, fuel oil equipment, ii, 477
Sugar Run oil field, Pa., i, 110
Sullivan County, Ind., oil, i, 88
Sullivan oil field, Ind., i, 89
Sulphate of lime, specific heat, ii, 811
Sulphur, boiling point, ii, 806, 807
determination by bomb method, i, 588, 709
heat of combustion, ii, 809
in fuel oil, ii, 340
in oil, i, 458, 525
in saline domes, i, 61
indication of oil, i, 3, 8
lamp test, i, 588, 707
latent heat of fusion, ii, 823
melting point, ii, 807
refinery use, ii, 368
specific heat, ii, 811, 812
Sulphur bearing asphalts, i, 527
Sulphur dioxide, expansion by heat, ii, 812
refinery use, ii, 392
specific gravity, ii, 832
Sulphur trioxide in fuming sulphuric acid, table, ii, 845
Sulphur water, associated with oil, i, 67
Sulphuric acid, boiling point, ii, 806
expansion by heat, ii, 812
fuming, content, table, ii, 842
gravity and strength, table, ii, 841–844
lubricant, ii, 643
recovery, ii, 402
refinery use, ii, 368, 392
specific heat, ii, 812
sweetening, ii, 336
temperature allowance, table, ii, 841
use, ii, 334
Sulphuric anhydride, ii, 411
refinery use, ii, 368
Sulphurous acid, ii, 411
melting point, ii, 806
refinery use, ii, 368
Sumatra, oil, i, 161, 162
analyses, i, 585
Summerfield gas pool, Ohio, i, 100
Summerland oil field, Calif., i, 16, 81, 82, 540
nitrogen in, i, 530
sulphur in, i, 526
Summit County, Utah, oil, i, 115
Sumner County, Tenn., oil, i, 111
Sump holes in oil drilling, i, 248

Sunbury shale, relation to Berea sand, i, 32
Sundance formation, i, 12, 20, 119
Sunnybrook oil field, Ky., i, 93
Sunset oil field, Calif., i, 82, 83, 542
Superheated steam, properties, table, ii, 795, 796
specific heat, ii, 811
Super heater, refinery, ii, 292
Surakhany, Russia, oil analysis, i, 581
Surakhany oil field, Russia, i, 27, 50
Surface indications of petroleum, i, 3, 167
Surface mapping, i. 172
Surface structures, petroleum geology, i, 174
Surveys, pipe line, i, 406
Sviatoi oil field, Russia, i, 127
Swab, oil well, i, 313
Swabbing, oil well, i, 313
Sweating, refinery, ii, 373
wax plant, ii, 199
Sweating building, refinery, ii, 231
Sweating pan, wax plant, ii, 199
Sweden, oil and gas, i, 136
oil shale, i, 848
Swedge, casing, i, 288
Sweet distillate, ii, 369
Sweetening, ii, 323, 324, 334
Sweetwater County, Wyo., oil, i, 573
Swing pipe, bleachers, ii, 38, 72, 73
tanks, ii, 32–37, 66–75, 107
Swing-pipe winch, steel tanks, ii, 107
"Swinging nipple," use with gushers, i, 312
Switzerland, bitumen, i, 792
oil and gas, i, 136
Swivel, rotary, i, 261
Swivel joints, oil burners, ii, 480
Sycamore oil field, Ohio, i, 100
Sylvestrine, i, 510
Sym-diphenyl ethylene, i, 516
Synclinal structures, oil geology, i, 45, 47
Synclines, United States oil fields, i, 79
Syria, oil, i, 144

T

Tabasco oil field, Mexico, i, 124
Tachira formation, Venezuela, i, 152
Taft oil field, Calif., standard casing requirements, i, 221
Tag closed tester, flash point determination, oil, i, 624, 631
Tagliabue, Baumé gravity, i, 595
Tagliabue open tester, for oil, i, 623
Tabliabue temperature control apparatus, ii, 374
Talc, grease filler, ii, 674
Tallow, melting point, ii, 806
saponification number, i, 711
Tallow oil, acidless, ii, 658
Tamaulipas, Mexico, bitumen, i, 792
Tamasopo limestone, Mexico, i, 12, 20, 63, 124, 126
Tampalachi, Mexico, oil seepage, i, 64
Tampico oil field, Mexico, i, 124
Tamping, in shooting well, i, 297
Taneha-Jenks-Red Fork-Glenn Pool oil field, Okla., i, 104
Tanhuijo oil field, Mexico, i, 124, 125, 574
Tank, acid, refinery, ii, 36, 76
air, Diesel engine, ii, 576
air, high pressure, ii, 78
air, refinery, ii, 78
capacity formula, ii, 99
capacity table, ii, 770, 772
filling, refinery, ii, 75
fire protection of, ii, 316
lye, refinery, ii, 40, 77
oil, locomotive, ii, 487
refinery, ii, 91, 212, 239
run-down, refinery, ii, 74
six-ring steel, ii, 102, 104, 105
steel, refinery, standard dimensions, ii, 92
steel roof, specifications and contract, ii, 101
storage, concrete, ii, 127
storage, refinery, ii, 31, 32, 34, 35, 65
working, refinery, ii, 70, 71, 72
Tank car shipments, natural-gas gasoline, i, 757
Tank cars, asphalt, i, 814
specifications, i, 359, 361
Tank heads, tank cars, i, 362
Tank measurements, ii, 134, 136
Tank spacing, refinery storage, insurance rates, ii, 85
Tank steamer, i, 371
Tank steel specifications, standard, ii, 116
Tank wagons, i, 359
Tankage, concrete, refinery, ii, 86
skimming plant, ii, 19
standard steel, ii, 87
underground, refinery, ii, 86
Tape, steel, i, 181
Tar, in cracking, ii, 432
in lubricants, ii, 661
precipitation test, petroleum products, i, 588
refinery, ii, 341
Tar agitators, ii, 186, 191
Tar plug, crude stills, ii, 43
Tar sand, Kentucky, i, 94
Tar sands, Canada, i, 5, 123
Tar springs, Athabaska, Can., i, 121
Tar stills, ii, 329
Tar tests, for petroleum products, i, 714
Tar treating, refinery, ii, 371

Tarentum oil field, Pa., i, 110
Tarry solids, in cylinder oils, ii, 662
Tasmania, bitumen, i, 793
oil shale, i, 166, 834, 847
Taylor County, Tex., Canyon formation, i, 28
Cisco formation, i, 28
Taylor County, W. Va., oil, i, 116
Taylorstown oil field, Pa., i, 110
Tcheleken oil field, Russia, i, 127
Teeter oil field, Kans., i, 91
Tehuantepec oil field, Mexico, i, 124, 125
Tejon formation, i, 12, 17, 83
Telescopic hand level, i, 179, 182
Telfair County, Ga., oil, i, 84
Temperanceville oil field, Ohio, i, 100
Temperature, adsorption, relation to pressure and boiling point, i, 748
bearing lubrication, ii, 640
conversion tables, ii, 801
cylinders, lubrication, ii, 653
determination by color, tables, ii, 804, 806
lubricants, ii, 632, 646
lubricants, relation to viscosity, ii, 632
observed, i, 598
steam, ii, 794
Temperature control, sweating, ii, 374
"cracking" oil shale, i, 864
Temperature corrections, Baumé gravity, petroleum, table, ii, 836, 838
specific gravity, table, ii, 835, 837
Temperature curves, viscosity steam cylinder oils, ii, 654
Tempering, filters, ii, 378
Tenacity, metals, ii, 815
Tennessee, Chattanooga shale, i, 34
Mississippian series, i, 29
natural gas, consumption and production statistics, i, 352, 354
oil, i, 111
oil production statistics, i, 331, 356
oil shale, i, 834, 839
Tensile strength, iron and steel, ii, 816
pipes, ii, 527
Tension test, boiler steel, ii, 144
pipe, ii, 551
structural steel, ii, 116, 122
Tensleep sandstone, i, 13, 22, 98
Tepetate oil field, Mexico, i, 124, 125
Terpene isomers, i, 514
Terpene (monocyclic), i, 510
Terpenes, i, 514
oxidation, i, 458
Terpinene, i, 510
Terpinolene, i, 510
Terraces, structural, i, 55, 73
United States oil fields, i, 79
Terrebonne gas field, La., i, 95, 96
Terre Haute oil field, Ind., i, 89, 544
Tertiary sediments, contact **with igneous** rocks, i, 64
volcanic plugs, i, 63
Tertiary system, Alaska, i, 80
Borneo, i, 162
California, i, 82, 83
geographic occurrence, i, 11
Greece, i, 137
Idaho, i, 86
Louisiana, i, 95
Mexico, i, 124
New South Wales, i, 164
oil shale, i, 873, 890
oil strata, i, 12, 14, 16
Persia, i, 142
Porto Rico, i, 161
Queensland, i, 166
Switzerland, i, 136
Tennessee, i, 111
Texas, i, 114
Trinidad, i, 159, 160
United States oil fields, i, 79
Venezuela, i, 152
Test, absorption, gasoline in natural gas, i, 761
charcoal, gasoline in natural gas, i, 763
compression, gasoline in natural gas, i, 760
internal combustion engine, ii, 548
pipe, ii, 526
pipe, flattening, ii, 551
pipe, hydrostatic, ii, 551
pipe, tension, ii, 551
tank cars, i, 364
tank steel, ii, 116
Vickers submarine Diesel engine, ii, 604
Testing methods, petroleum, i, 587
petroleum products, i, 589
specific gravity, i, 594
Testing, oil tankers, i, 376
Tetracosane, i, 470
Tetradecanaphthene, i, 500
Tetradecane, i, 468, 470, 533
Tetradecylene, i, 476
Tetraethyl benzol (sym 1, 2, 4, 5), i, 488
Tetrahydro-cardinene, i, 514
Tetrahydroferulene, i, 514
Tetrahydronaphthalene, i, 508
Tetramethyl benzol, i, 484
Tetramethyl-1, 2, 3, 4 benzol, i, 484
Tetramethyl butane, i, 466
Tetramethyl ethylene, i, 474
B-tetramethyl menthene, i, 478
Tetramethyl methane, i, 460
Tetramethyl propane, i, 464
Tetraphenyl ethane, i, 518
Tetraphenyl ethylene, i, 518
Tetratriacontane, i, 472

Texas, alluvium, i, 12
Amarillo gas fields, i, 21
Annona chalk, i, 19
Anticlinal and synclinal structures, i, 47
Austin group, i, 19
Bend limestone, i, 13, 31
bitumen, i, 791
Breckenridge oil field, i, 31
Burkburnett oil field, i, 28, 320
Burkburnett oil field, drilling system, i, 202
Caddo (Stephens County) oil field, i, 31
Cambrian system, i, 40
Canyon formation, i, 28
casing (rotary drilling), requirements, i, 221
Cisco formation, i, 28
coastal, oil production statistics, i, 337, 356
Desdemona oil field, i, 31
Duke sand, i, 31
Edwards limestone, i, 12
Electra oil field, i, 21, 22, 28, 47
Gordon sand, i, 31
Jones sand, i, 31
Marble Falls limestone, i, 31
Mexia oil field, i, 20, 60
monoclinal noses, i, 55, 60
mud volcanoes, i, 7
Nacatoch sand, i, 12
natural gas consumption and production statistics, i, 352, 354
natural-gas gasoline, statistics, i, 786
northern and central, oil production statistics, i, 336, 356
oil, i, 114
oil, nitrogen in, i, 530
oil production, per well per day, i, 349
oil production statistics, i, 341, 356
oil, refining, ii, 79
oil shale, i, 848
oil strata, i, 12
oil, sulphur in, i, 526
oil, sulphur with, i, 8
oil, typical distillation, ii, 355
oil, typical gravities, continuous stills, ii, 328
oil wells, number of, i, 349
Pennsylvanian series, i, 22, 28
Permian series, i, 21
Petrolia oil field, i, 22, 59
Quaternary system, i, 12, 16
Ranger oil field, i, 31
rotary drilling system, i, 251
saline domes, i, 59
salt with oil, i, 8
Saratoga chalk, i, 19
Spindle top well, i, 252
Strawn formation, i, 29
Texas, terraces, structural, i, 57
Trinity sand, i, 90
Veale sand, i, 31
Wichita County oil fields, i, 21, 22
Wichita formation, i, 22
Woodbine sand, i, 12, 20
Texas canvas packer, i, 293
Thamesville oil field, Ontario, Can., i, 122, 123
Thayer oil field, Kans., i, 91, 546
Thermal capacity, ii, 810
Thermal efficiency, ii, 530
Diesel engine, ii, 560, 582
Thermal units, ii, 808
Thermo dynamics, ii, 523
Thermometer, scales, comparison, ii, 832
Cleveland open cup tester, i, 627
flash and fire test, oil, i, 626
specifications, Pensky-Martens flash tester, i, 634
specifications, Saybolt Universal Viscosimeter, i, 644
specifications, Tag closed tester, i, 630
testing oil, i, 592
viscosity determination, oil, i, 626, 643
Thief, oil, i, 424
Thiophanes, in petroleum, i, 527
Third Mountain sand, i, 33
"Third" sand, i, 35
Thirty-foot sand, i, 33
Thirty oil, ii, 357
Thomas electric natural-gas meter, i, 435
Thomas Fork oil field, Ohio, i, 100
Thomas Morse Liberty gasoline engine, ii, 618
Thorn Creek oil field, Pa., i, 110
Thrall oil field, Tex., i, 113, 114, 563
Threading, pipe, ii, 550, 552
Threads, dies, ii, 544
standard, for tubular goods, i, 301
Three-four oil field, Okla., i, 104
Three-throw pumps, ii, 207
Throat, threading die, ii, 539
Thurston gas field, Ohio, i, 100
Tidioute oil field, Pa., i, 110
Tiffin oil field, Ohio, i, 100
Tiger Flats oil field, Okla., i, 104
Tiger Mountain gas field, Okla., i, 104
Tilbury oil field, Ontario, Can., i, 122, 123
Timor, bitumen, i, 793
oil, i, 163
Tin, latent heat of fusion, ii, 823
melting point, ii, 807
radiation and reflection of heat, ii, 824
specific heat, ii, 811, 812
Tin plate, weights and gages, table, ii, 779
Tiona sand, i, 36, 108
Titicaca oil field, Peru, i, 148

Titusville oil field, Pa., i, 110
Tolane, i, 516
Toluol (toluene), i, 480
hydrocarbon solvent, i, 536
Tonnage, oil tankers, i, 374
Tool joints, rotary, i, 259, 262
standard, i, 225, 230
Tools, fishing, i, 274
pipe line, i, 409
Topila oil field, Mexico, i, 124, 125, 574
drilling methods, i, 202
Top-slicing caving method, oil shale mining, i, 856
Topography, oil geology, i, 169
Topping, refinery, ii, 323, 324, 326, 328
Topping plants, ii, 82, 83
Tops, refinery, ii, 324, 338
Torchlight sandstone, i, 98
Toronto gas pool, Ohio, i, 100
Toronto oil field, Kansas, i, 91
Total immersion thermometers, i, 592
Tower bridge, condenser boxes, ii, 52, 54, 56
Tower stills, ii, 213, 325, 333
Towers, refinery, ii, 92, 184, 219
steam still, ii, 49
Towl, Forrest M., i, 359
Toyah oil field, Tex., i, 113
Tractor engines, horse-power formula, ii, 622
Trammel Creek oil field, Ky., i, 93
Transformer oils, sludge tests, i, 705
Transmission machinery, oil shale plant, i, 861
Transmission of heat, air to water, ii, 829
steam to water, ii, 828, 829
Transmitted light for petroleum examination, i, 662
Transportation, methods for petroleum, i, 359
natural gas, i, 432
natural-gas gasoline, i, 757, 759
pipe-line regulations, i, 428
wax, ii, 375
Transportation losses, pipe line, i, 425
Transverse framing system, tank vessel construction, i, 373
Transylvania, bitumen, i, 792
oil, gas, and mud volcanoes, i, 131
Trap rock, oil in, i, 66
Traps, oil well, i, 313
refinery, ii, 242
Traverses, i, 177, 198
Traversing, in mapping, i, 195
Travis County, Tex., oil, i, 562
Treating, fire precautions, ii, 409
refinery, ii, 367
typical time schedule, ii, 370
Trenton limestone, i, 15, 39, 40, 79, 86, 88, 102, 103, 121
sulphur in, i, 526
Trenton series, i, 13, 15, 94
Triacontane, i, 472
Triangulation, mapping, i, 197
Triassic system, geographic distribution, i, 11
Jugo-Slavia, i, 134
New South Wales, i, 164
North Carolina, i, 102
oil shale, i, 834, 846
oil strata, i, 12, 14
Queensland, i, 166
South Africa, i, 158
Wyoming, i, 119
Trickham oil field, Tex., i, 113
Tricosane, i, 470
a-Tricyclodecane, i, 522
b-Tricyclodecane, i, 522
Tridecane, i, 468
Tridecylene, i, 476
Triethyl benzol (sym), 1-, 3-, 5-, i, 488
Triethyl methane, i, 464
Trimethyl, 1-, 1-, 5, i, 478
Trimethyl, anthracene, 1, 3, 6, i, 514, 516
Trimethyl, anthracene, 1, 2, 4, i, 516
Trimethyl, anthracene, 1, 4, 6, i, 514
Trimethyl benzol, i, 482
Trimethyl 1-7-7-bicyclo-1-2-2-heptane, i, 522
Trimethyl 2, 7-, 7 bi-cyclo 1-, 1-heptane, i, 520
Trimethyl 1-1-3-butadiëne, i, 478
Trimethyl 2, butene, i, 474
Trimethyl-1, 1,3-cyclohexane, i, 496
Trimethyl-1, 2-, 4-cyclohexane, i, 498
Trimethyl-1-, 3-, 5-cyclohexane, i, 498
Trimethyl 1-, 3-, 4-cyclohexane, i, 498
Trimethyl-1, 2-, 3-cyclopropane, i, 490
Trimethyl 1-1-2 dimethylene, i, 490
Trimethyl, 1-, 2-, 5-, 4-ethyl benzol, i, 486
Trimethyl, 1-, 3-, 5-, 2-ethyl benzol, i, 486
Trimethyl ethyl methane, i, 462
Trimethyl ethylene, i, 472
Trimethyl-1, 2-, 5- hexamethylene, i, 498
Trimethylmethane, i, 460
Trimethyl 2-, 2-, 3-, pentane, i, 466
Trimethyl propyl methane, i, 464
Trimethylene, i, 490
Trinidad, asphalt, i, 158, 160, 796
bitumen, i, 792
bituminous lakes, i, 5
burnt clays, i, 7
oil, i, 158, 160
oil analysis, i, 575
oil, average specific gravity, i, 322
oil production statistics, i, 159, 325
oil seepages, i, 4
pitch lake, i, 68

Trinidad, refined lake asphalt, i, 794
rotary drilling system, i, 251
stratigraphic occurrence of oil, i, 17
Trinity sand, i, 12, 20
Tri-olefine, i, 478
Triphenyl 1-3-5-, benzol, i, 518
Triphenyl ethane, i, 516
Triphenyl methane, i, 516, 536
Triplex pumps, ii, 207
Tripper, filter plant, ii, 202
Tripping, fishing, i, 285
Trip spear, i, 285
Tritriacontane, i, 472
Trumble cracking process, ii, 439
Trumble dephlegmating towers, ii, 219
Trumble pipe still, ii, 213
Trumble refining process, ii, 385
Trunk pipe lines, i, 428
Trunk pistons, Diesel engine, ii, 566
Tubes, boiler, standard dimensions, ii, 292
iron and steel, thermal expansion, ii, 814
Tube stills, refinery, ii, 146, 238
Tubing, fishing for, i, 288
oil well, i, 301, 501
oil well, weights and dimensions, table, i, 306
seamless steel, weights, table, ii, 785
Tubing catcher, i, 289
Tubular boiler, oil fired, ii, 464
Tubular goods for oil country use, i, 300
Tubular stills, refinery, ii, 140
Tucker sand, i, 30, 106
Tulsa County, Okla., Claremore sand, i, 34
Mississippi lime, i, 29, 30
Mounds sand, i, 30
oil, i, 103, 558
Oswego lime, i, 27
Red Fork sand, i, 28
Second Booch sand, i, 30
Tungsten, melting point, ii, 807
Tunis, oil, i, 157
Turbine oils, characteristics, ii, 677
lubricants, ii, 645
Turbines, steam, refinery, ii, 294
Turbulence, spraying, Diesel engine, ii, 575
Turbulent flow, pipe lines, i, 381
Turkestan, oil analyses, i, 582
Turkey, oil, i, 134
oil shale, i, 848
Turkeyfoot oil field, Pa., i, 110
Turkeyfoot oil field, W. Va., i, 117
Turnback casing shoe, i, 237
Turntable i, 254
Turpentine, i, 512
boiling point, ii, 806
expansion by heat, ii, 812
Turpentine, melting point, ii, 806
specific heat, ii, 812
Tuxpam oil field, Mexico, i, 124, 574
28 oil, ii, 357
Two-cycle engines, ii, 529, 537, 558, 560
Two-plug method, cementing oil wells, i, 295
Tyler County, W. Va., oil, i, 116, 569
Tyrol, Austria, bitumen, i, 792

U

Ubbelohde viscosimeter, i, 648
Uinta County, Wyo., oil, i, 573
Uintah County, Utah, oil, i, 115, 567
Uintaite, i, 71, 794
Ukraine, oil, i, 11, 131, 132
Undecanaphthene, i, 500
Undecane, i, 468
Undecylene, i, 474
Undecylthiophane, i, 528
Undecylthiophane-sulphone, i, 528
Underdeck tonnage register, oil tankers, i, 376
Underground glory hole method, oil shale mining, i, 856
Underground tankage, refinery, ii, 86
Underreaming, standard drilling, i, 218, 246
Unguentum petrolei, ii, 380
Union County, Ark., oil and gas, i, 80, 539
Union Petroleum Company colorimeter, i, 662, 665
Unit of heat, ii, 808
Unit of viscosity, ii, 627
United States, asphalt and bituminous rock, i, 796
beginning of oil industry in, i, 107
bitumens, i, 790
mud volcanoes, i, 7
natural gas production and consumption statistics, i, 352, 354
natural-gas gasoline, statistics, i, 351, 786
oil, i, 79
oil analyses, i, 538
oil, average specific gravity, i, 322
oil production statistics, i, 324, 356
oil production, per well per day, i, 349
oil, quantity and value produced since 1859, i, 328, 356
oil shale, i, 836
oil shale areas, i, 871
oil wells, number of, i, 349
pipe line rates, i, 379
productive oil wells, i, 349
stocks of petroleum, i, 347, 357
stratigraphic occurrence of petroleum, i, 11
United States Committee on the Standardization of Petroleum products, carbon residue test, i, 704
colorimeter, i, 664

United States Committee on the Standardization of Petroleum products, evaporation tests, i, 716
fatty oil, i, 711
floc test, i, 691
United States Navy Dept., specifications for illuminating oil, wax, and asphaltic products, i, 715
United States Office Public Roads, melting point, asphalt, i, 802
solubility test, asphalt, i, 802
volatilization test, asphalt, i, 802
United States standard gage, sheets, ii, 773, 774
Unloading tank cars, rules for, i, 367
Upper Coroni formation, Venezuela, i, 152
Upper Cretaceous series, California, i, 83
Egypt, i, 156
Florida, i, 84
Galicia, i, 132
Louisiana, i, 95
Montana, i, 97
oil strata, i, 12, 14, 17
Syria, i, 144
Texas, i, 114
Venezuela, i, 152
Wyoming, i, 118, 119
Upper Devonian, i, 34
Upper Devonian sands, New Brunswick, Can., i, 123
Upper Devonian shales, England, i, 135
Upper Freeport, coal, i, 24
Upper Jurassic system, Germany, i, 134
Wyoming, i, 119
Upper Medina rocks, i, 39
Upper Miocene series, oil sands, i, 68
California, i, 83
Egypt, i, 156
Java, i, 162
Upper Oligocene series, Russia, i, 65
Upper Sandusky oil field, Ohio, i, 100
Upper Silurian rocks, South Africa, i, 158
Upper Sunnybrook sand, Ky., i, 39, 94
Upper Trenton sand, i, 94
Upset tubing, ii, 562, 563
Upshur County, W. Va., gas., i, 116
Upton-Thornton oil field, Wyo., i, 118, 120, 571
Uintah County, Utah, bitumen, i, 791
Utah, bitumen, i, 791
gilsonite, i, 71
Green River formation, i, 885
natural gas production and consumption statistics, i, 352, 354
oil, i, 115
oil analyses, i, 567
oil shale, i, 834, 839, 873, 875, 885, 888
Utah, uintaite, i, 71
Utica shale, i, 39, 42
Utopia oil field, Kans., i, 91
Uvalde County, Tex., bitumen, i, 791
oil, i, 113, 114

V

Vacuum method, natural-gas gasoline, production by states, i, 786
Valerylene, i, 478
Vallecitos oil field, Calif., i, 81
Valves, Diesel engine, ii, 562
internal combustion engines, ii, 544, 554
pressure relief, ii, 461
safety, Diesel engine, ii, 576
safety, steam, ii, 798
Valylene, i, 478
Van Dyke and Irish tower still, ii, 325
Van Dyke tower, ii, 216, 220
Vapor, coil condenser data, ii, 164–179
volume, natural-gas gasoline, i, 752
Vapor exchangers, condensers, ii, 181
Vapor-phase process, cracking, ii, 429, 436
Vapor pressure, gasoline, method of determination, i, 726
relation to hydrocarbon condensation temperatures, i, 736
Vaporization point, oils, ii, 497
Vaporizers, internal combustion engines, ii, 530
Vaqueros formation, i, 17, 83
Vaseline, ii, 360, 380
Veale oil field, Tex., i, 113
Veale sand, Tex., i, 31
Velma oil field, Okla., i, 104
Velocity, pipe line, i, 403
piston, lubrication, ii, 653, 665
lubricants, relation to viscosity, ii, 634, 636
Velocity head, pumps, ii, 210
Venango County, Pa., oil, i, 35, 107, 569
Venango First sand, i, 34
Venango oil field, Pa., i, 110
Venango sand group, i, 34
Venango Second sand, i, 35
Venango Third sand, i, 36
Venedocia oil field, Ohio, i, 100
Venezuela, asphalt, i, 150, 797
bitumen, i, 792
bituminous lakes, i, 5
mud volcanoes, i, 6
oil, i, 150
oil analysis, i, 578
oil, average specific gravity, i, 322
oil production statistics, i, 151, 325
oil, unit of measure, ii, 761
pitch lakes, i, 68
stratigraphic occurrence of oil, i, 11

Venice oil field, Pa., i, 110
Vent tanks, gasoline extraction plant, i, 775
Ventilation, oil tankers, i, 377
Vento condenser, ii, 216
Vents, concrete tanks, ii, 133
Ventura County, Calif., oil, i, 16, 80, 83, 541
oil, nitrogen in, i, 530
oil, sulphur in, i, 526
Ventura oil field, Calif., i, 81
Venturi meter, ii, 369
Vera Cruz, Mexico, bitumen, i, 792
oil, i, 20, 124, 126
Vera oil field, Okla., i, 104
Vernon County, Mo., oil, i, 97
Vernon oil field, Kans., i, 91
Vernon oil field, Tex., i, 113
Vertical Diesel engine, ii, 561
Vertical retort, oil shale, i, 866
Vicker's marine Diesel engine, ii, 599, 602
Vicker's solid injection sprayer, ii, 572
Victoria, bitumen and oil seepages, i, 166, 793
Vigo County, Ind., oil, i, 88, 544
Vinton County, Ohio, oil, i, 553
Vinton oil field, La., i, 96, 549
Virgelle sandstone, i, 12, 98
Virgil oil field, Kans., i, 91
Virginia, petroliferous deposits, i, 115
Viscosimeter, conversion table, i, 642
types, i, 643
Viscosity, i, 640
absolute, i, 640
coefficient, i, 393
comparison of viscosimeters, i, 641
conversion curves, steam cylinder, ii, 656
Engler viscosimeter, i, 588, 655, 656
fuel oil, ii, 340
hydrocarbons, ii, 647
kinematic, i, 640
lubricants, ii, 626, 360
oil, i, 641
oil temperatures of measurement, i, 643
oil in pipes, i, 389
Redwood Admiralty viscosimeter, i, 655
Redwood viscosimeter, i, 654
relation to density, i, 391
relation to gravity, i, 391
relation to temperature, i, 390
Saybolt Furol viscosimeter, i, 588, 648
Saybolt Thermo viscosimeter, i, 588, 648
Saybolt Universal viscosimeter, i, 588, 644
temperature curves, ii, 632
temperature curves, steam cylinder oils, ii, 654
Vitreosil basins, acid recovery, ii, 403
Vitriol, refinery use, ii, 368
Vivian gas sand, i, 18
Vogt chilling machine, wax plant, ii, 194
Vogt soft wax press, ii, 199
Volatile matter in shale, relation to coal, i, 835
Volatility, gasoline, ii, 346, 352
lubricants, ii, 644
Volatilization, in lubrication, ii, 663
Volatilization test, asphalt, i, 807
Volcanic neck structure, i, 63
Volcanic plugs, i, 63, 64
Volcano oil field, W. Va., i, 117
Volume, air, ii, 786
compressed air, relation to pressure, ii, 298
compressed air, relation to pipe diameter, ii, 300
measures, ii, 762, 766
pipe line flow, table, i, 395
weight and energy equivalents, ii, 803
Volume-air burner, ii, 452
Volumetric composition, natural gas, i, 734
Volumetric efficiency, ii, 530

W

Wabash County, Ill., oil, i, 86, 89
Wager sand, i, 92
Wagoner County, Okla., second Booch sand, i, 30
Wagoner oil field, Okla, i, 103, 104
Wahl oil field, Ohio, i, 100
Wales, oil and shales, i, 135
Walk out, geologic field work, i, 177
Walker's formula, fuel oil, ii, 341
Walkway, condenser boxes, ii, 53, 55, 57, 58
Wall Creek sand, i, 14, 17, 118, 119
Walnut-Hepler oil field, Kans. i, 91
Walters oil field, Okla., i, 104
Wann oil field, Okla., i, 104
Warm Springs oil field, Wyo., i, 120, 572
Warping, monoclinal, i, 53
Warren County, Ky., bitumen, i, 791
oil, i, 92, 94, 547
Warren County, Pa., oil sands, i, 35
Warren First sand, i, 36, 108
Warren group of oil sands, i, 36, 110
Warren group, relation to Bradford sand, i, 37
Warren Second sand, Pa., i, 36, 108
Wasatch County, Utah, bitumen, i, 791
Wasatch formation, i, 874
Wash still condenser box, refinery, ii, 59
Washers, refinery, ii, 221, 367
Washington, natural gas, production and consumption statistics, i, 352, 354
oil and gas, i, 115
Washington County, Ala., oil and gas, i, 80
Washington County, Ark., oil, i, 80
Washington County, Fla., oil, i, 84
Washington County, Ind., oil, i, 88
Washington County, Ohio, oil, i, 102, 553

Washington County, Ohio, Cow Run sand, i, 23
Keener sand, i, 31
Maxton sand, i, 25
Second Cow Run sand, i, 25
Washington County, Okla., Hogshooter sand, i, 29
oil, i, 103, 558
Peru sand, i, 27
Wheeler sand, i, 28
Washington County, Pa., oil, i, 558
Washington County, Utah, oil, i, 115, 568
Washington oil field, Pa., standard casing requirements, i, 220
Washington-Taylorstown oil field, Pa., i, 110
Waste heat, internal-combustion engine, ii, 554
Water, ii, 790–792
boiling point, ii, 806, 807
expansion by heat, ii, 812
freezing point, ii, 807
in monoclinal bulges, i, 59
latent heat of evaporation, ii, 823
radiation and reflection of heat, ii, 824
relation to oil occurrence, i, 72
specific heat, ii, 812
Water acid, ii, 369
Water ballast pipes, oil tankers, i, 377
Water cooling, double spray system, ii, 315
volumes of air required, ii, 312
Water evaporation, air required, tables, ii, 313, 314
Water exclusion from oil sands, i, 290
Water in oil, i, 531
centrifuge method of determination, i, 670
determination by distillation, i, 588, 668
test, i, 662
Water level, relation to petroleum occurrence, i, 74
Water lubrication, pipe line, i, 405
Water pipe, dimensions and capacity, tables, ii, 771
standard dimensions, table, ii, 782–784
Water supply, pump stations, i, 424
refinery, ii, 3, 10, 239
Water-top roof, tanks, ii, 95
Water vapor, air saturation table, ii, 314
specific gravity, ii, 832
Water white distillate, ii, 339
condenser coil data, ii, 168
Water white oil, i, 663
Waters' oxidation test, i, 705
Watsonville oil field, Calif., i, 82
Waverly group, oil sands, i, 33, 94
Wax, amorphous, ii, 359
paraffin, ii, 375
sweating, ii, 373
Wax chipping machine, ii, 222
Wax cooling machine, ii, 222
Wax distillate, ii, 355, 364
condenser coil data, ii, 170
Wax molding press, ii, 223
Wax oil, ii, 364, 365
condenser coil data, ii, 178
Wax plant, ii, 92, 194, 222
Wax pumps, ii, 194, 196
Wax separation, fire hazards, ii, 413
Wax tailings, ii, 329
Wayne County, Ky., "deep" sand, i, 94
Knox dolomite, i, 40, 94
oil, i, 92, 94, 547
Trenton limestone, i, 40
Upper Sunnybrook sand, i, 39, 94
Waverly group, i, 34, 94
Wayne County, Ohio, "Clinton" sand, i, 39
oil, i, 102
Wayne County, Utah, oil, i, 115
Wayne County, W. Va., oil, i, 116
Waynesburg oil field, Pa., i, 110
Webb County, Tex., oil and gas field, i, 114
Web frame system, tank vessel construction, i, 373
Web reinforcement, ii, 252
Weighing method, gravity determination, i, 618, 621
Weight of air, ii, 786, 788
Weight of gases, at atmospheric pressure, ii, 832
Weight of water, table, ii, 790
Weights and measures, ii, 760
Weights, volume and energy equivalents, ii, 803
Weimer oil field, Okla., i, 104
Weir sand, relation to Squaw sand, i, 32
Weld strength, tubes and pipes, ii, 530
Welded pipe, specifications, ii, 549
Well for filter wash still, ii, 51
Well logs, i, 238
Well records, subsurface mapping, i, 173
Well shaft, ii, 241
Welland County, Ontario, Can., oil and gas, i, 121, 122, 123
Wellman cracking process, ii, 436
Wells, heat-exchanging stills, ii, 50
Wells County, Ind., oil, i, 88
Wells County, Ohio, oil, nitrogen in, i, 530
Wellsburg oil field, W. Va., i, 117
Wellsville oil field, Kans., i, 91
Wellsville oil field, Ohio, i, 100
Welsh oil field, La., i, 96, 550
West Africa, bituminous lakes, i, 5
oil, i, 157
West Australia, oil, i, 166
West Hickory oil field, Pa., i, 110

West Indies, bitumen, i, 792
oil, i, 158
West Union gas field, W. Va., i, 117
West Virginia, anticlinal and synclinal structures, i, 47
anticlinal bulges, i, 59
application of anticlinal theory to pools, i, 44
Bayard sand, i, 36
Berea sand, i, 32
Big Injun sand, i, 32
Big Lime (Greenbrier limestone), i, 31
bitumen, i, 791
Cairo oil sand, i, 68
Cow Run sand, i, 24
domes in, i, 59
Dunkard sand, i, 24
Eureka-Volcano-Burning Springs anticline, i, 47
Fifth sand, i, 36
grahamite, i, 6, 68, 71,
Keener sand, i, 31
limestones, i, 29
Maxon sand, i, 31
Maxton sand, i, 25, 38
Mississippian series, i, 15, 29
natural-gas gasoline statistics, i, 786
natural gas production and consumption statistics, i, 352, 354
natural gas springs, i, 3
oil, i, 116
oil analyses, i, 568
oil, nitrogen in, i, 529
oil, oxygen in, i, 524
oil production, per well per day, i, 349
oil production statistics, i, 330, 356
oil, sulphur in, i, 526
oil and gas seepages, i, 3
oil strata, i, 15
oil wells, number of, i, 349
Pennsylvanian series, i, 22
Permian rocks, i, 21
pipe line regulations, i, 428
Pittsburgh coal bed, i, 23
quaquaversal structure, i, 59
Ritchie mines, i, 6, 68
salt brine, occurrence with oil, i, 8
salt sands, i, 25
sandstones, i, 29
Sixth sand, i, 36
Squaw sand, i, 32
synclines, i, 73
Upper Freeport coal, i, 24
West View oil field, Pa., i, 110
Western Argentina, oil district, South America, i, 146, 148
Westfork oil field, Tex., i, 113
Westmoreland County, Pa., Speechley sand, i, 36
Tiona sand, i, 36
Weston County, Wyo., oil, i, 570
Weston gas field, W. Va., i, 117
Wet-back Scotch marine boilers, oil fired, ii, 480
Wet natural gas, i, 731
Wetzel County, W. Va., oil, i, 116
Wexford oil field, Pa., i, 110
Wheeler oil field, Okla., i, 59, 104
Wheeler sand, i, 28, 106
Whipstock, i, 288
Whipstock grab, fishing tool, i, 282
Whitacre oil field, Ohio, i, 100
White Cottage oil field, Ohio, i, 100
White Medina sand, i, 39, 121
White petrolatum ii, 380
White Point gas field, Tex., i, 113
Whiteley Creek oil field, Pa., i, 110
Whitterbrook oil field, Ohio, i, 100
Whittier oil field, Calif., i, 16, 68, 81, 82, 83, 540
Wichita County, Tex., Canyon formation, i, 28
Cisco formation, i, 28
oil field, i, 114, 562
Pennsylvanian series, i, 28
Permian series, i, 21
Wier sand, Ky., i, 94
Wilbarger County, Tex., oil field, i, 114
Permian series, i, 21
Wilcox sand, i, 13, 34, 103, 106
Wiley Canyon oil field, Calif., i, 81
Wildwood oil field, Ky., i, 93
Wildwood oil field, Pa., i, 110
Willetts oil field, Pa., i, 110
Williamson County Tex., oil, i, 114, 563
Wilson County, Kans., oil, i, 90, 546
Wilson Run oil field, Ohio, i, 100
Winchester County, Ky., oil, i, 548
Windlass, tank, ii, 101
Winfield oil field, Kans., i, 91
Winfield oil field, Pa., i, 110
Wingett, P. O., oil field, Ohio, i, 100
Wire, refinery use, ii, 271
Wire drilling lines, i, 231, 233
Wirt County, W. Va., oil, i, 116, 117
Wolfe County, Ky., Caney sand, i, 39
oil, i, 92, 548
Wood, fuel oil equivalent, ii, 477
Wood County, Ohio, oil, i, 102
oil, nitrogen in, i, 530
oil, sulphur in, i, 525
Wood County, W. Va., oil, i, 116, 117, 570
Wood pulp, grease filler, ii, 674
Wood spirit, boiling point, ii, 806

Wood tanks, refinery, ii, 213, 214
Woodbine sand, i, 12, 20, 94, 95
relation to Blossom sand, i, 19
Wood-roof, tanks, standard specifications, ii, 95
Woods, specific heat of, ii, 811
Woodsfield oil field, Ohio, i, 100
Woodson County, Kans., i, 90
Woodville oil field, Pa., i, 110
Wool, refinery insulating, ii, 282
Wooster gas pool, Ohio, i, 100
Wooster oil fields, Ohio, i, 59, 100
Working stresses, reinforced concrete, ii, 257
Working tanks, refinery, ii, 70, 78
Working temperature, lubricants, ii, 644
World's production of petroleum, i, 78, 324
Worm, cast-iron still, specifications, ii, 163
Worm design, condenser, ii, 180
Worthington internal combustion engine, ii, 541
Worthington marine Diesel engine, ii, 608
Wrecks, tank cars, i, 368
Writing paper, radiation and reflection of heat, ii, 824
Wrought iron, radiation and reflection of heat, ii, 824
Wrought pipe, standard dimensions, table, ii, 782
Wurtzilite, i, 791,793, 794
Wyandot County, Ohio, oil, i, 102
Wynona oil field, Okla., i, 104
Wyoming, anticlinal and synclinal structures, i, 47, 59
asphalt, i, 67
Benton shale, i, 18
brea, i, 67
Cambrian system, i, 40
Colorado group, i, 17
crevice oil, i, 67
Dakota sand, i, 20
Embar sand, i, 22, 67
Frontier group, i, 17
Green River formation, i, 887
Chugwater red beds, i, 22
Jurassic system, i, 12, 20, 119
Lower Wall-Creek sandstone, i, 18
Montana group, i, 17
Morrison formation, i, 20
natural-gas gasoline statistics, i, 786
natural gas production and consumption statistics, i, 352, 354
Wyoming, oil, i, 118
oil in granite, i, 67
oil production, per well per day, i, 349
oil production statistics, i, 338, 356
oil shale, i, 834, 840, 873, 887, 888
oil springs, i, 68
oil tar, i, 67
oil wells, number of, i, 349
Peay sands, i, 18
Permian system, i, 22, 67
Pierre formation, i, 17
Powder River field, i, 20, 21
producing sands, i, 119
Salt Creek oil field, i, 18, 356
Shannon oil field, i, 17
Sundance formation, i, 12, 20

X

Xylene (xylol), i, 480

Y

Yale oil field, Okla. i, 104
Yampa oil field, Colo, i, 84, 85
Yard storage, refinery, ii, 84
Yarhola oil field, Okla. i, 104
Yates Center oil field, Kans., i, 91
Yeager oil field, Okla., i, 104
Yenangyat, Burma, India, oil analysis, i, 582
oil field, i, 47, 50, 140
Yenangyaung, Burma, India, oil field, symmetrical anticlinal structure of, i, 47
Yield point, tests of commercial pipe, ii, 526
Yields, refinery operation, ii, 421
Young's cracking process, ii, 428
Young County, Tex., Bend limestone, i, 31
oil, i, 563
Youngstown oil field, Okla. i, 34, 104

Z

Zacamixtle oil field, Mexico, i, 124, 125
Zimar oil field, Pa., i, 110
Zinc, latent heat of fusion, ii, 823
melting point, ii, 806, 807
radiation and reflection of heat, ii, 824
refinery use, ii, 279
specific heat, ii, 811
Zollarsville oil field, Pa., i, 110
Zorritos oil field, Peru, i, 148, 578

Lightning Source UK Ltd.
Milton Keynes UK
UKHW040159060119
334991UK00027B/2681/P

9 781527 879768